Proceedings of the Fourth International Congress on Mathematical Education

Edited by
Marilyn Zweng
Thomas Green
Jeremy Kilpatrick
Henry Pollak
Marilyn Suydam

Birkhäuser

Editor (for the ICME IV editing committee):
M. Zweng
Department of Secondary Education
N297 Lindquist Center
University of Iowa
Iowa City, Iowa 52242

This book was produced from the camera-ready manuscript provided by the organizers of the conference. Birkhäuser Boston, Inc., would like to thank Professor Marilyn Zweng who supervised the production of this manuscript.

Library of Congress Cataloging in Publication Data

International Congress on Mathematical Education
(4th : 1980 : Berkeley, Calif.)
Proceedings of the 4th International Congress on Mathematical Education.

"The Fourth International Congress on Mathematics Education was held in Berkeley, California, USA, August 10-16, 1980"—Introd.
1. Mathematics—Study and teaching—Congresses.
I. Zweng, Marilyn, 1932- . II. Title.
QA11.A1I46 1980a 510'.7'1 83-2695
ISBN 3-7643-3082-1

CIP—Kurztitelaufnahme der Deutschen Bibliothek

International Congress on Mathematical Education [04, 1980, Berkeley, Calif.]:
Proceedings of the 4th International Congress on Mathematical Education / [ed. by Marilyn Zweng...]. — Boston ; Basel ; Stuttgart : Birkhäuser, 1983.
ISBN 3-7643-3082-1
NE:Zweng, Marilyn [Hrsg.]

ISBN 3-7643-3082-1

Printed in USA

ACKNOWLEDGMENTS

This is the first time in the history of the International Congress on Mathematics Education that all submitted papers have been published in the Proceedings. Over half of the presenters at the Congress are represented in this volume; a total of almost 3000 pages of manuscript was submitted. Although not comprehensive, the totality of the articles presents a rather clear picture of the state-of-the-art world wide in mathematics education at the time the Congress was held in August of 1980. The Proceedings document the common concerns of mathematics educators around the globe and future directions for mathematics education.

Because of the enormity of the task, some difficult decisions had to be made. To stay within page limits the editors often had to delete several paragraphs, and at times, several pages from the submitted articles. It was not possible to return manuscripts to authors for their approval. We hope that our editing efforts maintained the essence of the presentations but accept the responsibility for any changes which may have altered the author's intent. We were unable to publish some very fine articles from projects and from working groups. The budget simply would not stretch to cover the expenses.

Many people have contributed freely of their time and talents to produce these Proceedings. The members of the editorial committee, Thomas Green, Jeremy Kilpatrick, Henry Pollak and Marilyn Suydam each read and edited almost 800 pages of manuscript. Their flair for clarity of expression and their concern for maintaining the integrity of an author's work provided the polish for the contents of the Proceedings.

The cost of the publication was underwritten by the National Academy of Sciences. Dr. John Goldhaber, Executive Secretary of the National Research Council, operating agency for the Academy, was most helpful throughout the production. Additional support was provided by the University of Iowa's Graduate College.

Douglas Dalman preserved the sanity of all of us throughout this past year. Doug, working as a graduate assistant, organized, supervised, edited, proof-read, arranged all details of the production and kept me on task. Doug was determined that the ICME IV Proceedings be a quality production.

The camera-ready copy for the Proceedings was produced by Words Worth - Word Processing Service of Iowa City. JoAnn Peiffer and Phyllis Stiefel are Words Worth. They provided far more than their word processing skills. They contributed ideas to improve the format, they edited when necessary and they offered intelligent guidance throughout. All this was accompanied by an unfailing sense of humor.

Finally, we extend our thanks and gratitude to the authors whose extra effort in committing their presentations to paper have made this volume possible.

Marilyn J. Zweng,
for the ICME IV editing committee:
Thomas Green, Jeremy Kilpatrick, Henry Pollak and Marilyn Suydam.

Iowa City, Iowa
September 1982

INTRODUCTION

Henry O. Pollak
Chairman of the International Program Committee
Bell Laboratories
Murray Hill, New Jersey, USA

The Fourth International Congress on Mathematics Education was held in Berkeley, California, USA, August 10-16, 1980. Previous Congresses were held in Lyons in 1969, Exeter in 1972, and Karlsruhe in 1976. Attendance at Berkeley was about 1800 full and 500 associate members from about 90 countries; at least half of these came from outside of North America. About 450 persons participated in the program either as speakers or as presiders; approximately 40 percent of these came from the U.S. or Canada.

There were four plenary addresses; they were delivered by Hans Freudenthal on major problems of mathematics education, Hermina Sinclair on the relationship between the learning of language and of mathematics, Seymour Papert on the computer as carrier of mathematical culture, and Hua Loo-Keng on popularising and applying mathematical methods. George Polya was the honorary president of the Congress; illness prevented his planned attendence but he sent a brief presentation entitled, "Mathematics Improves the Mind".

There was a full program of speakers, panelists, debates, miniconferences, and meetings of working and study groups. In addition, 18 major projects from around the world were invited to make presentations, and various groups representing special areas of concern had the opportunity to meet and to plan their future activities.

Innovations relative to previous Congresses included the following:

1. A series of miniconferences, in memory of Edward G. Begle, devoted to critical variables in mathematical education;

2. A series of miniconferences on topics in the mathematical sciences which deserve serious consideration, in all countries, for inclusion in the curriculum: algorithms, operations research, combinatorics, data analysis, algebraic coding theory, extrema without calculus;

3. Simultaneous translation among English, French, and Spanish, of one session at all times;

4. Daily informal gatherings with coffee and pastry in the morning, wine and cheese in the late afternoon.

There were also special meetings of the Inter-American Committee on Mathematical Education of the African Mathematical Union, of leaders of Mathematics Teaching Associations, of journal editors, of supervisors of mathematics instruction, and of those particularly concerned with planning the International Mathematical Olympiads. The final plenary session, devoted to the business of the International Commissions on Mathematical Instruction, expressed its concern for the future of IOWO in the Netherlands and the IREMs in France, and the welfare of J.L. Massera and A. Sharansky.

The short communications, in the form of several hundred posters, were especially popular.

The good will of many organizations, and the heroic efforts of many individuals, were necessary to make the Congress happen. It is obviously impossible to thank all of these individually; however, the outstanding cooperation of the host institution, the University of California at Berkeley, must be singled out for a special expression of gratitude. The National Academy of Sciences, U.S.A., the National Council of Teachers of Mathematics, and the Mathematical Association of America, along with their California affiliated and member groups, were most helpful. The Local Organizing Commttee under the co-chairmanship of Floyd Downs, Leon Henkin, and John Kelley put in many months of diligent labor, as did the various program committees. Finally, no person worked harder on a very difficult job, with a low budget and innumerable unexpected difficulties, than the chairman of the Editorial Committee of these Proceedings, Marilyn Zweng of The University of Iowa. She has carried off this task with unfailing good humor and true professionalism, and the entire mathematics education community in the world owes her a debt of gratitude.

TABLE OF CONTENTS

CHAPTER I - Plenary Session Addresses

I.1 MATHEMATICS PROMOTES THE MIND

MATHEMATICS PROMOTES THE MIND

George Polya
Honorary President of the Congress

delivered by
Gerald L. Alexanderson
University of Santa Clara
Santa Clara, California U.S.A.

First I may say that I regret very much not being able to be with you today and participate in the Congress.

The title of this short talk is Mathematics Promotes the Mind. I repeat with conviction that mathematics promotes the mind, but in fact this is not unconditionally correct. There is a condition: Mathematics promotes the mind provided that it is taught and learned appropriately.

I could quote to you examples of mathematics teaching which is quite good to transmit the intended mathematicalf acts and their proofs but does nothing to promote the mind. Certainly you had yourselves several occasions to listen to such teaching and you could quote examples yourselves. Hence the question: What can the mathematics teacher do in order that his teaching improves the mind. It would have been a worthy topic to discuss in this conference. Yet it is too late. It should have been proposed months ago. Therefore I propose it for the next Congress. I hope that there are some people in the audience who will think about my question, remember it and propose it to the next Congress. I have some ideas about if, if someone is interested we might discuss them privately.

I wish the congress much success.

I.2 MAJOR PROBLEMS OF MATHEMATICS EDUCATION*

MAJOR PROBLEMS OF MATHEMATICS EDUCATION*

Hans Freudenthal
Institut Ontwikkeling Wiskunde Onderwijs
The Netherlands

Forgive me, it was not I who chose the theme, though when it was first suggested to me, I experienced it as a challenge. A challenge, indeed, but to be sure not as one to emulate Hilbert, who at the Paris International Congress of Mathematicians in 1900 pronounced his celebrated 23 mathematical problems, which were to profoundly influence, nay presage, the course of mathematics for almost a century. If it were not modesty that prevents me from even trying it, it should not be the obvious fact that - problems, problem solving, and problem solvers - mean different things in mathematics education from what they mean in mathematics.

But first let us look at the other noun in the title: education. It can mean roughly three things: the educational process taking place in the family, at school, on street, and everywhere, an administrative establishment, a theoretical activity, called educational research.

The major problems I am going to summarize, are of the first kind, partly related to the second, while I will disregard the third kind. What I am interested in is problems in mathematics education as a social activity rather than problems as an entrance to educational research as one of the major problems of mathematics education.

Yet let me come back to what I announced as a difference between problems, problem solving, and problem solvers in mathematics on the one hand and mathematics education on the other.

Mathematical problems are problems within a science arising for a large part from this science itself arising from changing needs, moods and whims of a changing society. Hilbert's problems have been seminal for a century. The address I deliver today may be forgotten never to be remembered ten years from now.

From olden times mathematicians have posed problems for each other, both major and minor ones: here is the problem, solve it; if you can, tell me; I will listen to check whether it is a solution. In education problem solving is not a discourse but an educational process. In mathematics the problem solvers are mathematicians. In education problem solving is not a discourse but an educational process. In mathematics the problem solvers are mathematicians. In education problems are properly solved by the participants in the educational process, by those who educate and by those who are being educated.

Moreover in mathematics you can choose one major problem, say from Hilbert's catalogue, to solve it, and disregard the remainder. In education all major problems, and in particular those I am going to speak about, are strongly interdependent. As a matter of fact major problems of education are characterised by the fact that none can properly be isolated from the others. The best you can do at a given moment is focusing on one of them without disregarding the others, and this is in fact what I am going to do here.

In a sense the title of my address is wrong: all major problems of mathematics education are problems of education as such. In another sense it is exactly right: if you look for major problems the best paradigm of cognitive education is mathematics.

In order to terminate this introduction, I should add two points. First, there is no authority in mathematics education comparable to those in mathematics. The problems which I think are major ones, have been chosen according to my philosophy of learning and teaching mathematics, which I will not recapitulate here since it will be implicit in the problems and in the way they are submitted.

Second, although the problems have been inspired by my own experience and philosophy, I do not claim to have invented them. Originality too, has another meaning in mathematics education from that which it has in mathematics. My ideas have been anticipated not once but many times in the past. This then is the reason why I will refrain from quoting and citing, if some problem I mention has successfully been tackled before or not.

1. Allow me to start with the most earthly problem I can think of. Among the major ones it is the most urgent. What is even a problem is how to formulate it correctly and unmistakenly. Let us try a preformulation. It runs: Why can Johnny not do arithmetic? Does it sound sexist? I would not change it into - Why can Mary not do arithmetic?, least it may sound even more sexist, suggesting that girls are less able than boys. As a matter of fact both are wrong. My problem is not John Roe and Mary Doe. It is a problem, indeed, why many children do not learn arithmetic as they should, and it is a major one because more than anything else, failure in arithmetic may mean failure at school and in life. My concern, however, is not, or not primarily, what is wrong in classrooms and textbooks - today - that creates a host of underachievers.

Let me change the question. I now ask: Why can Jennifer not do arithmetic? Rather than an abstraction like John and Mary, Jennifer is a living child (though I have changed her name) whom I can describe in all detail. The two details that matter here, are that she was eight years old and could not do arithmetic. Meanwhile the question - Why can Jennifer not do arithmetic? - is no question any more, because today Jennifer is eleven and excels in arithmetic. Yet when she was eight, somebody observing her stumbling with numbers succeeded in answering the question and after ten minutes of remedial teaching, the problem Jennifer had, ceased to exist. Was it a miracle? Not at all. It was just an easy case. But so there are many. Noticed and unnoticed, cared for and uncared. But what about the less easy cases? THey have grown out of those easy ones that remained unnoticed and uncared for. Diagnosis and prescription are terms borrowed from medicine by educationalists who pretend to emulate medical doctors. What they do emulate is medicine of a forgone period, which is the quackery of today. Medical diagnosis iin former times aimed at stating what is wrong, as do the so-called diagnostic tests in education. True diagnosis tells you why something went wrong. The only way to know this is by observing the child's failure and trying to understand it. An expensive way. Would it be cheaper by computer? No, because in fact observing and understanding the individual child is not expensive. What is really expensive is wasting the vast resources of human experience.

Let me explain what I mean. By chance I know why Jennifer failed and I know about quite a number of other children because there were people who observed their failures and analysed them. All were different, and nevertheless they were an infinitesimal minority of those who need help. On the other hand I am sure that my own experience is only an infinitesimal part of a vast amount of knowledge, which has never been recorded nor even made conscious. Would we be better off if this bulk of knowledge had actually been recorded and reported? Not at all. Useful educational theory does not arise from blind generalising. What we need are paradigmatic cases, paradigms of diagnosis and prescription, for the benefit of practicioners and as bricks for theory builders.

Let me add two examples, two paradigmatic cases though not enough for theory building. To test underachievers it is useful asking them what is 2 + 9. Rather than looking to the result look to how the task is performed. Is it done by counting nine forward from 2, that is: 3, 4, . . .? Whether you call it a diagnosis or not, if it happens this way, you can be sure about at least two sources of failure: no awareness of commutativity nor of the use of the positional system.

Another example: a twelve-year old girl whom I taught understanding fractions simplified 16/24 to 3/8 - an unexpected failure. She explained it by 16 = 2 x 8, 24 = 3 x 8, thus 3/8. When I dug more deeply, the source appeared to be a failure of short term memory, that is, an error in storing or retrieving intermediate results. This experience led me to better understanding failures of hers and other children which in the past I had wrongly interpreted, for instance attributed to a lack of concentration. I started remedial work to improve short term memory, which proved successful, even in transfer: factorising mentally whole numbers below 100 with at least prime factors. This exercise draws heavily on short memory, indeed. Say 48 = 6 x 8, 6 = 2 x 3, 8 = 2 x 4, and now you can imagine what happens, the badly stored 2 x 3 cannot be retrieved. A fortnight after this exercise the girl performed the same task with no difficulty at all, and short term memory, in general, had greatly improved.

2. <u>The Learning Process</u>. In former times medical diagnosis started with anamnesis. Anamnesis of classroom experiences can be a ticklish affair. In most cases of arithmetic disability I came across, the case history was clear by circumstantial evidence, such as in the example of ignorance of commutativity: the subjects had never received any instruction, or at least any decent instruction, in arithmetic - I do not blame their teachers, who obviously had never learned what and how to teach. Studying illness is easier than studying health. Human biology indeed, started with medicine and medicine grew from cure to prevention. THis would suggest my second major problem, as it were an extension of the first: How should children learn?, in particular mathematics, which I immediately change into: How do people learn? which is the proper question, and the way to answer it would be: By observing learning processes, analysing them and reporting paradigms; learning processes within the total educational system, learning processes of pupils, groups, classes, teachers, school teams, councillors, teacher students, teacher trainers, and of the observer himself. Learning to observe learning processes, this is my second major problem in mathematics education. In mathematics education because I believe that learning to observe mathematical learning processes is the easiest approach to the problem of learning to observe learning processes in general. Observing involves analysing, by which I don't mean averaging or applying other statistical procedures nor fitting the observational data into preconceived patterns of developmental psychology. Grasping how people do learn would be a first step towards solving the every day problems of practicioners how to teach learning and towards building a learning theory, which should be based on evidence, rather than on preconceived ideas.

3. Progressive schematisation and formalisation. In the preceding I stressed observing learning processes against testing learning products. Amongst the learners I mentioned I forgot one, the biggest one. Mankind too is a learner. Observing its learning process is what we call history. How can the individual learner profit from knowledge about the big learning process of mankind? Rather than from its detail he can profit from the fact as such. Each stage in the growth of mathematics meant: knowledge acquired by insight transformed by schematising and memorising (or call it, codifying) into skills and insight of a higher order.

Let me explain by a few examples what I mean by schematising. A problem of long standing: A farm with chickens and rabbits, 20 heads and 56 legs; how many chickens and how many rabbits? I am sure you will solve it by insight. But as early as Babylonian antiquity one knew the schematism of a pair of linear equations with two unknowns to solve this kind of problems, and in more recent times this schematic insight has been anew schematised by the rule: Put the unknown numbers x and y, write down the connecting algebraic relations, and solve them by algebra. The most modern schematisation of this idea is vector space.

Clumsy algebra by insight was schematised by Vieta and Descartes' formal methods, and this process of schematising still continues in modern algebra. Calculation of areas, volumes, gravity centres, and moments, which once required the genius of an Archimedes - and even harder ones - are today within the grasp of our freshmen, thanks to Newton and Leibniz' schematising infinitesimal methods in what is known as Calculus.

But let us explain it in an even more elementary way. As early as written sources can remember, counting was schematised by introducing higher level units such as 5, 10, 100, 1000, and as early as reckoning was invented, it was schematised by a positional idea: higher level units materialised by counters on the abacus. Schematising arithmetic proceeded one step further by transforming sets of counters into digits, and the hardware abacus into a written one, written into sand or on paper, from which by a progression of schematising our present schematism of columns arithmetic arose.

Let me illustrate this process by a modern didactical version:

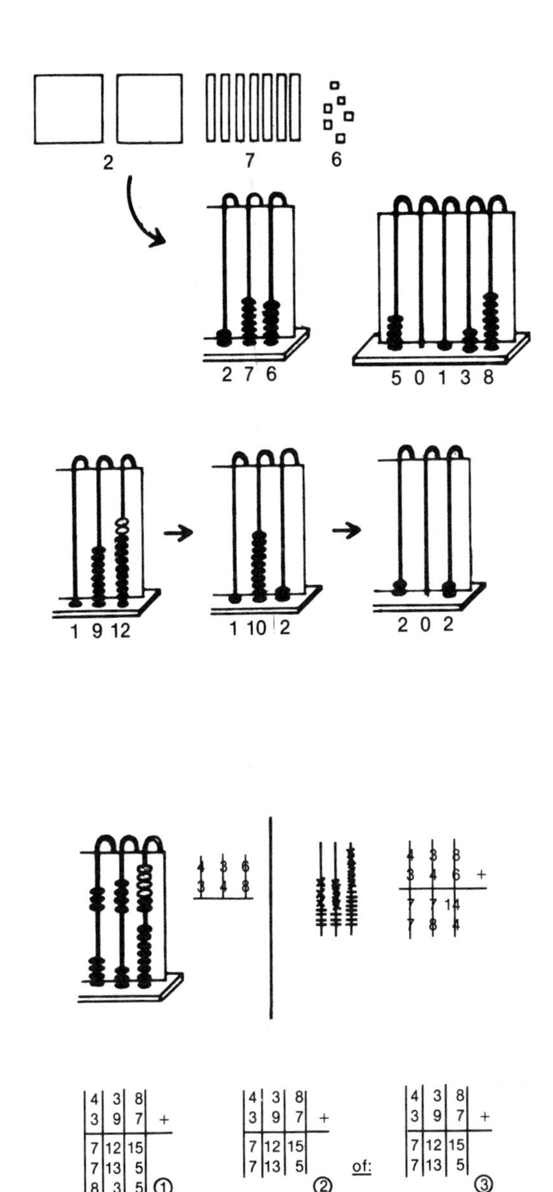

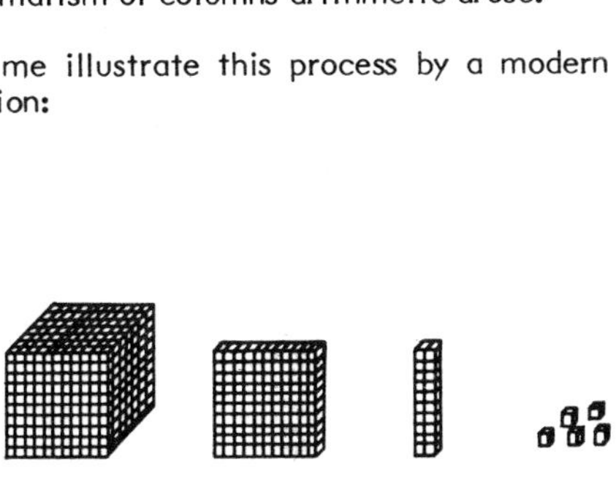

Figures 1, 2, 3, 4 & 5

History of mathematics has been a learning process of progressive schematising. Youngsters need not repeat the history of mankind but they should not be expected either to start at the very point where the preceding generation stopped. In a sense youngsters should repeat history though not the one that actually took place but the one that would have taken place if our ancestors had know what we are fortunate enough to know.

Schematising should be seen as a psychological rather than a historical progression. I think that in mental arithmetic of whole numbers we can fairly well describe schematising as a psychological progression or rather as a network of possible progressions where each learner chooses his own path or all are conducted the same way. Quite a few textbooks witness efforts to teach learning the traditional algorithms of column arithmetics of whole numbers by a progression of schematising steps though I am not sure whether their ideas are supported by actual learning and teaching. In teaching fractions, decimal numbers, algebra, and calculus I see little if any attempts at progressive schematising. The idea that mathematical language can and should be learned in such a way - by progressive formalising - seems even entirely absent in the whole didactical literature.

This then is my third major problem of mathematics education: how to use progressive schematisation and formalisation in teaching any mathematical subject what ever?

4. The role of insight. A cherished antagonism in teaching and learning mathematics is putting on one side of a deep gorge such noble ideas as insight, understanding, thinking, and on the other side such base things as are rote, routine, drill, memorising, algorithms. If I were malicious, I would add another pair of opposites - - theory versus practice, suggesting that learning by insight and understanding be a noble theory while the base practice is learning by rote and memorising. However, it is not that simple, and it has never been so. Even in our computerized age children shall memorise tables of addition and multiplication and acquire certain skills by rote, though one might argue that by the rise of the computer the balance has shifted in favour of the nobler activities.

It is not that simple, firstly because the option is not one between both sides of a gorge but rather bridging it by the learning process I called schematising. Secondly, I do believe that at any time more mathematics has been taught from the viewpoint of insight and more has been learned by insight than we are aware of. All agree and textbook writers witness that elementary arithmetic cannot be learned in any other way than by insight whether it is taught that way or not. But is also true that as things go on, as teaching proceeds to higher grades, to columns addition and multiplication, to long division, to fractions, ordinary and decimal ones, to algebra, to learning mathematical language, the part played by insight changes. Not just be diminishing, but its character changes. There is a tendency that the learner's insight is superseded by the teacher's, the textbook writer's, and finally by that of the adult mathematician. And the same holds on the long winding road, starting with concretely understood word problems and leading to highly formalised and badly understood applied mathematics.

This is why people who advocate learning by insight, disagree about what is insight. The wrong perspective of the so-called New Math was that of replacing the learner's by the adult mathematician's insight.

Yet this is not my main point. I have still to explain why we are not aware of how much is nevertheless being learned by insight. It is a most natural thing that once an idea has been learned, the learner forgets about his learning process, once a goal has been reached, the trail is blotted out. Skills acquired by insight are exercised and perfected by - - intentional and untentional - - training. This is a good thing. What is bad, is sources of insight clogged by acquired routines, to be never reopened, and this is what usually happens. It explains why teachers at higher grades so often complain about teaching habits at lower grades. If it is restricted to the first acquisition of some idea, learning by insight does not deserve this name. What is crucial, is retention of insight, which is gravely endangered by premature training, too much training, training as such. This then is my fourth major problem in mathematics education: How to keep open the sources of insight during the training process, how to stimulate retention of insight, in particular in the process of schematising?

5. How can this goal be pursued? The solution I propose is having the learner reflect on his learning processes. To a large degree, mathematics is reflecting on one's own and others' physical, mental and mathematical activity. The origin of proving theorems is arguing what looks obvious. Nobody tries to prove a thing unless he knows it is true. This he knows by intuition, and the way to prove it is by reflecting on one's intuition. Successful learning processes, if observed, should be made conscious to the learner in order to be reinforced and in order to be recalled if needed. This, however, is not what usually happens. Let us illustrate this abstract exposition by an example: Many children and adults can tell you that in order to multiply by 100 you have 'to add two zeroes' (which just holds for whole numbers) and most of them cannot explain why. It is even worse: most of them do not even understand that you can argue it and why you should do so. Did they learn such rules by rote? I do not believe it. I have observed too many children applying such rules intuitively before they were verbalized and formally taught at school. Rather than being taught the rules, they should have been taught to argue their intuitions, reflecting on what looks obvious. But this requires more patience than teachers can afford.

Let me add a personal experience. It happened with a sixth grade girl, a serious underachiever. When I started teaching her, she did not know anything about fractions, decimal fractions, percentage, the metric system. Fortunately - - I should add. (The only thing she know was: area equals twice length plus width.) So it was a fresh start. As I taught her mathematics, she taught me patience. By patience I don't mean, not getting angry, but rather abstaining from teaching her any rules. She learned by insight and schematising, and she now performs fairly well without knowing any rules or schematisms. At one moment at school when they noticed she had made progress, they tried to teach her rules for fractions. It was a catastrophe. I needed weeks to restore what had been destroyed. Now one year later she performs quite decently in mathematics. She does not master rules, she heavily depends on insight and half conscious schematisms. I have not yet succeeded in getting her to reflect on her mental activities. Is it too early? I am afraid it is too late. I guess that such an attitude should be acquired early. This then is my fifth major problem of mathematics education: How to stimulate reflecting on one's own phpysical, mental and mathematical activities?

6. Reflecting on one's own physical, mental and mathematical activities is an important component of what is called a mathematical attitude. I have hesitated whether I should include in the list of my

major problems how to develop a mathematical attitude? My own problem is that while we fairly well know what we understand under a mathematical attitude, we can describe it only by a long catalogue of examples and counter-examples - - too many to be dealt with in one hour or even in one week. I once tried to identify a few components of mathematical attitude. I will copy this list here, while omitting examples:

> developing language above the demonstrative and linguistically relative level, in particular on the level of conventional variables,
>
> change of perspective, a complex field of strategies with the common feature that the position of data and unknowns in a problem or field of knowledge is - partially - interchanged; including the recognition of wrong changes of perspective,
>
> grasping the degree of precision which is adequate to a given problem,
>
> identifying the mathematical structure within a context if there is any, and keeping off mathematics where it does not apply,
>
> dealing with one's own mathematical activity as a subject matter in order to reach a higher level.

Without examples this is a meaningless catalogue. The reason why I mentioned mathematical attitude is that you would have rightly objected if I had not, and the reason of showing a few components of mathematical attitude was to fight the usual misunderstanding of testing a mathematical attitude by asking questions about an attitude towards mathematics.

7. Let us consider one feature of progressive schematising that I did not yet consider, I mean that in such a progression not all learners progress at the same pace and reach the same goals. Progressive schematising is a way to account for natural differences in aptitudes and abilities. Differentiation is a general problem of education. In spite of the wide variability of linguistic mastery, it is a fact that people can communicate with each other in their mother tongue on a broad scale of subjects. Why is mathematics different, and should it really be different? There are reasons - - social reasons - - why in spite of their diverging learning abilities learners should learn together as they are expected to work together in the society. Cooperation involves levels of work. Cooperative learning presupposes levels of learning. It is a fact that mathematics lends itself as no other subject does, to distinguishing levels, in mathematics and in learning mathematics. My seventh major problem of mathematics education is: How is mathematical learning structured according to levels and can this structure be used in attempts a differentiation?

8. Perhaps you will compain that up to now I have paid almost no attention to subject matter and its didactics. If by subject matter you mean some chapter of a textbook, you will be disappointed. These are no major problems. But I agree that teaching is always teaching something. Something rather than anything. Something worth being taught. But what is worth being taught? In order to be taught it should be applicable, in some sense, in any sense, in any sense whatever. What does this mean? Teaching as much mathematics as the science teachers pretend they need? Or after a block of compulsory algebra and calculus a few choice subjects like probability, numerical methods, linear programming, or mechanics? Everybody knows that it does not work. Educationally viewed, application is a wrong perspective cherished by old math and even more by New Math. The right perspective is primarily from environment towards mathematics rather than the other way round. Not: first mathematics and then back to the real world but first the real world and then mathematising.

The real world - - what does it mean? Forgive this careless expression. In teaching mathematising 'the real world' is represented by a meaningful context involving a mathematical problem. 'Meaningful' of course means: meaningful to the learners. Mathematics should be taught within contexts, and I would like the most abstract mathematics taught within the most concrete context.

Let me tell a little story about what context can mean, not only for learning mathematising but even for learning mathematical skills. In a context of sharing and leasing gardens, fourth graders had to figure out, among other things, the rent of a plot like

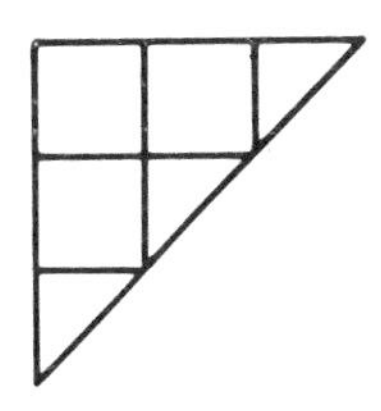

where each square pays five florins. All children who stayed within the context of squares and land and of florins to be paid got the correct answer of 22 1/2 florins, whereas all the others who prematurely divorced the problem from its context and schematised it as the numerical multiplication problem 4 1/2 x 5 got the wrong solution of 20 1/2.

What is this little story to suggest? That in the learning process mathematising a situation deserves priority on solving word problems by schematisms.

After this interruption let me formulate my eighth major problem of mathematics education: How to create suitable contexts in order to teach mathematizing?

9. Environment involves space, objects in space and happenings in space. The mathematised spatial environment is geometry, the most neglected subject of mathematics teaching today. For centuries geometry in the English terminology was synonymous with Euclid. But in history geometry started long before Euclid, and in children's life it starts even before kindergarten. Or shouldn't it? In any case, what starts, is the grasp of space and the relations in space, the grasp of space by seeing, by listening, by moving in space. When can this rightly be called geometry? The traditional answer is: when it can be verbalised in terms of definitions, theorems, and proofs. This then meant that geometry education started at an age when children were pronounced able to speak the conventional language of geometry, either Euclid's or the textbook's. Unfortunately you cannot learn the language in which a subject matter is expressed if you have not experienced

the subject matter itself. How do you learn the subject matter itself? The way is by becoming conscious about one's intuitive grasp of space. Let us take an example. Everybody knows that the diagonals of a parallellogram bisect each other. How do we know it? Well, you can prove it, by Euclid, by congruent triangles, and so on. But you knew in long before you learned formal geometry. How did you know it then? You do not remember? Perhaps you did not care. Why didn't you. Because your teachers did not care. But why should teachers not care? Ask a child: how do you know that these lines in that figure bisect each other? Every answer is welcome, even a wrong one, if it shows the child reflecting on his spatial intuitions. This is my ninth problem of mathematics education: Can you teach geometry by having the learner reflect on his spatial intuitions?

10. I am obliged to say something about calculators and computers. You would protest if I did not. I could refuse because I can prove I am incompetent. I know almost nothing about calculators and computers. It is a lack of knowledge that prevents me from tackling any minor problem of calculators and computers in mathematics education. It does not prevent me from indicating what in my view is a major problem. Technology influences education. The ballpoint, Xerox, and the overhead projector have fundamentally changed instruction. But this is as it were unintentionally educational technology. Programmed instruction, teaching machines, language laboratories, which were intentional education technology, founded on big theory, did not fare as well, to say the least of it. Calculators are being used at school and they will be used even more in the future. Computer science is taught and will be taught even more. How to do it - - these are minor questions. Computer assisted instruction has still a long way to go even in the few cases where it looks feasible. What I seek is neither calculators and computers as educational technology nor as technological education but as a powerful tool to arouse and increase mathematical understanding. Let me illustrate what I mean by a few examples. If you find they are trivial, or silly, please look for better ones. John and Mary are playing with their calculators. John starts at 0, Mary starts at 100. Alternately John adds 2 while Mary subtracts 3. Where will they meet? Or another: John starts at 0 and Mary at 100. Alternately John adds 3 while Mary adds 2. Where will John catch up? Or still another: John and Mary are asked to share 100 (say marbles) in the ratio 2 : 3. They will do it by alternating subtractions of 2 and 3 or multiples of them, while using their calculators. I hope you understand what I mean: discovering the laws governing ratio by numerical experiments, facilitated by calculators. It would be marvellous, indeed, if calculators, which know neither ratios nor fractions, could be helpful or could be even a key to understanding these fundamental mathematical concepts. With the provisos I have made before I would formulate my tenth major problem of mathematics education as: How can calculators and computers be used to around and increase mathematical understanding?

11. Is this not a marvellous display of problems of mathematics education? Show me a wonderful world where these problems might be solved, would be solved! Rather than in the armchair or in the laboratory, educational problems are solved in the educational process - - I claimed half an hour ago. If this is true, then in our real world solving them will be a slow process, a social process, a long learning process of the society. Can it be steered, can it be guided? Can there possibly be a strategy of change? I said change, rather than innovation. I do not like the term innovation. Innovation means newness. It suggests news as a necessary and sufficient condition for higher quality. In the sixties people believed in curriculum development as a strategy for change: curriculum prescribed by governmental decree or dropped as a new subject matter condensed in colorful textbooks on schools and classrooms, sold as easily in highly industrialized as in developing countries where for the majority of youth school life may end at the age of ten. Though I know almost nothing about developing countries, there are few things that shocked me more than the curricula sold to them. Curriculum development viewed as a strategy for change is a wrong perspective. My own view, now shared by many people, is educational development. This means an educationally integrated activity, aiming totally at education, rather than at details. Totally means: longitudinal, simultaneously at all levels, and viewing every subject area in its larger context. Isn't it an illusion, aiming at total education rather than at a number of aspects? Isn't is too many irons in the fire? no, aiming at education totally just means not overwhelming one's modest forces, not dissipating one's feeble energy. Only with a view of the whole can you discern the salient points, the nerve-fibres to influence education. My eleventh major problem of mathematics education is: How to design educational development as a strategy for change?

12. Where can you find the nerve-fibres to influence education? I will choose two extremes, the most conservative and the most progressive medium, the most powerful determinant of present and the most powerful of future education, that is textbooks, and teacher training. Let us start with textbooks. Teachers, most often, heavily depend on textbooks. I do not blame them for this lack of self-reliance. After three or four years of inadequate training, textbooks might be their last, their only resource.

You have been confronted here with ten problems of mathematics education that I thought of as major ones, and with the eleventh, of how to solve them. But from the start onwards I have warned you that problem solving in education is not a job for theoreticians but for the participants in the educational process. Textbook authors are participants, who in turn depend on the presumptive users of their production. Should I ask textbook authors to solve my problems? Of course not, even not to try it. The least and the most I can expect them to do is to ponder my problems. Perhaps they already did - - indeed it was not me who invented my problems. If they did, and if in some respect the succeeded, let it be known, not by guidelines and teacher manuals, which are often belied by the textbook, but by the textbook itself, by its built-in features. For instance, progressively schematised subject matter would be a good case, provided it is leading the teacher and learner, not along firmly preprogrammed paths but along reflection and retention of insight. Mathematising the environment is a hard thing to be taught by textbooks but just for this reason worth being tried. And so I could comment on quite a few among my problems.

I now turn to teacher training. Should I duplicate what I said about textbooks and ask teacher trainers as participants in the educational process to contribute to solving my major problems? Yes, I will, but that is not enough. Teacher students learning mathematics are expected to learn it as a didactical feature. In teacher training each of my major problems of mathematics education has its didactical counterpart, from why can Jennifer not teach arithmetic? via reflecting on one's own teaching, and developing a didactical attitude, to how can calcultors and computers be used to arouse didactical understanding?

Things are even more involved. Teacher students, in general, belong to the large group of adults where the sources of what they once learned by insight, have been clogged by the knowledge and skills they acquired meanwhile. To say it more concretely: they care neither about why multiplication by 100 is carried out by 'adding two zeroes' nor about that you can or why you should argue such piece of knowledge. So they have to undergo remediation: relearning such facts while teaching children and observing their learning processes. The higher the level of learning, the more paradoxical this conclusion may sound. Knowing a piece of mathematics too well may be a serious impediment to teaching it decently as long as the teacher is unconscious about the learning process that produced his excellence.

So he needs relearning by observing learning processes of less skilled people, of children. But now we are faced with one of the big problems in teacher training. Whereas one can easily arrange for observing short term learning processes, it is impracticable and hence impossible to do the like for long term learning processes in the school environment. Thought experiments, such as undertaken by textbook authors, cannot fill this gap if undertaken by unskilled people. Lack of experience in long term learning processes is the proper cause of the teacher student's future dependence on textbooks as their only sources for long term learning processes. How to solve this dilemma is a problem worth studying. Let me restrict myself to this one. Teacher training as a whole should be rethought and reshaped.

13. Educational development includes, amongst other things, educational research. Though I will not deal with any problems of educational research I promised you I would tackle educational research itself as a major problem of mathematics education. Perhaps you know that in the past I have severely criticised, if not castigated, educational research because of its irrelevance, and as a danger for mathematics education. What is called educational research represents an enormous production, still expanding both as to volume and variety. I admitted, and I still admit, that my knowledge, even if I include the most superficial kind, is restricted to a diminuitive part of this field. Although amongst the work I do know, there is little of good quality, it is still so large an amount - - and I should add that during the last few years it has increased by one order of magnitude - - that by extrapolating to the enormous quantity I am not acquanted with, I would guess that there ought to be a large amount of high quality educational research. But how to retrieve it, buried as it is under mountains of irrelevant and worthless production? Both in mathematical and educational research, production is high. The difference is that the retrieval of good and relevant research is easy in mathematics and almost impossible in education. I will not repeat any details of my well-known criticism of actual educational research, but for honesty's sake I feel obliged to say that since I first uttered it, the quantity of relevant research has sensibly increased, at least with respect to mathematics. Good research is, however, still in danger of being suffocated by the mass production of worthless, irrelevant educational research, which in addition is a danger to education itself.

There are people who claim physics is the curse of nature and biology is the curse of life. I do not believe they are right. Science is a good thing, for education, too, provided we see to it that educational research does not become the curse of education. I trust we will succeed. My hope is to set on educational research as part of educational development.

(I am indebted to J. Adda, H. Bauersfeld, A.J. Bishop, F. Goffree, A.Z. Krygowska, A. Treffers for critical remarks they made on the first version of this address even in those cases where I did not follow their advices.)

A version of this paper has appeared in Educational Studies in Mathematics, Vol. 12, No. 2, May 1981 pp. 133-150, Copyright 1981 by D. Reidel Publishing Co., Dordrecht, Holland.

1.3 YOUNG CHILDREN'S ACQUISITION OF LANGUAGE AND UNDERSTANDING OF MATHEMATICS

YOUNG CHILDREN'S ACQUISITION OF LANGUAGE AND UNDERSTANDING OF MATHEMATICS

Hermina Sinclair
University of Geneva
Geneva Switzerland

After a good deal of thought I decided to organize my paper as a function of a simple fact: children everywhere, at least in societies where they go to school, start learning to read and write and do arithmetic on paper around the age of six. How do children deal with this task, and what previous knowledge do they bring to it? Schools generally seem to regard the two tasks as independent one of the other; little attention is paid to possible confusions between the different systems that underlie spoken language, written language, spoken numerals and counting, and written numerals and computations.

My approach to the theme imposes limitations: in the first place, I will focus on research concerning young children. In the second place, I will deal only with that part of mathematics that is taught to children of, say, the first three grades. Speaking as a developmental psychologist, I will take the liberty of talking about the child" - a fictitious entity, or an epistemic subject, knowing full well that children are different in motivation, background knowledge, and - why not? - talent and interest; which makes a teacher's life rather more difficult than that of a researcher. I will also take the liberty of talking about "school" as an institution, knowing again full well that teachers are as different one from the other as the pupils they have to teach. Last but not least, the research I will refer to concerns only the <u>alphabetic</u> writing system, and only the nowadays <u>universal</u> number system with base 10 and place value.

Reading and writing on the one hand and written arithmetic on the other can both be considered to use particular notational systems. As such, they share some features, but they also present many differences.

Generally, it appears that educators see a big difference between written language and written arithmetic. Children can talk before they come to school, and it is throught that all they have to learn is to transpose spoken language into written language, and vice-versa. This transposition is generally regarded as a skill to be acquired by learning to spell. On the other hand, it is thought that children cannot add, subtract, multiply, or divide: this is what they have to learn, and the notation of these operations is not supposed to create any particular difficulties - if difficulties there are, they exist at the level of the operations themselves.

In other words, neither for literacy nor for written arithmetic are any conceptual difficulties pertaining to the notational system supposed to exist; in the case of literacy, phoneme-grapheme correspondences and their exceptions have to be learned in an associationist manner; in the case of paper-and-pencil computation it is the operations that have to be learned - a conceptual achievement. What is similar to both literacy and written arithmetic is thought to be simply perceptual skills of discrimination and motor skills of graphic accuracy.

These are the main assumptions that lead to the idea that the two subjects, literacy and arithmetic, can be taught simultaneously but independently one from the other, and that no conceptual confusions will arise, such as that between the number "5" and the shape of the letter "s" or the number "3" and the shape of the capital letter "E". And there are also the main assumptions I want to question, first as regards reading and writing, and then as regards arithmetic.

<u>Development of Literacy</u>

Several tacit assumptions that appear, in a general way, to underlie the teaching of reading and writing have to be questioned in the light of recent research carried out by Emelia Ferreiro (1980), with whom I have been associated for a long time.

The first of these assumptions is the following:

1. Children do not think about written language before they are taught to "read and write", either in school or by their family.

The second assumption is linked to the first, and has to do with the supposed "naturalness" or "inevitability" of alphabetic writing. It can be expressed as follows:

2. Children of school age already know how to talk, and thus the difficulty of learning to write and read resides in learning the various roles (and their many exceptions) concerning phoneme-grapheme correspondences, and in learning to produce letters graphically, in acceptable form (correct shape, orientation, and so forth).

To a developmentalist of the Piagetian school, the first assumption is highly improbable. Pre-schoolers have ideas about most of the objects and events they encounter habitually in their environment. Most children who are potential school-pupils live in an urbanized environment where they cannot but see much printed language (and numbers as well, but more about that later). Words appear on cans, posters, tee-shirts, street-signs, etc. Adults and older siblings provide scraps of information about the squiggles. Written language (and numbers) must be, in varying degrees according the the child's circumstances, an intriguing part of their familiar environmental. It would be surprising indeed if they did not have ideas about them.

It is, however, not so easy to find out what these ideas are. In the first place, children themselves cannot really ask questions about the writing system - generally speaking, one has to know quite a lot before being able to ask "an expert" a comprehensible question. And the "expert" does not know what questions to ask the child, since he does not know what the child's ideas are, and also since the child's reasoning is often not what we adults would call logical, even though it may be quite consistent within the child's own system.

The second hidden assumption, about the "naturalness" of alphabetic writing is its basic principles, and its difficulty in details of discriminating and producing shapes, is unacceptable for different reasons. Humanity took a very long time to work out the various writing systems, and apart from pictograms they all imply a certain degree of reflexion on language as an object of knowledge in itself. Most symbolic systems of whatever kind do not represent all the properties of whatever it is they symbolize and at the same time they have properties of their own. Our written system does <u>not</u> represent intonation, emphasis, speed of speaking, <u>etc.</u> It is based on a mental construct that has no objective existence: i.e., the phoneme. Neither auditive nor articulatory segmentations into single sounds exist; the only more or less natural unit in spoken language is the syllable. At the same time, texts in the alphabetic system have properties that do not exist in spoken language such as left-right, top-to-bottom (or vice-versa) order, and they introduce blank spaces between the words, without any regard for the degree of closeness of the link between the different words that follow one another in an utterance.

Let me give you some idea of the kind of facts Ferreiro uncovered and which contradict the tacit assumptions I mentioned. Already at the age of 3 or 4 many children have singled out letters (and numbers) from the other squiggles they see around them, such as decorations on wallpaper, patterns on cloth, etc. Some children may even have names for the particular squiggles that for us are letters; whether they do not not does not seem to change the developmental seqence. A two-and-a-half year-old I recently interviewed called all letters "aybeecees". forhim, these aybeecees do not mean anything, they are simply what they are, just as for adults lines and dots on a dress are simply lines and dots. By the age of four, many children, whether they use the adult term "letters" or not, have taken a big step: they now think that letters have a meaning. However, this meaning isnot determined by the sounds the letters represent for us, but by the nature of the object they appear on, or by the picture that accompanies them. A series of letters changes its meaning, and the way it is read, according to where it appears, or which picture accompanies it. If no such connection exists, the letters represent the intentions of the person who wrote them: only <u>he</u> can read them.

Indeed, we found to our surprise that many four-year-olds think that it is the postman who writes the addresses on the envelopes - - otherwise, how would he know where to bring them?

Thus, the first idea of letters carrying a meaning has nothing to do with letters as representing sounds. Curiously enough, children think at first that the meaning of the squiggles has to do with some quantitative property of what is symbolized: when there are three dogs in a picture, they think that there should be three squiggles in the writing underneath it, each squiggle representing one dog; they also may think that elephant should be written with more letters than butterfly, and that the name of a child three years old should have three letters. Thus, in the search for the meaning of written texts, the first link children think of is the domain of number and measurement, not in that of sound!

It is only after many further steps of reasoning about written material that children come to think in terms of a link between squiggles and sounds. At first, however, this link is thought to be syllabic: the child establishes a one-to-one correspondence between syllables and letters. For Spanish and French this proved to be a very tenacious theory, though it is continually bombarded by information from the milieu: conflicts arise especially in the case of proper names. In view of the fact that syllables are natural units of speech production, the syllabic theory is highly plausible; yet it should be emphasized that in an environment where writing is alphabetic, this idea, just as the quantitative ideas that precede it, is constructed by the children themselves - it does not come from information given by adults or older children.

This brief and partial account of an intricate developmental sequence clearly contradicts the common assumptions I mentioned in the beginning. Ferreiro's research leads to the following conclusions:

1. The writing system itself is, for children, an object to be conceptualized. For an early age on, children have theories about what letters symbolize and how the system functions.

2. There is nothing "natural" about alphabetic writing; children's theories about the alphabetic system start from something that is partly ideographic and proceed to syllabic and from their to alphabetic theories. The phoneme-grapheme correspondence is a mental construct which cannot be "simply" transmitted and taught in an associationist way. Children actively reconstruct the system, and in this reconstruction both endogenous ideas and information from the milieu play a part.

3. At some stage in their conceptualization of the system, children search for quantitative properties, such as number or size, in order to give meaning to the squiggles that to those who can read symbolize only sounds. In other words, what they think of as being present in alphabetic writing is in fact what belongs to the other notational system, that of numbers. It should be emphasized that this is a conceptual difficulty, and not one of perceptual discrimination.

Written Arithmetic

How about the teaching of arithmetic? Do schools make similar tacit assumptions about the child's previously acquired knowledge, and about his views on the fundamental nature of the notational system?

The answer seems to be yes and no. The idea that children have not thought about written numbers before they are taught, either at home or in school, is widespread, just as for written texts. The assumption is no more acceptable for numbers than it is for letters. Many very young children consistently differentiate numbers from letters: The same two-and-a-half-year-old who calls letters "aybeecees" calls numbers, when they appear in a book or a newspaper, "dollars". He does not attribute any meaning (such as "price" or "money") to the numbers, not does he link them to the real objects he is familiar with and which he calls "pennies" or "dollars" or "money". Numbers are just called "dollars". The number on his house is "my house" and a big number printed on a tee-shirt seems to have something to do with running or games. It only needs a quick look around for us to realize that information from the environnment seems to induce the idea that there are numbers and numbers, with different meanings, determined by the nature of the objects to which they are linked. On some packages the number is a cardinal number: for example, 50, on a package of coffee filters. On shoes, and articles of clothing, number indicates size. On cans, number indicates weight or volume. And what about clocks, calendars, petrol pumps, and so on and so forth? For those who know about numbers, these different uses pass unnoticed: for young children, it is confusing information, and they try to make sense of it as best they can. They do not decide to ignore it all until they get older! Moreover, there is evidence that the confusing use of numbers as cardinal or ordinal numbers or in measurement in urban environments appears to continue to influence children's thinking even at school age. In Geneva, urban bus-lines are numbered 1 to 9. Asked about the meaning of these numbers, we found several 6- to 7- year-olds who explained the number 2 bus in the following ways: "It's called 2 because it always has a trailer, so it's really two buses stuck together"; and "It's called 2 because there are two sorts, some run in town and others go out to the beach". Both explanations are based on fact.

On the other hand, the second assumption make for reading, that is, that the child knows how to talk and merely has to learn to put speech down on paper, has no counterpart in mathematics. Nobody seems to think that children already know how to add, subtract, multiply, and divide before they come to school, and that all they have to learn is to do pencil-and-paper sums. On the contrary - in most countries arithmetic is taught as if the conceptualization of arithmetic operations were the same as their written symbolization. Schools do not seem to envisage that the conceptualization of addition, subtraction, etc., may be a cognitive task separate from that of writing equations, and that the letter may present difficulties of its own.

Consequently, there are two still very widespread tacit assumptions that I think should be questioned, and that can be formulated as follows:

1. Before the age at which children generally are taught arithmetic at school, they have no capacity for dealing with numerical operations.

2. Being able to use the notational system for arithmetic operations is no different from understanding them.

It may surprise you that I want to question the first assumption as it is often thought to be supported by, or even to stem from, Piaget's number conservation task. The argument goes: children who do not know that the cardinality of a collection of discrete objects is independent of spatial arrangement cannot do arithmetic. Such children have no "number concept" - - Piaget dixit. Though numerosity-conservation is an important acquisition, it is certainly not the conceptual watershed it is often made out to be. I will first quote some evidence showing that certain important ingredients of number concepts are present before conservation, and, second, that long after conservation is acquired, some apparently equally important ingredients are still not mastered. These were the experimental results that led Piaget and his colleagues to speak about the "slow arithmetization" of number - a process that takes many years.

Eivdence for pre-conservation number concepts comes from a series of experiments Piaget and Inhelder published in 1963 (in Greco, 1963). The child is asked to put one bead in one container and another bead in another container, doing both actions at the same time, using both hands. After a while (well beyond the point where the child might actually have counted), he is asked whether the two containers have the same number of beads in them. This basic situation can be varied: the two containers may be made of glass, and may have different diameters (so that the beads will reach a higher level in the narrower one). Instead of beads for each hand, the child can use beads in one hand, paperclips or pebbles in the other, etc. The experimenter can also, instead of beginning with empty containers, put a few beads into one of the containers before the child begins his actions - - in this case, the correct answer is, of course, that the number will always be unequal. The remarkable finding of these experiments is not only that five-year-old children draw the correct inferences from the iterated actions, but that they spontaneously explain, for example: "if we had put 2 beads in one glass, and I in the other, it would be different; but because I always put in one and one, it's the same." In other words, the children are capable of constructive generalizations: having seen the begining - - one and one - - and maybe also two and two or three and three, they can state that if the equality (or the inequality) is true for a number N, it is also true for N'; but, even more remarkable, they are also capable of seeing that what is true for a number N is necessarily true for N + 1. Many subjects are capable of extending this reasoning to a quantity they have not constructed in action: when asked what would happen if one went on putting beads in the 2 glasses for a very long time, "the whole day long", they say that it does not make any difference, that the quantitites will always be the same as long as one carefully puts one bead here, and one bead there, and no "mistakes" are made. But surprises are not excluded - - one child added, "but if you did it also the whole night in the dark, then you can't know, because nobody would see it."

The precocity of this type of reasoning concerning numerical quantities appears to be due to the fact that in this particular case a coordination between actions and results is easily made. The union between successive actions and their cumulative effect prefigures the numerical synthesis between a cardinal union of equivalent elements and an ordinal succession of adding actions which allows the distinction between elements according to their rank-order in the succession.

By contrast, other experiments show the difficulties even 9- or 10-year-old children have with problems that bear directly on the properties of the series of natural numbers. Matalon (1963) asked the following questions in an experiment on children's understanding of odd and even numbers:

1. I think of a number, I won't tell you which, I add one to this number. Can I divide what I've got now into two?

2. I add one once again (to what I got when I added one). Can I divide what I've got now into two?

For these questions almost all the subjects (until the age of 9 or 10) take an example and generalize immediately: "Maybe you took three, add one, makes four, yes, you can divide. Sure, you can always divide into two." Sometimes the children themselves think of a second example, which may or may not be a counter-example; if it is not, the experimenter suggests a counter-example. Until the age of 9, the subjects usually simply change their minds: "Oh no, it doesn't go", or "Oh yes, I made a mistake, it does". The 9- and 10-year-olds conclude that one cannot know, sometimes it works, and sometimes it doesn't. The 10- to 12-year-olds either discover after some examples that it works if one starts with an odd number, or they spontaneously announce right away that it depends on whether one starts with an even or an odd number.

In other words, a good proportion of the 10- and 11-year-olds, who, by the way, were all doing quite complex arithmetical work in school, could not yet base their reasoning on the apparently simple notion of a regular alternation between even and odd numbers, and could not deduce its implications for divisibility into two.

Many other experiments yield equally surprising results. Clearly, the number concept is gradually constructed: the process takes many years, but it starts very early. By only focusing on certain properties of the number system, such as counting and conservation, schools both overestimate and underestimate children's conceptual capacities for dealing with numerical operations. Matalon's research also showed that the absence of a conceptual grasp of the implications of the even-odd alternation is no hindrance to success in quite complex school work.

What evidence is there to reject the second assumption, which tacitly supposes the existence of psychologically immediate logical and transparent links between arithmetical operations and their notation?

I have discussed some of the unexpected ways, discovered by Ferreiro, in which young children interpret the alphabetic system. On the numerical side, unfortunately, no study comparable to Ferreiro's is as yet available. Some recent research on this question strongly suggests, however, that the assumed obviousness of the link between operations and notations is quite unwarranted.

A study to Sastre and Moreno on children's spontaneous ways of symbolizing numerical quantities brought to light several different levels of representation. The children were interviewed on various tasks, in pairs. In one of these tasks, the experimenter placed a number of pieces of candy on the table in front of the children, and asked them to use paper and pencil to show how much candy there was "so that your friend will be able to use your paper to put out just as many". The following levels were found:

1. any drawing, having nothing to do with the numbe of objects; a house, a tree, a flower, a person, etc.

2. the drawing of one object which has number chanracteristics similar to that of the number of elements presented: for example a hand for 5 elements presented, an octupus for 8, a flower with 6 petals for 6, etc.

3. the same number of drawings as elements: 5 people or 5 trees, for example, for 5 elements

4. a drawing of the objects themselves, in their correct number: five pieces of candy

5. numerals 1, 2, 3, 4, 5, 6, etc., as counting numbers; i.e., 1, 2, 3, 4, 5, 6, written down for 6 elements

6. a single numeral, 5, 6, etc., written to represent the number of elements as a cardinal number.

When children were explicitly asked to use numbers, most of the 6- and 7-year-olds used representation level 5, and most of the older children gave a single number (level 6).

When the experimenter performed actions like giving 3 pieces of candy to a doll, then adding two more, or giving 6 and then taking away 4 pieces, and the children were asked to use paper and pencil so that the other child could give candies to the dolls in the same way, very few of the children used plus or minus signs. The invented other methods, such as crossing out, drawing a hand to symbolize adding or subtracting, or they used words like add, take away, and combined these methods with the representation levels described above. They u.sed tally-marks, crosses, etc.: when looking at their productions in this task and others proposed by Asstre and Moreno, one is struck by the resemblance to what we know about the history of written number systems: though many of the children had been working for years with the modern system, they were busy re-inventing it. All these children were perfectly capable of doing sums in school - some of them were already working at long division. But using numbers and plus and minus signs to symbolize actions with objects certainly did not seem to be "natural" to them.

Similar phenomena have been uncovered by Constance Kamii (1980) working on children's interpretations of written additions and subtractions. What do 6- or 7-year-olds make of the horizontal way of writing equations, and how do they interpret the "+", "-", and "=" signs? Kamii showed that the "=" sign is by far the most difficult to interpret, followed by the "-" sign; the "+" sign is apparently mastered before either of the others. However, as long as the "=" sign is not understood, a good intuitive interpretation of the sign "+" is no help in understanding addition equations.

Another study on the conceptualization of the notational system is being carried out by Mieko Kamii on the subject of children's understanding of place-value. All teachers know, of course, of the difficulties in teaching place value. However, until Kamii started her study, no systematic research was carried out on the way children interpret the spatial order of written numerals beyond 10. Kamii's results show the many curious ways children interpret the relative position of grouped numerals, some influenced by spoken numerals, others, unfortunately, by teaching methods such as Venn-diagrams and set-theory, still others, apparently, by systems worked out by the children themselves.

To sum up, just for the teaching of reading and writing, the largely tacit and widespread assumptions on which the teaching of written arithmetic is based appear to be incorrect. Available research suggests the following conclusions:

1. Children's acquisition of number concepts and numerical operations (as apart from the notational system) is a lengthy process, stretching over many years, of which many details are still unknown.

2. Just as for written language, young children have theories about written numerals and equations. These theories may not fit the symbolical nature of the systems elaborated by humanity over many centuries. In both domains children seem to re-invent or to re-construct the systems their societies have adopted.

If, as I have argued, children actively reconstruct the notational systems that symbolize spoken language and numerical operations, we should compare the two notational systems as to their differences and similarities, with a view to helping children avoid unnecessary confusions and so as to profit from progress in one system when building up the other.

Similarities, some not at all trivial, exist between natural language, both written and spoken, and mathematics. It has often been argued, for example, that in language we can distinguish elements (such as nouns) which often, though not always, symbolize persons or objects, and elements (such as verbs) that often (though not always) symbolize actions or relationships. In arithmetic, there are numbers: 3, or 27, that are like nouns; and signs: +, -, =, that are like verbs. Rules exist for the combination of elements in a sentence, just as they exist for writing equations: "Peter the slowly house" is no more a sentence than 5 = 7 x - is an equation. Some sentences are well-formed but incorrect: "San Francisco is the federal capital of the U.S.A."; some equations are well-formed but wrong: 7 + 4 = 13. And, maybe more important, there are an infinite number of well-formed sentences, just as there are an infinite number of equations; competent speakers and mathematicians can understand and produce sentences and equations they have never heard or seen before,. Joining nouns with the conjunction and can be repeated indefinitely, and so can adding a number N to a previous sum. There is no maximum number of words in a sentence, and there is no limit to the length of an equation - theoretically speaking.

However, the differences between arithmetical operations and verbal utterances are also very striking. spoken language in an everyday setting is highly context-bound, full of ambiguities and redundancies, as well as elliptical and pre-suppositional. Written texts are less context-bound, and for certain styles and subject matters ambiguities and ellipses are avoided. However, in structure, spoken and written language are basically identical, and their structural properties make them our supreme tool for their main purpose, namely communication of very varied subject matter. Possibly, it is precisely this communicative purpose which determines their structure. Mathematics, on the other hand, though it is sometimes called the language of number and size, is mostly explicit and unambiguous, non-elliptical and non-redundant. Much of its structure is very different from linguistic structure, and though it transmits information, its main function is not communicative in the large sense. Moreover, through written numerals are all around those of us who live in an urbanized environment, just as letters are, equations are not - - theyonly figure in specialized texts.

However, natural languages also have means of quantifying, which do not need numerals: one can speak (or write) of many, a few, a lot, more, less, etc. Numerals can be inserted into these quantifying procedures in the spoken form, in the numerical notation, or with letters, just like any other word.

Different languages may treat their quantifying (measuring and counting) elements in various ways, but in all languages linguistic constraints lead to curious, often ambiguous expressions. For speakers of Indo-European languages, the so-called numerical classifying languages are curiosities as far as their quantifying means are concerned. These languages appear not to make a difference between measuring and counting, and to treat all quantification in the way Indo-European languages treat measuring. For example, in Japanese, which is such a language, one does not simply say two books, two persons, two dogs, but for the latter, dog-two-non-human. In many Indo-European languages, something like this is done in special cases: three slices of bread, three pieces of cheese, three members of the committee. However, many inconsistencies exist: in English, for example, if one wants to order three cups of tea, "three teas" will not do" but "three cokes" is just as good as "three bottles of coke" (in French, e.g., "trois vins", en constraste avec "trois bieres"). Moreover, though precise, or approximately precise measures are used in natural language "three gallons of petrol", "three yards of material", etc., other, linguistically identical expressions have no precision at all: a glass of beer, a glass of wine or a glass of brandy do not contain the same amount of liquid. When talking about collections or groups most languages have specific expressions according to what is grouped: a herd of sheep (troupeau) but not a herd of fish; a group of people, but not a group of eggs.

Though for adults and older children these inconsistencies raise no particular problems, they must be very confusing for young children who try to make senseof what they see and hear.

Consider, for example, one particular problem: that of the different role played by order in spoken and written language on the one hand, and spoken and written numerals on the other. Already at the most superficial level, that of the way one says numbers above ten and the order of the written digits, there is a lack of correspondence in many languages. One cannot recognize ten-one in eleven (or in onze, or elf). In English, seventeen, where seven is recognizable, has the opposite order compared to twenty-three; in many languages this continues for higher numbers, as in Dutch or German where 43 is spoken and three-and-forty. Other languages (French, for example) make use of twenty in the spoken numerals: quatre-vingt, four-score or four-twenty, meaning eighty.

What is the child to make of all this? As we have seen, children between 4 and 6 or so are still busy working out the alphabetic writing rule that left-to-right spatial order corresponds to temporal order of utterance at the phoneme level as well as at the word-level. The order of letters and words in a written text does not introduce any other meaning than just that. However, in written numerals it does: thanks to the place-value idea, the 3 in 324 "means" three hundreds, whereas the 3 in 34 means three tens.

In spoken language, the problems of order and its function in determining meaning have been roughly worked out by children well before they start being taught writing and arithmetic. In languages such as English or French, where word order indicates functional roles such as agent and patient, children of two-and-a-half already know that "Peter hit Paul" means something different from "Paul hit Peter". Other problems, such as that in English the word hammer has a different meaning when it follows an article than when it follows the functor to (a hammer/to hammer) or examples in French such as un sourire/souriere or le diner/diner, are probably solved as early, though this has not be substantiated by research as it has been for the agent-patient roles.

In alphabetic writing, the function of order has to be worked out anew. Ferreiro showed that for young children order does not seem necessary, neither for the letters inside a word nor for words inside a sentence: papa is thought of as two p's and two a's, and "paap", "ppaa", or "apap" are all pronounced "papa"; an utterance such as "the duck swims in the water" may well be thought to be written down as duck water swim.

For the writing of numerals, the problem of order is made even more complex by the fact that about at the age where children are introduced to written numerals above 9, they have just come to the conviction that different spatial orderings of objects do not change the numerosity of the collection. So what is so important about writing sixteen as 16 rather than as 61? Sixteen blocks remain sixteen, whether they are lined up, spread out, or built into a tower; as children have just discovered, it does not matter in what order you count them as long as you count each block only once and do not skip any.

From the point of view of order, equations raise particularly difficult problems. The importance of order is not the same in addition as in subtraction, or in multiplication and division. I do not see how equations can be "spoken" in any consistent way, keeping order not only constant, but also consonant with word-order rules of natural languages. Saying something like "two plus three equals five" is no more than an oralization of the equation, and does not give a meaning of adding or of a separation; and an interpretation which seems sensible, such as "if you count to two, and then go on counting three more, then you get to five" introduces an if . . . then structure which is absent from the numerically written form, and, moreover, stays inside the counting system which many educators nowadays do not think appropriate for young children.

So, it appears that, theoretically speaking, there are similarities but also important differences between spoken and written numerals and equations. Moreover, from a practical point of view, quantifying expressions with or without numerals are part of language in the simple sense that they can be talked about. For young children, who are simultaneously taught to read and write and to do arithmetic, this mixture raises problems, and they have difficulty in understanding the difference in the essential nature of the link between the symbols and what is symbolized in the two systems. Clearly, in the process of re-constructing the link between alphabetic writing and spoken language, children look for meaning, that is to say, for some reference outside the notation itself. What happens in their efforts to re-construct numerical notations? For simple numerals, the question is not so very difficult: at an early age children look for some kind of quantifying meaning, even though theyhave trouble separating the different uses of numerals (cardinal, ordinal, etc.). But what about equations? Do they have a "meaning"? I realize that with these last remarks I am - almost - entering an epistemological controversy about the nature of mathematics. However, I will avoid this issue by speaking only as a psychologist, even if I do not think that epistemology and psychology should be separated in principle.

It seems to me that young children can only learn arithmetic if they can attach meaning to numerals and equations. Arithmetic, like reading and writing, has to do with the extraction and construction of meanings - at least for children. The difficulty lies in deciding what meaning equations can have for young children. A simple translation into words is no help. From all we know about children as constructors of knowledge, mathematical meanings are constructed as action-patterns, first on real objects and later interiorized. However, much research and much careful observation is still necessary on this last point.

Concluding Remarks

If we accept that very young children, already at the age of three or so, have ideas about the letters and numbers they cannot help but see around them and that fascinate them, and that they try to make sense of their observations, we have to ask ourselves how we can help them in this difficult task. It seems to me that in the first place we ourselves should become more aware of the confusing nature of the various kinds of information the child encounters in his everyday life. Instead of letting confusions arise, we should make both systems support one another. Secondly, much more research should be devoted to the theories the child elaborates about the notational systems. Thirdly, especially for arithmetic, it should be realized that the conventional notational system of written numerals and equations has to be conceptualized by the child as well as the nature of the quantifying operations it symbolizes. Most importantly, the question has to be asked: "What can you do with numbers and letters; how can youa make them work?" Unless the child can bring together his ideas on numerical operations and his ideas on the notation, he cannot make sense of the latter, and will be reduced to either learning to perform tricks correctly, or to becoming despondent and hostile.

Despite all the difficulties and obstacles, many children happily learn their three R's anyway. But there will also be many for whom the world of books and communicating by writing will never unfold, and who at any suggestion of mathematics will ever turn a deaf ear. It is my hope that a better understanding of what the learning process entails will help us to increase the contribution teaching can make towards a creative adult world.

References

Ferreiro, E. What is written in a written sequence? Journal of Education, Boston University, Vol. 4, No. 4, 1978.

Ferreiro, E. & Sinclair, H. L'enfant et l'ecrit. Medecine et Hygiene, 37, No 1350, 3530-3536, 1979.

Ferreiro, E. & Teberosky, A. Los systemas de escritura en el dessarollo del nino, Mexico, Siglo XXI, 1979.

Greco, P., Inhelder, B., Matalon, B. & Piaget, J. La formation de raisonnements recurentiels. Paris, PUF, 1963.

Kamii, C. Equations in first grade arithmetic: A problem for the "disadvantaged" or for first graders in general: Paper presented at the Annual Meeting of the American Research Association, Boston (Ma.), April 1980.

Kamii, M. Place value: Children's efforts tofind a correspondence between digits and numbers of objects. Paper presented at the Tenth Annual Symposium of the Jean Piaget Society, Philadelphia (Pa.), May 1980.

Sastre, G. & Moreno, M. Representation graphique de la quantite. Bulletin de psychologie de l'Universite de Paris, 30, 346-355, 1976.

THREEKS, RAINBRELLAS AND STUNKS

Reaction to Hermina Sinclair's
Plenary Lecture

William Higginson
Queen's University
Kingston, Ontario
Canada

I find myself with very little to oppose and a great deal to support in Professor Sinclair's stimulating presentation (1980). She is a psycholinguist and I am a mathematics educator. Because of this difference in academic orientation we construct slightly different meanings for the same phenomena. A reasonably accurate characterization of this paper would be that of a short complementary statement which reflects a shift in professional orientation and concomitant change in emphasis.

The issues and obervations Dr. Sinclair has brought to us are numerous and significant. The constraints of space are such that comments can be made on only a few of these. First, I wish to make some remarks about some general context of her paper, the relationship between mathematics and language, and its role in human culture. Second, I want to underline what I feel is a particularly important theme running through her remarks and, finally, I wish to note a few of the implications of her thesis for mathematics education.

Mathematics educators would seem to have come to a new awareness of the significance of language in the creation and transmission of their discipline. There would appear to be, for instance, a much stronger linguistic component in the programme of this conference then in any of the previous three. As is often the case, it seems somewhat surprising, now that we have arrived, that we didn't come sooner. Seen from close at hand the importance of the area is quite staggering, for a strong case can be made that in mathematics and language we have the two outstanding products of that ability which characterizes our species better than any other. Homo Sapiens is, as Cassirer (1970) argues, more accurately, Animal Symbolicum, the beast whose uniqueness resides largely in her ability to create and utilize symbols. Furthermore, in the opinion of some contemporary thinkers (such as Waddington, 1977), the only possibility for escape from the threatening situation our world finds itself in at the moment lies in a rigorous application of this ability which has been fundamental in creating the threat in the first place. In Dawkins' (1978) terms, only our memes (or conceptual units of cultural transmission) can save us from our genes. If these views have even the most modest of claims to validity surely we need look no further for a rationale for a focus on mathematical and linguistic activity as a cornerstone for any educational endeavor worthy of the name.

In looking for insights on this issue we can turn to the work of some of the best mathematical minds. Descartes had ambitious plans for a universal method (mathesis universalis) as did Leibniz for a universal language (characteristica universalis). In more recent times this search has been continued by Freudenthal, among others. In the nineteenth and twentieth centuries logic has frequently provided a bridge between mathematics and language (Higginson, 1978) as is evidenced in the writing of Boole, Peirce, Frege (Dummett, 1973) and Quine. The recent revival of interest in Wittgenstein's philosophy of mathematics (Wright, 1980) with its links to his work on language games and the social negotiation of meaning is very promising.

Mathematical terms and symbols have provided a rich base for the structural, historical and geographical aspects of the study of language (Cajori, 1974; Menninger, 1969). The activity of young children in creating numerical representations provides, as Dr. Sinclair skillfully documents, a compelling case for the validity of the controversial recapitulation thesis of Haeckel which contends that ontogeny recapitulates phylogeny, that is, that the growth experience of the individual retraces the growth experience of the species. Poised as we are on the edge of an information revolution (Toffler, 1980), it seems clear that much of the impact of the computer society will have to do with significant shifts in our attitude toward the use of mathematical symbolism and imagery (Iverson, 1972; Papert, 1980a, b).

A second point to which I wish to draw attention is the image, latent in Professor Sinclair's paper, of the child as an active constructor of knowledge. Given her prominent role in the work of the Genuvan school, this view is perhaps not one which should surprise us, but its significance is major. For, contrary to popular belief, young children do not arrive at school as mathematical 'blank slates.' As with other parts of their world, children struggle, from a very early age, to create meaning for mathematical concepts when they encounter them in their day-to-day lives. In addition, and perhaps even more significantly, this struggle to create meaning is a slow and gradual one which no one ever completes. Adults, and teachers are a particularly important subgroup here, are continually engaged in this same exercise. The traditional concept of mathematical knowledge, that of inert, 'basic,' impersonal and universal 'facts,' would now seem to be mistaken in some important ways.

Not only have we underestimated and failed to utilize the informal mathematical experience of young children (Gelman, 1979; Higginson, 1980a) but we have been insufficiently sensitive to the highly arbitrary nature of many conventions of formal mathematics and the difficulties that these present to young learners (Ginsburg, 1977; Higginson, 1980b). We really have no excuse, for instance, for accepting 'problems' such as "___ - 1 = 1". Accustomed as we are to the symbols and trappings of this game we may feel that the answer, '2,' is quite obvious but a seven-year old who sees only six sticks (used in four different ways) deserves our sympathy.

The commonalities and differences between the informal and formal stages of mathematics and language are informative. The role which numerical features can play in object-definition is indicative of their primal importance. Consider, for example, the child who exclaimed upon seeing a fish fork (which has three tines) for the first time, "see the threek." The construction of novel words is perhaps our best clue to the fact that, intellectually, children are active generators rather than passive imitators. Accordingly, we have the ingenious and amusing results of the wresting of sense from situations. Skunk is transformed to 'stunk' and umbrella to 'rainbrella'; from the verbal patterns perceived by the child we get 'fiveteen', twenty-ten' (for thirty), and 'fivety'.

I wish to conclude this brief reaction with some observations on the implications of Professor Sinclair's paper for mathematics education. The foremost challenge she delivers to teachers, teacher educators, researchers and curriculum developers is that of extrapolation. Dr. Sinclair's research has led her to some very valuable insights about the relation between language and mathematics. She has, however, concerned herself almost exclusively with young children and with arithmetical (almost pre-arithmetical) ideas. When we as mathematics educators turn to a consideration of these issues in the context of other branches of mathematics or of older children (who constitute the very great majority of our clients), we find ourselves with a very thinly-stocked research cupboard. The classical psychometric ("dipstick") model of research has provided very little which will help us understand the interactions between mathematics and language in the mind of the older learner. The recent shift toward clinical studies is a move in the right direction, but a great deal of work remains to be done.

What seems likely is that affective considerations become very important for many mathematics learners after experience in school mathematics programmes. With so much emphasis placed on manipulation and on the production of 'right' answers (Stake and Easley, 1978), it is understandable that many school children come to associate mathematics with failure, frustration, and irrationality (Erlwanger, 1973). Perhaps we can learn from our knowledge of the natural cognitive patterns of younger children how to structure more effective learning experiences in mathematics for older children. Certainly we will have to give much more prominance to linguistic factors than we have in the past. At the very least we must make teachers aware of the problems created by the unconscious use of ambiguous terms. Used carefully, ambiguity can be a powerful pedagogic tool, but to use 14 different terms of addition (Preston, 1978) can be seen as nothing other than an invitation to confusion. It seems likely that children who make responses like, 'twelve take away two equals one,' have more of a local logic operating than we have previously given them credit for. The model we should move toward in this respect has been outlined by Easley and Zwoyer (1975) and"teaching by listening-toward."

Professor Sinclair has pointed out some of the common misconceptions about informal and formal mathematics as far as young children are concerned. These misconceptions and the pedagogic problems which result from them have analogs at several other levels, for there has been a long-time tendency to link pedagogy to product rather than to process. Polya has been among the distinguished mathematicians to criticize this tendency. We might hope that the recent influence of Lakatos (1976) might be sufficient to help reverse this trend.

References

Cajori, Florian. A History of Mathematical Notations, Volume I: Notations in Elementary Mathematics. LaSalle, Illinois: Open Court, 1974.

Cassirer, E. An Essay on Man: An Introduction to a Philosophy of Human Culture. New York: Bantam, 1970.

Dawkins, Richard. The Selfish Gene. London: Paladin, 1978.

Dummett, Michael. Frege: Philosophy of Language. London: Duckworth, 1973.

Easley, J.A. and Zwoyer, Russell. Teaching by Listening-Toward: A New Day in Math Classes. Comtemporary Education, 1975, 47, 19-25.

Erlwanger, H. Benny's Conception of Rules and Answers in IPI Mathematics. Journal of Children's Mathematical Behavior, 1973, 1, 7-26.

Gelman, R. Preschool Thought. American Psychologist, 1979, 34, 900-905.

Ginsburg, H. Children's Arithmetic: The Learning Process. New York: Van Nostrand, 1977.

Higginson, William. Language Logic and Mathematics: Reflections on Aspects of Research and Education. Revue de Phonetique Appliquee, 1978, 46-47, 101-132.

Higginson, William. Berry Undecided: A Digital Dialogue. Mathematics Teachng, 1980 (a), 91, 8-13.

Higginson, William. Upstaging Piaget: The Child as Dialectical Constructivist (231-236). Proceedings of the Fourth International Conference for the Psychology of Mathematics Education, (R. Karplus, Ed.). Berkeley, CA: Lawrence Hall of Science, University of California, Berkeley, 1980 (b).

Iverson, Kenneth E. Algebra as a Language. Appendix A (325-337) in Algebra: An Algorithmic Treatment. Menlo Park, CA: Addison, Wesley, 1972.

Lakatos, Imre. Proofs and Refutations. Cambridge: Cambridge University Press, 1976.

Menninger, Karl. Number Words and Number Symbols: A Cultural History of Numbers. Cambridge, MA: MIT Press, 1969.

Papert, Seymour. The Computer as Carrier of Mathematical Culture. Plenary Address, Fourth International Congress on Mathematical Education. Berkeley, California, August 1980 (a).

Papert, Seymour. Mindstorms. New York: Basic, 1980 (b). To appear.

Preston, Mike. The Language of Early Mathematical Experience. Mathematics in School, 1978, 7, 31-32.

Sinclair, Hermina. Young Children's Acquisition of Language and Understanding of Mathematics. Plenary Address, Fourth International Congress on Mathematical Education, Berkeley, California, 1980.

Stake, Robert E. and Easley, Jack A. (Co-Directors). Case Studies in Science Education: Volume I: The Case Reports. Urbana, Illinois: Center for Instructional Research and Curriculum Evaluation, 1978.

Toffler, Alvin. The Third Wave. New York: Morrow, 1980.

Waddington, C.H. Tools for Thought. London: Paladin, 1977.

Wright, Crispin. Wittgenstein on the Foundations of Mathematics. Cambridge, MA: Harvard University Press, 1980.

1.4 SOME EXPERIENCES IN POPULARISING MATHEMATICAL METHODS

SOME EXPERIENCES IN POPULARISING MATHEMATICAL METHODS

L.K. Hua
Academia Sinica
Peking, China
H. Tong
University of Manchester
Manchester, U.K.

Summary

This report is a summing-up of some personal experiences so far accumulated in the on-going course of the popularisation of mathematical methods in the People's Republic of China. Some techniques and methodology currently in use both inside and outside China are critically reviewed.

Introduction

Starting from the mid-60's, apart from continuing my theoretical research, I was actively involved in popularising mathematical methods to the general public in China. Altogether, my assistants and I have visited twenty-two provinces, hundreds of cities, and thousands of factories, meeting millions of people. Such was the magnitude of my personal involvement. To sum up experiences accumulated in this still on-goping course, the following principles will form the basis of the discussion: 1. "To Whom?" or "For What Purpose?", 2. "Which Technique?", 3. "How to Popularize?" Some elaboration of these three principles is now in order.

1. To Whom?

Specialists or experts and factory workers do not often share a common language. My experiences have shown me that in order to achieve a common language, these two groups must look for their common need. Needless to say, it is unrealistic to expect the factory workers to be interested in infinite dimensional spaces, although these objects are important for the mathematicians.

2. Which Technique?

It has been my consistent aim to present users with the most efficient technique. Towards this end, my experiences have shown me that a deep understanding of the theory behind each technique is an absolute prerequisite, lest one might be completely misled.

3. How to Popularize?

My experiences have shown me that it is essential to start work on a small scale, for example, in a workshop. If my suggestions turned out to be effective they would naturally attract wider attention and might soon spread to other workshops, and then to the whole factory or even to a whole city or a whole province. In this way, my assistants and I sometimes had more than a hundred thousand people in the audience. (Of course, they had to be divided into smaller groups and duplicated lectures were given simultaneously by members of our team.)

In the course of my endeavour to select appropriate mathematical techniques for popularisation, I have discovered that it is not uncommon that an established technique is either misused or its fundamental assumptions are conveniently forgotten. I shall mention some such examples later in this paper.

Suggestions from other Branches of Science

The problem of how to find the surface area of a mountain from a contour map has been of considerable interest to geologists and geographers. A method often used by the former is the so-called Bauman method and that by the latter the so-called Boakob method (1). We may briefly describe these two methods in the following way.

Let ℓ_r denote the contour of height $r\Delta h$, $r = 0, 1, \ldots, n$, Δh being a positive constant and ℓ_n corresponding to the highest point of the mountain of height h, say. Let W_i denote the area on the map between ℓ_i and ℓ_{i+1}.

a. Bauman's Method

Surface area = B_n, where

$$B_n = \sum_{i=0}^{n-1} \sqrt{W_i^2 + c_i^2},$$

$$c_i = \tfrac{1}{2}(|\ell_i| + |\ell_{i+1}|)\Delta h,$$

$|\ell_i|$ denoting the length of the contour ℓ_i.

b. Boakob's Method

Surface area = $V_n = W \sec\alpha = \sqrt{W^2 + (\Delta h \,.\ell)^2}$, where

$$W = \sum_{i=0}^{n-1} W_i, \qquad \ell = \sum_{i=1}^{n} |\ell_i|, \qquad \text{and} \qquad \tan a = \frac{\Delta h.\ell}{W}.$$

We may raise two obvious questions, namely (i) do they converge to the real surface area, A, say, and (ii) if so, which method is better?

It turns out that

$$A \geqslant \lim_{n\to\infty} B_n \geqslant \lim_{n\to\infty} V_n.$$

The proof seems to be interesting although quite simple. Using cylindrical polar coordinates, (ρ, θ, z) where z denotes the contour height and the highest point is chosen as the centre, we have, from the equation

$$\rho = (z, \theta)$$

for the surface, the area of the surface,

$$A = \int_0^h \int_0^{2\pi} \sqrt{p^2 + \left(\frac{\partial p}{\partial \theta}\right)^2 + \left(p\,\frac{\partial p}{\partial z}\right)^2}\; d\theta\, dz$$

Introducing a complex valued function

$$f(z,\theta) = -p\,\frac{\partial p}{\partial \theta} + i\sqrt{p^2 + \left(\frac{\partial p}{\partial z}\right)^2},$$

we may write

$$\lim_{n\to\infty} V_n = \left|\int_0^h \int_0^{2\pi} f(z,\theta)d\theta\, dz\right| \leq \lim_{n\to\infty} B_n = \int_0^h \left|\int_0^{2\pi} f(z,\theta)d\theta\right| dz$$

$$\leq A = \int_0^h \int_0^{2\pi} \left| f(z\theta)\right| d\theta\, dz\,.$$

This result shows that neither method can give a good approximation of the surface area and it also points the way towards a more satisfactory approximation.

This example clearly demonstrates the importance of a sound theoretical analysis. However, in view of the fact that the above problem is of interest to only a limited audience, it is certainly not a subject for popularisation.

Optimisation

1. Meshing Gear-Pairs

It was in 1973 when my assistants and I visited the city of Loyang in central China on one of our popularisation missions. A worker in a tractor factory there presented us with the following problem.

"In designing two gear-pairs, we have to find to optimal gear ratios with the numbers of teeth chosen from the integers 20, 21, . . . , 100, so that the output speed is as close to a given number, ξ say, as possible."

Mathematically, the problem is to find integers, a, b, c, d from (20, . . . , 100) such that for a given real number, ξ,

$$\left|\xi - \frac{a.b}{c.d}\right|$$

is a minimum.

As an illustration, he also pointed out to us that, for the case of ξ being the number π, (68 x 62)/(22 x 61) is a better approximation than 377/120 = (52 x 29)/20 x 24), which is recommended in a handbook. He then raised the question "Are there better approximations?" It is not difficult to recognise that this is a problem of diophantine approximation with restricted numerators and denominators.

Now it is well known from the theory of continued fractions that, corresponding to every real number , there is a sequence of fractions, (p_n/q_n), the so-called convergents, which has the property that the n-th convergent, p_n/q_n, is the fraction that, amongst all fractions with denominators less than or equal to q_n, is the nearest to ξ. Our present problem is more complex in that p_n/q_n is restricted to the form ab/cd. For π, the convergents are

$$\frac{3}{1},\quad \frac{22}{7},\quad \frac{333}{106},\quad \frac{355}{113},\quad \frac{103993}{33102},\ldots$$

Because 113 is a prime number, 355/113 is ruled out. The next convergent is clearly inappropriate. Thus, it is clear that a direct application of the continued fraction approximation is futile.

I was given this problem on the day of my departure from Loyang and I gave my assistant a piece of paper, on it the expression

$$\frac{52 \times 29}{20 \times 24} = \qquad \frac{377}{120} = \qquad \frac{22 \times 355}{7 \times 113}.$$

He understood my suggestion of attacking the problem with the method of Farey means. In this way, two approximations better than those mentioned by the worker were found, namely,

$$\frac{20\times(355)+29\times(22)+2\times(3)}{20\times(113)+29\times\ (7)+2\times(1)} = \frac{7744}{2465} = \frac{88\times 88}{85\times 29},$$

$$\frac{10\times(355)+1\times(333)+2\times(22)}{10\times(113)+1\times(106)+2\times\ (7)} = \frac{3927}{1250} = \frac{51\times 77}{50\times 25}.$$

Note that the bracketed numbers are members of the convergents of π. It may be shown that the latter is the best choice. Of course, this method is applicable for a general ξ. My assistant and the factory worker re-examined the handbook carefully and replaced a number of recommended ratios by the best possible ones; they even filled in a few cases left blank in the handbook.

Although this encounter was interesting and led, in fact, to the formulation of the following problem in number theory = "given a real number ξ, find p, q belonging to restricted class of integers less than M such that $|\xi - p/q|$ is minimised" it was decided that this technique was not sufficiently general for popularisation.

2. 0.618 Method

I now describe a method of optimisation which turns out to be well suited for popularisation; I shall also describe how my team has popularised it in China. Further details are given in my forthcoming book "Theory of Optimisation" (2).

For simplicity, consider a continuous function with a single maximum, whose functional form is unknown. What is the most efficient way of determining the maximum of f by evaluating the function experimentally with as few experiments as possible, over an interval (a, b), say?

Our simple minded approach, sometimes called the "method of trials by shifting to and fro", is to perform an experiment at, say, x_0 and obtain a value y_0. Then I may take another point x_1 and obtain y_1. If $y_1 \quad y_0$, then I exclude the interval (a, x_0) and consider only the interval (x_0, b). From the third experiment at x_2, I get y_2. I then compare y_2 and y_1 and eliminate the part that gives a smaller value, etc., until the maximum or its nearby point is found.

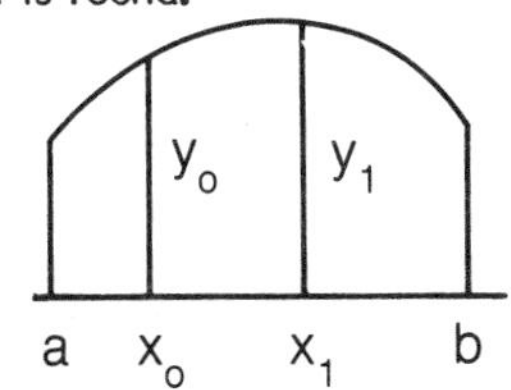

A natural question to ask is whether there is any rule with which I may select the "trial points" $x_0, x_1, x_2, x_3, \ldots$ so as to find the maximum as quickly as possible.

The Golden Section Method due to Kiefer (1953) of the United States turns out to be rather useful. Let

$$\theta = (\sqrt{5} - 1)/2$$

He suggested that one should set

$$x_0 = a + (b - a)\theta$$
$$x_1 = b - (b - a)\theta$$

where (a, b) is the interval chosen for experimentation.

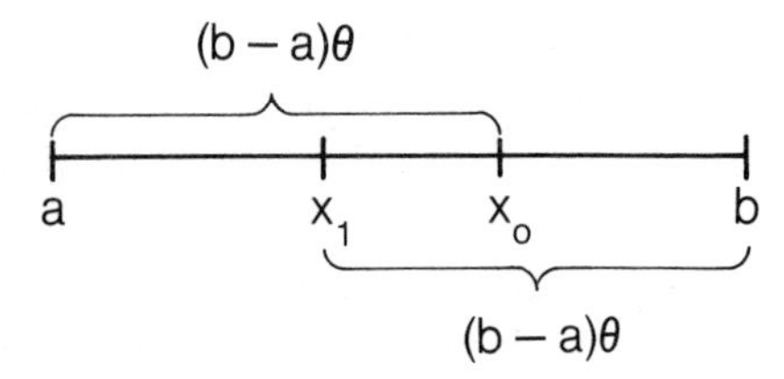

Note that (a, x_0) and (x_1, b) have equal length, $(b - a)\theta$. There are now two possibilities for the selection of x_2:

i) if $f(x_1) > f(x_0)$, then remove (x_0, b) and select x_2 from (a, x_0) such that

$$x_0 - x_2 = x_1 - a;$$

ii) if $f(x_0) > f(x_1)$, then remove (a, x_1) and select x_2 from (x_1, b) such that

$$b - x_0 = x_2 = x_1.$$

$x_3, x_4, \ldots$ may be selected in a similar way.

How can this method be popularised?

I now describe some personal experiences in this respect.

First, I ask the worker to memorise the number 0.618, but not the exact value of $(\sqrt{5} - 1)/2$! The first trial point is taken at 0.618 of the whole range. I often scorch a hole on a strip of paper with a cigarette to denote the point. To obtain the second trial point, I simply fold the strip in half and scorch the opposite side against the first hole. Results at these two trial points are compared. I then tear the strip at the worst point and keep that part which contains the better point. Next, I fold the shortened strip again in half and scorch the opposite side against the better point, thereby obtaining the next trial point. This process is then repeated a sufficient number of times and the maximum, i.e. the best (or near best), point is located.

I also have some satisfying response in popularising the discrete version of the above technique. Now, recall that $(\sqrt{5} - 1)/2$ has the convergents

$$2/3,\ 3/5,\ 5/8,\ 8/13,\ 13/21, \ldots F_n/F_{n+1}, \ldots,$$

where F_n is the n-th Fibonacci number. I sometimes illustrate the discrete method with matches. Twelve matches are arranged in a row to denote twelve different possible trial points. The first trial is made at the eighth counting from the left, and the second trial at the eighth counting from the right. If the "left" trial gives worse results, then I remove five matches starting from the left. Otherwise, I remove five matches starting from the right. In either case, seven matches remain. In this way, I have exploited the convergent 5/8! I next exploit the convergent 3/5 in exactly the same way, and so on.

I have found that by adopting the more visual and intuitive approach, the message is quickly conveyed to the general public with the gratifying result that many ordinary Chinese workers and farmers, usually with rather inadequate formal education, can master techniques of optimisation and operations research and increase their efficiency in production.

Workers of different industries, by using these simplified and visual methods in their own production processes, have achieved numerous fruitful results by choosing optimal parameters. By the way, these methods are widely popularized in China in the chemical, electronic, textile, machine-building, power supply, coal mining, metallurgy, construction, and other industries.

3. The Golden Number and Numerical Integration

So as to emphasize the fact that rather than hindering my theoretical research my involvement in popularising mathematical methods in fact often stimulates the former, Dr. Wang Yuan (of the Institute of Mathematics, Peking) and I have exploited the properties of the golden number for numerical integration. For example, we have found that the approximation

$$\int_0^1\int_0^1 f(x,y)\,dx\,dy \sim \frac{1}{F_{n+1}} \sum_{t=1}^{F_{n+1}} f\left(\frac{t}{F_{n+1}}, \left\{\frac{tF_n}{F_{n+1}}\right\}\right),$$

where F_n is the n-th Fibonacci number associated with the golden number, yields a very efficient numerical method. Note that the right-hand side involves a single summation only.

To extend the method to the s-dimensional case (s 2), we need only extend the F_n numbers. First, note that the golden number $(\sqrt{5} - 1)/2$ may be obtained by dividing the circle into five equal parts, i.e. by solving the equation

$$x^5 - 1 = 0,$$

or

$$x^4 + x^3 + x^2 + x + 1 = 0, \quad (x \neq 1).$$

Writing

$$y = x + 1/x,$$

we have

$$y^2 + y + 1 = 0,$$

giving

$$y = (\sqrt{5} - 1)/2.$$

Therefore, $y = 2\cos(2\pi/5)$, a cyclotomic number.

Now, what we have done in the two-dimensional case is to approximate $2\cos(2\pi/5)$ by F_n/F_{n+1}. It turns out that, for the general s-dimensional case, we need only consider approximating

$$2\cos\frac{2\pi\ell}{p}, \qquad 1\leqslant \ell < \frac{p-1}{2} = s$$

simultaneously. By Minkowski's theorem, we may confirm the existence of the simultaneous approximation of $2\cos(2\pi\ell/p)$ by x_ℓ/y for some $x_1, x_2, \ldots, x_{s-1}, y$. To obtain these s numbers, we consider the number field $R(2\cos(2\pi/p))$, obtained by adjoining $2\cos(2\pi/p)$ to the rationals. In this field, it is known that there exists a system of independent units,

$$\rho_\ell = \frac{\sin(\pi g^{\ell+1}/p)}{\sin(\pi g^\ell/p)}, \quad 1\leqslant \ell \leqslant s-1$$

where g is the primitive root (mod p). Let

$$\xi^{(i)} = (\rho_1^{(i)})^{x_1} \ldots (\rho_{s-1}^{(i)})^{x_{s-1}}, \quad 2\leqslant i\leqslant s-1,$$

where we have adopted the convention of using the superscript (i) to denote a conjugate of number. Also, $\xi^{(1)}$ is reserved for the number ξ itself. Consider the system of equations in $x_1, \ldots, x_{s-1}$ given by

$$\text{(1)} \qquad \log\left|\xi^{(2)}\right| = \ldots = \log\left|\xi^{(s-1)}\right|,$$

which is a system of s - 2 linear homogeneous equations in s - 1 unknowns. Therefore, there exist real solutions

$$x_1^{(o)}, x_2^{(o)}, \ldots, x_{s-1}^{(o)}, \text{ say}.$$

let

$$\ell_1, \ell_2, \ldots, \ell_{s-1}$$

denote an integral approximation of

$$x_1^{(o)}, x_2^{(o)}, \ldots, x_{s-1}^{(o)},$$

respectively. Let

$$\eta = \left|\rho_1^{\ell_1}\rho_2^{\ell_2}\ldots\rho_{s-1}^{\ell_{s-1}}\right|$$

and

$$\eta = \eta + \eta^{(2)} + \ldots + \eta^{(s)},$$

which is integral. From equation (1),

$$\eta^{(2)}, \eta^{(3)}, \ldots, \eta^{(s)}$$

are almost equal, and

$$\left|\eta\,\eta^{(2)}\ldots\eta^{(s)}\right| = 1$$

$$\text{Therefore,} \quad \left|\eta^{(i)}\right| = 0\left(|\eta|^{-\frac{1}{s-1}}\right), \; i=2,\ldots,s,$$

$$\text{and} \quad n = \eta + 0\left(\eta^{-\frac{1}{s-1}}\right).$$

Now, write

$$h_\ell \text{ for } \operatorname{tr}(\eta(2\cos 2\pi\ell/p)).$$

Therefore,

$$h_\ell = n\left(2\cos\frac{2\pi\ell}{p}\right) + 0\left(\eta^{-\frac{1}{s-1}}\right)$$

Therefore,

$$\left|2\cos\frac{2\pi\ell}{p} - \frac{h_\ell}{n}\right| = 0\left(\eta^{-1\frac{1}{s-1}}\right)$$

Hence, we conclude that the point set

$$\left\{\left(\left\{\frac{th_1}{n}\right\}, \left\{\frac{th_2}{n}\right\}, \ldots, \left\{\frac{th_s}{n}\right\}\right); t=1,2,\ldots,n\right\}$$

may be regarded in some sense as the generalisation of the point set

$$\left\{\left(\frac{t}{F_{n+1}}, \left\{\frac{tF_n}{F_{n+1}}\right\}\right); t=1,2,\ldots,n\right\}$$

used for the double integral, although we note that we can only have the big "0" result which is not so sharp as

$$\left|2\cos\frac{2\pi}{5} - \frac{F_n}{F_{n+1}}\right| \leqslant \frac{1}{\sqrt{5}\,F_{n+1}}$$

This generalisation leads to the efficient approximation

$$\int_0^1\int_0^1\ldots\int_0^1 f(x_1, x_2, \ldots, x_s)dx_1\ldots dx_s \sim \frac{1}{n}\sum_{t=1}^{n} f\left(\left\{\frac{th_1}{n}\right\}\left\{\frac{th_2}{n}\right\}, \ldots, \left\{\frac{th_s}{n}\right\}\right)$$

where the right-hand side is the single summation only.

5. Up-to-Date Technique

It seems to me that the technique developed by Powell, Davidson and Fletcher (PDF) is also suitable for popularisation. Incidentally, it is not difficult to prove that their technique is, in fact, better than what they originally claimed, i.e.

$$\|x_o - x_{n+s}\| = O(\| x_o - x_n\|^2),$$

rather than

$$\|x_o - x_{n+1}\| = o(\|x_o - x_n\|),$$

where x_n denotes the n-th iterate for seeking x_o, and $\|\ \|$ is the usual Euclidean norm. For further discussion of this technique, see my book "Theory of Optimisation" (Science Press of the Chinese Academy).

Overall Planning Method

1. Critical Path Method

As a result of extensive contacts with people engaged in management and organisation, I have found that the critical path method (CPM) is one of the most useful tools which can also be easily popularised. I have found that one hour of explanation is usually sufficient for the audience to come to grips with the essential points. Rather than describe the details of the method, I now summarise experiences gained by out team in the course of its popularisation.

i. Investigation: Items for investigation are: (a) a list of all relevant activities, (b) the connective relations between activities, and (c) the duration of each activity. Our method of investigation is by relying on the masses, i.e. the grassroots: the operator for each activity is asked to provide information about the duration of his activity and to name related activities which take place immediately before and immediately after his. In this way, a fairly complete list for (a) is usually determined.

ii. Critical Path Diagram: A draft CP diagram specifying all the critical paths is passed to the operators for general discussion, who will examine whether the draft diagram is reasonably complete. They are requested to reconsider whether some parallel operations, interacting operations and specialised technology, could be shortened. Our principle is to save time for each activity in the CP and to minimise the total time duration of the whole CP diagram. The aim is, of course, to speed up the completion date and to lower the cost.

iii. Execution of CP Diagram: During its execution, the CP diagram is constantly reviewed so as to direct it in a more favourable direction. In the light of new information, or if new development is about to take place, affecting the CP, the CP diagram is quickly adjusted.

iv. Summing-Up: At the completion of the project, the actual stage-by-stage progress is recorded in the form of CP diagrams so that this information may be utilised for future projects of a similar type.

I have found that CPM can be effectively applied two a wide range of activities, such as from the maintenance of a truck to the exploitation of a large oil field, the planning of "cutting", "transporting", "processing", and "replanting" in forestry as well as the planning and coordination of highly specialised developments.

2. Analysis Based on Ordered Samples

Another technique I have found useful for overall planning is based on ordering the sample in a sequence.

It is well known in mathematics that given non-negative numbers a_i and b_i ($1 \le i \le \ell$), the minimiser of

$$\sum_{i=1}^{\infty} a_{s_i} b_{t_i},$$

where

$$\{a_{s_i}\} \text{ and } \{b_{t_i}\}$$

are permutations of $\{a_i\}$ and $\{b_i\}$ respectively, over all possible permutations, is given by setting

$$a_{s_i} = a_{(i)} \qquad (i = 1, \ldots, \ell),$$

and

$$b_{t_i} = b_{(i)} \qquad (i = 1, \ldots, \ell),$$

where $\{a_{(1)}, a_{(2)}, \ldots, a_{(\ell)}\}$ is an ascendingly ordered sample of $\{a_i\}$ i.e. $a_{(1)} \le a_{(2)} \le \cdots a_{(\ell)}$ and $\{b_{(1)}, \ldots, b_{(\ell)}\}$ is a descendingly ordered sample of $\{b_i\}$ i.e. $b_{(1)} \ge b_{(2)} \ge \cdots b_{(\ell)}$.

To illustrate the above "prinicple", I would often use the following simple examples. A water tap is used to fill n buckets of capacities $a_1, a_2, \ldots, a_n$, say. How can the buckets be arranged so that the total waiting time is minimized? Without loss of generality, it may be assumed that it takes a_i units of time to fill a bucket of capacity a_i. It is clear that the first bucket has to wait for time a_1 for it to be filled. The second bucket has to wait for time $a_1 + a_2$ for it to be filled and the waiting times for the others are accordingly $a_1 + a_2 + a_3$, $a_1 + a_2 + a_3 + a_4$ and so on. The total waiting time T is given by

$$\begin{aligned} T &= a_1 + (a_1 + a_2) + \cdots + (a_1 + \cdots + a_n) \\ &= na_1 + (n-1)a_2 + M \cdots + a_n. \end{aligned}$$

Set $\qquad b_1 = n, b_2 = n - 1, b_n = 1.$

Hence, T is minimised on setting

$$a_i = a_{(i)} \quad (i = 1, \ldots, n)$$

where $a_{(1)} \le a_{(2)} \le \cdots \le a_{(n)}$ is an ascendingly ordered sample of $a_1, \ldots, a_n$. That is to say the smaller buckets should be filled first.

The above example may be easily extended to the case of s taps. Without loss of generality, let

$$n = sm,$$

i.e., there are m buckets of capacities

$$a_1^{(1)}, \ldots, a_m^{(1)}$$

for the first tap, m buckets of capacities

$$a_1^{(2)}, \ldots, a_m^{(2)}$$

for the second tap, etc. (If the taps are assigned different numbers of buckets, the numbers can still be assumed to be equal on introducing appropriate numbers of buckets of zero capacity.) The total waiting time T is now given by

$$T = \sum_{i=1}^{s} \left(ma_1^{(i)} + (m-1)a_2^{(i)} + \dots + a_m^{(i)} \right).$$

$$b_1 = b_2 = \dots = b_s = m$$

$$b_{s+1} = n_{s+2} = \dots = b_{2s} = m - 1, \text{ etc.},$$

the same result is obtained, i.e. by starting with the smaller buckets, T is minimised.

Sometimes, by substituting a lathe for a tap and an article for a bucket, a simple example is constructed for the benefit of the factory worker.

On the Use of Statistics

1. Empirical Formulae

Sometimes an empirical formula is derived from a set of statistical data. The meaning of such a formula can often be recognised by scientists with good common sense and some mathematical training. Let me give an example. R.C. Bose, an Indian mathematical statistician, obtained the empirical formula

A = length x width / 1.2,

for the area A of a leaf of rice grain, from a large sample of such leaves in India. I had no reason to question its reliability. However, some agriculturists in China applied this same formula to the harvest on their rice farm. On seeing the shape of the leaves on their farm, I pointed out to them that this formula could not be suitable for their leaves. They took some samples of their leaves and discovered that the formula tended to overestimate their areas. After their expression of surprise, I gave them a simple explanation by drawing the following picture, where the shaded area represents the area of a leaf.

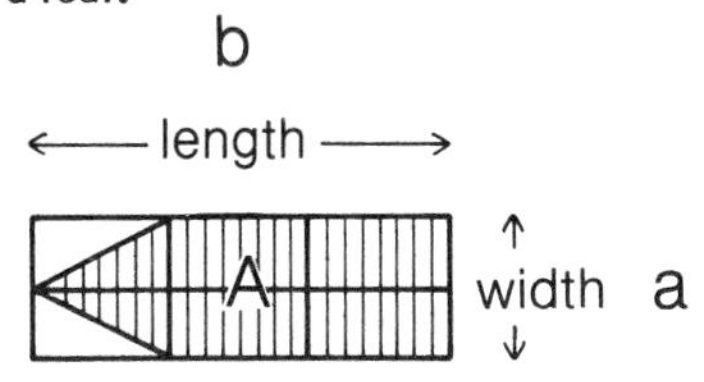

In this case, the ratio of the area of the rectangle to A is 6/5, i.e. 1.2. In contrast to this, a typical leaf in their farm has a more elongated tip, and I drew another picture. Thus, in this case, the ratio is closer to 4/3, i.e.

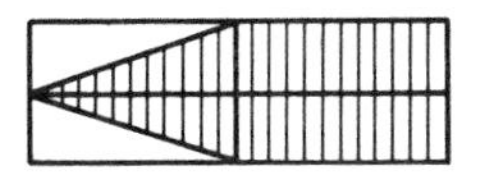

1.33, which explains the overestimation of A by using Bose's formula for their farm.

We all learned the lesson of the importance of seeking insight beneath an empirical formula.

2. A Simple Statistical Method

In the experimental sciences we frequently use statistical methods. Needless to say, their importance must not be denied. However, it is my view that some methods tend to be too sophisticated, and can easily be misused. I give a few examples first.

Example 1: Let $x_1, \dots, x_{20}$ denote twenty observations from independent repetitions of an experiment. Let

$$\bar{x} = (x_1 + \dots + x_{20})/20 \text{ (the sample mean)}$$

and

$$s = \sqrt{\sum_{i=1}^{20} (x_i - \bar{x})^2 / 19} \quad \text{(the sample standard deviation)}$$

The experimenter might then claim that the interval $(x - 1.96s/\sqrt{20},\ x + 1.96s/\sqrt{20})$ is a "reasonable" estimate of some underlying "average". Such a sophisticated method does not seem to be easily understood by an ordinary worker in China. Besides, the underlying Gaussian assumption might not be valid in his case!

In practical terms, I tend to prefer the following approach. Let

$$x_{(1)} < x_{(2)} < \dots < x_{(20)}$$

denote an ordered sample of observations. Intervals such as

$$\left(\frac{x_{(1)} + x_{(2)}}{2}, \frac{x_{(19)} + x_{(20)}}{2} \right)$$

and $(x_{(r)}, x_{(21-r)})$ for some r from (1, 2, . . . , 20), may be more readily accepted by our factory workers and farmers.

Example 2: Suppose that it is required to test which of two methods of production is better on the basis of five observations from each method. Let $(a_1, \dots, a_5)$ and $(b_1, \dots, b_5)$ denote the samples of the first and second methods respectively. It may be tempting simply to compare the means of the two samples by referring to the usual Student's t distribution. Underlying such a sophisticated method is a number of assumptions, such as normality, equal variance and independence, which are not easily understood by an ordinary factory worker.

It seems to be that a more robust and simple method based on the ordered samples $a_{(1)} > a_{(2)} > \dots > a_{(5)}$ and $b_{(1)} > b_{(2)} > \dots > b_{(5)}$ may be more suitable for popularisation in China. For example, an ordinary factory worker can accept it as being sensible to reject eh hypothesis H_0 that the two methods are equally good in favour of the first method if the mixed ordered sample is, say, either

$$a_{(1)} > a_{(2)} > a_{(3)} > a_{(4)} > b_{(1)} > a_{(5)} > b_{(2)} > b_{(3)} > b_{(4)} > b_{(5)}$$

or

$$a_{(1)} > a_{(2)} > a_{(3)} > a_{(4)} > a_{(5)} > b_{(1)} > b_{(2)} > b_{(3)} > b_{(4)} > b_{(5)}$$

I would usually demonstrate the former by holding my two hands up with the two thumbs crossing each other. The probability of rejecting H_o when it is true it less than 0.01.

2. PERT

Consider the so-called Program Evaluation Review Technique (PERT). Suppose that there are N activities in a network representing a project and there are three basic parameters describing the duration of the i-th activity. Thus, let a_i, b_i and c_i, denote the "optimistic duration", the "most probable duration" and the "pessimistic duration" respectively. The duration of the i-th activity is then usually assumed to follow a beta distribution over (a_i, c_i) and to have a mean duration m_i, where

$$m_i = (a_i + 4\,b_i + c_i)/6,$$

and variance

$$((m_i - a_i)^2 + 4(m_i - b_i)^2 + (m_i - c_i)^2).$$

It is often argued that the probability distribution of the total duration of the whole project may be approximated by a Gaussian distribution, presumably by appealing to the Central Limit Theorem (CLT). Even ignoring the fact that the assumption of beta distribution is by no means indisputable, I would question the wisdom of a hasty use of the CLT. It is obvious that if a fair proportion of the N activities are in series, then a careful analysis is needed to check whether a Gaussian conclusion is tenable.

Despite these reservations, PERT is basically a good topic for popularisation provided that some care is taken to avoid the above pitfalls, because

a. it is easy for the workers and the management to accept;

b. it is quite well suited for large projects which involve a multitide of people to participate;

c. it is essentially a good scientific method for organisation exercises involving several different industries.

I would argue that for a country like China with central planning, its application can be profitable.

4. Experimental Design

Up to now, it seems to be that insufficient attention is paid to non-linear designs. Past pre-occupation with linear models still seems to mask the important fact that these models are often unrealistic. I will not develop my point any further.

5. Types of Distributions

It has been suggested that the Pearson's Type III distribution be used to model the distribution of the waiting time for an "exceptionally big" flood (appropriately quantified). I would question the validity and wisdom of this approach in view of the scarcity of data which is inherent in this problem. It is even more unwise to expect that a sensible forecast for the next big flood may be obtainable from such a fitted distribution! The modelling and forecasting of such phenomenon is probably more appropriately handled by a "point-process" approach, in which the point events are epochs of flood. I understand that the problem is currently challenging the best brains in that field.

Mathematical Model

1. Generalised Inverse of Matrices

Consider a general regression model of y on $x_1, \ldots, x_n$,

$$y = f(x_1, \ldots, x_p) + e,$$

e denoting the random "error" term. Let $y^{(i)}$ denote the observed value of y at $x_1^{(i)}, x_2^{(i)}, \ldots p_p^{(i)}$. If f is assumed linear and there are n ($\geq$ p) observations, then

$$y^{(i)} = \sum_{j=1}^{p} \theta_j x_j^{(i)} + e^{(i)} \qquad i = 1, \ldots, n .$$

A general method for estimating $\Theta_1, \ldots, \Theta_p$ is by minimising

$$\sum_{i=1}^{n} (e^{(i)})^2$$

with respect to Θ_j's. That is, it is required to minimise with respect to

$$Q(\underset{\sim}{\theta}) = (\underset{\sim}{y} - \underset{\sim}{M}\underset{\sim}{\theta})'(\underset{\sim}{y} - \underset{\sim}{M}\underset{\sim}{\theta}),$$

where

$$\underset{\sim}{y} = (y^{(1)}, \ldots, y^{(n)})',$$

$$\underset{\sim}{M} = \begin{bmatrix} x_1^{(1)} x_2^{(1)} \ldots x_p^{(1)} \\ \vdots \\ x_1^{(n)} x_2^{(n)} \ldots x_p^{(n)} \end{bmatrix}$$

and

$$\underset{\sim}{\theta} = (\theta_1, \ldots, \theta_p)'$$

Now, for simplicity it may be assumed that $\underset{\sim}{M}$ is of rank p. Then

$$\begin{aligned} Q(\underset{\sim}{\theta}) &= [\underset{\sim}{\theta} - (\underset{\sim}{M}'\underset{\sim}{M})^{-1}\underset{\sim}{M}'\underset{\sim}{y}]'\underset{\sim}{M}'\underset{\sim}{M}[\underset{\sim}{\theta} - (\underset{\sim}{M}'\underset{\sim}{M})^{-1}\underset{\sim}{M}'\underset{\sim}{y}] \\ &+ \underset{\sim}{y}'[\underset{\sim}{I} - \underset{\sim}{M}(\underset{\sim}{M}'\underset{\sim}{M})^{-1}\underset{\sim}{M}']\underset{\sim}{y} \\ &= S_1 + S_2, \text{ say} \end{aligned}$$

Because $S_1 \geq 0$, therefore $Q(\underset{\sim}{\theta})$ is minimised on setting

$$\underset{\sim}{\theta} = (\underset{\sim}{M}'\underset{\sim}{M})^{-1}\underset{\sim}{M}'\underset{\sim}{y} \qquad (2)$$

which is sometimes regarded as a generalised solution of the equation

(3) $$\underset{\sim}{y} = \underset{\sim}{M}\underset{\sim}{\theta}$$

For this reason

$$(\underset{\sim}{M}'\underset{\sim}{M})^{-1}\underset{\sim}{M}'$$

is called a generalised inverse of $\underset{\sim}{M}$.

Of course, if the true model is adequately given by equation (1), then solution (2) is valid. Otherwise, if by substituting solution (2) in equation (3), substantial discrepancies are observed between the "observed" values and the "predicted" values of y_i's, then the assumption of linearity should be abandoned.

The use of generalised inverse is only one of the many examples of solving an essentially non-linear problem by posing it in a linear form. Other examples are linear programming and orthogonal design, to which the above remarks still apply.

2. Matrices with Non-Negative Elements

Since many economic variables are typically non-negative, the theory of matrices with non-negative elements is rather suitable for the analysis of economic relation. I believe that this theory has much to offer for the modelling of the Chinese economy.

Computation

I would just like to say a few words about modern computation. Although modern computers have enabled us to perform much more complicated calculations at very high speed, it still remains imperative to have a well-posed problem. It seems to me that the importance of numerical stability is not widely appreciated in China.

Concluding Remarks

If I were asked to say in a few words what I have learned in these last fifteen years of popularising mathematical methods, I would without hesitation say that they have enabled me to appreciate the importance of the dictum:

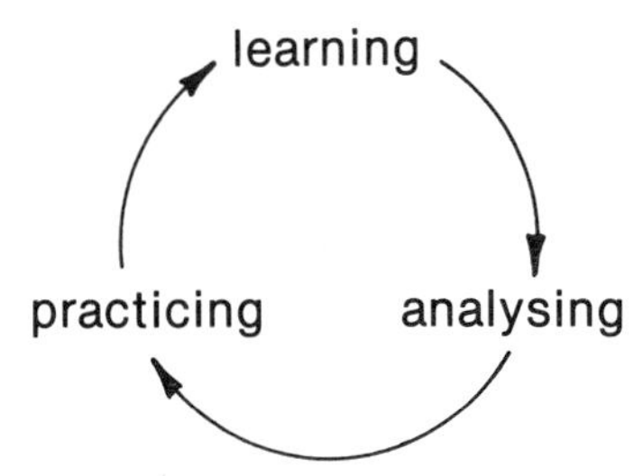

References

1. Hua, L.K. (1974) "Introduction to Higher Mathematics", Vol. I. (In Chinese, Science Press of Academia Sinica.)

2. Hua, L.K. (1980) "Theory of Optimisations". (In Chinese, Science Press of Academia Sinica.)

REACTION TO LOO-KENG HUA'S PLENARY LECTURE

Dorothy L. Bernstein
Brown University
Providence, Rhode Island

The manuscript of Prof. Hua's talk which was sent to me was rather different from what he said this morning, and therefore I will be changing some of my remarks. First, I would like to finish his account of the area of rice leaves. As he said, the formula which Dr. Bose used could be obtained as follows: Given a rectangle of dimension a and b, divide the length in thirds and the width in half, making 6 rectangles as the diagram shows, and draw the diagonals of 2 end rectangles; the shaded region approximates the leaf and hence its area is 5/6 (a x b) = (a x b)/1.2.

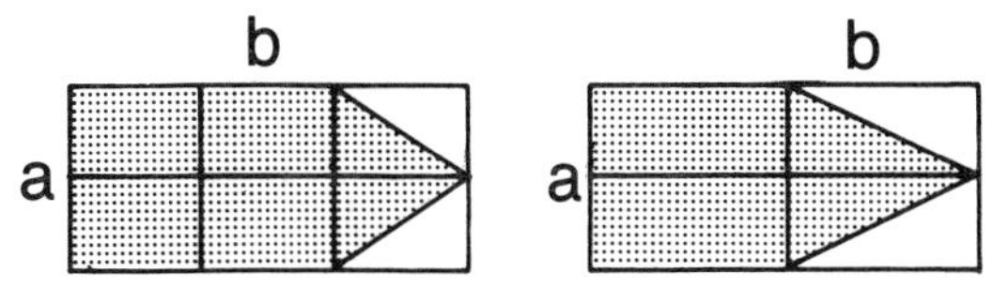

But the leaves at the experiment station in China were better approximated as follows: Divide the length and width in half, thus obtaining four rectangles; draw diagonals of 2 end rectangles. The shaded region, which represents the leaf, has area 3/4 (a x b) = (a x b)/1.33. This was the formula he recommended they use and it was satisfactory. This is, in my opinion, a good example, since the users can be given a fairly complete explanation of the formula.

But the example he gave of approximating the maximum of a continuous function f of an interval $[a,b]$ by calculating a sequence of points $x_0, x_1, x_2...$ and then determining $y_i = f(x_i)$ experimentally, is not so good, I believe. It involves $\theta = (\sqrt{5} - 1)/2 = .618$ and the mathematical explanation of how x_i is obtained from x_{i-1}, and why it gives the result is quite involved notationally. His clever trick with the cigarettes gives no idea of the reasoning and conveys the idea that mathematics is magic.

In fact, and here I quote from page 1 of Hua's manuscript, "Towards this end, my experiences have shown me that a deep understanding of the theory behind each technique is an absolute prerequisite, lest one might be completely misled." If this means the user must have a deep understanding, this is not true in his examples. If he means that the mathematician who wishes to popularize a method must have a deep understanding, this is true, but it still leaves the user with no idea of why the formula works.

In many of the examples in the manuscript, where he says he was talking to factory workers, I was puzzled since the tasks he described were certainly beyond the grasp of the ordinary U.S. factory worker. But when I read that they were in the electronic, chemical, power-supply, and construction industries, I realized that he was describing fairly skilled workers. And this morning he did add that his audiences included technicians and supervisors. Certainly the Critical Path Method, which he describes, is one for which a high degree of experience is necessary.

I will not go into a discussion of his statistical examples, since he did not get to them, except to remark that the formulas he uses to replace the usual formulas for a confidence interval have no justification; it seems to me that some place he must bring in the notion of a quantitative probability - - i.e. "it is 90% certain that..." Even unsophisticated persons can get some notion of probability. I was especially amused when in his discussion of PERT, which is based on highly sophisticated probability notions, he says this is good for popularization.

In general I do not think that in the United States today it is necessary to popularize mathematics. In the last two years students have increasingly enrolled in introductory mathematics courses in college. I would like to believe it is because we have been doing a good job, but I don't really think so. Students have an idea that "mathematics is a good thing" for them to take, although they are not sure just why. Many persons need a review of their beginning algebra, but they are not about to do that. However, one can give a computer-based problem-solving course to such people or an introductory computer-based statistics course, and in working on such a course they will review the needed algebra.

In his concluding remarks Hua says something about high speed computing that no one can object to, but I would like to say something about computation and of teaching undergraduate mathematics in this country which is not fully appreciated.

Anyone who looks at a group of college catalogues and compares the undergraduate offerings in mathematics in 1980 with the offerings at the same colleges in 1970 is immediately struck by the great increase in the number of new courses in applied mathematics and the emphasis on applications in regular math courses. There are courses in linear programming, applied statistics, mathematical modeling, numerical analysis, and applied differential equations. In addition, percolating down from the graduate schools, we are beginning to see courses in mathematical history and mathematical sociology as well as the older mathematical physics and mathematical economics. It is the widespread use of computers in undergraduate instruction which is largely responsible for all of this abundance.

Ever since World War II there has been an increasing interest in the applications of mathematics; because of the speed of calculations using computers, methods that were formerly only of theoretical interest became very practical and opened up entire new fields. For example, the theory of differential equations is an old standard subject and problems involving mechanical or electrical forces can be expressed by means of such equations. But the actual solution of a differential equation in a neat formula is only possible for a simple class known as linear equations, and even then, it is only first and second order equations and a few special higher order ones that allows an answer in closed form. So about all you could do in an undergraduate course was to give some problems in exponential growth or radioactive decay, involving first order equations, or simple spring-mass systems or electrical circuits using second order equations. However, since the time of Cauchy and Kowalevski in the last century, mathematicians have known how to solve a very large class of differential equations in terms of infinite series. The work of Picard, Peano, Runge-Kutta and others furnished new methods for numerical solutions of such equations. These methods generally arrive at the solution by an iterative process - a sequence of steps which can be performed over and over to get answers approximating the true one with increasing accuracy, and at each stage one has an error bound to tell you how far off you might be. The only trouble was that it required a huge number of calculations to make this feasible, so that, as physicists were fond of scornfully pointing out, these new methods were of no practical use.

The introduction of high-speed computers changed all this. It was not hard to write a program to perform one of these iterative procedures over and over and include an instruction to continue until the error bound was small enough to satisfy the accuracy requirements of the user. Once the program was stored, its application to a given differential equation could be had in a spectacularly short time. As a result, not only were non-linear problems in the physical sciences solved, but there was a great movement in the biological and social sciences to set up problems in terms of differential equations and solve them. Consequently, one can now give an undergraduate class a complicated predator-prey problem or a formidable problem in non-linear vibrations, and know they can handle it.

Or take the case of linear programming problems. One can write a set of linear equations or inequalities representing the constraints which will maximize or minimize certain variables, like quantities and prices in an economic situation, and ask for a solution of them. When the number of equations is small, say 2 or 3 or 4. the algebra involved in solving the set is not bad for a class, and simple applications are therefore found in many older texts on finite mathematics. But the interesting problems arise when you have 20 or 30 or 40 variables; here one needs to use methods of iteration and elimination or relaxation, which have been developed. They too are easily programmed for a computer, and appropriate error-bounds determined. So one can spend class time on setting up the problem or interpreting the results, which may be difficult still, but is more rewarding than calculating the solution.

And for anyone who wants to teach probability and statistics, the calculation of the various parameters like means, standard deviations, and coefficients of correlation, even for a very large data-base, can be done so quickly that one does not have to restrict oneself to artificially contructed examples (the programs are very easy to write also). In fact, students can be encouraged to seek out real-life problems in their own areas of interest which are challenging and intellectually rewarding.

Many other things, like computer simulation and computer graphics have affected and will affect the mathematics curriculum, but the above will serve to illustrate my point that the computer is an excellent tool in the teaching of mathematics and its applications.

REACTION TO PLENARY ADDRESS ON POPULARIZING MATHEMATICAL METHODS

J.S. Gyakye Jackson
University of Ghana
Legon

In this morning's talk we had a survey of the diverse ways in which the distinguished speaker had been trying to popularize mathematical methods in China.

The aim, I believe, was to make the factory worker become conscious of the power and beauty of mathematics and the role it plays in our everyday life.

In this age of rapid technological advance, automation, the computer, etc., there is an increasing demand for more mathematics to help keep up with the pace of advancement and it thus devolves on us who are mathematicians and mathematics educators to bring home to the general public the importance of mathematics and the role it plays in our everyday life and how an appreciation and ability to use it even to a small extent can help us understand a lot of the life around us and help us to improve our performance at work.

In the talk we had earlier this morning there is not much I can disagree with. However, my first reaction is that some of the examples cited require a very deep understanding of mathematics; also some of the mathematical principles involved are so sophisticated that even though there might be simple applications which can be popularised, the factory or workshop worker whose general education might be limited will have to be content with being told what basic principles to apply in a particular field to achieve better results in his job.

The example of the meshing gear pairs is very interesting but as it has been pointed out it is not sufficiently general for popularisation. The worker in the factory who posed the problem of finding two optimal gear ratios from the given set of numbers 20, 21, . . . , 99, 100, so that the output speed is as close to a give number, X , say, as possible, I consider to be a person who is mathematically literate and thus appreciates the mathematical principles involved in trying to determine the best approximation to his given number.

It is my belief that the majority of the audience in these popularisation lectures do not have a deep knowledge and understanding of the basic principles and sometimes sophisticated mathematics involved. They are thus content with visual aids as well as being presented with results which will help them in their various factories to achieve better results either in the form of more efficient running of machines, reduced production time or better products, etc.

The learned lecturer mentioned that in (some of) these popularisation lectures they sometimes had more than a hundred thousand people.

I know that the population of China is large and there is always a large turnout for any meeting. However, I wonder if this turnout is out of a genuine love of mathematics and desire to learn some simple ways of improving their methods of doing their work so as to achieve better results or is it because these lectures are compulsory for the workers and thus they have to turn up whether they are interested or not.

It would be interesting to know how popular mathematics is among the population in China. In a developing country like Ghana the usefulness and importance of mathematics in this scientific and technological age is widely appreciated. Mathematics is compulsory in both elementary and secondary schools, i.e., not only does everybody have to study mathematics up to that level (High School) but they must pass the examinations which are taken at the end of these grades. Entry into many jobs and many areas of study in higher institutions require a pass in mathematics at this level.

However I can say with certainty that mathematics is not popular. To many students mathematics is a foreign body which must be avoided by all means and they develop a hatred for it before they leave secondary school.

We are now trying, by adopting new and better methods of teaching to modernise the contents of syllabi and write new books to make the subject more intresting to children and get them to appreciate its usefulness and develop a liking for it. However, we are a long way from succeeding in this task.

As such I am sure that our factory workers would not attend lectures on popularisation of mathematical methods. At best if they are forced to attend, they would just sit around and probably get bored stiff.

However if they are to implement any new methods which incorporate new mathematical ideas to help improve their work, I am certain that will be done without question, so long as the order comes from above.

Another method which the lecturer was to talk about, but did not have time, is the Critical Path Method. This idea has a great future and is a method which can be adopted by developing countries. In a country like Ghana where there is a lot of manual labour there is a great tendency to waste labour and man-hours and thus spend a lot of time on production, making nearly all home produced goods very expensive. This wasteful nature does not only apply in factories but in nearly all facets of life.

By adopting the Critical Path Method, examining the activities which take place in the execution of any project, and finding ways of adopting parallel operations which save time and/or produce better results it should be possible to eliminate waste, cut down on production time and thereby reduce costs.

As I understand, the CPM has been effectively applied to a wide range of activities and thus has great potential in developing countries. It is a technique which can be applied in many developing countries with reasonable success even at the present level of education and mathematical understanding of the workers.

The technique based on ordering the sample in a sequence is another good point. This technique, as has been pointed out, is useful for overall planning. Again in activities where lots of manpower is involved this method can be adopted so as to reduce overall time for any activity and thus reduce cost.

This method also has a great future and needs to be adopted in the planning of developing nations.

In his comment on Experimental Design he stated that insufficient attention is paid to non-linear design. Even though linear models are sometimes unrealistic it is a fact that they are good approximations of the real situation. Non-linear models bring a great deal of complications into the mathematics of the problem that sometimes they become virtually insoluble.

In the theory of fluid dynamics the equations governing the motion of real fluids are non-linear and they are generally difficult to solve. It is only by making simplifying assumptions that they can be solved. This is also true of the notion of winds and electric currents within the ionosphere. The general equations governing the motions are not easy to solve unless we consider a one-dimensional model. Other non-linear models require other simplifying assumptions.

This year at the annual conference of the Mathematical Association of Ghana the theme for the conference was 'Mathematics in Industry.' Attention was focused on the various ways in which mathematics finds application in industry. Lectures were given on various topics both in mathematics and statistics to bring out to the members the various applications of mathematics in industry. They were thus made aware of the fact that mathematics can be used elsewhere than in the classroom.

One of the reasons why mathematics is not popular at home especially at the tertiary level is that many people believe if you study mathematics at that level there is nothing to be done with it other than to teach.

Mathematical methods are also useful in farming, etc., as has been pointed out. Some years ago I was approached by two men who had resigned their job with the ministry of agriculture and had acquired a piece of land to farm on their own. They were going into foodstuffs, especially corn.

They wanted to know if they could study Operational Research at Legon so that they could apply it to their farming. They wanted to know the maximum number of people they should employ and the area (region) within which they should be employed; whether they should provide accommodation on the farm for them or let them live at home and come to work. They were interested in all this so as to minimise their labour as well as other production costs.

I did tell them we were not offering the course but I would get an expert to get in touch with them. I did write to them that the expert was prepared to act as a consultant for them. However I did not see them or hear from them again. I do hope thay had an alternative solution.

The point I am making is that for once we could find individuals who were prepared to acquire mathematical knowledge and apply it to run their job more efficiently.

Mathematics is a language by itself which is understood by people who work and use mathematics. The results of our work and research are usually in the mathematical language and for these results to be applied they have to be put in a language which will be understood by the people who are going to apply it.

This is very essential for the popularisation of mathematical methods since the factory and other workers will in most cases have limited education and not be familiar with most of the mathematics we will be dealing with in our work.

Like extension work in agriculture, the agriculturist has to get his research and other results into a simple and applicable language which can be understood and appreciated by the farmer. This is the same situation with the popularisation of mathematical methods.

I borrow from Professor Sir James Lighthill in his Presidential Address to the 2nd ICME in Exeter in 1972, he stated that the slogan for Mathematicians and Mathematics Educators should be to learn the art of teaching the art of applying mathematics.

I sometimes give a course in elementary mathematics to our first year science students (all science students not reading mathematics have to take this course). They learn among other things calculus, differentiation, integration and differential equations. They learn to solve problems in these areas very well indeed. However when it comes to applying this knowledge to solving problems in say growth of plants in biological or chemical reactions then there is a great difficulty. The student is not able to apply the knowledge he has acquired. He cannot see how to translate equations in x and y to other variables.

This deficiency and difficulty is the fault of the teacher. We have not taught the student how the mathematics she is learning can be applied to the real life situation which is in their work in Biology. This has been corrected in that we try to take examples from the students' own field so that they see how the topic can be applied.

Finally, as I said earlier, mathematics is not popular even among highly educated Ghanians whose bent is along the humanities. In our efforts to modernise the teaching of mathematics attention has been focused on ways in which children can discover mathematical ideas for themselves, instead of being taught mechnical procedures, which they often master without understanding.

This we hope will make the subject more meaningful for many of them and they will get to like it and be able to master the basic principles up to at least secondary level. So that in whatever work they are engaged in after school, if they have to apply mathematics or use some ideas to achieve better results they will not shy away from it.

It is only when we get to this stage that we can try to popularise mathematical methods as is being done in Ghana.

In the meantime we have to be content with management making use of mathematical ideals to improve performance either through consultancy or by employing a mathematician to cater to the mathematical needs of the job.

CHAPTER 2 - Universal Basic Education

2.1 MATHEMATICS IN GENERAL PRIMARY EDUCATION

MATHEMATICS IN GENERAL PRIMARY EDUCATION

Romabus Ogbanna Ohuche,
University of Nigeria

Primary Education of Africa

About the time of the Addis Ababa Conference of African Ministers of Education in 1961 many African countries had their first commissions on education as independent nations. On the whole these reports indicated that primary education would be used to establish literacy, impart knowledge and skills, develop the personality of the individual and produce useful citizens.

Thus, it was the spirit of Addis Ababa which dominated policy on elementary education in African countries during the 1960s. That spirit was the idea of making that level of education which now lasts between six and seven years available to as many children as possible. In all African countries the rate of increase in primary school enrollment was significantly higher than the population growth. The 1960's also saw the initiation of major activities in curriculum development at the primary school level. A pan-African curriculum agency set up in 1961 defined for itself the main task of developing curricula appropriate for the African child in mathematics, science and social science. The mathematics component of the agency is the African Mathematics Program. It has had tremendous impact on the study of primary school mathematics in Africa.

The growth of elementary education which started in the 1960's continued in the 1970's at an even faster pace. Yet, it is in the area of policy and the quality of the education offered that the distinction between events in the 1960's and 1970's may best be made. The educational policies and structures of the immediate post-independence era were patterned after the educational policies and structures of the erstwhile colonial governments. Modifications in curriculum and other aspects of the educational delivery systems were taken.

It was in the 1970's that a call made by Lewis et. al. (1967) for links to be established among African psychologists, curriculum specialists, teachers and other educationists began to be taken seriously by African scholars, ministers and institutes of education and international agencies. Again, in 1968 the Nairobi Conference of African Ministers of Education pinpointed certain weaknesses in the educational systems, observed that qualitative improvement was necessarily a pre-requisite of quantitative expansion and went on to make recommendations regarding the reform of elementary education. Implementation of the recommendations started in the 1970's.

Then came the Lagos Conference of Ministers of Education of African member states of UNESCO in 1976. That Conference reviewed the trends in education in Africa during the period 1968 to 1976. It was revealed that, on the whole, African countries were now devising educational strategies to achieve the cultural, economic, political and social objectives stipulated in their national policies.

African countries have started seeking appropriate solutions to many of the problems which beset elementary education. These include the problems of quantitative expansion, teacher effectiveness, teacher shortages and physical inadequacies. Another problem arises from the inequitable distribution of opportunities in favour of the urban areas where only a small proportion of the people live. Another challenge is devising curricula which will serve the best interests of all primary school pupils, especially since in the majority of the countries primary school is likely to be terminal for a significant population of the children for sometime to come.

Objectives of Primary School Mathematics

Current development is motivated by two desires which are complementary but different in character:

i) to enhance the content of education by promoting the standardization and simplifying power of mathematical thought, with the object of improving each individual's level of understanding and grasp of an environment full of mathematical situations;

ii) to improve the learning process of every child and to introduce the study of mathematical ideas at the most appropriate moment.

In all countries, primary school mathematics is expected to contribute to the objectives of elementary education.

The contributions which primary school mathematics can make were examined thoroughly at The International Seminar On Developing Mathematics In Third World Countries in Khartoum in March, 1978. On that platform, debate on the objectives of teaching mathematics at the elementary school level was lively but the participants finally accepted five broad ones.

They were:

1. functional numeracy which was seen as knowing how and when to use arithmetical operations on both whole and decimal fractional numbers;

2. acquisition of certain mental attitudes which facilitate the development of problem solving attitudes and strategies;

3. acquisition of techniques of representation and interpretation of numerical and other data;

4. indication of abilities in measurement, approximation and estimation of number and quantity;

5. development of spatial concepts and the ability to represent them using such tools as maps and scale drawings.

These represented the concensus of the participants who met in Khartoum.

Individual countries have taken slightly different approaches to the indication of the objectives of teaching mathematics at the primary school level. Two examples may be used to illustrate this point. Thailand has specified five objectives similar to those stated in the Khartoum accord while Nigeria has specified objectives in terms of pupil behaviour for each content area at each grade level.

Content, Grade Placement, Language, and Methodologies of Primary School Mathematics

Table I is a simple version of the common core of the intended curriculum in Nigeria. It indicates that number and numeration, basic operations in mathematics and measurement are started in Elementary One and studied all through the six years of elementary school.

Table I.

Main Concepts and Their Placement

Concept	Year Covered					
	1	2	3	4	5	6
Number and numeration	✓	✓	✓	✓	✓	✓
Basic Operations in Mathematics	✓	✓	✓	✓	✓	✓
Measurement	✓	✓	✓	✓	✓	✓
Practical and descriptive geometry		✓	✓	✓	✓	✓
Everyday Statistics				✓	✓	✓

On the other hand, practical and descriptive geometry and everyday statistics are started in Elementary Two and Elementary Four, respectively. The former is studied for five years and the latter for three years.

A distinctive feature of this intended curriculum is that statistics is one of its components. In each of the last three years some work is outlined for students under the topic of everyday statistics. The main elements of the specified content are pictograms, bar charts, simple experiments, frequency tables, mean, median and mode. Since many introductory undergraduate courses in statistics do not specify much more than this, the essential difference must therefore, not be sought in content but elsewhere.

This brings us naturally to methodologies in elementary school mathematics. Here, Britain (HSI; 1979, p.1) provides a useful statement.

> "The belief that children should be enabled to discover important mathematical ideas for themselves has been developed over many years . . . Discovery methods are a sound approach when they are used to lead children to acquire a deeper understanding of the processes involved and more enthusiasm for the subject. Of course, neither teachers nor children have the time and skill to ensure that children discover everything, but, if their future attitudes to the subject are to be positive, it is important that each should have sufficient experience of personal discovery."

This is buttressed from Thailand where it is indicated that elementary mathematics emphasizes thoughts and understanding of numbers, algebra, measurement, geometry and statistics. Pupils are taught how to think about problems and thus gain confidence to confront new learning situations.

In several schools in Britain and New Zealand teachers of primary school mathematics still put most of their stress on computational efficiency. As for the developing countries, the Khartoum conference recommended that any major development of school mathematics curricula must be preceded by adequate preparation of teachers. The professional development of the teacher and the improvement of classroom conditions are higher priorities than further changes of syllabus.

The Accra Seminar added that the resolution of the 1976 Lagos Conference that

> "it is advisable for primary education to be given in the vernacular" was "consistent with what is known about how children learn mathematics".

Reflection and Conclusion

There is a major point of concern. It is that much more is expected from the elementary school teacher than he is trained to deliver. The expected level of proficiency in mathematics is usually not beyond that of the high school graduate. The actual level for most primary school teachers in Africa is lower.

There is a confounding factor. Not only is the level of mathematics preparedness inadequate, but the student teacher is not trained to relate well with children.

It is our view that specialization is called for. The primary school teacher in Africa should be trained to teach one of four blocks of subjects; arts and crafts, mathematics and science, language arts, and social studies. This specialization should enable him to acquire good depth in his subject of concentration and at the same time learn to understand and relate with the children he teaches. Only in this type of arrangement can the present-day teacher of elementary school mathematics cope with the challenges of the mathematics curriculum of this age.

It is perhaps necessary to counter one argument which may be made against this proposal. The Western argument in favour of the omnibus primary school teacher is that it is easier for the young child to relate with one teacher than with several. This is at best a questionable point of view in Afirca. Born into an extended family the child in Africa begins relating with many people from birth. By the time he is of school age, he would have so related with people that having four or five persons as teachers in his class with be nothing unusual.

Before concluding, it is necessary to say something about assessment. The Khartoum Conference recognized the importance of evaluation of both individual achievement and curriculum performance. It recommended the convening of a conference to refine the objectives of primary school mathematics outlined earlier and then work out a basis for curriculum evaluation. It then emphasized

(Final Report; p. 2) that

> "A competitive secondary selection examination is not adequate as an assessment mechanism for the attainment of the objectives of primary education.
>
> We recommend that further research needs to be undertaken, country by country, into the problems of the validity of various methods of pupil assessment."

The situation in Africa must improve in at least two dimensions. One is that the movement toward formative evaluation as now strongly advocated by Nigeria, must continue to gain momentum. The other is that we must design instruments which are appropriate for measuring achievement in the new mathematics curriculums. The traditional instruments now in use still emphasize computation.

Fortunately, foresight led to the establishment in 1972 of the International Centre in Educational Evaluation (ICEE) of the University of Ibadan as a result of the combined efforts of the University, the Science Education Programme for Africa (SEPA) and the Carnegie Corporation. The Centre is now training evaluators from different parts of Africa.

By way of conclusion, a historical review of the status of pre-college mathematics in the United States of America during 1955 to 1975 stated in part

> "Regarding practices in the schools: Too little is known about what happens in the typical classroom.
> Too little is known about the extent to which teachers differentiate instruction.
> To little is known about the extent and materials and instructional tools. The extent of teachers' dependence on drill-and-practice teaching strategies is not known."
> (Suydam and Osborne; 1977, pp. 10-11).

The authors might as well be discussing the case in most developing countries.

References

African Commission on Mathematical Education. A brief report of the inter-African seminar on teaching in of mathematics in African primary and secondary schools, Accra, Ghana: 16-19th May, 1979.

Cox, J.N. "Mathematics in New Zealand primary schools 1965-1975", mimeo.

Department of Education and Science; HMI Series: Matters for Discussion 9. Mathematics 5 to 11; handbook of suggestions, Her Majesty's Stationery Office, London, 1979.

Department of Mathematics, University of Khartoum. The Final Report, International Conference On Developing Mathematics In Third World Countries, Khartoum, 6-9 March, 1978.

Kirtikara, C. "Stability and change in primary school mathematics 1961-80", private communication.

Ohuche, R.O. "Recent attempts at mathematics curriuclum renewal in English-speaking West Africa", ABACUS, Journal of the Mathematical Association of Nigeria, Vol. 12, 1978.

Ohuche, R.O. and B. Otaala. The Child in the African Environment UNEP (in print).

Suydam, N.M. and A. Osborne. The status of pre-college science, mathematics and social science education: 1955-1975, Volume II. mathematics education, exective summary, Centre for Science And Mathematics Education, the Ohio State University, 1977.

2.2 BACK TO BASICS: PAST, PRESENT, FUTURE

BACK TO BASICS: PAST, PRESENT, FUTURE

Max A. Sobel
Montclair State College
New Jersey

"Back-to-the-basics" is a cry that is currently being heard throughout the United State—if not throughout the entire world. It appears that the entire world population has climbed aboard a bandwagon to condemn the teaching of mathematics and the level of student learning. It is very popular today for the general public, aided and abetted by the media, to deplore the level of mathematical competency of our youth--and of course to blame the teachers of mathematics for such student incompetency.

Such an approach is not new. It has always been popular to use teachers as scapegoats for the failures of society. Now I do not wish to imply that we are completely blameless; but on the other hand, the problem extends far beyond the four walls of the classroom.

My assignment in this paper is to explore the meaning of the "back-to-basics" movement—past, present, and future. I will try to review the past and discuss the present in some detail, but then I invite the reader's help in charting the course for the future.

Certainly the meaning of basic skills has never been defined in a universally acceptable manner. In the early 1950s the NCTM Commission on Post-War Plans recommended that schools "should guarantee functional competence in mathematics to all who can possibly achieve it." They listed 29 items that should be mastered by the mathematically literate person. This was their set of basic skills for the 1950s:

> Computation with whole numbers, fractions, decimals, percents, and signed numbers
> Estimation and significant figures
> Measurements, the metric system, and conversion
> Tables, graphs, maps, and blueprints

Basic geometric definitions and constructions
Sine, cosine, tangent, and the 3-4-5 relation
Square root
Addition of vectors
Basic algebraic symbolism and common formulas
Ratio, proportion, and similar triangles
Axioms, hypotheses, and conclusions
Elementary statistics, business and consumer arithmetic

These items were the ones designed to bring our youth forth into a bright new post-war world, but quite obviously by the end of the decade of the 1950s serious questions were being raised. Our curriculum in the United State had been changed to include courses such as general mathematics to provide better preparation for everyday life. But somehow the emphasis on the basic skills of the 1950s failed to provide the necessary scientific power to keep up with the race to space. It was inadequate for the role that President John F. Kennedy saw for our country in the decade of the 1960s.

The decade of the 60s was an exciting one for teachers of mathematics. The concept of basic skills changed quite drastically. Certainly we expected our youth to compute. But rather than see numerous problems such as

$$\begin{array}{r} 5278 \\ \underline{X\ 343} \end{array}$$

in the textbooks, we were far more likely to see somthing like this:

5728 X 343 = 343 X 5728   True or false?

The emphasis was focused on "mathematical structure." Meaning was emphasized as opposed to routine meaningless drill. This was the question that was being asked: "What is more important--to understand what one is doing, or to compute the answer correctly?" Of course we now recognize this as a false dichotomy. Evidence that dates back at least 50 years or more clearly indicates that successful learning of skills definitely requires basic understanding.

Of course there were excesses in the 1960s. Certainly no stretch of the imagination can make one declare that modular arithmetic, sets, or other number bases are topics that should be considered as "basic." And although these topics were really very small portions of what came to be known as "the new math," the general public and media seized upon these as examples of our corruption of the elementary curriculum. Many headlines stated that our youth could cite the computative property for multiplication, but could not multiply; that youngsters could add 5 + 6 in base eight, but did not know their sums in base 10.

The direction of school mathematics in the 1960s can readily be assessed by just reading the prefaces to some of the textbooks in common usage during that period of time. Invariably each one stated that they incorporated the recommendations of the School Mathematics Study Group. Other statements, frequently found, indicated that attentions was given to the structure of mathematics, the discovery of patterns, and to the fundamental ideas underlying the familiar practices and procedures of arithmetic. Emphasis on computational skills was definately not a feature that received much attention as a course objective as listed in textbook prefaces.

The start of the 1970s again brought indications of changing directions. Now our textbooks, in the preface, spoke of such points of emphasis as flow charting, mappings, computer language, functions, attention to explorations and discovery--but again scant mention of what we are still loosely calling "basic skills."

We must be careful to note that the development of arithmetic skills was not ignored during this period of time. As reported by the National Advisory Committee on Mathematics Education in 1975--the NACOME Report--teachers "sought improved skill performance through deeper student understanding of the structures underlying computational methods. Though the goal of increasing computational competence has not been reached on any massive national level, this failure does not invalidate the "understanding leads to skill" hypothesis."

In 1972-73, the National Assessment of Educational Progress explored the mathematical achievement of 9, 13, and 17 year olds in the United States. The news was either good or bad depending upon one's frame of reference. For example, 84% of the 13 year olds could compute the following:

Add: $ 3.09
10.00
9.14
5.10

On the other hand, 16% could not obtain the correct sum.

The NACOME report examined the data obtained in this assessment and felt that the results on computation did not confirm charges that basic skills seriously deteriorated during the "new math" era. It is interesting to note here that the report recognized the general interest in calculational proficiency, but felt "that the importance of computational skill is diminishing in the modern world." They went on to point out that there had been a general decline in all basic scholastic skills in the 1960s, and that mathematics achievement shared in this decline.

The failure of our youth to perform computational skills on a level that the general public felt to be acceptable caused a great deal of attention in the press. Textbook publishers are very much attuned to such public pressure and concern, and a definite change was seen in the books that were brought forth in the late 70s and that are being published today. Indeed, the prefaces of contemporary textbooks now list such features of their programs as:

Emphasis on basic skills
Real world applications
Problem-solving taught simply

It is evident that current programs of study devote extensive time to developing skills in arithmetic computation. Unfortunately, from my point of view, there are those who would make the total program of mathematics devoted exclusively to the development of such skills. To quote NACOME again:

> The members of NACOME view with dismay the great portion of children's school lives spent in pursuing a working facility in the fundamental arithmetic operations. For those who have been

unsuccessful in acquiring functional levels of arithmetic computation by the end of eighth grade, pursuing these skills as a sine qua non through further programs seems neither productive nor humane. We feel that providing such students with electronic calculators to meet their arithmetic needs and allowing them to proceed to other mathematical experience in appropriately designed curricula is the wisest policy.

In the late 1970s our professional groups issued statements and warnings to the profession concerning basic skills. For example, the National Council of Teachers in Mathematics wrote a position statement to note that computational competence is a major objective of our school programs. Thus they wrote:

> The National Council of Teachers in Mathematics is encouraged by the current public concern for universal competence in the basic computational skills. The Council supports strong school programs that promote computational competence within a good mathematics program and urges all teachers of mathematics to respond to this concern in positive ways.
>
> We are deeply distressed, however, by the danger that a "back to basics" movement might eliminate teaching for mathematical understanding. It will do citizens no good to have the ability to compute if they do not know what computations to perform when they meet a problem. The use of the hand-held calculator emphasized this need for understanding: one must know when to push what button.
>
> In a total mathematics program, students need more than arithmetic skill and understanding. They need to develop geometric intuition as an aid to problem solving. They must be able to interpret data. Without these and many other mathematical understandings, citizens are not mathematically functional.
>
> Yes, let us stress basics, but let us stress them in the context of total mathematics instruction.

The National Council of Supervisors in Mathematics made a distinct contribution to the profession by developing a position paper that sought to define basic skills beyiond the very narrow viewpoint of computation. Their listing of the following ten basic skills areas is currently serving as a guide to numerous curriculum groups and textbook authors:

Problem solving
Applying mathematics to everyday situations
Alertness to the reasonableness of results
Estimation and approximation
Appropriate computational skills
Geometry
Measurement
Reading, interpreting, and constructing tables, charts and graphs
Using mathematics to predict
Computer literacy

The warnings were clearly stated by many groups and individual educators as well: attention to computation to the neglect of other fundamental basic skills could lead to disaster.

During the 1977-78 school year the National Assessment of Educational Progress completed its second mathematics assessment. The results were released last year and are still being analyzed. It is important to examine some of the results, noting that this assessment came after a number of years of subjecting our youngsters to increased attention to basic skills in its very narrow sense.

At age 13 it appeared that most students could add, subtract, and multiply whole numbers. But by age 17, only about one-half had mastered whole number division and operation with fractions and decimals. Thus, only half could solve this problem at age 13 or at age 17:

$$28\overline{)3052}$$

It is interesting to note that with almost no formal instruction, over 80% of the 13 year olds and 90% of the 17 year olds could do this and other exercises with a calculator!

About two-thirds of the 13 year olds and three-fourths of the 17 year olds had learned elementary fraction skills. About 50% of both groups could add, subtract, and multiply decimals, but their powers of estimation were far lower than for computational skills. Note also that although 50% of the 13 year olds could multiply two decimals, only 18% could estimate their product. As stated in the May 1980 issue of the Mathematics Teacher:

> Students appear to be learning many mathematical skills at a rote manipulation level and do not understand the concepts underlying the computation.

I see this as a danger sign for our programs for the 1980s.

In general, the NAEP results showed that our students have a reasonable mastery of computational skills, especially those that involve whole numbers. However, the vast majority of students at all age levels demonstrated severe deficiencies in all the other basic skill areas. Performance was especially low in such areas as measurement, estimation, probability, and statistics.

It was in the area of problem solving that the most disastrous results were found. Drill on computational skills in our schools had been promoted to the virtual exclusion of attention to problem solving. But it is obviously the case that knowledge of computation does not insure ability to solve problems.

So where do we go from here as we head into the 1980s? Fortunately, the NCTM has charted a suggested direction with their document An Agenda for Action. Let me review several of their basic recommendations at this time:

1. Problem solving should be the focus of school mathematics in the 1980s.

2. Basic skills in mathematics should be defined to encompass more than computational facility.

3. Mathematics programs should take full advantage of the power of calculators and computers at all grade levels.

I think it is important to note that both the NACOME report and An Agenda for Action question the extensive amount of time being spent in our classrooms on manipulative skills. Thus NACOME said:

> It appears to us that the case for decreased emphasis on manipulative skills is stronger now than ever before.

An Agenda for Action includes this statement:

> Insisting that students become highly facile in paper-and-pencil compuations such as 3841 X 937 or 72,509 ÷ 29.3 is time-consuming and costly. For most students, much of a full year of instruction in mathematics is spent on the division of whole numbers--a massive investment with increasingly limited productive return.

I suggest we now consider a few recommendations for future action and discussion.

1. The NAEP results clearly indicate that many youngsters improve basic computational skills as they progress from age 9 to age 13 to age 17. Thus it appears that minimum competency programs that hold children back until they demonstrate proficiency of selected skills deprive them of the experiences they need to develop mastery.
2. Too many of our students conclude their formal study of arithmetic skills in grade 8. Those who attend college preparatory programs of algebra, geometry, and pre-calculus are deprived of many of the basic skills that we have defined here. Concepts of estimation, measurement, probability, statistics, reading of tables and graphs, etc., do not often appear in such academic courses and they should. In other words, we must pay careful attention to basic skills, defined in the broadest sense, in all of our courses. Problem solving must not cease once the student enters the exalted world of study of algebra and geometry. This suggests a careful revision of many of our current courses of study and textbooks at the secondary level of instruction.
3. An Agenda for Action calls for active participation by the student in the learning process: experimentation, exploration, use of manipulatives, use of references outside of the classroom, etc. Yet all studies completed to date indicate that most elementary and secondary mathematics instruction consists of whole class instruction.
4. It is important to emphasize the importance of in-service education of teachers to develop essential pedagogical skills. Our teachers ask for this, as indicated by a NCTM study conducted by their Commission on the Education of Teachers. But teachers want meaningful in-service programs, and high on their list of requests is instruction on appropriate methods of motivation.
5. We need public support. An Agenda for Action is concluded with this urgent request:

> Public support for mathematics instruction be raised to a level commensurate with the importance of mathematical understanding to individuals and society.

Back in the 1960s the teachers of mathematics were in the spotlight--we were very important in the eyes of the general public. Today the spotlight has dimmed:

> There is a serious shortage of qualified mathematics teachers.
>
> The media are extremely critical of the so-called failures of our programs of instruction.
>
> Young people are not entering the teaching profession.

It is time to have the spotlight come back on once again in a favorable manner. We must let the public know and force them to realize that we are dealing with the most precious commodity in our society: their children, the future citizens of the 21st century. Sure, they have the right and the obligation to criticize--but they also have the need to be supportive in our efforts for the days and years ahead.

2.3 SUGGESTED MATHEMATICS CURRICULA FOR STUDENTS WHO LEAVE SCHOOL AT EARLY AGES

SUGGESTED MATHEMATICS CURRICULA FOR STUDENTS WHO LEAVE SCHOOL AT EARLY AGES

Shirley Frye
Scottsdale School District
Arizona

In the United States attendance laws and child labor laws are influential in keeping young people in public schools until their mid-teens. The requirement to remain in school affords the unique opportunity for providing a common general education for all. But, as is well known, an outstanding curriculum needs to be coupled with a willing learner.

Although the vast majority of students complete high school, about 900,000 students drop out of high school annually. Many of them leave without salable skills. Educators are constantly assessing their programs to determine what minimum essential skills should be presented to the captive audience. The schools have a multitude of responsibilities, but the prime one is to prepare the student to be a functional citizen.

The particular goals of a minimal mathematics education include having:

1. a sense of number
2. the ability to quantify and estimate
3. skills in measuring
4. usable knowledge of the basic facts
5. the ability to select the appropriate operation to find a solution
6. the ability to use a calculator to perform operations
7. a "money-wise" sense.

The last skill relating to being "money-wise" is most important since an individual should have the ability to decide whether wages are being paid correctly and if purchasing transactions are fair. Minority groups stress this objective as a vital outcome of an education in the public schools.

Educators must be constantly alert to the need for including salable type skills, such as making-change, in school programs. Schools should be reponsive to the need and requirements of the business and industries in their locale and to the current technology that enhances arithmetic skills.

Throughout the United States, the outline of minimal skills will vary, but the general goal should be to challenge each individual student to his/her maximum.

SUGESTIVO CURRICULUM DE EDUCACION MATEMATICA PARA ESCUELAS ELEMENTALES CAMPESINAS DE 4 ANOS

Alonso Viteri Garrido
Quito, Ecuador

Factores Nagativos de la Educacion Elemental de los Grupos Sociales que viven en al campo

Los que se relacionan con el profesor:

- Formacion para educar otros y diferentes grupos sociales.
- Procedencia y/p representacion de clases explotadoras politicas, sociales y economicas.
- Burocratizacion de dirigentes y educadores.
- Abandono consuetudinario del trabajo docente en la escuela del campo.
- Negligencia en la tarea del desarrollo de la comunidad campesina.
- Desconocimiento del medio ambiente donde pretende educar, de las caracteristicas psico-sociales del sujeto a quien proyecta educar y del idioma materno de quienes se propone educar imponiendo la lengua oficial del pais, ignorada por los alumnos.

Los que se relacionan con el nino campesino:

- Desercion escolar elevada como producto de la pobreza, de la ocupacion en faenas domesticas y de campo, de la emigracion, de la desadaptacion al clima escolar, de la frustracion en el aprendizaje, etc.
- Desnutricion y enfermedades.
- Desmotivacion del aprendizaje ocasionada por la implantacion de un sistema educativo postizo.

Los que se relacionan con la infraestructura social del campesinado:

- Regresion del aprendizaje promovida por el escaso desarrollo ambiental, por la vigencia de un paternalismo dictatorial y sostumbres ancestrales y por una relativa inactividad mental.
- Desasimilacion de la comunidad al educando de corta edad, impidiendo su integracion socio-economica-cultural y, en consecuencia, creando una fuente mas de analfabetismo.
- Utlitarismo excesivo y exigente en el ambito socio-economico que restringe la proyeccion del conocimiento en el comprotameinto y a la functionalidad de la educacion.

Los que se relacionan con la estructura del sistema educativo:

- Esqualas de 3 grados servidas por un solo profesor.
- Implantacion de un curriculum elaborado para que su desarrollo dure 6 anos.
- Interrupcion del proceso educativo por la suspension del desarrollo curricular a los 3 anos.
- Rigidez del regimen escolar formal que promociona la desercion e imposibilita la rehabilitacion del desertor temporal.
- Contenidos programaticos con poca esencia, utilidad y funcionalidad por estar afectados de ampulosidad, curiosidades y sutilezas.
- Metas extenses y de largo alcance que distorsionan realidades educativas y confunden necesidades prioritarias de la educacion del campesino.
- Sistema educativo creado para otros grupos sociales.

Sugerencias para el Curriculum

Objetivos:

El nino sera capaz de

Adquirir habilidades, conocimientos y razonamientos matematicos.

Aplicar habilidades, conocimientos y razonamientos matematicos.

Crear aptitudes para adaptarse a situaciones nuevas.

Desarrollar actitudes de investigacion mediante la curiosidad, la comprension y el uso de sus propias iniciativas.

Continuar estudios matematicos.

Intelectualizar la Matematica.

Crear Matematica.

Contenidos:

Primer nivel: Conjunto y elemento. Correspondencia: conjuntos grandes y pequenos, conjuntos con pocos, muchos y ningun elemento, conjuntos con mas y menos elementos. Concepto de numero natural. Conjuntos con mayor y menor numero de elementos. Coordinabilidad y equipotencia. Operaciones con conjuntos: union y particion. Numeros naturales hasta 100 y el cero. Relaciones entre numeros naturales: mayor que, menor que, e igual a. Operaciones en el conjunto de numeros naturales: adicion, concepto, practica con graduacion de dificultades, tables, propiedades conmutativa y asociativa, algoritmo y problemas de aplicacion al medio campesino. Sistemas de numeracion: base de un sistema, sistemas de base 3, 4, 2, 5 y 10, sistema de numeracion decimal, conceptos de decena y centena.

Segundo nivel: Complementacion de conjuntos. Concepto de completacion. Resta de numeros naturales: concpeto en base a la completacion, quitar elementos y descubrir el sumando desconocido, pratica graduada, propiedad pseudoasociativa, algoritmo y problemas de aplicacion al medio rural. Localizacion en el espacio: arriba, abajo, delante, detras, ezquierda, derecha, etc. Nocion de volumen: delgado, grueso. Nocion de area: exteno, angosto y ancho. Nocion de longitud: largo, mediano y corto. Cuerpos: formas conocidas. Plano: formas mas usuales. Recta: abierta y cerrada. Nociones de dentro, fuera y en el borde de. La moneda nacional: conocimiento y usos de ma moneda metalica y de llos billetes. Problemas. Numeros naturales: extension del circulo hasta 1.000.

Tercer nivel: Producto cartesiano y par ordenado. Multiplicacion de numeros naturales: concepto como suma abreviada y como cardinal del producto cartesiano, tablas, practica graduada, propiedades modulatia, cancelativa, conmutativa, asociativa y distributiva con respecto a la adicion, algoritmo, multiplicacion por 10, 100 y 1.000 y problemas de apliacion al medio. Extension del circulo de numeros naturales. Reparticion: practica graduada. Division: concepto en base a la reparticion, a la busqueda del factor desconocido y a la resta reitarada, division con residuo igual y distinto de cero, practica graduada, algoritmo de la division hasta por 2 cifras y problemas de aplicacion. Equiforma y equitamana. Distinta forma y tamano. Segmentos congruentes e incongruentes. Longitud: equilingitud, segmentos mas y menos largos, suma de longitudes, relaciones del doble, triple, mitad, tercia, etc. Longitud, unidad de medida de longitud: el metro y el centrimetro, medicion de segmentos. Figuras geometircas mas usuales: el triangulo y el rectangulo.

Cuarto nivel: Numeros decimales: concepto en base a longitudes, decimos, centesimos, lectura, escritura, aplicacion a los precios de las cosas, orden en el conjunto de los decimales, operaciones: adicion, resta, multiplication y division. Problemas de aplicacion al comercio. Perimetro: concepto, calculo del permetro de figuras mas usuales y problemas. Area: concepto, calculo y problemas. Triangulacion y medicion de superficies, aplicaciones en la agrimensura y problemas. Medidas de peso, capacidad, tiempo, longitud y superficie, problemas. Numeros racionales: 1/2, 1/3, 1/4, 1/5, 1/6, 1/8, 1/10, 1/100, etc.

Volumen: concepto, calculo de volumenes de cuerpos conocidos y de aplicacion en la construccion. Operadores numericos fraccionarios y decimales, aplicaciones. Tanto por ciento e interes simple: conceptos, calculo y problemas sobre porcentajes, tasas, intereses, reditos, etc. El numero aproximado: estimacion de medidas.

Extension del conocimiento de numeros racionales: simplificacion, amplificacion, propiedades y operaciones, calculo y problemas. Operaciones con medidas de superficie y de volumen: calculo y problemas. Estadistica: muestreo aplicado a la agricultura y a la poblacion humana y ganadera.

Problemas.

Metodos:

Ensenanza indivudualizada y grupal.
La heuristica partiendo y regresando a la realidad.

El neo-objetivismo.
La presentacion multiple.

Utilizar medios de comunicacion y materiales educativos.

Actividades:

Onticas: Manipular, coleccionar, enlazar y esparcir objetos, etc.

Quebrar palitos, estirar y encoger objetos, recortar y dibujar, etc.

Colorear, modelar, etc.

Ontologicas: Unir y partir conjuntos, quitar y parear objetos. Configurar pilas, columnas, filas y series de objetos. Medir y contar objetos. Distinguir formas, tamanos y colores.

Caracterizar objetos. Diferenciar caracteristicas de objetos, etc.

Clasificar caracteristicas. Conversar y comunicar caracteristicas, etc.

Logicas: Construir y completar series de objetos con distintas caracteristicas. Clasificar medidas, formas, tamanos y colores. Cumplir ordenes. Contestar preguntas. Elaborar preguntas. Identificar muneros y bases de sistemas de numeracion. Consturir figuras. Realizar juegos logicos. Graficar soluciones. etc.

Simbolizar objetos y acciones. Analizar y comparar respuestas. Crear alternativas de solution. Esquematizar soluciones, etc.

Extraer y generalizar conceptos. Extraer y elaborar reglas de juegos. Inferir situaciones, etc.

Matematicas: Sumar, restar, multiplicar y dividir numeros. Traducir problemas a formulas y formulas a problemas. Resolver formulas y problemas. Probar calculos y soluciones. Utilizar propiedades de operaciones. Esquematizar problemas. Leer y construir tablas, etc.

Deducir respuestas. Estimar respuestas. Generalizar leyes. Extraer tecnicas. Simplificar algoritmos, etc.

Crear conceptos. Mostrar propiedades. Elaborar sencillos modelos matematicos. Descubrir matematica en juegos, materiales y herrmientas, etc.

Estratagias:

De la estructura del sistema educativo:

Educacion no formal de 4 niveles. Aulas abiertas todo el ano comercial laborable. Duracion de los niveles en funcion de las necesidades, intereses, aptitudes y actitudes de cada alumno. El ingreso al nivel correspondiente, el alumno hace en cualquier epoca del ano y egresa ocasionalmente al termino del respectivo nivel. Las sesiones de trabajo escolar son matutinas y vespertinas. La ensenanza de los 3 primeros niveles es en el idioma nativo y la del 4°, en el idioma oficial. La aprobacion de un nivel consiste en que el alumno meustra que sabe aceptablemente los contenidos del citado nivel.

Funcionamiento de un 5° nivel de equiparacion de estudios de la escuela primaria de 6 anos, conforme al programa de estudios oficial vigente.

Functionamiento de cursos de educacion para la comunidad: salud, crianza de ninos, manualidades femeninas, carpinteria, sastreria, albanileria, tejidos, etc., segun las necesidades comunales, y con estudios de Matematica Aplicada a las mencionadas ramas artesanales.

Del alumno:

Ingreso a la escuala o centro educativo desde los 6 anos de edad hasta los 16 o 17. Los que ingresan a los 12 o mas anos de edad terminan los 4 niveles en 2 o 3 anos, a lo sumo, para inmediatemante integrarse a la comunidad, iniciando el trabajo y vida independientes, o emigrando a lugares populosos en busca de trabajo. Al regresar ocasionalmente a la comunidad, continuar los estudios del 5° nivel y/o los artesanales que tienen una duracion aproximada de 3 meses. El ingreso tardio favorece el apredizaje, la integracion comunal y el cambio de actitud.

El ingreso a los 8 anos de edad hace que el aprendixaje sea mas lento, por lo que los 4 niveles son terminados en 4 anos. El aprendisaje lento se debe al la poca madures mental producida por la pobre actividad. El peligro de esta alternitiva radica en que el alumno egresado a los 12 anos de edad retorna al pasotreo y se produce la regresion del aprendizaje que culmina en el analfabetismo. Sin embargo, mediante acciones de convenciamiento a la comunidad y el uso de sesiones especiales de trabajo escolar: las ultimas horas de la tarde y las primeras de la noche, el alumno puede continuar sus estudios del 5° nivel y los de artesania, alternando con el pastoreo ocasinal y llegar, entonces, a la edad en que se libera. Asi, a una edad mas temprana, se halla capacitado para enfrentar su propia vida, hacer mejor uso de su libertad y rescatarse del analfabetsimo.

El ingreso a los 6 anos de edad causa problemas en el aprendizaje que se manifiesta demasiado lento y dificil, motivado por la inmadurez mental como fruto de su inactividad, por su desadaptacion a la escuela provocada por la ruptura violenta del maternalismo excesivo. Al egresar vuelve al pastorea y se producen los malos efectos ya mencionados.

Del profesor:

Un equiro de maestros altamente calificados y especializados planifica, dirige, orienta, sigue y supervisa las tareas de esta educacion no formal. El mismo equpo prepara informacion y material del trabajo docente y, ademas, forma, en pocas semanas, a los mointores indigenas seleccionados por la Comuna, a los que somete a un intenso entrenamiento teroico-practico sobre tecnicas educativas y practica docente. Los cursos de reciclaje, perfeccionamiento, acutalization y capacitacion se repiten peridicamente. El seguimiento, supervision y provision de trabajo y materiales funcionan perennemente, asi como la consultoria. Los monitores indigenas que sienten orgullo de ensenar a sus semejantes cobran poco por su trabajo docente.

Institucionalizar la formacion de monitores o instructores indigenas.

Instrumentar acciones para realizar investigaciones sociales, economicas, educativas, etc.

Aumentar la preparacion de los monitores y educarlos permanentemente.

Evaluar el sistema y el aprendizaje.

Extender el radio de accion del sistema evaluado.

Informar a los maestros de otros grupos sociales sobre los resultados obtenidos, con el fin de intercambiar conocimientos e ideas.

CHAPTER 3 - Elementary Education

3.1 ROOTS OF FAILURE IN PRIMARY SCHOOL ARITHMETIC

WHEN CHILDREN FAIL TO LEARN ARITHMETIC IN PRIMARY SCHOOL, WHAT ARE THE ROOTS OF THE FAILURE? WHAT APPROACHES CAN BE TAKEN TOWARDS A CURE?

Frederique Papy
Brussels, Belgium

Does the teaching of mathematics in primary school respect the child's natural way of thinking? If not, might that not be seen as one of the major causes of failure?

It is striking to note the ease with which little toddlers learn their mother tongue through day-to-day contact within the family circle, without adults using any teaching technique other than the spoken work, pronounced with tenderness and modulated in accordance with the intended meaning. As Venn diagrams, colored arrow graphs, Papy's Minicomputer, and electronic calculators are essential instruments of mathematical thinking, it is not surprising that they can be used with the youngest of children in the same way as the language by means of which they learned to speak.

From as early as four years old, when pupils obviously know neither how to write nor how to read, and are still very clumsy at the techniques of drawing, it is possible to introduce them straightaway to some of the important themes of mathematical thinking such as relations and sets. By means of a little story translated into the language of strings and colored arrows, the child learns to understand an abstract, ideogrammatic, and symbolic drawing, an ability that will continue to be useful to him throughout his studies and his life. At a time when he finds it difficult to draw a line on the board going from one dot to another and when the drawing of an arrowhead is utterly beyond his capabilities, he can follow the growth of a nice, big, abstract, colorful drawing which occupies his entire field of vision, and can answer questions by tracing along previously-drawn arrows with his hands, or by indicating where to draw strings around certain dots.

Passages from the story to the graph is made directly, without useless transitional stages. We have here a genuine translation from the mother tongue (which, it must not be forgotten, is conventional) into the ideogrammatic language (which in certain respects is less conventional) of Venn diagrams and arrow graphs. Figures, drawn bit by bit tell the story. The actors are represented by dots and the actions by arrows. A sober introduction of the tale sets up the bare canvas of the situation. The burden of verbal ornamentation falls on the children, who do not stint themselves, thereby showing the intensity with which they are caught up in it.

As every attempt at short-term evaluation shows us to be so much wasted effort, we will show the effects of such teaching three years later when the same children are confronted with numerical situations that are comparable with problems that are traditionally taught. (Second grade class: 7-8 years old).

> Our friend Nabu has 34¢. At the general store he sees some lovely green marbles for 3¢ each and some even nicer red ones for 4¢ each. He chooses among these marbles and spends all his money. If you had been in Nabu's place, what would you have bought? How many red marbles, how many green ones?
>
> The discussion of this problem is equivalent to looking for the solution set of a Diophantine equation.

In a single drawing, a student presented an array of solutions which allowed one to follow the train of thought. A succinct note on the page immediately interpreted that data in the language of arrows. Even though the first road enabled the student to achieve the goal, this student preferred to start again and provided three correct answers. Unlike some of the other classmates the usual mathematical expression "7 greens plus 2 reds" was not used, but the conjunction and was timidly substituted.

> Last Wednesday, our friend Christian played dominoes with his grandmother. Each game he won, he received 4¢; each game he lost, he paid 2¢. He won 30¢. Imagine a way in which this series of games could have developed.
>
> Another Diophantine equation.

Using numbers, colored arrows and calculations, a student provided several solutions in a graph indicating that for this student, mathematical writing was a real language, allowing transmission of pieces of information selected at will. The student spontaneously used negative numbers which were introduced at age 6, showing that the student had become familiar with them.

Graphs, which are thought of and then drawn by the children, structure the situation. As tools for work and discovery, they favor the child's imagination and enrich his graphical expression without imprisoning him in a narrow systematism.

In answer to the question posed at the outset, we have given priority to "preventive" mathematics. Experiments carried out since 1970 by Robert Dieschbourg in the Grand Duchy of Luxembourg, by Francis Lowenthal, Christiane Vandeputte, and ourselves in Belgium and the USA, show that these techniques may be adapted for 6- to 15-year-old pupils who are either mentally handicapped (IQ between 50 and 70) or who have serious behavior problems. This seems to indicate that the suggested activities are profoundly involving for the child and address themselves to the most basic mental structures.

Bibliographie

Comprehensive School Mathematics Program (CSMP), 1978. CSMP in action. CEMREL Inc., St. Louis, USA. (Trad. francais, CBPM, Bruxelles 1979.)

Dieschbourg, Robert. Un enseignement de la mathematique a des enfants mentalement handicapes. Nico 10, CBPM, Bruxelles (1971), pp. 34-63; Nico 13, CBPM, Bruxelles (1973), pp. 53-97; Nico 16, CBPM, Bruxelles (1974), pp. 129-152; Nico 19, CBPM, Bruxelles (1975), pp. 39-57.

Frederique et Papy. 1968. L'enfant et les graphes. Didier, Bruxelles-Montreal-Paris. (Trad. en anglais, Algonquin, Montreal; en italien, Societa Editrice Internazionale, Torino; en croate, Skolska Knjiga, Zagreb; en russe, Pedagogika, Moscou.)

Frederique. Les enfants et la mathematique. Didier, Bruxelles. Vol. 1, 1970; Vol. 2, 1971; Vol. 3, 1972; Vol. 4, 1973. (Trad. en anglais, Algonquin, Montreal; en italien, Societa Editrice Internazionale, Torino; en allemand, Ernst Klett Verlag, Stuttgart.

Frederique. 1977-80. Math Play Therapy. Vol. 1 et 2. CEMREL Inc., St. Louis, USA.

Frederique. 1975-75. Stories of Frederique. CEMREL Inc., St. Louis, USA.

Frederique. Contes mathematiques. Collection PAPY, Hachette, Paris. (Trad. en anglais, Thomas y Crowel Cy, New York; en italien, Societa Editrice Internazionale, Torino; en allemand, Ernst Klett Verlag, Stuttgart.

Frederique et Papy. 1973. Les enfants de 4 ans et le langage des graphes. Hachette, Paris.

Lowenthal, Francis. La mathematique peut-elle etre therapeutique? Nico 10, CBPM, Bruxelles (1971), pp. 69-86; Nico 13, CBPM, Bruxelles (1973), pp. 98-104.

Papy. 1968. Minicomputer. IVAC, Bruxelles.

Vandeputte, Christiane. Un enseignement moderne de la mathematique a des enfants paralyses cerebraux. Nico 13, CBPM, Bruxelles (1973), pp. 105-139; Nico 16, CBPM, Bruxelles (1974), pp. 86-121.

Van Halteren-Dieudonne, Anne. 1971. A propos de l'utilisation du langage mathematique des papygrammes. Nico 10, CBPM, Bruxelles, pp. 129-146.

3.2 DO WE STILL NEED TO TEACH FRACTIONS IN THE ELEMENTARY CURRICULUM?

DO WE STILL NEED TO TEACH FRACTIONS?

Peter Hilton
Case Western Reserve University
Cleveland, Ohio, U.S.A.

1. Introduction

Of course, the question should not be taken too literally - - certainly we should teach fractions as part of the elementary curriculum. But it is my contention that we should not teach fractions in the way they have been taught and still are being taught. Indeed, if the questions were 'Do we still need to teach fractions as they are taught today in the major elementary programs?', then the question could be taken literally and my answer would be 'No; in fact, we should never have taught fractions this way'.

I will point in the next section to five fundamental defects of the traditional approach to the teaching of fractions. Not all programs may exhibit all these defects, but I will testify from a painstaking scrutiny of texts in common use today that all exhibit some of these defects, and some exhibit all. I will propose, in Section 3, a different approach to the teaching of fractions. Naturally, I do not claim originality for all my suggestions - - especially do I acknowledge that the use of probability concepts to motivate the arithmatic of fractions is now recommended almost universally, though I find that some texts pay the merest lip-service to this recommendation.

The approach I am proposing can, of course, be inferred, to a considerable extent, from the description of the defects which precedes it. However, I do make, at the end of the article, some new suggestions for teaching fractions as a part of mathematics and these suggestions are not readily deducible from the content of Section 2. I claim that, through fear, distaste or perhaps wrong-headed principle, we tend to teach the arithmetic of fractions exclusively as a skill, without giving any attention to its rich mathematical content. A good curriculum would certainly place more stress on the mathematics and utility of the arithmetic of fractions and less on mere technical skill.

2. The defects of the present curriculum relating to fractions.

a. Phony applications. The multiplication of fractions lends itself readily to sensible applications; it is more difficult with the other arithmetical operations, but since division does not play so prominent a role in the curriculum, the proliferation of spurious applications is most conspicuous where addition and subtraction of fractions are concerned. 'Tommy works 2 7/12 hours hours in the morning and 1 2/3 hours in the afternoon. How long does he work altogether? A recipe calls for 7/8 a cup of sugar. Anita has already poured out 1/3 cup. How much more sugar does she need?' It should not be necessary to be explicit about the respects in which such examples, drawn from popular texts, are phony. But there is also a subtler form of phoniness in which the applications, which are perfectly genuine as real-life problems, do not begin to justify the attention given to the arithmetic process. Thus, in one text, there are some 40 drill problems devoted to the subtraction of mixed numbers involving the finding of least common denominator, regrouping and simplifying (thus, as an easy example, 6 1/6 - 4 2/3 = 1 1/2). To make this gruesome exercise more palatable, presumably, the drill problems are followed by word problems which, being honest, involve only the subtraction of fractions having, 2, 4 or 8 as denominator.

b. Confusion with the role of decimals. This confusion appears at two levels: there is a confusion between fractions and decimals as mathematical entities, and a confusion between their utilitarian roles in applications. Many texts claim to include lessons designed to show how to convert fractions to decimals. In fact, of course, such a conversion is rarely possible, requiring that the fraction be equivalent to one whose denominator has 2's and 5's as factors. This fundamental mathematical fact is glossed over by first concentrating (but not openly) on such fractions; and then saying that, in some cases, the resulting decimal 'does not terminate'. Up to this point, all decimals have terminated - - and so, of course, they should in the light of their definition. Decimals, the student has already learned, can be added, subtracted and multiplied by certain elementary finite algorithms. Infinite decimals cannot. Thus 'infinite decimals' are entirely new animals, if they exist at all. (The belief that infinite decimals are decimals is akin to the belief than an average height is a height or than an average family is a family.) The student is, however encouraged to write 1/3 = 0.333 . . . If he (or she) asks what the dots stand for he is presumably told that they stand for 333 and three more dots!

With respect to applications we find a similar confusion of roles. It is unfortunately true that, owing to the retention in the U.S.A. of a lunatic system of weights and measures, it is still sometimes the case that measurements are made using fractions. However, it is not at all natural to measure in fractions to prescribed degrees of accuracy, except in special cases (1/2 inch, 1/4 inch, 1/8 inch, 1/16 inch). Measurements are naturally effected in decimals; and particular absurd are measurements announced in fractions with utterly different denominators. 'To get to school, Johnny walks 3/10 mile and rides the bus for 2 1/3 mile. How far is it to school?' Not far enough, I would answer, if that's the rubbish taught there. Of course, the availability of hand-calculator further emphasizes the appropriateness of decimals in measurement and calculation.

There is a related tendency, too, where an implicit decimal problem is presented as a fraction problem because we're currently 'doing fractions'. ' 4 7/10 + 5 3/100' is a trivial problem if presented, honestly and correctly, as '4.70 + 5.03'.

Of course, there is an interaction between fractions and decimals, an important one, but this is largely ignored because of the absurd compartmentalization to which mathematics instruction is subjected. We do not refer to an artificial interaction to be found quite often, as in problems like 'Insert the correct sign (< , > , =) in the following '1/3 0.31' . . . ' But it is perfectly natural to be asked to take 1/2 of 5.74 or 3/4 of 9.27 (the inequivalence of fractions and decimals is adequately attested by the fact that it is highly unnatural to take 0.5 of 5 74/100!). Moreover, many 'shopping' problems lead naturally to the division of decimals by fractions. 'The cost of an article is made up of the price, together with a 7% sales tax. If the cost is $13.27, what was the price?' We must divide 13.27 by 107/100.

c. Absence of care if definition and explanation. Fractions always start life as parts of wholes. At this stage they are certainly not numbers, they are things - 'half a cake', 'three quarters of the pie'. Moreover, they are, of necessity, proper parts. At a certain stage we pass from the things themselves to amounts, or measures of things. At this stage, we are entitled to say that 1/2 = 2/4, and to introduce fractions greater than 1. However, none of the standard texts makes this transition explicit. Let us give one example. One text, in its initial explanation of fractions, never subsequently repudiated or even modified, says that the statement '3/8 of the apples are red' means that 'there are 8 apples and 3 of them are red'. So if I say 'a half of the applies are red' I mean there are 2 apples, of which one is red! Later, and without warning in the Teacher's Guide, the fraction 3/2 is introduced - - as if it had always been there. What is the student to make of this?

Most texts (in my experience, all but two!), seeking to show that 1/2 = 2/4, demonstrate this by exhibiting the same portion of the same region. But what is asserts is the equality of certain portions of equal regions - - it is, in other words, a statement about amounts and not about things. Try convincing a hamster that 5/5 of a hamster is the same as one hamster! The difficulty that all these texts have in explaining equality of fractions is that none is explicit about how a fraction becomes a number. Not one text I have seen (except the CSMP material and Real Math, Open Court Publishing Co.) ever introduces the term 'rational number'. It is not a difficult idea - - certainly much easier than 'repeating decimal, a common favorite - - and it is fundamental. With its help we can say that 1/2, 2/4 represent the same number but are different fractions. Without the notion of rational number it seems to me impossible to properly explain ratios. Thus a standard text says that a ratio 'is a pair of numbers'. If 6 items cost 89¢ then 'the ratio 6 to 89 tells the price'. There is a discussion of 'equal ratios' - - presumably, equal pairs of numbers. Nowhere in this text are ratios related to fractions, and certainly not, of course, to rational numbers. So there can be no 'simplification' of ratios.

A great advantage about introducing rational numbers explicitly and distinguishing them from fractions is that one can then discuss operations on fractions which are not operations on rational numbers. Perhaps the most important example is the operation

$$\frac{a}{b} \circ \frac{c}{d} = \frac{a+c}{b+d}$$

The traditional treatment of this operation is simply to tell the student this is not the rule of addition. It is, however, a perfectly good, and very important, operation on fractions, as anyone who has studied baseball players' averages would know. Averages are treated in most texts, of course; but their treatment is usually vitiated by the device, to which we will allude below, of 'fixing up the numbers' so that the average of a set of whole numbers is always a whole number! This, however, trespasses on the next defect we identify.

d. Dishonesty of presentation. This defect is to be distinguished from the phoniness of applications, though there is some overlap. We refer here to such devices as grandiose statements of objective. 'To multiply a whole number by a fraction with numerator 1' - - but we find the student is learning here to take a (unit) fraction of a whole number and can only take the fraction if the result is also a whole number - - a rare case in fact but a common device in curricular fiction. Nowhere in the text is it mentioned that we are dealing with a very restricted case of the general problem. Elsewhere in the text, the objective 'to write fractions as percents' turns out to refer to fractions whose denominator is 100!

Another example of dishonesty is simply writing down '1/3 x 12' below '1/3 of 12' and thus claiming we are learning to multiply by fractions. There is no explanatin of why we call the process of finding 1/3 of 12 multiplication. Another text adopts a different but even more surreptitious device - - having demonstrated that 12 x 1/3 = 4 by repeated addition, it produces as the next line '1/3 x 12 = 4' with no explanation.

The complementary disingenuity is also frequently found. The text demonstrates, by crosspatching a square that 1/2 x 1/3 = 1/6. It then writes '1/2 x 1/3 = (1 x 1)/(2 x 3) = 1/6., so in multiplying fractions we simply multiply numerators and multiply denominators'. It seems reasonable to doubt whether the student, in observing that 1/2 x 1/3 = 1/6, ever realized he was multiplying 1 x 1 to get 1!

A final example of dishonesty must - - and should - - suffice. To add two fractions, we are told to find the lowest common denominator (a silly phrase, since we only have to find the lowest common denominator when the fractions involved don't have a common denominator). This is routinely presented in three stages. First, if the fractions have the same denominator this is the l.c.d. Second, if one denominator is a multiple of the other, use it as the l.c.d. It is the third stage which is far less explicit. Most texts proceed immediately to examples like 2/7 x 5/12 and say 'use 7 x 12'. Others give examples like 3/4 x 5/6 and say "try 12'. There is, however, no mention of any systematic procedure, no hint even as to when you multiply the two denominators. The student is led to believe he (or she) has been taught a procedure and has learnt it, but this is not the true case.

e. Passion for orthodoxy. This malaise, which afflicts the entire traditional mathematical curriculum, is pecularily fatal to the effective teaching of fractions. We find in one text that improper fractions 'should be rewritten as mixed numbers'. Why? If I have to calculate 3/4 of (2/3 + 2/3) is it a progressive step to rewrite 3/4 of 4/3 as 3/4 of 1 1/3? Another text insists that fractions must be 'simplified'. Of course, what is meant is 'reduced', since it is a highly subjective viewpoint that 46/100 is simplified when written as 23/50. But is it always sensible to reduce? There is a ludicrous example in this same text where we find a whole partitioned into three percentages, namely, 45%, 40% and 15%. The instruction is, first, to express these percentages as fractions - - of course 'simplified' - - and then to add the fractions. There is no mention that this is an absolutely crazy way to verify that the percentages are compatible with a partitioning; and, likewise, it is not stated, as honesty should have dictated, that it is far easier to add the fractions if they are not first 'simplified'.

In concluding this section, let me make some points. Plainly the defects I referred to have their counterparts - - and even their exemplification - - in other domains of elementary mathematics. Plainly, too, there are other defects in the curriculum which we haven't mentioned even though they inflict grevious harm on the teaching of fractions. These defects have not seemed to us to be related to the teaching of fractions as closely as the five we have discussed. However they are serious too - - and done is especially serious but, so far as I know, nowhere explicitly referred to, so that I make no apology for mentioning it here. I will, however, be brief.

A substantial proportion of elementary teaching of mathematics is taken up with review, and it is proper that it should be so. My readers will surely be as familiar with the manifold reasons for this as I. Yet review (with the two exceptions quoted above) seems always to consist of a review of rules and skills and never of explanations. Thus is the idea insidiously instilled in the child that memory and accuracy are the principal components of mathematical success; and thus does the study of mathematics, and of a subtle area like the arithmetic of fractions, suffer, in many cases beyond hope of repair.

3. Ingredients for a good course in fractions.

As pointed out in the introduction, much that we might have said under this head is easily deducible from the content of the previous section. We will only repeat here our belief that we should quite explicitly introduce rational numbers (we refer here to positive rational numbers; the whole field of rational numbers comes into play, of course, once negative numbers have been introduced) - - and that there would be no difficulty in doing so. Let us then talk in this section of rational numbers where appropriate (this does not mean we avoid all mention herafter of fractions!).

We should present multiplication as the basic operation with rational numbers. This is correct mathematically, since rational numbers arise out of the desire, or need, to invert multiplication; and it is correct in practice because it is easiest to find real-life applications of the multiplication of fractions. Division by rational numbers presents no real problem mathematically, but it is harder to find convincing applications. There are questions of the 'how many?' type which call for division by fractions but in which the answer is obtained by rounding the quotient up or down to a whole number. There are other questions (e.g., price before tax or price before percentage reduction) which call for the division of decimals by fractions. Thus we should honestly admit that the precise division of rational numbers by rational numbers is not a very important practical tool, certainly not to be compared with multiplication.

The addition of rational numbers should be treated after their multiplication, being a far less important operation on fractions. The best exemplification comes from elementary probability theory, where the probability of a complete disjunction of two (mutually exclusive) events is the sum of the probabilities of the two events. It is to be noted, however, that in this application the two probabilities tend to be naturally represented by fractions with the same denominator. Thus, if I throw two dice and add the numbers showing up, the probability that I get 4 is 3/36 and the probability that I get 8 is 5/365, so that the probability that I get 4 or 8 is 8/36. It would be perverse to express the first probability as 1/12 and thus complicate the task of adding the fractions - - simplicity does not reside in the fraction but in the arithmetical process which the fraction serves. Even to present the answer as 2/9 is unnecessary and perhaps even unwise, if we want to compare the probability of getting 4 or 8 with the probability of getting 7, say.

A great advantage of typical probability calculations is that they often involve, simultaneously, the addition and multiplication of rational numbers. However, we should not be giving the (false) impression that we should teach probability theory to provide an outlet for our work with fractions. We should teach probability theory because it's useful and important to today's citizens; and we need the arithmetic of fractions to do probability theory. (It is our experience that adults, having grappled disconsolately and unsuccessfuly with fractions as children can be restored to competence, confidence and interest by relearning fractions in the context of probability theory.) (1)

The subtraction of fractions is of far less importance in practice than the addition of fractions, and this again should be honestly admitted. There are natural applications for the subtraction of simple fractions but the principal justification for giving time to this topic is that it is mathematical importance. I believe such justification is valid - - but it does not excuse over-emphasis on the topic, nor the dreary inculcation of dependability in the execution of an unexciting algorithm.

I should dispose here of an argument advanced for teaching arithmetic of fractions - - that it constitutes the prototype for the algebra of rational functions. If this argument (advanced, for example, by Morris Kline in his book, Why the Professor Can't Teach) merely means that mastering the arithmetic of fractions will stand the student in good stead if he or she ever reaches the level at which the algebra of rational functions in in question, then it cannot be gainsaid. But if it means either (a) that this should constitute explicit motivation for the study of the arithmetic of fractions, or (b) that prior study of the arithemetic of fractions is essential to the later study of rational functions, then I must disagree. (Morris Kline appeared to be advocating the argument in both senses a and b.) As to (a) I do not at all believe in the 'pie-in-the-sky' approach to motivation. To ask a student patiently to learn some meaningless and dreary skill, in the expectation of some reward at an unspecified future date, is to adopt a teaching strategy at once immoral and ineffective. As to (b), there seems to me to be no need for a long, detailed and painstaking study of the addition and subtraction of fractions as a preparation for the study of rational functions some years later. All that is necessary is exposure to the ideas; the techniques, together with the explanation of their validity, should be discussed in a review preceding the introduction of the corresponding operations in the field of rational functions.

There are two further points to be made on this question. First, intuition should play a far stronger role in the arithmetic of fractions than in the algebra of rational functions. One does not want a student to compute $1/2 + 1/2$, or even $1/2 + 3/4$, by appeal to the rule for calculating $a/b + c/d$. Thus, in the most important cases, the algorithm will not be used. Second, there are important differences between the two topics. A very important aspect of the algebra of rational numbers is the partial fraction expansion. The analogous process for rational numbers is far less important, and is not even unique. Thus $1/6 = 2/3 - 1/2 = 1/2 - 1/3$. Of course, the time to take up this question, and to consider the reason for the different behaviors in the two cases, in when one is studying rational functions.

There is a further feature of the set of rational numbers which is commonly treated in the textbooks - - and should be. I refer to the fact that rational numbers are ordered by magnitude. However, decimal notation is far better suited to comparisons of this kind than fractional notation; it is obvious by inspection that 26.524 is bigger than 26.518, but far from obvious that 60/19 is bigger than 22/7. Here we perceive an irony. There is a method, called cross-multiplying, which is an entirely reliable algorithm for determining which of two rational numbers is bigger. It is based on the principle

$$(3.1) \qquad a/b > c/d \text{ if and only if } a \times d > b \times c.$$

Yet this principle is not used in today's texts; instead the principle

$$(3.2) \qquad a/b = c/d \text{ if and only if } a \times d = b \times c$$

is used to test whether two fractions are 'equal'. (Actually, the situation is worse. Typically, we read 'if two fractions are equal, their cross products are equal', and then find it is the unstated converse which is being used. The quoted sentence is also reprehensibly because it very much appears to be saying that a fraction has a cross-product.) Now it is practically never the case that the quickest and easiest way to test whether two fractions are equal is to apply (3.2); for two equal fractions are liable to have denominators which are 'closely related', multiplicatively (that is, which are far from being mutually prime), so that it should be relatively easy to reduce them to the same fraction. On the other hand, it will be very often the case that (3.1) provides the quickest test of whether $a/b > c/d$, especially if a hand-calculator is available.

There is much to be said; but, in order to keep this article down to reasonable length, and to provide others with something that they may strongly disagree with, let me close by recommending that there are important mathematical features of the set of rational numbers which should be displayed, discussed and exploited in the curriculum. I do not, of course, claim that they are suitable topics in the first six grades, indeed, I will begin my own discussion by assuming that students have all become familiar with negative numbers. This I do largely to simplify my own exposition, and the reader will, I hope, readily see that some of my remarks would still be meaningful if the students were unfamiliar with negative numbers.

First, then I believe it important to point out that the set **Q** of rational numbers is closed under the four basic operations (addition, subtraction, multiplication, and division by non-zero divisor) and is the smallest set containing I with this property. It thus marks a sort of mathematical consummation - - but a consummation rudely shattered by the Pythagorean discovery of irrational square roots (of course, geometry must come into all mathematics!). It is important to make the distinction here between the fact that we can perform any of the operations on elements of **Q** and say inside **Q**, and any particular technique for actually calculating and expressing the result of such an operation.

Second, I suggest we should look at certain subrings of $\mathbf{Q}$, in particular, the subrings Z_P, where P is a family of prime numbers and Z_P is the set of rational numbers expressible as fractions a/b with b prime to P. It is interesting to do 'factorizatin into products of primes' in Z_P, instead of only in the ring of integers $\mathbf{Q}$. Moreover, there is a very attractive extension then possible of the beautiful 'casting out 9's'. This device involves the ring-homomorphism Θ:Z $\rightarrow$ Z/9, where Z/9 is the ring of intergers modulo 9. We use Z/9 because our base 10 enumeration system makes it particularly easy to calculate 'by adding the digits'. If F is a formula in Z, then Θ (F) is a formula in Z/9 and Θ(F) is true if F is true. On the other hand, formula in Z/9 are far easier to verify than formula in Z, so the ease of calculating and the ease of checking formulae in Z/9 makes 'casting out 9's' a very valuable tool in verifying calculations: if the formula Θ(F) is false, so must the formula F be false.

My point is that Θ extends naturally to a ring-homomorphism Θ : Z_3 --- Z/9; here (if P consists of a single prime p, we write Z_p for Z_P) Z_3 is the ring of rational numbers expressible as fractions a/b where b is not divisible by 3. This means that we can use the casting out 9's to check any calculation involving the addition, subtraction and multiplication of fractions, provided no fraction in our calculation has a denoninator divisible by 3. Let us give an example

Example 1 Check 5 13/28 + 14 71/220 = 19 303/385. We cast out 9's. Now 5 13/28 $\equiv$ 5 + 4/1 $\equiv$ 0; 14 71/220 $\equiv$ 5 + 8/4 = 7; 19 303/385 $\equiv$ 1 + 6/7 $\equiv$ 1 + (6 x 4) $\equiv$ 7; and 0 + 7 = 0. Note that we use the fact that 4 x 7 $\equiv$ 1 to infer that 1/4 $\equiv$ 7 and 1/7 $\equiv$ 4.

Almost as easy as casting out 9's is casting out 11's (for details see (2)). Now we have a ring-homomorphism Θ : $Z_{11} \rightarrow$ Z/11. We lose in casting out 11's rather than 9's in simplicity; but we gain because fractions encountered at random are more likely to have denominators divisible by 3 than by 11. Again we give an example to show the method.

Example 2 Check 6 12/29 x 2 3/71 = 13 7/71. We cast out 11's. Now 6 12/29 $\equiv$ 6 + 1/7 $\equiv$ 6 + 8 $\equiv$ 3; 2 3/71 $\equiv$ 2 - 1/2 $\equiv$ 2 - 6 $\equiv$ 7; 13 7/71 $\equiv$ 2 - 7/6 $\equiv$ 2 - (7 x 2) $\equiv$ 10; and 3 x 7 $\equiv$ 10.

Note that one casts out 11's by taking the alternating sum of the digits (starting at the right); and that we have used the congruence 2 x 6 $\equiv$ 1 to infer that 1/2 $\equiv$ 6, 1/6 $\equiv$ 2.

It may be objected that, having inveighed against pointless calculations involving fractions, we are now taking calculations as a pretext for discussing casting out 9's and 11's. Our reply is that we are here exemplifying a practical rule of great importance in mathematics - - try to solve the problem by simplifying the situation. (We could also claim that there's a lot of fun in these calculations; and fun is an essential element in the successful mathematics curriculum.) This, after all, lies at the heart of doing effective applied mathematics, and we - - and the students - - are here seeing the principle in action within mathematics itself.

References

1. Peter Hilton and Jean Pederson, Fear No More: An Adult Approach to Mathematics, Addison-Wesley (1981).

2. Peter Hilton and Jean Pederson, 'Casting out 9's revisited' (to appear).

THE FUTURE OF FRACTIONS

Mary Laycock
Nueva Learning Center
Hillsborough, California

My position as a teacher of elementary children and their teachers is that for the 80's fractions are a part of the elementary school mathematics. Whether fractions should be eliminated or not is irrelevant to my presentation. I will address three topics:

1. Developmental levels of elementary children.

2. Uses for fractions.

3. A demonstration of one of the effective sets of manipulatives for presenting meaningful fractions.

Research has established that elementary children are in the conrete stage of learning, yet we are teaching them with meaningless rules. Children believe and learn only what they experience. Many forces such as the coming of the Metric System, calculators, and computers simplify and alleviate the complicated forms of fractions and long computations with whole numbers and decimals. What is of great importance is the development of meaning into whatever we teach with emphasis on estimation and applications of mathematics to the real world.

Fractions are essential to the ratios and percents we accept as important to understanding the world. Because proportional reasoning is now expected at age 16 rather than 12, it is not reasonable to think fourth graders would understand 3/4 = 6/8 because 3 x 8 = 4 x 6; yet that is presented in currently used fourth grade texts. Through many different experiences with concrete objects children do understand that 3/4 and 6/8 are two ways to make the same number. If we do not give children this hands-on experience during their concrete learning stage to build the understanding, skill in the use of percent can only be acquired through meaningless rules.

I would like to recommend Z.P. Usiskin's article on "The Future of Fractions". As he cites it, there are important practical uses for fractions: Splitting a pie or a candy bar or an inheritance; rates like 80 km/2 hrs = 40 km/hr; averages; probability, etc. Much of the arithmetic curriculum is sterile with dull memorized drill and practice. Applications, especially probability experiments and the fractions they generate are important. Since calculators do not as effectively work with fractions, whatever fractional concepts we present must be well taught. That is the core of the problem as I see it. Because teachers learned by senseless rules, too often they teach that way. Much of my energy goes into teaching teachers to use many alternatives for presenting mathematics meaningfully, fractions among those ideas. The microcomputers are making it possible for students to learn about many things, including fractions, in spite of the teacher. The program must interact with excellent graphics. Such do exist, ask me! There are many good ways to make fractions make sense: Paper folding, cutting circles or squares, Cuisenaire Rods, tangrams, Fraction Bars, Pattern Blocks, and my favorite Fraction Tiles.

The remainder of my presentation is to demonstrate on the overhead how I present the meaning, comparison, addition, subtraction, division, and multiplication of fractions to children. These manipulatives are fascinating to see on the overhead. However, children do not understand the fractions by seeing them. They learn only by touching them. As I watch them pick up the pieces and try out their thoughts on a solution, I am excited as a teacher because it is like having a window into their brains - - I can see the way they are reasoning and quickly respond with a question or suggestion for more productive thinking. When the children understand, they do the problems without touching the tiles though often looking at them and starting to reach for them. When that occurs I say, "Great! How about doing the remainder of your problems without the tiles?" When they have finished, I say, "Can you prove your answers with the tiles?"

Let's take 1/2 and 1/3; if we examine this whole square first, why is this 1/2? (Because it takes two to cover.)

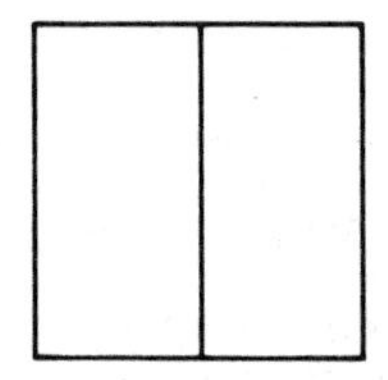

Why is this 1/3? (Because it takes 3 to cover.)

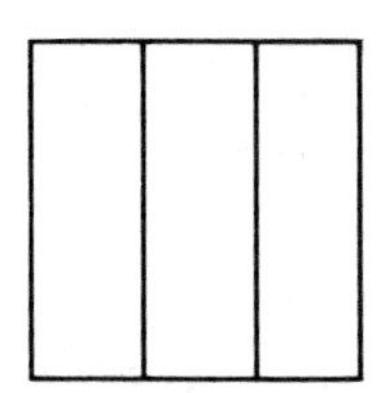

Which is larger? (We place the smaller on top.)

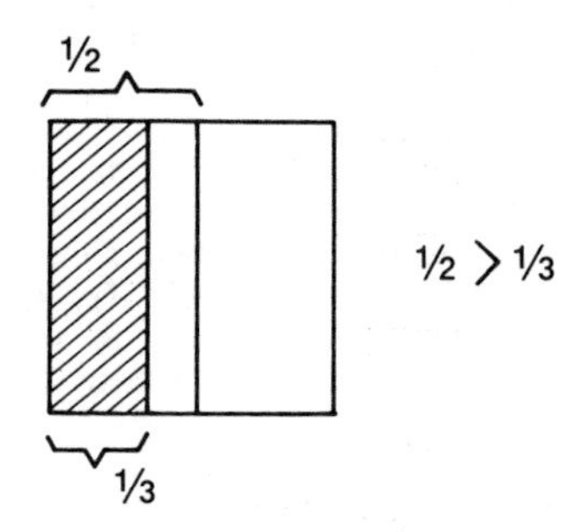

Take three units and two units, how do you add? (Place them together.)

We place 1/2 and 1/3 on the whole. (How much is covered?) We find one kind of piece to cover.

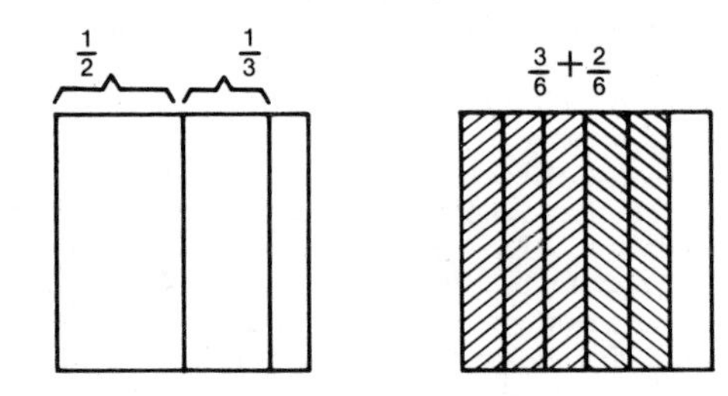

To subtract we refer again to our units. Get three units. Either take away two or cover three with two. What is either left or not covered? To act that out with 1/2 and 1/3 cover the 1/2 with the 1/3 and find what it takes to cover the rest.

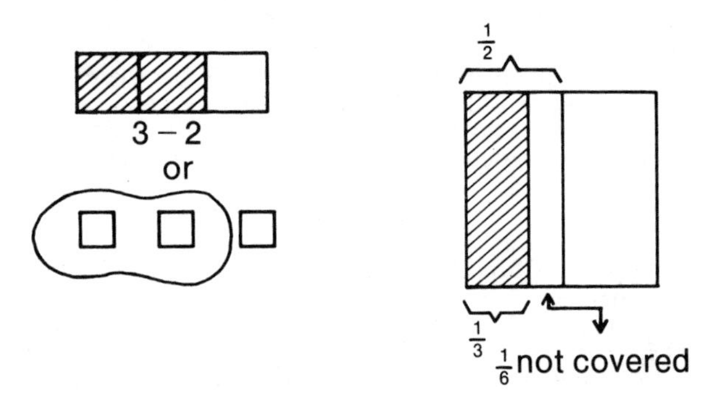

Or measure each with the same kind of piece and "take away" the third in that form.

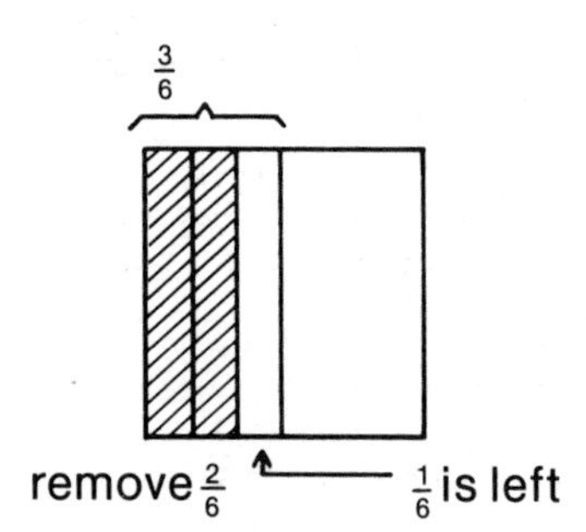

Often children comprehend for the first time the difference between subtracting by comparison and subtraction by "take away". They really were too young in the pre-operative stages to internalize these concepts.

Dividing has two ways of asking the question. First I have the student look at six units. How many twos are there is the six? We measure out two at a time and count three groups. Or if you separate the six into two equal parts, how many in each part? If you are dividing 1/2 by 1/2 ask: Using the 1/3 as a measuring unit, how many 1/3's are needed to cover 1/2?

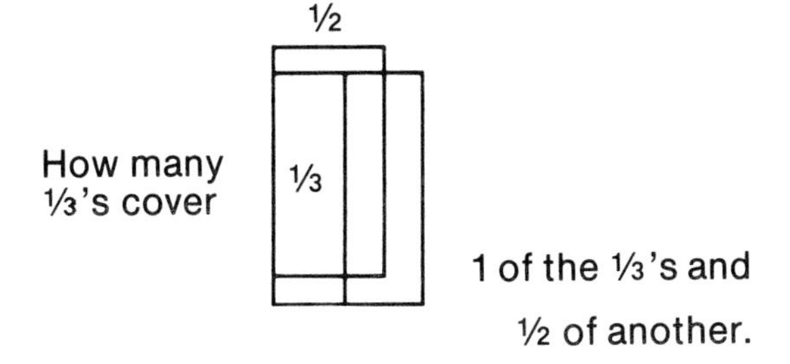

Or, if you want to divide the 1/3 by the 1/2 ask: Using the 1/2 as a measuring unit, how much of it is needed to cover the 1/3? All of this is at the intuitive or manipulative level. The student sees from many renaming activities that division, like addition and subtraction is easily done if the two are measured with the same unit. Then, division is just a matter of dividing the numerators.

How much of the ½ is needed to cover ⅓?

Multiplication makes beautiful sense IF - - IF whole number multiplication has been meaningfully taught so the learner knows that the factors are the length and width of a rectangle and the product is the area. For example, using our six units again: Three times two is a rectangle with an area of six.

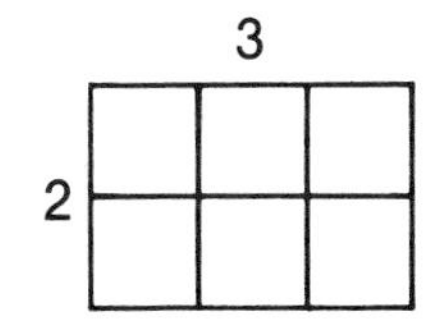

$2 \times 3 = 6$

To form 1/2 x 1/3 on the whole, form a rectangle with width of 1/3 and by overlapping a length of 1/2. The overlapped part is 1/6 of the whole. Children do this, record the result and make the rule themselves.

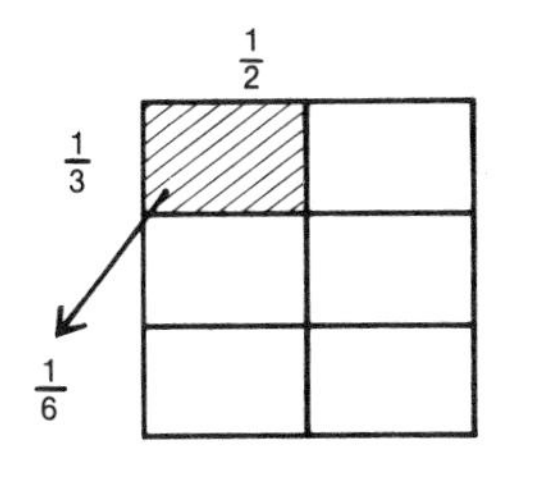

$\frac{1}{2} \times \frac{1}{3} = \frac{1}{6}$

These techniques have been received with enthusiam by boys and girls because they made fractions meaningful. If fractions remain as a part of the curriculum my premise is: (1) Teachers must understand the developmental learning stages of children. (2) Teachers must learn to present fractions concretely through a variety of embodiments. Therefore, the real burden of change depends upon those of us who are involved in teacher training.

References

Firl, Donald H. "Fractions, decimals and their future", The Arithmetic Teacher, March, 1977, 238.

Inhelder, B. and Piaget, J. The Growth of Logical Thinking from Childhood to Adolescence. New York: Basic books, 1954.

Jenkins, Lee and McLean, Peggy, Fraction Tiles. Hayward, CA; 1972.

Karplus, R. and Peterson, R. "Intellectual development beyond elementary school II, ratio, survey." School Science and Mathematics, 70, 1970, 873-820.

Payne, Joseph N. "Sense and Nonsense about Fractions and Decimals", The Arithmetic Teacher, January, 1980, 5.

Usiskin, Z.P. "The future of fractions', The Arithmetic Teacher, January, 1979, 18.

CHAPTER 4 - Post-Secondary Education

4.1 DECLINE IN POST SECONDARY STUDENTS CONTINUING IN MATHEMATICS: THE UNITED STATES EXPERIENCE

DECLINE IN POST SECONDARY STUDENTS CONTINUING IN MATHEMATICS: THE UNITED STATES EXPERIENCE

James T. Fey
The University of Maryland
College Park, Maryland, USA

Since 1970 the number of U.S. college and university students choosing to major in mathematics has declined sharply. There are several sources of data describing this decline and many conjectures about the causes. The situation is summarized in two sections of the paper below.

Data on Enrollments of Post-Secondary Mathematics Students

Since 1960 the U.S. Conference Board of the Mathematical Sciences (CBMS) has regularly surveyed course offerings and enrollments in undergraduate mathematical science courses - - mathematics, statistics, and computer science. The National Center for Education Statistics surveys general student enrollment trends and degrees granted, and several other agencies monitor students' secondary school mathematics study and indicate collegiate major intentions. Tables 1 - 5 summarize statistical trends in post-secondary mathematics from these various perspectives.

Table 1

Percent Distribution of Earned Bachelor's Degrees by Field*

	1964-65	1975-76
Biological Sciences	5.0%	5.0%
Computer/Information Science	-	0.6
Engineering	7.7	5.0
Mathematics/Statistics	3.9	1.7
Physical Science	3.6	2.3
Psychology	3.9	5.4
Social Science	16.4	14.0
Non-science	60.6	65.1

*Source: National Science Foundation, Science Education Databook, 1980, pp. 106.

The data in Table 1 show immediately the declining fraction of undergraduates attracted by mathematics as a major. In absolute numbers, mathematics majors increased throughout the 1960's to a peak in 1969-70; the decline has been steady and substantial since that time. Some of that decline is compensated by increases in computer and information science, but information on the intended majors of entering college freshmen suggests that the decline is not likely to end soon.

Table 2

Earned Bachelor's Degress in Mathematics*
(in thousands)

	Number of Degrees
1957 - 58	6,905
1963 - 64	18,624
1960 - 70	27,442
1973 - 74	21,635
1976 - 77	14,196

*Source: NSF, Science Education Databook

Table 3

Probable Majors of Entering Freshmen in U.S Higher Education*

	1965	1970	1975
	(percent)		
Biological Science	10.9	12.9	17.5
Computer/Information Science	NA	NA	NA
Engineering	9.8	8.6	7.9
Mathematics/ Statistics	4.5	3.2	1.1
Physical Science	3.3	2.3	2.7

*Source: American Council on Education, The American Freshman: National Norms for Fall, 1967, 1971, 1976.

Given the trends in earned degrees and probable majors, the trends in course enrollments since 1960 (see Table 4) are not surprising. The student population in typical "major" courses increased through the 1960's at a faster rate than overall college enrollments. But after peaking in about 1970, these course registrations have dropped sharply. The data in Table 4 show that overall mathematics enrollments have grown slightly faster than undergraduate enrollments in all fields. However, the growth has been concentrated in lower division mathematics, computer science, and statistics. The upper level courses commonly taken by mathematics, physical science, and engineering majors have actually declined in absolute enrollment numbers (see Table 5).

Table 4

Undergraduate Mathematical Science Enrollments*
(in thousands)

Level	1965-66	1970-71	1975-76
Below Calculus	812	1107	1417
Calculus	342	417	480
Upper Division	179	230	157
Computer Science	30	103	151
Statistics	48	108	168
Total Number of Undergraduate Students	4045	5703	6619

*Source: Fey, Albers, Jewett. Undergraduate Mathematical Sciences in Universities. Four-year Colleges and Two-Year Colleges, 1975-76, pp. 21, 27, 74, 76.

Table 5

Enrollments in Advanced Mathematics Courses*
(in thousands)

	1965	1970	1975
Linear Algebra	19	47	28
Modern Algebra	20	23	13
Theory of Numbers	3	4	1
Advanced Calculus	20	20	14
Mathematics for Physicists and Engineers	12	12	9
History/Logic/ Foundations	7	18	5
Advanced Geometry	12	13	5
Topology	3	5	1
Real and Complex Variables	9	18	10

*Source: Fey, Albers Jewett, pp. 28

The statistical picture of decline in U.S. students continuing mathematics as a major subject after secondary school in unequivocal. On a related issue, the reported decline in ability of those who do continue, evidence is more limited and opinion divided. In a 1975 survey of college and university mathematics departments, 75% of the reporting department heads said that in the preceding five year period the training and ability of their entering students had declined. Other data showed a sharp increase of 'remedial' level mathematics enrollments, apparently confirming the decline of student ability. But the entire issue is a complex mixture of causes and perceptions, with little clear evidence. In any event, the relation of the general quality issue to declining numbers of post-secondary mathematics majors is conjectural at best.

Causes of Decline in Post-Secondary Mathematics Majors

The phenomenon of declining numbers is clear, but the causes of this trend are less certain. Conjectures are numerous and based on a broad array of changes in school, societal, and economic factors.

The most plausible explanations, and those most clearly supported by data, suggest that student choices of an academic major are strongly influenced by post-college employment prospects. For the last decade opportunities in the computer industry have grown rapidly and those jobs pay well. Computer science has only recently emerged as an independent college major (separate from mathematics or electrical engineering) and it is clearly attracting many students with mathematical talent and aptitude.

While the 1960's were a period of strong economic growth, the United States economy grew more slowly and faced more signs of insecurity in the 1970's. This insecurity has led students to seek college preparation for a specific post-college job, rather than liberal education in an academic field leading to graduate school specialization. The sharp increases in business and engineering enrollments since 1970 seem to reflect this practical inclincation that is a strong American tradition. From this perspective the great burst of mathematics and physical science majors in the 1960's appear to be a unique occurrence in the history of U.S. higher education.

Signs that potential mathematics majors are being attracted by more practical majors can also be found in the characteristics of programs offered by mathematics departments that have not lost the great numbers of majors. Those departments have introduced applied mathematics and statistics courses and, frequently, industrial internships. At the same time the number of mathematics majors has declined, the service load of mathematics departments has boomed - many other disciplines are using more and more mathematics. Further, among the diminished number of mathematics majors, an increasing number are double majors - actually majors in another field who find it easy to earn a mathematics major with only a few courses beyond what they take in their first major field.

Many potential mathematics majors are attracted by opportunities elsewhere; they are also discouraged by impressions that the college and secondary teaching markets offer no opportunities. At the Ph.D. level there are not enough college faculty positions to absorb each year's graduates; but the secondary school market has never been oversupplied, and the last several years have seen emergence of a mathematics teacher shortage somewhat caused by attraction of experienced teachers to high paying computer jobs. The general impression of a teacher oversupply is not easily overcome.

Economic factors seem the clearest causes of declines in students continuing mathematics. However, some college faculty identify characteristics of the students themselves and their pre-college preparation as causes behind the decline. Some argue that secondary school efforts to provide improved minimal competence education for less able has deprived the talented of necessary challenge and stimulation. Others claim that contemporary youth are self-indulgent, unwilling to work hard and discipline their minds in the way that mathematics requires. Some blame secondary counselors for discouraging students from the 'hard' mathematics courses needed as preparation for college major work, and there is some evidence that students have recently elected less advanced mathematics in high school. However, it is also possible that rising expectations of college and university faculty - many with recent research training - have made the demands of a mathematics major more and more difficult to attain, thus discouraging potential students.

Responses to the Decline

Whatever the cause or causes for declining numbers of post-secondary mathematics majors, leaders in the mathematical community have begun to respond in several ways. A sub-committee of the Mathematical Association of America (MAA) Undergraduate Program Committee has, for several years, worked on alternative definitions of the undergraduate major including computer and application emphases. The MAA also sponsored a 1979 conference to outline Prospects in Mathematics Education in the 1980's, including efforts to attract and better serve majors. The results of these efforts are yet to be realized; there does not seem to be a broad based sense of urgency about responding to the current state of affairs.

WHO WANTS TO STUDY FOR A MATHS DEGREE?

R.R. McLone
University of Southampton
Southampton, England

There has been some concern amongst university mathematicians and others in the UK over the apparent inability to recruit a respectable level of intake to degree courses in mathematics and related subjects. This concern can be presented in terms of two observations, one factual and the other more a matter of conjecture. The first relates simply to the observed sharp decline in the number of applications for a mathematics degree which began in the early 1970's, and the second to the level of mathematical knowledge and ability of those who do gain admission. The second of these may well be largely generated by the first, but factual evidence is hard to find. This article attempts to set out some of the published data and to comment on the background to it.

Summary of Data

In attempting to set out the facts concerning the numbers applying to read for a mathematics degree in the UK, there are a number of complications. The largest of these relates to the position of Computer Studies. UCCA (the central administering body for applications to a British University) formally considered computer studies as part of mathematics and has displayed the two as separate subjects in its publications only from 1978. Earlier data can be ascertained, but in the early 1970's they are not complete. Sufficient data is available (1), however, to enable Figures 1 and 2 to be drawn.

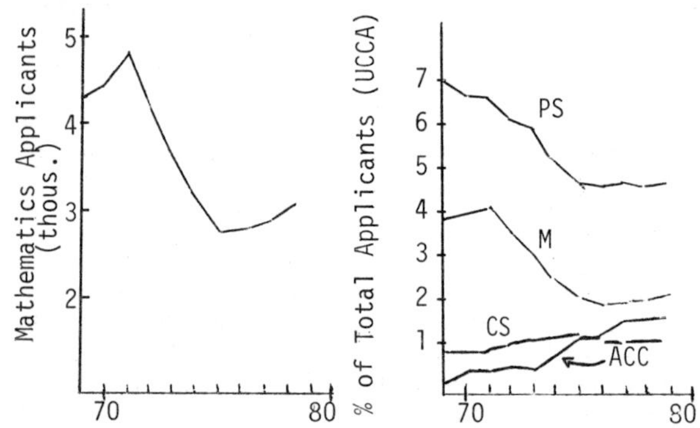

Figure 1 represents the total number of applicants and Figure 2 the percentage of all applicants through UCCA to read for a degree. The latter also shows details for Physical Science and Accountancy, as well as Mathematics and Computer Studies. The steep decline in the years 1971-75 is evident from both figures; the first, however, appears to show a recovry in numbers from 1975, albeit rather slight. It is only when a comparison is made with the total number of applications that we see that the position for mathematics is at best one of stabilisation after a sharp fall, with little real sign of recovery. (The total number of applications to UK universities has risen by approximately 30% between 1969 and 1979, largely in keeping with the total number of 18 year olds in the UK population (2). This latter number peaks in 1982-83, thereafter following a sharp decline to a value two-thirds of the peak value of 1996. The international reader might recall that entry to a UK university is generally highly selective, with a high proportion of those admitted successfully completing their course for an honours degree. For example, in 1978, 74,308 students were admitted through UCCA from a total 18 year old population of 854,900 or approximately 8.7%.)

The steep decline referred to in the last paragraph was also evident in the application rate for physical science and engineering degrees, although each of these groups reached a minimum in 1974, a year earlier than mathematics. Again there has been a recovery in total numbers. But whereas the relative position for physical science is comparable to that of mathematics, that of engineering has recovered almost to the level achieved at the beginning of the 1970's. Some fluctuation in demand can be expected and there exist "fashions" in acceptable subjects amongst university applicants as in other areas, but the decline in mathematics and physical science is generally not matched in other disciplines and certainly not in any other science or technological subject.

A notable increase during the same period occurred in the numbers applying to study for a degree course in accountancy (see Figure 2) and business studies. The application rate here has increased ten-fold in the period 1969-79. Even this increase is, however, overshadowed by that recorded in the application rate for computer studies (again see Figure 2). The rate (as a proportion of all applications) has changed from 0.6% in 1969 to 2.6% in 1979, an increase in the number of applications from approximately 650 to 3750. Indeed, although the final figures for 1980 are not yet available it appears that the number of applicants to read computer studies will for the first time exceed that for mathematics (including statistics and other mathematically related degree courses).

It is easy to speculate that the rise in applicants for computer studies accounts in large measure for the decline in mathematics applicants. However there are other factors which make this appear less likely. From the numbers themselves it is apparent that the sharp increase in computer studies applicants has occurred in the last three years or so and did not occur at the time of the decline in mathematics applicants. Moreover it appears from the sixth form study patterns of the computer studies applicants that a large number would have been unlikely, or indeed ineligible, to apply for a mathematics course. A numbr of courses in computer studies do not require study of mathematics beyond GCE Ordinary level (i.e. 16 years; eleventh grade). It is therefore debateable whether more than a small part of the mathematics decline is due to the advent of computer studies.

The situation relating to the level of mathematical knowledge and skill of prospective university students is much more confused. There are many assertions concerning "falling" entry standards for current undergraduates reading mathematics; usually emanating from college lecturers for whom "things are not what they used to be". Details are published annually (1, 3) of the performance of GCE Advanced-level (the public examination generally taken in the final school year and on which the major emphasis is placed in selection for a university place). These show that there does appear to be a steady (though rather small) decline in the average A-level score of mathematics entrants, but this change is overshadowed by variations within the intake and may be attributed to the change over the same period from a school study pattern dominated by mathematics in the sixth form (16-18 yrs.) to one in which mathematics occupies only one-third of the class time allotted to academic subjects. Certainly there is some evidence of a deterioration in algebraic manipulation, but it is not clear how this may have arisen.

Generally then there is a lack of interest in mathematics degree courses, and some doubt about the level of ability of those who do apply. (Although the data referred to in this article relates to university courses only, this constitutes a very high proportion (over 90%) of the total mathematics graduate population. Degree courses in mathematics and related subjects are offered at many Polytechnics and similar institutions, but the number of students on such courses are very small.)

Factors From Society Influencing Change

Perhaps one should ask the question "what has a mathematics degree course got to offer a lively intelligent 18 year old with some mathematical ability?" Indeed, does such a student see a bright, interesting future ahead on the basis of such a course? If it appears that the society into which the student graduates on completion of his course does not wish to avail itself of the expertise offered, then can we expect that prospective students will take note? Certainly such information, if correct, could be expected quickly to find its way to schools' careers departments and staff who might advise accordingly.

That there is some hearsay which gives rise to such a belief is true, but investigation of the career destination (4) and ability to obtain employment of mathematics graduates shows that even in the current difficult economic climate, these graduates have one of the highest success rates in finding employment, and other investigations (5, 6) have shown that such graduates are largely employed in the computing, statistical and operational research areas, with a large growing number interested in accountancy and other financial work, and a declining proportion entering the teaching profession.

Table I

First Destination of Mathematics Graduates

	1975	1978
Further academic study	15.7	14.0
Teaching	17.6	10.9
Management services (including computing)	21.3	32.4
Financial work	17.7	16.2
Management	2.4	2.1
Research and Development	3.6	5.2
Other Employment (including Overseas and Temporary)	12.7	9.8
Unemployed	4.1	2.4
Miscellaneous	4.9	7.0

(Figures are expressed as percentages of total graduate mathematics population of that year. Data supplied by University Grants Commission.)

One might note that the same investigations and subsequent local ones have shown that in general the mathematics graduate is employed for certain skills that seem to be developed by a mathematical training (e.g. logical approach, quick incisive understanding of essentials) rather than any particular mathematical expertise. Indeed, 30% of all mathematics graduates were shown never to use mathematics again in their careers, and the majority of the remainder used no more than first year undergraduate mathematics. This certainly does have implications for the style and content of the mathematics course which has been discussed elsewhere (e.g. in 5).

However we cannot conclude from job availability that mathematicians are difficult to employ. Rather the opposite appears to be the case. Detailed statistics relating to remuneration are more difficult to obtain, but those collected by such bodies as the Institute of Mathematics and its Applications (7) suggest at least comparability with graduates in other disciplines, with perhaps a few well recognised exceptions.

There are always "fashions" in desirability of certain careers in the minds of mid/late teenagers, which apply equally well to those in the highest academic bracket. These do not always correlate well with job availability in these careers! Currently there is considerable pressure in UK from suitably prepared students to enter the established professions; law and medicine have always featured strongly, but accountancy has recently joined this group and shows considerable growth over the past three or four years. In fact there has been a general increase over longer than this, and with little sign of any fall either in interest by prospective students, or in the demands of industry, commerce and public service for even greater numbers of qualified accountants.

Factors From Within Education Influencing Change

As it appears that mathematics graduates are generally sought after by employers and do not appear as a group to suffer unduly in salary comparisons (with perhaps a few reservations), we must look elsewhere for possible reasons for the dearth of applicants. The previous education of prospective entrants is a natural area for investigation.

We recall that our concern is a lack of applicants, not simply entrants. If it were a matter of low acceptance rate by university admissions tutors from those applying, we might have to look no farther than poor academic performance in the final school examinations. However, although we have noted a slight decline in the ability of entrants as measured by GCE Advanced-level performance, statistics published by UCCA (1) show that approximately 75-80% of applicants achieve a place on a mathematics degree course and that, of those who gain admission, approximately 80% succeed in obtaining a mathematics degree. There has in fact been some evidence that the failure rate in honours mathematics and related degrees has been declining since the mid-seventies with an 86.7% success rate in 1978 of 1975 entrants (later figures are not yet available). (What "success" means in this case is the award of a degree - whether "standards" are maintained is a perennial chestnut as much chewed over by university academics as by others.)

We return to the declining application rate. In England and Wales (Scotland has a rather different pattern of education in late secondary and early university years - whilst our discussion does not specifically refer to this, the overall conclusions are much the same) students in the sixth-form normally take a course in three subjects for the GCE Advanced-level. The position of mathematics in this context has been somewhat anomalous to that of other subjects, in that it is possible (with a number of variants whose enumeration would merely add to confusion) to take Mathematics and Further Mathematics as separate subjects. Thus the tradition, certainly throughout the 1950's and much of the 1960's has been for mathematically able students to study both these subjects in the sixth form (what came to be called "Double mathematics"). It was further traditional for university mathematics departments to expect a good Standard in Double mathematics in order to offer a place on a mathematics degree course. This situation prevailed for some time; however with the change from a grammar/secondary modern to a Comprehensive System of education, some schools have found themselves unable to offer Further Mathematics due to staffing problems; on the other hand, a much broader pattern of sixth-form study has been developed with greater choice of subjects. This has occurred especially in local authority areas which have developed a sixth-form college system where all 16 yr. and above are separated into a single institution. These changes have led to the proportion of sixth-form pupils taking "Double mathematics" to drop (Table 2) (8) from 13.0% in 1969 to 8.6% in 1977.

Table 2

Proportion of Advanced Level Students
Taking Mathematics

	1969	1971	1974	1977
"Double mathematics"	13.0	12.0	9.4	8.6
"Single mathematics"	30.1	31.9	30.2	30.4

(Figures are expressed in percentages. Data supplied by Department of Education and Science.)

On the other hand the proportion of Advanced-level students taking just one mathematics course has maintained a steady 30-31% over many years. It may be concluded that the number of sixth-form students following courses including sufficient mathematics as a preparation for a traditional single honours mathematics degree declined and that without a corresponding change in the initial level of the university courses, fewer sixth-formers will consider applying for such a course.

However there is another consequence of the changes in sixth-form structure and curriculum over the past decades. In the 1950's almost all the students studying mathematics in the sixth-form could be guaranteed to be doing so as part of a physical science package - either "Double Maths" and one other (almost always Physics), or "Single Maths" and two others, usually physics and chemistry. This ensured that any applicants for a university place were restricted to mathematics, a physical science or engineering as a degree course. Over the past ten years especially this has changed considerably. Not only is there now more opportunity for various combinations of sixth-form subjects including mathematics (such as economics, geography, mathematics, etc.) but also the degree course packages offered by many universities have also become less dominated by the traditional British single honours degree. This freedom of choice may have developed for good educational reasons, but there is no doubt that it has had a considerable effect on the willingness of school students to consider either mathematics or physical sciences as suitable disciplines for further study. It therefore becomes clear that the steady level of "Single mathematics" A-level students as shown in Table 3 obscures a general and significant change in emphasis on the role of mathematics as a sixth-form subject. The traditional patterns of study might be called "mathematics led" in that the mathematics studied was an essential prerequisite, or at least highly desirable for subsequent degree courses. These patterns of course still exist, but are now accompanied by an increasing number of "mathematics support" combinations - that is subject combinations in which mathematics plays a support role, often to areas such as biology, geography, economics, etc., which would not in previous decades have been seen to have much relation to mathematics. This involvement of mathematics in non-traditional areas is of value both to mathematics and the other subjects; however the changes appear to have occurred not in addition to the traditional study patterns involving mathematics but as a replacement for them.

Thus the real pool of potential mathematics applicants is declining at sixth-form level also. There is an important area for debate - in a situation of limited resources (and a limited supply of suitably qualified mathematics teachers) where should the emphasis lie? Do we redouble our efforts in re-establishing the previous level of "mathematics led" student patterns of study, or do we recognise the trend and make a virtue out of necessity by developing a more positive "mathematics support" role? The latter would have profound consequences for university mathematics departments in Britain, to say the least!

What steps are being taken amongst university departments in the UK to meet this problem? I am bound to say not many, although the conferences organised at Nottingham University on University Teaching Methods in Mathematics (9) have at least created a forum for discussion. Some University mathematics departments have been accepting applicants whose sixth form study contained just "Single mathematics" for a single honours mathematics degree course; others have introduced combined honours courses which allow further study of mathematics with a second subject to a higher level. Elsewhere unit course degree structures nearer the American model have been introduced. However, if we return to the application rate these moves do not appear to have produced a significant change. We have not discussed any particular factors from within the subject itself - e.g. the mathematics syllabus, teaching method, standard of teaching, etc., which may all play a part. One of these - the mathematics teachers themselves - is touched on in the next section.

Implications for Numbers of Mathematics Teachers

Earlier we referred to a particular issue of some importance arising from the trends we have noted above; namely the now quite rapid decline in the proportion of mathematics graduates willing to enter the teaching profession. Table 1 shows that by 1978 a mere 10.9% were entering the profession (or training for it in a one year postgraduate course) of those graduating with a mathematics degree. This compared with around 25% in the mid 1960's and even higher percentages in the 1950's. The single honours degree plus one year postgraduate training is the traditional route for a prospective secondary school teacher intending to teach the academically able mathematics pupils. The only other recognised route (that is a four year course at a college of education for a Bachelor of Education degree) attracts very few students wishing to specialise in mathematics.

The reasons for this decline are complex, but revolve predominantly around remuneration and career satisfaction. Certainly it is true that mathematics graduates can claim higher salaries in industry and commerce, especially in the computing field. However it is not clear that this is the sole reason for the decline which has taken place through a period when, at least for a time, teachers' salaries were improved substantially in relation to other jobs. Early surveys show that career satisfaction may be at least as important a factor for mathematics graduates. In particular graduates with a higher final classification find teaching especially unattractive. This can only be detrimental in the long run not just to the profession but also to any moves initiated to overcome the difficulties described in this article in recruiting mathematics students. It may well be that solution of the problem of finding sufficient able and committed teachers of mathematics will go a large part of the way to encouraging more advanced secondary students to seriously consider mathematics as a rewarding and interesting degree subject and career.

References

1. Universities Central Council on Admissions. Annual reports and statistical summaries.

2. Annual abstract of statistics. HMSO.

3. Annual reports of the various examining boards for the Certificate of Education (Covering England and Wales):
   University of London School Examinations Council,
   Oxford and Cambridge Schools Examination Board,
   Oxford Local Examinations Delegacy,
   University of Cambridge Local Examinations Syndicate,
   Joint Matriculation Board (Northern Universities),
   Southern Universities Joint Board,
   Associated Examining Board,
   Welsh Joint Education Committee.

4. University Grants Committee. "First Destination of University Graduates" (published annually).

5. McLone R.R. "The Training of Mathematicians". Social Science Research Council, 1973.

6. McLone, R.R. "The Employment of Mathematicians", Chap. of "Mathematical Education", ed. Wain, G.T. 1976.

7. Institute of Mathematics and its Applications. Statistics are published annually in its Bulletin.

8. Department of Education and Science. Statistics of Education (published annually in several volumes; Vol. 1 refers to Schools, Vol. 2 to School Leavers, Vol. 6 to Universities).

9. Proceedings of the Nottingham conferences on University Teaching Methods in Mathematics (1975 to 1978 are available).

DECLINE IN POSTSECONDARY STUDENTS CONTINUING IN MATHEMATICS: THE SOUTHEAST ASIAN EXPERIENCE

Bienvenide F. Nebres
Ateneo de Manila University
Manila, Philippines

This report summarizes the experience of the member countries of the Southeast Asian Mathematical Society (SEAMS): Hong Kong, Indonesia, Malaysia, Philippines, Singapore and Thailand (South Vietnam was also a member until 1975). The report is based, not on statistical data, but on analyses made by SEAMS over the past eight years. I believe that this is sufficiently scientific, because while some of the SEAMS countries are large, there are in each country only two or three universities, where serious mathematical training is being done. Moreover, as SEAMS was established in 1972, precisely to try to improve the state of mathematics in Southeast Asia, we have tried to study the matter with some care.

Analysis of the Situation

The SEAMS countries may be roughly divided into two: a relatively stronger group composed of Hong Kong, Malaysia, Singapore; and a relatively weaker group composed of Indonesia, the Philippines, Thailand.

Among the stronger countries, the situation appears quite good. There has not been a serious decline in either quality or quantity of students doing a mathematics major. There has also been an increase of students going into related fields: computer science, statistics. It should be noted, of course, that not all (not even a majority) of these undergraduates go on to graduate work as professional mathematicians, but this is simply because of the lack of career opportunities in these countries. In any case, many of the most talented students continue to do an undergraduate degree in mathematics and this gives promise for the future development of mathematics in these countries.

Among the weaker countries, the situation is poor, even dismaying. There have always been only a few students going into mathematics. They appear to be getting even fewer and the quality also appears to be declining. Many of those who start doing a mathematics degree shift to math-related subjects, such as computer science, actuarial science, etc. The best students usually go to medicine or engineering. Even more problematic, due to the paucity of career opportunities for mathematicians, many of those who go on to graduate work in math either remain abroad or go on to careers in business and industry. It has been extremely difficult, therefore, to get any kind of serious mathematics going in these countries.

Causes

Reflection on the situation suggests that we should divide the causes into what we might call internal and external factors. Internal factors are a cultural tradition that places high value on the intellectual life, for learning beyond utilitarianism. Such a cultural tradition is usually a product of an older culture and, among other things, manifests itself in the emergence of what we might call the "force of genius": talented young people doing mathematics (or philosophy) because of an inner drive, relatively independent of external support from society. External factors would be support from society: career opportunities, salaries and working conditions in academic life, respect (if not understanding) for intellectual work.

In the three stronger countries, both internal and external factors are present to a substantial degree. There is a respect for learning and an inner drive towards doing mathematics stemming from ancient Chinese cultural tradition. Similarly, Vietnam (both North and South) continued to do good mathematics despite a practically impossible environment, because of an ancient tradition of respect for learning. Moreover, in Hong Kong, Malaysia and Singapore the British left a legacy of substantial support for a few elite universities. There are, therefore, better career opportunities and support for academic life in these countries. Salaries in the universities are good and encourage the best to continue research and teaching.

On the other hand, both factors are weak in the Philippines, Indonesia, Thailand. There is not the same cultural tradition that produces mathematicians and philosophers with little or no support from society. There are more universities, but they are poorly funded. Salaries are one-third to one fifth of what one can get in business and industry. It is easy to understand, therefore, why few students go to mathematics and why among these few, even fewer continue to become professional mathematicians.

What Can Be Done?

Among the stronger countries much still has to be done to bring up mathematics research to world standards. Efforts are being made in this regard, but as they concern graduate training and research, they are not our concern here.

Among the weaker countries, the challenge is how to build a mathematics tradition. This means detecting mathematical talent early and fostering it. It is not realistic to simply hope that the best will rise to the top. They are too few and the environment much to unfavorable. It also means marshalling support from society: To improve career opportunities in academic life in general and in mathematics in particular. This is not easy, because the more immediate pressures on our developing countries to take care of their poor and destitute make it difficult even for us mathematicians to argue for the priority of not-immediately-useful learning. The different National Mathematical Societies and the Southeast Asian Mathematical Society are trying to find solutions to our present situation. In particular, there will be a Southeast Asian Mathematics Education meeting in Kuala Lumpur, Malaysia next April. Since this meeting will focus on mathematics education for 16 year olds and older, I am sure that the question of the numbers and quality of post-secondary students will be carefully discussed.

4.2 IS CALCULUS ESSENTIAL?

IS CALCULUS ESSENTIAL?

Margaret E. Rayner
St. Hilda's College
Oxford, England

There are two questions we are discussing today. One is the obvious one about the role of calculus. The second is the implied question as to the feasibiity or even desirability of a core syllabus in mathematics courses at universities and other insitutions of higher education. Let us take the second issue first.

In 1976, at the International Congress on Mathematical Education at Karlsruhe (3rd International Congress of Mathematics, 1976, Programme Part II p.25), Professor van Lint led a very general discussion on mathematics education at university (bachelor's) level and commented that, from the survey he had made, it was apparent that 'old-fashioned' calculus courses were being replaced by courses of a more abstract nature; he mentioned abstract integration theory, general topology, metric spaces. He also said that some universities were then moving back from the more abstract courses; I hope that in the subsequent discussion we may hear comments on the present situation in a number of universities. I think we may reasonably suppose that there is less common ground between first year university courses for mathematicians than there used to be (even in universities within the same country) and we should therefore examine what the implications are. Does it question the whole concept of 'a degree in mathematics'? I believe that it does and it could reduce the possibility of students moving to graduate programmes or employment in other universities and other countries. Such a consequence would be deplorable.

This problem of diverse backgrounds has confronted the universities in Great Britain for an earlier age group--the young people coming from schools into university courses which depend on prior mathematical knowledge. About four years ago, university departments expressed considerable alarm at the decreasing overlap between syllabuses used in schools for the 16-18 year olds and claimed that it was becoming almost impossible to run first year courses economically, i.e. without splitting audiences into small groups to take account of students' previous training. There have been consultations amongst teachers, examining boards and universities and a minimum core syllabus has emerged which is receiving substantial support (A minimal core syllabus for A-level mathematics. A joint statement by the Standing Conference on University Entrance and the Council for National Academic Awards; December 1978).

Has the situation yet arisen--or is it likely to arise--that some 'core' of mathematics would be advantageous at university level? This is not an occasion when Professor Roberts and I battle to the death; but is is, I hope, an occasion on which teachers from many universities and countries can give information (and so bring the rest of us up-to-date) and also contribute comments on the idea of a core syllabus.

Now let me return to the topic of this debate: Is calculus essential? If we accept that some core is required, should calculus be in it? This question must itself be interpreted in an elastic way to include the possibility that calculus may be a pre-requisite.

If any topic is to be included in a core, it must satisfy the following requirement:

The topic should introduce to students those ideas which are important to mathematics itself, are of very wide application in mathematics and elsewhere, and give rise to problems which are of interest to professional mathematicians.

I wonder if this criterion is universally acceptable?

A second criterion might be:

The topic should have applications which are of interest to students because it relates to other courses they may be studying at college or because they believe it might be needed in subsequent careers.

This second criterion appears to me to be an obvious one for including a topic within the mathematics course, but I think it is not itself an adequate reason for including a topic in the core. I am increasingly aware of students who enter university courses with no clear idea about careers or who have firm ideas which subsequently change as graduation approaches. I see therefore, enormous difficulties in finding, as a core subject, a topic which is especially career related, which interests all students for all of the time. I think therefore we must treat the second criterion less seriously than the first—I will repeat it:

The topic should introduce to students those ideas which are important to mathematics itself, are of very wide application in mathematics and elsewhere, and give rise to problems which are of interest to mathematicians.

On this criterion, I believe that calculus is assured of a place in the core.

By calculus, I mean the study of elementary functions of one variable, differentiation, integration and limits, in an intuitive way, with plenty of applications. It is the ideas here that are important, not the proofs. It is true that a few well posed questions can show the need for precision and for proof but these best come after some experience which will give a student confidence and will encourage him to be adventurous.

Dr. Roberts has already mentioned may areas of applicability of calculus to problems which arise outside mathematics, but of course it is the starting point for investigations in analysis of many sorts, in differential equations, in the calculus of variations, in geometry, all areas of very great concern to present mathematicians. I think I need not elaborate this point to this audience.

Over the past year I have asked many mathematicians if calculus is essential: most of them looked at me as if I was joking or mad. But just because a topic has been icluded in syllabuses for many years, it cannot continue to be included by tradition alone. Think of what has happened to geometry over the last fifty years.

I contend that calculus has, as yet, no rival for richness of ideas and applications and is therefore essential in a mathematics course.

There are two main themes in mathematics; the mathematics of finite processes and the mathematics of infinite processes and I think that a graduate of mathematics should have an appreciation of both, even if he becomes an expert in only one. The two themes are like twins, closely related, having a great deal in common, but each having its own existence. But unlike some human twins, they are also of the greatest assistance to each other by exchanging ideas for by-passing difficulties. To have a mathematics course which ignores either fininte or infinite processes is missing half the story.

The question is, what topic shall represent infinite processes? Shall it be convergence? From my experience, that is a subject which does not often generate enthusiam amongst students. Shall it be analysis without a preliminary course in calculus? This appears to be a curious idea because it suggests that students should be introduced to the remedies for pathological cases before they are really familiar with the well-behaved ones. So, they have no framework within which to work. Analysis has, undeniably, as important role as introduction to mathematical rigour, but is is nothing if it does not grow out of the problems raised by calculus. It is a good working principle: Don't prove a theorem until you need it. Analysis also lacks one other feature: scope for intuition. Here I would like to quote from the preface of one of the books of Richard Courant:

> The presentation of analysis as a closed systsem of truths without reference to their origins and purpose has, it is true, an aesthetic charm and satisfies a deep philosophical need. But the attitude of those who consider analysis solely as an abstractly logical, introverted science is not only highly unsuitable for beginners but endangers the future of the subject; for to pursue mathematical analysis while at the same time turning one's back on its applications and on intuition is to condemn it to hopeless atrophy. To me it seems extremeely important that the student should be weaned from the very beginning against a smug and presumptious purism; this is not the least of my purposes in writing this book.

And the book he wrote was Differential and Integral Calculus!

I claim, therefore, that calculus, with its ample fund of graphical illustrations, its concern with familiar concepts like area, volume and curve length provides the most appropriate introduction to infinite processes. I will therfore summarize my case for the inclusion of calculus in a core syllabus as follows:

1. It has applications in many different areas of study outside mathematics: physical science, statistics, numerical solutions, and so forth.
2. It provides the best starting point for a study of infinite processes because there is great scope for demonstration of its ideas.
3. Without its technique, whole areas of mathematics (for example, differential equations) which are of very considerable interest to mathemticians are out of reach.

4. It is a subject which gives great scope, even at an early stage, for inventiveness and original application.

I submit therefore that calculus is essential in any mathematics core which leads to a degree.

IS CALCULUS NECESSARY?

Fred S. Roberts
Rutgers University
New Brunswick, NJ

Calculus is not necessary. That will be my premise in this debate.

Let me explain myself. For me, the calculus course has five major rules to play.

(1) It illustrates a great intellectual accomplishment.
(2) It introduces the student to precise ways of mathematical reasoning, gives the student lots of experience with problem solving, and exposes the student to the beginnings of abstraction and rigor.
(3) It gives the student an introduction to modern mathematical tools, techniques, and applications, and a feeling for the types of things he or she will encounter in college mathematics. In short, it serves as an "entry" course to the math major and as a broad-based "survey" course for the non-major.
(4) It introduces concepts and techniques which will be used in later math courses.
(5) It introduces concepts and techniques which will be applied in courses in other disciplines; it serves as a "service" course for students in other disciplines.

I will argue that the first two roles can be equally well accomplished by using topics other than those of the caluculus. I will also argue that the third role is being carried out rather poorly. I will assert that the fourth role is being performed adequately for only some of the students taking calculus, and for many math majors, it could be accomplished better by replacing calculus. Finally, I will claim that the fifth role could also be accomplished better for many students by replacing at least parts of the calculus.

Perhaps the fastest growing area of modern mathematics is discrete mathematics. This has been spurred by the explosive growth of computer science, which has its own basic mathematical problems, but which also has made it possible to consider hitherto untouchable problems in industry and government and the social sciences which are inherently discrete in nature. Mathematical tools and topics such as linear programming and linear inequalities, networks and graphs, combinatorial algorithms, and probabilistic and statistical techniques, are being developed rapidly, and are of practical and theoretical importance. Many math majors now take a program of courses which, after calculus, emphasized such tools (more on this below). It is my feeling that an entry course or survey course on modern mathematics which does not expose the student to these topics is biased and misleading. The calculus does not begin to introduce the student to the broader variety of ideas which make up modern mathematics. It also probably causes mathematics departments to lose a large portion of its potential clientele, both majors and non-majors, to other disciplines such as Computer Science or Business Administration, which seem (after exposure to the calculus) more relevant to the kinds of problems and applications many students are interested in. Thus, calculus, as it is presently taught, is not serving very well to carry out role (3) above.

To what extent does calculus fulfill its fourth role, i.e., to introduce concepts and techniques to be used in later courses in mathematics? Obviously, this it does well for the traditional, theoretically-oriented math major and for the major interested in applications to the physical sciences and engineering, to the biological sciences, or in numerical methods. But, consider this: at my own institution many math majors take a program similar to the following, which I shall call a math major with emphasis on "modern" (as opposed to "classical") applied math. This program consists of two years of calculus, a course called Introduction to (classical) Applied Mathematics, applied linear algebra, probability, linear optimization, operations research, mathematical modeling and two courses in computer science or combinatorics. Such students. in my opinion, get a rigorous education in how to think and reason, which is, after all, what mathematics education is all about. They get their share of theory. They master certain subjects in depth. For instance, the operations research course has as a prerequisite, the probability course and also the linear optimization course, which in turn has as a prerequisite the applied linear algebra course. These students are well-prepared for the job market, and also for graduate school in applied math, operations research, statistics, business administration, etc. Many math majors around the country take a similar program. Is calculus necessary for such a major? It is certainly a prerequisite for some of these courses, such as the course Introduction to Applied Mathematics, which covers mathematical modeling in mechanical vibrations, population dynamics, and traffic flow, and involves ordinary differential equations and nonlinear first order partial differential equations. It is also a prerequisite for the probability course, which uses calculus briefly in the treatment of probability generating functions, though it does not cover continuous probability. It is a prerequisite for combinatorics, where ideas of sequences and series are used in the material on generating functions and on recurrence relations. It is a prerequisite for the mathematical modeling course, where differential equations models of population dynamics are discussed. However, with the exception of the Introduction to Applied Mathematics course, none of these courses uses calculus heavily, and the calculus topics required could easily be included in these other courses themselves. Calculus is required primarily because it demonstrates a certain level of mathematical maturity. Other courses could serve this purpose equally well. Thus, with the exception of one course, calculus is really not necessary for such a major in modern applied math. The Introduction to Applied Mathematics course is introduced to replace the traditional advanced calculus course. It does not play a central role for this type of major, and so it could easily be taken later on, or even omitted entirely, or perhaps be replaced by a course in statistics or numerical methods. Thus, at least as a prerequisite course, calculus is not necessary for a major in modern applied math, and is not necessary until very late in the major. Many of the concepts of calculus are not really used by this kind of math major. Thus, role (4) is not being fulfilled very well, at least for a certain percentage of students. Other topics in an entry course would be preferable for these students.

I am not trying to argue that modern applied math majors need not be exposed to the calculus. I am simply trying to argue that for some students, calculus is not so important as to be the primary subject of study for the first two years. A more appropriate entry course for such students would still have many topics from the traditional calculus sequence, but would also emphasize ideas of linear programming and linear inequalities, probability and statistics, graphs and networks, combinatorial algorithms, and so on. I believe that such an entry course can fulfill roles (1) and (2) that have heretofore been expected only of a calculus course. In sum, for all of roles (1) through (4), at least for a certain percentage of math majors, calculus is not necessary, and could be replaced by an alternate entry course.

Before turning to role (5) let me ask this question: Should there be more than one type of entry couse into the math major, one aimed for students that will amphasize modern applied math and the other for students that will emphasize more traditional topics? This might make sense if the entry courses start out the same, and only diverge after a semester or after a year. I think it would be a mistake to force a difficult, and probably irrevocable, choice of "track" for a math major sooner than that, before a student has had the exposure or the maturity to make a choice. Thus, it is probably that the beginning of the entry sequence will still have to emphsize calculus. However, before reaching the "fork in the road," an exposure to other topics is necessary, even for those who plan to go on eventually to more traditional math majors. In sum, the beginning of the entry sequence should include a taste of non-calculus topics for everyone. Moreover, more emphasis should be placed on applications from the non-physical sciences, and there should be considerable emphasis on the computer and computer-related mathematics.

The entry course can provide an alternate "track" as I have said, after the first semester or after the first year. Perhaps an alternate choice after the first semester is too soon. An alternative is to introduce an additional non-calculus type course for the first year, second semester, and urge students who get "turned on" to non-calculus type topics to enroll in such a course, perhaps even at the expense of putting off until the sophomore year the second semester of the more calculus-oriented entry course.

Let me turn to role (5) of the calculus course next, its roles as a service course. What happens to this role if we "fiddle" with the syllabus? I am afraid that many of the traditional "customers" will yell very loudly. We certainly want to keep these customers happy, for it is the service role of calculus (and other math courses) which justifies the size of mathematics faculties at many institutions. Also, students in the engineering and physical sciences need to learn certain things by a certain time in their student careers. I see no way to avoid this requirement, and so I see no easy way to tamper with the entry course in mathematics in the way I have suggested without creating a separate and more traditional calculus track for students interested in majoring in areas of the physical sciences and engineering. If such students decide later to major in mathematics, they could do it, but with some catching up required. I realize that creating different calculus or entry course tracks is easy for a large institution such as mine, but hard for a small institution. For a small institution, some compromises all around will be necessary.

What about students in the biological sciences, economics, business administration, etc.? These students are taking more and more calculus courses now, indeed, they are required to. What do we do about them; create still another track? (At Rutgers, we do; we have a separate bio-sci track and a separate economics/business administration track, but that, again, is the luxury of a large institution.) I believe that with the increased emphasis on non-calculus techniques in the biological sciences, economics, and business administration, the modified entry course I perceive would be appropriate as a service course for these students, indeed, more appropriate than the traditional calculus course.

The problem would be to convince faculty in these other disciplines to accept other mathematical topics in place of some calculus topics. My experience has been that such faculty often oppose deletion of some traditional calculus topics, not because these specific topics are required for students majoring in their disciplines, but because they feel their students are being short-changed and not being given as extensive or complete a mathematical preparation as possible. This prejudice can be overcome by careful, face-to-face discussion, convincing examples (of useful non-calculus topics), and a dedication to identify those calculus topics which are really useful.

There is one additional advantage of creating separate tracks for these students. The specific calculus topics chosen can be modified to be more appropriate for the areas of application. (At Rutgers, we introduce elementary differential equations and their biological applications in the second semester of calculus for bio-sci, but not in the calculus for engineering/physical sciences, and we introduce analytic geometry of three dimensions, partial derivatives, and optimization techniques in the second semester of calculus for economics and business.)

Finally, it goes almost without saying that for students in the computer and decision sciences, a modified entry course would be more appropriate with regard to role (5), and, indeed, such a course is already offered in a number of computer science departments.

Before closing, let me turn to the inevitable question: If calculus isn't (all) necessary, what topics can go? This is a question which requires considerably more space and time than I have been allotted, and will require considerable thought and experimentation. However, let me simply give two examples of topics which could easily be deleted from first year calculus: differentiation and integration of transcendental functions and techniques of integration.

It should be clear that my position isn't quite as extreme as my original premise suggested. I just feel that it is very important for us to modify the calculus course so as to expose a large number of students as early as possible to the increasingly broad variety of ideas which make up modern mathematics.

4.3 MATHEMATICS AND THE PHYSICAL SCIENCES AND ENGINEERING

MATHEMATICS AND THE PHYSICAL SCIENCES AND ENGINEERING

Gerhard Becker
Universitat Bremen
Bremen, Federal Republic of Germany

Daniela Gori-Giorgi, Roma, gives a presentation of a paper, worked out by herself, Emma Castelnuovo, and Claudio Gori-Giorgi: Integrated Teaching of Mathematics and Physics for Students from 15 to 19 years of age: Experience Report. Then, Jean-Pierre Provost, Nice, presents his paper: The Conceptual Nature of the Relationship between Physics and Mathematics, and its Consequences in the Teaching of both Sciences.

Besides technical remarks, the panelists' contributions, discussions and remarks of session participants cover a wide range of problems and questions, referring to attempts to incorporate applications of mathematics into instruction. They concern three main fields: 1. content problems, 2. goals, 3. the place and function of applications of mathematics within a mathematics course.

1. Content Problems

1.1 The extension of a problem: one or a small number of subjects applying mathematics - several local examples, in which only one aspect might be important, and which therefore shall not be treated completely.

1.2 The place of a problem with respect to the curriculum: obligatory - supplementary.

1.3 The attachment of an application and its special accessibiilty: math instruction (problem accessible perhaps without detailed special knowledge) - physics or science instruction - interdisciplinary instruction.

2. Possible Goals

2.1 Understanding mathematics as an applicable discipline, as an instrument to describe, to comprehend, and to improve real situations - understanding everyday-life situations by the help of mathematics.

2.2 Skills further needed in math education - skills to be used in other subjects - understanding everyday-life situations by the help of mathematics.

2.3 Mathematizing and modelling real situations (understanding one "exemplary" case - ability of mathematizing real situations, or certain types of real situations).

3. Organisation, Place, and Function of Applications of Mathematics within a Mathematics Course.

3.1 Coordinating "pure" parts and relevant information about specific fields of application.

3.2 Structuring a course by leading mathematical ideas - by a sequence of applications - by an alternating change of pure parts and applications - by a sequence of applications with increasing complexity or difficulty.

3.3 Continuation or endings (of pure parts, of applications).

3.4 Already present specific knowledge and special experiences - gathering relevant information (experiments, questionnaires, experts-interviewing).

3.5 Applications as a field of exercise, with many variations of the same type of situations.

3.6 Motivating pupils by applications of mathematics, and facilitating learning processes by applications - repulsing pupils without any interest in a special area of application.

3.7 Difficulties in transfer of training by reducing the portion of "identical elements".

INTEGRATED TEACHING OF MATHEMATICS AND PHYSICS FOR STUDENTS FROM 15 TO 19 YEARS OF AGE: EXPERIENCE REPORT

Daniela Gori-Giorgi
Universita di Roma
Roma, Italy

1. Some Characteristics of the Organization of the Italian High School

The main characteristics of the Italian High School are the following ones:

a. the same professor teaches both mathematics and physics to the same student for three years; the students study a total of six hours a week in mathematics and physics;

b. the mathematics and physics programs are established by the Ministry of Public Education; in these official programs, mathematics and physics are taught separately.

c. However, in Italy we are given much freedom in teaching, so professors are often motivated towards educational research.

In this frame, we have been working on a three year curriculum for the integrated teaching of mathematics and physics. Among the different subjects we have experimented with, we shall present two arguments in particular. The first one is usually developed at the very beginning of the course; it is very very simple but useful in giving the course a good direction. This argument is the motion of a heavy point particle, but we often call it "the launching of projectiles".

2. The Launching of Projectiles (students 15-16 years old)

We begin in a rather classical way: we and the students face the problem of finding the trajectory of a projectile launched by a cannon with known initial velocity. For instance:

Initial speed $V_o = 5$ m/sec, with components $V_x = 3$, $V_y = 4$.

Students obtain:

horizontal displacement $X = 3t$

vertical displacement $Y = 4t - 9.8/2\, t^2$

cartesian equation of the trajectory:

$$Y = -4.9/9\, X^2 + 4/3X$$

The students, who have already studied the parabola and its cartesian equation, immediately recognize that the trajectory is a parabola and so, easily find the range of the launching.

At this point we propose to exchange the two components, that is $V_x = 4$, $V_y = 3$, so that $V_o = 5$ always remains thus. All the students say: "It is clear, this time the range will be longer". They are very stupified when, by calculating the new range, they find the same number as above. From this fact, they are led to find the general formula for calculating the range, $X_r = 2V_xV_y/9.8$, in a motivated way. This formula immediately explains why the range remains the same by changing V_x for V_y and suggest this intuitive geometrical interpretation: V_xV_y represents the area of all rectangles having V_x and V_y as variable sides and V_o as the fixed diagonal. Since we have the maximum of this area in the case of the square, we can say that the launching will arrive at the maximum range when $V_x = V_y = V_o/\sqrt{2}$ and the angle of elevation is 45°.

The teaching method is becoming clear; the students are motivated towards scientific study by concrete, interesting problems. Then spontaneous discussion and active research lead them to the mathematical formulas as a good way to answer to open questions.

Now, can we know the velocity of the projectile as well as its trajectory? Let us face this problem with some graphical reasoning.

If V_o is the initial velocity, after one second the velocity changes because of the gravity and becomes V_1, after two seconds V_2, and so on. After a little discussion, the students themselves trace many drawings of the forces on the projectile. Now, as the vector velocity is always tangent to the trajectory, the parabola will touch all the drawn vectors.

Thus, let us observe the direction of these vectors; we discover an interesting property. This property suggests a well known construction of a parabola as an envelope of lines. Much experience has shown us that the students are fascinated by this kind of geometrical construction and gain a sure knowledge of the "two ways of drawing a curve", by uniting several points or by touching several lines.

As we have seen, the interaction between mathematics and physics is always very strong; in the first part of the experiment, mathematics helps the development of physics, while in the second part physics leads to some mathematical discoveries. Furthermore, the above experiment gives the students a good base for their further studies. In fact, it becomes very easy for them to calculate the trajectory of an electron in a uniform electrostatic field, and so on. Moreover, the physical meaning of the construction of a curve as an envelope of lines will be a good base for solving a more involved problem: how to calculate the possible trajectories in a central gravitational field.

3. Vibrating Strings and Fourier Series

The second and more advanced subject we present is usually developed at the end of the second year and accomplished in the third year (students 17 and 18 years old). We usually call it "Vibrating strings and Fourier series".

As it is well known, this study is generally reserved for an advanced University level, because of the differential equations present in the classical treatment. But here differential equations are replaced by physical models, difference equations and appropriate reasonings, so that all the students can work with this subject.

We begin by letting the students observe the string of a guitar being played. For a good observation multiflash photos are taken, so that the students can see some unexpected shapes of the vibrating string.

It is possible to give an elementary explanation of the different shapes; the string is thought to be composed of an infinite number of molecules. These molecules, held together by elastic forces, behave as small balls fixed on an elastic string.

This interpretation suggests a simple model for the study of the behaviour of a string: a very thin elastic, fixed at its ends, with one small ball fixed in the middle. We pull the ball up, release it and observe. In this case, we see only one shape with different amplitudes during the oscillation. Now it is easy to calculate the magnitude of the driving force F; we find $F = 4Ts/L$, that is $F = ks$, where k is constant.

The students, who have already studied the simple harmonic motion by means of graphical derivatives, immediately realize that the displacement s of the ball must follow the temporal law

$$s = \cos\, w\, t$$

when the initial conditions are $s(0) = 1$, $v(0) = 0$.

Students can also calculate the frequency of oscillation by using the dimensional analysis method; they obtain

$$w = \sqrt{k/m}, \text{ that is, } w = 2\sqrt{T/mL}.$$

Of course, this is a very rough model; if we want a better model, we can consider an elastic with two balls fixed at equal distance. In this case, we have many ways in which to pull the two balls up and so many possibilities of oscillations.

Multiflash photos of several cases during the motion show that the first and the second cases appear more regular than the others. Namely, in the first case, when ball A arrives at the middle of the initial displacement s_A, ball B also arrives at the middle of s_B. This happens because the same force acts upon the two balls. Analogous observations hold in the second case and, for these two special cases, it is easy to repeat the reasonings followed in the case of the elastic with only one ball. So, for every case we find the driving force F and the frequency of oscillation.

Examining some cases and discussing them, the students are led to discover that all the photographed shapes can be obtained by superimposing the first two in a suitable manner. For instance, at a given instant, say at the time $t = 0$, one has:

$a_o = m_1 + m_2$ $\qquad$ $m_1 = 1/2(a_o + b_o)$

hence

$b_o = m_1 - m_2$ $\qquad$ $m_2 = 1/2(a_o - b_o)$

and thus at all times

$s_A = m_1 \cos w_1 t + m_2 \cos w_2 t$

$s_B = m_1 \cos w_1 t - m_2 \cos w_2 t$

It is even rather easy to show that the first two shapes are the only ones so regular; the proof is based on the irrationality of the ratio ω_2/ω_1.

Now, let us summarize the properties for the elastic with two balls:

a. there are two privileged shapes recognizable by their regularity during the motion;

b. in these two shapes; the forces acting on the balls are proportional to their displacement at all times; thus the motion of each ball is harmonic with the same frequency;

c. by superimposing these two privileged shapes in a suitable manner, we can obtain all the shapes of the elastic at all times.

For these reasons, these shapes are often called "the two fundamental modes".

The above results have been generalized in two ways, according to the students' level:

a. At a more elementary level, we show many multiflash photos of an elastic with three balls, and then of an elastic with four balls. By the use of imagination and graphical intuition, the students are able to grasp that an elastic with n balls has n fundamental modes with the same properties as above.

b. At a more advanced level, we show many multiflash photos of an elastic with three balls and so the students themselves find the three fundamental modes. Then, they actually calculate the driving forces in every fundamental case and find that the formula $F = ks$ always holds. This formula allows them to calculate the shape and the frequency of each fundamental mode. Afterwards, the amplitude of each fundamental mode compounding some general shapes are found. This also leads to the discovery of a general criterion for calculating the amplitude of each fundamental mode compounding a general shape. The following criterion is found: If $s_1, s_2, \ldots$ are the displacements at a given instant of a general shape for the elastic with n balls, and if the displacements of the i-th fundamental mode are $x_{i1}, x_{i2}, \ldots$, then the general shape can be obtained by superimposing the n fundamental modes, each with amplitide m_i, where m_i is given by:

$$m_i = \frac{\sum_{j=1}^{n} s_j x_{ij}}{\sum_{j=1}^{n} x_{ij}^2}$$

This is the well known expression arising in the least squares method, already known to the students. By this superimposition, we are able to know the shape of the elastic at all times.

Now, let us go back to the study of the guitar string. We think of our string as being composed of an infinite number of molecules; hence infinite fundamental modes, recognizable for their regularity during the motion, are expected. As the point R is fixed, the arc PR in the second mode looks like the first mode for a string L/2 long and so on. So we shall know everything about the fundamental modes when we know the first.

To this purpose, let us extend the above property $F = ks$ and begin to calculate the force acting on the molecule B, placed between the molecules A and C. We obtain, at a given instant:

$$F_B = T\sin\alpha - T\sin\beta$$

but if α and β are very small then $\sin\alpha \simeq \mathrm{tg}\,\alpha = (s_B - s_A)/h$ and $\sin\beta \simeq \mathrm{tg}\,\beta = (s_C - s_B)/h$, that is

$$F_B = ((s_B - s_A) - (s_C - s_B))T/h$$

where $((s_B - s_A) - (s_C - s_B))$ is a "difference between the differences", or simply a second difference, and is indicated by Δ_2. Thus, the force F_B is proportional to S_2. Recalling that $F_B = ks_B$, we can also deduce that Δ_2 is proportional to s_B, and this property characterizes the shape of the string in the fundamental modes.

Since the students are already familiar with an analogous property concering the sine curve as the position function of a simple harmonic oscillator, they realize that even in this case, the shape of the string in the fundamental mode at a given instant is sinusoidal. Thus we obtain the first important result that the shape of the string in the fundamental mode at a given instant is sinusoidal.

Afterwards, the dimensional analysis allows the students to calculate the frequency w_1 of the first fundamental mode. They find

$$w_1 = \frac{1}{2L}\sqrt{\frac{T}{d}}$$

where L is the length of the string, T is the tension and d is the mass per unity volume. Considering the second mode as the first mode of a string L/2 long, we immediately obtain $w_2 = 2 w_1$, and so on. These formulas are very interesting for the students, because they give a scientific reason for the practical rules, followed by everyone when tuning a guitar.

Furthermore, these results lead the students to some interesting disucssions about the behaviour of the string in the first fundamental mode: the motion of each molecule during the time is sinusoidal with the same frequency w_1 and, furthermore, at each instant all the molecules lie on a sinusoidal curve. Thus, if we assume $z = \sin x$ as the initial shape of the string, each molecule has an harmonic motion with frequency w_1 and initial displacement sin x. Hence

$$Z(t) = \sin x \cos w_1 t$$

Of course, this result takes an unusual effort from the students and the teacher, but, in our opinion, it is worthwhile working in this direction. The students see a significant example of a function of two variables, discuss a physical phenomenon which evolves with the same mathematical law in space and in time and can experience the power of the mathematical and physical instruments.

Now we remember that, in the very beginning of our study, we saw many photos of a string. Some of these photos showed some nonsinusoidal shapes of the string; however, at this point the results above allow the students to realize that each shape of the string can be obtained by a suitable superimposition of the fundamental modes. This result immediately leads to Fourier's theorem. In the following year, when the students have some knowledge of integral calculus, they can also grasp the expression of the coefficients of the Fourier's series. In fact, if $z = f(x)$, $0 \_ x \_$, gives the shape of the string at the time $t = 0$ and $z_1 = \sin(ix)$ is the shape of the i-th fundamental mode, the formula

$$m_i = \frac{\sum_{j=1}^{n} s_j x_{ij}}{\sum_{j=1}^{n} x_{ij}^2}$$

becomes

$$m_i = \frac{\int_0^T f(x)\sin(ix)dx}{\int^T \sin^2(ix)dx} = \frac{1}{\Pi}\int_0^T f(x)\sin(ix)dx$$

So the students can also complete the study with the expression of the general shape of the string:

$$z(x,t) = m_1 \sin x \cos w_1 t + m_2 \sin 2x \cos 2 w_1 t + \ldots$$

The study of the vibrating string accomplished, we often tell the students, a significant episode of Enrico Fermi's life: to enter the University, Fermi had, as an examination, to write about vibrating strings. He wrote a relation about the classical treatment of the vibrating strings in terms of differential equations with partial derivatives. This relation had been judged absolutely exceptional, according to the usual preparation of a student at the high school level, and so it qualified him as a "scientific genius". The students are very satisfied that a subject generally reserved for the genius was intelligible for all of them.

As it has been shown, with this second argument one can go very far, always integrating mathematics and physics, the one motivating the other. It is now too long to list the subjects connected to this: one can study the 1, 2, . . . , n dimensional vector spaces by observing the set of the shapes of a string with 1, 2, . . . , n balls and Hilbert spaces by observing the string with an infinite number of molecules, or one can use some analogous concepts for the study of the mechanics of the waves, quantum mechanics, and so on.

In conclusion, integrated teaching can begin with very simple subjects and, gradually, advance into very involved theories, always giving the students an occasion for serious, but active and enjoyable work. We think that with this method it will be possible to organize a three year curriculum and in the next section we are listing the subjects prepared and experimented with for this purpose. Of course, there are still questions which will arise with these experiences; for example is it useful or interesting for students to look for the boundary between mathematics and physics? Should a person have a "pure mathematical point of view" or a "pure physical point of view"? Do these boundaries and these points of view exist?

4. Some Ideas for a Three Year Curriculum

Here we present a list of the subjects prepared and experimented with in the same spirit and with the same method as above. This list will give an idea of a three year curriculum, even if it is still incomplete.

a. Launching of projectiles (students 15-16 years old), published in "Educational Studies in Mathematics", vol. 10 (1979).

b. Simple harmonic motion and graphical derivatives (students 16 years old).

Subjects involved:

Math: trigonomitric functions, graphical derivatives, difference equations.

Phys: simple harmonic motion and its displacement, velocity, acceleration as a function of time.

c. Least square method and elasticity (students 16-17 years old) presented at the XXXI rencontre CIEAEM, 1979.

Subjects involved:

Math: first ideas on statistics, the least square method.

Phys: some ideas on elasticity, Hooke's law, fitting of data, measurements and errors.

d. Solar energy (students 16-17 years old), published in La Fisica Nella Scuola, March 1980.

Subjects involved:

Math: exponential and logarithmic functions, simple ideas on probability and statistics.

Phys: temperature, heat, heat flow transmission, absorption and emission, principles of thermodynamics, entropy, efficient use of energy.

e. Conics and universal gravitation (students 17 years old) published in Educational Studies in Mathematics, vol. 10 (1979)

Subjects involved:

Math: conics as envelopes of lines, focal properties, podaria of the conics with reference to the focus, pedal equation of the conics as first example of intrinsic representation of a curve.

Phys: Newton's law of gravitation, gravitational field and its potential, conservation of energy and of angular momentum, the trajectories in a gravitational field are only conics, escape velocity, black holes.

f. Vibrating strings and Fourier's series (students 17-18 years old) to appear.

g. Electrical networks and linear algebra (developed during different years with students from 15 to 17-18 years of age) presented at the 1980 IEEE International Conference on Cybernetics and Society, Boston.

Subjects involved:

Math: vector spaces, linear operators, eigenvalues, eigenvectors, Jordan's canonical form of a matrix, systems represented by linear difference or differential equations.

Phys: RC, RL, RLC and more complex networks and their time constants.

THE CONCEPTUAL NATURE OF THE RELATIONSHIP BETWEEN PHYSICS AND MATHEMATICS AND ITS CONSEQUENCES FOR THE TEACHING OF BOTH SCIENCES

Jean-Pierre Provost
Universite de Nice
Nice, France

The above title has a philosophical or epistemological character. However I shall not specialize my talk to questions of philosophy in which I am not an expert. The main problem which concerns me (and which I shall not solve!) is: "How to improve the collaboration between physics teachers (P.T.) and math teachers (M.T.)" or "How to build a real coordination between the corresponding teachings". Philosophical problems appear because, to my opinion, the major obstacle towards a solution of this problem is of an epistemological nature. Let me sum up this point of view:

As long as the connection between Physics (P) and Mathematics (M) is not considered to be essentially a connection at the conceptual level, i.e. as long as one reduces the usefulness of M in P to the supply of tools or to that of a rigorous language, the collaboration between teachers will keep its present character of "mutual assistance" and there will be no possibility of serious coordination between curricula. Mutual assistance means that although P.T. and M.T. may be ready to help their colleague, they do not in general expect any benefit for their own teaching. In general, a P.T. does not think that mathematical considerations may help to understand more deeply a physical concept and a M.T. does not think that a mathematical concept can be entirely grasped within a physical context.

In the first part of this talk (sections 1 and 2), I briefly present those ideas which either undervalue ("dangerous ideas") or favour ("good ideas") the above quoted conceptual relationship. These ideas have been developed in more detail by Provost (1). In the second part, I shall claim for and give examples of a new type of pedagogical research in the common field of M and P teaching.

1. Spontaneous Philosophy of Physicists and Mathematicians.

Under this name I describe some well accepted ideas which do not seem to be questioned but are dangerous since they undervalue the above quoted conceptual connection.

A. Mathematics as a Reservoir of Tools (typical physicist's point of view). According to this opinion, "three is a clear distinction between physical concepts (mass, force, atoms . . .) which are necessary to understand the physical world and mathematical tools which one cannot avoid in order to make these (abstract or qualitative) concepts come into contact (through calculus) with (concepts or quantitative) reality". This opinion may be summed up in the sentence "Mathematics apply to Physics" (whereas it would be more appropriate according to our opinion to replace the verb apply by are implied by or are constituitive of (1, 2).

As a consequence Mathematics in a Physics course are reduced to a series of automatisms (addition, multiplication, differentiation, integration . . .) which lead to numerical applications. Let us remark that this typical physicist's point of view is also shared by most mathematicians.

B. Mathematics as the (rigorous) Language needed by Physicists. (Typical mathematician's point of view). According to this opinion, "a physical concept contains a scientific core which can be formulated in a mathematical language and a "dirty" envelope associated with the use of the common language".

This opinion is dangerous because it can lead very quickly to a sterile axiomatization of P. teaching. The reason is that any physical situation is so mathematically rich that one is tempted to fill pages and pages of definitions in order to define everything in a proper manner, and so doing, one forgets about the main mathematical concepts which are present in the physical situation.

Therefore Mathematicians should remember that axiomatizing the teaching of Physics in order to make it "rigorous" is as much dangerous as it is for Mathematics teaching. On the other hand, Physicists should not fix their attention on the normative and syntactic aspects of the mathematical language, but on its semantic aspect.

Let us add that an exaggeration of the role of language may lead to sterile fights between M.T. and P.T. about words (such as vectors, angles . . .) instead of discussing and comparing ideas.

Finally let us remark that no theoretical consideration allows one to identify Math^{ics} and Logics (Logics being just a part of Math^{ics}; cf. the mathematical theory of models). Fortunately the role of Math^{ics} in Physics is richer than that of Logics.

C. Where should one place the division between Mathematicians and Physicists? The thought that Mathematical models is the natural place where M.T. and P.T. can meet may accredit the idea that Mathematicians should not have a look at physical reality and Physicists are unable to approach the mathematical theory. If so, the above mentioned meeting place becomes a demarcation line. The dangerous fact is that this compromise, although well accepted by both teachers since it respects their independence, impedes any possibility of a deeper cooperation.

I would prefer the theory which shows that the domains of work of physicists and mathematicians are the same (it extends from the real world to the mathematical theory) but that there exists a specialization in the type of work in the spirit of working. We should not forget that M and P were introduced and still are developed in order to apprehend more precisely our environment.

2. Towards a Better Understanding of the Conceptual Relationship between Physics and Mathematics.

The following ideas may help to a better understanding of the conceptual role played by Mathematics in Physics.

A. A look at the historical development of Mathematics and Physics shows that physical and mathematical concepts have grown up in parallel. Unfortunately all the historical discussions on, and the dialectic development of knowledge usually disappear in teaching.

The best example deals with the mathematical concepts which are present in the (physical!) concept of space: these concepts have been entirely introduced by mathematicians working for more than 2000 years in the spirit of present theoretician physicists. This historical fact explains why physicists are so much reluctant to consider the concept of space as a physical one, in spite of the severe lessons which "Nature" has given them. I refer of course to the Theory of Relativity at the beginning of this century, but to my opinion, the recent development of Gauge theories will show us that we have not paid enough attention for more than 50 years to the influence of quantum mechanical ideas on the concept of space (3).

B. The autonomy of development of mathematical concepts is quite relative: if most important mathematical concepts had not originated in the study of physical situations, one could never understand why very abstract mathematical concepts may be fundamental in Physics.

C. The migration of some teachings (such as Mechanics) from Mathematics to Physics should not be considered as a source of conflict, but as exemplifying the conceptual connection between the two sciences (as well as their evolution). Maybe in the future, this will be the case of elementary Geometry!

D. In the technological work (which may be considered as part of the real world!) the presence of Math is more apparent than that of Physics: it is easier to explain from Boolean algebra the automatization of an engine than to explain from electromagnetism how one's own razor works! Technology is for an important part a materialization of mathematical ideas. Let us add that through technology Mathematics have strongly influenced Physics at the conceptual level (concepts of coding, filtering, optimization, which are useful in Electronics, Optics, and also in theoretical Physics.) I think that the existence of both a mathematical approach and a physical approach to Technology can be the source of a deeper connection between P and M teachings.

E. The division between mathematicians and physicists is a particular case of the general technical division of labour. One should not introduce in teaching a division which is not clear at the level of research (cf. the division between mathematicians themselves or between physicists themselves).

3. Some Examples

A. Ohm's Law: An Example of Undermathematization. The physical concepts entering Ohm's law $V = RI$ are those of voltage and intensity (and their additive character); one remembers also the laws:

$$R = R_1 + R_2; \quad R^{-1} = R_1^{-1} + R_2^{-1}$$

relative to the resistance R of a system of two resistors (R_1 and R_2) set respectively in series and in parallel.

The mathematical concepts involved (vector space and manipulation of ordinary numbers) are very poor and this poorness in fact deserves Physics itself. This is the reason why I speak of undermathematization of the corresponding physical domain.

It would be more interesting for P and M to consider the relation between V and I as a functional relation $V = r(I)$ (r: resistance function) because:

1. in reality for a general dipole, the relation is a non-linear one;

2. the above laws can be formulated in the same simple way:

$$r = r_1 + r_2; \quad r^{-1} = r_1^{-1} + r_2^{-1}$$

But now, the concepts have changes: r is a function, not a number! r^{-1} is the inverse function with respect to the composition law of functions.

3. Many physical situations can be considered. Let us take for instance the example of generators: $V = E - RI$ (r and r^{-1} are affine functions). For two generators in parallel, one obtains (from the relation
$r^{-1}(v) = E/r - (1/R)\,V$:

$E/R = E_1/R_1 + E_2/R_2$; $R^{-1} = R_1^{-1} + R_2^{-1}$.

4. One gets a non-trivial example exemplifying the vector structure of a space of functions and the importance of the inverse function concept (4).

Remark: Since Galilea introduced velocity through the relation $L_2/L_1 = T_2/E_1$ (in Euclid's manner) any physical quantity has been introduced by a proportionality relation through the example of linear laws. However as everybody knows most physical laws are in fact non-linear ones. This leads to the important following question: What is fundamentally (in essence) linear and what is practically (within approximation) linear? The answer to this apparently naive question implies a serious reflexion on the conceptual aspects of the relationship between Physics and Mathematics.

B. Newton's Law: An Example of Overmathematization. The physical concepts entering Newton's law are those of inertia and force. Unfortunately the concept of force necessarily implies in dynamics that of differential equation and these equations cannot be solved (excepted when the force is constant) at an elementary level. Therefore there can be no interesting physical application of Newton's law, such as, Kepler motion, rule of air friction, harmonic motion of a puck on an air table . . ..

The solution to this problem is to go from the concept of differential equation to that of equation with finite differences (which one can solve) and, at the same time, from the concept of force to the concept of impulse (mass times a change of velocity). This figure constitutes a revival of Newton's approach.

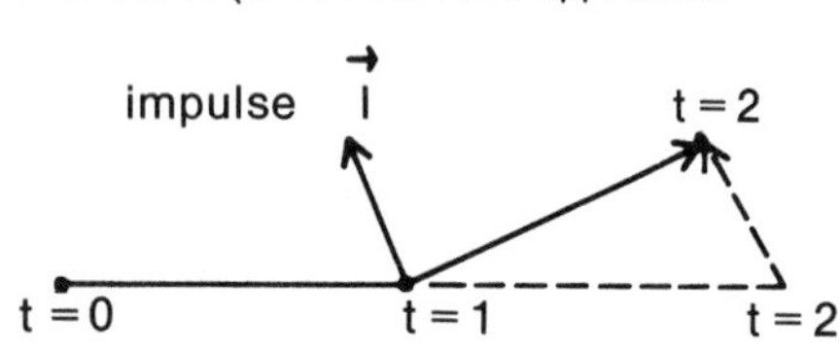

In that way 15 year old students can study the trajectories (and some of their properties) corresponding to the cases:

$\vec{I} = \vec{a} - k\vec{v}$ (gravity + air friction);

$\vec{I} = -k\vec{r}$ (harmonic motion);

$\vec{I} = k\vec{r}/r^3$ (Rutheford diffusion or Kepler motion)

$I = q(E + \vec{v} \times \vec{B})$ (motion in a constant electromagnetic field) (5).

The interest of this change of (Mathematical-physical) concept is:

1. the possibility of a great number of non-trivial physical applications at different levels (2nd School, university . . .).

2. The analogy with what is really observed in an air table experiment with a stroboscope.

3. The analogy with what a computer really does.

4. The motivation of the students to learn the concepts of vectorial derivatives and differential equations (in this connection it might be interesting in Mathematics to introduce differential equations from equations with differences; in that way differentiation and integration would be naturally connected).

Remark: the concept of inertia also has a deep mathematical content; it can be shown (although it is not obvious at first sight) that this concept is strongly connected to the concept of group (3, 6).

C. Geometrical/Metrical/Variational Optics: An Example of "Mathematical Polymorphisms" of Physics. The usual teaching of elementary geometrical optics is a rather dusty one. With some rare exceptions, it has not changed for about 400 years, dealing at length with the properties of similar triangles. I want to indicate now how the same subject can be taught with the help of almost no geometry in two other ways: use of matrices (linear algebra) or use of second order polynomials (variational calculus). These changes of "mathematical tools" do imply a deep change in the physical approach to Optics (7).

1. Why should be abandon the geometrical approach to Optics? There are many practical as well as theoretical reasons. (a) It is difficult to prove from pure geometry that any object has an image: for the simplest lens, there are two circular surfaces to consider; moreover Descartes-Snell's laws cannot be formulated in a simple geometrical way and approximations are necessary. As a consequence any elementary course admits the existence of images and is limited to the description of the rules of construction of rays. (b) This approach is not adapted to the present technology of optical systems (a zoom for instance has more than 5 lenses). (c) It does not explain the analogy of electronic Optics with ordinary (photon) Optics. From a geometrical point of view, a magnetic lens is very different from an ordinary lens. (d) The main concepts (vergence, magnification) do not appear as primary concepts. (e) Geometry is not involved in the generalization of geometrical Optics to wave optics (of optical systems).

2. Matrical Optics (centered systems). In that one considers a centered optical system as a black box which takes any incident ray of the object space into an emergeny ray of the image space. Since these rays are straight lines, they can be described by vectors. As coordinates of the rats in the planes O and O' (perpendicular to the optical axis in O and O'), one can take the heights y and y'and the slopes α and α' (or for photon Optics for the products $n\alpha$ and $n'\alpha'$ where n and n' are the indices respectively of the object space and of the image space. The zero vector corresponds to the optical axis. Since, due to the symmetry of the system, the optical axis is a possible path of light, it is natural to suppose that the optical system is a linear one. This hypothesis of course will be tested from experiment (either directly or through its consequences: existence of images . . .).

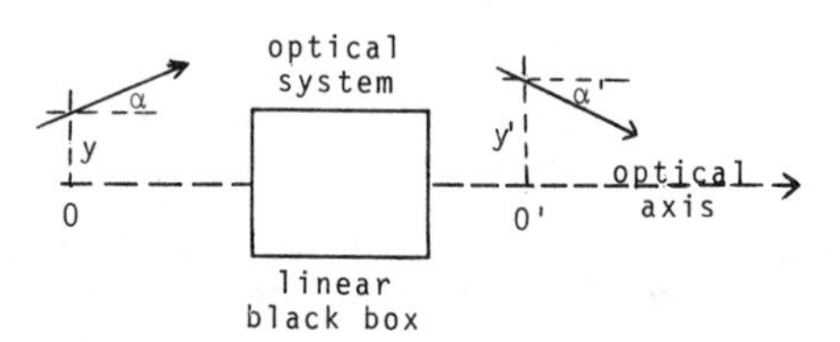

The linear relation between the coordinates may be written a priori:

$$\begin{vmatrix} y' \\ n'\alpha' \end{vmatrix} = \begin{vmatrix} a & b \\ -C & d \end{vmatrix} \begin{vmatrix} y \\ n\alpha \end{vmatrix} = M_{O',O} \begin{vmatrix} y \\ n\alpha \end{vmatrix}$$

The coefficient C is easily seen to be independent of the origins O and O' (C is related to the slope of the emergent ray associated with an incident ray parallel to the optical axis). It appears to be the vergence (power) of the system. It is a very simple exercise to prove that the above matrix is

$$\begin{vmatrix} 1 & x/n \\ 0 & 1 \end{vmatrix}$$

($x = \overline{OO'}$) when here is no optical system and is of the type

$$\begin{vmatrix} 1 & 0 \\ -C & 1 \end{vmatrix}$$

for a diopter (O and O' being the intersection point of the diopter and the optical axis). Since all matrices $M_{O',O}$ are the products of such particular matrices, their determinant is one. When C = O (afocal system), the matrix reads:

$$\begin{vmatrix} r & b \\ 0 & r^{-1} \end{vmatrix}$$

r clearly does not depend on O and O'; it is the magnification of the system.

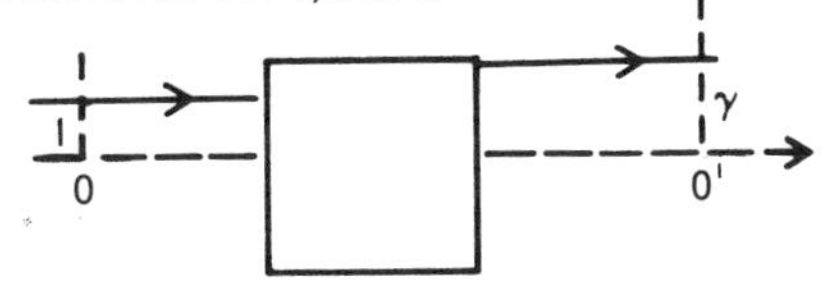

We have therefore described the two different types (focal and afocal) systems. The existence of images is a consequence of the hypothesis of linearity: if O and O' are such that b = O, the matrix of the type

$$\begin{vmatrix} r & 0 \\ -C & r^{-1} \end{vmatrix}$$

One gets $y' = ry$ whatever the slope of the incident ray is.

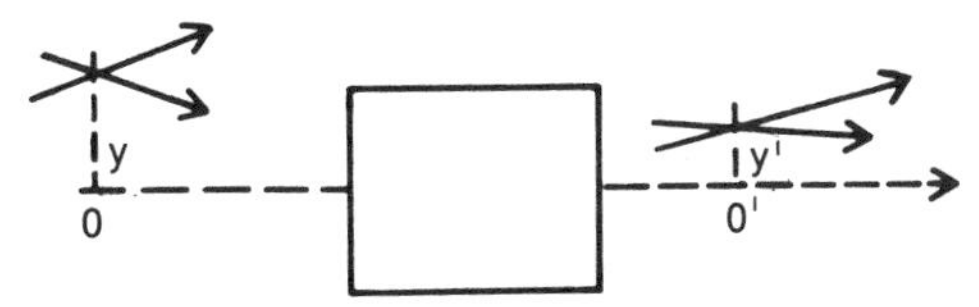

We stop here the matricial approach to Optics since this talk is not a lecture on Optics. Let us simply remark that we have answered the objections formulated in a), b), c), d). Moreover, although I cannot prove it here, this approach can be appropriate as soon as pupils have heard of matrices (indeed this example of the use of matrices has the advantage of lying outside traditional geometry). Finally, at a first year university level, this approach is the most efficient one; conjugation formulas (with any origin), or formulas of association of optical systems for instance are obtained within less than two lines; practical problems (with or without the help of computers) can be solved in a straightforward way.

3. Variational Optics (centered systems). According to Fermat, the path which is followed by light between two given points corresponds to an extremum of the time which is needed (or an extremum of the optical path). In all lectures on elementary Optics, Fermat's principle is abandoned as soon as Descartes-Snell's laws have been derived from it; it often appears as a homage paid by Physics to Philosophy. Indeed there is a straightforward (pedagogical) path which leads from this principle to the study of the most general centered system without using Descartes-Snell's laws.

Let us take the case of a thin optical system; such a system can be considered geometrically as a plane (no thickness); however, optically the optical path associated with a ray passing through the system depends on the height y of the ray (the thickness of glass depends on y).

1 2 y optical axis

The difference between paths 1 and 2 is an even function of y (symmetry of the system). For small y, we can write it as: $- 1/2\, Cy^2$ (this formula appears to yield a definition of the power C of the system from the very beginning of our approach).

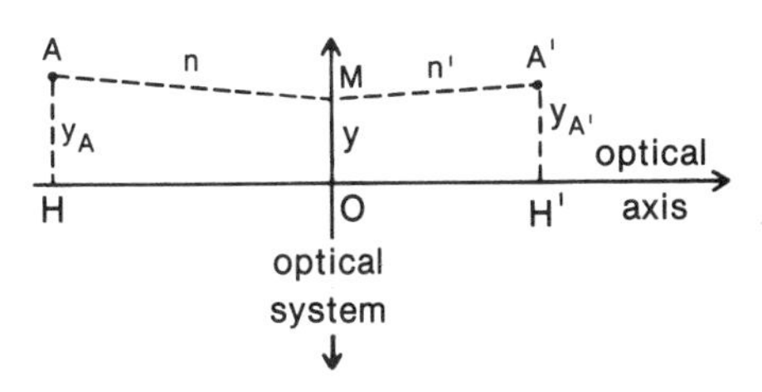

Let us now determine the shortest path from the point A to the point A'. We compare the paths AMA' and HOH' (OH = x, OH' = x'). With the use of Pythagoras' theorem, one gets for small y, y_A, $y_{A'}$:

$$MA' = [x'^2 + (y_{A'} - y)^2]^{1/2} \cong x' + \frac{1}{2x'}(y - y_{A'})^2$$

Therefore the difference between the optical paths reads:

$$(AMA') - (HOH') = - 1/2\, v\, (y - y_A)^2 - 1/2\, Cy^2 + 1/2\, v' (y - y_{A'})^2 \quad (v = n\, x^{-1};\ v' = n'\, x'^{-1}).$$

All main results of Geometrical Optics now come from the discussion of a polynomial of second order.

If $v' - v \neq C$, there is an extremum (maximum or minimum). The corresponding value of y is such that

$$\underset{\text{slope of the emergent ray}}{v' (y_A - y)} = \underset{\text{slope of the incident ray}}{- Cy + v\, (y - y_A)}$$

This relation establishes the equivalence between the matricial and the variational approach.

If $v' - v = C$ and $v\, y_A = v'\, y_{A'}$, the polynomial is a constant; any ray satisfies Fermat's principle; the points A and A' are conjugated and the above formulas are nothing else than the conjugation formulas.

Once more I recall that this talk is not a substitute for a course on Optics. My aim is to show that different mathematical approaches to the same physical subject involve (or are implied by) different physical concepts. The main advantages of the variational approach are the following.

It is a non-trivial example of a principle of extremum (without the necessity of highbrow mathematics). - The concept of power is formulated from the beginning - The existence of image is proved - It works for electron Optics (because in electron Optics, there is also a principle of extremum) - Last but not least, this approach allows for a short transition (at an advanced level) to Wave and Fourier Optics (8).

4. Conclusion

My conclusion will be short: there is no miraculous remedy to the problem raised at the beginning of this talk. Epistemological obstacles (as well as sociological obstacles: cf. the talk of Professor Revuz at the I.C.M.I. - C.T.S. - U.N.E.S.C.O. Conference in Bielefeld 1978) are very serious ones. To shed light on them is, of course, not sufficient. Conditions for a better situation are so numerous that their enumeration might let the reader think that my point of view is deeply pessimistic. Since I am optimistic I shall insist on one unique point: It is necessary to develop new directions for pedagogical research (besides and not instead of the present ones); these directions are those which exemplify the conceptual relationship between Physics and Mathematics. Teachers (but also pupils or students) are so much convinced that P and M are separated by a clear line of demarcation that the first things we need are examples of the falseness (and danger) of this opinion. This research should be made at any level; I insisted much on the university and secondary school levels which are more familiar to me. However the tradition of separating Physics (let us say more generally Science) and Mathematics begins at the elementary school level, the general feeling being that Sciences which involve Mathematics should be learnt once a minimal knowledge of M has been acquired; this might be the major defect of our teaching. Therefore this research should also concern the elementary school level (9). The subject of research, which to my opinion, deserves priority is the teaching of Geometry as a theory of Space: most of the mathematical concepts and of the physical concepts are related to the study of Space: Space (and time) are those physical concepts which have the deepest mathematical content, Geometry is an important source of intuition for Mathematics, moreover the concept of space is worthy of study from the elementary school level to the university level; any coordination between curricula should therefore be founded on its teaching.

References

1. J.P. Provost "Mathematical Concepts and Physical Concepts" I.C.M.I. - C.T.S. - U.N.E.S.C.O. meeting in Bielefeld (1978) published by the Institut fur Didaktik der Mathematik der Universitat Bielefeld.

2. J.M. Levy-Leblond in "Mathematiques et Physique" Encyclopedia Universalis.

3. J.P. Provost "Temps et Mouvement les Fantomes de la Geometrie, Colloque I.R.E.M. of Nice, 1974.

4. H. Nagerl and J.P. Provost "The physical concept of resistance and the mathematical concept of a function", Bielefeld, 1978.

5. J.P. Provost "Sur les traces de Newton" publication of the I.R.E.M. of Nice, 1974.

6. The equivalence of the inertia and group concepts is implicit in the book of Yaglom entitled "A simmple non-Euclidean Geometry and its physical basis" Springer Verlag ed., 1979.

7. J.P. Provost and P. Provost "Principe de Fermat; cours d'Optique Geometrique" C.E.D.I.C. Nathen ed. France, 1980.

8. P. Provost and J.P. Provost "Coherence; cours d'Optique ondulatoire" C.E.D.I.C. Nathan ed. France (to appear).

9. "Activites d'eveil a l'ecole elementaire. Mathematiques et Physique". Publication of the I.R.E.M. of Nice.

4.4 WHY WE MUST AND HOW WE CAN IMPROVE THE TEACHING OF POST-SECONDARY MATHEMATICS

WHY WE MUST AND HOW WE CAN IMPROVE THE TEACHING OF POST-SECONDARY MATHEMATICS

Henry L. Alder
University of California
Davis, California

1. Why We Must

A few years ago, the Editor of a major mathematical publication in the United States began a banquet address as follows: "Professor are untrained for much of what they do - - teach, administer and edit. Work experinece is the learning process for editing." He could have said the same for teaching, since work experience has traditionally been the learning process for teaching in college. Is this desirable? Emphatically not, many would argue, but why?

No doubt there has been and is a great deal of excellent teaching in mathematics at the college level. All of us can readily think of superb teachers of mathematics who have influenced us and may have been primarily responsible for our having become mathematicians. Few of us, however, will deny that there are others whose teaching could have greatly benfited from some program designed to improve their teaching ability.

In the past we may have been able to afford the luxury of having college mathematics faculty who are not good teachers. But there are many indications that these days are over and the time is ripe for us to do everything possible to elevate our teaching to the highest possible level. We as a profession have a responsibility to assure that the highest standards are maintained for the two main services we provide:

research and teaching. By a well-organized and widely-accepted process of refereeing papers we have established such standards for our research, but have we set similar standards for our teaching? And have we devoted the same efforts to maintaining quality in the teaching of mathematics courses?

With the predicted decline in college enrollments, we can maintain enrollments in mathematics courses only if we make them so attractive that more students will want to take them. For this we need good teachers. For the same reason, we can no longer have courses in a mathematical science taught outside the mathematics (or mathematical science) department. It would be unheard of for a course in French to be taught outside the French Department, or for a course in physics to be taught outside the Physics Department, yet we all know of courses in mathematics or a mathematical science, in particular, courses in probability and statistics, even linear algebra and calculus, that are taught in many departments. We shall not investigate the reasons for this phenomenon, almost unique to mathematics. There is, however, substantial evidence available that such proliferation of courses in the mathematical sciences outside a mathematical science department can be prevented, halted and even reversed if the mathematical science department offers the appropriate courses in a way that satisfied the needs of the students and presents them as an enjoyable, exciting and stimulating experience.

2. How We Can

About three years ago, the Mathematical Association of America appointed a Committee on the Teaching of Undergraduate Mathematics, which was charged to take a number of steps to help improve the teaching of mathematics and, expecially, to assist beginning teachers of post-secondary mathematics. I would like to describe briefly two actions which were taken by this Committee:

(a) This Committee has prepared and published last year a brochure entitled COLLEGE MATHEMATICS: SUGGESTIONS ON HOW TO TEACH IT which contains a wealth of suggestions for the beginning and the practicing college teacher of post-secondary mathematics. This brochure was distributed to every one of the approximately 19,000 members of the Mathematical Association of American and has been extremely well received.

(b) The same Committe also prepared and published last year a brochure entitled TRAINING PROGRAMS FOR TEACHING ASSISTANTS IN MATHEMATICS. This brochure gives a description of a basic Teaching Assistant Training Program which provides the bare minimum of guidance that should be provided to teaching apprentices. In view of the many other responsibilities both teaching assistants and regular faculty members have and in order to make the program acceptable, it is essential that it can be carried out with very little effort, in a small amount of time and at practically no cost. The basic ingredients of the program are the following:

1. A regular faculty program coordinator. This person takes responsibility for the mechanics of the program, and gives two sessions (before the first day of classes) for those teaching assistants who have been newly appointed as teaching assistants.

2. Each teaching assistant is assigned to a faculty supervisor. The responsibilities of the supervisor include monitoring the teaching assistant's classroom performance, checking on the construction of examinations and the assignment of grades, and serving as a general resource to be consulted whenever necessary. Experience indicates that, over the course of a term, the faculty supervisor spends an average of no more than two hours per week in his or her supervisory duties.

For the last five years I have had the privilege of serving as Teaching Assistant Supervisor for the Department of Mathematics of the Davis campus of the University of California. This program begins with two training sessions during the week preceding the start of instruction in the fall. At the first session I discuss the responsibilities of being a teaching assistant with the new trainees and give some helpful guidelines for beginning teachers.

In this session considerable time is spent on what to do at the first lecture which is obviously of prime concern to a new teaching assistant. Here is what I typically start out with: "You are walking to your first class. You should expect to be very nervous. If you are not, it is a bad sign. It is an indication that you are not sufficiently conscious of your great responsibility. Next you announce the number and title of your course to be given in the room...etc."

Particular stress is always paid in these sessions to ways and means of involving students in the teaching process, for example, by encouraging them to ask questions and by having the instructor direct questions to the students. To get student involvment is not at all easy; there are a number of helpful tricks which we discuss. Frequently in this training session I discuss a situation which might occur in class and ask the new teaching assistants how they would handle it. Let me give an example:

> A student raises his hand and asks a question which is so dumb that the whole class starts laughing. What do you as an instructor do? - - - -
>
> Here is a possible reaction from the instructor: "Now I know that many of you had the same question on your mind, but you were just too afraid to ask it. Mr. Smith deserves a lot of credit for having had the courage to ask that question...", and then proceed to answer it.

A very valuable technique to get the students involved in a mathematics class is to ask them to guess the answer to a problem. By asking the students to guess, they automatically are committed to the problem, that is, have a stake in its solution. As I always tell my students: "When you make a guess, you really never lose. There are only two possibilities: either your guess is right or your guess is wrong. If your guess is right and you find out you were right, you have something to be very happy about. If your guess was wrong and you find out that you were wrong, then you have something to be amazed about."

To conclude this first session we have an experienced teaching assistant come before the group and present a sample leclture of about 10 to 15 minutes on a topic chosen from freshman mathematics, usually calculus. The new teaching assistants then are asked to critique that lecture in as much detail as possible, which usually takes about an hour.

In the second session of our teaching assistant training program, normally held two days after the first, each new teaching assistant gives a lecture of about 15 minutes on a topic assigned to him or her at the first session. The teaching assistant gives the lecture to all the other new teaching assistants and three or four members of the faculty who are also present. Each of the lectures is carefully critiqued in a discussion of the assembled teaching assistants and faculty.

What have been the results of this program? We have had it evaluated and recieved, essentially, only positive comments from all involved in the process, the teaching assistants, the faculty, and the students in the classes taught by the teaching assistants. The program has had some rather surprising results. For example, when I asked a few of our new graduate students recently why they had come to our campus, one of the reasons given was that the student wanted to become a college teacher of mathematics and he knew that on our campus he would receive good training toward that goal.

The only criticism we received from the teaching assistants was that their faculty supervisors were sometimes too soft on them. They knew that they were not as good as their superviosrs led them to believe in the reviews of their lectures. They wanted us to be much more critical.

Now what have I learned from these experiments?

New teaching assistants--and this obviously applies equally to all new faculty members--can be taught successfully how to teach, and, what is not generally realized, it takes very little time, effort and money to do it.

Any such program must make due allowance for the fact that a new teaching assistant has a very full program of study and many other responsibilities. Therefore, the program should take only very little time, and that time needs to be used as efficiently and effectively as possible.

New teaching assistants, generally, lack the time to pursue the theories of teaching and learning. They want to know as many tricks of the trade as possible, but not necessarily why they work.

A training program for new teaching assistants can be viewed much like a driver education program. The learner wants to know what he has to do to get the car started, to get it to drive, etc. He is usually not at all interested to know why the turning of the key starts the motor. Similarly, a new teaching assistant does not need to care about why a particular trick of the trade makes the student learn the material and why another method does not work. What he or she wants to know is as many of the tricks of the trade as possible, and I believe we owe it to our new teaching assistants to supply them with these. If we give them more than that, it generally will—to use the current favorite expression—turn them off.

Another temptation which needs to be resisted is to spend undue time on the newer educational media, such as television, computer-aided instruction, etc. They are important—don't misunderstand me—but their use should be taught only to those who are likely to need them and this is clearly still only a small minority. From all indications, the lecture and discussion type of instruction will continue to dominate atleast 90% of all teaching and, presumably much more. This is not only because of the positive effect of the involvement of the human element in the teaching process, but also because it generally costs much less to have a live teacher than some mechanical substitute.

It is equally important to resist the temptation to stress innovative teaching methods in these training programs. For a beginning college teacher it is a big enough task to adapt to methods of teaching which have proved successful. There is little justification to encourage beginners to consider teaching methods, for which, the best that can be said of them is that they are innovative.

I want to emphasize again that I have described here a program which must be considered the bare minimum for training prospective post-secondary teachers of mathematics. It clearly would be highly desirable to have a more extensive program, but it should also be recongnized that initiation of even a bare minimum program is vastly better than no program at all, which, unfortunately until recently, has been the typical situation in most institutions in the United States. We are greatly encouraged that many institutions have recently instituted programs of this type, and we are confident that this will result in a substantial improvement in the teaching of undergraduate mathematics.

IMPROVEMENT OF THE TEACHING OF POST-SECONDARY MATHEMATICS

Detlef Laugwitz
Technische Hochschule Darmstadt
Darmstadt, Germany

There can be no doubt at all that we must improve the teaching of post-secondary mathematics. So let me proceed to my own remarks which will cover two aspects: Firstly, models which we have had and have in some parts of the European Continent, and secondly, some aspects of the pedagogy of mathematics teaching as developed by mathematicians.

I have to dwell for a little time on a description of the educational system in West Germany, or rather, in some parts of Continental Europe.

University study is preceded by 13 years at school. From the age of 6 to 10 there is the general elementary school, and gymnasium (high school) starts at the age of 10. Secondary phase I terminates at the age of 16. The last 3 years of the gymnasium (secondary phase II) have quite recently been transformed into something quite similar to a college, and in some parts of the region are actually named "college phase" (Kollegstufe). In general, gymnasium teachers have to be able to instruct all age levels, and in at least two subjects, like mathematics and physics. The training system of these teachers has, as I think, a very good tradition and reputation. They have a university education of about 5 years in their two subjects (and in addition some pedagogy and psychology). After their "first examination" they go for about 2 years to a so-called seminary of studies (Studienseminar) which is attached to a gymnasium. There they are supervised by one or two mentors. These should be experienced gymnasium teachers of some scientific level and should have good knowledges of modern education insights.

One of the disadvantages is that only in few cases the final qualification is obtained at an earlier age than 27. On the other hand, this training of teachers at the college level is satisfactory with regard to knowledge and experience.

Let me now speak of university teachers. Though there has been some discussion during the last decade or so we still have the so-called "habilitation" which leads to the degree of a "Privatdozent". In mathematics and related fields the usual prerequisites are a PhD plus about 5 years of successful work which includes published papers and, as a rule, assistantship at a university department. You might feel that this must be a very good training for the professor or a university teacher, and that other countries should copy the system.

I think that the system is not so bad, but needs essential improvements. Unfortunately it favours the publish-or-perish mentality since the scientific results are considered decisive, whereas only little or no teaching experience is expected. Actually during his or her years as an assistant the applicant had not much opportunity to teach classes independently. Only recently some positions have been called into being which are comparable to the teaching assistant (Hochschulassistent). More and more of us are becoming convinced that the "habilitation" should continue to exist but that improvements are urgently needed. I think there should be more independent teaching accompanied by some guidance by an experienced faculty member, perhaps in the manner that Professor Alder has proposed at this panel. As everything in Germany our universities are governed by state laws; even the new University Laws provide for some scientific guidance of the (teaching) assistant but not for any instruction how to teach.

Let me add some remarks on the mathematics instruction of students of technology, sociology and the sciences. As Professor Alder has pointed out we observe that many non-mathematical departments try to or even have succeeded in annexing mathematical lectures. I agree with Professor Alder's opinion that this development is deplorable, and that we should stop it soon. Please do not misunderstand me. Scientific level is certainly one important component when a person is considered to become a mathematics teacher in post-secondary instruction. But it is vital for both the future of the mathematics community and a good teaching that we are very careful about the contents and the presentation of mathematics instruction for non-mathematicians.

After 25 years of teaching experience at technical universities I am convinced that teaching assistants and lecturers and young professors should see the specific problems of the mathematics instruction for non-mathematicians. There must be sufficient motivation from their own fields of interest; we should keep in mind that many of them are primarily not interested in mathematical thinking in our sense but rather in methods which are useful in their fields. Of course, a mathematics instructor should not teach them more recipes and procedure (which will possibly be of minor interest a few years later) but general mathematics. The faculty must take care that constructions are more important than deductions, in this type of teaching. Abstract vector spaces are unimportant and useless if only rows and columns and matrices are needed, to mention only one example.

It is vital that good texts are available for these purposes, and I find it helpful if they contain some material from the student's realm of interest and experience.

To improve our teaching, more suitable texts are needed.

Can pedagogy and/or psychology help to improve mathematics instruction at the post-secondary level? I think that this question deserves serious consideration. Many non-mathematicians emphatically answer: "Yes". On the other hand most professor of mathematics are convinced that no essential help can be expected from psychology and pedagogy.

There are two aspects which I shall consider separately: General pedagogy (and general psychology) and, on the other hand, results which are relevant for the specific situation, contents and methods of mathematics instruction.

I agree that we can expect virtually nothing from the general theories, and I am able to report on negative experiences which underline that. Some ten years ago there was a Center of University Pedagogy (Hochschuldidaktisches-Zentrum) established at my university. The majority of the people working at this center were young and had little or no experience of their own in the fields of science or technology or mathematics. You will expect that this Center was bound to fail, even though some cooperation was offered by the departments of the university. I think that the practical experiences and the final aims of instruction are of a different kind, and that scientists speak languages which are foreign to the general psychologist. Thus I have no great hope in any future help from general pedagogy.

On the other hand I am convinced that we can get help for the improvement of our instruction from experienced mathematicians who have done research in mathematical instruction. There are many of them in North America, but let me mention at least one outstanding example from the European continent, the work of Professor Hans Freudenthal of Utrecht who is a mathematician doing important and influential work in mathematics instruction. The names of Wittenberg and Wagenschein may stand for all those whose roots are in pedagogy but who speak our language and have thorough knowledges of mathematics.

They agree in their well-founded recommendations which also apply to post-secondary instruction. To put it in a nutshell: Though completed mathematics is (often) printed in a deductive manner, this is not the right way to teach it, at least to beginners. (And beginners include all undergraduates and even graduates who want to become acquainted with a special field new to them.) The alternative which is offered is described by concepts like genesis and induction, or by more practical advices to the teachers (including teaching assistants and professors): Start from concrete problems which are inside the domains of interests and personal experience of the students. Be careful that these examples motivate the students, but that the problems are of some mathematical importance so as to give access to concepts and methods which lead on to more general mathematics.

During the limited time I have here I cannot dwell on these considerations on mathematics teaching. I am certain that many of you will know about them. Practical hints can be found in the publications of the MAA which were already mentioned today.

My point is: We should find ways to make those considerations better known to all teachers of mathematics, including the post-secondary level. Even if we should agree that this is an important aim, it is an open question how to realize it. In those countries where something like "habilitation" exists I suppose that we could achieve some progress in this direction. For instance, the applicant might be expected to include some reflections on methods and manners of mathematics instruction in a (public) lecture in front of the faculty. Please do not misunderstand me: I am not a friend of an introduction of such exercises as a necessary prerequisite of teachers at the post-secondary level, but I plead emphatically for recommendations. Perhaps it may help to tell teaching assistants to have a look at, say, Professor Freudenthal's mathematical work to convince them that he is both an outstanding mathematician researcher and teacher, and that his work on mathematics instruction may be worthwhile reading.

The last two (of thirty) pages of "College Mathematics: Suggestions on How to Teach It" (MAA) are dedicated to "Reading and Seminars Related to Teaching". You will find some useful hints there.

4.5 ALTERNATE APPROACHES TO BEGINNING THE TEACHING OF CALCULUS AND THE EFFECTIVENESS OF THESE METHODS

INTRODUCTION TO THE BASIC CONCEPTS OF INFINITESIMAL ANALYSIS

Georges L. Papy
Universite Libre de Bruxelles
Bruxelles, Belgium

Strange as it may seem, I have been asked to outline once again the topological introduction to infinitesimal analysis that I proposed about twenty years ago, the implications of which could affect the whole of mathematics education from kindergarten to university. I shall attempt to comply with this invitation, while at the same time adding a few comments with a view to answering some of the questions people are fond of asking nowadays.

If one of the necessary characteristics of a science consists in regarding facts as highly as Lord Mayors, mathematics has little right to that title. Parallel lines, which in the past never met, then met at infinity; and finally it even came to be considered quite natural that each straight line should be viewed as being parallel to itself. Originally without a solution, the equation $x^2 + 1 = 0$ was later endowed with the roots i and -i. The mathematician is a poet, and mathematics is his dream. When this dream has some connection with reality, it gives a version of that reality which is fanciful, ideal, harmonious, and solidly structured. Mathematics springs from man himself, and concerns him in the innermost reaches of his subconscious, which is just asking to be excited by appropriate stimuli. Now, there can be no true education which is not in large part mathematical education. It is undoubtedly possible to study dancing without learning how to dance, or to study music without learning how to play it. But any teaching of mathematics that is in the fullest sense educational cannot be restricted to playing with it as a prefabricated instrument, a gift from our ancestors, fallen from the heavens. For the deeply human character of mathematics is to be seen in a special way precisely in the activity of mathematisation, in the proper sense of the word, and in particular in the imaginative effort that gives birth to fundamental concepts. "The solemn moments of mathematical creation", said Francesco Severi, "are found in the definitions". Notably, might I add, in those of the grand structures and of their distinguishing features.

'Infinitesimal analysis is built on sand", complained Abel in a letter to Galois, and he added: "it is time to ensure that it has solid foundations". Such foundations were found in the shape of point set topology in what Hilbert called the "paradise created for us by Cantor". Even long after the discovery of non-Euclidead geometries and the bursting asunder of the Kantian yoke, it will readily be conceded that Kant had some excuse for asserting the impossibiilty of developing mathematics outside the immutable framework of space. The spectacle of today's mathematics seems to argue persuasively that there is little mathematics outside geometry. In particular, topology, with its topological spaces, would find it difficult to conceal its geometrical affinities. Bringing its topological concepts out into the open is a real geometrisation of infinitesimal analysis. Thus the introduction of infinitesimal analysis becomes one of the natural objectives of geometrical education, and most especially of the teaching of Euclidean plane geometry, which is infinitesimalised by a set-theoretic way of thinking.

This set-theoretic point of view both refines geometry and enriches it. Faced with a square cut into four, the situation itself inhibits the raising of the problem of the cutting-points, yet such a question is typical of infinitesimal analysis. A non-set-theoretic teaching of elementary geometry atrophies the turn of mind required by infinitesimal analysis. The introduction of this subject is facilitated by adopting the set-theoretic view of Euclidean plane geometry.

Cardboard geometrical shapes, with all their drawbacks, used to offer an intuitive support to a rather crude situation. But once the situation is infinitesimalised by a set-theoretic way of thinking, it is advisable to find a different intuitive support. This may be achieved by means of the green-red convention, which is inspired by traffic lights. A uniformly green disc on a white page, depicting the plane, represents the closed disc, that is, the set of points of the disc, including its boundary. The corresponding open disc, that is, the set of points of the disc, excluding the boundary, is represented by colouring the forbidden boundary-circle red. A line segment drawn entirely in green represents the closed segment, including its endpoints, either of which may be excluded by being coloured red. An open segment with its endpoints excluded is represented by a green stroke with two red ends. The same convention applies to closed, open, and half-open curved arcs. Conversely, a green surface (or its boundary) may have one of its points, or an open, closed, or half-open arc, removed by colouring it red. The only claim made by this modest convention is that it permits a clear, graphical, nonverbal presentation of a sufficient variety of infinitesimal situations. A personal touch may be added by refining it to a greater or lesser extent, but in no way will it ever be able to represent all the subsets of the Euclidean plane, such as the set of the rational points of a real number line or a real number plane, for example. One should not gain the false impression that this convention consists in colouring the points of a subset of the plane green and colouring those of the complement red. Using the green-red convention, one can draw a rather complete set of posters, each representing an infinitesimal situation. Comments are made about them, and then the pupils are encouraged to get their feet wet by being asked to suggest new posters of infinitesimal situations. The open nature of the question allows a variety of worthwhile answers, which are immediately commented upon. Particularly noteworthy, and hence important to be brought to the attention of the class, are the interior points, which the pupils sometimes spontaneously call safe points, thus attesting to the role played by affectivity in this enterprise. These are the points about which may be drawn a "safety circle"; if the point is question is allowed to move about a little, one may rest assured that it remains an element of the set, that is, it does not break out of the "safety circle". On the other hand, the slightest displacement of a noninterior point runs the risk of making it fly out of the set. Despite the haziness of this discussion, the concept of interior points very soon seems to have been acquired in the sense that the pupils are able, for every point of an infinitesimal situation, to say whether it is interior or not. A new challenge and the next step forwards consists in formulating a mathematical definition of interior points in terms of previously acquired concepts. It might be said, for example, that a point of a subset of the plane is interior to that subset if and only if it belongs to an open disc contained within the subset. The open sets of the plane are those of its subsets all of whose points are interior.

From a set-theoretic perspective it is now natural to ask what can be said about unions and intersections of open sets. The union of every set of open sets is open. The intersection of every finite set of open sets is open. Which, by a joyful leap with many unexpected side-effects, leads straightaway to the definition of topologies.

According to the neo-Kantian code of modern mathematics, topological spaces are the objects (or the points) of the category TOP of topological spaces. Its morphisms (or arrows) are the everywhere continuous functions. Continuity is the second basic concept of topology, and in particular of infinitesimal analysis. Let us restrict ourselves to outlining one way of reachng it.

As a starting-point a new set of posters is drawn, each representing a transformation of the Euclidean plane, henceforth endowed with a topological structure: translation: half-turn symmetry; line symmetry; constant transformation; homothety; orthogonal projection onto a straight line; descending orthogonal projection onto this infinite staircase:

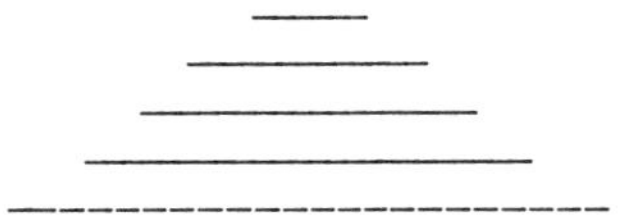

the transformation defined by a plane grid with square "tiles", each of which is mapped onto its lower left corner.

On each of these posters the teacher draws a continuous, meandering and winding road, and asks a pupil to draw its image on the same poster. On being asked to classify the transformations represented by the posters, the pupil's reaction is explosive. Some are spontaneously described as continuous, others being characterised by the fact that they cause jumps. These latter are examined more closely and the pupils are asked to mark those of their points that are responsible for the discontinuity. Once again it is remarked that the concept of continuity at a point has been acquried in the sense that the pupils soon show themselves capable of indicating the points of discontinuity of a transformation of the plane that is not everywhere continuous. The only improvement left to be made is to arrive at the mathematical formulation of the continuity - or the discontinuity - of a transformation of a plane at a point of that plane.

For this purpose attention is directed to a point of discontinuity x of a transformation of of the plane and the question is asked "where the line should be drawn so that it can certainly be crossed". In other words, determine an open disc D with a centre f(x) such that it is possible to make the image of a line through x jump out of D. Getting away from "lines through x", the definition of the discontinuity of f at x is eventually expressed in terms of the existence of nonempty open disc D with center f(x) such that no nonempty open disc X with centre x has its image fX included in D: $fX \not\subset D$. The definition of continuity is obtained by contraposition: The transformation f of the plane is continuous at the point x if and only if, for each nonempty open disc D with centre f(x), there exists a nonempty open disc X with centre x such that $fX \subset D$.

The definition of continuity at a point x of its domain of a function f from a topological space into a topological space is obtained by that poetic extension which transforms discs into open sets. The function f is continuous at x if and only if, for each open set Y containing f(x), there exists an open set X containing x such that $fX \subset Y$.

This leap from the continuity at a point of a transformation of the plane to that of a topological space function is considerable. Nevertheless, it is possible to regard the latter in such a way as to preserve our original belief in the continuity or jump alternative. In fact, by contraposition, discontinuity at a point x of its domain of a function f from a topological space into a topological space may be expressed in terms of the existence of an open set Y containing f(x) and satisfying the conditions: every open set X containing x contains at least one point y such that going from x to y makes the image jump from $f(x) \in Y$ out of Y.

Reference

Mathematique Moderne 1 - 2 - 3. (Didier, Bruxelles - Paris, 1963 - 1965 - 1967) Le Premier Enseignement de l'Analyse. (Presses Universitaires de Bruxelles, 1968).

APPROXIMATION ET INTERPOLATION EN TANT QU'ACTIVITES MATHEMATIQUES AU SEIN D'UN CURSUS D'ANALYSE ELEMENTAIRE

Daniel Reisz
Université de Dijon
Dijon, France

It s'agit de présenter ici les idées directrices et un exemple de travaux réalisés par le groupe d'Analyse des IREM français. Dans ce groupe, qui fonctionne depuis plusieurs années, nous sommes partis d'un certain nombre de constats critiques de la situation actuelle de l'enseignement de l'analyse dans les lycées français pour proposer quelques issues possibles aux difficultés que rencontre cet enseignement en France.

Nous avons d'abord fait notre définition de A. Krygowska: "l'éducation mathématique n'est rien d'autre que le développement de l'activité mathématique, et il n'y a pas d'activités sans problèmes". Ce truisme est trop souvent oublié dans l'enseignement actuel qui reste très inspiré d'un bourbakisme simpliste dans sa progression:

1. Ensembles, relations binaires, relations fonctionelles, applications, injections, surjections, bijections
2. $N \rightarrow Z \rightarrow D \rightarrow Q \rightarrow R\ (\rightarrow C)$
3. Topologie sur R
4. Fonctions numériques à variable réelle, domaine de définition, continuité et limites, dérivabilité, variations, branches infinies et points pathologiques, représentations graphiques
5. Intégration

et, en caricaturant à peine, ce n'est qu'à la fin du 4 qu'on peut réellement envisager de faire fonctionner cela au sein de situations non triviales et artificielles.

Les défauts majeurs d'une telle situation nous paraissent être:

- l'introduction des notions de base sans problématique sousjacente, ou avec une problématique très élaborée mathématiquement mais trop éloignée de l'intérêt de l'élève et de ses possibilités de compréhension.
- l'emploi dès l'abord d'un langage trop formalisé et souvent hermétique qui réduit l'activité mathématique à des acrobaties gratuites, voires factices, sur les symboles.
- un enseignement qui se ramène trop souvent à un discours du maître, bien au point, présentant les mathématiques comme un monde clos, tel un objet d'art, on propose à l'admiration béate des élèves. L'enseignant donne alors l'impression de trouver sa recompense quand l'élève parvient à reproduire - tel quel - ce discours ésotérique.
- une construction linéaire des concepts, bien hiérarchisée, n'amenant (et pas toujours) qu'en fin de construction des algorithmes et des méthodes opératoires. Les applications intéressantes arrivent trop tard, voire jamais, et les notions ne sont pas perçues dès l'abord comme étant efficaces pour la résolution des problemes, les problèmes posés n'ayant en outre que trop rarement un caractère quantitatif.
- un intérêt parfois trop précoce pour le pathologique, alors que le normal et l'usuel ne sont pas assimilés ou suffisamment manipulés (la recherche d'un exemple a une valeur didactique, mais sûrement pas le contre exemple donné d'entrée de jeu et sans problématique).

Face à ces écueils, une démarche sans doute plus fructueuse nous est suggérée par l'histoire même de la pensée mathématique. Bien peu de théories mathématiques ont été élaborées en partant des fondements et en allant vers les applications ou les procédures algorithmiques. La plupart des concepts ont mûri petit à petit, par des usages répétés dans des situations diverses. Une longue pratique est souvent nécessaire avant que les différents aspects d'un concept se clarifient.

- C'est pourquoi il nous faut organiser l'enseignement de l'analyse autour de quelques <u>grands problèmes</u> conduisant à des <u>situation riches</u> et <u>liées aux autres disciplines</u>.
- Pour contribuer efficacement à la formation scientifique, le choix des problèmes à étudier et des concepts qui leur sont liés, ainsi que celui des stratégies didactiques, doivent conduire à des activités mathématiques permettant notamment de développer les capacités suivantes: analyser une situation en dégager des hypothèses théoriques, découvrir et mettre en oeuvre des concepts propres à l'étudier, préciser les moyens expérimentaux propres à contrôler les hypothèses précédentes, analyser les résultats obtenus au regard des problèmes posés, et analyser la pertinence des moyens théoriques et expérimentaux ainsi construits.
- Dans ce but, les dialectiques suivantes jouent un rôle essentiel:
    - acquisition de connaissances et analyse de la pertinence des connaissances
    - maîtrise de l'acquis (entraînement, mémorisation) et exploration de nouveaux problèmes ou concepts (débroussaillage, conjectures)
    - approfondissement des exemples et approfondissement de la généralité
    - conjectures et démonstrations
    - stratégie de démonstration et rédaction de ces démonstrations

- construction d'objets complexes à partir d'objets simples et décomposition d'objets complexes en objects simples

- construction de méthodes variées d'attaque d'un problème ou d'un concept et analyse comparative de leurs performances.

- Les comparaisons entre problèmes voisins, les remarques sur plusieurs séries de résultats ou sur les qualités de certaines procédures permettent de créer des conditions favorables, sinon à la découverte d'une notion mathématique, du moins à une meilleure saisie de cette notion et de ses champs d'intervention. En outre, dans ce cheminement, il ne faut conceptualiser que ce qui demande à l'être au fur et à mesure des besoins. L'étude des structures n'est pas une fin en soi, elle doit être au service d'une maîtrise plus efficace de problèmes compliqués.

A travers l'enseignement de l'analyse, on doit s'efforcer avant tout de mettre l'élève en état de "comprendre les idées essentielles préalablement à toute formalisation". Maîtriser un concept, ce n'est pas seulement en connaître la définition formelle et les théorèmes qui l'accompagnent, c'est aussi être capable de le faire fonctionner, de le faire agir et réagir en liaison avec d'autres concepts dans la recherche de solutions à des problèmes issus de situations variées.

Bref, nous souhaitons que l'enseignement de l'analyse soit centré autour "d'activités significatives", de résolutions de problèmes, et que dans cette démarche, on situe théorisation et axiomatisation à leur juste place. Il faut amener l'élève à agir, à construire lui-même son univers mathématique - certes en toute modestie - mais au contact des grands problèmes des sciences mathématiques.

On trouvera dans Bulletin inter-IREM le développement de ces idées ainsi qu'un choix de grands problèmes mathématiques qui nous paraissent être des terrains privilégiés pour l'activité des élèves. Parmi eux, un des plus importants nous semble être constitué par les problèmes d'approximation et d'interpolation. Dans Bulletin inter-IREM une présentation substantielle de ce travail, assorti d'un choix important d'activités pour les élèves, est présenté. Je n'en présenterai ici que les lignes directrices.

I - Critères de choix

Pourquoi les problèmes d'interpolation et d'approximation nous paraissent-ils un secteur d'activités important?

1. c'est un ensemble de situations pouvant donner lieu à une étude suivie et pouvant être reprise à différents niveaux.

2. les situations rencontrées sont génératrices de problématiques pour l'approfondissement des concepts.

3. c'est un terrain privilégié de mise à l'épreuve des outils théoriques.

4. présence simultanée du qualitatif et du quantitatif.

5. importantes possibilités d'utilisation de calculatrices de poche ou de microordinateurs qui apparaissent ici non comme de simples outils mais comme participant à la dialectique entre le champ conceptuel et le champ des problèmes.

II - Analyse sommaire des problemes d'interpolation et d'approximation

La présentation schématique ci-dessous ne doit pas faire croire à deux problèmes distincts. Ce n'est le cas ni au niveau des problématiques, ni au niveau des méthodes.

A. Interpolation

Pour une fonction f on connaît $b_1 = f(a_1),\dots, b_n = f(a_n)$ et on désire pour une valeur a, avoir une valeur approchée de $b = f(a)$. La qualité de l'interpolation dépend de plusieurs facteurs dont les plus significatifs sont:

- le nombre et la répartition des a_i

- l'allure de f et de la fonction interpolatoire sur chacun des segments $(a_i, a_i + 1)$

- la position de a par rapport aux valeurs a_i

B. Approximation

Idée générale: trouver une fonction φ, telle que, pour tout x d'un intervalle, $\varphi(x)$ soit une "bonne" approximation de f (x).

Pour aborder les problèmes d'approximation, il faut regarder les "parametres" qui interviennent:

a. la façon dont est connue la fonction f (ensemble fini de valeurs $f(x_i)$, enregistrement expérimental discret ou continu, formulation mathématique explicite, ...)

b. le type de la fonction approximante φ

c. la définition de la qualité de l'approximation.

III Quelques exemples d'activités proposeés aux élèves

Faute de place, on ne trouvera ici que les sujets des exemples envisagés. On trouvera dans (Bulletin inter-IREM) une rédaction complete de ces exemples.

A. Interpolation linéaire

Interpolation linéaire de $x \mapsto y = x^2$

Interpolation linéaire de $x \mapsto y = \sqrt{x}$

Interpolation linéaire d'une fonction inconnue des élèves mais disponible sur une calculatrice.

Interpolation linéaire d'une fonction uniquement connue par des données numériques discrètes.

B. Interpolation parabolique

Determination d'une parabole par 3 points (cas particulier de deux points très proches)

Comparaison entre interpolation linéaire et interpolation parabolique (precision, performance).

C. Methode de Lagrange

Bases de l'espace vectoriel des polynômes

Choix d'une base adaptée au problème de l'interpolation (base de Newton, de Lagrange)

D. Interpolation comme méthode d'approximation

Approximation linéaire de la fonction sinus (majoration de l'erreur)

Approximation parabolique de la fonction sinus (majoration de l'erreur)

Expression du majorant de l'erreur dans la méthode de Lagrange.

E. Interpolation de Tchebychev

Analyse de la formule d'erreur de Lagrange

Polynôme s'écartant le moins possible de zéro sur un segment

F. Indications sommaires sur quelques autres types d'approximations

Approximations tayloriennes (établissement et utilisation de la formule de Taylor avec reste intégral pour quelques fonctions élémentaires majoration classique du reste)

Présentation géométrique de la méthode des moindres carrés

Series de Fourier (quelques exemples)

Approximations rationnelles (étude expérimentale de quelques exemples)

Lissage par méthode spline cubique (aspects mathématiques, pratiques, physiques).

References

Bulletin inter-IREM: Analyse (sous presse, à paraître debut 1981.

4.6 IN WHAT WAYS HAVE THE MATHEMATICAL PREPARATION OF STUDENTS FOR POST-SECONDARY MATHEMATICS CHANGED?

SIXTH FORM MATHEMATICS - CHANGES IN THE CURRICULUM AND ITS EFFECT ON PREPARATION FOR HIGHER EDUCATION

Kathleen Cross
Rosendale College
Accrington, England

Concern has been expressed that sixth formers who have taken 'A' level mathematics (the 18+ examination in England and Wales) do not seem to cope as easily as they used to in Higher Education Courses involving mathematics. If this is true, then why? Can the situation easily be remedied?

It is true that fewer students will be arriving at University and Polytechnic having studied two mathematical subjects at school. Higher Education staff are critical of the large number of mathematical syllabuses, but we have always had a large number. Examinations in England and Wales are administered by Examination Boards based in different parts of the Country. They devise their own syllabuses and set their own examination papers. Traditionally they all had syllabuses in Pure Mathematics, Applied Mathematics (Newtonian Mechanics) and Statistics, though many of these had changed to Mathematics and Further Mathematics where each syllabus contains both Pure and Applied topics. This automatically produces a large number of syllabuses but, in practice, there was a large measure of overlap.

The sixties brought a wave of curriculum development because of genuine concern in the schools about mathematical education. This led to new material being introduced into sixth form mathematics. The Examination Boards developed new syllabuses and ran these parallel to the old ones, thus allowing schools the choice. This meant that students were arriving at University with more varied backgrounds than before and lecturers met students who could not do certain topics in which they had come to expect mastery. This was because the student had either not met the topic at school or had met it but with a lesser degree of emphasis. Staff of Mathematics Departments at University have held several conferences to see how they can adapt their courses for the future.

This diversity of content and style of 'A' level courses (which seems to cause most problems to Physicists and Engineers) has led to SCUE and CNAA (representing Higher Education Institutions responsible for awarding degrees) jointly issuing a proposal for a minimal core syllabus for 'A' level mathematics (1). They proposed a core which is about two thirds of the present 'A' level which would be common to all syllabuses as a basis for discussion. It is said to be minimal because it sets out the least content "...which is required as the foundation for a wide range of degree courses". SCUE and CNAA stress that students should be well versed in the whole of the core which puts great emphasis on the "acquisition and retention of manipulative skills of the traditional kind" and suggests that examination assessment methods should be designed to ensure that the students have mastery of the core. This could lead to many teachers concentrating almost wholly on these manipulative skills at the expense of areas such as applications. It is a criticism now made by people in Higher Education that some 'A' level syllabuses produce students who can perform well on examination papers but cannot extend their ideas, apply their mathematics or show any awareness of its uses. A student cannot begin to apply his mathematics unless he has an understanding of the underlying principles.

The case for a core of material common to all 'A' level syllabuses is strong, but it is dangerous to allow the request for a high level of skills of a limited kind to determine the core. Also, mathematics is used by many different subject areas and it is not hard to envisage a situation where their varying demands determine the whole syllabus.

During the last decade more students have continued in full-time education up to the age of 18 and are taking 'A' level courses. Mathematics is now being studied alongside many different subjects. It not only supports physical and biological sciences, but others, including geography, sociology and economics. The course is taken by students going on to study

(i) mathematics, physics and engineering (who are the major users)

(ii) biological and social sciences (who will mainly use statistical techniques)

(iii) arts based subjects (who will probably not use mathematics at all)

In addition, many students will leave school and not continue any areas of further study but their view of the subject and attitude towards it will be determined for life by their experience of the course. For this reason the course must stand in its own right and not just be seen as a preparation for Higher Education.

But is it possible to devise a single course that satisfies all these students? Some schools do offer alternative courses, but this is not possible in smaller shcools, even if the staff wanted to. Many teachers find it beneficial to have students with varied subject interests in one class; efforts have to be made to link mathematics with a broader range of subjects than was traditional.

Most teachers would readily agree that the course should include some applications and traditionally this has been Newtonian Mechanics. But the majority of students in an 'A' level class will never require any mechanics; probability and statistics is probably more useful to them. This has resulted in many syllabuses, including some probability and statistics at the expense of the mechanics which Engineers say they require for their courses. Unlike other European countries the Engineering degree takes 3 years (rather than 4 years elsewhere) and Engineering Departments had come to rely on students having some knowledge of mechanics. (In England and Wales mechanics was part of both Mathematics and Physics course, so students had often studied it twice.) Concern has also been expressed that in courses dealing with these two areas of applications it is difficult to do justice to either. But a case can be made for also including other newer areas of applications and for using mathematics to model a variety of situations. Professor G.G. Hall wrote, in 1972, of the concern by many professors of Applied Mathematics that applied mathematics was losing its share of the school mathematics syllabus. He wrote of tensions between the older and newer branches such as operations research and computer science and argued that "all applied mathematics may be described as mathematical model building and that, by describing our activities in this way, we can eliminate our present confusion and lack of purpose."(2)

So the debate goes on. Do we have a CORE and, if so, what should it be and how big should it be in relation to the whole course" What about applications? Should there be a choice (i.e. have alternative courses like Pure Mathematics with Mechanics and Pure Mathematics with Statistics) and would it be the student's choice or the teacher's (or school's)? Many students have not made definite career choices at the age of 16 and so may easily make the wrong choice.

Most of the debate has centered on content because of the demands of the users. These cannot be ignored and there is obviously a need for students to be able to perform manipulative techniques accurately and confidently. But is this enough? SCUE/CNAA state in the introduction to their core syllabus that "students should be given more encouragement than is now common to strengthen their geometrical insight in space as well as the plane, they should understand clearly what is meant by a proof and should be numerate to the extent of having a real sense of orders of magnitude and degrees of error and approximation." They go on to say that "any course of Mathematics...should demonstrate the economy and beauty, as well as the discipline, of the great unifying concepts of Mathematics".

McLone, in a report for the Social Science Research Council, on the relevance of university mathematics courses to subsequent employment also identified a need for the ability to recognise how to apply mathematics to raw situations and to be able to interpret the results. Universities often say that they want students who can think for themselves; students who have a positive attitude towards mathematics and a keen interest in the subject; students who have the ability to generalise; to communicate their results; and so on.

Can we also develop these qualities through our 16 - 19 courses? I believe we can, but not by having a syllabus crammed full of unconnected topics and techniques and an examination geared to this. It is not enough to hope that the teacher will incorporate these more general strategies; these need to be thought out and carefully built into the teaching syllabus. The way people teach is heavily determined by the examination syllabus and the examination questions set. If the syllabuses are well written in content terms (with a few pious hopes in the introduction) we tend to lose the importance of the general skills and strategies a student needs in order to usefully apply his mathematics in the future.

Some attempts were made to look at sixth form mathematics differently in the Schools Council research programme into a new system of examining at 18+ (i.e. N & F level examinations as possible replacements for 'A' level). In the A.T.M. Commissioned Group a syllabus was devised with clearly defined process as well as content aims (3). Skills (numerical, graphical, spatial and algebraic) and general strategies (modelling, generalising, proving) were defined which would operate within the content syllabus framework.

This course has enormous implications for teaching method and commitment from teachers, but we cannot ignore teaching method when discussing aims of syllabuses (i.e. different aims are likely to be achieved in different ways). H.M.I., in a recently published survey of secondary schools (11 - 16) stated that "whatever the syllabus, and whatever the label under which it is classified, it is ultimately the interpretation which the teacher gives to the subject matter and the teaching approach which is crucial."(4)

If a core of content common to all syllabuses can be found which is helpful to engineers and other major users and also allows schools to teach a mathematics course satisfying the needs of all their students, maybe some of the present problems will be eased. Some Examination Boards have already incorporated the suggested SCUE/CNAA Core into their 'A' level examination packages, but the relative ease with which they have achieved this almost certainly means that criticisms levelled at many present courses in relation to the student's lack of ability to apply their mathematics will continue. I do not believe we will see any long term improvement unless we tackle more fundamental issues. Simply changing content will achieve very little. The impact of calculators and micro-processors in recent years has questioned the necessity of certain skills once considered basic. Do we honestly know what content is required by our students in the 1980s and 90s? Much more developmental work needs to be done into the area of effective teaching method so that our students can acquire necessary skills and strategies and be able to adapt and apply their mathematics to whatever situation they find themselves in the future.

References

1. A minimal Core Syllabus for 'A' level Mathematics (A joint Statement by the Standing Conference on University Entrance and the Council for National Academic Awards. December 1978).

2. G.G. Hall "Modelling - A Philosophy for Applied Mathematicians", published in Bulletin of Institute of Mathematics and its Applications (August, 1972).

3. For further details see "Mathematics for Sixth Formers' (A.T.M. 1978).

4. H.M.I. "Aspects of Secondary Education in England". (H.M.S.O. 1979).

STUDENTS FOR POST SECONDARY MATHEMATICS AND REACTIONS TO THESE REFORMS

S.M. Sharfuddin
Institute for Advancement of Science
and Technology Teaching
Dacca, Bangladesh

My dear colleagues, allow me to say that I greatly appreciate that I have been asked to speak at this unique gathering.

In a gathering like this with international audiences, it is very important to recognise that, as well as sharing common experiences, there are many differences among us. In particular what may be the case of one country may be different elsewhere.

In all developing countries like ours the impetus for changing the curriculum of Mathematics education came from international movements which disseminate new points of view through conferences and publications. Local professionals who have the opportunites of fimiliarising themselves with the new trends and events, through their experiences mainly from abroad, are the prime factors for the enforcement of the same in their own country. As a result many parts of it are incompatible with national needs and resources. Some resistance came from secondary teachers, but it gradually faded away because they recognised their gaps and dared not enter into controversy. Resistance from students was almost nil because they knew that the changes were made under the pressure of university professors and they were to go for higher study in the university as it was not easy to get a job after secondary education. Most of the parents never received a secondary education and such changes in syllabuses had no effect on them.

The new curriculum certainly is the product of prolonged hard work by different groups of both mathematics and teaching workers all over the world. Of course we cannot believe that anybody has found the final answer to any of our problems; we cannot even believe that a final answer exists. We are very much convinced that a better work may come.

Though the analysis of the mathematics edifice shows that there is no such natural division as elementary mathematics and advanced mathematics, yet there is a dogmatic separation between the two. We, however, felt that many of the notions which are normally treated at secondary levels contain the essentials of some of the most modern mathematical notions. We tried to establish a close collaboration between the universities, colleges and secondary schools in order to ensure the harmonious development of mathematics teaching and put an end to the detestable dogmatic quarrels.

In Bangladesh, in the new curriculum, mathematics has been made a compulsory subject for students of all disciplines up to class X (secondary school) and for all science groups up to class XII (higher secondary school). Our overall objective is to help students develop an ability to think mathemtically and logically by identifying and recognising various phenomena and by considering and treating these phenomena in a unified progressive way and to get them to realise the role of mathematics in a society. The specific objectives are: to have the students deepen their understanding of basic mathemtical concepts, principles and laws, to have them deepen their understanding of the significance of mathematical terms and symbols and thereby creating and fostering an ability of expressing simply and precisely the properties of terms and symbols, to develop in them an ability to percieve things with proper insight and the habit of examining the results according to the purpose or objectives, to have the students understand the mathematical way of approaching things systematically and to learn the significance of and methods of doing so, to help them to develop the ability to apply mathematical skills accurately and efficiently.

In the teaching and learning of mathematics language plays a vitally important role. The problem encountered in developing countries is all the more multiplied because the learner has to learn mathematics in a language other than his mother tongue. In a new curriculum abstractions, axioms, rigour, symbols and terminology need some basic minimum mathematical literature. Developing countries have now started teaching mathematics in their own languages. The deficiencies in mathematica vocabulary are being made good sometimes by keeping the English term with its spelling in mother tongue or by creating a matching term in the vocabulary.

Whatever may be the new automobile model, it also contains the same basic parts as the old, so our school mathematics still consists of arithmatic, algebra, geometry and trigonometry. But the responsibility of the teachers of mathematics has been increased immensely. He is to emphasise understanding, to build on a logical structure, to introduce new vocabulary, to give new presentations of old ideas and to show the use of mathematics in a scientific society.

As regards some of the contents of the curriculum, in Bangladesh we have kept a good part of Euclidean geometry of the plane as we feel that it is a valuable tool of intuition. Concrete geometrical intuition and formalisation is certainly a good combination of teaching and learning mathematics. Geometry with co-ordinates is designed. This relies more heavily on algebra. The fusing of geometry with algebra erases the boundary lines between the branches. We feel that this gives analytical geometry its proper place in the organic structure of mathematics. Free use of calculus is also made. Number with different bases, descriptive statistics, probability and some transformation geometry have been introduced.

Application of mathematics to our everyday life, its relation to science and society has a great formative value in the process of mathematics teaching and learning. Today our everyday life is not confined to supermarkets, banks and stores only. Today it also includes the world of sciences, engineering and computers. So we are faced with the challenging question of how problems involving scientific ideas can be presented without having to teach, not only the mathematics involved, bu the science as well.

The population changes, the increasing interdependence of people throughout the world, and the biological evolution has placed those phases of the mathematics curriculum that deal with human relationships in a position of major importance.

In countries like ours "change in curriculum" and "change in textbooks" are synonymous. Curricula are framed by the National Curriculum Committee. Production of textbooks is a centralised system. A single textbook is followed in all schools. The Bangladesh Text Book Board selects a group of authors who are entrusted to write a textbook for a particular class. This book is then edited by a group of selected editors. Sometimes textbooks for a particular class are invited from intending authors. Several books are generally submitted by different authors. These books are then given to a group of reviewers. The best one marked by them (with changes if they feel necessary) is selected.

We do not have enough trained teachers to cope with the new curriculum programme. So we have undertaken a gigantic teacher training programme.

Like all other countries of the world we are deeply cncerned with (as Professor Hassler Whitney, President of ICMI has put it in his circular of May 11, 1979, and February 21, 1980) "the failure of large numbers of students in schools throughout the world to absorb mathematics as a working tool and the common failure of students in mathematics in schools." In this connection we read with great interest th suggestion put forward by him in his circular of February 21, 1980: "the greatest need is to help the students to take more responsibility for doing their own thinking rather than just trying to learn what we teach; and this really requires facing all social, political pressures on them and on their teachers." We do not know whether any experiment has yet been made with Professor Whitney's suggestion. The idea is certainly revolutionary.

Changes in mathematics teaching in schools demand that a pupil has the ability to read with understanding. In a teacher directed class the teacher will learn to play the game in the teacher's language and in a student directed class unorthodox languages will be used and will present the teacher both with new opportunities and new problems and challenges.

4.7 CURRICULUM FOR A MATHEMATICS SCIENCE MAJOR

CURRICULUM FOR A MATHEMATICAL SCIENCES MAJOR

Alan Tucker
State University of New York
Stony Brook, New York

I. Background

This paper discusses a college-level mathematical sciences curriculum project of the Mathematical Association of America's Committee for the Undergraduate Program in mathematics (CUPM).

In 1977 CUPM decided that a major re-examination of the mathematics major was needed. The CUPM model for the math major contained in the 1965 CUPM reports on Pregraduate Training in Mathematics and a General Curriculum in Mathematics in Colleges (revised in 1972) was felt to be out of date. Following a six month study, it was decided that the CUPM math major curriculum should be substantially revised and broadened to define a mathematical sciences major. For this purpose, the CUPM Panel on a General Mathematical Sciences Program (MSP) was formed. The members of the CUPM Panel on a General Mathematical Sciences Program are: Richard Alo (Lamar Univ.), Winifred Asprey (Vassar), Peter Hilton (Case Western Reserve/Batelle), Don Kreider (Dartmouth), Bill Lucas (Cornell), Fred Roberts (Rutgers), Gail Young (Case Western Reserve), and Alan Tucker (SUNY-Stony Brook), chair.

The MSP panel coordinated its work with the National Research Council's Panel on Training in Applied Mathematics (chaired by P. Hilton, a member of MSP). The Hilton panel had a much broader mandate than the MSP panel. Its report addressed the unification of the mathematical sciences, the attitudes of mathematicians, academic-industrial linkages, and society's image of the mathematical sciences, as well as curriculum. The MSP panel strongly endorses the Hilton report's emphasis on the importance within mathematics departments of proper attitudes towards the uses and users of mathematics and of a unified view that respects the content and teaching of pure and applied mathematics equally.

In 1979, five subpanels were created with the aid of a Sloan Foundation grant, to develop course recommendations in: (i) the calculus sequence, (ii) computer science, (iii) modeling and operations research, (iv) statistics, and (v) upper-level core mathematics. Final curriculum recommendations and course syllabi are scheduled for early 1981.

2. Curriculum Background

The uncertainties created by recent upper-level enrollment shifts and by the explosion of new theory and applications in all the mathematical sciences have led the MSP panel to look for guidance from past CUPM curriculum development experiences, and, farther back, from the traditional goals of the mathematics major before CUPM's creation. No matter how great the advances in the past generation, the fundamental objectives of training in mathematics, defined over scores of years, should be the basis of any mathematical sciences program.

Until the 1950's, mathematics departments were primarily service departments, teaching necessary skills to science and engineering students and teaching mathematics to most students solely for its liberal-arts role as a valuable intellectual training of the mind. The average student majoring in mathematics at a better college in the 1930's took courses in higher (college) algebra, trigonometry, and analytic geometry in the freshman year, followed by two years of calculus. The mathematics major was filled out with five or six electives in subjects such as differential equations (a second course), projective geometry, theory of equations, vector analysis, mathematics of finance, history of mathematics, probability and statistics, (applied) complex analysis, and advanced calculus. Most mathematics majors also took a substantial amount of physics. The training of secondary school mathematics teachers rarely included more than a year of calculus. In the early 1950's, twenty years later, the situation had changed only a little; top schools did now offer modern algebra and abstract analysis.

In 1953, amid reports of widespread dissatisfaction with the undergraduate program, the Mathematical Association of Amerca formed the Committee on the Undergraduate Program (CUP, later to be renamed CUPM). CUPM concentrated initially on a unified introductory mathematics sequence called Universal Mathematics, consisting of a first semester analysis/college algebra course (finishing with some calculus) followed by a semester of "mathematics of sets" (discrete mathematics). CUPM hoped its Universal Mathematics would "halt the pessimistic retreat to remedial mathematics...(and)...modernize and upgrade the curriculum". The first comprehensive curriculum report of CUPM, Pregraduate Training for Research Mathematicians (1963), outlined a model program designed to prepare outstanding undergraduates for PhD studies in mathematics. Emphasis on PhD preparation represented a major departure from the traditional mathematics program and was the source of continuing controversy. A more standard curriculum for the mathematics major was published in 1965, but it also reflected the point of view of research mathematics.

3. Current problems facing the Mathematics Major

The unprecedented growth of computer science as a major new college subject parallels the theoretical growth of the discipline and its ever-expanding impact on business and day-to-day living. The number of computer science majors now exceeds the number of mathematics majors at most schools offering programs in both subjects. Enrolment in many upper-level pure mathematics courses have declined dramatically in the 1970's as students turned to applied and computer-related courses. Yet, while the number of mathematics majors is decreasing, the demand for broadly trained mathematics graduates is increasing in government and industry. The mathematical problems inherent in scientific computations and in projects to optimize the use of scarce resources and, more generally, to make industry and government operations more efficient, guarantee a strong future demand for mathematicians. These problems require people trained in disciplined logical reasoning and versed in the basic techniques and models of applied mathematics. If mathematics departments do not train these quantitative problem-solvers, then departments in engineering and management science will.

The shortage of secondary school mathematics teachers nationwide is worse than ever before. This shortage appears to be due in large measure to the greater attractiveness of computing careers to college math students. On another front, pre-calculus enrolments have soared as the mathematical skills of incoming freshmen have been declining (a problem that concerned CUP in its first year).

University mathematics departments, faced with a heavy pre-calculus workload, shrinking graduate programs, and competition from other mathematical sciences departments, appear less able to broaden and restructure the math major than do most liberal-arts college mathematics departments. Many university mathematicians seem to prefer to retain their current pure mathematics major for a small number of talented students.

On the positive side, there appears to be a natural evolution in the math major taking place at many schools. Students and faculty have developed an informal "contract" for a major that includes traditional core coursework in calculus and algebra along with electives weighted towards computing and applied mathematics. Complementing this trend, the ACM Curriculum 78 Report implicitly encourages students interested in computer-based mathematical problem-solving to be mathematical sciences majors, as the computer science curriculum concentrates on software and system design.

4. Principles for the Mathematical Sciences Curriculum

The MSP panel's goal is to produce a flexible set of recommendations for a mathematical sciences major, a major with a broad, historically rooted foundation for dealing with current and future changes in the mathematical sciences. The panel seeks a unifying philosophy for diverse coursework in analysis, algebra, computer science, applied mathematics, statistics, and operations research.

The following principles should guide the design of a mathematical sciences curriculum. THESE PRINCIPLES SHOULD SERVE AS A STARTING POINT FOR LOCAL DISCUSSIONS TO DEVELOP A PROGRAM APPROPRIATE TO THE INTERESTS OF A PARTICULAR SCHOOL'S FACULTY AND STUDENTS. TO BUILD A GOOD PROGRAM, FACULTY AND STUDENTS MUST ALSO UNDERSTAND AND RESPECT EACH OTHER'S EXPECTATIONS FOR A MATHEMATICAL SCIENCES CURRICULUM.

Program Philosophy

1. The curriculum should have a primary goal of developing the attitudes of mind and analytical skills required for efficient use and understanding of mathematics. The development of rigorous mathematical reasoning and abstraction from the particular to the general are two themes that should unify the curriculum.

2. The mathematical sciences curriculum should be designed around the abilities and academic needs of the average mathematical sciences student (with supplementary work to attract and challenge talented students).

3. A mathematical sciences program should use interactive classroom teaching to involve students actively in the development of new material. Whenever possible, the teacher should guide students to discover new mathematics for themselves rather than present students with concisely sculptured theories.

4. Applications should be used to illustrate and motivate material in pure and applied courses. The development of most topics should follow the paradigm: applications--- mathematical problem solving--- theory--- applications. Theory should be seen as useful and necessary for all mathematical sciences.

5. First courses in a subject should be designed to appeal to as broad an audience as is academically reasonable. Most mathematics majors do not enter college planning to be math majors, but rather are attracted by beginning mathematics courses.

Coursework

6. The first two years of the curriculum should be broadened to cover more than the traditional four semesters of calculus--linear algebra--differential equations. Calculus courses should include more numerical methods and non-physical-sciences applications. Also, other mathematical sciences courses, such as computing and applied probability/statistics, should be an integral part of the first two years of study.

7. All students should take a set of two upper-division courses leading to the study of some subject in depth. Rigorous, proof-like arguments are used throughout the mathematical sciences, and so all students should have some proof-oriented courses. Real analysis and/or algebra are natural choices but need not be the only possibilities.

8. Every mathematical sciences student should have some coursework in the less theoretically structured, more combinatorially complex mathematics associated with computer and decision sciences.

9. Students should have the opportunity to undertake "real-world" mathematical modeling projects, either as term projects in an operations research or modeling course, or as independent study, or as an internship in industry.

10. Students should have a minor in a discipline using mathematics, such as physics, computer science, or economics. In addition, there should be a sensible breadth in the physical and social sciences. For example, a student interested in statistics might minor in psychology but also take beginning courses in economics and biology (heavy users of statistics).

This mathematical sciences curriculum would model the historical evolution of mathematical subjects: some problems are introduced, formulas and techniques are developed for solving the problems (usually with heuristic explanations), then common aspects of the problem are examined and abstracted with the purpose of better understanding "what is really going on". The difference in this scheme between beginning calculus and upper-division probability theory would be primarily a matter of the difficulty of the problems and techniques and the speed with which the material is covered and generalized, i.e., a matter of mathematical maturity of the audience. In the course of the two or three years of such coursework, there would be a steady increase in sophistication of the materials and more importantly, an increase in the student's ability to learn and organize the ideas of a new mathematical subject. The MSP panel feels that such maturity is a function of how a subject is learned as much as what is learned.

All courses should have some proofs in class and, as the maturity of students increases, occasional proofs as homework exercises. In particular, students should acquire facility with induction arguments, the basic method of proof in the mathematical sciences. In surveying both current student performances and the work in typical mathematics programs 30 years ago, the MSP panel is led to the conclusion that many able students do not have the mathematical maturity to take theoretical courses before their senior year. On the other hand, in the senior year students should be ready for some proof-oriented courses that show the power of mathematical abstraction in analyzing concepts that have been seen to underlie a variety of concrete problems. For example, part of a flowchart of courses leading to a senior-year real analysis would contain

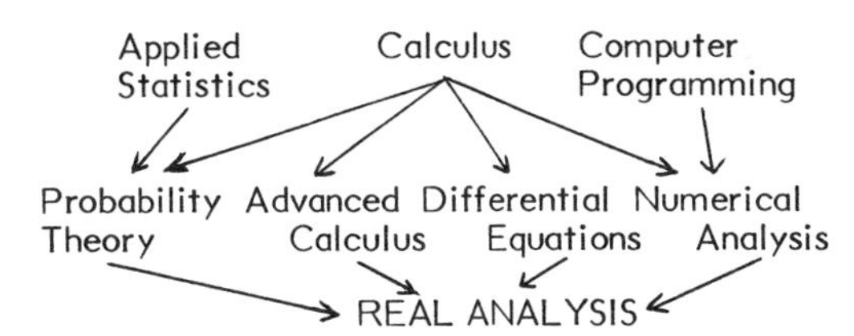

The panel has found the question of whether to require courses in algebra and analysis its most controversial problem. In light of the strongly differing opinions received on this subject, the MSP panel is making only a minimal recommendation (Principle VII) that it feels is reasonable for all students. Possible two course sequences besides analysis and/or algebra are: analysis and proof-oriented probability theory, analysis and differential equations, algebra and proof-oriented combinatorics, or algebra and theory of computation. Some departments will want to make stronger requirements.

Students should not be required to study a subject with an approach whose rationale depends on material in later courses nor should they be required to memorize (blindly) proofs or formulas. Some upper-level elective courses should always be taught as mathematics-for-its-own-sake, but an instructor should be very careful not to skip the roots and historical development of a subject in order to delve further into its modern theory.

The recommendation for interactive teaching seeks to encourage student participation in developing new mathematical ideas. It constrains an instructor to teach at a level that students can reasonably follow. Interactive teaching implicitly says that mathematics is learned by actively doing mathematics, not by passively studying lecture notes and mimicking methods in a book. To most students, the most useful thing learned in their mathematics major will be simply the ability to reason in a logical, analytical fashion.

5. Teaching Mathematical Reasoning

Because a mathematical sciences major must include a broader range of courses than a standard (pure) mathematics major, many mathematicians have expressed concern that it will be harder to teach the average mathematics student rigorous mathematical reasoning in a math sci major. They believe that the major will develop problem-solving skills but that without more abstract pure mathematics, students will never develop a true sense of rigorous mathematical reasoning. The MSP panel thinks that a mathematical sciences major with primary emphasis on problem solving is in keeping with time-tested ways of developing "mathematical reasoning".

Historically (before 1950), the main thrust of the mathematics major was problem-solving. Most courses in the major could be classed as mathematics for the physical sciences: trigonometry, analytic geometry, calculus (first year and advanced), differential equations, and vector analysis. Courses such as number theory and geometry had more "problem-solving" theory than abstract theory. Proofs in advanced calculus were computational. Theory of real variables or modern algebra was delayed until graduate school, or senior year for honor students. Proofs in number theory were, and still are, usually combinatorial problems. The one abstract "pure" course in the curriculum was logic. A "rigorous" course did not mean an abstract course, "mathematics done right". A rigorous course used to mean a demanding, more in-depth treatment that required more technical skill and ingenuity from the student. That past curriculum surely had some faults, but its problem-solving and close ties to physics came from traditions that go back to the roots of mathematics.

While problem-solving may be the truly traditional way of teaching mathematical reasoning to undergraduates, the complexity and breadth of modern mathematics and mathematical sciences require theory to help organize and simplify learning. Rigorous problem-solving will lead students to appreciate theory and formal proofs. In a mathematical sciences major, theory should be primarily theory for a purpose, theory born from necessity (of course, this is also the historical motivation of most theory). Students may find theory difficult, but they should never find it irrelevant.

Virtually all courses in a mathematical sciences major should be case studies in the pedagogical paradigm of real world questions leading to mathematical problem-solving of increasing difficulty that forces some abstraction and theory. Instructors should resist pressures to survey fully fields such as numerical analysis, probability, statistics, combinatorics, or operations research in the one course a department may offer in the field. The instructor of such a course should give students a sense of the problems and modes of reasoning in the subject, but general pedagogical goals should always take precedence over the demands of individual course syllabi.

6. Sample Mathematical Sciences Majors

In this section we present two quite different 12 semester-course mathematical sciences majors. Many other sample majors could be given. The MSP panel believes that most majors should be a "convex combination" of the two majors given here. Major A contains much of a standard mathematics major, while major B is a broader program designed for students interested in problem-solving. Both majors should be accompanied by a minor in a related subject

SAMPLE MATHEMATICAL SCIENCES MAJOR A

* 4 semesters of calculus sequence (with linear algebra, differential equations, some numerical calculus, and applications)
* probability and statistics
* differential equations (with computing)
* 2 semesters of advanced calculus/real analysis
* 1 semester of modern algebra (one-half linear algebra)
* 2 courses from the following set: modern algebra (second course), geometry, topology, complex analysis, combinatorics
* mathematical modeling

PLUS (related coursework): 2 semesters of computer sciences and 2 semesters of physics, to be taken in the first two years.

SAMPLE MATHEMATICAL SCIENCES MAJOR B

* 4 semesters of calculus sequence (as above)
* introduction to computer science
* numerical analysis OR second course in computer science
* probability and statistics
* advanced calculus OR modern algebra
* applied combinatorial mathematics OR differential equations
* mathematical modeling/operations research
* 2 electives (continuing a subject in theoretical depth)

The common required core of all tracks would be 4 semesters of calculus, two courses in computer science (or one formal course and further computing experience acquired in several other courses), a course in probability and statistics, the equivalent of a course in combinatorial problem-solving, modeling experience, and 2 theoretical courses of continuing depth.

Major B is for a liberal-arts college where computer science is part of Mathematics. Major A is meant to be close to the spirit of the major suggested by the NRC Panel on Training in Applied Mathematics. That panel viewed differential equations as a unifying theme in the major. The proper mixture of Majors A and B (with appropriate electives) would also allow students to make statistics or operations research or computer science a unifying theme. The program provides excellent secondary mathematics teacher preparation.

The one "new" course in these sample majors is applied combinatorial mathematics. As mentioned in principle 8, the MSP panel feels that the central role of combinatorial reasoning in computer and decision science requires that some combinatorial problem-solving should be taught in light of the four semesters devoted to analysis-related problem-solving in the calculus sequence. To this end, the modeling course should be heavily combinatorial if students have not taken a formal combinatorics course.

4.8 UNIVERSITY PROGRAMS WITH AN INDUSTRIAL PROBLEM FOCUS

CLAREMONT'S MATHEMATICS CLINIC AND APPLIED MATHEMATICS PROGRAM

Jerome Spanier
Claremont Graduate School
Claremont, California, U.S.A.

The Claremont Colleges consist of six small private institutions which retain autonomy in most respects but which agree to cooperate in the conduct of some of their programs. Five of the colleges (Claremont Men's, Harvey Mudd (HMC), Pitzer, Pomona, and Scripps) enroll only undergraduates while the sixth, Claremont Graduate School (CGS), is a self-standing graduate institution.

The impetus to create a graduate program in mathematics in the sixties stemmed mainly from the undergraduate departments' feelings that such a program would both enrich the professional life of the undergraduate faculty and improve the curriculum available to undergraduate students. With initialgrowth supported almost entirely by a National Science Foundation grant, the mathematics program at CGS was initiated in 1968.

Founded as a small, traditional PhD program, the new department appeared to meet its objectives well in its first few years of existence. Each of its entering classes contained one or two outstanding students who would have been a credit to any mathematics department, as well as a reasonable number who were good if not excellent. A colloquium series, involving a different speaker each week and a special annual lecture series were initiated (and have continued without interruption), undergraduates were strongly attracted to the new graduate courses, all of the department's graduates were placed after graduating, and a healthy cooperation with the undergraduate departments was begun. With the help of the NSF grant and Ford Foundation funds, a rudimentary collection of books and journals was transformed into a first rate library. By any measure, the mathematical life of the Claremont community seemed considerably enhanced.

While the new programs seemed to be expanding at a normal rate, the now-familiar nationwide problems that beset such programs were emerging clearly by 1971. With the expected decline in federal support of graduate students and with the undergraduate colleges willing to provide paid teaching positions to graduate students only in a rather haphazard way, it was recognized that the new program would be unable to compete with the widely available teaching assistantships offered even to mediocre students at most schools.

Because of these factors, in 1972-73 the department and the administration at CGS made the decision to develop a Master's program based in the application of mathematics, with several novel features. At the same time dialogues with the colleges began which led to the formation of joint degree programs in three new applied math concentrations. The new MA concentrations focused attention on practical training for immediate employment in industry and government. However, it was deemed essential to continue to serve the needs of bright undergraduates as well as those wishing to continue to a Ph.D., either in Claremont or elsewhere.

Substantial curricular changes were made in order to serve the new program. Traditional courses were modified to receive a new emphasis on problem-solving. Along with these more or less conventional courses, a unique practicum called the Mathematics Clinic, and its theoretical counterpart, a course in mathematical modeling, were initiated.

The mathematical model course, taught for the first time in 1976 at CGS, was developed to stress problem formulation, derivation and simplification of equations, dimensional analysis, perturbation theory, and other techniques essential to the modeling process. Traditional mathematical methods for solving equations, while discussed, are de-emphasized to permit most of the course to be taken up with material not normally encountered in other courses. Students must also work, either individually or in small teams, on projects which emphasize the modeling tools they are attempting to acquire.

The Mathematics Clinic, patterned after a highly successful model in the HMC Engineering Department, and a year's experience with a Clinic in the HMC Mathematics Department, was established jointly with HMC in 1974-75. This novel course affords students an opportunity to work in teams on real problems originating in industry or government. Fixed-fee research agreements are negotiated with the sponsoring firms before projects are undertaken. Funds from these provide some stipends for graduate student support, partial release for faculty project supervision, and support of some overhead costs.

Since its inception in 1973 the Mathematics Clinic has attracted projects from 19 different sponsors, 10 of whom have supported more than one project. These 19 clients have sponsored nearly 35 project-years of work in a wide variety of problem areas. It has sometimes been helpful to classify these into categories which indicate their modeling content:

Math Modeling and Analysis (Hard)

Math Modeling and Analysis (Soft)

Analysis

Economics

Computer - Related

The modeling categories are used only if modeling is a significant component of the project, and the qualifiers "hard" and "soft" are used (respectively) to indicate that the modeling process uses continuous techniques (e.g., differential equations) or discrete techniques (e.g., discrete probability theory, game theory, programming).

A classification of projects according to subject matter reveals the great diversity of Mathematics Clinic work. In the general area of Environmental Studies, a Clinic team has developed (for the Pomona Valley Municipal Water District) a mathematical model to predict groundwater nitrate concentrations based upon land use patterns and subsurface geologic structures. Other teams have been involved in the assessment strategies for fire control and prevention in Southern California chaparral lands through modeling of existing and potential environmental conditions. The latter projects have been supported by funds from the U.S. Forestry Fire Service Laboratory in Riverside, California. In the area of Economic Planning, a Clinic team spent one semester desigining a Life Cycle Costing model to estimate, for the Office of the State Architect in California, the true cost of a structure over its life. This model is used to make decisions about the purchase, design, and refurbishment of buildings. In another year-long study, an empirical mathematical model was designed for use by Becton, Dickinson and Company to forecast economic activity levels over a five year period within a corporate division. Under the broad category of Strategic and Social Policy Analysis, a model was developed for the Rand Corporation for optimal strategies to use in the event of a surprise attach by enemy submarines on U.S. bomber bases. For the Los Angeles County Superior Court, a one-semester project team developed a predictive model to improve the efficiency of the juror selection process. Many other projects dealing with the Modeling and Analysis of Physical Systems, Aerospace Problems, and Computer Science, have been successfully addressed in the Clinic.

Two large NSF grants, totalling $364,142, have helped the Mathematics Clinic and related curriculum grow to a stable, self-sustaining size during the period 1976-80. These grants have helped establish a post doctoral program based in the Clinic, have brought senior mathematicians to Claremont for work and study in the Clinic, have assisted in the development of the modeling course, have provided some support for administration of the Clinic, and have sponsored a highly successful Conference on "Experimental Applied Mathematics Education" in the early summer of 1979. During the past three academic years a total of fifteen post-doctoral mathematicians have visited Claremont and become deeply involved in Clinic work for periods ranging from one semester to two years. Five of the long-term visitors have served as faculty supervisors of two projects each: the remaining ten have served as consultants on one or more projects. Of the fifteen visitors, ten have either returned to their home universities or have accepted new university positions which use their Clinic experience in some nontrivial way. The remaining five have accepted excellent industrial positions. Three of the fifteen visitors have used mainly their own support to finance their visits to Claremont.

At the student level, a total of 173 have registered for 248 student-semesters of Clinic work in 69 project-semesters since 1973. Of the 173 students, 48 participated as graduate students, the rest as undergraduates. Of the 48 graudate student registrants, 22 continue to work in advanced degree programs in mathematics, operations research, and other fields, 14 have positions in industry which make good use of their training and education, 3 are teaching at various levels and one manages a major university computer center. The statistics for undergraduates are not too different. Of the 125 former participants, 51% are still in school, 31% have taken good industrial jobs, and 18% are presently unaccounted for.

There is universal agreement that Clinic experience has been of great value to students as well as being a cost-effective investment for project sponsors. Students receive practical training simply not available in standard courses and often establish excellent employment opportunities through their Clinic work. In addition to receiving valuable help with the formulation and solution of some of their pressing problems, Clinic sponsors are helping to support a novel educational experiment and are often able to recruit talented students and thus to reduce costs associated with training employees in key technical positions.

Since 1973 there has been a dramatic increase in the MA component of the graduate mathematics program, an increase which parallels the evolution and success of the Mathematics Clinic program. Project fees totaled $139,500 in the past academic year alone, with a total fo $28,350 of this amount helping to support graduate student "team leaders". Involvement of students and faculty has grown nearly monotonically since 1973, and graduate enrollments, which had fallen to 17 students in 1975-76, have topped 30 each of the past three academic years.

The Claremont program has attracted considerable attention, mainly because of the unique Mathematics Clinic. Additional visibility has been created through lectures and published papers (1 - 13), as well as through the visiting program and confernece mentioned earlier. External evaluations, produced as a requirement of the NSF grants, have been favorable. Placement of graduates has continued to be a strong asset.

With all of its positive features, the Claremont program is not without its defects. The Clinic remains a moderately expensive and labor-intensive component which retains considerable risk factors because of its perennial need to attract new paying sponsors. It also requires a steady, predictable supply of students. This has sometimes been a problem because the Clinic is not universally required of undergraduate math majors. Nevertheless, the benefits are perceived by most to outweight the disadvantages and it is likely that the Clinic will be an important part of the Claremont curriculum for many years to come.

References

1. Robert Borrelli and Jerome Spanier, The Mathematics Clinic: A Review of its First Seven Years, to appear in the UMAP Journal.

2. Robert Borrelli and Stavros Busenberg, Undergraduate Classroom Experiences in Applied Mathematics, to appear in the UMAP Journal.

3. Stavros Busenberg and Wing Tam, An Academic Program Providing Realistic Training in Software Engineering, Proceedings of the ACM, May 1979.

4. Practical Experience in Top-Down Structured Software Production in an Academic Setting, SIGCSE Bull., Vol. 9, No. 1 (February 1977), pp. 31-36.

5. Lee Harrisberger, Richard Heydiner, John Seeley, and Margaret Talburrt, Experimental Learning in

Engineering Education, Report to the Exxon Foundation, (August 1975).

6. Melvin Henriksen, Applying Mathematics Without a License, American Mathematical Monthly, October 1977, pp. 648-650.

7. On the Juror Utilization Problem, Jurimetrics Journal, Summer 1976.

8. Jerome Spanier, Academic Realism: An Innovative Approach, SIAM News, Vol. 9, No. 3 (June 1976).

9. Education in Applied Mathematics: The Claremont Mathematics Clinic, SIAM Review, Vol. 19, No. 3 (July 1977).

10. The Mathematics Clinic: An Innovative Approach to Realism Within an Academic Environment, American Mathematical Monthly, Vol. 83, No. 10 (December 1976).

11. The Education of a Mathematical Modeler, Proc. Second Int. Conf. on Math. Modeling. St. Louis, July 1979.

12. Thoughts about the Essential of Mathematical Modeling, to appear in Int. J. Math. Modeling.

13. Thomas Woodson, The Harvey Mudd Experience, Engineering Education (February 1973), pp. 345-348.

THE SWEDISH INSTITUTE OF APPLIED MATHEMATICS: A LINK BETWEEN INDUSTRY, SOCIETY AND UNIVERSITY

Germund Dahlquist
Royal Institute of Technology
Stockholm, Sweden

The Swedish Institute of Applied Mathematics, the ITM, was formed in 1971 to strengthen the Swedish efforts to develop applicable mathematics important to industry and government. ITM is an acronym for the Swedish name "Instituted for Tillampad Mathematik". The word "institute" is perhaps a misleading name for the strongly decentralized activity of the ITM. The institute operates mainly as a link for information and project cooperation between researchers, chiefly graduate students, in different specialties at the eleven universities of our country. (We include here the five Institutes of Technology.) The population of Sweden is approximately 8,000,000.

In the sixties, some Swedish industrialists realized that there was much to be gained from increased research in applicable mathematics with a strong link to real world problems in the operating, planning and construction departments of their companies. A great deal had recently happened, with vast potentialities for the future: think of computers, control theory, operations research etc. They felt that problems existed, for which no single company could establish resources sufficient for the development of the necessary new methods.

At the same time, several university mathematicians also felt the need for improving the existing occasional contacts between university and industry. In particular, Professors Ulf Grenander (mathematical statistics) and Goran Borg (mathematical analysis) were instrumental in the formation of the ITM. Grenander is nowadays active at Brown University, but he still stimulates the ITM with his ideas and comments on its activities. Borg is a member of the board of the Institute. A group of young mathematicians around Lars-Erik Zachrisson, Professor of Optimization and Systems Theory in Stockholm, was well prepared for industrial applications of new techniques of applied mathematics, sometimes brought together under the term "operations research". Zachrisson had about twenty years of experience in industry and in military research and development, before he became a university professor. He and his students have indeed been of central importance to the ITM from the beginning.

In 1971, contacts were established between about 15 companies. The decided to set up an industrial foundation for applied mathematics to contribute financially to a basic research program, to begin with for the period 1971-1974. They formulated it with the help of university people. The resources available at the universities were graduate students working towards their Ph.D. It was important to find a form of organization that fitted within the framework of our academic life.

The industrial foundation then turned to the Swedish Board for Technical Development, which is the government establishment for supporting research and development of importance to our industry. It decided to contribute with about the same amount of money as the industrial foundation. The board of the ITM consists of representatives of these two parties.

The basic research program is renewed every three years. Annual adjustments of the program are decided by the board of the ITM. The annual budget for the basic research program is about $300,000. In additin to the basic research program, the ITM can also accept contracts for proprietary consultancy work, preferably in branches of applied mathematics which are not yet adequately covered by firms of consulting engineers in our country. The ITM can also receive additional funds from the national resarch councils for special projects, for example if they are more interesting to the public sector than to industry.

The basic research program consists of a number of project areas. The headings for the fourth period (1980-1983) are as follows. Note that they deviate a little from the traditional concept of "applied mathematics".

1. Mathematical planning
2. Control of processes and systems
3. Computational problems of science and technology
4. Optimization techniques
5. Simulation techniques
6. Techniques for the efficient use of computers.

These headings cover two types of projects, which are carried on in close contact with each other.

A. Applications-oriented projects having a direct connection with practical cases. Here new methods and mathematical software are treated in a realistic environment, and feedback experience is obtained for the continued development of methods (primarily project areas 1 - 3).

B. Projects where new ideas for computational methods are developed, with impulses from the general development of applied and applicable mathematics inside or outside Sweden as well as from the Institute's own applications-oriented projects (primarily projects 4 - 6).

We mentioned earlier that most of the research work is performed by graduate students at the universities of Sweden involving on the average 25 people, with a certain natural turnover every 3 - 4 years. Indirectly, of course, their supervisors are also involved. The permanent staff is very small: the executive officer Daniel Sundstrom and three to four part-time employees. There is a "collegium" of approximately six professors, selected in rotation so that every region, geographical as well as scientific, will be represented for a time. It has the role of combining the distributed scientific competence and interests together with the needs of industry, which are expressed in general terms by the basic research program. The collegium, in which the executive officer and a representative from industry also participate, transforms the program into suggestions for concrete research projects, which have to be approved by the board. The collegium has certainly improved contact between the different departments and disciplines of applied mathematics in our country. The first chairman of the collegium was Ulf Grenander. He was succeeded in 1972 by Germund Dahlquist (numerical analysis), and since 1977 Karl-Johan Astrom (control theory) has been the chairman.

In a way the ITM functions as a specialized national research council but, since the basic research program is formulated by the industrial foundation, the ITM probably emphasizes those areas of mathematics where our industry recognizes the need for increased knowledge much more than a traditional research council for the mathematical sciences would have done. The closeness to practical demands strengthens the motivation for many students. As a rule it is not difficult for them to find stimulating employment after having worked with ITM's projects for a few years.

The ITM also serves as a channel for short-contract jobs. The students sometimes take leave from their work within the basic research program in order to work on a perhaps even more motivating contract job during a few months. Before the ITM existed it was sometimes difficult to handle the offers for contract jobs, which arrived rather stochastically and at short notice to the departments. Typically, when the jobs came in, the students were heavily involved in courses, either as teachers or as students. And when students were available, there were no jobs. The graduate students working in the basic research program form a reserve, and the ITM can handle such situations more efficiently through its nation-wide connections.

Frequent direct contacts with industry in the applications-oriented projects are important for the success of the ITM idea. Other forms of contact have also been arranged:

1. Contact days, where the students meet people from industry working on related problems, and both parties present their work. A contact day is centered around one or two areas of high activity within the ITM, e.g. optimization techniques, simulation, fluid mechanics, microprogramming or microcomputers.

2. Informal international workshops, where the students and researchers on related problems from abroad can exchange ideas. It goes without saying that this broadens the perspective of the graduate students. They also participate in the practical arrangement of the meeting, a valuable experience. In 1979 a workshop was held on "stiff systems of differential equations", and in September 1980 there will be one on numerical methods in automatic control.

3. The students sponsored by the ITM also present their work in ordinary seminars. In this way the experiences attained in the ITM projects are spread to university people, who are not directly involved. They become aware of new types of applications, and their concept of applicable and applied mathematics is gradually transformed.

All this is on the graduate or post-doctoral level, but what about the effect on undergraduates? This is more indirect. One of the least indirect effects is through term projects and diploma projects which emerge as spin-off effects, with the ITM-sponsored students as advisers. These students are often working part-time as teaching assistants. However, we must not forget the indirect effect on the composition of exercises, illustrative examples in the lectures, etc., which results from the altered view on applied and applicable mathematics. This is first evident in the teaching given by those involved in the ITM work, but gradually percolates through a large part of the department concerned.

The informal set-up of the ITM is very relaxing for university departments, which are otherwise hampered by the increasing bureaucracy of our academic life. However, the fact that the ITM places strong emphasis on its role in supporting graduate students, who wish to obtain applied experiences also means that most of the researchers involved are fairly inexperienced or have many other duties. This makes it difficult for the Institute to strive for the position and authority of "qualified consultant". There may be difficulties e.g. accepting large projects, where it is important to deliver the results before a deadline.

The executive officer also has to work hard to prolong the Institute for new three-year periods. Some new industries join the foundation, and others leave. When the ITM started, our industry felt that they were not able to utilize the new trends in applicable mathematics by in-house resources. The situation is different now. Some companies have acquired adequately trained personnel, or are satisfied with methods which are at present in routine use. In both ways the ITM has contributed to reducing its own role in some problem areas.

Our industry has, however, encountered new problems during the seventies, which demand a new mathematical approach. For example, in recent years there has been increased attention to the importance of environmental control, i.e. to the conditions of work as well as to the impact on the external environment. The possibility of applying mathematical models in the study of ecological systems and biochemical processes is therefore of growing interest.

Also, the shortage of energy and natural resources and the changes in relative prices increase the importance of methods for planning and controlling production processes with the purpose of saving material and energy. New optimization methods may also be required in the effort to design products and systems so that material and energy are not wasted in their use, while due consideration is given to safety factors, technological restrictions and the impact on the environment, formulated as constraints.

Having described how the ITM works, we shall now provide a few examples of what they are doing.

The ITM is very active in applied mathematical optimization. In fact, this takes up about 50% of the basic research program, and an even larger fraction of the contract jobs. There are many projects in logistics, concerned with .the collection of raw materials (for example in forestry) or with the distribution of products for customers. One of the first tasks for the ITM was to work on improved methods for food distribution. Actually, a renowned Swedish industrialist's interest in this application was a great help at the founding of the Institute. There have been numerous other applications of this type, both for our basic industries, e.g. steel and forestry, and for the public sector, e.g. a model of mail distribution, (1), and optimization of local telephone networks (2).

Another application, in which large savings have been obtained, is the recipe optimization, with integer constraints, occurring in the composition of batches for high quality steel production (3). The problem is how to mix available raw material resources, including scrap pieces with known composition from earlier production, in order to achieve the least expensive production plan to meet the customers' orders.

We have also worked with the well-known trim scheduling problem (4 p. 182). The manufacturer makes rolls of newsprint paper to meet customers' specifications as to width and diameter. In cutting these customer rolls from larger reels of paper, trim losses are incurred. The manufacturer must determine on which machines and in what combinations the orders should be cut. In the original problem, the purpose was simply to minimize the over-all trim loss. It turned out, however, that in the practical situation the frequency of changes of the setting of the knives must also be taken into account. We now have a very efficient program (5), for this modified paper-trim problem, after a very long process of successive modifications and experiments in cooperation with industry. The total process of development, including all these successive modifications and experiments as well as the marketing of the final applications software in order to have it accepted as a natural tool in the day-to-day work in our paper mills, has extended over a long time, perhaps 10-13 years.

In a project of a different type, L. Edsberg had, for a couple of years, a grant for developing methods to simulate chemical reactions on a computer. Within this project he designed a program package (6, 7), for a homogeneous batch reactor, where the user tells the computer the structure of the reactions and the numerical values of the rate constants. Consider for example an autocatalytic system, where three species A, B, C, are involved in three simultaneous reactions, with rate constants k_1, k_2, k_3,

$$A \xrightarrow{k_1} B$$

$$B + C \xrightarrow{k_2} A + C$$

$$2B \xrightarrow{k_3} C + B$$

The program sets up automatically the system of three non-linear ordinary differential equations describing how the concentrations of the species vary with time, and calls a subroutine for the step-by-step solution of the differential system. The concentrations of interest are then plotted versus time on a screen. The programm is designed to allow the user to interact with the computer and to modify the model with ease or change the scales in the output, etc. The package has been used as a research tool by graduate students in chemical engineering, and it has also been used in the undergraduate training there. One undergraduate improved the package as a diploma project under the supervision of Edsberg (8). Now students can follow the simulation of the famous Belousov-Zhabatinskii reaction on a screen in the classroom simultaneously with a beautiful demonstration of the reaction itself, where colour shifts show the oscillations characteristic for this reaction. Edsberg is now a lecturer at the Royal Institute of Technology, and he is involved in making the theoretical subjects in the training of metallurgy more tangible with the aid of computers. From the educational point of view this is a beautiful sequel to our project.

Edsberg's work has also been very positively received in the international scientific literature.

It is less encouraging, however, that the simulation technique has not yet become a hit in our chemical industry, and we have not been able to discover how to make it more attractive to the chemical engineers. Perhaps we did not find the right contact persons in industry at the right moments. (This is often an important problem.) The project was started because it was known that mathematical models were used in chemicalenginnering in the U.S. and U.K. Moreover, chemical reactions offer fascinating examples of "stiff" differential systems, which was the main research area at the department, where Edsberg was a student. Hopefully, things will change, when the new generation of chemical engineers, which has been brought up with this technology, goes out into industry. Once again, perhaps a delay of 10-13 years.

The final product of most of our projects is a computer program, but as a rule a large amount of model construction, theoretical analysis and algorithmic development precedes the coding. It seems likely that in an increasing number of industral applications the final product of the mathematician's work will become an algorithm stored in a microprocessor, more or less considered to be part of the hardware.

The ITM is therefore interested in methods of making the use of computers of all sizes more convenient for mathematicians. The interactive use of computers has been emphasized in the past. Two projects have been concerned with the development of languages for interactive computing in the design and analysis of control systems. Another project was concerned with computer graphics as a tool for technological computing.

Several years ago, we designed (as a contract job) an algorithm for identity checking in a system of unattended cash dispenser terminals (9), where ever terminal contains a microcomputer which is remotely connected to a large central computer. Our solution, which is now used in a nation-wide network of bank offices, is based on a functional relationship between a number on the customer's magnetic card and a 4-digit number which he inputs at the keyboard of the terminal. This function is defined in a very compliated way, where a table of random numbers, stored in the microcomputer and never seen by a human eye, is involved. Thus, the function is secret even to the designers! This project initiated much subsequent activity both inside and outside the ITM. Simulators for several microcomputers were written, and further work is being done in mathematical cryptography.

The computer is such a basic tool for the ITM that, in 1974, we even dared to accept a contract job for a government establishment to "forecast" the development of the computer field during a ten-year period. A difficult and stimulating task for three graduate students and a great contrast to all well-defined exercises in the traditional training of a mathematician! The response to their report on this tricky question was very positive, and a verification in 1979 showed good agreement with the actual development so far.

On these pages we have tried to summarize nine years with the ITM, with particular emphasis on its role in university education. It is clear that a large amount of enthusiasm, persistence and capacity is required from all parties in this activity. Although our construction of a link between the university and the "outside world" is in many respects influenced by the university traditions and the mixed economic system of our country, we hope that both the good and bad news we have related can be of some use also outside Sweden.

References

1. Bjorklund, B. and Lundgren, G., A model of mail distribution of Sweden. (Presented at the Tenth International Symposium on Mathematical Programming, Montreal, 1979, UTM Report 42, 1979).

2. Holmberg, G. and Lindberg, P., Optimization of digital local telephone networks. (Presented at the International Teletraffic Congress, Madrid 1979).

3. Westerberg, C-H, Bjorklund, B., and Hultman, E., An application of mixed interger programming in a Swedish steel mill. Interfaces (ORSA/TIMS), Vol. 7, 1977.

4. Gass, S.I., Linear programming, methods and applications. 1st edition, McGraw Hill, New York, 1958.

5. Holmberg, G., Plans for the development of software for trim scheduling for a paper mill. UTM Report 45, 1977, (In Swedish).

6. Edsberg, L., Integration package for chemical kinetics. in Stiff Differential Systems, ed. R.A. Willoughby, Plenum Press, New York 1974.

7. Edsberg, L., Numerical methods in mass action kinetics. in Numerical Methods for Differential Systems, eds. Lapidus, L., and Schiesser, W.E., 1976.

8. Uhlen, M., Kinrate and Kinbox, two program packages for interactive simulation of chemical systems, examination project. Supervisor L. Edsberg, Report TRITA-NA-7912, Department of Numerical Analysis and Computing Science, Royal Institute of Technology, 100 44 Stockholm, Sweden.

9. Dahlquist, G., Ingemarsson, I., and Riesel, H., A randomly generated program for automatic identity checking. BIT 15 (1975), 381-384.

AN INTEGRATED APPROACH TO GRADUATE EDUCATION IN THE MATHEMATICAL SCIENCES

Clayton Aucoin
Clemson University
Clemson, South Carolina

In the early seventies, the Department of Mathematical Sciences at Clemson University began to question how the Department could best train its graduate students for the changing market for such students. One result of this questining was to begin advising all of the students, even those whose primary interests were in pure mathematics, to take a broader range of courses. In order to include a broader range and still maintain quality, most of the students at the M.S. level began taking more courses than were actually required at that time. The rationale for taking these additional courses was that the graduates would have the option of pursuing non-academic careers or would be able to teach courses in applied areas. (It was predicted that the academic market for mathematical sciences graduates would be much stronger in applied areas.)

When N.S.F. began making funds available for grants to explore alternatives in higher education the department realized that such a grant would provide the opportunity to expand its existing ideas and experiment with new ideas for a significantly different graduate program. In July, 1975, N.S.F. awarded a three year grant (extended to four years) to Clemson. This grant was part of a national effort, which included four institutions (Claremont Graduate School, Clemson University, Rensselaer Polytechnic Institute, and Washington State University) to develop, test, and report alternative programs to current M.S. and Ph.D. programs in the mathematical sciences. Clemson's grant was a companion grant with one to Washington State University with specific provision for collaboration between the two institutions. The objective of these programs is to produce graduates who are better equipped than graduates of traditional programs to satisfy contemporary and future needs for mathematical scientists.

The facility at Clemson was immeasurably aided in its grant implementation by input to the project from many outside professionals from both academia and industry. The official board of advisors for the grant was composed of six individuals, distinguished representatives of both academia and industry. Also, the grant provided for various visiting professors and other professionals. The types of visits varies: many visitors were on campus for one or two-day colloquia, five each visited ten days, two each visited for five days, and many spent one or two semesters on campus. Also, in September 1977, Clemson hosted a conference on graduate programs in the applied mathematical sciences which was attended by twenty-eight applied mathematicians from industry and government and thirty academic mathematicians.

The development and implementation of the M.S. program received major emphasis under the grant. The resulting program was based on the following premises:

1. The major source of employment for mathematical scientists in the future will be non-academic agencies.

2. Most such employers will require more than a B.S. degree but less than a Ph.D. degree in the mathematical sciences.

3. Employers will prefer personnel who possess not only a concentration in a particular area of the mathematical sciences, but also diversified training in most of the other areas.

4. Graduates should have had more than superficial education in applying mathematical techniques to solve problems in areas other than the mathematical sciences. Inherent in such training is the ability to communicate, both orally and in writing, with persons from these application areas.

5. It is desirable to obtain such broad-based education in the mathematical sciences prior to specializing for the Ph.D. degree.

The M.S. program requires all students to have a minimum competency in each of these areas: computing science, core mathematics, operations research, and probability and statistics. Thus, all students are required to earn (or to have earned prior to entry into the program) credit in the following courses:

1. Computing sciences: An undergraduate course in a high level programming language, a numerical computing science course, a computing course emphasizing discrete models.

2. Core: One undergraduate course in each of linear algebra, abstract algebra, and advanced calculus, two core courses beyond advanced calculus and modern algebra.

3. Operations research: One course in deterministic optimization.

4. Probability and statistics: A course in basic statistics, a senior-level course in probability, an advanced course in applied statistics, and an additional course in statistics or operations research.

The requirement that all students take a models course is a major feature of the M.S. program. The models courses are taught by someone in the application area or are team-taught and are designed specifically for students in the mathematical sciences. The principal goal of the models course is to present a preview of the experiences which confront mathematicians who are consulted for help in the mathematical formulations and solutions of problems from such diverse areas as biology, environmental engineering, and management, to name a few. Three models courses were designed and offered as specified in the grant proposal. Some of the students satisfy their modeling requirements by taking regularly scheduled courses in other departments. Such courses are carefully screened to ensure that prerequisite training in "outside" terminology and concepts is not excessive and that the emphasis is on initial problem formulation and subsequent refinements rather than elementary mathematical concepts. Also available to the student are a two-semester course sequence in stochastic models in operations research and a two-semester course sequence in digital models.

In order to attain sufficient depth in a concentration area, each student is required to take six additional courses to be selected from or to complement the concentration areas.

Every student is also required to complete a project (either in connection with some course or independently) and to present the result of the project to his/her committee. Many of these projects originate with a local industry or government agency.

The following is a sample program for a typical student entering the M.S. program at Clemson University. Most students will take two academic years and a summer to complete the program (42 semester hours total).

Digital Models
Numerical Processes
Core
Applied Models
Mathematical Programming
Data Analysis
Operations Research or Statistics
Concentration Area (including electives)
Project

The concentration areas are:

Algebra
Analysis
Probability and Statistics
Operations Research
Computing Science

Ninety-five students have received M.S. degress in the mathematical sciences at Clemson since the beginning of the grant. All of them have obtained employment which utilizes their interests and training.

The following indicates the types of activities which these students entered immediately after graduation:

62% - industrial and governmental positions
24% - Ph.D. programs
9% - high school and college teaching

4% - miscellaneous (military, married, etc.)

The activities of the project with respect to the Ph.D. program culminated with the approval in Spring, 1979, of a new set of requirements for the Ph.D. program, which is also designed for industrially as well as academically oriented students. In addition to being required to specialize in one of the mathematical sciences, students completing the Ph.D. program are expected to have obtained breadth in the mathematical sciences, as is the case in Clemson's M.S. program, and to have completed at least three or four advanced courses outside the area of specialization. To insure depth in the specialty area, students begin no later than the third year to participate in formal seminars and to pursue independent research.

During the grant period several students completed Ph.D. program which were much broader and required less core mathematics than traditional Ph.D. programs. Some of these students prepared dissertations which combined their interests in an outside area with the appropriate areas of the mathematical sciences. Of the fifteen students receiving Ph.D. degrees during 1975-79, seven accepted non-academic positions.

For a more detailed account of Clemson's programs, see New Opportunities in Applied Mathematics, a report from Clemson University and Washington State University.

MATHEMATICAL MODELING: INTERACTION OF EDUCATION WITH INDUSTRY

William E. Boyce
Rensselaer Polytechnic Institute
Troy, New York, U.S.A.

Several years ago Rensselaer Polytechnic Institute, with assistance from the National Science Foundation, established a curriculum leading to a Master of Science degree in Applied Mathematics as an alternative to the trditional master's degree program in Mathematics. This new curricolm was designed primarily for students who wish to prepare themselves for careers as applied mathematicians in business, industry, or government at the M.S. level.

Prior to launching the program we spend considerable time discussing with representatives of various industries the kinds of qualifications that are sought by industrial employers of applied mathematicians. There was remarkable agreement in the responses from a wide variety of companies and laboratories. All indicated a need for

problem solving skills

ability to communicate effectively with non-mathematicians;

a broad base of knowledge, both in mathematics and in areas of application.

The problems that are encountered by industrial aplied mathematicians sometimes are already in mathematical form, but usually appear in the context of some other field. Occasionally problems are precisely formulated, but usually they are not, and sometimes it may not even be clear what th eproblem really is. How then does a mathematically trained individual approach a poorly or partially formulated problem from a more or less unfamiliar field? What sort of training is appropriate for persons who will have to deal with such problems regularly?

Although some applied mathematicians like to refine the process further, there is general agreement that using mathematics to solve problems in the real world involves at least the following three steps:

Formulate a mathematical model, that is, express the problem in mathematical terms, usually a system of equations or inequalities involving apropriate variables, parameters, or functions.

Determine the logical consequences of the mathematical model through analysis or computation.

Compare these consequences with experience or observation to determine the validity of the model and its analysis.

If the comparison of results with observation is unsatisfactory, then it may be necessary either to modify the mathematical model, or to carry out a more accurate or detailed analysis. Or, of course, it may also be necessary to make more or sharper observations, or to modify the experiment or circumstances in which they are made. It is not necessarily the mathematical mode or analysis that is to blame for discrepancies.

Mathematical modeling - the formulation of a problem in mathematical terms - is the crucial first step in mathematical problem solving. It is frequently the most difficult step because it requires some insight into the phenomenon under investigation, and often entails a judgment as to which influences are of primary, and which are of secondary importance. In constructing a mathematical model of a complex situation there is an inherent conflict between what is realistic (and therefore usually mathematically complicated) and what is amenable to analysis (and therefore possibly oversimplified). Of course, what is realistic may depend strongly on the purpose of the model, and what is amenable to analysis certainly depends on the resources, both intellectual and computational, of those doing the analysis. In addition, there are often constraints of time and money.

Mathematical modeling is an art as well as a science. Although there are some recurring themes or ideas that are repeatedly useful, modeling resists codification in the same sense that mathematical methods can often be classified and their scope delineated. Consequently, mathematical modeling is difficult to learn, and therefore difficult to teach effectively. Nevertheless, we must try, for as we have noted, modeling is indispensable to problem solving by mathematical means.

Whatever else may be involved in teaching and learning the skill of mathematical modeling, it seems clear that it requires at the very lest a good deal of time, persistent and active participation by the student, and an instructor who is both adept in mathematical modeling and sympathetic to the difficulties faced by the learner. Unfortunately, all of these factors tend to be in short supply.

Even under the best of circumstances, most instructors find that it is easier to teach mathematical techniques than mathematical modeling. Most of us are more comfortable with the precision of mathematical methods and theories than with the uncertainties and ambiguities of modeling. There is seldom enough time to cover the purely mathematical topics of a course as well as we would like. It is hard to find texts that deal consistently and realistically with modeling; on the contrary, in textbooks the discussion is polished and problems are usually formulated precisely and contain exactly the right amount of data. Good homework assignments on mathematical modeling are not easy to construct. It is also not easy to evaluate a student's progress, or lack of it. Given all of these difficulties, it is no wonder that most of us, myself certainly included, spend most of our time teaching techniques and theory rather than modeling.

Nevertheless, mathematical modeling should form a larger part of the mathematics curriculum that it now does. It should also probably start earlier in the educational process. Certainly it should be an object of emphasis in secondary school, and possibly even sooner. One problem is that effective modeling requires at least some familiarity with other fields of knowledge, and consequently there must be some coordination of instruction in mathematical modeling and in the other fields that are involved. It is possible to do quite a bit, however, simply by drawing on the everyday experience of students.

The Department of Mathematical Sciences at Rensselaer offers several courses in which modeling is a major component. I would like to focus on only one of them, a new course called Advanced Mathematical Modeling that is a requirement in the Master of Science in Applied Mathematics curriculum, and which is based largely on academic-industrial interaction. This course is offered at the first year graduate level, and has the following principal goals:

> To broaden the student's outlook as to what constitutes applied mathematics.
>
> To provide the experience of dealing with loosely formulated problems and lack of data in relatively unfamiliar areas, and attempting to bring a measure of order to such problems.
>
> To simulate to some extent the experience of working in industry, for example, by encouraging cooperative rather than solely individual effort, and by spending time on problems of demonstrable current interest in the real world.
>
> To improve the student's skills in oral and written presentations.

The course has two largely separate parts that proceed in parallel. The first consists of several series of lectures on various current topics in applied mathematics. Each series usually occupies from nine to twelve class hours, and there are typically three series during each semester. Although some of the lectures are given by members of the Rensselaer faculty, we attempt in the main to arrange for lectures by individuals who are employed in industrial or government research installations. In general, in any given year we seek to have represented a variety of fields of application and as great a range of mathematical methods as possible. Guest lecturers are urged not to dwell unduly on details of mathematical methods and analysis, but to discuss fully how and why the problem was formulated as it was, what methods of solution were chosen and why, and what insights were gained as a result of the mathematical investigation. A discussion of false starts and unexpected difficulties can also be extremely illuminating.

The other major component of the course is semi-independent investigation by the students of selected research projects. These investigations are the subject of a seminar which meets roughly once a week and is an integral part of the modeling course. At the beginning of the year the class is divided into groups of two or three students and a project is assigned to or chosen by each group.

For example, typical projects have stemmed from questions such as the following:

> How does one decide whether to launch a large scale public immunization program, for example, against swine flu?
>
> How does one set standards for allowable exposures to known or suspected carcinogens?
>
> How does one develop al algorithm for assignment of membershp in a legislative body such as the House of Representatives following a periodic census?

Questions such as these have many ramifications, some technical and some non-technical, and can be approached in various ways and at various levels of sophistication. In some cases one needs models of phenomena that are imperfectly understood. The acquisition of reliable data may also present a major difficulty. The object is not so much for the students to solve the problem as to come to grips with the difficulties that it presents, to consider various possible ways of dealing with them, and to experience both the frustrations and benefits of a mathematical approach to such questions. Members of the class present oral progress reports on a rotating basis at each meeting of the seminar, and each individual student writes a term paper summarizing his or her investigations at the end of the term.

During the second semester the groups are formed again, possibly with some changes in membership, but the projects for investigation now come from local industrial laboratories or government agencies. Care must be taken to choose projects of a proper magnitude, that is, difficult enough to be challenging, but of such a nature that reasonable progress can be anticipated within a few months. It is easy to overestimate what can be accomplished.

A liaison person from the company proposing the problem meets with the student group from time to time to explain the background of the problem and to assist in the consideration of alternative methods. A faculty member monitors progress on a frequent basis, assists with logistical arrangements, provides moral support, and lends an occasional scienific hand. To the maximum extent possible, however, the students are responsible for generating and executing their own plan of investigation.

In evaluating our experiences with this course during the past four years, we conclude that it has accomplished each of its goals, at least in part. Students certainly have had their horizons broadened both as to areas of application and mathematical techniques, and some have become extremely interested in subjects hat came up either in lectures or as a part of their project work. The experience of talking and writing about their work is certainly good, and some students have improved markedly in their abilities along these lines. Perhaps most important, especially with respect to the industrial projects, is the effect of the course as a confidence-builder. Most students have been thoroughly intimidated when first faced with the projects they were expected to work on, and felt that it would be out of the question for them to achieve any significant progress within the allotted time. Nevertheless, in almost all cases worthwhile progress has been achieved, and this as resulted in an unmeasurable, but very noticeable, gain in self-confidence for many students. For these reasons we believe that this rather novel course has been of substantial educational value to the great majority of the participants. We encourage others to undertake innovations along similar lines.

INDUSTRY IN THE CLASSROOM

J.L. Agnew
Oklahoma State University
Stillwater, Oklahoma, U.S.A.

1. The Use of Industry in a Case-Study Course in Applied Mathematics.

During the past five years, we have been engaged in a project designed to use the expertise of regional industry to enrich the teaching of mathematics at the college level. Because of the limited time available, I want to talk for the most part about our industry-related case-study course in applied mathematics which is designed to be the capstone of the undergraduate experience. Its main goal is to provide experience in (a) modeling a real world problem and interpreting its solution in the actual setting in which the problem arose, (b) integrating the mathematics learned in various courses and extracting new techniques from the literature, (c) communicating in writing about a problem, its solution, and its interpretation. These are skills which are needed in nonacademic employment and which are given little emphasis in traditional courses.

The basic ingredients of our case-study approach are: (1) industrial resource persons, that is, individuals working in nonacademic situations who are willing and able to discuss their mathematical needs; (2) problems encountered by these people, and which use undergraduate mathematics; and (3) a mechanism for interaction between our students and the industrial resource person.

The first two of these ingredients are identified by personal visits to regional industry. We go to visit with persons in industry who use mathematics, these persons having been identified through graduates, friends, colleagues, etc. We are received with a surprising degree of enthusiasm. These people are glad to help an educational institution, and glad to talk about their work. Almost always in the course of two or three visits a suitable problem is identified and a plan for using it formulated.

One of the unique aspects of our case-study technique is that the industrial resource person comes to campus and describes the problem to the students. They must determine or refine the model, seek out appropriate reference material, suggest approaches to solving the problem, and discuss the relative merits of these approaches. During this time, the instructor gives support and a minimum of direction. A particular method of approach must be chosen by a student or group of students who must then solve the problem, working individually or in groups. Finally, the students are required to discuss their solutions individually in a written technical report. The industrial resource person returns to campus for a second visit with the class. Frequently, he reads the reports and discusses them with the students. He also discusses a variety of methods of solving the problem, describes any further considerations which must be taken into account in obtaining a final solution, and answers any questions the students may have. This second visit, usually held in an informal situation, is a very meaningful exchange.

2. A Typical Semester of Problems.

Perhaps the best way to answer questions about how long, what problems, how many, what is suitable, etc., is to describe a typical semester. As a rule, we begin with a problem that we hope will not be too difficult - - a "confidence builder". The problem provided by Richard Greenhaw, American Fidelity Assurance Company, concerns insurance to cover a loan. A combination is being considered made up of regular life insurance which repays a fixed amount, and Businessman's Protector, a term insurance which decreases linearly. What combination of such insurance will cover the loan while yielding the minimum premium? Since the students know surprisingly little about insurance, they are asked to read about several insurance plans before the first presentation. Mr. Greenhaw provides additional pamphlets and information about costs.

For this problem, the students must create the model. First, they realize that they need an expression for the principal outstanding after s months. They can either calculate this or find an appropriate formula in a business text. Once this is done, they must complete the model by realizing that the amount covered by the combined policy at any time will be represented by a straight line, that the y-intercept of this line will represent the amount of whole life plus the amount of BP that must be bought, and that the slope of the line will be related to the proportion of each involved. The creation of the model is complete when someone realizes that the straight line should be tangent to the curve. It is now a simple matter to produce the cost function, express everything in terms of a chosen variable, differentiate to locate the minimum cost, then calculate the corresponding premium. The emphasis in this problem is not so much on the mathematics used as on the creation of the model. On his return visit, Mr. Greenhaw helps the students compare the premium for this plan with the cost of plans offered by competing companies. This problem requires three to four weeks.

The second problem is chosen from a different area of application and involves a different type of mathematics. A good choice is the problem provided by Mr, Kenneth Roger from the Boeing Company, in Wichita. The response of a plane to gusts is expressed in terms of departures from a mean position and is in general a linear combination of a number of responses:

$$x = x_{mean} + \Delta x = \Sigma c_i[L_{i(mean)} + \Delta L_i]$$

The gust-loading region consists of n-tuples

$$(\Delta L_1, \Delta L_2, \ldots, \Delta L_n)$$

which satisfy a certain design criterion for all sets of constants c_i. The particular criterion considered is

$$(\Delta x)^2 \leq U_\sigma \sigma_x^2$$

or, in matrix notation,

$$(c_1)^T(\Delta L_i)^2 - U_\sigma^2(c_i)T(\sigma_{ij})(c_j) \leq 0$$

where σ_x^2 is the variance of x, U_σ is a given constant, and

$$\sigma_x^2 \qquad U_o \qquad [\sigma_{ij}] = \mathrm{cov}(\Delta L_i, \Delta L_j)$$

The design loads envelope is the boundary of the gust-loading region. It consists of all potentially critical loading combinations which must be investigated by the stress analyst. The problem is to devise a way of determining and describing the design loads envelope, given the covariance matrix for the loads. Mr. Rogers provides numerical data and asks the students to solve particular cases with two or three variables, and also to devise a technique for the general case. As preparation, the students are asked to review linear algebra and find the mathematical meaning of the word "envelope". In this problem, the mathematical model is relatively accessible, since it is provided by the criterion and the definition of the quantities involved. By considering the two-dimensional case the students are able to find two approaches. One involves thinking of the envelope of the hyperplanes

$$(\Delta x)^2 = U_\sigma^2 \sigma_x^2$$

Some students, have difficulty with the concept of envelopes but recognize that the criterion requires that the quadratic form

$$(\Delta x)^2 - U_\sigma^2 \sigma_x^2$$

in the variables $c_1\ c_2\ \ldots, c_n$ be negative definite. From this requirement, the equation of the set of boundary points is determined:

$$[\Delta L_i]^T[\sigma_{ij}]^{-1}[\Delta L_i] = U_\sigma^2$$

In both methods, the final identification of the surface requires orthogonal diagonalization of a matrix. This problem can be done in about five weeks.

Since the first two problems have not involved either differential equations or numerical methods, it is natural to try to choose a third problem using these. It is also valuable if this problem is open-ended, one that can be pursued longer if time permits, or shortened a little if the first two problems have required more time than planned. A good third problem for our typical semester is the one brought by Dr. Dennis Zigrang, from Rockwell International's Tulsa Division. It concerns an analysis of the moisture content in the lip of the payload bay doors of the space shuttle. These doors are designed and constructed at Tulsa and are made of graphite epoxy. Moisture is absorbed directly from the atmosphere by diffusion through the epoxy matrix, and too much moisture affects the material in an adverse manner. Initially, the doors are perfectly dry, but prior to the first flight, they are exposed to a variety of storage and shipping environments. These conditions are given to the students, and they are asked to determine the amount of moisture in the 0.1 inch thick lip of the payload bay door at launch, and to report inermediate results in percent moisture by weight. Here the model isgiven by the classical differential equation

$$\frac{\delta m}{\delta t} = \Delta(T)\frac{\delta^2 m}{\delta x^2}$$

where M is the weight percentage of moisture, D is the diffusion coefficient, T is the temperature, t is the time, and x is the depth in the lip of the door. Although the students do not need to create the model, they must understand it and do considerable work to identify the appropriate value of D and the boundary conditions from the data given. The students usually find an appropriate numerical method from the literature. An exact solution in series form can be found by standard methods. If time permits, the students find the exact solution, at least for the first time period, and compare the results with the numerical solution. It is instructive for them to discover that the calcultion of numerical values from the exact solution presents some difficulty, especially for small values of t. During his return visit Dr. Zigrang discusses the relative merits of different solution techniques.

Each semester, a different set of problems is used. Three each semester gives a variety along with time to assimilate each problem. Fortunately, our supply of problems was increased with the help of a CAUSE grant in which we were joined by Professors James Yates and Ray Beasley from Central State University (Oklahoma), and Professors Lysle Mason and Philip Ames from Phillips University (Oklahoma). They participated in making contacts, collecting and writing up problems, teaching the course in their institutions, and making the material available around Oklahoma and neighboring states.

3. Problems Available

The industrial problems we have identified are suitable for a variety of purposes. Some are quite simple and can be used effectively as enrichment in college algebra or elementary calculus. Others, like the ones described here, are longer and can be used for independent study or in a case-study course. A catalog listing the problems and describing them briefly can be obtained from Dr. Marvin Keener, Department of Mathematics, Oklahoma State University, Stillwater, OK 74077. If in the catalog you identify a problem which you would like to use, you can obtain a write-up of that problem on request for a small charge to cover cost of reproduction. We hope that some of you will find it possible to make use of this material, or to experiment with the technique in your own region, and that you will find it as exciting and rewarding as we have found it.

CHAPTER 5 - The Profession of Teaching

5.1 CURRENT STATUS AND TRENDS IN TEACHER EDUCATION

CURRENT STATUS AND TRENDS IN TEACHER EDUCATION (U.S.A. AND CANADA) 1980

D. W. Alexander
University of Toronto
Toronto, Canada

I. Status

The 1975 report of the National Advisory Committee on Mathematical Education (NACOME) identified the lack of data regarding teacher education programs for mathematics teachers, particularly secondary teachers (NACOME, 1975: 144, 145). In response to this need for baseline data the Commisssion on the Education of Teachers of Mathematics (CETM) undertook a survey of preservice mathematics teacher education programs (elementary and secondary). Questionnaires were prepared, tested, and revised. In 1977 the 535 institutions accredited by the National Council for Accreditation of Teacher Education for the preparation of teachers in the U.S. and 48 Canadian institutions involved in such preparations were sent questionnaires. Fifty-six percent of the forms were completed. An analysis of the responses on the basis of size of institution and geographic region satisfied the members of CETM that the survey was representative. The data collected is available through ERIC (Sherrill, James).

An analysis of the data suggests the following descriptions of "typical" programs offered by U.S. and Canadian institutions. These programs differ significantly in some respects. Reference to the actual data (Sherrill, 1978) would provide those interested with a much better concept of the programs.

The "Typical" Program of Preservice Education of Secondary School Mathematics Teachers, 1978

In the United States admission to the program depends upon an overall grade-point average (GPA) of 2 (on a 4 point scale) and a GPA of 2 in mathematics, together with a demonstrated ability to speak clearly, and recommendations by members of the education faculty.

The program graduates five students per year.

Students are required to take at least five mathematics courses beyond calculus including algebra and geometry. Courses in statistics and analysis are strongly recommended. One course in mathematics education is required.

The mathematics education course emphasized lesson planning, exposure to the activities and journals of the NCTM and the local mathematics education organization, test construction, micro-teaching, individualized instruction, the study of learning theories related to mathematics, and the use of manipulatives. Some attention is also given to the use of media, the use of computing devices, and problems of teaching slow or gifted students. Prior to a practicum of ten weeks duration the student is required to observe in a mathematics class (Grades 7 to 12) for eight hours and engage in tutoring or small group instruction for five hours.

In Canada admission to the overall program depends primarily on a personal interview, although overall grade point average is also used as a criterion. The program graduates 20 students per year.

Students are required to take 2 to 4 mathematics courses beyond calculus with the same requirements of algebra and geometry. Statistics and analysis are also recommended. A mathematics education course is required.

This mathematics education course emphasizes lesson planning, exposure to the activities and journals of NCTM and local mathematics education organizations, the use of manipulatives, individualized instruction, the use of computing devices and micro-teaching to peers. In contrast to the U.S. programs, the study of learning theories related to mathematics, test construction, the use of media, and the problems of teaching the slow or gifted students receive comparatively little emphasis.

Prior to a practicum of eight to ten weeks the student has little prior field experience, limited in most cases to observation, if it exists at all.

The "Typical" Program of Preservice Education of Elementary School Teachers, 1978

In the United State admission to the program depends upon an overall grade point average (GPA) of 2 (on a 4 point scale) together with recommendations by members of the education faculty.

Students take the equivalent of two full year courses in mathematics and one course in mathematics methods. The methods course emphasized the use of manipulatives lesson planning, the study of learning theories related to mathematics, the NCTM journals, the use of media, and problems in employing individualized instruction and in teaching slow or gifted students.

The manipulatives involved are geoboards, colored rods, geometric shapes, place value materials, attribute blocks, computation games and the abacus. The use of the hand-held calculator is given little emphasis.

Prior to their practicum some students observe in an elementary school classroom and engage in tutoring and small group instruction but neither the practicum nor the previous experiences need involve an mathematics teaching.

There is the opportunity (usually taken by less than five students in a given year) to major in the teaching of mathematics in the elementary school. These students are required to take one or two additional mathematics courses.

In Canada, the majority of institutions have no special criteria for admission for the elementary teacher education program. Students take no specific mathematics or mathematics education courses.

Where a methods course is offered it streeses the use of manipulatives, lesson planning, the use of media, exposure to the NCTM journals, the study of learning theories related to mathematics, and individualized instruction. The manipulatives employed include place value materials, counting materials, geometric shapes, geoboards, computation games, the abacus, the hand-held calculator, and attribute blocks.

Prior to their practicum the students engage in one-to-one tutoring and some other field experience, but neither the practicum nor the field experience need involve any mathematics teaching.

There is little opportunity for an elementary education student in Canada to take a major or concentration in mathematics.

Inservice Courses

The National Council of Teachers of Mathematics reported on inservice courses in 1977 (NCTM, 1977). Three styles of inservice education were identified: (i) state sponsored programs related to certification and salary, (ii) degree related programs, (iii) local programs related to locally identified instructinal and curricular problems (NCTM, 1977: 16).

In a report of a survey concerned with inservice education the following characteristics of inservice programs were identified:

The programs are conducted by personnel within the school systems and are concerned more with teaching methods than mathematical content; however, many teachers felt that the programs did not fit their needs in the classroom and were too theoretical. Very few teachers reported that their school district had an inservice program specifically for teaching mathematics.

The role of degree related programs has declined as the age of the teacher population has increased. The diversity of local programs makes it difficult to identify their features and establish common elements. It can be noted that teachers surveyed generally felt that more emphasis should be given in inservice to: Motivation, applications of mathematics, students with learning difficulties, remediation, improving student attitudes about mathematics, metrication, computational skills. For elementary teachers the use of the calculator was seen as receiving (and deserving) little emphasis.

Both secondary and elementary teachers saw little need for emphasis on transformational geometry.

2. Trends

The trends in North America are:

1. Declining numbers of students entering programs for preparing mathematics teachers.
2. Pressures within school systems to provide mathematics teachers by "retraining" certified teachers with experience in teaching other subjects and threatened by declining enrolments.
3. Increasing emphasis on special education reflected by the increasing emphasis on diagnostic and prescriptive mathematics both in preservice and inservice courses.
4. Increasing emphasis on preparing teachers to use calculators in teaching mathematics.
5. Concern over how to prepare teachers for teaching applications problem solving and statistics. A concern not yet reflected in a major way in course requirements or design.

References

National Advisory Committe on Mathematica Education. Overview and Analysis of School Mathematics - Grades K-12. Washington, D.C.: Conference board of the Mathematical Sciences, 1975.

National Council of Teachers in Mathematics. An Inservice Handbook for Mathematics Education. Reston, Virginia, 1978.

Sherrill, James M. NCTM Commission on the Education of Teachers of Mathematics Survey of Pre-Service Secondary Mathematics Teacher Education Programmes. Reston, Va.: National Council of Teachers of Mathematics, 1978. (ERIC Document Reproduction Service No. ED 162 878).

NCTM Commission on the Education of Teachers of Mathematics Survey of Pre-Service Secondary Mathematics Teacher Education Programmes in Canada. Reston, Va.: National Council of Teachers of Mathematics, 1978. (ERIC Document Reproduction Service No. ED 162 878).

STATUS AND TRENDS IN MATHEMATICS TEACHER EDUCATION (AUSTRALIA)

Jeff Baxter
Sturt College of Advanced Education
Bedford Park, South Australia

Introduction

Whilst this is a particular study, limited to Australia, it is potentially useful in providing an example of attempts to blend influences from both American and European approaches to teacher education. As a 'developed' country, Australia provides an example of teacher education experience that borrows from both approaches to provide its own unique structures and practices.

Education in Australia is funded partly by the Federal Government and partly by provincial authorities (called 'States') and is administered independently by the States, which seem to be more aware of their differences than their similarities across the range of educational practice. Pre-service teacher education follows two basic models found elsewhere in the world, namely 'concurrent' (usually at Colleges of Advanced Education, where academic and professional studies are taken jointly) and 'consecutive' (usually at universities where professional study follows a three- or four-year general degree). The majority of primary teachers train in colleges for three years, and currently from one third to one half of beginning secondary mathematics teachers enter the profession with a four-year degree from a college. Pre-service teacher education is usually in one of three distinct awards directed at ages 5 to 7, ages 7 to 12, or ages 12 to 18, graduates being trained and authorised to teach only at one of these three levels.

In recent years there has been a series of official inquiries in the post-secondary education sector, all touching in some way on teacher education. Despite this, no broad national survey with unequivocal findings has emerged: controversy continues about the efficacy of the mathematics education currently provided in our schools, and current information and detail on mathematics teacher preparation is almost non-existent, save for some specific reports commissioned by the Australian Association of Mathematics Teachers. (It is from these sources that any data in this paper are quoted.)

Status of Pre-Service Teacher Education

a. Primary

Currently in all institutions responsible for the education of primary teachers, there is at least one compulsory course on the teaching of mathematics, but this varies from a total instruction time of 10 hours to about 150 hours, with a mode close to 60 hours. Mathematics educators (some 70% of them) are generally dissatisfied with the amount of time pre-service trainees spend in mathematics courses, both professional and academic, given that a primary teacher in Australia almost certainly will be expected to be able to teach mathematics at any level either in K-2 or in 3-6 for as much as 20% of the instructional time available each day.

In compulsory courses emphasis is focused on three main categories of the mathematics of primary schooling:

increased understanding - - i.e., improving trainees' comprehension of relationships, purposes of basic mathematical concepts

skills - - ensuring trainees can themselves perform all relevant processes and operations they will be expected to teach

attitude improvement - - nearly all institutions assume students generally enter primary teacher education with poor or negative attitudes towards mathematics, as few have been very successful in secondary school mathematics.

Approximately two-thirds of teacher educators express dissatisfaction with existing pre-service mathematical requirements for intending 'primary' teachers. Two factors contributing to this disquiet are that there is no formal requirement in all awards that students either (i) demonstrate personal basic mathematics competence, or (ii) successfully display competence in teaching mathematics to children.

b. Secondary

In most consecutive teacher education courses, there is little formal or even informal communication between mathematics and education faculties; in addition, rarely do universities offer courses designed specifically for intending teachers, as this is not perceived as a task relevant to a general undergraduate degree.

Only in the last decade have college graduates been qualified to enter the profession to teach the full range of secondary mathematics. Their courses, usually 'concurrent', are generally less rigorous mathematically, but are specifically designed for intending teachers. However, mathematics educators in both kinds of institutions have expressed a need for programs to devote more time and emphasis to helping students acquire skill in teaching mathematics in a variety of modes and organisations, and for training programs to reflect the slowly growing inclusion of practical experiences in early secondary school mathematics education.

The Status of In-Service Education

Following a general cutback in funding for education in Australia, money for in-service work has been severely limited since 1978. The Federal Government provides funds to the States to conduct in-service education through a specially constituted committee representing government and non-government school authorities. Tertiary institutions are not directly funded to support any kind of formal in-service program, teacher participation is voluntary, and each State's work is independent of the others.

A dramatic reduction in the demand for newly qualified teachers over the last three years has also seen an increased stabilisation of staffing for schools. It was not uncommon for schools to have a 30-40% staff turnover annually. These two factors, together with a policy shift to school-based curriculum development, have meant the involvement of teachers in the consequences and results of their work in a more direct context than before. The implication of this is to change the focus of teacher education from pre-service to in-service functions.

In the primary education sector there has been careful re-examination of current curriculum practice, and guidelines ranging over years K (Kindergarten) to year 12 are being, or have been, developed in some States. Such guidelines are likely to have an increasingly important impact on mathematics education in Australia in the future.

At the secondary level there are two dominant concerns in in-service work. Firstly, the complex and broad issues of coping with mixed-ability mathematics classes in lower secondary are critical, and secondly, problems resulting from public examinations involve all teachers of secondary mathematics in some way. As such examinations inevitably constrain syllabus, methodology, and purpose they are dominant; about 45% of all students entering secondary school attempt such an external examination in their final year of schooling (except in one State in which the summary examination system no longer exists).

Trends in Mathematics Teacher Education

Currently, a national inquiry into the whole area of teacher education is continuing and the trends identified here reflect some of the desired changes submitted in evidence together with ideas and opinions from colleagues around the country.

Pre-Service

Undoubtedly, the next four years are critical ones for many Australian institutions and their programs. College-based awards which allow flexibility of specialisation after an introductory year of general tertiary study are likely to become more popular. In primary teaching awards, continuing pressure from concerned groups should result in an extension of compulsory hours spent in the study of mathematics and its teaching, as existing awards undergo their regular reviews and revisions.

In addition, a formal 'numeracy' component seems likely to become mandatory. A campaign for extending the basic primary training course from three to four years of full-time study will gather momentum, with part of this extra time being spent in mathematics and certainly 'practice teaching' experiences in schools. Indeed, exhibiting competence in the teaching of mathematics in particular may be the next formal component in assessment.

Secondary pre-service programs are already reacting informally to the implications of a K to 12 Curriculum, considered as a continuum of experience for each child. Also the involvement of many Australian teacher educators in 'problem solving' and 'applicable' mathematics is reflected by such emphases in methodology courses - - particularly as resource materials are gradually becoming available. In general, it is likely that a political move to encourage an upgrading of qualifications and updating of expertise of mathematics teachers, will include making in-service work mandatory - - or a criterion for promotion anyway. This would give otherwise expensively underemployed college staff a formal role within the in-service realm.

In-Service

At present there is no indication that this will change from its current fragmented, voluntary, short term, and parochial nature, with its consequential partial success and short term effects. Possible systematic developments like the 'master-teacher' idea (to retain expertise in the classroom), formal involvement of tertiary institutions in in-service education and the emphasis on the schools', rather than a central, authority and responsibility for the education children receive, are all critical.

Specifically in mathematics, one major influence over the next four year period is certain to be the recently funded Australian Mathematics Education Program. Planning for this project is well under way, and one of its areas of focus must be the identification and development of support services for teachers if they are to implement changes in the teaching of mathematics in schools.

As it is the first national program with support in every State, it is uniquely an opportunity to begin a viable long-term, specific, and coordinated assault on the critical issue of mathematics education in Australia.

CURRENT STATUS AND TRENDS IN TEACHER EDUCATION (SOUTHEAST ASIA)

Sr. Iluminada C. Coronel, f.m.m.
Ateneo de Manila University
Manila, Philippines

INTRODUCTION

In the past three decades, population explosion, policies of equal access to education and greater interest in education among the masses of the region who consider education as the only source of hope for the future, have caused a burgeoning increase in school population in the member countries (Hong Kong, Malaysia, Indonesia, Singapore, Thailand and the Philippines) of the Southeast Asian Mathematical Society (SEAMS). The rapid increase in school population, which in Singapore has reached a peak, has taxed the resources of these countries and obliged school and education authorities to recruit unqualified teachers, mostly from secondary graduates and holders of non-teaching degrees. In Indonesia, for example, "to reach an enrolment target of 85% of the primary school-age children by 1979, the Government has built (by 1978) 201,100 classrooms, rehabilitated 56,000 old classrooms, recruited 290,000 teachers and other staff, and distributed large numbers of books and other materials" (1, p. 73) while in Thailand, the "Fourth Development Plan aims toward producing 296,565 qualified teachers at all levels" (1, p. 251) and in the Philippines, the steady increase in student enrolment "created by the year end (1977) a backlog of 45,911 classrooms in the elementary schools" (2, p. 1). It is against this background that teacher education in these countries has to be viewed.

Pre-Service Training

In the countries under consideration, teaching requires a professional degree: a certificate or degree in teaching for the first level, and a degree in mathematics for the second and third levels. Teacher Training Colleges and Faculties or Colleges of Education attached to universities exist for the training of prospective teachers.

Secondary school graduates qualify for two to four years of training for the first level. In the Philippines, since 1950, a B.S. in Elementary Education degree has been offered (3, p. 92), though previously the two-year Elementary Teachers Certificate was sufficient. For secondary mathematics, a B.S.E., major in Mathematics is required. In Hong Kong and Singapore, a two-year training qualified trainees for the elementary level and an additional year qualified them to teach the first two or three years of the second level.

Prospective elementary teachers are trained by experienced first and/or second degree holders and their training prepares them to teach any subject in elementary school. There is a discernable trend to give some specialized training in a specific field even to teachers for this level. For secondary teaching, specialization is a prerequisite and in the apprenticeship period, the trainee learns to teach one or more mathematics subjects.

A trend in teacher education which is a response to a felt need is the "integrated teacher training approach". In Malaysia, since 1973, "whether a student is a prospective primary or secondary teacher, the training is similar," (1, p. 128). In the 36 teachers colleges of Thailand, which have been established to meet local needs, the integrated teacher training approach prevails. The same trend is gaining ground in the Philippines, where each village wants a high school of its own. The objective of the integrated training is to produce teachers who can teach at the first and the second level.

Acceptance of the "Field-Based Teacher Education" principle has effected the lengthening of the practicum from one semester to two with the first phase of teaching usually being in a laboratory school attached to the teaching college and the second, in an off-campus school where teachers of the host school cooperate in supervising the trainees. In fact, in Hong Kong, there is a shifting away from pre-service towards in-service training (4).

The indiviudalized or pupil-centered approach has found its way into several private and public schools in the Philippines, though sometimes in a modified form. To prepare prospective teachers to cope with this trend, Colleges of Education train their students to use both the class and the individualized technique. In addition, the use of the discovery, the laboratory and other approaches are taught. Classroom management by objectives, especially behavioral ones, is stressed. With the emphasis now being placed on activities and games in the lesson, the prospective teacher has to be trained for these.

Governments encourage the upgrading of the teaching of mathematics and science. Among others, the region has the UNESCO-sponsored Southeast Asian Regional Center for Science and Mathematics in Malaysia, the Institute for the Promotion of Teaching Science and Technology in Thailand, and the Science Education Center in the Philippines.

Teacher training for teaching mathematics in tertiary schools is essentially an apprenticeship. Officially, the required qualification is a second degree in Mathematics. In practice, due to a shortage of teachers for the tertiary level, first degree holders teach the basic mathematics courses. This is true in the Philippines where "an 8.29% yearly increase has brought the tertiary enrolment to nearly a million students", (2, p. 1). The relatively much smaller tertiary enrolment in the other five countries has enabled them to minimize this problem.

Third degrees in mathematics are obtainable in leading local universities, but most of the Ph.D.'s in Mathematics in the region have been trained abroad. Difficulty of adjustment by returning Ph.D.'s has recently brought about a marked determination by local mathematicians to train their own faculty and send them abroad only for post-doctoral work. Close cooperation among leading members of the Southeast Asian Mathematical Society is making this decision a viable stance.

In-Service Training

The recruitment of unqualified teachers to staff newly-organized classes each year, the accelerated production of poorly qualified teachers to cope with short-term deficiency, and the reforms in the mathematics curricula in the past two decades have made in-service training an enormous problem. Singapore and Hong Kong with populations below five million and more adequate financial and/or personnel resources have systematic programs of in-service education in the form of part-time courses, which basically reflect the content of pre-service courses and lead to a Certificate in Education. In addition, Singapore has introduced "residential weekend seminars for teachers, principals, and Education Ministry personnel", (1, p. 215).

Indonesia and the Philippines have similar in-service training problems and both employ a two or three-tiered system of training. Every summer, in the Philippines, thirteen Regional Staff Development Centers offer Government-funded 6-week training in mathematics for elementary and secondary teachers. Participants conduct echo seminars in their respective districts. Another in-service training program exists as a component of the textbook scheme of the Ministry of Education and Culture. At the release of a new book (distributed free to all public schools) for the first or second level, a one-week training on the use of the book is given to supervisors and trainers who in turn train the classroom teachers for a period of one to two weeks, in the use of the book.

Graduate programs leading to different levels of masteral degrees are offered in several universities in the Philippines. Courses offered in these programs are also availed of by those who simply want some knowledge in a particular field or earn the units required for a promotion. The Government funds some of these programs.

One faculty development program for teachers of private colleges and universities based at the Ateneo-FAPE Graduate Center for Mathematics of the Ateneo de Manila University, and funded by the Fund for Assistance to Private Education, gives promise of becoming a good model for an effective upgrading of mathematics teachers as well as mathematics education. Begun in 1973, the two-year program leading to an M.S. in Mathematics degree originally aimed at producing graduates who would upgrade mathematics education in their sponsoring schools. Deep involvement of teachers and fellows of the Center in preparation for the 1978 ICMI-sponsored First Southeast Asian Conference on Mathematical Education, which included their conducting a national survey of Philippine undergraduate-graduate mathematics and working for the revision of the different mathematics in curricula, has enriched the program. The Center has, as a consequence, broadened its horizons and taken upon itself the leadership in mathematics, both in education and in research. A government-funded Ph.D. program, in consortium with two universities started in June, 1977, and the Center is now the headquarters of the Mathematical Society of the Philippines (MSP) and the Mathematics Teachers Association of the Philippines (MTAP). A Center Staff composed of teachers and fellows conduct seminars all over the Philippines for teachers at the three levels in simultaneous parallel sessions. During these seminars, discussions on curricular and professional problems encountered by teachers are held to provide an opportunity for a dialogue among the participants. These often end in the organization of a provincial/city chapter of the MTAP.

Conducting seminars is now an integral part of the training of the fellows and demands for their services even while in training is increasing. The heaviest demand is from elementary teachers who are currently in the process of implementing a "back-to-basics" curriculum and have found out that they have forgotten their arithmetic through disuse. Several of the graduates of the Center have taken up leadership roles upon graduation and return to their sponsoring schools.

Based on this model, there is a tabled proposal to the Ministry of Education and Culture for the training of public school teachers along similar lines and who upon graduation will have the in-service training of the teachers of their districts as their major commitment.

PROBLEMS OF MATHEMATICS TEACHER EDUCATION

A. Pre-service training: The main problems are:

The low status of teachers and the consequent difficulty in getting students to take up a teaching career. The best students flock to more prestigious and/or financially promising courses like medicine, law and economics;

Lack of qualified teacher educators. There is a short supply of these as there is no training course for them. Research in mathematics education is negligible;

The great tension between quality training and mass producing them to teach at either the first and/or second levels; and

Lack of proper textbooks and reference materials. This is especially true where the national language has newly become the medium of instruction and there are not sufficient locally-authored books.

B. In-service training: The main problems are:

Persuading teachers that training is necessary and worth the effort. Even when teachers are convinced to participate, they keep demanding for help on how to make mathematics more interesting. There are strong indications, however, from the experience of the Ateneo Graduate Center for Mathematics and the two national mathematics societies that the main problem of teachers is content and not methodology but they are unaware of, or refuse to acknowledge, this weakness;

Financing. The cost of re-training such a massive number of teachers is beyond the resources of developing countries; and

Lack of qualified trainers. To make up for this, leaders of national mathematical societies have taken it upon themselves to be concerned with mathematics education and conduct seminars for teachers, but their number is woefully inadequate.

CONCLUSION

In brief, pre-service education is patterned after that of western nations. However, it has an element of its own - - the integrated approach which developed out of an urgent need to train teachers capable of teaching both the first and second levels. In-service training, at the moment, seems like a bottomless pit - - financial and/or personnel resources are inadequate to meet the great need for it.

References

1. "Education in Asia and Oceania," Bulletin of the Unesco Regional Office for Education in Asia and Oceania, June, 1979.

2. "Education in Asia and Oceania," Reviews, Reports and Notes, November, 1979.

3. Carson, L.C. The Story of Philippine Education, New Day Publishers, Philippines.

4. SEAMS members supplied information for this paper. These are Dr. Lee Peng Yee (Singapore), Dr. R. Turner-Smith (Hong Kong), Dr. Mokhtar Bidin (Malaysia), Dr. Milagros Ibe, Ms. Carmen de la Pena (Philippines).

TEACHER EDUCATION IN MATHEMATICS IN ENGLAND AND WALES

Hilary Shuard
Homerton College
Cambridge, England

Teacher education in England and Wales at present is much affected by external forces. A falling birth-rate is producing a substantial fall in school rolls. The economic situation means that falling rolls will not be an occasion to improve the quality of education; rather, teachers who leave a school are often not replaced, tattered text-books and equipment continue in use, the release of teachers for in-service education is curtailed, and teacher morale is suffering. In the early 1970s, initial teacher education was greatly reduced, so that the college of education system is now only one-third of its 1973 size, and about 20% of the time of college of education lecturers is now devoted to in-service education for teachers, whereas there was formerly only a minimal commitment. Despite the reduction of teacher output, many young teachers have difficulty in finding jobs, especially in primary schools. Another background factor is that responsibility for the curriculum in England and Wales rests with each local education authority, which normally delegates responsibility to each individual school. Recently, there has been much criticism of education, and mathematics teaching has come under particular pressure. Employers of youngsters who leave school at 16 have been vocal in criticism of their innumeracy. These criticisms are difficult to pin down to hard facts, but they have led to the setting up of a government Committee of Inquiry, the Cockcroft Committee, into the teaching of mathematics in primary and secondary schools. I am a member of that committee, which has not yet reported, so this paper represents my personal views.

Much informed comment on education is contained in recent major surveys of English primary and secondary schools by HM Inspectorate. HMI found much attention to basic skills in both primary and secondary schools, but in many cases the work in mathematics was narrowly based. In primary schools, where HMI were able to test the 11-year-olds, better results were found in classes where mathematical relationships were sought and practical activity and visual presentations used as well as practice of basic written and mental skills. We need to ask why all teachers do not make use of these teaching techniques. In secondary schools, the isolation of mathematics from the rest of the curriculum was a cause of concern to HMI, as was the leadership of mathematics departments. HMI recommended that one-third of heads of mathematics departments should undergo in-service education, and that a majority of schools should devise a more appropriate curriculum for less able pupils. Another problem is that in spite of total over-supply of teachers, there is an acute shortage of qualified mathematics teachers, so that much mathematics teaching is undertaken by non-specialists.

These problems all concern the ways in which teachers do their work. Many teachers of mathematics need to change their teaching in style or content. This can only be tackled through in-service education, or by providing new teachers. Few mathematics graduates choose to enter secondary teaching, and few new primary teachers can find employment. Thus, our major concern must be with helping serving teachers to become more effective. At present, in-service education has to compete for scarce funds against keeping teachers in post, and against textbooks, equipment or computers. In-service education must be seen to be effective, or it will not be funded.

In preparation for this Congress, about forty teachers of mathematics and providers of in-service education were asked to write about the professional life of mathematics teachers. Their book will shortly be published (1), and some of the things they say are very illuminating. Here are three new teachers in their first weeks of teaching in secondary schools:

> I don't suppose I could be more unprepared - - not anyone's fault in particular; teething troubles of two schools merging and a new head of department trying to organise 'order out of chaos'...I though it would help to go into school the day before anything happened...but no one else appeared...I found some notes suggesting what books I might use with some classes.
>
> *   *   *
>
> When I hear myself shouting in class (to keep the noise down) I feel I have seen it all somewhere before and it quite horrifies me to see myself slipping so easily into what could be a predictable and somewhat boring teacher.
>
> *   *   *
>
> I was not well prepared by the head of department and soon found out the law of the jungle concerning equipment and books. I will start homework and more defined lessons instead of working straight from the books.

These are the realities of life in secondary schools, not only for new teachers; time and money would be well spent in preventing new teachers from being socialised into an easy acceptance of the present state of mathematics teaching. Pilot schemes for the induction of new teachers, which started in the mid-70s, are faltering, and although probably more new mathematics teachers fail the probationary first year of teaching than in other subjects, little is being done to ensure that mathematics teachers start in the right direction.

In primary schools, many authorities are trying to ensure that at least one teacher in each school has some expertise in mathematics, so that his colleagues can turn to him for support. These 'mathematics coordinators' have an uphill job, and there is a shortage of suitable teachers. One of them writes that he thought he would only have to keep his colleagues informed about in-service courses and new developments in mathematics teaching, but

> I realised rather quickly that mathematics courses were the last thing they would want to attend - - 'That's your job!'. And "New methods?" No, thank you! Haven't you heard? There's a big move back to formal methods as the government is very concerned about the drop in standards since the new methods came in.'

Thus, a major task is to encourage teachers to evaluate their own teaching and to use knowledge of mathematics teaching and learning to improve it. What do we know about the effectiveness of in-service education in doing this? It is easy for a teacher to come back from a course intending to put what he has learnt into practice. But this may involve difficult rethinking, and the other teachers in the school may not want to change their ways. If two or more teachers from a school attend the course, they may support each other. Perhaps the course should take place in the school, so that all the teachers can take part and we can reach the many teachers who, to quote Edith Biggs, "know they are not teaching mathematics very well, but don't want to change the way they do it". School-based in-service work is very expensive in manpower. Ideally, the school's own mathematics leader should provide in-service education, but it takes much confidence to develop an advisory role with colleagues.

Increased knowledge can boost confidence, and the Mathematical Association has recently started to run a Diploma in Mathematical Education for teachers of the 5-13 age range. This two-year part-time course of 200 hours is being taken by over 1000 teachers in 50 colleges and is being very well received. Its curriculum contains mathematics and the study of its learning and teaching. How much is put into classroom practice, I do not know, and it does not cover leadership skills and ways of dealing tactfully with anxious or hostile colleagues.

Too many in-service activities regard teachers as passive recipients rather than active participants in their development. Each primary school is responsible for its own curriculum, but the authority often provides guidelines which its schools can adapt. These guidelines are usually written by groups of teachers, together with the mathematics adviser. Guidelines may not solve all the problems of the schools, but they certainly help the thinking of the teachers who take part in writing them. At the secondary level, a similar means of teacher development has been the Certificate of Secondary Education, a 16+ examination for pupils of average ability. This examination can be devised by the teachers in the school, and moderated by teachers from other schools. It has provided a very useful forum for the sharing of ideas between schools. Although only able teachers take part in these activities, and so have these opportunities for growth, they gain confidence in providing leadership for their colleagues.

Another secondary school problem is the shortage of mathematics teachers. A recent scheme provided one-year retraining courses for teachers of other subjects who are willing to become mathematics teachers. These teachers need concentrated mathematical study at their own (not very advanced) level, closely related to the classroom. They also need experience of teaching mathematics, as this may be very different from their previous teaching style. The retraining of teachers will be very important while few new teachers enter schools, and experience shows that while teachers may learn to deal with new subject-matter, they often interpret this in terms of their old teaching style.

More research is needed in this area. We know something about conceptual development in children's mathematics, but we know very little about why many teachers do not apply that knowledge. We need to explore the interactions of

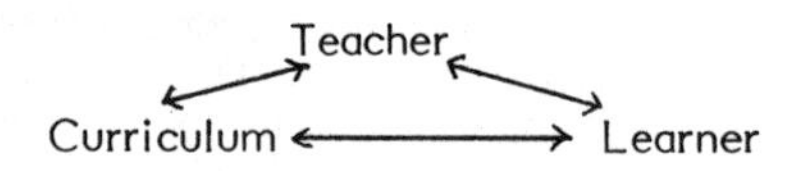

as they actually happen in real classrooms. In in-service education, there is another dimension:

In-service classroom

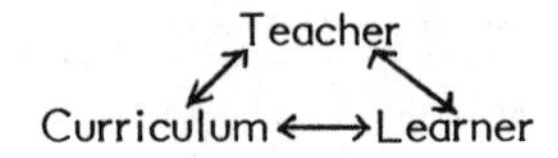

School classroom

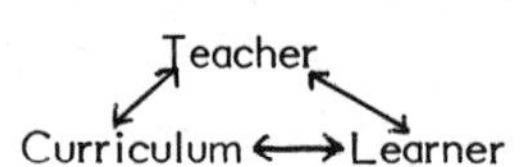

We need to understand more about transfer from the in-service classroom to the teacher's own classroom. We need to study the in-service classroom as well as the school classroom.

Reference

1. Shuard, H.D. and Quadling, D.A. (Eds.). Teachers of Mathematics: Some Aspects of Professional Life, Harper and Row, 1980.

5.2 INTEGRATION OF CONTENT AND PEDATOGY IN PRE-SERVICE TEACHER EDUCATION

INTEGRATION OF CONTENT AND PEDAGOGY IN PRE-SERVICE TRAINING OF MATHEMATICS TEACHERS

Zbigniew Semadini
Institute of Mathematics
Warsaw, Poland

I will speak on mathematical education of teachers-generalists in primary grades. Some ideas apply also to the case of teachers-specialists in higher grades (with change of emphasis and with different allocation of time, effort etc.), but I restrict myself to elementary teachers who are in charge of several subjects, including reading, writing, crafts, etc.

Frankly, I do not believe in full integration of content and pedagogy. It is a very attractive idea and many nice examples can be found in the literature (1), but it is very hard to find instructors with suitable attitudes, capable of teaching teachers this way. Yet it is is possible and highly desirable. I will try to describe it as I see it.

The worst (though quite common) solution is to have a course of traditional-style content course in math followed by a method-oriented course. Whatever the topics of such a content course are it will not give satisfactory results because the student teachers are usually poor in mathematics and fear it, they lack confidence in their ability to learn, and they switch off whenever exposed to mathematics of this kind. Let me recall the significant article by Arthur Morley (4): "to completely dissociate discussion of aspects of teaching mathematics, and, vitally, work with children, from teaching the mathematical content is to invite trouble, because it cuts these students off from their main source of motivation." I would add that it is the student who is then blamed for the failure.

Thus, I would propose two separate courses (or two sequences of courses). The first will be labelled "content-oriented" and the second "method-oriented". These terms may deceive somebody accustomed to traditional boundary line between "content" and "methods", so let me explain what I mean. The content-oriented course may include mathematically exploring topics using Cuisennaire rods or geoboards. The student teachers learn what can be done with the rods, what mathematical problems can be raised, solved, or illustrated. During the method-oriented course the student teachers should deal with another question: how to use the rods in the classroom. In the former course they think of the material used and of mathematics; for future teaching is somehow in mind but not explicitly involved. The child's learning is the core of the latter course.

The former course should be taught by a competent mathematician (willing to explore this kind of mathematization-concretization problems), not necessarily familiar with real school problems. The latter course requires an educationist, experienced in primary school teaching.

This way of separating content and methods, which indicates to what extent content and pedagogy should be integrated, seems most feasible. It believe it is possible to find capable instructors for the content course and to convince them hat they should try to teach concrete-based mathematics. Many working mathematicians are likely to accept laboratory or workshop approaches to mathematics for elementary teachers (at least as a principle) and, if given good examples, may be willing to teach this way.

On the other hand, those mathematicians and educationists who are able only to reproduce what they have been taught often resist the approach to a content course outlined above, finding various arguments against it. My experience is that such people cannot simply see mathematics embodied in the concrete situations and therefore reject it or call it "methods".

Such a tendency should be overcome, and this is where the help of research mathematicians with sufficient authority is needed.

The question is whether the method-oriented course should follow the content one. Ths would be the easiest to organize: yet, there are voices that some preliminary or interwining contacts with children may help the student teachers to get interested in the subject.

Let me give now a few further hints on the proposed way of partially integrating content and pedagogy.

The content course should be structured. Yet, the order of topics need not be the logical order or their purely mathematical order. It should rather be some spiral structure of learning. We may assume certain knowledge on the part of the teacher (e.g. of arithmetic) and use this knowledge for other topics, broadening and deepening it successively. The new topics should enrich the previous ones. Natural motivation for a problem may be more important than logical or even competency-based order.

When the student teachers begin to deal with a concept, especially a new or difficult one, they should follow what I laconically call the child's way; they are to enact what is devised for the children (at a different pace, of course) or to perform them in mind. This should be augmented with partial mathematization and symbolization with comments, and should be systematized. Thus, the student teachers are first to do child's mathematics, and then to reflect on what they have done.

Bear in mind that teachers in the classroom are likely to follow the way they learned the subject. If now they are taught it in an "adult way" later they will also reproduce it in an adult way.

The new ideas should be introduced by abstracting a given situation, by paradigmatic examples rather than by definitions. For example, it is important that elementary teachers know that a square is a rectangle with equal sides; yet, it is pointless to distinguish whether this statement is a definition or a theorem.

If a concept is to be generalized, particularly if the generalization is hard to understand, the following approach, which I shall refer to as scheme permanency, is often the best way: Formulate the concept in a case which is already familiar to the student, using a suitable chosen paradigmatic example. Keep the example and devise a strategy of changing a variable in it. The students will be likely to accept the extended concept if the scheme holds the same. ("Student" here means: school pupil or student teacher.)

The following example will help understand this principle. We shall consider the question: What does it mean to divide fractions? In traditional teaching one spoke of how to perform the division. It is equally important that the student understand what it actually means to divide a number by a number; the way we handle the division $2/3 \div 5/8 = 2/3 \cdot 8/5 = (2 \times 8)/(3 \times 5)$ should be the result of understanding the meaning of division, not the starting point.

Currently it is popular to explain that dividing means finding an unknown factor; the above quotient is such a fraction which multiplied by 5/8 gives 2/3. Such an approach would be perfect if it were really understood by the student. It is more likely to be a mechanical extension of the procedure. ("Mechanical" may be a merit in case of a formal theory; in case of a student it is not.)

In order to make use of scheme permanency, let us think of an adequate paradigmatic example. I am choosing bags of sand. Conceivably, the children would play with real bags and sand. The teachers just imagine such actions, peform them in mind, or use graphical representation.

We should start with the familiar range: "6 kilograms of sand is put in bags of 2 kg each. How many bags will be filled?" The students explain it, make simplified pictures, and write formulas. We then keep constant capacity of a bag (only 2 kg bags are available now) and change the amount of sand.

After a few examples, when the students become familiar with the scheme, we pass to fractions. This requires a change of wording; instead of "how many bags" we ask "what portion of a bag" - this is an example of how language difficulties may break the scheme (the word "filled" helps to minimize the change). Hopefully, it is the concrete situation rather than a word problem which is to hold permanent. "I kg of sand is put in a 2 kg bag. What portion of a bag is filled?" The common-sense answer "half" should be followed by explanationn what we are really doing with these numbers and by formula like $1 \div 2 = 1/2$. After a sequence of such questions, answers and comments, when the students grasp the division by 2, we change the size of the bag (to 3 kg say) in order to see whether the students can transfer the ideas.

The crucial point is when we pass to fractional bag. "6 kg of sand is put into half-kilogram bags. How many bags will be filled?" Again we start with the common-sense answer: 12; the formula is $6 \div 1/2 = 12$ is to follow this. The student is to accept that in this case "to divide" means to find an unknown factor (the unknown number of bags) and that the quotient may be greater than the divident. After a sequence of questions with 1/2 kg bags, we pass successively to more difficult cases, the help of iconic representation will be indispensable. Eventually, we reach cases like $2/3 \div 5/8$. If this is mastered, we check the understanding by modifying the concretization, e.g. to pouring (in mind) milk into bottles, etc.

The above example gives an idea of possible integration of content and pedagogy in preservice teacher education. I believe that such a sequence of problems will give the student teacher a sound basis for teaching the given concept. It goes without saying that they are the students who are to perform the activities, guided by the instructor. The same sequence presented as a lecture will loose much of its value.

Varying constants is a significant method. Instead of solving unrelated problems, we take one and change the data. I believe this should give better results. Moreover, it is a basis for a future concept of a function.

Let me comment on what is the core of mathematical thinking: proving conjectures. It is well known that proofs are particularly hard for young pupils and for their teachers. Yet, learning statements without justification is to invite rote learning. The question is: what should be meant by a proof in primary mathematics teaching? Of course, it cannot be deductive proof. This should be a way the student can convince himself (or herself) that his/her observation is a logical necessity rather than an experimental fact or a statement taken for granted.

In my opinion, both in case of pupils and student teachers, the base is what I shall call an action proof, or premathematical proof (3, 5, 2). Take an enactive and/or iconic representation of a paradigmatic example (not too simple, not too complicated), and perform certain actions (manipulations, drawing pictures) to get the desired results. Do the same with other examples, keeping the scheme permanent and varying constants. When you no longer need concrete physical actions, perform them in mind until you see that the same can be done for a whole class of cases and therefore the statement is surely true. Examples of such proofs can be found in (5, 6, 2).

What is the main difference between an action proof and a proof is an axiomatic theory (such as Euclidean geometry)? In the latter one abstracts (from reality) primitive notions and some properties of them formulated as axioms; other notions are defined and other properties are deducted without reference to the concrete. An action proof is a result of abstracting the action rather than logically interferring from given premises.

Of course, an action proof must be correct in the sense that it is possible to mathematize it as to become a formal proof. Some version of how to translate such a proof to ordinary mathematical language may (but does not have to) be shown to the teachers after they have understood it.

The concrete example and the mode of representation should be properly changed to avoid misinterpreting the concepts.

In my opinion, the above program of partial integration of content and pedagogy is the best and most practical form of such integration, provided there are enough laboratory activities and inter-student discussions (9, 10). Lectures should be to sum up previous experiences and to structure them rather than to present the subject as an introduction to follow-up activities. This will most likely guarantee that the teachers gain acceptable understanding of the subject, deep in the sense (8).

References

1. Fred Goffree, John - a teacher training freshman studying mathematics and didactics, Educational Studies in Math 8(1977), 117-152.

2. Arnold Kirsch, Beispiele fur "premathematische" Beweise, In: Beweisen im Mathematikunterricht, W. Dorfler, R. Fischer, editors, Holder-Pichler-Tempsky, Wien 1979; 261-274.

3. Arthur Morley, Changes in primary school mathematics - are they complete, Mathematics teaching (1969) 20-24.

4. Arthur Morely, Goals for the mathematical education of elementary school teachers, Arithmetic Teachers 16 (1969) 59-62.

5. Zbigniew Semandeni, What should be the mathematical education of primary-school teachers (in Polish), Matematkya (Warsaw) 26(1973), 299-309.

6. Zbigniew Semandeni, The concept of premathematics as a theoretical background of primary mathematics teaching, Institute of Math. Polish Acad. of Sciences, preprint, 1976.

7. Zbigniew Semandini, The mathematical training of elementary teachers, Proc. Conf. on Math. Education, Manila, 1978.

8. Tamas Varga, On primary school teachers, Mathematics, Educational Studies in Math. (Utrecht) Vol. 7 (1976) 171-177.

9. Julian Weissglass, Exploring elementary mathematics: A small-group laboratory approach, W.H. Freeman, 1979.

10. James E. Schultz, Joan R. Leitzel, Arthur L. White, Alternatives in the mathematics preparation of elementary teachers, The American Math Monthly (to appear).

INTRODUCING PEDAGOGY INFORMALLY INTO A PRE-SERVICE MATHEMATICS COURSE FOR ELEMENTARY TEACHERS

Julian Weissglass
University of California
Santa Barbara, California

In considering the question of combining mathematical content and pedagogy in preservice courses for elementary teachers I find it useful to introduce the idea of <u>quality</u> for mathematical learning experiences. A learning experience has <u>quality</u> if (i) the learner enjoys the experience, (ii) the learner is more interested in the subject after the experience than she or he was before, and (iii) the learner has a sense of pride or accomplishment. Although this is an imprecise concept it is a useful one. If elementary teachers are to change the conditions that lead most upper level students to regard mathematics as drudgery, then they must learn to set up learning experiences for these pupils which have quality. This requires that they have such experiences themselves. Unless a prospective teacher develops a positive attitude toward and enthusiasm for mathematics they will find it difficult to not pass on their negative feelings about mathematics to their pupils.

When I teach prospective elementary school teachers I try to remember that whether or not their learning experiences have quality will probably have as great an effect on their teaching as the amount of mathematical or pedagogical theory that is assimilated. For this reason I use a method of teaching which incorporates pedagogical experiences in an informal way while attempting to provide each student with a mathematical learning experience which has quality in the sense defined above. There are many ways for students to learn pedagogical principles. Before beginning a formal study of pedagogy prospective elementary teachers need to (i) observe how they learn mathematics themselves, (ii) have a learning experience in which the instructor encourages free exploration and question-asking, (iii) think for themselves about how they might help young people learn particular mathematical concepts, and (iv) observe children in the process of learning.

The method I use has students learning mathematics while working together in small groups. The students use concrete materials to investigate mathematical concepts whenever possible. I call this method a small-group laboratory approach. Since the course I teach is offered through a mathematics department and since students are required to also take a course in the procedures of teaching mathematics, I do not emphasize pedagogy as much as I would like. The basic method, however, is flexible enough to include pedagogy to any extent desired.

The essential characteristics of the small-group laboratory approach are:

(i) Students working together in groups of 4, 5 or 6.

(ii) Written exercises guide the students to learn mathematical concepts and skills with minimal assistance from the instructor.

(iii) Students are directed and encouraged by both the written material and the instructor, to help each other learn.

(iv) Whenever possible students explore theoretical concepts using concrete material (e.g. colored rods used to investigate properties whole numbers and fractions and their operations, multi-base blocks used to learn the algorithms for the operation of arithmetic, geoboards and tangrams used to investigate geometry.

(v) Students discuss what young people might do with equipment on their own and how teachers might use the equipment to teach mathematics.

(vi) Mathematical theory is presented so as to challenge the intelligence of college-level students.

(vii) Students are given information about how people learn, what interferes with learning mathematics and why the small-group laboratory approach is being used.

(viii) Students are asked to think about why certain groups of people (for example, women and some minority groups) have not succeeded in mathematics in proportion to their numbers in the general population.

(ix) Students are encouraged to stop blaming themselves for not having learned mathematics, but to instead view this particular class as an opportunity to regain their power over this part of the world.

(x) The instructor regards himself or herself as a facilitator of learning than as an imparter of information.

(xi) The instructor circulates from group to group, observes when learning is not taking place, determines the difficulties and attempts to resolve them by asking questions and/or providing information.

(xii) The instructor educates herself or himself about the mathematically related activities of the groups of people represented in the class and provides that information to the class. For example, information about women mathematicians, the accomplishments of the Mayan civilization, African number systems and symmetry design.

More detailed information (1, 2, 3) about using the small-group laboratory approach is available and several textbooks are available which incorporate group work and use of concrete materials to varying degrees. See Davidson's article (1) for bibliography of articles and books.

There are many advantages to using this approach. Some of the ones I have noticed are:

(i) Resources are increased by having the students help each other learn.

(ii) Communication of ideas by the students helps them learn.

(iii) The likelihood of students being overwhelmed by too much new information is reduced (not eliminated!) by students having more control over the rate of presentation of new concepts.

(iv) It is easier for students to ask questions of their peers than an instructor.

(v) Students become friends with their classmates. This reduces feelings of loneliness that interfere with learning.

(vi) The use of concrete material provides students experiences with the physical world which they may have missed. Their concrete experiences form a foundation for more abstract reasoning.

(vii) The use of concrete material helps the students see the relevancy of what they are learning to their career as teachers.

(viii) The students begin thinking for themselves about how to help young people learn rather than being told from the start what experts think.

(ix) The students see in operation a model of teaching that encourages free exploration, discussin and question asking. This will be helpful to them when they begin to teach.

(x) The instructor has the opportunity to observe the thinking processes of students - - how they learn, how they misinterpret definitions and explanations, and where they have difficulty learning.

(xi) Mathematicians who have little experience or information about pedagogy can incorporate it informally into the course to any degree desired. Even if the course is taught in a mathematics department the method leads the students to think about teaching.

In summary, the small-group laboratory approach increases the likelihood that the students learning experience will have quality without increasing the teaching resources used for the course. It is a particularly appropriate method for prospective teachers since it allows them to learn mathematics, experience peer teaching and begin to think about how to teach their students. This combination increases their understanding of both mathematics and pedagogy. It helps both subjects be seen as open and exciting fields of investigation and not as a body of theory or techniques which they must memorize.

References

1. Davidson, N., "Small Group Learning and Teaching in Mathematics: An Introduction for Non-Mathematicians" in Cooperation and Education (Sharan, Hare, Webb and Lazarowitz) Brigham Young University Press (1980).

2. Weissglass, J., Instructors Manual for Exploring Elementary Mathematics, San Francisco, W.H. Freeman and Co., 1980.

3. Weisglass, J., Mathematics for Elementary Teaching: A Small-Group Laboratory Approach, American Mathematical Monthly, Vol. 84, No. 5, May 1977, pp. 377-382.

5.3 PREPARATION IN MATHEMATICS OF A PROSPECTIVE ELEMENTARY TEACHER TODAY, IN VIEW OF THE CURRENT TRENDS IN MATHEMATICS IN SCHOOLS AND IN SOCIETY

PREPARATION IN MATHEMATICS OF A PROSPECTIVE ELEMENTARY TEACHER TODAY, IN VIEW OF THE CURRENT TRENDS IN MATHEMATICS IN SCHOOLS AND IN SOCIETY

James E. Schulz
Ohio State University
Columbus, Ohio

The thrust of this presentation is that "The mathematical preparation of elementary school teachers should emphasize a problem solving approach supported by the use of calculators and computers within a framework of mathematical structure." Intended in this statement are three themes:

1. We should emphasize a problem solving approach;

2. This approach should be supported by the use of calculators and computers; and

3. This approach should be within a framework of mathematical structure.

Each of these themes will be discussed and illustrated in view of current trends in mathematics, in schools, and in society.

By a "problem solving approach" we mean one in which a student learns the material primarily by doing problems. Thus problems serve as the main vehicle for developing material as opposed to, for example, being a check on whether certain material from a given lecture or reading is comprehended. Thus, instead of the problem "Show that a set with n members has 2^n subsets", a series of problems such as "How many versions of a medium-sized pizza are possible if we have 2 toppings available and can use one, some or all of the toppings? If we have 3 toppings? If we have four toppings? If we have 10 toppings? Develop a formula for n toppings and explain why it works."

There is currently widespread support for emphasizing problem solving. The recommendations of the National Advisory Committee on Mathematical Education of the Conference Board of the Mathematical Sciences (NACOME) state (p. 141) that problem solving and the use of the calculators and computers are areas which should receive emphasis in teacher education. The National Council of Teachers in Mathematics (NCTM) in its Recommendations for School Mathematics of the 1980's entitled An Agenda for Action urges (p. 1) that "problem solving be the focus of school mathematics in the 1980's" and it devoted its 1980 yearbook to problem solving.

We have found in a study of alternatives in the mathematics preparation of elementary school teachers at Ohio State University that a laboratory component is highly desirable. Besides being an effective way for our pre-service teachers to learn the mathematics, the laboratory approach provides an appropriate example of how elementary mathematics can be taught in the schools and it provides experience in communicating about the mathematics as groups work together solving problems in the laboratory setting. The approach also bolsters student attitudes, since it is perceived as (and is!) relevant to their needs. Of course, the mathematics which took centuries to develop cannot be rediscovered in a year's coursework, but perhaps an environment can be created where a few ideas can be legitimately rediscovered by students. We must provide reasonable time for such activity to avoid what I saw in a Chicago school where the teacher said to a second grader, "Quit playing with those rods and discover something."

In view of current trends in mathematics, schools, and society a problem solving approach is a sensible one. It reflects the trend to apply mathematics to other disiciplines; by providing a wide range of problems it allows for meeting the needs of students in a class operating (perhaps temporarily) at vastly different levels because of desegregation or mainstreaming; it provides an interesting setting for learning basics; it builds the much-needed skill of deciding which computational strategy to apply in a world where the computations themselves usually will be done by machines; and finally, it prepares students for a rapidly changing world, for problem solving skills are important to many existing fields of study and are likely to be important to fields of study to be created in the near future.

Unlike passing fads in education, electronic calculating devices are sure to have a lasting influence. Calculators and computers are now almost commonplace as they become more powerful, yet more compact and less expensive. Considering the rapid growth of technology, the potential for electronics staggers the imagination.

The current support for incorporating calculators and computers in teacher training comes as no surprise and, for lack of time, need not be mentioned here. More interesting perhaps is that this was recognized much earlier, for example by the Committee on the Undergraduate Program in Mathematics (CUPM) in its

1971 recommendations and by H. Freudenthal, who in Mathematics as an Educational Task (p. 74) in 1973, referred to "the miracle of computers" as "the most sensational subject mathematics can offer."

Examples of how calculators and computers are useful beyond mere calculation are well known and include introducing multiplication (or division) using the repeat add (or subtract) feature of a calculator and reinforcing concepts such as place value and functions through calculator games. (See Judd, 1976.) When, in a recent course for prospective elementary school teachers, we investigated repeating decimals, first with 4-function calculators and then with programmable calculators, students were excited about discovering and analyzing patterns and were not instead spending critical time doing tedious computations.

Calculators and computers have permeated society, including the fields of education and mathematics. In view of this trend they cannot be ignored in programs which prepare elementary teachers.

Perhaps the most controversial part of the idea that "the mathematical preparation of elementary school teachers should emphasize a problem solving approach supported by the use of calculators and computers within a framework of mathematical structure" is the reference ot mathematical structure. In his famous indictment of the modern math, Why Johnny Can't Add, Morris Kline argues (p. 100) against teaching structure to young people, mentioning (p. 51) the centipede who could no longer move once he began to think too much about which foot to use next. Yet even Kline concedes (p. 98), "there is nothing intrinsically wrong with the goal of teaching structure, though one might question its importance at the elementary state of mathematical learning." I maintain that mathematical structure, appropriately taught, is important for elementary teachers whether or not it is taught in the schools.

Some aspects of mathematical structure which I consider appropriate are the following:

1. Teachers should encounter mathematical proof to be aware of its beauty and power.

2. Teachers should be aware of how sequencing of topics is important in mathematics.

3. Teachers should be aware of the structure which unifies the treatment of number concepts.

4. Teachers should be aware of the structure which relates various topics within mathematics, such as number with measurement.

5. Teachers should understand why fractions and negative integers are necessary, and they should understand what motivates us to work with the rational numbers and the real numbers.

To illustrate these points, we have for example (1) the beauty of Euclid's proof that there is no largest prime or the necessity of proof upon examining "false patterns", such as when considering the maximum number of regions formed by chords connecting n points of a circle; (2) how 3 1/2 -- 5 1/4, a problem involving division of mixed numbers, is reduced through a series of steps to7 x 4/2 x 21, a problem involving multiplication of whole numbers; (3) the analogy between dividing fractions and subtracting integers; (4) how addition of whole numbers can be modeled on the number line; and, (5) how the whole numbers are extended to the integers to accomodate subtraction.

Yet these points should not be overemphasized. I have seen attempts to rigorously develop everything which is covered. I have seen preoccupation with sequencing which rejects the number line model for addition of whole numbers until the entire real number system is built up. I was distressed by a professor of education from another university who boasted that his greatest contribution to teacher training at his institution was to reduce the number of hours that teachers were exposed to formal mathematics courses which he felt were detrimental. I witnessed the effect on a classroom teacher of stressing formal vocabulary, when the teacher demanded of a pupil that instead of the word "number" he use another word ("numeral") which, said the teacher, "means the same thing."

My concern is that certain trends may present threats to the understanding of appropriate mathematical structure by teachers:

1. The trend toward problem solving and applications might ignore building the understanding of mathematical structure.

2. The trend toward integrating content and methods instruction might be at the expense of understanding relationships within the mathematics.

3. The trend toward increasing the school experience component of teacher training programs might reduce the time devoted to the study of mathematics.

I wish to illustrate each of these points. 1. I agree with the NCTM position, again quoting An Agenda for Action (p. 2) that, "Structural unity and the interrelationships of the whole should not be sacrificed" (when focusing on problem solving). Certain books featuring a problem solving approach claim that the material can be taught in any order. I confess immediate skepticism of such approaches. 2. Approaches which integrate content and methods by relating a mathematical concept to its application in the classroom might ignore building the relationships between one mathematical concept and another in the process. Organizing books or courses around certain teaching materials instead of concepts also poses a danger. The question should not be. "What can I do with a geoboard?", but rather "How can I best teach a given mathematical topic?" 3. Though school experience related specifically to mathematics is highly desirable, certification requirements which increase the school experience at the expense of content are threatening. A teacher gets school experience on the job, but further work in content may not be readily available.

As we revise programs for teachers to focus on problem solving, on the integration of methods and content, or on school experience, we must not ignore the structure which is the backbone of mathematics.

References

Committee on the Undergraduate Program in Mathematics: Recommendations on Course Content for the Training of Teachers of Mathematics. CUPM, Berkeley, California, 1971.

Freudenthal, H.: Mathematics as an Educational Task. D. Reidel Publishing Company, Dordrecht-Holland, 1973.

Judd, Wallace: "Instructional Games with Calculators" Arithmetic Teacher. (Vol. 23, No. 7), November, 1976.

Kline, Morris: Why Johnny Can't Add. Random House, New York, 1973.

National Advisory Committee on Mathematical Education: Overview and Analysis of School Mathematics Grades K-12. Conference Board of the Mathematical Sciences, Washington, D.C., 1975.

National Council of Teacher of Mathematics: An Agenda for Action - Recommendations for School Mathematics of the 1980's. NCTM, Reston, Virginia, 1980.

National Council of Teachers of Mathematics: Problem Solving. 1980 Yearbook of the NCTM, Reston, Virginia, 1980.

Schultz, James E.; Leitzel, Joan R.; White, Arthur L. "A Comparison of Alternatives and an Implementation of a Program in the Mathematics Preparation of Elementary School Teachers" (Final Report). Department of Mathematics, The Ohio State University, Columbus, Ohio, 1979.

5.4 EVALUATION OF TEACHERS AND THEIR TRAINING

THE EDUCATION OF MATHEMATICS TEACHERS

Thomas J. Cooney
University of Georgia
Athens, Georgia

This essay will focus on issues related to the education of mathematics teachers. A basic premise of the essay is that there is a body of knowledge specific to mathematics education, albeit eclectic, which deserves inclusion in mathematics teacher education programs. The argument that mathematics teachers have knowledge specific to their field is not a trivial one. Popham (1971) found, for example, that experienced teachers were not particularly skilled at bringing about prescribed behavior changes in learners and, in fact, seemed to do no better than those knowledgeable about the content but with no training or experience in education. Popham's study did not involve the teaching of mathematics. Still the question remains, "Do those trained in the teaching of mathematic outperform those not so trained?" Or to put it another way, "Do mathematics teachers have knowledge specific to their profession which distinguishes them from the mathematician or from the teacher of other subject areas?" The next section addresses the question of such knowledge.

The Need for Pedagogical Knowledge

To emphasize the need for pedagogical knowledge several episodes will be presented which I have observed. While the episodes are not all from secondary school classrooms, they do highlight the need for teachers to generate alternatives - - alternatives based on knowledge accumulated through years of experience and/or training.

Episode 1.

A seventh grade mathematics teacher was teaching youngsters how to factor whole numbers into their prime factors. The lesson began with a quickly stated definition of prime number followed by two examples of prime numbers. No nonexamples were given. Two demonstrations of how to obtain the prime factorization of a whole number were given. Students had obvious difficulties including the following mistakes as alleged prime factorizations of the numbers on the left.

$8 = 5 + 3 \quad 12 = 4 \times 3 \quad 40 = 4 \times 10 \quad 24 = 16 + 8$

The teacher recognized there was a problem, repeated the definitions, and gave one more demonstration. Students returned to the worksheets but few corrections were made as they were still quite confused.

I contend that the teacher's inability to cope with the difficulty stemmed far more from her lack of pedagogical knowledge than from her lack of mathematical knowledge. (The young lady was, in fact, an honor student in mathematics.) She seemed to have no instructional alternatives save repeat what had been said before. Strategies for teaching the prerequisite concepts were essentially nonexistent save for a quick "define, example" strategy. In my view the teacher could have prospered from a study of ways to present mathematical knowledge.

The ability to present mathematical concepts through the use of examples and nonexamples is one kind of "special knowledge" teachers should have. School mathematics textbooks in the United States provide virtually no nonexamples of concepts. Thus, if you believe, as I do, that the use of nonexamples is an important pedagogical technique in teaching concepts, then teachers must be educated about their use.

Episode 2

The following dialogue was observed in a beginning algebra class.

T: O.K. Now what do we mean by linear function? How do we define it? Jake?

Jake: I don't know. I forgot.

T: Elsie?

Elsie: Well, it has something to do with a straight line.

T: That's true. But we need more. Todd?

Todd: Things like f (X) = 2X + 3 and f (X) = 4X - 10. These are linear functions aren't they?

T: Yes. That's good. O.K. now let's see how we can graph some linear functions. Look at this problem.

It appeared the teacher was asking for the definition of linear function but when the student produced two examples the teacher seemed satisfied. This may have been a case of a teacher being unclear as to the nature of the content being taught, and, hence, instruction suffered from vagueness. Or it may have been that the teacher was willing to accept examples as evidence of understanding rather than press for a correct definition. Did the teacher make a conscious decision to accept examples rather than a definition? What alternatives did the teacher generate? The point is that an instructional decision decision was made, if only passively. It seems highly desirable that such decisions should be based on a number of alternatives, alternatives available because of the teacher's special knowledge of the teaching of mathematics.

Episode 3

The following episode occurred in a mathematics course in which the Pythagorean Theorem was being discussed.

> Students had constructed squares on the legs of right triangles. The unit squares on the legs were then rearranged to form a larger square of unit squares on the hypotenuse. Following this activity the following dialogue transpired.

T: Now consider the right triangle with legs a and b and hypotenuse c. What does the theorem say about this triangle?

Jane: $a^2 + b^2 = c^2$

T: O.K. Very good. Now suppose we have this right triangle with legs a and c and hypotenuse b. What does the theorem say about this triangle?

Karen: The theorem won't work for that triangle. It doesn't apply.

It appeared that Karen's conception of mathematics was largely synonymous with symbol manipulation. How can a teacher encourage and promote meanings rather than manipulations of symbols? Answers to this question further constitute the type of knowledge that ought to permeate teachers' knowledge about teaching mathematics. Based on that knowledge, alternative actions can be considered and selected including the following: call on another student, tell Karen that the theorem does apply and state the correct relationship, ask Karen to clarify her statement or to state the conditions under which the theorem does or does not apply. The question is not which alternative is better. Rather, the focus is on the identification of possible alternatives - - a task facilitated by the existence of an explicit knowledge base.

Teaching the Teacher Pedagogical Knowledge

Brownell (1956) once emphasized that mathematics teachers should maintain a balance between understanding and skill. In a sense I think his admonition also applied to mathematics teacher educators. That is, knowledge must be provided to help teachers reflect upon and understand the dynamic flow of classroom interaction. This implies providing the intern with certain constructs which can be used as a focus for considering instruction. While providing those constructs may be necessary for their use, the provision is not sufficient for skilled use. That is, practice must be provided in order for interns to become skilled in applying whatever knowledge they are taught.

The task, then, of teacher education programs is twofold. First, pedagogical knowledge must be taught. This can be accomplished through the use of lecture/discussions, or the use of videotapes, role playing, etc. Second, practice must be provided for using the knowledge. Practice can occur in the context of microteaching, tutoring, clinical interviews, small group instruction or large group instruction. In short, mathematics teacher education programs should focus on imparting the best of what we know about teaching and learning and allow prospective teachers to practice their creativity in using this knowledge in classrooms. Teaching is a skill in the sense that practice is necessary (but, I would argue, not sufficient) for the development of teaching competence. Practice can be enhanced by knowledge of the teaching act, that is, pedagogical knowledge, in much the same way students' understanding of mathematical algorithms can facilitate the development of skills in using the algorithms.

Research on Teaching the Teacher

At the present time we have a dearth of information about mathematics teacher education programs. It is nearly impossible to describe the typical graduate of a preservice program much less that same individual after possible exposure to a wide variety of inservice training experiences (NACOME Report, p. 81). In terms of research, I submit that there exists a considerable imbalance between our efforts to study how students learn mathematics and our efforts to learn how the students' teachers learn to teach mathematics.

What might a possible research agenda be for mathematics teacher education? Consider an analogy. Mathematics educators spend considerable energy to learn how children learn mathematics. In doing such research the following general questions often predominate.

1. How does the child think about mathematics?

2. How can we take advantage of the child's natural thought processes to teach him mathematics?

3. How does the child process information?

But I contend that these questions can serve also as a general guideline for research on teacher education. That is, we ought to consider the following questions.

1. How does the teacher think about teaching mathematics?

2. How is the role of a mathematics teacher perceived by teachers?

3. How can we take advantage of the teachers' perceptions and conceptions about teaching to teach them knowledge about teaching mathematics?

4. How do teachers process information?

5. What cues are attended to when instructional decisions are made? What characterizes teachers' belief systems? What characterizes their conception of mathematics?

Magoon (1977) has argued that educational research should adapt a constructivist approach. Basic to his position is the view that the teacher is a knowing being and operates in a purposeful manner. Research questions consistent with a constructivist approach would focus on how mathematics teachers develop their "professional knowledge," what factors contribute to the acquisition of the knowledge, and how that knowledge affects instructional decisions. To Magoon, the value derived from such research lies in a heavy emphasis on construct validity, that is, the meaning of events or situations to participants and also in the related hypotheses generated from the findings. The undeniable impact of Magoon's position is a shift away from a product orientation toward a process orientation. That is, we ought to be seeking ways of finding out how teachers acquire pedagogical knowledge and what the impediments for learning and using that knowledge are.

Otte (1976) at ICME III, underscored the lack of research in mathematics teacher education. He made the point that teacher education is basically reactionary to whatever school reform exists at a given point of time. Hence, according to Otte, teacher education is moved by forces better characterized as whimsical than as rational. If we wish to take seriously the education of mathematics teachers, we have a responsibility to seriously reflect upon what that special knowledge is which permits the development of professional mathematics teachers and to consider what processes exist when teachers acquire that knowledge. The process of educating the professional mathematics teacher is too important to allow ourselves to be moved by whimsical forces.

References

Brownell, W.A. Meaning and skill - - Maintaining the balance. Arithmetic Teacher, 1956, 15, 129-136.

Magoon, A.J. Constructivist approaches in educational research. Review of Educational Research, 1977, 17, 651-693.

National Advisory Committee on Mathematics Education NACOME). Overview and analysis of school mathematics, grades K-12. Washington, D.C.: Conference Board of the Mathematical Sciences, 1975.

Otte, M. The education and professional life of mathematics teachers. Paper presented at the Third International Congress on Mathematics Education, Karlsruhe, Germany, August 1976.

Popham, W.J. Performance tests of teaching proficiency: Rationale, development, and validation. American Educational Research Journal, 1971, 8, 105-117.

EVALUATION OF TEACHERS AND THEIR TEACHING

Edward C. Jacobsen
Unesco
Paris, France

Introduction

There was general agreement amongst those I contacted from different parts of the world that this panel should concentrate on the evaluation of teaching, not of teachers. However, this would mean refraining from evaluating specific teachers for the purposes of certification, promotion, improvement of instruction and research - - activities which are carried out throughout the world. The broader interpretation of evaluation is that it is an activity necessary to gain overall national knowledge of current classroom practices.

We will look at each of these purposes in turn, then discuss evaluation methods and instruments used in evaluating teaching. Then we turn to the reaction of teachers to their being evaluated, ending by looking at two recent evaluation exercises.

The Evaluation of Individual Teachers

Certification of teachers takes place at two levels - - the first one following their initial training, and then at various points of their career. It is the responsibility of the teacher training institution to determine if the teacher entering the profession possesses sufficient knowledge of the subject and is proficient in teaching. While it is true that a study done by the School Mathematics Study Group showed that qualifications greater than those 'required' did not make any 'difference' as measured by student learning, most would agree that a teacher without 'sufficient' knowledge would not be able to teach mathematics in a way that we would advocate today.

Continuing certification is a problem in all professions. It is usually solved by requiring the professional to undergo periodic examinations and scrutiny of his work by colleagues. This is badly lacking in the teaching profession and is not a problem only of mathematics teaching. Some consider that periodic testing should be done to see if a teacher possesses at least minimum knowledge in the subject he is teaching.

Initial certification also requires a certain proficiency in the art of teaching, which is usually determined by teaching practice supervision. Admittedly most of my professional life has been in the developing countries, but my observations of teaching practice are not very favourable. Students from the teachers' colleges are put out into schools throughout the country and when the college staff goes out to observe their teaching they often find a changed timetable or the school dismissed for the day. Also, the teacher under whom the student teacher is working is often using methods which are not in agreement with those advocated by the college. The practicing mathematics teacher is seldom visited by tutors specialized in mathematics. Tutors have found observation schedules useful for improving the feedback by tutors after the observed lesson, providing the teacher with comments and suggestions of behaviours which might be adopted. I would encourage those of you operating in situations such as I have just described to consider replacing the first part of teaching practice with micro-teaching. In the four countries in Africa where I have done this, the improvement of the student teacher has been striking, and after ten years there is every indication that it will continue.

The World Confederation of Organizations of the Teaching Profession (WCOTP), in response to my request for its position on the evaluation of teachers, has indicated that the present framework is too narrow, that we have concentrated on those tasks which are easily measured, usually based on operational objectives, while neglecting the broader objectives of education. The concept of external evaluation should be replaced by the concept of self-evaluation, designed to focus on professional autonomy in order to develop a truly professional teacher - - self-evaluation meaning evaluation by the teacher himself or by collaborating professional colleagues whose presence in the classroom would not destroy the natural climate. These colleagues might exchange classroom visits and exchange their impressions more openly.

So far we have been discussing the observation and feedback of individual teachers in order to improve their performance, or to certify them. Let us look at evaluation for the purpose of improving the overall system.

Evaluation of Teaching for Systems Improvement

Frequently we want to know the 'state of the nation' with regard to classroom practices, especially before planning curriculum changes or in identifying weaknesses in the teaching force in order to take corrective action through in-service training or modifying pre-service education. This type of evaluation is usually carried out by the inspectorate, an arm in most countries of the ministry of education. The role of the inspectorate has changed over the past several decades, from being a police force to becoming an information service, with the goal of encouraging good educational practices in the schools. Generalizations are made by the inspectorate for those who take decisions and make policy. When an individual teacher's career is not influenced by the inspectorate, inspectors are more likely to be welcome into classrooms. However, in many countries the inspectorate is involved in the promotion or appointment of teachers, and many a mathematics teacher dreads the visit of the inspector, for fear of finding himself posted in some remote village should his performance be displeasing.

A more acceptable inspectorate does not inspect teachers, but rather teaching; does not examine pupils, but rather looks at the overall effects; does not decide upon syllabuses, but rather advises bodies and committees as to current practices. By reporting back confidentially to schools visited, giving information as to what they saw duly interpreted, it is hoped that schools will take corrective actions. This type of evaluation was done in the National Secondary Survey carried out in England in 1979 and is well documented in two reports by the Department of Education and Science, entitled Aspects of Secondary Education in England: A Survey by H.M. Inspectors of Schools and Aspects of Secondary Education in England: Supplementary Information on Mathematics, available from HMSO.

Research plays a part in the improvement of mathematics teaching, a role increasingly being requested especially by developing countries as they modify their school systems in accordance with their rapidly changing societies. Here the evaluation of teaching is done to identify factors which affect learning. Researchers also use the evaluation of teaching to test the superiority of a particular method or programme over another.

Methods of Evaluating Teaching

Teaching is evaluated through observation of classrooms, scrutiny of recorded work of students, questionnaires to teachers and students, testing gains in student achievement and interviews with teachers, students and community leaders. Classroom observations are expensive, time-consuming and the coverage is limited by the number of inspectors or trained observers. There also exists a problem of observer reliability (especially important in research related observations), as observer's preferred methodology and other biases may influence the objectivity of the feedback or report. This is especially true in cross-cultural observations, a situation still found in many regions. However, inspectors, supervisors, headmasters, headteachers and researchers all use this method as their primary source of information about mathematics teaching, and most teachers would be unhappy if their teaching was never observed directly but rather evaluated through the achievement of their students. Suggestions made by the WCOTP in the last section are relevant here.

Linked to classroom observation is scrutiny of the work of the pupils. An example from the HMI Report mentioned earlier will show the value of this method: "Exercise books showed that although a great deal of time was spent on algebraic manipulation, only a little time was spent applying techniques to problems". It is also possible to tell the usual teaching approaches (those when the observer is not in the classroom) by looking at pupils' workbooks: whether the class is teacher controlled, leading through small steps with closed questions or with a more open approach. While at school, it is possible to look for specialist rooms and mathematics teaching equipment. Modern observational schedules focus as much attention on the activities of the pupils as on the teacher, and how these two interact. Professor Freudenthal pointed out that we must learn how to observe the learning processes rather than testing what has been learned. This contrasts sharply with McNeil and Popham's statement in their Handbook that "A focus on pupils reveals far more about the effectiveness of teachers than does direct study of teachers themselves". Many large-scale evaluations utilize as their dependent variable the gain in student achievement. Teachers and schools are often rated on the basis of results on certification examinations. Whether they are called primary leaving exam, school certificate, O-level, baccalaureat, abitur, studenten, etc, the results are often published in the newspapers, by school, resulting in pressure on the teachers and schools. This is especially serious if what is tested is out of line with the objectives of teaching...a situation which seems true of the world. Let us examine here some difficulties of measuring student achievement which are due to teaching alone. It is difficult to statistically factor out family factors of student achievement (such as economic status, education of parents, ethnic background which includes language differences). Some might argue that these should not be entirely factored out, as the teacher should have considered these factors in teaching. Also teachers react differently to various teaching materials and their implied teaching methods. A teacher whose students do poorly with open-ended or self-study materials might have much different results with more tightly structured materials. The Second Mathematics Survey, which is being carried out by the IEA in attempting to overcome some of these objections first considers what is intended to be taught, then what is actually taught, and then compares these with what is learned by the students (as measured by tests).

If we use pupil gain scores for measuring effectiveness of teaching, are we to use simple gross score gains from pre-test to post-test, or residual gains, or proportional gains (since higher pre-test scores should result in lower gain scores). In any case, pupil gain scores are unfair for evaluating individual teachers, as pupils are not normally randomly assigned to classes. If such scores are used, the results will probably be that the teacher will teach for the test, i.e. he will concentrate on those aspects of mathematics which are easiest to examine rather than on objectives of higher priority. Also research has shown that a teacher who produces a large gain in class A may not do so in class B. However, pupil gain scores remain useful in uncovering deficiencies in the overall system and in pinpointing differences between systems.

Large scale evaluations of mathematics teaching usually depend more upon questionnaires than direct observations. Teachers are asked to make self-evaluations which force them to reflect on their methods. It is thought that if the teacher is asked to consider alternative methods of presenting a topic, this act alone might make him a better teacher. Perhaps we should not ask questions about methods to find out what methods are used, but should observe them.

Students are less often given questionnaires to find out about teaching. Yet it is common knowledge in a school who are the good teachers. Studies have shown that evaluations by students corroborate with outside ratings like those of inspectors, especially concerning the environment of the classroom, e.g., intimacy, friction, satisfaction, difficulty, apathy. Research has shown that if both students and teachers are questioned, there is even closer agreement, especially if each knows the other is responding.

Community involvement in evaluation can be especially useful with regard to validity of goals, this being more true today with the trend to make education more responsive to the needs of society. In Bangladesh, by promoting greater involvement of parents in the affairs of the school, and by increasing the level of supervisors, it is hoped to raise the quality of instruction to obtain higher and more lasting pupil achievement, and to reduce teacher absenteeism, a major educational problem there.

By neglecting community reactions over the past two decades, mathematics education has been placed on the defensive in many, indeed most, countries of the world. Pressures are brought to bear on individual teachers who use methods or materials different from the norm. For this reason an independent evaluation of teachers is important: to counter or underscore community demands.

There is much variance in the degree to which school officials and the community interact with the teacher in his classroom, and hence differing teacher reactions to evaluators. For instance, in Sweden teachers are paid to administer evaluation achievement tests to the pupils, and the teacher unions must agree to the tests being given. In the United Kingdom, because of the independent characteristic of the inspectorate they are welcomed by the teachers. Responding to my question about removing a teacher because of poor teaching, a teacher's union official from the Republic of Ireland told me that the teacher would always be defended.

We might consider the WOTP's suggestion for self-evaluation which leads to teacher reflection on goals, attitudes and methods utilized.

Instruments for Evaluation

I intend to say very little about evaluation instruments. My experience has been that instruments which have been developed and validated for a particular situation do not transfer well to another situation or another country. It has been well argued by evaluation specialists from Africa (who have seen more than their share of the application of evaluation instruments developed elsewhere) that one should not even try to adapt an outside instrument, but rather start from the basic assumptions and develop instruments entirely within the culture. Since those who have done this work know how time-consuming it may be, evaluators in some countries will probably have to be satisfied with adapting instruments. I can only refer to ERIC for information about a large collection of evaluation instruments.

The process which I have used in adapting and validating instruments has been described in my paper, "An Evaluation of the First Two Years of Individualized Learning of Mathematics by Mathcards in Swaziland, Lesotho and Botswana", published in Kwaluseni (Swaziland) in 1976.

Examples of the Evaluation of Teaching

Both examples come from studies being planned or carried out by the International Association for the Evaluation of Educational Achievement (IEA). Both of these IEA projects concern the evaluation of teaching, each attempting to identify variables which affect the outcomes of learning. The first, with which many of you are already familar, is the second Mathematics Survey which is being carried out in 26 countries throughout the world. The variable is the curriculum. Curriculum is divided into three aspects: the intended curriculum, the implemented curriculum, and the attained curriculum.

The intended curriculum internationally is obtained from an analysis of all participating countries' curricula and from the construction of a content and behavior grid. Each country reports if the item to be taught in each cell is important, very important, or not important. These international grids provide blueprints for constructing tests, with the 'importance' ratings providing the weighting. Thus the international test is a slice of the intended curriculum, with each country deciding upon the goodness-of-it of the test to its intended curriculum.

In the first IEA Mathematics Survey, carried out at the onset of the mathematics curriculum reforms, the implemented curriculum was identified by asking teachers to rate each test item as to whether or not it was taught or reviewed this year, and if not, why not. Detailed questions are asked to identify approaches used to teach various topics, giving a national profile.

There may exist a strong congruence between the intended and implemented curriculum. It is now possible to prepare a profile of student outcomes: those expected based upon the above-mentioned procedures and those obtained based upon student performance on the tests. What will be interesting are departures from the expected, with each country being encouraged to pursue the reasons for the discrepancies. Some expected possibilities for discrepancies are (i) unrealistic expectations of students, (ii) test items that are too abstract or complex, (iii) curriculum materials that are inadequate for the topic in question, and (iv) instructional strategies inappropriate for the topic or age-level.

Professor Ken Travers of the University of Illinois, the chairman of the survey, would be able to provide further information on this interesting example of evaluating teaching.

The second example of an evaluation study on teaching which I wish to present is another IEA project, the Classroom Environment Study, which is in an advanced planning stage. Its objective is to obtain knowledge concerning improved teaching practices which will lead to a better utilization of resources. It is not exclusively evaluating the teaching of mathematics but rather all school subjects. The study will contain two aspects, a correlational study which will reveal relationships between specific teaching practices done under normal teaching conditions, and student achievement and attitudes. This will be followed by an experimental phase, training these teachers to increase the use of teaching practices identified above which seem to improve learning, and testing to see if this training has actually improved students' achievements and attitudes.

On the basis of previous research, factors which have been shown to affect achievement will be chosen from the educational system and environment. These factors will be correlated with the mean student achievement and attitude in related classes. So far, nothing new! Then a training programme will be prepared for the participating teachers, aimed at improving the teaching practices which were found to correlate well with achievement. Then, these teachers, along with a control group not specifically trained in this programme, will be observed in their classrooms to determine the effectiveness of the training. Finally, the achievement and attitudes of students being taught by two groups of teachers will be compared.

Thus, teaching practice is expected to be the independent variable. These are divided into two main factors: managerial tasks and instructional process variables. Briefly the, managerial variables are time-on-task (including students as well as the teacher, e.g., looking at the rules of conduct, disciplinary action taken, monitoring of seat work, etc.) and task orientation, considering the activity of the class focusing on the accomplishment of specific academic objectives. The instructional process variables seem to me similar to those emphasized in micro-teaching programmes, e.g. instructional cues (which link content and teaching skills such as emphasizing main points, outlining objectives, etc.), questioning skills (frequency, cognitive level, mode, reactions to answers, etc.), and feed-back given.

This Study which is just now starting to work on the instruments, will include teacher and student questionnaires, observation schedules, achievement and attitude tests. Because they are aiming at multi-national use of the instruments one might keep this in mind for future reference, or obtain more information for the IEA Secretariat in Stockholm.

5.5 HAND-HELD CALCULATORS AND TEACHER EDUCATION

L'EXISTENCE DES CALCULATRICES INFLUENCE-T-ELLE L'ENSEIGNEMENT DES MATHEMATIQUES?

Willy Vannhamme
Universite Catholique de Louvain
Louvain-la-Neuve, Belgium

Je n'ai pas les informations pour répondre à cette question de manière internationale. Je ne puis que donner mon avis sur ce que je constate autour de moi en exercant mon métier qui consiste à m'occuper de la formation de futurs enseignants en mathématiques. Et cet avis est négatif: la possibilité de la présence de calculatrices dans la classe n'a pas, en général, influencé à ce jour la manière dont s'enseigne la mathématique.

Peut-être convient-il de dire brièvement comment, à mon point de vue, elle s'enseigne. Le professeur est tenu de voir un "programme". Ce programme est interprété, la plupart du temps, par un "manuel". Ce manuel présente la mathématique sous une forme "élaborée", "structurée". La matière se déroule devant les élèves, au tableau noir, très linéairement, en partant de définitions, débouchant sur des théorèmes et aboutissant, parfois, à des exercices d'application faisant intervenir uniquement la matière que l'on vient d'étudier. Ceci se passe souvent dans une ambiance qui fait participer les meilleurs élèves à l'élaboration des notions par d'astucieuses questions qui demande réponses immédiates. Un tel enseignement crée, chez l'élève, une image tres figee de la mathematique et la calculatrice n'y trouve place que pour traiter certains types d'exercices. Les professeurs sont d'ailleurs unanimes pour admettre que, grâce à la calculatrice ils peuvent aborder avec les élèves des exercices numériques qu'il était impensable de traiter, faute de temps, lorsqu'elles n'existaient pas.

J'ignore quelle est la place faite à la calculatrice dans les programmes de mathématiques dans différents pays. En Belgique francophone, les programmes actuellement en vigueur parlent de la "règle à calcul" et des "tables de logarithmes" mais ne disent rien dês calculatrices.

Un programme expérimental qui touchera l'an prochain la deuxième année du cycle secondaire (12-13 ans) dit ceci:

"Utiliser des machines à calculer pour résoudre des problèmes conduisant à des calculs pratiquement irréalisables à la main".

Les auteurs de ces programmes, eux aussi, percoivent d'abord cet aspect de la calculatrice: celui qui consiste à diminuer l'effort, aspect, somme toute, assez négatif!

La littérature pédagogique qui traite de l'enseignement mathématique abonde en excellents articles quie proposent, pour former les élèves, des attitudes de celles décrites plus haut. Il y est fait une part importante à la réflexion; l'élève y est invité à construire la mathematique et non a la recevoir toute faite. Encore faudrait-il que les messages que nous proposent les spécialistes de l'enseignement passent de la littérature à la pratique de tous les jours, dans les classes.

Comment faire adopter par les maîtres les méthodes résultant de recherches pédagogiques concluantes?

L'introduction de nouveaux programmes qualifiés à l'époque (il y a 20 ans de cela) de "mathématique moderne" avait soulevé chez certains, un espoir de renouvellement pédagogique. Nous savons maintenant que ce ne fut pas souvent le cas et nous en sommes très décus.

Il y a quatre ans, à Karlsruhe, des avis divers sur l'utilisation des calculatrices avaient été presentés lors d'une table ronde. Une majorité voyait, me semble-t-il, dans leur emploi un moyen de renouveler la pédagogie de l'enseignement. Un séminaire, tenu à Luxembourg, avait confirmé ces tendances. Il me semble que le chemin parcouru depuis lors, du moins dans mon pays, est peu important.

Les étudiants dont je m'occupe sont des universitaires qui, en quatre ans d'études, obtiennent un diplôme de licencié en mathématiques. Parallèlement à ces études ils recoivent une formation pédagogique et méthodologique. Une part importante de cette dernière formation consiste en un stage. Le peu de temps dont ils disposent pour cette partie de leur formation les conduit à enseigner une vingtaine d'heures dans des classes. Bien sûr, ce contact avec des élèves est preparé, contrôlé et discuté.

Une preoccupation importante consiste à faire passer ces future maîtres de la mentalité d'enseigné à celle d'enseignant.

Dans cet esprit, j'avais, cette année conseillé à ces futurs enseignants de la mathématique une réflexion personelle sur l'usage qu'ils pourraient faire d'une calculatrice dans des classes et leur avais suggéré de choisir l'un ou l'autre thème du programme. Ils devaient proposer une approche de la notion, principalement l'introduction du concept, en faisant intervenir la calculatrice. Une cinquantaine d'étudiants étaient répartis en quatorze équipes. J'insiste sur le fait que ces étudiants en étaient à leur troisième année d'études universitaires en mathématiques pures ou appliquées et que la formation pédagogique venait en complément. L'activité de travail d'équipe que je leur proposais pouvait, pour certains, se limiter à quelques heures (3 ou 4 h) et pour d'autre prendre beaucoup plus de temps (15 a 20 h) en fonction de leur intérèt pour l'enseignement.

Première remarque: ce travail leur a semblé difficile au départ. Ils manifestaient peu d'enthousiasme et deux équipes ont passé pas mal de temps à rédiger un document dans lequel ils nient la possibilité d'utiliser la calculatrice pour aider à la création d'un concept, lui déniant tout intérêt pédagogique. Tout au plus admettaient-ils que, pour effectuer des calculs, elle est parfois utile, mais qu'il faut s'en servir avec prudence. Une des équipes insiste sur le fait qu'il ne faut pas négliger les tables numériques qui sont importantes, disent-ils, pour traiter les problèmes d'interpolation. Ils ne réalisent donc pas qu'une calculatrice est notamment une table numérique et que l'interpolation (moins nécessaire qu'avec une table) est également possible.

Deux autres équipes traitent de problèmes que l'on rencontre dans le cycle inférieur des humanités (12-15 ans).

- Utilisation d'une calculatrice élémentaire (avec constant) pour l'étude des opérateurs. Dans des graphes fléchés sorte de jeu motivant un calcul: comment passer de n_1 à n_2 en utilisant + a et x b.
- Prise de conscience et découverte du théorème de Pythagore: lecture sur des dessins de longueurs de côtés de triangles et calcul de $a^2 + b^2$ et de c^2.
- Une des équipes ayant dénié tout intérêt aux calculatrices dans l'enseignement fait cependant un essai pour aborder la résolution d'un système d'équations mais avec un grand manque d'imagination. Il est vrai qu'ils voulaient montrer qu'on pouvait se passer de la calculatrice.

Les autres équipes ont signalé ou abordé des thèmes du cycle supérieur (16-18 ans).

- Calcul d'erreurs et erreurs de calcul: ils utilisent les erreurs d'arrondi des calculatrices pour mettre en evidence l'influence de ces erreurs.
- Etude de la notion de distance dans R^2.
- Convergence de suites et de séries et calculs de limites.
- Développements en séries - calcul de π .
- Etudes de coniques.
- Etudes de fonctions.
- Calculs d'intégrales définies.

Ce sont effectivement des thèmes importants et la manière dont ils les abordent est très variée. Pour l'étude de fonctions c'est la possibilité de construire, rapidement, des graphes point par point qui est la plus souvent utilisée. A partir de nombreux graphes ainsi construits on cherche à découvrir des proprietes. Dans l'étude des limites et de la convergence c'est la possiblité d'itérer rapidement (avec une calculatrice scientifique ou même programmable) qui est utilisée. La plupart des étudiants qui ont traité du calcul de π découvraient personnellement les détails historiques des différents procédés utilisés pour approcher ce nombre et ceci leur plaisait assez.

Le travail proposé par les étudiants ne manquait pas d'intérêt; la difficulté est ailleurs: ils ne voient pas comment procéder concrètement dans la classe. Ils ont, pour la plupart, cherché à discuter avec des maitres en fonction et ceux-ci leur ont fait, en général, remarquer que de telles méthodes sont peu compatibles avec le temps dont ils disposent pour traiter le "programme". La seule manière d'arriver au bout, disent-ils, c'est exposer la matière.

Je suis en effet frappé par le fait que, pour un enseignant, "voir son programme" est plus important que de se poser la question: "mes élèves ont-ils compris?" En accordant la priorité à ce deuxième point, ils accepteraient, peut-être, de prendre le temps nécessaire pour faire passer l'essentiel, laissant agir l'élève en l'aidant à apprendre.

Les maîtres doivent être soutenus et motivés dans leur effort de renouvellement pedagogique

- par le "systeme" qui sanctionne les études,
- par les inspecteurs et les autorités pédagogiques,
- par les responsables d'examens d'entrée.

La difficulté que le maître éprouve à équilibrer son enseignement a également été mise en évidence lors de séminaires qui regroupaient, une fois par semaine, une vingtaine de professeurs du secondaire pour discuter des calculatrices.

En effet, s'ils trouvaient plaisir à s'initier personnellement à une utilisation de calculatrice peu d'entre eux envisageaient de traiter de ces problèmes dans leurs classes. Par contre, plusieurs ont créé dans leur école, en dehors du temps prévu à l'horaire, des activités au sein de "clubs calculatrices" fréquentés librement mais assidûment par certain élèves. Bien sûr, ils en arrivent alors rapidement a utiliser des calculatrices programmables et il s'agit avant tout d'un jeu pour lequel les élèves acceptent de maîtriser des notions parfois difficiles, ceci vers 14-15 ans mais rarement avant.

La calculatrice doit, avant tout, être considérée comme une occasion de se poser des problèmes.

5.6 COMPUTERS IN MATHEMATICS TEACHER EDUCATION

MICROCOMPUTERS AND TEACHER EDUCATION

Rosemary Fraser
College of St Mark and St John
Plymouth, England

"The most critical problems hampering the spread of comper use are money and teacher education" (Engel 1976) (1).

In this paper we assume that with the advent of the microcomputer, money for machines is no longer a serious problem but teacher education is a huge challenge which will need ideas, money and organisation.

Teacher education faces two major tasks. The first is to help teachers gain confidence in their ability to handle the technology and the second is to help them to adapt to the changes in the curriculum which the availability of cheap calculating and computer power will promote.

As we can only suggest what these changes may be and as we can only roughly predict what future technology will offer, how can we effect a programme of teacher education that is not out of date before it starts to be useful? Using a new classroom observation system, (SCAN Beeby, Burkhardt, Fraser 1979) (2) we have made detailed records of teachers' lessons and have noted the wide variety of teaching styles and that significant changes in teaching style can occur when a computer program teaching unit is integrated into a class activity. The lessons used a single microcomputer with a large V.D.U. screen and exploratory situations were created both by the program and the teachers.

These experiences led us to suggest that the most vital ingredient in a teacher education program is good flexible material which illustrates the potential of the microcomputer as an aid to teaching.

Before considering how flexible material may be created it is useful to define three major categories of computer use in education.

Categories of Computer Use

Category One - Tutor

The use of the computer as a tutor is often with material that attempts to be independent of the teacher. PLATO (3) and TICCIT (4) are two projects that used this approach and a great deal of experience has been gained since 1960. This experience will be valuable as the microcomputer can now be used to make such packages widely available at a reasonable cost.

Category Two - Pupil

This is using the computer as a problem solving tool i.e. the user teaches the computer by presenting it with an algorithm, to "solve" certain problems. Thus the computer is in the role of a pupil. An example of this approach is the LOGO (5) project.

Category Three - Partnership

Here the computer is in partnership with the teacher and the pupils in the creation of learning environments in the classroom. Our work has its centre in this area.

For all three categories of use, it is at present necessary for the teacher to have some training in setting up the most commonly available microcomputers. Many systems still require the operator to load control programs before loading the actual teaching program and it is essential to know how to do this and often reassuring and helpful to understand the level of the procedures as well. Although this is an interim problem it is very important to cope with it successfully so that teachers with little or no experience of computers have the opportunity to join in the classroom trials of computer material now. It is pointless to wait until all systems are designed so that it is possible just to plug in the teaching module and press the start button! We need to draw on the teaching experience of a wide range of teachers so that the design of the teaching modules for the future can be made to meet their educational objectives.

For teaching programs to be written to support categories one (tutor) and three (partnership) and for the computer to be used as described in category two (pupil), a large number of teachers will actually need to learn to write programs. We believe that in addition to mounting an IN-SERVICE programme of teacher training, all mathematics students on initial teacher training courses for secondary schools and some training to teach in primary schools should be given programming courses as part of their education. These courses should also consider algorithmic approaches in the teaching of mathematics. The computer language taught should be widely available on the microcomputers being used in schools - - at present this means BASIC which may become a lingua franca that is hard to displace.

In the ITMA (6) project we have development teams working together to produce curriculum material that illustrates how a computer program can be used as an integrated part of the lesson. Each curriculum group needs a programmer who is an experienced teacher and a classroom observer who is also an experienced teacher to support and guide the group not only in the design of the material but also in the use of the material.

A Multidimensional Model for Teacher Education

From our experience we suggest a model for teacher education as shown in Figure I. The model is such that it should be possible for each teacher or trainee teacher to find a participating role that suits them. The required involvement may only be the introductory course in operating a microcomputer, followed by the course on using prepared computer-based materials; on the other hand trainee teachers and experienced teachers could work well alongside each other to the mutual benefit, on the development of teaching units using the computer or in the development of the use of such units in the classroom.

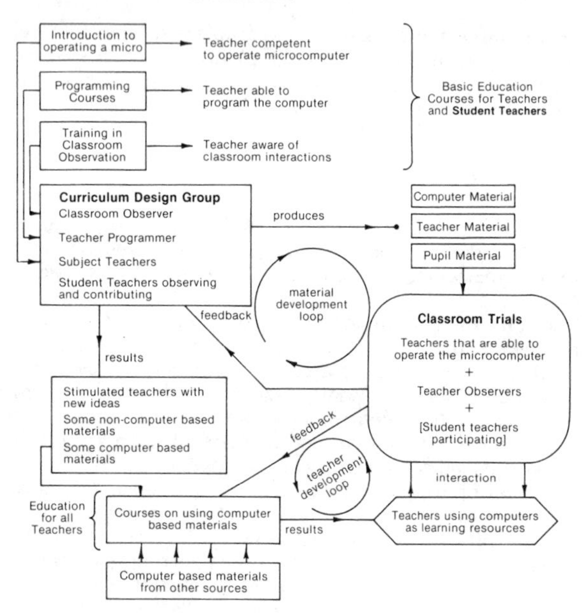

Figure I

The model must be well supported by course and computer materials. The centre of the model shows curriculum development and teacher training as parallel activities with two possible development loops. The top material development loop includes three activities:

1. Design and production of new computer based curriculum material

2. Development and experimentation of the different ways in which the teacher can use such material

3. Curriculum changes through these new experiences

The teacher development loop does not involve the teachers in creating new material but involves activities (2) and (3).

Over the past decade most of the computer based materials that have been produced use the computer in the first two categories i.e. either as tutor or pupil. The possibility of using a microcomputer in active partnership with the teacher, the third category, opens up a considerable range of new activities and presents some very different problems to the program/curriculum designers. We will now concentrate on this area as it has great significance for teacher training courses.

Driving a Teaching Program

For the microcomputer to play an effective role in an interactive discussion session it must offer the teacher a fast and flexible method of control which will provide quick responses to the demands of the situation. The programs must be easy to 'drive' - - learning to drive the teaching program is educationally much more important than learning to switch on the computer. Because the microcomputer offers us powerful methods of display with colour, animation and graphics, we must learn to write materials that will use this power to the classroom. The microcomputer is focussing teachers' attention to the question "What are the pedagogical difficulties in teaching and how can calculators and computers help with these problems" (POLLAK 76) (7).

In our own curriculum development groups many pedagogical ideas are exchanged - - the idea of using a microcomputer stimulates our thoughts and brings us together in a creative environment from which our teaching can be enhanced. We have established that the classroom use of a microcomputer demands that a program must be written such that the teacher is not constrained by long fixed sequences; he must be able to draw in displays, information, problems, etc. relevant to the situations that arise with his pupils. The methods of program control in many previous CAL applications are not suitable for this new use. James and Irland (8) state "the basic strategy (of previous CAL) is to obtain information from the user by a series of questions and answers which enable the user to reply in free format and do not require access to any external written documents". No teacher in the middle of a demonstration will wish to stop to answer a long string of questions posed by a computer program wanting to know what to do next! The teacher will wish to tell the computer what to do next. Thus he must have a clear picture of the program control in his mind and be able to communicate his decisions quickly and efficiently to the program.

Elsewhere a system to build such a picture of the teacher is described (Fraser, Burkhardt and Wells 1979) (9). It is based on the assumption that only two different types of decision are necessary to control the program: one is of the A or B or C type and the other is a menu selection such as

A	D	G
or	or	or
B and/or	E and/or	H
or	or	or
C	F	I

The program is presented to the teacher as being made up of these types of decisions connected by routes that show the flow paths that are possible. As a mnemonic the teacher uses a DRIVECHART which shows clearly at which points he may communicate with the program and indicates to what signals the program will respond and how. Where necessary question/answer sequences may be employed but mainly the teacher is able directly to indicate his decision without being interrogated.

The current DRIVECHARTS are 'externally written documents' to which the user refers - - this may appear a disadvantage but in practice does not prove so. Most school microcomputers only have one V.D.U. available and only a keyboard for communication, thus it is not yet possible to display the DRIVECHART under computer control; it is also questionable as to whether it is desirable to use an expensive medium to display a simple diagram. It is already possible to use the keyboard under computer control such that each message that the user wishes to communicate can be entered by pressing just a single key.

Another feature of DRIVECHART design is that it is carefully constructed so that only a few items need to be remembered at any point. These items aim to be a natural part of the teachers thoughts and the transfer of these thoughts to the computer keyboard is aided by making the choice keys link mnemonically with the program flow. In this way most DRIVECHARTS can be memorised without much effort once the program has been used a few times. As the user becomes acclimatised it is possible to allow for more sophisticated control - - this is a move towards the creation of a more sensitively responding interface.

This digression into the discussion of the design of responsive interfaces highlights the different levels of teacher training needed to help teachers to use the computer to add a new dimension to their teaching. The teacher must be part of the development and not just observers. We must not teach teachers methods that are already out of date.

Changes in the mathematics curriculum will emerge as we learn to present information with a speed and clarity that has never before been possible. Children and teachers will be able continually to explore and investigate. "What would happen if...?" will be the common phrase that indicates most clearly that a change has taken place. We have witnessed children's obvious enjoyment in the experimental stages of this work which is a major encouragement for further development.

The microcomputer serves as a catalyst to our discussions in mathematical education as well as an aid to the teacher in the classroom.

References

1. Engel, A. (1976) The Role of Algorithms and Computers in Teaching Mathematics at School, Proceedings of the Third International Congress on Mathematical Education.
2. Beeby, T., Burkhardt, H. and Fraser, R. (1979) SCAN - a Systematic Classroom Analysis Notation for Mathematics Lessons. Shell Centre for Mathematical Education, Nottingham.
3. PLATO Project. Programmed Logic for Automatic Teaching Operation. Computer-based Education Research Laboratory, University of Illinois, Illinois 61801.
4. TICCIT Project. Time Shared Interactive Computer Controlled Information Television. Mitre Corporation, USA.
5. Papert, S. LOGO Project, Mass. Institute of Technology, USA.
6. ITMA Project. College of St. Mark & St. John, Plymouth, England.
7. Pollak, H. (1976) Panel Discussion: what may in the future Computers and Calculators mean in Mathematical Education. Proceedings on the Third International Congress on Mathematical Education.
8. James, E. B., Irland, D. (1980) Microcomputers as Protective Interfaced in Computing Networks. Imperial College, London University.
9. Fraser, R, Burkhardt, H. and Wells, C. How DRIVECHARTS Work. College of St. Mark & St. John, Plymouth, England.

5.7 THE MATHEMATICAL PREPARATION OF SECONDARY TEACHERS - CONTENT AND METHOD

THE MATHEMATICAL PREPARATION OF SECONDARY TEACHERS - CONTENT AND METHOD

Trevor J. Fletcher
Staff Inspector
Durham, England

The question is:

> What fundamental changes should there be in the preparation in mathematics of a prospective secondary teacher today, in view of the current trends in mathematics, in schools and in society?

We found it necessary to broaden the question. Mathematics continues to grow in scope and complexity and it is applied to an increasing range of social and technological problems. As mathematics advances there is not infrequently feed-back to the more elementary parts of it, and ideas on how to sequence well-established material are sometimes modified. In consequence a teacher needs a general idea of the range of the subject, including new developments, as well as detailed familiarity with a modest area which he (throughout, he denotes he/she) can use with competence. Furthermore, he needs a foundation which he can continue to expand and modify throughout his working life.

In schools we are now trying to teach mathematics (as distinct from arithmetic) to all pupils and not only to a selected few. Teaching the subject to this wider audience presents substantial problems, and in most schools (I would guess) satisfactory solutions to many of these problems have still to be found. (For detailed evidence and a discussion of the problems in England for the age group 14-16 see Aspects of Secondary Education in England, a survey by HM Inspectors of Schools, HMSO 1979, and the accompanying Supplementary Information on Mathematics, HMSO 1980.) Pupils' needs, capabilities and expectations require analysis , and methods of presenting the subject matter require reappraisal. To match the increased range of applications of mathematics pupils at all levels of ability need to meet far more applications of mathematics to problems in other areas of the school curriculum and in the world outside.

In society at large, although mathematics enjoys a certain prestige, there is incomprehension of what it can offer and bewilderment over what schools have been teaching. Mathematicians need to communicate more effectively with their fellow men, and ways have to be found for their knowledge to be applied more perceptively. A much more profound analysis might be made of current trends, but these simple considerations are sufficient to identify more than enough problems for a short talk.

Twenty years ago there was a strong feeling that the content of school mathematics needed changing, and there was a surprising degree of unanimity among mathematicians over the mathematical preparation which secondary teachers needed in order to bring the desired change about. The Royaumont seminar of 1959 and the subsequent publications (Synopses for Modern Secondary School Mathematics, OEEC 1961; New Thinking in School Mathematics, OEEC 1961) presented a clear programme of content-oriented reform. Few people at the time could have predicted how extensive and worldwide the resulting movement was to be. The key was abstract algebra, with arithmetic and geometry formulated compatibly. The programme also included some probabiltiy and statistics - these topics being innovations for many countries concerned. I still think there is much to be said for this programme, but it became widely discredited because it was so often implemented with minimal feeling for the requirements of the classroom and with complete disregard both for the historical purposes of the new ideas and their current applications. Recent publications (see for example Proceedings of the Third International Congress on Mathematical Education, Karlsruhe 1976 and New Trends in Mathematics Teaching IV, Unesco) confirm that the programme retains nothing of its original momentum and the teaching world is looking elsewhere for solutions to its problems.

I have asked many mathematicians, and still more people in the business of mathematical education, the question to which this seminar is addressed. These people were mainly from England - but a number were from other countries. My enquiries were in no way a scientifically controlled survey, and I present no more than personal views formulated after consultation. In general, I found little consensus on desirable changes in content; but I found considerable consensus on changes in approach and I will return to this later.

Thinking only of mathematical content, a majority of my respondents might agree that the two most dominant topics in a course for prospective teachers should be, as they now are in many institutions, analysis (calculus) and linear algebra. A sizeable number say that courses should also contain a strong element of geometry. It is clearly the opinion of many that the reduction of geometry in courses has gone so far that the restoration of some traditional material could properly be called a change, although there is little agreement on what geometry should be put in. Personally, I would advocate a course that included an historical perspective of geometry leading through to an appreciation of Hilbert's axiomatics (without bothering with the finer details). Groups of transformations should be considered, but for a long initiatory part of the course transformations should be a means rather than an end. Important configurations (Pappus, Desargues, etc.) should get due attention, and geometrical constructions should be a major motivation.

There was wide agreement among my respondents that numerical analysis and a computer-oriented approach to problems should be a major component of any course. To think in terms of numbers should be a natural, first approach to problems, but students need to appreciate the limitations of such methods and see the need to validate solutions by critical analysis.

Statistics and probability are important studies and the key ideas of hypothesis testing should be part of every mathematician's general knowledge. Courses should include realistic case studies with the emphasis on data analysis rather than on the applications of routine procedures.

Mathematics is a growing subject and some areas of recent growth should be studied by intending teachers. (These are not necessarily areas of major research interest to University mathematicians.) It is salutary from time to time during a school course for the teacher to ask, "How much of what I have taught in the past year is knowledge which the community acquired in my own lifetime?" In most subjects of the curriculum it would not be acceptable for the answer to be "Nothing". It is acceptable for the answer to be "Nothing" in mathematics, and does this properly reflect the position of mathematics in society? If this is a valid line of thought then the teacher's preparatory course should certainly include some mathematical knowledge developed in his own lifetime. Exactly what may not matter, but there should certainly be some.

A number of areas offer opportunities of this kind; for example, recent advances in combinatorics, coding theory, number theory, graph theory and the design of algorithms. The student should know what these things are, and come to appreciate their flavour through selected examples, perhaps at a "Scientific American" level. There is no time to study all of these topics in any depth, but a choice should be offered of deeper studies in one or two. On one matter there is great agreement. Whatever is studied the applications should be included. Applications motivate and give life.

But the intending teacher requires more than a good knowledge of a number of parts of mathematics. He requires an overview of how the parts fit together; he needs a perspective into which he can fit fresh knowledge as he acquires it. Many of those I questioned thought current courses weak in these respects. Such an overview involves the history of the subject, the philosophical foundations and a knowledge of where the current expanding frontiers are to be found. To see these things as still more separate topics would defeat the very need we should be trying to meet. The history and the foundations of the subject need to be incorporated in courses in such a way that their social relevance and their relevance to the school classroom are clearly seen. The history is to enable today's ideas to be seen more clearly. The foundations might be approached by a critical study of the idea of 'proof' - what are mathematical proofs (in a variety of circumstances) and how does this kind of analysis enable us to explain mathematical ideas to other people?

A number of my respondents advocated mathematical modelling as a component of courses. This means more than a study of the conventional areas of application where the appropriate mathematical formalisations of the elements of the problem are already well established; it means also a study of less well-charted areas where the identification of the issues and the appraisal of the validity of the solutions are all part of the problem. It is certain that potential users of mathematics in business and industry require this as part of their studies, and the intending teacher needs some familiarity with the activity if only to understand better why society employs so many mathematicians. (A course for teachers of children aged 9-13 (approximately) is currently offered by the Open University in England - Mathematics across the Curriculum, Course PME 233. This course involves television programmes, tapes and printed materials.) Whether the teacher will later be able to develop model building skills in his pupils is currently a matter of controversy. Exceptional teachers may be doing a certain amount of successful work of this kind in schools already and where this is so it deserves to be more widely disseminated, but the feasibility of expecting large numbers of teachers to do work of this kind is certainly not proven. Such work is hard to do well, and there is no advantage in doing it badly.

Where my correspondents were most in agreement was in the view that changes in content were not enough, and furthermore they agreed strongly on the changes which they thought desirable. One striking comment came to me from a University teacher (Professor H. Halberstam of Nottingham University) - "I would like teachers to be much more concerned with encouraging use of the richness of language, rather than stunting it with premature and excessive use of symbolism. It seems to me that the mathematics teacher and the English teacher have a common task to show how language can be used to express and to illustrate difficult thoughts."

Students need to gain confidence in their ability to pick up fresh mathematics by their own efforts. Courses should "teach for success" more than they do. As things are there is a tendency for many students to gain no more than a precarious grasp of ideas which are presented at a level as sophisticated as the lecturers can make it. The abstraction and generality of existing courses could be reduced. It would be preferable for students to attempt rather less (in terms of depth and technical difficulties) but to acquire greater confidence with the work they do cover. The intending teacher needs the opportunity to learn (at his own level) by methods which he could later adopt in his teaching. Whatever mathematical content is involved the student should meet the ideas in forms which respect his previous knowledge and his methods of thinking as far as they have evolved. Definition-theorem-proof is a form of mathematical exposition with which students should become familiar, but it is not suitable as the mainstay of the course. The student needs a variety of teaching approaches - problem solving, investigations, work in groups, reading contemporary sources - which he might take as examples for his subsequent work. (A series of conferences at Nottingham, for University teachers and others, has considered matters of this kind. The proceedings of the conferences are obtainable from the Shell Centre for Mathematical Education, Nottingham University, England.)

In my experience there are two very common complaints which young teachers make about their courses of preparation:

1. the course did not help the student to deal with the day to day problems of classroom management.

2. the course was too much concerned with developing the mathematical knowledge of the student, and too little with showing him worthwhile methods of teaching.

Perhaps we cannot tackle the first problem in this seminar, but let us try to contribute to the second.

When we are teaching mathematics to prospective secondary teachers teaching method is not a subject apart. In many training institutions these two components are treated separately, but good mathematics and good methods can be studied simultaneously to the benefit of both. Let us make this our major fundamental change.

5.8 SPECIAL ASSISTANCE FOR THE BEGINNING TEACHER

SPECIAL ASSISTANCE FOR THE BEGINNING TEACHER OF MATHEMATICS AT SCHOOL LEVEL

Edith Biggs
London, England

During the past five years I have been engaged full-time on an in-service project in mathematics. In this research, which took place in fourteen schools in an outer London borough with a high proportion of immigrants from various countries of the British Commonwealth, I tried to assess the relative effects (in terms of changes in the classroom) of providing not only courses of working sessions in mathematics but support in the classroom while teachers were implementing changes in their teaching styles. Many of the teachers involved were in their first or second year of teaching. Some of these were the most responsive of all the 150 teachers concerned in the project.

I must make certain assumptions about the beginning teacher's knowledge of mathematics and her attitude to the subject. I shall assume that she enjoys mathematics and has at least a fair knowledge of the subject - - in other words that she is confident in what she does know. If this is a false assumption, and more often than not it is false, then something needs to be done beforehand in college to bring the new teacher to this relative state of bliss! I shall asssume that you all know ways of deepening and extending pre-service education in mathematics.

Now there are different ways of teaching mathematics successfully; in other words there is a variety of successful teaching styles appropriate to different situations and different children. A judicious mixture of more than one style, formal or informal, would probably be the most stimulating and successful. But the large majority of student teachers lack the carefully graded experiences which the successful use of an active teaching style requires. The first essential is that beginning teachers should have the experience of learning mathematics through investigations themselves, if they have not already done so. They need to sample, at first hand, the advantages of peer exchange - - working discussions with other young teachers as they undertake practical activities. Until they are convinced that mathematial concepts can be acquired much more easily and permanently through carefully planned activities than through instruction and that this way of working is far more exciting and beneficial, they may not be willing to expend the extra effort required to organise their own classes in this way.

Working sessions - - which could be arranged by mathematics consultant, adviser, master-teacher or teacher-tutor - - for beginning teachers from a number of schools (elementary as well as high schools) could be organised to meet these and other requirements. The purposes of all aspects of the programme should eventually be made explicit to the participants. Chief among the objectives is the need to give these teachers experience of an informal teaching style. During these sessions the teachers should work in groups on investigations which can be solved in more than one way. They could compare and appraise the various methods suggested. The following example at child-level illustrates the importance of collecting and discussing a variety of methods of solving a problem.

Eight seven and eight-year olds from the abler half of the class were asked to take a handful of unit squares and, without counting, to make the largest square they could. Three of them made square frames, one with six squares along the edge. I asked how many unit squares he had used. "24" was the prompt answer. "Are you sure?" I asked. Peter counted and found that he had used 20 units. Joanne said, "That's because you counted the corners twice." "So how many units would you need to make a square frame of edge 10 units?" I asked. "36 - - that's four tens subtract four", came the answer. "How else could you find this?" I urged. "Four 4s add 4", was the next suggestion. "How did you get that?" I enquired. The girl placed her squares four along each edge and completed the '6' square by placing a square at each empty corner. Two further suggestions were made: "Two sixes and two fours," was the next, then four fives. The final one was harder to obtain. It was only when I asked the group to find the smallest square frame of all that Lisa said: "3 sets of 3 take away I square, that's 8" "What would that be for a 10 by 10 frame?" I asked. "Ten tens subtract nine ones - - no, it's ten tens take away eight eights," was the final answer. At each suggestion I asked the children to describe their methods in words. For example, the first was given as "Four sets of the number of units in the edge subtract four" We were approaching algebra!

Later on, when a group of 10 and 11-year olds were discussing the smallest square frame, a boy said it was a two by two with a zero middle!

Perhaps the greatest value of comparing methods is that children learn to listen to each other and to appreciate that there may be many interesting ways of arriving at a solution of the same problem.

Sequences of activities to assist concept formation and games devised to help children to memorise number facts should also be included in the working sessions. In playing games teachers should come to realise the importance of playing with a group of children themselves - - to ensure that the rules are kept and that useful strategies are developed. Children, too, should understand the purpose of each game and appeciate that games are to help learning as well as for pleasure. Games sessions should be followed by strenuous questioning to see whether learning has taken place. The following account illustrates the importance of this follow-up.

Brian was a nine-year-old in a remedial group who had not begun to memorise essential number facts. Using a set of playing cards with the court cards removed he was playing a version of 'elevens' Patience. He placed 12 cards face up; when he noticed a pair of cards adding up to eleven he covered these cards with the next two from the pack. He continued, with the over-eager help of the group around him, until the pack was finished. I then asked him to 'pair' piles of cards whose top cards had a sum of 11. That done, I asked him to tell me the pairs of numbers whose sum was 11. Hesiatingly he gave 10 and 1, 1 and 10 and finally 9 and 2. I suggested he should play again. This time he played alone. At the second questioning he managed 10 and 1, 9 and 2 and 6 and 5 - - and their reverses. I left him to continue but asked him to find me at break. By then he had forgotten 6 and 5, but gave 8 and 3. At the dinner break he managed all except 7 and 4 but at the end of the afternoon he gave the whole set although he had not played again. I asked Brian's teacher to continue to question him until he was 'number perfect'. Next day, flushed with excitement because he still remembered the whole set of numbers whose sum was 11, he decided to work on pairs of numbers whose sum was 10, throwing out the 'tens' from the pack. He also invented a variation: if there were no pairs whose sum was 10, he covered three cards adding up to ten. It was not long before he had mastered the addition and subtraction facts he needed to know. For the first time he had made the effort required and been successful. He then asked his teacher if he could help others in the group, promising, "I shall not tell them the pairs. I'll let them find out as I had to." Brian proved to be a good teacher; all the children in that class learnt their addition and subtraction facts.

Another purpose of the working sessions is to encourage teachers to look for good starting points, preferably from everyday life, as an introduction for new topics. Once teachers become accustomed to looking for suitable examples they can persuade their children to take part. For exampe, the pattern of the tiles on my doorstep provided an interesting problem in area. Advertisements, physiology books, current events and journeys afford many other starting points.

There are other advantages of working sessions for beginning teachers: the consultant could observe, in an informal way, which of the teachers require a greater background knowledge of mathematics and which are already developing mathematical 'imagination'. She could also give encouragement for the efforts teachers make and avoid the criticism implied by comments such as, "That's wrong." Later on, she needs to make the purpose of a positive approach to the teaching of mathematics explicit. Children, like adults, require encouragement and appreciation. Teachers should avoid discouraging comments, asking children instead to describe the methods they used in reaching their answers. Furthermore, the contact between elementary and high school teachers is valuable in laying the foundation for a greater understanding of the problems at different phases and for encouraging continuity in the teaching of mathematics.

At the conclusion of the working sessions teachers should be asked to cooperate with the consultant by experimenting with active learning situations with the children they teach. She needs to explain that each teacher who volunteers will receive support in her classroom during the experiment. Such support could be supplied by the consultant herself or by a head of department of mathematics at the school (provided that he is a skilled exponent of active styles of teaching). There are stages of development here, too. If a beginning teacher has not yet worked with a small group of children (four to six) providing them with an activity in problem form, observing what they do and listening to their discussion, she should begin here. Teachers need to convince themselves that children can learn from carefully planned activities and questioning without prior instruction. Moreover, this type of experience gives teachers the opportunity to develop their questioning technique: trying to provide open questions which will stretch able children and smaller steps (more directly questioning) for those who have difficulty. Some teachers will require repeated experiences, particularly in the planning of activities and the sequencing of questions. Sometimes it may not be possible to plan questions in advance as these will depend on the pupils approach to a problem, as the account which follows illustrates. The children were an able group of eleven and twelve-years who had been investigating areas of rectangles with the same perimeter. At the next session they were asked to study triangles with a constant perimeter; loops of string were available. One of two girls working together placed two fingers on the string to steady it, so fixing the base. Keeping the string taut and with a pencil in the loop, the other girl traced an outline with the pencil. "It's a circle", she said. I suggested, perhaps too soon, that they should experiment with different 'bases' to find out whether the tracings really were circles. After many changes of base they decided that the outlines were 'ovals', but that when the base was zero, the pencil did trace a circle. We returned to the original problem of the triangles made as the pencil moved from one position to another. As they experimented the pupils said that, as with the set of rectangles, the smallest area would be zero. "Which triangle has the largest area?" I asked. Moving the pencil once more they said, "At the highest point." "Why is the area greatest at the highest point?" I enquired. David, who had been listening, replied that the area of a triangle was half the base times the height. "How do you know?" I questioned. David drew a rectangle to enclose the triangle, and on the same base. The others agreed that he was right. "Then why is the triangle at the top the greatest in area?" I pressed. After some thought, the youngest girl in the group said that since all the triangles had the same base, the one with the greatest height must have the greatest area. "And it is an isosceles triangle," she added. Because the original problem was open-ended, another group made an entirely different start; they investigated the changing areas of isosceles triangles. The question sequence had to be matched to the reactions of the pupils.

Some teachers soon gain confidence and are ready for the next stage. At this point the consultant should discuss with the teachers concerned the planning of future sessions and subsequently help with the appraisal of these. The teacher need to accept that, initially, group work in her classroom is a joint undertaking. First she should take complete control of one group of children (maximum 8) while the consultant takes responsibility for the rest. Next the teacher should work with two groups, dividing her time between them. Some of the sessions could be introduced by the teacher, some by the consultant. Occasionally the head of department should ask the teachers to help him with his own class so that she can see how he deals with group organisation. Finally, the beginning teacher should take the entire responsibility for all the groups (perhaps 4 or 5) in her class, moving from group to group to observe and discuss the suggestions which have been put forward. It is only when the teacher has the confidence to do this that the consultant should withdraw her help gradually and leave the teacher on her own. Even so, she may still require access to the consultant and to colleagues at the same stage from time to time. She will by then have had experience of working both with the whole class and with groups, introducing new and carefully planned topics to the children. She should be in a position to vary her teaching according to the needs of a particular group of children working on a specific topic. I hope that she will use many mixtures of formal and informal teaching styles, remembering how important it is 'to teach guessing' as advocated by Professor Polya. Above all I hope that she will be able to impart her own enthusiasm for mathematics to the children she teaches.

In conclusion, what objectives should the beginning teacher try to achieve? First of all her children should enjoy mathematics. Secondly, they should be confident in their own ability to achieve success in the subject. This can be furthered by the teacher's own positive approach - - but the children themselves should be helped to master essential facts, including number facts. Thirdly, they should understand that mathematics is much more than number facts. It is fundamentally concerned with pattern. Fourthly, they should be able to talk about what they have done and to listen to their peers and compare the methods they have used. Fifthly, they should appreciate the importance of mathematics in the electronic revolution and that "It has transformed the way we think. Neither radio, television, telephone, satellites, calculators nor computers would be possible were it not for numerous results of 'pure' mathematics." (1) Finally they should be able to use mathematics to deal with situations they encounter outside mathematics lessons, whether in other aspects of the curriculum or in everyday life.

Reference

1. Mathematics Today. Ed. Steen, L.A. Springer Verlag. New York, 1978.

SPECIAL ASSISTANCE FOR THE BEGINNING TEACHER

M. E. Dunkley
Macquarie University
North Ryde, Australia

My comments are of a general nature and apply equally to all teachers, not only to beginning teachers of mathematics. They are based on the assumption that no teacher is a finished product at the end of a three or four year course of professional training. Today, most educators would agree that the neophyte teacher is not a finished product but is only at the readiness stage to begin teaching: the gap between the ideal situation, as studied during pre-service training, and the real situation in the schools has still to be bridged. Rubin (1969) in "A Study on Continuing Education of Teachers" suggests that in-service training is probably more important than pre-service preparation because in the pre-service program the prospective teacher learns about teaching, whilst in the classroom the begnning teacher learns how to teach. Further, Rubin makes the point that a school could and should provide for the professional development of its teachers as teaching weaknesses can now be more readily diagnosed and developmental practices instigated. In general, teacher education is increasingly seen as a continuum of which pre-service and in-service programs form integral related parts.

Undoubtedly, there is considerable variation in the type and extent of the problems and needs of beginning teachers. Since they have different backgrounds and enter schools of different kinds, it is not surprising to find that the difficulties mentioned by newly-qualified teachers show a divergence. If the beginning teachers are to complete their adjustment to teaching, including their socialisation into the role of full-time teachers, then they will need the opportunity to relate the theory they have mastered to the practice in which they are now involved, and to do this in a supportive environment.

A necessary condition for a supportive environment is one in which there is a good two-way communication. Within the Australian context, I'm not convinced that this always exists. As an illustration I would like to contrast two of the findings found and reported by Tisher (1978). A large number of principals involved in this study reported that "beginning teachers display a confidence that their competence does not justify" while a considerable proportion (more than 40%) of beginning teachers feel that they are not fully accepted as members of the school with their own contribution to make.

Before looking at the type of assistance that beginning teachers should be given, I would like to report on the type of assistance beginning teachers in the Australian Capital Territory (ACT) and the Northern Territory (NT) were receiving and their evaluation of this assistance. The table below indicates aspects of support received and its helpfulness.

During this year did you...	Received		Recommend as particularly helpful	
	ACT	NT	ACT	NT
	%	%	%	%
- receive written materials of appointment - e.g. leave, salaries?	71	87	81	81
- receive written materials on school matters e.g. curriculum, rules, duties?	68	63	87	83
- accept advice in class-room management and/or help in producing pro-grams of work?	75	66	87	83
- accept any evaluation of own teaching?	67	64	79	80
- take part in organised consultations with experienced school personnel?	48	46	72	59
- attend group meetings for beginning teachers at school?	19	27	60	61
- attend group meeting for beginning teachers else-where?	35	24	61	53
- observe other teachers' methods of teaching?	52	41	79	79
- visit other schools for observation/ consultation?	28	19	63	57
- confer informally with beginning teachers from other schools?	48	47	74	57
- look at local education resources e.g. teachers' centers, museums, parks, etc.?	63	43	73	70

Educational systems have a responsibility to provide support and assistance to enable beginning reachers to develop towards their maximum potential as quickly as possible.

These support programs should be directed towards meeting the needs and concerns of the beginning teachers. Such support programs would have four aspects:

1. General pastoral support. All senior and experienced staff in the school give this support as opportunities occur. Specific provision is made when, for example, the in-school adviser to a group of probationers, arranges meetings attended by several probationers together during school time. Other teachers, new to the school, will also need this type of support.

2. Classroom support. This support involves advice given to each probationer individually upon detailed content, method and skills, and can also, at times, take the form of direct assistance in the classroom. It is at this level that assistance with lesson preparation becomes a major element in the support program. So also does observation of the probationer's lessons with follow-up advice. It is equally important for the beginning teacher to have opportunities to watch experienced techers at work and the chance to talk with them about their methods.

3. Peer group support. Beginning teachers learn a great deal from their peers, for example the teacher at the desk next to them, other members of their teaching team, or the teacher working in the room next door. There are many things they do not ask supervising officers. It is important to recognise this peer group influence and to provide for it. "Buddy" systems capitalise on this and can be a vital part of the school support program. Efforts can also be made to ensure that probationers have the opportunity to associate with other young teachers with one or two years experience of teaching and who can assist them with practical advice upon every day matters. Peer support can be seen as non-threatening and may be better as an informal arrangement, not a formalised part of the school organisation.

4. External support. This fourth level of support is provided through resources and out-of-school in-service activities arranged by a centralised body. The activities can include general in-service courses which beginning teachers attend in company with other teachers and courses specially provided for beginners. External support should provide such training and experience as cannot normally be provided within the schools.

The specific content of induction programs should reflect the major concerns of beginning teachers. What are the main areas of concern to beginning teachers? For about two out of three beginning teachers, the tasks of teaching groups with wide ability ranges and teaching slow learners are a source of worry. Within the Australian Capital Territory and the Northern Territory, the two territories in which I was involved in studying the induction of beginning teachers, we found the main areas of concern to these teachers were as follows:

	ACT % concerned	ACT % concerned
Teaching slow learners	55	78
Teaching over a wide range of abilities	67	70
Motivating pupils	59	70
Evaluating their own teaching	62	61

In summary, induction programs should be available to all begining teachers as an indespensable step in their continuing professional development and should include:

1. Induction programs conducted mainly within the begining teacher's own school.

2. School-based programs in operation from the time a teacher takes up duty in a school.

3. School-based induction programs which are the responsibility of a specifically designated experienced member of staff in each school who should himself have appropriate special training for this role.

4. School-based induction programs that concentrate on areas of concern as perceived by beginning teachers and senior staff members.

Finally, we should not lose sight of the fact that many new entrants to the teaching profession are of first-rate calibre, well-motivated towards their work and capable of reaching a high level of performance in their first year. In developing an induction program, it is easy to overlook or forget their particular problems and needs and to devote attention to the average or unsuccessful beginning teacher. Only by careful planning can we develop induction programs which identify and satisfy all these varied individual needs and only then will the initial year become meaningful and relevant to every new teacher.

References

Dunkley, M.E. et al., (1978), Report of the Enquiry into the Induction of Beginning Teachers in the ACT and the NT, Teacher Education Program, Macquarie University.

Tisher, R.P., Fyfield, J.A., and Taylor, S.M., Beginning to Teach Vol. I, ERDC Report No. 15 (1978) and Beginning to Teach Vol. II, ERDC Report No. 20, (1979), Australian Government Publishing Service, Canberra.

Rubin, L.J., (1969), A Study on the Continuing Education of Teachers, Santa Barbara: University of California, Center for Co-ordinated Education.

5.9 THE MAKING OF A PROFESSIONAL MATHEMATICS TEACHER

THE MAKING OF A PROFESSIONAL MATHEMATICS TEACHER

Gerald R. Rising
State University of New York
Buffalo, New York

Current Status of U.S. Mathematics Teaching and Teacher Preparation

In the United States today we face the problem of extinction of mathematics teaching tradition. Seven years ago in another paper I estimated our distribution of teaching quality as ten percent excellent, seventy percent pedestrian, and twenty percent unsatisfactory. I believe that our situation is markedly worse today. In this country salary and achievement have come to be too tightly associated and low salaries are attracting few talented students to teaching. In self defense against this and uncaring school boards whose concerns seem always for low tax rates and against teacher and often even student perquisites, our teachers have turned to unionism; and unions have in turn further drained the teacher's energy and interest from the classroom. We have passed through one of the most permissive decades in our history when drugs have reached out from the cities into suburb and countryside to strangle young minds, when television has hypnotized even more adults than children sending millions of brains into a state of Laverne and Shirley induced lethargy, when broken families and working mothers have further reduced an already low level of discipline in the home, and when the schools themselves have turned from educational institutions into entertainment centers in competition with the so-called other media. But worst of all from the point of view of this audience, the faculties of our schools of higher education have fled from their responsibility for the preparation of teachers to what is very loosely designated educational research, an activity whose relation to the climb up the professional career ladder of the so-called researcher is inversely proportional to its relation to classroom realities.

It would be quite unfair to say that none in this country have taken seriously the problems of teacher preparation over the past decade. In fact the National Council of Teachers in Mathematics formed its Commission on Mathematics Teacher Education about ten years ago and a number of states including my own New York have mounted similar if less continuous efforts.

The recommendations of these groups and others in the teacher education community have tended to focus on what are called alternatively competency criteria or performance criteria. Students must reach an acceptable level of performance or competence on each of these specific criteria in order to be certified as teachers.

I have written elsewhere of my opposition to this simplistic approach to teacher training. (See, for example, "Teacher Education and Teacher Training in Perspective," Educational Technology 13, November 1973, 53-59.) That opposition is based on several concerns: (1) teaching is a global task not subject to such atomization, (2) the listing process invites misuses such as infinite lengthening and focus on low level activities to the almost complete exclusion of those more sophisticated and harder to measure, (3) their conservative influence, (4) the illusion that they give of providing a basis for evaluation when in reality they turn our attention away from critical evaluation, and (5) the mechanizing carry-over of the approach into classroom instruction.

But quite frankly my opposition to this movement has softened over the past few years. Although most of my reservations remain, I am willing to concede that this approach to the methods course component of teacher preparation may make a positive contribution to at least movement on my scale from unsatisfactory to pedestrian teaching.

One other component of teacher preparation in our schools of education: If you ask a U.S. classroom teacher to describe his college preparation, he or she will almost without exception descredit all but what we call student teaching, the one to six month internship in a school setting. Here is where the student first confronts the realities of teaching and learns survival skills in the classroom.

Yet it is exactly here that my problem of critical mass in the extinction process applies. The idea that the internship will result in a good new teacher is a xerographic model. It calls for a master cooperating classroom teacher to copy. Not only do we not have enough of these teachers to play this role, but the dynamics of our school-college relationships tend to attract to the role of cooperating teacher the pedestrian classroom teacher or worse.

Characteristics of a Quality Mathematics Classroom Teacher

In the best of all possible worlds we would certainly like to have classroom teachers who know mathematics in considerable depth, who are so well prepared academically that they are comfortable with the role of mathematics in society, who understand students and can communicate with them well, who can and will ably perform the tasks of the classroom - - organizing and teaching lessons, correcting student papers, providing individual assistance, etc. - - who will lead students to the powerful goals of independence and self-esteem, who will provide moral and intellectual models for their students. Quite a bit to ask. When we examine it in this form it seems clearer why we have so few master teachers.

A major problem for an individual teacher in climbing to these laudable heights is the movement of most societies toward increasing individual freedom from constraint. In those older days - - which in this country lasted until almost 1950 - - when the strictures were tighter, when Victorian laws so narrowed our perceptions as well as our actual choices, such expectations were more easily achieved. Today it is a much harder task for any individual - - student or teacher - - to delay gratification, to act altruistically, to make the unselfish sacrifices necessary to reach these humane goals. I note this not as a criticism of democracy to which I remain toatally committed, but only as a recognition of a fact of the modern world. As some things get easier, others get tougher.

What is the lesson in all of this for those with concerns for mathematics teaching and mathematics teacher preparation? I suggest that it is one of scale. I see our task as teachers and teacher trainers not as being or creating optimum teachers, but rather as raising ourselves and our students up the personal scale within which we each function and which is essentially fixed for us by the end of adolescence. Most of us cannot be master teachers; what we should aim for is the best our individual personality allows us to achieve.

What Are Our Problems?

Before I turn to specific recommendations for pre- and in-service teacher education, I list here what I believe are the most serious practical problems we face in the United States.

The Problem of Resources

For us this is clearly the most serious. In the United States we have, for essentially snobbish reasons, distanced ourselved in teacher education from the normal school whose sole goals was to train teachers. This distancing has been more than in renaming our institutions from normal schools to colleges. It has been in resources as well. Today our undergraduate colleges and universities providing teacher preparation programs are most often schools of liberal arts with activities centered around substantive departments. Education has remained or slid in only at the bottom of this university hierarchy well below its predecesssors medicine, engineering, and agriculture. Thus in a modern United States college or university, as in ICME, a mathematics educator must first overcome the negative image of his academic home.

Now any reasonable program of teacher preparation makes heavy demands on the time and energy of faculty; much more, for example, than a content lecture which may be given to a hundred or more students at the same time. Because of resource allocations that are associated most often with the lecture model, individual attention in a low priority activity is almost impossible to achieve. Thus much of our teacher preparation is either in large classes or is assigned to overworked graduate students, to part-time in-service teachers, or as partial load to a mathematics faculty member.

The Problem of Time Allocation

As our priorities have changed, so too has our total program. Today most U.S. colleges no longer offer a major in secondary education and the movement is away from the major in elementary education as well. Students major instead in a content field. Once general degree requirements and the major are fulfilled, what is left for pedagogical training? At my institution one semester, sixteen weeks, which must include not only all education courses but the internship as well. Since the mathematics methods course is only one part of this program, we see students in this setting for a total of less than sixty class hours.

So we in mathematics education are assigned the responsibility of providing pedagogical resources to a prospective teacher in what amounts to less than four percent of the college program. The more general resources of the education faculty - - educational psychology, educational sociology, curriculum and media - - have another ten percent, but that ten percent includes the essentially extracurricular internship.

In recording these data about time in program, I do not wish to communicate the idea that program quality is necessarily positively related to program time. If the college time share devoted to pedagogical preparation were increased from fifteen to thirty percent, the resulting program might be twice as good or twice as bad.

The Problem of Choice

We do have a few hours assigned to us to provide pedagogical background for mathematics teachers. What do we do with them? If the majority of so-called methods texts, the books most often used in these courses, are any indication, we reteach the algorithms of school mathematics. I have not said that we teach how to teach the algorithms; I have said that we teach the algorithms again. Thus elementary teachers, for example, are retaught how to carry (or to use one of my favorite pieces of educational jargon, to decompose) in adding.

The choice comes down to just this: Do we in the small amount of time left to us to teach mathematical pedegogy instead teach more math? In most cases in my country the answer is yes despite the fact that our students have all studied mathematics for ten or more years in school and college before they came to us. To me this choice is unacceptable.

The Problem of the Internship

My experience with prospective teachers has been that before they student teach they are too idealistic and theory oriented to the point of impractibility. During this period I find myself trying to balance this with some of the practical concerns they will face. But once they cross the school threshold and they face their first discipline problem it becomes a much more difficult fight to get them to retain any of their earlier ideals. Far from balancing their views of the real world of the schools, student teaching adds a crushing weight to the practical end of the see-saw. It doesn't just bring the students down to earth, it buries them.

The Problem of Supervision

I have already indicated when considering resource problems, that university supervision is quite often in the hands of a much overworked graduate student. It is quite reasonable, however, to include in the training of a teacher at least the pre-tenure years in the classroom. Should not preparation continue here? Of course it should. But in United States schools almost all of that preparation is self-preparation without any assistance. If a beginning teacher in this country is observed twice in any one year he or she is lucky. Despite a few healthy exceptions, my U.S. classroom teaching colleagues will agree with me when I say that teacher evaluation by administrators is largely a matter of accumulated parent complaints (unevaluated), a count of students sent to the office for disciplinary reasons, and drinking fountain gossip.

The current training of principals and their job perception has led them to focus their attention on everything but staff supervision. A turn-around here would make a major change in the national quality of instruction. That this key figure in the educational hierarchy is lost to us is an unfortunate but well established fact. That our university departments of school administration are contributing to rather than countering this refocus is unconcsionable.

Meanwhile mathematics supervisors and department chairmen are being eliminated in a major share of our schools. The teacher is left to his or her own resourcs.

Recommendations for Pre-Service Programs

We in the United State cannot mount a national response. Instead each college must reassess its resources and muster them to do the best job possible within local constraints. Local impacts can be made in direct proportion to the intelligence and the energy investment of the teacher preparation staff involved.

Mathematics Content Recommendations

1. All teachers should be provided the strongest possible mathematics background. For secondary school teachers this should be a mathematics major (which translates to about a year and a half of full time study of college level mathematical content or about a third of a university undergraduate program).

2. We should stop fooling ourselves into thinking that recommendation (1) carries any value to prospective teachers whatsoever without suitable translation and interpretation.

3. To respond to recommendation (2) what Peter Braunfeld of the University of Illinois has aptly designated pedagogical shadow courses should be introduced. Such courses seek answers to the dual and interactive questions, How should you learn this mathematical content? and How should you teach this mathematical content?

4. We should not attempt to itemize the mathematics content required of all teachers. Instead what we should be after for our teachers is a general understanding of what mathematics and mathematization are about.

5. Instead in what may seem a contradiction of recommendation (4), we should seek to expose prospective teachers to the general concepts of a few basic areas, most notably analysis, algebra, finite mathematics, geometry, and (surprise!) topology. Note that by finite mathematics I mean here the study of algorithms, errors, logic, etc., especially as they relate to computer science.

General Education Recommendations

6. Of course prospective teachers should have the broadest and deepest possible exposure to the humane tradition through literature and philosophy, to the human condition through psychology and sociology, to civilization through history and political science.

7. I also recommend as much science as possible should be taught.

8. Teacher educators should identify prospective teachers early in order to provide them with all the assistance they can in the never ending battle for the best course instructors.

Mathematics Pedagogy Recommendations

9. Pedagogical preparation should be viewed as a global experience extending over at least seven years, the fourth years of undergraduate studies and the pre-tenure years in the school classroom.

10. Mathematics teaching candidates should be identified early and provided appropriate counsel, coursework, and exposure to and even possibly early participation in the best area schools

11. The program of mathematical pedagogy should be closely related to both the content training (as in the recommended shadow courses) and general pedagogy. We find, for example, that it is helpful to our students to interpret, "make sense of", and occasionally even contradict the ideas they meet in their educational psychology and sociology courses.

12. As I recommended at some length in an earlier paper ("Minimum Content for a one Semester Mathematics Methods Course" in John A. Shumaker and Meredith W. Potter, Designing Methods Courses for Secondary School Mathematics Teachers, ERIC, 1977, 34-49, available from ERIC SMEE Clearinghouse, 1200 Chambers Road, Columbus, OH 43212 USA) the methods course organization and approach of my colleague at Buffalo, Stephen Brown, should be employed. The focus of this course is on the learning process and its cognitive and affective components. But into this process framework are then placed products and techniques of school mathematics instruction. I consider this my most important recommendation.

13. The methods course should include opportunities for real (with school students) or pseudo (with peers) microteaching opportunites which are discussed in detail and if possible retaught in new settings.

14. The methods class should overlap with the internship, if possible spanning before, during, and after that practical experience.

15. The internship should be intimately related to the methods course. The school cooperating teacher should be a program disciple, whenever possible a person who has not only taken the course but also helped to teach it.

16. The internship should be preceded by as much clinical practice such as tutoring and teaching small groups as possible.

17. Opportunities to share and compare experience should be provided not only with supervisors and mathematics education staff but also with other student teachers. Under skillful guidance such seminars encourage these beginners to think of their own solutions to problems and to generate new and highly personal teaching techniques.

18. Insofar as it is possible your goals and techniques should be sold to the principal of the school in which interns serve so that he can play a supportive and non-contradictory role

Recommendations for In-Service Programs

In New York and many other states in this country a fifth year of college preparation or a master's degree is required for permanent certification as a teacher. Some states also or alternatively require continuing accumulation of graduate coursework or within school in-service credits. These programs provide further opportunities for university faculties to influence classroom teachers. In this regard the fact that classroom teachers have organized teacher centers in which they serve themselves speaks directly to their misunderstanding, distrust, and outright rejection of what schools of education have to offer. We are rightly, I believe, placed on the defensive, but at the same time our teacher center activities have little more claim to being an answer to our problems than have some of our misguided graduate courses. Sharing dittos is not the avenue to teacher professionalism.

Some of the pre-service recommendations apply here with equal force, most especially (9) the need to consider teacher preparation as a global experience. Here are a few more:

19. Despite the strong pull to respond to practical classroom problems in these courses, graduate coursework should move rapidly to general constructs and deeper thinking. This does not mean that the practical problems should be dismissed; quite the contrary they may be used variously to build the higher conceptualizations or a examplars of these contructs.

20. Students who first enter your program at the graduate level should be provided a replication of most of the methods experience so that they "speak the same language" as other graduate students.

21. All graduate students and especially those who will serve as school cooperating teachers should participate in a practicum in which they assist in teaching the undergraduate methods course

22. At least part of the graduate program of an in-service teacher should be cooperatively arranged between college and school.

23. In-service courses are often mounted to answer specific needs like providing background for teaching transformational geometry or for using calculators in the instructional program. While a major focus of these courses may be on content, they provide an excellent opportunity to address pedagogical problems in a new setting. These opportunities should not be missed.

24. No part of the educational program for teachers through the master's degree should address educational research in its current form. I detect only zero or even negative effects on classroom instruction from the carry-over of such courses.

25. Curriculum development should not be identified with the classroom teacher's role. Their job is something quite different: curriculum interpretation.

What of the Future?

I have made it clear that my prognosis for the future of mathematics teaching in the United States is not a happy one. Despite this I close on a hopeful note. Each teacher should seek to create a strand of professionalism in his or her clasroom by doing the best job possible in the face of all the many problems, seeking out in the process help wherever it may be found. And each mathematics educator should start in a small way to create an island of quality teaching whose area is ever expanding. From just such small beginnings we can begin to recreate locally a mathematics teaching tradition.

Count yourselves lucky, those of you from other countries whose classrooms are not so ill served. But beware. My observation is that you criticize us, even deride us, but despite this too soon you copy our worst ways. My characterization, recommendations, and prognosis have been for a country whose educational system is deeply troubled. Since that characterization could apply to you in the near future, you would do well to seek now the means to head off our predicament.

(Copies of the full text of this Bowdlerized paper are available from the author at 580 Baldy Hall, SUNY Buffalo, Buffalo, NY 14260.)

THE MAKING OF A PROFESSIONAL TEACHER

A. G. Howson
University of Southampton
Southampton, England

I am reminded of an article I read recently:

> "Perhaps the most decisive criterion we can take of the real mind of society on the subject of education, is the estimation in which the educators are held. If their work is really honoured, they will themselves be honoured for its sake....."
>
> "Now, judging by this test, we are compelled to conclude that the general estimate of the value of education is still very low among us. The profession of educator is not honoured... It is not generally felt, that those who are devoted to the office are the greatest benefactors of society. Their labours are ill remunerated, and often grudgingly. Many of them could earn higher wages by a handicraft trade; and most of them would have thriven better in the world than they do, if they had applied the same amount of mental power and activity to the details of business.... They are not socially recognised as equals, by those to whom they may be, intellectually and morally, far superior. The very parents of the children upon whom they are conferring, we may say, an intellectual existence, consider themselves, generally speaking, as the benefactors and patrons of the teachers, rather than as benefited by them in the persons of their children."
>
> "Symptoms such as these are strong evidence, that the worth of education is not so thoroughly and heartily appreciated as might, to a superficial view, appear to be the case; and the want of a deeper sense of its dignity and value is the greatest obstacle conceivable to the real progress of education itself."

The passage was written in 1839 and comes from an essay on 'The expediency and means of elevating the profession of the educator in public estimation'. The author, the Reverend Edward Higginson, saw two main causes for the low esteem in which educators were held. First, the attitude of the general public towards education and secondly the 'inefficiency of the educators'. As he wrote, 'It is to be admitted and regretted most deeply, that the actual state of the educational profession, if it does not justify contempt, at least requires great internal improvement'.

It is possible when considering this quotation alongside Professor Rising's address to be filled with despair. Here we are, one hundred and forty years later, saying the same things. Undue pessimism would, however, be as out of place as witless optimism. Enormous progress has been made in the past century - - but expectations have advanced even further. I suspect, however, that some of the excellent teachers who taught me would be far less successful in the present day comprehensive school where the demands made on them would be much greater, even, perhaps, excessive. Teaching is now, in England at any rate, a vastly more complex and demanding task than formerly. Exactly what it is reasonable to expect of a professional teacher is a point to which I shall return. Meanwhile, I should like us to look again at the way in which the general public views our work.

Teachers are inefficient. We spend a considerable part of our time setting examinations, the results of which demonstrate our inefficiency. Just to ensure that the point is not lost on the general public, national surveys, professional institutes, industrialists and individual authors write to tell us the 'Johnny can't add' or, to quote a recent example, only 21 per cent of 11 year olds in England and Wales could order in size, smallest first, the decimals 0.7, 0.23 and 0.1.

What the press and others involved forget - - or more likely never even knew - - was that forty years ago such work was for the few. The 1938 government report on education in grammar schools (attended by less than 21 per cent of the country's children) set out as the first objective of grammar school mathematics the consolidation of such knowledge of decimal fractions.

This, of course, is not to deny that we as teachers must increase our efficiency. We must, however, insist on being judged according to realistic standards.

As a body we have done little to give the public a true impression of what we can and cannot do. Teacher professionalism has become associated more with union wrangles, strikes and discontent, than with argued statements about aims and possibilities and with raising the profession in the general public's esteem. Professionalism depends not only upon the teacher, it also lies in the eyes of the beholder. In England a government committee is currently inquiring into school mathematics. One hopes that in its report it will suggest not only how teachers can become more successful at their job, but also how the public can be educated better to appreciate teachers' problems and more fairly to judge and to value their work. Mention of the Cockcroft Committee reminds me that in this short time available to me I must draw to your attention recent developments in England which should help in the provision of better teachers. There are three enterprises I should particularly like to mention - - each concerned with a different sector of teacher training. I shall not offer detailed descriptions, for representatives of all three projects are present here at Berkeley and are prepared to discuss their work with anyone interested.

The Mathematics Teacher Education Project has prepared materials - - based in the main on possible student activities - - for use in the preservice training of teachers: these are on view at the Congress.

The Mathematical Association has recently insituted a Diploma in Mathematical Education intended primarily for teachers in the 7-13 age range. To obtain the diploma the serving teacher must attend a recognised institution - - usually, but not always, a college of education - - for part-time education (most often in the evenings) for a period varying from fifteen months to two years. The teacher will follow a course specially devised by the institution concerned but which conforms to guidelines laid down by the Association and which has the Association's approval. Assessment is moderated by individuals nominated by the institutions and appointed by the Association. On succesful completion of the course the teachers obtain diplomas which will not only help them gain promotion, but in some cases will immediately confer an increased salary.

The third initiative is sponsored by the School Mathematics Project and consists of in-service training materials designed for use within schools. I was disturbed to hear that 'department chairmen' were disappearing in the U.S.A. The role of Chairman or Head of Department in raising the quality of teaching in our schools would seem to be a key one - - I cannot envisage how countries in which such a post does not exist can conceivably achieve large-scale improvements. The SMP through its packages provides the Head of Department with a resource which can be used to improve the level of mathematics teaching within that particular school.

Planned in-school, in-service training is still fairly novel - - although the unplanned, unstructured kind has for many years been the most potent way of increasing the professionalism of the teacher. It is a form of professional education which demands more attention. Unfortunately, at the moment it does not provide 'tangible' incentives - - there is no diploma at the end, no letters to put after one's name.

I now wish to return to the question of what it is reasonable to demand of a teacher - - what does professionalism entail? In England the teacher is allowed more freedom than in most other countries. Yet greater freedom, also means greater responsibilities. Recently, one has begun to wonder whether the freedom which the better teacher has rightly earned and which he intelligently uses, has not placed too great a burden on weaker colleagues.

The traditional British 1930s arithmetic or algebra textbook usually consisted of a collection of summaries followed by hosts of carefully graded exercises. They provided a framework for teachers, but left it to them to determine how to introduce and motivate the topic. The teacher was assumed (perhaps incorrectly) to be capable of doing this and in that sense was a "developer of the curriculum". Recently, there have been great changes. Certain teachers banded together to devise completely new curricula, thus becoming 'developers' in a wider sense. Other teachers felt it necessary to imitate these innovators. Additional pressures on them to do so arose following the establishment of Mode 3 CSE examinations which encouraged schools to devise their own syllabuses. Yet this movement towards a greater taking of responsibility was in some ways counter-balanced. For a variety of reasons, the textbook style changed. New textbooks frequently attempted to inform both pupil and teacher; and to influence, and usually to change, styles of teaching. Weaker teachers lacking sufficient knowledge and confidence to set the book temporarily aside, were now more inhibited: their influence on the curriculum lessened rather than increased.

The teachers who wished to adopt the Nuffield Mathematics Primary Project or the secondary school Mathematics for the Majority Project had not even textbooks to fall back on. They, with the help of guides and paradigm lessons, had truly to develop a mathematical curriculum - - often simultaneously with ones in several other subjects. It was an impossible burden to place on any teacher. Recently, teachers have been offered more limited and almost certainly more profitable ways for involvement in curriculum development. For example, groups of primary teachers in many areas have come together to draw up guidelines which they and their colleagues might follow in their mathematics teaching. Here we have 'professional' involvement on what would seem a more appropriate scale.

Does it still make sense then to speak of "the teacher as a curriculum developer" or "the teacher as an educational researcher"? As we have observed, there are certain tasks which teachers cannot fairly be asked to undertake (without enormous changes in the time they are allowed for preparation and in-service education). Yet, what are the alternatives?

In the USA, it would seem, the teacher is seen as a consumer of curriculum materials which are carefully constructed away from the classroom. In such a climate the production of would-be 'teacher-free' packages is natural. We have seen, however, the limitations - - practical and educational - - of such an approach.

These lead us inescapably to the view that the teacher must play a part in determining the curriculum of his class - - in making those minor adjustments that are necessary to cope with individual personalities, with local interests and conditions, etc. The teacher who says "tell me what to teach and then I'll do it" is the one who can be safely and economically exchanged for minicomputers plus software.

Yet making such adjustments is not to be confused with preparing a whole course from scratch. If one has such large-scale developments in view, then it is presumptuous to ask for the "teacher as a curriculum developer", and problems have probably resulted as a result of such calls. Surely, what must be regarded as our goal is that teachers should be prepared (in both senses - - ready and equipped) to exercise choice and to make decisions from a position of knowledge and self-confidence.

Whilst still discussing aspects of professionalism, may I mention one which seems to have been given little attention. Earlier, I referred to things which teachers cannot and shold not be expected to do: large-scale curriculum development is one; ensureing that every child can achieve all desirable goals is another. It is a mark of a professional lawyer or doctor that he is prepared to say that a case cannot be defended, or that he cannot provide a cure for certain conditions. Yet teachers recently would seem to have accepted impossible tasks, and have aroused ridiculous expectations - - here mathematics teachers are not alone; one also thinks, for example, of what has happened in England in the field of modern language teaching. Such over-ambitiousness leads not only to disappointment, but tends to divert us from that we can do well. Let us always remember that the doctor who pretends that he has the cure for all complaints is regarded not as a professional, but as a quack.

Finally, and more contentiously, may I turn to another point - - that of the part of pre-service training that U.S. students value most is their period of practice teaching in schools. I am certain that this would be the response of many newly qualified teachers. It is but one reason which makes me wonder if the whole of professional education needs restructuring. The present system has arisen over many decades and was designed to answer demands that have now changed. In England, for example, the postgraduate year of professional training was originally designed to produce teachers in grammar schools. Now our students may elect, after their training, to go into 11-18 grammar schools, 11-16 comprehensives, 11-18 comprehensives, or what in effect are 16-18 comprehensives - - to mention but four possibilities. The mathematics tutor has approximately 100 hours with a group of such students. Can a teacher trainer do a 'professional' job of preparation in such a time? Can any quantity of materials such as those of MTEP to which I referred above do more than ameliorate an unsatisfactory state of affairs?

Perhaps, a restructured professional education could commence with a two-year school-based first state of which the major part, but not the whole, should be concerned with the gaining of practical competence and confidence. This would be followed three or so years later by a more theoretical course of study, possibly on a part-time basis, aimed at widening the young teacher's horizons. Such a structure would also help to avoid the present teacher/teacher-trainer dichotomy and could provide outstanding heads of departments with extra promotion prospects within schools which did not automatically mean their giving up mathematics for administration. In the coming years the effects of population changes on many educational systems will be great, and the pressure to produce new teachers will be considerably eased. It would seem, then, to be an excellent time to conduct limited experiments with new structures of pre-service education. I hope that there will be some countries who will be prepared to experiment with alternatives to the present patterns.

5.10 THE DILEMMA OF TEACHERS BETWEEN TEACHING WHAT THEY LIKE AND TEACHING WHAT THE PUPILS NEED TO KNOW: HOW MUCH FREEDOM SHOULD TEACHERS HAVE TO ADD MATERIALS, HOW MUCH MATERIAL, WHICH TEACHERS?

THE CONTENT OF ELEMENTARY SCHOOL MATHEMATICS IN THE UNITED STATES

Andrew C. Porter
Michigan State University
East Lansing, Michigan, U.S.A.

In 1963 the International Association for the Evaluation of Educational Achievement (IEA) conducted a survey of mathematics achievement in 12 countries (Husen, 1967). The IEA is now in the process of conducting a second international survey of mathematics achievement. As was true for the earlier survey, it seems almost certain that the second survey will reveal several substantial inter-country differences in mathematics achievement. These differences in achievement will lead to speculations about their origins. One logical explanation is that the content of instruction provided students varies from country to country and that these differences in what is taught account for much of the differences in achievement. This speculation, in turn, leads to questions about how the content of mathematics instruction is decided and the extent to which the content of instruction is common across students even within a country.

The Content Determinants project within the Institute for Research on Teaching at Michigan State University is conducting a series of studies to describe the mathematics content decisions of elementary school teachers in the United States (Porter et al., 1979). The purpose of this paper is to provide an overview of the findings thus far from that work.

Our research can be viewed in terms of three sources of influence on teacher decision making: (1) a formal hierarchical component which transmits the policy decisions of higher authorities and which, therefore, reflects political processes bearing on a school system as a whole; (2) other influences from inside or outside the agency which are brought directly to bear on the teacher; and (3) the teacher's own conceptions of what outputs are desirable and feasible. Our eventual aim is to estimate the strength of these influences on teachers using measures of content as the dependent variable.

Factors Teachers Say Influence their Content Decisions

Our first study investigated teacher perceptions of the effects of six sources of pressure to change the content of fourth-grade mathematics (Floden, et al., 1978). Our approach, following other research on human judgement (Slovic & Licktenstein, 1973), used written descriptions of hypothetical schools. The factors considered in these descriptions included pressures from parents, teachers, and the school principal as well as district instructional objective, textbooks supplied to the teacher, and test results reported in the local newspaper. By systematically varying the presence or absence of these factors, sixty-three vignettes were created. In each case, the pressure advocated the addition of five new topics and provided no support for the teaching of five topics that the teacher ordinarly covered. The pressures were always consistent in that the topics supported by each source were identical. After reading a vignette, teachers were asked, once in terms of core topics and once in terms of peripheral topics, whether they would teach the five new topics and whether they would continue to teach the five old topics.

Sixty-six teachers were recruited for the study from five diverse Michigan cities and surrounding areas. The most notable aspect of their response to the vignettes was an expressed willingness to change instructional content, whatever the pressure for change. Differences between pressures in degree of expressed acquiescence were generally slight. The most endorsed pressure was not textbooks, but district objectives. For example, when we looked at the core topic question about whether the teacher would add five topics usually covered in other school systems, the mean response for vignettes with each of the following pressures standing alone was as follows: objectives, 1.67; tests, 2.06; textbooks, 2.47; principal, 2.49; other teachers, 2.52; and parents, 2.73, where a response of 1 is "virtually certain to teach these topics," 2 "fairly certain to teach these topics," and 3 "more inclined to teach these topics than not."

When we took the same question and looked at the means for all the vignettes with one source of pressure, two sources, etc., as the number of pressures increased, the teachers' expressed willingness to change approached virtual certainty: one-pressure vignettes, 2.35; two-pressure vignettes, 1.75; three-pressure vignettes, 1.59; four-pressure vignettes, 1.39; five-pressure vignettes, 1.35; and six-pressure vignettes, 1.39.

Content of Textbooks and Tests

While the policy capturing results may not accurately reflect what teachers do, since they are based on self-report data, they do suggest a certain lack of teacher autonomy in content selection. If teachers do look to others for advice on what should be taught, it is also important to understand whether the advice received is generally consistent. Put another way, to what extent is there a standard content for elementary school mathematics in the United States?

One approach to this question is through examination of the mathematics materials provided to teachers. Since textbooks and standardized tests represent a large part of these materials, we have subjected several of each to careful content analyses.

Our content analyses of textbooks and tests were guided by the classification manual that describes a three dimensional taxonomy developed for this purpose (Kuhs, et al., 1979). The three dimensins of the taxonomy describe: the general intent of the item (e.g., conceptual understanding or application), the nature of material presented to students (e.g., measurement of decimals), and the operation the student must perform (e.g., estimate or multiply).

Four widely used standardized tests and three commonly adopted textbooks were selected for these comparisons. The tests were:

MAT - Metropolitan Achievement Tests (Elementary Level/Grades 3.5 - 4.9), Harcourt Brace Javonovich, Inc., 1978;

SAT - Stanford Achievement Test (Intermediate Level/Grades 4.5 - 5.6), Harcourt Brace Javonovich, Inc., 1973;

CTBS - Comprehensive Tests of Basic Skills (Level I/Grades 2.5 - 4.9 and Level II/Grades 4.5 - 6.9), McGraw-Hill, 1976;

IOWA - Iowa Test of Basic Skills (Level 10/Grade 4), Houghton-Mifflin, 1978.

The three textbooks of fourth grade mathematics that were selected are:

Mathematics in Our World, Addison-Wesley Publishing Co., 1978;

Mathematics, Houghton-Mifflin Co., 1978;

Mathematics Around Us, Scott-Foresman and Co., 1978.

Content analyses of textbooks were limited to the student exercises.

The five tests were reasonably consistent in their treatment of whole number computational skills, but there were striking differences in their treatment of such general topics as applications involving graphs or tables and number sentences. Further, there were only nine specific topics that were common to all five tests (porter, Kuhs, & Freeman, 1979). All three books reflected the current "back to basics" and "minimal competency trends" in their common emphasis on computational skills. However, more than one-half of the specific topics that were covered across the three books were unique to the book in which they appeared (Freeman, Kuhs, Knappen, & Porter, 1980).

From these analyses, there is strong reason to discount the possibility of a common content of mathematics instruction for elementary students in the United States. But what about the consistency of content messages that teachers receive from the particular textbook and test with which they work? Is there the possibility that these messages are consistent, as in our policy capturing study, or are teachers necessarily confronted with content choices between textbooks and tests?

To answer these question, pairs of textbooks and tests were compared both by describing the percent of tested topics that are covered in a textbook and be describing the percent of topics in a book that are also covered on a test. For the textbooks and tests analyzed, the percentages of test topics in a book ranged from 52 to 74 percent while the percentages of textbook topics in a test ranged from 15 to 45 percent. Regardless of the particular test and textbook a teacher uses, a large percentage of the content in one is not considered in the other. Thus, teachers are forced to make some further choices.

Seven Case Studies

We have just finished data collection on case studies of 7 teachers in 6 elementary schools in 3 districts. Each teacher was studied for a full school year to document, through teacher logs, the content of their mathematics instruction, to describe the advice they received on what to teach, and to assess their reactions to that advice. The school districts were selected to vary on the extent to which they attempted to standardize mathematics content. In each district, one school was selected where teachers functioned in a fairly isolated fashion (e.g., self-contained classrooms) and one school was selected for the opposite (e.g., open spacing, team teaching, instructional aides). The selection of schools represents our hypothesis that teachers may be more responsive to external pressure for particular content when they work independently of each other. In one school, we studied two teachers in order to get some idea about content variance within a school.

While data analysis is far from complete, we already know that the seven teachers varied widely in the content of their mathematics instruction even when they taught at the same grade level in the same district. For example, the content of one teacher was exactly as prescribed by the district's instructional management system. Another teacher in the same district ran two parallel mathematics curricula, one of which was based on the management system and the other on a textbook. At this point it would appear that each of the seven teachers arrived at their content decisions in somewhat different ways and that the content of their mathematics instruction differed. The extent to which these differences were reflected in student achievement has not yet been analyzed, though at this point the relationship seems almost tautological.

The implications of our research for mathematics education are still emerging. Only when we know how teachers make content decisions, what external factors they take into account, and what the consequences are for content covered, will we be in a position to be prescriptive. If teachers have little freedom, we might want to change their school system, not the teacher. If they have much freedom, we might want to help teachers use this freedom wisely or we might restrict freedom used with harmful consequences for students. We might also make teachers more knowledgeable about the external factors to be taken into account to give them more skill as political brokers. Teachers already have the freedom to refuse to respond to external factors, but can they adequately judge the benefits and costs of this refusal?

References

Floden, R.E., Porter, A., Schmidt, W., Freeman, D., & Schwille, J. A policy-capturing study of teacher decisions about content. Paper presented at the meetings of the American Educational Research Association, Toronto, 1978.

Freeman, D.J., Kuhs, T.M., Knappen, L., & Porter, A.C. A closer look at standardized tests. The Arithmetic Teacher (in press).

Husen, T. International study of achievement in mathematics, vols. I and 2, Stockholm: Almqvist and Wiksell, and New York: John Wiley, 1967.

Kuhs, T., Schmitd, W., Porter, A., Floden, R., Freeman, D., & Schwille, J. Taxonomy for the classification of elementary schools mathematics content. (Res. Ser. No. 4). East Lansing, Michigan: Institute for Research on Teaching, Michigan State University, 1979.

Porter, A.C., Schwille, J.R., Floden, R.E., Freeman, D.J., Knappen, L.B., Kuhs, T.M., Schmidt, W.H. Teacher autonomy and the control of content taught (Res. Ser. No. 24) East Lansing, Michigan: Institute for Research on Teaching, Michigan State University, 1979.

Porter, A.C., Kuhs, T.M., & Freeman, D.J. Is there a core curriculum in elementary school mathematics? Invited paper presented at the Textbook in American Education Conference co-sponsored by the Center for the Book in the Library of Congress and the U.S. National Institute of Education, May 2-3, 1979. (The paper will be published in a book of the proceedings.)

Slovic, P., & Lichtenstein, S. Comparison of Bayesian and regression approaches to the study of information processing in judgement. In L. Rappaport and D.A. Summers (Eds.), Human judgement and social interaction. New York: Holt, Rinehart and Winston, 1973.

5.11 INTEGRATION OF MATHEMATICAL AND PEDAGOGICAL CONTENT IN-SERVICE TEACHER EDUCATION: SUCCESSFUL AND UNSUCCESSFUL ATTEMPTS

THE INTEGRATION OF MATHEMATICAL AND PEDAGOGICAL CONTENT IN THE IN-SERVICE EDUCATION: SUCCESSFUL AND UNSUCCESSFUL ATTEMPTS

David A. Sturgess
University of Nottingham
Nottingham, England

Preamble: The Aims of In-Service Activities

I think most people would agree that all in-service activities have two main aims:

1. to change the content of the mathematics being taught in schools

2. to change the teaching methods used by teachers.

What I am going to propose to you is that it is these very aims that have caused many in-service ventures to be ineffective because one or other of them dominated to the extent of taking over the course.

For instance one author (1) who discusses various modes of in-service work, talks of them all in terms of innovation; another (2) who is highly critical of present structures for in-service work states that only training of the teachers in their academic subject will improve matters. I want to propose another aim, which I am going to label the "professional growth" of the teacher, which succeeds in integrating mathematical and pedagogical content by not attempting to do either of them. I have chosen the word "professional" because I believe that we have to treat teachers as professionals capable of solving their own problems and not merely as employees who need to be told what to do. I have chosen the word "growth" to indicate that everyone has to start where they are and gradually grow into something different. Teachers are not empty vessels to be filled with knowledge by "experts", nor are they straws to be blown in the wind of each fashion, nor is it necessary for more than a few in-service activities to be innovatory.

I am suggesting the concept of professional growth as a way forward that avoids the Scylla and Charybdis of mathematical or pedagogical content by sailing through the middle of them.

The Failures of Past In-Service Activities

Let me look first at why I think the aims I stated first have led to activities that have not succeeded.

Projects worldwide have set out to change or modernize the content of the mathematics being taught in schools. The main model for a project has been to produce text books for children which include the new material and to give various amounts of training for teachers in the mathematics associated with the new material. In England projects have succeeded in a wholesale change in the content of mathematics taught in schools, but failed to produce a similar change in the way in which things were taught. Teachers now teach sets instead of fractions, but teach them in the same didactic way that they had taught fractions, in spite of the attempts of the authors to promote ways of teaching that involved the conceptual understanding of the topic. Such an approach is based on the curriculum innovation model of in-service educatin which talks of curriculum as being an "agent of change" (3, 1).

The second of the two aims is less often the main aim of a project but is certainly an aim of all in-service work, and this is to change the ways in which teachers teach. Now it is perfectly possible to change content, if you can decide what it is you want to change, all you have to do is to write texts for children, write books for teachers, change the public examination syllabuses, and the content has been changed. The assumption that many in-service activities and projects seem to have made is that you can change teachers in the same way,and this is demonstrably false. I have only to ask you to examine your own experience and try to think of any case where someone has changed you without your cooperating in the change. One fundamental piece of knowledge that has been ignored in the writing of in-service activities that I have found is that it is not possible to change anyone else, people can only change themselves. All we can do is to create an environment in which change is possible.

The Initial Training Model

In most countries in the world in-service education is a relatively new venture. In England large scale in-service activities only started in the late 1960's and in many countries it is more recent than this. There were no bodies of 'good practice' established and when in-service training first started nearly all the methods of working, ways of designing courses, and content were based on initial training methods. And yet initial and in-service are so different that it is a great mistake to use one as the model for the other.

The Importance of Experience

The factor that creates this great difference is that on an in-service course you have teachers with experience. They will have different kinds of experience depending on how long they have taught, what ages they have taught, in what kinds of school they have taught, but they all have experience in teaching. This experience is absolutely priceless. Any of you who may be engaged in initial training of teachers will know how impossible it is to convey to students who have little or no knowledge of the actual teaching situation what it is like to work with a class of children.

The "Professional Growth" Model

It is this very important factor which is my main reason for putting before you a professional growth approach to the planning of in-service activities. We have positively to make use of this experience so that it becomes, not an incidental factor, but the absolutely key factor in our design and planning.

Not all experience is beneficial, we all have had experiences that we would rather be without; some experiences may confirm mistakes, for instance a teacher who is teaching didactically very able children, who are able to make generalisations very easily, may not believe that this is a skill that other abilities of children do not possess. But this is the experience those teachers have had, this is what we must work with, because it is what is there and it must be used positively and developed. It is no good telling someone that the conclusions their experience has led them to are wrong, you have to give them more experiences that enable them to see that they are wrong.

It is providing the machinery for developing courses from what teachers already have experienced and know, rather than assuming that they have to be told what they should change to, that I call "professional growth". We assume that teachers are professional people and that they can use their experience to help them to grow in the exercise of that profession.

There are disadvantages of course. Growth is slow, it is sometimes painful, but is is organic. One great advantage, is that people who have grown into ideas only adopt those things that they really believe; there is no adopting the latest fashion without understanding what it is about.

Considerations in Planning Professional Growth Courses

I can only briefly mention the different facets of a professional growth course, these have been dealth with in greater detail in articles (4, 5, 6, 7), but I like to think of the structure of an in-service activity as consisting of a number of bricks which will vary in importance and position for different activities but which will all contribute to the final structure.

Previous Experience

The first brick is previous experience. If you have a group of thirty teachers on an in-service course it is very unlikely that any two will have the same experiences. If they have an average of 4 years teaching experience each, there will be 120 years of experience to be shared. One of the first things I discovered about courses is that this experience will not be brought forward and shared unless the sharing is made a central feature of the course. If the style of the course is entirely one that involves imparting information to teachers and asking them to learn new or different things, then there is no opportunity to share; if the style of the course is such as to recognise the teachers' professional expertise and make use of it, then it will be a resource available for the course.

Shared Experience

My next brick is shared experience. Previous experience has to be related, but an experience on the course such as a simulation or workshop is something that is immediately available for discussion and comparison. People experience the same event in different ways and it is important to bring this into the open and discuss it. The style of many in-service courses is to tell teachers about "discovery methods" or "activity methods" or"new maths", whereas what is needed is to be able to experience these things and then to share that experience with others.

The Space in Which to Grow

My next brick is the "space in which to grow". People can only grow if there is room for them to grow into. If a course is packed with content, lectures, events, the teacher becomes punch drunk and cannot assimilate what is going on. What is needed is the space to be able to compare the ideas being presented against previous experience, to be able to assimilate them and accomodate to them. It is quite wrong to expect a professional experienced teacher to be a sponge and just absorb new ideas that are presented by those who "know". Teaching experience is without price, it cannot be bought, it has to be acquired with hard sweat. An essential part of the "professional growth" course is to create the space within a course for new ideas to be compared and contrasted with previous experience.

Time

Creating the space needs time and my next brick is time. It is incredible to me how much in-service trainers expect to achieve in remarkably short space of time. Any yet growth is slow, it needs time.

I believe teaching to be a very demanding task, and in order to feel reasonably secure every teacher develops a collection of tricks of the trade that get him successfully through the day. If we attempt to alter any of these behaviours we are attacking the security of that teacher and this is something that has to be done with the greatest of care and sympathy. If the teacher feels too much under pressure he will merely put up his defences and reject everything being offered. If there is time however to rebuild the security after taking on new ideas, then those ideas will be absorbed as part of the professional growth. There must only be as much disturbance to the teachers equilibrium as there is time to allow for a return to stability. The more time we have the more disturbance we can risk.

Size of an Activity

What I have said so far implies that we are working with relatively small groups of teachers on any particular activity. A small group to me would be the staff of a primary (elementary) school, or the members of a department in a secondary school, or a group of teachers performing a similar role within different schools. This means a group will consist of not less than six teachers and not more than thirty. Where the normal mode of in-service work is by lectures to a hundred or more teachers the "bricks" I have talked about can not be used to build the structure. Clearly, working with small groups increases the difficulties of organising in-service activities, which, in those countries with few trained teachers, is a very acute problem, but I suggest that even if it is going to take longer, we must make what we do effective, and the "professional growth" model does help teachers to change.

Content and Focus

My last brick is the one that most people start with but I have left to last because if you are planning a professional growth course it is not the first consideration. This is the content and focus of the course. I believe that many courses are far too vague in their aims. It is not uncommon to find courses for Teachers labelled "teaching Primary Mathematics" or "teaching Secondary Mathematics" or just "Modern Mathematics". One must ask "What mathematics?" "What teachers?". In a single school there will be teachers with different amounts of experience, different mathematical backgrounds, and it is important to make sure that we define much more closely the kind of teachers for which the course is suitable. The actual mathematical content of the course has to be chosen so that it exemplifies ideas that the teacher can generalize and should not be too specific, for instance, there could be a session or sessions on a course talking about how to teach number bases, but this is all the teacher will take away from the course. Unless we are to be faced with the impossible task of covering the whole syllabus, then we have to choose a topic that contains within it explicit ideas that can then be used with other content. Examples are the teaching of computation through a study of number patterns or the use of geoboards/nail boards to investigate the properties of shape. In these instances there are basic ideas, number pattern and investigation of geometrical shapes, which can be applied to a wide range of content.

I am concerned with developing the awareness of the teacher about what it means to function as a mathematician (8), at a level which will have meaning for the children being taught. I would not regard an in-service activity as worthwhile if it did not have this as one if its major aims. It has nothing to do with teachers learning specific mathematical content, but with attitudes to the learning of mathematics that have within them the seeds of growth. It should not be too difficult for any teacher to gather factual knowledge from the literature; the sense of working as a mathematician at the level of children (which does not necessarily, or even often, mean a low level), is something that has to be experienced and therefore shared on a course.

The Course Structure

My experience of running courses along the lines that I have suggested and using the bricks that I have put forward, is that this automatically integrated pedagogy and content because all the time one is saying "what do the teachers know?" "How am I going to make use of this?" "What methods of working are available to me?" "What is the suitable content to demonstrate these ideas?"

The present trends are towards more in-service work that is based in the school rather than taking teachers to courses outside the school (9). What I have been saying is extremely important for those who are planning in-school courses and particularly for those teachers with responsibility for developing work within schools. I have been running courses for teachers, who, whilst remaining class teachers, take on the responsibility for the mathematics in 5-11 age schools (7). These courses, which make explicit the ideas I have put before you, have proved very successful in helping these teachers to develop the work of their colleagues. I call teachers with this kind of responsibility consultants, and the ideas of tackling the vast problem of involving all teachers in in-service activities through a structure of consultants has been explored in different ways by others (1, 2).

Another example of putting these principles into practice is an activity I organised which required autonomous groups of teachers to identify their own problems in teaching mathematics (10). The striking aspect of this experiment was the way in which the problems they chose to work at all moved from content to pedagogy and then ended somewhere inbetween the two.

I fear that the time available is too short to do more than present the ideas but these are based on many years work and I am very willing to pursue them with anyone who wishes to do so.

References

1. M. Erant - In-service education for innovation, occasional paper No. 4, National Council for Educational Technology (England) 1972.

2. P. Greer - Education's stepchild - In-service Training occasional paper 28 - Council for Basic Education, Washington, D.C. (1979).

3. E. Hoyle - The role of the teacher - Routledge (1969).

4. D.A. Sturgess - An experiment in In-service training - Mathematics Teaching No. 65, December 1973.

5. D.A. Sturgess - A Framework for Mathematics Teaching in the Primary School - University of Nottingham, School of Educatin (1973).

6. D.A. Sturgess - Teachers can do it - Mathematics in Schools Vol. 3, No. 1, January 1974.

7. D.A. Sturgess - Humanizing in-service education in Mathematics - Mathematics Teaching, No. 91, June 1980.

8. C. Gattegno - The commonsense of teaching mathematics - Educational Solutions, Inc.

9. Making INSET work - Department of Education and Science - H.M. Stationary Office (England) (1978).

10. D.A. Sturgess - The problems of Teachers of Mathematics - Nottingham University, School of Education, (1980).

INTEGRATION OF MATHEMATICAL AND PEDAGOGICAL CONTENT IN IN-SERVICE TEACHER EDUCATION

E. Glenadine Gibb
The University of Texas
Austin, Texas

Introduction

In-service education is a generic term with different definitions depending largely on who defines it and for what purpose. There is general agreement, however, that it means all activities for employed teachers that contribute to their professional growth and development. Massanari (1977) has identified five modes and the related roles of teachers. They are:

Mode	Role of Teacher
Job embedded	Employee of the school
Job related	Colleague of other teachers
Credential-oriented	Student of higher education
Professional organization related	Member of a profession
Self-directed	Individual craftsman

Furthermore, in-service education for maintaining and improving teacher competency both in knowledge of mathematics and in teaching mathematics may be oriented towards different objectives. A program may be

1. An "on the job" program to do that which could not get done in the initial preservice preparation and directed towards the specific job responsibilities in a particular school or school system. Unfortunately, this has been viewed negatively as a remedial program beginning with the idea that "something is wrong";

2. An upgrading program for keeping teachers abreast of new advances in knowledge (this may include opportunities for higher degrees);

3. A program of special needs for considering special population groups not considered adequately in the preservice programs; and

4. A curriculum development program for introducing new curriculum and methods of teaching to implement that curriculum.

In considering the integration of mathematical and pedagogical content, my remarks are limited to job embedded, staff development, for primarily "on the job" programs. What then are some ways of integrating mathematical content and pedagogy in this mode and program?

Lesser Successful Efforts

Based on personal experience and reports from others, programs that have failed or at least have not been too successful have not always reflected the true concerns of teachers. Teachers have been overwhelmed by theoretical treatments of both mathematics and pedagogy although they seek each. They have been unable to relate theory to realistic, manageable classroom practices. Efforts attempted were perceived by the participants as irrelevant to their needs in their own classrooms. Often, teachers participting in in-service programs did so as individuals. Any motivation or enthusiasm for implementing suggested ideas for the classroom met with opposition, lack of commitment, or no support from fellow teachers and administrative staff when they returned to their teaching assignments. Adequate follow-up and reinforcement had not been a part of planning those in-service programs and thus participants did not sense accountability for their in-service education.

In a survey by the National Council of Teachers of Mathematics (1977) in the mid-70s, over half of the elementary (800) and secondary (1400) teachers of mathematics participating expressed dissatisfaction with their experiences in in-service education programs in mathematics. Ranking of reasons for dissatisfaction were:

Reason for Dissatisfaction	Ranking Elementary	Secondary
Did not fit needs in the classroom	1	1
Too theoretical	2	3
Did not help in my teaching of mathematics	3	2
Did not help to select topics	4	5
Leaders have not taught classes like mine		4
Materials too expensive for practical classroom use	5	

We should be reminded that evaluation of many in-service activities has relied predominately upon opinion with little attention to tangible evidence of teacher growth, student achievement and side effects.

More Successful Experience

Let us now consider some more positive kinds of experiences. Three plans for integrating content and teaching are described based on personal, direct experience. I admit that I cannot share with you evaluation data based on evidence of change. Evidence of success has come from survey-type information from participants reflecting their perceptions of benefits.

1. Partnership approach

This approach provided two-part sessions, using at least a two-member team as resource leaders. For example, if the topic requested by the teachers was help with teaching fractions, one member of the team assumed responsibility for refreshing the participants' knowledge and understanding of rational numbers. Based upon this mathematical foundation, the second member of the team directed the discussion to the teaching of fractions - - for the purposes of developing understanding and skill on the part of the participants' students. Success of this approach was deemed dependent on team effort to inform, teach and harmonize with each other in planning, setting forth criteria and developing materials. Both looked at what they did, what each other did and what they wanted teachers to do.

Participants came together from different school buildings to meet in one location. Efforts also were made to provide for individualization according to the specific classroom responsibilities of each teacher. In fact, participants made plans for teaching and then reported back to the group on the success of their experiences. Sharing of experiences with others sparked enthusiasm and also motivated responsibility for change in the classroom. Demonstrations by the second member of the team also were helpful. Seeing students in action seemed to provide a model to be used in their own classrooms. At least it stimulated the spirit to try.

2. Field-based Approach

Rubin (1971) has recommended that most in-service education activities should be carried on within the setting in which the participants normally work together. Resources often were limited and thus it was not feasible to use a partnership approach. A two-part session still was possible. The mathematical background and theoretical aspects of teaching were provided by a 30-minute television presentation where teachers in each single school building in the TV viewing area had the opportunity to view the presentation together. This activity was followed by a discussion of the presentation directed by a leader or leaders from among their own group. The discussion then extended to the implications for pedagogical considerations in that particular school and the needs of students in that school. In small working groups, teachers made plans for trying alternative strategies in their classrooms. Each teacher then assumed responsibility for implementation in his/her teaching.

This approach was enhanced by follow-up visits to individual schools where possible, and communications between leaders in each school building and the person making the television presentation. Certainly television has yet to fulfill its rightful place in programs for change but it may when video cassettes, video disks and feedback systems contribute more flexibility and more individualization.

3. Supporting Background Approach

Any of us who have worked with teachers in a school setting are well aware of their interest to have something that is practical and that can be used in the classroom immediately. In this approach, the focus is on the teaching concerns with the mathematical content providing the supporting background as needed in considering alternative strategies deemed to be effective in the participants' classrooms. In attempting to provide for a wide range of needs, this program usually has been planned across the school system or school systems in an area rather than in a particular school building. Predictability of success is high since the intent is focused on that which is perceived by participants as useful in the classroom. Care must be taken, however, to make a distinction between the content considerations for teachers and the applications of this knowledge in making instructional decisions. Also the success of this approach seems highly dependent on how well the resource leader can work within this integrated schema with clarity as well as the leader's competence both in mathematics and familiarity with teaching mathematics in the participants' particular school settings.

Teacher centers should not be overlooked as still another approach. Limited to only indirect experiences with this approach, I am reluctant to assess its merits. Yet, I feel that teacher centers have significant promise for improving the quality of instruction in our schools. Certainly they can provide a much wider selection of in-service activities in mathematics, including special sessions, a resource center for materials and ideas and interactions with other teachers. As with the other approaches, the integration of mathematics content and the teaching of mathematics must be planned.

Some General Comments

Regardless of what plan for integrating content and pedagogy may seem appropriate for any type of in-service mode and program, we should remember that success of efforts is enhanced if the program is relevant, need-oriented, well-conceived and well-implemented. A program should provide opportunity for knowledge, competency in teaching, and commitment for change on the part of the participants. Bishop (1976) has suggeted the following plan to effect these three components of staff-development.

Component	Individual	Small Group	Large Group
Knowledge	Reading Modules Audio Tape Mediated Programmed Materials	Study Group Case Study	Lecture Film-TV
Competency	Directed Practice	Simulation Laboratory Exercises "Training" Session	Demonstration
Commitment	Visitation Internship Interview Research Utilization	Discussion Gaming Real Situations	Field Trip Feedback groups

For us in the United States, there is a clear trend towards locally-based decision-making authority. Teachers must have a part in identifying needs and selecting topics for in-service education programs. Perhaps this is the major key to successful programs, regardless of their purposes.

As a mathematical educator, I personally subscribe to in-service education in mathematics for both elementary and secondary teachers as a school-faculty enterprise if the mathematics programs of the schools are to be improved. The ultimate beneficiaries of such efforts should be the participants' students. Thus, how successful or unsuccessful any efforts to integrate mathematics content and pedagogy must be measured in terms of the achievement and attitudes of children and youth in the classroom.

References

Bishop, Leslie J., Staff Development and Instructional Improvement. Boston: Allyn and Bacon, Inc., 1976.

Massanari, Karl. What we don't know about inservice education: An agenda for action. Journal of Teacher Education, 1977 28, 41-43.

National Council of Teachers in Mathematics. An In-Service Handbook for Mathematics Education. Edited by Alan Osborne. Reston, Virginia:

National Council of Teachers of Mathematics, 1977.

Rubin, Louis, Ed. Improving In-Service Education: Proposals and Procedures for Change. Boston: Allyn and Bacon, Inc., 1971.

IN SERVICE TRAINING TEACHERS OF SECONDARY SCHOOL

Martin Barner
Deutsches Institut fur Fernstudien
Freiburg, Germany

Since 1967 the German Institute for Distance Studies (DIFF), affiliated with the University of Tubingen, developed distance study courses in mathematics for further training and refresher courses for teachers.

During 1967-76 the "Basic Mathematics Course" was produced.

In 1970-71 DIFF and Zweites Deutsches Fernsehen developed a multi-media distance study block "Introduction to Integral Calculus".

The distance study course "Mathematics for Elementary School Teachers" was planned in 1970/71 and produced between 1972 and 1976.

For detailed information see the report in the "Proceedings of the Third International Congress on Mathematical Education".

At present DIFF develops two further distance study courses in mathematics the planning of which was reported as well in the Proceedings":

1. Mathematics for Teachers at Lower Secondary Level/General High Schools

Planning of this distance study course started in 1976, it will be completely developed in 1980.

The course is designed for teachers at the lower secondary level and at General High Schools. It may also be used in all stages of teacher training.

The course tries to treat General High School maths mainly as applied maths. The applicability of maths should not only be evident within the discipline of mathematics or of natural sciences, but must also include the students' everyday life and the world of their later careers. So the students have to learn the convenient use of pocket calculators. Another example is statistics which they often meet and which they should be able to understand and interpret.

General High Schools maths must gain its own characteristic features. These characteristic features should include topical applications and a mathematical approach. Together with these comes the acquisition of techniques, including automatic processes and calculating skills, not neglecting modern aids.

This distance study course is a continuation of the course "Mathematics for Elementary School Teachers".

It includes 12 units: Percentages, Interest, Ratios; Calculations in economy, Increase and Disintegration Processes; Plane Figures; Three-dimensional Figures; Isometries and Similarity Transformations; Caluclations on Plane and Three-dimensional Figures; Electronic Pocket Calculators; Decimals; Equations and Inequalities, Part 1 and 2; Descriptive Statistics; Calculation of Probability.

These units together with 8 units of the course "Mathematics for Elementary School Teachers" may be grouped in five study packages: Practical mathematics (word problems, real-life applications): Quantities, elementary functions, percentages, interest, ratio, calculations in economy, increase and desintegration processes; Geometry: Constructive and descriptive aspects of plane and three-dimensional figures, isometries and similarity transformations; Arithmetic/Algebra: Pocket calculators, decimal (e.g. approximations, approach to real numbers), equations and inequalities; Statistics/Calculation of Probability: Concept of descriptive statistics, calculation of probability, some stochastic distributions; Number Systems: Theory of divisibility, prime numbers, factorisation, integers, fractional numbers (positive rational numbers), rational numbers.

Each unit is arranged in three parts: the F-part (mathematical theory) supplies the teacher with the background knowledge which is indispensible for modern mathematics instruction, independent of special methodical approaches. After this the D-part (methodology) points out the significance of the treated subjects for school lessons as well as offering motivational elements, illustrations and types of exercises. The exercises of the A-part (worksheets) are intended to encourage the teacher to work actively on the treated mathematical subjects.

2. Mathematics for Teachers at Upper Secondary Level

This course has been planned since 1976. Its development will be finished in 1980.

The course is designed for teachers at the upper secondary level (grades 11-13) in all schools of the Federal Republic of Germany which lead to Abitur. (This is the final examination which entitles those whoo pass to attend the university).

The course can also be used for the education of teachers at universities and other institutions for teacher training. There are three main mathematical topics in the upper secondary level: Calculus, Probability and Statistics; Analytic Geometry and Linear Algebra. Calculus and Probability are treated in this project. Geometry will follow in a new project.

While calculus and geometry are classical fields in the upper secondary level, probability and statistics have entered the curriculum only recently. THe majority of teachers in charge therefore have insufficient qualifcations in this field. Consequently there is a great demand of adequate literature especially in probability and statistics.

The course consists of the following study units:

Calculus:

Approaches to Differential Calculus: Different aspects of differentiation are presented, for instance the derivative as rate of change, the linear approximation aspect and the aspect of tangent corresponding geometrically to the linear approximation. They lead to different classroom approaches which are described and compared.

Differentiable Functions, Critiera of Monotony, Mean Value Theorem: It is shown that these "global" theorems are nearly equivalent aids in analysing differentiable functions, in approximating functions by Taylor-polynomials and in solving simple differential equations. Several classroom approaches are given and compared.

Approaches to Integral Calculus: Three different ways are discussed: the approach of Riemann and Darboux by sums of products, an axiomatic approach based on the notion of area and a way beginning with the problem of finding a primitive (antiderivative) of a given function. The different approaches are compared and prepared for the classroom.

Exponential and Logarithmic Functions, Increase Functions: Three different approaches to the exponential functions are given and compared: Continuous extension of a function which is given by rational powers of a fixed basis, solution of the functional equation $f(x+y) = f(x)*f(y)$ under several additional conditions, solution of the differential equation $f' = f$ with $f(0) = 1$.

Trigonometric Functions and Oscillation Processes: Three approaches to trigonometric functions are given and compared: A way which is based on the geometrical definition of sin and cos, the trigonometric functions sin and cos as solutions of a linear system of first order differential equations, the functions sin and cos and solution of a linear second order differential equation.

Probability/Statistics

Descriptive Statistics: Statistical notions as relative frequency, mean, median and standard deviation are introduced. Ways of demonstrating statistical data are given and prepared for the classroom.

Approaches to Probability: Random experiments and probability spaces, conditional probabilities and independence of events are defined, classroom approaches are designed. Bernoulli's law of large numbers is discussed.

Random Variables and Distributions: The notions of random variables and distribution functions are defined. Expected value and variance are introduced for discrete and continuous random variables. The concept of independence of random variables is discussed. Various distributions are given. Limit theorems as the Poisson approximation to the binomial distribution and the De Moivre-Laplace limit theorem are demonstrated. A classroom approach is described.

Introduction to Statistical Inference: Hypotheses testing, estimation and confidence intervals are introduced. At an elementary level tests are compared on the basis of their risk functions. An estimation problem in which there is no straight forward estimate is discussed and different estimates are examined. For classroom demonstrations the importance of simulations is emphasized.

The study units of this project generally show several classroom approaches to important topics in calculus and probability. Up to now teachers and teacher students were thrown more or less on university text books in their preparation for the lesson. But these books are not designed for the classroom situation. On the other hand school books do not supply the teacher with sufficient information about the variety of possible approaches.

Here the project tries to fill the gap. To this end a special structure of the units has been chosen.

The first part (part A) contains a rigorous mathematical description of the topic. The election and formulation of the facts however has been done with a view on possibilities of realisation in the classroom. Several mathematical aspects, possible approaches and developments are taken into consideration.

The second part (part B) presents a didactic reflexion of the subjects given in part A. Different approaches which are the result of the analysis in part A are compared. Advantages and disadvantages of the approaches when applied in the lessons are discussed. Knowledge supposed to be present or to be previously taught is mentioned. Alternatives of argumentation on different levels of stringency are demonstrated as well as possible branchings of the instructional ways. Possibilities of motivation and application are given.

The third part (part C) contains proposals of instructional realisations of the different approaches on the basis of the analysis in part A and the results of part B. Sections have been described more in detail when this has been found to be necessary. In other cases the description has been given in rough outlines. As the instructional situation can never be described in a comprehensive manner the instructional ways shown here can only be models which are supposed to give some orientation to the teacher.

The study units in the field of stochastics follow a slightly different line. The middle part containing didactic relexions in the units on calculus is absent in these books. The didactic discussion in the field of stochastic is obviously not in an advanced stage as for instance in calculus. The main reason, however, why these study units have only two parts is that in general German teachers are not sufficiently educated in the field of probability and statistics. So they have to be taught themselves. Therefore it appeared to be natural to describe only one approach. The comparative and reflective part is inapplicable in this situation.

Final Remark:

Both courses which have been introduced here intend to help teachers adequately in the preparation of their lessons. As already mentioned University text books are inadequate because they give no hint how to adapt the subject to a given classroom situation. High school text books mainly intend to be exercise books for the students. They show only one way through the subject and give no analyses and didactic reflexions. For the preparation however of a qualified instruction it appears necessary to have at one's disposal a larger variety of approaches and reflections. Therefore high school text books are inadequate too. The courses of DIFF shall lead out of this dilemma between university text books and high school text books.

UNE FORMATION CONTINUE INTEGREE A L'ACTIVITE PROFESSIONELLE

Michel Darche
IREM d'Orleans

Je decrirai rapidement ce qu'ont ete et ce que sont les IREM, puis je decrirai quatre realisations qui me paraissent exemplaires des activites des IREM pour developper et faire evoluer a la fois la formation continue et la pratique enseignante malgre les difficultes qui appraissent depuis deux ans.

25 instituts de Recherche sur l'Enseignement des Mathematiques ont ete crees de 1968 a 1974.

Leurs missions:

- mettre en place une formation continue des enseignants;
- developper la recherche et l'experimentation sur l'enseignement;
- contribuer au developpement de l'enseignement en liaison avec l'evolution des idees, des sciences et des techniques;
- participer a la formation initiale des futurs enseignants.

Leurs exigences:

- la formation continue doit faire partie integrante de l'activite professionnelle. Elle ne doit pas etre "autre" et se detacher de la pratique;
- la formation continue doit etre prise en charge par le systeme de production des connaissances scientifiques et techniques (l'universite);
- la formation continue doit favoriser la prise en charge de la formation par les interesses eux-memes;
- la formation continue doit se trouver de lieux de validation au contact des autres systemes de formation (formation des adultes par exemple).

Les IREM doivent favoriser les interactions entre formation continue, formation initiale, recherches et pratiques enseignantes.

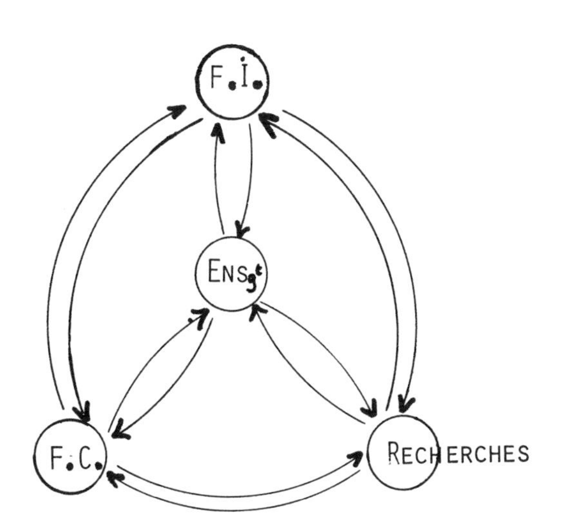

LES MOYENS ET METHODES.

- Les IREM sont rattaches a des universites, ce qui peut favoriser la liaison formation-recherche et le contact rapide aux nouvelles connaissances.

- Tout le personnel (animateurs et stagiaires) est volontaire et les activites IREM sont prises en compte dans leur temps de service (pour les stagiaires cela fut vrai jusqu'en 1978).

- Tous les enseignants sont a l'IREM a temps partiel (de 10% a 50% du service) et pour une duree limitee (2 a 5 ans). Ce sont donc des praticiens.

- Le travail s'effectue au sein d'equipes d'enseignants de statuts differents, de niveaux et de matieres d'enseignement differents.

- La formation est prise en charge par les enseignants eux-memes, les animateurs ayant un role d'appui technique.

- le systeme de formation doit etre souple et decentralise donc EVOLUTIF.

LES GRANDES PERIODES DES IREM.

1ere etape 1968 - 1972: Centree essentiellement sur les contenus mathematiques les universitaires diffusent la connaissance aux enseignants des Lycees et Colleges. Cette periode correspond a celle du mathematisme triomphant qui favorise l'immobilisme face a un savoir puissant et sclerosant.

2eme etape 1972 - 1978: Les experimentations. Le groupe. La reflexion sur la matiere fait place a la reflexion sur l'aspect groupe-communication-relationnel. C'est la multiplication des recherches-actions ou la psycholosociologie tient une grande place. La formation est centree sur le groupe, l'equipe enseignant qui prend en charge ses propres problemes.

3eme etape 1978 - 19??: La periode predidactique. En meme temps que s'elevent les premieres critiques et que les premieres reductions de moyens sont annoncees le courant didactique fait son apparition avec comme principe, la rupture: la connaissance des phenomenes d'appropriation ne peut se faire a l'interieur de l'enseignement mais en dehors.

Des UER de didactiques sont creees a Paris, Bordeaux/Marseille, Strasbourg/Nancy.

Simultanement apparaissent une rupture entre recherche-action et recherche en didactique et une rupture entre chercheurs nobles et innovateurs en formation.

LES QUESTIONS ACTUELLES:

Actuellement les questions son de 3 types:

- Comment poursuivre la formation continue maigre la modestie des moyens.

- Comment eviter le cloisonnement Recherche/Innovation/Formation.

- Comment impliquer, sans moyens, les enseignants dans une formation continue.

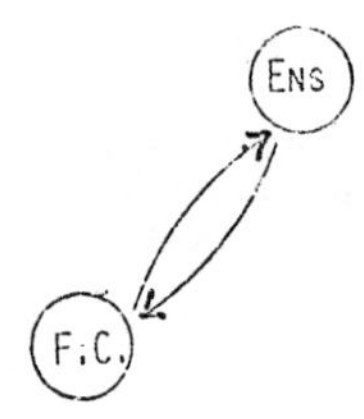

1er exemple: LE GROUPE - L'EQUIPE (I) vs. Interaction Formation Continue/Enseignement.

Suivant les IREM les stagiaires dand les premieres annees etaient regroupes par niveau d'enseignement. Ces groupes cloissonnes ont bientot fait place dans de mobreaux IREM a des groupes d'interets de 5 a 15 personnes reunis autour d'un theme ou d'un projet. L'animateur, s'il n'est pas issu du groupe, a alors des taches bien definies:

- faire formuler et analyser par les stagiares leurs besoins en formation,
- amener les stagiares a elaborer un projet de formation, a le realiser et a l'evaluer.

Pendant la realisation du projet l'animateur est le lien entre le groupe et l'IREM, c-a-d les autres groupes, les autres animateurs.

Cette valorisation du groupe, de l'equipe n'a pas uniquement pour but l'une des missions des IREM:

La prise en charge par l'enseignant de sa propre formation, elle vise aussi:

- a partir des problemes soulives par la pratique enseignante et a y revinir rapidement en peermettant des experimentations en cours de formation.

elle vise surtout

- l'apprentissage de methodes, de techniques que les enseignants maitrisent peu et leur transfert a la classe.

- l'animation d'un groupe de travail ------ l'animation pedagogique.

- l'elaboration/la realisation d'un projet d'actions ---- -- la planification pedagogique et l'evaluation.

Le premier vecteur de transformation de la pratique enseignante doit en effet se situer dans la facon de travailler d'abord pour rompre les multiples cloisonnements qui existent entre enseignants mais surtout pour pratiquer une pedagogie vivante centree sur l'activite individuelle et collective des eleves et l'intercommunication.

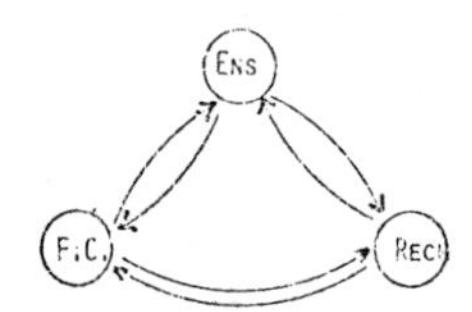

2eme exemple: ATELIERS/PROJETS E'EQUIPE/EQUIPE DE RECHERCHE vs. Interaction Formation Continue/Enseignement/Recherche.

Pour eviter la coupure Formation Continue/Recherche l'IREM d'Orleans a propose de 1976 a 1978 un ensemble d'activites qui se voulait complet et en constante evolution, assez structure pour permettre une planification et assez souple pour permettre une adaptation constante.

3 types d'activites etaient proposes aux enseignants.

- Des ateliers de 3h/semaine pendant un trimestre. Ces ateliers prepares par des animateurs permettaient aux stagiares d'approfondir un probleme particulier (pedagogique, didactique, psychologique, sociologique . . .)

- des equipes de recherche sur des problemes d'apprentissage soutenues par l'IREM mais financees, si possible, par des organismes exterieurs (INRP, DGRST, CNRS . . .)

- des projets d'equipe, directement inspires de nos amis quebecois de PERMANA, permettent aux enseignants au sein de leurs etablissements de constituer une equipe autour d'un projet d'innovation ou de reflexion et de suivre une demarche pre-experimentale avec planification, realisation, evaluation et bilan.

Chaque type d'activites est en lisison avec les 2 autres:

- les ateliers repondant a des besoins precis et nourris par l'experience acquise dans la realisation de projets ou de recherche et permettant d'amorcer un nouveau projet ou une nouvelle recherche.
- les projets permettant d'analyser des problems pedagogiques vecus et de faire emerger de nouveaux terrains de recherche, de constituer des equipes interetablissements et/ou interdisciplinaires.

- les recherches finalisant certains travaux IREM et alimentant les autres types d'activites en introduisant de nouvelles connaissances sur l'apprentissage.

L'ensiegnant, agent de ces liaisons, s'organisant et gerant sa formation dans l'ensemble de ces activites peut suivre des trajectoires variees: atelier-projet-recherches ou atelier-projet-animation d'atelier ou . . .

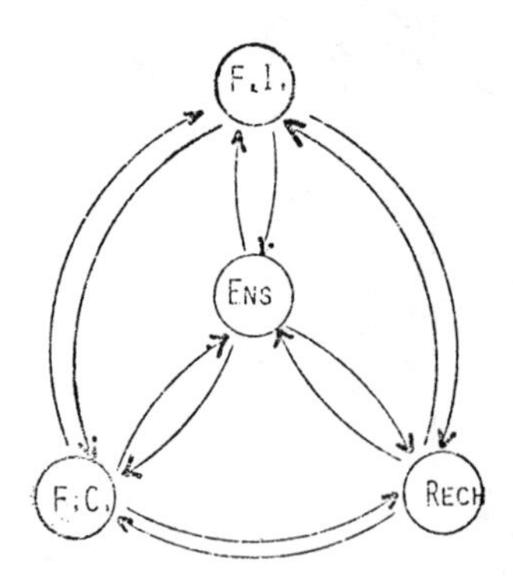

3eme exemple: LES COLLOQUES INTER-IREM vs. Interaction Formation/Recherche.

Tous les IREM ont la meme structure mais une totale autonomie dans l'organisation de leurs activites. Ils ont cependant besoin d'une validation externe de ces activites. L'un des outils privilegies de cette validation est realise par la dizaine de colloques nationaux annuels qui permettent la communication et la confrontation des travaux realises dans les IREM sur differents themes.

Ces themes sont choisis en fonction de leurs interets intrinseque soit en fonction des travaux realises.

Ils servent de bilan d'activites, de mise au point dans un parcourt de recherche ou de lancement de nouvelles activites dans certains IREMS.

Ils permettent aussi les contacts directs entre animateurs et stagiaires de differents IREMS.

Ces recontres durent chacune deux jours, regroupent entre 50 et plus de 100 participants chercheurs, innovateurs et praticiens, et donnent lieu le plus souvent a la publication de documents et travaux de synthese.

Parallelement a ces colloques se rencontrent a un rythme plus soutenus (de 3 a 6 reunions annuelles) des groupes plus restreints de travail inter-IREM qui prolongent leurs activites sur le terrain IREM par une reflexion commune.

Dans les deux cas ces rencontres permettent aux chercheurs de communiquer aux pacticiens leurs travaux, aux innovateurs de valider leurs experiences avec les chercheurs, aux practiciens enfin de recueillir une masse importante d'informations. C'est pour chacun, et pour le practicien en particulier un moment fort dans l'annee et dans la periode de formation ou de recherche.

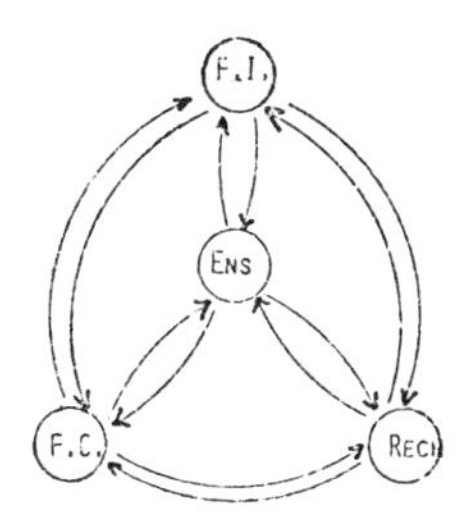

4eme exemple: ELABORATION ET DIFFUSION DES DOCUMENTS vs Interactions:

L'elaboration et la diffusion des documents ont ete les elements les plus observables et les plus caracteristiques des activites iremiques. C'est le haut de l'iceberg mais c'est celui qui touche le plus de monde, gens de IREM mais aussi personnes exterieures, enseignants ne participant pas ou plus aux activites IREM.

Rediger un document, faire connaitre son travail c'est aussi un objectif motivant et valorisant pour de nombreux groupes, c'est enfin et surtout un element de validation de l'action d'un groupe, d'un IREM. De nombreux types de documents ont ete et sont encore diffuses, allant du document de travail pour communication au sein d'un IREM au rapport de recherche faisant le bilan de 3 ou 4 ans de recherche en passant par des publications d'innovation, d'experimentation, par des descriptifs de mateiels, de jeux pour la classe. De nombreuses maisons d'edition allechees par ces travaux trouvent dans ces documents un pool d'auteurs de manuels scolaires et de publications.

Ces documents sont envoyes par l'IREM diffuseur dans tous les autres IREM et dans les centres de recherche, de formation ou de documentation francais et etrangers qui le desirent.

Par ailleurs des bulletins synthetiques sont diffuses par une commission inter-IREM sur les problemes brulants et les themes les plus preoccupants du moment.

Apre une periode de foisonnement et de profusion vecue jusqu'en 1978, dans la periode de limitation des moyens que nous vivons actuellement un travail de synthese de ces documents est en train de se faire qui devrait deboucher sur des publications inter-IREM de documents et materiels directement utilisables dans la classe de la maternelle a l'universite.

Malgre les difficultes que nous rencontrons actuellement: 1978 - 1979: - 20% des credits de fonctionnement, 1979 - 1980: - 20% des credits et suppression des stages sur le temps de service, 1980 - 1981: suppression des 3 emes cycles de didactique des maths a Bordeaux, Paris et Nancy-Strasbourg . . . je voudrais conclure sur une note optimiste et une note d'espoir qui soulignerons l'interet des exemples presentes:

La note optimiste:

En depit des restrictions tres importantes des moyens de fonctionnement chacune des actions de formation citees continue d'exister:

- une partie des equipes qui se sont constituees precedement poursuivent leurs travaux benevolement,
- les colloques inter-IREM se deroulent avec le meme frequence et un interet accru: ce sont les derniers lieux de rencontre et de confrontation,
- les publications IREM continuent deparaitre et un projet de large diffusion aupres des enseignants est en preparation,
- la recherche en didactique se structure et se developpe:

  seminaires trimestriels a Paris,

  naissance d'une revue trilingue sur les reserches en cours (2 numeros deja parus avant le congres ICME IV),

  creation d'une ecole d'ete de didactique.

La note d'espoir:

Nous esperons que ces derniers mais important moyens nous permettrons de relancer le militantisme pedagogique, de contrecarrer les projets ministeriels de formation delivree par ceux qui "savent" aux enseignments qui "ne savant pas" et de faire obstacle aux ambitions das editeurs qui accompagnent leurs manuels scolaires de "livre du maitre" qui sont un succedene sclerosant de la formation continue.

Footnote

(1) voir groupes "nouvelle formule" Paris-Sud; groupe "formation continue articulee sur la pratique" Rowen; grupes autogeres, Caen; projets d'equipe, Orleans.

Bibliographie

Colloque de TOURS. 1977. Formation continue des enseignants (IREM - Poitiers)

Bulletin inter-IREM n° 17: les IREM missions et activites. 1979. (IREM - LYON)

On acheve bien les IREM. Editions SALIN 1979.

Mathematique, enseignement, societe: La politique de l'ignorance. Editions RECHERCHE 1980.

"Formation continue et recherche". Projet IREM-FRANCE Berkeley - aout 1980, avec bibliographie plus complete pages 8, 9, 10.

PERMAMA: UN PROGRAMME INNOVATEUR DANS LE PERFECIONNEMENT DES PROFESSEURS DE MATHEMATIQUES

Richard Pallascio
Teleuniversite du Quebec
Montreal, Quebec, Canada

Introduction

L'exposé qui va suivre va comporter essentiellement deux parties, la première exposant ce qu'est PERMAMA, sigle qui signifie PERfectionnement des Maîtres en MAthématiques, ce qu'il est devenu et les étapes de son evolution; la seconde partie visera à expliquer les causes du succès de ce programme auprès des professeurs de mathématiques.

1. Qu'est-ce que PERMAMA?

PERMAMA est un programme de perfectionnement des professeurs de mathématiques du niveau secondaire (i.e. enseignant à des élèves de 11 à 16 ans), en cours de service. A son terme, le projet aura fait de plus de 1000 professeurs de mathématiques (sur 3500 au Québec), des gradués deténteurs d'un diplome de premier cycle universitaire en enseignement des mathematiques. Ces professeurs se réunissent pour travailler par petites équipes d'une dizaine de personnes, disséminées à travers le Québec (plus de trois fois la superficie de la France à couvrir) en une centaine de "centres PERMAMA". Une équipe centrale hétérogène composée de mathématiciens, de professeurs chevronnés du secondaire, de pédagogues, de psychologues, de physiciens et d'informaticiens, assure la conception des activités de perfectionnement (soit des cours de mathématiques (v.g. langage logique, theorie de la probabilité, algèbre linéaire, théorie des graphes, relations et fonctions, algèbre de Boole, ...), soit des cours portant sur l'activité mathématique (projets de mathématiques appliqués à l'étude de l'espace, résolution de problèmes, modèles mathématiques et réalite, ...), soit des cours portant sur l'enseignement ou l'apprentissage des maths (v.g. relations élèves-professeur, simulations et jeux, la fiche dans l'unité d'apprentissage, fonctionnement cognitif, conception d'un plan d'études, apprentissage par projets, divers atelier de lectures, ...), de même que l'encadrement des projects d'intervention définis par les enseignants en rapport avec des projets pédagogiques dans la classe (v.g. fiches de travail, banque d'applications, programme de voie allégée, ...) et des plans de perfectionnement de chaque enseignant inscrit au programme.

Tous ces modules (maintenant au-delà de 75) peuvent être regroupés de différentes facons pour former des ensembles plus vastes, centrés autour de thèmes, par exemple:

Exemple 1

Une équipe de trois (3) enseignants s'inscrit à un projet d'intervention qui consiste à expérimenter une nouvelle approche visant l'apprentissage des premiers concepts de la probabilité en secondaire II.

Ils compléteront leur plan de perfectionnement par les modules suivants (en moyenne, un module par mois, 10 mois par année):

- statistiques descriptives,
- probabilité I,
- modèles mathématiques et réalite,
- la fiche de travail dans l'unité d'apprentissage (module double),
- simulations et jeux.

Enfin, un module de gestion en vue d'élaborer leur plan de perfectionnement de l'année suivante, complétera leur plan d'activités de perfectionnement pour l'année en cours. Les autres professeurs du centre peuvent ou non se joindre a eux pour certains modules.

Exemple 2

Les professeurs d'un centre s'entendent entre eux pour se perfectionner au niveau du contenu mathématique ayant pour thème "hasard et stratégie". Ainsi, ils s'inscriront à des modules:

- mathématiques:
  - statistique descriptive,
  - probabilité I,
  - inférence statistique,
  - phénomènes linéaires;
- d'activite mathematique:
  - modeles mathématiques et realite;
- et d'enseignement des mathematiques:
  - simulations et jeux,
  - apprentissage par projets,
  - enseignement de la probabilité,

- enseignement de la statistique.

Tout comme leurs collègues de l'exemple I, ces professeurs s'arrêteront un moment au printemps pour élaborer leur plan de perfectionnement pour l'année suivante, le module de gestion étant le seul module obligatoire de l'année.

C'est donc dire que le perfectionnement, tout en étant offert à l'intérieur d'une diplomation à durée forcément limitée, est de type continu, et de plus, la gestion de ce perfectionnement continu se réalise conjointement en impliquant et les professeurs dans les centres et l'équipe de conception des activités de perfectionnement.

D'une part, les professeurs dans les centres

- acquièrent des connaissances mathématiques et didactiques,
- réalisent des projets pédagogiques dans l'école,
- planifient chaque phase de perfectionnement,
- organisent leur centre, et
- orientent le projet en fonction de leurs besoin personnels et professionnels.

D'autre part, l'équipe de conception des activites de perfectionnement

- crée progressivement une banque de cours modulaires,
- fournit des conseillers de projets d'intervention aux centres,
- coordonne les plans annuels de perfectionnement,
- facilite l'organisation des centres, et
- administre et anime le projet global.

D'une facon générale, PERMAMA vise donc

- à développer la formation mathématique de l'enseignant selon ses besoins,
- à développer la formation pédagogique de l'enseignement selon ses besoins,
- une personnalisation du processus enseignement-apprentissage, et
- une meilleure influence sur l'évolution de l'enseignment de la mathématique au secondaire.

2. <u>Quelles sont les principales caractéristiques de PERMAMA expliquant son succès auprès des professeurs de mathématiques?</u>

Quatre caractéristiques importantes permettent d'expliquer le succès du programme PERMAMA auprès des professeurs de mathématiques.

1°) PERMAMA est un programme de perfectionnement pour des enseignants en cours de service nettement distinct des autres programmes de formation initiale. D'une part, il n'est ouvert qu'à des professeurs en exercice; d'autre part, il fait appel à l'expérience même des enseignants auxquels il s'adresse, par des réflexions sur leur situation professionnelle, des experiences a réaliser avec leurs élèves, des projets personnels ou d'équipe permettant d'intervenir sur le milieu et de le transformer petit à petit, ...

2°) Plutôt que de centrer le perfectionnement uniquement sur l'apprentissage de mathématiques dites supérieures, plusieurs modules ont porté sur des concepts mathématiques qui sont sujets d'étude au secondaire, mais en poussant leur étude dans un sens vertical tout en examinant ses composantes didactiques (v.g. les fractions, les relations et les fonctions, etc.). De plus, la possibilité d'acquérir un savoir pratique par la mise en chantier de projets d'intervention permet aux enseignants de se concentrer sur les veritables problemes de la classe et d'essayer de les resoudre.

Plusieurs instruments élaborés pour les cours de PERMAMA sont d'ailleurs utilises par la suite par les professeurs dans leur classe.

3°) La possibilité de se perfectionner sur les lieux même du travail, à savoir l'école. D'une part, le Québec étant un pays géographiquement très vaste, les distances a parcourir pour se rendre à l'université la plus proche sont parfois très grandes (jusqu'a 1500 km). De plus, PERMAMA est parti du postulat ecologique suivant:

> "Si un changement mésologique est souhaitable, il est préférable que les principaux acteurs de ce changement soient sur place".

C'est pourquoi PERMAMA a adopté une formule alliant l'enseignement à distance avec le travail en équipe, l'animation des centres étant assurée par des responsables élus par leurs collègues et recevant une formation spéciale leur permettant d'assurer un dépannage léger. Evidemment, les conseillers pédagogiques des centres, membres de l'équipe centrale, assurent un suivi, d'une part par l'usage occasionnel de la conférence téléphonique et d'autre part par une participation à une rencontre de chaque centre, environ trois fois par année, i.e. une fois par session universitaire en moyenne.

4°) Enfin, la philosophie de l'éducation sous-jacente aux messages contenus dans les activités de perfectionnement à l'endroit des élèves des professeurs-étudiants, est celle-la meme qui constitue la philosophie du perfectionnement offert aux enseignants. Voici un extrait du rapport de révision, datant de mars 1974, qui inaugurait, en quelque sorte, la seconde carrière de PERMAMA:

> "PERMAMA n'est pas simplement une machine à produire des detenteurs de baccalaureats. PERMAMA a un rôle social: offrir aux enseignants de la mathématique au secondaire un perfectionnement qui leur permette de transformer leur enseignement.
>
> Or, réformer l'enseignement de la mathématique est une tâche sociale essentielle, si on songe a l'immense majorité de la population qui sort de l'école dégoutée des mathématiques, peut-être parce que tout ce qu'on y enseigne est l'art de se conformer aux bonnes méthodes, aux bonnes définitions, aux bons théorèmes démontrés infailliblement dans le bon ordre.

Nous preparons la société de demain, et pour la préparer avec comme objectif de rendre l'étudiant au point ou il pourra se développer de façon autonome, ne faut-il d'abord que le maître fasse lui aussi l'apprentissage de l'autonomie, de l'ouverture d'esprit et de la confiance face à la mathématique et son enseignement?

PERMAMA doit créer des compétences dans le milieu en termes d'équipes de professeurs ayant acquis les attitudes et les ressources nécessaires pour continuer à se développer et à développer l'enseignement par eux-mêmes.

PERMAMA doit également laisser à ces équipes du materiel didactique et des éléments de méthode utilisables dans leur enseignement.

PERMAMA ne peut donc ressembler à un programme de baccalauréat traditionnel. Il s'agit de tenir compte à la fois de la faible préparation mathematique des maitres inscrits a PERMAMA et de partir de leurs expériences et de leurs besoins d'enseignants".

(Tiré de: Rapport du Comité PERMAMA à la commission de la TELUQ, 15 mars 1974, pages 5 et 6.)

En outre, le Comité énumérait les critiques suivantes qui s'adressaient au modele initial: taux élevé des abandons, inadaptation du contenu, caractere lineaire du programme, mécontentement des étudiants et manque d'effet sur la pratique pédagogique.

Sur un plan plus positif, le Comité recommandait que le modèle soit revisé en tenant compte des elements suivants:

1. Atténuer, dans la mesure du possible, le caractère lineaire du programme et modifier les conditions d'admission pour favoriser l'ouverture et la souplesse.

2. Modifier le programme en fonction des caractéristiques réelles des élèves afin qu'il influe sur l'enseignement donné en classe.

3. S'attacher à changer les attitudes des maîtres à l'egard du processus d'enseignement au lieu de mettre l'accent sur l'acquisition des connaissances mathématiques.

4. Etablir une relation plus étroite entre les cours de PERMAMA et les programmes de mathématiques en usage à l'école secondaire.

5. Donner plus d'importance aux problèmes d'épistémologie et à l'apprentissage de la mathématique.

6. Orienter le contenu mathématique de PERMAMA pour qu'il puisse servir à résoudre les problèmes concrets.

Voici un extrait d'un avis d'un expert québécois indépendant du programme lors d'une rencontre de l'OCDE à Philadelphie en 1976:

"L'elaboration de la nouvelle version de PERMAMA fait ressortir l'opposition entre deux fasons de concevoir l'acquisition des connaissances. Dans la conception traditionnelle, le "système d'enseignement", par l'intermédiaire de ses structures pédagogiques et administratives et du choix de ses programmes, sélectionne un échantillon de connaissances tiré de l'univers du savoir (et filtré par le système d'enseignement). Dans la version traditionelle, la relation entre ces quatre éléments (l'univers du savoir, le systeme d'enseignement, le maître et l'étudiant) est linéaire et univoque. Dans ce contexte, le maître est le principal agent de la "transmission de savoir", tandis que le rôle de l'étudiant consiste à le recevoir. Le premier modèle PERMAMA, en dépit de ses variations structurelles, reposait encore, pour l'essentiel, sur cette conception traditionnelle dont on peut affirmer, sans risque d'erreur, qu'elle prévaut encore chez la majorite des universitaires.

Au contraire, le personnel de PERMAMA preconisait une autre conception de l'apprentissage considéré comme un processus d'échange réciproque dynamique entre l'élève et son environnement, processus par lequel l'élément essentiel et actif est l'élève. C'est lui l'agent principal de la formation, et celle-ci à son tour, tire son origine de la dynamique interne de l'élève. Dans cette optique, l'agent principal n'est pas le maître qui transmet le savoir mais l'élève qui le construit.

Le personnel de PERMAMA a poussé beaucoup plus avant cette notion de l'acquisition des connaissances. Elle n'est évidemment pas nouvelle. En fait, la notion du "sujet-connaissant" est l'une des idées maîtresses de l'épistémologie génétique de Piaget. Elle n'est pas nouvelle non plus pour ceux qui connaissent les méthodes informelles appliquées dans les écoles primaires britanniques du premier et du second cycles, ou dans l'"open education" américain. Si l'exploitation de cette notion pour l'apprentissage des enfants a fait l'objet de nombreux débats, il est assez reconfortant de trouver dans les propositions de PERMAMA cette même idée appliquée aux adultes, et de plus, à la formation en cours de service! Mais ici encore, cette application aux adultes n'a pas de quoi surprendre ceux qui s'interessent à certains types de centres pédagogiques où cette idee fait souvent partie d'un présupposé implicite ou explicite. Le personnel de PERMAMA, quant à lui, a preferé faire de cette notion la base conceptuelle explicite - implicite - du modele revisé. Il convient de faire renarquer et non que beaucoup de ceux qui proposent des modèles de formation des enseignants en cours de service (FECS) ont tendance à dissimuler leurs conceptions de base tout en édifiant une superstructure très élaborée. Parmi les grands problèmes soulevés par les activites de FECS, il en est beaucoup qui semblent se situer, non pas au niveau des structures, mais plutôt à celui des bases conceptuelles, c'est-a-dire, au niveau de la "philosphie" ou de "l'idéologie".

En exposant sa conception de l'apprentissage et de la connaissance, le personnel de PERMAMA prenait un certain risque, car ce ne sont pas ces idées qui l'emportent dans le milieu universitaire. En effet, si l'on va plus loin dans ce sens, il faut, pour rester logique, repenser le rôle de l'enseignant et celui de l'élève. En allant plus loin encore, il faudra réexaminer l'ensemble des structures et des fonctions de l'enseigment dans les societes modernes, tout au moins en ce qui concerne la pédagogie, les programmes et l'appareil administratif. Le modèle PERMAMA s'adresse à un public restreint, composé de professeurs de mathématiques de l'enseignemnt secondaire, mais ses bases conceptuelles sous-jacentes, sous leur forme revisée, débordent de loin ces limites.

En revisant le modèle, le personnel de PERMAMA a été très influencé par un document qui fait partie du rapport annuel pour 1969-1970 du Conseil supérieur de l'éducation du Québec. Rédigé par l'éducateur québécois Pierre Angers et intitulé "L'activité éducative", ce document a influencé nombre de ceux qui, au Québec, s'intéressent à la réforme de l'enseignement. Angers soumet les notions traditionnelles d'apprentissage - ce qu'il appelle les notions "mécaniste" - à une analyse approfondie. Il etudie egalement ce qu'il appelle la notion "organique", c'est-à-dire une serie d'idées qui sont à l'origine de nombre d'innovations éducatives adoptées dans différentes parties du monde. Dans ce sens, on peut dire que la nouvelle version de PERMAMA est un modèle de FECS qui s'inspire d'un point de vue de plus en plus largement répandu sur la nature du processus éducatif.

L'un des objectifs fondamentaux du nouveau projet PERMAMA consiste a tenter de modifier l'idée que les profeseurs de mathématiques se font du processus éducatif. Il s'agit là d'une entreprise bien plus abitieuse et plus ardue que la poursuite de l'objectif essentiel du premier modele qui consistait à perfectionner et à mettre à jour les connaissances théoriques et pratiques relatives à l'enseignement des mathématiques à partir d'une conception traditionelle du processus éducatif".

(Tiré de: Innovation dans la formation en cours de service des enseignants, par Maurice Belanger, OCDE, 1976, pages 23 et 24.)

Conclusion

En terminant, voici quelques-unes des multiples facettes de l'experience de perfectionnement vécue à PERMAMA:

- PERMAMA s'appuie sur le dynamisme de ses responsables de centre pour assurer son insertion dans le milieu scolaire et sa communication avec les étudiants.

- Par le développement des "matériathèques", PERMAMA devient un service accessible et utile à tout enseignant desireux de profiter de son matériel didactique.

- Les manuels, vidéogrammes, diaporamas, cartes, graphiques, affiches et autre matériel produits par PERMAMA, bien que prioritairement au service de leur prefectionnement, sont, dans certains cas, utilisés par les enseignants comme outils dans leur propre enseignement.

- PERMAMA a gagné le respect et la confiance de ses étudiants par l'exercice de la cogestion pédagogique.

- La composition diversifiée de l'équipe pédagogique réunissant des mathématiciens, des didacticiens et des enseignants chevronnés assure aux cours offerts par PERMAMA un contenu et une pédagogie mieux adaptés.

- Le recours continu à des professeurs invités assure la mise en disponibilité des meilleures ressources pédagogiques du milieu québécois ou de l'étranger.

- PERMAMA a fait un pas significatif dans la voie de l'intégration du perfectionnement dans la tâche et dans celle de la recherche-action, en créant le projet d'intervention comme outil de perfectionnement.

- La présence active de PERMAMA dans les organismes de promotion de l'enseignement de la mathématique (A.M.Q., G.R.M.S., C.I.E.A.E.M., IREM) tant au Québec qu'ailleurs, lui a assuré une solide implantation dans le milieu et a favorise sa participation à l'évolution de l'enseignement de la mathématique aux plans national et international.

- La notion de programme évolutif dont les cours nouveaux sont choisis annuellement dote PERMAMA d'une flexibité spéciale en congestion avec les etudiants.

- PERMAMA se nourrit et se regénère d'une facon continue par son interaction avec les enseignants.

- Les étudiants participent directement à la conception des nouveaux cours de PERMAMA par une interaction planifiee avec l'equipe pédagogique.

Références

Belanger, Maurice, Innovation dans la formation en cours de service des enseignants, OCDE, 1976.

Daniel, John S., New Models for In-Service Teacher Training in Quebec: the PERMAMA and PERMAFRA programmes, OECD, 1976.

Filion, Paul et Bédard, Roger, Mémoire à la commission d'étude sur les universites, Tele-universite, juillet 1978.

Gaulin, Claude, Innovations in Teacher Education Programmes, Conseil des Sciences du Canada, février 1978.

Pallascio, Richard, Etude de l'evolution des interactions élèves-professeur due a l'implantation d'un nouveau modele de perfectionnement des professeurs de mathématiques, Universite de Montreal, these de doctorat inédite, 1978.

5.13 SUPPORT SERVICES FOR TEACHERS OF MATHEMATICS

SUPPORT SERVICES FOR TEACHERS OF MATHEMATICS

Michael R. Silbert
Hamilton Board of Education
Hamilton, Ontario, Canada

What is Meant by 'Support Services for Teachers of Mathematics'?

Let me start by defining support services for teachers of mathematics to include those services provided to the teacher of mathematics in order to facilitate learning on the part of the learners in his/her charge. Such services may be tangible, material services (such as provision of textbooks or geoboards) or theoretical services (as is the case in the provision of alternative instructional strategies). These services may be direct or indirect.

In North America, the major focus, the 'preoccupation' tends to fall on the material aspect of support services. Yet, increasingly, the major need tends to be ideational in nature. Currently, this point is exemplified by the number of schools and school districts that acquire computer hardware for instructional purposes only to find out later that the required instructional software is either unavailable, inadequate or not suited to the needs of the learners. In a survey of 61 Ontario school boards released in June, 1980 (1), Didur has observed that "most concerns were about the hardware innovatins - - and little concern was expressed about software innovations!" Consequently, due to the constraints of time, I will be confining my remarks to the ideational aspect of support services.

Looking at the question from an operational point of view, one is faced with considerations of both development and delivery.

What Support Services Currently Exist?

As an individual intimately involved with both the development and delivery of support services, I have observed, over a period of years, the tendency for statutory organizations (such as state education departments and school districts) to put a much greater emphasis on the development of support services relative to teachers' needs and, conversely, the voluntary professional organizations such as state mathematics associations have found themselves primarily active in delivery of support services through the organization of conferences, professional meetings and the publication of professional periodicals. Unfortunately, there appears to me to be a rather serious mismatch of activities and resources.

The effective delivery of support services is undoubtedly a far more difficult and resource-consuming activity than the development of such services. Yet, the statutory institutions (which undoubtedly have the more stable and extended financial base) tend to allocate relatively more resources - - both human and material - - to the development of such things as curricular guidelines and policies. Both at the provincial and the local level, the development process (which tends to be finite and more clearly defined) receives, generally, appropriate funding; whereas, the delivery process (which tends to be more open ended and abstruse and, which by the nature of the process must be evolutionary) tends to be very seriously 'under-resourced' if supported at all!

Let me give you an example. In the fall of 1976, the Minister of Education, Mr.Thomas Wells, announced a tightening up of the secondary school program and indicated that students entering secondary schools would be required to take "an expanded core of mandatory subjects" in order to qualify for a Secondary School Graduation Diploma.

Consider the following scenario. On Friday, November 12, 1976 the Minister announced that five curriculum guidelines (including the mathematics guideline for all grade seven to ten courses) were to be revised and expanded by the spring in order that teachers might make use of them in preparing courses for September. A provincial committee was struck the week of November 15, and, in just under three months, developed a new curriculum guideline to replace courses that were developed ten to fifteen years previously.

On Thursday, February 10, 1977 the last page of Preliminary Draft I was typed. The following day, Friday, this document (of approximately thirty-five pages) was to be copied and sent to each of the school boards in the province for reaction in less than two weeks (by February 24, 1977)! Well, needles to say, there was a tremendous uproar 'from the field'. The net result was that attitudes towards any substantive departure in the curriculum guideline from accepted practice were almost exclusively negative. To make a long story short, there was a subsequent draft copy prepared which, today, is still not released in its finalized form.

If this case is exceptionally pathological, it is, unfortunately, not unrepresentative of many of the efforts put forth in the development and delivery of curriculum at both state and local levels.

Future Directions

Let me be so bold as to propose an alternative model for the development and, more importantly, the delivery of support services to teachers.

At present, there appears to be somewhat of a division between the development and delivery of services. While there may be consultation or consultative involvement between, for example, the professional teacher organizations and the state educatin department, there appears to be the need for a more collegial approach to the development of support services (including the sharing of material as well as ideational resources). In terms of the delivery of support services, there is the definite need for the development of some guerrilla-type strategies for use in the delivery process.

I have deliberately chosen to loosely use the metaphor of guerrilla warfare. Before you recoil in horror, let me reassure you with some disclaimers and an explanation.

First of all, I do mean warfare - - and I make no bones about it. However, this warfare is not directed at people but at the more abstract notions of inertia (or apathy), ignorance (or misunderstanding) and lack of priority (or lack of time) that infect us all in different ways and to differing degrees. I use the metaphor very loosely for I do not include terrorism as a strategy or demoralization as an objective.

Guerrilla warfare is a strategy for the morally strong and the materially weak. The young T.E. Lawrence, archeologist-turned soldier, in describing the process put it most succinctly: "Guerrilla war is far more intellectual than a bayonet charge." He went on to indicate that it requires no more than "a friendly population, not actively friendly, but sympathetic to the point of not betraying rebel movements to the enemy . . ."

We have entered an era in which we must face up to the fact that both the institutions and the professional organizations, even when working together, do not have sufficient resources to effectively deliver the variety and extent of ideational support services required by teachers now or in the foreseeable future. I strongly believe that we must take a page out of the guerrilla's handbook and work towards more flexible, cost-effective approaches in the delivery of support services and be willing to pay with the hard currency of time. (Time is, needless to say, the one resource where everyone gets exactly the same daily allotment!)

Let me give two examples from my own experience of the conscious application of this general principle. In 1976, our jurisdication initiated the "Applications of Mathematics Development Project" with the assistance of a pair of grants from the Ministry of Education. The stated goal of the project was to develop materials in order:

> to provide a new focus on problem-solving, a focus that is rooted in the view of mathematics as a useful tool in many areas of human endeavour. Through the use of real-world applications, an approach to problem-posing and problem-solving can be developed that is highly motivating and that provides students with a better understanding of the world in which they live.

The project team consisted of four teachers and four students in a consortium which collected applications and developed pseudoapplications for student use. While the teacher members were responsible for the suitability of content and the potential integration of these ideas into the program, the students were responsible for assessing potential learner/material interaction and for enhancing motivational impact on learners.

Approximately three hundred applications and pseudoapplications have been developed since the project's inception and, after field testing, much of the material has been diseminated to our schools. A new feature will be appearing in the "Mathematics Teacher" in September, 1980, edited by the project leader, Mr. George Knill, which focuses on these materials. But something far more important came from this project. The eight team members gradually developed an acute sensitivity to mathematics potentialities in the environment in which they worked and lived. They started using mathematical modeling as a more natural and more significant part of their everyday experiences. It was a process which they described as mathematization.

Arising out of this project were two distinct delivery problems - - the delivery of an ideational product and an ideational process. As the two were closely interwoven, we opted for a unified delivery approach.

Rather than focusing on a traditional implementation approach using formal communication networks on a restricted timeline, it was decided to deply a small cadre committed to both process and product which would use informal networks for the disseminatin of the product and the stimulation of the process.

Let me share several observations with you. First of all, while the effective dissemination of the product is not as extensive as what we ultimately hope it will be, it is beyond our initial expectations. This appears to be due predominantly to the fact that both the product and the delivery are strongly identified with peers and, to a much lesser extent, that informal student networks appear to have propagated positive interest in the product. Our assessment of the effectiveness of the delivery, I might add, comes through observations of reference to the product in individual and group discussions, in the minutes of departmental meetings and in courses of study. Official support through formal channels has been significant but very low key.

As far as effective delivery of the process is concerned, out impact has been, as we had expected, not as great, yet by no means unsatisfactory. The effectiveness of process delivery is strongly associated with the relational proximity of a member of the cadre. The cadre is gradually expanding and, certainly, process devliery has been effective at least to the extent that I would have expected as a result of substantial inservice commitment.

A second example is provided by the Mathematics Heads Groups in our jurisdiction. The Mathematics Heads Group is a formal group which has met, on a regular basis, monthly for over fifteen years. I recall that, when I first joined the group as a Mathematics Department Chariman nine years ago, the meetings tended to centre on discussion which did not lead to action, appeared to lack a clear purpose and direction and, tended to be dominated by a small number of individuals. All items for consideration were put on the floor and discussed by the group as a whole. Very little seemed to result from these meetings.

When I assumed my present position, I did two things in the hope of altering the existing situation. As an ex-officio member of the group, I asked that a committee be struck to look at the structure/function and, indeed, the existence of the group and that this committee report back to the group with specific recommendations. As well, I asked that the members of the group advise me, collectively, what a Head of Department does as a Head of Department. (After all, if we shouldn't "have meetings for meetings' sake" should we "have Heads of Departments just for the sake of having Heads of Departments?")

I should say from the outset that I did not consciously contemplate the potential interaction of these two considerations but, it was considerable.

The committee considering the group's organizatinal structure recommended that items for group consideration needed to be pre-screened by a small committee. The group, as a whole, accepted this recommendation for a variety of reasons (some members felt that "the length of these boring meetings could be cut down" and others saw "the opportunity to focus on the important items of business") and a Steering Committee consisting of the group's chairman, secretary and one member-at-large was set up.

At the same time, the Heads discovered that each Head had a different interpretation of his role and this caused considerable discomfort. Members of the Steering Committee, sensing potential in the situation, suggested the formation of a committee to consider the role of the Head of Department.

I have worked with the Steering Committee in an advisory/supportive capacity and they were starting to provide conscious active leadership (as if their future depended on it!)

In summary, the Committee to Consider the Role of the Head conducted an extensive survey (seven provinces, twenty-nine states) as well as a literature search and discovered a great deal in the process. The ultimate product was an agreed upon outline of the Duties and Responsibilities of the Mathematics Department Head.

This very rapid initial development has provided considerable momentum which led to a number of other interesting developments about which I do not have sufficient time to elaborate.

One interesting sidelight: The Committee to Consider the Role of the Head was deliberately comprised of three reasonable liberal individuals and one staunch conservative (who gained considerable in the process).

I am deeply concerned about the future of mathematics education as a result of certain decisions made in the light of the decline in enrollment resulting from several significant demographic and social factors.

Our jursidiction, like many others, has chosen to reduce teaching staff on the basis of seniority rather than ability.

I want to share with you briefly the potential impact of such a decision on the delivery of mathematics education in one component of our jurisduction - - our secondary schools.

During this past school year, 15,843 (equivalent) students were enrolled in mathematics courses taught by 97 full-time (equivalent) teachers. (In fact, approximately 12 percent of the mathematics timetable lines were taught by individuals whose major assignment was not in the mathematics department.)

From 1980 to 1985, our student population will decline by approximately 30 percent. On a purely probabilistic basis one would expect to lose 30 percent of the mathematics timetable lines and 30 percent of the teachers assuming no other adjustments are made. However, 19.3 percent of all mathematics teachers were hired from 1974 to the present yet, of all teachers (the total population including mathematics teachers), only 12.5 percent were hired from 1974 to the present. As a consequence, a policy based on 'total years of continuous Service to the Board' will potentially lead to the decimation of the present mathematics staff and severely affect the quality fo the program. Already we are involved in providing retraining for teachers who are entering mathematics teaching from other disciplines and the situation will likely accelerate.

The other issue that is starting to make itself felt is the educational and social impact of the new technology on the school.

To the present, we have spent a great deal of time establishing tools which, once established, have not often enough been applied and, hence, have tended to lie around and rust. To my mind, the cognitive end-product of mathematics is problem-solving - - the ability, having obtained the tools, to know which ones to use and to use them. We, unfortunately, have too often paid only lip service to problem-solving and have not used it as a whetstone to keep our tools and minds sharp. I say both tools and minds for it is the mind which dictates which tools to use and in what order.

There are many tasks which the human mind can perform exceptionally well and many others which efficient machines were appropriately designed. The ideal combination of mind and machine - - using the best features of both - - provides a vast new horizon for learning. In this our teachers need help for it is their responsibility to prepare the young people in their change to face an uncertain and rapidly changing future with a confidence that is built on understanding. It is only when teachers can boldly confront new ideas that they will be able to prepare their students to confidently face the future.

Mathematics, as presented in our schools, is on the move from being a rather deterministic-continuous oriented sytem towards one which is more discrete and, to some extent, heuristic.

School will be, I believe, strongly affected by the impact of social changes due to an increasingly distributed technology and, if society is not to pass them by, teachers will find themselves thrust into the roles of both agents for change and victims of change.

Ladies and gentlemen: these are the challenges that face us. While the challenges are great, the rewards of satisfactorily comquering them are great as well.

Thank you for your consideration.

Reference

1. Didur, D., A report on Computer Science in Ontario Secondary Schools, Educational Computing Organization of Ontario (ECOO), June, 1980.

SCHOOL-BASED SUPPORT SERVICES FOR TEACHERS OF MATHEMATICS

Max Stephens
Education Department of Australia
Victoria, Australia

In proposing to give special consideration to school-based support for teachers of mathematics, I do not wish to underestimate the practical support which comes from their professional associations. It is clear that these associations, operating at regional, state and national levels, provide important services for teachers. Conferences have tended in recent years to steer away from the "speaker-and-those-spoken-to" model, and have encouraged greater participation through workshops and materials developed. Journals, newsletters and a wide variety of informal publications have taken on a practical note of communicating to teachers information about successful programs and teaching approaches. Local branches and teachers' centers have also given teachers opportunities to discuss particular school programs and to observe them in operation.

As vehicles of support for teachers, professional associations and teachers' centers have as their weakness, as well as their strength, the voluntary nature of their membership. In the United Kingdom it is estimated that about one in five of all secondary full-time teachers of mathematics belongs to the Association of Teachers in Mathematics or the Mathematical Association. This pattern of membership supports my argument that support services for teachers of mathematics will be inadequate unless such services reach and involve a teacher in his/her own school. This is the essence of school-based support and will be the recurring theme of this presentation.

Special support must be provided for teachers of mathematics in elementary schools. Their needs are the least well addressed by our professional associations. Elementary/primary school teachers are generally not specialist teachers and cannot be expected to belong to a professinal association in every discipline they teach. However, their need for support, especially at school and local levels, is critical to the development of good teaching and learning. Too often services for elementary teachers tend to be limited to an exchange of worksheets, and do not address more fundamental issues of curriculum and teaching in the elementary school.

Three principal considerations have led me to argue for an extension of school-based support for teachers of mathematics. The first is a practical consideration, and stems from reduced financial support, in many countries, for in-service education of teachers and for teachers' center. An extension of school-based support may be the best means of compensating for these diminished services. The second reason is pedagogical: regardless of whether there is a well articulated syllabus for the state or region, or whether a school has adopted a particular textbook or program, it is clear that "what happens when the classroom door is closed" will differ from the pattern of implementation one would expect given only the syllabus or textbook at hand. The third consideration is in response to continuing pressure for schools to be accountable, and will be taken up in the final part of this presentation.

Although teachers do welcome the kind of support which clear guidelines bring to the teaching of mathematics, one cannot assume that the guidelines will be implemented as recommended. Guidelines need also to foster confidence among teachers. However, even where good initiatives exist at the systemic level, there is no substitute for sound co-ordination within the school itself. Co-ordination of mathematics teaching is an indispensible condition of any school-based support for teachers. Whether this is done through a designated position of head of department, or through some alternative structure, it is the role which is critically important. How schools organize that role will vary according to the size of the school and the country in question.

In co-ordinating the teaching of mathematics within a school, one of the key functions is that of maintaining links between teachers and those support services external to the school. This will include maintaining school memberships of regional, state and national associations of teachers of mathematics; it will usually entail liaison with subject specialists employed by the local authority; and equally important will be the maintenance of informal local networks.

A Gene Maier has pointed out, informal networks can also be powerful forms of support for teachers; their strength resides in their ability to respond to the expressed needs of teachers in a direct yet flexible manner. In Madison, Wisconsin, for example, a small group of elementary teachers has been meeting to share teaching ideas and to assist one another in the assessment of students' learning. Tests developed by this group have been adopted by teachers in other city schools, including many teachers who have not been part of the initiating group. In this way, a small group of teachers has been able to influence both the content and methodology of mathematics teaching in their local area.

The question, however, remains: how can these informal/voluntary groups and networks be influential among teachers who are not part of them? In almost all cases, the effectiveness of these informal networks requires a regular and clear pattern of communication among teachers within a school.

Informal networks can also benefit from connections with specialists and others who might have an official responsibility for supporting teachers of mathematics. Simple matters such as sending out notices and a degree of recognition can be fostered by connections with sympathetic supervisors and subject specialists. In addition, these networks can provide opportunities for teachers to have a measure of political influence.

Political influence is necessary at a time when mathematics teaching is exposed to the activities of various "predators", an apt term used by Professor H. B. Griffiths. He was referring to the danger of corruption and/or abolition of mathematics teaching by predators outside the profession. He may have had in mind the danger that what and how we teach will be influenced unduly by the media, politicians, and misinformed academicians, or by the decisions of school administrators who, faced with declining numbers of mathematics teachers, may decide either to reduce the time allocated to mathematics learning or to deploy into mathematics classes teachers whose training in the subject is barely adequate. On the other hand, Brian Griffiths was equally insistent in referring to another sort of predator: that which threatens corruption by incompetence from within the teaching force.

Because there is a need to guard against incompetent teaching of mathematics or mediocre teaching, one of the crucial tasks for the school is to review and evaluate its own curriculum and instruction. School accreditation visits and other formal reviews and inspections can assist in this task, but these are too infrequent and sometimes superficial; but this external form of accountability does not always assist in promoting evaluation from within a school, nor does it ensure that evaluation is closely attuned to the resources, capacities and obligations of a particular school.

Continuous evaluation by teachers of their own mathematics programs is an integral part of a larger process by which a school gives an account of its teaching to is own community and to the profession. Should the employing authority and/or the government require that schools undergo formal evaluation on a regular basis, the pedagogical (as opposed to the political) value of this kind of evaluation will depend upon the ability of teachers as a group to have reflected critically, and on their own terms, about their implementation of the school's mathematics program. Their critical insights can be enriched by contacts with teachers' centers and professional associations.

A strong pressure for accountability is likely to remain with us throughout the 1980s, as an inevitable concomitant of the financial stringency affecting education in so many countries. Its predatory aspect is that it sets the criteria by which teachers shall be accountable in their teaching; and while an accountability movement has a powerful influence on what and how teachers teach, its direction is seldom open to input from teachers themselves.

For these reasons it is vital that teachers, supported by administrators, mathematics specialists and college/university teachers, establish and direct their own accountability procedures; procedures by which schools will be seen to be accountable themselves and to their community. This does not necessarily call for an elaborately structured program of evaluation, but it does require that the first initiatives be taken by the schools themselves. The fact that this form of accountability is public and assumed to be effective reduces the possibility that less welcome forms of accountability will be imposed from the outside.

Unless school-based support can be organized to assist teachers in being seen to be responsible, teachers are likely to be made accountable in ways which are possibly less responsive and probably more arbitrary. I believe that Brian Griffiths was not speaking extravagantly when he referred to these forces as threatening corruption and possibly the abolition of mathematics teaching as we know it. In this context, support services for teachers of mathematics - - from state and district employing authorities, from professional associations, teachers' centers and informal networks, and from within the school itself - - will continue to be a major priority throughout this decade.

5.14 WHAT IS A PROFESSIONAL TEACHER OF MATHEMATICS

WHAT IS A PROFESSIONAL TEACHER OF MATHEMATICS

John C. Egsgard
Twin Lakes Secondary School
Orillia, Canada

Teachers of mathematics can be found in every country of the world teaching mathematics to children and adults, to the illiterate and the learned, to the eager and the reluctant. Indeed, the most important person in all of mathematics education is the teacher of mathematics. If there were no teachers of mathemtics at the elementary and secondary school level, there would be none at the post-secondary level. There would be no need for schools of education to instruct people how to teach mathematics. No one would be able to do research in mathematics education. Thus, it is fitting at this Fourth International Congress on Mathematical Education that some time is being taken to appreciate the basic element in mathematics education - - the teacher. In past Congresses there has been the tendency to look at the more esoteric elements in mathematics education; to try to discover the theoretical principles that effect the teaching of mathematics. The teacher of mathematics per se has been largely ignored.

In this session, and the next to follow, we are going to concentrate on this forgotten but important person. We will examine the elements that make a professional teacher of mathematics. All of this is being done in the hope that professional teachers of mathematics will be encouraged by the fact that they will recognize themselves in this; that other teachers of mathematics will have a model they can strive to imitate; that young men and women who are thinking of becoming teachers of mathematics will know what will be expected of them.

What is a professional teacher of mathematics? Stated simply, a professional teacher of mathematics is one who is a good teacher of mathematics. But what is a good teacher of mathematics?

> A good teacher of mathematics is one who uses his knowledge and love of mathematics as well as his love and respect for his students to lead these students to enjoy the study of mathematics.

Notice that I have said nothing about the student being successful, for I believe that a student who enjoys the study of mathematics will learn mathematics, will be successful. Observe also, that this definition makes it very simple to determine whether or not someone is a good teacher of mathematics. It is only necessary to discover the attitude toward mathematics of the students of that person.

Let us examine the definition more closely. This definition of a good mathematics teacher indicates that there are two key elements that will lead a student to enjoy the study of mathematics. One is the love and knowledge the teacher has of mathematics. The other is the love and respect the teacher shows his students. In other words, a very personal relationship must exist between teacher and students, founded on the mutual love and respect they will show towards each other.

In what manner does a good teacher show his love and respect for his students? The answer to this question will depend somewhat on the individual teacher and upon the cultural background of this teacher and his students. I will indicate some of the methods that I use that are common to good mathematics teachers in Canada and the United States.

I teach high school students who range in age from 14 to 19 years. Daily for five days each week, I meet six classes of mathematics students in periods that last forty minutes. This past year the total number of students I taught each day was 172.

In order to show students that I do love and respect them, I try to give each one as much individual attention as possible. To solve the problem of doing this in classes of varying size, I use the socractic method of teaching that involves questions and answers - - the questions being asked by both teacher and student and answered wherever possible by the students. This enables me to ask most if not all of the students one question each day. This question provides daily personal contact with each student. Ordinarily also twenty of the forty minutes of class time are given over to students working alone on problems. Again this allows me the opportunity of personal contact as I go from desk to desk answering questions, praising good work, encouraging the weak and giving hints. In addition, I am available before and after school to help students who have been away or who have difficulties that have not been cleared up in class time. In all of this I find that praise and encouragement tend to help the student to try harder whereas sarcasm and sharp criticism cause them to become discouraged and sullen so that they stop making an affort.

All of us will agree that students must practice, must work if they are to learn mathematics. This frequently involves doing problems at home. I find it more effective in aiding the growth of self-discipline in my students to assume that they will do their work without being checked. Rather than examining carefully the homework of all students, I concentrate on the ones who have shown by lack of knowledge on tests or attitudes in class that they are not doing their work.

Students want us to think that they are more than just students of mathematics so that I find that I can increase my personal relationship with students if I show a genuine interest in their other activities - - academic, cultural, athletic, etc.

In all of this effort on the part of the teacher to show love and respect the primary quality that must be evident is kindness. Kindness is easy to give to someone who shows us love and respect. But kindness demands a true self-discipline on the part of the teacher as he tries to encourage students who appear to be unaffected by his efforts. Meanness, lack of work, rowdiness, etc. by the students must all be treated firmly, but kindly. It is a slow and sometimes discouraging process trying to change a student who dislikes mathematics and all teachers into one who is happy to be doing mathematics in your class. Each year I find students in my classes who begin to enjoy mathematics as the year is drawing to a close. The fact that these students do change gives me the confidence I need to continue working with all students no matter how hopeless it may seem.

Yet all of the kindness, love and respect for students will not help unless the students find that they are learning mathematics, sometimes in spite of themselves. Indeed, successful learning of mathematics is an absolute necessity for any student who is going to enjoy mathemtaics. For this to happen the teacher has to teach so that the student can learn. Too many teachers of mathematics - - and these cannot be called good mathematics teachers - - think that mathematics should be a difficult subject. Rather than teaching mathematics so that it is as easy as possible as a good teacher does, such poor teachers of mathematics seem to take delight in making the simple things appear to be complex. There are many ideas in mathematics that are difficult in themselves, but the good teacher of mathematics, the truly professional teacher, is able to peel away the unnecessary complications and to present the core ideas so that these ideas can be understood. This is one of the key qualities that the good mathematics teacher has that enables him to lead his students to the enjoyment of mathematics. A good mathematics teacher, through his questions and explanations, is able to help all of his students to understand and successfully do mathematics. Every time a student does a mathematics problem successfully, especially if success seemed impossible to the student, that student has made a giant step on the road to the enjoyment of mathematics.

Each student must attain some level of success in problem solving if he is to experience the joy of studying mathematics. The good mathematics teacher continually indicates the importance of academic excellence to his students but must be ready to accept it in his students in varying degrees. In Canada and the United States we have the phenomenon of the slow learner - - a student who is only supposed to be able to learn mathematics with much difficulty. Izaak Wirszup, in a recent report to the National Science Foundation of the United States pointed out indirectly that there are no slow learners in the Soviet Union. In the Soviet Union, all students go to school for ten years and all students study mathematics each year up to and including the calculus. In the United States and Canada many students drop out of school before the twelve years of elementary and secondary school are finished and of those who remain only a select few take the calculus. Also in Canada and the United States the slow learner is ordinarily a student who has the ability to learn but who has decided that he does not enjoy mathematics - - and/or school itself - - and makes little or no attempt to learn. If every teacher of mathematics in both elementary and secondary school in Canada and the United States was a good teacher of mathematics, as I have defined one here, then there would be no slow learners of mathematics in these countries. The same can be said of any other country that can claim the slow learner among its students.

All of the previous discussion of a good mathematics teacher leading students to enjoy the study of mathematics makes two assumptions. One that the teacher has the necessary knowledge of mathematics; the other that the teacher enjoys mathematics himself.

What is the necessary academic knowledge for a good teacher of mathematics? The answer is simple - - the best possible academic knowledge of mathematics.

In general, as a minimum, a teacher of mathematics must know all of the material that he will need to teach as well as the place of this material in the spectrum of the mathematics curriculum. The extent of this knowledge should be determined for each country or part thereof by a committee consisting primarily of elementary and secondary school teachers, assisted by current teacher educators and a few post-secondary mathematicians who have demonstrated an interest in elementary and secondary mathematics education.

This process of learning mathematics is never finished for a good mathematics teacher. For, no matter how well a teacher is prepared, he must continually learn; for example, by reading the national and local mathematics journals such as The Mathematics Teacher or The Arithmetic Teacher of the NCTM; by attending meetings of national or local organizations of mathematics teachers; by taking courses to keep his knowledge up to date.

In addition, good mathematics teachers need other good mathematics teachers who will be leaders of their profession. Much is demanded of a good mathematics teacher. It is the obligation of the profession itself to see that mathematics teachers get as much assistance as possible in their growth to become better mathematics teachers. Professional organizations, both at the national and international level like the NCTM and local organizations of cities and states are an essential part of the growth of the good mathematics teachers. These organizations, if they are to be helpful to teachers at all levels, elementary, secondary and post-secondary, must be led by teachers from all levels. One of the gravest mistakes that organizations like this can make is to think that such representation is not needed. Mathematics organizations will do more good if the articcles in their journals and the talks at their meetings, are given by teacher experts who are communicating with people at their own level be it elementary, secondary or post-secondary. For teachers can learn best from teachers who are experts at the same level. All of the good will in the world cannot make up for the fundamental error made in assuming that the elementary teacher learns best about teaching elementary mathematics from secondary or post-secondary teachers in their classroom practice. Teachers at the post-secondary level can and must assist elementary and secondary teachers to increase their knowledge of mathematics. It is only rarely that post-secondary school teachers are able to give worthwhile practical help, useful in the classroom, to elementary and secondary school teachers. The good secondary school teacher must be willing to share with other secondary school teachers. The good post-secondary school teacher must find a way to show other post-secondary school teachers that good teaching is important and possible at that level.

Now that we know what is meant by a good teacher of mathematics, it is reasonable to ask the question, Do such teachers exist? From my own contacts made at the previous three International Congresses on Mathematical Education and also during my two years as President of the National Council of Teachers of Mathematics I can assure you they do. Yet the dislike for mathematics found among the non-mathematically oriented public seems to indicate that such teachers are in short supply.

I wonder how many people were taught by a teacher whom they loved and respected; one who helped them in many areas; but, one who unfortunately was a person who feared mathematics and/or disliked it? Until we get a generation of children who are happy in mathematics, with teachers who are happy teachers of mathematics, then there is little likelihood that we can change the public attitude towards mathematics.

The question school teachers that good teaching is important and possible at that level.

Now that we know what is meant by a good teacher of mathematics, it is reasonable to ask the question, Do such teachers exist? From my own contacts made at the previous three International Congresses on Mathematical Education and also during my two years as President of the National Council of Teachers of Mathematics I can assure you they do. Yet the dislike for mathematics found among the non-mathematically oriented public seems to indicate that such teachers are in short supply.

I wonder how many people were taught by a teacher whom they loved and respected; one who helped them in many areas; but, one who unfortunately was a person who feared mathematics and/or disliked it? Until we get a generation of children who are happy in mathematics, with teachers who are happy teachers of mathematics, then there is little likelihood that we can change the public attitude towards mathematics.

The question now remains, Can one decide whether or not a given individual should become a teacher of mathematics?

First let me make some statements that follow logically from my definition and my discussion on it.

1. Only a person with the required knowledge of mathematics for level X should become a teacher at level X.

2. Only happy persons should become teachers of mathematics.

3. Only persons who enjoy other people and want to help other people should become teachers of mathematics.

4. Only teachers who want to become mathematics teachers should become teachers of mathematics.

5. Only teachers who love mathematics should become teachers of mathematics.

Note that this is a list of necessary conditions, not of sufficient conditions for determining whether or not an individual should bcome a teacher of mathematics. In other words, no one lacking one of the five qualities should become a teacher of mathematics. But it does not mean that everyone who has these five qualities will become a good teacher of mathematics.

I believe that the best way to determine whether or not a person with the necessary conditions to become a mathematics teacher also has the sufficient conditions to assign that person assign that person as an apprentice to an outstanding teacher.

Experience has shown us that many students who raduate from teacher training institutions in mathematics are not ready to become good teachers of mathematics. Indeed, many of these teachers give up after a short time in the classroom. Those who do go on to become good teachers of mathematics usually have had much assistance from their peers who are good teachers of mathematics.

Much time and effort in all countries of the world has gone into the training of teachers of mathematics by dedicated teacher educators. During the summer of 1976, the International Congress for Mathematics Education met in Karlsruhe. One section was devoted to the preparation of teachers of mathematics throughout the world. Much research went into the development of a document called "The Education and Professional Life of the Mathematics Teacher". The people who wrote this document were unable to come up with an effective example of teacher education that they felt could be used universally. Notice that I did not say that they felt they were unable to find examples of effective teacher education. The researchers thought that there was enough variation throughout the world that no one program would be useful for all. Yet my personal belief is that the effectiveness of a program for the preparation of mathematics teachers is directly proportional to one thing and inversely proportional to another. First, I believe that the effectiveness of a program varies directly as the amount of time the teacher trainee spends teaching in a classroom in an apprenticeship role under the direct guidance of an outstanding teacher. Second, I believe that the effectiveness of a program in mathematics teacher education is inversely proportional to a time t. I define "t" as the number of years that have elapsed since the teacher educators have spent time effectively teaching at the level at which they are attempting to prepare their teacher trainees. In other words, the closer a teacher educator is to his or her classroom teaching experience, the more he or she will be able to assist the teacher trainees.

One of the difficulties of the present system of teacher training in colleges of education throughout the world is that there are colleges of education where some of the faculty has lost touch with what is going on in the classroom. There are people expounding ideas on how to teach that have never tried these ideas in a class of children for any length of time. Indeed, there are some mathematics educators trying to explain how to teach at the secondary or elementary level who have never had a full time job in either an elementary or secondary school. As Confucius said, "It is not what teacher educators do not know about teaching that hurts their teacher trainees, but what they do know for sure that turns out to be wrong".

The real learning about teaching and about whether or not one should become a teacher of mathematics comes in the classroom. I firmly believe that one learns to teach by doing not by listening. An apprenticeship type of training for beginning teachers would require "experts" to whom they would be assigned in elementary or secondary schools. These "experts" would need to be successful teachers who are currently teaching in their own classrooms each day. A maximum of three teacher-trainees would be apprenticed to each of these teachers for a minimum of one year. As in other apprenticeship programs, the trainees would be paid a nominal salary. Under this alternative, there would be not need for preservice courses in the philosophy or psychology of education, classroom practice, and so on, since the important ideas come from these courses would arise in the practical sphere during the year that the trainee spent with the cooperating teacher. Only experience would tell us if these ideas would need to be reinforced in courses given in the summer after the apprenticeship.

As I envision the program, the "expert" teacher and the trainee would be responsible for the same classes. Ample time would be set aside in the school day for them to prepare classes together, to decide what was to be taught, how it was to be taught, the questions that would be asked of the pupils, and so forth. Some days the teacher would teach all the classes and on other days the teaching would be shared with the trainee. Time would be taken daily to discuss both the teacher's and the trainee's classes so that the trainee could understand the why as well as the how.

Every country, every city and town, every school, every classroom, every pupil has a great need for good mathematics teachers. I think that apprenticeship as I have described it is our best way of assuring these teachers will be developed.

I have been a teacher of mathematics now for nearly thirty years. During that time I have had many students ask me why I have been content to teach the simple mathematics of the high school level when I could be challenged more by university level mathematics. They openly wondered why it was not boring to teach the same mathematics year in and year out. My reply has been a simple one. "I teach students rather than mathematics". Even though the mathematics does not change from year to year, students do. Each year I have a new group of individuals to be with in my classroom. My thirty years in the classroom have been happy ones. I have treated my students with love and respect and they have reciprocated. We all have much fun in my classroom, yet we all learn. But my attitude of love and respect for the students is not enough for them. They must also be successful. Somehow in each of my classes I have been able to find something that every student can do well. I have been fortunate in being able to do this in classes with four students and in classes with forty-five students. The fact that my students have learned and that they have enjoyed themselves learning, has brought me great joy.

Teaching children is difficult. Many people try it and cannot handle the one to one and many to one relationships that are a part of teaching. Yet, too much stress is given to the difficulties of teaching. We must encourage our young teachers to look for more joy in their teaching. We must assure them that it is there. One of the reasons I think that apprenticeship is an essential part of teacher preparation is that it gives the teacher trainees an opportunity to share in the joy of teaching - - a joy that many may believe does not exist.

We must continue to find good teachers of mathematics, professional teachers of mathematics. I hope this paper helps to understand better whom we are looking for and how we can find them.

WHAT IS A PROFESSIONAL TEACHER OF MATHEMATICS

Jacques Nimier
IREM de Reims
Reims, France

Je remercie beaucoup le professor John C. Egsgard d'avoir autant insisté sur les phénomènes affectifs. Différentes recherches menées dans plusieurs pays confirment leur importance dans l'enseignement. Mais je voudrais plus particulièrement insister sur l'attitude du professeur à l'égard des mathématiques, car finalement: "Pourquoi devient-on professeur de mathématiques?" - Qu'est-c que fait qu'un professeur a telle ou telle attitude à l'égard des mathématiques? - Comment expliquer qu'un même professeur peut "réussir" auprès de certains élèves et pas auprès d'autres?" Autant de questions qui, à mon avis, ne peuvent avoir un début de réponse que par la prise en compte de la dimension inconsciente de toute attitude.

Une première recherche portant sur près de 1,500 élèves de 3 pays différents m'avait premis d'étudier les roles que ceux-ci faisaient jouer inconsciemment aux mathématiques dans la problématique de leur personnalité, et actuellement j'essaie de savoir ce qu'il en est pour les professeurs de mathématiques, grâce à 900 professeurs de mathématiques francais qui ont bien voulu répondre à mon enquête.

Certains d'entre eux, qui aiment les mathématiques sont particulièrement attirés par leur unité, leur cohérence, leur capacité à aller de la diversité à l'unité; ils ont l'impression que dans un problème, là où il y a apparence de chaos, il y a en définitive une organisation à retrouver. On peut émettre l'hypothèse que pour ces professeurs, l'objet mathématique symbolise l'unité de leur personne qu'ils cherchent à acquérir. Ils seront sans doute particulièrement portés dans leur enseignement à montrer aux élèves cette unité, cette cohérence. Ils seront peut-être aussi particulierement sensibles à l'unité de leur classe.

D'autres professeurs aiment les mathématiques pour leur puissance, soit comme instrument au service des autres sciences, soit comme moyen permettant une maîtrise totale d'une situation: "En mathématiques on a toutes les données en main".

Ils pensent acquérir cette puissance par appropriation de cette discipline un peu comme certains peuples pensaient s'approprier la force de leurs adversaires en mangeant leurs cervelles. Ces professeurs ne seront-ils pas portés à insister sur l'aide qu'ils peuvent apporter à l'élève "qui ne sait pas", ce non-savoir étant le fondement et la preuve de leur propre savoir. Peut-être que certains succomberont alors, comme l'a si bien dit le professeur Egsgard, au plaisir de rendre difficile ce qui peut être expliqué simplement? Par contre, d'autres se sentiront fiers d'aider les élèves à s'affronter au défi que leur opposent ces maths puissantes. Ils présenteront alors les mathématiques comme un véritable combat où toute décourverte de solution est une victoire.

Pour d'autres professeurs, les mathématiques sont un autre monde, un univers dans lequel on peut pratiquer un jeu parfois solitaire et où l'on peut s'évader et se retrouver à l'aise loin des soucis, des déceptions et des contraintes du monde quotidien. Les mathématiques sont alors "une drogue dure qui n'en a pas les inconvénients" comme disait l'un d'entre eux. Ces professeurs aimeraient montrer à leurs élèves comment les mathematiques sont "libératrices", ils insistent sur leur aspect "ludique, non dangereux, sans affrontement inquiétant, sans conséquences graves".

Par contre, pour certains les mathématiques sont avant tout un objet rigoureux. Ils aiment l'ordre, la précision et ont le souci de former leurs eleves a un raisonnement correct. Mais ce souci de raisonner juste ne couvre-t-il pas parfois une peur inconsciente de derâisonner? En tout cas, cet ordre et cette rigueur leur paraissent importants comme si la fonction qu'ils font jouer aux mathématiques était celle de barrière, de limite et d'obstacle à des tendances ou pulsions répréhensibles. Ces professeurs aimeront que l'ordre règne dans leur classe, que chacun y parle à son tour, qu'il y règne une bonne ambiance de travail, que les élèves travaillent beaucoup.

Enfin, nous pensons aussi à ces professeurs pour lesquels l'amour des mathématiques est amour de la verité. Amour de ce qui est beau, harmonieux, amour de la structure qui régit le monde ou encore forme unique et parfaite du fonctionnement de la pensée humaine. Cette idéalisation des mathématiques provient d'un clivage permettant de maintenir loin de soi la faille des échecs, des approximations, des inconnues... Elle peut être source d'éblouissement et d'emerveillement propres à rentrer parfois en raisonnance avec certains eleves, mais cause parfois de réveils douloureux.

Il existe encore bien "visions" des mathématiques qui entrainent des attitudes très différentes à l'égard de cette discipline. La question que se pose au professeur de mathématiques qui désire prendre une certaine distance vis à vis de ses propres désirs afin de garder plus de souplesse dans son enseignement, pourrait etre celle-ci: quelle fonction suis-je tenté de faire jouer à l'objet mathematique dans la dynamique de ma personnalité: Les mathématiques sont-elles un objet paré de qualités? Ou bien, les mathématiques sont-elles un objet revêtu de propriétés idéales qui me fascinent et qui répresentent pour moi un idéal à atteindre? Ou alors, les mathématiques sont-elles pour moi une loi rigoureuse gardienne de l'ordre face à l'exhubérance du désir et au service d'une conscience surmoique que certaines pulsions inquiètent?

Il ne s'agit là que d'éléments à prendre en compte pour comprendre la situation didactique. Cette dernière n'est pas seulement faite d'élèves en apprentissage conformément à certains programmes et de maitres ayant choisi certaines pédagogies facilitant cet apprentissage. La situation didactique comprend une relation maître-élèves dans laquelle se fait, au travers de gestes, d'attitudes, de silences, de réflexions, une communication des inconscients du professeur et de l'élève où se transmettent l'attente du professeur et de l'élève, leurs réponses et, ce qui se joue justement à ce niveau fantasmatique, ce que répresentent pour chacun d'eux les mathématiques. Le discours du professeur sur sa pédagogie ne sera, bien souvent, qu'une "rationalisation" lui donnant une justification de son action bien plus profondément enracinée dans son désire et sa fantasmatique.

Quand on parle des mathématiques et de leur enseignement, on pense souvent ou bien à leur enracinement dans la réalité, l'expérience, la vie, ou bien à leur abstraction, a leur aspect symbolique et leur aspect linguistique. Mais dans là plupart des cas, on "court-circuite" inconsciemment leur aspect imaginaire et fantasmatique. C'est pourtant là que s'enracinent les investissements tant des élèves, comme je l'ai montre dans mes travaux, que des professeurs. Il parait urgent que cette dimension soit prise en compte dans toute formation de professeurs de mathématiques.

J'appuierai volontiers les dires du professeur Egsgard au sujet de l'importance d'une formation à partir de stages en situation. Ces stages sont à mon avis plus importants que tous les cours théoriques de psychologie ou de pedagogie. Je proposerai qu'une partie du temps de ces stages soit consacrée à des échanges entre stagiaires sur leur vécu affectif des mathématiques à l'occasion d'évènements survenus en classe. Cela donnerait l'occasion aux stagiaires de prendre conscience, en partie, de la fonction que jouent les mathématiques pour eux et pour les autres stagiaires, et assouplirait leur perception des mathématiques. Cela leur permettrait également de découvrir la dimension imaginaire de cette discipline et les aiderait à accepter les différences de vécu chez leur élèves.

Pour illustrer ce propos et mettre en evidence cette dimension imaginaire, je voudrais vous rapporter un passage d'entretien d'un mathématicien qui m'explique ce que représente pour lui le fait de faire des mathématiques.

"P.- Oui, mais pratiquement si vous voulez, quand on fait de la recherche c'est quand même pas tout à fait comme ça que ça se présente parce que, si vous voulez...c'est une comparaison que j'emploie souvent avec mes étudiants, les mathématiques c'est une espèce de forêt vierge, n'est-ce pas, enfin les mathématiques de la recherche; alors plutôt quand vous faites de la recherche c'est vraiment une forêt vierge que vous avez à...non pas tellement à explorer, mais à...il faut percer: vous avez un certain point qui est à atteindre, vous savez que c'est a peu près par là, et il faut percer, et si vous percez comme ça tout droit...il y a des gens qui font ça,...ça permet même je crois de diversifier un petit peu les races de mathématiciens: c'est qu'il y a des gens qui ont...une force extraordinaire...n'est-ce pas ils y vont au coupe-coupe comme ça, ils scient tous les arbres qui sont devant eux au bulldozer...et ils ont une telle force qui ils arrivent...ils arrivent à faire des choses intéressantes. Mais je crois que ce n'est pas ca qu'il faut faire, enfin ce qu'il faut faire c'est un petit peu tâter le terrain, se dire ben (!), puis alors là il y a évidemment le flair, et se dire bon (!) si je vais tout droit, non c'est tres brousailleux par-là, il vaut peut-être mieux obliquer un peu par-la, et puis alors la, ben la, ca a l'air plus clair, bon, etc. et puis alors à chaque fois que vous avez devant vous vraiment un gros obstacle, une montagne, un roc, un...il ne faut jamais y aller devant comme ça, il faut toujours essayer de tourner...

"P.- Alors finalement je crois que l'important...c'est pas tellement d'avoir trouvé un chemin qui, partant du point A, aille vers le point B qui était, qui constituait le but; ça a été de trouver des chemins, j'allais presque dire des chemins royaux...vous savez, il y a...je crois que c'est peut-être Ptolemée qui disait qu'il n'y avait pas de voie royale en mathématique, c'est vrai et c'est pas vrai, c'est-à-dire que dans cette forêt vierge des mathématiques ...il y a quand même des chemins, n'est-ce pas, et puis, bon, les "mutter structur" de Bourbaki, n'est-ce pas les structure-mères, bien finalement ces structure-mères elles se sont dégagées peu à peu et ce sont des chemins royaux, n'est-ce pas, ce sont des chemins, . . . c'est les grandes autoroutes à l'interieur de la forêt vierge; donc il y a quand même...des chemins privilégiés et c'est une forêt vierge qui n'est pas homogènement embrousaillée, si vous voulez,...il y a quand même des points forts, des points de résistance, des points faibles, des points...et tout le travail du mathématicien au fond, ça consiste à éprouver...la resistance...de ces murailles...qui sont devant lui...et a flairer quels vont être les points plus faibles que les autres, les points de passage..."

Enfin, il termine par cette phrase:

"M.- C'est important pour vous, ce problème de l'unité?"

"P.- Ah! oui, oui parce que, alors là...c'est toujours cette espèce de rêve que j'ai en moi...de condenser, d'avoir un instrument, qui me permette de condenser au maximum, qui me permette d'avoir prise sur...un immense empire, si vous voulez, au moyen justement...de tout petits germes, de tout petits embryons...n'est-ce pas?"

Ceci n'est qu'un exemple, mais il peut permettre de comprendre combien le travail de recherche en mathématique comme celui de l'enseignant de mathématiques, comme celui de nos élèves se fonde, s'enracine, trouve sa sève dans une fantasmatique personnelle, qui en assurant sa motivation, sa dynamique en fait son originalité. Un professeur de mathématiques expert sera un professeur qui saura prendre en compte cette dimension imaginaire des mathématiques dans son ensignement; il comprendra alors pourquoi il n'est pas évident que plus on explique à un élève, mieux il comprend. L'explication n'etant recevable par lui que dans la mesure où elle est fondée sur une demande de l'élève, sur un désir de l'eleve. C'est alors que ce professeur saura, comme l'a dit le professeur Egsgard, amener ses élèves sur la route du plaisir des maths.

QUE ES UN EDUCADOR PROFESIONAL DE MATEMATICA?

Leopoldo N. Varela
Buenos Aires, Argentina

Como Director del Proyecto de Matematica, del Programa para el Mejoramiento de la Ensenanza de la Ciencia, que en Argentina auspicia el Consejo Nacional de Invetigaciones Cientificas y Tecnicas, me vi hace dos anos en la necesidad de responder a la pregunta que plantea esta conferencia.

Hay en mi pais catorce mil profesores de matematica en el nivel scundario a quienes pensamos llegar con nuestro curso.

Un censo general y una prueba de nivel inicial nos respondieron a la pregunta, como es el profesor actual de matematica? El segundo paso era tratar de determinar el perfil profesional del profesor de matematica. Debiamos, entonces, responder a la pregunta del titulo, para poder encaminar nuestro trabajo.

Pensamos que debiamos tomar en cuenta, al menos, tres protagonistas:

- el alumno al cual se consagraria el trabajo del docente,
- el docente mismo,
- la matematica

y las necesidades que surgian de cada uno de ellos.

Por supuesto que sabiamos que con ello no agotabamos las variables a tener en cuenta. Como telon de fondo tendriamos en mente la estructura de nuestro sistema educativo, las condiciones ambientales en que se desarrollan las clases, las necesidades de la sociedad en la que vivimos, etc., etc.

Como se ve, para iniciar nuestro trabajo habiamos elegido la misma terna que aparece en la definicionque recien nos propuso Mr. Egsgard.

Veremos que esta no es la unica coincidencia.

Tampoco nos asombra demasiado que asi sea. Salvando las distancias, respecto del valor de los conceptos involucrados, recordamos las palabras de Wofgang Bolyai a su hijo:

> "... porque hay tambien algo de verdad en esto: que muchas cosas tienen una epoca, en la cual son descubiertas al mismo tiempo en muchos lugares, precisamente como en primavera brotan las violetas en todas partes..."

Mr. Egsgard nos ha invitado a que manifestemos como tratamos de mostrar nuestro amor y respeto por los estudiantes.

Consideramos que el amor y respeto hacia los estudiantes consiste en ayudarlos a que:

- desarrollen y perfeccionen sus capacidades potenciales relacionadas con las operationes mentales ligadas a la matematica,

- logren su autoafirmacion personal mediante el conocimiento de sus capacidades y limitaciones,

- comprendan y utilicen las ideas de la matematica de su epoca,

- vean que esta ciencia es un objeto cultural valioso, obra del espiritu creativo, y que ellos mismos pueden obtener placer intelectual si adoptan actitudes y realizan acciones creativas o recreativas ante los problemas que plantea la matematica,

- aprecien los valores esteticos de esta ciencia,

y tomando un concepto del Dr. Polya:

- Desarrollen actitudes eticas que les lleven a estar dispuestos a revisar cualesquiera de sus creencias, cambiandolas si existen razones para ello o manteniendolas en caso contrario.

Quiza todo lo anterior pueda resumirse diciendo que, a nuestro juicio, mostraremos amor y respeto por nuestros alumnos si les permitimos y ayudamos a ser personas autonomas, gozosas con su actividad intelectual y coherentes con los resultados de sus pensamientos.

Como lograrlo? se pregunto Egsgard. Nos dijo que estaba convencido que solo es posible si se parte de una profunda relacion persoal con cada uno de los alumnos y nos ha contado como lo hace el, al mismo tiempo que nos ha pedido que expliquemos como tratamos de hacerlo nosotros. Tambien senalo que creia que seguramenta ello dependera de la idiosincracia del profesor y de sus alumnos. Otra vez coincidimos. Permitaseme, entonces, narrar suscintamente mi propia experiencia para fundamentar esta opinion.

Tambien, llevo alrededor de treinta anos en la tarea de ensenar.

Comence como maestro en una Escula Unitaria. Se llaman asi, en mi pais, a escuelas - generalmenta ubicadas en zonas rurales - donde veinte a veinticinco alumnos, con edades que oscilan de los seis a los catorce o quince anos, estan a cargo de un maestro unico. En estas escualas el maestro debe repartir su atencion entre alumnos que van desde aquellos que estan aprendiendo sus primeras letras hasta esos otros que preparan su ingreso a la escuela secundaria.

La heterogeneidad del alumnado me llevo obligatoriamente a una ensenanza totalmente individualizada en la que cada alumno debe trabajar en su propio tema: este luchando por dibujar bien su primera letra, aquel discutiendo un problema de regla de tres, el de mas alla ubicando en el mapa un pais europeo.

Esta situacion obliga al maestro a conocer profundamente a cada alumno, sus capacidades y habilidades, sus dificultades en el aprendizaje, las posibles causes de las mismas... Acercarse a cada uno, conversar, estudiarlo, estimularlo,... Prepararle trabajos especiales: fichas, lecturas, cuestionarios, guias... Es claro que entonces el alumno comprueba dia a dia, momento a momento, la atencion personal que el maestro le presta.

Esto tipo de trabajo obliga a lograr la tarea responsable del alumno, y aqui otra vez coincidimos, el medio fundamental es el carino, el respeto y la confianza.

Aprendi, en esos anos, cuan cuidadoso y cuan eficas - a veces mas que nosotros mismos - puede ser un alumno aventajado ensenandole a otro que se inicia. Aprendi como la libertad responsable y la imposibilidad de apoyarse demasiado en el maestro ayuda a los alumnos a lograr autonomia y a buscar y brindar cooperacion. Aprendi lo importante que resulta que el alumno goce con sus labores.

Anos mas tarde tuve a mi cargo un curso con alumnos de los dos ultimos grados de la escuela elementa: algo mas de veinte alumnos entre once y trece anos. Aqui cambie el metodo en lo formal pero siguio basandose en la confianza y el respeto hacia los educandos.

El aula se transformo en una empresa periodistica. Editariamos una revista. Como el alumnado provenia de familias de escasos recursos economicos y la misma escuela era pobre, la revista debia financiarse con la venta de avisos (estaba prohibido solicitar ayudas o contribuciones desinterasadas). Esto obligaba a los alumnos, convertidos para ello en administradores dentro de la escuela y promotores fuera de ella, a vender avisos, calcular, discutir, decidier... Cuando nos transformabamos en periodistas habia que entrevistar a personas, redactar criticas, buscar datos... la entrevista a un vecino extranjero obligaba a verificar sobre el mapa la informacion que nos suministrara, calcular las distancias con escalas, calcular superficies aproximadas...

Todos, ellos y yo, compariamos la tarea trbajando codo a codo. Se dicutian y analizaban las ideas de todos y los alumnos proponian la resolucion de tantos problemas cuantos proponia el maestro. Era habitual que fueran los alumnos los que dijesen: "debemos calcular esto", "ahora deberiamos discutir aquello",...

Todos, ellos y yo, aprendimos juntos gran cantidad de coses. La tarea les interesaba enormemente y tuve la dicha de que algunos padres (felizmente pocos) viniesen a quejarse proque sus hijos dedicban demasiado tiempo, en sus cases, a las tareas escolares. (He aqui en ejemplo de por que la terna alumno, maestro, ciencia es insuficiente para el analisis de la tarea educativa). Lo que se logro, estoy seguro, se debio otra vez a la confianza dispensda, al respeto brindado a los trabajos de los alumnos y al proponerles labores en las que se sentian suficientamente libres y totalmente responsables.

Con esto aprendi que la clase debia ser un taller alegre y que la mejor manera de lograr un esfuerzo continuado era proporcionandoles tareas que les interesasen profundamente y a las que les encontrasen sentido.

Luego tuve a mi cargo distintos cursos de ensenanza secundaria suyas caracteristicas en horario, numero de cursos y de alumnos son similares a los que menciono Mr. Egsgard. En esta tarea ensenaba exclusivemente matematica.

Nuevamente hubo que cambiar de metodo. Diversos ensayos poco exitosos me llevaron a volver a una variante del que habia utilizado anteriormente como maestro de escuela elemental

A partir de alli los alumnos trabajaron con fichas, guias, etc., pero lo hacian en pequenos grupos a los cuales sucesivamenta iba atendiendo. De vez en cuando una interrupcion para alguna explicacion colectiva, un resumen de resultados, una discusion ante propuestas distintas... Agui debo reconocer que la atencion personal, pese a mi empeno, me resulto siempre mas dificil que en la escuala primaria o que en el nivel terciario. Esta dificultad se debe, en parte, a que no solamente el tiempo pasado con los alumnos es menor simo que ademas, donde resido, el profesor dificilmente puede concentrar sus horas en una unica escuela lo cual dificulta aun mas el conocimiento personal de los alumnos.

Desde hace diez anos mi tarea docente se divide entre la formacion de futuros profesores y el perfeccionamiento de los que estan en ejercicio, especialmente del nivel secundario, en la cual trato de volcar las experiencias anteriores.

Cual es al perfil del docente que hay que formar?

Que es un educador profesional de Matematica?

Pensamos que el profesor de Matematica es un docente que:

1. Posee un conocimiento suficientemente amplio de la Matematica que le permite:

   1.1. Jerarquiizar y seleccioner los contenidos de esta ciencia de modo tal que puede propner actividades convenientes a sus alumnos.

   1.2. Ordenar coherentemente los distintos temas.

   1.3. Conducir el aprendizaje de las ideas y conceptos matematicos graduando el rigor cientifico de modo adecuado al desarrollo psicologico de sus alumnos.

   1.4. Reconcocer la proyeccion de los contenidos y metodos que transmite tanto en la formacion de sus alumnos como en el valor social, cientifico, tecnologico, estetico y etico de esta ciencia.

   1.5. Reconocer las aplicaciones de la matematica a otras actividades.

   1.6. Elegir criticamente los medios auxiliares.

   1.7. Orientar vocacionalmente a sus alumnos.

2. Posee habilidades que le permiten:

   2.1. Generar y mantener en sus alumnos interes y aprecio por las actividades metematicas.

   2.2. Transmitir conceptos y metodos actualizados de la Matematica.

   2.3. Idear situaciones motivadoras.

   2.4. Evaluar el aprendizaje de sus alumnos.

Para precisar el conocimiento al que se refiere el primer punto indiquemos que debe incluir:

a) Estructura y lenguaje actual Ide la Matematica
b) Desarrollo histrico y fundamentacion epistemologica

c) Aplicaciones

todo ello atendiendo a las necesidades del alumno que antes hemos mencionado.

Un parrafo aparte creo que merece el problema de la responsabilidad del profesor de hoy y especialmente de quienes se ocupan del area cientifica.

Ya es un lugar comun el problema de la contaminacion ambiental que preocupa a todo el mundo. Tambien es unanimemente reconocido que la polucion no se da solo en el medio fisico sino tambien en el espiritual. Nuestros alumnos, como nosotros, nos movemos en un espacio cultural en el que nos vemos sometidos a un verdadero bombardeo, especialmente de los medios de comunicacion masiva, de falacias, de trampas, de trivialidades y de consignas denigrantes. Es comun que la llamada opinion publica sea formada y modelada por individuos mediocres que desde una pantella de TV, desde la radion portatil o desde el periodoco actuan constatnemente sobre las personas, durante much mas tiempo que la escuala y con majores recursos a su alcance, sin mahor preocupacion etica y, generalmente, proponiendo arquetipos de escasos valores positivos, cuando no de valores francamente negativos.

El docente, especialmente el de ciendias y en particular el de Matematica esta obligado a ayudar a sus alumnos a adquirir los anticuerpos que le ayuden a sobrevivir espiritualmente en esta situacion. La formacion de una personalidad autonoma critica, sensible, con coraje y honestidad intelectual y con sabia contencion puede intensarse muy eficazmente por intermedio de la educacion cientifica.

Quiza logremos un viejo ideal de la humanidad expresado en todos los idiomas. Al venir hacia aqui, al pasar por Mexico, desde uno de los muros del Museo de Antropoligia, los indigenas americanos proponen su invocacion:

Que aclare!

Que amanezca en el ciolo y en la tierra

No habra gloria ni grandeza

mientras no exista la criatura humana:

el hombre formado.

El educador profesional de Matematica es uno de quienes mayor responsabilidad tiene para llevar a cabo este ideal.

Por ultimo permitaseme advertir que el perfil del educador profesional de la Matematica no se agota, hoy, en el del maestro que dirije el aprendizaje cara a cara con sus alumnos.

Nuestro futuro profesor, en formacion, debera manana escribir un libro, filmar una pelicula, crear una maquina de ensenanza, grabar un video-tape, particpar o dirigir alguna investigacion sobre educacion matematica, colaborar en la elaboracion de un curriculum. TOdas estas tareas estan suficientemente ejecmplificadas en esta IV Conferencia que esta concluyendo... Muchas otras, hoy inimaginables, apareceran... Los avances tecnicos iran introduciendo mas y mas actividades nuevas para eses profesional. Como consequir preparlo para ello.?

Es mi firme creencia que el unico modo de lograrlo es asegurando que mantenga durante toda su vida un espiritu plastico capaz de adaptarse a las nuevas exigencias, amplio pero a la vez critico y cauto ante las innovaciones...

Nuevamente hemos vuelto al principio. Solamente lo lograremos cuando hayamos humnizado suficientemente a la ensenanza de la Matematica. Cuando hayamos convencido al futuro profesor, y al profesor mismo, que el objeto ultimo de nuestra tarea es el hombre.

CHAPTER 6 - Geometry

6.1 GEOMETRY IN THE SECONDARY SCHOOL

GEOMETRY IN THE SECONDARY SCHOOL: SOME CONSIDERATIONS

Eric Gower
Fareham Park School
Fareham, England

For a long time now, there has been a question mark over geometry in the secondary school. The struggle to find a suitable alternative to the ELements of Euclid as a school text or certainly as a basic approach to geometry teaching has occupied the minds of mathemics educators for well over a century. In the preface to an edition of the Elements in 1862, the following was written:

> "Numerous attempts have been made to find an appropriate substitute for the Elements of Euclid; but such attempts, fortunately, have hitherto been made in vain...it is extremely improbable, if Euclid were once abandoned, that any agreement would exist as to the author who should replace him".(1)

Whilst not wishing to be quite so enthusiastic about the suitability of Euclid as a geometry text book for the secondary school, one detects a certain prophetic ring to these words. The reforms of the sixties and seventies went a long way in developing alternative approaches and in Britain, the geometry has been characterised to a large extent by Transformation geometry in one form or another. There were many good reasons for this, including a genuine desire to give a consistent overall view of mathematics to the learner. However, the type of pupil for whom this must be an over-riding consideration cannot now be said to be in the majority. The resulting geometry work does not seem to meet the needs of youngsters of all abilities almost all of whom now have to study mathematics until they are sixteen.

History should teach us to be circumspect about achieving a solution to the problem, so I propose undertaking a simpler task initially. In trying to establish the lines along which a geometry course for all abilitites should develop, it is perhaps helpful to try to identify where current courses appear to be failing. This list is not intended to be exhaustive, but to provide the backdrop for the type of development I see as both necessary and possible.

Firstly, there has been a tendency to gear the geometry to the needs of the most able pupils. I suggest that providing a unified view of mathematics can only be significant to fairly sophisticated youngsters and this offers little to the majority. The dominance of standard symbolisation, algebraic description and matrix manipulation in such an attempt to show unity, serves to underline the inappropriateness of such an overall aim for the majority.

Secondly, the continuing need of many pupils to have genuine concrete experience on which to base their concept formation in geometry tends to be neglected. At best, this is seen as a requirement in the very early stages of development to be abandoned as soon as possible. In turn, without this concrete base, there is evidence that other areas of mathematics are deprived of their vitality.

Thirdly, the relation between geometry and the physical space in which our pupils live is neither explored nor exploited sufficiently. A goal such as "Geometry teaching shall enable the students to structure the real space and to explore the utilization of this structure" (2) should have some place in a course, but this is certainly not apparent at present. Nowhere is this more obvious than in the field of measurement, which to a lot of pupils consists of selecting formulae and algorithms unthinkingly. The consequence of this is poor appreciation of ideas such as units, approximations and error.

Fourthly, whilst the hallmark of approaches to geometry derived from Euclid was proof and deductivity, this aspect has become very much underplayed in the British approach to Transformation geometry. Understandably, a thoroughgoing axiomatic approach has never found general acceptance in Britain, but the consequence has been to remove deductivity as an essential component of geometry study. Even the most able pupils are not exposed consistently to the logical interdependence of geometrical ideas. This tends to weaken the structure and reduce geometry to a collection of isolated facts and procedures.

Fifthly, and this is a consequence to a large extent of the others, the geometry that most children experience at secondary level has little genuine problem-solving capability. It is true that geometry problems are encountered, but often these are little more than opportunities to rehearse set routines. Without problems which will catch the imagination of the solver geometry is dead. This is true for pupils of all abilitities.

With these shortcomings in mind we should turn our attention to the main problem of how to construct a geometry course which will have something to offer to all our pupils. This is by no means a simple task. From what I have indicated already, there are a number of areas of concern which I do not believe will be solved by adopting any one dominant geometry theory, be it isometries, vectors or congruence. Little has changed since Trevor Fletcher wrote in 1971, "Progress is more likely to come from the effective mobilisation of existing resources than from any new, previously untried, revolutionary remedy" (3). The richness of geometry and the variety of approach available must be reflected in the curriculum. However, the result may not be mathematically 'tidy' and it will almost certainly require a broadening of understanding on the part of teachers as to what constitutes geometry in secondary school.

What I want to suggest are some lines of development based on a broad definition of geometry. The course I have in mind, which would be entirely embedded within the mathematics programme in the British context, has much in sympathy with what Freudenthal has written in his book, Weeding and Sowing, "I imagined a course of geometric instruciton . . . which depending on the ability and bent of the pupil moves more quickly or more slowly, or stays exclusively at the lowest level, possibly proceeding to local, or even global, organising and which for a few may end with the incorporation of geometry into a system of mathematics" (4).

Such an idea depends on an enlarged view of geometry which is typical of his work, but it is this enlarged view which we should explore. His definition of geometry as 'organising space' or 'becoming conscious about one's intuitive grasp of space' demands considerable work to translate it into genuine classroom practice. However, since it is all too easy to trivialise discussion on geometry to the level of the traditional/modern debate on the merits of one set of axioms over another, I feel we have a responsibility to tackle the more difficult task of developing geometry teaching based on such a definition.

What follows is by no means worked out in detail, but rather lays down possible classes of activity which should contribute more profitable attitudes to geometry at this level.

The first of these is what I am terming Geometry which responds to perceptions and/intuitions about the real world. The concepts necessary for a development of geometry have a vital relation with the perceptions and intuitions about the physical space in which our pupils exist. One of the foundational tasks of geometry would seem to be to establish that relation and create a well-formed model of that space. This requires observing and handling real objects, examining various aspects of their form to help towards the formation of essential concepts. This is especially necessary with respect to features of three dimensions where many important properties cannot be grasped simply through the use of 2 dimensional diagrams. Indeed one of the types of abstraction which requires specific exploration is the relation between this pair of dimensional frameworks. Both representation in 2 and visualisation of 3 dimensions must play an important part. Development of intuitions about, for example, viewpoints, orientation and connectivity is important and I would suggest that the rich potential of photographs and diagrams needs to be more explicitly exploited here. Questions like "Where was the photographer standing?", "Give instructions to get from point A to point B on the map", "Which view would you see at X?" seem to be in the spirit of this level of enquiry and should form a continuing strand of a course. Characteristics of mathematical abstractions such as lines, planes, spheres and so on can only be fairly discussed once their corresponding approximations in the real world have been thoroughly experienced. I reject the view that this is only relevant to early development and that a once-for-all experience is what is to be aimed at. Abstraction and concretisation are important at any level; witness the degree of involvement when adult mathematicians explore a problem based, say, on the cube when they are actually allowed to handle the framework model as opposed to simply drawing diagrams or manipulating symbols.

Further examples of concept formation in this way and the sophistication to which it can be taken can be found elsewhere (see, for example, (2)). A continuing explicit development of a satisfactory model of real space appears to be a prerequisite if youngsters are to make any sense of higher abstractions to which they are necessarily exposed both in geometry and other aspects of mathematics.

However, it is not sufficient to encounter this model as a series of isolated concepts and features. Part of the work of forming the various inter-relations comes under my second heading, Geometry as a consequence of transformation and invariance. These twin ideas are central to any genuine study of mathematics and geometry is the natural place for pupils to encounter them. It follows that attempts must be made to give them a significant place in every course. However, formal approaches have been remote and frequently expected a pupil to be able to operate at a high level of abstraction. The ideas should permeate all the work and the questions "What has changed?", "What has caused the change?" and "What is unchanged?" should be asked at every level and refined as youngsters gain experience. Work with, say, mirrors, is common enough, but pupils need also to be exposed to situations where it is vital to ask these sorts of questions in order to understand what is happening. For example, ordering the views of a moving observer or the inverse of this activity implies the attitude to transformation that I am looking for. Although this is not set out as one of its primary objectives, the example of the IOWO work on the "Rescue of the Bermuda" demonstrates some of the possibilities.

In a similar exploratory way, aspects of invariance can be experienced in the context of 2 dimensional representation of 3 dimensions, in work on shadows and optical projection and enlargement.

Although a formal study of say, the plane isometries, may be an appropriate endpoint for some pupils, experience of many types of transformations in their own right together with notions of mappings, inverses and combinations should be available to all. A fruitful approach could be via a study of linkages and mechanisms. Although useful work was brought together about 10 years ago in Britian in the book Machines, Mechanisms and Mathematics (6), little appears to have been done with it. Perhaps now is the time to revive interest in its scope. At every stage of these various approaches there is scope for further levels of abstraction and in this way sophisticated aspects of transformation and invariance can be experienced.

Geometry solidly based in the development of the perception and intuition of youngsters in the real world coupled with a sound idea of transformation and invariance goes some way to puttng a course within the grasp of most pupils. However, little has been gained unless there is a major component, that of Geometry as a response to problem-solving needs. There is a pressing need to find problems which can be seen to be worthwhile and genuine to the pupils as well as establishing relevant problem solving strategies. Doubtless the selection of these will be highly personal, but there do seem to be some areas which deserve re-examination. Possibly as a consequence of the direction which reforms in mathematics teaching in Britain have taken the importance of measurement, drawing and deductivity seems to have declined. As far as measurement is concerned, with a geometry more closely related to the real world, it should follow that it should provide a vast field of problems as well and important set of strategies. As an example, without going into details, I would mention problems related to area measurement where it should be a priority for pupils to experience a range of measurement strategies and where, all too often, approaches are sterile and unproductive.

Secondly, the elevated place which ruler-and-compass constructions held in the Elements and its derivatives, has no counterpart in the approaches to geometry adopted in the past 15 to 20 years. Consequently the significance of drawing in problems has not always received due recognition. This deprived youngsters of a vital area which is accessible at some level to most. However, I am not advocating a return to the rigidity of ruler-and-compass only. Various alternatives exist which have been found to be fruitful - not the least of which are problems in perspective. (See, for example, (3) and (7)).

Finally we come to a class of problem which demands more than measurement or drawing in their solution. In addition to the standard type, Freudenthal's so-called 'haphazard' list of problems (for example, "Why does a piece of paper fold along a straight line?", "Why does a tied paper ribbon show a regular pentagon?", "Why can the radius of a circle be transferred six times around its periphery?", "Why can a table with four legs wobble, and what is the difference with a table with three legs?") shows how many of these readily stated geometric problems are at hand. What is disturbing about such a list is how seldom our pupils are exposed to such problems and, if they are, how ill-fitted they are in tackling them. Part of the difficulty lies in the feel that they have - or rather do not have - in putting forward a logical case, in other words deductivity and proof. Certainly during a course of geometry in the secondary school pupils should find themselves in positions where the need for these processes is felt at first hand. I am in sympathy with William Wynne Willson when he says, "To achieve this requires careful choice of subject matter; situations are needed in which the pupil is likely to be able to make a plausible conjecture and also to argue in defense of this statement . . . Geometry is fertile ground for such topics" (7). With appropriate background experience it is probable that the results and consequences of, say, Pythagoras, the circle theorems of concurrence features of triangles will yield sufficiently surprising and interesting results for the most able youngsters to ask "How do you know?". It is not appropriate to be dogmatic as to how and when other pupils will experience the same urge to ask "Why?" but it can only be fostered in a climate where the teacher is asking the child "How do you know?" when dealing with some of the basic intuitive responses mentioned previously.

Whatever context is chosen we must give every child the opportunity to expand his ability to handle this vital aspect of problem-solving.

In conclusion, the significance of geometry in the secondary school mathematics programme cannot be overestimated. The subject was at the forefront of the reforms of the sixties and seventies which catapulted school mathematics belatedly into the twentieth century, putting geometry into a far wider and healther context than was formerly the case. However, some of these reforms have conspired to put it beyond the grasp of our pupils and a radical overhaul of attitudes is now needed if the eighties are to put it in the context of the reality they experience. I hope that some of the ideas indicated in this paper give some clue as to the manner in which this overhaul should be undertaken.

References

1. Todhunter, I.: 1879, The Elements of Euclid, Macmillan and Co., London, p. vii.
2. Bender, P. and Schreiber, A.: 1980, "The principle of operative concept formation in geometry teaching." Educational Studies in Mathematics 11, pp. 59-90.
3. Fletcher, T.J.: 1971, "The teaching of geometry, present problems and future aims", Educational Studies in Mathematics 3, pp. 395-412.
4. Freudenthal, H.: 1978, Weeding and Sowing, D. Reidel, Dordrecht, p. 277.
5. IOWO: 1976, "Five years IOWO", Educational Studies in Mathematics 7, pp. 183-367.
6. Mathematics for the Majority: 1970, Machines, Mechanisms and Mathematics, Chatto and Windus, St. Albans and London.
7. Wynne Willson, W.: 1977, The Mathematics Curriculum: Geometry, Blackie and Son Ltd., Glasgow and London, p. 121-122.
8. Freudenthal, H.: 1973, Mathematics as an Educational Task, D. Reidel, Dordrecht, pp. 403-405.

PROCESS OBJECTIVES IN THE GEOMETRY CURRICULUM OF SECONDARY SCHOOL

Gerhard Holland
Universitat Geissen
Geissen, West Germany

1. Introduction

Geometry as a subject of teaching in secondary school is - more than other subjects - only justified if it aims at process objectives in addition to the learning of special concepts, theorems and procedures. Process objectives are found by analyzing the specific activities and performances of the active mathematician. They are related to mathematics as a process - to mathematics as a prefabricated product. In contrast to content objectives, process objectives can only be realized in the long run. They also presuppose instructional methods, which give the students large scale opportunities to carry out their own mathematical activities.

In the following, three different aspects of geometry are discerned and analyzed with regard to a possible realisation of process objectives (1).

a. Geometry as the "theory of space";

b. Geometry as a "logical structure capable of axiomatization";

c. Geometry as a "training area for problem solving".

2. Geometry as the Theory of Space

The subject matter of geometry from this point of view is the three-dimensional Euclidean space of intuition - the idealized model of the surrounding physical space. The predominant task of geometry instruction at the secondary school is to qualify the students to apply geometry within geometry and other mathematical fields as well as in real life situations. The application of geometry presupposes a solid knowledge of geometric concepts, relationships and techniques. Therefore the instruction of geometry at school must ensure the building up and refinement of the conceptual framework of three-dimensional space within the student's cognitive structure as well as its enrichment with content. Geometry as the "theory of space" does not presuppose a high level of formalizatin. Concepts are learned without formal definition in the strict sense of mathematical theory. The use of language is restricted to verbal circumscription of objects which exemplify the concept or to the description how to produce or draw those objects. To realize the generality of a geometric relationship, a sophisticated and detailed formal proof is neither needed nor desirable. The only - yet important - aim in proving a theorem is to create understanding and to provide insight into the theorem's general validity. All means of visualization should be permitted in pursuing this aim, as long as they lead to desirable simplification (2). In consequence, it makes no sense to prove a theorem which is already evident by intuition.

We now turn to the question of instructional strategies which are suitable for the realization of process objectives besides the desired learning of concepts relationships and skills. (Some valuable research work has been done by A.W. Bell (3)). We have to agree with J. Bruner that some kind of discovery learning is necessary if we strive both for the realization of mathematical process objectives and the meaningful learning of geometric content. In the following I want to point to some instructional strategies which enable students to find and prove a generalization by guided discovery learning (4). Each strategy requires specific mathematical activities and consequently contributes to the corresponding process objectives.

2.1. Inductive Method

This well known and often used method consists of two phases: First a hypothesis is formed and tested by experimentation (drawing and measuring), subsequently the hyposthesis is proved. But not every theorem is accessible to this strategy. The preconditions for the application of the inductive method strategy of guided discovery learning to a special theorem are:

a. the teacher has to find an appropriate formulation of the presented problem to initiate the empirical investigation. If the question is too wide, it will be difficult or impossible for the student to find the unknown relationship. If on the other hand, it is too narrow it may look very artificial.

b. The proof of the generalization should be accessible by the application of the usual heuristic strategies. A theorem for which only a tricky proof is known is, for this reason, unsuitable for the inductive method.

The inductive method contributes to the realization of such process objectives as

- the ability to generate examples
- the ability to find and formulate a generalization
- the ability to solve a proof problem.

2.2. Analyzing a given configuration

A suitable geometric figuration is given by the teacher. The analysis of this configuration by the students, under a special aspect hinted by the teacher, leads to the discovery of the theorem and of its proof. For example, a quadrilateral inscribed in a circle and radii drawn to the the vertices is given. By colouring congruent angles with the same colour, the student finds that opposite angles of the quadrilateral add up to 180°.

The range of application of this special strategy of guided discovery learning is rather broad. It depends on the teacher's wealth of ideas whether he finds a suitable geometric configuration of the given theorem. The method of analyzing a given configuration contributes to the realization of such process objectives as

- the ability to analyze a given geometric configuration
- the ability to find and formulate a generalization
- the ability to specify the idea of a proof to a complete work

2.3. Solving a construction problem

Many theorems in plane geometry are narrowly related to a specific geometric construction. This gives rise to a further strategy of guided discovery learning. In a first phase the students solve the construction problem connected with the theorem. In a second phase of analysis, they discover the theorem in question together with its proof. For example, the problem of constructing a circle through the three vertices of a given triangle leads in a second phase to the discovery and proof of the theorem that the three perpendicular bisectors of the triangle meet at one point, the circumcenter of the triangle. The problem of constructing to a given triangle a second triangle whose midpoints of the three sides are the vertices of the given triangle leads to the discovery and proof of the theorem that the three altitudes of a triangle meet at one point. Regarding this strategy the main part of the proof is already done during the first phase of solving the construction problem, namely in proving the validity of the construction. The method of solving a construction problem contributes to the realization of such process objectives as

- the ability to solve a construction problem
- the ability to prove the validity of a construction
- the ability to specify the idea of a proof to a complete proof.

2.4. Solving a computation problem

The students start with the solution of a suitable geometric computation problem which, in a second phase, will be generalized by the introduction of variables. The solution of the generalized problem supplies the theorem in question together with its proof. For example, a geometric configuration consisting of a square inscribed in another square is given. The problem is to calculate the length of the side of the inscribed square from the given lengths of the legs of the four congruent right triangles within the configuration. The solution of the generalized problem provides the Pythagorean Theorem. With the same strategy of guided discovery learning nearly all relationships between angles in triangles, quadrilaterals and circles can be discovered and proved. Naturally this far-reaching strategy is restricted to theorems which express a relationship between measures (of angles, length, area and volume). It contributes to the realization of such process objectives as

- the ability to solve a computation problem with specific measures
- the ability to generalize a computation problem and its solution
- the ability to improve and curtail a proof.

3. Geometry as a logical structure capable of axiomatization

This aspect of geometry concerns the logical network of geometric concepts and theorems. The mathematical activities are defining, deducing and axiomatizing. It starts with the local ordering of some separate and limited regions and leads finally to a global axiomatic theory of Euclidean Geometry. Because the objects of axiomatization are concepts already known and theorems already proved (as far as a proof was necessary to convince students of the theorem's general validity), this new aspect of geometry presupposes geometry as the "theory of space". The aim of defining or re-defining a concept, proving or re-proving a theorem is now the logical reduction to other given concepts or theorems. Only from this point of view is it meaningful to prove a theorem whose general validity is already evident by intuition. Defining and proving is now a pure formal performance which permits no reference to a geometric figure. Although this extreme position is surely not practicable at the school level, geometry as a "logical structure capable of axiomatization" is accessible to the most gifted secondary school students if the level of exactness is adapted to the capabilities of the students. Instructional objectives are

- the ability to deduce a theorem from given theorems or axioms
- the ability to analyze a given proof with regard to completeness and correctness
- the ability to define a concept (in different ways) by logical reduction to other concepts
- the ability to demonstrate the logical equivalence of different definitions of the same concept
- the ability to axiomatize a limited area of connected theorems
- the ability to use a language which involves formalized elements and logical terms.

All these objectives are process objectives which may be realized only on a long term basis. As Freudenthal has repeatedly pointed out (5) the presentation of a prefabricated axiomatic theory is by no means an effective method for the realization of these aims. Only if the students are given opportunities to perform these activities of proving, deducing, defining, formalizing and axiomatizing by themselves is there a realistic chance of increasing the corresponding abilities. While there are many opportunities for formalizing, analyzing and deducing, it is not so easy to find suitable examples for defining and axiomatizing. Some examples are the following:

1. Define the concept of a rhombus (or of a half-turn of the plane) in different ways. Prove the equivalence of the definitions.

2. Consider the following four theorems referring to the angles on two parallel lines which are intersected by a third line:

(T1) Two vertical angles have the same measurement.

(T2) Two adjacent angles together measure 180°.

(T3) Two alternate angles have the same measurement.

(T4) Two corresponding angles have the same measurement.

Find a subset of (T1,T2,T3,T4) (as small as possible) from which you can deduce the remaining theorems.

4. Geometry as a training area for problem solving

The subject matter of geometry as a "training area for problem solving" is the nearly inexhaustable supply of geometric problems one can find and invent in relation to the topics of traditional geometry curriculum, and which can be classified as proof problems. The predominant instructional objective from this point of view is the improvement of the ability to solve geometric problems, in expectation that there will be an observable transfer to non-geometrical problems. For the realization of this aim a systematic training in problem solving activities is required. In my own reserach work I have started to investigate problem solving training with well defined problem classes (6).

Each problem class is defined by its special problem type (proof problem, construction problem or computation problem) and by the set of operations (applications of theorems and rules) which are sufficient to solve the problems of the class. It is supposed that the students are well acquainted with these operations before the problem solving training starts. In this way the barrier in solving a problem of the class is reduced to the correct choice and ordering of the required operations. During the problem solving training phase the students learn to apply heuristic strategies appropriate to the problems of this class. To achieve a continuous improvement in problem solving activities, problem solving phases with well defined problem classes should be added to each suitable area of the geometry curriculum. In the same measure as the student's knowledge of concepts, theorems and rules accumulates, the range of the problem solving operations may be amplified. The degree of difficulty of the problems belonging to the same problem class is usually widely variable. Therefore it is possible to adapt the degree of difficulty to the capabilities of the individual student or learning group. Examples of the problem classes are:

1. The solving of proof problems by applying theorems regarding similar triangles.

2. The solving of construction problems by applying an enlargement from a centre.

3. The solving of computational problems by applying calculation rules for the areas of triangles, rectangles and trapezoids.

5. Summary

With respect to process objectives, three aspects of geometry as a subject in secondary school have been discussed. Geometry as the "theory of space" aims at the learning of concepts, relationships and procedures. But suitable instructional methods of guided discovery learning also permits the realization of process objectives as a desirable by-product. Geometry as a "logical structure capable of axiomatization" is directed to the realization of those process objectives which correspond to such mathematical activities as defining, deducing and axiomatizing. This aspect is only relevant for the most gifted students. Geometry as a "training area for problem solving" strives for the enhancement of general problem solving activities. Geometry in secondary school may contribute to this ambitious aim if suitable methods are invented. A problem solving training with well defined classes of geometric problems is proposed.

References

1. Holland, G.: Das Beweisen geometrischer Satze in der Sekundarstufe I unter verschiedenen Aspekten von Geometrie. In: Didaktik der Mathematik 7 (1979), 104-116.

2. Kirsch, A.: Aspects of Simplification in Mathematics Teaching. In: Proceedigs of the Third International Congress on Mathematical Education. Karlsruhe, 1977.

3. Bell, A.W.: The Learning of General Mathematical Strategies. Shell Centre for Mathematical Education, University of Nottingham, 1976.

4. Holland, G.: Unterrichtsstrategien zum Gewinnen und Beweisen geometrischer Satze. In: Praxis der Mathematik 22 (1980), 97-113.

5. Freudenthal, H.: Mathematik als padegogische Aufgabe Bd.2, Stuttgart 1973.

6. Holland, G.: Geometrische Berechnunsprobleme in der Sek. I. In: Beitrage zum Mathematikungericht 1980. Hannover 1980.

WHY WE STILL NEED TO TEACH GEOMETRY!

Jean Pedersen
University of Santa Clara
Santa Clara, California, U.S.A.

Today I have three things to say. First I would like to talk about the present status of geometry, at the secondary level. Second I want to discuss the role of geometry in mathematics. Finally, on the basis of the first two topics I will give some suggestions for improving mathematics education.

I. The Current Status of Geometry at the Secondary Level.

My observations deal mostly with the situation as I see it in the United States. This summer while visiting parts of Southeast Asia and Europe I discussed this topic with many University and Pre-college mathematics teachers. It was clear to me that in general the problems surrounding the teaching of mathematics in various countries are quite varied. However, my impression is that in all countries the status of geometry is very much the same. To put it briefly: Geometry is underemployed! This seems to be true at every level of pre-college mathematics.

Although we were not asked to discuss the status of geometry in the grade-schools I believe that aspect deserves a brief mention. At that level whatever geometry exists is often omitted when there is pressure to finish the syllabus by a given time. Worse, when geometry is included in grade-school it often consists of having students simply learn the names of certain geometric objects. The children exposed to this kind of 'geometry' see it as a useless taxonomy whose only purpose is to enable them to answer the sometimes ridiculous questions posed on tests. Imagine, after learning the definitions of several special polygons, encountering a test on which you must 'classify each of the following polygons as exactly one of the following

square
rectangle
parallelogram
equilateral triangle
isosceles triangle
obtuse triangle'.

Even if the question is re-stated so that it becomes logically reasonable, the approach is wrong. It is static, sterile and dull! To see the real point of this example consider the following questions:

What are students supposed to learn from this exericse?

What will students be able to do after they learn this material that they were unable to do before they learned it?

How does this material help students develop their ability to think?

In what way can this material be expected to increase the student's interest in mathematics?

These are important questions to ask about everything we teach in mathematics. Much of what is currently taught in the grade-schools fails to answer these questions. (I have recently received copies of the new elementary series of textbooks published by Open Court Publishing Company and the final experimental version of CEMREL's CSMP elementary textbooks. Both of these series contain a less traditional and more dynamic treatment of geometry.)

Before I discuss the status of geometry at the secondary level I should make mention of several new geometry textbooks that are now available. I have recently been sent books by Harold Jacobs; O'Daffer and Clemens; Clemens, O'Daffer and Cooney; Goodwin, Vannatta and Crosswhite; Ken Seydel; and Alan Hoffer, just to mention a few. I salute these authors for having the courage to try to do something better than the usual treatment. But despite what would seem to be an outpouring of interest in geometry, in terms of new textbooks, most schools in this country do not require geometry for high school graduation. And, when geometry is taught, the usual approach is to teach classical Euclidean plane geometry. The rationale for this approach is that geometry is an ideal medium for teaching students about abstract reasoning and provides an example of an axiomatic system.

I believe that the traditional approach is a mistake and I expect that many of you will disagree with what I am about to say.

> It is my opinion that for most students the classical axiomatic approach to plane geometry is not the best way to introduce either geometry - - or systems of abstract reasoning!

I do not deny that it is possible to teach geometry this way - - in fact, the evidence of that is overwhelming. What I wish to emphasize is

a. that for most students abstract mathematical systems could be better exemplified by using algebra or analysis, most especially, elementary number theory; and

b. that geometry could be made a good deal more exciting and useful to students if it were free from the constraints imposed by using the axiomatic approach.

If we try to teach both geometric facts and abstract reasoning at the same time we are violating Polya's second rule of style - - " . . control your self, say first one and then the other'. I believe that, when presented too early, the classical treatment, particularly if it is rigorous, is extremely detrimental to both of these important subjects.

2. <u>The Role of Geometry in Mathematics.</u>

In addressing the International Congress of Mathematicians in 1900 David Hilbert said, "Mathematical science is, in my opinion, an indivisible whole, an organism whose vitality is conditioned upon the connection of its parts'.

Hilbert is, of course, correct! Geometry is woven so tightly into the mathematical fabric that we dare not ignore it. It is important for us to observe exactly how geometry fits into the mathematical picture, for then we will have a better idea of how it should be taught.

So, just what are the important roles that geometry plays in mathematics? I list some of them

A. Geometry provides an economical and mnemonically useful way of presenting information acquired in the real world - - and sometimes that information involved non-geometrical facts.

Most of our students will eventually have a need to express their ideas by means of some sort of geometric symbols. We should prepare them to do so. We should also prepare them to recognize when geometric symbols are being used to distort the truth.

B. Geometry provides a vital link between the various parts of what Hilbert called mathematical science. It also provides a connection between the real world and the mathematics we use to solve problems about that world.

This aspect should be emphasized in all of the mathematics that we teach, especially at the pre-college level. Notice that when I say this I am not asking for the introduction of any new geometry courses in the curriculum, rather what I am advocating is that we include, within the existing courses, the geometry that helps to clarify the ideas at hand.

C. Geometry is a rich source of inspiration, of ideas and of questions!

It seems foolish to deprive the next generation of the motivation and inspiration that geometry offers. Therefore the teaching of mathematics should include the history of how ideas are born and some discussion of the 'intuitive' geometric questions that led, for example, to irrational numbers, calculus and topology.

3. <u>Suggestions for Improving Mathematics Education.</u>

First I believe we need to remember the following facts:

A. The geometry of three dimensions is closer to the experience of our students than the geometry of the plane.

B. We live on a 'sphere'.

C. Geometry is a skill of the eyes and the hands as well as the mind.

D. Students need concrete experiences with geometry and a certain mathematical maturity before they can appreciate the axiomatic aspect of geometry.

Here is a partial list of suggestions.

1. When selecting, or writing, material think about the four questions raised in Section 1.

2. Whenever possible select material that has some element of surprise to it, or

   that leads to an important concept in mathematics, or

   that connects with other parts of mathematics, or

   that is useful.

3. Treat geometrically those many parts of mathematics that connect with geometry (e.g., linear equations, quadratic equations, probability as measure of length, area, volume).

4. Use as many concrete examples as possible.

5. Whenever possible use geometry as a mnemonic device.

<u>Post Script</u> Many purely geometrical proofs and facts are psychologically satisfying. They appears to our natural tendency to want to conserve our own energy, to our sense of balance, to our preception of beauty and to our enjoyment of surprises. These are powerful motivational tools and we ought to make the most of them in our teaching.

References

1. George Polya, How to Solve It, Second Edition, Princeton 1973 (p. 172).

2. Harold Jacobs, Mathematics a Human Endeavor, Freeman 1970 (p. 427).

3. G. Polya, Mathematical Discovery, Vol. 2, John Wiley and Sons, Inc. 1960 (p. 124).

4. Phares G. O'Daffer and Stanley R. Clemens, Geometry. An Investigative Approach, Addison-Wesley 1976 (p. 270).

EN DEFENSA DE LA GEOMETRIA

Julio Castineira Merino
INB de Cuellar
Spain

La ensenanza de la Matematica, en los ultimos anos, ha estado caracterizada, en parte, por un aumento del rigor, entendido como:

a) aumento del formalismo y del discurso deductivo

b) fundamentando la ensenanza en el lenguaje de la teoria de conjuntos.

c) resltando las propiedades estructurales.

Esto motivo en gran medida el menosprecio de la geometria elemental clasica, pues en efecto:

a) Ciertas nociones como angulo, medida de angulos, son dificiles de formalizar.

b) La introduccion de las nociones conjuntistas se realizo desplazando gran parte de la geometria. Si la geometria habia dejado de ser fundamento logico de la matematica por que ensenarla?

c) En muchos casos se reduce la ensenanza de la geomtria al estudio del algebra lineal, olvidando o postergando a cursos superiores problemas topologicos (teorema de Euler sobre los poliedros, etc.).

Necesidad de la Geometria Elemental

Las nciones espaciales (recta, circunferencia, curva, angulo, plano) y las relaciones entre ellas (paralelismo, perpendicularidad, distancia, tangencia) no son innateas sino consecuencia de un aprendizaje que realiza el hombre desde su infancia y que la humanidad ha elaborado en sucesivas culturas. Ciertas nociones geomericas son tan usuales que su desconocimiento o conocimiento defectuosos raya en el analfabetismo. Todas estas razones aconsejan su ensenanza en la escuela.

Aclaraciones Sobre el Termino Geometria

Entiendo por geometria elemental la disciplina de las nociones espacielaes y no el mero estudio de la geometria de los Elementos de Euclides o la geometria del triangulo; y en cuanto a la forma de ensenarlas opino que debe ser descriptiva en el nivel primario y posteriormente, al tiempo que se aumente el discurso deductivo, ena mezcla de geometria sintetica y analitica.

Utilidad del Estudio de la Geometria

a) El estudio de la geometria proporciona una base solida de conocimientos en otras ramas de la ceincia y de la matematica.

b) Pedagogicamente conviene utilizar nociones geometricas en la introduccion y desarrollo de conceptos matematicos. Analizare este punto a continuacion.

Aritmetica y ALgebra

a) El concepto de numero

La didactica tradicional del numero racional es, a mi juicio, superior a las innovaciones basadas en operadores, relatcion de equivalencia, etc. El numero racional se aprende facilmente a partir de las partes de una figura (su fundamentacion logica puede esperar a niveles superiores), a continuacion la notacion decimal y su utilization practica como medida de longitudes y otras magnitudes. Surgen entonces, de una forma natural, cuestiones sensillas pero importantes:

- El desarrollo decimal indefinido (que prepara el camino al numero real)

- Ofrece ejemplos sencillos de las paradojas de Zenon, por ejemplo 0/9 = 0 se puede interpretar como la paradoja de Aquiles y la tortuga.

La introduccion al numero real considero conveniente realizarla sobre tres ideas basicas:

- La medida de magnitudes

- Los numeros decimales

- El axioma de Cantor

Los conceptos de raiz cuadrade y de raiz cubica se deben introducir ligados a los conceptos geometircos correspondientes, por dos razones: es mas intuitivo que el concepto formal de operacion inversa y es inmediata su aplicacion practica.

Respecto al numero complejo basta recordar la importancia del diagrama de Argand y de las superficies de Riemann en la matematica superior y un hecho historico: los numeros complejos no fueron plenamente aceptados hasta que se dio de ellos un modelo geometrico. La interpretacion geometirca de las operaciones es un hecho natural, pensemos en el isomorfismo entre el grupo S y el grupo ortogonal, la relacion entre grupos ciclicos y raices de la unidad.

b) ecuaciones e inecuaciones

El estudio de las ecuaciones e inecuaciones debe ir ligado a su modelo geometrico. Si bien los metodos graficos o geometricos ofrecen ciertas deficiencias en la resolucion de ecuaciones, el estudion geometrico es importante en el problema de discusion. Los metodos geoetricos son, algunas veces, el mejor metodo de resolucion de inecuaciones.

La difusion de las calculadoras manuales, esta permitiendo ensenar algoritmos y otros temas antes reservados a los profesinales. Soy un partidario aferrimo de la introduccion de las calculadoras en la escuale y no para realizar operationes sencillas sino como procedimiento para acercar al alumno los metodos de calculo.

La introduccion a los procesos interativos, como el de Newton, se puede realizar a una edad muy temprana, por ejemplo a los 14 anos, con una herramienta sencilla, desplazando temas caducos como el calculo logaritmico o la regla de calculo. Un buen ejemplo es el calculo de la rais cubica con una maquina con las cuatro operaciones basicas, el planteamiento geometrico del problem sugiere el metodo y resuelve intuitivamente el problema de convergencia.

c) Estructuras algebraicas

La geometria proporciona la mejor introduccion al algebra lineal. Los movimientos euclideos, a parte de su valor formativo intrinsico constituyen, a mi juicio, la mejor introduccion a la teoria de grupos por las siguientes razones:

- Presenta la estructura de grupo de forma natural, evitando una introduccion demasiado abstracta o a traves de conjuntos susceptibles de una estructura rica como Z, Q, R, etc.

- Ofrece bellos modelos de ciertos tipos de grupos: los grupos de movimientos directos, que dejan invariante un poligono regular de $\underline{n}$ lados, los grupos de los poliedros, los grupos diedrales.

- Ciertos conceptos como generadores, relaciones entre subgrupos invarintes, suma directa, isomorfismo, orbitas, etc. admiten ejemplos geometricos muy interesantes.

Analisis

Las interrelaciones entre geometria y analisis son constatnes en la historia de la matematica y convenientes en la ensenanza. Las nociones de function, function creceinte convexidad, deben de ir unidas a su representacion grafica y a su interpretacion geometrica.

Las functiones trigonometricas deben defnirse a partir de las razones de un angulo, dicho metodo no es "muy riguroso" pero tiene la ventaja de una mayor visualizacion y de su utilidad practica. Las functiones hiperbolicas presentan un aspecto geometrico interesante. En cursos preuniversitarios se puede culminar el estudio con las functiones seno y coseno lemniscatico como ejecmplo de trascendentes no elementales, siendo a su vez una aproximacion elemental a las funciones elipticas.

La nocion de integral definida conviene hacerle a partir del problema del area de un recinto plano pues:

- Es facil de comprender y asimilar

- En el sentido de Lebesgue la integral es una medid como lo es el area.

- Las sucesivas generalizaciones del concepto de integral: integral de linea, integral multiple, integral de superficie, se adquieren facilmente a traves de su interpretacion geometrica.

- Los conceptos geometircos visualizan diversos teoremas como por ejemplo el teorema fundamental del calculo.

Otras Aplicaciones

La geometrica es fuente de problemas curiosos y estimulantes, como por ejemplo:

a) La forma de los panales de las abejas, el mosaico hexagonal y la forma romboidal del fondo.

b) La razon aurea y sus relaciones con los numeros de Fibonacci y la filotaxis.

En el Calculo de Probabilidades y en la Estadistica es menos importante la utilizacion de tecnicas geometricas. Cierto es que no faltan aplicaciones geometricas interesantes, aunque algo exoticas, como la iteracion de Fejes Thoh sobre un poligono y la marcha aleatoria en un grupo abeliano finito. Sin embargo las probabilidades geometrcas y en especial la Paradoja de Bertrand manidiestan la necesidad de axiomatizacion del Calculo de Probabilidades.

6.2 GEOMETRIC ACTIVITIES IN THE ELEMENTARY SCHOOL

GEOMETRIC ACTIVITIES IN THE ELEMENTARY SCHOOL

Koichi Abe
Osaka Kyoiku University
Tennoji, Japan

I shall begin with the present status of our program in the elementary school, which has been in force since this school year. The mathematic content consists of four domains, one of which is geometrical figures. The improved program is based on the reflection that the hasty and too formalized modernization of school mathematics has been imposing heavier burdens upon pupils, and it seems to have been influenced by the public opinion "back to the basics." For example, the concept of sets, together with some of the symbols, was introduced but could find no place where it proved to be significant and indispensable, at least in geometry. To identify the squares among various quadrilaterals was sometimes called the lesson based on set-theoretical thinking, but it could be done without any reference to sets, and it would give rise to some confusion by pupils to use sophisticated set-theoretical language and symbols.

Nevertheless, our modernization in school geometry was not so radical because of a great diversity of opinion on what it meant that the new program has not so many modifications in geometry. Of course, some modifications emerge. For example, the emphasis on transformations has become modest. I think it is partly because most of the geometry course at the lower secondary in Japan is devoted to the Euclidean demonstrative geometry, and transformational aspects of geometry will not be expected to play an important role. Whereas in many countries many debates are available on the place of traditional geometry as school mathematics, in Japan there are comparatively many educators who take the affirmative point of view about the pedagogical value of Euclidean demonstrative geometry. I think in this respect I belong to the minority in Japan.

One observation: as you know well, the reform movement in school mathematics, including geometry, started at the beginning of this century. The textbooks published by the Ministry of Education in Japan from 1935 to 1945, which were multi-colored and stimulating, contained various topics in geometry: observing natural things around children such as leaves, flowers, butterflies, snails, and so on; folding and cutting paper; using colored boards to make geometrical figures; tracing the spiral; making the envelope; and experimenting with Moebius bands. The leading idea was to give pupils spatial awareness. Considering which was more important in teaching of geometry - - the global, synthetic, and intuitive understanding or the local, analytic, and logical understanding - - the authors of these textbooks gave the former priority and intended to develop and elaborate the latter on the basis of the former. It is regrettable to me that such pedagogically precious achievements in the pre-war period have neither been inherited nor been verified in the post-war period.

Now let us come back to the present situation. An Agenda for Action published by NCTM points out that basic skills in mathematics should be defined to encompass more than computational facility, and that geometry should be contained in the basic skill areas. This is true, and then we should ask ourselves, what are the basics in geometry - - especially at the elementary levels? If our aim is to make pupils familiar with geometrical figures, both in the plane and in space, which can often be seen in daily life situations, it is sufficient to give them the names of those figures and to make clear how they function in daily life situations. On the other hand, if we want to systematize the contents of geometry, we should not teach the geometrical figures in isolation, but should consider in what context they must be treated. Geometrical figures taught now are almost the same as from the elementary through the upper secondary school, and the systematization of school geometry will be quite different from the systematization of number systems. The expansion of the concept of number in itself must play the leading part in the systematization of number systems. On the other hand, in the systematization of geometry the expansion of the concept of figures is almost meaningless, and what must play the leading part is the gradually deepening understanding of figures already familiar to pupils.

When we look for such a systematization in relation to our framework of Euclidean demonstrative geometry and to the future perspective for vector space structure, the dominant and central position should be given to the parallelogram. It has been tried out in our project. It is well known that the parallelogram has the dominant position in the vector space as well as in traditional geometry, though it is hardly seen in daily life situations. A parallelogram is divided by one of the diagonals into two congruent triangles, and almost all the properties of the parallelogram are proved by using this property. Such specialized parallelograms as square, rectangle, and rhombus can respectively be characterized by specialized triangles divided by a diagonal. Therefore this property is called the unifying property in our project. There is no need to say that in the elementary school we do not intend to teach the properties formally, nor to deduce the other properties from the unifying property, but we teach them informally and give an implicit awareness. The formal treatment belongs to secondary geometry.

As to transformational geometry, our official program suggests almost nothing. Although symmetric figures are introduced in the sixth grade, symmetrical transformation as an isometry is treated only in the eighth grade. But I think it is quite reasonable to expect that introduction of various (not necessarily isometric) transformations through operational activities will contribute to developing the geometrical imagination or intuition of pupils. As Piaget pointed out, geometrical intuition is far more than a system of perceptions and is the basic awareness of space, at a level not yet formalized. Geometrical intuition is essentially active in character and it consists primarily of virtual actions.

Connected with spatial awareness, I shall refer to two points. One is the location of a point in two- or three-dimensional space, leading to the idea of coordinates. The other is concerned with tessellations. The introduction of tessellations has two objectives. On one hand, it leads to a recognition of the infinite extension of the plane, and on the other, it helps pupils consider what figures can be tiled around a point on the plane. These considerations can be developed further to the discovery of relationships between mutually adjacent congruent figures, which also leads to the consideration of transformations.

Operational activities also play an important role in measuring areas and volumes. Interpreting transformations in a wider sense, we can include as kinds of plane transformations:

a. transformations which preserve the area,

b. transformations which double the area,

c. transformations which halve the area.

Operational activities of type (a) can be seen in transformating the parallelogram to the rectangle and the triangle to the parallelogram. Similar activities of successive combinations of type (b) and type (c) yield type (a). Pupils are eagerly involved in these activities and find the above-mentioned transformations easily by using, for instance, geoboards.

All of us would agree with the importance of operational activities in elementary school geometry. And I think one of the most urgent problems is how to organize and systematize operational activities in geometry lessons so as to increase and deepen cumulatively the pupils' understandings of geometrical space. This problem is not yet solved.

SPACE AS A MODEL FOR ELEMENTARY SCHOOL GEOMETRY

John Del Grande
Newmarket, Canada

Elementary Geometry relies heavily on what we know about children, in particular how children learn geometry. Traditional courses are of no help becuase they are limited to: measurement, terminology, facts about familiar geometric figures, definitions, theorems, proof.

Such courses do not provide the teacher with what is needed. No coherent view of the nature of geometry has ever been given to teachers in a traditional setting.

What do teachers need? Firstly, teachers should know: what geometry is about, where it leads, what it is good for, and why anyone - particularly children - should study it. Since teacher background is inadequate to answer these questions, it becomes imperative to give a functional and appropriate view of the nature of geometry. Geometry for children must not be preoccupied with structure and proof as presented secondary schools. Children' geometry must start on an intuitive basis and should build on objects and experiences that are part of a child's environment. To provide children with informal environmental experiences, the teacher needs experiences that emphasize the relevance of geometry to one's environment and to the real world. The teacher also needs experiences in geometry like those that would be relevant and interesting for children. Many such activities do exist and can be found in the North York Guidelines (1,2,13,14).

Geometry, at the elementary level, consists of two components - Transformation and Shape - which appear separately at first but become interralted as one studies the properties of shapes, symmetry, and the properties of transformations (1,2).

Geometry cannot be taught in isolation. It is related to the real world not only by objects and motions of objects in space but is also related to art, physical movement, science, and music.

Geometry can be characterized as the study of space experiences. The entire geometry program can be built around these experiences that set an informal tone relating geometry to the child's environment. Thus, space is the model we use for geometry in Elementary Schools.

A child's view of Space

When a child is born the physical apparatus for seeing is fairly well-developed, and the visual processes are about the same as in an adult. But much of vision is like a language that has to be learned, word by word and sentence by sentence.

- Human vision begins with a response to intensity of light.
- Small babies are interested in faces and in actual round objects rather than in 2-D pictures of these objects.
- By experimenting, touching, reaching, and tasting, infants build a store of experiences and information that will one day allow them to make judgements on the basis of their eyes alone.
- By the end of the first year of life the child begins to grasp the idea of identity and permanence. Things that were square and solid yesterday are square and solid today.
- A child learns about size and distance and understands that a large toy across the room is not the same as a small toy nearby although their physical images are almost the same.
- One of a child's first discoveries is that certain things are solid, and that some are moveable while others are not.
- The child learns to recognize objects backwards or upside down just as readily as in their "proper" position.
- The child learns that a red ball and a white ball are not two utterly different kinds of objects.

Man's capability of visual organization in his highly developed brain sets man apart from all other species. His brain is capable of screening the mass of information, selecting what is needed, and using the rest for comparisons or background, making decisions, and achieving a stable, consistent and meaningful vision of the universe in which he lives.

How do people learn Geometry?

Educators of the world ask this question: Why is it that so many students who master most other subjects, get nowhere in their study of geometry?

To answer this question we must first know how children learn geometry. There are two psychological school that have much to say on this topic.

1. Piaget and Inhelder (1959) (4,5) asserted that traditional geometry starts too late and then takes up the concept of measurement thus omitting the qualitative phase of transforming spatial operations into logical ones - that is, geometry developed from measurement to shape. The natural development of geometric operations in children actually proceeds the other way - from the qualitative to the quantitative.

2. The work of Dutch teacher P. M. van Heile and his wife was brought to light by Professor Hans Freudenthal. Van Heile was the author of a paper on The Thought of the Child and Geometry (1959) which suggests five levels of though development in geometry (6,7). Between 1960 and 1964 the Soviet Academy of Pedagogical Sciences found, through intensive research, that the statements and principles expounded by the van Heiles were all valid.

A Russian description, in 1968, of the five van Heile Levels (7)

Level I

Student
- perceives geometric figures as a whole
- does not see parts of figures nor components of figures
- does not recognize a square as a rhombus or a rhombus as a parallelogram

Level II

Student - begins to discern components of a figure, relationships among components and relationships between individual figures.

- establishes the properties of a figure experimentally.

- tells a figure by its properties, e.g. a rectangle has 4 right angles, diagonals congruent, opposite sides equal.

- notices that parallelograms and rectangles have opposite sides congruent but does not yet conclude that a rectangle is a parallelogram.

Level III

Student - logically connects figures and their properties by definitions.

- does not grasp the meaning of deduction as a whole - the teacher or text gives the order of the argument or logic.

- does not understand the role of axioms and cannot yet see the logical connection of statements.

- reasons using experimentally obtained properties.

- recognizes a square as a rectangle and a parallelogram.

Level IV

Student - grasps the significance of deduction.

- understands the essence of axioms, definitions and theorems.

- understands the logical structure of proof.

Level V

Student - appreciates Hilbert's approach to geometry.

- develops a theory without the use of concrete objects.

The research of the van Heiles and the Soviets has shown that the passage from one level to another is not dependent on the student's maturity or his age. many adults are still at Level I as they have never had the experiences necessary to proceed to Level II. A teacher through content and method of instruction, can encourage and hasten the movement from one level to another.

Van Heile's research makes it possible to identify some of the faults with current geometry programs.

- Elementary Geometry begins with measurements and concepts at Levels II and III.
- When an explanation is given at the teacher's level and the student is a different level - the result is nothing more than a monologue.
- Most high school students are still at Level II or III.
- Secondary courses demand Level IV thinking.

No wonder our secondary school geometry has been a total disaster!

Brain Theory

Recent developments on brain theory rely on two important findings (8,9,10).

- Man has two brains instead of one.
- Each hemisphere is specialized for different modes of thought.

Left Hemisphere	Right Hemisphere
- Processes language	- Spatial/musical
- Thinks in words	- Artistic, symbolic
- Orders sequentially	- Intuitive, creative
- Orders comples motor sequences	- Thinks in images
- Sees figures in parts	- Does non-verbal tasks
- Analyzes a problem	- Grasps a problem as whole and solves it at once

Most infants are right brain oriented. When young children enter school they have good intuitive ideas about space but current mathematics programs do little to nurture and develop space concepts. There seems to be a great haste to develop number and number concepts in the formative years of a child. But, for such children, space concepts are intuitive whereas number has to be taught.

Using what we know about how children learn geometry and starting with their intuitive notions of space, it is now possible to design an elementary geometry program that will prepare the student for future work in geometry. An elementary program should give children an opportunity to learn the following

1. Simple properties of the isometries
2. Simple properties of geometric figures
3. Applications of the isometrics to symmetry, pattern design and the art of Escher (11,12).

The geometry of secondary school should rely on these notions that children develop in elementary school. The properties, obtained through concrete and manipulative experiences, can be used as the axioms for secondary geometry. Students can accept and use only those axioms that they understand and have experienced. This approach leads to an intuitive approach to geometry which has more meaning for students and gives them a geometry they can apply to their real world.

The North York Board of Education aggressively encourages their schools to implement the following objectives (13,14).

Objectives for a K-3 Geometry Program

1. Makes and draws congruent figures on a geoboard.
2. Creates and identifies figures and designs that have symmetry.
3. Relates solids and shapes to his/her own environment.
4. Identifies common solids and plane figures by name and characteristics.
5. Illustrates slides, flips, and turns using a cutout of a plane figure or a geoboard.
6. Identifies congruent figures.
7. Identifies similar figures.
8. Identifies sides and corners of plane figures and faces, edges and corners of solids.
9. Constructs solids and skeletons of solids using appropriate materials.

Objectives for a Grades 4-6 Geometry Program

1. Identifies solids and plane figures that have symmetric and identifies the symmetries.
2. Identifies parallel and perpendicular lines.
3. Uses and describes slides, flips, and turns to map one figure onto its congruent image.
4. Identifies and draws symmetric figures and patterns.
5. Uses one, two or three figures to tile a plane.
6. Identifies and draws right angles.
7. Makes and measures angles.
8. Uses properties of figures as those characteristics that enable one to define the figure.
9. Makes and draws similar figures and scale drawings.
10. Uses ordered pairs of whole numbers to locate points in the first quadrant of a Cartesian coordinate system.
11. Relates all geometric experiences to his/her own environment whenever possible.
12. Makes solids and skeletons of solids using appropriate materials.

References

1. Shape K-9, J. Del Grande, North York Board of Education.
2. Transformations K-9, J. Del Grande, North York Board of Education.
3. Life Science Library - Light and Vision (p. 9-15).
4. The Child's Conception of Space, Piaget and Inhelder, Routledge and Kegan Paul, London 1956.
5. The Child's Conception of Geometry, Piaget, Inhelder and Szeminska, Harper and Row, New York 1964.
6. Geometry and Visualization, Creative Publications, Palo Alto, CA 1977.
7. Space and Geometry, Research Workshop, Eric Center for Science, Mathematics and Environmental Education, Columbus, Ohio 1976.
8. Right Brained Kids in Left-Brained Schools, Madeline Hunter, Today's Education, Vol. 65, No. 4.
9. The Right Hemisphere's Role in Problem Solving, Grayson Wheatley, Arithmetic Teacher, Nov. 1977.
10. How the Brain Works, Basic Books, New York 1975.
11. Tessellations the Geometry of Pattern, Bezuszka, Kenney, Silvey, Creative Publications 1977.
12. Creating Escher - Type Drawings, Ranucci and Teeters, Creative Publications, Palo Alto, CA 1977.
13. North York Guidelines for Mathematics K-6.
14. North York Geometry Units K-6, North York Board of Education, c/o Coordinator of Mathematics, 5050 Yonge Street, Willowdale, Ontario, Canada M2N 5N8.

6.3 THE DEATH OF GEOMETRY AT THE POST-SECONDARY LEVEL

SHOULDN'T WE TEACH GEOMETRY?

Branko Grunbaum
University of Washington
Seattle, Washington, U.S.A.

The point of view stressed earlier in the discussion of Diudonne sees geometry in almost all currently active and important branches of mathematics. Naturally, from such a point of view, we teach geometry whenever we teach mathematics. This view goes far back - - was it not customary to call every mathematician "geometer"? - - but somehow, I cannot find much solace or joy in this view. It seems to be that to call all the activities described by Dieudonne by the name "geometry" is rather analagous to - - and as justified as - - calling the activities of a French chef "agriculture". Indeed, in both cases much love,thought and effort is given to specially chosen and suitable objects (meats, vegetables, theorems) while converting them into delicacies for the true connoisseurs. And in both cases, many of the "raw materials" must be wasted if the only destiny deemed worthwhile for them is the transformation into gourmet dishes, or highly abstract mathematics.

Dropping the simile, it seems to me that the situation of geometry in colleges and universities is much less satisfactory than the opinions of my colleagues might have led you to believe. To justify this assertion, I should probably first give a brief explanation of what I mean by "geometry". Geometry is a special way of replacing objects of the real world by "simpler", "idealized" "figures and shapes", and then investigating the mutual relations of these. This, I believe, is what geometry - - "intuitive geometry" - - is, and should be. Geometry arose in such manner and - - as I hope to elaborate shortly - - this is how meaningful and interesting questions continue to arise. In Greek times this process of abstraction led to the points, segments, circles, triangles, and all the other objects of Euclidean geometry, and in the course of its development gave the ancients the opportunity to create the deductive system we know as Euclidean geometry (1).

Unfortunately, in later times this great achievement of the ancients was perverted to a religion. And even the Age of Reason did not dispel the superstition of the divine or a priori root of Euclidean geometry until about 150 years ago for the most "radical" among mathematicians, - - and not at all, to this day, for most mathematicians and laymen alike. Very few people are conscious of the fact that all geometry - - and Euclidean geometry in particular - - is a product of our thinking and represents just one of the ways in which we try to communicate about our surroundings and understand certain aspects of reality. Anybody who thinks he can "see", or find in nature, a Euclidean point (or a straight line, or a differentiable curve) needs urgent medical attention!

Geometry is one of the few approaches our senses give us to the outside world. Every since Euclid (or even before) some "experimental" discovery - - when idealized, generalized and digested - - led to certain geometric results or problems which, in turn, often led to far-reaching developments in geometric or other directions. Whether it is the diagonal of the square leading to irrationals; or the smoothness and continuity of curves to derivatives; or the renewed interest in "foundations" to "non-Euclidean geometries" and to the axiomatic method; or - - more recently - - properties of convex polyhedra (arising in linear programming) to a vast field stretching from combinatorics to commutative algebra; or any other of the myriad of analogous cases - - it is always some relatively simple, easily imaginable, intuitive context which generates the whole field, often so great in extent that the roots become hard to discern.

Most mathematicians are totally unaware of the fact that the elementary, intuitive approach to geometry continues (and will continue) to generate mathematically profound and interesting problems and results. There is no argument against the assertion that all questions in the geometry of points and straight lines in the place are reducible to (and possible even solvable by) linear algebra. But many attractive and genuine problems become, under such a transformation, artless conglomerations of symbols and provide no hint as to why one would ever want to spend even a moment thinking about them. A simple example will surely persuade you of the jurisdiction of this claim, if you attempt to carry out the translation into linear algebra: If S is a finite set of points in the (real Euclidean) plane, then there is a straight line containing all points of S or there is a straight line containing two and only two points of S. There is no completely trivial geometric proof of this fact (which is known as Sylvester's theorem, although Sylvester did not manage to prove it); but I am certain that you will not be able to come up with an algebraic proof which is simpler than the geometric ones (2).

The opinion that the interest in visual-geometric aspects of a topic is lost or negligible just because the topic can be translated into algebraic language is as fallacious as the (long prevailing) view that if a problem is solvable in a finite number of steps, further consideration must be boring - - the whole challenging field of computational complexity refutes such a stance.

It is a rather unfortunate fact that much of the creative introduction of new geometric ideas is done by non-mathematicians, who encounter geometric problems in the course of their professional activities. Not finding the solution in the mathematical literature, and often not finding even a sympathetic ear among mathematicians, the proceed to develop their solutions as best they can and publish their results in the journals of their disciplines. Crystallography is a prime example of this phenomenon, but other examples range from physics and biology to geography and anthropology, from industrial design to art. All these lead to very challenging mathematical problems - - more specifically, to intuitively accessible geometric problems - - which could provide a fertile ground for the development of mathematical theories (3).

But what is happening instead, and why do I see geometry in colleges and universities in a very gloomy light? The simplest answer is that we teach very little geometry, and that what we do teach is rather misleading. On the one hand, we often follow "tradition" and pretend to be teaching the "classical" Euclidean geometry, holding it up as an example of a logically perfect deductive science. But we all know that this is intellectually dishonest and mathematically impractical. This led to Dieudonne's famous slogan "Down with Euclid". Unfortunately, this slogan was generally understood as a call for eliminating all elementary geometry, and Dieudonne's insistence on teaching linear algebra in high school as a substitute for geometry did not help matters, either.

On the other hand, many of the courses we teach and which have "geometry" in their name, are good examples of the "rule of the tool" syndrome: A certain technique is found to be applicable to geometric problems of some kind; soon the interest switches to development of the technique, and then - - by actions if not by words - - only those geometric problems are deemed important which are tractable by the (by now well-developed) technique. For example, the applicability of calculus to some questions in the theory of curves and surfaces led to differential geometry, with all its modern ramifications; simple obervations concerning the numbers of points in which lines, circles, ellipses and other simple curves in the plane intersect each other led to the towering generalizations of contempory algebraic geometry. Now, I have nothing against p-adic fields, or any other developments in these areas - - provided the narrow applicability of these disciplines and their very special preconditions are acknowledged and admitted, and the viability and validity of other points of view is granted. But this is where, in practice, the system breaks down. Even if (as is often the case) the teacher is aware of various other aspects of the geometric objects he is exploring, the students in his corse on, say, manifolds learn only very few (usually rather trivial) background facts and examples. The results are similar to those that would be obtained in studying the Iliad in the original by somebody who knows only 200 words of Greek, and learns no more. Possibly, he could get the general drift of the action; certainly, he could master all the details of the meter, and other superficial aspects. Would it then be very surprising if - - when he became a teacher - - the erstwhile student were to put undue stress on the meter in the Iliad?

What is missing from our curricula in colleges (as well as in high schools) is the accent on the many aspects of reality which are susceptible of meaningful geometric interpretation. Equally absent are the elementary geometric considerations which - - besides intrinsic interest - - can contribute in very significant ways to the students' comprehension of topics in other sciences, in engineering and in art, as much as help guide future progress in mathematics. As Dieudonne has said, geometry is a great tool, a powerful stimulus, and a trustworthy guide. However, geometry can render all these services only to people who know some geometry. Disparaging the importance of the visual, intuitive - - even tactile - - aspects of geometry and urging their replacement by tool-oriented techniques certainly will not make the future role of geometry any easier. Such an attitude is inherently as absurd as the promotion of soundless music, or verbal rendering of paintings. Just as nobody argues against writing music - - and a few fortunate individuals can even enjoy and "hear" music in this form - - would the act of writing music not be utterly ridiculous if the composer were oblivious to (or even ignorant of) sounds? A "score" for some movements in a ballet; how much of the excitement of the real thing could it convey to you? But such replacement of visually accessible information by formal symbols is precisely what most mathematicians are inflicting upon themselves and upon their students regarding geometry.

I believe that - - at least in the United States - - we are at the moment caught in a quandary from which there is no easy exit and for which the very easy exit and for which the very existence of a satisfactory solution is rather uncertain.

On the one side it must be noted by anybody willing to give the matter even minimal attention, that an understanding of the spatial figures and shapes, and their mutual relationships, is becoming increasingly important to scientists as well as to engineers, and certainly to mathematicians. Even the three-dimensional space of everyday experience is much more complicated than we mostly give it credit for; we hardly stand a chance of misunderstanding any of the more complicated "spaces" needed in fields ranging from astronomy to nuclear physics, from biology to art, from space travel to highway design, - - if we are as ignorant of the ordinary geometry as most of us are.

I believe that geometry "works" so well for the rest of mathematics and for other fields because the "seeing" aspect of it coevolved with our other mental capabilities ever since we descended from the trees (or possibly even longer). But whether this is so or not - - why deny to our students both the pleasure and the utility of exploring the figures and shapes of our space?

On the other side, I am afraid that the proponents of "intuitive geometry" among mathematicians are so few in number and so dispersed and unorganized that they fall well below the "critical mass" needed to sustain any development and changes in attitudes. There seems to be no doubt that intuitive geometry will continue to develop - - the question is whether it will find a place among mathematical disciplines, or will it continue to flourish mainly among non-mathematicians.

Even the well-intentioned college professor can hardly be expected to be able (or willing) to develop by himself a suitable course on a topic with which he is largely unfamiliar and for which hardly any texts are available; therefore additional generations of students grow up with the same distorted perspectives. And it has to be admitted that there is very little motivation for potential authors to write (and for publishers to publish) books for which at the moment there seems to be almost no demand.

Clearly, if any solutions are to develop they can come about only through the dedicated work of many individuals. Let's hope that the discussions of the panel have contributed to the understanding of the problem, and will help motivate people to an increased effort towards its solution.

Notes

1. For a very lively discussion of the nature of Greek geometry see the articles by S. Unguru ("On the need to rewrite the history of Greek mathematics", Archive for History of Exact Sciences 15(1975/76), pp. 67-114, and "Some reflections on the state of the art", Isis 70(1979), pp. 555 - 565), B.L. van der Waerden ("Defiance of a 'shocking' point of view", Archive for History of Exact Sciences 15(1975/76), pp. 199-210), H. Freudenthal (ibid. 16(1976/77), pp. 189-200). and A. Well ("Who betrayed Euclid", ibid. 19(1978), pp. 91-93.

2. Relatively simple proofs of Sylvester's theorem are presented in H.S.M. Coxeter's book "Introduction to Geometry" (Wiley, New York 1969). For the history and for references to the abundant literature, as well as for other topics of a similar nature, see the author's "Arrangements and Spreads" (CBMS Regional Conference Series in Mathematics, Number 10, American Mathematical Society 1972).

3. Prof. G.C. Shephard and the author have written several survey articles and a book illustrating these facts. The two most recent papers are at the moment being published; they are "A hierarchy of classification methods for patterns" (to appear in Zeitschrift fur Kristallographie) and "Tilings with congruent tiles" (to appear in Bulletin of the American Mathematicsl Society). The book "Tilings and Patterns" is being published by W.H. Freeman and Co., San Francisco.

STRUCTURE VS. SUBSTANCE: THE FALL AND RISE OF GEOMETRY

Robert Osserman
Stanford University
Stanford, California

To the question put before our panel: "Is geometry dead?" the answer of Dieudonne would appear to be:

"Geometry is alive and well and living in Paris under an assumed name." (For an earlier discussion of this question, see the interview with H.S.M. Coxeter in the Two Year College Mathematics Journal, Vol. 11, No. 1, January, 1980, pp. 2-19). As Dieudonne convincingly documents, geometrical ideas and terminology have penetrated the most diverse and seemingly distant parts of mathematics. But what of geometry as a subject in its own right? What I propose to argue is that although we can all agree that to speak of the "death of geometry" at the post-secondary or any other level is clearly an exaggeration, it nevertheless reflects a reality. I would describe that reality as a serious decline in the position of geometry relative to other branches of mathematics during a period that might roughly be described as the middle third of this century. My purpose here is to analyze in some detail one factor that may have been of singular importance in contributing to that decline. The factor I have in mind is the predominant role that has been played by one special approach to mathematics, which, as would be the case with any particular approach, has the effect of shifting the emphasis toward certain areas of mathematics, and away from others.

The approach I refer to is the one that asserts the primacy of the abstract, the general, and the theoretical, over the concrete, the particular, and the applied. For simplicity, I shall refer to it as the "Bourbaki approach," although Bourbaki is undoubtedly a follower as well as a leader of a general trend. Also, whether Bourbaki specifically espouses that approach, he or she clearly epitomizes it. I think as good a formulation as any can be found in the introduction to the recent Abrege d'histoire des mathematiques, 1700-1900, written by Jean Dieudonne with a number of collaborators: "One sees thus why current mathematics has inevitably become the study of very general abstract structures, such as groups, rings, topological spaces, operators, sheaves, schemes, etc." (1). (This same phrase was singled out in the review of the book in 1979 Advances in Mathematics (Vol. 34, pp. 185-194), where it is also remarked (p. 187) that "On the whole, Geometry is slighted in the Abrege d'histoire.")

Is that, in fact the case? Apparent counterexamples spring to mind. There is the fascinating and extensive work done over the past decade on the Korteweg de Vries equation, known affectionately as KdV. That equation is certainly neither abstract nor general, but it has been the subject of extensive study by Peter Lax, Jurgen Moser, Henry McKean, and many others. (See, for example, the series of lectures entitled "Integrable systems and algebraic curves" by H. P. McKean in Global Analysis, Vol. 755 of Lecture Notes in Mathematics, Springer-Verlag, New York 1979, pp. 83-200. Note in particular the historical remarks on pp. 94-96.)

Another example is the map of the interval $[-1, 1]$ into itself defined by the quadratic polynomial

$$x \to 1 - cx^2$$

for a fixed c, $0 \leq c \leq 1$. The study of the iterates of this map has yielded the most unexpected properties as one takes different values of the constant c (see P. Collet, J. -P Eckmann, and O. E. Langford III, "Universal Properties of Maps on an Interval," Communications of Math. Physics, to appear).

An interesting feature of both the above examples is that some of the most important discoveries that inspired later theoretical work were originally made using computer experimentation: for example, the idea of "solitons," along with their basic properties, in the case of KdV.

But the specific question I want to address is whether the Bourbaki approach is fundamentally antithetical to geometry. I would like to start by quoting the most outrageous statement on the subject that I know. It can be found in the marvelous annotated bibliography at the end of the fifth and final volume of Michal Spivak's Comprehensive Introduction to Differential Geometry. Under the listing of Bourbaki's Varietees Differentielles et Analytiques, Fascicule de Resultats, Paragraphes 1 a 15, Spivak comments:

"Bourbaki is the originator of that famous pedagogical method whereby one begins with the general and proceeds to the particular only after the student is too confused to understand even that any more. His influence is to be seen everywhere, probably in these volumes too. Bourbaki has apparently decided that the theory of manifolds has now entered the domain of "dead" mathematics to which he hopes to give definitive form. In this summary of results the corpse is laid out to public view; the complete autopsy is eagerly awaited."(2)

There you have an appropriately morbid metaphor to match the subject of this panel. Before commenting on its main thrust, I would like to call attention in passing to the opening sentence, describing a certain approach to presenting or teaching a mathematical subject: starting from the abstract and narrowing down gradually. Spivak rightly observes that he himself has been strongly influenced by the approach, to the extent that the first of his five volumes is entirely devoted to manifolds, differentiable structures, etc., and not until Volume 3 does he get to the topic of surfaces in Euclidean 3-space. There are, I think, important pedagogical questions concerning the advantages and disadvantages of such an approach, rather than the reverse one of starting with the easily visualized and understood surfaces in 3-space, and gradually generalizaing to higher dimension and codimension, Riemannian metrics, connections, etc. The latter approach is less efficient, but has the advantage of providing clear motivation for each of the progressively more abstract concepts introduced. However, that is a different (although not totally unrelated) matter (the same dilemma is in fact referred to in the opening sentence of the Preface to Dieudonne's Abrege d'histoire des mathematiques mentioned above), and there is no space to pursue it further at this point. I would like, rather, to focus on the latter part of Spivak's statement. But in doing so, I propose to start with a detour, leading away from geometry, and from mathematics altogether, to an area that may seem initially of no great relevance. I hope that by putting the whole discussion in a broader context, some significant light may be shed on our central topic.

What I wish to do is to review briefly the movement, if that is indeed what it is, referred to as structuralism. More specifically, I want to compare the "structuralist approach" to various fields with the "Bourbaki approach" to mathematics.

I will make no attempt to give a detailed account of what I mean by structuralism. Basically it seems to me no easier or harder to define "structuralism" than, say, "mathematics," or, for that matter, "geometry." In each case there is a core of activity, where nearly everyone would agree to the terms, and then as we move more toward the fringes, we find less and less accord. I am interested in the core area of structuralism, where I believe one can discern some typical features. Among those features I would include, first, a tendency toward the abstract, the general, and the theoretical, and second, a method of approach that consists of searching for "structures" and viewing the subject through the perspective of those structures.

Now the analogy between this approach and that of Bourbaki seems clear enough superficially, so that it is not surprising to find in a book devoted to structuralism (Michael Lane, Introduction to Introduction to Structuralism, Basic Books, New York 1970, p. 20. Compare also Wolfgang Stegmuler, The Structuralist View of Theories: A Possible Analogue of the Bourbaki Programme in Physical Science, Springer, New York 1979.) a reference to "the Bourbaki," followed by a parenthetical explanation: "the pseudonym of a group of French structuralist mathematicians." I am not sure how the Bourbaki would view that characterization, but let me put that aside and explore a bit further the notion of "structuralist."

A structuralist approach to a given subject generally involves an attempt to organize the subject on a large scale, to provide a broad theoretical framework and a scientific foundation, and if possible to mathematize it. Claude Levi-Strauss, for example, asked Andre Weil to formulate an algebraic structure for his theory of kinships, and Weil's contribution appears as an appendix in Levi-Strauss' book (3). One also finds frequent use of mathematical terminology, as in Roman Jacobson's often-quoted definition of the poetic function: "The poetic function projects the principle of equivalence from the axis of selection into the axis of combination." (4) (Upon explanation of the terms, it appears that they are not used with their standard mathematical meanings. One could indeed speculate on the relative values of the ratios mathematical content/mathematical terminology in structuralism and geometrical content/geometrical terminology in various parts of mathematics. Although further afield from our central concerns, it may be worth noting that the first book of Jaques Derrida was a translation and introduction to Edmund Husserl's Die Frage nach dem Ursprung der Geometrie als intentional-historisches Problem. An English translation has appeared as Edmund Husserls' "Origins of Geometry": An Introduction, by Jaques Derrida, translated, with a preface by John P. Leavey, Jr., New York, Nicholas Hays, 1978.)

A further feature of the structuralist approach is the emphasis on relationships between various entities in a subject under study rather than on the entities themselves. In part, this feature stems from Saussurean linguistics, with its "binary opposition."

As a counterpart to these facets of the structuralist approach, it may be of interest to compare the excellent analysis by Jahnke (5) of the mathematization of physics. Jahnke draws special attention to a paper of Guass (6) which asks the question, "What is the essential condition in order that a combination of concepts may be thought of as referring to a quantity?" Gauss writes that "Mathematics is in its most general sense the science of relationships, in which one abstracts from any content of the relationships," and later, "It will be of particular importance to bring clarity to the theory of oppositions without quantities." (This very brief set of observations by Gauss contains two further remarks, both of which refer to the use of geometric representation by points in the plane. Again I should like to refer to detailed discussion of Gauss' paper and related matter in the presention of Jahnke.)

These sketchy and unelaborated statements leave considerable room for interpretation, but it seems clear (and not surprising) that in pondering some of the more philosophical aspects of mathematics and the ways that it may be applied to the "real world," Gauss was led to principles clearly related to those later taken up by the structuralists. However, as these ideas were pursued by the structuralists and their successors, a curious thing happened. (It is amusing to note that while we are here debating the "death of geometry," "the death of structuralism" seems to be widely accepted as a *fait accompli*. For example, it is referred to without further comment in the opening sentence of Leavay's preface to Derrida's book on the Origin of Geometry, referred to above: "On the present French intellectual scene, the advent and demise of structuralism.") What they found when they reached the end of the road was that, just as with Gertrude Stein's Oakland, there was no "there" there: the original object of study seemed to fade away in the process of analyzing it. Among the many statements to that effect, one could cite Levi-Strauss, "I believe the ultimate goal of the human sciences to be not to constitute, but to dissolve man" (7). Roland Barthes refers to both literature and fashion as "systems whose function is not to communicate an objective, exterior *signified* which pre-exists the system, but merely to create an equilibrium of operations, if you will, they signify 'nothing' " (8). The most dramatic version may be, appropriately enough, a (post-(?)) structuralist view of drama, "Modern theater does not exist" (9).

I grant that these statements taken out of context may easily be misconstrued. (In fact, the structuralist view would likely be that they are designed more to call attention to themselves than to be construed any which way.) Nevertheless, they point to a certain movement away from that which a naive approach to the subject would likely consider to be of principal interest and concern.

But what connection, if any, is there between all this and Bourbaki? I suppose that on a superficial level, one could say that any approach, no matter how soundly based, when pushed to an extreme, can lead to distortions, if not total self-destruction. In the case of Bourbaki, problems arise from the attempt to subdue all mathematical activity (some of it quite unruly) in the straitjacket of a single approach. On a deeper level, given that a central goal of the structuralist approach is to provide a scientific or mathematical foundation for a subject, one might conclude that such a goal becomes vacuous when the subject in question is itself mathematics. But does it? The answer is not so clear if what we mean by "mathematizing" is abstracting the purely axiomatic and deductive aspects, and avoiding the more human activities of creating, learning, and teaching mathematics. All of those are notably "unscientific" activities, linked to the fuzzy notions of intuition and heuristics. Polya, in his wisdom, developed an art of plausible reasoning, rather than a science of problem solving. Certainly mathematical creativity (like literary creativity) involves highly "irrational" functions. I should like to note here the beautiful discussion by David Pimm (10) on whether the history of mathematics should be rated X; that is, whether there is a danger to the eager and innocent minds of the young in presenting them with a picture of how mathematics is really done, with all the false starts, the wrong-headed approaches, the stupid mistakes, and the frustrating pursuits down blind alleys, to say nothing of the occasional bitter feuds over priorities. (Pimm draws attention also to the paper of Rene Thom in the Proceedings of ICME II at Exeter in 1972. Many of the remarks in Thom's paper are highly pertinent to our discussion here. The exact reference to Thom is (shades of Kristseva) "Modern mathematics" Does it exist?" in Developments in Mathematical Education; Proceedings of the Second International Congress on Mathematical Education, Cambridge University Press, 1973, pp. 194-209. The same volume also contains the paper "Geometry as a gateway to mathematics" by Bruce E. Meserve. That paper contains a good deal of interest in connection with our present panel. In particular, I should like to note two quotations that Meserve cites from T. J. Willmore's "Whither Geometry?" (Mathematical Gazette 54 (1970), pp. 216-224.) "Geometry as such is no longer a subject. What is important is a geometrical way of looking at a mathematical situation." and "I predict that geometry as a self-contained body of knowledge will become less important, while the geometrical attitude towards mathematics will become increasingly important.") There are so many human sides to mathematics that are rarely discussed. For example, there are the highly emotional as well as intellectual responses that we may have to mathematical books, papers, or lectures. Speaking personally, I would say that my responses to mathematical works are comparable in range and intensity to my responses to music or literature, and certainly broader and deeper than my responses to painting. I may find myself attracted or repelled, I may be full of admiration but uninterested, I may be angered and irritated by either style or content. It seems quite clear, for example, that whatever the intellectual content of Spivak's critique quoted above, his fundamental response is one of anger.

Let me now return to Spivak and to the specific question of how geometry fits into the world view of Bourbaki. I should like to suggest a different metaphor in place of that used by Spivak. I would compare the Fascicule de resultats to the framework for a structure which the completed volume on Varietes differentielles would represent. I do not know why that framework, erected in 1971, has been left to stand unfinished, or what obstacles have prevented the completion of the structure in its entirety. However, that is secondary. It is clear from the existing framework that were the work completed as planned, it would still have a serious shortcoming. It would provide a structure or an edifice that one may well admire, but what would still be missing is the contents. After all, what is the theory of differential manifolds? It is a framework that was invented in order to do differential geometry. It is a stage on which the action can take place. But where is the action? To my mind, the action is in all the intriguing results that have been discovered, and are still being discovered today, concerning a welter of specific geometric objects and classes of objects: the 3-dimensional hyperbolic manifolds whose properties have been uncovered by Milnor and Thurston and many others; minimal surfaces, which in the hands of Yau and his co-workers have revealed such unexpected power, playing a role in the solution of the positive mass conjecture in relativity and the Smith conjecture in topology. Note that these examples involve the central concept of curvature, and "curvature" is a word not to be found anywhere in the fascicule de resultats. But curvature is perhaps the central notion of differential geometry. In short, whatever else it is, differentiable manifold theory is not all of differential geometry. The best one could say is that it is an interesting subject in its own right. Some would see it, as we have said earlier, more as a setting for carrying on the real business of geometry.

Note also that the "real busines of geometry," as in the examples I have given, tends to be a rather ragged and uneven affair, full of loose ends, unfinished business, and decorative detail. It is a kind of antithesis to the neat, elegant, and rigid structure which may or may not succeed in housing and containing it.

Lest there be any misunderstanding, I should like to emphasize that none of what I have said is meant to belittle the value of Bourbaki's approach, which I consider to be an important and beautiful part of contemporary mathematics. My only quarrel is with the unfortunate tendency in many circles to accept it as all of mathematics - - to de-emphasize if not deny the other face of mathematics: the special, the concrete, the unstructured. And my concern is specifically that much of what I value in geometry falls in that category.

Let me conclude by addressing the second question put to our panel. The first was, "Is geometry 'dead' or irrelevant?" and the second: "Can or should anything be done about it?" I have said earlier that I believe that geometry has in fact suffered a serious decline. It has gone through a period of neglect, while the arbiters of mathematical taste and values were generally of the Bourbaki persuasion. On the other hand, I would say that there is nothing we need do about it, because that period is already drawing to a close. The recent rise of such unabashed geometers as Thurston and Yau is a sign that the low ebb in the fortunes of geometry has passed. I would predict that with no effort on any of our parts, we will witness a rebirth of geometry in the coming years, as the pendulum swings back from the extreme devotion to structure, abstraction, and generality. Perhaps the most important role we can play is to shorten the inevitable time-lag between changing currents at the research level and corresponding adjustments in our college curricula. We can initiate and revitalize courses in which students become familiar and comfortable with geometric insights and methods. Perhaps most important and difficult of all is to develop courses where the fragile but vital ability to invoke geometric intuition will be fostered and nurtured. In that connection I should like to refer you in closing to a remarkable passage from Andre Weil's introduction to the recent volume of papers by S.-S. Chern:

> The psychological aspects of true geometric intuition will perhaps never be clear up. At one time it implied primarily the power of visualization in three-dimensional space. Now that higher-dimensional spaces have mostly driven out the more elementary problems, visualization can at best be partial or symbolic. Some degree of tactile imagination seems also to be involved. Whatever the truth of the matter, mathematics in our century would not have made such impressive progress without the geometric sense of Elie Cartan, Heinz Hopf, Chern, and a very few more.

It seems safe to predict that such men will always be needed if mathematics is to go on as before.

References

1. Jean Dieudonne, Abrege d'histoire des mathematiques, 1700-1900; Paris, Hermann 1978, Vol. I, p. 11.

2. Michael Spivak, A Comprehensive Introduction to Differential Geometry, Vols, 1-5, 2nd ed., Publish or Perish Press, Berkeley 1979, Vol. 5, p. 608.

3. Andre Weil, "Sur l'etude algebrique de certains types de lois de mariage (Systeme Murngin)," Appendix to Part I of Claude Levi-Strauss, Les structures elementaires de la parente, Presses Universitaires Francaises, Paris 1949, pp. 278-285. (Reprinted in Weil's Collected Works, Vol. I, pp. 390-397, with commentary on pp. 567-568.)

4. Roman Jakobson, "Linguistics and Poetics," in Style in Language, ed. T. Sebeck, MIT Press, Cambridge 1960, p. 358.

5. Hans Niels Jahnke, "Numbers and quantities: Pedagogical, philosophical and historical remarks," Proceedings of ICME IV, Berkeley 1980.

6. Carl Friederich Gauss, "Fragen zur Metaphysik der Mathematik," in Werke, Vol. X/1, pp. 396-397.

7. Claude Levi-Strauss, The Savage Mind, University of Chicago Press, 1966, p. 247.

8. Roland Barthes, Critical Essays, translated from the French by Richard Howard, Northwestern University Press, Evanston 1972, p. 152.

9. Julia Kristeva, "Modern Theater Does Not Take (A) Place," Sub-Stance N, 18/19, 1977, p. 131.

10. David Pimm, "Why the History of Mathematics should not be rated X--The need for an appropriate epistemology of mathematics for mathematics education," Proceedings of ICME IV, Berkeley 1980.

11. Shiing-shen Chern, Selected Papers, Springer-Verlag, New York 1978, p. 12.

6.4 THE DEVELOPMENT OF CHILDREN'S SPATIAL IDEAS

THE REPRESENTATION OF PARALLELS AND PERPENDICULARS IN CHILDREN'S DRAWINGS

M.C. Mitchelmore
University of West Indies
Kingston, Jamaica

When young children draw 3D objects, they frequently represent perpendiculars with considerable accuracy, distorting the parallels if necessary. Since this is opposite the usual convention, their drawings are regarded as poor representations.

The usual explanations for this phenomenon is that young children's conception of space is dominated by topological and local relationships (which includes perpendicularity); that they are unable to coordinate perspectives, so that faces are drawn as seen straight on; and that faithful representation marks the development of a fully operational, euclidean frame of reference (Piaget & Inhelder, 1967).

However, it may also be that poor 3D drawings are the result of "lack of drawing skills." In other words, it may be that young children simply cannot draw angles and parallels accurately in 2D. There are very little data on this question. Campbell (1969) found that a 45 degree angle was copied as a 90 degree angle up to the age of 7 years, with a sudden increase in accuracy up to age 9. Wursten (1947) found that drawings of oblique parallels showed a definite improvement from ages 5 to 7; vertical and horizontal parallels are drawn accurately from age 5. No studies have been found comparing the accuracy of drawing perpendiculars. Moreover, when drawing regular 3D figures, parallels must be drawn in the presence of distracting oblique lines; no study of errors on this type of task have been located.

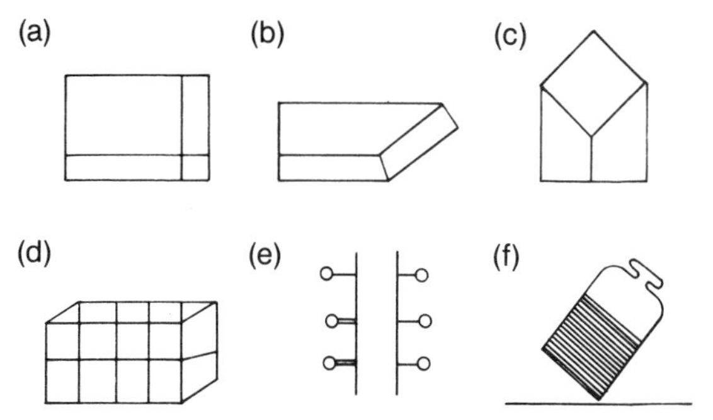

Examples of spurious perpendiculars in young children's 3D drawings.

The present study therefore set out to investigate age trends in 2D drawing skills and to study the relation between 2D and 3D drawing skills.

Method

Sixteen right-handed subjects (6 boys and 10 girls) were selected at random from a second stream in each of Grades 1, 3, 5, 7, and 9 in a primary school and an adjacent secondary school in a middle-income area of Kingston, Jamaica. Mean ages were 7.4, 9.4, 11.3, 13.4, and 15.4 years.

2D drawing was tested by freehand copying tasks. A figure was presented on the left-hand side and the same figure with one line missing was on the right-hand side; subjects were instructed to make the second figure look "exactly the same" as the first. Ten parallelogram tasks were chosen, in mirror-image pairs, to mimic the faces of a cuboid drawn in isometric projection (with sides horizontal or sloping at 30°). The 10 angle tasks and 6 parallel tasks were constructed from the parallelogram tasks by deleting redundant lines. Six perpendicular tasks were constructed by "straightening" the angles. The resulting 32 items were arranged in random order and printed in a small booklet with two tasks to an opening.

3D drawing was tested by three tasks: drawing telegraph poles alongside a given road winding into the distance, drawing the water level in tilted bottles, and drawing a given cardboard cube and cuboid.

Results

Perpendiculars and parallels were drawn accurately from grade 3 onward; more than 90% were within 5°. Angles were drawn somewhat less accurately (74% within 10°) and there was a consistent bias towards the perpendicular in most cases. Grade 1 children were considerably less accurate. The parallelogram tasks indicated a clear susceptibility to the distracting effect of the oblique line which depended on the shape of the parallelogram and decreased steadily with age. The reliability (Cronbach alpha) of the average error on the 10 parallelogram tasks was 0.90.

The decline in mean errors on the "poles" task matches the decline in error on the parallelogram tasks; correlations for the three sections were 0.40, 0.53, and 0.44. On the "bottles" task, many subjects tilted the water level away from the perpendicular, so correlations with errors on the parallelogram tasks were lower.

A total score on the 3D drawing tasks was obtained to categorising responses to each item as shown in Mitchelmore (1975); its reliability was 0.80. The correlation of this score with the average error on the parallelogram tasks was -0.66, reducing to -0.44 when the effect of age was partialled out.

Conclusions

The results suggest that there is a fairly strong relation between performance on 2D and 3D drawing tasks which involve the drawing of parallel lines at an angle to other lines in the figure. We cannot, of course, deduce that poor 3D drawing is the result of lack of 2D drawing skills, but some inferences are possible. For example, children who cannot copy a line parallel to a given line in the presence of an oblique line (i.e., most primary-age children in this study) cannot be expected to draw upright poles or horizontal water levels, even if they do understand the concepts involved. The 2D and 3D drawing skills develop together as a result of maturation, children eventually becoming conscious of the need to take care to draw accurate parallels in both 2D and 3D drawing.

The results also suggest that it is doubtful whether young children are in fact attempting to represent perpendiculars in their 3D drawings and ignoring the parallels. It seems more economical to ascribe the perpendiculars to the 2D effect revealed by the parallelogram tasks. But if this "pull" to the perpendicular is 2-dimensional rather than 3-dimensional, why does it happen?

References

Campbell, D.T. Parallelogram reproduction and the "carpentered world" hypothesis: Suggestions for cross-cultural research. Unpublished paper, Northwestern University, 1969.

Mitchelmore, M.C. The perceptual development of Jamaican students. (Unpublished doctoral dissertation, The Ohio State University, 1974.) Dissertation Abstracts International, 1975, 35, 7130A.

Mitchelmore, M.C. Predictions of developmental stages in the representation of regular space figures. Journal for Research in Mathematics Education, 1980, 11, 83-93.

Piaget, J., & Inhelder, B. The child's conception of space. New York: Norton, 1967.

Wursten, H. L'evolution des comparaisons de longueurs de l'enfant a l'adulte. Archives de Psychologie, 1947, 32, 1-144.

OBSERVATIONS CONCERNING THE CHILD'S CONCEPT OF SPACE AND ITS CONSEQUENCES FOR THE TEACHING OF GEOMETRY TO YOUNGER CHILDREN

Dieter Lunkenbein
Universite de Sherbrooke
Sherbrooke, Quebec, Canada

1. The Context

In teaching geometry to younger children, one often uses more-or-less concrete materials and situations in the hope of communicating to the learner the idea of what one "sees" in the given material or situation. All too often, one does not take into account what the child "sees" in the material. This can lead to a situation of conflict between the intention and the effect of a material which can be aggravating to the progress of learning. So it seems very important to us to get to know more about certain learning materials to determine what activities (mental or otherwise) they stimulate in the child, and to compare these reactions with the intentions of the given material and theoretical reflections that lead to its design.

In this context, we should like to report on some observations made during classroom experiments of geometry teaching units and on a number of subsequent laboratory interviews. These observations and experiments concern the child's conception of geometrical objects, the recognition of their properties and their variations, both regarding the actual physical object and various types of representations. The concrete learning material which we used is a set of polyhedra designed for a series of teaching units on polyhedra in general and on hexahedra, pyramids, prisms, and semi-regular polyhedra in particular. The intentions of the material (and of its accompanying learning situations) were to encourage children to engage into a variety of mental processes which characterize mathematical activity. In particular, we were aiming at the development of spatial representation, the formation of geometrical concepts, and the gradual structuring of the conceptual context of polyhedra. We thought of such processes as being constituted of the recognition of properties of geometrical objects and of their constituting elements, as well as of the conception of relations between geometrical objects and between their constituting elements. Such mental processes were to be stimulated by activities of manipulative exploration and construction as well as by description, discrimination, and classification activities. The material, concrete polyhedral objects made of cardboard or wood, contained examplifications of various kinds of pyramids, prisms, and regular and semi-regular polyhedra; it represented variations in the number and shape of faces and in the number of vertices and edges, as well as variations of convexity and parallelism of faces and edges. The gradual natural conception and the deliberate conceptionn of such properties and relations was the subject of our study.

2. Observations

Here are some observations made primarily with elementary school children of ages 10 and 11 who did not have much, if any, previous geometrical training. These observations are almost identical to those made with adults (pre- and inservice teachers) during teacher training courses if they did not have previous experience. Some small variations will be indicated as we proceed.

a. Global perception of geometrical objects.

The most striking and revealing observation was the fact that our students perceived the geometrical objects globally as entities. The objects were judged and described according to their appearance. The students did not see the parts of the objects, nor did they perceive the relationships among components of the objects and among the objects themselves. The polyhedral objects they were asked to explore and to describe appeared to them as different entitites, which they named according to real objects of which they were reminded by the polyhedral blocks: tent, chair, bird's house, ball, box, house, barn, etc. Even one and the same block appeared to be different to the same student when considered in different positions. Each global perception of geometrical objects may be the reason why most students consider the blocks they were presented with as being "all different objects" at an initial state of exploration. Their judging according to appearance indicates a surprising degree of absence of conservation of shape. The idea of shape as an invariant to a number of transformations, and describable by the recognition of its components and a collection of their properties, seems to be subject to a process of development based primarily on more-or-less concrete experience.

First signs of progress of such a process are indications of rather vague global properties common to some blocks: some regular or semi-regular polyhedra "look like balls", some pyramids are recognised as being "pointed", some concave polyhedra are preceived as having "cavities" or "grooves". Students who have already had some experience with plane geometrical figures indicate some blocks as "having triangles" or "squares" or as being triangular. The description of such properties is stimulated by the forming of collections of blocks which "look alike". Students realize the vagueness of these properties, when asked to verify their presence in other blocks, after which exercise they are rather eager gradually to specify these properties, in order to develop more precise criteria of discrimination.

b. Conception of Properties of Geometrical Objects.

The process of recognition and of specification of properties can be stimulated and directed by activities of comparison of concrete objects. Given a pair of polyhedral blocks, the students were asked to describe the resemblances and differences, by which activity we intended to learn which properties were easily conceivable by the students and in which mode (resemblance or difference) they tended to be recognised most frequently.

Generally, and for the beginning of such comparison activities, students find it much easier to identify differences between two polyhedral blocks than to conceive resemblances.

After some experience on the contrary, and particularly after having learned some properties relevant to polyhedra, they tend to apply these properties more easily in resemblance situations. Such observation leads us to think of the process of recognition and specification of properties as an interaction of difference and resemblance situations, where the role of difference situations is predominant in the original conception or recognition of a property, whereas the resemblance situation plays a more important part in the application and refinement process. The most frequently identified properties of polyhedra by the children we observed were

1) the convexity, or more precisely, the non-convexity of polyhedral shapes; this property is the most easily recognised, either when both objects are non-convex or when one is convex and the other is not.

2) the pyramidal shape of polyhedra; this property is indicated mainly by the particularity of one vertex or by the presence of several triangular faces; it is frequently named a triangle.

3) the shape of faces: this property is the most frequently realised when all faces have the same shape (triangles, squares, rectangles) or when two different blocks have faces of the same shape.

4) the number of facts: this property seems to emerge mainly out of difference situations, although it is quickly applied to resemblance situations.

Other properties, such as the number of vertices or edges, relations between facts (disjunctness or parallelism), the degree of vertices, etc., emerge out of such activities but much less frequently. The natural process of recognition and specification of geometrical properties tends to develop from global perceptions to the recognition of more and more detailed properties and relations. Thus, for example, the determination of the property of the number of faces leads to the activity of counting the faces of a given polyhedron, which activity turns out to be of rather geometrical character: in order to ensure a systematic counting process, students have to conceive an interior structure of the geometrical object. For that matter, certain hexahedra (those with 8 vertices and 12 edges) are conceived as triplets of pairs of opposite faces, pyramids are "seen" as a series of laterial triangles connecting to the base polygon with an opposite vertex, prisms are considered to be a pair of parallel faces connected by a belt of parallelograms, etc. The beginning of such structurings is normally expressed by sign-language when counting or justifying the number of faces of a polyhedron. Expressions of this sort indicate an important progress in the formation of spatial representation, particularly if they occur without the use of the concrete object.

c. Drawings of Geometrical Objects

In the attempt to know how children "see" a given geometrical object, we asked them to make drawings of blocks of their choice. Although most children enjoy such an activity, they are very much hampered by technical problems and in such a measure that it becomes extremely difficult to recognise what they want to express in a particular drawing. Nevertheless, we did recognise perspective, topological, and measure-conserving elements in these drawings. Therefore, we drew perspective, topological, and net-work pictures of several blocks and asked the children to associate these pictures with the block the most appropriately represent. With such an experiment, we wanted to know which of these features of two-dimensional representations of the concret objects are the most familiar to a child at a given age.

Children of age 10 to 11 have a net preference for perspective pictures and recognise more easily a given block in a net-work picture than in a topological picture (Schlegel-diagram). Adults, however, recognise the blocks in the net-work or the perspective picture with almost the same facility, but they tend to refute a Schlegel-diagram as a picture of a block. Although they also have a preference for perspective pictures, young children of age 5 to 7 recognise more easily a block in a topological than in a net-work picture. Experience seems to direct the learner's attention and his considerations of relevance away from topological features towards Euclidean properties. For younger children, such different features seem to be much more accessible than for adults.

3. Some Consequences for the Teaching of Geometry to Younger Children

Most importantly to us, geometry teaching, in particular to younger children, has to take into account the process character of the conception of shape, its components and properties, and of various relations between such elements. Such processes can be initiated from a very young age on by concrete experiences such as manipulation, construction, model-building, drawing, observation, and description activities as long as they are oriented by the natural pattern of such processes. This pattern can be described broadly by the gradual transition from global perception to the recognition of more and more detailed properties and relations. Observations of variations and resemblances of geometrical objects, constructions, or drawings, as well as their communication, seem to stimulate such processes in an effective way. Manipulative and constructive activities that favor the invention of some kind of procedure (such as the counting of the faces of a polyhedron) and their description appear to be beneficial to the development of spatial representation. Young children of school age do not seem to have a pronounced preference for either topological, perspective, or Euclidean properties, for which reason we tend to recommend that such properties should be considered more equally in order to develop a balanced geometrical conception.

4. Conclusion

In the course of exploration activities such as described above, the student's conception of geometrical objects changes gradually from a global perception determined by appearance to a more and more detailed visual description of its properties. If, originally, the geometrical objects appear to him as "all different", they may suddenly become "all alike" if he conceives a property common to some of them. Such a resemblance may disappear as abruptly as it appeared, when he realises another discriminating property; now they seem to be quite different again. The change of conception seems to take place through an oscillating process between "all different" and "all alike" which diminishes gradually in magnitude until geometrical objects are seen as bearers of properties and can be recognised and described by their components and their properties. Such change of conception enables the student to group or classify the objects more and more precisely and to differentiate in a more subtle manner. Relationships of resemblance and difference between the objects are established according to common or differentiating properties, although relationships between properties are not yet seen.

Such observations are in accordance with those made by P.M. Hiele (1) and seem to indicate some details concerning the transition between the first two Van Hiele levels of development in geometry.

The development of this material, as well as the research activities reported here, are part of a research project supported by a grant of the Minister of Education of the Province of Quebec in its programme F.C.A.C. (Ref. No. EQ-646).

Reference

1. Wirszup. Breakthroughs in the Psychology of Learning and Teaching Geometry. In L. Martin (Ed.), Space and Geometry. Columbus, Ohio: ERIC Clearinghouse for Science, Mathematics and Environmental Education, 1976. pp. 75-97.

THE CHILD'S CONCEPT OF SPACE, AND GEOMETRY EDUCATION FOR THE CHILDREN AT THE AGE FROM 4 TO 6

Kiyoshi Yokochi
Yamanashi University
Yamanashi, Japan

Introduction

Nowadays, in nursery schools and kindergartens, there should be geometry education. In this paper, I will look into the child's concept concerning the surface, position, and parallel lines, and moreover indicate certain activities which may be necessary in order for children aged 5 to realize those concepts at a higher level. These studies have been worked out by The Association of Kindergarten Teachers since 1960; under my leadership.

On the other hand, in Japan, many school children have never been through regular geometry education before they enter school. For those children, some projects of geometry education have been done at Osaka University of Education in the new semester of this year. The leader of the projects is Prof. Hirokazu Okamori. Some child's activities at the age of 6 in the projects are explained in this paper.

Position in Space

1. Experiment 1. Picture of a Woman's Face.

The experiment is to see how far a child can abstract the concept of figure and surface from a real object.

A. Procedure: In June 1980, the experiment was carried out in the following way. The subjects were 12 children aged 1, 19 children aged 2, 28 children aged 3, and 39 children aged 4 at Kyorei Daini Nursery School (Tokyo). The experimenter asked the children at age 1 and 2 to draw with crayons the picture of their mother's face, and one ones at ages 3 and 4 to draw the experimenter's face.

B. Results: The children's works were classified by the following points: Is their a contour line of the face: Are there eyes or a mouth? Are there cheeks as a part of the face? Is the inside of the contour painted with crayons - - in other words, is the child conscious of the surface?

Besides the type having no contour(a in the table), there are 4 types in their work. The first type has only a contour (b). The second type has eyes or a mouth inside the contour (c-e). The third type has cheeks or has the inside of the contour painted (f-i). The table shows the classification according to age.

Age	a X	b	c	d	e	f	g	h	i	Total
1	9	2	1							12
	75.0 %	16.7	8.3 (c–e)							100.0
2	3	7	5		4					19
	15.8	36.8	47.4 (c–e)							100.0
3		2	2	1	17	2			3	28
		7.1	71.5 (c–e)			21.4 (f–i)				100.0
4		1			1	1		26	10	39
		2.6	2.6 (c–e)			94 . 8 (f–i)				100.0

C. Conclusions: The child aged 2 begins to be conscious of there being eyes or a mouth inside contour of the face. The child aged 3 becomes conscious of part of a surface, that is, the cheeks, or the surface as a whole. At age 4, a child is certainly conscious of the surface.

These conclusions suggest the following: A child aged 2 attains the concept of a contour surrounding a real object and some parts of it being inside or outside of the curve. A 4-year-old child attains the concept of their being the surface of a real object.

2. Experiment 2. Picture of a Standing Woman Holding a Big Ball.

A. Procedure: In June 1980, the experiment was carried out with 34 children aged 3, 35 children aged 4, and 40 children aged 5 at Miyasaki Futaba Kindergarten (Kawasaki City). The experimenter stood erect holding a volley ball upon her breast with her right hand. Each subject was asked to draw a picture of the experimenter's posture with crayons.

B. Results: The children's works were classified by the following points: With which hand is the ball held? Is there a ball upon her breast (her body)? Are there fingers drawn upon the beforehand surface of the ball?

Besides the type having either no ball or two balls a, b), there are 4 types in the children's works. The first type has the ball held with her left hand (c). The second type has the ball held with her right hand (d). The third type has the ball upon her breast, but with no fingers (e, f). The fourth type has the ball upon her breast with fingers (g, h). The table shows the classification according to the age.

Age	a	b	c	d	e	f	g	h	Total
3	8	1	14	9		2			34
	23.5 %	2.9	67.7 (c–d)		5.9 (e–f)				100.0
4	1		1	18		13		2	35
	2.9		54.3 (c–d)		37.1 (e–f)		5.7 (g–h)		100.0
5				5	2	7	1	25	40
			12.5 (c–d)		22.5 (e–f)		65.0 (g–h)		100.0

C. Conclusions: The child aged 3 does not distinguish between the right and left hand. The child aged 4 not only does, but also begins to be conscious of the position of the ball upon the breast. The child aged 5 is conscious of the fingers holding the ball.

Those findings will suggest that the child at the age 4 is conscious of the right-left and front-rear position. At age 5, they do it more severely.

3. Experiment 3. Picture of Gathering Berries.

I wrote the paper on this experiment in 1978 (1), and so now summarize it briefly. The children at age 4 begin to be conscious of space surrounding real objects, so some of them added the ground and grass in their pictures. At age 5, they came to conceive the objects put in a more extended space, so we could see the river bed in their pictures.

4. Educational Activities of Drawing for the Children at the Age 5.

Here are some activities which are supposed to strengthen a child's concept of surface, position, and surrounding space: (a) sketching a kettle; (b) drawing pictures of a still life; (c) drawing pictures of outdoor scenery.

Parallel Lines and Direction

1. Experiment 1. Drawing the Route Map from Kindergarten to One's Home.

Last year, I wrote the paper about this experiment (2); the following is its brief summary. The children at age 3 begin to be conscious of a straight line and a right angle, so some of their maps were polygonal lines. At age 4, they become conscious of parallel lines, so they draw their route with parallel lines. At age 5, they come to conceive a plane in the sense of a great number of parallel lines consisting it.

2. Experiment 2. Picture of a Woman Standing Erect and Seen Sideways.

The experiment was carried out to know the child's concept of direction.

A. Procedure: In May 1980, the experiment was carried out with 33 children aged 4 and 37 aged 5 at Miyazaki Futaba Kindergarten. The children drew the picture of the experimenter standing erect, but they saw her sideways, so only one hand was visible, and one of her legs was hidden behind the other.

B. Results: Children's pictures were classified into 3 types under the following view points: Do the hands hang from the shoulder? Is the face seen sideways? Is one hand clearly visible? Do two legs make nearly one straight line?

The first type has the experimenter facing the children (a-c). The second type has the figure with the face turned sideways (d-f). The third shows the figure seen from its side (g-i). The table shows the classification according to the age.

Age	a	b	c	d	e	f	g	h	i	Total
4	1	6	7		3	3	9		4	33
	42.4%			18.2			39.4			100.0
5			3	1	1	6	9	1	16	37
	8.1			21.6			70.3			100.0

C. Conclusions: I would like to say one comment. When children draw a picture of a man standing erect and seen sideways, they have a strong feeling to communicate with the man, and this feeling makes them draw the man facing them. Those children who drew the experimenter as seen sideways can be considered to have the sure concept of direction.

From the table and their pictures, the following conclusions are drawn. The children at age 4 begin to draw the correct picture of the woman standing erect and seen sideways. The chldren at age 5 can do this much better. Those conclusions suggest that the children at age 4 have the concept of direction to some degree, and that the ones at age 5 have a much clearer concept.

3. Educational Activities of Drawing and Working with Papers for the Children at Ages 5 and 6.

Here are some activities which are supposed to strengthen children's concepts of the straight line, right angle, parallel lines, direction, and movement.

A. Age 5: (a) Making a polygon-hat. (b) Making a marker using compasses. (c) Drawing a table center or curtain.

B. Age 6: (In preschool time, no geometry education.) (a) Drawing pictures of a man standing erect and seen sideways, using a doll we call "Pinokio". (b) Making a car through its development.

References

1. Yokochi, K., (1978), Perspective Geometry and Its Meaning in Educatin. Bulletin for Math. Ed. Study, Vol. 19, No. 1-2, 10-21.

2. Yokochi, K., (1979), From Line to Plane - - Child's Map. Bulletin for Math. Ed. Study, Vol. 20, No. 1-2, 29-44.

SPATIAL ABILITIES AND MATHEMATICAL THINKING

Alan J. Bishop
University of Cambridge
Cambridge, England

There exist many research sources for ideas concerning spatial abilities. The factor analysts' tradition (Michael, 1957), the work of developmental psychologists (Piaget and Inhelder, 1956) and the researches of the individual-difference field (Anastasi, 1958) have all left us with many descriptions and theories. Mathematical learning and thinking have similarly attracted much research attention, and conferences such as ICME use much of the vocabulary and concepts developed by this research.

Of interest here, however, is the area of interface between these two families of constructs, and this area is not so well mapped (Bishop, 1980). There have been notable attempts to emplore it, for example MacFarlane Smith (1964) from the factor analysis standpoint, and Hadamard (1945) and from the individual difference perspective. Others consider the area of interface to be essentially geometry, and have therefore studied children's behaviour with respect to that (Werdelin, 1961 and Piaget et al., 1960). In the last decade there has, however, been an increasing interest in this interface, stimulated in part by a concern over sex differences in mathematical performance (Fennema, 1979). More fuel was added to the fire by the publication of the English translation of Krutetskii's work in 1976, which focused attention of 'mathematical abilities" and it is this construct which may offer the most promise yet for linking spatial and mathematical behaviours. I wish therefore to present in this paper a description and elaboration of two different abilities of crucial importance in this interface, which have not been clearly separated previously.

Much of my research has been guided by my concern with teaching - to understand more about how one can help children to improve their spatial abilities in a mathematical context. Moreover, when one works intimately in this field, one is always struck by the ease with which one can teach ideas about geometric properties, terminology, even relationships, compared with the difficulty of teaching anything to do with imagining and visualising. We can hypothesise many reasons for this, but the fact remains that those two areas of teaching and learning have a very different feel about them - they seem to involve a different type of ability.

Recently I was engaged in some fascinating work in Papua New Guinea, provoked by a concern expressed there that the indigenous students were very weak at visualisation and at geometric work generally. My research was concerned with trying to identify not just their weaknesses but also their spatial strengths which could perhaps be exploited through teaching. I used various test materials, including standardised spatial tests, and whilst superficially at least the students seemed to have difficulty with much of the spatial and visual material, when I probed further a different feeling emerged. Of crucial importance seemed to be their familiarity with, and knowledge of, the spatial forms used in the tests. For example one test (NFER Spatial Test 2) used the diagram of a matchbox with a dot on one corner. The student was required to place a dot on the corresponding corner of rotated versions of that matchbox. The test was performed much better than many others. It clearly involves the ability to imaging and visualise, but the visual conventions required to represent the matchbox are not complicated. As a result of more detailed testing and analysis, the feeling grew that, whereas those students' visualisation abilities had been previously assumed to be weak, in fact the problem really lay with their lack of knowledge of, and general unfamilarity with, the many spatial and visual conventions which we use in spatial and geometric test materials (Bishop, 1979). The tests may not have included many words but they were just as culturally biased! Indeed when one moved to a thoroughly familiar context (for example, locating promising fishing grounds) the students' visualising abilties seemed as good as anyone's.

What this example rather dramatically shows is the necessity of separating knowledge of the problem's context (and representation of it) from the abilities required to perform successfully in that context. It is this difference which I seek to elaborate here.

Two Kinds of Ability

I should therefore like to propose two very different kinds of ability in the spatial/mathematical interface:

1. The ability for interpreting figural information (IFI).

2. The ability for visual processing (VP).

IFI involves knowledge of the visual conventions and spatial "vocabulary" used in geometric work, graphs, charts, and diagrams of all types. Mathematics abounds with such forms and IFI includes the "reading" and interpreting of these. It is an ability of content and context, and it concerns the form of the stimulus material presented. VP, on the other hand, involves the ideas of visualisation, the translation of abstract relationships and non-figural data into visual terms, the manipulation and extrapolation of visual imagery, and the transformatin of one visual image into another. It is an ability of process.

Let us consider how this distinction helps to interpret the interface. Firstly concerning testing, we can see that part of our interest lies with visualising from non-figural data, and spatial ability testing, on which much factor analysis depends, tells us very little about that.

We must therefore separate clearly testing for IFI ability from testing for VP ability. With regard to the former there is much that has been achieved and the problems do not seem so great. In Bloom's (1956) terms IFI is a relatively low-level ability concerned with understanding and interpreting various stimuli. What does need to be emphasized is the very wide range of figural forms used in mathematics and in a previous paper (Bishop, 1974) I presented an analysis of these.

The testing of VP is a very different matter. I have already pointed to one deficiency of spatial testing in this regard. Another is the fact that most spatial tests are group tests, but I would argue that one can only assess an individual's ability for VP by individual testing. We know from many studies (Myers, 1958) that people solve spatial problems idiosyncratically. For instance, it is perfectly possible for Krutetskii's (1976) analytic pupuls to score highly on spatial tests by adopting analytico-deductive methods, but that would tell us nothing about their ability for VP. The only way to discover how an individual solves a problem is to test that individual alone and in depth. Krutetskii has shown the way by using introspective and retrospective techniques.

Developing and Teaching these Abilities.

Concerning IFI, once again the problems do not seem so great, although my evidence from various classrooms is that it is an ability which should perhaps be taken more seriously in view of the range and variety of figural forms in use. Certainly in any minimum curriculum this ability should be included even if extensive exposure to geometry itself should not be considered necessary. Figural forms are as much used and needed by society as are numerical and algebraic forms and, to make a different analogy, the ability to read and interpret figural forms is as important as the ability to read sentences.

Developing the ability for visual processing is, however, a very different order of problem. First of all, it is a matter of individual tendency and therefore one can ask whether it is necessary to develop it? Krutetskii is in no doubt and feels that his analytic pupils show "a certain one-sidedness in their mathematical development" (p. 321). I would agree and, in fact, I would argue further that we should emphasise more the ability for VP in our teaching precisely because it is a difficult ability to develop. The more verbal and sequential nature of the logico-analytic methodology makes it much easier to communicate and therefore to teach.

Given that developing pupils' VP ability is valuable, how can it be done? First of all there are many examples in the training literature of improvements in spatial test scores, but we must treat these with caution, of course, having exposed certain deficiencies of spatial testing (Lean, 1980). However, we can extrapolate some possibilities from that source, such as the use of manipulative apparatus which seems to have a beneficial effect (Bishop, 1973) and the use of image-provoking methods with film and stimulated recall (Kent and Hedger, 1980). The growing interest in intuition

(Fischbein et al., 1979) will hopefully generate some teaching ideas, and of course another surce is the classroom itself where some skilled teachers know the importance of VP and intuitively cultivate it already.

The teaching of IFI must also help by developing the pupil's visual vocabulary for use in later visual processing. Furthermore the teaching of other subjects such as art and poetry will be likely to encourage visualisation (McKim, 1972).

Finally, it may be provoking to consider whether the typical mathematics classroom is the best place to develop and teach VP ability, with its heavy reliance on textbook, boardwork, and exercises. I feel that the reason many children turn away from mathematics is their dissatisfaction, and unhappiness, with an excell of logico-analytic methods. I think that many of them are predisposed more towards VP but teachers and texts are not stimulating or rewarding that ability. I tested all the 14-year-old pupils at a local comprehensive school on a visualising test and amongst other analyses we contrasted their scores on this test with their school mathematical performance scores. Many pupils who were near the bottom of the year group on the mathematics test scored highly on the visualising test. Yet their mathematical "diet" had beome almost entirely arithmetic, because it was assumed that they could not do 'real' mathematics. If their mathematics teaching had emphasised and encouraged their use of VP, I am certain they could have performed at a higher level mathematically.

References

Anastasi, A. Differential Psychology. Macmillan, New York, 1958.

Bishop, A.J. The use of structural apparatus and spatial ability - a possible relationship. Research in Education, 1973, 9, 43-49.

Bishop, A.J. Visual Mathematics. Proceedings of the ICMI-IDM Regional Conference on the Teaching of Geometry. IDM, Bielefeld, West Germany, 1974, 165-189.

Bishop, A.J. Visualising and mathematics in a pre-technological culture. Educational Studies in Mathematics. 1979, 10, 135-146.

Bishop, A.J. Spatial abilities and mathematics education - a review. Educational Studies in Mathematics. 1980, 11, 257-269.

Bloom, B.S., Englehard, M.D., Furst, E.J., Hill, W.H. and Kratwohl, D.R. (Eds.) Taxomony of educational objectives: The classification of educational goals. Handbook I: Cognitive domain. David McKay Company, New York, 1956.

Fennema, E. Women and girls in mathematics - equity in mathematics education. Educational Studies in Mathematics. 1979, 10, 389-401.

Fischbein, E., Tirosh, D. and Hess, P. The intuition of infinity. Educational Studies in Mathematics. 1979, 10, 3-40.

Hadamard, J. The Psychology of Invention in the Mathematical Field. Dover, London, 1945.

Kent, D. and Hedger, K. Growing tall. Educational Studies in Mathematics. 1980, 11, 137-179.

Krutetskii, V.A. The Psychology of Mathematical Abilities in Schoolchildren. University of Chicago Press, 1976.

Lean, G.A. The training of spatial abilities: A bibliography. Mathematics Education Centre Report no. 8. Mathematics Education Centre, Papua New Guinea University of Technology, Lae, Papua New Guinea, 1980.

MacFarlane Smith, I. Spatial Ability: Its Educational and Social Significance. University of London Press, London, 1964.

McKim, R.H. Experiences in Visual Thinking. Wadsworth, California, 1972.

Michael, W.B., Guilford, J.P., Fruchter, B. and Zimmerman, W.S. The description of spatial-visualization abilities. Educational and Psychological Measurement. 1957, 17, 185-199.

Myers, C.T. Some observations of problem solving in spatial relations tests. College Entrance Examination Board Research Bulletin RB-58-16. Educational Testing Service, Princeton, New Jersey, 1958.

Piaget, J. and Inhelder, B. The Child's Conception of Space. Routledge and Kegan Paul, London, 1956.

Piaget, J., Inhelder, B. and Szeminska, A. The Child's Conception of Geometry. Routledge and Kegan Paul, London, 1980.

Werdelin, I. Geometrical ability and the space factor in boys and girls. University of Lund Press, Sweden, 1961.

CHAPTER 7 - Stochastics

7.1 STATISTICS: PROBABILITY: COMPUTER SCIENCE: MATHEMATICS. MANY PHASES OF ONE PROGRAM?

THE ALGORITHMIC STRAND IN THE MATH CURRICULUM OF UPPER SECONDARY EDUCATION

Leo H. Klingen
Helmholtz Gymnasium
Bonn, West Germany

This report refers to prerequisites relative to hardware, software, and operating systems in the scenario of the school. Then problem categories are described for computer-orientated solutions, namely exercises for beginners, normal algorithms in the lower and upper Secondary Education, subject-overlapping applications and complex projects. Finally conclusions are drawn for the curriculum of Mathematics in following years.

Some large school centers in Western-Germany now begin with modern time-sharing systems (cpu 128 K, harddisk 10 M, with 4-6 terminals and 2 monitors and printer or plotter, based on microprocessor, e.g. Z-80), other schools prefer independent systems with 1 terminal (each one with 64 K and 2 floppies). Many schools still work with desk-calculators, which cannot be considered as sufficient instruments.

Software in form of a primitive "language", as original BASIC, is not sufficient. Modern European development of structural programming needs languages of the ALGOL-family, such as ELAN (Elementary Language). Following is a short description of 3 structural elements of ELAN:

1) Structural element of "procedures" (as in all languages of the ALGOL family) with local variables and independence of surrounding context and possibility of recursions.

2) Structural element of "refinements" (similar in COBOL) which allow a top-down-structure of programs. Long names of variables, refinements etc. relieve the readability of foreign programs and therefore relieve teachability in a decisive manner too.

3) Structural element of "packets" for corresponding procedures with abstract datatypes and specific infix-operators. Abstractions generate a new simple level of operation and better understanding.

Such a language only needs three weeks of intensive instruction. Then targets of following teaching are defined by the nature of problems and projects and not by a systematic approach to the structure of the language. Some simple examples of these three elements of ELAN are necessary:

A well-known model of Samuelson simulates trends in a national economy by an additive formula

S_t (social product) = C_t (private consumption) + I_t (investment) + G_t (governmental expenditure)

based on three assumptions:

$$C_t = 0.7 \times S_{t-1}$$
$$I_t = 0.6\,(C_t - C_{t-1})$$
$$G_t = 1 \text{ for all } t$$

Instead of solving a difference-equation of 2nd order we work with a recursive computer-procedure:

```
PACKET          national economy
DEFINES         consumption, private investment,
                governmental  expenditure,  social
                product:
REAL PROC       consumption (INT CONST):
     0.7 * social product (t-1)
END PROC        consumption

REAL PROC       private investment (INT CONST t):
     0.6 * (consumption (t) - consumption (t-1))
END PROC        private investment;
REAL PROC       governmental expenditure: 1
END PROC        governmental expenditure;
REAL PROC       social product (INT CONST year):
     IF year  =  1
     THEN         2.0
     ELIF year = 2
     THEN         3.0
     ELSE       consumption (year) + private invest-
                ment (year)
     FI         + governmental investment (year)
END PROC        social product
END PACKET      national economy;
PROCEDURE       Samuelson
     INT VAR year: :  1  ;
     page;
     REPEAT
        draw histogram;
        year INCR  1
     UNTIL year > 10
     END REPEAT .
     draw historgram;
     put (5 + int (socialproduct (year)* "+"));
     line
END PROCEDURE  samuelson;
```

Refinements are of an immense value for large programs of more than 100 lines. (For instance, the program for an interactive production of a time-table for a large school contains 2500 ELAN-lines.) Here only a (too) little example of the idea can be presented: a stochastic integration of a circle is realized by a random-rain falling upon a square.

```
PROCEDURE archimedes
     initialize situation;
REPEAT
     generate a random drop;
     count the drop;
     IF drop inside
     THEN
        count success;
        print the drop
     FI
     UNTIL number of drops > 1000
     END REPEAT
     show total result.
```

```
initialize situation:
INT VAR number of drops:: 0, success:: 0;
REAL VAR  x, y; page
generate a random drop:
x: = random;
y: = random.

count the drop:
number of drops INCR 1.

count success:
success INCR 1.

drop inside:
x * x + y * y < 1 .

print the drop:
LET a = 40; LET b = 20;
plot (a * x, b * y, " * ").

show total result:
page;
put (4.0 * real (success)/ 1000.0)
put ("is approximation for π by 1000 randoms drops")

END PROCEDURE archimedes;
```

Packets are apt to generate a certain information-hiding: the pupils have not always to learn the same as the computer. For example a packet for arithmetic of long natural numbers is inserted, using a datatype LONGINT and specific infix-operators. Pupils can use the functions of this packet without knowledge of its detail structure; here is a recursive definition of faculties with 100 exact digits:

```
LONGINT PROC fak (INT CONST n):
     IF n = 0  OR n = 1   THEN
        longint  (1)
     ELSE
        longit (n) * fak (n-1)
     FI
     END PROC fak;
```

In another didactical situation, the teacher may decide to open the "glass-box" of this packet. Here is the head of the packet and as an example the definition of "+" between LONGINT - datatypes:

```
PACKET long natural numbers
DEFINES LONGINT, +, -, *, DIV, put, get:

LET max number of digits = 100,
     base = 10
TYPE LONGIT = ROW max number of digits INT;

LONGINT OP + (LONGINT CONST left, right):
     LONGINT VAR sum;
     calculate the sum;
     consider overflow if necessary;
     sum.

calculate the sum:
     INT VAR position; set carry to zero;
     FOR position FROM 1 UPTO max number of digits
               REP add for the actual position
     END REP.

consider the overflow if necessary:
IF sum overflow
THEN errostop ("LONGINT overflow").
FI.

set carry to zero:
INT VAR carry :: Ø.

add for the actual position:
sum (position) : = left (position) + right (position)
                     + carry;
IF sum position ≥ base
THEN              sum (position) DECRbase
                  carry : = 1
ELSE              carry : = Ø

Sum overflow:
Carry > Ø

END OP + ;
.
:
.
END PACKET long natural numbers;
```

The operating system (OS) for school-computers must be very simple. Any system should be rejected which needs more than 1 hour's explanation of its menu (including supervisor commands for time-sharing.) Editing, copying and archiving of programs must be supported by a good flexible editor. A normal keyboard of a typewriter without special keys is desirable.

The scenario of computer-teaching can be organized as follows: In the morning by development of fundamental ideas for algorithms etc. with paper, pencil (and rubber) accompanied by simultaneous editing with 1 terminal with 2 or 3 parallel monitors. Also exams are written in the same traditional manner. In the afternoon exercises and studies of variations etc. with groups of pupils are preferred in time-sharing work with 4 to 6 terminals. Free choices of additional projects are possible.

Problems for computer-work at school can be subsumed into 4 categories:

1) Exercises for beginners can be an imitation of functions implemented in pocket calculators (as $\sqrt{x}$, $|x|$, sign(x), max (a,b) etc.) or characteristic programs for desk calculators (as tables of values of functions, series etc. or descriptive statistics)

2) Normal algorithms for Lower Secondary Education are: recursive arithmetic, transformation between different representations of numbers, Euclid's algorithm for greatest common divisor, iterations for the solution of a linear equation, bisection of intervals for roots of functions, approximations for calculation of areas and volumes. Here some non-numerical problems, which increase motivation are important; examples are:
   a) automatic change of coins combined with a parking house with different decks and random access of automobiles
   b) coding and decoding by various methods
   c) interactive dialogue looking for successive refinements of terms (of animals, professions, etc.)
   d) simulation of collection of complete sets of advertising pictures, distributed by chance. Here is the last program, which corresponds to the so called "coupontest" for random-numbers:

```
PROC  coupontest  (INT  VAR  number  of
        simulations):
        INT VAR total number:: Ø, trial;
        FOR trial FROM 1 UPTO number of
```

```
simulation
REPEAT
  organize new simulation
END REPEAT;
put ("result:"); line;
put (real (total number), real (number of
  sim)).

organize new simulation:
initialize situation:
FOR picture nr FROM 1 UPTO 10 REPEAT
  buy packages if desirable
END REPEAT; strike balance;

initialize situation:
INT VAR i ; ROW 10 INT VAR duplicates;
INT VAR picture nr, number of packages::
Ø;
FOR i FROM 1 UPTO 10 REPEAT
  duplicates (i) : Ø
END REPEAT

buy packages if desirable:
  WHILE duplicates (picture nr) = Ø REP
      number of packages INCR 1;
      pictures nr : = random (1,10);
      duplicates (picture nr) : = dupl
    (picture nr) + 1
  END REP.
strike balance:
total number : = total number
        + number of packages.
END PROCEDURE coupontest
  Expectation: number of packages = 29
```

Transforming the uniform distribution of random numbers to an oblique distribution (e.g. by a parabola $y = x^2$) the expectation value for the number of packages increases considerably; but also the collectors (the boys and girls) are induced to be creative: they begin changing pictures between 2 partners and the average number of packages to be bought decreases.

Normal algorithms in the upper circle of Secondary Education (in America: the post-secondary section) refer to methods of Horner for calculation of polynomials, Newton for roots of functions, Simpson for integration of areas and volumes, Gauss for solution of a system of linear equations, Galton, Pearson, Buffon etc. for simulation with random-numbers and inductive statistics.

Excessive time and effort needed for numerical calculation has prohibited the general entry of these algorithms into school so far. If time allows for example methods of Gauss-Seidel for an interative solution of systems of linear equations or of Dantzig for linear optimization, these algorithms can be a supplement of the normal curriculum.

Here also subject-overlapping applications are very important. Some examples:

a) Reduction of statistical information by curves of regression.
b) Approximation of empirical functions by spline functions (e.g. partial volume of a tank with irregular measures, compared with indication of its volume).
c) Currents and loss of tension in an electrical network.
d) Score-statistics with efficient sorting-procedures, e.g. used in sport-meetings.
e) Kinematic laws governing the motion on tridimensional curves (method of small differences) e.g. for a roller-coaster.
f) Even partial differential equations, e.g. conduction of heat, of course, by difference equation and finite methods.

Some difficulties are involved in the teaching of applied Mathematics. Although German teachers have studied two subjects with university-level, there may exist only a restricted knowledge of model limits and of theoretical background. The propagation of numerical errors of different origin sometimes produces mistakes. Finally, the contiguity of subsequent applications cannot be realized in all cases.

4) The modern math curriculum gives a certain field for free teacher work. Here complex projects can be a good possibility. Projects can be complex by quantity of data (a) or by nested application of various algorithms (b).

   a) The school-typical problem of time-table needs 1200 decisions for a school of average size (1000 pupils).

   Another example: groups of pupils can construct with cheap little resistances networks of 2-3 meshes and solve the corresponding system of linear equations (Kirchhoff I and II) by Gauss elimination. The union of the small networks to a large network of 500 resistances and unknown currents is possible, the same method of solution by matrix impossible. A new idea is realized by an iterative method (Hardy Cross): by graph theory a trivial initial solution can be found, which only fulfills the laws of Kirchhoff I in each knot.

   Then one single mesh is corrected in the cpu considering Kirchhoff II. This correction does not touch the validity of Kirchhoff I in the corresponding knots of this mesh. But the correction will propagate to neighbor-meshes by common lines: an iteration begins through the whole network and it converges. Only the mainprogram can here be presented:

```
PROCEDURE hardy cross:

declaration of arrays; initialization;
initial approximation;
REPEAT
FOR meshes FROM 1 UPTO M REPEAT
  calculate correction of actual mesh;
  correct there all currents;
  calculate total of corrections
  UNTIL total of corrections < epsilon
  END REPEAT;
  print array of lines with final currents;
  print loss of tensions for all knots'
END PROCEDURE hardy cross
```

Here real industrial work was done by Helmholtz-Grammar-School (Gymnasium) in Bonn for losses of tensions in large meshed networks of gas pipelines, and also for meshed networks of steam tubes for nuclear industry.

b) Complex projects by nested application of various algorithms are for example:

Forecast of results of elections (statistical and graphical methods), Construction of highways in hilly country (in curves: integrals of Fresnel solved by finite methods), food chains for animals, problems with queues (difference equation, random numbers), interval-arithmetic (numerical methods), fully automatic discussion of functions (methods of calculus and graphical methods), net plans with critical paths (graph theory, theory of sets, graphical methods).

Complex projects support important educational aims, which cannot be realized in such perfection by other teaching-contents: Pupils learn that order of procedure, layout of description, problem-solving by breaking the problem down into subsections and by introducing new abstractions are decisive instruments for success. They learn too, that team-work in respect of neighbors is essential.

Finally some conclusions for the Math Curriculum in the 1980's should be drawn. There is not strict deterministic dependence between development of curriculum and development of instrumentation. But a certain interdependence must be seen, if teachers do not want to lose the motivation of the last pupils. As more perfect instruments enter in schools, there should result a good reaction - not to forget Euclid and Euler but to underline an algorithmic strand.

1) I support in the development of school mathematics some "mathematics en gros" instead of "mathematics en detail". Example: Pupils should not learn - better: not only learn the discussion of a concrete function $y = x^3 + 2x^2 - 5x + 7$ (because this will be a simple machine-work and machines do this work better with more precision and higher speed and without mistakes). But they should have the knowledge of and should be able to prove the theorem "Each rational function of degree 3 has exactly one point of inflextion".

2) For many applications mathematics of finite differences should be preferred to too sophisticated calculus. (The starting point for the solution of many problems in natural sciences and social sciences is as easy as a rule of three).

3) I wish an avoidance of over specialization in pupils' options in senior classes in favor of a more general education including related disciplines in the true sense of a Mathesis universalis (G.W. Leibniz).

STATISTICS: PROBABILITY: COMPUTER SCIENCE: MATHEMATICS - MANY PHASES OF ONE PROGRAM?

Richard S. Pieters
Moses Brown School
Providence R.I.

Changing fashions in curricula are part of the educational scene in many, if not all, countries of the world. But such changing fashions frequently have only limited and temporary effects on the offerings of most schools and colleges. Particularly this is true in those countries such as the United States (where I have spent my teaching career) where there is no central Ministry of Education to specify what fields or even what specific topics in certain fields should be in a curriculum and to determine when they should appear.

Nevertheless, changes do occur, perhaps more slowly than some of us would hope and more quickly that some of us would like. Looking back over the fifty years that I have been teaching, I can see signficant changes. However, many of the changes, that twenty-five years ago I thought were just around the corner, are still not evident on the current scene at any significant level.

My personal experience in the 20's and 30's as an undergraduate major in mathematics and in several years of graduate work, was that Probability and Statistics were not considered a necessary part of such a program in the mathematical sciences if indeed they were available at all, and, of course, Computer Science was not then in existence. But in time this began to change.

In 1955 the College Board Commission on Secondary School Mathematics began its work and in its recent report made a strong recommendation for the inclusion of a course in Probability and Statistics as part of the college preparatory program in mathematics. But little came of this. In spite of the prestige of the members of the Commission and the weight of the College Board's recommendation, the time was not ripe for such a significant change in the secondary school mathematics program. Even so, many people recognized that something needed to be done and shortly thereafter the School Mathematics Study Group (SMSG) was organized with substantial support from the National Science Foundation. Over the course of several years it proposed and developed curricula that soon became know as the "The New Math".

During the 60's corresponding groups, such as the Southampton Mathematics Project (SMP) in England and the African Education Program in East and West Africa, also developed new curricula which, it was hoped, would be widely adopted. But in the 70's it was realized that, at least in the United States, these hopes were largely being unfulfilled. Meanwhile in several countries, including for example, France, Germany, England, and Sweden, as well as the United States, many people began to feel that Statistics, at least its fundamental ideas and methods of inquiry, was important enough that it should be included in the mathematical training of all students, not just those who were college or university bound.

In an attempt to meet this feeling, the American Statistical Association (ASA) and the National Council of Teachers of Mathematics (NCTM) appointed in 1967 their Joint Committee on the Curriculum in Statistics and Probability. Its activities, such as preparing and publishing the books "Statistics By Example" and "Statistics: A Guide to the Unknown" and conducting over several years a dozen conferences on the teaching of statistics and its place in the curriculum, have fostered, or at least kept alive in some places, the belief that Statistics and Probability do have a rightful place in the education of most people.

As I have observed from my position as a school teacher, it seems to me the situation in many colleges and universities has also been changing. Having become convinced of the importance of statistics in the mathematics programs in secondary schools, I was at first gratified when statistics was introduced and integrated into mathematics departments, but I was soon disappointed when again statistics departments became separated from mathematics departments as more and more specialized fields in each discipline became important. Now, numerical anlysis and computer sciences are becoming fields so separate and distinct from "pure" mathematics that many times they are no longer even housed in the same buildings with mathematics.

Has not the time come for those of us interested not only in the education of specialist in each disparate field but also even more in the education of all future citizens of the world, to realize how important each of these fields of Probability, Statistics, Computer Science and Mathematics are to this education? Is it not possible that each can contribute to this education more effectively if they are interrelated and their separate ideas and methods brought to bear on each other? If so, by working towards such a union we may really build a whole greater than the sum of all its parts.

Let us consider some of the efforts in this country to build such an integrated curriculum that might begin in the late elementary grades and continue through the secondary schools and colleges.

The ideas and methods of probability and statistics have long been considered too abstract, subtle and complex to be introduced to students before their university days. But some recent investigators (1,2) have found that by the time youngsters are twelve or thirteen years old they are already familiar with such ideas as "equally likely events", perhaps not with such formal words themselves, but with the underlying concept. They know that in flipping a coin, heads are about as likely to come up as tails. Of course they may not always be able to tell whether two events are actually equally likely. For example, a youngster confronted with two bags, one containing one red and two white marbles and the other containing ten red and twenty white marbles and asked which bag he would rather draw from to get a red marble, may very well choose the second bag since it has many more red marbles. But he does have some recognition of the idea of equally or unequally likely events. What better time is there to actually try such an experiment and thus begin to develop this idea more clearly, than in those middle school years when experiments with coins, dice and marbles are still new, fresh and exciting.

Alan Hoffer at the University of Oregon in 1976-77 was director of a project funded by a grant from the National Science Foundation that developed and tried out in a sixth grade class in a public school in Eugene, Oregon many suggestions for class room discussions and experiments in statistics and probability. The results of this project were published by Creative Publications, Inc. of Palo Alto, California under the title "Statistics and Information Organization". Since I worked on this project and am, therefore, familiar with the material I shall refer to it frequently.

Why is the study of statistics and probability even in these very elementary aspects important for all of us? There are several reasons that seem to me to be valid. Among these are:

We all have to make decisions in the face of uncertainties. Some knowledge of how to select, classify and interpret relevant data may enable us to make such decisions with more confidence.

Statistics and probability are a natural link between arithmetic or elementary geometry and the phsycial or social sciences even at a beginning level.

Concrete and formal experience with these fields obtained by gathering, organizing and interpreting data about one's class or home, or by relating probability to games and school or professional sports will be a good introduction to more formal aspects and methods to be studied later.

The interesting applications of elementary statistics will give many opportunities to review and strengthen the basic skills of arithmetic and geometry whose study sometimes becomes so boring after endless repetitions. But some may complain that the arithmetic involved in working with a mass of data to find the representative numbers such as the mean, median and mode or the common measure of dispersion such as the range, the percentiles or even the variance and standard deviation are beyond the ability of these students or may even destroy their interest in the problems at hand. Obviously we should begin with a limited number of data items, say ten or a dozen where the ideas may be developed but the arithmetic is limited to a few operations. Then when we move to larger masses of data, the availability of electronic calculators and mini- or micro-computers can be of great help. Later I will give some simple programs in BASIC that show how such work can be integrated in these fields.

A very important objection to these suggestions may be raised. How can we find time to add to the curriculum in the face of all the demands already being made on our time and effort? It surely is a question of priorities. I believe there are appropriate answers both for the very able students of mathematics and also for the ones who must progress at a slower and more leisurely pace. For the first group there has been a growing tendency to push the introduction of algebra down into the eighth or seventh grades. My experience is that, except for the very exceptional student, the concreteness of the applications of arithmetic and geometric processes make for a much more satisfying and fulfilling course than moving too soon into the abstractions of algebra. For the other students who may have been subjected to endless and boring repetition of trying to learn multiplication tables, manipulation of fractions, long division, etc., to use in "consumer math" courses, the relief of being able to use calculators and/or micro-computers to ease the drudgery brings an added incentive to think about problems vitally related to their own lives.

All of us, students or adults, blue collar or white collar workers, professionals or executives, face such problems daily. We live in a world of statistics. We must be able to judge the claims of advertisers, pollsters, politicains and special interest pleaders as to the validity of their arguments and the reasonableness of their conclusions. What better preparation for making such judgements is there than some early experience with problems that can be attacked by collecting, organizing and studying appropriate statistical data and using probabilities, not only to make correct inferences from such data, but to estimate the reasonableness of the answers we get?

Let us now consider several examples of problems that have been used successfully to introduce some ideas of statistics and probability at the middle school level. A simple problem might be to choose a member of a class that most nearly is representative of that class with regard to some quality, such as weight. To do this a student may suggest that we try to find the average weight of the children in the given class and perhaps extend that to find the average weight of all the children in that grade in the school. Immediately the question arises as to what we mean by "average". We must think about and perhaps even define informally different kinds of averages and decide how to find them and when each is appropriate. Calculating the mean will provide opportunities to review addition and long division; finding the median will involve counting and arranging; finding the mode will involve constructing bar graphs. Suppose we want to choose a representative child from the room, how will we decide which average we should use? If we wanted to do this for the whole grade would it be better find the average of the averages of each class or the average of all the individuals in the grade? If the latter, we would probably move to a frequency distribution and thus review basic multiplication as well as addition and division. This problem has been successfully used in a fifth grade class in France as well as in this country and I would be surprised if not in many other countries as well.

Now the use of a calculator or a simple computer program would help. The opportunity to teach youngsters at this age how to write simple programs in a language such as BASIC, should not be ignored. Programs to find the mode and perhaps even the median are probably not ones that could be worked out except by the computer buffs of whom there are surprisngly many these days in the middle schools.

Another problem that seems to be successful is to have a class consider the number of M & M candies in each of several small packages. How does this number vary from package to package? What is the average number? How does the distribution of the various colors of M & M's vary from package to package? Are the proportions different if you consider a large family or party size package? Bar graphs come naturally as the students line up the candies of different colors and then disappear quickly as the class ends. (I wonder why?)

Now suppose a class has done quite a bit of work on experiments involving coins and dice and the probabilities of certain specified outcomes. They may be ready for a more sophisticated and interesting problem involving sampling from an infinite population and some of the probabilities involved.

Suppose the OHSOGOOD CEREAL COMPANY packs one free felt-tip pen in each box of its SOGOOD cereal. There are pens of six different colors and there are the same number of pens of each color distributed randomly in each day's output. Jill wants to get a complete set of pens. How many boxes of cereal will she have to buy to get a complete set? This question really involves three situations. What is the least number she may buy? How many may she expect to buy? How many might she have to buy? Here is a problem very different from an ordinary arithmetic problem. While there is an easy answer to the first question, there is no easy answer or solution to the second and no unique answer to the third. Posing the questions to a class should lead to an open ended discussion and perhaps to suggestion for different approaches to a solution of the second question. One suggestion might be for each member of the class to start buying cereal (if that were possible) but that would soon prove to be too time consuming and too expensive. Now what? If some simulations of problems have been tried before, some student may suggest such a smiluation using a die where each face is assigned to one of the colors. Counting the number of rolls it takes until all six faces have appeared would be one trial of a simulation of a series of actual purchases. I tried this a couple of times and it took thirteen rolls once and fourteen once to record all six faces. (These are surprisingly close to the theoretical value of 14.7.) But having a class full of children rolling dice on desks and under chairs may not be your idea of a peaceful and quiet activity. A better way might be to use a table of random numbers. If such a table is not available to you, a simple computer program will quickly print out 2500 random digits from one to six. Now start reading at any point in the list and count how many digits you have to look at before you get at least one of each kind. But even this will not quickly give us enough trials of the experiment to make us confident that we have a reasonably close approximation to the real answer to the question "What is the average numer of purchases for success?" We could resort to a computer run of the experiment. Figure I gives a program in BASIC to do this.

Colored Pens (Balpen)

```
100 REM This is a program to simulate how many purchases you need to
110 REM make to get all the pens if N differently colored pens are
120 REM packed one pen to a box of cereal
130 REM There may be up to 28 different kinds
140 DIM A (20)
150 Randomize
160 Print "How many different pens? (up to 20)"
170 Input N
180 Print "How many different trials desired? (up to 100)"
190 Input M
200 If M  100 then 220
210 Go to 240
220 Print "No more than 100 trials are allowed"
230 Go to 180
240 For I = 1 to M
250 For J = 1 to N
260 Let A(J) = 0
270 Next J
280 Let K = 0
290 Let Z = Int(RND(x)* N + 1)
300 Let K = K + 1
310 Let A(Z) = 1
320 For J = 1 to N
330 If A(J) = 0 then 290
340 Next J
350 Let S = S + K
360 Next I
370 Print
380 Print "The average number of purchases is "S/M" for "N" different pens."
390 End
```

Figure I
from Statistics and Information Organization

At STEP 340 we have completed one run of the experiment. This program sets the maximum number of runs allowed at 100 but by a simple change of the program at STEP 180 and omitting STEPS 200, 210, 220, and 230 you could allow any number of runs. This program does not give any idea of the probabilites for the different number of purchases required from the minimum of 6 to any large number. But a few steps added could do it. For instance

```
200   DIM T(100)
210   FOR I = 6 TO 100
220   LET T(I) = I
230   NEXT I
345   IF K = T(I) LET N(I) = N(I) + 1
385   FOR J = 6 TO 100
386   PRINT "P(J) = "N(J)/M
387   NEXT J
```

Now by allowing 1000 runs we could check the probabilities of the various number of purchases required. (Could there by a P(14.7)? Would you expect P(14) or P(15) to be the largest of all? Maybe we did not have to consider T(I) to be as high as 100.) Also note that this program could be used for different values of N (the number of different colors of pens).

This is getting rather sophisticated for a 6th or 7th grader, but my experience is that usually there is at least one student in the school if not in that class that could work out such a program. If some statistics, probability and computer work is introduced in the middle grades, then how much easier will be the introduction and integration of such ideas in the secondary program.

I must be frank to admit that although this introduction has been advocated in several reports and suggested curricula, it has not made a great deal of headway in the overall picture of American secondary education. The present availability of mini- and micro-computers in many schools to take over the task of "crunching" the large masses of data that are involved in really interesting problems will assuredly help this situation. The ability to use calculators and computers, at least to the extent of being able to write reasonably simple programs as well as to use those already prepared by others, is rapidly becoming as necessary for college students as the ability to use the library, and those of us in the elementary and secondary field should help our students by beginning this training at an early age. Even for non-college bound students the employment opportunities for young people with even elementary knowledge of progamming and computer use are so great that it is important we should prepare them for it as far as we can. What better opportunity is there than in some simple statistical problems?

Where can we begin? Zalman Usiskin's recent book "Algebra Through Applications?" published by the NCTM in 1978 is an example of the effort to integrate these fields. Using it should make it impossible to omit statistics by omitting that last chapter that so many books offer as a bow to the request to include statistics and probability in the standard course. New York's latest curriculum spirals statistics through their program all the way up. But many systems both there and in other states still use books in which little of this kind of material is available. What can be done by those who are convinced of its importance? The books "Statistics By Example" put out by the Joint Committee of the ASA and NCTM are resource books containing many examples, experiments and projects that can be used as bases for individual or classroom work of short or long duration. In fact, some of us have used these books as the basis of term long courses in statistics and probability although that is not what I am advocating or hoping for at this time. The Joint Committee under the chairmanship of James Swift is continuing to work to help all teachers move towards a unified curriculum that will prepare our students, both college bound and non-college bound, to better face this world of uncertainties, by making decisions in the light of available facts that may have been judged carefully as to their appropriateness and reliability.

May I quote in conclusion a few words from as long ago as 1956 and 57?

> "Uncertainty dogs our every step...We must act on incomplete and unsure knowledge...Our system of education tends to give children the impression that every question has a single answer. This is unfortunate because the problems they will encounter in later life will generally have an indefinite character. It seems important that during their years of schooling, children should be trained to recognize degrees of uncertainty, to compare their private guesses and extrapolations with what actually takes place, in short to interpret and become masters of their own uncertainties". (3)

Also from a report on the need for a new curriculum.

> "The notions of probability, correlations and sampling are among the fundamentals of modern social science...There is in all statistics a salutary concern for the uncertain and the incomplete - - for the gray that is real more than for the black and white that is abstraciton. It is well for the student to learn both that matheamtics has uncertainty and that uncertainty can be mathematically treated...If we had a curriculum to build from the ground up we cannot suppose that it would omit statistics from a general education". (4)

How much more powerful this argument becomes now that we have, through the use of calculators and micro-computers, the ability to handle masses of data as never before.

As we move into the last two decades of the twentieth century let us not be content with the curriculum fashioned for the fast disappearing world of the nineteenth and early twentieth centuries, but endeavor to build a curriculum for all our students that will help them face the problems and the uncertainties of the new world of the twenty-first century.

References

1. Cohen, J. and Hansel, C. "The Idea of Distribution". The British Journal of Psychology Vol. 46 Part 2 (May 1955a) pp. 111-121.

2. Piaget, J. and Inhelder, B. The Origin of the Idea of Chance in Children. New York: Norton and Company, 1975.

3. John Cohen "Subjective Probability" Scientific American Vol 197, No. 5, November 1957 p. 138.

4. General Education in School and College. Harvard University Press. Cambridge, Mass. 1956. p. 56.

7.2 VIGOR, VARIETY AND VISION - - THE VITALITY OF STATISTICS AND PROBABILITY

VIGOR, VARIETY AND VISION - - THE VITALITY OF STATISTICS AND PROBABILITY

I. J. Good
Virginia Polytechnic Institute and State University
Blacksburg, Virginia

Abstract

It is argued that elementary probability and statistics, taught in an appropriate manner, would increase the rationality of human thought and behavior.

Introduction

I was greatly honored when Jim Swift invited me to give the keynote speech for these sessions, and overwhelmed by the title of my talk that he selected. He asked me to present an overview of the case for greater emphasis of statistics and probability in school curriculum. I understand that "school" is used in the British sense of pre-college.

The field is one where I feared to tread, but I agreed to rush in.

My overview has to contain some recommendations. Most recommendations are partly predictions because there's no point in making recommendations that are very unlikely to be adopted. When I talk about the future I shall have in mind, say, the next ten or twenty years.

In a recent conference in a hotel in Detroit there was a session on parapsychology which appropriately enough was held in the Crystal Ballroom. I need a crystal ball again today.

The question of what probability and statistics should be taught in schools, from the kindergartens to the high schools, has often been considered, and a historical account was given by Pieters (1976). Many other articles on the subject can be quickly found under the heading "teaching" in the index of The American Statistician published in 1979 December, and in the Current Index to Statistics, Volumes 1 to 5 (1975 to 1979). There is even a journal called Teaching Statistics which started publication last year.

In view of all this literature it is unlikely that any of my main themes are new, but perhaps complete novelty is not expected in an overview.

Education in General

To make a case for a change in the curriculum it is necessary to hold in mind some philosophy of educationin general. Such a philosophy covers the purposes of education, its problems, its methods, and its compromises. I shall now say a little about these aspects. They all apply generally and in particular to statistics and probability, and I shall return to some of them when discussing those topics.

The purposes can be classified as personal, vocational, and social (including political), and I expect our sessions will touch on all three of these aspects in more or less detail.

Among the problems of education are the questions: Can we teach students how to learn? how to reason? and how to be creative?

I think it is possible to learn how to learn. There are books on the art of study, how to use a library, and how to read fast. The student can be taught the need to change gear when reading mathematics, and to ask questions of himself when reading. Seymour Papert at M.I.T. believes that very young children can be taught how to learn with the aid of computers. A child can have an active interaction with a computer, whereas a teacher hardly has time to attend to each child individually.

I think students can be taught how to reason, at least when they have the latent ability, but I'll return to that topic later.

I think students can be taught how to be creative and can also be given the opportunity to be creative. By suitable prompting, the student might, for example, conjecture Pythagoras's theorem, and perhaps conjecture the inverse square law of gravitation. We don't have to tell the students the solutions of such problems before they have tried to solve them themselves. Even in the soft sciences, such as political science, students can be asked to suggest explanations. For example, if they are told that the fine is a few thousand dollars for giving an illegal political contribution of a million dollars, they might come up with a reasonable hypothesis to explain why the fine is so small.

Regarding the methods of education, perhaps the main aim is to interest the student, holding in mind that he is probably asking himself "What's in it for me?" Also we need to decide how much use should be made of individual projects, books, films, audio cassettes, television, and microcomputers. Christopher Evans, in his book The Micro Millenium (p. 129) says of microcomputers "To the beginners, the backward, or the poorly motivated, they will be lucid, non-patronizing, and endlessly patient... To the brighter, more advanced child they will be challenging and demanding, but still endlessly patient."

Compromises arise in education when there are apparent conflicts. Let me mention some of these conflicts, each of which needs to be resolved by suitable compromise or synthesis.

(i) Choice of topics in crowded curricula

(ii) Theory versus practice

(iii) Teachers versus computers. One attempt at an integration was described by Wagner and Motaze (1972), who combined the Proctorial System of Instruction (the Keller system) with computer support. But the computer support was not essential to their experiment.

(iv) Authority versus the student's choice in the allocation of the student to various courses. Aptitude tests can be used, and to help the students to choose there could be very short courses about the courses, and about the "labor markets". One disadvantage of the authoritarian approach is that it can turn students off. If it is compulsory to learn poetry many students will never read poetry after they leave school. I am reminded of the Pink Floyd song "We don't want no education, We don't want no thought control".

(v) Technique versus understanding. The "New Math" perhaps put too much emphasis on understanding and not enough on technique. A difficulty here is that the optimal mixture depends on the student - - and on the teacher. In fact Zelinka (1980) argued that the main or only cause of failure of the New Math was that teachers were inadequately trained. Her paper is a good defense of the New Math.

(vi) So called "elitism" versus uniform mediocrity. I suppose most of us here are in favor of special instruction for the very brightest students. The teachers of very bright students migh need to learn more about the topics taught in which case they would become better teachers of the other students. It is always useful to know more than you have to teach.

Before leaving the matter of education in general, it seems appropriate to note how inefficient it is in schools, except in keeping children off the streets. Speaking for myself I think my elementary education was no more than 10% efficient. In most subjects I learned less than 10% of what I was told, and in mathematics I learned everything in 10% of the teaching time. The only solution to such problems is to match the curriculum to the student instead of putting him in a procrustean bed. Unfortunately, procrustean beds are less expensive but a solution might come from the Keller system, or perhaps from computers because computers keep going down in price, for any given computing power.

General Comments about Probability and Statistics

Cicero described probability as the guide to life. He didn't know that two thousand years later his comment would be used as an argument for studying probability instead of studying the writings of Cicero.

The twentieth-century prophet, H. G. Wells, predicted in 1903 that, not very long after that date, efficient citizens would need to be able to think in terms of averages and optimization. In 1931 he said "the movement of the last hundred years is all in favour of the statistician". (See Tankard, 1979 and Tee, 1980.) Now, as the ninth decade of the century approaches (it starts in 1981), we seem at last to be agreeing that statistics and probability should be taught in high schools.

The teaching of probability might do more than just produce efficient citizens because probability is basic for reasoning as a whole. We do not always consciously judge probabilities, but when we are rational we act as if we were judging them. I am here using "probability" in the sense of a "degree of belief". Bernard Shaw remarked that "It is not disbelief that is dangerous to our society, it is belief." Arthur Koestler, in The Ghost in the Machine (p. 266) says "... the damages wrought by individual violence for selfish motives are insignificant compared to the holocausts resulting from self-transcending devotion to collectively shared belief-systems." Perhaps if people thought more in terms of degrees of belief, instead of absolute belief, there would be fewer wars, holy or otherwise.

When H. G. Wells wrote in 1903, he had in mind the social and economic applications of statistics. The word "econometrics" had not yet been invented and the probabilistic nature of quantum theory was unknown although radioactivity had been discovered by Becquerel in 1896.

Probability and statistics now permeate all the sciences, both hard and soft, as well as engineering. Statisticians are finding it easy to get jobs in industry even without doctorates, and assistant professors, unsure of tenure, are being tempted away from university departments. Likewise the demand for statistical consulting within the universities is still rapidly increasing. In my university it has increased about tenfold in five years. So the value of statistics and probability from the social and vocational point of view, as prophesized by Wells, will be questioned only by those who believe that statistics are too often misused.

Of possibly even greater importance is Cicero's comment (that probability is the guide to life). It can be reexpressed by saying that probability is basic to rational thought and to rational decisions, at least if probability is interpreted in terms of degrees of belief. Probability judgements in this sense can apparently be improved by eduction though more evidence about this would be worthwhile. Sharper probability judgements would be of value both to the individual and, in the long run, to society. I think the hidden social value of forms of education is not well understood. It is possible, for example, that over thousand of years the study of mathematics has improved languages by making their grammars more logical. Whether this is so or not, it is often claimed that mathematics trains the mind in logical thinking and in objectivity. But most thought is probabilistic and this is true to some extent even within pure mathematics. The process of discovery in mathematics is not very different from that in experimental science; in fact pure mathematics is to some extent experimental. This has been recognized by several mathematicians (Polya, 1954; Good, 1977, where further references are given). A theorem has what I like to call an evolving or dynamic probability of being true, a probability that varies in accordance with the evidence generated by the mind of the statistician.

The fact that the thinking of even the pure mathematician is probabilistic shows that probability permeates all rational thinking. There is also a form of thinking that is less overtly rational; it is the thinking believed to be done largely by the right hemisphere of the brain, the kind of thinking that recognized patterns at a glance. When you recognize a face you do not consciously apply discriminant functions although the process is implicitly Bayesian (Good, 1950, p. 68). It is possible that many highly experienced medical doctors often recognize a disease state by this largely unconscious process. It's like the way we recognize words and phrases when reading without much conscious attention to the letters. Just as there are two schools of thought about how children should be taught to read, there might be some dispute about how rational the training of a doctor should be. Many doctors of the old "right-hemispherical" school have some difficulty in understanding the Bayesian approach to clinical decisions, whereas a "left-hemispherical" doctor might ask "What other approach is possible?" My own guess is that children should learn the alphabet, that medical students should be taught rational methods, and that musicians should learn to play the scales. Use both hemispheres of your brain.

By the same token, I think students should know how to think probabilistically and rationally, as well as intuitively and creatively. Statisticians of the subjectivist Bayesian philosophy accept the need for intuitive judgements of probabilities which have to be combined with the objective evidence by means of Bayes's theorem. Sooner or later this approach will be understood even by legal judges.

It has been a tradition of educational theory that mathematics and logic are useful for training the mind. Chess too has much merit for those who enjoy it; and I am now claiming that probabiity and statistics, taught with the right mixture of technique and understanding, may very well be intelligence-amplifiers.

H. V. Roberts, in a recent article, said " . . . we should introduce something of a 'cookbook' flavor into our textbooks. . . . It is good to encourage students to think but thought is costly . . .". In the discussion Winkler said "It seems much better to have a basic understanding ... than to have a superficial exposure to a large 'bag of tricks'." In his rejoinder Roberts said "I find little or nothing that I disagree with." If this interchange between Roberts and Winkler is smoothed out I think it agrees with my opinion that we need to look for a good mixture of cookbook and thought. It should also be held in mind that they were discussing college education. In high schools the students are in even greater need of learning how to think rationally. If this can be achieved, they will be able to think rationally in their adulthood also, and the social benefit might be immense. I know this is controversial; theories of education often are.

Generalities Concerning the Curriculum

We ought not to prejudge the issue of the curriculum too much, but we mut make some guesses or judgments. I once attended a colloquium concerning the activities of economic consultants. The speaker said that his main value in nearly every consultation was to talk about "the size of the cake" and to ask the client "What are you prepared to give up (in exchange for some new activity)?". We can ask the same question for anything new that is to be added to a crowded curriculum. I mentioned this when discussing education in general. Can we sacrifice some of the mathematics when some of it is needed anyway to understand elementary statistics and probability?

I think the answer is that the effect of teaching a little statistics and probability is to motivate some of the pure mathematics. The beginnings of probability is to motivate the manipulation of fractions rather than decimals (as I believe Dr. Pollock has mentioned elsewhere), and the beginnings of statistics motivate graphs and the extraction of square roots. Perhaps the best policy is therefore to integrate the most elementary probability and statistics into the mathematics course.

The very brightest students could do a little regression theory which could help to motivate differential calculus. The normal distribution exemplifies integration, the Poisson distribution exemplifies infinite series. Matrix theory and n-dimensional geometry are exemplified by the theories of the multivariate normal distribution and the general linear model, but only the most exceptional student would reach these topics at school. Weyl, in his Group Theory and Quantum Mechanics (p. xxii) says "It is somewhat distressing that the theory of linear algebras must again and again be developed from the beginning, for the fundamental concepts of this branch of mathematics crop up everywhere in mathematics and physics, and a knowledge of them should be as widely disseminated as the elements of differential calculus." We could now add statistics as a field of application of linear algebra. When I was at Cambridge the course on matrix theory was not motivated even by geometry.

Perhaps only one high-school student in a thousand should bother with the general linear model but that student is likely to be the one that will advance knowledge more than the other 999 laid end to end.

For the general mass of students I think the curriculum recomended by this Congress should be restricted to the very elementary level at least at first. The curriculum could be gradually made more advanced as the teachers gain more experience. One reason, but not the only one, for keeping to the elements at first is that not many teachers inhigh schools are likely to know very much statistics. Statisticians earn too much in industry at present for many of them to take up school teaching. Over the years it will be possible to organize courses for teachers to make possible the growth of the curriculum. Paradoxically enough, the teaching might be at its best when there is a glut of statisticians instead of a shortage. Eventually an ideal suggested by Allan Birnbuam (1971) might be achieved. He said "... it may well turn out that the familiar pedagogic difficulties, as well as the central educational purpose, can be met effectively just by establishing from the outset a realistic informed perspective on applied statistics, and a lively conceptual involvement, to be progressively challenged, broadened, and deepened through well-planned several-sided interdisciplinary study and experience." (His italics.)

There are some general comments about the curriculum that should be made. One of them concerns the question of cookbook techniques versus philosophy or understanding which I mentioned earlier. Children are usually better at rote learning than at logic and philosophy, but I believe we should encourage them to think. For example, perhaps some of them could guess Mendel's laws given the results of Mendel's experiments. They might even suspect that Mendel's statistics were too good to be true!

A useful method for teaching almost anything is by emphasizing how not to do it. One amusing book on this topic is Darrel Huff's How to Lie With Statistics. See also Good (1978) who, unlike Huff, gives detailed references. Teachers using this method of eliminating the negative would find it useful to have statistics of misuses of statistics. The misinterpretation of a tail-area probability is one of the common mistakes that merit increased attention in elementary teaching.

One technical concept that deserves to be more widely taught is that of weight of evidence in a sense introduced by the philosopher C. S. Peirce in 1878. (See also Good, 1950.) It captures the intuitive notion extremely well and I believe it would increase the apparent intelligence of doctors, lawyers, detectives ordinary men and potentional paranoiacs. When dealing with two simple statistical hypotheses the weight of evidence is the logarithm of the likelihood ratio, and, more generally it is the logarithm of the "Bayes factor". The Bayes factor is the ratio of the final odds of a hypothesis to its initial odds.

In combinatorial probability theory there are many fascinating propositions and problems some of which could well be stated without proof. Let me just mention a few.

(i) A coin has no memory, a point that many gamblers to not appreciate.

(ii) The law of large numbers (weak or strong).

(iii) Some aspects of gambler's ruin.

(iv) The arc-sine law. (See Feller, 1950, p. 252.)

(v) Applications of probability to football pools. If the numbers of goals scored by teams A and B have independent Poisson distributions with means a and b, then the probability of a draw is $e^{-a-b} \sum (ab)^n/(n!)^2$, which happens to be a Bessel function.

(vi) The birthday problem. For an application of this well know problem to the theory of fingerprints see Good (1978).

The readable book that might be at the right level is Warren Weaver's Lady Luck (1963).

In elementary statistics some topics that might appeal to high school students are

(i) Statistics of sport. This should go down especially well, at least for the boys.

(ii) Expectation of life.

(iii) Health statistics in relation to nutrition.

(iv) Accident statistics as a function of age.

(v) Accident statistics as a function of alcohol.

(vi) The use of statistical methods to decide whether marijuana is harmful. (Perhaps too controversial, as the statistics of sexual behavior might be.)

(vii) Murder statistics in various countries.

(viii) Statistical fallacies in newspapers.

(ix) Starting salaries of various classes of graduates.

(x) Use of statistics for assigning grades.

There are many othe suitable topics in Tanur et al (1972) though not all the articles in that anthology are good.

The first nine chapters of Kimble (1978) would be at about the right level for the top 5% of the students.

In decision theory, the concept of utility and its maximization are essential, a simple example being the question of whether it is rational to make a "fair bet".

For the uses of statistics in industry see Snee et al (1979).

For the potential use of statistics for today's social problems see Good (1972).

To sum up, my main theme is that the ideas of probability and statistics are basic not just to the tricks of the trade but to the whole of rational thought, and I believe they should be taught in that spirit some of the time. A little technique should be taught too, and the best mixture would depend on the student. Examples should be found that grab the student's attention but I have not thought it appropriate for me to try to give many detailed suggetsions as there are several other sessions to follow where such suggetions will presumably be made.

Let me wind up this overview with a fairy story. It is intended to be an example of an attempt to encourage the students to think.

Once upon a time there was a land far away where dwelt many a scientist and they all performed their experiments without consulting a statistician until the experiments were completed. Now the experiments were piddling in the sense that each of them provided only a negligible amount of evidence about whether one treatment was better or worse than another. On the other hand the scientists thought in advance that a specific one of the treatments was preferable and they were right four-fifths of the time, for in this respect they had good judgement.

It came to pass that each scientist, after completing his experiment, sought the advice of a statistician. Because there was always a dreaded deadline the clients demanded definite recommendations in every case. Now the statisticians were divided into two camps known as dogmatic Bayesians and pigheaded non-Bayesians. The dogmatic Bayesians ignored the results of the experiments and always recommended the treatments that the scientists had preferred initially. Therefore the dogmatic Bayesians were right four-fifths of the time. The pigheaded non-Bayesians also knew that the experiments were too small, but they were forced to recommend the treatments that appeared better judging by the experiments alone, and they were right just over half the time when there were just two treaments.

The moral of this story is left as an exercise for the student.

References

Birnbaum, Allan (1971). "A perspective for strengthening scholarships in statistic", The American Statistician 25, 14-17.

Evans, Christopher (1979). The Micro Millenium (New York: Viking Pres, 1980).

Feller, W. (1950). An Introduction to Probability Theory and its Applications, Vo. I (New York: Wiley).

Good, I. J. (1950). Probability and the Weighing of Evidence (London: Charles Griffin; New York: Hafners).

Good, I. J. (1972). "Statistics and today's problems", The American Statistician 26, 11-19.

Good, I. J. (1977). "Dynamic probability, computer chess, and the measurement of knowledge", Machine Intelligence 8 (eds. E. W. ELcock and D. Michie; Ellis Norwood Ltd. & Wylie), 137-150.

Good, I. J. (1978). "Fallacies, statistical", in The International Encyclopedia of Statistics (ed. William H. Kruskal and Judith M. Tanur; The Free Press), 337--349.

Huff, D. (1954). How to Lie with Statistics (London: Gollancz).

Kimble, Gregory A. (1978). How to Use (and Misuse) Statistics (Englewood Cliffs, N.J.: Prentice-Hall).

Peirce, Charles Saunders (1878). "The probability of induction", Popular Science Monthly, reprints in The World of Mathematics, 2 (ed. James R. Newman; New York: Simon & Schuster), 1341-1354.

Pieters, Richard S. (1976). "Statistics in the high-school curriculum", The American Statistician 30, 134-139.

Polya, G. (1954). Mathematics and Plausible Reasoning, two volumes (Princeton: University Press).

Roberts, Harry V. (1978). "Statisticians can matter", The American Statistician 32, 45,57 (with discussion).

Snee, R. D. et al (1979). "Preparing statisticians for careers in industry", American Statistical Association 1979 Proceedings of the Section on Statistical Education (Amer. Stat. Assoc., Washington, D.C.).

Tankard, J. W. Jr. (1979). "The H. G. Wells quote on statistics", Historia Mathematica 6, 30-33.

Tanur, Judith M., Mosteller, F., Kruskal, W. J., Link, R. F., Pieters, R. S., and Rising, G. R. (editors) (1972). Statistics: a Guide to the Unknown (San Francisco: Holden-Day).

Tee, G. J. (1980). "H. G. Wells and statistics", Historia Mathematica 6, 447-448.

Wagner, G. R. & Motazed, B. (1972). "The Proctorial System of instruction combined with computer pedagogy for teaching applied statistics", The American Statistician 26, 36-39.

Weaver, Warren (1963). Lady Luck (New York: Anchor Book, Doubleday and Co.).

Weyl, Hermann (1931). The Theory of Groups and Quantum Mechanics (London: Methuen: tr. by H. P. Robertson from the German of 1928).

Zelinka, Martha (1980). "The state of mathematics in our schools", American Mathematical Monthly 87, 428-432.

7.3 THE PLACE OF PROBABILITY IN THE CURRICULUM

PROBABILISTIC REASONING AS AN EXTENSION OF COMMONSENSE THINKING

Ruma Falk
The Hebrew University
Jerusalem, Israel

In recent years a growing need has been felt to improve people's judgemental abilities. An accumulating body of research on information processing, decision making, clinical judgement, and probability estimation has documented a substantial lack of ability across both individuals and situations (Slovic, 1972; Slovic, Fischoff & Lichtenstein, 1977). These studies challenge the previously prevalent assumption (influenced mainly by psychoanalytic theory) that the above shortcomings are necessarily due to interference from noncognitive sources. Dawes (1976) claims that rather than attribute cognitive dysfunction to motivational and emotional factors, one should acknowledge the fact that cognitive capacities are limited, and rationality is "bounded." Tversky and Kahneman (1974) describe several heuristics on which people rely when assessing the probability of an uncertain event or the value of an uncertain quantity. These heuristics reduce the task complexity, and in general they are quite useful, but sometimes they lead to severe and systematic biases. Intuition has been shown to lead us astray in judgements of randomness (Falk, 1975) and in prediection tasks (Goldberg, 1968). Predictive ability has been found to have low (and even zero) validity in clinical settings and neither professional training nor a large amount of available information are known to increase predictive accuracy. However, the fallibility of intuitive judgements does not seem to discourage the judges. Einhorn and Hogarth (1978) describe the "persistence of the illusion of validity." They emphasize their point by quoting La Rochefoucault: "Everyone complains of his memory and no one complains of his judgement."

Expanding Commonsense

Several attempts at devising corrective procedures and teaching programs for improving judgement were published recently (Slovic, in press; Fischhoff, Slovic & Lichtenstein, 1979; Kahneman and Tversky, 1979). Nisbett and Ross (1980, Chapter 12) discuss a few educational programs designed to improve inferential strategies. They advocate introducing elementary statistics and probability theory at least as early as secondary school. The statistics courses they have in mind should be mainly oriented to everyday problems of informal inference and judgement and should be taught in conjunction with material on intuitive strategies and inferential errors. (A statistics text emphasizing judgemental heuristics and the contest between intuitive and formal inferential strategies is reported to be currently in preparation by W. H. Dumouchel and D. Krantz of the University of Michigan.) However, Nisbett and Ross also mention a problem that may turn up if one starts by persuading students how prone they are to inferential errors. Students may become so convinced of their inferential incapacity that they may dispair of mastering more appropriate techniques. To avoid such a possibility, demonstration of students' misconceptions and fallacies should be accompanied by reassuring them of their exising capacity, and by retaining and encouraging the valid component of intuitive judgement (Kahneman & Tversky, 1979; Nisbett & Ross, 1980).

The best approach, in my opinion, would be to start the instruction of probability by building on the sound elements in students' intuitions rather than by making them doubt their own judgement. Many of the rules prescribed by probability theory are compatible with the conclusions derived from our daily commonsense. Teachers of mathematics may be familiar with the following statement made by students (which I encounter often when teaching probability): "I solved it just by commonsense. I don't understand the mathematics involved." Let us convice the students that probability theory is there not to extinguish there intuitive commonsense, but rather to strengthen and extend it.

Probability theory has got some inherent psychological difficulty - - on top of the notorious anxiety associated with symbols and abstractions - - due to the indeterministic nature of the solutions it offers. (Ray Hyman, of the University of Oregon, who had been teaching probability to various groups of students, told me that often engineering students found probability more difficult than social-studies students. The engineers, perhaps just because of their previous mathematical education, were trained to expect accurate deterministic answers.) On the other hand, probability, in contrast to other mathematical fields, often offers extremely commonsensical predictions. Students can very easily guess "directions" in probability problems, i.e., they can correctly tell which of two (or more) events would be more probable. If we focus on this intuitive ability, the teaching of probability could serve both to decrease math-anxiety and to create a bridge between mathematics and other subjects of science.

Educating for Judgement.

Ideally, "the place of probability in the curriculum" should be set in a program designed to enhance judgemental ability. Fred Hooper of Mercer Island, Washington, has circulated a proposal for a six-year course (extending through junior and senior high school) in Evaluating Evidence that should include, besides statistics, subjects like evaluation of documents and psychology of thinking. Some of his ideas are best represented by quoting T. H. Huxley: "Science is, I believe, nothing but trained and organized common sense." In a similar vein, Nisbett and Ross (1980) recommend a program of research methodology and philosophy of science, in addition to probability and statistics and they quote D. Hume: "Philosophical decisions are nothing but the reflections of common life, methodized and corrected."

I believe that a course in probability theory may tie very well to courses in logic and philosophy of science. Likewise, some probability issues may relate directly directly to problems of research methodology and statistics. The experience could be considerably enriched if taught in parallel to a course on the psychology of thinking. The latter may elaborate on our susceptibility to biases and distortions. One could demonstrate, in the context of the psychology of thinking, some of the fallacies and notorious pitfalls in probabilistic thinking. Thus, rather than get discouraged, the students will gain better insight both into their own thought processes and the correct probabilistic analysis. Class demonstrations of this kind may also alert the students to watch for their false intuitions.

Inductive Inference in a Nutshell.

I shall now demonstrate, via a few examples of the use of Bayes' Theorem, how one can base the teaching of probability on the student's genuine commonsense and I'll point out possible links to some of the other topics in the (desired) curriculum. The selection of Bayes' Theorem was not made at random. Typically, when solving problems with Bayes' Theorem, one infers backwards from data to possible underlying states. This thought process comprises an important part of inductive reasoning. Bayesian analysis may well illustrate our process of learning from experience in a variety of fields. Finally, the analysis offered by Bayes' rule is commonsensical, and coincides with our accepted "logic" in specific limiting cases. Thus, the Bayesian computation comprises an expansion of deductive inference (or logical implication).

The following examples should be best introduced after the students have mastered the concepts of conditional probabilities, probabilistic dependence versus independence, and formally understand the structure of Bayes' formula.

Example I
A man was arrested as a suspect of murder. Let us denote the event "the man is guilty" by G. The investigating officer collected all the relevant information, added his impressions of the suspect, and arrived at the conclusion that his probability of guilt was .60.

a. The investigation went on, and it was found (beyond any doubt) that the murderer's blood-type was O. The relative frequency of blood-type O in that population is .33 (i.e., this is the probability that a "random person" in that population has blood-type O). A blood test of the suspect was carried out and his blood-type turned out to be O.

Compute the posterior probability of the suspect's guilt (from the officer's point of view) considering all of the data. We start with the prior probabilites P(G) = .60; P(G) = .40.

Before proceeding to compute the required probability, the teacher may, as a general strategy, let the students guess how the evidence would affect the probability of G; would it increase that probability, decrease it, or leave it unchanged? The answer would be quite obvious even though most students wouldn't be able to use technical terms like statistical dependence or likelihood ratio to justify their guess. Some difficulty may arise with subsequent conversion of the informaiton into symbols and this should be introduced carefully. "O" should denote the observation that the suspect's blood type was found to be O. Consequently, P(O/G) = 1 and P(O/G) = .33. We are interested in the probability of guilt, given the evidence, namely P(G/O). Now, employing Bayes' formula, one obtains:

$$P(\frac{G}{O}) = \frac{P(\frac{O}{G})P(G)}{P(\frac{O}{G})\ P(G) + P(\frac{O}{G})\ P(G)}$$

$$= \frac{(1)(.60)}{(1)(.60) + (-.33)(.40)} = .82$$

b. Suppose the murderer's blood-type was found (with certainty) to be A, and the suspect's blood-type also turned out to be A. The relative frequency of blood-type A in the population is .42. How would the posterior probability of guilt in that case compare to the probability in section a? Would it be greater, smaller, or equal?

Students may be encouraged to answer even without relying on the formula. (While I was telling my husband about this problem, our 9-year old son (who was listening without my being aware of it) started laughing and said: "Of course blood-type A would make him less guilty, since there are more of that kind." This made me think the problem would be appropriate for school children.) Some students speculate about limiting cases and suggest the two extreme possibilities that either 100% of the population share the murderer's (and the suspect's) blood-type, or, there is known to be just one person in the population with the blood-type found to be the murderer's (and the suspect's). Such a discussion should best be summarized by the statement that the smaller the rate of that suspicious blood-type in the population, the more incriminating for the suspect is the evidence. Of course, there is nothing wrong with using the formula, and it would be nice to note it yields the same conclusion.

By the same token, one can now study the role of the prior probability as follows:
c. Suppose everything stays constant as in section a above, only the prior probability of guilt is now .40, instead of .60.

Here, again, one could predict the direction of the effect of this change on the posterior probability of guilt by "commonsense" and by subsitution in the formula.

Fischhoff et al. (1979) called that technique of asking subjects to make the same judgment several times while varying the value imputed to one variable, "subjective sensitivity analysis." This method may sensitize the student to the role of each input variable. Specifically, prior probabilities and base-rate information were shown to be ignored in intuitive judgements (Kahneman & Tversky, 1972; Lyon & Slovic, 1976). Thus, section c of Example I would alert the student to the need to attend to the parameter of the data. The conclusion would also be better verbalized explicitly.

The experience of being able to correctly guess directions of change of probabilities, and having correct expectations concerning more or less probably events, may prepare the ground for the acceptance of probabilities as numerical values measuring the degree of uncertainty. The advantage of the formula over our commonsense lies in its ability to supply not only ordinal comparisons, but also exact numerical values.

Example 2

A doctor is called to see a sick child. The doctor knows (prior to the visit) that 90% of the children in that neighborhood ar sick with flu, denoted F, while 10% are sick with measles, denoted M. Let us assume for simplicity's sake that M and F are complementary events.

A well-known sympton of measles is rash, denoted R. The probability for a child sick with measles is .95. However, occasionally children with flu will also develop rash, the probability for that is .08.

Upon examining the child, the doctor finds a rash. What is now the probability of measles?

The prior probabilities of the two competing hypotheses are clearly given here as base rates (in relative frequency terms). In order to test whether the students would apply the lesson from Example I about the need to attend to priors, to that problem, one may start by asking for a guess or an estimate of the required probability. Complete failure to consider the low base rate of measles (.10) will result in the answer .95. Otherwise, the answer would be somewhat lower, but higher than .10. Sensitivity analysis, i.e., changing only base rates and asking for new judgments, may be employed again.

The Bayesian solution is straightforward:

$$P(\frac{M}{R}) = \frac{P(\frac{R}{M})P(M)}{P(\frac{R}{M})P(M) + P(\frac{R}{F})P(F)} = \frac{(.95)(.10)}{(.95)(.10) + (.08)(.90)}$$

$$= .57$$

Such a medical example may highlight the inductive nature of our probabilistic inference. "Disease" versus "symptom" represents "underlying state" (or hypothesis) versus "data", respectively. "Disease" represents the unobservable of the "genotypic" event, whereas "symptoms" stand for observations or "phenotypic" phenomena.

We may ask, first in medical terms, when a symptom is diagnostic. Both the formula and intuitive considerations may lead students to the answer that in order for a symptom to cause us to modify our diagnosis (i.e., change our probabilities) the conditional probabilites of that symptom under the alternative states should differ. Rash turned out to be diagnositc because $P(R/M) \neq P(R/F)$. Statistically speaking, in order for data, D, to be informative with respect to a hypothesis H, D and H should be dependent, otherwise we'll end up with $P(H/D) = P(H)$ and make no progress. Thus, an intelligent investigator should collect data that is relevant to the hypothesis in question, namely, statistically dependent on it. The doctor would be incompetent if s/he would measure the sick child's height instead of his/her temperature.

Suppose the doctor visits that child the following day and tests for some other symptom. The prior for tomorrow's computation should be today's posterior probability. Thus, when successive symptoms are registered, one changes the probability of the disease accordingly. This process exemplifies the continuity of inductive inference. Our prior, at each point, is the sum total of our knowledge up to date, and new data is assimilated to this knowledge via Bayes' rule to revise our states of uncertainty.

Example 3

Assume that people can be sorted unequivocally into two distinct groups according to their hair color: dark -- denoted D, and blond -- denoted B.

A person's hair color is determined by two genes, each transmitted at random by one parent. The gene for dark hair, denoted d, is dominant over that for blond hair, denoted b. Hence, the first two out the three genotypes dd, db and bb would result phenotypically in D, only the third would be B. One can ascertain that a blond person is homozygous, i.e., bb. However, upon observing a dark-haired person, one cannot know whether that person is genotypically homozygous (dd), or heterozygous (bd).

Consider a couple with both mates dark haired and heterozygous for hair color. The genotypes and phenotypes of the potential offspring of that couple are given in the table below:

		Mother D	
		d	b
Father D	d	dd D	db D
	b	bd D	bb B

The probability of such a couple giving birth to a dark-haired child is 3/4 and to a blond child is 1/4. The probability of a random D child of such parents being heterzygous is 2/3.

A couple is interested to know whether any of their future children could be blond. Both husband and wife are dark haired. Their phenotypes, along with those of their parents and siblings, are shown in Figure 1. Note that each of these prospective parents has got a couple of dark-haired parents and a blond brother.

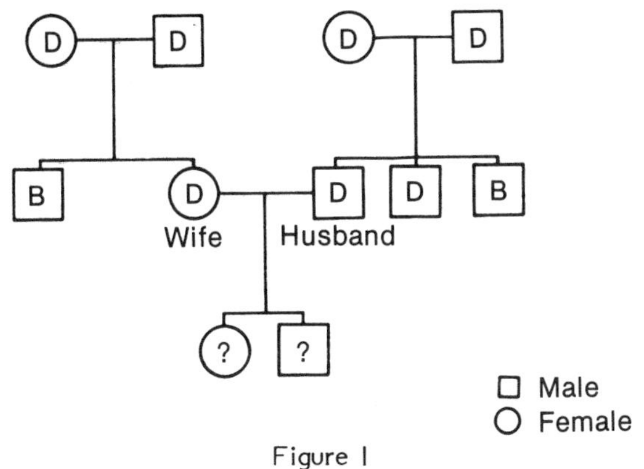

Figure 1

Let H denote the genotype event that the couple has got the potential to produce a blond child. H is equal to the intersection of the events that husband and wife are heterzygous (i.e., that each is a carrier of a recessive gene b). $\overline{H}$ would mean that the couple is incapable of producing a blond child (i.e., at least one spouse is homozygous dd). If H is true, then the probability of that couple getting a B child is 1/4.

a. Find the prior probability of H (before any children are born in that family). Denote that prior $P_0(H)$.

b. A dark haired baby was born to that couple. Denote that event D_1 (first born child dark). D_1 did not change the possibility that H is true. Did the probability of H change? Denote the probability of H after the birth of the first child by $P_1(H)$. What is the value of $P_1(H)$?

c. After a few years the couple had another baby. The secod child was blond, denote this event B_2. What is the posterior probability of H in the light of that information? In line with the previous notations we are looking now for $P_2(H)$.

Let us outline the solutions to these questions briefly and comment on their significance.

a. Since both husband and wife have a blond brother and D parents, these four parents must all be carriers of a b gene, i.e., they are genetically bd. Consequently the probability of the D husband being a carrier is 2/3 and so is the probability of his wife. Because of the independence of these two events, the probability of their intersection is:

$$P_0(H) = \frac{2}{3} \times \frac{2}{3} = \frac{4}{9}\text{; and } P_0(\overline{H}) = \frac{5}{9}$$

b. The probability of obtaining D_1 under the two competing hypotheses are:

$$P(D_1|H) = \frac{3}{4} \text{ and } P(D_1|\overline{H}) = 1$$

(In fact, these probabilities are independent of birth order.) Therefore,

$$P_1(H) = P(H|D_1) = \frac{P(D_1|H)\, P_0(H)}{P(D_1|H)\, P_0(H) + P(D_1|\overline{H})\, P_0(\overline{H})} =$$

$$\frac{\frac{3}{4} \times \frac{4}{9}}{\frac{3}{4} \times \frac{4}{9} + 1 \times \frac{5}{9}} = \frac{3}{8}$$

The probability of H did change, it decreased a little.

c. $$P_2(H) = P(H|B_2) = \frac{P(B_2|H)\, P_1(H)}{P(B_2|H)\, P_1(H) + P(B_2|\overline{H})\, P_1(\overline{H})} =$$

$$\frac{\frac{1}{4} \times \frac{3}{8}}{\frac{1}{4} \times \frac{3}{8} + 0 \times \frac{5}{8}} = 1$$

The possibility that the couple is capable of producing a B child jumped to certainty. For next time, their probability of a blond child is 1/4, and this would not change whatever the hair color of their subsequent children.

The following points are worth comment:

1. Again, the focus of our discussion was an unobservable combination or a genotypic event. The data we used to infer the probability of that underlying state were observable, or phenotypes.

2. Whereas in Example 1 the student got the prior probability of guilt ready-made without any insight as to how it was assessed, here one can easily

utilize all the phenotypic data in the pedigree and come up with a reliable (objective) prior for H.

3. It may seem that the birth of a D child should not change the probability of H, since it does not rule out the possibility of the couple being capable of producing a B offspring. Indeed, students often argue vehemently that this piece of evidence is worthless, since it is inconclusive. The lesson to be learned from the Bayesian analysis in this case is that each item of evidence could be informative once it is statistically dependent on the hypothesis in question. One agrees readily that had the couple produced 10 successive dark-haired children, this should have significantly reduced the possibility of H. Why, then, should not the birth of one D child have its due impact on that probability, however small it may be?

   Thus, the probabilistic inference via Bayes' Theorem offers an extension of "common logic" as it allows incorporating even inconclusive evidence to revise our views. Inferential progress is accomplished although certainty is not achieved.

4. The birth of a blond child, however, is conclusive evidence. We do not need Bayes' formula to logically deduce the couple is capable of producing a B child, but it is good to realize that the formula yields the same conclusion. The extended probabilistic model includes the extreme instances as specific cases.

5. The evolutionary nature of probabilistic inference was well illustrated by using the posterior probability of stage i, namely $P_i(H)$, as the prior stage i + 1.

6. Note that although $P(D_1/\bar{H}) = 1$, the observation D_1 did not prove $\bar{H}$, it only slightly increased its probability. An item of evidence predicted with certainty by one of the hypotheses was still not conclusive. On the other hand, because $P(B_2/\bar{H}) = 0$, obtaining B_2 proved H.

Some Links to the Curriculum.

Negative evidence (B_2) turned out (in Example 3) to be more powerful than positive evidence (D_1). This should come as no surprise to students of philosophy of science who have read Popper's (1959) views that hypotheses can only be disconfirmed by evidence but never confirmed. Indeed, the supporting evidence, D_1, only strengthened $\bar{H}$ but did not prove it, whereas the disconfirming evidence, B_2, ruled out $\bar{H}$, In Example 3, the two hypotheses were mutally exclusive and exhaustive, therefore, disproving $\bar{H}$ resulted in proving H. Usually in science, one has more than just two complementary alternatives, hence ruling out one of them does not resolve all the uncertainty.

The above discussion has an associative link to the theory of logic. We know that $A \rightarrow B$ does not imply $B \rightarrow A$. Observing a datum that is predicted with certainty by a theory does not prove that theory, maybe this datum could be accounted for by some other theory. However, if $A \rightarrow B$ and we learned that B is not true, this rules out A. In logical symbols:
$A \rightarrow B \Rightarrow \bar{B} \rightarrow \bar{A}$. The course on the psychology of thinking may examine to what extent people are utilizing disconfirming evidence. A class demonstration of people's failure to do so, via the well-known experiment of Wason and Johnson-Laird (1972, p. 173), is bound to work. The implications for research methodology are clear: One has actively to search for negative evidence also. Don't be content with the experimental group exhibiting the predicted effect. Design a control group and test for no effect in the absence of the experimental manipulation.

In conclusion, probability theory is fruitful both in uncovering and extending the commonsensical nature of our inferential process, and in generating solutions to problems from various facets of real life.

References

Dawes, R.M. Shallow psychology. In J.S. Carrol & UJ.W. Payne (Eds.), Cognition and social behavior. Potomac, Md.: Erlbaum, 1976.

Einhorn, H.J. * Hogarth, R.M. Confidence in judgment: Persistence of the illusion of validity. Phychological Review , 1978, 85 , 395-416.

Falk, R. The perception of randomness. Doctoral dissertation (in Hebrew). Jerusalem: The Hebrew University, 1975.

Fischoff, B., Slovic, P. & Lichtenstein, S. Subjective sensitivity analysis. Organizational Behavior and Human Performance, 1979, 23, 339-359.

Goldberg, L.R. Simple models or simple processes? Some research on clinical subjects. American Psychologist, 1968, 23, 483-496.

Kahneman, D. & Tversky, A. Subjective probability: A judgment of representativeness. Cognitive Psychology, 1972, 3, 430-454.

Kahneman, D. & Tversky, A. Intuitive prediction: Biases and corrective procedures. TIMS Studies in Management Science, 1979, 12, 313-327.

Lyon, D. & Slovic, P. Dominance of accuracy information and neglect of base rates in probability estimation. Acta Psychologica, 1976, 40, 287-298.

Nisbett, R. & Ross, L. Human inference: Strategies and shortcomings of social judgment. Englewood Cliffs: Prentice-Hall, 1980.

Popper, K.R. The logic of scientific discovery. London: Hutchinson, 1959.

Slovic, P. From Shakespeare to Simon: Speculations - - and some evidence - - about man's ability to process information. ORI Research Monograph, 1972 12(2).

Slovic, P. Toward understanding and improving decisions. In W. Howell (Ed.), Human performance and productivity. Hillsdale, N.J.: Erlbaum, in press.

Slovic, P., Fischoff, B. & Lichtenstein, S. Behavioral decision theory. Annual Review of Psychology, 1977, 28, 1-39.

Tversky, A. & Kahneman, D. Judgement under uncertainty: Heuristics and biases. Science, 1974 185, 1124-1131.

Wason, P.C. & Johnson-Laird, P.N. Psychology of reasoning. Cambridge: Harvard University Press, 1972.

THE PLACE OF PROBABILITY IN THE CURRICULUM

Tibor Nemetz
Hungarian Academy of Sciences
Budapest, Hungary

I propose to proceed with the panel discussion along the following flow-chart:

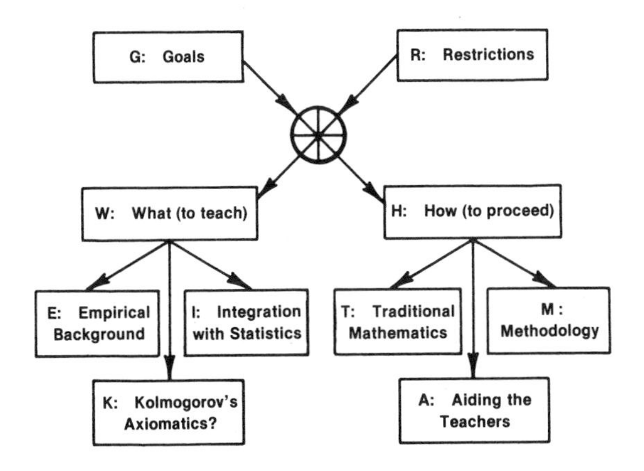

We have to start with listing our GOALS and all barriers in the way of attaining these goals. Taking into consideration all RESTRICTIONS, and only then, can we formulate WHAT can be taught, showing, at the same time, HOW. I will comment on the entries of the first two levels separately, although the strong connection among them is obvious.

As to the details, we consider the questions in the bottom-level of central interest. Besides remarks, we bring some examples illustrating our standpoint in this respect.

G) Goals:

In many countries stochastics is just being introduced into the pre-university education. At this stage it is better to formulate approximate goals rather than fully specialized ones. Never-the-less, there is a minimum, on which one has to insist. Here we try to list these minimum-goals in a flexible form.

a) Get the pupils acquainted with the notion of "randomness". Make them able to view stochastically the random events.

b) Teach them to evaluate statistical data. They have to be able to understand the characters of the statistical decisions, limitations of stochastical arguments.

c) The stability of relative frequencies should be made as clear as possible.

d) Get the pupils acquainted with the basic laws of the probability theory and some main ideas of mathematical statistics. (In this respect one must insist on teaching the weak law of the large numbers, as a minimum.)

e) Show them how to construct stochastic models in real-life problems. At least, do the first steps in this direction.

R) Restrictions:

Many countries share the following difficulties during the introduction of stochastics into the curriculum:

a) The introduction of a new subject, or just new material, requires a redistribution of the time at disposal, which means a resistance on behalf of the representatives of "well-established" areas. ("Time limitation")

b) Teachers are reluctant to teach stochastics. This is mainly due to the lack of experience and their poor education. Also, they do not like "being tricked" by the random.

c) Conducting statistical experiments is very time-consuming. There is a limited range of statistical "populations", which can be easily inspected and processed by the students. We need populations with such stochastic behaviour, which is almost independent of the place and time.

d) Stochastics teaching as part of the mathematical education means a compromise.

e) Evaluating the knowledge and progress of the students is a difficult problem, especially during the phase of providing them the empirical background.

f) In many cases the national school system means additional difficulties.

W) What to teach:

a) A real comprehension of the main notions of the probability theory and statistics supposes a sound empirical background.

b) We have to demonstrate that the theory matches the practice. This calls for the rudiments of descriptive statistics.

c) Goals c) and d) can be obtained by restricting ourselves to discrete (or finite?) random variables. The concept of dependence - independence, characteristics of random variables like location parameters and measures for dispersion, Markov's and Chebisev's inequality, the weak law of large numbers should be included in all curriculum.

d) Probability theory and the corresponding statistical items should proceed simultaneously.

e) The subject matter has to balance among problems and situations, where

- only purely statistical arguments can be applied (see experiments 2, 5, 6, 9, 10 in the Appendix);

- easy combinatorial reasoning can help (experiments 1, 3, 4, 8);
- there are exact solutions, which are, however, hard to compute (experiments 7, 12, 13).

H) How (to attain the goals):

I am convinced that the introduction of the probability has to be a gradual one and should be preceeded by classroom-trials with "average" teachers. The following guidelines, intended for both projects and the classroom-work are motivated by our experiences:

a) Activity approach is usually preferred to formal lecturing. Games and statistical experiments are to be followed regularly by classorm discussions, where the pupils can argue with and convince one another.

b) The statistical experiments have to be carefully planned. It is a vital point in fighting the "time-factor", that the data-collection be as fast as possible (see item M).

c) A better utilization of the traditional mathematics is needed: wherever it is possible, problems should be smuggled into the "classical disciplines", which are fruitful in attacking problems in stochastics (see item T).

d) Exactness is desired whenever it is possible. Nevertheless, one should not disregard heuristics with merely statistical demonstrations.

e) The teachers have to be aided in ever respect of their work (see item A).

The teaching of stochastics should start with a complex introduction via games and statistical experiments, where the pupils can meet quite a number of problems with simple random structure. Our teaching experiments during the last decade allow us to conclude, that adequate empirical background can be provided in some 15 - 20 lessons for 14 - 15 years old students.

E) Empirical background:

In stochastics, unlike in geometry, the pupils do not have sufficient prior experiences for accepting the main notions and axioms. The following examples illustrate some typical false reasoning:

a) Throwing 2 coins we can observe 3 outcomes: 0 - 1 - 2 heads. These 3 outcomes are frequently claimed of equal probability.

b) When a coin is tossed repeatedly, then throwing 3 heads in a row increases the probability of throwing a tail the next time.

c) Excerpt from a newspaper: "The difference between the number of heads and that of tails, when a coin is flipped 2n times, is about zero, provided n is large. E.g. we would be surprised to see how small this difference is when the coin is flipped one trillion times."

d) When a dice is cast till the first 5, the most likely number of cases needed is one. This simple fact is quite hard to accept.

Obviously, only experimenting can help us to overcome such difficulties. As an example in this respect, we bring a set of 13 experiments used to introduce the location-parameters via 3 contests. This set meets both points R/c and W/e. As to point R/c written languages proved to be very useful "populations".

I) Integration with statistics:

In the every-day-practice we have problems similar to the experiments 1, 5, 6, 9, 10. Probability theory can work in these cases only after establishing an appropriate model. This means clearly a need for a statistical concept, namely the principle of maximum-likelihood estimation.

Investigating point E/a we need to test a simple hypothesis against another simple one, while in E/b a test of independence is needed.

Characteristics of statistical populations and those of random variables can be explained along the same line. If location-parameters are introduced by extreme-value problems, then a short time suffices to discuss the method of least-square.

K) Kolmogorov's axiomatics?

Here the interrogation-mark summarizes our opinion. We can restrict ourselves to discrete random variables. In this case there are no measure-theoretic problems, and the concept of a σ-algebra is not needed. On the analogy to the case of the distance, area, volume and the relative frequency we can collect the basic properties of probability as a measure (and we may even call them axioms). What we need is the "art of counting" rather than "set-theoretic considerations".

T) Traditional mathematics - a more effective utilization:

Generally one has to solve many problems in pure mathematics during stochastic lessons. These problems should be shifted to the "traditional" mathematics lessons. By this, we need less time, and, hopefully, stochastics becomes a bit more attractive for the teachers. The following examples illustrate this possibility.

a) The location parameters can be introduced by some extreme-value problems (see contests 2 and 3 in the Appendix). In this case one has to find the minimum points of function of the types

$$f_1(x) = \Sigma a_i \cdot |x - b_i|$$
$$f_2(x) = \Sigma a_i (x - b_i)^2$$

Both problems belong to the traditional subject mattter.

b) The expected value of the difference in the "newspaper" problem of point E/c is easily seen to be

$$E_n = \frac{1}{2^{2n}} \cdot 2 \cdot \Sigma_{k=0}^{n} 2k \cdot \binom{2n}{n+k}$$

Applying

$$\sum_{k=0}^{m} \binom{m}{k} = 2^m$$

repeatedly, one gets

$$E_n = \frac{2n}{2^{2n}} \cdot \binom{2n}{n}$$

The usual manipulation can be carried out in lessons on combinatorics.

c) In n trials an event E is observed k times. For estimating the probability of E one needs the maximum point of the function

$$f_3(x) = x^k \cdot (1 - x)^{n-k}, \; 0 \leq x \leq 1$$

A nice solution to this problem can be given, when the inequality between arithmetic and geometry means is discussed. Similarly, one can find the maximal element of the "hypergeometric probability"

$$\frac{\binom{M}{m}\binom{N-M}{n-m}}{\binom{N}{n}}$$

when 3 of the 4 parameters are given.

M) Methodology:

It is of vital importance to speed up the work with the statistical experiments, expecialy with recording and data-collection. We will concentrate on this question, and show that it can be performed really fast. The 13 experiments in the appendix were carried out in different classes by the following method

- 1 - 3 "responsible" pupils were chosen from the class for all experiments
- Work-sheet A was given to all of the students. They had to perform all experiments 10 times, and record the outcomes on the work-sheet. They were free to do them in any order. (The total time of execution varied between 60-90 minutes.) They were allowed to fix their tips afterwards.
- The class was divided into groups of 10-15 pupils. Each group had received data collecting sheets B for all experiments. These were filled in by the students in a rotating way. (Only 25-30 minutes were required.)
- The data collecting sheets were used by the "responsibles" to determine the frequency distribution of the given experiments. (This can be done as home-work. In class it takes about 30 minutes.) They were provided a pattern, see Figure 1, to ease and unify their work.
- Finally, the "responsibles" read the frequency-distributions aloud so that all students could fill in their own work-sheet C.

A) Aiding the teachers:

In point R/b we had comments on the teachers' negative attitudes. This can be changed by carefully prepared manuals. These guide-books should contain

a) information about the "exact" part of the manual;

b) list of experiments tested in classrooms;

c) a number of samples with respect to all experiments, together with the graph of the empirical distributions;

d) a collection of the "tricks" of the random, observed in the classroom work, false ideas of the students and methods of overcoming these difficulties;

e) detailed description of methodological questions (see item M);

f) experiences of teachers during a trial-teaching of the material.

There is usually sufficient information about points a) and b), and as to c), the teachers can perform the experiments for themselves. As to the last 3 points, however, we are convinced that really detailed material is needed if we wish to bring the pre-university stochastics teaching into existence.

Appendix

Contests:

1) HIT OR MISS: Guess the outcome of the experiment. You will get the point whenever an observation in the class coincides with your guess. (Only 1 guess is allowed.) The player with the most points will win.

2) APPROXIMATE: In this and the following contest you have to give a prediction for the outcome, and a penalty is calculated according to the rule:
Penalty = / Prediction - Observation /
The player with the least sum of penalties for all observations in the class will win.

3) LEAST SQUARE: A different prediction can be given, and the rule:
Penalty = (Prediction - Observation)^2

PATTERN for the class: how to collect the frequencies

Outcome	Frequencies	
1	卌 卌 |||	13
2	卌 卌 卌 卌 |||	23
3	卌 卌 卌 卌 ||	22
4	卌 卌 卌 卌 卌 |	26
5	卌 卌 卌 卌 卌 卌 |	31
6	卌 卌 卌 卌 卌 卌 卌 卌 卌	45
7	卌 卌 卌 卌 卌 卌 卌 ||||	39
8	卌 卌 卌 卌 卌 卌 卌 卌 |||	43
9	卌 卌 卌 卌 卌 |||	28
10	卌 卌 卌 卌 卌 卌 卌 ||	37
11	卌 卌 卌 卌 ||	22
12	卌 卌 |	11
13	卌 卌 卌	15
14	卌 ||||	9
15	卌 |	6
16	卌 ||	7
17	|	1
other :	20, 22	2

Figure 1: Observed data in experiment 6

Experiments:

1. Throw 20 coins and count the number of head.

2. Drop 10 thumbtacks from 1 meter to the floor, and count how many of them point upwards. This number is the outcome.

3. Cast 5 dice, and count the number of sixes.

4. Cast 5 dice, and count the number of sixes and threes.

5. Choose "randomly" a letter in an old newspaper, and count the number of vowels in the segment of the following 40 consecutive letters. Use adjacent segments for the 10 experiments!

6. Choose "randomly" a page in a thick book. The number of letters in the tenth word on this page will be the outcome.

7. Cast 5 dice. The outcome of the experiment is the sum of the numbers cast.

8. Cast 5 dice. Outcome is the smallest number cast.

9. Choose "randomly" a page in a thick book, and consider the 10th letter on that page. Starting with this letter, count the letters preceeding the following first vowel. This number is the outcome of the experiment.

10. Find the vowel following the 10th letter of a "randomly" chosen page of a book, as in the 9th experiment. Starting with this vowel, count the letters preceeding the following vowel. This number is the outcome of the experiment.

11. Cast a die till you hit the first 6, and count how many casts you needed.

12. Cast a die till one of the numbers shows up a second time. The number of casts gives the outcome.

13. Shuffle a pack of cards, then check them one by one till you find the first ace. Outcome = number of cards checked.

7.4 THE NATURE OF STATISTICS TO BE TAUGHT IN SCHOOLS

THE VITALITY OF STATISTICS

Jim Swift
Nanaimo Senior Secondary School
Nanaimo, British Columbia

Introduction

The Fourth Congress on Mathematical Education is being held at the dawn of a new decade, a dawn that puts a new light on our old institutions, on the needs of the children we teach and on the mathematics curriculum we use.

Today I wish to consider one aspect of that mathematics curriculum, and ask "Does the mathematics curriculum reflect the world in which we live?". Consider the task of a photographer focussing the camera on the reflection of a tree in a pool of water. If she focusses on the water, the tree may be out of focus. So rather than begin this talk by focussing on the curriculum, I would rather direct your attention to the world around and remember that we now live in an "information society". In North America, over half the labor force have jobs that are related to the processing of information, and much of that information is statistical. And this proportion is increasing!!!

It is said that humans require three kinds of food, the food we eat, the air we breathe and the impressions we get. It would appear that the public's impressions of Statistics gives them indigestion. But it only takes a slight alteration of perception to change those impressions. But impressions are subject to change. You can percieve Statistics as a dry collection of calculations or as an enjoyable way of shedding illumination on the world and society in which we live. It is simply a matter of looking at something familiar in a different way.

So I would like to do three things in this talk. First, examine examples of the use of Statistics in newspapers and magazines. Second, look at the use of such examples in the classroom, and finally show two examples that have been developed into deeper investigations.

Examine Examples of the Use of Statistics in Newspapers and Magazines.

To begin with, a few examples of the frequent use of statistical language.

- In a recent letter to an Editor, the writer used the words "signficant adverse effects were unlikely", to justify the safety of the herbicide 2-4-D.
- We read of probability and chance in many contexts, from the probability of flood damage on the River Thames, to the probability of being hit by a piece of Skylab.
- In an article on dating, you are asked to believe that "your chances of finding potential lovers increase in direct proportion to the number of men you meet".
- We read of the importance of the interpretation of statistics in and editorial entitled "A Statistical Canada" - (when did you last read an editorial on the joys of solving quadratics???)

Computers have made it possible to produce quite sophisticated graphics. We see with increasing frequency, coloured maps, the shades of which represent levels of two variables. Reading such graphs is no trivial exercise, and is a long way from the graphics chapter at the beginning of most first statistics books.

Perhaps the most dramatic use of the language of probability and statistics appears in the energy debate. Are we to use more nuclear energy? Such decisions are rooted in the use of statistics and probability and, indeed, make professional gamblers of the politicians but for very much higher stakes. The widespread use of nucelar energy give cause for concern. Companies such as Atomic Energy of Canada produce high-quality magazines for sale on newstands, and expect the average person to follow statistical arguments on such questions as the effects that followed the accident at Three Mile Island. These arguments can be quite sophisticated and difficult to follow. An article in Scientific American on the safety of fission reactors contained this statement about a procedure used in the Rasmussen report: "There may somewhere be a statistican who believes that this is valid but he has yet to make himself known"!!

With such widespread use of statistical arguments, it is not surprising that data analysis is being given increasing emphasis in our schools. Encouragement of this trend appears frequently in the NCTM report "Agenda for Action" and in a new journal devoted to the "Teaching of Statistics" in schools. But what really happens in a Statistics class? As I hope to show, a student can do much more than sit at a desk and work only from a textbook, calculating means and standard deviations.

Using Newspaper Clippings in the Classroom.

A newspaper item about Canada's increasing population is contrasted with an item about declining school population in Toronto. Students explore the contrast by studying populations pyramids from Statistics Canada publications and using a computer simulation of the growth of such pyramids.

A Salvation Army advertisement concerning income distribution is related to a study of Lorenz curves for different countries.

The existence of health hazards provokes arguments that are entirely based on statistical methods. I am sure that, in particular, we shall hear a lot more about the increased incidence of hyperthyroidism following the accident at the Three Mile Island nuclear power station. In class we look at the knowledge required to decide if this increase is significant, and if it is, does this mean that the accident was the cause.

Reports of opinion polls are studied, some of which may contain statements like:- "if you would like to participate in this random survey, please call me at ..."!! Such statements, when connected to practical survey exercises, reveal many of the difficulties involved in using information from polls. A ban on polls has been proposed on the grounds that they are an attempt to manipulate public opinion. In class, we consider this argument and look for informative articles on polls. One such article asked that we treat polls like a tool and beware of headlines that talk about gains and losses. Why? The idea of sampling error is relevant. It is unusual to see a report that gives a range of possible values of the population percentage. More common is the statement that "this poll is accurate within 4 percentage points 19 times out of 20". When that is seen next to a statement that an increase of 2% has occurred, one is perhaps entitled to wonder at the lack of understanding of polls.

A well-known example of the misuse of probabilistic arguments comes from Time magazine and was described in "Statistics Made Relevant" (Baum and Schauer 1976). Janet and Malcolm Collins were convicted of robbing an elderly woman in San Pedro. The evidence was circumstantial. Witnesses saw a white girl with a blonde ponytail run to a yellow car driven by a bearded negro. The Collins couple fitted this description and were arrested. The prosecution, aided by a college math professor, argued that the probability of a couple possessing those characteristics was 1 in 12, 000,000. Since the population of San Pedro was less than 12,000,000, the jury was satisfied that the Collins were the guilty couple. The Appeal Court overturned the verdict, giving reasons that used conditional probability:- given that there is only one couple in the area fitting that description, what is the probability that they are the only couple. This turned out to be about 0.4 - reasonable doubt that the Collins were the only couple fitting that description.

Newspaper Clippings that Led to Further Investigations.

In conclusion, two examples that stimulated further investigation.

The first example concerns a proposal to replace the ferry system between Vancouver and Vancouver Island with a 40 km. tunnel or bridge. A local Newspaper reported a statement by a Cabinet Minister that a survey found a 2 to 1 majority in favor of this proposal. Refusing to accept this statement, three students undertook a phone survey to check his facts, and plotted their data on a zoning map of the city. They found a 2 to 1 majority opposed to the link, together with some interesting differences in the responses from various areas of the town. Their work was written up in the same Newspaper, providing an item for another class.

The second example concerns the opening of a shopping centre in Nanaimo. A student read that it provided 253,000 sq. ft. of new shopping space. Shopping centres were being built simultaneously with many shops closing down. A visit to the City Planning Dept. provided him with a copy of Profile Nanaimo, which contained the startling statistic that 2,200,000 sq. ft. of new retail shopping space was either constructed, under construction or proposed. So how big an increase is that? Nanaimo now has 900,000 sq. ft. of retail businesses. What revenue will be needed if all these shops are to stay in business? The 900,000 sq. ft. generated $174,000,000 in 1979, or about $200 per sq. ft. per year. So, to keep all the new shopping space in business, the people of Nanaimo will need to spend and extra $440 million per year. Even if everyone in a 60 km radius of Nanaimo shopped there, it would still require an extra $2,000 per person above the present expenditure of $3,400 per person per year!! As the report was completed, a Canadian Press article gave similar findings. The conclusion of the student's report - members of City Council, who approved the large increase, would do well to read such publications as Statistics Canada's "How a Retailer can Profit from Facts".

Conclusion

Earlier, it was said that the photographer might put the tree out of focus if she focused on the surface of the water, but she knows that can be prevented. With sufficient depth of field, you can have the water, the surrounding world and the students all in focus at the same time. At the dawn of the new decade, let us aim for such a perspective.

THE VEHICLE OF STATISTICS

Albert P. Shulte
Pontiac, Michigan

Introduction

Statistics interacts with us in many ways. We are all confronted with masses of data, out of which we must try to make some sense. We are continually presented with results of polls or surveys, showing preferences ranging from presidential candidates to laundry detergents. We are faced with generalizations based on statistics, whose truth we must attempt to assess. Much material is presented to us in the form of tables or graphs, sometimes set up to be deceptive or misleading.

This constant interaction with statistics provides a strong argument for adults being skilled in dealing with statistics. However, the question in this session is "Why teach statistics in the elementary schools and the middle schools?" Why shouldn't statistical instruction wait until students are capable of formal reasoning?

Some answers to this question are: (1) Young students are exposed to uses of statistics in the same way adults are. They need to become knowledgable and critical consumers at their level. (2) The ability to work effectively with statistical information develops gradually. To achieve the desired ease and familiarity with statistical concepts that adults need, it is most effective to start the process early. (3) Statistics provides a chance to provide real-life applications of many of the mathematics skills taught in the elementary and middle school. (4) Some statistics is already being taught in the schools. We are suggesting that the amount and the emphasis on statistics be increased.

If more statistics is to be taught to young students, how should it be taught? Students need to be actively involved in the learning process. They should be involved in collecting and organizing information. They need to know how to tally, how to arrange information in a table and how to use a graph to present information visually. They need to understand such ideas as "the average score on the test." They need to learn that variability is a part of the real world. They need to see statistics applied to science, in laboratory activities in mathematics, in measurement situations and in their own games. Students need to work on projects where data are collected and presented in sound ways to other students.

Examples at the Elementary Level

Imaginative teachers have for years involved students in graphing in the early elementary years. A sound sequence for developing the ability to graph is the following: (1) Start with object graphs. If the students wish to compare preferences for chewing gum, a stick of the appropriate gum is attached to the graph for each student that prefers that type of gum. (2) Move to picture graphs. If children wish to graph the types of pets in their families, they cannot attach the actual pets to the graph. However, they can use pictures to present the pets. (3) As information is collected, tally graphs may be used to record and display data. (4) In the upper elementary grades, students can be taught the techniques of properly constructing bar graphs and line graphs, and where to use each type. Reading and interpreting the graphs should also be stressed.

In the middle elementary grades, students may be exposed to ideas of sampling. For example, they may draw colored popsicle sticks from a bag and make inferences about the distribution of popsicle sticks in the bag. Or, for a tastier example, they may explore the proportions of different colors in a package of M and M candies.

At all levels, students should be involved in data collection. As examples, students could record the favorite TV shows of members of their class, the number of missing teeth for each child or the total number of children in each family. Once the data are collected, then discussion about displaying the information can lead the students back to graphing.

Examples at the Middle School/Junior High Level

At this level, students should have experiences where they describe data using a single number to represent the middle, typical or most common value. That is, they should learn to use the mean, the median and the mode, to distinguish among these measures and to learn the advantages and disadvantages of each measure.

Middle school or junior high students should be actively

involved in doing experiments and recording data. For example, they could play games to determine experimentally whether the games seem fair or unfair. As another example, they could perform an experiment to determine the fastest skateboarder, as illustrated in the recent statistics film Hyposthesis Testing.

Certainly the junior high or middle school is an appropriate level for students to collect samples and carry out surveys. This activity can lead to an understanding of random sampling, to analyzing surveys for flaws (for example, the famous Thomas Paine survey from Statistics By Example) and to a better appreciation of how public opinion polls are carried out and how their accuracy is assured.

At this level, students can use more elaborate methods of organizing and displaying data. The stem-and-leaf display can be used effectively to present a picture of a distribution, and to aid in finding the median of the distribution. Work with graphs can deal with bar and line graphs of greater complexity. As students learn to measure angles and to use per cent, circle graphs may be introduced.

Statistical Projects

Once students know a variety of statistical techniques, they should be given the opportunity of putting these techniques to practical use by carrying out a statistical project. They should go through the steps of formulating the project, checking with the teacher for additional suggestions, for sharpening the focus of the project or for revision. Once the project is approved, the student (or several students working together as a team) should gather the data, organize the data appropriately and draw conclusions from the data.

The final step in the project should be presenting the project and its results to the class. In preparation for this presentation, the students should illustrate their work with appropriate visual displays - - tables, charts or graphs. In the presentation itself, they should show how the data support their conclusions.

Working on a statistical project has many advantages. It brings into play all the statistical topics the student has learned. It demonstrates to the student that these principles and techniques are not just found in textbooks, but that they have practical use in the student's own world. It thus involves the student actively in using statistics at his/her level.

Summary

Students at the elementary or middle school level can learn statistical concepts and techniques presented at their level. They can also apply these concepts and techniques in studying problems of interest to them. What is needed is more imaginative teachers who are willing to introduce statistical topics, and more curriculum materials which contain sections on statistics for the young student.

THE VIGOR OF STATISTICS

Peter Holmes
University of Sheffield
Sheffield, England

The general educational reasons for including probability and statistics in the school curriculum have been well made by Jim Swift and Al Shulte in earlier talks of this mini-conference. The aims of such material are that children should become aware of and appreciate both the role of statistics in society (the breadth of statistical applications) and the power and limitations of statistical thought (the depth of statistical thinking). In this talk I shall be exploring the possibilities and problems posed by statistics being an interdisciplinary subject.

At present statistical ideas are beginning to appear in many subjects in the main school curriculum. We can distinguish between the statistics that occurs in mathematics courses and that which occurs in non-mathematical courses. In constrast to the general educational reasons propounded by Swift and Shulte, the main reason for incorporating statistics in the mathematics curriculum (judging by the type of statistics that appears) is that it is seen as an important component of present day mathematics. In other subjects statistical ideas are introduced becasue they are seen as giving useful insights into these subjects.

In mathematics courses this attitude usually leads to an overemphasis on techniques such as drawing histograms and calculating medians and means. Less emphasis is placed on the problems of collecting accurate data and on the drawing of inferences. Occasionally the dilemma between statistics for general education and statistics for mathematics education can appear quite sharply. In Modern Mathematics for Schools (Blackie/Chambers 1973) the teachers' notes talk of the importance of questionnaire design, opinion polls and obtaining representative samples as well as the difficulties of forecasting. On the other hand,none of these occur in the pupils' texts which concentrate on establishing tecniques. Such a dichotomy is not unique to this particular course.

An initial analysis of the work of a statistician indicates two broad areas in which he might work. He may be concerned with work which uses surveys (either complete or sample) or he may be concerned with experimentation. In survey work he has to take the results that are available, though he will be able to influence the size and method of choosing the sample. In experimentation he will be able to influence the design of the experiment and the variables will be much more under his direct control. Both of these aspects can be seen in the use made of statistics by school subjects other than mathematics. Here the emphasis is on the ability to read and use statistical information rather than on the ability to carry out specific statistical calculations (though this also might be required). Many subjects in the school curriculum are becoming more quantitative and so are making greater use of statistics. Examples can be found in both good and bad uses. One of the major users of statistics in U.K. schools are the Geography departments. Currently the British Geographical Association and Mathematical Association have a joint committee considering the uses being made of mathematics in geography. Many of these mathematical ideas are statistical and the list is long. There are simple ideas such as re-drawing the map of Great Britain with distance representing the time it takes by public transport to travel from London, or with area representing population density. The contrast between these two maps is very illuminating. Scatter diagrams are used as are many other techniques in a practical context. Some of the items - e.g. Lorenz curve, Gini coefficient, triangular graphs - are fairly specific to geographers and it is reasonable to expect the teaching ofsuch topics to be done there. Greater interdisciplinary cooperation may mean that more statistical ideas could be introduced practically in the context of some other subject's teaching.

Some syllabuses in Religious Education require knowledge of statistical sources and the limitations of such ideas as correlation. This is particularly true of syllabuses which are oriented to social problems such as crime rates, quality of housing and the inner-relationships between the two.

Many of the more recent courses in Science make extensive use of statistics and refer to statistics being an important component in modern scientific thinking. Examples are of the use of the mean of several readings in estimating a mass, of the effect of fluoride on tooth decay, of weight gained when on a milk diet compared with not being gained on a milk free diet, as well as many others. Mistakes occur as in the textbook which said the distribution of pupils with different colour eyes would not be normal since the sample size was too small.

Environmental Science also uses many statistical examples such as birth and death rates and the interpretation of sample survey data of different types of plants found in a particular region. It is interesting to note that questions asked in such courses can be very demanding in requiring statistical insight even though the calculations to be done may be trivial.

Recently developed courses in Social Science are very concerned with the problems of carrying out surveys, and the accuracy and implications of such data as may be obtained. General Studies courses may well refer to such problems as smoking and heath for which some of the information is statistical. As History becomes more quantitative so it, too, makes more use of data with the added complications of having to use such data as may be available and which may have been collected for an altogether different purpose.

Such a widespread use of statistics across subject disciplines poses many problems as well as giving great opportunities for realistic teaching. The problems can be summarised as those of coherence, inadequacy, repetition, omission, timing and communication. The appearance of bits of statistics across the curriculum, with different notations and different expectations, means that students do not see statistics as a coherent subject. Teachers feel inadequate in having to use and teach a subject for which they have not been trained. Simple ideas, such as a bar chart and a mean, may be repeated many times sometimes with slightly different definitions and so confuse the student. There may be an omission of areas of statistical application which are important for general education since each teacher considers what is useful from the point of view of his own particular discipline. There may also be an omission of particular techniques and concepts which the student could find useful since the teacher does not recognise that these may be used, indeed he may be ignorant of their existance. There is a real problem of timing in that the teacher may be requiring a level of understanding for which the student has not been adequately prepared. The major problem in many schools is one of communication. Teachers of different subjects just donot know what use of statistics is being made by their colleagues in other disciplines. Within each school it would be most useful to have one member of staff responsible for coordinating the statistics teaching to overcome the problems.

With this background of statistics across the curriculum there are clearly many opportunities for cooperation to make the teaching of statistics more relevant and realistic. Cooperation may take the form of using data gathered in one subject in the teaching of another. It will include decisions as to who shall be responsible, and when, for instructing particular statistical concepts and techniques. It may also take the form of having project work done with links across the disciplines. Evidence from England at the 16-18 year old level shows that, when given a choice, students will come up with a variety of projects and will develop an understanding of statistics which is not easily tested by the more standard forms of test and examination.

Statistics is a vigourous subject fighting its way into all parts of the school curriculum. We should use this vigour as a positive means of improving the quality of statistical education.

7.5 STATISTICS AND PROBABILITY IN TEACHER EDUCATION

IN-SERVICE COURSES FOR TEACHERS OF STATISTICS

Peter Holmes
University of Sheffield
Sheffield, England

In this talk I want to focus attention on a number of questions relating to in-service courses in statistics for teachers and illustrate them from the experience of the Schools Council Project on Statistical Education. When planning in-service courses there are several variables which we have to consider. There are different types of teachers and these all have different needs, they are from different backgrounds, they have different levels of understanding of statistics, they occupy different positions in the school hierarchy. There are also different courses that can be run. They may differ in terms of time available; for example from one 2-hour period, 2 hours per week for 5 weeks, full courses lasting one or two days, a week's residential course, a 30 hour summer school or as part of a full time one or two semester course leading to a higher qualification. They may also be on a part-time continuing basis with a group of teachers working together over a long period of time.

They will differ in content and approach. They may involve listening to lectures, working with some previously developed material, developing new material, learning more statistics or learning how to coordinate the teaching of statistics across disciplines and so on.

They will differ in terms of geographical region from which the participants are drawn. They may be national courses, regional courses or even courses developed in a single school. Course content and approach have to match up with all these various backgrounds. In this talk I only have time to look briefly at some of the more important types of course.

Statistics is still a relatively new subject in the school curriculum, particularly for pupils ages 16 and under. This means that there are many teachers having to teach the subject who have not had experience with statistics in their initial training courses. Such statistical instruction as they have received will usually be subject oriented rather than classroom oriented. Emphasis will have been on statistical techniques and their mathematical justification rather than on the nature of statistical thought, the wide range of applications of statistics in society and the implications that these have for the appropriate methods of teaching and content in the classroom.

A. Courses for School Principals, Head Teachers and those in charge of the overall school Curriculum

I start here because if statistics is to be included effectively in the school curriculum then these are the people who need to be convinced that it is an importnat subject. It is unlikely that such people would attend a long course specifically on "Statistics in the School Curriculum". We are therefore, from the School Council Project on Statistical Education, adopting two approaches. The first is to try to get a 2 hour session on statistics in other longer courses which are already being organised for such members of staff. In the U.K. such courses are run by the regional education authorities, usually on administrative matters, and it is occasionally possible to persuade these organising authorities to schedule a 2 hour session provided that personnel is available to run it. The second approach is to run a half day couse specifically on this topic.

In each case the aims are the same. They include

1. to show participants the many ways in which statistics is used in everyday life,

2. to convince participants that all pupils need appropriate statistical understanding in order to cope with the world in which they live,

3. to raise the problems posed by incorporating statistics in the school curriculum,

4. to encourage participants to appoint someone within their school to be responsible for coordinating the statistics being taught across the curriculum.

At present the Schools Council Project on Statistical Education is setting up a network of Regional centres which can initiate and staff these short sessions. We had a tapeslide sequence which indicates the role of statistics in everyday life, discusses the reasons for incorporating statistics into the school curriculum and shows how our teaching material can be used to help grapple with these problems. It is essential that statistics be seen as generally useful since this is the main reason for including it in the main school curriculum. Since statistics is seen as being useful in many subject disciplines as well as being an important part of present day mathematics this means that it is already to be found in many schools to a greater or lesser extent. But the problems posed by this multidisciplinary use and the fact that statistics becomes seen as a subject tool rather than an essential part of education mean that there is a need for establishing within each school someone to oversee all the statistics that is taught and act as statistics coordinator.

B. Course for Statistical Coordinators within a School

The word "course" is perhaps not appropriate for this group of teachers. They need help in trying to identify the current uses being made of statistics within the school, in matching this up with the general needs of the pupil, in establishing a working group of teachers within the school to improve cooperation and develop appropriate teaching materials (perhaps adopting some that are presently available). It is useful if such coordinators can meet from time to time to discuss mutual problems and receive advice and help from outside experts. There is also a place for a two day course or a 2 hours a week for 5 weeks course for them to see the breadth of statistical applications, learn more statistics and consider the hierarchies of statistical concepts and techniques so that they can more readily identify and help solve problems of interdisiciplinary cooperation within their own schools. The Schools Council Project on Statistical Education is currently preparing material for use by such coordinators. It will be administered and tested through our Regional Centres. It includes such items as a Topic Analysis sheet which helps each department head analyse the statistical techniques and concepts he is using, identify others that he might be using and decide whether or not what he is doing is appropriate to the age and ability of the pupil. It also helps decide who should introduce a given technique and see the contexts in which it might be used by looking at subject themes (such as "the population explosion" already in the curriculum.

The type of in-service course which takes place within the school and involves many, if not all, of the staff is very important here.

C. Courses for Mathematics Teachers

These can be long or short depending on the needs of the individual. One problem is how to encourage teachers to come on these courses - - often it is the people who don't come who most need to come. One answer is incentives, either as a credit to build up to a higher degree or diploma or by extra recognition within the school. Another answer is to have school based courses in school time to which members of staff are obliged to come - - but this can also cause resentment and give course organisers problems with motivation. Whether long or short, there are a number of common aims and principles behind such courses. Much statistics teaching in current mathematics courses is technique oriented. It fails to take into account the nature of data, its inherent inaccuracy and the problems of data collection and of drawing meaningful inferences at all stages. In many cases it is not seen as useful since mythical data is invented to illustrate a technique and no useful inferences can be drawn. Aims 1 to 3 of section A of this paper are appropriate here. We would also expect:

1. to match up the statistics being taught with the statistical need of the pupil,

2. to consider hierarchies in statistical concepts and techniques and match up with the age and the ability of the pupil,

3. to consider how to get the pupils to appreciate the breadth of statistical applications, the essential usefulness of statistics,

4. to consider the relationships between statistics, mathematics and the other subject disciplines,

5. to work with currently available material adapting, where necessary, for use in one's own school,

6. to consider different methods of assessment, including examinations, appropriate for statistics and the effect they have on teaching.

With the growing use of desk top computers one aim of a longer course may be to consider the use of such machines in teaching statistics and to develop appropriate teaching material. I take it as important that people learn more by doing than by listening so that much of the course time should be spent doing activities related to the above aims. These activities may include working through teaching material, carrying out simulations, collecting data from various sources, analysing articles from newspapers and journals, reading textbooks from other subjects that use statistics, taking part in discussions, testing ideas in the classroom and reporting back later with evaluative comments and so on.

D. Courses for Teachers of Non-Mathematical Subjects

It may not be appropriate at a conference on mathematical education to consider courses for such teachers. Nevertheless I have included this category for completeness and to ensure that we do not forget that statistics is an interdisciplinary subject and that cooperation is an essential ingredient for successful statistics teaching.

ESTADISTICA Y PROBABILIDAD EN LA FORMACION DE PROFESORES: ESTUDIOS PREVIOS A LA ENTRADA EN SERVICO

L.A. Santalo
Buenos Aires, Argentina

1. Introduccion

La Estadistica y la Probabilidad ha sido uno de los temas que mas ha preocupado a los responsables de disenar planes de estudio para la ensenanza de la Matematica, desde la revolucion iniciada en el Seminario de Royaumont en 1959. "Parece evidente - se dijo entonces que la introduccion de la Probabilidad y la Estatistica en el curriculum de la ensenanza media es una tendencia importante por su utilidad" y en la primera reunion de especialistas en Dubrovnik (1960) se recomendaron los primeros programas para esas disciplinas (15).

Despues de varios anos de experiencia, en 1972, al senalar la UNESCO las "Nuevas tendencias en la ensenanza de la Matematica" vol. III (11) dedicaba un capitulo a la Probabilidad y Estadistica, en el cual decia "En la ensenanza de las ciencias, ya no es suficiente desarrollar el pensar determinista. Debe merecer mayor atencion en nuestra educadion escolar el pensar probabilistico, que domina los fenomenos de la herencia, procesos radioactivos y astrofisicos, y distingue la astronomia estelar de la vieja astronomia. En la vida diaria se encuentran numerosos fenomenos aleatorios (problemas de trafico y de contagion de enfermedades) y la lectura de diarios y revistas requiere un conocimiento minimo de Probabilidad y Estadistica para su correcta interpretacion. Graficos, histogramas, indices y procentajes, se utilizan en muchos dominios (seguros, impuestos, accidentes de vehiculos, datos demograficos, etc.) y problemas de muestras al azar, juegos y sondeos de opinion, son del mayor interes para el publico, sin que muchas veces comprenda los metodos utilizados ni el fundamento de los mismos".

En el ICME III de Karlsruhe (1976) se repitio lo mismo: "Uno de los temas mas importantes que ha empezado a ser introducido en la ensenanza elemental es el de la Probabilidad y Estadistica" (F. Colmez (11)) y "La Estadistica descriptiva y la probabilidad estan adquiriendo pueden ilustrar conceptos matematicos y motivar la computacion. Respecto de la probabilidad no se trata, a ese nivel, de aprender una teoria, sino esencialmente de acostumbrar al alumno con el pensamiento probabilistico en accion, que es muy diferente del pensar determinista (A.Z. Krygowska (12)).

Al llegar a 1980 se ha avanzado en extension (numero de escuelas en que se ensenan Probabilidad y Estadistica) pero poco en intensidad pues son muchos los problemas de ordenacion de contenidos y metodologia que quedan por resolver. Se ha llegado mas o menos a un acuerdo sobre los contenidos de Probabilidad y Estadistica que se pueden ensenar a cada nivel, pero se concoe poco acerca de la ordenacion mas conveniente de los mismos y sobre la metodologia y didactica para esa ensenanza. No se trata de ensenar unicamente algunas reglas practicas para calcular probabilidades o para aplicar rutinariamente algunas reglas de inferencia estadistica, si no que el ideal es desarrollar el "pensar probabilistico", haciendo que muchos resultados, en su aspecto cualitativo, lleguen a ser intuitivos, y el alumno comprenda de manera natural el comportamiento de la Probabilidad y tambien el peligro de muestras sesgadas o razonamientos falaces que pueden conducir a paradojas o al "mentir de la Estadistica".

Hay muchos problemas para resolver: a) Vinculaciones y ordenamiento raltivo entre la ensenanza de la Estadistica y la ensenanza de la Probabilidad; b) Didactica de esa ensenanza en cada nivel; c) La Estadistica y la Probabilidad en la formacion de profesores.

Aunque estos problemas estan muy vinculados entre si, aqui nos vamos a referir esencialmente al ultimo de ellos.

2. Objetivos y Metodologia Para la Probabilidad y Estadistica en la Formacion de Profesores

Puesto que la ensenanza de la Probabilidad y Estadistica en las escuelas elemental y media es relativamente reciente y, por tanto, no hay abundantes libros de texto para poder comparar y elegir segun las necesidades de cada escuela, el problema fundamental para el profesores el de la transferencia de sus estudios en la universidad o en el instituto en que se ha formado, a la escuela. Debe saber elegir los topicos apropiados y, sobre todo, aplicar una metodolgia sobre la cual hay poca experiencia. Por esto, en estas disciplinas, estimamos como mas recomendable la aplicacion del "principio de congruencia" segun el cual los futures profesores deben ser ensenados tal como ellos deberan ensenar (12). Tambien debe darse primordial importancia a la ensenanza orientada hacia la solucion de problemas y hacia la practica.

Hay que tener en cuenta los siguientes objetivos:

1. En el aspecto teorico, desarrollar el pensar probabilistico y la manera de usar la informacion como base para deducir consecuencias y tomar decisiones;

2. En el aspecto practico, ejemplificar los conocimientos de Probabilidad y Estadistica con temas de otras materias (Goegrafia, Biologia, Fisica, Sociologia, Idioma) y con sucesos de la vida diaria (seguros, rifas, trafico, fenomenos meteorologicos, deportes, . . .);

3. En el aspecto de investigacion, hay que lograr que los futuros profesores sepan buscar por su cuenta en el futuro toda la informacion que puedan necesitar sobre nuevos programas, textos y experiencias realizadas en otros lugares. Deben estar tambien preparados para realizar experiencias e investigaciones didacticas propias, como manera de contribuir al esfuerzo general para consequir una buena ensenanza de la Probabilidad y Estadistica desde las primeras etapas de la escala educativa.

Esta ultima parte es muy importante. El futuro profesor debe ser entranado a manejar las principales fuentes de informacion (libros, revistas, insitutos de investigacion) para poder tomar parte activa en las discusiones que suscitan muchos puntos de la didactica y del aprendizaje de la Probabilidad y Estadistica. Debe tambien conocer las tecnicas estadisticas para evaluar experiencias didacticas (analisis factorial). Hace falta mucha experimentacion para comprender las dificultades que aparecen en el alumno en el aprendizaje de la Probabilidad y Estadistica y el futuro profesor debe conocer las tecnicas para ponerlas de manifiesto (17). Tambien debe conocer los elementos de la teoria del muestreo. Ver la Bibliografia al final, especialmente (16).

Las operaciones de sacar promedios, calculo de variaciones y otras medidas de dispersion resultan engorrosas si se hacen con muchos datos y poco instructivas si se practican con pocos datos. De aqui la importancia del uso de las computadoras de bolsillo, cuyo uso se hace casi indespensable para la ensenanza de la Probabilidad y Estadistica. Los profesores deben ser instruidos en el manejo eficaz de esas calculadoras y sus tecnicas computacionales.

3. Experiencia Reales y Experiencias Ideales: Tablas de Numeros al Azar

Los problemas de probabilidad pueden ser ejem;lificados con experiencias reales, en base a lanzamientos de dado o monedas, extraccion de bolillas, juegos de naipes, ruletas, etc. Tambien la Estadistica puede ejemplificarse con datos reales: alturas de los alumnos, marcas meteorologicas, gastos diarios, . . . Pero algunas de estas experiencias llevan mucho tiempo y hay que realizarlas en horas fuera de clase. El "metodo de los proyectos" para el aprendizaje de la Estadistica a partir de problemas experimentales, es muy recomendable. Como trabajo previo hay que aprender a coleccionar datos, recolectar muestras, elaborar cuestionarios y hacer encuestas, analizando las posibilidades y los peligros de sesgo.

Tanto en los ejercicios hechos en clase, como en los proyectos fuera de ella, las experiencias reales pueden ser engorrosas y dificiles de realizar, siendo preferible la simulacion por experiencias ideales. Por otra parte, parece que esta sustitucion no perjudica el aprendizaje. .D. Austin (1), despues de analizar los resultados de una cuidadosa experiencia, llega a la conclusion de que "los resultados finales no indican ninguna diferencia significativa entre alumnos que realizaron manipulaciones o experimentos y alumnos a quienes se explicaron los resultados de estas manipulaciones o experimentos."

Para simular experiencias es muy conveniente el uso de Tablas de numeros al azar. Cada alumno debe tener una tabla propia, de unos 500 digitos, distinta para cada alumno, de manera que en total se dispone en las clase de un numero de digitos al azar suficientemente grande para obtener resultados confiables. Estas tablas pueden ser consetruidas directamente por cada alumno, o copiadas de algun texto. Lo importante es que sean distintas de un alumno a otro, para comparar resultados.

En el libro de Glaymann-Varga (8) se encuentran interesantes ejemplos sobre el uso de Tablas de numeros al azar, muy instructivos para cualquier curso distinado a futuros profesores.

Vamos a mencionar, como ejemplo, tres problemas que no son inmediatos teoricamente y que se prestan a ser simulados mediante tablas de numeros al azar.

1. Un pasajero llega en un instante al azar a una parada de omnibus. Si los omnibus pasan exactemente cada T minutos, el tiempo medio de espera es T/2. Si los omnibus pasan al azar con un promedio de un omnibus cada tiempo T, el tiempo medio de espera es T. Comprobar este hecho, que parece paradojico, con una tabla de numeros al azar. Cada alumno debe elegir para T un valor particular, por ejemplo, 6, 7, 8, . . . minutos.

2. Un servico de omnibus esta programado para que por la parada A paso de un omnibus cada 8 minutos. Por inconvenientes de trafico, se supone que cada omnibus puede adalenatarse o retrasarse hasta 4 minutos, con probabilidad uniforme. Un pasajero llega a la estacion A en un momento al azar. Cual es el tiempo medio que tendra que esperar hasta que llegue el primer omnibus?

Indicacion: Se toma una sucesion de ternas (x_0, x_1, x_2) de numeros al azar, excluyendo el 9. Supongamos que los omnibus debieran pasar en los minutos 4, 12, 20, . . . y que por tanto pueden pasar en los intervalos (0,8), (8,16), (16,24), . . . Se conviene en que x_1 indica el minuto en que pasa el omnibus 0_1; se supone que el omnibus 0_2 llega en el minuto $8 + x_2$ y que x_0 indica el minuto en que llega el pasajero. La diferencia $x_1 - x_0$, si x_1 x_0, o la diferencia $8 + x_2 - x_0$ si x_0 x_1 es el tiempo de espera en cada caso. Con esto, la Tabla de numeros al azar permite hallar la solucion. El valor teorico es 28/6 = 4,66 minutos.

3. Problema del estacionamiento de Reny. Sobre una calle de longitud x se estacionan autos de longitud unidad. La razon entre el numero de autos que han podido estacionar (suponiendo que cada uno estaciona al azar) y la longitud total x tiende a C = 0,747 . . . para x tendiendo a infinito.

Para la simulacion, tomese por ejemplo x = 1000 decimetros y supondase que cada automovil tiene 30 decimetros de longitud. Se toman ternas de numeros al azar que indiquen la distancia, en decimetros, del origen al comienzo del auto que estaciona. Se procede hasta que no quede lugar para ningun automovil. En este caso el valor teorico, no facil de calcular, es que pueden estacionar, en promedio, 24 automoviles.

4. Aplicaciones de la Estadistica

La Estadistica debe ir siempre acompanada de ejemplos, que pueden tomarse de la vida diaria o de las otras materias que figuran en los planes de estudio de la ensenanza media. El futuro profesor debe familiarizarse con estas aplicaciones e ir formaando una "bolsa de ejemplos" para su futura vida profesional. He aqui algunos ejemplos:

La Estadistica y la Escuala. Alturas y pesos de los alumnos, numeros de hermanos y de tios, longitudes del paso y del palmo, fechas de nacimiento, distancias de la casa a la escuela, calificaciones, deportes preferidos.

La Estadistica y la Ciudad. Medios de transporte: frecuencia y uso por los alumnos. Diarios y revistas de mayor circulacion: distribucion de su contenido en noticias, articulos literarios, avisos, graficos. Estadisticas demograficas. La guia telefonica como posible tabla de numeros al azar y como muestra de letras iniciales de los apellicos. Sondeos de opinion.

La Estadistica y el Pais. Los habitantes y su distribucion en ciudades y provincias o estados. Importaciones y exportaciones. Productos industriales y agropecuarios. Precios y salarios. Consumos de alimentos y energia.

La Estadistica y la Geografia. Longitud y caudal de rios. Alturas de montanas. Poblaciones de cuidades y paises. Estadisticas meteorologicas. Productividad agricola y ganadera.

La Estadistica y las Ciencias Naturales. Especies de animales y plantas mas frecuentes. Crecimiento y decrecimiento de poblaciones. Influencia de los abanos en el crecimiento de plantas. Duracion de la vida en distintas especies.

La Estadistica y el Idioma. Frecuencia de las distintas letras en el idioma: aplicacion a la criptografia. Frecuencia de las primeras letras de las palabras (numero de paginas de cada letra en el diccionario) y de las primeras letras de los apellidos (numero de paginas de cada letra en la guia telefonica). Longitudes de las frases para distintos autores. Frecuencia de sustantivos y adjectivos.

La Estadistica y la Matematica. Frecuencia de los numeros primos (hasta 500 o 1000). Frecuencia de los numeros naturales que son suma de dos cuadrados. Distintos caminos para ir de un lugar a otro en una ciudad cuadriculada: probabilidad de encuentro. Camino al azar.

Bibliografia

1. Austin, Joe Dan, Experimental study of the effects of three intructional methods in basic probability and statistics, J. for research in Math. Education, 5, 1974, 146-154.

2. Bognar, K. y Nemetz, T. On teaching statistics in secondary level, Educational Studies in Mathematics, 8, 1977, 399-404.

3. Engel, A. Teaching probability and statistics in intermediate grades, Intern. Journal of Math Educ. 2, 1971.

4. Engel, A. Wahrsecheinlichkeit and Statistik, Klett, Stuttgart, vol. 1, 1975, vol. 2, 1976.

5. Engel, A. The Probalistic Abacus, Educational Studies in Math., 6, 1975, 1-22.

6. Fischbein, E. The intuitive sources of probalistic thinking in children, Reidel, 1975.

7. Freudenthal, H. Probability and Statistics, Amsterdam, 1965.

8. Glaymann, M. y Varga, T. Les probabilies a l'Ecole, CEDIC, 1973.

9. Las Aplicaciones en la Ensenanza y el Aprendizaje de la Matematica en la Escuela Secundaria, Oficina de Ciencias, UNESCO, Montevideo, 1974.

10. Mies, T., Otto, M., Reiss, V., Steinbring, H. y Vogel, D. Tendencies and problems of the Training of Mathematical Teachers, Institut fur Didaktik der Mathematik, Universitat Bieldfield, 1975.

11. Nuevas Tendencias en la Ensenanza de la Matematica III, 1973; Nuevas Tendencias en la Ensenanza de la Matematica IV, 1979 (Traducciones castellanas de los New Trends III and IV) UNESCO.

12. Otte, Michael, Formacion y vida profesional de los profesores de Matematica, Nuevas Tendencias IV, UNESCO, Montevideo 1979.

13. Rade, L. The Teaching of Probability and Statistics, Stockholm, Almquist et Wiksell, 1970.

14. Symposium on Combinarorics and Probability in Primary Schools, Warsaw, 15-18 Agosto 1975.

15. Synopses for modern secondary school mathematics, OECD, 1961.

16. Teaching Statistics, Revista publicada por Dep. of Probability and Statistics, Sheffield University, Sheffield, England.

17. Wood, R. y Brown, M. Mastery of simple probability ideas among G.C.E. ordinary level mathematics candidates, International J. Math. Ed. Sc. Techn. 7, 1976, 297-306.

CHAPTER 8 - Applications

8.1 MATHEMATICS AND THE BIOLOGICAL SCIENCES -- IMPLICATIONS FOR TEACHING

THE SCHOOL MATHEMATICS - BIOLOGY RELATIONSHIP

Sam O. Ale
Ahmadu Bello University
Zaria, Nigeria

Preamble

In recent years there has been a considerable development of emphasis on the interplay between mathematics and biology, espeically at the tertiaral levels of education. These can be seen in the established areas of Biostatistics, Catastrophe theory, Population Biology, Mathematical Genetics and Cybernetics. Prominant journals produce research works of high standard on these interplays. These recent developments have forced some curriculum developers to contemplate the introduction of the subject of "Biomathematics" into the school curriculum. Such a move, however, may be open for a debate.

Though the much celebrated interplay and extensive research continues at a high level, it is disheartening tonote that these have not been reflected at the lower levels of education. Most school pupils still think a great gulf exists between mathematics and biology. Unfortunately some have, from the onset, aligned themselves to one with the hope that they might avoid the other.

My contention here is that these two subjects are related in many aspects even at the school level. There are mutual benefits for the biology pupils and teachers when the obvious and hidden relationships are investigated and integrated in a meaningful way.

The Objectives

The mathematics - biology relationship at school level should be based on certain objectives that contribute to the general education of pupils and teachers of the two disciplines. As Dudley (3) rightly emphasized, each has to gain from the other. The following eight objectives are here identified.

1. The Appreciation of the Notable Links Between Mathematics and Biology.

As far as many pupils are concerned, no links exist between the two. In my survey (1) carried out among Nigerian pupils of mathematics and biology, 61% said no to the question of if mathematics and biology are related, 32% answered yes, while 7% admitted that they don't really know. The result is even worse among girls as over 70% think they are not related. However, among students who have admiration for both subjects only 45% think they are not related. But the worse situation is that only 25% of those who like biology, but dislike mathematics said they are related. Also 20% of those who like mathematics, but hate biology, said they are related. Many other interesting statistics of students' views on the links between mathematics and biology are reported in that survey (1).

It is of real importance that pupils should begin to know at the school stage that there are links between the two subjects. For if interdisciplinary studies are to occur at the high levels, they must be reinforced at the lower levels. If not we may be in danger of having only very few being interested in the links.

2. A Gradual Conception of the Fundamental Uses of Mathematics in Other Disciplines.

There is a general lack of interest or hatred for mathematics in our schools. They do it because it has been made compulsory or it is a requirement for certain courses. One of the reasons for this, is that many students fail to see the ses of mathematics in the various disciplines. This is particularly true among those who offer biology. Many students complain of mathematics being theoretical with no practical utility. Even, when the application is obvious, some students fail to grasp it. Nonetheless students of various disciplines should understand that mathematics is a core subject to every area of knowledge. Also, students of mathematics should not that the end product of mathematics is in its application. And in particular, biology provides a wide area where mathematics is applicable.

3. The Development of a Logical Approach to Problem Solving.

Mathematics has been described as the single subject that provides the most logical activity of our mind in its purest and most perfect form. A mathematical demonstration is a scientific logic which contributes to the foundation of scientific education. If the aims of teaching biology to pupils in school as enumerated by Steentoft (14), (which includes the acquisition of knowledge of the interactions between organisms and their physical environment) are to be achieved, some scientific logic must be developed. And the ability to analyze data and draw reasonable conclusions is a unique function that training in mathematics offers.

4. Ability to Apply Mathematical Knowledge to Biological Situations.

Some students will not like to see mathematics being taught during a biology class, however relevant it may be. The survey (2) shows that about two-fifths of the students who even like both subjects fall into such a group. However, students should know how mathematics fits into biology.

5. A Competence in the Mathematics Underlying Biology.

It is essential to acquaint students of biology with the mathematical topics of magnitude, measures, proportions, ratios, relations, numerical data and graphing. A mathematical competency in these areas is most valuable if a student biologist is to make future progress.

6. The Formation and Use of Mathematical Models.

The Pythagorians had a common view that any system of the real world can be represented by a mathematical model. This has come to be the common view of many mathematicians and biologists today (14). However, H.G. Flegg has pointed to the crisis in the mathematical education of scientists and technologists today. This he said, is partly due to the fact that the modern mathematics syllabuses being taught in the schools do not provide a suitable foundation upon which traditional 'service' courses can be built. He, therefore, suggested the replacement of such service courses by integrated courses that are based on mathematical modelling (9). This is certainly of particular interest in the school mathematics - biology relationship.

7. The Development of Positive Attitude Towards Mathematics and Biology.

If any link between mathematics and biology is to be taught, pupils must first have a positive attitude toward both subjects. As shown in (2), only a third of the students have such an attitude. So a major objective for establishing a mathematics - biology relationship should be to help pupils have admiration for both subjects. If not, pupils will think that any relationship being taught is extra burden on the other.

8. The Relevance of Mathematics and Biology in Everyday Life.

The outstanding attribute of the two disciplines is the direct applicabiity to the everyday lives of people, whether they live in cities or jungles. Our lives are governed by many mathematical and biological processes. Biology student enthusiasts say biology teaches them about themselves and other living things, while mathematics student enthusiasts say mathematics teaches them how to live in their world. Any school mathematics-biology relationship should endeavor to emphasize these aspects.

The Problems

One must draw attention to the variety of problems involved in a school mathematics - biology relationship if any significant progress is to be made. One of the problems involves the mental disposition of the students and teachers of the two subjects. Generally, biology students and teachers have no liking for mathematics. Essentially, biology is assumed descriptive, while mathematics is but quantitative. And, as some pupils complained, they take to studying biology only becase they want to avoid calculations. Such views are the result of the lack of communication between mathematicians and biologists. It is worth noting that about 50% of pupils who like biology, but not mathematics, will come to like mathematics if they knonw it is related to biology, see (2). Most pupils have not come to realize the relationships between disciplines, since most educational systems have been structured in bits and pieces right from the beginning. This is where mathematics suffers most. Right from the primary school, mathematics is taught as a single subject on its own, since it is considered to have wide application and be of great importance, which is rightly so. However, the deficiency comes in the syllabuses and teaching methods, which are full of abstract phenomena and separated from our everyday life. Mathematics educators often argue the diversity of users of mathematics, for example biology is a discipline among many. Therefore who should we have in mind when teaching mathematics, especially at the school level? A suggested principle is, to be broad, only avoiding the abstract and the incomrehensible. When the abstract can be associated with the living, it certainly helps in the education the the child.

The relevance of mathematics to biology has not been felt so much in the past in our schools because the occurence of biology examination papers requiring mathematics is rare. As noted by Dudley (4) this situation is changing as more mathematics is being required in biology examinations. Also, mathematics has not usually been a prerequisite course for students going into courses involving biology. This has made mathematics irrelevant to these students. This situation is also changing, for example, in Nigeria an ordinary - level requirement is a must for all candidates entering into all scientific disciplines like Agriculture, Medicine, Pharmacy, Environmental Design, etc.

Another major factor that has suppressed the mathematics - biology relationship is the pre-University combinations or courses that students take. The traditional combinations are being changed with the modern combinations. These modern combination systems allow relevant mathematical topics of biological nature to be taught during biology classes and vice versa. This has provided a good way of integrating the teaching of mathematics and biology.

There is also the problem of the mathematics and biology curriculum being already overloaded. Therefore the idea of adding more mathematics to the biology syllabus is unacceptable to many. However there is a need for the review of both syllabi so that certain topics in mathematics could be infused into the biology syllabus and also in mathematics that its applications to biology could also be taught. This inclusion of topics from one to the other must imply the dropping of certain topics in the syllabi. And those to be dropped should be such that are not easily comprehensible, and too technical for the students. Without any doubt, mathematics and biology contain many such topics.

The Kindergarten

Education in the kindergarten is primarily what is seen, touched and played with, that is, an education of the basic world around infants. And it is right at this level we begin to see the fundamental relationship between mathematics and biology which the infant can grasp.

Of primary importance is geometric conceptions. The world around the child is full of geometric shapes which can be recognised and identified with familiar objects and living things. The infant can recognise and associate circular, rectangular or triangular objects, which are often common with what they play with or even eat. It is worth noting that the rudiments of biology that could be taught to infants surrounds foods. It is amazing to see how different food materials can display some different geometric shapes.

The idea of building blocks is common with infants, and most often they use squares, triangles and other polygons to model different objects. Such models result in some kinds of animals and plants. They therefore learn the parts as contained in what they model. Since simple geometric shapes like triangles, squares or pentagons can be drawn by infants, they can begin their drawing of biological objects using these.

The drawings also enable infants to gain some familiarity with numbers as they add blocks one to another or even take away and yet form objects. This has in many cases provided an interaction between mathematics and biology right from the infant school. The basic notion of numbers comes from counting, but most often infants count using animals and food items and even perform some arithmetical operations using these objects, providing the link between mathematics and biology right from the kindergarten.

The Primary School

In most countries, biology is taught under the integrated science or primary science programme at this level. The basic programme as it concerns biology includes hygiene and nature study. The mathematical programme is generally arithmetic and some geometry. The biology is related to mathematics in the sense that the questions pupils ask about biology concern how much, how many, how large, how big, how fast, etc. Fundamentally the answers to such questions involve the pupils being able to measure, find volumes, read meters and calculate some simple ratios and proportions. These are involved in estimating the size of animals and plants; drinking and eating; reading body temperatures by using thermometers; and finding growths in animals and plants respectively.

The knowledge of basic operations in arithmetic can help pupils understand the primary science course. Counting the teeth, for example, is the most common part of the body to all human beings, which we are not born with. Infants and pupils often add one at a time. Also there comes a time when they get removed one by one in children while others spring up. Since the maximum can only be 32, this provides a finite number in the body that first begins with zero. The process of addition and subtraction are involved in practically every area. The pupils know that yeast +flour + sugar + butter = bread or beans + oil + onion + salt = akara (bean cake) although these have not involved numbers, but the operation addition. Thus they begin to know about "additions" in different forms and even in bases. Animal reproduction provides a multiplication example and also flow diagrams. Fractions are learned from division of an entity like one orange for 4 boys. Divisions are not often clear to pupils, like $0/1 = 0/2 = 0/n = 0$, are best illuminated with objects involving biology. In this manner, biological examples can be drawn for every mathemtical topic.

The teaching of biological classification can also begin in the primary school. The importanct of classifactory systems can not be over emphasized. As Hutchinson (11) stated, they are not only used for purposes of communication but also as a basis for generalizations and predictions. By and large, the best way to introduce classification in biology is via set theory. Pupils recognise first that a set is a well defined collection of objects, e.g. the set of insects in the garden = (butterflies, ants, grasshoppers, flies, mosquitoes). The pupils can then learn about different living organisms and put them in their different families, but before this, they have learned about the set of other familiar objects. Empty sets are easily grasped as they think of the impossible, e.g., the set of pupils in their class with two heads. Subsets, intersection, and union of sets also form a central part in a classificatory system, this is developed in the next section.

The Secondary School

The first two years of the secondary school curriculum, especially in mathematics and biology, is generally a development of the upper primary school programme. The development of set theory in mathematics and classificatory systems in biology are a simple example.

Many simple examples can be drawn of objects in biology that display some mathematical features. The mathematics itself may be difficult to grasp by secondary school students, but the picture is beheld in the biological objects. We give some few other examples here:

A vector being a magnitude that has magnitude that has direction is best illustrated in many animals' hair and even hairy plants like tridax procumbens, oil palms and clematis. In the study of polar coordinates, the equation which is of the form $r = a(1 + \cos\theta)$, shows an example of fruits like tomatoes and seed germination; and some animal kidney, and the equation $r = a \text{ sine } 2\theta$, shows a flower structure like Tribulus. If the area between the graph $y = x^2$, the line $y = 4$ and the y-axis is rotated about the y-axis, and the figure produced is like some caryopsis of maize or the flower Aristolochia Ductchman's pipe.

The system of cultivation used by many farmers in ridges display parallel lines, rectangles and other geometric shapes. Thus, in this way the theoretical mathematics is made practical in biology for school pupils. Many books have appeared recently (see 10), seeking to introduce mathematical works into school biology so that teachers and pupils can become familiar with the language of mathematics without leaving the realm of biology. Such books often provide examples of mathematics - biology relationships. A list of other relationships as given by students are contained in my survey "Student Views on Mathematics - Biology Relationships" (1).

Considerable work has also been done on the interplay between mathematics and biology at school level, especially by B.A.C. Dudley. The report on Mathematics for Biologists (12) provides the links between the two at school level. From that report and other investigations, it seems that the following mathematical topics are inevitable for the biologists at school level: ratios, proportions, fractions, elementary statistics, numerical data, graphs and their interpretations and measurements in metric units. These mathematical topics are applied throughout the whole area of biology. Other mathematical topics are often only relevant to certain biological topics, like set theory for classification and permutations and combinations for genetics. Thus, the biology student certainly needs mathematics.

Conclusion

The argument then is that right from the onset of education, i.e. from kindergarten, to the primary, through the secondary schools, many relationships exist between biology and mathematics. It is no exaggeration to say that every field of school mathematics is applicable to school biology. And also that school biology provides the practicalities of any abstract mathematics. However, some areas are more prominent than others and at each level. Some mathematics is so important at the school level that it needs to be included in the biology curriculum, e.g. simple number theory and elementary statistics.

Teachers and pupils also need to be informed of the interplays between mathematics and biology, many are simply ignorant as my surveys (1) and (2) show.

Furthermore the following innovations:

1. the inclusion of some mathematical topics into the primary science/biology syllabus;

2. the closer interraction between mathematics and biology teachers and students through lectures, seminars and conferences;

3. further research into the interplay between the two subjects at the school level;

4. development of more materials for the teaching of mathematics to biologists;

5. adequate backing in mathematics for biologists;

6. mathematics courses for biologists;

are things to be encouraged in the school mathematics biology relationship.

Acknowledgement

In the preparation of this lecture and other papers on biology the author has been working in collaboration with some colleagues in the biology section, in particular, he is grateful to C.O. Abah, S.E. Yakubu, and also, his colleagues in the Mathematics Department.

References

1. Ale, S.O. Students perceptions on mathematics-biology relationships. Ahmadu Bello University Research Project, 1979/80.

2. Ale, S.O. Mathematics or biology - A survey of student options. Ahmadu Bello University Research Project, 1979/80.

3. Dudley, B.A.C. Bringing mathematics to life. Int. J. Math Educ. Sci. Technol., 6 (1) 1975.

4. Dudley, B.A.C. The mathematics of school biology exam papers. Jour. of Bio. Educ., 11 1 1977.

5. Dudley, B.A.C. Investigating the effect of size upon metabolic rate of mammals. Jour. of Bio. Educ., 7 (3) 1973.

6. Dudley, B.A.C. Some mathematical relationship of bilogical value. Jour. of Bio. Educ., 5 1971.

7. Dudley, B.A.C. The maths of whether an animals legs will hold it up. The Teacher Jan. 1972.

8. Dudley, B.A.C. The mathematicsl bases of phenotype ratios. Int. J. Math. Educ. Sci. Technol., 4, 1973.

9. Flegg, H.G. To be published.

10. Gibbons, F. & Blofield, B. A Mathematical Approach to Biology.

11. Hutchinson, C.S. Biological classification. The Jour. of the Ass. for Sci. Educ. 61 (215) 1979.

12. Report, Mathematics for Biologists. Jour. of Bio. Educ. 8 (5) 1974.

13. Stone, R. & Oozens, A. New Biology for Tropical Schools, Longman, 1972.

14. Thom, Rane Structural Stability and Morphogenesis, Benjamin, Inc. 1975.

THE STRUCTURE OF HOMEOSTASIS MODELS

Diego Bricio Hernandez
Universidad Autonoma Metropolitana
Mexico City, Mexico

A homeostasis is intended to portrait the biological processes making up the regulatory actions that take place in living organisms, and which are essential for the persistence of life. A case is made for explicitly incorporating such features as feedback loops into these models, as well as for proving the stability properties implying the good regulation actually observed in living beings.

This exercise is intended to illustrate on how mathematics can aid the biological sciences in describing a most fundamental phenomenon taking place in living matter.

Introduction

This presentation responds to a desire to illustrate the role of mathematics in biology. Of the various aspects normally emphasized when dealing with mathematical treatments of biological phenomena (see last paper in (BNL)) only the construction and evaluation of models will be considered, on account of its being the most fundamental one. On the other hand, many biological phenomena could be selected for the purpose of illustration. However, homeostasis is known to play an essential role in the sustainment of life itself, hence its fundamental character and hence our choice of subject matter.

A basic requirement on any mathematical model of a real situation in that it should relfect observed phenomena. The kind of observations normally made in physiology include subjecting the system being considered to suitable inputs and observing the corresponding response. This "black-box" type of description can in turn be explained by more than one set of assumptions concerning internal structure, i.e. by more than one model of the type

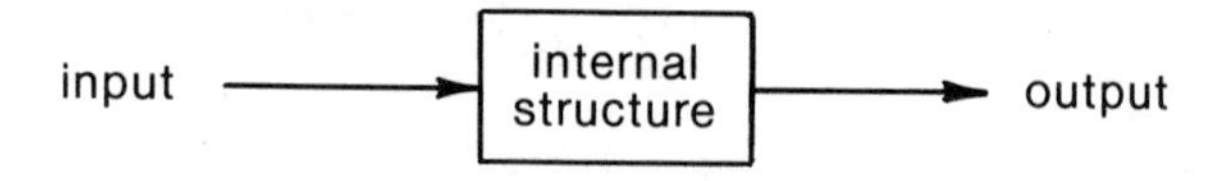

Now, which internal structure is chosen for the black box in order to explain observed behaviour depends on many factors, the current state of physiological research (i.e. current paradigm) playing no minor role in these matters.

We shall be interested in satisfying a basic requirement: a model must be isomorphic to underlying physiological structure. Hence a model intended to portray a homeostatic mechanism must reflect the way homeostasis actually works, besides reflecting accurately the basic facts of homeostasis, namely

a) existence and uniqueness of steady states
b) asympotic stability of steady state under perturbations.

<u>Modelling homeostasis</u>

Quoting from a well known text in physiology (Ganong, p. 17), homeostasis consists of the various physiological arrangements that collaborate towards restoring normal conditions after a perturbation. The physiology of living organisms is full of examples of homeostatic mechanism (Ganong; Talbot-Gessner; Banks; BNL) in charge of regulating

a) the level of blood impurities
b) blood pressure
c) pH in blood
d) body temperature
e) tissue metabolism

and many others.

Consider, for instance, body temperature regulation. Simplifying matters somewhat, the goal is to maintain body temperature around 36.5°C, with fluctuations not exceeding 0.7°C, although there are of course variations from one individual to another. A slight perturbation in environmental temperature unleashes both internal and external reactions that tend to counteract the effect of that external variation upon body temperature. This action is accomplished through appropriately balancing the rates of heat production and dissipation. Thus temperature signals reach the hypothalamus from various body and surface sources, this organ acting then as a signal-processing unit generating commands that constitute homeostatic action (Gannong, pp. 169-173).

All homeostatic mechanisms achieve their goal in a fashion similar to the one just described. Regulatory action involves:

1. Signal of interest (<u>observation</u>),
2. Processing of this information into the appropriate regulation commands (<u>control</u>), and
3. Transforming these commands into direct action (<u>transduction</u>).

In turn, ths three-step action results in

4. Improving the performance of signal of interest, which will continue to be the subject of observation (<u>plant dynamics</u>).

This overall picture of homeostasis can be expressed in terms of the basic concepts of <u>feedback</u> (Wiener), and is summarized in a closed loop system as a basic model of regulatory action:

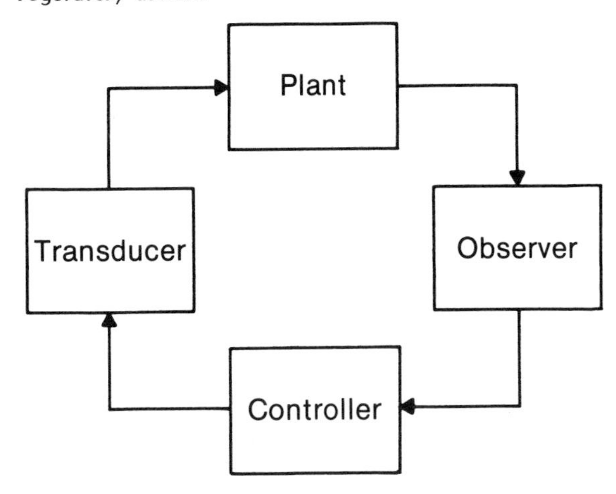

Any model of hemeostatic action should incorporate these four elements if it is to be "isomorphic to underlying physiological structure".

As to actual regulatory action, it is embodied in the mathematical concept of <u>asymptotic stability</u> to be formulated in the next section.

<u>Feedback and Stability</u>

Each block in the feedback loop is an instance of what is normally called an input-output system in the literature (Zeigler). For example, let signal to be controlled take values in a set X, so that its time evolution can be represented by a function

$$x : [0,\infty) \rightarrow X, \text{ with}$$

x(t) : = value of signal to be controlled at time t.

The observer processes x into another signal taking values in say set Y and evolving in time according to a function $y : [0,\infty) \rightarrow Y$. A mathematical model of the observer should consist of

a) The class $\mathcal{X}$ of functions X it can accept,
b) The class $\mathcal{Y}$ of functions Y it can accept,
c) The rule ω for transforming input into output, namely a function $\omega : \mathcal{X} \rightarrow \mathcal{Y}$ subject to the intrepretation

ω(x) : = Observer output when input is x.

Letting $y = \omega(x)$, it is natural to restrict the class of acceptable ω's to those being <u>non-anticipative</u>, in the sense that

$$x', x'' \in \mathcal{X}, \; y' = \omega(x'), \; y'' = \omega(x'')$$

$$x'(s) = x''(s), \; s \leq t \Rightarrow y'(t) = y''(t), \text{ all } t \geq 0$$

Otherwise future values of input with influences present values of output, contrary to causality. Let Ω be the class of all admissible observers.

Analogously, can give set $\mathcal{K}$ of all admissible controllers, set $\mathcal{T}$ of all admissible transducers and set Π of all admissible plants. A controller $k \in K$ will be a nonanticipative mapping $k : \mathcal{Y} \rightarrow \mathcal{V}$, $\mathcal{V}$ being the class of all possible time evolutions of controller output. And analogously for the remaining elements of the loop.

A model of homeostasis will then be a 4-tuple

$$< \pi, w, k, t >$$

with

$$\pi:\mathcal{U}\to\mathcal{X},\ w:\mathcal{X}\to\mathcal{Y},\ k:\mathcal{Y}\to\mathcal{V},\ t:\mathcal{V}\to\mathcal{U}$$

Summing up, the closed loop behaviour can be characterized by the system of equations

$$x = \pi(u),\ y = w(x),\ v = k(y),\ u = t(v)$$

This system of equations should have a unique solution for each initial condition (xo, yo, vo, uo) $\in X \times Y \times V \times U$. Transient behaviour is represented by nonconstant solutions, whereas a constant solution such as

$$t \to (\bar{x}, \bar{y}, \bar{v}, \bar{u})$$

represents steady state.

Let

$$B = X \times Y \times V \times U$$

and let

$$\sigma : B \to B$$

be given by

$$\sigma(x,y,v,u) = (\pi(u), w(x), k(y), t(v))$$

Then closed loop behaviour can be characterized by the condition

$$\sigma b = b$$

with b = (x,y,v,u). In other words, possible closed loop dynamic behaviours should be sought among fixed points of operator σ, and there are as many such fixed points as there are "initial conditions"

$$(x_0, y_0, v_0, u_0) \in X \times Y \times V \times U \equiv B$$

For each $t \geq 0$, a transformation $T_t : B \dashrightarrow B$ can be given, with $T_t b_o :=$ closed loop behaviour at time t if initial condition was b_o.

Clearly

$$T_0 b_0 = b_0, \text{ all } b_0 \in B$$

and $\bar{b} \in B$ represents equilibrium if and only if

$$T_t \bar{b} = \bar{b}, \text{ all } t \geq 0$$

On the other hand, uniqueness of closed loop behaviour is easily seen to translate into

$$T_t T_s = T_{t+s}, \text{ all } t, s \geq 0$$

thus showing that $\{ T_t,\ t \geq 0 \}$ constitutes a semigroup (Bhatia-Szego, Chapter I).

Now, suppose B is equipped with a norm (Riesz-Nagy, p. 211). An initial departure from equilibrium will become after a time t a deviation $\|T_t b_o - \bar{b}\|$. Stability means that future deviations can be kept as small as desired provided initial departures from equilibrium are suitably restricted. In other words, $\bar{b}$ will be a stable equilibrium point for closed loop system if given $\varepsilon>0$, there exists $\delta>0$ such that

$$\| b_0 - \bar{b} \| < \delta \Rightarrow \| T_t b_0 - \bar{b} \| < \varepsilon, \text{ all } t \geq 0$$

If, moreover

$$\| T_t b_0 - \bar{b} \| \to 0 \text{ as } t \to \infty$$

stability is said to be <u>asymptotic</u> (Bhatia-Szego).

This is the kind of dynamic behaviour that should e predicted by any model of an effective regulatory system. In particular, homeostatic mechanisms are known to be highly effective, as required for the preservation of living condition in organisms. Hence a model of homeostasis should be accompanied by proofs of asymptotic stability of equilibrium states, besides clearly indicating the closed loop components taking part in the regulation process.

Conclusions

The various component elements taking part in regulatory action achieved by homeostatic mechanisms have been identified as plant, observer, controller and transducer. Each of these elements has been modelled as a deterministic input-output system, overall closed loop behaviour being in turn characterized as the fixed points of a certain operator. To be precise, that operator acts on an appropriate function space which contains constants, constant fixed points being candidates for representing equilibirum conditions.

A semigroup of operators was in turn constructed fromthese elements, successful regulatory action being translated as asymptotic stability of steady states for the dynamical system defined by that semigroup.

These are the structural and functional properties that have to be guaranteed whenever a model of homeostasis (or regulatory action, in general) is being put forward. Needless to say, it cannot be done without quantitative agreement with experiment, same as in any other modelling situation. See the author's reference in the bibliography for a detailed example illustrating validation of homeostasis models (body temperature regulation) from this perspective.

References

Banks, H. T., "Modelling and control in the biomedical sciences", Springer-Verlag, Berlin, 1975.

Bhatia, N. P. & Szego, G., "Dynamical Systems: Stability theory and applications", Springer Lecture Notes in Mathematics, No. 35, Berlin, 1967.

Brookhaven National Laboratory, "Homeostatic Mechanisms", Brookhaven Symposia in Biology, No. 10, Upton, N.Y., 1957.

Ganong, W. R., "Review of Medical Physiology", Lange Medical Publications, Los Altos, Calif., 1973.

Hernandez, D. B., "Sobre la validacion de Modelos de Homeostasis", Rep. Inv. UAMI-CBI-20, Universidad Autonoma Metropolitana-Iztapalapa, Mexico, 1979.

Riesz, F. & Nagy, B. S., "Functional Analysis", Frederick Ungar Publishing Co., New York, 1965.

Talbot, S. A. & Gessner, V., "Systems Physiology", John Wiley & Sons Inc., New York, 1973.

Wiener, N., "Cybernetics", The MIT Press, Cambridge, Mass, 1961.

Zeigler, B. P., "Theory of modelling and simulation", John Wiley & Sons Inc., New York, 1975.

MATHEMATICS AND THE BIOLOGICAL SCIENCES IMPLICATIONS FOR TEACHING

Lilia Del Riego
Universidad Autonoma Metropolitana
Mexico City, Mexico

Dr. Ale has taken up several biology problems which involve mathematics, but which appear generally hidden to biology students. In this connection, I always recall the great surprise these students receive the first day of classes, when they are faced with me and my mathematics! Indeed, their weak background in mathematics, together wih the general distaste they exhibit for their study makes it difficult for them to believe that mathematics will serve for something practical.

I will now tell you about my own experience organising and teaching mathematics to biology undergraduates at the Universidad Autonoma Metrolitana-Iztapalapa. The contents of the courses were selected by the Biology Department, who outlined the basic needs they had, while we mathematicians were left in charge of acutally organizing the various themes. We worked with biologists in a seminar in order to work out examples and provide biology problems which would be presented during the first week of classes in order to build a mathematical need. One could in this way catch the student's interest, and the teaching of the mathematics was reduced to an easier task.

The courses were presented in lectures twice a week together with tutorials once a week. We have four calculus courses (named Mathematics A through D), and two statistics courses (named Biostatistics I, Biostatistics II).

In designing these courses we had in mind two objectives. Firstly, they would have to include relevant biology problems. Secondly, they would have to provide a student with an acceptable mathematics course (from a mathematician's point of view), but one which was free from "epsilonitis". Because it is only after actually teaching the course that you realize whether the bilogy undergraduates are really receiving the mathematical ideas they should, the process of designing new courses to test afterwards doesn't really end - we are in a process of correction at this moment. The last programs we used are the following.

Mathematics "A"

Real functions of one real variable

1. The Straight Line
2. Exponential and Logarithmic Functions
3. Power Functions, Rational and Trigonometric Functions.

Because in calculus courses, applications don't seem so natural, we have tried to develop mathematical concepts with realistic examples. Mathematics A is a pre-calculus course. The main objective is to get the student to recognize an equation in order to sketch the curve and vice versa. One can start by describing a process of population growth to introduce the very important exponential functin and then pass to discuss the logarithmic function and the straight line by use of a semi-logarithmic plot. The other functions are described in a similar way. The textbook which approaches these themes best is Hadley's "Elementary Calculus". Afterwards, in Mathematics B, they study:

Differential Calculus

1. Growth Rates, Limits, The Derivative
2. The Derivative as Slope of the Tangent Line
3. The Chain Rule
4. Maximum and Minimum Values, Curve Sketching.

One defines rates of growth (of a population perhaps) in order to discuss differential calculus and then in Mathematics C one may reverse the process to recover the actual population to introduce the following:

1. The Antiderivative
2. Integration Techniques
3. The Definite Integral as an Area, The Fundamental Theorem of Calculus
4. Differential Equations.

Mathematics D consists of:

Real Functions of Several Real Variables

1. Definition, Graphs and Contour Lines
2. Partial Derivatives, The Chain Rule
3. Directional Derivatives, The Gradient, The Differential

4. The Double Integral as a Volume, Iterated Integrals.

Mathematics D is not a "must' for all biology students: Those who need more training take this course to get acquainted with the calculus in several real variables.

My own experience from teaching these courses shows that students are prepared to receive the mathematical concepts once they see their applications. So by always presenting biological examples, they not only have been very receptive, but also they worked very enthusiastically. They also work harder than calculus students in the engineering branches, perhaps because most themes are new to them. We always encourage biology students to look for applications of the mathematics they learn in the other biology courses, in order to relate more closely the mathematics to their interests. The topics covered in the Mathematics B, C and D are almost all dealt with in Rodolfo de Sapio's excellent book: "Calculus for the Life Sciences".

While in calculus courses a biology student finds applicability difficult to see, in contrast, the statistical courses are accepted very well. Realistic biology examples are easy to find and the student naturally sees whyhe must organize and compare his data more efficiently. They apply directly what they are learning. The statistical description leads naturally to the problem of generalization and to decision-making and several statistical tests are studied. The topics covered in Biostatistics I include:

1. Descriptive Statistics

2. Probability, Discrete and Continuous Probability Functions

3. Statistical Inference, Estimation and Hypothesis Tests

4. Significancy Tests (parametric tests)

Finally, in Biostatistics II, the hypothesis cases more used by biologists are analyzed such as:

1. Least Square Methods

2. Variance and Regression Analysis

   a. Media Comparison

   b. Block Design

3. Contingency Tables

   a. X^2 Distribution

   b. 2 x 2 Tables

   c. R x c Tables

A study was made to try to discover up to what point the mathematics received was of use in biology and to what extent it was actually applied. Two groups were tested: one which consisted of biology students who had not finished their degree but who had finished all required mathematics courses, the other was formed of biology graduates.

The results are as follows: The former group consisting of 122 biology students out of 600 showed that for 102 students (83.6%), the mathematics learned were already in use in the other biology courses.

As to the latter group, the questionnaire showed that 87.5% of the biology graduates were using mathematics in several branches including:

Dynamics of Population,

Phytopathology,

Ecology,

Experimental Sciences,

Biochemistry, and

Research in Health Sciences.

8.2 THE RELATIONSHIP OF MATHEMATICS AND THE TEACHING OF MATHEMATICS WITH THE SOCIAL SCIENCES

RECENT DEVELOPMENTS IN GEOGRAPHY TEACHING AT THE SECONDARY LEVEL AND THEIR RELATIONSHIP WITH MATHEMATICS

J.F. Ling
School Mathematics Project
London, England

Personal Background

I first became interested in the subject of this talk when I was working with a project set up by the Schools Council to produce a critical review of the mathematics curriculum of 11-16 year olds. This review has now been published as a series of books on various different topic areas (1). One of the books reviews the use of mathematics in other subjects. In order to collect the necessary information a special group was set up to examine each subject area. One of these groups was concerned with mathematics and geography.

My own function was to write the book on the basis of the findings of the groups. Before work started I expected the chapter on geography to be fairly short. I imagined that it would cover at the more elementary level such items as map scales, compass bearings, contours, gradients, percentages and various ways of presenting data - graphs, charts, and so on. At a more advanced level I expected some geometry of the sphere, latitude, longitude and possibly some work on map projections. But once the work had started I soon became aware that there had been an explosion of quantitative work in geography starting at univrsity level but now taking root in secondary schools.

What I have to offer in this talk is based partly on the work of the project and partly on what I have learned afterwards through following up the interest in geography which started with the project. My approach is that of a person engaged in mathematics education at secondary school level who is an interested observer in current trends in geography but not qualified in geography himself.

The General Character of Recent Ideas in Geography Teaching

The situation in geography teaching resembles in some ways the situation in mathematics teaching at the time when so-called 'modern' and 'new'mathematics was emerging. In both cases there are teachers who have seen that there is an increasing gap separating the content of the subject at university level and its content at school level. Teachers leaving university have been unwilling to abandon the view of their subject which the university has given them. They have been unwilling to go back to what seems to them to be out of date and increasingly unrelated to what is going on at the creative edge of their subject.

There is another point of resemblance in the ideas of the internal structure of the two subjects. The advocates of 'modern mathematics' have held that mathematics is based upon certain fundamental notions such as set, relation, function. It is said that a student who grasps these ideas at an early stage will be able to appreciate an ordered and unified presentation of mathematics and it has been held that the fundamental notions of mathematics can be grasped through quite everyday examples. For instance, function can be exemplified by the relationship between countries and capital cities, where to each country there is assigned one capital city. Similarly in geography certain fundamental ideas can be approached through familiar examples. One such idea is what geographers call the 'friction of distance'. Stated abstractly this says that the influence of a place upon surrounding places falls off as the other places get more and more remote. This principle can easily be illustrated by considering, for example, the frequency of visits to relatives living near and far away.

Generally speaking, what characterises the newer approaches to geography is a concern with spatial arrangements which recur in one form or another all over the globe and can be abstracted from particular context. The understanding of such spatial arrangements can be advanced, so it is held, by quantitative methods of analysis and/or the building of theoretical models. Examples of such spatial arrangements (usually called 'spatial distributions' by geographers, a term with confusing statistical overtones) are: transportation networks (arrangements of links connecting places); settlement patterns (arrangements of points in a region); land use patterns (arrangements of regions).

Mathematics in Geography: Examples

In the remainder of this talk I will be describing and commenting on some specific examples of the use of mathematics in geography at secondary school level. All of the examples I quote are taken from books written for teachers of geography or from textbooks for use in schools. In some cases I shall be very critical, but I have decided not to give sources for my examples. It is not part of my purpose to commend or criticise this or that particular contribution. What I am trying to do is to convey through the examples I describe, a general impression of the 'scene' in this area of geography. Many of the examples I choose are to be found in the same or in equivalent forms in a number of publications.

A Theoretical Model: Von Thumen's Land Use Model

Von Thumen's model of land use round a market centre is a classic example of the use of a theoretical model. It is hardly 'new geography', dating as it does from the early nineteenth century, but it has been incorporated recently into some secondary school courses, in some cases modified to deal with urban rather than agricultural land use.

The assumptions of the model are

(a) There is an isolated state with a single market centre.

(b) Land is of uniform productivity.

(c) The cost of transportation is proportional to distance.

For each crop grown, the graph of profitability against distance from the market centre would slope downwards. Some crops will be highly profitable near the centre, but with profitability very sensitive to distance. (For example, highly perishable fruit and vegetables.) Other crops will be less profitable but not so sensitive. Suppose for example there are three crops with the graphs shown here.

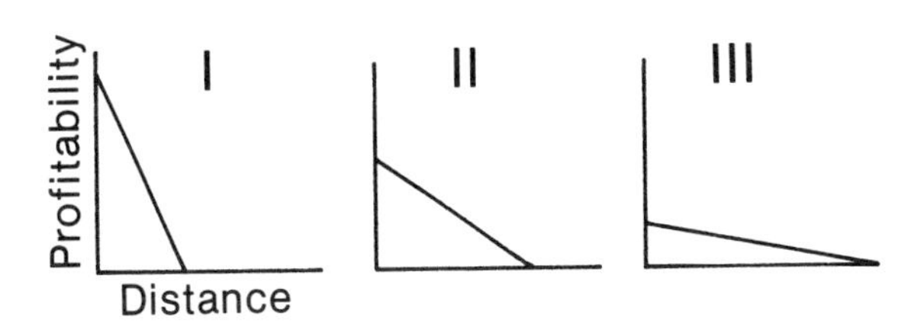

At any point farmers who grow the most profitable crop will be able to bid the highest price for land. So at a given distance, the most profitable crop will be grown. When the 3 graphs are superimposed we get a composite graph showing the most profitable crop for each range of distances.

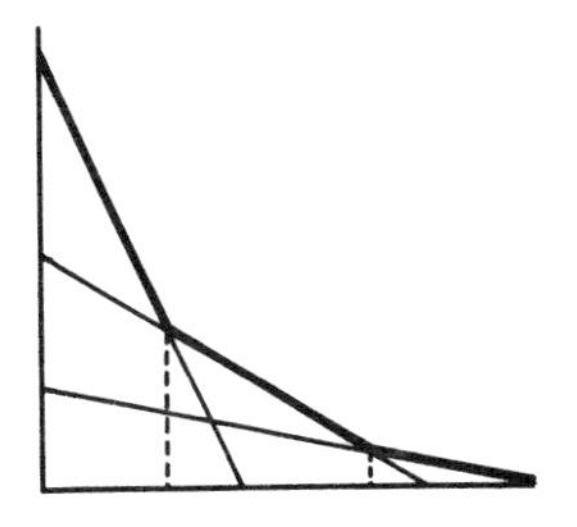

Translating this into a geometrical arrangement we get this pattern of land use round the centre.

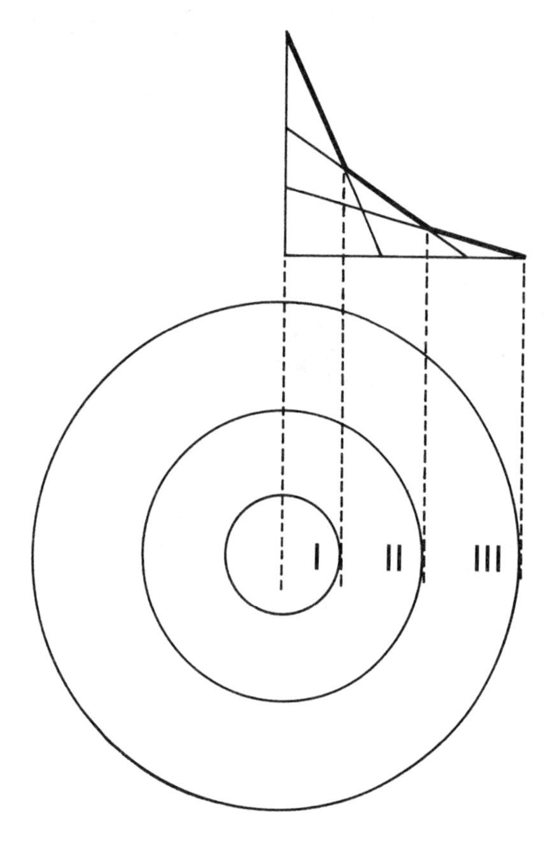

We can predict the effect on this arrangement of the building of a superior line of communication, thus modifying assumption (c).

The purpose in setting up such a model is to examine real-life instances to see how far the predictions of the model are borne out in practice. If they are not, the question arises as to which of the assumptions of the model are not applicable.

Probabilistic Models and Simulations

In addition to the three stated assumptions of the von Thumen model certain other assumptions were made. One was that at each place the most profitable crop will be grown. A weaker form of this assumption is that at each place the most profitable crop will tend to be grown, or, put another way, that at each place it is more likely that the most profitable crop will be grown than that other crops will be grown. Probabilistic models attempt to quantify tendencies and recognise the inherent chanciness of processes in geography.

In the next example I describe a probabilistic model for the growth of a town.

(a) To start with, a physical map of the area around the town is divided into a grid of squares, each of which is given a reference number.

(b) Certain physical features are selected as being factors which will affect development, either positively or negatively. Such factors might be, for example, slop of ground, nearness to a river and hence liability of flooding, etc.

(c) Each factor is now considered in turn.
Suppose we start with slope.
Each grid square is given a desirability score on the basis of slope.
Then we turn to the next factor and allocate desirability scores. When all factors have been considered, an overall desirability score is allocated to each grid square, so that each square is not assessed as to the likelihood of its being settled, or developed.

(d) It is recognized that at any particular time during the process of growth existing settlement patterns affect subsequent settlement. For example, a square which is next to a settled square increases its probability of being settled. This dynamic element is then built into the model. It may be decided that when a square is settled, all neighbouring squares to up by 1 in their desirability scores.

(e) The process of settlement is now simulated by using a random number generator. The probabilities of settlement are made proportional to desirability scores. The results of some long-run simulations are then compared with the known pattern of settlement.

Of course there is a large degree of arbitrariness in the selection of factors and in the allocation of desirability scores on the basis of them.

Indices

The tradition of geography up until fairly recently has been largely descriptive, with explanations expressed in qualitative rather than quantitative terms. The newer quantitative techniques in school geography courses are demanding a level of mathematical understanding greater than that possessed by many of the teachers. The 'cookbook' approach to statistics has been with us for many years. It is now unfortunately paralleled by an uncritical use of other branches of mathematics.

The first step in applying a quantitative method to a feature of interest is to find a method of measuring the feature in question. Such a measure is often referred to as an index, and modern geography abounds in such indices. The search for a suitable index can be an interesting activity, if engaged in critically, with an awareness of the general properties an index should have in order to be useful.

The examples of indices which I shall give come from the study of networks. This topic too is a popular one in the new geography courses, so I will examine it next. Some indices will arise in the course of this discussion.

Networks

Geographers have adopted the terminology of graph theory (vertices, edges, regions) to describe transportation networks. One aspect of a network which they wish to measure is its degree of connectedness, or connectivity. Here are two networks. Which of them is more connected up?

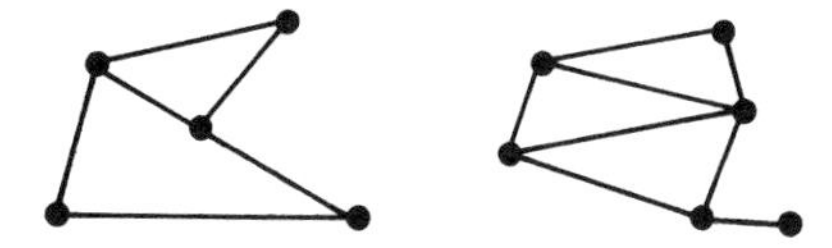

Several indices of connectivity have been used. Ideally we might think it would be a good idea if students first thought about the problem of measuring connectivity and devised methods of their own. At very least we would expect that the thinking behind each of the 'standard' indices would be explained to them. Unfortunately this is not always the case and the formulae for the indices are often just stated and used.

The simplest measure of connectivity is the β - index, defined by E/V. According to the values of this index, this network

$$\beta = \frac{1}{2}$$

is less connected up than this one,

$$\beta = \frac{3}{3} = 1$$

even though in both cases every edge which could be in the network is present.

It would be better to measure connectivity by comparing the acutal number of edges with the maximum number which could be present, for a given number of vertices.

Given V vertices, the maximum number of edges is V(V-1)/2. (For a planar network, i.e. one with no edges crossing one another, the maximum number is 3(V-2).) The index E/1/2 V (V-1) is called the γ-index.

Indices of connectivity are used in geography to compare rail or road networks. It would appear to me that there must be severe limitations to this approach. By treating a rail or road network as a graph in the mathematical sense we are declaring ourselves interested only in the number of vertices, the number of edges, and whether or not a given pair of vertices is linked by an edge. All considerations of distance and relative location do not apply.

The first problem is to identify the edges and vertices of the real network. The representations of networks on fairly small scale maps are very different from their representations on large scale maps. A process of smoothing, or elimination of detail, takes place as the scale is reduced. Not only are minor links eliminated but also vertices. A small scale map might show a single direct link between two towns. A larger scale map shows smaller towns also served by the route. Are they vertices as well? What counts as a vertex? How can we compare the connectivities of networks unless the methods of deciding what counts as a vertex and what counts as an edge are clear? So what happens to objective comparison of networks by means of quantitative measures if there is this much doubt about how the numbers should be arrived at?

Now look at these two hypothetical rail networks? Both are situated on islands.

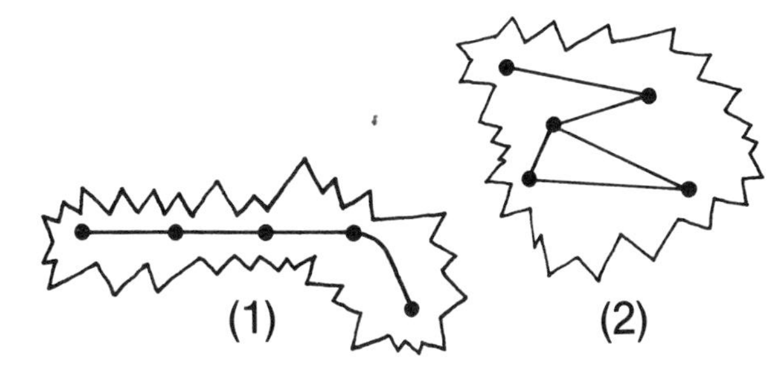

Network (2) has a higher index of connectivity than network (1). Yet (1) is as connected up as one could reasonably expect it to be, but (2) lacks several connections which would improve it. In the mathematical sense of connectedness, (2) is more connected up than (1). So is this idea of connectedness the most relevant one?

My impression is that pupils are not generally being encouraged to think critically about this kind of problem.

Use of Statistics

The quantitative approaches I have just describe sought to measure features of networks so that corresponding features of different networks can be compared in a more precise way than mere qualitative comparison will allow.

The two examples I now describe illustrate the use of statistics. Correlation is the topic most frequently found at the elementary level. Both the exercises I describe involve correlation.

i) The question is posed: How is the value of farming land related to slope? The data consists of the values per hectare for each of a number of fields on a map, and the average angle of slope for each field. The data are plotted on a scatter diagram. It appears that there is a strong tendency for steep slopes to be associated with low land values. There turn out to be some exceptions: there are certain fields with small angles of slope which nevertheless have low values. Reference to the map shows them to be situated in marshy ground.

ii) The question is posed: Why are some market towns larger than others? A hypothesis is formulated: the size of a market town is related to the size of the area it serves and thus to accessibility. To test this hypothesis, data have to be collected. The region chosen was East Anglia. The size of a town was measured by its population. The accessibility was measured by the number of roads entering the town. These were weighted: 2 for a 1st class road, 1 for a second class road. Now the data has to be processed; a scatter diagram was drawn. From the graph it appeared that population is positively correlated with accessibility. A more precise measure of the degree of relationship was wanted, so Spearman's rank correlation coefficient was calculated. The value obtained, 0·67, indicated strong evidence of positive correlation. By calculating $R^2 = 0{\cdot}45$ it could be seen that 45% of the variation in population size can be explained by accessibility.

At this point one should go back and see where this started from. One of the variables was accessibility, defined by counting roads and applying a weighting procedure, and then converted to a rank order. The other was population sizes also reduced to rank orders. Yet the conclusion is that 45% of the variation in population size can be explained by accessibility!

Conclusion

In this talk I have attempted to show you something of what is going on in geography teaching at secondary level, how mathematics enters into it, and how it is being used and misused. It may be that my observations do not apply outside the confines of my own country, so I hope you will use the opportunity afforded by the time remaining to tell us whether the picture is very different in other parts of the world.

Footnote

1. The Mathematics Curriculum: Series published for the Schools Council by Blackie. The volume referred to is entitled Mathematics Across the Curriculum.

LESS PROOFS AND MORE MODELS

Ivo W. Molenaar
University of Groningen
Groningen, Netherlands

Summary

In the Netherlands, passing a final exam of a so-called VWO school (ages 12 to 17/18) is a necessary and sufficient condition for admission to a university. Until 1970, there were two standard forms of such an exam. Since then, the student may almost freely choose seven subjects from a long list that will make up his/her exam program, dropping all other subjects. The various disciplines at the University may require one or two such subjects as compulsory for admission.

Many disciplines, ranging from Mathematics and Physics to Economics and Social Sciences, decided to make Math I such a compulsory subject. This created a serious problem for the VWO schools, as the rather formal contents of Math I are far from ideal for students who only have to see certain mathematical models and tools in their university career. Moreover, the Math I and Math II programs failed to teach enough spatial insight.

The paper discusses two programs Math A and Math B recently proposed by an advisory committee to the Dutch government, of which the former emphasizes the application of mathematics to real world problems as contrasted to formal derivation within an axiomatic system. The success of this program will largely depend on the answer to two "burning questions":

1. How much, and what, mathematics is needed for a major in one of the behavioral and social sciences?

2. Is model-oriented math accessible for people scared of pure math?

In the discussion of these questions, believed to be of common interest to many participants of the ICME Conference, attention is paid to the various abilities playing a role in mathematics, to the "translation" of a real world problem into a mathematical model, to the recruitment conditions under which students decide to study one of the behavioral or social sciences, and to the role of computers and pocket calculators.

1. Some Data on the Dutch Situation

A brief sketch of the school system in the Netherlands will be helpful for appreciating the context in which a revised mathematics curriculum was proposed. It is essential to note the following two features:

a. at the age of 11, a choice is made by the school and the parents jointly which assigns some 10 to 20 percent of all pupils to the school type VWO (since 1970, a later switch to some higher grade within VWO is also possible);

b. the final exam of the VWO, with nationally uniform difficulty level, is necessary and sufficient for admission to any university (with exception of a fixed quota selection for the medical professions).

Until 1970, there were basically only two types of VWO final exams, each entitling admission to certain university careers coupled to them. Since 1970, each VWO pupil is, apart from minor restrictions, free to choose seven subjects to be intensively taught in the last two or three VWO years in which he/she will take a final exam. Each "faculteit" (college, set of departments) of the unviersities is invited by the legislators to name one or at most two of the subjects that would be compulsory for entrance to that "faculteit". As an exception, students defecting this requirement could take a special admission exam.

It is important to note that the new subject Math I was also made compulsory for studying Economics or Social Sciences (which include Psychology, Sociology and Education and Social Work), reflecting the growing use of mathematics in these areas which had admitted students without a math final exam until 1970. Math II was not compulsory for any study, and partly for that reason not frequently chosen as one of the seven subjects.

The requirement of Math I by the "faculteiten" of Economics and Social Sciences was greeted with fierce criticism. Some critics claimed that it was purely a "status affair", others argued that there should be some university studies left for their children withlow mathematical ability. Within the social sciences themselves, the growing mathematization was opposed both by conservative elderly scholars and by marxist oriented students and young teachers, fearing de-humanization and "social engineering" by leading powers. In the VWO schools, the math teachers were faced with growing numbers of students of very models mathematical ability who insisted on taking Math I in order to keep more choices open for a university career. Moreover, there were controversies as to who should organize courses for the so-called "deficient" students who missed a subject compulsory for the study they desired to follow, and as to what topics an exam for the deficient students should consist of.

Several departments within the Natural Sciences and Engineering & Technology had a different complaint about the new Math I and Math II programs. They admitted students with Math I or with both Math I and Math II, but even the latter were found to exhibit a serious lack of spatial insight. The almost complete replacement of Euclidean geometry, both in the plane and in the space, by analytical geometry and linear algebra led to a strong tendency to solve all problems by manipulating coordinates without visualizing the geometrical aspects.

Between 1973 and 1976, the "Math I problem" was intensively debated between VWO teachers, university administrators and politicians, and several internal reports were brought out suggesting solutions. finally in May 1978, the Dutch Government nominated a committee of ten "wise men", both from VWO schools and from universities, who should propose a "reallocation of topics" between Math I and Math II, including new suggestions for making them compulsory for certain university studies. A draft proposal was intensively discussed in schools and universities in the spring and summer of 1979, and a final report appeared at the end of that year. Although the committee had advised to start the retraining of teachers and the development of revised text books in 1980, it is to be feared that the government budget cuts forced by the economic crises will lead to a serious delay.

This author was both one of the ten committee members and president of the earlier report writing groups. The general philosophy behind the revised curriculum will be outlined in this paper, because the problem of modernizing a math program is believed to be of common interest to participants from many countries at this ICME conference.

2. The Revised Program: Some Burning Questions

After considering several alternatives like abolishing the Math I requirement for Social Sciences, or the installation of a new Math III subject at the same or a lower difficulty level, possibly combined with an abolition of Math I or Math II, the committee has decided to propose a replacement of the two existing programs in the VWO by two new ones, Math A (applied) and Math B (fundamental).

The committee proposes that Math A will be compulsory for studying Economics (for Economics and Econometrics, Math B will probably also be a valid entry ticket), Social Sciences (except possibly Education and Social Work, but including Social Geography), while it might also be useful for Biology, Medicine and Agriculture. On the other hand Math B should be compulsory for studying Mathematics, Astronomy, Physics, Chemistry, Geology, Physical Geography and all Engineering and Technology studies. Within each of the four rather different topics included in Math A, the emphasis should be on the application of mathematics to real world problems, as contrasted to formal derivation within an axiomatic system.

The committee expects that most Math B students will be taking a substantial amount of mathematics courses during their later university career, in which they will be taught most of the specific mathematics or computer science topics that they need. It is essential that their capacities for formal reasoning and spatial insight be developed during their VWO school years.

Most Math A students, on the contrary, will use some mathematical models in their university study, but hardly be taught math there. It seems desirable that their VWO program helps them to

- become familiar with the language of math, with expression in formulas, with graphical representations.
- work with mathematical models
- judge the relevance of such a model (H. Freudenthal, Weeding and Sowing, Preface to a Science of Mathematical Education, Reidel, Dordrecht/Boston, 1978).

The committee has spent much time debating whether a student of Economics or Social Sciences needs all the mathematics involved in such a program, and whether he/she can successfully complete it. This brings us to the following two "burning questions":

1. How much mathematics is needed for a major in Economics, Econometrics, Psychology, Sociology, Education, Social Work, Social Geography?
2. Is model-oriented math accessible for people scared of pure math?

Let us examine these two questions in more detail.

The first question is immediately seen to be falsely formulated, even though it was formulated in exactly these terms by almost all government officials and university administrators who met the "Math I problem". It is illustrative to compare it to the question: how much money is needed for staying 24 hours in San Francisco? Among the immediate answers are:

the more the better,

depends on who you are,

depends on the season,

depends on who you know there,

paid by your boss or yourself?

Coming back to the amount of math rather than money, it is clear that the time spent on teaching math should be compared to the time spent on teaching other useful skills like English, history, politics, report writing, etc. Moreover, in each of the fields mentioned in question 1 one may distinguish a variety of sub-fields of which some use mathematics intensively and at an advanced level, whereas other subfields avoid it to a large extent. Even within a subfield like business economics, clinical psychology or family sociology there will be considerable divergence of opinions on the importance of mathematics for that subfield.

Things get worse when question 1 is pinned down to "exactly what mathematics is needed for a major in . . .". The question may seemingly be evaded by taking the position: "It does not matter that much exactly what, as long as the students develop some general mathematical maturity and mathematical ability". This position is appealing at first sight, but it presupposes that such a maturity and ability is a one-dimensional trait which can be transferred without loss from one part of math to another. There is a parallel with intelligence, which on closer looking can be diversified into verbal, spatial and arithmetical intelligence, and even further.

Mathematics distinguishes itself from many other disciplines by the relatively minor role of the ability to reproduce knowledge previously learnt by heart - an ability which in general is less cherished in today's schools as compared to fifty years ago. Seeing somebody else's finished solution of a math problem is of limited help when it comes to actively solving a problem: the ingenuity of a mathematician is closely tied to his/her ability to creatively decide which of the many possible roads within the system will quickly lead to a solution, and which roads will turn out to be dead alleys.

The role of creativity and imagination becomes still more dominant when one agrees on the intricate role of mathematical methods in solving a "real world problem" in some other discipline.

3. Math and Social Sciences: The Second Burning Question

The following quotation from "Making Artists out of Pedants" by James G. March (The Process of Model-Building in the Behavioral Sciences, Edited by Ralph M. Stogdill, 1970, ch. 4) is relevant to us:

> I consider it a self-evident proposition that an increase in the amount of creative model building carried on within the social and behavioral sciences would be desirable. Patently, the proposition is not self-evident except by proclamation. The point to the proclamation is simply to avoid the necessity of arguing the case and to turn directly to the question of means.
>
> We wish to modify our program so that the next generation of scholars will be radically less incompetent than we are. We can accomplish this objective through some mix of two basic strategies:
>
> 1. We can change our style of teaching and the content of what we teach.
>
> 2. We can change the raw material (i.e., the kind of students) that we teach.
>
> These are not independent strategies; generally one reinforces (or interferes with) the other. A change in curriculum, teaching style, or course content changes the decisions of students with respect to interests. A change in the kind of students creates pressure for us to make our program consistent with student interests and capabilities."

March next argues that recruitment for the behavioral and social sciences suffers from three major disabilities:

1. Virtually nothing from those disciplines is taught in the first twelve years of school.

2. The skills required for them are far more unique to them.

3. "The social norm press towards social science tends to be antianalytical. The behavioral sciences are associated (quite appropriately) with human beings and social problems. As a result, they are associated (quite inappropriately) with a rejection of things, quantities, abstractions. They tend (except for economics and political science) to be relatively 'feminine'".

The major emphasis of the program proposed by March is to develop the artistry of thinking analytically about social sciences.

This brings us back to our second burning question: may we expect that a student obtaining moderate or low grades in a traditional purely formal math high school program will be more successful in a revised program in which model building and applications are the central theme? This is a real world problem that can definitely not be solved by formal reasoning, but it may clarify our thoughts to distinguish in more detail some abilities having to do with mathematics.

The present author believes that the impact of the advent of computers and calculators, as well as the use of math in previously non-mathematical fields, will strongly influence the way in which math is taught to the next generation. Let us also note in passing the importance of early childhood experiences for the development of spatial abilities, as argued by Piaget and other developmental psychologists.

A mathematical problem as formulated in a traditional textbook is a purely formal puzzle, which is a source of excitement and an incitement to students fond of such puzzles. The majority of the present student generation, however, experiences little motivation or satisfaction in the solution of such puzzles, and this generation demands that their teachers provide motivation because authority and tradition are not compelling for them. For at least some students now scared of math, stressing its usefulness and its logic in use rather than its formal logic in reconstruction may provide such a motivating factor.

Let us take the exponential and logarithmic functions as an example. They are introduced in a popular Math I textbook via an ingenious argument about the Riemann sums of the function $f(x) = 1/x$, leading via $\log(ab) = \log a + \log b$ to $\exp(a)\exp(b) = \exp(a + b)$ and to the derivatives of $\log x$ and $\exp(x)$. It may keep more pupils awake, however, to imagine what happens when the growth of a plant (a biological or a business one) in time is proportional to the size at the moment. Measured reaction times in a psychological task are almost invariably analyzed via an additive analysis of variance model, but is not it more plausible that an experimental manipulation will produce a decrease of the previous reaction time by a fixed factor, rather than by a fixed amount? If a new method of teaching geography is expected to increase performance as compared to a control group, surely there is not the same additive increase in the scores on a 40-item multiple choice test, but such scores must be linked to the latent ability via a logistic model (a detailed explanation of such a Rasch model would be the subject for another talk).

Students lacking creativity, tidiness, geometrical insight or logical thinking will fail on the Math A program as well as they did on the Math I one. Maybe they should not try to obtain a university degree in one of the social sciences. Brilliant mathematicians will hopefully go on to show increased interest in the social sciences, and provide them with brilliantly new models and analysis techniques. There is a large class of students between the two extremes, however, for which I can only hope that they will become competent social scientists by using less proofs and more models in their mathematics curriculum.

SOCIAL SCIENCE APPLICATIONS IN THE UNDERGRADUATE MATHEMATICS CURRICULUM

Samuel Goldberg
Oberlin College
Oberlin, Ohio, U.S.A.

1. Introduction

The worth or utility of mathematical models and methods for the study and understanding of social pressures and human behavior is best left for social and behavioral scientists, not mathematicians, to judge. But the record is clear that social scientists are in fact making substantial and ever-increasing use of the theories and techniques of the mathematical sciences. (When speaking of the social sciences, we mean to include anthropology, economics, demography, history, linguistics, political science, psychology, and sociology. Applied and even basic social science research is also done under such headings as management science, operations research, systems analysis, and decision analysis. Boundaries between fields are blurred and much current work is interdisciplinary.) Our topic is very broad indeed, for it deals with the interaction between two extraordinarily broad fields. We limit ourselves in this paper to (1) a brief survey of the literature of mathematical social science, emphasizing books and journal articles of special value to teachers of mathematics, (2) a presentation of a decision problem arising in genetic counseling to illustrate the use of mathematics in the social sciences, and (3) some comments on pedagogical implications for teachers of mathematics and for the mathematics curriculum.

2. Background Studies

Advantages, difficulties, and abuses involved in the interaction between mathematical and social sciences have been the subjects of many authors. The Gibbs lecture delivered by Fisher (1930) and the paper titled "Mathematical Adventures in Social Science" by Davis (1938) now have only historical interest. More noteworthy are the contributions of Arrow (1952) and Samuelson (1952), the remarks by Lazarsfeld (1955) made at the end of a symposium on mathematical models of human behavior, by Miller (1957) on "mathematics and thinking" in his brief summary of mathematics in social psychological research, and by Sibley (1963) on the need for mathematical training of sociologists.

As part of the 1967-68 Survey of the Behavioral and Social Sciences organized jointly by the National Academy of Sciences and the Social Science Research Council, a report (Kruskal, 1970) issued by the Mathematical Sciences Panel discusses the relevance of mathematics, statistics, and computation for the social sciences, the problem of data collection, the relation of statistics to public policy, the importance of modern computing facilities, and the problems associated with cooperation, in both teaching and research, between social scientists and mathematicians. The critique of some mathematical work in the social sciences made by Berlinski (1976), as well as the cautionary comments by Coser (1975) in his American Sociological Association Presidential Address show that in spite of many successes, the use of mathematics in the social sciences continues as a source of some anxiety.

Of greater value and worthy of being much better known by teachers of mathematics are the many monographs and textbooks on mathematical social science. A very small sample would include works in anthropology (Ballonoff, 1976; Kay, 1971), in demography (Keyfitz, 1968; Pollard, 1973), in economics (Allen, 1966; Benavic, 1972; Franklin, 1980; Gandolfo, 1971), in history and archaeology (Hodson et al. 1971; Swieringa, 1970), in linguistics (Brainerd, 1971; Partee, 1976), in political science (Alker et al., 1973; Brams, 1975; Riker and Ordeshook, 1973), in psychology (Coombs et al., 1970; Krantz et al., 1974), in sociology (Beauchamp, 1970; Coleman, 1964; Fararo, 1973; Leik and Meeker, 1975; Mapes, 1971), and a number of more general books (Kemeny and Snell, 1962; Kim and Roush, 1980; Lave and March, 1975; Lazarsfeld and Henry, 1966; Maki and Thompson, 1973; Olinick, 1978; Roberts, 1976; Selby, 1973) that cut across a number of fields.

Also noteworthy is the guide to the literature of mathematical sociology at the end of Fararo's textbook, and the review articles on mathematics in anthropology by White (1972), in geography by King (1970), in political science by Fagen (1961), Fiorina (1975), and Taylor (1971), in psychology by Luce (1964) and Rapoport (1976), and in sociometry by Roistacher (1974).

3. An Illustrative Example: A Decision Problem

A thirty-five year old woman is pregnant for the second time. Her first child is a healthy six year old, but she and her husband are concerned about the possibility of this pregnancy resulting in a mentally retarded newborn with Down's syndrome ("mongolism"). Their obstetrician tells them that the likelihood of having a child affected by Down's syndrome increases with maternal age, is not negligible at age 35, and so they should consider amniocentesis. If a positive diagnosis could be made prenatally, then the couple would have the option of electing to have a therapeutic abortion.

Amniocentesis involves inserting a needle into the uterus, then withdrawing, culturing, and examining a small amount of fluid surrounding the fetus. Unfortunately, this diagnostic test is not perfect: affected fetuses are sometimes erroneously classified as healthy and unaffected fetuses as affected. The procedure also slightly increases the risk of a miscarriage.

What medical information should the couple seek from the obstetrician before making the decision to have or reject amniocentesis? How do their values and preferences enter the analysis? Is the decision properly the couple's or the doctor's? These are the kinds of questions discussed by Pauker and Pauker (1977) and this is the way we have presented this example to students.

It is natural to summarize the situation in the form of a decision tree. There is only a single decision node at the start of the tree and then all subsequent branching occurs at chance nodes with probabilities specified as in Figure 1. The error probabilities e_1 and e_2 are known for the laboratory doing the test, typical values being approximately 0.005. The doctor is able to make some judgements about the values of s, a, and p for this particular patient.

Figure 1. Decision Tree for Amniocentesis Example.

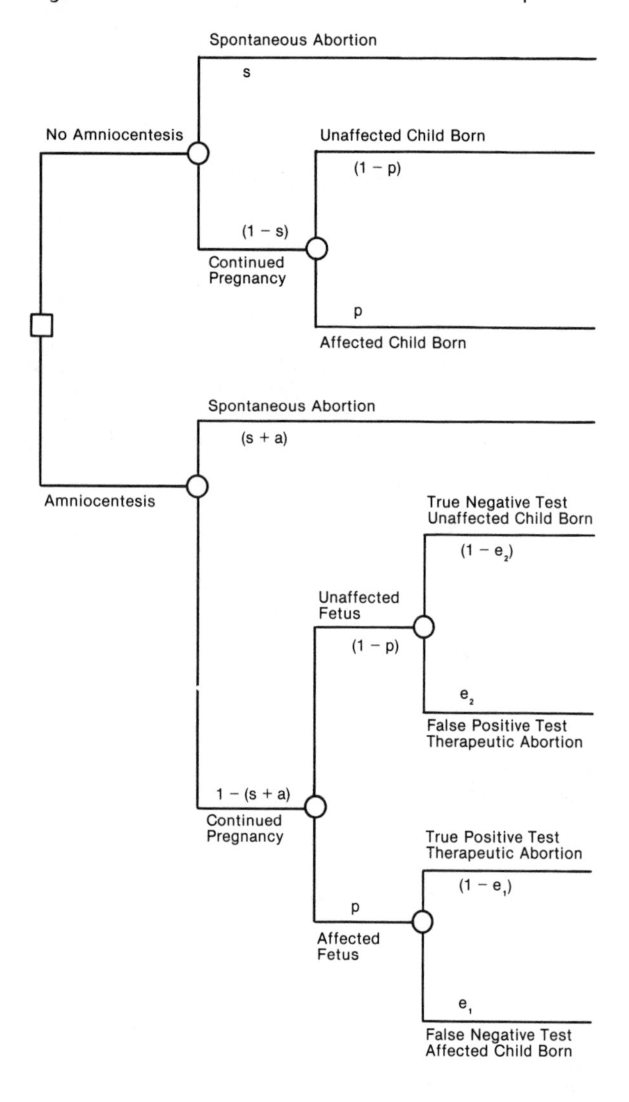

Students are quick to observe that these probabilities are only able to be approximated and thus that some sort of sensitivity analysis is needed. They also realize that the probabilities are not sufficient to determine the optimal decision. It is also necessary for the doctor to develop some means of assessing the parents' attitudes, expressed quantitatively by utilities or relative costs, toward the five possible outcomes: birth of an unaffected child, spontaneous abortion, therapeutic abortion of an affected fetus, therapeutic abortion of an unaffected fetus, and birth of an affected child. It is clear the amniocentesis becomes less preferred as an option as the parents' attitude toward having a therapeutic abortion becomes more negative.

Assessing parental preferences is a difficult task, but follows the usual protocol for determining a decision-maker's utility function. (For details, see Holloway, 1979; Raiffa, 1968.) Specifying the best and the worst outcome is relatively easy for the parents and one can arbitrarily assign a cost of 0 to the best (say, unaffected child) and a cost of 1 to the worst outcome (say, affected child). Obtaining the parents' answers to a sequence of questions concerning their preferences in lotteries involving pairs of possible outcomes allows the assignment of relative costs to the remaining three outcomes. (See Figure 2.) Then a straightforward computation of expected costs associated with each of the two branches of the decision node leads to the determination of a threshold probability T such that the optimal decision is to have amniocentesis if and only if the probability p exceeds T. Letting $r = a/(1 - s)$, one finds

$$T = \frac{(1 - r)e_2C_{TU} + rC_s}{1 + (1 - r)[e_2C_{TU} - (1 - e_1)C_{TA} - e_1]}$$

Figure 2. Utilities and Relative Costs.

A sample set of utilities of the five possible outcomes is indicated on the horizontal axis, with U_A and U_U arbitrarily assigned values 0 and 1, respectively. The relative cost of any outcome X is the difference between $U_U = 1$, the utility of the best outcome, and the utility of X, i.e., $C_X = 1 - U_X$. Outcomes: U = unaffected child born, S = spontaneous abortion, TA = therapeutic abortion of affected fetus, TU = therapeutic abortion of unaffected fetus, A = affected child born.

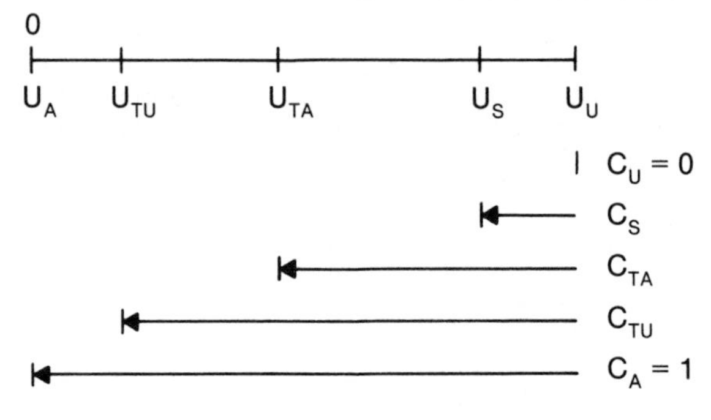

In the Pauker and Pauker article already cited, one can find an interpretation of this solution as well as an analysis of the sensitivity of the optimal decision to the various probabilities and relative costs appearing in the formula for T. There is also a brief discussion of the legal and social burden of informed consent in genetic counseling and the following mention of a very important policy question:

> . . . if facilities for processing amniotic fluid continue to be expensive and of limited availability, then resource constraints will limit the total number of amniocenteses which can be performed. In such circumstances it would not be appropriate to allow the parents' attitudes to be the only utilities considered; overall societal good should also be considered explicitly.

Thus, this problem leads naturally to the consideration of deep social, economic, and ethical issues for which the simple decision tree in Figure 1 is quite inadequate. A similar situation, in which optimal medical care for an individual patient may not remain optimal from the point of view of society, arises in a decision analysis of coronary artery bypass surgery (Weinstein et al., 1977).

Applications of expected value from probability and of utility theory to such decision problems have many features that commend their use in the classroom. These problems arise not only in health care (Krischer, 1980), but also in business management (Brown et al, 1974) and in many areas of public policy (Keeney ad Raiffa, 1976). They can be presented to the student initially in nonmathematical prose form and so offer useful practice in structuring of a problem in the form of a decision tree. Students easily understand that probabilities and utilities need to be assessed and since this leads only to approximate values, the important idea of sensitivity analysis arises naturally. Finally, the limitations of the model and the need to perform a more complex analysis become apparent when deeper questions, involving multiattribute objectives or group rather than individual decision making, are raised.

4. Pedagogical Remarks

Mathematics courses, textbooks, and teachers tend still to be oriented mainly to the physical sciences and engineering. However, there have been recent signs that the mathematics community is paying more attention to the curriculum needs of social science students. (The role of the Social Science Research Council in the advance of mathematics in the social sciences is discussed by Mosteller, 1974.)

The Committee on the Undergraduate Program in Mathematics (CUPM, 1964) of the Mathematical Association of American recommended that the first two years of college work for students of the biological, management, and social sciences include a mix of calculus, linear algebra, differential and difference equations, probability, and the rudiments of computer programming. The committee also recommended additional work in probability and statistics.

Another CUPM Panel (1972) made two important recommendations.

> The first recommendation of the Panel is that each department should offer a course or two in applied mathematics which treat some realistic situations completely, beginnng with a careful analysis of the nonmathematical origin of the problem; giving extremely careful consideration to the formulation of a mathematical model, solution of the mathematical problem, and relevant computations; and presenting thoughtful interpretations of the theoretical results to the original problem. In other words, there should be a few courses which give students the experience of grappling with a entire problem from beginning to end.

The Panel gave detailed syllabi for three such applied courses. The first two, an optimization and on graphs and combinatorics, are of special interest and value for social science students. Such courses are becoming more popular, but they are still relatively rare in undergraduate programs of study.

> The second recommendation of the Panel is that a greater number of realistic applications from a greater variety of fields be introduced into the mathematics courses of the first two years.

Some recent writing efforts make this advice that can be easily implemented. Sixty short educational modules, about a third of which deal with social science topics, were produced by a College Faculty Workshop entitled Modules in Applied Mathematics, held at Cornell University during the summer of 1976 under the directorship of W.F. Lucas. Similar modules were prepared in another National Science Foundation supported project in 1977 at the University of Illinois. Many useful teaching materials have been produced by UMAP (Modules and Monographs in Undergraduate Mathematics and its Applications Project.) The new UMAP Journal and many lesson-length and self-contained modules in mathematics and statistics and their applications are now published by Birkhauser Boston, Inc. A joint project of CUPM and the Mathematical Social Science Board (Goldberg, 1977) has produced fifteen expository essays illustrating the use of undergraduate mathematics in the social sciences and supplying fairly complete bibliographies. These are only a small sample of the materials recently produced in order to make it easier and hence more commonplace for writers of mathematics textbooks to draw on the social sciences for applications and to encourage teachers to use such examples in their classrooms. But the cooperation of social science departments is also needed. Courses and course-sequences need to be developed to show beginning students that the mathematical ideas and techniques they learn are actually used in the social science curriculum

But students need even more realistic exposure to applications than can be included as part of traditional mathematics courses. The Ad Hoc Committee on Applied Mathematics Training of the National Research Council in its recent report (Hilton, 1979) recommended:

> That experience in applications (including internships, where feasible; courses of case studies; and senior research project) be an integral part of the program for the major in the mathematical sciences.

The Claremont Mathematics Clinic (Spanier, 1976, 1977) is the prototype for this type of applications-experience program.

An interdisciplinary effort is required that will rethink the nature of an undergraduate course and allow experimentation designed to train social science and mathematics students to deal with real-world problems. Beginning graduate students in operations research departments and in schools of business and public administration regularly join faculty in real applied social research, often sponsored by government agencies, municipalities, and industrial companies. Why is it not possible in a much more modest fashion to duplicate this effort for undergraduates as a culmination of their work in mathematics and social sciences? Faculty from a number of departments, including persons who can contribute mathematical, statistical, computing, and social science expertise, can join to offer a workshop course devoted to such real-world problem solving. Students would get valuable experience in the art of selecting a manageable part from a problem that is clearly unmanageably large. They would learn that real data is messy, often unavailable or only partly available, and expensive in time and money to collect reliably. They would surely have to use the computer and thus give some thought to the trustworthiness of their computer output. They would get experience in developing a suitable mathematical model for the situation under study or in adapting a textbook technique for their purposes. They would learn that often simple mathematics is all one needs or can use since the problem is too complex and only a part of it can be approached in a reasonable period of time. They would recognize that some progress in solving even part of a problem and making appropriate policy suggestions is better than no progress on the large unwieldly problem with which they began. And finally, the would realize that all their effort is for naught unless they carefully consider the client, the user of their analysis and receiver of their recommendations. For they will want to conduct themselves during the study and when they prepare their report, whether written or oral, in such a way as to make implementation a reality. These are important goals and not easy to realize. We do not really attempt to teach them as the curriculum is presently structured. But they are worthy of our efforts, mathematicians and social scientists alike, as we think about the education of the next generation of students.

Finally, we note that social science problems are difficult. In a symposium on the future of applied mathematics, this point was made by Cohen (1972):

> . . . we have so far mostly dealt with relatively simple systems. When we turn to sciences others than the physical sciences we are confronted with greater complexity in structure and relationships, a larger number of variables, and often, it seems, a higher degree of interdependence and hence nonlinearity.

Because of this, we need to be especially careful not to be carried away by enthusiasm for the mathematical method. After all, it is good sociology, good psychology, etc., that is needed and the use of existing mathematics or the invention of new mathematics by researchers doing social science needs to be judged with that in mind. This point allows me to conclude with a literary flourish. According to his biographer (Hone, 1962), the poet W.B. Yeats in an imaginary letter to a school master wrote as follows concerning his son:

> Teach him mathematics as thoroughly as his capacity permits. I know that Bertrand Russell must . . . be wrong about almost everything else, but as I have no mathematics I cannot prove it. I do not want my son to be as helpless. (p. 420)

A slight paraphrasing: we need sons and daughters, men and women social scientists, well trained in mathematics (and mathematicians well trained in social science), so they can not only advance their subject, but also make wise and knowledgeable judgements about the worth of the more mathematical research in the social sciences - so they will not feel helpless like Yeats.

Bibliography

Aiker, H.R. Jr., K. Deutsch, and A.H. Stoetzel, eds. Mathematical Approaches to Politics. San Francisco: Jossey Bass, 1973.

Allen, R.G.D. Mathematical Economics. 2nd ed., London: Macmillan, 1966.

Arrow, K.J. "Mathematical Models in the Social Sciences." pp. 139-154 in Lerner, D. and H.D. Lasswell, eds., The Policy Sciences. Stanford: Stanford University Press, 1952.

Ballonoff, P.A. Mathematical Foundations of Social Anthropology. Paris: Mouton, 1976.

Beauchamp, M.A. Elements of Mathematical Sociology. New York: Random House, 1970.

Benavie, A. Mathematical Techniques for Economic Analysis. Englewood Cliffs: Prentice Hall, 1972.

Berlinski, D. On Systems Analysis: An Essay Concerning the Limitations of Some Mathematical Methods in the Social, Political, and Biological Sciences. Cambridge: MIT Press, 1976.

Brainerd, B. Introduction to the Mathematics of Language Study. New York: American Elsevier, 1971.

Brams, S.J. Game Theory and Politics. New York: Free Press (Macmillan), 1975.

Brown, R.V., A.S. Kahr, and C. Peterson. Decision Analysis for the Manager. New York: Holt, Rinehart and Winston, 1974.

Cohen, H. "Mathematical Applications, Computation, and Complexity," Quarterly of Applied Mathematics, Vol. XXX (1972), pp. 109-121.

Coleman, J.S. Introduction to Mathematical Sociology. New York: Free Press, 1964.

Committee on the Undergraduate Program in Mathematics. Tentative Recommendations for the Undergraduate Mathematics Program of Students in the Biological, Management and Social Sciences. Berkeley, 1964.

Committee on the Undergraduate Program in Mathematics. Applied Mathematics in the Undergraduate Curriculum. Berkeley, 1972.

Coombs, C.H., R.M. Dawes, and A. Tversky. Mathematical Psychology, An Elementary Introduction. Englewood Cliffs: Prentice-Hall, 1970.

Coser, L. Presidential Address. American Sociological Review, Vol. 40, (1975), pp. 691-700.

Davis, H.T. "Mathematical Adventures in Social Science." American Mathematical Monthly, Vol. 45 (1938), pp. 93-104.

Fagen, R.R. "Some Contributions of Mathematical Reasoning to the Study of Politics." American Political Science Review, Vol. 55 (1961), pp. 888-900.

Fararo, T.J. Mathematical Sociology. New York: Wiley, 1973.

Fisher, I. "The Applications of Mathematics to the Social Sciences," Bulletin of the A.M.S., Vol. 36 (1930), pp. 225-243.

Franklin, J. Methods of Mathematical Economics (Linear and Nonlinear Programming, Fixed-Point Theories). New York: Springer-Verlag, 1980.

Gandolfo, G. Mathematical Methods and Models in Economic Dynamics. Amsterdam: North-Holland, 1971.

Goldberg, S., ed. Some Illustrative Examples of Undergraduate Mathematics in the Social Sciences. Haward, California: Mathematical Association of American, Special Projects Office, 1977.

Hilton, P. The Role of Applications in the Undergraduate Mathematics Curriculum. Report of the Ad Hoc Committee on Applied Mathematics Training, National Research Council. Washington, D.C.: National Academy of Sciences, 1979.

Hodson, F.R., D.G. Kendall, and P. Tautu, eds. Mathematics in the Archaeological and Historical Sciences. Edinburgh: Edinburgh University Press, 1971.

Holloway, C.A. Decision Making Under Uncertainty: Model and Choices. Englewood Cliffs: Prentice-Hall, 1979.

Hone, Joseph. W.B. Yeats 1865-1939. 2nd ed., London: Macmillan, 1962.

Kay, P. Explorations in Mathematical Anthropology. Cambridge: MIT Press 1971.

Keeney, R.L. and H. Raiffa. Decisions & Multiple Objectives: Preferences and Value Tradeoffs. New York: Wiley, 1976.

Kemeny, J.G. and J.L. Snell. Mathematical Models in the Social Sciences. Boston: Ginn, 1962.

Keyfitz, N. Introduction to the Mathematics of Population. Reading: Addison-Wesley, 1968.

Kim, K.H. and F.W. Roush. Mathematics for Social Scientists. New York: Elsevier North Holland, 1980.

King, C.A.M. "Mathematics in Geography." International Journal of Mathematics Education, Science, and Technology, Vol. 1 (1970), pp. 185-205.

Krantz, D.H., R.D. Luce, R.C. Atkinson, P. Suppes, eds. Contemporary Developments in Mathematical Psychology, Vol. 1, Learning, Memory, and Thinking, Vol. 2, Measurement, Psychophysics, and Neural Information Processing. San Francisco: Freeman, 1974.

Krischer, J.P. An Annotated Bibliography of Decision Analytic Applications to Health Care," Operations Research, Vol. 28 (Jan. - Feb. 1980), pp. 97-113.

Kruskal, W., ed. Mathematical Sciences and Social Sciences. Englewood Cliffs: Prentice-Hall, 1970.

Lave, C.A. and J.G. March. An Introduction to Models in the Social Sciences. New York: Harper & Rown, 1975.

Lazarsfeld, P. "Concluding Remarks." pp. 97-103, in Mathematical Models of Human Behavior, Proceedings of a Symposium. Standfor, Conn.: Dunlap and Associates, 1955.

Lasarsfeld, P.F. and N.W. Henry, eds. Readings in Mathematical Social Science. Cambridge: MIT Press, 1966.

Leik, R.K. and B.F. Meeker. Mathematical Sociology. Englewood Cliffs: Prentice-Hall, 1975.

Luce, R.D. "The Mathematics Used in Mathematical Psychology." American Mathematical Monthly, Vol. 71 (1964), pp. 364-378.

Maki, D.P. and M. Thompson. Mathematical Models and Applications. Englewood Cliffs: Prenticee-Hall, 1973.

Mapes, R. Mathematics and Sociology. London: Batsford, 1971.

Miller, G.A. "Applications of Mathematics in Social Psychological Research." SSRC Items, Vol. 11 (1957), pp. 41-44.

Mosteller, F.. "The Role of the Social Science Research Council in the Advance of Mathematics in the Social Sciences." SSRC Items, Vol. 28 (1974), pp. 17-24.

Olinick, M. An Introduction to Mathematical Models in the Social and Life Sciences. Reading: Addison-Wesley, 1978.

Partee, B.H. Mathematical Fundamentals in Linguistics. Stanford, Conn.: Greylock, 1976.

Pauker, S.P. and S.G. Pauker. "Prenatal Diagnosis: A Directive Approach to Genetic Counseling Using Decision Analysis," The Yale Journal of Biology and Medicine, Vol. 50 (1977), pp. 275-289.

Pollard, J.H. Mathematical Models for the Growth of Human Populations. Cambridge: Cambridge University Press, 1973.

Raiffa, H. Decision Analysis, Introductory Lectures on Choices and Uncertainty. Reading, Mass.: Addison-Wesley, 1968.

Rapoport, A.. "Directions in Mathematical Psychology I." and "Directions in Mathematical Psychology, II." American Mathematical Monthly. Vol. 83 (1976), pp. 85-106 and pp. 153-172.

Riker, W.H. and P.C. Ordeshook. An Introduction to Positive Political Theory. Englewood Cliffs: Prentice-Hall, 1973.

Roberts, F.S. Discrete Mathematical Models with Applications to Social, Biological, and Environmental Problems. Englewood Cliffs: Prentice-Hall, 1976.

Roistacher, R.C. "A Review of Mathematical Models in Sociometry." Sociological Methods & Research, Vol. 3 (1974), pp. 123-171.

Samuelson, P.A. "Economic Theory and Mathematics - An Appraisal." Papers and Proceedings, American Economic Review, Vol. 42 (1952), pp. 56-66.

Selby, H.A. Notes of Lectures on Mathematics in the Behavioral Sciences. Washington, D.C.: Mathematical Association of American, 1973.

Sibley, E. The Education of Sociologists in the United States. New York: Russell Sage Foundation, 1963, especially pp. 134-145.

Spanier, J. "The Mathematics Clinic: An Innovative Approach to Realism Within an Academic Environment," American Mathematical Monthly, Vol. 83 (1976), pp. 771-775.

Spanier, J. "Education in Applied Mathematics, The Claremont Mathematics Clinic," SIAM Review, Vol. 19 (1977), pp. 536-549.

Swierenga, R.P., ed. Quantification in American History, New York: Atheneum, 1970.

Taylor, M. "Review Article: Mathematical Political Theory." British Journal of Politcal Science. Vol. I (1971), pp. 339-382.

Weinstein, W.D., J.S. Pliskin, and W.B. Stason. "Coronary Artery Bypass Surgery: Decision and Policy Analysis," pp. 342-3371 in Bunker, J.P., B.A. Barnes and F. Mosteller, eds. Costs, Risks, and Benefits of Surgery. New York: Oxford University Press, 1977.

White, D.R. "Mathematical Anthropology," in J.J. Honigmann, ed. Handbook of Social and Cultural Anthropology. Chicago: Rand-McNally, 1972.

8.3 APPLICATIONS, MODELING AND TEACHER EDUCATION

TEACHING AS BASED ON INTERDISCIPLINARY MATHEMATICAL MODELS

Aristides Camaragos Barreto
Pontificia Universidade Catolica do Rio de Janeiro, Brazil

The start of this decade saw mankind surprised and astonished in the face of his impotence to face basic problems of survival. The energy crisis provoked a discontinuity that changes the flow of history in a still unknown way. We do not have any realistic program to face the greatest challenges by the end of our century: the sun and the cell, that is, energy and biology.

Paradoxically, while this is ocurring, the scientific knowledge grows exponentially (de Solla Price, 1963). In particular, biology is the fastest developing science.

How are we to take advantage of this progress? How could both the available pure and applied sciences become applicable science?

Only through education, in its widest sense, as UNESCO has recognized in 1976.

Also according to T.W. Schultz, the 1979 Economics Nobel Prize Winner, education is the main investment in human capital (Schultz, 1963).

Teachers must prepare their students to live in the next decade. This obvious fact becomes more and more challenging in periods of discontinuity, as now. Education - and particularly mathematical education - must turn itself into a global multidiscipline. It must be characterized by the feature of anticipation, which means sensitivity to possible future events. Besides arts, both education and mathematical education must become definitively applicable sciences.

Nowadays it is commonplace to say that the objective of mathematical education at all levels is to develop in the learner the triad: knowledge-ability-attitude, with special emphasis on the formation of attitude. This triad, knowledge-ability-attitude, also dominates the technological processes. This is certainly not by chance.

Knowledge is the full set of concepts, tools and methods the student must acquire (what?); ability is the capacity of applying knowledge in a efficient way (how?); and attitude is the critical sense, both of analysis and synthesis (when, where, why and for what?).

The knowledge to be gained must result from a judicious selection of their interests, necessities and social and professional priorities, in their time and social environment.

The improvement of abilities constitutes a step inseparable from the acquirement of knowlege. The failure to appreciate this important issue was one of the distortions committed by the so-called "new mathematics".

Educational psychologists agree on the major importance of the phenomenon called "transfer of learning". This consists of learning in a certain situation in order to use the eventually adapted or generalized outcomes in new situations later. As such, transfer is the basis to both effective problem solving capacity and, in a wider context, improvement of creativeness.

There are two main barriers against transfer:

a. A certain inertia that comes from the original association of every knowledge with the specific motivation received when learned.

b. The fact that transfer from a past situation to a new one is not in general immediate, but depends strongly on the pedagogic means.

"The best way to teach any activity is by showing it" (Comenius, a seventeenth century pedagogist). Furthermore, according to H. Freudenthal, "the best way to learn any activity is doing it".

Let us consider as activity a significant application of mathematics, in the sense of mathematization of a given situation: situation - abstraction.

In this sense the strategy of teaching based on interdisciplinary mathematical models aims at enabling the learner to apply mathematics at all areas and levels. This strategy consists firstly of anticipating situations, chosen from the field of the student's interests. These situations will be passed through a process of mathematization in parallel with the theory approached along the course and with their effective participation, either individually or in small groups.

Hence this strategy is a dynamic one, fitted in the category of mechanical learning, as opposed to rote learning in Ausubel's classification of types of learning

(Ausubel, 1968). Furthermore it satisfies Ausubel's principle of integrative reconciliation, because of requiring a step of analysis of different knowledges before the final synthesis in the model is performed.

A model can be either local or global, depending on whether it embraces only a few or many topics in the course. Undoubtedly, the choice of a global model excludes neither local models nor submodels of the original one.

Formulating a mathematical model is, in general, an open answer problem. As such, it helps to contradict the usual prejudice among students that the solution to a problem of mathematics must be unique and well determined. On the other hand, looking for solutions can suggest, naturally, some kind of either algorithms or statistical and numerical methods. Hence in some way there exists the possibility of interaction among subareas of mathematics.

The more inclusive the model, the more transfer is reinforced.

An inseparable step in constructing a model is its validation, that is, to come back to the situation: situation $\leftarrow$- abstraction. When the fit of the outcomes of the model to the available data is unsatisfactory, we must adjust or perhaps remake the model, through a sequence of better and better approximations. In the case that there exists more than one solution, we must discuss them comparatively and finally make a decision. Both discussing and making a decision belong to the main goal of mathematical education, that is to say the formation of attitude.

Constructing good models for relevant situations can depend on assistance by experts in the corresponding areas, since the most difficult is the formulation.

This strategy presumes on the part of the teacher some interdisciplinary interests and tendencies, which can be improved through courses of training. In my experience both future and in-service teachers can acquire a familiarity in models from interdisciplinary seminars, attended by professionals of different areas.

In 1976 C.B. Wilmer, in his master thesis under my guidance conciliated the strategy of models with the main ideas of Piaget, Polya and Dienes, on the psychopedagogic level (Wilmer, 1976).

In 1979, following the same lines, Pardo Sanchez looked for theoretical foundations of a combined strategy of both instructional modules and interdisciplinary mathematical models. A module is a unit of teaching, which consists of a self-contained approach to a subject, with clear indication of goals, pre-requisites, alternatives of learning, and tests for self-evaluation. Consequently, it permits directed individual study. In the context of global theories of learning, it followed that the double strategy grasps in an unified way the so-called six Bloom's categories: knowledge, understanding, application, analysis, synthesis and evaluation.

In the second part of his thesis, for some validation at high-school level (age range: 15 - 17), he prepared several modules conjugated with models, using quantitative information from newspapers and magazines as a source of material. The outcomes of final evaluation were the following: 40% of the control group scored the minimum of 80% in the post-test (Pardo Sanchez, 1979).

It is to be expected that inability to apply mathematics in construction of models or even in interpretation of proposed models is not necessarily symptomatic of a lack of the understanding of the subject being tested. For example, in a first course on differential and integral calculus of one variable last semester at Pontificia Universidade Catolica do Rio de Janeiro, the average marks of my class (about 150 students) rose by more than 100% after evaluations based on models were replaced for a totally conventional one.

Furthermore, there is no doubt that the strategy of models requires on the part of teachers a great deal more in effort and time outside the classrooms. In compensation, there exists a reward: the teacher will be enlarging his own field of knowledge, that is, education could reach its wide bilateral role - DOCENS DISCET.

References

Ausubel, D.P., Educational Psychology: A Cognitive View, Holt, 1968.

Barreto, A.C., Interdisciplinary models of teaching geometry, III International Congress on Mathematical Education, Programme Part III, Abstracts of the Short Communications, Section A4, Karlsruhe, 1976.

Baretto, A.C., Ensino a partir de modelos (in Portuguese), Boletin Informativo, ICMI-CIAEM, no. 5, Septiembre 1977, IMECC-UNICAMP, Campinas, 8-15.

Baretto, A.C., Propuesta de estrategia: la ensenanza partiendo de modelos, Actas del V Congreso de la Agrupacion de Mathematicos de Expresion Latina (Palma de Mallorca, 19-23 septiembre 1977), Consejo Superior de Investigaciones Cientificas, Instituto "Jorge Juan" de Matematicas, Matrid, 1978, 482-484.

Baretto, A.C., Practical Geometry, THE MATYC JOURNAL - a Journal of Mathematics and Computer Education. Vol. 13, No. 1 (1979), 33-36.

Hunter, M., Teach for Transfer, TIP Publications, El Segundo, California, 1971.

Botkin, J.W., Elmandjra, M. and Malitza, M. (Ed.), No Limits to Learning: Bridging the Human Gap, Club of Rome, Pergamon, 1979.

Pardo, A., J.E., Estrategia Combinada de Modulos Instrucionais e Modelos Matematicos Interdisciplinares para Ensino - - Aprendizagem de Matematica a Nivel de Segundo Grau: Um Estudo Exploratorio, master thesis, Departamento de Educacao, Pontificia Universidade Catoloca do Rio de Janeiro, 1979.

Pollak, H.O., La Interaccion entre la Matematica y Otras Disciplinas Escolares, Nuevas tendencias en la ensenanza de la matematica, Volumen IV, Capitulo XII, UNESCO, 1979.

de Solla Price, D.J., Little Science, Big Science, Columbia University Press, 1963.

Schultz, T.W., The Economic Value of Education, Columbia University Press, 1963.

Wilmer, C.B., Modules na Aprendizagem da Matematica, master thesis, Departamento de Matematica, Pontifica Universidade Catolica do Rio de Janeiro, 1976.

APPLICATIONS, MODELLING AND TEACHER EDUCATION

Hugh Burkhardt
Shell Centre for Mathematical Education
Nottingham, England

Three major reasons for giving importance to applications have been recognised by many people for a long time:

1. Mathematics gets a lot of curriculum time, largely because it is thought to be useful in tackling practical problems, but this usefulness can only be realised with specific training in applications.

2. Applications give practice in the execution, and also in the selection, of mathematical techniques; these must be fluent and accurate to be of much use.

3. Applications reinforce conceptual understanding by providing multiple concrete illustrations of the principles involved.

Nonetheless, the study of practical situations in mathematics plays almost no part in the mathematics curriculum of most European children from 11-16 years and the situation is not very different elsewhere. There is now a strong movement for change but the difficulties should not be underrated - - the desire has been there before but with little effort.

It is my aim here to establish a framework for considering the implications for teacher education of the renewed emphasis on applications in mathematics, and of the relatively recent introduction of realistic problems of practical concern to the pupils themselves into the mathematics curriculum, in a small proportion of schools. There are two main aspects of the change:

new curriculum content and materials, and

changes of teaching and learning style that they imply.

The first area is referred to continually - - interesting new material at various levels is emerging from a number of centres. Jack Downes and some of the other speakers in this session will briefly describe their own work and I shall mention some of the things we are doing at the Shell Centre. John Baker's section on real problem solving is devoted to this important new area and Max Bell has compiled a review of available material. This is only a begining and we have much to discover as to what are realistic challenges for pupils at a given stage, but I have no doubt that enough good material can and will be produced. However, the implications of these developments for teaching style are profound and they will be more difficult to achieve. It is these that pose the challenge to teacher education; I shall return to them in introducing the second part of the session. It should be emphasised that everything in this area is still experimental so that we must strive to build into our work ways of finding quickly when we are failing.

Tackling Real Problems

How do real problem situations differ from those we usually train our students on in applied mathematics? Real problems are not well phrased, often occurring first as vague 'troubles'; it is a long way from "the Thames stinks and you can't catch any fish" to the partial differential equation for oxygen concentration (1), and even from "I can't find enough time for TV, my boyfriend, and my homework" to recording how the time is spent and trying to establish some sort of timetable. Real problems are usually new - - you have not met them before with slightly different numbers. This makes life much harder because you are now exploring without a guide, so that at any given level of skill the problems that may be tackled must be much easier. There are no right answers to real problems, though there are wrong ones; one seeks useful understanding and the criteria for deciding whether this is adequate must come from the needs of the situation. The primary objective is always to understand the situation, and not to practice mathematics.

I should like to make a few definitions and distinctions that are useful. I start by classifying the interest levels of problems:

Action Problems concern decisions that will affect your own life; an example for most of us is:

"How shall I choose which of the parallel sessions to attend?"

Believable Problems are action problems for your future or for somebody you care about, perhaps:

"How should a student organise his travel?"

Curious Problems simply intrigue you - - they may have no practical relevance, for example:

"Why can some birds hover and others not?"

Dubious Problems are there to make you practice mathematical techniques:

"Two hamburgers and one milkshake cost $1.70, while one hamburger and three milkshakes cost $2.60. What is the price of each?"

Educational Problems are special - - though dubious, they illuminate some mathematical insights so well that no one would throw them out:

"If inflation continues to give a 20% annual increase of salaries and costs, a teacher in the year 2000 could be earning over $1m a year and a cup of coffee costing $50".

This rating is, of course, partly subjective. Mathematics teaching has rarely aimed higher than the Curious and for many students it has been almost entirely Dubious; we are now aiming to include in the curriculum the tackling of Action and Believable problems on a sigificant scale.

A distinction between standard and new situations is useful; both are needed. I shall comment further on their roles. We also need to be aware that the prime purpose in studying a given application may either lie in understanding the situation being studied or in illustrating the value of the particular mathematical technique being used. Again both are improtant but the presentation needs to be quite different according to which has the priority.

What methods have we for realising these aims in the classroom? The didactic teaching of standard models of important situations has a fine tradition; it is now being revitalised with the introduction of much more realism and a wider range of interesting situations from different fields of human activity. Max Bell's review gives some picture of what is going on. On the other side, the USMES project staff were pioneers in developing the tackling of new, real action problems on a class project basis, first in elementary schools but later with all ages; everyone should be aware of their work and the material is of great value for teacher education courses (4). In England the Open University (5) has started an in-service course for teachers based on this approach. Susie Groves and Kaye Stacey at Burwood State College for teachers in Australia have been running a problem-solving course including real problems for the last 5 years (6). We, at the Shell Centre, have developed material at several levels - - an idea book (7) to explain the approach, review the material available, and provide enough examples to help the independently innovative teacher in his own explorations; a 'starter pack' (8) giving teachers fairly explicit material and directions for a 5-10 lesson course; a problem collection PAMELA (9); and video-tapes aimed at teaching modelling skills (10). The early results of the research programme aimed at finding more about how pupils tackle real problems and how they can be made more effective at it are also of interest to those developing this approach. Next Professor Downes will discribe his work in finding interesting situations with models accessible to student teachers. Other sources mentioned in Max Bell's review include the NCTM 1979 Yearbook on applications (11); most of them emphasise models to be learnt rather than modelling by the student. Mention should also be made of Chris Ormell's Mathematics Applicable series of textbooks for pupils of about 17 years of age which aim to teach mathematics itself through a modelling approach (12).

Real Problems in the Classroom

The teaching of modelling in the classroom involves the tackling by the student of real problems, the study of standard models of important or interesting situations, exercises on minor variants of standard situations, exercises to develop particular skills (in modelling processes as well as in mathematics), the validation by observation or experiment of the predictions of the model, and the explanation of the results of the analysis. It should include the 'descriptive modelling' of a situation on the basis of given or collected data, summarising its content as a prelude to the search for underlying causes which is 'analytic modelling'.

To those who would like first to try this approach for themselves without having to study any of the sources quoted, I offer some key elements for a small-scale, informal approach; if you are happy to encourage pupils to take the initiative in their work, it will probably go well. You will be dissatisfied at the mathematical level - - the level that they can operate at independently is much lower than one might think.

1. Find problems - by asking the students what has concerned them in the last day or so, or going to one of the sources quoted above. The problems must look much simpler mathematically than the ones you are currently showing them how to do because they are going to have to solve them.

2. Let the students have a go:
   - in small groups (3-5), with a recorder taking notes
   - spread over more than one session
   - in a consultant role; solving a problem for someone else requires a more complete analysis and explanation than for oneself
   - provide heuristic hints of a general kind when they are stuck, based on a model of problem solving such as in reference (7), (2)
   - perhaps exercise them in relevant skills such as listing factors, generating or choosing between relationships, estimating, posing specific 'questions' to answer
   - if they are really stuck, don't solve the problem for them, find them an easier one, perhaps a sub-problem of the first but still realistic, and return to the original one later as their strength grows
   - emphasise practical usefulness as the measure of the value of the analysis - - they will know how good it is.

"Beginning to tackle real problems" (8) aims to provide ingredients for such an approach for the teacher (most of us, I think) who would like some help with detailed suggestions and examples, and who enjoys some didactic teaching on testable skills.

Teaching Methods, Style and Teacher Education

The last 20 years have seen the gradual development of teaching methods and materials that can help pupils to acquire the wider range of skills, knowledge, and experience that enables them to use mathematics effectively in tackling real problems of concern to them. However, these new developments, like most others in mathematical education, ask the teacher to go beyond the straightforward traditional and didactic approach (the exposition of ideas and techniques with illustrative examples leading to illustrative exercises by the students). This approach still seems, however, to be used in the overwhelming majority of classrooms, paticularly with pupils over the age of 9 or 10. A more investigative approach in which the teacher encourages the pupil to find and pursue his own path through problems of a more open kind has undoubted benefits but places extra demands on the teacher in at least three ways:

1. mathematically, by asking him to discern the pupil's line of argument diagnose its possibilities and errors, and direct the next stage in a way that builds on the pupil's achievements so far;

2. pedagogically, by asking him to accept and handle the much wider range of problems, and progress through them, that differnt approaches and abilities of individual pupils will produce; and, above all,

3. personally, by requiring of the teacher the confidence to follow pupils into areas that he has not himself explored, to admit that there are many things that he does not know, to accept that some pupils will often see points before he does, and to live with less clear-cut measures of pupil achievement.

These demands clearly go on well beyond the didactic approach with its statement, illustration and restatement of facts, ideas and techniques; it only requires of the teacher a thorough understanding of a single 'track' through the problem and little diagnostic ability beyond recognition that the pupil has left that track. The teacher is also then able easily to assess his pupils' and therefore his own success, since there are well-defined right answers. The rewards of the investigative approach in transferring responsibility and initiative to the pupils, and in thus preparing him to function mathematically without the teacher's constant support over a wider range of challenges, are overwhelmingly clear to those who have successfully tried it. However, they are still very few, even in countries where such a style has been widely advocated for at least 20 years. This suggests that it is likely that adopting profound style changes is difficult. More ideas and experiments (13) are needed, including:

a. experiments on ways to help teachers on initial and in-service training adopt investigative methods as at least part of their style;

b. adaption of new curriculum material for use by teachers who will operate in a largely exposition-exercise mode, recognising the results will not be ideal but optimising them under this constraint; and

c. exploring style specialisation (as opposed to subject content specialisation) as a way of ensuring that each pupil receives regular teaching from those teachers in the school who can teach in an investigative way.

References

1. Sir James Lighthill (Editor), "Newer uses of mathematics", Penguin, 1978.

2. Treilibs, Vern, Hugh Burkhardt and Brian Low, "Formulation Processes in Mathematical Modelling", Shell Centre for Mathematical Education, Nottingham, 1980.

3. Burkhardt, Hugh, "Temporary Traffic Lights", Journal of Mathematical Modelling for Teachers I II (1978).

4. Unified Sciences and Mathematics in Elementary Schools: List of publications from Moore Publishing Company Limited, Box 3036, Durham, North Carolina 27705, U.S.A.

5. Open University, "Mathematics Across the Curriculum", course PME 233.

6. Groves, Susie, and Kaye Stacey, "The Burwood Box - a course in problem solving", Burwood State College, Victoria, Australia, 1980.

7. Burkhardt, Hugh, "The Real World and Mathematics", Blackie and Son, Glasgow, 1981.

8. Treilibs, Vern, Hugh Burkhardt, Kaye Stacey and Malcolm Swan, "Beginning to tackle real problems", Shell Centre for Mathematical Education, Nottingham, 1980.

9. PAMELA, Shell Centre for Mathematical Education, Nottingham, 1980.

10. Diana Burkhardt and Hugh Burkahrdt, "From Problem to Program - a multimedia introductory course", Computer Science Department, University of Birmingham, 1981.

11. S. Sharron (Editor), "Applications in School Mathematics", National Council for Teachers of Mathematics, 1979.

12. C. P. Ormall and others, "Mathematics Applicable", a series of tests, Schools Council/Hienemann (1976).

13. T. Beeby, H. Burkhardt and R. Fraser, "SCAN - a systematic classroom analysis notation for mathematics lessons", Shell Centre for Mathematical Education, Nottingham 1979.

8.4 THE USE OF MODULES TO INTRODUCE APPLIED MATHEMATICS INTO THE CURRICULUM

THE USE OF MODULES TO INTRODUCE APPLIED MATHEMATICS INTO THE CURRICULUM

R. John Gaffney
Wattle Park Teachers Centre
Wattle Park, South Australia, Australia

Since 1974 I have been Director of a project sponsored by the Australian Academy of Science. Our aim has been to prepare mathematical materials to be used by some of the high school students studying mathematics in their eleventh year of schooling. In particular, we were interested in those students who intended this to be the last course in mathematics they would ever study.

For such a large area, the population of Australia is small. There are currently about 50,000 students throughout Australia taking mathematics courses of the type which interested us. I should also point out that most Australian students commencing school in year one remain at school until at least the completion of their eleventh year. Almost all study a subject called mathematics until the end of year 10, and, of those who remain at school in year 11, about 85% study some kind of mathematics course in year 11.

It is unsurprising, if disappointing, that many of those students are motivated by factors other than a love for mathematics. Parents, teachers and employers all make demands for mathematics to be included, and many post-high school educational institutions list mathematics as a desirable, if not pre-requisite study. A significant proportion of the students had not been successful in previous mathematical studies, but found themselves in year 11 on an aggregate performance across all arts of the curriculum.

What mathematics should be studied by this group? What process would be most likely to bring some success?

We accepted a fundamental principle which was to be used to guide our development of the materials. We would aim to develop generalized problem solving skills, rather than directly applicable vocational skills. The applications of mathematics to be included should demonstrate the inter-disciplinary nature of useful mathematics, and should encourage students to make mathematics work for them when they attempt to solve real problems.

Our first approach was similar to that used by Prof. Finney. Unfortunately it was not so successful. I decided to use the contacts so readildy available through the Australian Academy of Science. I asked some applied mathematicians if they could write up a problem which involved mathematical modelling, using mathematics accessible to my group of students. In each case, a case-study with examples — similar to those produced by UMAP -- was the outcome. Each would have been useful to our year 12 most able high school students, but for my group the gulf was too wide.

It seemed that it might be more practicable to ask some high school teachers to write some materials. We created some (non-mathematical) themes which would use several areas of mathematics. From the several such themes we developed and tried, two examples reveal typical problems which we were not able to solve.

One teacher-writer from a large city high school developed a module on designing car parks. Many of the students drove a car, or expected to drive one shortly. Nearby was a large shopping centre with a variety of parks holding hundreds of cars. To develop a good design, the teacher and students discovered they needed to know about:

"average" car size -- and this "average" was not of a type they were familiar with

the geometry of turning cars

the trigonometry of stacking parking stalls at an angle

some good measuring and counting techniques

compromising — accepting many possible solutions

changing your assumptions -- getting cars in and out easily is more important than fitting in the greatest possible number.

Some drawbacks:

this problem is not interesting to everyone

not all students do research at the same rate

some of the activities must be done outside the classroom

when to teach new mathematical material; (e.g. "average" = 95th percentile car)

when to fit this kind of activity into the program

A second teacher prepared a module in card form, based on planning on overseas trip. It included:

reading (complex) airline timetables

currency conversions

planning a budguet

understanding baggage regulations

time zones

speed and distance calculations

insurance

climate - temperature conversions

equivalencies in clothing sizes

There was also supplementary material on:
aircraft capacity and fares

great circles

vectors

The draw backs were:

the only process workable was individual progression

the material was difficult to manage

the mathematics remained implicit, which worried the teachers

the teachers believed the students were being kept occup ied, but weren't learning much mathematics -- a view shared by the students!

the teachers couldn't fit this activity into the traditional syllabus.

As the Mathematics for the Majority Continuation Project had encountered similar problems, we knew we were not alone.

A Change of Direction

In the face of the problems which we were not able to solve, we decided to abondon the non-mathematics theme approach, but to incorporate some of the ideas and examples into differently conceived modules.

During 1980, six modules, each containing enough materials to support approximately 20 hours of instruction, plus some supplementary problems, are on trial in some Australian schools. Their titles, and an outline of their contents are as follows:

MAKING THE BEST OF THINGS; Some simple optimization strategies readily transferable to everyday problems — deliveries, placement of phone boxes, school zones, choosing a home, routing taxicabs and the like.

TAKING YOUR CHANCES: We are trying to give examples of many real-life situations where a stochastic model is appropriate. We have included modelling queues, an analysis of some popular gambling games and the counting of runs. We have not included permutations, combinations or the binomial theorem. Despite this, we believe it important that these students should encounter problems for which probable -- rather than certain — answers can be given. In many mathematics texts, this kind of problem only occurs after the normal distribution has been studied. It is unlikely that many students in our group would study mathematics at that level.

PEOPLE COUNT: In this module, which deals with interpretation of statistical data, we are trying to develop a critical appreciation of arguments based on statistics. Most students enjoy making comparisons and judgements, so problems involving scoring and combining scores for consumer group tests, talent quests, gymnastic and athletic competitions, taste tests and the like are included.

SHAPE, SIZE AND PLACE: Developed from a previous attempt to integrate several areas of mathematics around the theme "Design for Living". Ideas about measurement, ratio, proportion, enlargement, trigonometry and maps are included. There were several anguished debates about the place (if any) of trigonometry in a contemporary mathematics course. The main argument which forced the inclusion of some trigonometry was its presence in the core of almost every prescribed high school course in Australia.

UNDERSTANDING CHANGE: in which the concept of "variable" is considered in real life contexts, and the commonly encountered forms of functional dependence that relate variables are investigated. Deciding how much (if any) algebra these students should (and could) know and use is still to be resolved. We agreed that we would try to find ways to show that tables, graphs, flowcharts and formulas can each be used to describe a relationship, and often one is more useful than the others.

MATHEMATICS AND YOUR MONEY: Unlike the others, this module consists of many self-contained sections with no mathematical sequencing from one to the next. The mathematical context would be found in many texts written for students in years 6 to 9. The context is directed towards young people about to leave school and undertake the associated responsibilities and commitments.

Thematic Treatment

There has been some retention. In UNDERSTANDING CHANGE, four young people decide to earn some extra money by setting up a delivery service. They estimate trip times, maintenance costs, profit margins and charges using each of the four techniques mentioned previously. They discover step functions. For a while, all relationships are linear.

The distribution of advertising leaflets to a specified area of the town introduces the first non-linear relation. Others follow -- depreciation, speed zoning and finally some graphs for which no simple formula, table or flow chart exist.

The calculator is assumed to be continously available to the students for all the modules. We have yet to find a school where this has been unacceptable to the teacher. Generally the students have owned a calculator for at least two years. It greatly aids the yreatment of real-life data. We would not be game to ask these students to solve $(0.775)^t = 0.2$ other than by estimation and testing.

There are opportunities to remind the students that errors can creep into the use of calculators. Most students simplified

$C = (t/60) \times 2.05 + e$ — C = charge (dollars), t = time (minutes)

to $C = 0.034t + e$ — e = entry cost (dollars)

by rounding off $2.05/60 = 0.034166\ldots$

For small values of t, there was no problem. When a journey of 18 hours was considered, the round-off error appeared - - and puzzled most of the students, who couldn't account for it.

The booklets are still being revised as a result of the trials, but a few observations from our trials-and-tribulations are worth noting.

1. Teachers generally have extreme difficulty if new material demands any major changes to their teaching style.

2. Students are mistrustful of mathematics which is left implicit.

3. Most teachers asked for student materials with little descriptive material ("they can't read anyway") and lots of examples. We did not satisfy these requirements.

4. Including examples derived from the Social Sciences is difficult. They generally require longer introductory descriptions and provide less quick repetition of standard algorithms.

5. Encouraging students to develop their own problems and propose solutions to them requires patience, adaptability, confidence and resourcefulness from the teacher. Much less content will be covered. There will not be answers in the back of the book, or in a Teachers Guide, and a variety of answers to match the variety of assumptions must be expected - - and discussed.

6. A modular format is useful to the extent that it can promote flexibility - - it can accomodate a wide network of authors and styles, and can be adopted to a wide variety of learning environments. Flexibility decreases with size, but economy may increase. This can lead to unconfortable compromises with publishers.

7. Most teachers in Australian high schools have little work experience other than being a student until becoming a teacher. Very few have worked as applied mathematicians; not many more have undertaken undergraduate programs containing significant amounts of mathematic modelling. It

is not surprising that most of us find difficulty in helping our students model aspects of the real world when we have had so little training and experience in it ourselves.

It is in this last area that the progress report by my colleagues on this panel gives some hope for the future. If more of the flavour of the models of the real world can continue to be infused into undergraduate programs and refresher courses undertaken by present and future teachers, then some of the problems to which we have provided partial solution may be more greatly reduced.

8.5 TEACHING APPLICATIONS OF MATHEMATICS

APPLICATIONS OF MATHEMATICS FOR NON-MATHEMATICIANS

F. van der Blij
Rijksuniversiteit
Utrecht, Netherlands

We discuss some aspects of mathematics courses for students in other disciplines, e.g., chemistry, biology, social sciences and so on. We restrict ourselves to students in last years of secondary and first years of university education. Teaching such courses one must realize that the major aims of this type of education cannot be:

memorisation of isolated facts.

acquisition of abstract theories.

ability in logical deductions in axiomatic systems.

problem solving of pure mathematical problems (in number theory or geometry or . . .)

Only common-sense problems from real-life situations and problems proposed in descriptions of real situations in their own major field of study must be used as motivation and stimulus for mathematical education.

Examples

1. The teaching of the theory of logarithms and the exponential function cannot be introduced from an abstract point of view as an isomorph between multiplicative and additive groups. Neither can the logarithm any longer be introduced as a necessary aid for calculations, we live in the era of hand-held calculators. Even problems about $\log_7 x = \log_{0.7}(x + 1)$ are not so important for every day (scientific) life! So one of the most appropriate ways to introduce exponential functions and logarithms seems to be the study of growth processes, as well in biology as in economics as in physics, chemistry and so on. Logarithms can be interpreted as time-intervals in this view.

2. The teaching of skills for "calculating" definite integrals is not so wise. It may be an interesting mathematical problem to characterize the functions that can be integrated in elementary functions (= finite terms). But it does not make much sense to calculate

$$\int \frac{x^2 + x + 1}{(x^2 + 1)\sqrt{x^2 - 1}}\, dx$$

expressing it in log, arctg if one is asked to calculate this integral between 2.864 and 5.975 to four decimal places. Instead of using tables one shall prefer the computer or calculator. And in that case it can be easy to use direct methods of numerical integration. Although it can be nice and useful to deal with $\ln x\, dx$ and such simple problems one should be aware of the fact that from a point of view of practical applications there is not so much difference between the integrals

$$\int \frac{\sin x}{e^x}\, dx \text{ and } \int \frac{\sin x}{x^2}\, dx$$

3. Far more important than "explicit" solutions in terms of elementary functions is a study of qualitative nature of local and global properties of the solution. Stabiity , attractors and such notions are more important than solving special equations such as Clairaut with special methods. Even in those cases in which one can obtain an explicit or implicit solution it is not always easy to get a good impression of the behavior of the solution. We refer to the equation of logistic and restricted logistic growth:

$$\frac{dx}{dt} = \alpha x(1 - \frac{x}{N}) \text{ and } \frac{dx}{dt} = \alpha x(1 - \frac{x}{N})(\frac{x}{M} - 1)$$

These equations can be solved in elementary function but if you look at these solutions you still have to do some thinking or some calculations to remark the essentials of the theory. One look at the picture of the line-elements is enough, however! So please more graphical display and less partial fractions, splitting of variables for our non-mathematicians. They ought to be interested in stable or non-stable equilibrium, in asymptotic behavior of the solution.

4. The mathematical models that are used in psychology, in sociology and in many other recent fields of applied mathematics use quite other topics than we were teaching in traditional courses in applied mathematics. The theory of graphs, of Boolean algebra, of difference equations. of sequences defined by recurrance, of branching processes seems to be important. I don't speak about statistics, all of you know enough of these problems.

But alas, many of the simplest problems in the "new" fields of applied mathematics are from a mathematical point of view rather difficult. And we poor, pure mathematicians fall many times in the temptation to go over to quite non-realistic simplifications, that we just can handle with our mathematical apparatus. This is rather dangerous if we don't underline ceaselessly the dangers of two strong a simplification and of pseudo-mathematical proofs of scientific statements in several natural and social sciences.

TEACHING APPLICATIONS OF MATHEMATICS - TO WHOM AND TO WHAT PURPOSE?

Douglas A. Quadling
Cambridge Institute of Education
England

In a Congress such as this it is all too easy to concentrate our attention on "teaching" and to forget the needs and motivation of the students for whom our courses are designed. In this paper I hope to examine critically some of the ways in which we associate the word "mathematics" with some of the words derived from the verb "to apply": "applied", "applicable", "application", "applications", "for applications", etc. etc.

Those of us whose work is directed primarily at mathematics in comprehensive schools are unlikely to ignore the fact that mathematics has its roots in man's need to understand and to extend his control over his environment. To teach arithmetic without reference to quantitative measures of time and money, geometry without reference to machinery and the solar system, calculus without reference to problems of optimisation, would be to withhold from our students the very things which give our subject its justification and meaning. There is a sense in which the whole of school mathematics is "applicable" or "for application", area if not "applied". But these are not the issues with which this panel is concerned.

In England, and a number of other countries, there is a long-standing tradition of presenting, as part of the mathematical education of older students in schools and of university students, a course in "applied mathematics". In school (for ages 16-18) this takes the form of mechanics - the statics and dynamics of particles and of rigid bodies moving in a plane - whilst in university it may be extended to subjects such as hydrodynamics, electromagnetism, mathematical economics or theory of relativity. Such courses have come under increasing attack in recent years, particularly in schools. It is argued that:

i) Taught as they are by mathematicians, they place little if any emphasis on the experimental aspects of the subject and on the process of modeling the real-life situation, and concentrate on problems which lead to interesting mathematical solutions.

ii) The courses deal largely with closed problems, and rarely discuss the range of applicability of the theory, or the degree of reliability in the final solutions.

iii) The subjects treated tend to be the historical fields of applied mathematics - note the past participle! - rather than those in which the students may find themselves applying mathematics in their future careers. That is, the choice of course tends to narrow students' horizons.

iv) The courses have little attraction for students who have already decided that they wish to pursue careers in other fields; in fact, the direct relevance of the school mechanics course is largely restricted to those who intend to become engineers.

The most common reaction to these arguments in English schools has been to substitute for the mechanics course a study of statistical methods. This has the advantage of a wider range of applicability - problems can be drawn from a variety of disciplines in the social, biological and (to a lesser extent) physical sciences - though it is unlikely that at this stage the students will themselves see the need to apply statistics for their own purposes; its use is mainly as a research tool, rather than being an integral part of the discipline in which it is applied (as mechanics is to engineering). Moreover, it retains some of the drawbacks referred to above in relation to mechanics: if it is taught by mathemeticians who have no personal experience in using statistics but who know it only as a set of theoretical procedures, then the collection and analysis of untidy and ill-organised data is too often ignored and replaced by the solution of closed problems more amenable to mathematical treatment.

I have presented my arguments so far in the context of school mathematics, since that is the scene which I know best. But the problems are no easier to solve - indeed, perhaps some are more difficult - at college or university level. It seems to me that they call for discussion of a more fundamental kind than they often receive. For teaching applications of mathematics - I would rather myself talk of "teaching application of mathematics" - raises in an acute form all the questions of interdisciplinary education; and there is an obvious danger in even discussing it here, in a Congress devoted to just the one discipline of mathematics.

We must, I think, distinguish two classes of student: the one whose first interest is in mathematics, but who wishes to apply his mathematical expertise in some other discipline; and the intending biologist, town planner, economist or industrial chemist who will need to use mathematics in his work. (Of course, this is a crude dichotomy, and some students will come between these extremes; but the organisation of university education tends to produce these types - and perhaps rightly so.)

For the mathematical student we have to ask, do there exist skills of mathematical application which are teachable and which are independent of the particular fields in which the student will ultimately work? There are of course some general principles of mathematical modeling which can be exhibited at work in a variety of particular examples; and a few university courses have been developed in England along these lines. It is probably too early to say whether they are more successful than conventional courses in applied mathematics in producing adaptable, usable mathematicians for work in other disciplines. Apart from these courses, what is needed by such students is a wide knowledge of the more applicable parts of mathematics which they can bring to bear when required within the field of application.

For students of other disciplines the primary need from mathematics is not a detailed knowledge of particular processes so much as the ability to handle the language, symbols and concepts of mathematics appropriate to their field of study with meaning and confidence. There can be too much emphasis on looking at ways in which mathematics has been used - this often applies only in particular areas of research, and need not be of concern to the general student. But in every area of pure or applied science mathematics offers an important reserve of power. The implication of this for education is that mathematics should enter as far as possible in a form where it is integrated into the main subject matter; the role of the mathematical community lies more in the provision of suitable resources to support such courses than in setting up seperate courses of instruction in mathematics. A recent example from England is a set of materials covering a range of mathematical topics designed for individual instruction which can be brought into use as and when they are needed to support the student's work within his own discipline. Another is the establishment in some colleges and universities of units for "engineering mathematics" or "business mathematics", for example, within departments of engineering or management studies, staffed by mathematicians committed to a resource role in the particular field of application. But here we are moving away from questions of didactics and student motivation to the even more intractable problems of institutional organization and career development. Where does the "applied mathematician" belong?

THE PRACTICAL PROBLEMS OF TEACHING THE APPLICATIONS OF MATHEMATICS

P.C. Rosenbloom
Teachers College
Columbia University

Leading mathematicians and mathematical educators have been advocating teaching the applications of mathematics for a long time. I may mention, among others, Felix Klein, John Perry, E.H. Moore, and Moris Kline. Some textbooks and other curricular materials have been provided by individuals and groups, yet applications are hardly mentioned in schools. In American schools the most widely taught "applications" of mathematics are the so-called "social appliations" - buying and selling, profit and loss, interest, mortgages, and the like. Even though mathematics is an integral part of most vocational school programs, the role of mathematics in the skilled trades is not mentioned in school texts. Certainly nothing is said about the uses of mathematics in the natural and social sciences.

The advocates of the applications stress the importance of relating mathematics to the real world. But they will never accomplish their purpose until they themselves grapple with the real world of the educational system and its structure. Look at how difficult it has been to get an adequate treatment of applications at the college level, where the teachers have better mathematical backgrounds and enjoy greater freedom of choice. How much more difficult it is to get applications into the school curriculum.

At the school level, we must face the problem of teacher education. Furthermore we must cope with the way schools are restricted by official syllabi and constrained by external examinations such as College Boards, baccalaureate, and certification examinations. The system whereby textbooks are chosen in many places also poses a serious problem to innovation in curriculum.

Let us consider several possible ways of introducing applications into curriculum. One is to incorporate them in the existing curriculum, that is, to teach the mathematical ideas and their applications in close proximity to each other. This has the advantage that a topic is better motivated if it is applied immediately. It also often corresponds to the historical development. This solution also avoids the problems of deviating from the official syllabus. We simply accept the topics and their sequence as given, but insert applications wherever appropriate.

A major difficulty is that in order to discuss properly applications of mathematics to physics, biology, or economics, one must know something about these fields. A college teacher, who is supposed to be a professional scholar, often feel uncomfortable in discussing subjects he knows only superficially. How much more difficult it is for most school teachers, who are not accustomed to studying new topics independently. (I may mention that in the '60's the Committee on the Undergraduate Program recommended that all high school mathematics teachers be required to take at least one course with mathematical content in some other field, but I do not believe that this has been implemented in certification requirements anywhere.)

A second difficulty is the time problem. For example, consider the application to moments, work and water pressure which on finds in many standard calculus texts. In most of the textbooks the treatment of the physical background is quite inadequate, and the 'applications' are merely excuses for stating exercises on definite integrals in a picturesque language. If one tries to explain the physical meaning of these topics, one finds that it takes more time than one would wish to allot in a calculus course. (Here we encounter the question of whether there is any value in merely stating mathematical problems in applied language in order to suggest that mathematics is useful. Of course, a real understanding only comes from seeing how a mathematical model is set up for a real world problem, and how it is analysed.)

An alternative approach is to replace some topics in the curriculum by applications. This has the disadvantages of raising the political and public relations problems associated with changing the official syllabi and external examinations.

At the college level the third alternative of introducing separate courses on the applications has been used at many institutions. This has the advantages that it requires no changes in the rest of the curriculum, that these courses are taught by people with interest and background in the applications, and that the decisions on offering such a course and on its content are made by the mathematics department. It has the disadvantage that students can still take a full sequence of courses in algebra, analysis, and geometry without hearing a word about their applications. At the school level the tme alloted to mathematics is much more rigidly prescribed, as are also the contents of the courses, and decisions about additions or substitutions involve others besides the mathematics staff.

Certainly, whichever approach is adopted it must be implemented by practical teaching materials. Some textbooks were produced by the School Mathematics Study Group. They were tried out with moderate success, at least when the teachers got inservice education and consultant help. The same is true of the Minnemast materials for elementary schools. There was also an Australian project for coordination of mathematics and science at the secondary school level. The UMAP project of the Educational Development Center has produced quite a number of modules on applications for the undergraduate level. The Mathematical Association of America has published some excellent source material on applications, which needs to be translated into practical teaching materials. The National Council of Teachers of Mathematics has also published material on applications.

I do not think any approach to the teaching of applications in the schools will be successful without a program of inservice education. This will cost money. It will be necessary for the mathematical community to persuade the appropriate agencies to support such a program. The advocates of the applications have a responsibility to wrestle with these practical problems of implementing their objectives in the real world.

8.6 THE INTERFACE BETWEEN MATHEMATICS AND EMPLOYMENT

THE INTERFACE BETWEEN MATHEMATICS AND EMPLOYMENT

N. Connie Knox
Sheffield, England

In our education and employment session there are six speakers. The first two, Peter Price and I, cover the mathematical needs of employees starting work. Mine is for the 16 year old school leavers, while Peter Price looks at the needs of graduates starting work.

The second session had two speakers from West Germany. Rudolf Strasse shows techniques in technical and vocational education for 16 - 18 year olds in a sandwich system, using both college and work. Werner Blum was to have looked at the relations between mathematics at full time technical school level and employment. He was ill, unfortunately, but I am pleased to say he became well enough to come, and will speak and join in the session in questions and discussion. Trevor Crudge, from Scotland, has been researching for the past three years into the mathematical content of work done in factories in the Lowlands of Scotland. He will talk on the common ground of applying mathematics learnt in school onto the job.

The third session has speakers who move us to the planning of courses. David Matthews looks at engineering craftsmen or tradesmen and technicians, and finally Robert Lindsay treads a hopeful path for a high school syllabus, 12 -16 year olds, which will reconcile the expectations of both employers and teachers.

What are the mathematical needs of the young worker? and how is the best way to train workers? are problems throughout the world. Somehow in the United Kingdom we have not been training children in the way employers had hoped. Employers know that a good general education is a fine start. Further education or training can be provided either at work or in a Further Education College. What employers say is that we have failed to keep up the solid constant practice in basic arithmetic. Children are no longer drilled to get the right answer in everyday arithmetic problems. Also, employers found school leavers with the wrong attitude to work as well as the wrong attitude to being consistently accurate. This concern was nationwide.

Many causes can be cited. From 1944 the children in Grammar schools (11 - 16 years) followed an examination syllabus with almost one-third of it arithmetic. The Secondary Modern Schools, containing 70% of all children, had an age range of 11- 15 years and had a syllabus approximately two-thirds arithmetic. Then the changes really came in the 1960s. New syllabi with little arithmetic were taught in the new Comprehensive Schools, which combined all secondary children into one school. Real slogging practice in the four rules of number did not take a prominent part.

The government listened to the public concern and called for an enquiry into the teaching of mathematics in schools. Professor Cockcroft chaired the enquiry. Bob Lindsay, who speaks later, is engaged at the moment in writing in support of this enquiry. Their report should be out next year.

The report "Numeracy and School Leavers" (Knox) was made in 1977. It is a survey of the mathematical needs of school leavers aged 16 in their first year of work. As such, it should give a minimal to useful content of mathematics for the young school leavers.

What was the best way to constructively help teachers to be aware of the frustrated expectations of the training officers in employment? One fact stood out. Although there were meeting points in the Training Boards - -where, say, engineering knew its own needs, the construction industry knew theirs, and the distributive trades knew theirs - - they each had no knowledge of the other employers' needs. So situations arose like this: A local firm, say engineering, would visit a nearby school and show the teachers what mathematics they hoped the children might know as they start work. The teachers did not pay much attention. They looked at their classes. Perhaps there were a few budding engineers, but alongside them were budding butchers, hairdressers, clerks, bankers, and so on.

It was clear that to help the employers in a way which was respected by both teachers and employers, a broad survey should be made. The results of the survey are in Knox (1977). It is not possible in a short paper to go through the elaborate stages of compilation. However, it is always interesting to find out why the research takes the form it does.

For instance, the mathematical needs of school leavers are not found in the examination which are sat. These reflect the school syllabus. Employers set selection tests in mathematics. We found these did in part reflect what the firm thought it needed. They often reflected the mathematics taught in school twenty years before, when the training officer was at school.

So it was essential to reach the school leavers in their first year at work, and find out what mathematics they used on the job. Who can estimate the mathematics used on the job without actually watching the work? This would be a mammoth task, with so many categories of work to cover. For the first step in such research, we used the next best method, which was to ask the training officer. It was vital we covered all employment. To find a list of all employers in the area, there is only the Government Employment Services

Agency. Their list is confidential. However, their statisticians were kind and gave us the 8% random sample firms' names. The size of area used was the large urban unit covering Sheffield, Rotherham, Doncaster, Barnsley and Chesterfield. The report (Knox, 1977) shows how we got the evidence in a form suitable for computerisation - - and by fairly painless methods.

I would like to show you just one part of the work. It is the extraction of a core of useful mathematics by a fairly objective method. The twenty eight topics used in the survey are listed in Knox (1977, pp. 23-24). For each topic, for each school leaver, the training officer had to answer what it was in one of four categories: "Very important," "Useful," "Used occasionally," or "Not needed."

Putting the first two categories together gave a measure of good usefulness. This total score made the x coordinate in a series of scattergrams (Knox, 1977, pp. 13-17). The y coordinate was the total of "Not needed" replies for that topic. There were 1427 children. Should we have included the "used occasionally" with the "not needed" to make y, we would have for every topic, x plus y equals 1427. This would make all the topics, when plotted, fall onto a straight line. By omitting the "Used occasionally" total, the topic points scattered. This made the points drift down vertically from x plus y equals 1427 by a measure of the response "Used occasionally." The first scattergram accumulates scores from all 1427 children. It gives a spread of need of the topics across all employment. The topics with a placement well down to the bottom right hand corner are those with a high count for usefulness and little evidence of being not needed.

This is how the topics emerged:

1st: Topic 2. Four operations with whole numbers

2nd: Topic 1. Mental arithmetic

3rd: Topic 4. Length

4th: Topic 3. Metrication

This actual process of changing from one set of units to another came high. The schools had moved more quickly to full decimalisation in metric units. Industry still had much machinery still in the old units of measurement.

5th: Topic 10. Decimals

6th: Topic 7. Weight

7th: Topic 13. Use of simple directive tables

The reading of tyre pressures opposite the model of a car would be an example of this.

8th: Topic 5. Area

As the results were all assembled onto computer cards it was not difficult, either by program or manually, to get many groups of results. An interesting pair of tables are the scattergrams on pp. 14-15 of Knox (1977) for all topics except the first eight, giving boys separate from girls. The scattergrams on pp. 16-17 show which topics are important for craft apprentices and for commercial and sales workers.

Such results as these have been put before the teachers in the United Kingdom. Now they await the results of the government inquiry. Remembering that half of all children finish full time education at 16 years of age, do teachers give much attention to their needs? Another sizeable group leave at 16 but go on to apprenticeship or full time Further Education. How much attention should teachers give to their needs? Only 12% of the 16 year olds go on to take A level examinations.

References

Knox, N.C. Numeracy and school leavers: A survey of employers' needs (Paper no. 12). Sheffield: Sheffield Region Centre for Science and Technology, May 1977.

Knox, N.C. Changing needs for mathematics in employment (Report of a pilot survey). Sheffield: Sheffield Region Centre for Science and Technology, September 1979.

MATHEMATICS IN THE PLANNING AND DIAGNOSTIC SKILLS USED BY ENGINEERING CRAFTSMEN AND TECHNICIANS

David Matthews
Engineering Industry Training Board
Watford, England

In the United Kingdom engineering industry, the system of apprenticeship incorporates the training of those young school leavers (mostly 16 years old) who are selected as future craftsmen and technicians. The early stages of training for the two categories of employee have much in common and, indeed, the division between the two sets of jobs is sometimes a hazy one.

Cognitive skills have been recognised for a long time as being important for the technician, but the role of such skills in the work of the craftsman has been less well appreciated. Prominence was, however, given to this aspect of craft occupations by research carried out in the engineering industry in the late 1960's (1). Skills of 'planning' were identified as central to many craftsmen's jobs; appropriate methods of training these skills were developed. While these ideas have not been implemented as one would wish, there has been a growing acceptance in other industries of the importance to craftsmen of the skills involved in planning and diagnosis. In this context, the term "planning" is defined as follows: 'the selection, review and control of methods which are appropriate to the successful completion of a given task'. Planning is not to be seen, therefore, as a process which takes place only before commencing operation. It is an intrinsic part of operating, and takes place continuously throughout the task.

In the variety of engineering tasks which involve planning to some significant extent, certain characteristics are common. Of these the most important are that:

1. choices have to be made and evaluated,
2. there is a requirement to optimise some criterion or criteria,

3. there is an element of uncertainty about the result of the choices which are made.

It must be recognised, however, that the existence of characteristics common to planning in many different contexts does not imply that there is a single skill of planning (as it is defined above) and there is no evidence to suggest that such a skill exists.

Craftsmen plan in a variety of tasks. In machining, for example, they may choose the operations to be performed, the sequence of those operations, work-holding methods and measuring and test equipment. Review and control of operating methods follows the initial choice as part of the planning process. We find craftsmen and technicians involved in fault diagnosis for systems as varied as railway locomotives, electronic control system and machine tools.

Research carried out at Chelsea College, London, from 1971 (2) was based on the importance of planning such tasks, and examined the relevance of school experience in subjects such as science, mathematics and craft and design to the development of planning skills. Some relationship was established, and the results led to work which is now taking place at the Engineering Industry Training Board. The programme is concerned with training in planning skills and, consequently, involved identifying related skills and abilities. It is recognised that planning is but a part of the craftsman's or technician's job and that there is a danger in atomising even further by looking at constituents of planning, but there are two considerations of particular importance to any examination of the mathematical element of planning.

It may be that it is possible to identify skills and abilities <u>required</u> for planning, but these may become confused with 'skills' which are merely artefacts of the system of analysis. If, however, we look for abilities and skills which may be <u>exploited</u> in the act of planning we will be less in sympathy with the selector, but more with the trainer, and we will be more able to make use of such models of the real task as may be devised.

The mathematical skills which employers value in their recruits are by no means the only ones which may be of use to the trainee considered as a planner of engineering tasks. They may well be necessary, but they reflect a narrow view of the potential of young people who have had some years of mathematical experience. One of the problems is that few employers have carried out job analysis of a form likely to identify the mathematical characteristics of planning. The following, however, are some areas of mathematics which might usefully be exploited by craftsmen and technicians in a wide range of tasks involving planning:

1. averaging,
2. numerical progression,
3. optimisation,
4. probability and relative probability,
5. two- and three-dimensional shapes and their relationships,
6. approximation,
7. simple linear programming.

There is also a case for the involvement of mathematics teaching in providing experience of planning through models of more technical tasks. Some such tasks were developed in the course of the research at Chelsea (op. cit.) and, while they were developed as test instruments, they have merit as learning devices.

Some of the tests involved subjects in planning engineering tasks - - in one case removing a delicate pump impeller from a shaft, in another planning a sequence of operations for manufacturing an article on a lathe and a milling machine. The subjects did not carry out the tasks, but had to devise a methods for doing so, thus limiting the planning involved.

In other tasks subjects were a little more actively involved, but the tasks which they performed were less obviously related to engineering. The first of these was termed the Route Test. It satisfied the requirement for a task involving the making of decisions about a sequence of simple operations, with the optimisation of one criterion as an objective. The task itself involved plotting a 'route' consisting of certain chess-like moves along a twisting track made up of squares; in certain respects the operation was little like that of solving a maze. Subjects were encouraged to revise and adapt their routes with optimisation of distance achieved in set moves as the objective. In a great variety of approaches to this task, successful outcomes depended on subjects recognising the interaction between 'moves' which were spatially separated on the track.

Sounding a little more startling than the Route Test was the Stabbing Test. This was a search task in which success was determined by the efficient use of information and the efficient selection and testing of hypotheses. As a good search task it had the capacity to induce the same sorts of good and bad characteristics as are to be found in fault diagnosticians. In the test subjects were required to find the exact position of a hidden shape by stabbing through the concealing paper with a tapered needle; they were required to find the position in as few stabs as possible.

It has since been found that, while there is still the incidence of some quite extraordinary strategies, if subjects are given time to think about the task (that is, are prevented from active search) within the normal time limit they perform markedly better than those subjects who commence active search from the start. If merely orienting subjects towards planning produces such effects, one might wonder what experience of exploitable skills might do.

In the United Kingdom the findings of the research into planning skills have been received particularly well by teachers of craft, design and technology. They have at various times provided encouragment in the face of criticism by employers, and they have enabled the discussion of the interface between school and work to be broadened. There is, I suggest, every reason to hope that such a broadening of the arguments can be brought about in the area of mathematics.

<u>References</u>

1. EITB (1971) <u>The analysis and training of cerain engineering craft occupations.</u> Research Report No. 2. Engineering Industry Training Board, Watford, U.K.

2. Mathews, D. (1977) The relevance of school learning experience to performance in industry. Research Report No. 6. EITB, Watford, UK.

APPRENTICESHIP AND COLLEGE—MATHEMATICS IN TECHNICAL AND VOCATIONAL EDUCATION ("TVE")

Rudolf Straesser
Universitat Bielefeld
Bielefeld, Federal Republic of Germany

Technical and Vocational Education in the FRG

The Federal Republic of Germany has a complicated system of part-time and full-time courses in technical and vocational colleges. These schemes prepare students for their entry into the labor market and aim at rather different qualification levels, providing preparatory education for very different vocations. Bluntly stated, this education is structured as follows: The lower the qualification level to be attained, the higher the proportion of on-the-job-training paralleled by part-time schooling. In my contribution, I will concentrate on the approximately 50% of the youth who attend these part-time college/part-time on-the-job-training schemes which lead to a certificate of a qualified worker. For these schemes I will use the word "dual system" of TVE (for details cf. Blum, 1979; Straesser, 1980).

During the normal three years of this "dual system" scheme, the student or apprentice attends college courses for one to two days a week; the rest of the week he will be in private firms and will be trained on-the-job to a varied extent. During these three years, one or two of his approximately eight hours of (weekly) college attendance will be devoted to lessons which offer mathematical topics. These are meant to be taught in close relations to situations of the vocation the apprentice is trained for.

Mathematics in TVE for Lower Qualification Levels

Entrance Tests Set by Employers

Comparable to other industrialized countries, there is a shortage of places in the dual system in the FRG. So the employers set entrance tests in order to select the applicants for these trainee posts. In these tests, mathematics plays a very important role. Research about the qualifications tested shows that the tests concentrate on numeracy (arithmetic with decimals and to a varying extent fractions; proportions; change of measurements, rule of three), percentages, and some geometry. There are no standardized tests offered for this special purpose. Sometimes employers even set tests with apparently no predictive value nor mathematical or psychological background (for all these findings cf. Loercher, 1980).

The Implemented Mathematics Curriculum in West German TVE

As there is no empirical study analyzing the mathematics lessons really taught in West German TVE, I will give a brief survey of what ought to be the contents of the lessons in the dual system—judging from syllabi and textbooks.

Mathematics lessons center around such topics as basic arithmetic, percentages, and proportions (rule of three, etc.). In some vocations, these techniques already seem to be all the mathematics needed, and they are applied in these lessons to a great variety of situations taken from the vocation the student prepares for. In addition to this "core of topics" which is present in all vocations, there are some specific topics which depend on the vocation the student is trained for: basic algebra (especially the manipulation of formulae);
table-, diagram-, and graph-reading and interpreting; and some more advanced calculating techniques and problems as logarithms, exponential functions, and trigonometry. They are added to the curriculum for metal workers, and apprentices in the electrotechnical and electronics field, as well as for those going into careers in the construction vocations. Commerce apprentices are being trained for business calculations, including calculation of interests, discount rates, and securities. Some basic descriptive statistics is now entered in these commerce curricula. Geometry and technical drawing play an important role in the construction vocations (metal and wood workers, bricklayers).

Most of these topics may have been touched upon or learnt in compulsory education before. The greatest difficulty for the students (apprentices) seems to be the application of these techniques to vocational situations, as can be seen in the results of the final examinations at the end of this vocational training. The search for and identification of the right way to solve a vocational problem - - the "mathematization of the vocational situation" - - turns out to be twice as difficult as the calculation itself (cf. Ploghaus, 1967, p. 524). Apprentices often fail because they cannot correctly relate mathematics and the vocational situation; they have not learned to apply mathematics.

Criticism Concerning the Topic-Oriented Approach

Arguments from Research in Mathematics in TVE.

Going through the reports published on research i the field and talking with people doing the research work, one often hears about the problem of really determining the needs of industry and commerce: German researchers complained explicitly about the problem of finding the right person to indicate the needs. The same statement seems to be true for the English research work, as was said in discussions with English researchers. Entrance tests at the beginning of vocational education obviously only test skills of applicants that have been trained in compulsory schooling beforehand. These tests, however, cannot be used as indicators of the mathematics needed in the jobs. The great variety of different entrance tests even points at the possbility that employers do not know the qualifications they want from their applicants when they apply, without taking into account the qualifications the future craftsman needs in her or his job.

Looking into the results of the research work, "confusion" seems to be the right description: Answers even for the same profession are contradictory, even when companies of the same size and production area are asked, or as Dawes and Jesson (1979) put it: "The analysis (of employers' answers as to what mathematics they need) provided some indication that the mathematical needs of the young employee are not . . . clearly defined by their employers. Even the 'common core' identified by Knox can hardly be said to be 'common'"(p. 399).

Do these contradictions only indicate the deficiencies of our research methodology and questions, or do they reflect a more deeply rooted mistake we make when only searching for the mathematics needs of industry and commerce.

The Process of Mathematization as a Topic in TVE

As already pointed out, apprentices have a lot of difficulties in finding the appropriate mathematics for the solution of their vocational problems. Why is this true, and how can we change this state of the art?

One simply stated problem may stem from the organization of this vocational education: Having separated college and on-the-job training, the college seems to have problems in keeping step with the evolution of technology and thus in teaching the apprentice the application of mathematics to real working life situations. The vocational problems which are to be solved in college mathematics may tend to become artifical or even old-fashioned and irrelevant to the real working life.

Besides this problem of keeping step with the evolution of technology, if not being ahead of technological progress, mathematics in TVE has to teach not only applicable mathematics, but also the process of the application of mathematics, the "mathematization process." West German research done in the field of application of mathematics to vocational situations (Damerow et al., 1974; Jaeger & Schupp, 1979; Schotz, 1980) shows that this process cannot be described or understood merely in terms of the mathematical discipline. Concepts such as modelling, communication skills, and cooperation with experts in other fields than mathematics are needed to understand what is going on in such a process of mathematization. I am not so fond of the often quoted circle describing the application of mathematics as the process of

> problem definition - model building - building of a mathematical model - solving the mathematical model - interpreting the mathematical results in terms of the problem definition

and maybe reiterating this circle. This description, however, shows one aspect of the problem very clearly: the process of application of mathematics is not only a process inside mathematics. Mathematics plays a minor role in this business, which is determined by facts and motives outside mathematics. Consequently, mathematics in TVE cannot concentrate on the teaching and learning of special mathematical topics but also has to provide the knowledge and skill of the mathematization of a given or future vocational situation.

"Mathematical Thinking," not Mathematics, is Needed in TVE

In the FRG, two studies have been published concerning the way and extent probability and statistics are being used in commerce and industry (Jaeger & Schupp, 1979; Scholz, 1980). One of them was done by interviewing workers and employees about the statistical knowledge they really needed in their jobs. The jobs to be analyzed were chosen because of their numerical importance for apprentices. The other study used the case study method and analyzed how and to what extent stochastical phenomena are noticed and handled by physicians and managers of smaller branches of banks. Both studies came to the identical conclusion that it was not probability and statistics as such that was used in the vocational situations. The workers and employees used statistical techniques and concepts in a way quite different from how the discipline of mathematics presents them. What they need is the overall concept of stochastical information and decision making on the basis of uncertainty, instead of special frequency distributions, probability theory, or Kolmogorov's axioms.

A second piece of evidence comes from a broader discussion about the organization topics, and methods in TVE in the FRG: The trend seems to favor not a narrow qualification for just one special activity or job, but preparation for whole areas of working life in order to enable workers to change their occupation, their job, or even their vocation. Such a broadening of the field the apprentics is trained for may induce a different mathematics education than a disciplinary one.

I have the impression that mathematics in TVE should stress the overall concepts of mathematics such as numbers, functions, geometry, and stochastic thinking. Mathematics should offer basic ways of thinking such as the idea of algorithm, logical reasoning, and sequential analysis. Offering these "topics" together with a deep understanding of the methods, the possibilities, and the problems when applying mathematics to vocational situations could offer a link between the mathematics to be taught and the working situations it is needed for.

References

Blum, Werner: Stichwort "Berufliches Schulwesen." In: Volk, Dieter (ed.), Kritische Stichworter Mathematik. Munchen (Fink) 1979, pp. 15-32.

Damerow, P., Elwitz, U., Keitel, Ch., Zimmer, J.: Elementarmathematik: Lernen fur die Praxis? Ein exemplarischer Versuch zur Bestimmung fachuberschreitender Curriculumziele. Stuttgart (Klett) 1974.

Dawes, W.G., Jesson, D.S.: Is There a Basis of Specifying the Mathematical Requirements of the 16 Year Old Entering Employment? In: International Journal of Mathematics Education in Science and Technology, vol. 10, no. 3, 1979, pp. 391-400.

Jaeger, J., Schupp, H.: Mathematikunterricht und Arbeitswelt - Zwischenbericht. Sarrbrucken (University) 1979, typescript with appendix.

Knox, C.: Numeracy & School Leavers - A Survey of Employers' Needs (Sheffield Regional Centre for Science and Technology, Paper no. 12), May 1977.

Loercher, G.A.: Allgemeinbildender und berufsbildender Mathematikunterricht. Diskrepanzen und Koordinationsmoglichkeiten. In: ZDM, vol. 12, no. 4., 1980.

Ploghaus, G.: Die Fehlerformen in Metallgewerblichen Fachrechnen und unterrichtliche. Mabnahmen zur Bekampfung der Fehler. In: Die berufsbildende Schule, vol. 19, no. 7/8, 1967, pp. 519-531.

Scholz, R.W.: Berufliche Fallstudien zum stochastischen Denken. In: Beitrage zum Mathematikunterricht 1980. Hannover (Schroedel) 1980, pp. 298-301.

Straesser, R.: Mathematik in der (Teilzeit-) Berufsschule. In: ZDM, vol. 12, no. 3, 1980, pp. 76-84.

A MODEL FOR A MATHEMATICAL SYLLABUS WHICH WILL RECONCILE THE EXPECTATIONS OF BOTH EMPLOYERS AND TEACHERS

Robert L. Lindsay
Shell Centre for Mathematical Education
Nottingham, England

Currently in the UK there is much activity in reviewing the present state of school mathematics teaching. After twenty years of experimentation in methods, in content, and in school organisation, amid fierce criticism during the last decade by industry and commerce, two successive political governments have appointed, or supported, the (Cockcroft) Committee of Inquiry into the Teaching of Mathematics in Schools, which will report in 1981.

Part of this task has been the collation of case-studies of people in a variety of employments, particularly those in which the lower sixty percentiles of the population use modest mathematical techniques.

An extreme, but not unusual quotation from these case-studies has been: "At School I liked arithmetic, and I like it now, and I find it useful, but the rest of maths is rubbish". Such remarks beg the question of teachers: "In what way is the maths we teach irrelevant to the requirements of the early school-leavers and injurious to their life-long attitudes to mathematics?"

The problem for the teacher is how to maintain balance between the cultural aspects of mathematics and its utilitarian content - i.e., between mathematics as "queen or servant". But what constitutes a utilitarian content will differ according to the employment, and this first needs to be expounded.

At the age of 16, in England and Wales, there are Public Examinations for some school leavers. The top 20% are entered for the General Certificate of Education (GCE) which is generally acceptable to employers. The next lower 40% are entered for the Certificate of Secondary Education, which is not always so acceptable to employers, while the remaining 40% have no Public Examinations, except the inappropriate CSE. For these, and for half or more of the CSE candidates - for at least the lower 60% of the school population - a more satisfying programme still awaits discovery.

A model of the syllabus priorities would be of help: both to those who, as heads of departments, ordain what subordinate staff must teach; and to those same subordinate staff who often might not be trained as mathematicians but who find themselves time-tabled to teach mathematics, both because of the national shortage of trained mathematicians in schools and because of a surplus of teachers of certain other subjects. Their problem is one of knowing the priorities within the syllabus, especially when dealing, as they mostly do, with the less able who are unlikely to master or even cover the entire syllabus.

The usual model of syllabus priorities is the Apple:

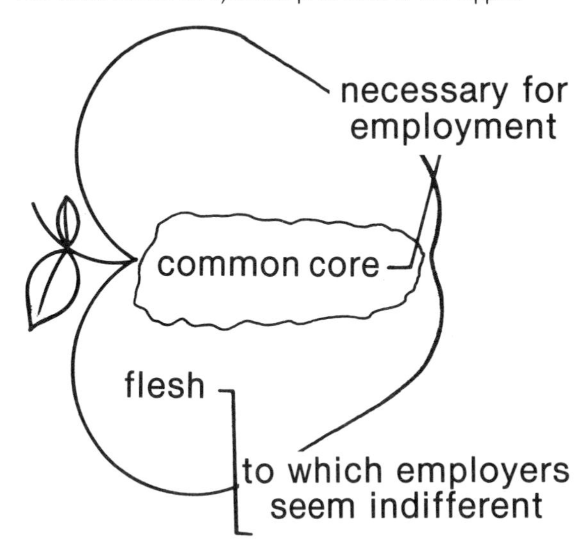

But this model fails when, in collaboration with teachers and employers, an attempt is made to define its Common Core. The engineers will rightly demand inclusion of Pythagoras' theorem and the trigonometry of the right-angled triangle, but the nurses will have none of this. Yet some sort of priority should be assigned to these topics.

In practice, a tripartite model has been found to be more practicable, nor has any further subdivision been found to be necessary to establish useful priorities. This model is a Peach.

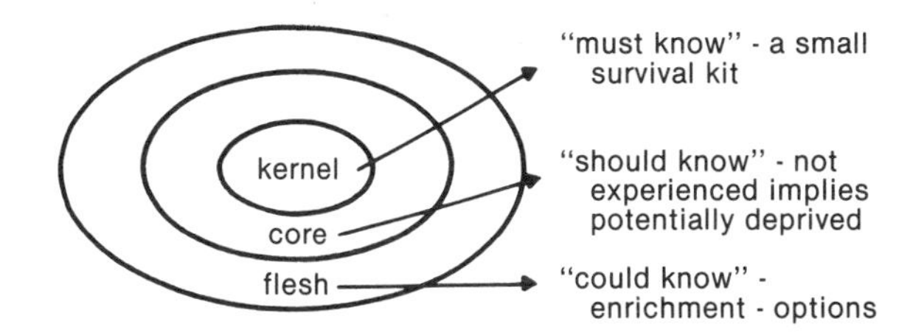

In this model, the Pythagoras and trigonometry are property allocated to the Core (no longer the Common Core) since school children are mostly not classified as potential (male or female) nurses or craftsmen when these topics are normally introduced. Therefore, if they are not made familiar with them, these children could be deprived of the opportunity of certain careers.

When this tripartite analysis has been completed, it is then possible, using this criterion, to assess the many available syllabuses:-

If the Union of all syllabuses is:

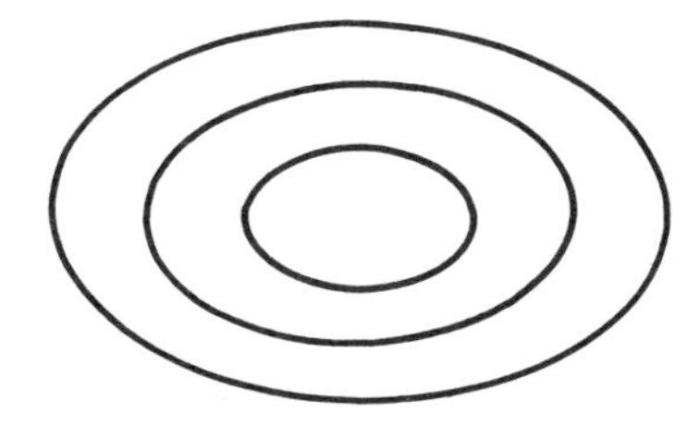

then one subset, the hypothetical syllabus "Brand X" might be:

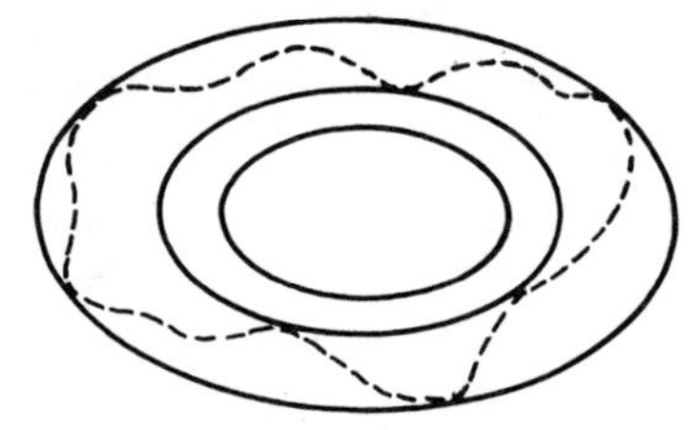

and is acceptable to the extent that it respects the absolute priorities of kernel and core.

But the following syllabuses may be deemed effective:

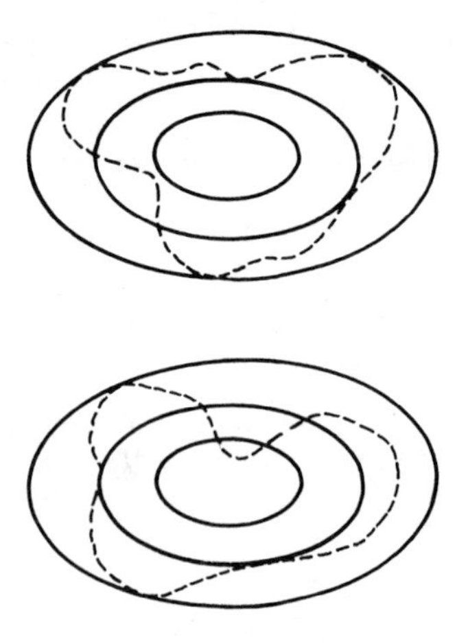

It is furthermore possible to classify the careers available to 16 year olds into four classes. The four entry-levels into engineering are well-defined in the UK, and many other careers conform to a similar structure. These levels are for those who become (1) full-time students at advanced, or "A", level; (2) "Technicians" with GCE qualifications; (3) "Craftsmen" with CSE qualifications; (4) "Operators" with little or no mathematical testing by their employers.

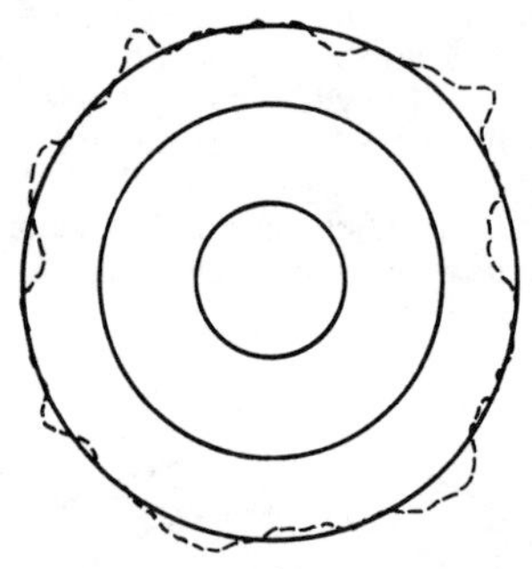

Good potential A level mathematics pupils will ideally master all the syllabus with some slight omissions and some excursions beyond.

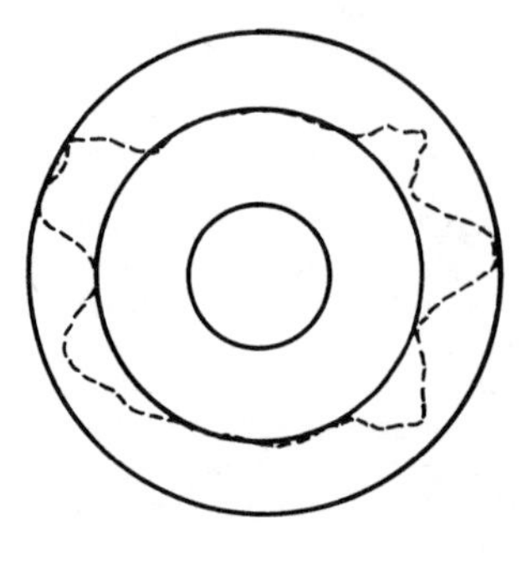

A 16+ leaver at "Technician" level or equivalent will ideally extend mastery beyond the kernal and core into areas relevant to the intended profession.

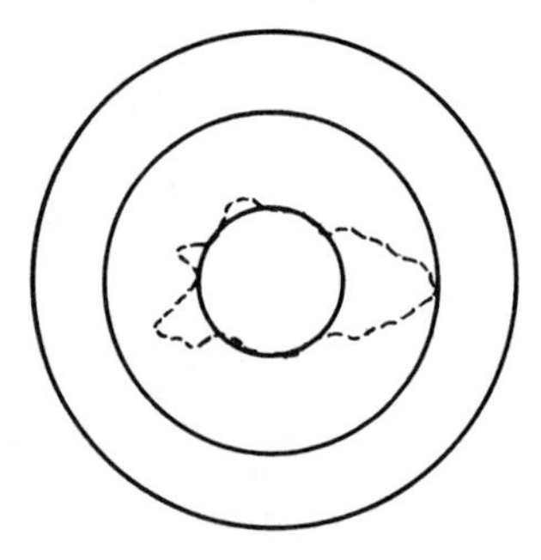

A 16+ leaver at "Craft" level or equivalent will ideally extend mastery into those parts of the core which are relevant to the intended craft.

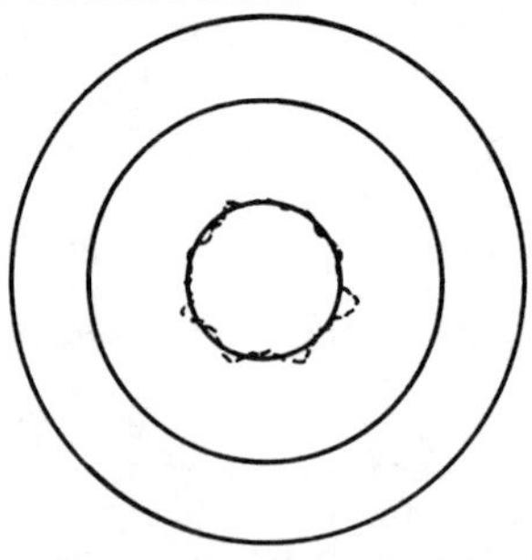

A 16+ leaver at "Operative" level will ideally conform as closely as possible to the pattern of the peach model.

Nothing said here should be taken to imply that we are defining a time sequential teaching programme. We do not support teaching the kernel, the core and then the flesh consecutively in that order; it is probably impossible to do so without dire results. Work can be spread over each of the three zones in any convenient order once a minimal working basis has been established. Indeed, the most fruitful zone for practicinig the techniques of the core and kernel is the flesh, where there will naturally arise many opportunities to practice division, for example, in a wide variety of contexts.

The naive model for this recommended process is, therefore, nearer to that of successively carving out crude slices of a cake:

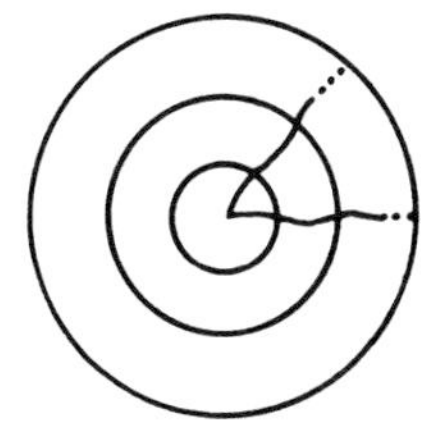

but the notion of overlapping sectors better demonstrates the processes of forging lateral connections and of recalling to mind previous forays.

THE MATHEMATICAL KNOWLEDGE AND SKILLS NEEDED BY GRADUATE ENTRANTS TO INDUSTRY AND COMMERCE IN GREAT BRITAIN

Peter Price
Stockton-on-Tees, England

1. Introduction

This paper is essentially a review of a Symposium (1) called by the Institute of Mathematics and its Applications (IMA) in May, 1977. The Symposium attracted about two hundred participants and was addressed by a distinguished list of speakers from a wide range of employing organisations. Contributors included Sir Hermann Bondi (Chief Scientist at the Department of Engergy), Sir James Lighthill (then the Lucasian Professor of Mathematics at Cambridge University) and Mr. E. A. Johnston (the Government Actuary). The present writer contributed a paper on the requirements of the chemical industry. Copies of the Proceedings of the Symposium are obtainable from the IMA, Maitland House, Warrior Square, Southend-on-Sea, Essex, England.

Two preliminary points need to be made in this Introduction:

The first is that the Symposium was intended to study the mathematical qualities needed from all graduate entrants, not exclusively mathematical specialists. The second is that the Symposium contributions can be properly understood only by reference to the character f British university and polytechnic education. This has inherited a tradition of selective entry, high specialisation, and high academic standards for any given age-group. Thirty years ago only about three per cent of each year-group entered university: the majority of professional men, engineers and technologists were educated outside the universities in the further education (technical college) system. Even today the proportion entering universities and polytechnics is only about ten per cent. So in discussing the "graduate entry", the Symposium contributors were essentially discussing the education of the top academic decile, who would be expected to make a professional-level contribution to their specialist discipline, and/or proceed to middle or senior management positions.

2. Technical Mathematics

It might be expected that the need for graduates to understand and be able to use the basic mathematics of their specialism would be fully met, and would not require detailed examination from a Symposium. While most speakers recognised that in many disciplines the mathematical education of graduate entrants in industry was satisfactory, certain areas of weakness were disclosed and discussed in detail by the contributors.

The most important was the inability of some technical graduates to formulate and solve practical mathematical problems in their field from first principles. In the words of one writer "There is a tendency for technical workers who are weak in fundamental knowledge and analytical skill to try to solve a problem by recalling an apparently similar example from their textbooks or lecture notes. On straightforward problems this approach should work, but in industry a substantial proportion of the problems that come up are not simple and straightforward. In these cases, attempts to construct the correct solution by imperfectly recollecting apparently similar examples can lead to results which may vary from the clumsy and laborious, but ultimately correct, to the completely absurd." (2)

The question of how far this weakness can be attributed to the British university system of assessment by annual formal examinations was explored in the paper by Sir James Lighthill. He emphasised the degree to which, in this system, examination questions define the curriculum. "Collectively" he stated "the (examination) questions set constitute the main channel for communicating to students the body of knowledge and skills which they should above all aim to acquire." (3)

He expressed concern about the consequences of the fact that while for obvious practical reasons any examination question must be capable of solution in about one hour, in industry or commerce even the simplest problem will require days or weeks of work in collecting and evaluating data, developing a mathematical model, and investigating the significance of its formal solutions to the practical problem. The technical graduate in the real world therefore needs personal qualities such as persistence, and mathematical skills such as a disciplined iterative approach to the formulation of a mathematical model, which cannot be tested in the examination room and which tend to be inadequately developed in the traditional examination-oriented university course.

He, therefore, recommended the inclusion within university courses of project work in depth on real-world practical problems. He was supported in this conclusion by Dr. Sherman (GEC Machines Ltd.), who also emphasised the value of experience in developing mathematical models. "It is necessary to apply mathematics: to know how to take a business problem and model it for mathematical solution. The technologist must be able to think about the right assumptions and approximations which can be made, what numbers can be given to parameters and how to interpret a solution back into real terms." (4) Several other speakers also gave support to the need for students to have more project-type experience and more training in the use of mathematical models.

In addition to this general problem, two special problem areas were noted by contributors: a general tendency for chemists to exhibit weakness in mathematics, relative to graduates in other disciplines, and a lacknd more training in the use of mathematical models.

In addition to this general problem, two special problem areas were noted by contributors: a general tendency for chemists to exhibit weakness in mathematics, relative to graduates in other disciplines, and a lack of formal training, among graduates in more than one subject, in the statistics of the design and analysis of experiments.

The mathematical weakness of chemists is probably attributable to the early specialisation in British schoplines, and a lack of formal training, among graduates in more than one subject, in the statistics of the design and analysis of experiments.

The mathematical weakness of chemists is probably attributable to the early specialisation in British schools and higher education. A pupil of poor mathematical ability may opt for an education on the science side and then find his mathematical weakness a handicap: chemistry, as the least mathematical of the physical sciences taught in school, is a natural destination for such a student. Unfortunately chemistry in industry, except in certain very specialised fields, is a highly numerate subject, and the chemist who has a pronounced allergy to mathematics can be a real liability.

Chemists, engineers and some other technical graduates are often found to be inadequately prepared with respect to the statistical treatment of experiments. Today students must be expected to employ the same rigorous standards in evaluating experimental results as are used throughout industry, commerce and government, and must be given access to the appropriate computer programmes and training in their use.

3. Management Mathematics

It is perhaps also necessary to record that nearly all the mathematical techniques of management are important to all graduate entrants to industry and commerce. Even the technical graduate, who is initially hired to work in his specialist area, is likely to find himself in middle of senior management later in his career. Although in large organisations there will be corporate planning and operational research teams able to provide specialist services in areas requiring advanced techniques, the general manager and the technical graduates in a multi-disciplinary team must have a good appreciation of management mathematical methods in order to be able to communicate with the expert and assess the practical value of their advice.

Technical graduates do not, in general, have much difficulty in understanding basic management mathematics. Their own grounding in technical mathematics provides the essential foundation. The only serious problem is that some specialists, particularly chemists and engineers, have little knowledge of statistics. Universities should appreciate that the administration of large production and marketing organisations is an essentially numerate activity, requiring a sound foundation in theoretical statistics and some practical experience in elementary applications, and should provide appropriate option courses and encourage their technical students to take them.

The situation is much more serious with respect to Arts graduates, who constitute a substantial part of the entry to industry and commerce in Britain, particularly in functions such as personnel and marketing. Some British Arts graduates have not studied mathematics beyond the age of sixteen, and cannot be expected to cope with, for instance, the details of interpretation of a linear programme or Box-Jenkins methods applied to market forecasting. It is difficult to avoid the conclusion that a substantially higher standard of mathematical sophistication is need for everyone likely to reach a managment position.

A special problem is the mathematical education of economists. Economists in industry are expected to make their professional contribution in multi-disciplinary teams in such areas as market forecasting and corporate planning. They are econometricians more than economists. Yet the tradition that economics is really a branch of the humanities, taught with a historic and literary emphasis with only a minimum of essential mathematics, is still very much alive. Many university departments of economics have recognised the problem and are increasing the mathematical content of their courses, but progress is very variable from university to university and is altogether too slow. The lack of mathematical preparation of many Arts-side school leavers acts as an obvious constraint on progress in this field.

4. Communication Skills

This subject might be regarded as outside the strict range of this paper, as it does not deal with a mathematical skill. But so many contributors emphasised its importance that it would be entirely wrong to omit it.

Mr. McDonnell (British Aerospace) set out the position as it appears to the industrial manager. "The most effective work in problem solving is almost certainly done in small teams ... Communication with the team is vital and the graduate should be able to produce written documentation in good English and good mathematics for his colleagues, as well as being able to discuss problems with them less formally. The ability to communicate with others outside the team, managers for instance, is also important. A team is likely to have to fight fairly hard for its budget. To do this one of its members at least must be able to describe the problem, its importance, its progress and justify its cost. This kind of communication requires an ability to see the essentials, to understand the other man's position, and may even require a touch of guile or showmanship". (5)

The same point was emphasised in his introduction by Sir Hermann Bondi: "Something that is needed in almost any type of work is the ability to communicate. However, it must be admitted that to date there are relatively few courses in mathematics at school or university level that put sufficient stress on this, if indeed they put any stress on it other than occasionally the ability to communicate with the examiner. And I think there is a real gap here... I think it is not just the skills in communication that it is so important to get across. It is even more important to teach the essential need for communication, the fact that every enterprise is a social enterprise, that the number of positions open for hermits is rather small... In many other subjects, on the arts or the social sciences side, people do get trained in debate, in writing essays. I do feel we should not allow this to go by default in mathematics". (6)

This last point was echoed again by Mr. Johnston, the Government Actuary. "Like other professional men, the actuary exists to communicate his results to his customers, so an inability to express himself is a bad fault, but, I am afraid, one which our Board of Examiners often finds in candidates. A degree course which never requires the student to write a sustained piece of English prose is not a help." (7)

5. Conclusion

It is hoped that members of this mini-conference on the interface between mathematics and employment will be interested in the points raised in the IMA Symposium. It would be of interest to learn how far the problems disclosed are peculiar to the British system of higher education, and how far they are paralleled by experience in the USA and other countries.

6. References

1. The Mathematical Skills and Qualities Needed in Graduate Entrants to Industry and Commerce, 1977, I.M.A. Symposium Proceedings No. 17.

2. P.C. Price, "The Mathematical Skills and Qualities Needed in Graduate Entrants to the Chemical Industry", ibid., pp. 62-63.

3. Sir James Lighthill, "Bridging the Chasm Separating Examination Questions from Real-World Problems", ibid., pp. 48-49.

4. W.G. Sherman, "The Needs of Technologists in Manufacturing Industry with Special Reference to Heavy Electrical Engineering", ibid., pp. 84.

5. J.H. McDonnell, "Mathematical Graduates in the Aircraft Industry", ibid., pp. 38-39.

6. Sir Hermann Bondi, "Introduction to the Symposium", ibid., pp. 2-3.

7. E.A. Johnston, "Actuary: A Commercial Profession", ibid., p. 17.

RELATIONS BETWEEN MATHEMATICS AND EMPLOYMENT IN MATHEMATICS EDUCATION IN FULL TIME TECHNICAL AND VOCATIONAL EDUCATION

Werner Blum
Gesamthochschule Kassel
Kassel, West Germany

1. Some Remarks on the West German System of Full Time Technical and Vocational Education

It would be much too complicated to explain in detail the organisation of the German full time Technical and Vocational Education (TVE) system. There are schools at the lower secondary level (pupils aged about 15 to 16 years) as well as colleges at the upper secondary level (students aged about 17 to 20 years) and colleges for further education (students aged over 20 years). For details about the German TVE system and mathematics education in it, see (8) and (3). All types of these full time schools and colleges provide the student with both a general certificate and a preparation for some vocational area. In all types of schools and colleges mathematics is an important subject. The contents of mathematics education are - - with some exceptions - - roughly comparable to the contents in those schools and colleges in the "general" education system that provide students with equivalent certificates. (In this paper, I cannot consider meaning of or problems in "general" education and mathematics education in it; see (8) and (3).) That means: algebra (equations and functions) and geometry in the lower secondary level, and in the upper secondary level mainly calculus, in addition to analytic geometry and some stochastics.

2. Important Goals of Mathematics Education in Full Time Technical and Vocations Education

I deal neither with "official" nor with "actual" goals of mathematics education in full time TVE. Instead, I propose three goals which in my opinion are all important for full time TVE. (The goals - - explicit or implicit - -actually aspired to in the classroom are very often the "vocational part" of goal 1 or goal 2 and very seldom both together. Roughly speaking, mathematics education in full time technical and vocational schools and colleges is often either an auxiliary subject for technology or business or a self-supporting subject oriented toward the "gymnasium".)

Goal 1:

Mathematics education should contribute to better describing, better understanding, and mastering present and not-too-distant future problems from the student's vocational and everyday environment. Goal 1 has a "vocational" and a "non-vocational" part. Taking into account the latter, preparation for vocations in one but not the only direct use that should be made of mathematics.

Moreover, more important than dealing with particular problems and particular mathematical topics is that the student learn by a few examples to translate between mathematics and the real world, and especially that he or she learns to mathematize meaningful contexts (7) and learn general strategies for tackling real problems, according to Figure 1 (see also 6, 11, 10, 5, 2).

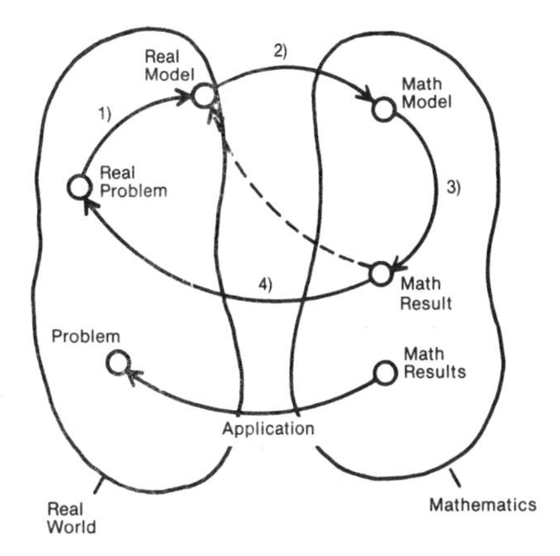

Figure 1

Goal 2:

Mathematics education should try to attain so called formal aims, that is (in accordance with Wittmann (13)): to advance cognitive strategies and intellectual techniques, such as mathematical reasoning and algorithmic thinking.

Goal 3:

Mathematics education should contribute to equalizing the opportunities of students by making possible further learning and advancement to higher education.

3. Implications for the Interface Between Mathematics and Employment in Full Time Technical and Vocational Education

The interface between mathematics and employment in mathematics education works mostly by dealing with situations and problems coming from the corresponding vocational area and containing or leading to suitable mathematical topics. It stands to reason that vocational situations and problems provide important hinds for the mathematics curriculum in order to achieve the "vocational part" of goal 1 (role a).

But, according to goal 1, vocational problems are not the only important ones in full time TVE. As, for example, Strasser (12) points out, the mathematical needs for various jobs and vocations are not as clearly defined as would be desirable. Therefore, vocational situations and problems cannot immediately or even exclusively determine the mathematics curriculum. Instead, such situations and problems (for criteria for suitable application problems, see 9, 10, or 2) play a methodological role as they show by example the interface between mathematics and real world situations (role b), motivate students to study mathematical topics that may have another justification, according to goals 2 and 3 (role c), and illustrate mathematical topics and thus help the student understand them better and retain them longer (role d).

4. An Example: The West German Income Tax Function

As an example of an application that can play all for roles identified in section 3, we look at the German income tax. This example yields a vocational problem from the economic and financial area. At the same time, and perhaps didactically even more important, it yields an everyday problem for all present or future taxpayers. I have had much instructional experience using this example with 17- to 19-year-old students (see also chapter 7 in textbook 1).

4.1 Information

The present West German income tax law runs as follows:

. 32a
Einkommensteuertarif

(1) Die tarifliche Einkommensteuer bemisst sich nach dem zu versteuernden Einkommen. Sie betragt vorbehaltlich der .. 32b, 34 un 34b jeweils in Deutsche Mark

1. fur zu versteuernde Einkommen bis 3719 Deutsche Mark: 0;

2. fur zu verteuernde Einkommen von 3720 Deutsche Mark bis 16019 Deutsche Mark: 0,22x - 812;

3. fur zu verteuernde Einkommen von 16020 Deutsche Mark bis 47999 Deutsche Mark: (((10,86y - 154,42)y + 929)y = 2200)y + 2708;

4. fur zu versteuernde Einkommen von 48000 Deutsche Mark bis 130019 Deutsche Mark: 109,95)z + 4800)z + 15298;

Deutsche Mark an:

"x" ist das abgerundete zu verteuernde Einkommen. "y" ist ein Zehntausendstel des 16000 Deutsche Mark ubersteigenden Teils des abgerundeten zu versteuernden Einkommens. "z" ist ein Aehntausendstel des 48000 Deutsche Mark ubersteigenden Teils des abgerundeten zu versteuernden Einkommens.

Let, as in this law, x be the annual income (in DM), and let s(x) be the income tax in DM) that has to be paid by an unmarried peson earning x. Idealizing, we allow x to be any element of IR . Disregarding rounding off, the German income tax function $s : x \to s(x);\ x \in \mathbb{R}^+$ is a piecewise polynomial function. The graph f this is shown in Figure 2.

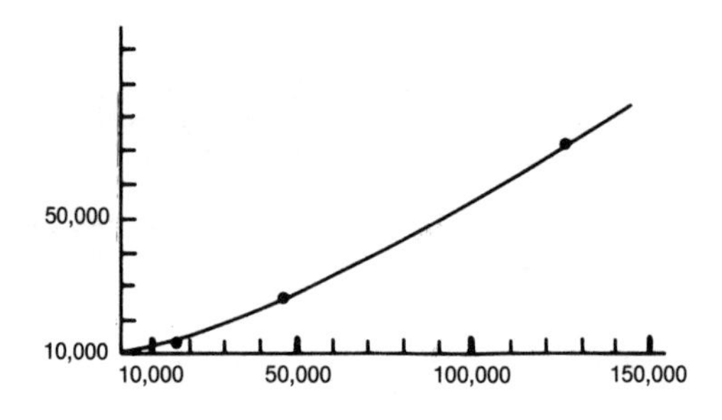

Figure 2

Let $d(x) = s(x)/x$ be the average tax rate for income $x \in \mathbb{R}^+$ and $d : x \to d(x);\ x \in \mathbb{R}^+$ be the average tax rate function. If, as in the German law, the income tax function is almost everywhere differentiable and everywhere differentiable from the right, let s'(x) be the (right) derivative at $x \in \mathbb{R}_0^+$, that is, the instantaneous rate of change of s at x, or the marginal tax rate. Let $s' : x \to s'(x);\ x \in \mathbb{R}_0^+$ be the marginal tax rate function. The two functions d and s' are more important in political and public discussions that the function s itself. The graphs of d and s' for Germany are shown in Figures 3 and 4 respectively.

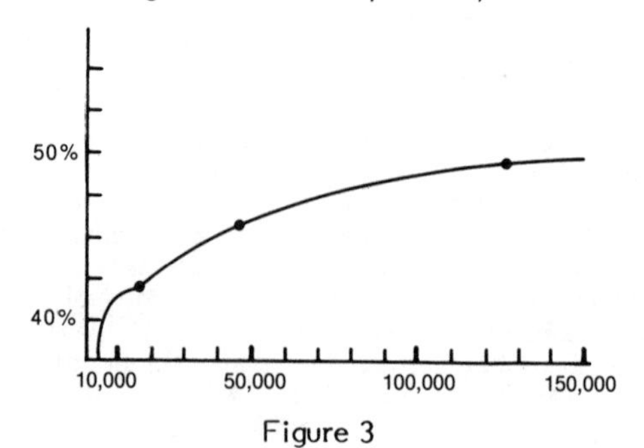

Figure 3

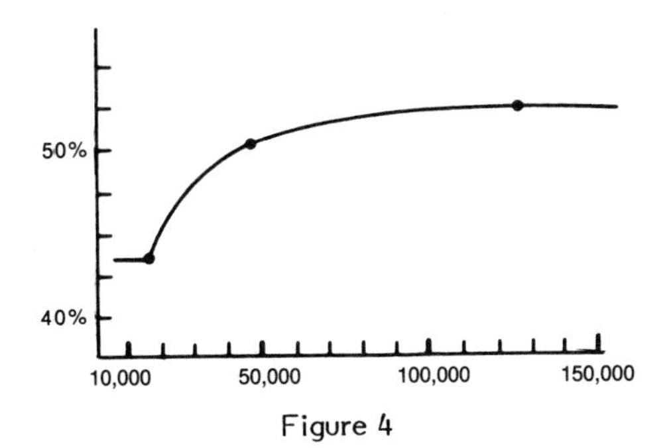

Figure 4

4.2 Income Tax as a Subject in Mathematics Education in Full Time Technical and Vocational Education

Firstly, this example can show - - according to role a in section 3 - - how mathematical topics (like functions and their properties, secants, tangents, derivatives, etc.) can contribute to achieving goal I (with regard to some economical problems). For example, mathematics education can help students understand the idea of tax rates. This is important not only for future tax consultants but for all citizens. Therefore, this example is important not only for vocational colleges in the business area but also for all other types of colleges, where it plays an analogous role to a, according to the "non-vocational" part of goal I. The appropriate mathematical topics (like derivatives) need not necessarily be introduced in a formal and rigorous way. The aim is that students may be able to use these topics intelligently as a tool. Simplifications which are not falsifications are allowed or even necessary (see 4).

Secondly, this example can demonstrate - - according to role b - - the interface between mathematics and real world situations shown in Figure 1 (see 2).

Thirdly, this example yields - - according to roles c and d - - motivations and illustrations for a great many mathematical topics, especially for the uppe secondary level. For example, it is easy to show the importance of determining by means of calculus the regions of increase and of convexity of the polynomial functions that compose the German income tax function (see again 2).

References

1. Athen, H., & Griesel, H. (Eds.): Mathematik heute, Grundkurs Analysis I. Hannover, 1978.

2. Blum, W.: Einkommensteuern als Thema des Analysisunterrichts in der beruflichen Oberstufe. Die berufsbildende Schule, 1978, 30 (11), 642-651.

3. Blum, W.: Berufliches Schulwesen. In: Kritische Stichworter zum Mathematikunterricht (Ed.: Volk, D.), Munchen, 1979, pp. 15-32.

4. Blum, W., & Kirsch, A.: Zur Konzeption des Analysisunterrichts in Grundkursen. Der Mathematikunterricht 1979, 25 (3), 6-24.

5. Fischer, R., & Malle, G.: Fachdidaktik Mathematik, Lehrbrief fur das Fernstudium Padagogik fur Lehrer an hoheren Schulen. Klagenfurt, 1978.

6. Ford, B. & Hall, G. G.: Model Building - An Educational Philosophy for Applied Mathematics. International Journal of Mathematical Education in Sciences and Technology, 1970, 1 (1), 77-83.

7. Freudenthal, H.: Mathematics as a Pedagogical Task. Dordrecht, 1973.

8. Kell, A. & Lipsmeier, A.: Berufsbildung in der Bundesrepublik Deutschland - Analyse und Kritik. Schriften zur Berufsbildungsforschung, Vol. 38, Hannover, 1976.

9. Pollak, H. O.: How Can We Teach Applications of Mathematics? Educational Studies in Mathematics, 1969, 2 (3), 393-404.

10. Pollak, H. O.: The Interaction betwen Mathematics and Other School Subjects. In: New Trends in Mathematical Teaching (Ed.: UNESCO), Vol IV, Paris, 1979, pp. 232-248.

11. Steiner, H. G.: What is Applied Mathematics? Indian Journal of Mathematics Teaching, 1976, 3 (1), 1-18.

12. Strasser, R.: Apprenticeship and College Mathematics in Technical and Vocational Education. In these proceedings.

13. Wittmann, E.: Grundfragen des Mathematikunterrichts. Braunschweig, 1976.

8.7 HOW EFFECTIVE ARE INTEGRATED COURSES IN MATHEMATICS AND SCIENCE FOR THE TEACHING OF MATHEMATICS

CONSIDERATIONS AND EXPERIENCES CONCERNING INTEGRATED COURSES IN MATHEMATICS AND OTHER SUBJECTS

Mogens Niss
Universitetcenter Roskilde
Roskilde, Denmark

The increasing interest within the community of people involved in mathematics education in proposing and discussing courses where mathematics is taught integratedly with other subjects, seems to stem primarily from two different types of concern, which may be phrased as follows:

I. Mathematics teaching does not function too well as it has developed. It is probably necessary to recognize, even if only with uneasiness, that an increasing number of students do not feel sufficiently motivated to work with abstract mathematical topics or constructed, simplified application examples. Is it possible to restore students' motivation for the study of mathematics as a subject by demonstrating its role in tackling genuine, extra-mathematical problems, for instance by letting it be taught in integration with one or more other subjects, probably most adequately science subjects?

To this concern the idea of integrated courses represents a (new) strategy for pursuing the ultimate end: success for the teaching of mathematics as a subject in itself.

2. Mathematics teaching does not function too well as it has developed. One major reason for this is that it displays mathematics as being a purpose in itself, as a formal game, too far removed from the rest of the world. We should admit that the teaching of mathematics cannot serve as its own end. An important ingredient of the rationale for teaching mathematics is its role as a powerful tool in tackling problems outside mathematics itself. And this aspect has not gained a position which corresponds to its importance, a point which is also reflected in the often reported fact that students have major difficulties in applying the mathematics they have been taught, to "real life situations" at and after school. Is it possible to improve this state of affairs by letting mathematics be taught in direct contact with subjects for which it is a fundamental tool?

To this concern the ultimate interest in mathematics teaching is based on the view that mathematics is valuable in a broader context than just as a subject. This concern aims at improving the understanding and utilization of mathematics in extra-mathematical situations. The idea of integrated courses represents a (new) strategy to this end.

* * *

How are these concerns present as regards our session? Its title is "How Effective are Integrated Courses in Mathematics and Science for the Teaching of Mathematics?" This formulation seems to presuppose implicitly that the final goal of our endeavours is success for the teaching of mathematics as a subject in itself. According to this interpretation this session is devoted, then, to discussing to what extent integrated courses provide a useful means for pursuing this goal, how the integration should be carried out if the case should arise, and what problems are likely to be encountered in the implementation of it. This suggests that we are dealing with the first-mentioned concern.

The subtitle asks the question: "How to tackle the dilemma between the need for preparing children/students to deal with useful mathematics in natural situations, and the need for teaching them mathematics as a systematic structure." It takes the need of teaching students mathematics as a systematic structure for granted but introduces an additional need, with which the first might be in potential conflict and with which it hence should be balanced: the need of preparing students to use mathematical tools in coping with natural situations. This brings into the discussion elements of the second-mentioned concern. So, we are supposed not only to treat the theme of our session under the perspective of the first concern but also under the perspective of the second one.

* * *

In my view we should, however, not continue without having examined a little bit the purpose of giving mathematics education to large groups of children and students, since any answer to the questions of our session is bound to reflect somehow a basic conception of the purpose of mathematics education. If this purpose were taken for granted without further comment, the discussion might end up with talking at cross-purposes.

I agree with those who reject that mathematics education should be given for its own sake. (Moreover, I consider it over-optimistic to believe societies to accept "its own sake" as a sufficient argument for assigning to mathematics education the dominant position it actually possesses in almost any country. For further comments on this matter, cfr. (1).) The reason for giving mathematics education to a majority of children and students at every level must refer to regards outside mathematics itself.

Having not enough space to go into detailed analyses of these issues, I shall restrict myself to pointing out what to me is the main justification for giving extensive mathematics education to everybody.

Mathematics plays a central and increasing role in the function and development of fundamental aspects of our societies. It is of democratic importance, to the individual as well as to society at large, that any citizen is provided with instruments for understanding, judging and utilizing this role of mathematics. Anyone not in possession of such instruments becomes a "victim" of societal processes in which mathematics is a component. So, the purpose of mathematics education should be

> to enable students to realize, understand, judge, utilize, and sometimes also perform, the application of mathematics in society, in particular to situations which are of significance to their private, social and professional lives.

Two goals can be derived from this purpose:

a. students should acquire understanding of those factors within mathematics (such as ideas, concepts, edifices of theory, methods, etc.), as well as those outside mathematics, which are of importance to the applicability of mathematics, its potentials and limitations;

b. students should themselves acquire experiences with applying, independently and in a non-mechanical manner, mathematics in treating extra-mathematical situations.

* * *

Let us, on the background of these premises, rephrase the question of our session as follows: Are integrated courses in mathematics and science the right vehicle for the pursuit of these aims?

My answer to this question is "yes" and "no": intergrated courses are necessary but not sufficient. The arguments for this answer are partly based on concrete experiences in Denmark with integrated courses on different levels, partly on theoretical considerations of a more principal nature.

* * *

So far, we have not tried to define what to mean by "integrated courses in mathematics and other subjects". At this stage a few comments on terminology might be adequate.

By an integrated course in mathematics and other subjects (one or more) we shall understand an educational course in which the non-mathematical subjects occupy a salient position and are taken seriously, i.e. are not serving as only an excuse for doing mathematics, or as a source of illustrations to it, but their integrity and particular requirements are respected.

Using this definition, we can distinguish between two principally different types of integrated courses:

1. integrated courses where the main interests lie with the non-mathematical subjects and where mathematics is a service subject introduced only when and only to the extent that it serves these main interests.

2. integrated courses intending to serve also purposes paying particular regard to aspects of mathematics.

* * *

At the Natural Science Basic Education Programme at Roskilde University Centre in Denmark (my own university) students are engaged in half of the time in two years with performing problem-oriented projects on complex situations, which are not from the outset defined by subject categories but in which mathematics and/or different natural sciences become involved under various perspectives, according to the situation. These projects may be considered as (extreme) instances of integrated courses of the first kind.

One marked experience which we have gained from these projects, is that students do not, in general, feel much inclined to introduce mathematical consideration, unless this was explicitly planned from the very beginning. In projects where mathematics, or matters closely related to it, do not form a determined part of the problem in focus of attention, students tend to avoid mathematics, and even if they don't the involvement tends to be narrowly directed towards exploiting methods or results as recipes, without much regard being paid to their justification, let alone their embedding in a systematic theoretical context.

If, on the other hand, we are talking about projects, where the applicational role of mathematics is part of the initial problem, for instance through explicit intentions to construct or examine a mathematical model, things are markedly different. Students usually treat the issue of the project and the mathematics to be learned in connection with it, with a maturity, competence and sense of proportions, which are not often seen with students who learn mathematics and applications separately.

Experiences similar to those gained from the Natural Science Basic Education Programme are reported from various experimental courses of the same character but of a far smaller scale, at upper secondary level (16-19 years) in Denmark.

In the programme leading to a final degree in mathematics at Roskilde University Centre (a programme presupposing and continuing the basic education programme mentioned above) all activities have, of course, mathematics as a subject (considered in a broad context) in focus of attention. Within a complicated organisational framework, which I shall abstain from describing here (for a reference, cfr. (2)), courses where mathematics is integrated with other subjects form part of the programme. Also in this programme we have obtained encouraging experiences with projects where mathematical models in biology, economics, and physics are constructed or investigated, encouraging also as regards the learning of the associated mathematics, which proceeds faster and with stronger motivation than is usually seen with a similar mathematical content. However, there seems to be a tendency to a weakened long-term fastening of the mathematics learned in this way, unless it is supported by additional "fastening activities", like traditional exercise sessions.

* * *

The experiences reported show that if mathematics is left alone to integrated courses of the first kind it tends to reach an only marginal position, and a position which is insufficient for a successful pursuit of the purposes for mathematics education which I am advocating. In contradistinction, integrated courses of the second kind seem to be most valuable in this respect. Moreover, it is difficult to imagine how students could acquire the sort of understanding and personal experiences with mathematics in extra-mathematical situation, required by the goals stated, if not from integrated courses (in the sense here defined) forming part of their mathematical education.

This poses a new question: Couldn't we content ourselves with giving mathematics education alone in the form of integrated courses (of the second kind)? My reasons for answering "no" to this question are the following:

The capacity and power of mathematics in treating problems of "the real world" is not an illusion. Mathematics indeed offers remarkably powerful instruments for understanding and treating a large variety of complex situations outside mathematics itself. Any mathematics education pursuing the goals put forward here must aim at understanding the origin and character of this capacity. It is probably well justified to claim that the determining factors in this capacity are the abstract, deductive structures forming the edifice of mathematics. Therefore this edifice as such should be an object of study in mathematics teaching. This is why the dilemma mentioned in the subtitle of our session is a genuine and not an artificial one. It is necessary for preparing children/students properly to deal with useful mathematics in natural situations that they obtain insight in and knowledge about mathematics as a systematic structure. Such insight and knowledge do not automatically result from integrated courses. Actually our experience suggests that this is not the case. But balanced with courses on mathematics as a systematic structure they constitute a most valuable, and in my view even indispensible, component of any mathematics teaching aiming at giving not only mathematics instruction but mathematics education.

References

1. Mogens Niss: Goals as a reflection of the needs of society. To appear in "Studies in Mathematics Education II", Unesco.

2. Mogens Niss: The crisis in mathematics instruction and a new teacher education at grammar school level. Int. J. Math. Educ. Sci. Technol., 1977, vol 8, No. 3 (303-321).

AN INTEGRATED APPROACH: THE INTERRELATIONS OF MATHEMATICAL DISCIPLINES FROM A PEDAGOGICAL VIEW

Helmut Siemon
Pedagogische Hochscule I
Ludwigsburg, West Germany

1. Historical Remarks

I am going to use the term integration in mathematics teaching in a fairly comprehensive fashion, and propose to treat school mathematics in its entirety from the very beginning. My chief aim is to emphasize the interrelatedness of the different mathematical disciplines and to propound that greater attention be given to an interdisciplinary approach within the mathematical field where one offers itself. There are many areas of mathematics where such an approach commends itself and I intend in this lecture to point some of them out. Applied problems are especially useful in this respect to promote in students an interest in mathematics.

The idea of integration in mathematics teaching was probably first taken up in the middle of the last century in Austria and Germany and at the beginning of this century also in France (1). The name then was "Fusion". However, the considerations were confined to an integrated treatment of plane and solid geometry. At that time it was realistic to confine activities to this fairly moderate objective, for some of the curricula made a separate treatment of plane and solid geometry compulsory. It was difficult enough to convince school administrations of this innovation. The International Congress of Mathematics Teaching in Milan in 1911 (18-20 Sept.) discussed the matter thoroughly (2) and school authorities in Italy were prompted to introduce integration of plane and solid geometry as compulsory. But this innovation was only temporary. The programs developed did not meet the necessary requirements and the whole idea had to be abandoned two years after it had been introduced.

Interesting are some of the arguments in support of this "Fusion" idea, which were taken from experimental psychology (3): To comprehend plane figures a greater extent of abstraction is required than is necessary to grasp solid figures. Solid structures can be experienced immediately. Motivation to work with plane figures could then easily be arrived at in a natural way. The mutual interaction between plane and solid structures would produce adequate associations and would foster and strengthen the acquisition of the subject.

It was Felix Klein (4) who first suggested publicly that integration should not be completely confined to geometric subject matters. He had in mind "Fusion" of arithmetic, geometry and analysis first of all in the training of mathematics teachers and then also with respect to the teaching of mathematics at school level.

Most certainly the reason for his convictions may be traced back to his successes as a research worker in the field of elliptic modular functions, where the integration aspect comes through most conspicuously in his book "Vorlesunge uber die Theorie der elliptischen Modulfunctionen", vol. I, Leipzig 1899.

The "Arbeitsschule" movement, the idea, that is, of the movement of creative schooling took up Klein's idea of generalized "Fusion". One of Klein's main concerns was to stress the role of application in mathematics teaching. An idea subsequently adopted by well known educational theorists. Jungbluth (5), for example, says: "Most of all it should be emphasized that fruitful possibilities for practical application arise if a further demand is recognized, that is, the demand for greater integration of separate mathematical disciplines".

Although integration in school mathematics is scarcely found in practical school work, it is aimed at in the programmes of textbook authors.

2. Arguments for integration

As I mentioned above, the argument for integrating plane and solid geometry came from experimental psychology. As far as the more comprehensive integration of mathematical subjects is concerned, concrete experimental psychological support for this idea is scanty. One can say that the suggestions for this type of integration resulted essentially from scientific experience research workers had while pursuing their research. There are many examples where in quite a natural way geometrical, algebraical, number theoretical, and analytical considerations can be usefully combined, particularly at an advanced level in order to effect deeper insight into the context of the problem at issue. The network of relationships expands and the subject matter will become more familiar through having been illuminated from different points of view.

In mathematics teaching it is a particular problem to awaken an interest in the subject in students. How can that be effected? It is necessary to take as a basis the creative drive every human being is endowed with and to be conscious of the relationship between practical and intellectual work. Especially with young children there is quite a considerable desire for physical activity which can easily be satisfied in teaching mathematics by including such activities as drawing or constructing solids. At the intellectual level the creative drive can be developed by establishing various relationships between different subjects. Here an integrative treatment of school mathematics is most helpful because the different views of the same problem which are thereby disclosed, not only make the problem itself more perspicuous, but lend mathematics, as a discipline, greater comprehensibility.

Example 1: Introduction to divisibility

In order to achieve integration of number theoretical, geometrical, and analytical aspects from the very beginning one has to modify slightly the definition of divisibility.

First we introduce a-curves. That is, the graphs of the function $x \to a/x$, the geometric locus of the upper right vertex of rectangles with given area a cm^2.

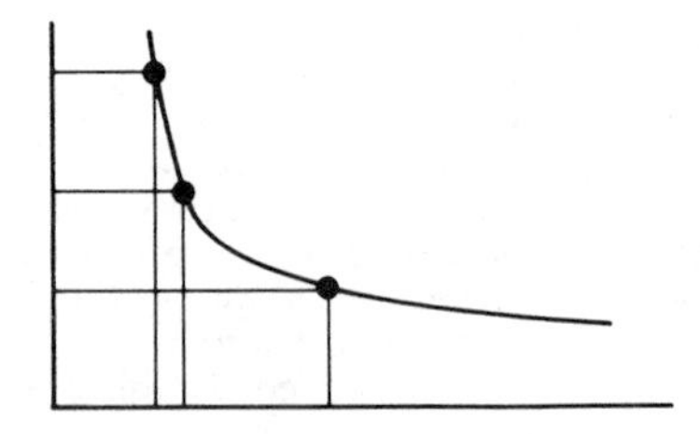

We define for natural numbers (0 included) a and b:

b divides a ($b \mid a$) if the curve a has at b a lattice point.

From this definition we arrive at a two-dimensional interpretation of elementary situations in divisibility. The proofs of elementary lemmas (for example concerning the division algorithm, the least common multiple, and the greatest common divisor) can be made, in a sense, concretely visible. At the same time this procedure yields the possibility of introducing the gnomon construction which allows the construction of an instrument, such that the curves a can be drawn the same way as a circle can be plotted by a pair of compasses. A collaboration with the teacher of handicraft is possible at this stage. If integers have been introduced already, number theoretical considerations can be further developed. One would, for example, consider the presentation of the greatest common divisor d of a and b as linear from $d = ax + by$. The set of solutions $L_{ab}^{(d)} = \{(x,y) \quad d = ax + by\}$ can be interpreted as a line through the lattice points (x,y). We speak of the contour line L_{ab}. At this point a wealth of algebraic ideas comes into play if one considers, for example, the parallel class P(a,b) which contains all contour lines $L_{ab}^{(d)}$. We refer here to the paper by Gunther Pickert: Teilbarkeitslehre unter Benutzung des ebenen Punktgitters (Divisibility by applying the lattice plane), Bweitrage zum Mathematikunterricht 1969, vol. 2, Schrodel Verlag - proceedings of the Ludwigsburg conference of teaching mathematics 1969, and further to a recent paper by A. M. Freadrich: Mathematische Aktivitaten am ebenen Gitter, MU 26, 3, 1980 (Mathematical activities at the lattice plane), and a forthcoming book by the present author: Anwendung der elementaren Gruppentheorie in Kombinatorik un Zahlentheorie, Klett Verlag Stuttgart (Application of elementary group theory to combinatorics and number theory).

Example 2 Finite euclidean planes

We turn now to an example from the mathematical training of teachers, where geometry, algebra, number theory and combinatorics are integrated. This belongs in the field of finite euclidean geometries.

Let $F = GF(p^{\alpha})$ be a Galois field with p^{α} elements and A(2,F) be a two-dimensional affine plane over F such that in A(2,F) an orthogonality relation is defined which satisfies:

i) The theorem of the orthocenter (the altitudes of a triangle are concurrent).

ii) There exists no pencil of parallel lines which are self-perpendicular.

From (i) it follows that there exists a constant of orthogonality c such that two lines $(y = a_1x + b_1)$, $(y = a_2x + b_2)$ are orthogonal if and only if $a_1a_2 = c$. And from (ii) it follows that line reflections can be introduced. We impose on A(2,F) the further condition, that the x-axis be perpendicular to the y-axis. Then the following conditions are equivalent:

1. -1 is not a square in $GF(p^{\alpha})$.
2. In A(2,F) there exists a square (a rectangle with perpendicular diagonals).
3. There exists a bisector of the angle the sides of which are the x-axis and y-axis.
4. c can be normed to -1.

Let us call a plane euclidean if in it one, and thus all, of the conditions (1) - (4) are satisfied.

Now we have a number theoretical problem: Which condition for the prime number p and for the exponent must be assumed so that $A(2,GF(p^{\alpha}))$ is euclidean? We find $p \equiv 3(4)$ from the first supplementary theorem of the quadratic reciprocity law for the prime field (or some other argument) and α odd. We consider now the group which is generated by line reflections. The subgroup of translations is elementary abelian of order $p^{2\alpha}$. Is the group R(X) of rotations with a given center X cyclic? We consider the isomorphic image R(0),0 being the origin of the coordinate system. R(0) is abelian and of order $p^{\alpha} + 1$ which follows from the theorem of the three line reflections. Is R(0) cyclic? To decide this question we make use of the matrix representation MR of R(0) and map MR into the multiplicative group of the quadratic extension field $GF(p^{2\alpha})$. We find that MR is isomorphic to some subgroup of $GF(p^{2\alpha})$. Now we know $GF(p^{2\alpha})^*$ is cyclic, so are MR and R(0). The procedure we applied can be compared with mapping the group of rotations of an $(p^{\alpha} + 1)$-gon with center 0 of the real plane into the multiplicative group of the quadratic extension field of complex numbers. The isomorphic image being in the group of the $(p^{\alpha} + 1)$-roots of unity.

The fact that R(0) and hence R(X) for every X is cyclic has a remarkable geometric implication. We are capable of defining circles and regular polygons in the following fashion:

$C(X,A) = \{ A^{\delta} \mid \delta \in R(X)\}$ is called circle with center X through the point A.

$[A_1, \ldots, A_n]$ is called a regular n-gon, if there is a center X, such that $A_1 \in C(X,A_1)$ and if there is σ^i $R(X) = \langle\sigma\rangle$ with $(A_1,\ldots,A_n)$ being a cycle of σ^i when R(X) is considered a permutation group of the points of $C(X,A_1)$ - actually the group R(X) operates on all points of the plane, the orbits being the circles with center X. However the consituent $\langle\sigma\rangle^{C(X,A_1)}$ operates faithfully, so we identify $\langle\sigma\rangle$ with $\langle\sigma\rangle^{C(X,A_1)}$.

Now from our assumption $p \equiv 3(4)$ and α odd we obtain $p^{\alpha} \equiv -1(4)$, that is, $4 \mid p^{\alpha} + 1$. This is equivalent of saying in $\langle\sigma\rangle$ there exists an element of order 4. $\sigma^{(p^{\alpha}+1)/4}$ is this element which consists of $(p^{\alpha}+1)/4$ cycles of length 4 (4-cycles). We can show that the 4 cycles are squares in the former sense. We can also invert the argument. If squares exist, then the group $\langle\sigma\rangle$ must have elements of order 4, so $p^{\alpha} = -1(4)$ which implies $p \equiv -1(4)$, α odd. The motion group (translations and rotations) is the semidirect product of the group T of translations by the group R(0) and has order $p^{2\alpha} \cdot (p^{\alpha} + 1)$. M is a Frobenius group with T as the Frobenius kernel. The full group of congruence Γ is a semidirect product of M by a line reflection $\langle u\rangle$: $\Gamma = M \times_s \langle u\rangle$. It is not a Frobenius group. The order of Γ is $2 \cdot p^{2\alpha} \cdot (p^{\alpha} + 1)$. We now define the congruence of figures. Let U, be figures, that is, sets of points (or lines), then $U \cong V$ if and only if there is $\alpha \in \Gamma$ such that $U^{\alpha} = V$. If we conceive Γ as a permutation group, then $U \cong V$ belong to the same orbit of Γ. Now we can count congruence classes or orbits of figures by means of the Burnside lemma: With N as the number of orbits, $\chi(\alpha)$ as the number of fixpoints, we have

$$N = \frac{1}{|G|} \cdot \sum_{\alpha \in G} X(\alpha)$$

G being the group that operates on the set of figures under consideration.

There are, for example, $p^{\alpha} - 1$ congruence classes of line segments, the number of congruence classes of lines through a point is two. To determine the number of classes of congruent triangles we have to distinguish the cases $p \equiv -1(12)$ and $p \not\equiv -1(12)$ and obtain, respectively,

$$D_{[p \equiv -1(12)]} = (p^{\alpha} - 1)(p^{\alpha} + 1)^2/12$$
$$D_{[p \not\equiv -1(12)]} = (p^{\alpha} - 1)(p^{\alpha} + 3)/12$$

Here we will break off the discussion. Many interesting details can be uncovered by pursuing this type of investigation. What matters in the present context is not so much the results, however interesting they may be, but the interplay of different fields. The results being the links which hold together the network of the relationship.

It becomes quite clear that integration in teaching mathematics requires a teacher with a general mathematical background of high level rather than one which is specialized in a very small subject. The teacher himself must be conscious of the connection between the different fields of mathematics so that he is capable of exploiting these relations to enrich his own teaching.

References

1. C. A. Bretschneider: Lehrgebaude der niederen Geometrie, Jena 1844. J. H. T. Muller: Lehrbuch der Geometrie, 1844. Firschauf (Osterreich): Elemente der Geometrie, Leipz 1877. E. Merary: Nouveaux Elements de Geometrie, Dijon 1906.

2. W. Lietzmann: Der Kongress in Mailand vom 18 - 20 Sept. 1911, Heft VIII der Ber. u. Mittl., vernalasst durch die IMUK, Leipzig 1912.

3. D. Kratz: Psychologie u. math. Unterricht, Deutsche IMUK - Abhandl. Bd. III, Heft 8 Leipzig 1913.

4. F. Klein: Elementarmathematik vom hoherne Standpunkt aus, II. Geometrie, Leipzig 1914, S.4 u.5.

5. Franz A Jungbluth: Mathematischer Arbeitsunterricht, in: Handbuch des Arbeitsunterrichts fur hohere Schulen, 9. Heft, Frankfurt/M 1927. Translation of the quotation by the present author.

8.8 MATERIALS AVAILABLE WORLDWIDE FOR TEACHING APPLICATIONS OF MATHEMATICS AT THE SCHOOL LEVEL

MATERIALS AVAILABLE WORLDWIDE FOR TEACHING APPLICATIONS OF MATHEMATICS AT THE SCHOOL LEVEL

Max S. Bell
The University of Chicago
Chicago, Illinois

(Some of the background research for this was done as part of the Arithmetic and Its Applications project, supported by NSF grant SED 79-19065, but the opinions expressed herein do not necessarily reflect those of the National Science Foundation. Mary Page was an invaluable assistant in gathering materials for the survey.)

Introduction

My title reflects the assignment given me by the International Congress program committee: to assess not merely what is being said about teaching applications but what actual materials are available to support such teaching. For that, I have attempted a worldwide survey of such materials and to the extent it has succeeded, the results are given in the Appendix. When I did not find people to correspond within a given country, I turned to published curriculum outlines or scholarly studies. For some countries I found no information. Hence the survey is not truly comprehensive. Readers can help me make this more complete by telling me of things I may have missed. (A form for that purpose is at the end of the Appendix.)

I will be urging in this paper that school mathematics must become much more closely linked to its actual uses, but I want to say here that I do not urge that applications become the main or only motivation for the learning of mathematics. There is no need to choose between applications or theory, between concrete work or work with symbols, between playful or the disciplined inquiry, between skills or concepts, or any other of the false dichotomies that too often dominate discussion about school mathematics. There will be plenty of room for new emphases in merely getting rid of the useless, the ugly, and the essentially dreary aspects of the school mathematics experience, and by the learning efficiencies that come from capturing the commitment and imagination of children and adolescents instead of merely their compelled and dutiful attention.

It is, of course, very difficult to change school instruction, and this is nowhere better illustrated than in the repeated failures to meet demands that the teaching of school mathematics be better linked to the actual uses of mathematics. In the Unites States such recommendations have been a feature of essentially every proposal for the reform of school mathematics since at least 1900, and I have often seen similar recommendations in reform-minded documents from Europe and elsewhere. That same concern has been a prominent feature of each International Congress and of many other national and international meetings, including the one in 1960 that moved Freudenthal to the memorable phrase. "The teaching of mathematics so as to be useful" (Appendix, part IV, item 8) hereafter, such references to material in the Appendix will be given in the form - (IV, 8). All that earnest recommending and discussion however, has up to now had very little impact on the actual school mathematics experience.

At the same time, a lot happened in the 1970's that did not reach classrooms but that has provided a solid basis for efforts now to achieve an applications emphasis in mathematics teaching. To discuss this basis and survey the various sorts of available materials, some structure is needed. An analysis of our past difficulties and our failures points the way to such a structure by subdividing the enormous problem of teaching applications into smaller pieces, each of which can then be systematically attacked.

First, there must be good raw materials from which to build applications for school use. Expositions about the uses of mathematics as well as good sources of data must exist at a level accessible to teachers and curriculum builders who are not themselves specialists in applications of mathematics. This is not a problem - the raw materials from which to build applications exist in rich and lovely abundance. Even as I first worked on this matter in 1967 (with SMSG), it was easy to select from hundreds of readily available articles about applications more than one hundred excellent expositions accessible with only school mathematics. It was also easy to collect more than fifty useful books containing nice material about applications and also accessible with only school mathematics. I found myself nearly buried in interesting numerical and other information from which to build school problems. Since then many additional articles, books, and sources of information have been published. In short, I can assure the reader that this first part of the problem of creating applications for use in school teaching presents no difficulties.

Second, with assurance of ample resources from which applications materials can be created, there is the task of transforming these raw materials into actual school-usable problems, problem sets, or units of work. These must be related to specific school topics, must use realistic data in realistic ways, and must be interesting to school students, although not every problem need be interesting to every student. Furthermore, large numbers of such problems and units must be created in order to help authors, curriculum planners, and teachers meet a wide range of interests and abilities of particular students. In 1967, little of this work of translation of good raw materials into good problems had been done, although there were a few nice problem collections even then. That situation moved Henry Pollak to an appeal:

> The big unfinished task is to collect appropriate examples of honest applied mathematics . . . I know it is possible to bring real applications into the secondary, and even the elementary, school and to motivate and illustrate much mathematics by such examples. A major, probably international, effort is needed to collect a sufficient variety of examples to fit all our different school situations. (IV, 8)

Since then much work has been done. More is needed, of course, but we seem to have learned how to accomplish this transformation. Hence, we may be on our way to solution of this part of the problem.

Third, even given this new abundance of problem material, one of the clearest lessons we have learned in mathematics education is that the mere existence of even the best of problems and units does not assure that these materials will be used in school work. In order to become a part of school experience, those problems, problem sets, or units must be worked into the curriculum guides, examination syllabuses, and textbook materials which guide most teachers in what they do with their students. Like the proverbial engineer who promised to learn the theory of relativity as soon as it was printed as a table in his engineering manual, many teachers say they will try to use applications in teaching when they appear in textbooks and examinations. But most widely used school textbooks remain completely inadequate in their attention to applications, and this is one of two places where we must focus our attention next if school mathematics is ever to be linked to its uses.

Fourth, even if fine problems and units are created and then worked into school books and examinations, the difficulties of implementation into actual school practice are still formidable and are seldom solved. To begin with, teachers must be made aware of the existence of such materials, and (at least in the United States) we do not communicate well with our teachers. When teachers are aware of these materials, they must then be persuaded to use them. Once willing to use them, they must be made able to use them - - a significant problem, since very little in the training of secondary school mathematics teachers links mathematics to its uses and, at least in the United States elementary school teachers seldom have much feeling for mathematics itself, let alone for its uses. Furthermore, teaching methods appropriate for teaching applications of mathematics are often quite different from the expository methods that dominate most teaching of mathematical skills and theories. Essentially everyone worldwide who helped me in the survey of available materials spoke of apparent intractibility of the problem of training teachers to use applications, and that alone could defeat any efforts to link school mathematics to its uses.

The Survey of Materials Available for Teaching about Applications

The survey in the Appendix is organized into subdivisions correponding to the subproblem just discussed. As to the first subdivision, raw materials from which to fabricate problems are abundant everywhere and more appear each day. Rather than attempt a listing of those, part I of the Appendix briefly notes where bibliographies of such materials can be found in parts II, III and IV.

With respect to the second subdivision, individual problems and problem sets for school use have in the past tended very heavily to what Polya has called "one-rule-under-your-nose" word problems, often with cooked-up data, and usually concerned with a narrow range of money transactions or physical science problems. Generally I have omitted sources of such problems, though many are available and they are undeniably useful in teaching. The items listed in part II of the Appendix show that more substantial problem material is available; indeed, for each problem or brief unit easily available to me in 1967; I can today find ten to twenty within a few steps of my desk. Nearly all have been produced within the last ten years. For several reasons this abundance of problems for school use is unknown to most people. Problems tend to be produced by individuals or local writing groups for their own special purposes. They are seldom published for widespread use, and when they are, they seem to go out of print rapidly. Worldwide, there are language barriers; for example, I am excited by what I can tell of what is in the materials produced in the Netherlands by IOWO (II,16), but I cannot actually use them. Also, the coverage of many school topics and student interests is still inadequate. Hence, the international effort to produce problem collections that Pollak called for some time ago must surely continue, and we need to find ways to share the results of those efforts more widely. Problem sharing efforts such as PAMELA (IV,16) and several new journals (IV, 13-15) are useful beginnings, but more must be done.

Still it is very encouraging that we seem to have learned how to transform applications data and other raw materials into excellent problems and units, as indicated by the listings in part II of the Appendix. The transformation is not easy, but the results of the survey and first-hand experience have convinced me that putting capable people to the task virtually assures that useful problems and units will be produced.

Turning to the third subdivision of the problem, I found correspondence worldwide often drew comments such as these: "There is no program I know of that succeeds in including 'true' applications of mathematics in school textbooks." "It is only a slight exaggeration to say that mathematics education for most children beyond age 11 is completely devoid of applications except as some elementary school teachers try to show how arithmetic is used." Many of my correspondents added comments such as this: "However, there has been a series of attempts over the last ten to twenty years to remedy this situation. Much interesting material has been produced, but it has not yet found its way into common school practice." Some letters noted that experimental versions of textbooks sometimes contained strong applications, but many such textbooks did not survive, and for those that did survive, "dilution and corruption" of material difficult to teach had in revised editions eroded the applications emphasis. Respondents from the many countries where examinations play an important role in specifying the school syllabus noted that they expected little change in school emphases without changes in those examinations, but also noted that examining boards seemed willing to provide special examinations if helped to do so.

Some exceptions to the rule that textbooks do little with applications are listed in part III of the Appendix. I would be very pleased to be informed of other exceptions. I would also like to be informed of examination practices anywhere that encourages teachers to attempt to link mathematics to its uses. I know of none in the United States. If we cannot increase the number and quality of classroom-ready textbooks and other materials which genuinely link the content to its uses, we can expect to make no further progress in school teaching of mathematics so as to be useful.

With regard to the fourth subdivision of the problem of teaching applications, it seems clear that even if we had the best of school textbooks and examinations oriented to applications, many teachers would still be reluctant to exploit those materials or would find it difficult to do so. On the other hand, some teachers succeed in linking mathematics to its uses without such textbooks by making use of published raw materials, their own experience, and problem sets and units such as listed in part II of the Appendix. Many other teachers could learn to do so. Hence, materials for teaching teachers are at least as important as materials for school teaching.

Fortunately, much of what is available for students in any of the first three subdivisions of this problem can also be adapted for use in teacher training. I have seen considerable learning by teachers in workshops that merely extend an invitation to review and add to such materials written mainly for students. Also, it is well known that one of the most effective ways for teachers to learn new things is to use new materials with their own students, especially if capable consultants are available to them.

There is no shortage of good advice for teachers. On the contrary, there are by now hundreds of articles spanning more than sixty years that appeal to teachers to do more with applications and give suggestions on how to do so. Part IV of the Appendix lists some of these. But course outlines of classes and workshops for teachers are missing from this survey because I found very few. I hope readers will tell me of such courses for teachers for inclusion in any revision. Obviously, much more must be done with materials for helping teachers link mathematics to its uses, and this is the second place where major new efforts must be focused.

Some New Reasons to Believe Teaching Applications at the School Level May Be Possible

The Appendix notes the existence of many materials of certain sorts, but overall it gives us little reason to believe that applications are likely to become a strong emphasis in the mainstream of mathematics instruction. Yet there are once again many demands for another substantial reform of mathematics teaching, and these demands are even more insistent than usual that the uses of mathematics become a genuine part of teaching. In spite of the dismal history of such efforts, I find it possible to be cautiously optimistic that over the next decade applications actually will be worked into at least secondary school instruction. Applications in elementary schooling are quite another and more difficult problem, but even there some optimism is possible, especially if we can exploit the possibilities in application of such relatively simple things as the non-computational uses of numbers, of measure, and of geometry.

To believe that progress is now possible where very little progress has been made in eighty years or more, one must believe that there are powerful new factors in today's situation. Let me outline some of these.

First, the world of applications has itself been changing in the past few decades (say since 1940) in the direction of more and more utilization of mathematics in more and more places. Most people must cope with mathematical applications in their everyday work, and average citizens are confronted with considerable numerical and graphical information that they should cope with. Mathematics is now used in many fields other than physical sciences, and many of the new applications are accessible with only school mathematics content. Hence it is easier to find believable school level applications, and the greater variety of applications can appeal to more students.

A second optimistic indicator is that over the past few decades consistent and clear ways have evolved for talking about some of the processes by which mathematics is applied. Thus, for example, the phrase "mathematical models" conveys something useful to many interested non-specialists, and can help us think about what to do with teaching applications in school.

As a third optimistic factor, we observe that while every new reform wave since 1900 has earnestly demanded that applications be made a part of the school experience, this has never before been the central concern. For example, the reform recommendations of the 1960's strongly urged linking mathematics to its uses, but the actual reform textbooks were directed mainly at repair of conceptual gaps between school mathematics and mathematics itself. Today there are strong demands from every side (the public, specialists, educators) that most people should learn to use mathematics. Hence, in a new reform this demand may finally become a central one. (Of course, there are some dangers in making "usefulness" a central demand that we should be alert to.)

Fourth, there is substantial and quite recent support for applications in the rapid spread (since only about 1975) of small calculators and, even more recently, of microcomputers. Such sudden and widespread technological innovation is sure to have significant effects in our society outside of schools. Schools can, and often do, ignore what goes on in the outside world, but personal access by nearly everyone to cheap and powerful computation offers many new possibilities for making applications a part of the school experience. Even the simplest calculator makes it possible at last for textbook problems to include realistic data and realistic amounts of data. The solution methods we teach can be more general and can more closely approximate those actually used outside of school. The second grader who wants the solution to a problem and knows he should multiply, can multiply, even if he has not learned an algorithm. Older students can use iterative methods, simulations become possible, and so on. That is, calculators and microcomputers make things possible in teaching applications in schools that were just not possible only a few years ago.

A fifth optimistic factor is the new abundance of problem material developed for school uses. That is a main subject of the Appendix that follows. That abundance, along with the sudden transformation in cheap and widely available calculation power, surely enables substantial new progress in linking school mathematics to its uses.

Of course, there are also many barriers to more teaching of applications, and in education one can never be truly optimistic about reform. But there is a saying that optimism consists mainly in the belief that the future is not predictable from the past. Perhaps we can at least be optimists in that sense.

Some Things To Be Done

If we are to increase the likelihood that applications will be effectively taught in many school classrooms, say by ten years from now, certain things must happen.

First we need to be clear about what we want to accomplish in school mathematics and what that has to do with the linking of mathematics to its uses. That is another discussion for another time, but one attempt to outline what most people should get from school mathematics is given in my article, "Teaching mathematics as a tool for problem-solving" in Prospects, Vol. 9, No. 3, 1979. It is linked to a mathematical models point of view and is an example of an exercise that the reader might attempt. That is, if we expect to make further progress, we must clarify what we really want to accomplish with most people, not merely with a chosen few who survive our present courses.

Second, the transformation of vast amounts of raw materials about the real world into projects, problems, and units intended for school use must continue. We are still far from the variety that will be needed for the rich linking of school mathematics to the world outside of school.

Third, better ways of sharing materials for teaching applications must be devised; we who work on this are much too isolated from one another. This sharing and communication should cross national lines, hence translation services are needed. Of course some judgment and screening of materials should be built in since not everything is of a quality that should be translated and widely shared. This one place where such an organization as UNESCO could make a contribution.

Fourth, more textbooks with effective stress on applications simply must be produced. Here there is a peculiar circular bind that can probably be broken only by non-commerical sponsorship of some textbook writing. Publishers want assurance of existence of courses and teachers as markets for any new sort of textbook they are urged to print. But teachers say they cannot offer new emphases without supporting textbooks. Some publishers have taken risks to break this impasse, as indicated by entries in parts II, and II of the Appendix, but other ways must be found to produce the necessary seed materials. I believe that books for relatively advanced mathematics courses will soon be adapted to calculators and computers and hence also include richer applications. But similar adaptation of books for arithmetic, algebra, and geometry will probably require special efforts and non-commercial support.

Fifth, many new initiatives must be attempted to try to cope with the unsolved problems of helping teachers use applications in their classrooms. Here the effects of magnitude and scale are perhaps the most serious difficulties. Perhaps we must use the tools of mass communication, such as Poland has done with radio and TV channels dedicated to professional training. Programming for television, cheap video disk or videotape recordings, or cable television may provide ways around the problems of scale that we face. Perhaps producing certain sorts of software for micro-computers with visual displays would be helpful. We certainly must think about this problem in new ways since there are vast numbers of teachers that must be reached, and in the past we have never learned how to do that.

Sixth, wherever examinations are a barrier, that barrier must be removed. Few examiners are likely to maintain that linking mathematics to its uses is not important.

Seventh, we cannot ignore the fact that the school mathematics curriculum seems already too full. Teachers say there is no room for anything else and we should always listen seriously to teachers. Here are a few suggestions:

a. It is very clear (at least in the United States) that not much mathematics is taught in the first three or four years of school. Much of what would truly begin to link mathematics to its uses can be done in those now relatively empty years; for example, plain and fancy counting, measure, classification, locating things in reference frames, coding things with numbers, non-metric geometry, various kinds of relationships, and so on.

b. Many applications of mathematics are imbedded in some wonderful materials that exist for the early teaching of science (e.g., II, 33-37). But generally speaking these materials are just not used in primary schools, and correcting that would go far in the early linking of mathematics to its uses.

c. With calculators so available, the amount of time spent on teaching of calculation as such can surely be reduced to make more room for the applications that, after all, provide the only reason to want answers to such calculations.

Such remedies as these would, I believe, enlist new commitment and energy from students, so that what would seem to be an addition to an already crowded curriculum might really save time by increased efficiency. A series of developmental teaching experiments might show whether this is so and how it could be accomplished.

Eighth and finally, we must better confront the fact that for most people (perhaps ninety percent) the main uses of mathematics involve merely the flexible use of arithmetic. Hence, perhaps the central problem of teaching mathematics so as to be useful to most people is simply to handle better the teaching of arithmetic, especially non-computational uses of arithmetic. Here again, teaching experiments may show us ways to do this with children and with arithmetically inept adults. Perhaps we should also explore TV programming as a way of helping the countless adults handicapped by inability to use arithmetic, and also as a way of getting parents to demand that schools more effectively link arithmetic to its uses.

Conclusion

Countless papers similar to this one have appeared with little effect during the past eighty years or more. Yet some things today do seem to be different from times past, and it is at least conceivable that we can make progress in ways that have never been possible before. If so, materials such as are outlined in the Appendix give us some startling points on these perpetually difficult problems. In addition, there are other fairly obvious steps that should be taken, and I have outlined some of these. Perhaps we can now go to work on such tasks so that ten years from now we may be closer to what Freudenthal urged twelve years ago - - the teaching of mathematics so as to be useful.

APPENDIX

Survey Results

Max Bell
University of Chicago
Chicago, Illinois

This listing of materials for the teaching of applications is preliminary and incomplete. Additional information will be welcomed. Thanks are due to the following respondents for their contributions; I apologize for any omissions or errors:

Shmuel Avital	Murray S. Klamkin
Alan Bell	George Knill
Rolf Biehler	Marcia C. Linn
David Burghes	John Mack
Hugh Brukhardt	John Niman
Jerzy Cwirko-Godycki	E.J. Piel
Beatriz D'Ambrosio	Henry Pollak
Ubiritan D'Ambrosio	Kenneth Retzer
Mary Falk de Lozada	Fred Stephen Roberts
Jean Delerue	Thomas L. Saaty
Margaret Farrell	William Sacco
Walter Farmer	Hal Saunders
James Fey	George Scroggie
Alan J. Friedman	Janet Shroyer
John Gaffney	Michael Silbert
Claude Gaulin	Cliffort W. Sloyer
Peter Holmes	John C. Stone
Robert Ineichen	Frank Swetz
Robert Karplus	Zalman Usiskin

Category I: Raw Materials from which applications might be constructed.

This category is so rich in materials that no list would be adequate, so none has been provided. Good bibliographies can be found in many of the books listed in Category IV, especially the Sourcebook of Applications for School Mathematics and Cooperation Between Science Teachers and Mathematics Teachers.

Category II: Problems, problem sets, and units with genuine applications content but not yet worked into school courses, textbooks, examinations or syllabuses.

1. Ahrendt, M. The Mathematics of Space Exploration. (1965).

A book of problems or exercises (called Space Flights) relating the applications of mathematics to the space age. Written for studetns who have completed two or three years of high school mathematics. Chapter titles include: Motion; Mass, Weight, and Gravity; Measurements for the Space Age; The Nature of Space; Launch Vehicles; Bodies in Orbit; and Flights in Space. There is a summary, equations and formulas, answers to problems, tables, bibiliography, and index. 160 pp.

Holt, Rinehart and Winston.

2. Atzinger, E.M.: Copes, W.C., & Sacco, W.J.; eds. Trianalytics.

   Designed for high school students, these materials are being developed to promote applied mathematics, and come in three formats: (1) Journal, Roundtable, published three times a year for high school students which gives problems, games, math history and college information; (2) Math Kernels, enrichment sheets for math classes using applied math; and (3) TAMES, Tri-analytic math enrichment series with 20-30 pages devoted to one topic such as "Dynamic Programming."

   To order any or all of these materials, contact:
   Tri-Analytics, Inc.
   P.O. Box 55
   Churchville, MD 21208

3. Avant, J. (Director) Career Laboratory in Mathematics Education (CLIME).

   Creates a mathematics laboratory using student assistants that encourages creativity and ingenuity in solving problems and that stimulates real on-the-job occupational activities. Written for the student in grades 8-12 who might possess average or below average ability in mathematics. There are 32 activities, 23 corresponding "quizzes," 32 guide sheets for teachers. Activity topics include board foot, bridge, business cards, carpet, piston, sprinkler and pantograph. A teachers guide accompanies the book; it give "answers" to quiz problems and general ways of approaching problems, noting the answers and approaches are flexible.

   Project CLIME
   James Avant Director
   6325 W. 33rd St.
   St. Louis Park, MN 55426

4. Baumann, R. and a team of teachers (Gymnasium - Sedundarstufe II). Luneburger Projekt zur prazisnahen Entwicklung von Matierialien fur problemorientierten Unterricht in Analysis und Linearer Algebra (Development of material for problem-oriented instruction in calculus and linear algebra). (1978-present)

   Written for students in Sekundarstufe II (grades 11-13), there are materials for 45 lessons. Topics are taken from special literature in the field of biology, economy, technology, and demography and include discrete mechanics, evolution of systems feedback and equilibrium, development of ecosystems and populations, and economic systems.

   For futher information contact
   R. Baumann
   In den Stuken 16
   2120 Luneburg
   West Germany

5. Becker, G.; Henning, J.; Lindenau, V.; Schindler, M.; et al. Projekt Anwendungsorientierter Mathematicunterricht (Applications-oriented Mathematics Instruction Project). (January 1977-present)

   Incorporates more extended areas of application into mathematics instruction than presently used in secondary school, in order to increase interest in mathematics itself and in applying mathematics, as well as to demonstrate the process of mathematization. Materials are written for grades 7-10, especially in the Hauptschule and are also used as suggestions for teachers' in-service and for students' training. Six units, elaborated in great detail, have been published so far, and include additional worksheets for the classroom, a guideline for adapting given untis or tasks to a classroom situation. Published units include maps, queue problems, railway, air journey, calendar, elections; further units are in preparation, such as bridges, loan installments, and planning a camp.

   A booklet of six sample applications is available for DM 19.80 from:
   Gerhard Becker
   Modersohnweg 25
   2800 Bremen 33
   Germany

6. Bell, M.S. Mathematical Uses and Models in Our Everyday World. Studies in Mathematics, 20. (School Mathematics Study Group, 1972)

   Many problem sets, each usually built around a theme and introduced by a brief exposition. Mathematical demands are mainly from arithmetic, with a few sets requiring elementary algebra. However, even some of the sets that require just arithmetic are fairly challenging. Some problems are open-ended and some require finding additional information. The problems are written to be worked by students in, say, grades 7-9, but the book as a whole is intended as a sourcebook from which teachers can draw problems for use with their students. In practice its main use has been in teacher training. Several hundred problems are grouped into five chapters: uses of numbers for description and identification; uses of pairs of triples of numbers (coordinate systems, ratios, calculations); the role of measures in applications; formulas as mathematical models; and examples of problem collection themes. There is a bibliography.

   ERIC (#ED 143 557. $8.69 plus postage; $0.83 on microfiche)
   Computer Microfilm Corporation
   P.O. Box 190
   Alexandria, VA 22210

7. Bordier, J.; Sanche, J.G.; & Terrill, R. Enseignement de la Mathematique et Realite (Teaching of mathematics and reality). (Course PPM 5201)

   A course for pupils and teachers in early high school with materials for nine weeks given by three books:
   1. Guide d'Activites (37 pp.): Week by week suggestions for using the materials from the other two books.
   2. Document de Base (150 pp.): Background expositions about various applications, as

support for the problem book, Recueil de problemes.

3. Recueil de problemes (163 pp.): 86 diverse problems and solutions using arithmetic, algebra, geometry from real or simulated applications. It includes probability based problems, problems from statistics, decision making using linear programming-type graphs, and several problems about house construction - - building a stairway, estimating materials needed, etc.

Contact:
Permama
University du Quebec
Tele-universite
Quebec City, Canada

8. Bushaw, D.; Bell, M.; Pollak, H.O.; Thompson, M.; & Usikin, Z. A Sourcebook of Applications of School Mathematics. (NCTM, 1980)

A resource for teachers with 500 problems. Introductory essays on processes of applying mathematics (Henry Pollak) and using applications in school classrooms (Pamela Ames). Problems and solutions are in five areas: advanced arithmetic (118 problems), algebra (132), geometry (135), trigonometry and logarithms (55), combinatorics and probability (25), and "odds and ends" (20). A series of five slightly larger "projects" with problems follows; e.g., "fuel gauges" and "home heating." There is an extensive annotated bibliography of many books and more than 150 expository articles from U.S. journals from which applications can be drawn or about the teaching of mathematics.

National Council of Teachers in Mathematics
1906 Association Drive
Reston, VA 22091 $15.00

9. Burkhardt, H., & Treilibs, V. Mathematical Modeling Package.

Develops skills in formulating and handling mathematical models. Written for students 14 and over (and teachers of that age group), the material includes 12 lessons of student work. Each student activity is accompanied by a teacher's page suggesting questions and classroom techniques and noting resources required. Sample topics: "What factors influence a shopper seeking the best buy in instant coffee?" "What information would a clerk need to keep a ready supply of an item in a store?" "Is it worth buying a bike for travel to school if the bus costs 5p?"

For more information, contact:
Hugh Burkhardt
Vern Treilibs
The Shell Center for Mathematical Education
University Park
Nottingham NG7 2RD
England

10. Calculating Better Decisions Staff of Texas Instruments Learning Center. (1977)

Written for high school and above, the book contains four chapters arranged by "need" in a decision-making situation. Little justification for statistics given, but clear instruction on how and when to use the calculator.

Available from: Texas Instruments Learning Center

11. Friebel, A.C., & Gingrich, C.K. SRA Math Applications Kit. (1971)

Written for junior high school students, the kit includes a student handbook with explanatory information and some equivalence tables; "activity" cards which list a problem, materials needed and related activities; reference cards providing some materials needed for activities; teachers' handbook with a chart which correlates activity with mathematical topics. The problems are listed within six categories: appeteasers, science, sports and games, occupations, social studies, and everyday things. Questions are posed such as "How far away can you read a sign?" "How predictable is Slapjack?" "How does an engineer plot forces?"

For further information contact:
Science Research Associates
Chicago, Illinois

12. The Great International Math on Keys Book. Staff of Texas Instruments Learning Center. (1976)

Short one or two page expositions written for high school level and above on how to use calculators for sundry purposes. Chapters cover areas such as business, physics, probability, and trigonometry, and such specific uses as recipe conversions, interest on the interest, gas mileage, and universal gravitation.

Available from:
Texas Instruments Learning Center

13. Green, D.R., & Lewis, J. Science with Pocket Calculators. (1978)

This book is part of the Wykeham Series, which is intended to bring up to date accounts of scientific subjects to sixth-formers and first year undergraduates The majority of the book is concerned with applications and contains much valuable material which could be classified as numerical analysis, statistics, or mathematical modelling. An attempt is made to cater for most types of calculators since for all specimen calculations detailed key-stroke sequences are given both for machines with the more "algebraic" logic and for those with Reverse Polish logic. Most chapters contain a good selection of exercises. Some of the exercises are rather open-ended; for instance at the end of the chapter on calculus Exercise 13.24 can be summarised as "find out about Romberg Integration and use it to repeat some of the problems in this chapter."

(From R.J. Goult's review in the Journal of Mathematical Modelling for Teachers, Vol. 2, no.1, June 1979.)
Wykeham Publications
(England)

14. Hoffer, A.R. (project director). Mathematics Resource Project, the University of Oregon. (1978)

Intended as a teacher resource file with a large number of teacher materials and student worksheets in five main collections:

1) Number Sense and Arithmetic Skills (pp. 832)
2) Ratio, Proportion and Scaling (pp. 516)
3) Geometry and Visualization (pp. 830)
4) Mathematics in Science and Society (pp. 464)
5) Statistics and Information Organization (pp. 850)

In each collection there is a "didactics" section and a "classroom materials" section. The classroom materials include a number of lessons, each with some exposition of a given topic or the setting up of a problem, followed by number of exercises, nearly all with "problem solving" intent and most of those with links to the real world. Each collection also has an extensive annotated bibliography; the one for statistics and information organization, for example, annotates about 80 articles or books, many of which would provide useful problems or raw materials for teaching applications. This is rich resource for what in the U.S. would be grades 5-9.

About $30 for each collection:
Creative Publications
3977 East Bayshore Road
P.O. Box 10238
Palo Alto, California 84303

15. Holmes, P., et al. School Council Project on Statistical Educaton. (1975-present)

This project has produced 27 sixteen-page teaching units for students ages 11-16 and their teachers to be used with any course which includes some statistics, e.g. mathematics, sciences, humanities etc. The mathematical competence. The unit titles include: "Tidy tables," "Wheels and Meals," "Retail Price Index," "Figuring the Failure," "Smoking and Health," and "Phony Figures." There is also a teachers' book: Teaching Statistics, 11 to 1.

Materials available from:
Foulsham Educational
Yeouil Road
Slough, SL1 SBS
England

Other correspondence:
Peter Homes, Schools Council Project on Statistical Education
University of Sheffield
Sheffield, S10 2TW
England

16. I.O.W.O. Project. Utrecht, Holland. (1972-present)

I.O.W.O., Instituut Ontwikkeling Wiskunde Ondewijs, or Institute for Development of Mathematics Education was founded about 1971 and has engaged in short and long term curriculum development for children 4-12 by the Wiskobas project and for children 12-19 by the Wiskivon project. The latter was in turn divided into materials for ages 12-16 and 16-19. According to its former director Hans Freudenthal its philosophy is "Mathematics as a Human Activity." The idea is to show that mathematics is everywhere and use applications to introduce mathematical concepts. "Around the World in 80 Days" is a project for 13-14 year olds. Titles for the older age group include "Functions of Two Variables," "Exponents and Logarithms," "Differentiation and Periodic functions." Material for children age 4-12 are described in "Five Years IOWO," Educational Studies in Mathematics, vol. 7, no.3, August 1976. (Jan deLange in H.G. Steiner, ed., Cooperation Between Science Teachers and Mathematics Teachers. See IV, 10 below)

Contact:
Instituut Ontwikkeling Wiskunde Onderwijs
Tiberdreef 4 Utrecht (030) 611 611
Netherlands

17. Judd, W. SRA Problem Solving Kit for Use with a Calculator. (1977)

A collection of nearly 300 problems on individual cards for junior high school students on topics such as "What's the Car Worth Now?" (declining value of new car); "Bills in Stereo" (differences among payment plans); "Population Explosion." Problems ask for estimate before actual calculation; short answer questions related to original information are on back of card. Logic of problem solving emphasized. Explanatory guide to calculator usage is also included.
Science Research Associates
Chicago, Illinois

18. Knill, G. (Mathematics Consultant, Board of Education for the City of Hamilton, Ontario, Canada), and a team of four high school students and six teachers. Applications of Mathematics Development Project. (1977-present)

Applications and pseudo-applications that motivate interest and involvement with the process of mathematization and help to answer the question "Why do we have to do this stuff?" For use in grades 7-12, materials are designed for students but may also be used for in-service work with teachers. About 150 brief units, each consisting of a one- or two-page intorduction including data, diagrams, etc., and followed by questions or problems about the data. Topics were selected by student indication fo high interest: examples include music, crime detection, fast cars, art, spaceships, the environment, Olympic records, readability tests.

A booklet of six sample applications and pseudo-applications is available at no cost.
Michael R. Silbert
Supervisor of Curriculum & Instruction (Mathematics)
Board of Education for the City of Hamilton
P.O. Box 558
Hamilton, Ontario, Canada L8N 3L1

19. Larsen, R., & Stroup, D. Statistics in the Real World, A Book of Examples. (1976)

Written for the advanced high school or undergraduate level, the book conveys the relevance of statistics through the use of real data taken from real experiments. There are 50 such examples taken from research in anthropology, psychology, sociology, literature, history, etc. These are divided among six chapters: characterizing variability: the one sample problem: the two sample problem: the paired-data problem: the correlation problem: non-parametric methods. Examples are followed by statistical objective, procedure section, and questions to help students complete analysis. Some sample topics: handwriting analysis; bumper stickers; how to stop smoking; future shock.

Macmillan Publishing Company, Inc.

20. McFadden, S.; Anderson, K.; & Schaff, O. MathLab Junior High (Mathematical Experiments for Students). (1975)

Listed under nine categories including: calculator: shape: science and applications. There are over 70 exercises such as "Number Palindromes," "Planning a Vacation," and "Taking a Survey" designed to supplement the regular curriculum in grades 7 and 8. Although written for students, materials could be used in teacher training courses which emphasize the laboratory approach.

For further information:
Action Math Associates Incorporated
Eugene, Oregon

21. Mathematik-Unterrichts-Einheiten-Datei, MUED. Contributors: about 400 mathematics teachers. (1977-present)

Teachers of Sekundarstufe I/II (grades 5-13) have contributed about 140 units, some with additional background information, such as reports on class experience and working sheets. Material rests largely on applications with a stress on a particular pedagogical philosophy ("orientation towards the emancipation of pupils"). Topics of units include electricity in the home (UE120); calculators (UE115); pocket money (UE107); go carts (UE27); cybernetics (UE117).

All units available.
Heinz Boer, MUED
Bahnhofstr. 72
4405 Appelhulsen
West Germany

22. Mosteller, F.; Pieters, R.S.; Kruskal, W.H.; Rising, G.R.; Link, R.F.; Carlson, R.; & Zelinka, M. Statistics by Example (1972)

Presents "clear, interesting, and elementary written descriptions" for high school students of real life statistical problems. Four progressively more difficult books of essays, each presenting problems, discussing them, then giving exercises. Each essay is individually authored. The books are Exploring Data (14 essays; arithmentic, rates, percentages); Weighing Chances (14 essays; notion of probability, elementary algebra); Detecting Patterns (12 essays; elementary probability, intermediate algebra); and Finding Models (12 essays; elementary probability, intermediate algebra).

Addison Wesley Publishing Company
Reading, Maine 01867

23. Munzinger, W. (Hessian State Advisor for Mathematics in "Gesamtschulen," Ministry of Cultural Affairs, Wiesbaden.) Alternative Energy, Applications of Mathematics. (1979-present)

Projects materials and worksheets are designed for comprehensive school students (age 15-16). Project documentation also accompanies units. Topics include mathematics and environment problems; energy problems; form and function; and worksheets to build a sun-mirror.

A booklet for pupils ("How to build a sun-mirror") is available for 5,00 DM
Fachmoderator Mathematik fur hessische Gesamtschulen
Wolfgang Munzinger
6039 Florsheim am Main
Lassalle Str. 5
West Germany

24. Niman, J. Mathematics Teaching through Artistic Forms. (1979-present)

Seeks to integrate mathematical concepts of the middle and junior high school curriculum with works of art selected from various periods and civilizations. Created for grades 5-9 (and in-service courses for teachers), the project uses 35mm slides, overhead transparencies, lectures, and worksheets. Topics studied include symmetry, tessellation, number patterns, projective geometry.

For further information:
Professor John Niman
Mathematics Education
Hunter College of the City
University of New York
695 Park Avenue
New York, New York 10021

25. Price, S. & Price, M. The Primary Math Lab An Active Learning Approach. (1978)

A student workbook for grades 1-3, containing classroom suggestions for the teacher and worksheets for activities such as: body measurement, "kitchen math," graphing time periods, and throwing dice.
Goodyear Publishing Company
1640 Fifth Street
Santa Monica, California 90401

26. Saunders, H.M. (Mathematics Teacher, La Colina Junior High School, Santa Barbara, California) When Are We Ever Gonna Have To Use This? (1980)

Motivates the usefulness of topics in the secondary math curriculum by inverstigating their applications in 100 different occupations. Written for students in grades 7-12, there are several hundred word problems including the occupation from which they were obtained and the solutions. Also a poster sized chart cross-referencing math topics with occupations. For example, in the fractions chapter, vocational problems are included such as airplane mechanic cutting a piece of tubing; chiropractor setting an X-ray machine; social worker computing welfare payments. Most traditional secondary math topics from fractions to second-year algebra and trigonometry are covered. There are some topics, however, for which no respondent could suggest and application.

The charts ($3.00) and collections of word problems ($6.00) are available.
Hal M. Saunders
290 Hot Springs Road
Santa Barbara, California 9310827.

27. Schupp, H., et al. Stochastik in der Hauptschule (Stochastics in Upper Elementary School). (1976-81)

The project developed propositions for a project at the end of a curriculum in stochastics in order to show the significance of that discipline by treating problems of real life. Designed for the final grade in elementary school (pupils about 14-15 years old), materials also include curriculum papers or teachers who intend to start such a project, with an introduction to the theme and a frame of the project (arrangement, aims, methods, helps for difficult situations, exercises, test, etc.). Unit topics include problems of filling prepared packings; how signs are transmitted; the workless -- lazy or injured people?; price index -- how is it built and what does it say?

Curriculum papers are available from:
Professor Dr. Hans Schupp
Universitat des Saarlands
D-66 Saarbrucken (BRD)

28. Slesnick, T.; McHolm, L.; Luegrmann, A.; & Hakansson, J. The Lawrence Hall Science Shuttle. (1979-present)

Spreads computer awareness and literacy throughout the educational community - - to teachers, students, parents, and administrators. Aimed at children (ages 9-17), teachers, and administrators, it inculdes a teaching guide for a sequence of computer workshops (organized by day), and covers topics such as general computer programming concepts, graphics, and computer awareness. This is an ongoing project.

The Lawrence Hall of Science
The University of California
Berkeley, California

29. Sloyer, C.W. Fantastiks of Mathematiks. (1974)

For secondary school teachers, prospective teachers and students, with only precalculus mathematics required. Problems are chosen from real-world situations in areas such as medicine, card games, or business. The problem is posed by an expository section, which is followed by a step-by-step discussion of the solution. At the end of each, on or more related practice problems are given. Thirty-nine sections, for example: "Here Cums de Judge," the mathematics of determining an optimal jury size (algebra, geometry); "A Smug Drug," determining the amount of drug in the body after n injections at constant intervals (algebra, geometric sequences and series). An annotated bibliography is included.

C.W. Sloyer
University of Delaware
Newark, DE 19711

30. Space Mathematics: A Resource for Teachers. National Aeronautics and Space Administration. 1972.

Many problems with realistic formulas and data in ten chapters with titles that mirror standard school topics: e.g., elementary algebra, conic sections. Brief annotated bibliography.

Stock number 3300-0389. $2.00
Superintendent of Documents
U.S. Government Printing Office
Washington, D.C. 20402

31. Truxal, J.G.; Visch, M. Jr.; & Piel, E.J. National Coordinating Center for Curriculum Development (NCCCD) Minority in Engineering Project. (1978-present)

An ongoing project to promote better understanding by teachers and students of the nature of modern engineering work, and better preparation of college-bound students for entry into quantitatively-oriented higher education. There are 12 units under four categories in booklet form written for high school students, each with an extensive teacher's guide. By category they include:
Algebra: mixtures, water, glass, minimization, problem formulation, maximization, automobile production
Mathematics: problems in search of a calculator, fractions
Geometry: offshore limits, railroad crossings, bridging the gap (logic)
General Science: flywheels
Problems investigated include: choosing optimum signal types for railroad crossings; heat loss taken into account in building design; tendency of flywheel to fall apart while spinning at great speed.

For further information:
E.J. Piel
Department of Technology & Society
University of New York at Stony Brook
Stony Brook, NY 11794

32. Usher, M., & Bormuth, R. (Sharpe, G.H., consultant.) Experiencing Life through Mathematics. (1975)

Provides a workbook of problems in many different practical life experiences that shows why and how mathematics is really used. Written for high school students - - requires percent, basic operations, proportions; basic or general math background. Two volumes, each 122 pages in sections of occupation or situation - - e.g., vacation, planning a wedding, managing a restaurant, working in a hospital. Description of situation starts sections, followed by problems which often build on each other.

Michael A. Usher
Greeley Central High School
Greeley, CO

Note: We found in this search that applications materials that for elementary schools the best applications of mathematics were often in science textbooks and supplementary materials. We include here some of the examples we found especially useful, and that class of school materials should propably be closely examined as sources of applications, especially for the early school years.

33. Elementary Science Study (ESS) (Education Development Corporation)

A general science program for grades K-9 which provides materials that motivate students through "science experiences" that are fairly loosely structured with the intent that students and teachers will focus on what interests them most. Fifty-four units are included with such titles as these: light and shadows; animals in the classroom; mobiles; balancing; mapping; pendulums; mirror cards.

Webster Division, McGraw Hill Book Company
Manchester Road
Manchester, Massachusetts 63011

34. Ennever, L., et. al. Science 5-13

"The project aims at helping teachers to help children between the ages of five and thirteen years to learn science through first-hand experience using a variety of methods. The project has produced . . . units of work dealing with subject-areas in which children are likely to conduct investigations. These units are linked by objectives that we hope children will attain through their work." (From the Preface) Twenty-one books have been printed, including "With Objective in Mind," "Source Book," and units with such titles as: working with wood, science from toys, structures and forces, minibeasts, like and unlike.

Macdonald Educational
49-50 Poland Street
London W.1, England

35. Karplus, R.; Lawson, C.; Their, H.D.; et al. Science Curriculum Improvement Study (SCIS). (1962-76)

A complete curriculum for science teaching in grades K-6 plus in-service work for teachers. Contains 13 units - - one for kindergarten, six in life science and six in physical science. Also, teachers' guides, student manuals, and complete equipment kits including shipments of living organisms. Classroom topics include material objects; organisms; interaction and systems; life cycles, energy sources; communities.

Rand McNally & Co.
P.O. Box 7600
Chicago, IL 60680

36. Mayer, J.R.; Livermore, A.H.; & the Commission of Science Education of the AAAS. Science - A Process Approach (S-APA). (1967)

Undertakes to develop skills in grades K-6 that lead to mastery of the basic scientific processes, which are deemed more important than content as such. Units focus on these processes: i.e., observing, measuring, using space-time relationships, predicting, inferrng, formulating hypotheses, interpreting data, controlling variables, and experimenting. Teacher in-service course and manual available.

Xerox Education Division
600 Madison Avenue
New York, NY 10022

37. USMES (Unified Sciences & Mathematics for Elementary Schools). EDC. (1970-present)

Materials for grades K-8 to help teachers motivate mathematics students through a problem-solving approach. Materials are as follows: (1) Units (we have seen 26 such) based on real problems stated as "challenges"; e.g., classroom furntiture that doesn't "fit"; busy intersections; crowded playgrounds. Problems have no "right" solutions, but encourage a problem-solving approach. (2) "How to" cards and books for teaching the skills needed for a particular "challenge." (3) Design Lab: tools and supplies for constructing models or other apparatus. (4) Curriculum Correlation Guide: how activities can be related to other subjects. (5) Teacher Resource Guide: in-depth analysis of "challenge," other teachers' experiences with it, resources.

For a summary of evaluation results:
E.D.C.
55 Chapel St.
Newton, MA

Materials obtained from:
Moore Publishing
Box 3036
Durham, NC 27705

Category III: School course materials with considerable Category II type materials built in.

1. D'Abrosio, U. (project director); Soares, M.G. (coordinator); et al. New Materials for Mathematics Instruction. (1974-77)

The project's aim was to provide school teachers with mathematics laboratory and curriculum materials for developing experimentation associated with mathematization. Written for grades 3-8 (ages 9 to 14), materials include student texts and teacher's manuals in "Geometry," "Functions," "Equations and Inequations." A mathematics laboratory was also designed. Materials are now being used in the State of San Paulo school systems.

For reports from field testing and large scale use, contact:
Prof. Almerindo Marques Bastos
Coordenadaria do Ensino e Normas Pedagogicas
Rua Joao Ramahlo no. 1546
Sao Paulo, Brasil

2. David, E.E., Jr.; Piel, E.J.; & Truxal, J.G. The Man Made World - - Engineering Concepts Curriculum Project. (1966-71)

The aim of the text is to encourage "technical literacy" and logical thinking (not pure science methodology or mathematical problem solving), and is written for pre-calculus (high school or junior college) students. It consists of 15 chapters (e.g. decision making; systems; feedback) each with problem sets and laboratory projects. In general, problems are in terms of systems, (input-output) which are modeled with block diagrams with increasing complexity of these diagrams. Sample problems include: growth of an epidemic, spectograms for speech and music, water pollution, and traffic control, prosthetics, and much else. A wonderful textbook on technology but with no place to go in standard U.S. mathematics and science curricula.

McGraw-Hill Book Company

3. Foerster, P.A. Algebra I: Expressions, Equations, and Applications. (1980-82?)

   Companion volume to the author's Algebra and Trigonometry: Functions and Applications. Written for first-year high school algebra students, it will be 400-500 pages when finished. Materials will be organized to reach quadratics as soon as possible. Emphasis will be on "word problems" in which variables actually vary, rather than simply standing for unknown constants. Its philosophy: given an expression, (1) find value of expression when variable is given; (2) find value of variable when expression's value is given; (3) use these techniques in problems from the real world. Not additional information is available at the present time.

This is a good model for us all of interesting and believable applications worked into the standard format of a standard (U.S.) course. It has many very fine problems, many of them named and included in a separate index for easy reference. One hopes it will be widely used as a textbook; it is also useful as a resource or problem book.

Addison-Wesley Publishing Company

4. Goodstein, M. Sci Math. (1980)

   Nine chapters (200 pages) developing through clear exposition and many examples the mathematics helpful in high school science courses. (The author is a chemistry professor.) That means mainly a thorough treatment of rates, ratios, direct and inverse proportions, "units analysis," graphing data, and, implicitly, a lot about measure. It is about a ten-week course in advanced arithmetic without an obvious place to reside in the standard U.S. curriculum but useful to most middle school or high school students, not excluding many in "college preparatory" courses who have poor intuition about rates, ratio, and measure. Experiments in Sci Math has 30 projects for "hands on" experiences designed to accompany material in the t ext, but not expressly related to a certain section of the text. Examples: How thick is a page?; bicycle pedaling; comparing brands of vinegar.

   Preliminary editions are available:
   Madeline P. Goodstein
   Department of Chemistry
   Central Connecticut State College

5. Greenfield, L.; Noogueirs, M.F.; Eufinger, B.; & Montalto, B.J. Improving Math Competence. (1978)

   Adult education workbook. Each chapter is built around commonly confronted situations such as buying on credit, insurance protection, traveling from place to place. Most problems concern money. A "math skills" section provides arithmetic instruction.

   Cambridge Publishing
   488 Madison Avenue
   New York, NY 10022

6. Haber-Schaim; Skvarcius; & Hatch. Mathematics I and 2. (1980)

   Two "independent but reinforcing" one-year texts (10 chapters, 300 pages each) leading to algebra. A lot of work is done with mapping, graphs, etc., with exercises alternating between word problems and drill. Some units use more application than others, depending on mathematical subject matter. Topics include: How much can you change a map and still have it be useful?; When do we add and when do we subtract?; coded messages.

   Prentice-Hall, Inc.
   Englewood Cliffs, NJ

7. Koflia, A; Dwiggins, B.H.; & Luy, J.A. Practical Problems in Mathematics for Automotive Technicians/Carpenters/Electricians/Machinists/Masons/Printers/Sheet Metal Technicians (1972-73)

   Presents terms and practices of a particular occupation in problems following standard format of a basic mathematics course. Written for students with junior high math background interested in the particular occupation or actual employees needing basic math improvement. The course included Basic Math Simplified which teaches arithmetic beyond basic four operations. Seven 140-page booklets organized into 9 sections are to be done while studying the corresponding unit in Basic Math Simplified then doing review problems in the particular occupation. Topics include board measure, estimating sheathing and roof boards, discounts, horsepower, graphs.

   Editor-in-Chief, Delmar Publications
   Box 5087
   Albany, NY

8. Kollegschulversuch NW (BRD). KOSEK II, Landesinstitut fur Curriculumnentwicklung, Neuss, BRD (National Institute for Curriculum Development). (1976-present)

   Seeks to develop a complete curriculum for a new type of school at the level of Sedundarstufe II with the guideline of connecting aspects of general college preparatory and vocational education. Courses and related information on teaching guidelines are intended for teachers of pupils of grades 11-13. Type and amount of applications are chosen on the basis of the aspired qualifications; e.g., there are differences between courses for prospective biological-technical assistants, mechanical engineers, and economists.

Course materials and information about the whole system and the intention of the new school type are available.

D.r Wolfgang Humpert
Landesinstitute fur Curriculumentwicklung
Gorlitzer Str. 3
4040 Neuss
West Germany

9. Ormell, C., & project team. Mathematics Applicable.

These materials include: Teaching Mathematics Applicable - an introductory guide (48 pp); Understading Indices (144 pp); Geometry from Coordinates (96 pp); Introductory Probability (64 pp); Polynomial Models (128 pp); Vector Models (128 pp); Algebra with Applications (252 pp); Calculus Applicable (136 pp); Logarithmic/Exponential (180 pp); Mathematics Changes Gear (64 pp).

This series of textbooks casts mathematical modelling in a central role in the learning of mathematics itself. It is the main output of the Schools Council Sixth Form Mathematics Curriculum Project, which was started at Reading University in 1969, and provides the basis for a one or two year (A) level course; a related examination is provided through the London Board. The mathematics itself is the conventional tool-kit of the applied mathematics at this level, as the book's titles imply, but the approach is both original and thorough; the Director and the Project team have created a rich profusion of conceivable situations with which to reinforce the mathematical skills and concepts involved. The Project itself is now complete, but the Mathematics Applicable Group provides a vehicle for communication between teachers using the course, and for the development of the approach at other levels of age or ability. It can be reached at the Department of Education, Reading University, London Road, Reading. (From H. Burkhardt's review in the Journal of Mathematical Modelling for Teachers, vol. 2, no. 1, June 1979.)

Heinemann Educational Books, Lt.
22 Bedford Square
London WC1B 3HH

10. Steinbring, H., & von Harten, G. EPAS I (Development of practice-oriented materials for teachers of "Sekundarstufe I"). Stochastics in "Sekundarstufe I." (1980-present)

Seeks to develop guidelines and teacher material in stochastics at the Sekundarstufe I (grades 5-10) which should enable the teacher to handle the various existing stochastics courses in a critical and variable way. It is written for teachers with no knowledge in probability and statistics, especially teachers in German Sekundarstufe I. Chapters of the book include:

I. Remarks on the position of stochastics in the "S I" curriculum
II. Didactics of the concept of probability
II.1 What is probability - the various foundations of the concept of probability
II.2 A didactical characterization of the concept of probability
III. Application as a tool of the development of the notion of probability; concrete suggestions for a course
III.1 Probability theory as general theory of measurement
III.2 Discussion of important examples for teaching

H. Steinbring/G. von Harten
Universitate Bilefeld, IDM
Universitatsstr. 25
4800 Bielefeld 1
West Germany

11. Trotta, F.; Imenes, L.M.P.; & Jakubovic, J. Matematica Aplicada (Applied Mathematics). (In Portuguese.) (1980)

Designed to present a mathematics course, closer to reality, closer to student's everyday problems, and to be more useful to the majority that frequent our schools (Brazil). Three student textbooks and three teacher's manual were written for "Segundo Grau" - - ages 15-17. Topics were selected based on the curriculum of the "sedundo grau," and include applications related to percentage, progressions, functions, equations, inequalities, plane geometry, trigonometry, logarithms, analytic geometry, linear systems, probability, limits, derivatives, space geometry, and complex numbers.

Editora Moderna Ltda.
Rua Afonso Bras, 431
Sao Paulo - SP - Brasil 04511

12. Usiskin, Z. Algebra through Applications with Probability and Statistics. (1979)

A textbook for a first-year algebra course which develops algebra from real world applications and includes probability and statistics. (This is normally a 9th grade course in the U.S. but the book is usable earlier or with college students who need such a course.) The material is organized into 124 lessons, each followed by various types of exercises. Chapter titles include uses of numbers, patterns and variables, additionand subtraction, models for multiplication and division, sentence solving, linear expressions and distributivity, slopes and lines, powering, operations with powers squares and square roots, sets and events, linear systems, quadratic equations and functions. A number of statistical topics are integrated into the problem material of the text. These include sampling, mean, standard deviation, simple probability, maximum likelihood estimates, permutations, line-fitting, events and independent events, "law of averages," probability functions, and chi-square calculations.

Text $9.50 U.S. ($7.60 for NCTM members):
NCTM
1906 Association Drive
Reston, VA 22091

Notes for Teachers $14.00 U.S. prepaid:
Professor Zalman Usiskin
University of Chicago
5835 Kimbark Avenue
Chicago, IL 60637

Comment: In the early and mid 1970's a number of applications-oriented books were published by educational publishers in the United States (and perhaps elsewhere) that were intended for the U.S. 7th or 8th grades or for students beyond that not in the "college preparatory" sequence. Mostly the applications consisted of brief word problems closely linked to specific arithmetic topics, but some (such as the Ginn and Freeman books below) went beyond that, and all are interesting. Here is a necessarily incomplete sampling:

Applications in Mathematics (2 vol.). Johnson, D.; Hansen, V.; Peterson, W.; Rudnick, J.; Cleveland, R.; & Bolster, L. Scott Foresman and Company, 1972. (High school)

Activities in Mathematics (2 vol.). Johnson, D.; Hansen, V.; Peterson, W.; Rudnick, J.; Cleveland, R.; & Bolster, L. Scott Foresman and Company, 1973. (Junior high school)

Modern Applied Mathematics. Gold, M., & Carlberg, R. Houghton Mifflin Company, 1971. (High school)

Ginn Mathematics: An Applied Approach. Ames, R.; Immerzeel, G; Moulton, J; & Scott, L. Ginn and Company, 1975. (levels 7-8)

Mathematics: A Human Endeavor. Jacobs, H. W.H. Freeman and Company, 1970.

Category IV: Materials to help teachers learn about applications or about the teaching of applications.

1. Alberti, D., & Laycock, M. The Correlation of Activity - Centered Science and Mathematics. (1975)

   Correlates specific lessons in science projects (ESS, IDP, MINNEMAST, S-APA, SCIS) with concepts covered in The Fabric of Mathematics. For teachers of grades K-8. Contains 149 pages, with 12-15 activities for each of 8 concepts such as observing, classifying, measuring, collecting, and organizing data, or predicting and inferring. Activities are accompanied by specific references to science materials and to appropriate mathematical concepts.

   The authors
   Neuva Day School and Learning Center
   Hillsborough, California

2. Bell, M.S., & Usiskin, Z. Arithmetic and Its Applications Project. (ongoing)

   Develops one set of materials for teachers, supervisors, and curriculum determiners and anothe set of materials for students in grades 6, 7, and 8 which categorize and illustrate the ways in which arithmetic is used. The teacher material consists of a Handbook for Applications (approximately 225 pages) organized into these chapters: Numbers and Labels (uses without calculation; pairs and n-tuples; types of labels); Uses of numbers Fundamental Operations of Arithmetic (addition, subtractin, multiplication, division, and powers); and Adjustments of Numerical Information (choices of notation, estimation, rounding, and various modes of representing data). The student material consists of a collection of prototype classroom activities that illustrate the Handbook categories. Each begins with actual data which the students then analyze and work toward solutions.

   No materials are presently available. A preliminary version of the Handbook may be available in 1981. Student materials will be pilot tested in 1980-81.

   Professor Zalman Usiskin or Professor Max Bell
   Department of Education
   University of Chicago
   5835 Kimbark Avenue
   Chicago, IL 60637

3. Burkhardt, H. The Real World and Mathematics. (1978)

   A comprehensive exposition of available materials about teaching mathematical modeling. Written for teachers and educators the pre-calculus problems are directed twoard junior high or high school interests. The book (290 pages) is a combination of problems (situations) for student use, a guide to teaching modeling for teachers, and information on other available student materials of the same type and where to get them. See especially pp. 162-163 for PAMELA - - a store of problem situations for which additional contributions are solicited. Topics include allocations of resources: time, money wallpaper; movement situations: mechanics, athletics, traffic; personal relations problems: friendships, methods of communication; and calculators and computers.

   The Shell Centre for Mathematical Education
   University of Nottingham
   Nottingham NG7 2RD, England

4. Effe; Fuchs; Kaiser; Riedel; Schmid; Schmidt; Stein; Stumpf; Trierscheid; Volk; & Wunsche. Grenzen des Wachstums & Lohnerhohungen. (1974-1979)

   The project developed material for teachers and prospective teachers for a "problem-oriented teaching of mathematics with an emancipatory intention." Two booklets for teachers:

   a. "Lohnerhohungen in mathematischer Behandlung" (A mathematical treatment of rising wages): problems in connection with different models of the rise of wages.
   b. "Skizzen und Materialien fur facherubergreifende Unterrichtsreihen zu den GRENZEN DES WACHSTUMS anhand der studei von D. Meadows" (Outlines and materials for comprehensive instructional sequence for the "Limits of Growth" from the study of D. Meadows): analysis and problems in connection with D. Meadows' book on "Grenzen des Wachstums" (report by the Club of Rome). Mathematical content: descriptive statistics, geometric sequences and series, exponential function, difference equations, logistic models.

   The two booklets are available from:
   G. Kaiser
   Freiderich-Ebert-Str. 176 a
   3500 Kassel
   West Germany
   or D. Volk

Oswaldstr. 8
4650 Gelsenkirchen
West Germany

5. Farrell, M.A., & Farmer, W.A. Systematic Instruction in Mathematics for Middle and High School Teachers. (1980)

A text for undergraduate and graduate pre-service and in-service mathematics teachers. Can be used for methods courses, curriculum and materials courses. Topics include "Mathematics, Science, and the Everyday World," an annotated bibliography, and classroom illustrations. A parallel volume for science instruction exists.

Available at $13.95 from:
Addison-Wesley Publishing Co., Inc.
Reading, MA

6. Mathematiques Du Hasard Dans Les Lycees: Activites de calcul de prbabilites dans le second cycle. Institut National De Recherche Pedagogique. (France)

This project has aims similar to the one just below, but here the students are supplied with data sets and use more of the formal tools of statistics. As with the proceding project, it is the students that formulate the question and decide what to do with the data.

7. Mathematiques Du Quotidien Dans Les Colleges: Activites d'analyse de donnees dans le premier cycle. Institut National De Recherch Pedagogique. (France)

Projects in which children in the first four years of the secondary school (ages about 12-16?) collect numerical data and work with it in various ways. The report gives some of the projects and discusses the results of use with children. It is reported that by selecting the projects and questions coverage of most topics in the standard curriculum is possible.

8. Proceedings of the Colloquium How to Teach Mathematics so as to be Useful. Educational Studies in Mathematics, Volume 1, No. 1 and 2. May 1968.

This very useful volume has a number of contributions from people in a number of countries. The introductory essay by Hans Freudenthal on "Why to teach mathematics so as to be useful," and "Propositions on the Teaching of Mathematics from the Symposium at Lausanne' would be difficult to improve upon.

9. Sharron, S., ed. Applications in School Mathematics. 1979 Yearbook of the National Council of Teachers of Mathematics.

Twenty separate articles of about ten pages each, some with joint authorship. Some articles talk about the teaching of applications, but most talk about specific applications, some of the latter give classroom exercises. One of the articles is an extensive bibliography of sources and types of applications. There is too much variety here to describe briefly, but it is a quite useful resource.

The National Council of Teachers of Mathematics
1906 Association Drive
Reston, VA 22091

10. Steiner, H.G., ed. Cooperation Between Science Teachers and Mathematics Teachers. IDM Materialien und Studien Band 16.

A useful sorting out of the topic given by the title, with six plenary papers, twenty-four essays by conference participants, and reports from six working groups: science applications in mathematics classrooms; mathematics in physics classrooms; mathematics in biology classrooms; mathematics in chemistry classrooms; conditions for cooperation and innovation. There are several quite useful bibliographies of articles about applying mathematics in science.

Institute fur Didaktik der Mathematik (IDM)
Universitat Bielefeld
D4800 Bielefeld 1
Federal Republic of Germany

11. Tanur, J.M. Statistics: A Guide to the Unknown. (1972)

This consists of forty-five brief essays about specific uses of statistics. It is aimed at a lay audience, and almost no numerical information or formulas or computations are included. There are four main parts: man in his biologic world, political world, social world, and physical world. A few essay titles at random: safety of anesthetics; deathday and birthday: an unexpected connection; do speed limits reduce traffic accidents; the meanings of words; the consumer price index; statistics, the sun, and the stars.

Holden-Day, Inc.
500 Sansome St.
San Fransisco, California 94111

12. Wain, G.T.; Woodrow, D.; Brown, M.; Booth, L.; Orwell, C.P.; Dean, P.G.; Harper, J.; et al. Mathematics Teachers Education Project. (1974-80)

Material to train mathematics teachers at the secondary level to integrate mathematics with other subjects. Handbook for teachers and set of student exposition material published summer, 1980. Contact:

G.T. Wain
Centre for Studies in Science Education
Leeds University
Leeds, UK

or Margaret Brown
Centre for Science Education
Chelsea College
Bridges Mall
London SW6, UK

Note: Recently several groups have begun to publish journals about using applications in teaching mathematics or subscription series of problems and units. Here are some that have come to our attention.

13. Undergraduate Mathematics and its Applications Project (UMAP)

Units: Many topics have been undertaken by individual authors to show applications of mathematics (often elementary calculus) and statistics to biomedical sciences, business, economic, etc. Booklets are field tested, then revised and made available by subscription. Some topics: measuring cardiac output; food service management; interior design; preparing cost estimates.

Journal: The Journal of Undergraduate Mathematics and its Applications. R.L. Finney, ed. Quarterly. $15 or sFr. 25 or DM 27.

Both units and journal are published by:
Birkhauser Boston, Inc.
380 Green Street
Cambridge, Massachusetts 02139

14. Teaching Statistics. P. Holmes (ed.). Three issues per year. (1979, $4; 1980, $9; 1981, $10.50; or 2, 4, and 4.75 pounds sterling, respectively.)

"Teaching Statistics is a new journal for all who use statistics in their teaching of pupils and students aged 9-19 . . . (it) seeks to inform, enlighten, entertain, and encourage all teachers to make better use of statistical ideas."

Department of Probability and Statistics
University of Sheffield
Sheffield S10 2TN
England

15. Journal of Mathematical Modelling for Teachers. Burghes, D.N., & Read, G.A. (eds.)

"This journal promotes the study of applicable mathematics through the publication of articles related to the modelling process. Particular emphasis is placed on the translation of a real situatin to a mathematical model and on the verification and interpretation of the mathematical solution."

Department of Mathematics
Cranfield Institute of Technology
Cranfield, Bedford MK43 0AL
England

16. PAMELA . . .

"The problems in the collection all arise outside mathematics and, as usual, the process of modelling will begin and end there. They should come from everyday life (or every-year life at least) and should be recognisable as problems by the interested, though perhaps mythical man-in-the-street; this restriction is designed to rule out most of the myriad of problems arising from other school subjects or from particular jobs. Finally, PAMELA is primarily a collection of problem situations not of models for describing them (but) there are suggested solutions and references for many of the problems. Access to PAMELA is freely available to those involved in the development and teaching of mathematics and its applications. After 1 May 1979 listings from the collection of problems will be sent on request; equally contributions to the collection are solicited. In order to cover costs we are adopting a crude but simple system of charging. Please send £5 with your first request - - you will be notified of the charges made and of your remaining credit when the material you have asked for is sent.

Hugh Burkhardt
Shell Centre for Mathematical Education
Univeristy of Nottingham
Nottingham NG7 2RD
England

Here are additional materials which have come to our attention, with little additional detail, or well known things with present availability not known.

She Wan ge Weishenne (One Hundred Thousand Questions). (Shanghai: People's Republic Publishing House, 1970) Vol. I

Applications-based enrichment material with topics such as:
- How do we estimate the production of rice from a paddy?
- How does a computer work?
- Do you understand how landlords used interest to cruelly exploit the peasant?

Applications of Mathematics in Industry and Agriculture. (Peking: Education Department of the City of Peking, 1975)

Hua LoKeng "Applications of Mathematical Methods of Wheat Harvesting," Acta Mathematica Sinica 1961, 1:77-91.

(Information on these materials from Frank Swetz, "The 'Open-Door" Policy and the Reform of Mathematics Education in the People's Republic of China," Educational Studies in Mathematics 8, (1977).)

Madgett, A.C.; Madgett, P.; Thomas, J.F.S.; & Weyging. (Laurentian University) Applications of Mathematics: A Nationwide Survey (Canada). (1974-76)

The project sought to determine what kinds of mathematical techniques were currently in use in Canada, obtain a core section of sample problems, and present those in a form for classroom use. A useful interim report exists (inquire of the authors at Sudbury, Ontario, Canada).

8.9 MUTUALISM IN PURE AND APPLIED MATHEMATICS

MUTALISM IN PURE AND APPLIED MATHEMATICS

Maynard Thompson
Department of Mathematics
Indiana University
Bloomington, Indiana

Abstract

The complex interrelations between pure and applied mathematics are examined in the context of broadening mathematical sciences and expanding areas of application. Special attention is paid to the fact that many applications are currently discussed within science. Educational implications are considered.

Historically there have been deep and highly productive relations between mathematics, especially analysis, and the physical sciences and engineering. This is, of course, widely known and the implications of the relations understood. However, there is a commonly expressed feeling that the sort of mutual stimulation which existed previously - for instance between mathematics and physics during the birth of quantum mechanics - represents the heyday of interaction between pure mathematics and applications and is today relatively rare. It is certainly the case that in contrast to a period in which most researchers in mathematics were stimulated more or less directly by scientific problems, today most mathematical research is generated by questions which arise within mathematics. The rise in importance of such internal stimulation is a natural consequence of the growth of relatively specialized subfields and the extensive mathematical background necessary to make progress in them. Many people simply have no time to consider the possible relations between their work and problems which arise outside of mathematics.

The tendency toward reducing the links between mathematics and scientific problems has in some ways been intensified by the explosive growth of the use of mathematical ideas and techniques in other fields. Many of these fields lack the strong traditional bonds with mathematics that the physical sciences have, and may even have their own journals for the publication of mathematical results. For instance, much activity which in an earlier era would have been carried out by researchers primarily identified with mathematics is today carried out by scholars primarily identified with other disciplines. The results may be published in journals such as Journal of Mathematical Economics, The Mathematics of Operations Research, Mathematical Biosciences and Theoretical Population Biology, to name a few, which may not even be housed in a mathematics library (as is the case at my school). Yet these journals publish papers, reviewed in Mathematical Reviews, which contain precise definitions, clearly stated theorems and rigorous proofs. In short, legitimate mathematics.

These tendencies have intended to strain the ties between those who do mathematics and those who use it. However, there is risk of oversimplification and it is useful to be somewhat more careful in our use of terms. On one hand there is mathematics which is created for its own sake and with (at most) stimulation from another area of mathematics. We refer to the activity of creating this mathematics as "doing pure mathematics." Of course, it may happen (and it frequently does) that results obtained in this way are very useful outside of mathematics. However, any possible contribution to resolving a scientific problem played no role in their creation. Similarly, we refer to the activity of using mathematical ideas and techniques to aid in understanding a situation or solving a scientific problem arising outside mathematics as "doing applied mathematics." The mathematics used in this way may be new, invented by the user for the purposes, or it may be gleaned from the literature. There may be aspects of this activity which are locally identical to aspects of doing pure mathematics, discovering and proving a theorem for instance, but globally the two activities have quite different motivations and normally evolve in rather different ways. Whereas one who does pure mathematics usually strives for generality and elegance, one who does applied mathematics is usually primarily interested in resolving scientific questions. The generality and elegance of the mathematics is less an issue that its contribution to the scientific question.

The mutual stimulation between these two activities which previously took place to a large extent within mathematics now takes place more and more between mathematics and other disciplines. Nevertheless, the stimulation exists and to an extent that is perhaps underestimated. In fact, it can be argued that the long-term vitality of both science and mathematics requires that the interaction becomes less of a peripheral and more of a central concern. The relation between the two activities is truly one of mutualism.

Our goal is to examine this mutualism in a setting which emphasizes its various aspects, and since applications and science are so thoroughly intertwined, we pay special attention to the scientific perspective. Also, we are concerned with the implications of our analysis for the training of mathematical scientists and scientists in other fields. We conclude with some examples of recent efforts which illustrate our theme.

Mathematics in the Sciences

In order to investigate in more detail the relations described above it is helpful to have a broad classification of the ways mathematics is used in the sciences. We begin by citing a common - but by no means universal - pattern in the ways that mathematics is used in other discriplines. In most cases there has been a tendancy toward increased mathematization as a discipline evolves, but this view is conditioned by hindsight and there may be important differences in the future as the role of computers becomes more important.

We shall identify four stages which correspond roughly to the degree to which mathematics contributes to the discipline. We emphasize that from the scientific perspective it is the contribution of the mathematics that is important and not the sophistication or elegance of the mathematics. A relatively simple mathematical idea may have great impact when used ingeniously; and on the other hand a very elegant mathematical discussion may contribute little to our knowledge of the scientific problem.

Stage I. Mathematics frequently enters into scientific work in an area with the collection, organization and interpretation of data and information. In some instances there are large amounts of data collected from observations: Tycho Brahé's data on the motion of the planets, Gregor Mendel's data on plant reproduction, and the results of the numerous experiments in psychological learning beginning in the 1930's. In other instances such as relatively uncommon natural phenomena (major earthquakes, rare diseases, etc), there may be relatively few instances to examine. The task of deciding which data are needed and how to go about the collection process is frequently a difficult one. The task is, however, essential and the absence of good data can significantly retard the general development of an area, and in particular its quantitative development.

From a mathematical standpoint, the interaction primarily involves statistics and computing. The traditional influence of applications on the development of statistics remains large, and many prominent mathematical statisticians have also contributed to the solutions of problems of great applied interest. We return to this later.

Stage 2. Some collections of data display only the most general regularities - economic statistics frequently are of this sort. Other collections display remarkable patterns when examined closely. Two of the best examples of this are Brahe's data on the motion of the planets and Mendel's records of plant hybridization experiments. Sufficiently regular patterns in data or information may be summarized in empirical "laws of nature". These laws, which at this point are no more than convenient ways of summarizing observations, may be very precise or quite vague. In the physical sciences such laws are frequently quite precise. For instance, Kepler's laws of planetary motion, which are empirical laws deduced from Brahe's observations, and Snell's law of refraction are very precise statements. Empirical laws tend to be much less precise in the life and social sciences. For instance, Gause's law of competitive exclusion (two different species competing for exactly the same biological niche cannot coexist) and Gossen's law of marginal utility (the marginal utility of a good decreases as more of that good is consumed) are much less precise and their legitimacy can be debated in a way that Snell's law cannot be.

There are some caveats worth mentioning in discussions of empirical laws. First, such laws are frequently derived on the basis of aggregate behavior. That is, the data are smoothed or averaged and the law may be more applicable to a (perhaps nonexistent) "average" situation than to any specific situation. Also, there are frequently (implicit or explicit) assumptions underlying the statement, and the law may describe observations only to the extent that the assumptions are fulfilled. For instance, one of Mendel's law of genetics asserts that genes relating to different characteristics recombine at random. This law provides an accurate description of the behavior of genes which are not linked. If we are interested in studying characteristics associated with linked genes, then this law does not apply.

Empirical laws which serve as succinct and illuminating summaries of experiment and observation may be one of the goals of an experimentally inclined scientist. Such laws may be midway in the analysis of a situation by a theoretically oriented scientist. Indeed, it is uncommon for there to be much interaction between mathematics and science in this stage.

Stage 3. Following the identification of empirical laws, the next state in the mathematization of a science is the creation of a mathematical structure - sometimes referred to as a theory - which accounts for these laws. The expression "accounts for" requires elaboration and we shall return to it later. For the moment the usual intuitive meaning will vary from one discipline to another and from one time period to another. A mathematical structure together with an agreement as to the relations between the symbols and notation of the mathematics and the objects and actions of a real world situation is known as a mathematical model of that situation. There may be more than one mathematical structure which accounts for an empirical law, or there may be one structure which accounts for some of the laws of a situation and another structure which accounts for other laws. (An example of the latter from elementary physics is the dual wave and particle models for light. The wave model provides explanations for the phenomena of physical optics: reflection, refraction, dispersion, etc. However, the photo-electric effect which poses problems in a wave model is perfectly comprehensible in a particle model.)

An investigation of the mathematical structure leads to conclusions, usually called theorems, which are deduced by logical arguments from the assumptions and definitions. Using the agreements between the symbols and terms in the mathematics and the original setting, these theorems can be translated into assertions, usually called predictions, about the real world setting. We can now sharpen our meaning of "accounts for". We say that a mathematical model accounts for an empirical law if one of the theorems of the model translates into that law.

The task of determining whether a model accounts for a set of observations is frequently a very complex one. It is common for models to involve parameters, symbols which represent real world quantities such as the velocity of a particle, the cost of a commodity, the arrival rate of cars at a tollbooth, or the likelihood that a subject will remember a nonsense word after a single presentation. In order to compare predictions with observations one must know the values of the parameters which appear in the model. The estimation of parameters is a challenging mathematical and scientific task, and many otherwise apparently acceptable models have not been pursued because they relied on parameters which could not be estimated.

A mathematical model, as a mathematical structure, may of course raise questions which are interesting from a mathematical standpoint. Indeed, it is at this point that the stimulation on pure mathematics by applications may be greatest. Once a structure is identified or created and sharp mathematical questions posed, it may be clear that these questions are meaningful and perhaps intersting and important in a much more general setting. These new questions may, but need not, have significance for the situation which gave rise to the model. A recent example of this from my own experience involves memory and learning models in psychology. A psychologist and I were interested in developing an algorithm for determining moments of certain random variables (occupancy times) defined on Markov chains. While formulating the problem precisely we realized that it was an interesting one in a much more general setting. It is most unlikely that the general problem would have occurred to me - and it had apparently not occurred to others since it has not appeared in the literature - if it were not for the special case which arose in the learning model.

This example illustrates two aspects of the relation which are worth noting. First, in making the problem precise we were forced to be very clear about the connections between the symbols and relations in the mathematical model and the original setting. Although the burden of this work was carried by my colleague, it was important for me and provided intuition which was helpful. Second, the special case motivated work which went beyond the original problem, but in turn raised questions which are significant from the viewpoint of the original setting. This kind of activity is actually more illustrative of the next stage.

Stage 4. Suppose that we have created a mathematical model, developed it and derived a body of predictions (theorems) which account for some observations. The next stage is the examination of the model with a goal of obtaining **new** scientific insight. That is, we look for aspects of the science of the situation that are either uncovered or illuminated by the use of the mathematical model. Frequently this means that the mathematics leads to predictions of scientific phenomena which are subsequently observed. In Stage 3 we used existing data to validate the model - here we use the model to tell us where to look for data. Here, the stimulation is directed largely from pure mathematics to the application.

It may happen that in developing the mathematical structure we encounter mathematical relations or conclusions which can be interpreted in terms of the application. Also, it may be that the mathematical structure which arises in one situation may be very similar (or even identical) to the mathematics which arose in a very different situation. One can then utilize all the results derived in the second study to help understand the first. This may well lead to the recognition of aspects of the applied problem which were not previously recognized. A good example of this occurs in chemical kinetics. In particular, a type of chemical reaction known as the Belousov-Zhabotinskii reaction exhibits regular sustained oscillations between states (colors), and in some cases outward propagating circular waves. An analysis of a mathematical model for the situation resulted in a prediction of inward propagating waves as well. Waves of this type have been observed after their existence was predicted. A confirmation such as this greatly enhances our confidence in using a model as a description of the real world setting.

Sources of Stimulation

Let us now review these stages briefly with a view to identifying where those doing pure and applied mathematics may look for useful links. The stages can be summarized as follows:

Stage 1: Data and information collection, analysis and interpretation.

Stage 2: Quantitative formulation of scientific principles and empirical laws.

Stage 3: Formulation, study and validation of mathematical models.

Stage 4: Use of mathematical models to gain scientific insight.

As a part of the scientific activity of stage 1 there may well be statistical or computing problems which either a) cannot be resolved or b) are more general versions of special problems. These problems provide a continuing source of new material for the more mathematically oriented statistician and computer scientist. Also, deeper understanding of the statistics of a situation may enable an applied statistician to contribute more helpfully.

The mathematical problems generated in stage 3 are a common source of motivation for those doing pure mathematics, and on occasion are the germs for major developments. For instance, the active interest in conservation laws by researchers in differential equations has its origins in models which are closely tied to the physical world. Likewise, the applied mathematician may find among the mathematical results created for their own sake exactly the idea or method needed for the problem being studied, for instance, the results from finite stochastic processes which are so commonly used in the social sciences. Alternatively, it may be that a result created for one purpose turns out to be ideal for a quite different purpose. One of the best illustrations of this is a lemma derived by J. Farkas in 1902 in connection with an engineering problem. This lemma provides a key result in many proofs of the fundamental theorem of linear programming - a result in a theory which saw rapid development in the early 1950's.

The utility of mathematical results in suggesting new scientific insight (stage 4) may rest on the result being cast in a form that the connection between the mathematics and science is most revealing. Alternatively, it may be that the identification of an appropriate special case will provide the key.

The most well known examples of mutualism, interactions which are truly beneficial to both groups, came from the physical sciences and primarily before 1940. It may be useful to cite two more recent examples, one from communication theory and one from population dynamics.

Error correcting codes.

(2), (5),(8) The original problem concerns the reliable communication of digitally encoded information. A message is a string of 0's and 1's. When a message is transmitted over a noisy channel there is some likelihood that a digit which is transmitted as a 1 will be received as a 0, or vice versa. In order to overcome this problem additional digits are transmitted to provide redundancy. The message is not simply repeated, but rather check digits are added which enable the receiver to determine whether the received message is distorted, and if so to determine the message actually transmitted. Typical problems are to determine the number of check digits for a given message length to enable certain types of errors to be corrected, and how to use the check digits (encoding and decoding). The study of these problems has led to many interesting purely mathematical questions in a variety of fields: combinatorics, number theory, finite groups, etc. In the other direction, mathematical results have led to major advances in the design of efficient decoding methods.

Discrete population models.

The natural periodicity associated with many biological systems has led to the creation of several discrete time models for population dynamics. In some cases what appears to be a relatively simple model turns out to be quite complex. Indeed a system whose behavior through time is completely determined step-by-step by a simple equation can, (for certain parameter values) display global behavior which appears totally random. This phenomenon, known as "chaotic behavior", has generated a great deal of work on the mathematical structure of such systems. The predictions (theorems) have led to considerable biological speculations; the implications are by no means completely understood.

Maintaining Mutualism

Although it is easy to cite recent examples of strong mutual stimulation between pure and applied mathematics, it is undoubtedly true that such links are declining both in frequency and in strength. It seems to me that the efforts with the greatest potential for long term profit are those devoted to students. In particular, there are a number of specific actions which seem likely to enhance communication. It is important, however, to recognize that there are students who have almost no interest in the uses of mathematics just as there are students in other disciplines with great mathematical talent who have minimal interest in doing mathematics for its own sake. The extent to which individuals in either group should be "force fed" is a sensitive issue.

It is now common for mathematical problems which arise outside of mathematics to be studied by people trained primarily in other disciplines. In fact, today most people doing applied mathematics would not refer to themselves as mathematicians or applied mathematicians. As a result of natural feedback mechanisms, we find the teaching of mathematical ideas and techniques becoming a standard part of the curriculum of other disciplines. Also, there is diminishing awareness within the mathematical community of many important and interesting problems to which mathematics can be applied, and the isolation of mathematics from its uses is enhanced. This process is self perpetuating unless deliberate efforts are made to reverse it. An essential part of the reversal effort must be to expand the ability of those trained in mathematics to communicate with those who need to use mathematics. One method of developing communication skills is to proceed deeply enough into another field that important problems in the field can be phrased in mathematical terms. In most American universities this would normally be in the third year of undergraduate study in fields such as biology, economics and psychology. It is tempting to react that an undergraduate mathematics major has no time for such an effort. But the study of the substance of another discipline is essential to acquire the perspective from which others view mathematics. An alternative is an upper division mathematical modelling course in which problem settings are fully developed and there are opportunities to discuss the substance of the problems. Although this alternative is more time efficient, only in rare circumstances does it have the effect of a sequence of courses in another field. From an educational standpoint the choice of field is not nearly as important as the decision to take enough work to understand the origins of the mathematical problems that arise. Next, it is essential that students receive broad training in the mathematical sciences. The core areas (algebra, analysis, geometry) have not diminished in importance, but they are now forced to share their importance with probability and statistics, optimization and mathematical programming, and computer science. It appears inevitable that there will be compression and reorganization of undergraduate courses to accomodate these new arrivals. Just as many geometry, theory of equations, and similar courses disappeared to make room for linear algebra, modern algebra, and topology, there will be casualties among today's standard courses to provide for new offerings. This winnowing process is well underway at some schools. Indeed at one with which I am familiar more than half of the courses in the curriculum in 1980 are new or substantially revised since 1975. The image of mathematics as a single subject must be broadened into a cluster of subjects, and students need training in all of them.

As a result of the usual structure of the mathematics curriculum, students rarely have an opportunity to select among different techniques to solve a problem. In mathematics courses it is usually obvious that one uses a method from analysis (or algebra, or probability, or . . .) to attack a problem. On the other hand it rarely happens that a problem which arises outside of mathematics comes with a specified mathematical technique. Even if it does - say a method suggested by previous efforts on related problems - it may not be the most appropriate approach. Consequently, the selection of an appropriate mathematical method is usually a part of the problem solving process. The development of the ability to make a mathematical diagnosis should be a goals of mathematics instruction. This goal is certainly closely related to that of broad training. Clearly having a repertoire of potential methods is prerequisite to making a selection.

Finally, students need to be confronted with open-ended situations. We must provide experiences for students which begin with a loosely described situation and lead up to a precise mathematical problem or a family of such problems. It is essential to "convince them that the formulation of a problem in mathematical terms is an achievement and often the major step towards it solution" (7). Most traditional mathematics courses provide no experiences in dealing with situations in which many decisions are open. The process of applied mathematics is a complex one and there are no easy guidelines which are widely applicable. Consequently, there is a limit to what can be learned by observing, and students must get their hands dirty in posing problems, selecting methods, developing models and relating these results to the setting of the original problem. Students should be encouraged to pursue the approaches that seem natural to them. The usual results, that different students develop quite different models, can be evaluated and compared. Care should be taken to dispel "the model" syndrome (6),(10).

Conclusion

The mutual stimulatin between those doing pure and applied mathematics retains its role as more of the applications of mathematics shift to other disciplines. However, the relation between mathematics and its applications needs special attention to maintain its strength. A recognition of the role of mathematics from a scientific perspective, and efforts to train mathematics students to function within that role are essential to preserve this mutualism.

References

1. J. Guckenheimer, On the Bifurcation of Maps of the Interval, Inventiones Math. 39 (1977), 165-178.

2. N. Levinson, Coding Theory: A Counter Example to G.H. Hardy's Conception of Applied Mathematics, Am. Math. Monthly, 77 (1970), 249-258.

3. R.M. May, Biological Population with Nonoverlapping Generagions: Stable Points, Stable Cycles and Chaos, Science, 186 (1974), 645-647.

4. R. May and G. Oster, Bifurcation and Dynamic Complexity in Simple Ecological Models, Am. Naturalist, 110 (1976), 573-599.

5. V. Pless, Error Correcting Codes: Practical Origins and Mathematical Implications, Am. Math. Monthly, 85 (1978), 90-94.

6. S.M. Pollock, The Modelling Studio (Discouraging the Model Model), in Discrete Mathematics and its Applications, ed. by M. Thompson, Indiana University, Bloomington, 1977, pp. 131-154.

7. Report of the Committee on New Priorities for Undergraduate Education in the Mathematical Sciences, Am. Math. Monthly, 81 (1974), 984-988.

8. N.J.A. Sloane, A Short Course on Error Correcting Codes, CISM 188, New York: Springer-Verlag, 1975.

9. S. Smale and R.F. Williams. The Qualitative Analysis of a Difference Equation of Population Growth. J. Math. Biol., 3 (1976), 1-4.

10. M. Thompson, The Process of Applied Mathematics in Case Studies in Applied Mathematics, Mathematical Association of America, Washington, D.C., 1976, pp. 4-21.

MUTUALISM IN DECISION MATHEMATICS

D. Bushaw
Washington State University
Pullman, Washington

This will be a scholium on Maynard Thompson's account of "mutualism in pure and applied mathematics." I will emphasize an aspect of mathematical science he touched on only briefly, and to which his analysis applies better after some modification.

"Decision mathematics" may be characterized in various ways: as the mathematics of the decision sciences; as the mathematics of purposive action; as the mathematics that tries to help answer questions of the type "What should be done?" as contrasted with questions like "How (or why) does this happen?" It includes, but is by no means limited to, questions about the best way to do something, that is, optimization; but sometimes it is content merely to find satisfactory ways.

Like applied mathematics in general, decision mathematics cannot be characterized in terms of the traditional structuring of pure mathematics. It cuts right across algebra, analysis, geometry, probability, statistics, and combinatorics, not to mention such somewhat mathematical disciplines as computer science and logic.

This observation has an immediate implication for mathematics education: namely, that topics from decision mathematics are accessible in almost every mathematical classroom context. It is a matter of accent in the kinds of problems considered and solutions sought, not of the mathematical areas invoked. I shall mention a few specific representative possibilities later.

Professor Thompson has identified four major stages in any interaction between pure mathematics and a real-world problem. He had in mind mainly the needs of the descriptive, rather than the decision, sciences. I believe, however, that the same four stages apply also to activities in decision mathematics, but would like to suggest certain shifts of emphasis and interpretation.

Stage 1. The collection and organization of data are of course important for any problem whose solution is to have some bearing on the real world, decision problems included. Indeed, this stage tends to be especially difficult for many kinds of decision problems, because the facts are often elusive and fast action is often required. Effective methods for arriving mathematically at decisions in specific instances therefore tend to be, if not "quick and dirty," then at least unleisurely and unfastidious; and this applies particularly to the gathering and assimilation of data. Statistics and computing often play their expected essential roles at this stage.

Stage 2. Decision mathematics is less concerned with formulating empirical laws abstracted from the data than with the identification of performance criteria and constraints. This is often a straightforward matter - - after all, the problem itself usually specifies, if only crudely, what these entities are. Nevertheless, identification of significant variables and of important relations among them is a crucial and sometimes difficult task. It may require art and experience as well as a capacity for clear thinking and, more particularly, mathematical skill. For example, one needs to be able to exercise good judgment about the complexity of the representation of reality used. A common criticism of traditionally trained mathematicians is that they often lack this judgment - - they tend to indulge in the voluptuous intricacies of precise and elaborate representations when much simpler ones would yield perfectly satisfactory results with greater speed and lucidity. On the other hand, simplification can be overdone; every practitioner of decision mathematics knows how easily one can choose a "wrong" performance criterion or omit constraints needed for adequate realism.

Stage 3. In Professor Thompson's enumeration, this was the stage of formulating and analyzing a mathematical structure to account for the empirical laws identified in stage 2. In decision mathematics this may well take the form, as for the descriptive sciences, of adopting or devising and then analysing a suitable mathematical model. A distinctive feature of decision mathematics at this stage, however, is the emphatic interest in solution algorithms. The familiar processes of analogy and generalization inevitably lead to a desire for effective procedures for finding concrete solutions to problems constituting broad classes. The stress here should be on the word "effective," which represents both the typical element of urgency mentioned earlier and such desiderata as aptness for computer implementation. An excellent and familiar example is that of linear programming, where the subject really came into its own as a valid branch of decision mathematics only after the discovery, in about 1950, of the simplex method, the first effective technique for finding specific solutions to problems of a conceptually fairly transparent type (extremization of a linear form in variables subject to linear inequality constraints).

Stage 4. This is the stage of obtaining new insight into the phenomena under study. It might be supposed that this would be at best a minor aspect of decision mathematics, whose goal is to find specific directives for action in specific circumstances. In fact, however, general insights often emerge as by-products of the consideration of classes of similar decision problems.

Instead of obtaining specific recipes for behavior under specific conditions, for example, one might obtain general rules providing at least partial guidance for responding to problems of a broad class. These rules often take the form of statements that certain prima facie alternatives may be eliminataed. The observation in linear programming that one need consider only the vertices of the polytope defined by the constraints is one example; the "bang-bang principle" asserts that - - under conditions fairly easy to verify - - only extreme values of control parameters are needed to obtain optimal results. This observation, when it applies, considerably simplifies the basic task of finding the "best" control functions over time.)

In fact, optimal control theory provides a fine example of the mutual interaction of pure and applied mathematics that is the central theme of this session. The basic problem of optimal control, in several simple forms, was formulated about 1950 in connection with the steering of aircraft and submarines and the use of servomechanisms. Mathematically, the problem clearly belonged to the calculus of variations, but the established techniques of. that subject did not seem to fit. So the problem was attacked by ad hoc direct methods. The solutions obtained in this way, based in part on engineering insight, suggested some extensions of classical methods associated with such names as Hamilton, Jacobi, Weierstrass, and even Euler and Lagrange. The development of these extensions led to new techniques - - notably the Pontryagin Maximum Principle - - for treating the original problem. These ideas in turn raised further mathematical questions as well as problems of computer implementation, some of which are still not satisfactorily solved. In its present form, optimal control theory draws not only on the calculus of variations and its extensions but also on vector bundles, Lie groups, stochastic processes, and many other branches of mathematics which are themselves enriched thereby.

In the developments just sketched, which have involved engineers, physicists, economists, computer scientists, and many others besides mathematicians, you may find all four of the stages we have described, and not just once but repeatedly.

How do these remarks bear on mathematics education? I certainly would not propose that we try to indoctrinate all our students with an explicit account of these very general ideas. I do believe, however, that the point of view they express should influence the choice of topics we teach and the spirit in which much of our teaching is done. I offer you three justifications for this statement. There are surely others.

1. Decision mathematics is important. Decision mathematics has become a major and distinctive component of the cultural and specifically scientific scene for which our students should be prepared.

2. Decision mathematics is appealing. Many young people, even those who find mathematics for its own sake pointless and are lukewarm about mathematics as a handmaiden to the sciences, respond with interest and often with enthusiasm to suitable glimpses of decision mathematics. It speaks to the pragmatic strain in almost everyone, and, properly taught, entices students with the prospect of being able to use it in their own lives.

3. Decision mathematics is accessible. The flavor of decision mathematics can be conveyed in countless contexts of mathematical instruction. Many of the specific revelations of the subject can be taught to very young students. The leading ideas of optimal control theory are not beyond a good calculus student; linear programming may be introduced to students after a year or two of high school algebra. The key ideas of dynamic programming and of graph theory might well be taught in the grades.

What this all amounts to is another plea that some teachers, teachers of teachers, and authors of textbooks and examinations consider changing their ways somewhat. The task of working more decsion mathematics, and more of the spirit of decision mathematics, into the mathematics curriculum is large, but I would expect that our students will be grateful, in their way, for whatever we can do.

MUTUALISM IN PURE AND APPLIED MATHEMATICS

Candido Sitia
Centro U. Morin
Paderno del Grappa, Italy

My talk, too, will be a commentary on Prof. Thompson's account. I will look at the four stages presented in his account from the point of view of a teacher, i.e., from a didactical point of view.

In fact, I shall refer to my personal experience as a teacher of Mathematics in an Italian High School (Liceo Scientifico) and, more recently, to my experience as principle of this Liceo, i.e., responsible for the teaching activities of my colleagues.

The pupils we are concerned with are 16/18 years old. The official programmes of Mathematics are very traditional: analytical geometry, trigonometry, calculus (differentiation and only the basic ideas of integration). The final examinations that give access to our universities are only written for Mathematics and consist of some routine exercises: study of function, drawing its graph, etc.

The teaching of physics and chemistry starts at the age of sixteen and is very limited:

age	Mathematics (hours/week)	Physics (h/w)	Chemistry (h/w)
16	3	2	3
17	3	3	3
18	3	3	2

The teaching of Chemistry however, includes many other topics, such as: Biology, General Geography, Astronomy, Earth Sciences, . . .

It is clear that in this situation it is very difficult to give a real insight into the scientific method whatsoever and into the real nature of Mathematics. For purposes of comparison I will show the complete schedule of our "Liceo Scientifico":

ORARIO DI INSEGNAMENTO

MATERIA	Liceo scientifico				
	I	II	III	IV	V
Lingua e lettere italiane	4	4	4	3	4
Lingua e lettere latine	4	5	4	4	3
Lingua e letteratura stranfera	3	4	3	3	4
Storia	3	2	2	2	3
Geografia	2	-	-	-	-
Filosofia	-	-	2	3	3
Scienze naturali, chimica e geografia	-	2	3	3	2
Fisica	-	-	2	3	3
Matematica	5	4	3	3	3
Disegno	I	3	2	2	2
Religione	I	I	I	I	I
Educazione fisica	2	2	2	2	2
	25	27	28	29	30

However, even in this situation and at this level, I am convinced that the traditional methods of teaching are quite obsolete and that the starting points of the process of learning Mathematics (or any Science whatever) are real life situations. The conquest of an insight into Mathematics is based on discovering analogies betweeen many different real situations. The point of arrival, at this level (16/18 years), should be the creation of one (or more) mathematical model(s), and some formalisation. Last but not least, the students should acquire the consciousness of the actual scientific proceeding and methods.

To reach these objectives we have developed in our "Liceo" an alternative line of approach to the mathematical notions implied in the programme, aiming at the final target of giving a broad idea of the scientific thinking and process. The stages we pass through to arrive at a new mathematical project or idea or methodology, are the following (they fit well with the ideas of Prof. Thompson):

1. Starting from a wide interdisciplinary variety of situations (using historical problems preferentially);

2. Looking for some symbolisation of every situation we have studied;

3. Discovering the analogies and therefore the same mathematical structures in their background;

4. Developing the theoretical part of this model (generalisation);

5. Coming back to the real situations to be mastered and interpreting the "Theorems" of the theory as natural laws, acquiring thereby a better insight into sciences;

6. Coming to numerical applications to some concrete situation;

7. Criticism of the results through the study of

   (i) the limiting hypothesis made in relation to points (3) and (4);

   (ii) the discrepancies between the theoretically obtained results and the concrete situations.

Examples: Introduction to differentiation:

First of all, I discuss with my students some examples of simple intuitive ratios, e.g., the price of goods, the capacity of a glass, the thermal and electrical capacities, etc.

Then we move on to investigate a lot of different situations selected from very different fields. This investigation involves

(i) classroom discussion
(ii) physics or chemistry lab tests and activities
(iii) reading selected literature (suggested by the teacher).

Here are some possible situations that can be discussed and studied:

- Velocity: intuitive sense; its refinement through laboratory activities; the problem with which Newton was confronted.
- Slope of a line, of a graph, . . . the problem of Leibnitz.
- Marginal quantities in economics.
- Rate of growth of a population . . . and also its problems.
- Rate of radioactive decay.

And so on.

Of course all these items shall not be discussed equally in full extent, but only those which will arouse the interest of the students.

At this point the need for some symbolisation of all these analogous situations will arise and discussion will lead us to outlining the analogies existing and the possibility of finding a general symbolism for them.

It is possible then to arrive easily at the concept of a finite rate of change and at its symbolisation and evaluation. Coming back to different real situations in order to interpret the symbols and to apply them to the concrete world will be an interesting activity.

However, the students will soon understand that, in many cases, this model does not work. The need for a further more sophisticated model requires further investigation to pass from an "average" model to a "point" model.

Another example. Referring to one of the examples quoted by Prof. Thompson I can add that the problem of communication of digitally encoded information has been used by our group to introduce some interesting problems and the logarithmic function, problems of combinatorics, of finite group theory, etc. This work was published in the journal of our Centro Morin: "L'Insegnamento della Matematica e delle Science Integrate".

In this way real situation and mathematical models interact continuously with each other and the interest in one of them will generally excite the interest in the other. Coming back to the real situation is a very important step for two reasons (at least):

(i) The students will acquire the art of interpretation of a model in a given situation; particularly they will be able to "read" the theorems of a model as "natural laws".
(ii) Very often this return to the real situation and this interpretation will lead them to question the theorems and the natural laws, and, as a consequence, the students will be obliged to find a new model or to refine the old one (for ex.: the fall of mass in vacuum and the comparison with the reality; the limitations of classical mechanics; the inadequacy of the particle model of light in explaining interference phenomena, and of the wave model in explaining photoelectricity, etc.)

Given sufficient time, it is therefore possible to discuss with the students the very nature of science and of mathematics, the mutual interaction with each other and to give them a consistent consciousness of the scientific method.

Some final observations

a. This approach to science and to mathematics in a high school requires wide interdisciplinary preparation of the teachers or, at least, good cooperation between them.
b. as pointed out by Prof. Thompson, there is no doubt that the links between pure and applied Mathematics are declining both in frequency and in strength. But I agree that the efforts with the greatest potential for long-term profit are those devoted to students. The past, and also recent history, have accumulated a wealth of material to be used in this direction. However, I recognise that it is very difficult for a single teacher to profit from this wealth. Where can he find the resources of this work? I find it difficult to know such useful materials for a Mathematics teacher! It is a consequence of the isolation of Mathematics and of mathematicians from other disciplines and it enhances, in turn, this isolation. I suggest that we should produce more interdisciplinary materials to be used by the teachers of Mathematics in their classrooms and for the education of prospective teachers. In our journal, quoted above, I translated some papers of Euler, Newton and other prominent creative men and this work was widely appreciated in my country. The corresponding literature is very deficient too. A pointed out by Prof. Bushaw, however, the creation of this material is an urgent task, for which our colleagues and our students will be grateful.
c. This approach, however, required much time and it is therefore necessary to limit it to the fundamental ideas and to make some limiting choices.
d. In the Italian situation, and maybe in many others, this method is very hard to realise, but it gives satisfactory immediate results with the cleverest students, and, in long term, I found also good results in most of the students, even in those who avoided every difficulty. These results, however limited and often shifted in long periods, seem to me the best support for a constant and renewed endeavour in this direction.

CHAPTER 9 - Problem Solving

9.1 TEACHING FOR EFFECTIVE PROBLEM SOLVING: A CHALLENGING PROBLEM

WORTHWHILE PROBLEMS FOR THE MATHEMATICS CLASSROOM: A NON-EMPIRICAL CLASSIFICATORY ATTEMPT

Shmuel Avital
Israel Institute of Technology
Haifa, Israel

Problem Solving in the Mathematics Curriculum

An enormous number of studies about problem solving are being carried out at various centers of mathematics education. The reason for this fervent activity is quite clear to anybody involved in this area. It is generally agreed that to achieve the aims and goals of mathematic education the student must exercise and practice the direct application of learned algorithms, and then develop his ability to apply mathematics to the solution of problems encountered in the subject itself, in other sciences, and in everyday life. However, experience shows that the achievement of this two step goal is among the most enigmatic phenomena in education. Many students who show a reasonably good ability in the solution of exercises, that is, in the application of learned algorithms to straightforward tasks, fail completely when they are faced with a problem in which there is some gap between the problem and its solution; they fail when they have to organize the data to make a decision about what algorithm to apply. In general the ongoing research and activities are aimed at finding ways for the improvement of the student's ability to solve problems.

So far there seems to be no reason to expect a breakthrough in this enigmatic situation. As usual in education, when we don't know what is a good way to achieve some goal, we just go on exposing students to tasks related to this goal, hoping for the best. As Carl Bereiter once said: When you don't know how to teach something, just go and do it, practice and hope. Mathematics teachers all over the world continue to expose their students to various problems, hoping that the improvement in the ability to attack and solve these problems will come with the practice. It seems that experience supports this approach; all mathematics textbooks are stocked with various problems, and one must assume that at least a decent part of these problems is being successfully tackled, and solved, even by average students.

Lack of Classificatory Approaches

The compilation of exercises and problems in textbooks usually follows the habit of accumulating these at the end of each unit related to topics discussed in the same. Very rarely do we find in the literature an anlysis and suggestion for some specific classes of exercises and problems which should be included in texts. This lack can be explained as having its roots in the fact described in the previous section. The scarcity of knowledge of how to teach to improve problem solving compels the textbook writer to accumulate large numbers of exercises and problems and arrange these in order of increasing difficulty. The inclusion of a problem at the end of a unit is usually being decided on the reason that it requires an application of ideas discussed in the unit, and that previous exercises, and/or problems, included in the set, or some solved problems included in the text itself, should have prepared the student for the solution of this specific problem.

In some cases we find an exception to the above. We sometimes encounter special sets of more challenging problems aimed at the more able student. Again it is hoped that by exposing these students to the greater challenge will provide them with the needed practice to be able to cope with such problems.

However, a deeper analysis of accepted tenets, in education in general, and aims of mathematics education in particular, shows a dearth of exercises and problems which are needed to achieve broader aspects of these aims. We shall point out in the next secions what some of these aspects are, and list classes of exercises and problems as examples of what is needed to satisfy these aspects.

Some Basic Tenets About Objectives and Needs of Mathematics Education

Need for Review and Connections

Research about learning, remembering and applying what has been learned has generally shown that anything learned, that has not been retrieved and used for a longer period of time, undergoes attrition and becomes more difficult to be retrieved when needed. We also know that anything that has been learned without contiguity between it and other material tends to become compartmentalized and, therefore, more difficult to be applied when needed in connection with some other ideas. These results compel us:

(i) To find ways for continuous usage of concepts, processes and ideas developed in any unit in mathematics.
(ii) To establish an interwoven system of connections between various mathematical topics.

That this is generally not done can be seen, for instance, from the fact that the concept of writing integers or rational numbers in different bases, usually taught in grade 7 or 8, is rarely, if ever, touched upon in later units. Similarly, the concept of absolute value, usually discussed for the first time when operations with negative numbers are introduced, is generally not used when dealing with quadratic equations, graphing of quadratic functions, etc. Classroom teachers could suggest many more examples of this nature.

We claim therefore, that there is a need for problems that will purposely develop ties between processes and concepts encountered in the past and new topics presently introduced.

Need for Reflection and Reconsideration

The Luchins in their study of Einstellung have shown the general tendency of the human being to continue in the beaten path when solving problems. A student who has learned the use of an algorithm persists to use it in situations for which a much simpler approach can be found. He may persist in his attempts to use the given algorithm even in a situation in which this seemingly appropriate algorithm is not applicable any more. To break this tendency of cognitive rigidity we have to expose the student to many examples in which a well established algorithmic approach can be replaced by a much shorter and faster one. In mathematics education we have to care not only for the products, facts, and outcomes that students have learned, but also for a variation f processes, approaches and ways that lead to conclusion. To hope for the development of such ideational flexibility we have to expose the student to both types:

(i) Problems for which a shorter, faster algorithm can be found instead of a habitual one.
(ii) Problems for which a seemingly appropriate algorithm does not work.

Problems of these types are badly needed while they are rarely encountered in textbooks; the usual "fallacies" are often quoted in textbooks, but these have the aura of exotic examples, their influence on the students' performance seems to be minimal.

Mathematics abounds with examples of problems for which the answer is "the solution set is empty". These also should find their place in the sets of exercises and problems regularly listed in textbooks. The usual attitude of students is "if teacher asks, there must be an answer, and this answer must be positive".

Need for Exploration and Conjecturing

Most exercises and problems, as used in textbooks, atack only a single point, "solve this equation", "simplify this expression", "prove this theorem", etc. The student has been taught certain algorithms, has acquired certain skills, and is expected to apply these towards the solution of the exercise or problem. There is nothing beyond the exercise; "you have to practice by solving exercises, so that you improve your skills, you ameliorate your ability to use the algorithm, etc". "You would like perhaps to solve the problem, but when done you may forget it and go ont to solve another". If we want the student to get the feel of what mathematics is, we have to expose him not only to problems which are cooked and baked before by somebody else, but also to the exploration of mathematical phenomena in search of a pattern, the pattern should then lead to a conjecture which may eventually be proven by the student himself.

The first person who gave a description of an approach, similar to the one described here, is Archimedes in The Method, written in the form of a letter to Eratosthenes. Archimedes describes there the intuitive modes that led him to the conviction about the formula for the volume of a sphere, and the area of a parabolic segment. According to Archimedes' opinion only after one is convinced about the truth of certain statements, only then can one move to prove these in a rigorous way.

The exploratory appraoch is even more important in those examples in which the student is practicing the use of an algorithm while he explores. The student who explores sets of specific numerical examples, to find out which linear functions of the form $x - ax + b$ are commutative under the operation of substitution, will obtain outstanding practice in subsitution of linear functions before he formulates his conjecture. Similarly, a student who tried to discover, by actually marking the points with a compass, the pattern of points in the plane whose distances from two given points are in the ratio of 2:1, will obtain considerable practice in the use of compass, and in using the inequality of the triangle.

One can formulate exploratory problems in almost any branch of mathematics, starting with the curriculum of grade one (example: In how many different ways can you write a whole number as a sum of two whole numbers, without paying attention to order?). It seems definitely expedient that every set of problems, at the end of any unit, should contain a few exploratory problems related to that unit. It is not mandatory for the student to be able to prove, or refute, his conjecture. Even if the student does not have the means for such a proof, or refutation, his conjecture remains open until the time in the future when he acquires the necessary means.

The Need to Sow the Seed for the Future

Psychology has taught us that distribution of practice over a longer period of time improves learning. It is also well known that concentration of a large number of new concepts, to be learned in a short period of time, produces inhibition that diminishes greatly the amount of learning that takes place. It seems therefore imperative to spread out the amount of preparatory practice over longer periods of time. This can be attained by introducing into the curriculum of the lower grades, concepts and relations that shall serve in the future as an underpinning for the development of new ideas in the upper grades. It is not expected that students, when they reach the upper grade, will remember everything encountered before. However, we know from research that even if the student does not remember explicitly when, where, and what he has learned in the lower grades, the relearning of this material is nevertheless substantially faster. The modus operandi seems to be as if the learned material were embedded in some subconscious area and the new exposure to same brings it forth to the fringe of the conscious. Teachers aware of these needs and acquainted with the courses of study of the upper grades should then attempt to introduce exercises and problems that may facilitate future learning.

The Need for Problems Related to the Culture of the Subject

The culture of any subject encompasses all those processes, endeavours, strifes, struggles, contentions, controversies, successes, failures, sagas, etc. that have accompanied the development of mathematics from the first day of its beginning to the present. It seems to be true that the same way living organisms have developed the need to explore and learn to know their physical environment, homo sapiens has developed the need to explore and learn to know his intellectual environment. This is particularly true if instruction and exposure to ideas is carried out in ways that encourage the learner's interest in the cultural setting of the learned material. Such encouragement can be attained if the teacher will expose the student to problems that have attached to them rich historical connotations of particular interest. Such historical connotations enliven the subject, raise the interest of the student, and often open for him further vistas for new developments and deeper understanding. To give just one example: There is a good chance that a teacher's story of human opposition over the ages to recognize negative numbers with full right may help dialectically some youngsters to overcome their own misgivings about the operations with these numbers.

Some Catch Phrases for Listed Needs, with Mathematical Examples

Before moving on to provide examples for each of the listed classes one should be clear that this is only a classificatory attempt, and in no way, whatsoever, a categorical one. Almost any example one can produce can be related to more than one class. In many cases one can find a predeominent motive which is decisive under what class the example should be listed. However, this seems to be of very little importance. We should let the student reap the whole range of benefits an example may provide.

We shall now attempt to provide one or two examples for each class of needs. These should also help to clarify the idea behind the listed needs.

We shall list these examples under "catch phrases" that can serve the teacher to keep the classes in mind and have him check himself whether he has introduced some of these needed examples. The order here corresponds exactly to the order of the subheadings given in the previous sections.

Need for Review and Corrections
Connect - Review - Remember

Many curricula contain in courses of study for grade 7 a chapter, or two, on writing numbers in different bases. this topic is often completely forgotten at a later stage. It might be an interesting task for the student who later studies formulas for $(a + b)^2$ and $(a + b)^3$ to follow the pattern obtained from 121_b for integral $b > 2$ and try to give a reason why the pattern occurs. He can then explore, on his own, some other three digit numerals in various bases that produce a similar pattern. He can then go on to explore 1331_b.

The concept of $|x|$, the absolute value of a number, is extensively discussed when students study operations with "signed numbers". Almost any text requires the student to graph functions of the form $x \to |x - a|$ and solve equations $|x - b| = c$. There is no reason for the omission of similar tasks when the student studeis quadratic equations and functions. This can be a challenging task for the student to construct a "quadratic equation" which has four real roots.

Also the graphs of functions in which the variable x appears in absolute value, such as $x \to x^2 - 2|x| - 3$, or $x \to x^2 - 6|x| + 8$ or $x \to x^2 - 2|x| + 1$ will definitely improve the students grasp of the concept of absolute value.

Reflection and Reconsideration
First Think, Then Compute

It is very easy to produce exercises and problems in this class, nevertheless such exercises very rarely occur in textbooks. One of the nicer examples that have the flare of the culture of the subject is the following:

A car leaves town at 6 a.m. driving towards town B with an average speed of 40 m.p.h. A second car leaves town B at 7 a.m. and drives towards town A with an average speed of 50 m.p.h. Both cars drive along the same road, and for the sake fo this problem may be considered as geometrical points. Which of these cars will be nearer town A when they meet?

You can be sure that many of your students will start to set up an equation! To reflect here means to comprehend first what exaclty is wanted.

Students who are studying the solution of trigonometric equations might benefit from being exposed to the following:

Solve the equation $\cos^2\alpha + 8\cos\alpha = 10$.

In a recent text (published in 1979) we found a series of examples of the nature $\sqrt{x^2+3} + 3\sqrt{2} = 0$. The students for whom the text is designed know that $\sqrt{a^2} = |a|$. Nevertheless, the text suggests to the students to apply the usual algorithmic approach to establish that the solution set is empty, instead of encouraging them to see this just by looking at the exercise.

The following is an interesting problem in this class:

Find three consecutive terms of the Fibanacci sequence which are the measures of the lengths of the three sides of a right triangle.

Exploration and Conjecturing
Explore and Conjecture

Students studying the solution of sets of linear equations should be exposed to the following:

What can you say about the solution set of sets of two linear equations whose coefficients, including the free term, are consecutive integers? Generalize and expand your investigation.

The student who has not yet studied the solution of such sets with parametric coefficients will accumulate numerical examples, formulate a conjecture, and later use his new experience with parametric coefficients to prove his conjecture. He can then attempt to generalize to coefficients that form an arithmetical sequence, to sets of three equations with coefficients of the same nature, to coefficients taken from other sequences such as the Fibonacci, etc.

A student who studies arithmetic series can explore about the nature of integers that cannot be written as a sum of positive consecutive integers, or the number of ways that other integers can be written as such a sum.

A student who studies polynomial and rational functions can explore: for what values of the coefficients of the function $f: x \to (ax + b)/(cx + d)$ with a,b,c,d integers, and $ad - bc \neq 0$ do we have $f(f(x)) = x$ or $f(f(f(x))) = x$. To facilitate the first steps he can begin with a,b,c,d = 0, 1, or -1.

An interesting exploratory problem that connects combinatorics with geometry might be to explore how many different triangles with sides of integral measure are there whose perimeter is less than n units.

How to Seed the Future
Think Forward, Not Only Back

Students in Junior High who study numerical fractions should be exposed to the periodic decimal as an infinite geometric series and the intuitive concept of convergence. Their familiarity with these topics will be of much use in a later grade, when they are going to study infinite geometric series.

Many schools introduce an elective calculus course for students in grade twelve. A basic concept on which the major ideas of differential calculus are developed is the rate of change of a function. There is no reason why students shold not acquire this concept in the lower grades. The rate of change of any function, over a given interval, is essentially a measure of the change of the function over one unit of the independent variable in the given interval. Ths is a perfectly concrete concept and can be understood even by weaker students. Incidentally, the calcualtion of the rate of change of different functions provides a good opportunity for meaningful practice in factorization, cancellation of fractions, etc. Such an early preparation of the student might also help with the understanding of harder theorems as, for instance, about the product of two functions.

Problems Related to the Culture of the Subject Culture Your Subject

These are badly needed problems to bring into the classroom some of the flavour of the historical development of mathematics, or some of its open problems that attract the keenest minds to attack them.

Students studying the concept of a polynomial could be exposed to the following:

Given a polynominal $a_n x^n + a_{n-1} x^{n-1} + \ldots + a_1 x + a_0$ in which a_i are integers for all $i = 0, 1, 2 \ldots n$, $a_n > 0$, n - non negative integer. Define $h = a_n + n + |a_{n-1}| + |a_{n-2}| + \ldots + |a_1| + |a_0|$. How many different polynomials are there for $h = 1, 2, 3, 4$?

Experience of the author has shown that this problem provides an opportunity to deepen the understanding that the polynomial x has an $h = 2$, that both polynomials $x + 1$ and $x - 1$ have an $H = 3$, etc. This provides also an opportunity to expose students to the countability of algebraic numbers.

Students in grade 9 or 10 can easily handle the problem: Find three consecutive integers such that the sum of the cubes of the two smaller ones is equal to the cube of the larger. These students have the means to prove that the solution set is empty. This provides an opportunity to expose the students to the famous theorem of Format, a theorem that is open until now.

Students who study square roots can be exposed to the following approach to $\sqrt{2}$:

We know that $\sqrt{2} = 1 + x$ with $0 < x < 1$. From this we get $(1 + x)^2 = 2$ or $x^2 + 2x = 1$. Hence $x(2 + x) = 1$ or $x = 1/(2 + x)$.

Substituting this value for x into $\sqrt{2} = 1 + x$ leads to expressing $\sqrt{2}$ as an infinite continuous fraction

$$\sqrt{2} = 1 + \cfrac{1}{2 + \cfrac{1}{2 + 1}}$$

This is an opportunity to expose the students to the charming, puzzling, enigmatic topic of infinite continuous fractions about which so little is known from the time of Bombilli until today.

Summary

This is an attempt to bring together some ideas about worthwhile problems for the mathematics classroom, worthwhile for reasons other than practice in use of concepts of algorithms just learned in the unit. The classes listed here do not exhaust in any way the wealth of reasons that one can suggest for including various problems in the matheamtics classroom. We have purposely left out the class of problems of application of mathematics about which more is being said by others. Let us hope that others will come up with different reasons for other types of problems. Whatever the reasons, let us make an effort to amend the mathematics curriculum with enriching, inspiring, enlivening mathematical problems.

LOS PROBLEMAS DE MATEMATICAS EN LA ESCUELA SECUNDARIA. ANALISIS DE UNA EXPERIENCIA

Jose R. Pascual Ibarra
Madrid, Spain

El caracter esencialmente formativo de la escuael secundaria demanda para la matematica una ensenanza integral, que reponda tanto a los posibles futuros matematicos, los menos, como a los alumnos que no han de llegar a serlo, lo mas, pero debe suminstrar a todos, los instrumetos de pensamiento y accion que han de necesitar para poder enfrentarse con acierto y eficacia a sus problemas vitales.

En esta perspectiva, hay que tener en cuenta que si bien la matemtica es ciertament una ciencia logica, junto al frio razonamiento deductivo hay otro tipo de razonamiento anterior a el; razonamiento plausible, en denominacion de G. Polya, al que no puede ser ajena la ensenanza de la matematica. Aprender a formular conjeturas plausibles, a hacer su critica y a comprobar su valides, es la mejor forma, y la mas practica, de apreder a razonar, y estas actividades se realizan mejor con la ejercitacion en el planteo y resolucion de problemas, que con el estudion formal de teorias ya elaboradas.

La resolucion de problemas, como tarea pedagogica, no es nada facil; es mas comoda una labor expositiva, pero esta, pr ocultar al alumno el proceso de aprendizaje, es, sin duda, menos eficaz. La actividad despegada por los alumnos, cuando se enfrentan con un problema nuevo, al tiempo que les permite el desarrollo de la imaginacion creadora, el cultivo de la intuicion, el desarrollo de las facultades de observacion y analisis, etc., les inculca el amor por el trabajo bien hecho. Deben adquirir la conviccion de que nada es imposible si se trabaja con teson y con esfuerzo. La superacion por si mismos de las dificultades reales conduce a la noble alegria del exito, y el exito es para nuestros estudiantes como para todo el mundo, el factor condicionante del afianzamiento de su seguridad personal. Porque, como decia Rey pastor, "no es la posesion de los bienes lo que hace felices a los hombres, sino su adquisicion".

* * *

Veamos, pues, en un ejemplo, como entiendo yo que pueda ser tratada en la clase la resolucion de problemas. Se trata de una experiencia realizada con alumnos aspirantes al profesorado de escuela secundaria.

Les propuse como tema didactico la elaboracion de una leccion activa a partir de la situacion, bien clasica, de "dado un segmento AB y una recta r, pralela a el - no importa como haya sido dibujada -, encontrar, utilizando exculsivamente como unico instrumento la regla de un solo borde, el punto medio del segmento."

Despues de aclararles lo que es la regla de un solo borde (el serrucho), les pedi insistentemente que no intentaran aplicar ninguno de sus conocimientos matematicos, porque no los necesitaban, sino que trabajasen inteligentemente sobre el problema concreto que se les habia presentado. Una persona inteligente, en el sentido kantiano del termino, les dije, es aquella que ante una situacion caulquiera de la vida, primero examina lo que puede hacer, despues analiza lo que debe hacer . . . y finalmente lo hace.

Ante su actitud mas bien pasiva, todavia les tuve que ayudar. Siguiendo a Polya, les pregunte: cual es la situacion? cuales son los datos del problema? de que medios dispongo?

Los datos son: los puntos A y B, extremos del segmento y la recta r, paralela a el:

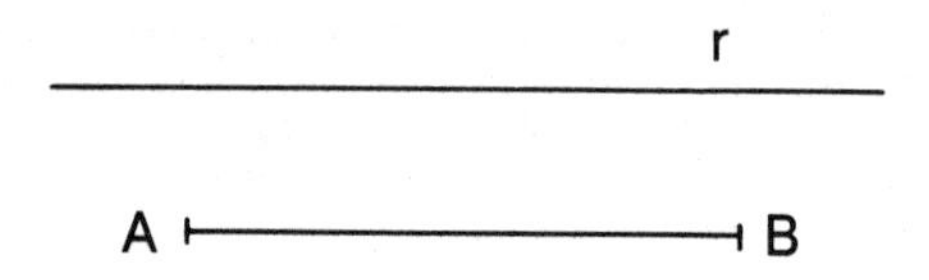

Nos preguntaremos: que es lo puedo hacer? Como solo dispongo de la regla de un borde (y, por supuesto, de un lapiz), la unica operacion posible es dibujar rectas. Nada consequire, evidentemente, trazando rectas "a barullo". Debo utilizar los datos del problema. Tendre, por tanto, que dibujar una recta que pase por el punto A:

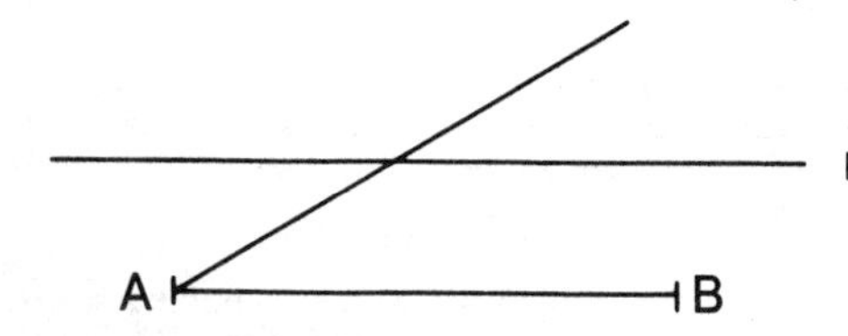

Y, ahora, oviamente, otra que pase por B

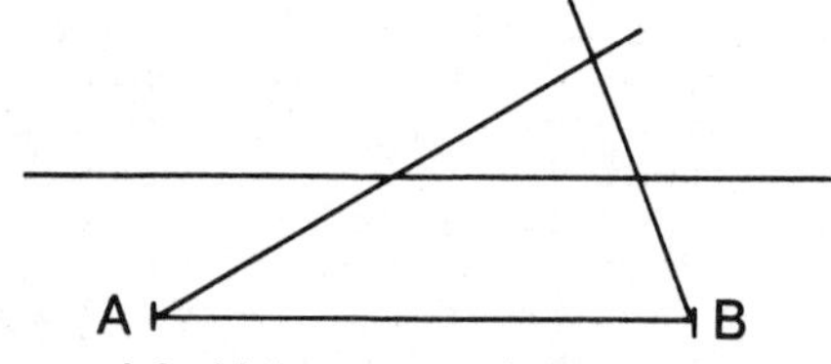

¿Como sequir? Habra que utilizar necesariamente el otro dato - la recto r - y, si quiero que lo que llevo hecho me sirva de algo, he de fijarme en los puntos A' y B' que las dos rectas ya dibujadas han determinado en r. Unire, pues A' conB, y B' con A:

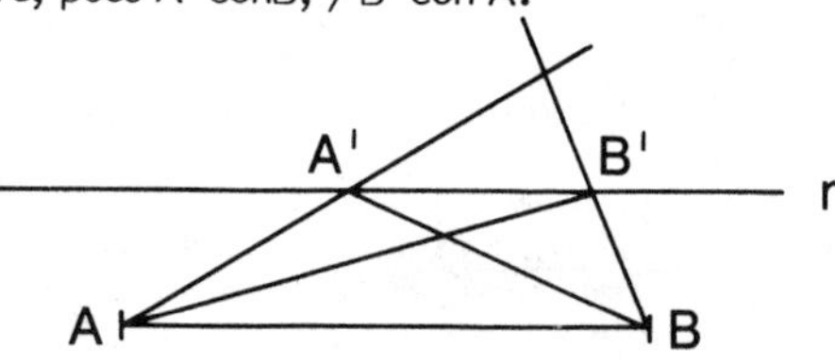

Y hecho esto - lo unico que realmente podia y debia hacer - ya "parece" que "sin querer" tengo resuelto el problema. La recta PQ me da el punto buscado; es M:

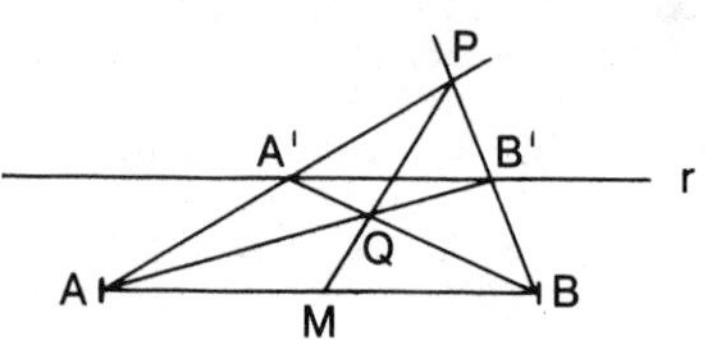

Al llegar aqui la pregunta surge espontanea en la clase: por que es M el punto medio de AB? Justamente ahora, les digo, es cuando comienza la matematica; en el memento que sentimos la necesidad de demostrar una proposicion cuya verdad, de alguna manera, hemos intuido. En nuestro caso, estamos cnvencidos de haber llegado a la solucion, y estamos convencidos de ello porque hemos trabajado inteligentemente, porque hemos realizado lo que podiamos y debiamos hacer. Pero, no obstante, meintras no construyamos una demostracion, hasta que no encontremos un proque, la proposicion para un matematico es solo una conjetura. Podriamos reforzar el valor de la misma, afianzarnos en ella, repitiendo cuidadosamente la construccion con otras rectas diferentes y comprobando que siempre se llega al mismo punto M.

Los alumnos parecen ya mas animados y les propongo que encuentren una demostracion. Que es una demostracion? Demostrar es convencer. Para muchos alumnos es inutil, y hasta absurdo, buscar "demostraciones" de verdades que para ellos son evidentes; mientras que para otros sera un agradable juego de la razon que les permitira deducir de proposiciones conocidas, otras desconocidas. Les indico, pues, que vean si pueden considerar la configuracion construida como caso particular de otro mas general: que secederia, les pregunto, si repetimos la misma construccion, pero sin ser r paralela a AB?

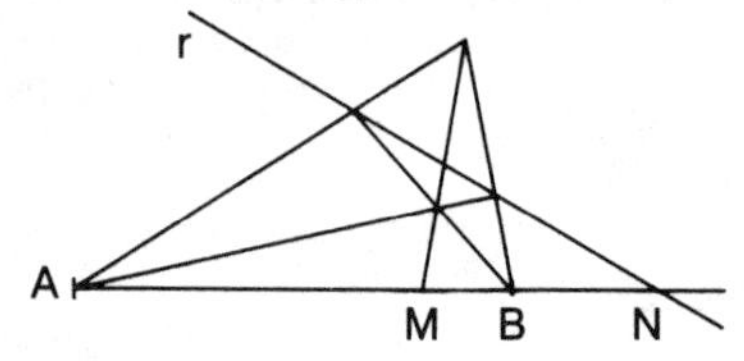

Comprueban que la recta r corta ahora a la recta que contiene a AB en un punto N, y observan que el punto M, debido a la inclinacion de r, se ha desplazado hacia la derecha.

Recuerdan ahora la propiedad del cuadrilatero completo y que la cuaterna (ABMN), por ser armonica, vale -1, y, por eso, en el caso particular de ser N el punto del infinito de la recta, M es el punto medio de AB.

Pero - me dicen - esta demostracion desborda los conocimientos de un alumno de bachillerato. Les invito, en consecuencia, a investigar otra de caracter mas elemental. No la encuentran. Les sugiero entonces que observen bien la configuracion final y que traten de expresar en terminos "mas geometricos" las construcciones realizadas (pensar, muchas veces, es decir lo mismo de otra manera). Lo que hemos hecho, en definitiva, ha sido: proyectar AB sobre r desde un punto P, arbitrario del plano, en A'B'; despues hemos proyectado A'B', en BA desde Q. Por tanto, por ser r paralela a AB, hemos realizado dos homotecias. Como las respectivas razones de estas homotecias son A'B'/AB Y BA/A'B', su producto es -1, y la composicion de ambas homotecias, la simetria central: M es el punto medio de AB.

Es conveniente aprovechar todo lo posible la riqueza de una situacion, y, en particular, las situaciones reversibles. Les propuse a continuacion enunciar los problemas inversos del que habiamos resuelto. Son dos:

I) Dados A, B y M, hallar r. Esto es, dados un segmento AB y su punto medio M, dibujar por un punto cualquiera del plano una paralela al segmento AB. Siempre, naturalmente, con el uso exclusivo de la regla de un solo borde. Ahora, todos, sin excepcion, encuentra la soluciòn reconstruyendo la figura.

b) Dados A, M y r, paralela a AM, hallar B. Es decir, duplicar un segmento. Todos, ahora, tambien sin excepcion, se empanan en resolver el problema siguiendo la misma via de reconstruir la figura, a partir de los nuevos datos. Ninguno lo consigue. Sin embargo, la solucion en este caso, es aparentemente mas facil: En r puedo tomar un segmento arbitrario A'B'; como se dispone de AM paralelo a r, es posible, por el teorema directo, hallar M', punto medio de A'B'. Uniendo A con A' y M con M', tengo P. La recta PB' nos da el punto B:

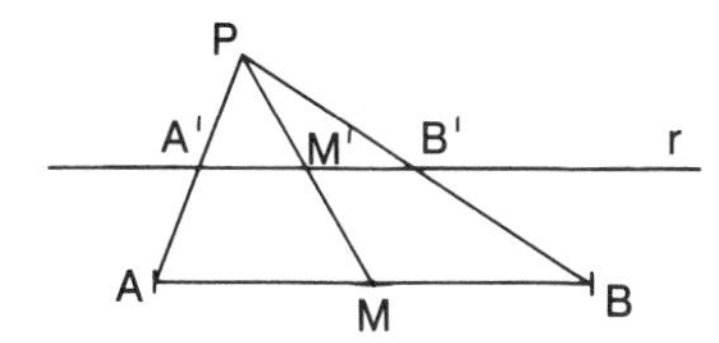

El comportamiento de los alumnos nos permite verificar, una vez mas, la tendencia que tienen, cuando ya han adquirido una cierta tecnica, esto es, cuando poseen un instrumento que convierte un problema en ejercicio, a aplicar esta tecnica de forma indiscirminada, sistematica y rutinaria, aun cuando no sea ya la adecuada, y, por otra parte, a comprobar tambien que la solucion de un problema que a nosotros, profesores, puede parecernos trivial, no lo es en ocasiones para los alumnos, y no lo es, desde luego, siempre que para ellos aparezca una situacion nueva. No olvidemos, pues, este hecho, y que lo sepan tambien ellos; que las "ideas felices" siempre leegan al final, como resultado de muchos intentos fallidos. Llamada a la constancia, a la tenacidad, al esfuerzo continuado!

Ahor ya se les puede hacer ver que por operationes reinteradas es posible construir sobre la recta AB la recta racional, esto es, encontrar cualquier punto P de ella tal que PA/PB = m/n; m, n N.

* * *

Un somero analisis de esta experiencia nos plantea una serie de consideraciones. Veamos solo algunas:

Si, como afirma el maestro Polya, la actividad mas caracteristica de la inteligencia, es decir, la mas especificamente humana, es la capacidad para plantearse y resolver problemas, en donde radica esta capacidad y como impulsarla?

¿Por que tantos alumnos, en mi opinion, encuentran hoy mayores dificultades que nunca para plantearse y resolver problemas?

¿Estan los profesores recien salidos de la universidad, en general bien provistos de las nociones mas abstractas de la matematica de hoy, suficientemente preparados para guiar los procesos de aprendizaje de la matematica elemental?

No se habra exagerado la finalidad de la educacion matematic, que se dice formadora de la inteligencia - y lo es - pero sin tener en cuenta la inteligencia practia, en un sentido mas estricto, como fermento de la aptitud para comportarse inteligentemente en situaciones concretas?

Como observa la profesora Krygowska, en las ultimas reformas de la ensenanza de la matematica no habra privado mas la idea de construir un edificio coherente y logico de las matematicas elementales, sin preocuparse tambien - lo que es imprescindible para el exito de la reforma - de los metodos de realizacion concreta en la clase de las nuevas ideas.

El ideal de Klein de ensenar la matematica elemental desde un punto de vista superior, no habra sido invertido por ciertos profesores haciendolo al reves: ensenando, o tratando de ensenar, matematicas superiores desde un punto de vista elemental, lo que, en una perspectiva psicologica, a mi modo de ver, es puro disparate, porque hace de la escuela secundaria una caricatura de la universidad?

El predominio del algebra en la construccion axiomatica de la matematica, no comportara una limitacion del alumno para un aprendizaje activo, aprendizaje mas ligado a procesos intuitivos que a los regidos metodos exclusivamente deductivos y demostrativos, culminacion de aquellos procesos?

Como conclucion final, creo que una didactica adecuada en la resolucion de problemas puede y debe contribuir a la implantacion de una nueva metodologia en la ensenanza de la matematica, que nos permita, no solamente contribuir a impulsar las vocaciones matematicas, sino tambien a proporcionar una mejor y mas eficaaz educacion matematica, en sentido amplio, a un mayor numero de alumnos.

Este es el desafio, creo yo, con que nos enfrentamos los profesores, y el esfuerzo que la sociedad, y nuestros alumnos, nos exigen.

Bibliografia

C.I.E.A.E.M. - "La problematique et l'enseignement de la Matheatique". Comptes rendus de la XXVIIIe rencontre. Louvain-la-neuve. 1976.

Glaeser. - "Matematica para el profesor en formacion". Eudeba, 1977.

I.R.E.M. de Strasbourg. - "Le livre du probleme". CEDIC, 1976.

G. Polya. - "Mathematics and Plausible Reaoning". Princeton University Press, 1954.

G. Polya. - "How to solve it". Princeton University Press, 1957.

G. Polya. - "Mathematical Discovery". John Wiley and Sons, Inc. 1962 y 1964.

P. Puig Adam. - "La matematica y su ensenanza actual". Ed. Revista de Ensenanza Media. Madrid, 1960.

Luis A. Santalo. - "La educacion matematica, hoy". Ed. Teide. Madrid, 1975.

TEACHING PROBLEM SOLVING IN A SIXTH FORM COLLEGE WITHIN THE CONFINES OF A PRESCRIBED SYLLABUS

Ian Isaacs
University of the West Indies
Mona Kingston, Jamaica

The Problem

Is it possible to improve the problem-solving abilities of mature high school students of average and above average mathematical ability by direct teaching? Furthermore, can this be done in the context of a regular classroom, following a prescribed syllabus? The two studies described below are part of an on-going series designed to explore the feasibility of teaching problem solving in the classroom.

The Background

High schools in Jamaica offer a seven-year course in mathematics. However, the majority of students take mathematics for five years, at the end of which they sit the General Certificate Examinations (G.C.E.) Ordinary Level or the Caribbean Examinations Council (CXC) General or Basic Proficiency papers in mathematics. Students who have done well in these examinations and wish to pursue mathematics or mathematically related subjects (such as physics or economics) continue for a further two years, at the end of which time they sit the University of Cambridge's Advances level papers in mathematics. Most of the students entering the two-year A level programme are competent at solving routine exercises based on the external syllabus but incapable of dealing with novel problems requiring the use of the same content.

The students in these studies were (and are) enrolled in the pre-university course at the Excelsior Community College in Kingston, Jamaica. These students do two or three subjects at A level over a two year period. They are recruited from high schools in and around the city of Kingston. Most of them failed to gain entry to the sixth Forms of their own schools because they did not have very good passes in the three subjects they wished to pursue at the Advanced Level. A few of them attended high schools which do not offer A level courses. The syllabus followed in these two studies is the "traditional" Cambridge A level mathematics syllabus covering topics in analytical geometry, calculus, mechanics, statistics and probability, and trigonometry.

The Studies

The first teaching project took place during the school year 1977-78. The major aim of that study was to develop flexibility and reflectiveness in the students during problem-solving episodes. The second project, which was started in September 1979 and will be completed in May 1981, has the same aims as the first study but has the additional aim of developing the student's ability to devise a plan.

The First Study

The first study (for a detailed report, see Isaac, "Teaching Students in Sixth Form to Use Some of the Heuristics for Solving Problems in Mathematics, available from the author, School of Education, UWI, Mona, Kingston 7, Jamaica) involved 21 students (17 males, 4 females), 16 of whom had received Grades A or B in their G.C.E. O level mathematics examination; the remaining five had C grades. Three members of the group reported at the beginning of the project that they were doing mathematics because it was their favourite subject. The others were taking mathematics as a tool subject or to satisfy university entrance requirements.

A problem-solving style of teaching was used throughout the project. This style is characterised by the following teacher behaviours:

i. Introduce topics by using examples which: (a) can be solved in more than one way, (b) embody some of the salient characteristics of the general problem and will encourage the students to inquire how the problem can be solved in general, or (c) embed the problem in a "real life" situation.

ii. When developing, reinforcing, or reviewing a topic, (a) organise the teaching-learning situation around questions or problems raised by the student or the teacher, (b) emphasise conflicting, difficult, or puzzling (seemingly paradoxical) aspects of the topic, and (c) pose questions which are not readily answered by recalling definitions, theorems, or illustrative examples done previously.

iii. In problem sessions (held once per week), (a) encourage multiple suggestions for tackling problems before trying any of them, (b) encourage students to use more than one method to solve a problem (when problems are solved on the chalkboard use at least two approaches to arrive at the solution), and (c) model problem-solving behaviours in front of the class.

Three problem-solving tests were administered during the school year. Five members of the class were audio-taped as they solved the problems on these tests. The tapes and their written solutions were analysed for changes in problem-solving behaviour. The written solutions of the other members of the class were also analysed for similar changes. In addition, the classwork and homework assignments were examined during the year for evidence of the development of problem-solving skills in the students.

Four of the students showed discernable growth in their problem-solving abilities during the year. The rest of the class showed no consistent changes in their problem-solving behaviours.

The results of this initial study suggested that the following modifications should be incorporated in the follow-up study:

i. When a problem-solving approach is used to introduce a new topic, then a substantial period of time should be allowed for the students to consolidate the concepts, nomenclature, and techniques by letting them work through graded exercises before presenting them with novel problems based on the new material.

ii. Novel problems discussed in the weekly problem sessions should be based, if possible, on content which is well learned.

iii. To demonstrate the power of a general heuristic, problems from different branches of the subject should be used to illustrate the use of the heuristic.

iv. Students should be supplied with an overall plan for tackling problems.

The Second Study

The second study, which began in September 1979, involved 26 students (20 males, 6 females), 10 of whom had received grades A or B in their G.C.E. O level mathamtics examination; the remainder had C grades. On paper they were an academically weaker group than the first group, and this was confirmed by their performance during the school year.

The teaching programme was modified to include practice time for more routine exercises, and all novel problems were clearly related to subject matter on the syllabus. At the end of the first term a general plan for tackling problems was introduced during a review session of the term's work. This plan is based on one developed by Schoenfeld (2). The students were encouraged to paste it in their notebooks and refer to it whenever they were faced with solving a problem. Three problem-solving tests were given, and 8 students were audio-taped solving some of the problems. A sample of the weekly problem-solving sessins were also audio-taped. The taped problem-solving session were analysed using the checklist prepared by Brown, Teaching Practices Observation Record (3), to determine if the investigator was using a problem-solving style of teaching. Some of the problem-solving protocols were analysed to determine the frequency of certain heuristic behaviours, namely flexibility and willingness to explore as well as use of diagrams/tables and "productive checks." In addition, a student's path through a given problem space was coded using the scheme developed by Hollowell (4) to determine what development, if any, had taken place in that student's ability to work through a problem in a systematic manner. (Schoenfeld's plan is very similar to Hollowell's coding scheme, both of which are derived from Polya's general model for solving problems.)

From this limited analysis it seems that the results are similar to those of the first study. Five students showed definite signs that they had developed problem-solving skills which they consistently used in regular classroom situations as well as under test conditions. The others sporadically showed some types of problem-solving behaviours in the problem-solving sessions which seemed to be heavily dependent upon the contextual framework of the problem. Under test conditions these behaviours were rarely displayed.

Conclusions

The following are some tentative conclusions based on these two studies.

i. Using a plan in conjunction with a checklist of related questions did help the majority of students in the weekly problem-solving sessions but had little carry-over to problem solving under test conditions.

ii. One year is probably too short a period in which to try to change the views of students of average mathematical ability that mathematics is a collection of specific techniques for finding right answers, especially after eleven years experience to the contrary.

iii. To develop flexibility of approach in school mathematics one might have to start much earlier in the school programme using discovery/inquiry methods so that pupils develop a dualistic view of mathematics, both as a prescribed body of knowledge with algorithms for applying it and as processes for inventing knowledge.

References

1. Schoenfeld, Alan H. "Can Heuristics be Taught?" (unpublished report), SESAME Project, Physics Department, University of California, Berkeley, 1976.

2. "Teacher Practices Observation Record", Mirrors for Behavior III: An Anthology of Observation Instruments, Anita Simon and E. Gil Boyer (eds.), Communication Mateirals Center, Wyncote, Penn. 1974, pp. 199-202.

3. Hollowell, Kathleen A. "A Flowchart Model of Cognitive Processes in Mathematical Problem Solving" (unpublished Ph.D. dissertation), Boston University, 1977, University Microfilms No. 77-11363.

9.2 REAL PROBLEM SOLVING

THE SCOPE OF REAL PROBLEM SOLVING

Diana Burkhardt
University of Birmingham
and
Shell Centre for Mathematical Education
University of Nottingham
England

Real problem solving is concerned specifically with the formulation and solution of real world problems that are of interest and immediate concern to the individual student pursuing this activity. This paper looks at how such problems may be identified and categorised in primary, secondary, and tertiary education. It discusses what should be looked for and how to go about looking for problems; it stresses the essentially pupil oriented approach that must be taken and the comparatively reduced role that the teaching and exercising of mathematics itself must take.

We are concerned that children should be educated to recognise that mathematics provides a set of tools whch can be of great use in helping the understanding of a very wide range of problem situations (1, 2, 3). It is the purpose of this paper to discuss the areas from which such problems are drawn and to provide some tentative guidelines whereby one can identify "good" and "bad" examples. However, we should immediately qualify the preceding statement by saying that a "good" problem in one situation with one set of pupils can be a "bad" one on another occasion.

It could be argued that Real Problem Solving (RPS) is a misleading name for this activity since there are many fascinating "real" problems in the world around us that would not qualify for inclusion in a RPS session in the classroom. We would most of us agree that organising one's bank account is a central problem in the real world. Recently a computer appreciation question in a public examination for average ability sixteen year olds asked about the use of magnetic ink to assist in the date processing of cheques and asked what other information appeared on a cheque. The first part of the question was answered tolerably well, but in the second part the majority of the pupils omitted to mention that all cheques showed the sum of money to be transferred. A very real problem for the teacher, but to the pupil one of utter irrelevence to his day to day life - - probably as unimportant to him as whether two submarines steaming across the Atlantic on converging courses will meet.

So we are looking for problems which are not only obviously relevant to the real world around us but are of concern to the student. Ideally we are looking for what we call Action problems (1) - - this is problems whose answers will directly affect decisions in our everyday lives. For example, the answer to "Should I cycle to my new school or go by bus" may determine whether I go out and buy a bicycle or not. Finding somewhere to live is an Action problem for undergraduates, whereas the pocket money dilemma "I got a pound pocket money on Monday morning and it just seems to disappear" occurs in various guises and is an Action problem for almost everyone.

At a lower interest level are those problems which the pupil recognises as Action problems for himself in the future or for someone he cares about. These we categorise as Believable problems. A teenager may be given the opportunity to design and equip his room as a study bedroom - - probably within a fixed budget! - - this is an Action problem, while designing a kitchen layout would be a Believable one for the school child.

There are many problems which are intriguing but worth pursuing despite their lack of Action content. These, using our ABC nomenclature, we dub Curious such as "Why are there two tides a day?" or "How far can I longjump on the moon?"

The next categories are Dubious - - they exist simply to provide exercise in a mathematical technique. These problems have no place in Real Problem Solving sessions; most textbook problems belong to this group.

Finally reaching E in our ABC of categories comes the Educational type; these are essentially Dubious but make an important point of mathematical principle so clearly and beautifully that no one would want to get rid of them - - "If I had invested 1p at 5% compounded interest in 512 AD, what would it be worth now?"

We would suggest that the most important criterion for RPS problems is that they should preferably be Action problems but, if not, at any rate Believable for the particular student involved. Having said this,do we discard all our traditional mathematical models - - exponential decay, Newtonian mechanics and so on? Indeed, this is not our intention. Standard models give insight into the use of mathematical techniques and should continue to exist alongside RPS - - both require the teaching of mathematical activity, although the estimates of what proportion of the curriculum should be devoted to it range from 10-50%.

Where do we look for Action problems? There exist some collections of tested problems which may help us (2, 3). Our approach should be child centered. We must see the world through the eyes of our students and gently probe to test our own view of their perceptions against their own. Suggesting that each child writes down the problem that concerns them at the present time can generate a lot of useful starting points. In most cases the problem which most attracts the child's interest and enthusiasm is one that grows naturally out of group discussion. With experience a teacher may be able to nudge the discussion into her preconceived problem area, but in RPS there is always an element of the unexpected, and the teacher has to learn to grapple with an entirely fresh problem. When a situation arises where both teacher and pupils are unsure how to proceed, this element of equal partnership can be most rewarding, but it also places new demands on the teacher. She has to have faith that some solution will emerge; she must continue to exude a positive approach and remain confident, buoyant, and encouraging. Training for such a style, which is new to many teachers, is not easy; we have found that tackling an Action problem oneself is a good way to approach teaching RPS. We have also done research to identify various skills important in RPS; they can be exercised individually, and we have developed a "Modelling Starter Pack" (4) to let the interested but uncommitted teacher try this approach without major changes of teaching style.

How much mathematics should a good problem include? This is one question that should be approached with caution. First and foremost we must recognise that considerable progress can usually be made with the simplist of skills - - the enumeration of alternative possibilities, simple arithmetic with a calculator, tabulating and plotting graphs, and for older students, simple algebra. Experience has shown that any mathematical techniques used must be very well assimilated; usually they must have been exercised over a period of at least two to three years. On the other hand, from time to time the problem may provide a setting to illustrate more recently acquired skills or even introduce new ones, but this must be thought of as a secondary effect, and great care taken not to try to steer the pupil's effort in order to achieve it.

In conclusion we are primarily aiming to show pupils how mathematics can help in their everyday life, hence we must search for problems that are part of their daily concern. We must avoid those, which although part of their daily life, are not seen by the children to be important to them at this moment in time. Experience has shown that there are very few problems that are not better tackled with some help from mathematical techniques and that looking at problems that seem to have obvious mathematical pay-off may well be a recipe for disaster.

References

1. Hugh Burkhardt, "The Real World and Mathematics", Shell Centre for Mathematical Education, Nottingham 1980 and Blackie and Son, Glasgow, 1981.

2. "The U.S.M.E.S. Guide". Unified Sciences and Mathematics in Elementary School, Moore Publishing Company, Durham, South Carolina, U.S.A., 1978.

3. "Mathematics Across the Curriculum", Open University In-Service Training Course PME 233, Open University, Milton Keynes.

4. Vern Treilibs, Hugh Burkhardt, and Brian Low, "Formulation Processes in Mathematical Modelling", Shell Centre for Mathematical Education, Nottingham 1980.

9.3 MATHEMATIZATION, ITS NATURE AND ITS USE IN EDUCATION

MATHEMATIZATION, ITS NATURE AND USE IN EDUCATION

Eric Love
University of Nottingham
Nottingham, England

Introduction

I propose to discuss some work done by students when they are encouraged to mathematize. It might be helpful to give you a little background about the kinds of lessons from which this work arose. They were all from classes I have taken, ones in which the students are encouraged to pursue a mathematical investigation. Sometimes the situation is presented to the whole class and is investigated communally, but more often a choice of starting points is given and the students work individually, or in small groups. From the age of 11 the students spend something between a third and a half of their time working in this way - although many of the other lessons have an exploratory orientation.

I will say more later about the characteristics of such lessons. For the moment I shall concentrate on the products. A typical investigation starting point is:

> The first 8 binary numbers may be arranged in order so that only 1 digit is changed at a time, and the last one can be changed back to the first in the same way. One order is:
> 000 → 001 → 011 → 010 → 110 → 111 → 101 → 100
>
> Find all such orders and investigate relations between them.

Here is what one student (aged 15) writes about the initial stages of his enquiry:

When I started this investigation the first thing I decided to do was to find out all the possible chains that can be made with the 8 numbers. After this decision I had to devise a method of finding out all the possible chains. The first method I tried was to try and make up the chains just by writing them down as they came out of my mind; this for obvious reasons was not very successful. By this method I only found about 2 complete chains. The next method I tried was to draw out each of the 8 binary numbers and then try and join them up, and in this way again try to find all the chains. This again did not get me very far. It was at this point, when I had the 8 numbers just drawn out randomly, that the 8 numbers could represent the 8 corners of a cube.

This is how I had my figures drawn out just before I found they could be represented by a cube:

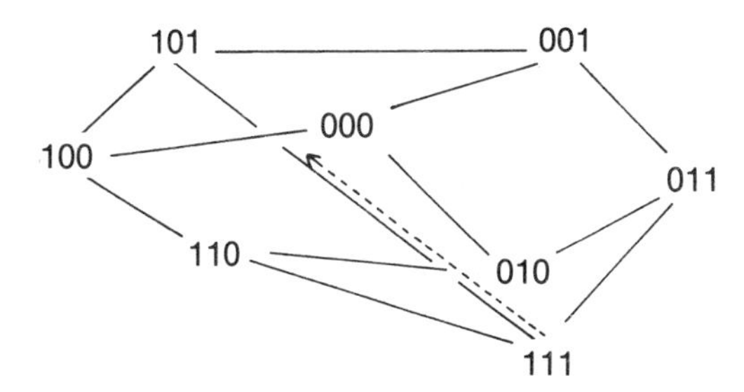

Here it can be seen that if the figure 111 is moved along the dotted line, a diagram of a rough cube can be made.

The number of possible chains that can therefore be made can also be said to be the number of routes along the edges of a cube, passing through each corner once and only once and ending up where you began.

By this stage I still hadn't found all the possible chains, so another method had to be found. The next method I came up with turned out to be the successful one.

What I decided to do was to write out the first number of the chain, i.e. 000. THen you put down each of the numbers that this one can go to, i.e., 100, 010, 001. Each of these then goes to each of the numbers that they can go to; this goes on and on not repeating numbers in the chains. This carries on until none of the chains can be added to.

Part of this large diagram is shown below. In this part 000 just goes to 001 just to illustrate the idea. Below one can see an example of the diagram.

The chains marked with the circle are chains that don't end, the chains marked with the square are those that go through all 8 numbers without going over the same one twice, but in these cases the last number in the chain will not return to 000.

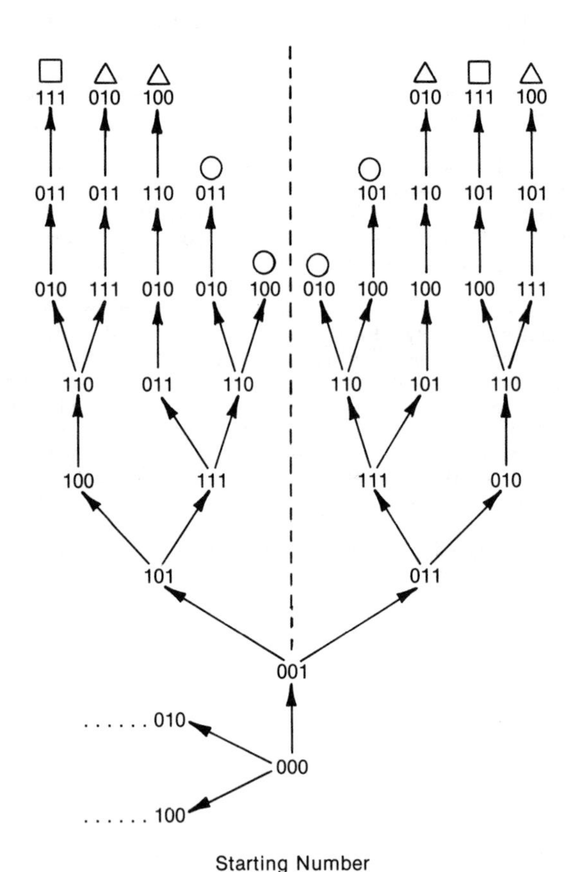

Starting Number

It can be seen from the dotted line inserted on the diagram that there are an equal number of each type of chain on each side of the line. By doing the whole diagram I was able to find that the number of chains that complete all 8 numbers and return to 000 is 12.

The number of chains that complete all 8 numbers but don't return to 000 is 6. The number of chains that don't go through all 8 numbers is also 6.

This can also be taken as being the number of routes around a cube. So there are 12 routes around a cube going through each corner once and ending up at the point at which you started.

Comments

a) It starts with a random "messing about", a stage in which something is tried to see what happens. This is a vital prelude, but also occurs again and again, when the students are unsure of the directions they wish to take.

b) The diagram that he drew he suddenly "saw" as a cube drawing: a moment of putting a structure onto something still very amorphous; and he was able to see that his problem could be transformed into finding the routes on the cube. It would have been nice to report that that structuring enabled him to solve his problem, but that didn't, in fact, happen.

c) He tries another way - a tree diagram. He has met these diagrams before, but not in this context. Fortunately, the diagram is manageable here, and he can see how it gives him certainty.

d) Notice that he only give part of the drawing, "just to illustrate the idea", with the implication that the other parts are similar: this is typical moment of abstraction - working with one instance that can stand for a number of others. He also notes the symmetry, but doesn't follow this up.

e) In the rest of his account he goes on to look at the relations between the chains - the symmetries, the transformations between them; and in a short section shows analogous results with 2-digit binary numbers.

Characteristics of the Work

Learners' control

It is quite clear from the example that I have given that the students know what they are trying to do, and this occurs because they have personal control over the direction and methods of the investigations. They were not simply following out something designed by the teacher. This control by the students seems to me to be essential if they are to be able to mathematize.

It is possible to offer this control by a range of means, but because the normal position of children in school is not one of being in control of the work they are expected to do, it is something that has to be worked at by the teacher. Three aspects which I want to raise briefly are:

a) Simple situations. The initial starting points must be simple, and accessible to students of a wide range of ability and achievement. It is often helpful if they use simple materials, or utilise simple imagery (I am thinking here of materials such a pegboards, geoboards, counters, and imagery involving elementary geometrical shapes - circles, grids, and the like).

b) Diversity of response. The situation offered to the students must not only be stimulating and accessible, they also need to give opportunities for diverse responses by the students. Only in this way can the student take control. If the teacher is trying to direct the student to a particular approach - or even to solve a particular problem - then the teacher will be in control The diversity must then be not only about methods but also about formulating the problem. The students need to formulate the problem for themselves.

c) The style of teaching. I have indicated that the teacher needs to operate in a different way to lessons where s/he is expositing. In this way of working, s/he is assisting students in carrying out their own plans. Mathematization is not a learning outcome - it is the way in which students learn what they do. The teacher does not have intended outcomes but needs to be opportunist, always seeking ways in which s/he can assist in the mathematizing. The style is one of supportive criticism in which, as Stephen Rowlands has said,

> The way that the teacher collaborates with the student in the execution of his plans is of fundamental importance. It involves helping him to clarify his ideas and offering guidance, and even instruction and criticism where appropriate. But this help is given in such a way that the control of the work - its objectives and method - rests firmly with the student.

d) The language used by the teacher. Much teacher conversation in the classroom takes the initiative from the students - and is intended to do so: it is usually concerned with trying to get the students to do with what the teacher wants. The teacher may often be quite unconscious of this, and it needs to become part of the teacher's awareness. There is a world of difference in these questions:

- What patterns have you found?
- What have you found out?
- Have you found anything?
- Tell me about what you have done.

It is at this level that the relationsips between the activity and the teacher and the learner will be determined.

References

This list of references is confined to items concerned with classroom activity leading to mathematization. In particular, all of the items contain either (i) records of conversations with children or (ii) extracts from children's writings.

1. From Mathematics Teaching, the journal of the Association of Teachers of Mathematics.

Billings, C., Problem solving in the primary school, M.T. 70 (1975).

Bird, M., Using Noughts and Crosses as a starting point, M.T. 90 (1980).

Clark, J., Elizabeth and the mystic rose, M.T. 73 (1975).

Dichmont, J., Balancing - a conversation, M.T. 59 (1972).

Hatch, G., Mathematics and ice-cream, M.T. 89 (1979).

Hedger, K., The importance of being Kevin, M.T. 88 (1979).

Hedger, K., Reflections on a question of method, M.T. 66 (1974).

Hedger, K. and Kent, D., Given two points, M.T. 84 (1978).

Hedger, K. and Kent, D., Armes' Theorem, M.T. 89 (1979).

Jeffrey, R., Happy numbers, sad number, M.T. 85 (1978).

Kent, D., The dynamics of put, M.T. 82 (1978).

Kent, D., Isobel, M.T. 82 (1978).

Middleton, G. et al, Out of School, M.T. 47 (1969).

Renwick, E.M., Three articles, M.T. 74 (1976).

2. From Educational Studies in Mathematics

Goffree, F., John - A teacher-training freshman studying mathematics and didactics, ESM Vol. 8 (1977).

IOWO Team, Eleven minutes group work - a transcript, ESM Vol. 8 (1977).

IOWO Team, Five plus four minutes class instruction - a transcript, ESM Vol, 9, (1978).

Hedger, J. and Kent, D., Growing Tall, ESM Vol. 11 (1980).

3. Some books

Armstrong, M., Closely Observed Children. Writers and Readers, London 1980.

Banwell, C. et al, Starting Points. Oxford U.P., 1972.

Bell, A.W., Rooke, D., and Wigley, A., JOurney into Maths. Blackie, 1977-9.

Goutard, M., Mathematics and Children. Educational Explorer, 1964.

Goutard, M., Mathematiques sur Measure. Classique Hachette, 1971.

Renwick, E., Children Learning Mathematics. Stockwell, 1963.

MATHEMATIZING WITH A PIECE OF PAPER - AN INTROSPECTION

Marion Walter
Department of Mathematics
University of Oregon
Eugene, OR

Having often claimed that one can mathematize with almost any material or situation as a starting point, I thought it might be useful to demonstrate how I go about this in the hope that: 1) it might serve as a starting point in discussions on what takes place while mathematizing, 2) it might give us clues as to what we might do to encourage students to mathematize, and 3) it will encourage others to present their own case histories.

Why would one want to mathematize from 'anything' besides for the pure fun of it? I believe that there are at least two reasons. First, mathematizing can lead, at all levels, to new and perhaps good and useful curriculum ideas and topics. Secondly, if students are exposed to such examples of mathematizing and are encouraged to carry out such activities themselves, they may realize that this natural activity IS valued though 'it is not in the text'.

I decided to use a piece of paer as a starting point since everybody is likely to have one.[1] I had two questions in mind. I wondered what problems and ideas I would generate from the piece of paper and whether I could catch and record my thoughts as I went along. If I could do that, I thought I might be able to begin to answer - What facilitated my asking the many questions, carrying out the various activities, and my going in the various directions? What hindered me?

After working one afternoon with the piece of paper and reflecting on the thoughts that I had while doing this I made the following list:

1. I believe the most important aspect was to LOOK and CONSIDER. This ties in with noticing attributes. Depending on what I 'saw' and focused on, I asked different questions and was reminded of yet other things. This ties in with the 'What-If-Not' method of problem posing.[2]

2. A liking of and facility of free association.

3. A willingness not to discard ideas as useless or silly until after some consideration - even if not pursued.

4. A liking for playing with materials and situations.

5. Past experience had certainly a bearing - even childhoold memories, past mathematical experiences and past experiences of doing such work.[3]

6. Knowing what the curriculum required probably influenced me in several but not in all the ideas with which I came up.

7. A tolerance for and even a liking for work where the goals are not specific - where the situation is quite open ended. Perhaps even a preference for work where there is no confined course of action.

8. Knowing that a particular problem suggested itself has been investigated or solved in some sense put me off.

I obtained this list only after examining ALL the material I developed while mathematizing with a piece of paper. What follows here is only ONE small part but I hope it is enought to indicate the flavor of activity. At the meeting I hope also to present a few of the other ideas and problems and to develop a few new ones with the group.

The bibliography contains some of the material which must have influenced by thinking and a few related articles of which I thought afterwards.

I FOLD THE PAPER RECTANGLE R

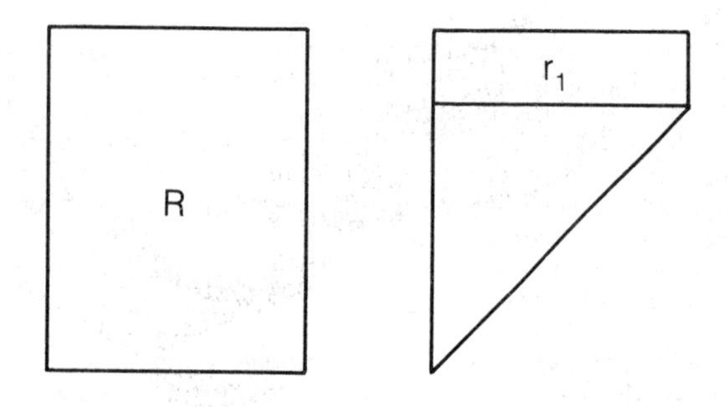

This was the first thing I did - I was thinking of making a square. I did this so automatically that I am pretty sure my action resulted from childhood memories of making squares as a starting point for many things!

I TEAR OFF THAT EXTRA RECTANGLE obtaining SQUARE S and small rectangle r_1

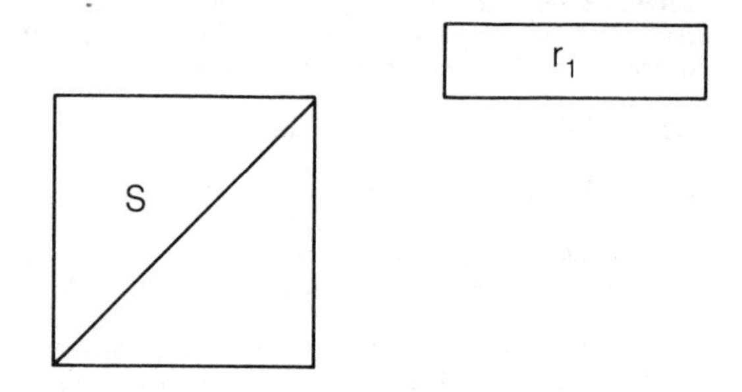

I am not sure if I tore off square S because I wanted the square or because I was thinking that extra bit (r_1) must be useful for something. I do have an attitude of "everything is useful for something."

I LOOK AT THE RECTANGLE r_1 TURNED LENGTHWISE. START FOLDING SQUARES AGAIN

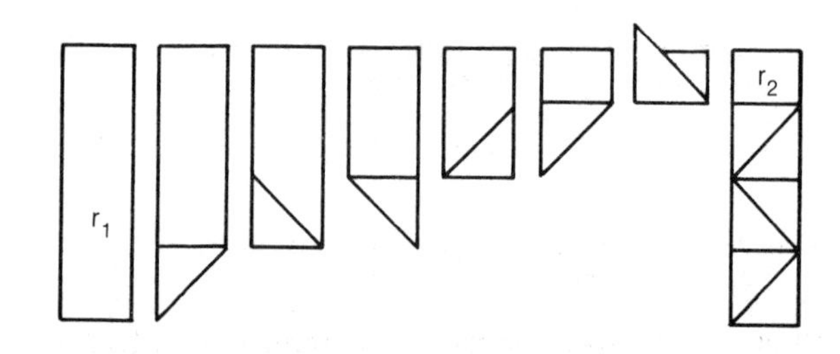

I wander how many squares can I get from this small rectangle. I get 3 squares and again have a small rectangle r_2 left over.

I ASK, WHEN WILL I GET EXACTLY k SMALL SQUARES AND NO RECTANGLE LEFT OVER AFTER I REMOVE THE SQUARE?

I began to like this problem. I realized that in the usual way of folding a piece of standard paper we always get ONE square and a piece left over.

WHEN WILL RECTANGLE R GIVE I SQUARE AND RECTANGLE r_1 WHICH GIVES EXACTLY k SQUARES?

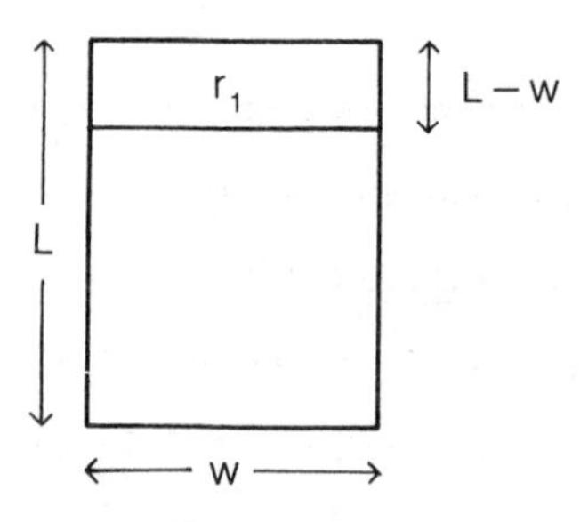

r_1 will give exactly k squares iff

$w = k(L - w) = kL - kw$

$w(k + 1) = kL$

$L/w = (k + 1)/k$

In the back of my mind, after drawing the diagram, lurked the Golden rectangle property: Cut a square off a golden rectangle and a smaller rectangle which is also golden remains. I purposefully and quite consciously tried NOT to think of this because I did not want to be influenced by it since I knew to where the golden rectangle led.

k	1	2	3	4	5
$\frac{k+1}{k}$	$\frac{2}{1}$	$\frac{3}{2}$	$\frac{4}{3}$	$\frac{5}{4}$	$\frac{6}{5}$

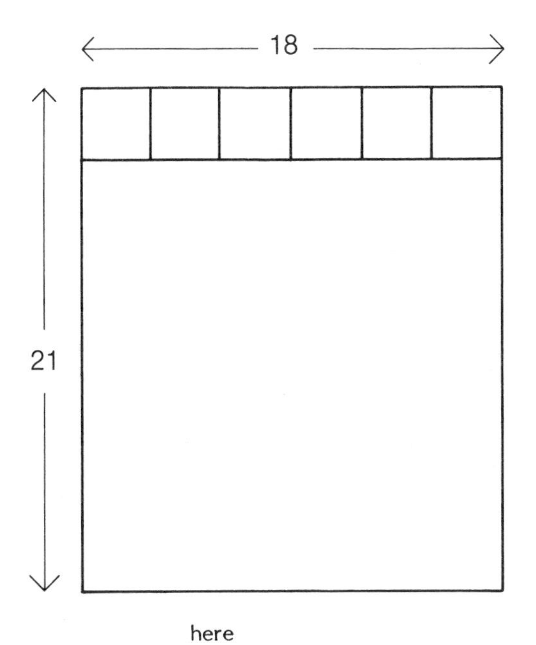

here

L/W = 21/18 = 7/6   k = 6

This surprised me and I substituted a few values of k. I tried a specific case (of course it is obvious if one LOOKS at a diagram) and it hits home that it is only the RATIO of the two sides that we are dealing with. It seems a nice application of ratio and proportion: Given any rectangle that splits into one square and a rectangle which in turn divides into exactly k squares, then the ratio of length to width of the original rectangle is $(k + 1) / k$. I like that! So I ask - What if the rectangle r_1 folds into k squares but has remainder of rectangle r_2?

WHEN WILL RECTANGLE r_2 CONSIST OF EXACTLY n SQUARES?

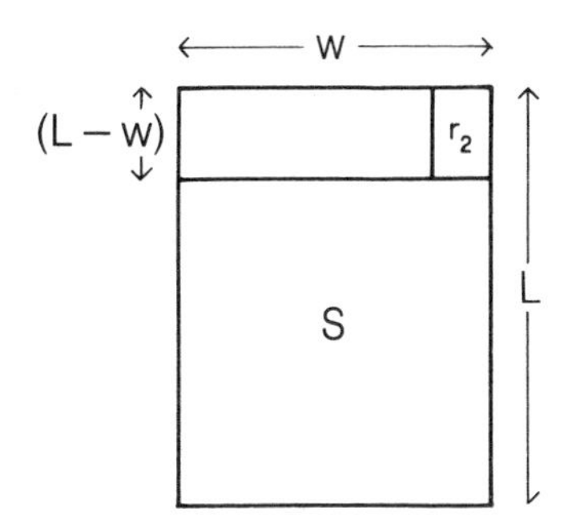

So, r_2 consists exactly of n squares iff:

$$L/w = (n + 1 + nk)/(1 + nk)$$

Rectangle r_1 has dimensions w by (L - w) and it forms k squares leaving rectangle r_2. Rectangle r_2 is (L - w) by $[w - k(L - w)]$. So r_2 forms exactly n squares iff:

$$n[w - k(L - w)] = L - w$$
$$nw - nkL + nkw = L - w$$
$$nw = L - w + nk(L - w)$$
$$(L - w)(1 + nk) = nw$$
$$(L - w)/w = n/(1 + nk) \text{ or}$$
$$L/w = (n + 1 + nk)/(1 + nk)$$

WHEN WILL r_2 CONSIST OF THE SAME NUMBER OF SQUARES AS r_1?

If n = k,
$L / w = (1 + k + k^2)/(1 + k^2)$, 3/2, 7/5, 13/10, 21/17...

here n = k = 3

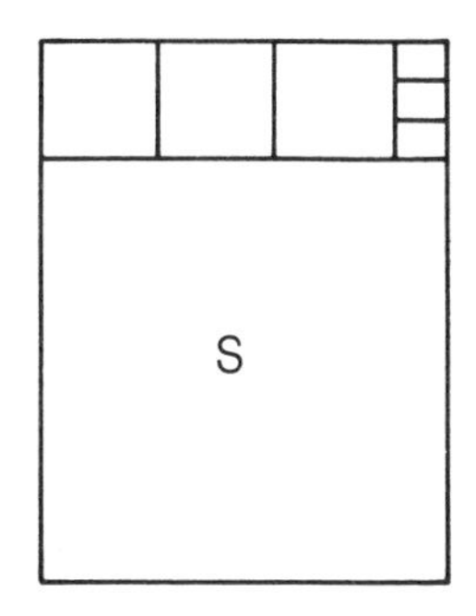

IF r_1 AND r_2 EACH HAVE k SQUARES AND r_3 HAS EXACTLY k SQUARES THEN

$L / w = 1 + 2k + k^2 + k^3/2k + k^3$ 5/2, 17/12. 43/33...

I thought this looked a bit messy and so asked what if n = k?

At this point I did not stop to ask what happens at the next stage. I did later. What is your guess for the ratio of L/w if r_1, r_2 and r_3 each have 3 squares and r_3 had 3 squares exactly?

Now generalize!

I went on to examining the folds in the paper as my second activity but I will stop here.

NOTES

1. Walter M. and Brown S. "What if Not?" Mathematics Teaching #46 pages 38-45. "Problem Posing and Problem Solving: An illustration of their Interdependence" Mathematics Teacher Jan. 1977 pages 4-12. Brown S. and Walter M. "What if Not? An elaboration and second illustration". Mathematics Teacher #51 Summer 1970 pages 9-17.

2. There are numerous places that give rich examples. See for example Banwell, Saunders and Tahta Starting Points Oxford University Press 1972, especially pages 67, 139, and 44-49 but all of the text is relevant.

BIBLIOGRAPHY

Fielker, D. "Notes from a Maths Centre" Mathematics Teaching #51 Summer 1970, pages 18-21.

Gardner, Martin "Mathematical Games: The combinatorial richness of folding a piece of paper." Scientific American May 1971.

Hole, David "One Thing Leads to Another" Mathematics Teaching #72 Sept. 1975, pages 18-21.

Steinhaus, H. Mathematical Snapshots. Chapter 2 "Rectangles, Numbers and Tunes."

Sturgess, David "An Afternoon with Five Cocktail Sticks" Mathematics Teaching #47 Summer 1969, pages 53-54.

Walter, Marion Boxes Squares and Other Things National Council of Teachers of Mathematics 1970.

Walter, Marion "Frame Geometry: An Example in Posing and Solving Problems" Arithmetic Teacher October 1980.

Walter, Marion "Two Problems from a Triangle" Mathematics Teaching #74 March 1976, p. 38.

MATHEMATIZATION: ITS NATURE AND USE IN EDUCATION

David Wheeler
Concordia University
Montreal, Canada

1. Mathematization refers to the process by which mathematics is brought into being. It is both entirely individual, because every act of mathematization is a mental act performed by one person, and entirely universal, because the ability to mathematize is a human power belonging to everyone (Wheeler, 1974).

We may (metaphorically) express the difference between mathematization and mathematics in the way we note the difference between number and numeral. Number is the "essence"; numeral is the "appearanec" or "representation". Number is the absolute, the "given"; numeral is the relative, socially-approved way of signifying number. The distinction is not always preserved: numbers and numerals are so intimately related that for most purposes we identify them and move from one to the other without noticing. Just so we may shift attention from the process (mathematizing) to the product (mathematics), and vice versa, without noticing.

The distinction is not important to the mathematician, but it is crucial to the educator. Awareness of the human power to mathematize is the only rational basis for the belief that anyone who does not yet know mathematics can come to know it.

The educator's awareness that mathematizing is a human possibility defines

(i) an attitude. The teacher can be optimistic since all students possess something that will enable them to gain entry into mathematics.

(ii) a task. The teacher's job is to trigger the mathematizing ability, not try to put mathematics into the learner.

(iii) a tool. By studying how mathematizing works, the teacher can know what he must do for the students and what they can do for themselves (Gattegno, 1974).

2. We can "know" mathematization without being able to define it, or account for it. We know it in the way we know gravity (in the external world) or perception (in our internal world) - i.e. by experiencing it and finding out how it functions for us. In this sense we all know it. But few of us know mathematization in the sense of a precise awareness of what constitutes this power.

To become aware of mathematization is to become aware of the way the mind works in certain situations.

3. If we accept that mathematization is a special activity of the mind then we shall attempt to become aware of it by watching our minds at work when we are actively occupied with mathematics - solving a mathematical problem or thinking about a mathematical question. We may find - I believe we will find - that our minds also work in similar ways in some apparently non-mathematical situations, i.e. the mathematizing power is not always engaged in producing the kind of mathematics that we know.

Those who would identify mathematization with "making a mathematical model" look at the relation between mathematization and mathematics in a different way. They say, in effect, the only when something has been converted into mathematics has mathematization taken place. This approach does not seem to tell us anything useful: first, it stops short of telling use anything about the processes that go on "in our heads"; second, it makes the ability to mathematize seem to be wholly a learned, cuture-specific skill (Davis and Anderson, 1979) - which is like believing that man's liguistic ability allows him to learn only the language of his particular society.

4. In order to help awareness of the activity of mathematization come to the surface, I propose the following clues to its presence:

(a) Structuration

"Searching for pattern" and "modelling a situation" are phrases which grope towards this aspect. But our perceptions and thoughts are already structures; reality never comes to us "raw". So mathematization is better seen as "putting structure onto structure". Existentially, however, it seems more like discovery or re-structuration since what we have brought into being seems new to us. The "eureka" feeling is an extreme case, marking the release of energy brought about by a new structuration.

Awareness of this aspect of mathematization can be elicited by considering simple examples where an "obvious" structure is concealed because one is not given the "rules";

e.g. continue the sequence 1 5 6 1 7 8 (Banwell et al, 1972).

(b) Dependence

Mathematization puts ideas into relation and coordinates them; in particular it seeks to establish the dependence of ideas on each other.

Socrates gets the slave boy to question the coordination between the areas and lengths of squares. Leibniz coordinates the sum of the terms of an infinite geometric progression with the sum of the differences between consecutive terms (Hofmann, 1974). The children in a recent account by Dick Tahta coordinate the movements of two rotating line segments by considering the motion of the midpoint of the segment joining their extremities (Tahta, 1980).

(c) Infinity

Poincare points out that all mathematical notions are implicitly or explicitly concerned with infinity. The search for generalizability, for universality, for what is true "in all cases", is part of this thrust.

The methods of exhaustion (Archimedes) and of infinite descent (Fermat) formalize common institutions of infinite processes. Indefinite iteration "and so on", is an insight of the mathematizing power.

Can we say we understand the notion of "midpoint" if we do not see that when we can produce the midpoint of one segment we can immediately conceive the midpoint of any of an infinite class of segments? If we think of the midpoint of a given segment, and then of the midpoint of one half of it, and then of the midpoint of one half of that, and so on, our act of mathematization relates the infinity of a class to an infinite process.

Consider the problem of cutting a square completely into a number of (not necessarily congruent) squares. One step in the mathematization of the problem is the awareness that what is done to the original square can be repeated with any sub-square; the infinite has entered and the problem is "solved".

Algebraic structures, functions, and infinite sets, are the mathematical entities created by these types of mathematizing awarenesses.

References

Banwell, C.S., Saunders, K.D., and Tahta, D.G., Starting points. London: Oxford University Press, 1972.

Davis, P.J., and Anderson, J.A., "Nonanalytic aspects of mathematics and their implications for research and education". SIAM Review 21, 1 (1979).

Freudenthal, H., Mathematics as an educational task, Dordrecht: D. Reidel and Co., 1973.

Gattegno, C., "Functioning as a mathematician". Mathematics Teaching 39 (1967).

Gattegno, C., The commonsense of teaching mathematics. New York: Educational Solutions, Inc., 1974.

Goutard, M., Mathematiques sur mesure. Paris: Classique Hachette, 1970.

Hofmann, J.E., Leibniz in Paris. Cambridge: The University Press, 1974.

Servais, W., "Mathematisation et problematique"

Steiner, H-Gl, "Examples of exercises in mathematisation at the secondary school level". Educational Studies in Mathematics 1, 1-2 (1968).

Tahta, D.G., "About geometry". For the Learning of Mathematics 1, 1 (1980).

Wheeler, D., "Mathematization". Notes of the Canadian Mathematical Society October, 1974.

Wheeler, D., "Humanising mathematical education". Mathematics Teaching 71 (1975).

9.4 THE MATHEMATIZATION OF SITUATIONS OUTSIDE MATHEMATICS FROM AN EDUCATIONAL POINT OF VIEW

THE MATHEMATIZATION OF SITUATIONS OUTSIDE MATHEMATICS FROM AN EDUCATIONAL POINT OF VIEW

Rolf Biehler
Institute fur Didaktik der Mathematik
Bielefeld, Federal Republic of Germany

1. Applications of mathematics and process orientation - - various approaches

In general, advocates of mathematization agree on two important aspects:

1. The intention of making school mathematics more applied: i.e., linking it to various contexts, such as

the everyday world of the students, other disciplines, and practical uses of mathematics in society.

2. The emphasis on the processes by which knowledge is developed as opposed to teaching prefabricated mathematics or prefabricated mathematical models.

But behind these common goals, we find quite a diversity of concretizations which depend on, among other things, the specific context of goals of (mathematics) education for particular populations of students. I think we can distinguish between different types of goals:

1. Neither of the two above described aspects is considered an end in itself but rather a means. Thus, the reference to non-mathematical contexts can merely be a solution to the problem of motivating the students to study particular mathematical subject matter. Or, the process can merely be a means of teaching, just as "discovery learning" is frequently rather interpreted as a way of learning a certain content more efficiently, and not so much as a way of "learning to discover."

2. Both aspects are an end in themselves. Students should learn to solve "real life problems" mathematically as a preparation for their future or present life, in which they frequently will have to be active problem solvers themselves.

3. Teaching mathematical modelling to students is considered as preparing them for quite different future situations in which critical understanding of other people's modelling or models, the reasonable application of known models, and communication and cooperation with "expert mathematical modellers" will be required rather than active modelling itself.

4. Modelling or mathematization has mainly the function of providing an opportunity to learn various things about modelling and mathematization in different domains of science or society. For instance, mathematical modelling can be an opportunity to show the usefulness of mathematics in an exemplary fashion, to discuss philosophical topics such as the relationship between mathematics and reality, to treat the motives and features of the historical development of mathematics in relation to its applications, etc.

Discussions on the relative importance and feasibility of these types of goals and their implications for teaching and curriculum are a problem I have to leave to further thoughts and investigations in mathematical didactics.

As to the design of curricula there seems to be almost general agreement that at least part of the curriculum should be devoted to discussing "genuine" applications of mathematics or to solving "genuine" modelling problems. But a main problem is also how to design the rest of the curriculum if it is to be devoted to a systematic application-oriented development of mathematical knowledge. The problem is that fundamental concepts such as the concept of "variable," "function," "probability," and "random variable" have quite a different meaning and are somewhat differently used in the context of applying mathematics. Thus, the question is how to teach these concepts in order to enable students to use them in the context of solving modelling problems.

2. Epistemological backgrounds

2.1 Mathematical model and mathematical modelling

It is very fashionable, and not only in didactics of mathematics, to represent the general features of how mathematics is applied by means of a universally applicable modelling diagram which shows several subprocesses and their interrelations. Besides these process descriptions, the concept of the model itself is considered important. (See, for example, the booklet Mathematics Changes Gear from the curriculum project by Ch. Ormell for an approach in mathematical didactics which puts an emphasis not only on modelling but also on the concept of model. It attempts to explain the concept of mathematical model to students by putting it in the broader context of physical models, scale models, etc.) Particularly, newer fields to which mathematics is being applied require an understanding of the model as a separate entity which has to be distinguished from the "system" or "process" it represents. This requirement is not a philosophical luxury but a necessary condition for reasonable application of models.

We can identify the following problems:

1. How can the ability to model be (perhaps stepwise) developed so the modelling diagrams can really fulfill their claimed heuristic function?

2. Does the current state of knowledge about mathematical modelling already provide a sufficient basis for such (meta-) knowledge required by students, teachers, or mathematics educators in their function of teacher trainers or curriculum designers?

3. How can we achieve a deeper understanding of mathematical modelling with respect to the structure of knowledge and activity required for users of mathematics?

2.2 Real problem solving

We find some approaches in didactics of mathematics describing processes of using mathematics as problem solving processes (see the materials of the USMES project in the USA or the course "Mathematics across the curriculum" developed at the Open University (U.K.)). Although applied mathematicians would also say they are also solving problems, they in their self-reflections usually do not apply the (theoretical) knowledge about problem-solving processes found in one of Polya's famous books or in psychological studies on the problem-solving process. Thus, some important aspects of problem solving such as looking for similar already solved problems, defining sub-problems, making and carrying out plans, do usually not appear in descriptions of mathematical modelling. On the other hand, the specific nature of mathematical modelling - - namely representing and exploring real systems by means of mathematical models - - is usually not well represented in descriptions of real problem solving.

An advantage of interpreting mathematical modelling as real problem solving can be that results and methods of the somewhat more elaborated empirical research on mathematical problem solving and on teaching mathematical problem solving becomes applicable, although modifications and further elaboration are certainly necessary.

Besides being faced with the need for exploiting the research context mentioned more systematically, we are confronted with the problem of relating and comparing the problem-solving approach and the mathematical modelling approach (see V. Treilibs, 1979, Formulation processes in mathematical modelling, Shell Centre for Mathematical Education, University of Nottingham, for an attempt in this direction). This also implies the difficult problem of coordinating the standpoint of psychology with the standpoint of methodology and epistemology of science.

2.3 The mathematization of situations

In the didactics of mathematics, the concept of mathematization is mostly used to describe the process of creating mathematics in relation to situations and problems inside or outside mathematics (see, for example, the use of the notion of mathematization in Freudenthal, H., Mathematics as an Educational Task, Dordrecht, 1973). As opposed to that, this concept is used in discussions on the development of scientific theories to describe one component of their development, namely the transformation of knowledge (theory, hypotheses) which are not yet mathematically expressed in the so-called mathematical language. Cf. Bunge, M., Scientific Research I, Springer, 1967, p. 446.) Because it refers also to situations inside mathematics, the notion of mathematization is a broader concept that the notion of mathematical modelling. But associated with using the expression of "mathematization of situations outside mathematics," we find a perspective on the processes involved different from those of mathematical modelling. The main emphasis is on the formation and development of mathematical concepts, structures, and theories in relation to non-mathematical situations. Hence, mathematization is frequently used as a didactic principle to teach and learn mathematics in the context of non-mathematical situation (see, for example, the curriculum material developed by IOWO in the Netherlands, where the concept of model and modelling play no significant part). It is often associated with the "genetic principle" of teaching.

2.4 Philosophy of mathematics

As classical philosophies of mathematics such as logicism, formalism, and intuitionism do not sufficiently elaborate the relationship between mathematics and its applications, other approaches have been chosen as a basis. In this connection, I should like to refer to the curriculum development project directed by Ch. Ormell, which is partly based on C.S. Pierce's pragmatic philosophy of mathematics. However, discussion on a broader philosophy of mathematics including dynamical, pragmatic, or social aspects of mathematical knowledge have been revived in recent years but have not yet been very much exploited for mathematical didactics so far.

2.5 The interdisciplinary nature of mathematical modelling

There is still another perspective from which the problem of mathematical modelling is sometimes approached, i.e., the role of mathematics in the development and representation of scientific theories and the interdisciplinary nature of mathematical modelling. The fact that an epistemological clarification of the relationship between, e.g., physics and mathematics is particularly important for the intention to foster cooperation between mathematics and physics teachers has been recognized, e.g., by A. Revuz (see Revuz, A. in: Cooperation between Science Teachers and Mathematics Teachers. IDM-Materialien und Studein, Bielefeld 1979, pp. 12-29.).

The relation of mathematical modelling to the development of scientific theories draws attention to the following problems:

1. If we take the "mathematical-modelling philosophy" of applied mathematics as a basis, we can interpet activities such as those of physicists just as a specific typ e of mathematical modelling. On the other hand, it cannot be the main objective of mathematical modelling as part of the mathematics curriculum to learn the activity of a physicist, for example. Thus, the nature of mathematical modelling and its relationship towards building scientific theories has to be more clearly described than has been done so far in the didactics of mathematics.

2. Although mathematical modelling is sometimes described as "the" scientific method where the hypotheses are formulated in mathematical terms, the discussion on scientific method and the fairly elaborated discussion on the development and structure of scientific theories have rarely been exploited by the didactics of mathematics. In particular, the discussion on general strategies and principles of modelling have direct correlates in the discussions on the possibility and practical value of a "logic of discovery." Also, the discussion in the didactics of the sciences on the problem of how scientific activity can be learned seems to be of great importance to discussions on teaching mathematical modelling.

I think it should have become plausible that choosing an adequate path between the extremes of teaching a body of knowledge and teaching "processes" or "methods" has to be based on a deeper understanding of science and scientific activity. In particular, the new developments in the reflections on science that try to reunite the traditional separation of the body of subject matter knowledge from the thought processes, value systems, and orientations of its practitioners and their communications and socializations seem to be of central importance for the development and communication of knowledge in school.

A NOTE ON THE MATHEMATIZATION OF SITUATIONS OUTSIDE MATHEMATICS FROM AN EDUCATIONAL POINT OF VIEW

Tatsuro Miwa
Osaka Kyoiku University
Osaka, Japan

1. The Aim of Mathematics

Regarding what the mathematization of situations outside mathematics aims at, there are two different standpoints:

(1) It aims at helping pupils to understand mathematical concepts, principles, and techniques.

(2) It aims at helping pupils to comprehend the situations better which deserve recognition and to gain insight into them.

In terms of aims, the Japanese mathematics course of study for the primary school says as follows:

> The aim is to develop in children the abilities and attitudes which will enable them to consider matters of daily life mathematically.

This aim may be interpreted as "to develop in the children the abilities and attitudes which will enable them (a) to create mathematics out of real situations, and (b) to analyze and deal with concrete matters, making use of mathematics." The above-mentioned standpoint (1) corresponds to (a), and (2) to (b).

As to the standpoint (1), I should like to take an example of how children form concepts of the rectangle as an abstract figure. Children observe from a certain viewpoint many rectangular objects familiar to them in their daily lives, such as sheets of paper, books, and desks, and learn to abstract the shape common to these concrete objects. Strictly speaking, the accurate forms of such objects in reality are not exactly rectangular. In the process of learning, however, they are idealized and abstracted to form an ideal image of rectangularity in the children's minds. Thus it is important to abstract the figure from concrete objects.

This is not limited to geometric figures only, but applied to all areas of mathematics. By considering situation of daily life in a broader sense, the above description is valid in secondary school.

The standpoint (2) is the essentials of the application of mathematics; as Professor Hall said, "The goal of applied mathematics is to understand reality mathematically." In the mathematization of situations it is necessary to establish conditions and assumptions and to describe them in terms of mathematics. This always should be done in the application of mathematics whether we are conscious of it or not.

Mathematization of what and how differs according to whether we stand on standpoint (1) or (2). On standpoint (1), it is desirable that situations are simple and do not contain impurities so that pupils can easily idealize and abstract the concepts and principles. On the other hand, situations need not be simple on standpoint (2). They should be interesting to pupils in the ordinary sense and as real as possible within their knowledge and abilities to deal with them.

2. The Merit and Role of Mathematization

I should like to focus my discussion on standpoint (2). First, I should like to take note of the merit derived from mathematization. This can be elucidated when various patterns of thinking used in the process of mathematization are examined: Scientific method consists of induction from factual observation, deduction from model, and verification of the results - - all of which are found in the process of mathematization. In more detail, idealizing, simplifying, and therefore, abstracting in model-making; examining whether assumptions are consistent and whether conditions are necessary and sufficient for attaining the results; and checking the results with the real situations and refining the model for better validation (successive approximation) are examples of patterns of thinking used in the process. These are fundamental elements of the mathematical way of thinking. To help pupils to foster these abilities is included in the aim of methematical education. Therefore the process of mathematization constitutes an important component in mathematical education.

Second, I should like to take not of the role that mathematization should play in mathematical pedagogy. A mathematical course often falls into the formalized teaching of the accumulation of highly abstract concepts, principles, and techniques in mathematics. However, it is necessary to bring the course back to the concrete and make it as vivid and interesting as possible in order to let pupils know the power of mathematics. Mathematization plays a significant role in this respect.

3. Some Problems Involved in Mathematization

a. The Aspect of the Pupils

Both the understanding of situations outside mathematics and the mathematical techniques used in working out a mathematical model depend heavily upon the maturity of the pupils. In particular, primary school children are so limited in experience and knowledge that the difficulty is great. But there have been many teachers who have contrived to overcome this difficulty and tried to let children make models commensurate with their maturity.

Among those innovators I should like to list the late Mr. Jingo Shimizu of Japan, who did an excellent practice of "problem making" about 50 years ago. Mr. Shimizu was a primary school teacher attached to the Nara Women's Higher Normal School. He said that it is better for children to study facts in daily life and elucidate the relationship between them, and then make up problems for themselves. By doing so, the children will gradually cultivate the attitudes and habits of finding relationships between the facts. At this juncture, the teacher's guidance is most important because without it children cannot make suitable problems, and the problems thus made tend to be valueless or invalid.

In problem making, it is important to describe a real state of a situation, and unnecessary to have children conform to the convention of written problems. Therefore, only requirements or objectives drawn from the situation may well be expressed. When children try to find the solution for such a problem, they should provide conditions completely by searching and collecting necessary data. From among children-made problems, appropriate ones are selected as "class problems," and the teacher guides them in their studies in a wider context.

Mr. Shimizu's spirit remains partially in my country, but problem making in a thorough form is not practiced now. I should also like to note the fact that pupils, especially in secondary school, are not familiar with multidisciplinary teaching. Often pupils in mathematics class learn only mathematics in a pure form and do not think of other subjects, although mathematization outside mathematics requires knowledge not only of mathematics but also of other subjects.

b. The Aspect of the Teacher

The most critical problem is found here. Initial teacher training is insufficient. It provides highly abstract mathematics courses in order to train specialists in mathematics, but it has little time for an education in mathematization outside mathematics. In fact, in teachers' colleges courses on algebra, geometry, analysis, probability, and statistics are emphaiszed, but those on applications of mathematics are not necessarily stressed. Therefore, a course on mathematization is established in very few colleges. Thus teachers are provided neither sufficient knowledge nor experience in mathematization. In inservice training, however, mathematization courses are established in a few institutions since their importance is gradually recognized. Nevertheless, it is insufficient now.

Moreover, cooperation between teachers of mathematics and teachers of other subjects is insufficient because they have little knowledge of other subjects and like to teach independently. The fact that the mode of using mathematics in other subjects is different from that of the mathematics class makes the above cooperation more difficult. Regarding the teaching of mathematics and science, one can find some discord, as shown in our report (Miwa et al., 1979); e.g., usage of letters in representing variables and constants, usage of concepts of function, mathematical inference in science lessons, spatial recognition, and graphics.

c. The Aspect of the Curriculum

The total process of mathematization is so big a learning task or learning unit that it takes a great amount of time. Moreover, it is difficult to construct the school curriculum on the basis of several learning units on mathematization of situations. Many teachers regard mathematization as important, but they often hesitate to do in class for this reason. In fact, if it is done within the hours of mathematics class, many things to be learned will probably be sacrificed. Thus on the premise that the subject matter of school mathematics is changed, pupils may as well learn one or two learning units of mathematization when they are in school, e.g., at the terminal grade of each school level.

The most suitable material for this unit and how the unit is developed are determined by the actual state of the pupils, including their interests, knowledge, and abilities. Valuable and interesting examples of these units on mathematization are found in many documents from various countries. But we must take national climate into consideration for adopting examples worked out in other countries.

Note

One example of Mr. Shimizu's practice:

On Vernal Equinox Day, the children in his class went up to Kasuga Hill in Nara with him and a science teacher before sunrise. They stayed there for a while. They observed and recorded the movement of the sun. Then the science teacher gave a lecture about the vernal equinox. Some problems made by the children are as follows:

1. Is daytime as long as nighttime?
2. Why does the sun look larger when it rises than when it is in the sky?
3. How far does the sun move in an hour?

References

Miwa, T., S. Shimada, & K. Kinoshita. Discords between mathematical instruction and science instruction. Proceedings of the Conference on Co-operation between Science Teachers and Mathematics Teachers. IDM, Bielefeld, 1979.

THE PERCEPTION OF MODELLING PROBLEMS AS 'REALISTIC'

Christopher Ormell
Reading University
Reading, England

Introduction

The mathematisation of real-life situations consists of a kind of crystallisation of these situations. The situation is represented by a mathematical model which then enables us to derive solutions to certain "real problems" arising in the original situation. We can undertake such mathematisation in many ways, on many different types of situation. The professional scientific or industrial modeller tackles pre-set questions (i.e. question usually chosen by others) and these are normally of a fairly technical nature. In education it is possible to adopt a different stance. In a word, we can choose our own problems and we can choose them with the object of undertaking a crystallising-and-modelling exercise of the greatest possible educational value. In education our modelling can be based on the principle that it is more interesting to travel than to arrive!

We shall call this educational modelling.

Educational modelling can be aimed at students of various levels of ability and career intention. It can also be aimed at providing a part or the whole of the student's course. I am concerned here mainly with the second case, viz. where educational modelling is used as a vehicle to carry the load of a complete course. Some readers may be surprised to learn that this is possible. It is of course only possible by exploiting the idea of projective modelling, that is, the application of mathematical modelling to explore the implications of "good ideas" for developments of situations slightly beyond the range of present experience. I am particularly concerned with the development of systematic educational modelling for use with students aged 16-18 of a wide ability spectrum.

To say that a course is based on projective modelling does not specify it completely. Onto what kinds of "good ideas for developments slightly beyond the range of present experience" does it project? How does it cope with the perennial problem that students seize up when they see a mass of words and no obvious cue about the type of mathematical trick required? I shall be concerned mainly with issues which have arisen as a result of work with Mathematics Applicable a course designed for non-specialist sixthformers in schools in England and Wales. It was written with two main emphases: (1) to project the mathematics onto situations of the broadest probable appeal, that is, situations in which the "problem" is defined in terms of mainstream or even cinematic values, (2) the material which we developed at Reading over ten years (1969-78) uses the device of semi-programming to give the student hints, answers, and background notes of various kinds. Without this provision of hints, answers, and notes the average or weak student would all too quickly seize up. With them, he or she can keep going, and after a period of time the "point of the activity begins to sink in.

Given such material, it is possible to solve the age-old problem of striking a balance between teaching too fast (with the aim of stimulating interest) and of teaching too slow (with the aim of consolidating students' ideas). The projective subject matter of such material enables the teacher to maintain interest almost indefinitely, whilst holding the mathematical content stationary. It is this capacity to hover on a given mathematical topic which gives the new material its educational potential, and one of the main tasks for mathematical education in the next twenty years will be to develoop courses spanning the full ability range and for all ages based on this principle.

But the construction of educational modelling material based on projective themes poses a general problem: that modelling-derived questions are not always perceived by teachers and students as being 'realistic'. What are the conditions under which modelling-derived problems appear to be 'realistic'? The immediate answer is that the perception of 'realism' depends on the preconceptions of the reader. Some teachers of a cautious disposition have convinced themselves that systematic modelling based courses are a pedagogic impossibility, and they have allowed the wish to become the father of the thought. The result is that they have sometimes applied the adjective 'unrealistic' to problems which are unquestionably derived from real problems. For example, several commentators have claimed in print that a problem concerned with the repeated reflection of light on the surface of mirrors possibly used in the painting of tombs in the Valley of Kings is 'unrealistic'. There is no way in which such a judgement will stand examination. The tombs are real; the paintings inside are real. The question "How were these painting executed?" is a real question. There is no sign of the extensive sooty deposits which would result from the use of candles or oil lamps over many hours. Given such facts, the most probably hypothesis must be that the ancient Egyptian painters used mirrors. It then becomes a real question whether the light in the inner chamber would be sufficiently bright to allow painting to proceed. When commentators casually apply the label 'unrealistic' to such well-founded problems, they only succeed in displaying the extent to which they are out-of-touch with what is 'real' and what is not.

Some commentators seem to apply the adjective 'unrealistic' to anything which is not of evident technical or commercial value. If we adopted this criterion, little that is taught in school in other subjects would count as 'realistic'. Is Hamlet realistic? Is The Fall of Constantinople realistic? One would not have thought that it was necessary to spell out to teachers of any subject that a question that is posed in terms of real elements and in relation to which someone (not necessarily a scientist, technologist, or industrialist) needs to know the answer is a 'real' and sometimes an educative question.

Is it a 'realistic' question to ask how much water an open-topped car would collect if driven for t hours through a rainstorm of known intensity? Open-topped cars are real; so are rainstorms. Some rainstorms are carefully monitored by amateur and professinal meteorologists, so that their intensity t hours after the first drops appear is subsequently known. It is of interest to a psychologist to know how much rain a disturbed individual ignored in his single-minded intention to get from A to B. It is of interest to the car designer to know how much rain the car absorbed, i.e. what proportion of the rain it gathered on its way. There are also real conditions under which a detective would be interested in such questions. Yet just this situation is another example of a state-of-affairs which has been claimed to be 'unrealistic".

Factors in the perception of realism

How is it possible to explain the repeated comment of some teachers that certain problems are 'unrealistic', when it is known that they began life as reflections on real situations. The immediate comment must be that realism, like beauty, is at least partly, "in the eye of the beholder". This is a paradox, since comments about realism do not sound at all like comments about beauty: they sound very positive and down-to-earth. People who look at modelling-derived material in the light of only a weak or partial understanding of the general applicability of mathematics often fail to bring to the interpretation of the material those "reconstitutive" elements presupposed by the new modelling approach. It is as if a housewife, on trying out dehydrated soup, should complain that it bears little resemblance to real soup! Obviously, in the interests of consision and intelligibility, the number of words in such modelling-derived material will have been reduced to a minimum. It is assumed that the teacher will remedy this situation with any student who is unable to recreate the full sense of the implied backgroung for himself or herself. What we have found is, however, that some teachers need a lot of help in recreating the backgroung too.

One reason for this response may be that there is a general presumption that almost all so-called "applications" found in mathematics texts are unrealistic. The majority of such "problems" show no sign of the writer having taken the most minimal account of the empirical background, let alone the motivational context. It will, no doubt, take time for this expectation to be erased in the average teacher's mind.

Some realistic problems, though basically sound, have been ruined by excessive repetition, e.g. filling-the-bath and the ladder-against-the-wall. One can hardly find more realistic problems than that of ensuring that a ladder which one is about to climb will not begin to slip when one reaches the nth rung.

Four types of complaint about the alleged lack-of-realism of modelling-derived material are: (a) the information is given in a form which one would rarely or never encounter in real life; (b) it is difficult to see any technical, industrial or commercial point to the question; (c) the empirical conditions of the problem have been over-simplified; and (d) the numbers quoted in the problem are far too neat and convenient to echo one's experience of such problems in real life.

The answer to (a) is that 'real life' is changing. We are entering an era of greatly increased telemetry and the public is getting used to new quantifications all the time. It is quite reasonable, in the view of the present author, to extrapolate this trend a little into the future.

The answer to (b) is to listen to the conversations of students and to take into account the kind of things which really interest them: these are by not means only, or even largely, things of "technical, industrial, or commercial" concern.

The answer to (c) is that it is alwyas highly desirable in material designed for a broad ability spectrum to find the simplest possible case of a process, relationship, or phenomenon. The same applies to (d). Here, too, it is essential that the figures students encounter in grappling with a problem for the first time should not cause needless dismay and difficulty. Once the basic pattern of the phenomenon has sunk in, it is possible to look at more complicated cases. And it is possible to repeat the processes used on more typically awkward numerical parameters - - especially now that the calculator removes most of the grind from handling numbers.

SOme teachers seem to feel that in breaking away from the lofty, idealistic, aesthetically pleasing perspectives of an ultra-pure, hyper-abstract conception of mathematics, they are required to undergo the penance of working on dull, formally unpleasing, complicated, unimaginative tasks. They are inclined to overdo the "rolling up sleeves" and "getting down to the nitty-gritty" bits, and to forget that mathematics, in its central essence, is an imaginatively and conceptually gymnastic activity. The fascination of applicable mathematics, like that of pure mathematics, is that it gives us an insight into the unexpected ifs of life.

[1]The series of ten books Mathematics Applicable is published by Heinemann Educational Books and is available in the United States from the following address: 4 Front Street, Exeter, New Hampshire. Other materials, produced by a voluntary group of teachers (the MAG) in England, are obtainable from The Mathematics Applicable Group, School of Education, London Road, Reading RG1 5AQ, England.

THE MATHEMATICAL MODELLING ABILITIES OF SIXTH FORM STUDENTS

Vern Treilibs
Para Hills High School
Warradale, Australia

Considerations about the nature of the mathematization process indicate some possible strategies for its teaching. Another important determinant of appropriate teaching methodology is the existing modelling capabilities of the students. This paper describes an exploratory study of the mathematical modelling abilities of untrained sixth form (17 year-old) mathematics students (1). The emphasis of the study is upon the formulation processes used by the students when solving realistic problems. The choice of relatively able students in the sample was intended to produce a stronger "signal" for the observers; in particular it was expected that these students would produce more analytical modelling solutions and that they would indulge in less descriptive modelling (2). The choice of realistic problems reflect the bias of the experimenters concerning the nature of the modelling process and its role in mathematics education.

Initially a test of general modelling ability was constructed, scoring schedules devised, and the test administered to a sample of 118 students. The results of this test indicated that modelling ability differs from, but is not independent of, ability in "conventional" mathematics. Next a set of component skills of model formulation was postulated, and tests of these skills were constructed and then administered to the student sample. The component skills tested were:

1. generating variables - - students were asked to identify as many as possible of the relevant variables in a given realistic problem.

2. selecting variables - - students were asked to rate, on a four-point scale, a given list of possible variables for the problem situation.

3. identifying specific questions - - students were given a real situation problem and were asked to identify precisely what specific questions had to be answered in order to resolve the problem.

4. generating relationships - - students were asked to graph the relationships between key variables in the given real problem.

5. selecting relationships - - students were required to select from a choice of given relationships (in graphical form) the one appropriate for the problem.

Analysis of the results showed that performance on all but the second (selecting variables) of the component skills tests correlated significantly with performance on the test of general modelling ability. The component skills 3, 4, and 5 (above) proved to be the abilities that differentiated "good modellers" from "poor modellers."

Another phase of the study involved the construction of a theoretical model, in flowchart form, of the modelling process followed by a series of case studies of groups of students formulating models of real problem situations. In addition to providing support for the theoretical model, the case studies revealed that "good modellers" were characterized by their greater ability to work at the operational level. Thus it was seen that:

1. Good modellers seemed to have a strong sense of direction when tackling problems; they appeared to foresee the underlying structure of the solution method.

2. Good modellers were much more likely to direct the organization of the solution than were poor modellers.

3. Good modellers were able to partition their treatment of the problem.

4. Good modellers showed a willingness to make assumptions.

5. Almost all students preferred to produce arithmetic modelling solutions rather than algebraic ones.

Other findings of the study included:

1. In formulating and solving models students rarely used mathematics that they had learned recently (i.e. in the last two years).

2. Students with genuine modelling ability were rare, but generally their teachers were able to identify them (despite teaching "conventional" mathematics courses).

3. The students, untrained in working co-operatively, performed relatively poorly in a group modelling exercise.

The general implications of the findings of the study for curriculum designers are that explicit steps must be taken to develop modelling ability; it is inefficient to merely present large numbers of standard models and hope that the students will learn modelling skills "by example." Such steps should include the practice of the component skills of modelling listed above (and others e.g. estimating), practice in organizing solution procedures, and the development of an awareness of the need for, and nature of, validation procedures.

The teaching of modelling presents particular problems for the teacher; the traditional approach based on teacher exposition of principles and rules with illustrated examples and similar pupil exercises is not suitable for developing the creative skills of modelling. A substantial change of style is required of the teacher; there must be a willingness to tackle open-ended, interdisciplinary problems with the students, to accept that frequently he or she will not know the solutions, and to follow students in whatever lines of argument they develop. It is certain that many teachers of mathematics would find it most difficult to make such a change.

Notes

1. The study was undertaken during 1978-79 while the author was on leave at the Shell Centre for Mathematics Education, University of Nottingham, U.K.

2. The term "descriptive modelling" refers to the creation of a model mainly for its ability to describe the given problem situation, i.e. to identify the overt structure of the problem. In "analytical modelling" the model created explores the underlying structure of the problem, i.e. the relationships between the variables in the problem situation are manipulated to provide more knowledge of the situation and, eventually, the answers to the problems posed.

CHAPTER 10 - Special Mathematical Topics

10.1 ALGEBRAIC CODING THEORY

ALGEBRAIC CODING THEORY

J.H. van Lint
Eindhoven University of Technology
Eindhoven, Netherlands

Introduction

One of the purposes of the mini-conferences at ths meeting is to introduce new areas of mathematics which have found their way into the curriculum at some universities and colleges but deserve more attention. At the same time one can consider the question whether it concerns a development in mathematics that could in some way be incorporated in the high school curriculum. In the present case the topic is algebraic coding theory, an area between information theory, combinatories and applied algebra, which has only been around for about 30 years. More than likely at most of the world's universities there has never been a course in this subject. On the other hand at a few it has been taught for at least 15 years and it is usually received with enthusiasm by the participating students.

I am going to assume that the reader has either no knowledge or only a vague idea of what coding theory is. We shall start by showing the origin of coding theory by using a few examples. This introduction will also serve as an example of how the subject can be introduced to students. The introduction is also designed to show that the subject can be introduced at the high school level.

By no means do I want to advocate the introduction as a subject at this level. However, I do believe that for little groups of students with an interest in mathematics, or for special projects, or even as a part of a general science course, certain parts of the subject are quite suitable. At a somewhat higher level I believe the subject is extremely suitable as an illustration of an area which is important for practical reasons, where all kinds of mathematics (which the students by then have learned in introductory courses) are applied. One of the things coding theory can illustrate is that a mathematician needs a very broad basic education and that he will often be surprised about the places where he can apply his knowledge.

The second part of the paper sketches a possible course content and serves as a continuation of the introduction. We stress the algebraic aspects.

Most teachers of combinatorics are well aware of the possibilities of coding theory and often have a section on the subject in their course. For this reason we shall hardly go into this part of the subject.

In the final part we look at the possibility of using certain tipics from coding theory in courses on other parts of mathematics, showing connections or applications.

The main reference is "The Theory of Error-Correcting Codes" by F.J. MacWilliams and N.J.A. Sloane, which contains just about everything there is to know about algebraic and combinatorial coding theory and has a list of nearly 1500 references. The authors are preparing a short introductory text book which will appear in 1981 (Springer-Verlag). For some subjects this book is given as a reference. The second part of the References will mention suitable introductions into this area.

Error-correcting codes

We illustrate error-correcting codes by means of a well known recent example. Many readers will have seen the excellent pictures which were taken of Mars, Saturn and other planets by satellites such as the Mariners, Voyager, etc. In order to transmit these pictures to earth a fine grid is placed on the picture and for each square of the grid the degree of blackness is measured in a scale of 0 to 63. These numbers are expressed in the binary system, i.e. each square produces a string of six 0's and 1's. The 0's and 1's are transmitted as two different signals to the receiver station on earth (the Jet Propulsion Laboratory of the California Institute of Technology). On arrival the signal is very weak and it must be amplified. Due to the effect of thermal noise it happens occasionally that a signal which was transmitted as a 0 is interpreted by the receiveer as a 1 and vice versa. If one simply transmits the sextuples mentioned above, the errors would have great effect on the quality of the pictures. In order to prevent this, so-called redundancy is built into the signal, i.e. the transmitted sequence consists of more than the necessary information. We are all familiar with the principle of redundancy from everyday language. The words of our language form a small part of all possible strings of letters (symbols). Consequently a misprint in a long (!) word is recognized because the word is changed into something which resembles the correct word more than it resembles any other word we know. This is the essence of the theory of error-correcting codes. In the example of the Mariner 1969 the sextuples of 0's and 1's were mapped into strings of 32 symbols. If a received string contained less than 8 errors the decoding device interpreted it correctly. We illustrate the idea by a smaller example.

0 = 000	000	00000
1 = 001	001	10110
2 = 010	010	10101
3 = 011	011	00011
4 = 100	100	10011
5 = 101	101	00101
6 = 110	110	00110
7 = 111	111	10000

Here the integers from 0 to 7 are written in the binary system and then mapped into codewords of length 8 (i.e. strings of eight 0's and 1's).

The reader can easily check that any two codewords differ in four positions (and that it is not difficult to add more codewords to the list which also differ in at least four positions from all the previous ones). If we just use the eight words given above we can drop the fifth symbol since it is zero for all of the words. In the example of the Mariner 1969 a codeword of length 32 consisted of six information symbols and the rest was redundancy. We say that this code has information rate 6/32.

It is not difficult to understand that if we are willing to use extremely low information rate then t is possible to correct many errors in each received word. E.g. if we repeat each symbol (0 or 1) a number of times, say $2n + 1$ times, then a received string of $2n + 1$ symbols may contain up to n errors and still be interpreted correctly on the basis of maximum likelihood. The information rate is now $1/(2n + 1)$. The following example illustrates in a simple way why coding theory has become so interesting and important.
We conduct an experiment in which a coin is tossed many times and the results heads (= 0) resp. tails (= 1) are transmitted over a so-called binary symmetric channel:

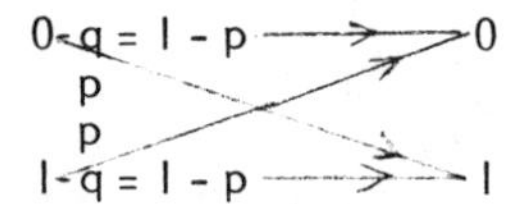

(The probability that a symbol is received incorrectly is p.) Let us assume that the channel is fairly good, say $p = 0.001$. If we simply transmit our results then the receiver will have a 0.1% incorrect information at the end of the experiment. Suppose the channel can handle two symbols in the time needed for a toss of the coin. In the terminology used earlier we can transmit with information rate 1/2. The old educational principle of attempting to be better understood by saying everything twice does not work in this example! If we transmit 00 resp. 11 instead of 0 rep. 1 then the receiver will know that a received sequence 01 is incorrect but he cannot correct the error. Now we wait a few seconds until ten tosses have taken place and then start transmitting twenty symbols for every ten tosses of the coin; (again the information rate is 1/2). We need a mapping f from $(0,1)^{10}$ to $(0,1)^{20}$ or in other words we need a subset of 2^{10} elements from $(0,1)^{20}$ which we call a code C consisting of s^{10} codewords of length 20. Since we have 2^{20} words from which to pick it is not so very surprising that it is possible to choose C in such a way that any two codewords from C differ in at least six places. This means that if a received word contains one or two errors these can be corrected. It is not difficult to calculate the probability that a received word contains more than two errors and thus discover that the receiver now has about 0.001% incorrect information.

Shannon's famous theorem which marks the beginning of this area of research states that in fact the probability of error can be made arbitrarily small (in our example with $p = 0.001$ and information rate 1/2). The important thing to note is that this is achieved without lowering the information rate! The proof of this theorem is based on a probabilistic argument. So the question remains, to this day, how to construct good codes.

This has been a short and sketchy introduction.

If the reader has become interested he can read more about what was explained above in the references (5), (7), (9), (10), (a), (b), (c), (d), (e), (f).

A course on coding theory

A course on coding theory should start with several examples of communication channels affected by some kind of noise, followed by a motivation for working in this area such as the previous paragraph.

Of course, at the university level one actually gives a rigorous proof (and a precise statement) of Shannon's theorem. Before embarking on general thoery it is good to illustrate that usually three different aspects can be distinguished. First of all there is the definition of the code C, i.e. the question of constructing in some systematic way the mapping f from information strings $(a_0, a_1, \ldots, a_{k-1})$ to codewords $(c_0, c_1, \ldots, c_{n-1})$. Then follows a detailed analysis of the properties of the code C, the most important being the so-called minimum distance. Finally one can go into the question of decoding algorithms of different kind, and even complexity questions can be considered. In many courses for mathematics students the third part, which is of a more practical nature, is often hardly treated at all.

Once the setting is clear the actual mathematics starts. We introduce structure. The first thing to do is to spend some time on the theory of finite fields. If the course is for mathematics majors one can go into this extensively or possibly it is already known to the students. I claim that even if one chooses coding theory as a special topic to show (better) high school students some interesting application of mathematics including several things they may have had to learn without realizing where it is used, then one can easily explain F_2, even F_p and maybe also the fields F_4, F_8, etc.

Let us introduce some terminology.

Let F be a finite field. We call F the alphabet. A code C is a subset of F^n. The distance $d(\underline{x}, \underline{y})$ of two codewords $\underline{x} = (x_1, x_2, \ldots, x_n)$ and $\underline{y} = (y_1, y_2, \ldots, y_n)$ is the number of places where they differ (i.e. $x_i \neq y_i$). The weight of $\underline{x}$ is $d(\underline{x}, \underline{0})$, where $\underline{0} := (0,0, \ldots, 0)$. The minimum distance d of the code C is

$$[d(\underline{x}, \underline{y}) \ : \ \underline{x} \in C, \underline{y} \in C, \underline{x} \neq \underline{y}]$$

Clearly d is a measure for the error-correcting capability of C. The information rate which we used in a previous section is defined as $n^{-1} \log|C|$, where the base of the logarithm is the size of the alphabet, i.e. $|F|$.

The next chapter in the course is a very nice application of linear algebra: one requires that C is a linear subspace of F^n, say of dimension k. Such a code is called an $[n,k]$ code. By just mentioning some key words we hope to give an impression of what comes up: basis, generator matrix, standard form, cosets, linear equations, inner product. A very important concept is the dual code $C^\perp$ defined by $C^\perp := \{\underline{x}\ F^n : (\underline{x}, \underline{y}) = 0$ for all $\underline{y} \in C\}$.

We illustrate this by treating the [7,4] binary Hamming code. We wish to construct a linear code over the alphabet $F_2 = \{0,1\}$ which can correct one error. Therefore d must be at least 3. We do this by constructing a matrix H which has as its rows a set of basis vectors for the dual code $C^\perp$. By definition every codeword $\underline{c}$ has inner product $\underline{0}$ with every row of H, i.e. $\underline{c}\ H^T = \underline{0}$. If we change the i-th symbol of $\underline{c}$ we obtain a word $\underline{c}'$ with one error. Clearly $\underline{c}'\ H^T$ is the transpose of the i-th column of H. Hence we can correct the error if all columns of H are different!

Now, let the i-th column of H be the binary representation of the integer i ($1 \le i \le 7$):

$$H = \begin{pmatrix} 0 & 0 & 0 & 1 & 1 & 1 & 1 \\ 0 & 1 & 1 & 0 & 0 & 1 & 1 \\ 1 & 0 & 1 & 0 & 1 & 0 & 1 \end{pmatrix}$$

We then have a very easy decoding algorithm. If $\underline{c}'$ contains one error in position i then $\underline{c}'H^T$ is the integer $\overline{i}$ in binary and $\underline{c}$ is obtained by changing the i-th received symbol back into what it should be. The three rows of H are linearly independent; so the code we have just defined is the set of solutions of three independent linear equation in seven variables, a [7,4] code with d = 3. Finally we could analyse this code and show connections to several interesting areas of combinatorics.

This simple but important example was chosen to make it clear that one can teach this subject quite clearly. To really get an idea of the most important codes used in practice today one must use more algebra. Again, we mention the minimum which is necessary to be able to treat much interesting material. We need the concept of a ring, ideals, principal ideals, polynomial rings. It is useful to know something about a group algebra but we can do without this for a while. A cyclic code is a linear code C such that for every $(c_o, c_1, \ldots, c_{n-1}) \in C$ the word $(c_{n-1}, c_o, c_1, \ldots, c_{n-2})$ is also in C. The mapping $(c_o, c_1, \ldots, c_{n-1}) \rightleftarrows c_o + c_1 x + \ldots + c_{n-1} x^{n-1}$ maps C 1 - 1 onto a principal ideal in the ring of polynomials mod $(x^n - 1)$. This opens a wealth of possibilities and an advanced course would now go into the connections between the two representations, zeros of polynomials in extension fields of F and a discussion of the important class of BCH-codes. If "applications of algebra" is the main theme one can include much more which we shall mention below. The reader mainly interested in high school or first-year college education should not stop reading! The subject of arithmetic with polynomials is certainly taught in high school. Multiplying polynomials is not so interesting, division even more distasteful. Now study the following example and think about its possibilities. We are going to work with polynomials $c_o = c_1 x + \ldots + c_6 x^6$ in which every c_i is 0 or 1. We first add, subtract, multiply in the normal way and then reduce mod 2 (so 1 + 1 = 0). Now some division: polynomials of degree greater than 6 are reduced by taking the remainder after dividing by $x^7 - 1$. Let C be the set of $(c_o, c_1, \ldots, c_6)$ such that $c_o + c_1 x + \ldots + c_6 x^6$ is divisible by $x^3 + x + 1$. Then C is a [7,4] cyclic code with minimum distance 3, in fact it is a cyclic representation of the Hamming code (easy to decode!) treated earlier. A lot of things that require proof, with motivation provided gratis. If one wants more: let (a_o, a_1, a_2, a_3) be a string of four information symbols; map this into $(c_o, \ldots, c_6)$ where $c_0 + c_1 x + \ldots + c_6 x^6 = (a_o + a_1 x + a_2 x + a_3 x^3)(x^3 + x + 1)$, an extremely simple encoding scheme for our code. My experience is that at this stage many students like to hear something about the actual circuitry (shift registers) which are used to do the encoding and decoding (see 1, 9). We now turn to a connectin with geometry. We introduce the idea of Reed-Muller codes by looking at a simple example. The code for Mariner 1969 which was mentioned in the introduction is of this type.

A simple approach with pictures is possible but it is better to use linear algebra since it is fairly common nowadays to use methods from linear algebra in geometry courses. We take F_2 as the alphabet. Consider the m-dimensional space V over this field (if necessary only take m = 3, i.e. look at the unit cube). Let $P_o, P_1, \ldots, P_{n-1}$ be the points of this space, numbered in some way ($n = 2^m$). Define the vector $\underline{v}^{(i)}$, (i = 1, 2, . . . , m), with n coordinates by $V_j^{(i)} = 1$ if the i-th coordinate of P_j is 1; = 0 otherwise.

So $v_j^{(i)}$ is simply the i-th coordinate of P_j and hence $v^{(i)}$ is the characteristic function of the hyperplane $\{x \in V: x_i = 1\}$. The vectors $\underline{v}^{(i)}$ are the basis of an [n,m] linear code. The code word $\sum a_i \underline{v}^{(i)}$ is clearly the characteristic function of the hyperplane $\{\underline{x}: a_i x_i = 1\}$. Observe that if we use only these hyperplanes then more of them will contain the origin; i.e. all words will have a 0 in the corresponding position, which is what happended in the example in the introduction. Therefore every nonzero codeword has 1/2n coordinates equal to 1. Since two hyperplanes intersect in a space of dimension m-2 we see that two different nonzero codewords have a 1 in common in 1/4n positions, i.e. their distance is 1/2n. The example with n = 8 was treated in the introduction. If we add the vector $\underline{v}^{(o)} := (1, 1, \ldots, 1)$ to the basis we increase the dimension of the code by 1. The new codewords are the characteristic functions of the whole space and all hyperplanes through the origin.

Starting from this example one can treat several codes which all show nice applications of both finite geometries and linear algebra.

The course which we are sketching could stop at this point. However, let us look at other topics which can be included. For the purpose of teaching, nonlinear codes are not so suitable. On the other hand, here is an area where the student who has become interested can do a lot of interesting little research projects on construction methods. Of course there is extensive literature on the subject (cf. 3, 9). A much better subject is "bounds on codes". Here the problem is to find inequalities involving $|F|$, n, $|C|$, and d.

Especially at the advanced level some fascinating mathematics comes into the picture (e.g. orthogonal polynomials, application of linear programming). For fixed n, combinatorial techniques of many kinds can be used. More important are the asymptotic results. For all these subjects we must refer to the literature.

To show that there are interesting problems even for a lower level course we consider the following problem, which is left as an exercise for the reader. We wish to construct a binary code C of length n = 6 with minimum distance d = 3. Show that C cannot have more than nine words. Then show that C cannot have nine words. An example with eight words is easily found. My experience with this exercise, with mathematics majors in their third year, is that the solutions which are given usually take a few pages although a few lines suffice. In fact, this kind of problem becomes hard extremely rapidly. If we replace the wordlength in the above problem by n = 8 then the maximal value of /C/ is 20. It is very unlikely that a reader who does not know coding theory will be able to prove this. For large values of n and d it is usually impossible to give exact values.

Special topics that can be included depending on the level of the students and the purpose of the course are e.g. arithmetic codes (used in computers, the alphabet is not a field but a set of integers, addition is normal addition), perfect codes, codes and combinatorics (cf. 3, 9), automorphism groups.

Topics from coding theory as applications

It is a good habit of many teachers of courses in some areas of "pure" mathematics to show that their subject can be applied either in other areas of mathematics of in some other science or in practice. With this purpose in mind we shall consider a number of areas of mathematics and mention possible applications in coding theory. This section is mainly intended to be a collection of suggestions for teachers of such courses. It is clearly impossible to treat the problems of coding theory which are mentioned. We give only references and some indication of the problems and their level. We do not include the obvious area of combinatorics (see 3, 9).

Linear algebra

Here there are many applications at the elementary level. We refer to the previous section and the standard text books on coding theory. As an application of systems of linear equations one can treat the concept of dual code $c^{\perp}$. Here it is worth pointing out that our intuitive idea that C and $C^{\perp}$ are complementary is not true for finite fields. In fact self-dual codes (i.e. $C = C^{\perp}$) are among the most intersting.

Algebra

We have mentioned rings and ideals in the previous section. The Goppa codes provide another nice example of the use of rings. There are several nice applications of the theory of group algebras in coding theory (cf. e.g. 5, 8, 9, 13). These usually also involve some character theory. In a course on semi-simple algebras one can include the idempotents of cyclic codes as an example (9) and for a very nice and important application the association schemes of coding theory are highly recommended (3, 9).

Finite fields

In our discussion of cyclic codes we have already seen that some theory of finite fields is necessary for coding theory. More advanced concepts such as traces turn out to be quite useful, e.g. to describe the Kerdock codes (cf. 9). Sometimes the description of a code makes use of the normal basis theorem (9). Coding theory presents some problems which one would not expect after having learned about finite fields. If C is a code over F_q where $q = p^r$ we can represent F_q as $(F_p)^r$. The field F_q is unique. However, if we substitute the representation for the symbols in the codewords of C then the minimum distance of the resulting code over F_p depends on which representation of F_q we choose. There are open problems in this area.

Group theory

In a course on permutation groups one can include several well-known codes as examples, e.g. the quadratic residue codes with groups PSL (2,n), again an area with open problems. Of course the Golay code is a must in such a course (cf. 3, 9).

Invariant theory

This topic has recently regained popularity. There are several very nice applications in coding theory of which Gleason's theorem on weight enumerators of self-dual codes is probably the most important one (see 9, Ch. 19).

Polynomials

The Reed-Muller codes which were treated in the previous section can also be defined using polynomials. This leads to theorems on the number of zeros of polynomials over finite fields (5). The reader who is familiar with the theorems of Chevalley, Warning and Ax in this area can find an application to RM codes in (6). Questions on irreducible polynomials clearly play a role in the description of finite fields. A very important application is connected to the Justesen codes (7).

Number theory

Elementary number theory is used in the treatment of arithmetic codes (7, 11). There is a proof of the distance properties of RM codes which uses Lucas' theorem (7). Several non-existence theorems for perfect codes are nice applications of number theoretic arguments (cf. 5, 6, 7, 8, Ch. 6). The Carlitz-Uchiyama bound shows an application of A. Weil's famous results on the Riemann Hypothesis in function fields (5, 9). For a course of modular functions the connection with self-dual codes as described by Sloane in (12) is highly recommended.

Orthogonal polynomials

There are several places in coding theory where the Krawtchouk polynomials turn up. Many theorems depend heavily on results which are usually treated in a standard course on orthogonal polynomials. Perhaps the best illustration is the proof by McEliece, Rodemich, Rumsey and Welch of what is the strongest upper bound presently known for codes with prescribed length and distance (cf. 9, Ch. 17). The strongest non-existence results for perfect codes also heavily rely on properties of Krawtchouk polynomials (2, 6, 7). An elementary proof of Lloyd's theorem (cf. 4) is another nice application, in this case of recurrence relations.

Fourier theory

As an application of general ideas from Fourier theory one can treat one of the most famous theorems of coding theory, namely MacWilliams' theorem on weight enumerators. Another nice example is the so-called Mattson-Solomon polynomial which provides very elegant proofs of a number of results on weights (e.g. Goppa codes). These results can be found in the standard text books on coding theory. In our introduction we mentioned the Mariner 1969 mission. The decoding of the RM code which was used shows a good application of the Fast Fourier Transform (Chapter by E.C. Posner in 10).

Finite geometries

Although finite geometries are certainly interesting in their own right I believe that one can say that the increasing popularity of the subject is partly due to many applications in coding theory. We briefly looked at RM codes. For several other applications we refer to (3) and (9). In these references one can also find examples of "applications" in the other direction!

Linear programming

This area clearly does not lack application. We mention it because it may be nice to show how this subject can be applied "theoretically", without any actual computing. For this we refer to the upper bound mentioned above under orthogonal polynomials.

Most of this section is rather sketchy but that probably does not matter. If I have succeeded in convincing the reader that coding theory is full of elegant applications of many areas of mathematics he will surely consult one of the references and find much more than could be discussed in these pages.

References

1. E.R. Berlekamp, Algebraic Coding Theory (McGraw-Hill, New York, 1968).

2. M.R. Best, On the existence of perfect codes, Thesis, University of Amsterdam, 1981.

3. P.J. Cameron and J.H. van Lint, Graphs, Codes and Designs, Cambridge University Press, 1980.

4. D.M. Cvetkovic and J.H. van Lint, An Elementary Proof of Lloyd's Theorem, Proc. Kon, Ned. Akad. v. Wet. A 80 (1977).

5. J.H. van Lint, Coding Theory (Springer, New York, 1971).

6. J.H. van Lint (ed.), Inleiding in de Coderingstheorie (M.C. Syllabus 31, Amsterdam 1976).

7. J.H. van Lint, Introduction to Coding Theory (Springer Verlag, 1981).

8. J.H. van Lint and F.J. MacWilliams, Generalized Quadratic Residue Codes, IEEE Trans. on Inf. Theory II 24 (1978), 730-737.

9. F.J. MacWilliams and N.J.A. Sloane, The Theory of Error-correcting codes (North Holland, Amsterdam 1977).

10. H.B. Mann (ed.), Error-correcting Codes (John WIley & Sons, New YUork, 1968).

11. W.W. Peterson, Error-correcting Codes (M.I.T. Press, Cambridge Press, 1961).

12. N.J.A. Sloane, Binary Codes, Lattices and sphere-packings, in Combinatorial Surveys (P.J. Cameron, ed.), (Academic Press, London, 1977).

13. H.C.A. van Tilborg, Uniformly packed Codes, Thesis, Eindhoven University of Technology (1976).

Introductions to Coding

a. E.F. Assmus, Jr., and H.F. Mattson, Jr., Coding and Combinatorics, SIAM Review 16 (1974), 349-388.

b. E.R. Berlekamp, A Survey of Algebraic Coding Theory, CISM Courses and Lectures 28, Springer Verlag, 1970.

c. N. Levinson, Coding Theory, A Counterexample to G.H. Hardy's Conception of Applied Mathematics, Am. Math. Monthly 77 (1970), 249-258.

d. J.H. van Lint, see References (5), (6), (7)

e. N.J.A. Sloane, A Short Course on Error Correcting Codes, CISM Courses and Lectures 188, Springer Verlag 1975.

f. N.J.A. Sloane, Error-correcting Codes and Invariant Theory: New Applications of a Nineteenth Century Technique, Am. Math. Monthly 84 (1977) 82-107.

g. J. Swoboda, Codierung zur Fehlerkorrektur und Fehlererkennung, Oldenbourg Verlag 1973.

10.2 COMBINATORICS

REFLEXIONS SUR L'INTRODUCTION DE LA COMBINATOIRE DANS L'ENSEIGNEMENT MATHEMATIQUE

Nicolas C. Balacheff
Equipe de Recherche Pedagogique en Mathematique
Grenoble, France

II - Introduction

La combinatoire connait depuis deux decennies une developpement particulierement important. Dans une acception naive elle regroupe les problemes d'enumeration, de denombrement, de classement.

En s'appuyant sur la notion de configuration (disposition d'objets en respectant certaines contraintes) C. Berge (1968) enumere les principaux aspects de la combinatoire: etude d'une configuration connue; recherche de configurations inconnues; enumeration de configurations; denombrement approximatif lorsque la complexite ou la nature des configurations ne permet pas leur denombrement exact (on cherche dans ce cas des inegalites ou des ordres de grandeur); optimisation.

Dans la premiere partie (II et III) de cet expose nous envisagerons le probleme general de l'introduction de la combinatoire dans l'enseignement des mathematiques.

Dans la seconde partie (IV) nous decrirons, a partir de recherches que nous conduisons actuellement, les comportments spontanes d'eleves de 10 et 16 ans pour resoudre des problemes de denombrements.

L'introduction d'un nouvel objet d'enseignement peut s'appuyer sur trois types de considerations: ses apports culturel, formatif, utilitaire a la formation des individues (Kuntzmann 1976).

L'importance croissante des problemes combinatoires dans le monde scientifique et industriel contemporain. Le developpement de la combinatoire abstrate en tant que science (theorie des graphes, recherche operationnelle) donnent un poids evident aux arguments utilitaires et meme culturels.

Dans cet expose, j'aborderai les arguemnts du second type en essayant de montrer ce que la combinatoire peut apporter a la formation mathematique de l'individu.

II - Combinatoire et Formation Mathematique

II.1 Si l'on decoupe l'enseignement francais en trois periodes, enseignement obligatoire (6 a 16 ans), enseignement secondaire du second degre (17 a 19 ans) et universitaire (au dela de 19 ans), on peut considerer que la combinatoire est absente des deux premieres.

Quelques problemes sont abordes a l'ecole elementaire (6 a 11 ans): problemes d'enumeration lors d'activites de codage et aussi quelques problemes de parcours de graphes. L'essentiel est la presentation d'un outil: l'arbre d'enumeration.

Dans l'enseignement secondaire la combinatoire est abordee, en tant que telle, pour preparer l'etude des probabilites. En fait elle est reduite a la presentation de formules fondamentales de denombrement.

A l'universite les grandes theories combinatoires font l'objet d'un enseignment, peu dans les filieres mathematiques, mais plutot dans les filieres informatiques en 3e et 4e annee d'etude. Il existe des Diplomes d'Etudes Approfondies, mais il s'agit moins la d'enseignement que de formation a la recherche.

II.2 L'enseignement des mathematiques sur les contenues classiques - - arithmetique, algebre, geometrie, analyse - - privilegie tres largement le raisonnement deductif. Au point que pour les eleves il apparait non seulement comme un mode d'exposition, mais aussi comme une demarche privilegiee en mathematique: resoudre un probleme c'est deduire la conclusion des hypotheses. Les phases de manipulation, les approches empirico-deductives ont un statut d'illustration et les meses en garde sont nombreuses pour qu'elles ne soient pas confondues avec l'activite mathematique. Aussi pendant une periode assez longue de la formation des individus, jusque vers 14 ans ou sont abordees les premieres demonstrations, l'activite proprement mathematique semble etre reduite a l'execution d'algorithmes, a la resolution de problemes pour lesquels l'essentiel est la determination de l'operation a effectuer.

Lors d'une etude sur la resolution de problemes de denombrement par des eleves de 10-11 ans (Balacheff, 1979) nous avons pu constater les consequences d'une telle concption. Certains eleves, apres avoir cherche la solution par une enumeration, cherchent une preuve mathematique (ils l'affirment eux-meme); c'est a dire une sequence d'operations arithmetiques qui, appliquee aux donnees numeriques du probleme, restitue le resultat qu'ils ont obtenu.

En voici deux exemples:

- Pour les escaliers le Lili, combien de facons de monter un escalier de 6 marches avec des enjambees de 1 ou 2 marches?

  Tel eleve ayant trouve 12 (il y a 13 facons) justifie sa reponse a posteriori par la multiplication 6 x 2 (6 marches, 2 facons de monter).

- Pour denombrer les promenades dans un parc partage par trois murs, chacun perce de 3 portes, "il ya 9 portes et 3 murs. 9 (portes) x 3 (murs) = 27 (facons de traverser le parc)".

  Ici le resultat est correcte mais pas l'explication.

Les eleves qui ont ce comportement ne sont justement pas les mauvais eleves au sens du maitre. Au contraire, et les solutions par enumeration qu'ils ont trouvees se fondent sur des procedure systmeatiques. Ce qui est en question eci c'est le statut de l'explication et de la preuve en mathematique.

L'un des apports de la combinatoire est de permettre d'elargir la notion de preuve mathematique. D'eviter qu'elle ne soit enfermee dans le seul modele de la demonstration hypothetico-deductive tel que le formalisme mathematique l'a fixe.

Les problemes peuvent concerner aussi bien une situation recouvrant un nombre fini et petit de configurations, que la generalisation d'une proposition a une classe denombrable de configurations. Prenons l'exemple du probleme de l'enumeration des mots de p lettres sur un alphabet de q lettres sans repetition. Il peut etre aborde pour des valeurs petites de p et q par des eleves assez jeunes et sans grands moyens mathematiques (Mendelsohn 1980), des valeurs assez grandes de p et q disqualifient les procedures empiriques et conduit a l'etude approfondie de l'ensemble des solutions. Par exemple, pour $p = 3$, $q = 6$ on peut remarque qu'il y a 6 classes qui chacun contient autant de mots qu'il y en a pour le probleme $p = 2$, $q = 5$. Dans le cas general l'etude d'un processus de production des mots, le choix d'une lettre quelconque dans des ensembles qui ont de q a q-p a chaque pas une lettre de moins, conduite a la formule bien connue $q!/(q - p)!$

A chaque niveau de complexite on peut aborder la question de la preuve de la solution obtenue, c'est-a-dire entreprendre une analyse qui remet en question la certitude des resultats. La discussion des arguments permet de degager la demarche fondamentale de la resolution de probleme qui est habituellement occultee par les presentations deductives - - elaborations de conjectures, distinction entre conjecture et theoreme, critique de la preuve a l'aide de contre-exemples, analyse et decomposition des procedures de resolution.

II.3 Un aspect interessant de la combinatoire est qu'elle peut etre consideree comme un champ d'application des autres secteurs de la mathematique. En particulier s'y rencontrent l'arithmetique, l'algebre et la geometrie. On en trouve de nombreux exemples dans deux articles (Kappur 1970; Engel 1971).

III - Quelques Remarques sur l'Introduction de la Combinatoire

Bien que le nom "Combinatoire" soit utilise dans les programmes, il est assez claire que la combinatoire elle-meme n'a jamais ete un objet d'enseignement en France.

Voici, pour le mettre en evidence, les aventures de la notion d'"Arrangement" a travers quelques ouvrages scolaires francais.

Dans Papelier (1916) l'analyse combinatoire est presentee comme un paragraphe du chapitre "Formule du binome". La notion d'arrangement est definie comme "disposition que l'on peut former en placant p de ces m lettres a la suite les unes des autres, ces dispositions differant soit par l'ordre des lettres, soit par la nature des lettres qui y figurent". Il s'agit d'un ouvrage destine aux classes preparatoires aux grandes ecoles.

Dans Wicker-Boursin (TG 1972) on a la presentation inverse: l'analyse combinatoire fait l'objet d'un chapitre autonome (Ch. 2) ou les arrangments sont presentes de facon analogue a Papelier; le principal resultat de ce chapitre est la formule du binome.

Jusque vers les anness 70, la presentation des formules de denombrement se fait dans une premiere partie du chapitre consacre aux probabilites. Elles sont enoncees dans l'ordre suivant: arrangement, permutation, combinaison. (Voir, par exemple, Lebosse-Hemery (TD 1966).

Dans certains ouvrages des annees 70 le mot arrangement disparait ainsi que les references a la combinatoire. C'est par exemple le cas de Queysanne-Revuz (TCE 1971) qui presente le nombre Card (E_p,F_n) des injections de E fini dans F fini dans un paragraphe "Entiers naturels et ensembles finis" du chapitre "Entiers naturels". Dans ce meme paragraphe les nombres C sont definis et calcules et il est indique en note qu'ils sont aussi appeles coeficients binomiaux (il y a plus loin le developpement du binome dans le chapitre sur les polynomes).

Il est claire qu'il y a eu une evlution de l'objet d'enseignement "Arrangement" en fonction de l'evolution des notions scientifiques qui se trouvent en amont. Dans les annees 70 sous l'influence du mouvement de renovation des mathematiques la presentation, en terme de configuration, des formules de denombrement est abandonnee pour une presentation ensembliste plus formaliste.

La notion meme de denombrement peut avoir disparue. Ainsi dans l'ouvrage de Queysanne-Revuz, la presentation de Card (E_p, E_n) est fondee sur un raisonnement par recurrence construit sur la notion de prolongement injectif. Bien sur, cette presentation est aussi sous-tendue par une enumeration, mais celle-ci reste implicite.

Ce n'est donc pas la combinatoire qui fait l'objet d'un enseignement ni meme la partie de cette discipline consacree aux denombrements, mais quelques formules qui seront indispensables dans d'autres chapitres mathematiques.

En fait jusqu'ici les combinatoriciens n'etaient pas concernes en tant que tel. Il leur appartient de mener, avec les specialistes de l'enseignement des mathematiques, une reflexion specifique sur l'introduction de la combinatoire dans les cursus mathematiques. Il est claire qu'elle ne peut se reduire a identifier un morceau de theorie pour l'inserer dnas les programmes de mathematique. A partir des objets, concepts et relations qui constituent la theorie scientifique, il faut elaborer les objets d'enseignement qui en sont une transposition en vue de l'apprentissage. Les contraintes essentielles sur cette transposition sont:

- la conservation du sens: la signification de l'objet enseigne ne doit pas etre en desaccord avec celle de l'objet scientifiques dont il est issu.

- la prise en compte des modeles spontanes que l'eleve se constitue, en particulier sous l'influence du milieu, des connaissances concernees. Il est clair qu'avant de subir un enseignement, les eleves ont des concpetions sur ce qu'est enumerer, denombre, classer, etc.

IV - Etude de la Resolution d'un Probleme de Denombrement

Nous nous proposons d'etudier la resolution, par des eleves de 11 et 16 ans, du probleme suivant: denombrer le nombre de rectangles dans la figure ci-contre.

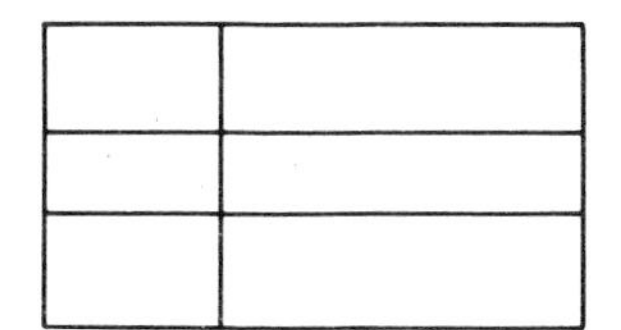

Il s'agissait pour nous d'examiner les moyens que ces eleves allaient mettre en oeuvre pou resoudre le probleme puis pour s'assurer du resultat.

IV.1 Comment resoudre ce probleme

Une premiere direction de solution consiste a se demander quels sont les objets a enumerer. Ice ce sont des rectangles, il faut alors les identifier non de facon ablue mais relativement au probleme: c'est-a-dire l'intersection de 2 bandes.

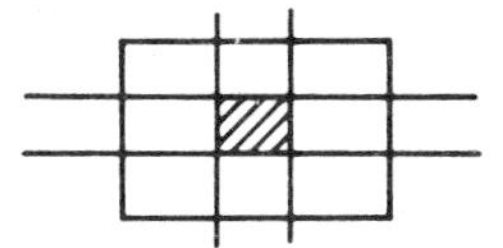

Puis reconnaitre qu'une bande est completement determinee par 2 poins sur le cote qu'elle coupe. S'il y a n points sur la longueur et m points sur la largeur, il y a C_n^2 bandes verticales et C_m^2 bandes horizontales; et donc il ya a $C_n^2 \times C_m^2$ rectangles dans une figure n,m.

Une autre direction de solution consiste a definir a priori un mode de structuration de l'ensemble des solutions qui permette de ramener le probleme a celui de l'enumeration dans des classes plus petites. Le critere le plus frequemment utilise est celui de la taille des rectangles en nombre de petits rectangles elementaires.

Une autre procedure enfin consiste a associer le denombrement a un processus de reconstruction de la figure.

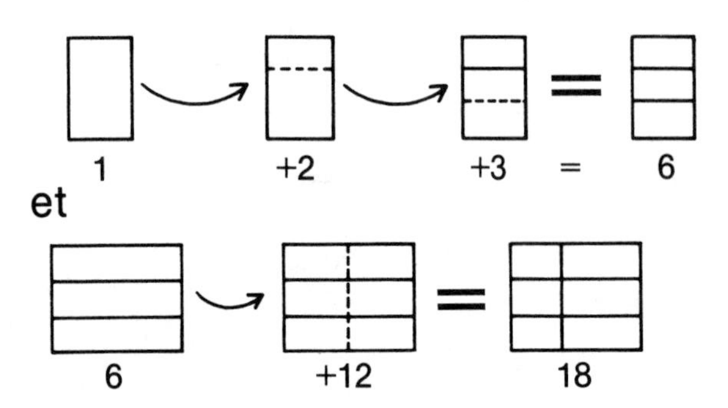

IV.2 Les observations dont nous allons rendre compte ici sont le resultat de l'analyse d'experience conduites dans le cadre de l'equipe de Recherche Pedagogique du Laboratoire d'Informatique et Mathematique Appliquee de Grenoble. (La description de ces experiences, les protocoles et leur analyse complete sont disponibles comme Rapports de Recherche de ce laboratoire.)

Pour les eleves de 11 ans (sixieme)

Les premieres tentatives pour resoudre le probleme sont d'abord empiriques, puis assez rapidement la procedure d'enumeration se stabilise. De facon implicige elle s'appuie sur une structuration a priori de l'ensemble des solutions a partir des traits perceptifs de la figure: les grands, les petits, les lignes, les colonnes, les lignes doubles, les petits doubles. L'incertitude sur le resultat de l'enumeration conduit les eleves a deux types de comportements:

- la recherche de nouveaux rectangles. Mais cette recherche reste empirique.
- la verification soit en reiterant, parfois plusieurs fois de suite la procedure, soit en utilisant un autre mode d'enumeration.

Dans ces deux cas les eleves reviennent sur la procedure utilisee, mais sans l'analyser comme articulation d'operations elementaires, ni expliciter la structuration de l'ensemble des solutions sur laquelle elle s'appuie. Deux eleves font des remarques qui montrent qu'ils percoivent qu'ils utilisent une procedure systematiqe, mais leur remarque reste sans consequence.

Dans quelques cas, ils reviennent sur les objets a enumerer mais sans chercher a les identifier. En tous cas, ils n'en font pas une analyse qui pourrait servir de point de depart a une solution.

Dans leur tentative d'explication du resultat qu'ils ont obtenu, il est clair qu'ils cherchent a indiquer comment ils ont resolu le probleme. Mais pour les raisons que nous avons donnees plus haut, ils ne parviennent pas a decrire la procedure qu'ils ont utilisee, et lui substitue l'enumeration des rectangles trouves en faisant apparaitre sa structure. Voici un exemple:

(B6):

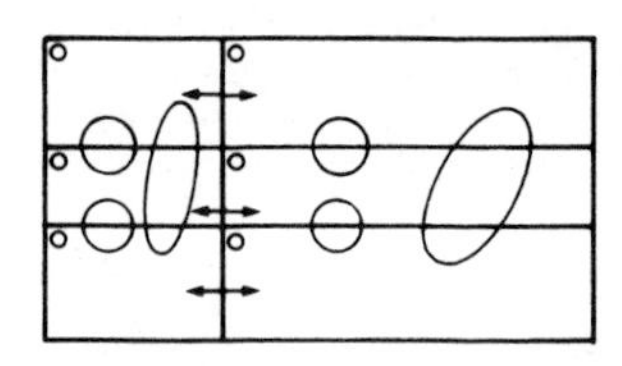

Petits ronds 1,2,3,4,5,6
Ronds moyens 1,2,3,4
Grands ronds 1,2
Fleches 1,2,3
Grand rectangle 1
6 + 4 + 2 + 3 + 1 = 16

Le nombre de rectangles qu'il y a dans la forme geometrique est 16.

Pour les eleves de 16 ans (troisieme)
L'approche est empirique, puis apparait une procedure qui s'appuie clairement sur une structuration a priori de l'ensemble des rectangles solutions. Les eleves explicitent leur souci d'une procedure "logique", c'est-a-dire systematique, allant du plus petit au plus grand; ou pour deux d'entre eux, que chaque rectangle enconce soit obtenu par decomposition ou reconstruction a partir des rectagles deja obtenus.

La description qu'ils donnent de la solution de ce probleme n'est pas l'enonce statistique de l'enumeration des rectangles, eventuellement structuree en classes, mais la description du processus qui permet de les obtenir. Les operations sont indiquees comme realisees par un operateur:

- "Maintenant a partir de ces 6 rectangles de base je compose une seconde serie formee de 2 rectangles de base" (3eA).
- "Au rectangle (246) on enleve le rectangle 2, cela donne le rectangle (46)" (3eB).

Neanmoins c'est bien le processus effectivement mis en oeuvre qui est decrit, et non pas la procedure d'enumeration en tant qu'objet. Dans le meme sens nous remarquons que l'incertitude sur le resultat obtenu conduit les eleves soit a reprendre des recherched empiriques, soit a faire des verifications. Mais il n'y a pas de remise en question et donc pas d'analyse de la procedure en tant qu'articulation d'operations elementaires.

Dans cette observation le case de l'eleve Bertrand est particulierement interessant. Apres une approche empirique du probleme il trouve un resultat (16) confirme par une verification. Il se demande alors comment il va "resoudre ca", c'est-a-dire trouver une preuve. Il essaie d'abord d'y parvenir par reconstruction, puis par decomposition de la figure. Cette approche parait particulierement difficile. Il tente alors de trouver une regle en examinant les figures, mais "ca fournit rien". Lorsque le resultat (18) est obtenu par d'autres eleves il est accepte, et Bertrand cherche a l'etablir en le "demontrant par des chiffres", puis en essayant d'utiliser un arbre d'enumeration. Il s'agit dans les deux cas d'investir des connaissances mathematiques deja acuises. Il echoue a utiliser l'arbre, puis cherche a donner un sens a l'expression 6 x 3 = 18; 18 est le resultat, 6 correspond au nombre de petits rectangles, mais 3 n'a pas de signification.

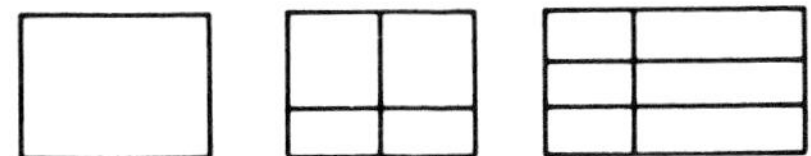

1	9	16	rectangles
4	12	17	segments

Il abandonne cette voie pour essayer de generaliser la situation. Il examine une autre configuration de 6 petits rectangles pour laquelle il trouve 21 solutions, mais qui est sans lien avec le probleme donne. Finalement il revient a l'enumeration a l'aide d'un arbre et l'associe a l'expression 6 x 3 = 18 de la facon suivante:

"On trouve en tout 18. C'est 3 fois 6, donc dans chaque machin on doit faire 3 rectangles differents".

Il dessine alors 6 graphes en etiquettant chaque racine par le nom donne a un rectangle elementaire.

Clairment cet eleve a cherche a elaborer une preuve qu'il reconnaisse comme etant une preuve mathematique. Lorsque son camarade lui propose de decrire l'enumeration, il refuse car:

"C'est pas une explication ecrite. C'est bon quand on fait a la main. C'est exactement ca qu'on fait".

References

N. Balacheff (1979) Quelques aspects du sens donne a l'explication mathematique par des eleves de 10-11 ans. Seminaire de Recherche Pedagogique, Grenoble USMG-LA7.

C. Berge (1968) Principes de combinatoire. DUNOD.

A. Engel (1971) Geometrical activities for the upper Elementary school. Educational Studies in Mathemtics 3 (1971) 353-394.

J. N. Kappur (1970) Combinatorial analysis and school mathematics. Education Studies in Mathematics 3 (1970) 111-127.

J. Kuntsmann (1976) Evolution et etude critique des enseignements. Mathematiques. CEDIC.

P. Mendelshon (1980) Pensee naturelle et logique combinatoire. Laboratoire de psychologie experimentale. Grenoble.

Ouvrages Scolaires

C. Lebosse, C. Hemery, P. Faure. Geometrie et elements de probabilites, classe terminale D. Fernand Nathan 1967.

G. Papelier. Precis d'algebre, d'analyse et de trigonometrie. Vuibert 1916.

M. Queysanne, A. Revuz. Mathematiques T1: Nombres, Probabilites. Terminale CE, Fernand Nathan 1971.

G. Wicker, J.L. Boursin. Mathematiques. Terminale G, Bordas Dunod 1972.

THE EDUCATIONAL VALUE OF THE HUNGARIAN "MAGIC CUBE"

David Singmaster
Polytechnic of the South Bank
London, England

"An example in the hand is worth two in the mind."

Ladies and Gentlemen, I feel a bit like Tom Lehrer's Old Dope Peddler. I am here today to peddle an insidious mathematical addiction which has already wasted the creative energies of tens of thousands of mathematicians, computer scientists, physicists, engineers, etc. and millions of students. My justification in spreading this addiction among you is that it is probably the most educational toy ever invented.

I am referring to the Magic Cube invented some six years ago by Professor Erno Rubik of Budapest. Production began about three years ago and since then the addiction has spread out from Hungary to the UK and Europe and now the world!

For those who have not seen a cube, it appears to be a 3 x 3 x 3 array of little cubes (or cubelets) which is most ingeniously held together so that each 3 x 3 face can be turned about its center piece. Further, once one face is turned by an integral multiple of 90 degrees, then any other face can be turned and so on. When purchased, the faces of the small cubes are colored with six colors so that the whole cube has six solid colored faces. However, after a few turns, the colors move about and become quite randomized. Indeed, only 3 turns will usually lose a novice and 5 turns will lose all but the most expert cubist.

The Basic Mechanical Problem is: how does the thing work? I will say no more than that there are no magnets and no rubber bands and that there are not 26 (or 13) Hungarian midgets inside (nor even one 26-armed one!). The question which should interest us is the Basic Mathematical Problem: how do you restore the cube back to its pristine state of monochromatic faces (variously called START, GO, Square One or Cube One)? This can take anywhere between a day and a year to solve, depending on your background and how hard you work. All solvers manage to construct an algorithm which works, but few manage to characterize the set of possible patterns, which is necessary to show that the algorithm is complete.

I have just mentioned a mathematical point which the cube exemplifies, and this brings me to the theme of this lecture - what mathematics does the cube teach us? The answer to this depends on the level of the student. There is material for all levels of student from about age 14 on and it can fascinate even much younger students.

I shall outline the mathematical topics which the cube develops under four headings and I will illustrate the mathematical idea sufficiently for you to see how to develop an algorithm. The four headings are: three-dimensional geometry; general mathematical thought; group theory; combinatories. As you can see, this is a hybrid topic for a combinatories session.

Three-Dimensional Geometry

It is reported that Rubik invented the cube because he wanted something to improve students' three-dimensional abilities. He began as a sculptor, then became an architect and now teaches 3-dimensional design. On his patent, he describes himself as an architectural engineer. With the decline of solid geometry in the schools, students in higher education are often deficient in spatial skills. The Magic Cube certainly develops such skills - try convincing yourself that you can move any 4 of the 8 corners into one face. Further, the basic properties, such as there being 6 faces, 12 edges and 8 corners, soon become second nature. The symmetries, the direct symmetries and even the even direct symmetries of the cube all arise naturally.

General Mathematical Thought

Under this, I include a number of ideas.

First, as already mentioned, is the need to prove a characterization of the achievable positions in order to show that the algorithm is complete.

Second, is that there is a mathematical approach to a real problem. An example, like the cube, where the mathematics 'discovers itself' is valuable in conveying to students the way in which mathematics is present in everyday situatios.

Third, is the point that this is not a completed and dead body of knowledge. There is opportunity for students to make original observations and to raise interesting problems. A great deal of cubism has been done precisely as the result of casual questions.

A fourth point is that it is possible to construct a large number of pretty patterns which are aesthetically pleasing, even fascinating. This ties in strongly with the symmetries of the cube.

The main point under this heading is the necessity for and the value of a good notation. Everyone who works on the cube soon discovers the need for an adequate notation. Further, there are several notations based on different coordinate systems - and some are more useful for some problems than are others. We first observe that turning a face does not move its center. The six face centers are physically attached to the hidden center cubelet. Hence there is only one possible START position and we can always recognize which face should be red by its red center piece.

John Conway used a notation based on the colors of his original cube: Red - Right; Blue - Bottom; Orange - Out; Yellow - Yonder; Uncolored (i.e. white) - Up; Leaf Green - Left. However this system became awkward when he saw a second cube, which had the colors differently arranged. (Combinatorial Exercise - how many ways can you put six colors on the faces of a cube?)

I independently devised a less Conwayesque notation which seems to be becoming standard. The six faces are denoted by the initials of six position names: Right, Left, Front, Back, Up, Down. The faces are identified by their centers within each process, but we may then change the position of the cube between processes. The initials are used to denote positions and the pieces which should be there, as in Figure 1.

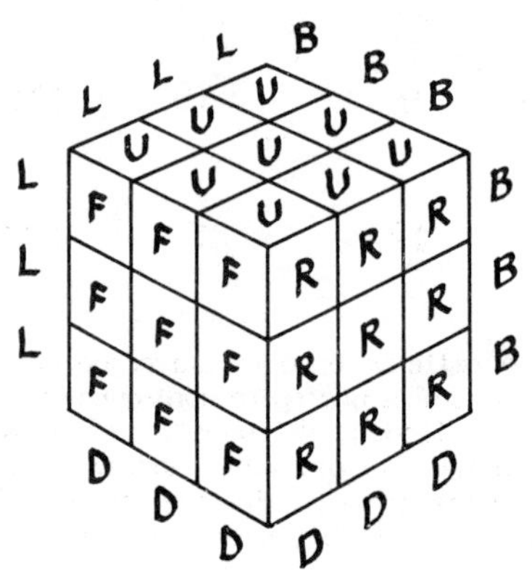

Figure 1.
E.g. UR (or RU) is the edge position between the U and R faces and also denotes the piece that should be there (i.e. that was there at START). Likewise URF (or RFU or FUR) denotes the corner of the U, R, F faces. (Conventionally, we always take the faces in clockwise order.)

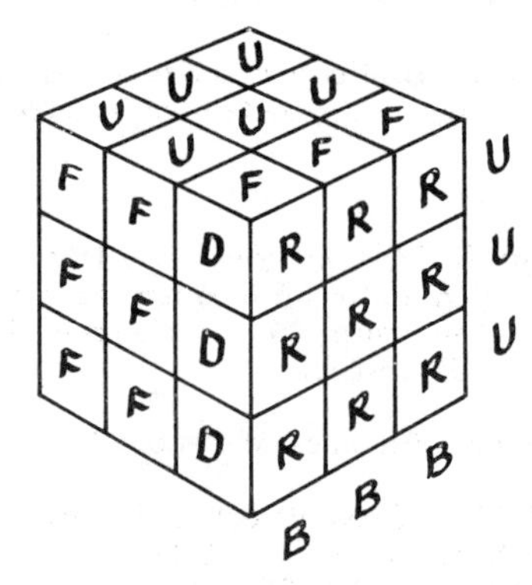

Figure 2.

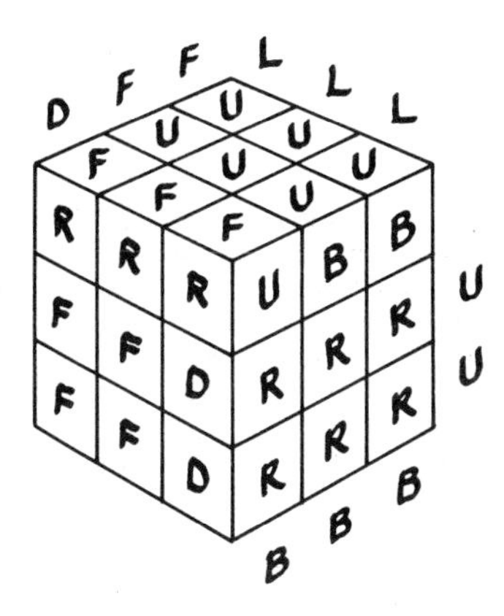

Figure 3.
In Figure 3, the UR piece is at BR and the URF piece is at RFU. (Here the distinction between URF and RFU is significant.)

The letters will also denote a 90 degree turn of the corresponding face, clockwise as seen from outside the cube. R^{-1} or R' denotes the 90 degree counter- (or anti-) clockwise turn and R^2 denotes the 180 degree turn. E.g. the move R is shown in Figure 2. A sequence of moves such as RBLF means to apply R first, then B, then L, then F. E.g. RU is shown in Figure 3. This system is called the center coordinate system.

The corner coordinate system is similar but based on fixing coordinates with respect to some corner rather than the centers. E.g. in Figure 4, it is easier to say the centers have moved with respect to the corners than vice versa.

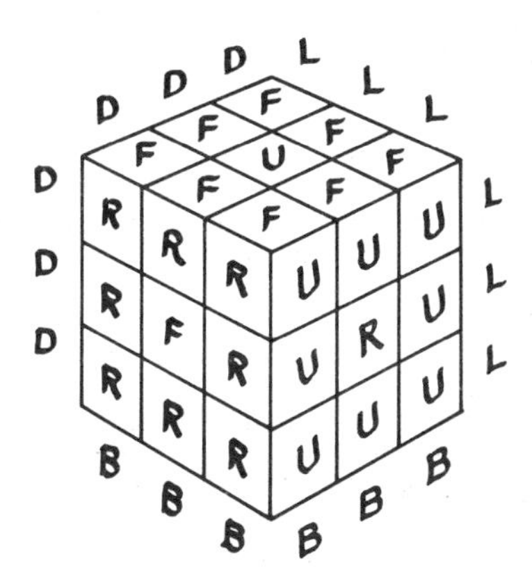

Figure 4.

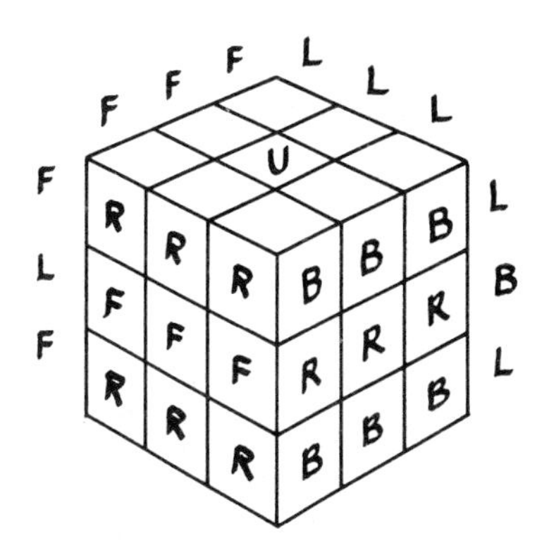

Figure 5.
If we move Y and D^{-1}, we get Figure 5. We can think of this as the middle 'slice' being moved with respect to U and D layers and we can think of the slice as being moved with respect to a system of fixed coordinates in space. Then the coordinate axes are not fixed to either the centers or a corner so it is easy to lose track of them. I call this the spatial coordinate system.

Though the spatial and corner systems are not as easy to see and to use, they do have uses and the fact that there are different systems with different uses is a valuable educational lesson.

Group Therapy

The majority of the specific ideas which the cube develops fall into the general topic of group theory but much of this is of an elementary nature which overlaps into combinatories.

The first concepts which I shall discuss are the understanding of permutations and their representation. The distinction between permutations acting on symbols (i.e. as functions) and acting on contents of positions (i.e. as motions of an object) must be clearly made here - as it is not usually done in elementary courses.

The notion of cycle structure is basically standard, but we need an extra feature to deal with the orientations of pieces. We write (UR, FU, . . .) or UR→FU→. . . to show that the U face of the piece at UR goes to the F face at the position FU. It can happen that the positions in a cycle repeat but with different orientation. E.g. we might have
UR→FU→RU→UF→UR→. . . We abbreviate this to $(UR, FU)_+$, where the $_+$ shows that the square of this cycle twists each of the pieces 120 degrees clockwise, i.e. the square has URF→RFU→FUR→URF→. . . We refer to corners being twisted + or - and to edges as being flipped and we call the above cycles twisted. As a specific example, we have:
RU = (FR,UF,UL,UB,UR,BR,DR) $(URF)_+$ (BRU,DRB, FRD,UFL,ULB)$_-$.

Once you have a notation for recording the effects of a process, you can explore the group of the cube, or, betterm some of its simpler subgroups. All of the basic notions of group theory now start to appear:
$(FR)^{-1} = R^{-1} F^{-1}$, order of an element, groups of permutations, subgroups, cyclic subgroups, order of a group, subgroup generated by a set, subgroup leaving

somethings fixed, etc., etc. (However, when P is a lengthy sequence of moves, then $PP^{-1} = I$ may be true theoretically, but is only 90% true in practice!)

Of course, the Basic Mathematical Problem is what absorbs attention at first. Anyone with determination will find some processes which can be used to restore the cube and it is a remarkable phenomenon that everyone seems to find a different combination of processes and strategy. Several further basic ideas of group theory arise naturally in trying to do this. One idea is that of the conjugate of a process, which appears very visually as the same process but acting on different positions. Thus the cycle structure of the conjugate is clearly seen to be the same as the cycle structure of the original. Another basic idea is that of the commutator, e.g. $FRF^{-1}R^{-1}$ - that this is not the identity is a visual illustration of noncommutativity. A third concept is that the 3-cycles or the pairs of 2-cycles generate the alternating group and the 2-cycles generate the symmetric group. A fourth concept is that the group is multiply transitive on edges and on corners. Using all these ideas, one can see that the following two processes are sufficient to restore the cube:

$$P_1\,(R,F) = F^2R^2F^2R^2F^2R^2 = (FU,FD)\,(RU,RD),$$

$$P_2\,(R,F) = FRF^{-1}R^{-1} = (FLU,FUR)_+(FRD,DRB)_-(FU,FR,DR).$$

The notion of even permutation has already been mentioned and it arises naturally because each face turn is a 4-cycle of edges and a 4-cycle of corners, hence is an even permutation of the set of all 20 pieces. Consequently, we cannot achieve all possible patterns of pieces by turning faces. This leads back to the problem of characterizing the group of achievable patterns. It is easiest to first consider the 'constructible' group of patterns obtainable by taking the cube apart. Then the 8 corners can be permuted in 8! ways and each corner has 3 orientations, while the 12 edges can be permuted in 12! ways and each edge has two orientations, giving

$$8!\ 12!\ 3^8\ 2^{12} = 5\ 19024\ 03929\ 38782\ 72000 = 2^{29}\cdot 3^{15}\cdot 5^3\cdot 7^2\cdot 11$$

constructible positions. However we already know that we must divide by 2 since an achievable permutation must always be even on the 20 edges and corners. The rest of the characterization is less easy to see - it depends on choosing some fixed orientation for each piece and each position. (There is a reasonably natural way to do this for edges, but no natural way for corners. Any arbitrary choice will serve here, but various systematic choices yield further insights.) We then compare the orientation of each piece with the orientation of the position it is in. It is convenient to choose the orientations so that every piece has the orientation of its position when the cube is at START. Examination shows that, no matter what the present orientation of the pieces in a face, a face turn will flip the orientation of an even number of edge pieces and will twise the orientations of corner pieces by a total amount which is an integral multiple of 360 degrees. E.g. we can flip two edges, but we cannot flip just one edge. We can twist two corners in opposite directions or three corners in the same direction, but we cannot twist just one corner nor two in the same direction.

(Note that mod 2 and mod 3 arithmetic arises here.) These restrictions introduce divisors of 2 and 3 yielding:

$$N = (8!\ 12!)/2 \times 3^8/3 \times 2^{12}/2 = 43252\ 00327\ 44898\ 56000 =$$

$2^{27}\cdot 3^{14}\cdot 5^3\cdot 7^2\cdot 11$ achievable positions. Thus the group of achievable positions has index 12 in the group of constructible positions. Conway describes this as meaning that if you take apart the cube and reassemble it at random, then you have only a 1 in 12 chance of being able to get back to START. (Of course your probability of getting back to START may be so small that it doesn't matter if you are in the group of achievable positions.)

The above value N seems large, but we can make it larger. If we look closely, we see that the center pieces do move - though they stay in place, they can rotate. This can be made visible by putting pictures on the faces. We get $4^6 = 4096$ times as many constructible positions but the total number of 90 degree center turns has the same parity as the permutation of corners or of edges. Thus we only get a factor of $4^6/2 = 2048$ and there are $2048\,N = 885\ 80102\ 70615\ 52250\ 88000 = 2^{38}\cdot 3^{14}\cdot 5^3\cdot 7^2\cdot 11$ positions in the 'supergroup'. At 10^6 patterns per second, it would take about 3 billion ($= 3 \times 10^9$) years to get through this number. This is about 1/3 of the current estimated age of the universe. It seems unlikely that all the patterns of even the achievable group will ever be examined, even though it seems that hundreds of computers are at work on it. This is a fine illustration of the 'combinatorial explosion' - a small number of objects can lead to a large number of patterns.

A great deal of more advanced group theory arises very naturally as one studies the cube. Some group theorists are now using the cube as a canonical example in teaching group theory. I shall mention a few of the topics which arise. Direct products arise clearly since edges and corners are separately permuted, but the rather obscure wreath product arises in a most concrete manner - the constructible group is a direct product of the wreath products Z_3 wr S_8 and Z_2 wr S_{12}. Transitivity, stabilizers, orbits, blocks of imprimitivity, presentations, representations and Jordan's theorem on A_n all arise and one can use the cube as a concrete example for such ideas as the center of a group, the commutator subgroup, composition series, solubility, Sylow theory, etc. Groups generated by elements of order 2 arise, e.g. F^2, R^2 which is the group of symmetries, D_6, of a regular hexagon, as do groups generated by two elements. I have recently analyzed $\langle F^2, R^2 \rangle$ and verified the assertion of several group theorists that the action of this group, on the six corners that it affects, is only S_5. This is a degree 6 representation of S_5, identical to the action of PGL (2,5) on the 6 point projective line over Z_5.

Combinatorics

We have already seen a good deal of combinatorial thinking interwoven with the above material on group theory. Basic enumerative techniques have been used several times and the notions of algorithms, permutation and symmetry are all basic in combinatorics. I will now mention some topics which are still interwoven with group theory but are more combinatorial.

The major topic is the problem of finding a best possible algorithm for restoring the cube to START (or equivalently, to any other desired destination). This is a minimax problem - we want to minimize the maximum number of moves required. Since R^2 takes about the same amount of work as R or R^{-1}, cubists have generally counted R^2 as a single move. However, one may prefer to count R^2 as two moves and then we shall refer to the 'length' of a process. At present, the best known algorithm is due to Morwen Thistlethwaite and requires at most 52 moves. This number has been decreasing steadily at the rate of about 2 moves per week, but Thistlethwaite says his present strategy probably cannot be brought below 45 moves, even with exhaustive calculation. So any further major improvement will require a new strategy.

A lower bound for the minimax is obtained by seeing that there are at most $1 + 18 + 18 \times 15 + 18 \times 15^2 + \ldots + 18 \times 15^{n-1}$ positions after n moves. Setting this equal to N yields n = 16.6, so some positions require at least 17 moves. Attempts to eliminate redundancy in the positions counted, e.g. from $FBF = F^2B$, have only managed to increase this lower bound to 18. I have rashly conjectured that the minimax, or the length of 'God's algorithm', is actually about 20 moves!

The minimax problem also can be considered on various subgroups. Only the simplest subgroups have been examined so far. One which could be done is the U group, i.e. the group of patterns in which only the U face is changed. There are only $(4!\ 4!)/2 \times 3^4/3 \times 2^4/2 = 62208 = 2^8 3^5$ patterns in this group and one can apply the symmetries of the square to get a much smaller set of 'distinct' patters. (This is a natural problem to use Brunside's lemma on, but I don't know that it has been done.) There is a conjecture that each pattern in the U group requires at most 18 moves, but I think this might be reducible to about 12.

A 2 x 3 x 3 Magic Domino is being produced in small quantities and Rubik's patent describes a 2 x 2 x 2 cube which has not yet been produced. The group of the 2 x 2 x 2 cube has not be minimaxed yet and this would be of interest.

Two other combinatorial problems relating to permutations have arisen. What is the maximum possible order of an element in the group? This is know to be 1260. What is the number of elements of a given cycle structure or of a given order? This can be done fairly straightforwardly, but the calculations are tedious and have not been done, except that I have found the number of elements of order 3. Thus we do not yet know the average order or the most popular cycle structure in this group.

Perhaps the most important point about the cube is that it is a truly fascinating puzzle and students cannot put it down. Any serious student will informally assimilate a vast amount of basic group and combinatorial theory by studying (playing ?) with the cube and will thus be prepared to appreciate and understand the formal development of these ideas.

In closing let me cite a few anecdotes to illustrate the popularity of the cube. One family had to buy a cube for each member of the family "in order to preserve domestic tranquility". After I lectured at Nottingham, one student wrote "I used to be a full time Ph.D. student in computer science until I discovered the Magic Cube". Finally, what is probably the first medical syndrome due to mathematics is due to excessive cubing. Dame Kathleen Deledicq and Touchard (4) are primarily concerned with the Basic Mathematical Problem but go on to many other things. I have seen about 25 algorithms but I list only those which are easily available:

1. Angevine, J. Solution for the Magic Cube Puzzle. 11 pp, 1979 (Available for $2 and SAE from Logical Games, 4509 Martinwood Drive, Haymarket, Virginia, 22069, USA.)

2. Berklekamp, E.R., J.H. Conway & R.K. Guy. Winning Ways. (In press ?)

3. Cairns, C. & D. Griffiths. Teach Yourself Cube-Bashing. 6 pp, Sep 1979. (Available from the authors, ICL Dataskil, 118-128 London St., Reading, RG1 4SU, Berkshire, UK.)

4. Deledicq, A. & J. -B. Touchard. Le cube Hongrois - Mode d'emploi. 73 pp, 1980. (Available from IREM, Paris VII, T. 56/55, 3eme etage, 2 Place Jussieu, 75005, Paris, France.)

5. Howlett, G.S. "Magic Cube" - A Guide to the Solution. 2 pp, 1979. (Available from Pentangle, Over Wallop, Stockbridge, Hampshire S020 8Ht, UK.)

6. Neumann, P.M. The Group Theory of the Hungarian Magic Cube. Chapter 19, pp 299-307 of: Neumann, et al. "Groups and Geometry". Lecture Notes, April 1980. (Available from Mathematical Institute, 24-29 St. Giles, Oxford, OX1 3LB, UK, for 2.00 plus 1.22 UK postage or 0.70 overseas postage.)

7. Ollerenshaw, K. The Hungarian Magic Cube. Bull. Inst. Math. Appl. 16 (Apr 1980) 86-92.

8. Singmaster, D. Notes on Rubik's 'Magic Cube'. Fifth ed.m 75 pp, Aug 1980. (Available from the author, Polytechnic of the South Bank, London, SE1 OAA, UK for 1.50 pounds (= $4.00) in the UK & Europe and 2.00 pounds (= $5.00) elsewhere. Also available from Logical Games - address under (1.). A US edition may appear soon.)

9. Taylor, D.E. The Magic Cube. 18 pp ms, Nov 1978.

10. Taylor, D.E. Secrets of the Rubik Cube - A Guided Tour of a Permutation Group. In preparation.

Ollerenshaw has been afflicted with 'cubist's thumb' - a form of tendonitis requiring minor but delicate surgery for its relief and otherwise mostly due to excessive finger snapping in disco freaks.

I hope this lecture indicates some of the many educational opportunities which this amazing toy presents. (If nothing else, you can give one to the bright disruptive student and it will keep him quiet for days!)

Postscript

The above lecture was given at the ICME in Berkeley on Monday morning, 11 August 1980. A considerable amount of new material came to me during the Congress. In the two weeks of 7 to 21 August, I received 27 typed pages and a 73 page booklet (referring to a 20 minute videofilm) (5.), I have seen another booklet and a computer program and I have heard about another booklet and another program, all on the Magic Cube! Tamas Varga showed me a number of cubes on which he has simplified the color patterns - e.g. one with just two colors, one with only the corner pieces colored (hence the same as the 2 x 2 x 2 cube), etc. He and his colleague Julienne Szendrei report that these are fascinating for younger children down to ages 5 or 6. I have also seen a 2 x 2 x 2 cube designed and made by Dan Sleator at Stanford.

I found cubes for sale at prices from $7.76 to $13.95 (+ 6% tax). Toys R Us was the cheapest and I am told that K Mart also has them at a cheap price.

REFERENCES

The most extensive material presently available is the fifth edition of my "Notes on Rubik's 'Magic Cube'" (8.). D.E. Taylor has written a technical group theoretic paper (9.) and is preparing a paper (10.) using the cube as a canonical example for studying group theory. Peter Neumann's chapter (6.) deals with the basic group theory. Berlekamp, Conway and Guy (2) and Deledicq and Touchard (4) are primarily concerned with the Basic Mathematical Problem but go on to many other things. I have seen about 25 algorithms but I list only those which are easily available.

1. Angevine, J. Solution for the Magic Cube Puzzle. 11 pp, 1979. (Available for $2 and SAE from Logical Games, 4509 Martinwood Drive, Haymarket, Virginia 22069, USA.)

2. Berlekamp, E.R., J.H. Conway & R.K. Guy. Winning Ways. (In Press).

3. Cairns, C. & D. Griffiths. Teach Yourself Cube-Bashing. 6 pp, Sep 1979. (Available from the authors, ICL Dataskil, 118-128 London St., Reading, RG1 4SU, Berkshire, UK.)

4. Deledicq, A. & J.-B. Touchard. Le cube Hongrois - Mode d'emploi. 73 pp, 1980. (Available from IREM, Paris VII, T. 56/55, 3eme etage, 2 Place Jussieu, 75005 Paris, France

5. Howlett, G.S. "Magic Cube" - A Guide to the Solution. 2 pp, 1979. (Available from Pentangle, Over Wallop, Stockbridge, Hampshire, S020 8HT, UK.)

6. Neumann, P.M. The Group Theory of the Hungarian Magic Cube. Chapter 19, pp 299-307 of: Neummann, et al. "Groups and Geometry". Lecture Notes, April 1980. (Available from Mathematical Institute, 24-29 St. Giles, Oxford, XO1 2LB, UK.)

7. Ollerenshaw, K. The Hungarian Magic Cube. Bull. Inst. Math. Appl. 16 (Apr 1980) 86-92.

8. Singmaster, D. Notes on Rubik's 'Magic Cube'. Fifth ed., 75 pp, Aug 1980. (Available from the author, Polytechnic of the South Bank, London, SE1 OAA. Also available from Logical Games - address under (1).

9. Taylor, D.E. The Magic Cube. 18 pp ms, Nov 1978.

10. Taylor, D.E. Secrets of the Rubik Cube - A Guided Tour of a Permutation Group. In preparation.

10.3 THE IMPACT OF ALGORITHMS ON MATHEMATICS TEACHING

ALGORITHMS

Arthur Engel
University of Frankfurt
Frankfurt, Federal Republic of Germany

Introduction

Because of the enormous proliferation of computers of all sizes the designing of algorithms has become an essential skill. It is a nontrivial skill and so it presents a formidable challenge to mathematics educators. If we do not succeed in integrating algorithmics into mathematics, our subject will lose much of its prestige and importance.

Our introduction to algorithmics consists of three lectures. The first lecture on numerical algorithms looks at three topics of classical school mathematics from an algorithmic point of view: computing functions, solving equations, and extrapolation. The aim of the lectures is to show that algorithmics has become a mature discipline, which is teachable. All the algorithms of this lecture require no more than a nonprogrammable pocket calculator.

The second lecture deals with seminumerical algorithms. Its aim is to show the deep influence of the algorithmic approach on three topics: number theory, probability, and statistics. Some of these topics become radically simplified.

The third lecture is devoted to combinatorial algorithms. Combinatorial algorithms are more interesting and also more important for applications. The major algorithmic progress of the last generation is in the area of combinatorial computing. Combinatorial problems show clearly the power and limitations of the computer. But content and methodology of combinatorial algorithms do not fit into traditional school mathematics. They are huge, challenging, and time consuming. They require much computing power and a language like Pascal (with recursion). Due to space limitations, the third lecture will be published elsewhere.

First Lecture: Numerical Algorithms

Algorithmics deals with the design and analysis of algorithms. Algorithms are abstract machines which solve problem classes. A programming language is a tool kit containing a small stock of standardized parts, from which you can fabricate algorithms. A computer is a gadget which executes algorithms.

I use Dijkstra's (1976) IF and DO control structures:

$\underline{\text{if}}\ B_1 \rightarrow S_1 \,[]\, B_2 \rightarrow S_2 \,[]\, \ldots \,[]\, B_n \rightarrow S_n\ \underline{\text{fi}}$

$\underline{\text{do}}\ B_1 \rightarrow S_1 \,[]\, B_2 \rightarrow S_2 \,[]\, \ldots \,[]\, B_n \rightarrow S_n\ \underline{\text{od}}$

where the B_i are Boolean variables (the guards) and the S_i are unordered statement lists. IF operates by executing exactly one S_i whose guard is true. DO operates by repeatedly finding a true guard B_i and executing its S_i until none of the B_i are true.

Algorithmics was an art. But now it is becoming a science. What makes a body of knowledge a science is a collection of coherent results and a small stock of paradigms (fundamental ideas and methods). You teach a science by making the student familiar with the pivotal results having many applications and with the paradigms.

If you teach geometry you do not prove a random sample of theorems. You select those theorems from which you can dominate the subject. And you teach some paradigms. The dominant paradigm of Euclidean geometry was the triangular paradigm. All theorems, big and small, are proved by identifying chains of congruent or similar triangles. The dominant paradigm of transformattion geometry is symmetry. All theorems, big and small, are proved by symmetry (isometry and similarity maps).

It is hard to change the paradigms you once learned. Take the rise and fall of the flowcharts. Once they were considered essential. Later they were called useful, still later useless, and today they are considered harmful. But it takes some effort to get rid of the flowchart paradigm if you were raised on it.

Most paradigms of algorithmic design are still either too general or too specific. I will concentrate on powerful paradigms that are specific enough to be used for solving many problems. In selecting the examples no entertainment was intended. All algorithms are important on their own, or they are building blocks for other important algorithms. The examples of the first lecture can be handled by a pocket calculator. They illustrate the paradigms of numerical mathematics. And they treat two related topics: computing functions and solving equations.

The dominant paradigm of algorithm design is the technique of structured programming, which is essential for constructing large programs. It plays no role in the small scale numerical examples of school mathematics. Instead we will stress the idea of invariance. It is a fundamental idea in all of mathematics. It is indispensible in analyzing processes, especially in designing and analyzing algorithms and games. In algorithms you start in a state S. You try to reach a terminal state T by a sequence of legal steps (moves, transformations) which leave some relations invariant.

Computing Functions

Example #1. Design of a square root procedure

For given $a > 1$, I want to find $z > 0$, such that

(1) $z \cdot z = a$

An important technique is to replace an unknown constant z by variables which are pushed toward z. So we replace z by the variables x and y such that

(2) $x > y$ (3) $xy = a$.

These relations are easy to establish initially by setting $x = a$, $y = 1$. Now I move toward my goal (1) by keeping (2) and (3) invariant. That is I must find a step which changes x,y so that the difference $x - y$ decreases but (2) and (3) remain valid. This would be a step toward (1). Now the step $(x,y) \leftarrow (x',y')$, defined by

$$a = x.y = ((x+y)/2)\cdot((2xy)/(x+y)) = x'.y'$$

leaves obviously (3) invariant, and also (2) since

$$0 < x' - y' = \frac{x+y}{2} - \frac{2xy}{x+y} = \frac{(x-y)^2}{2(x+y)} = \frac{x-y}{2} \cdot \frac{x-y}{x+y} < \frac{x-y}{2}$$

So one step reduces $x - y$ at least by the factor 1/2.

Do we gain just one bit per step?

No! this is a superfast procedure. Suppose $0 < x - y < 2^{-n}$. Then for the next difference $x' - y'$ we have

$$0 < x' - y' < 2^{-2n}/(4x).$$

One step about doubles the number of correct digits. So we have the $\sqrt{\ }$-procedure in Fig. 1 with the invariant relations in braces.

$x, y \leftarrow a, 1$ $[a > 1, xy = a]$

do $x > y \Rightarrow x, y \leftarrow \frac{x+y}{2}, \frac{2xy}{x+y}$ **od**

print x $[x \cdot x = a]$

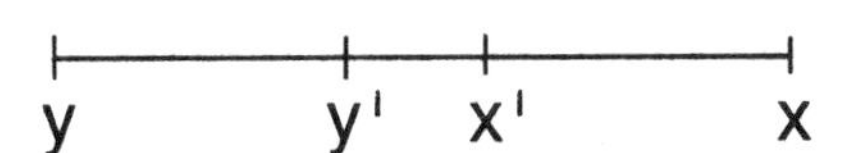

Fig. 1 & Fig. 2

If you know that the harmonic mean is less than the arithmetic mean, then Fig. 2 shows without computation that $x' - y' < (x - y)/2$.
In Fig. 1 we can simplify using $xy = a$, and replace the concurrent assignment by sequential assignments
$x \leftarrow (x + a/x)/2;\ y \leftarrow a/x$.

Then we get Fig. 3. But now the correctness proof is not so immediate. We can even eliminate the variable y by using $x \leftarrow (x + a/x)/2$. But then we do not even have a good stopping rule.

$$x, y \leftarrow a, 1$$

$$\underline{\text{do}}\ x > y \Rightarrow x \leftarrow \frac{x+y}{2};\ y \leftarrow a/x\ \underline{\text{od}}$$

$$\underline{\text{print}}\ x$$

Fig. 3

Example #2. An important algorithmic lemma

The tiny algorithm in Fig. 4 generates the sequence

$$x_0 > 0,\ x_1 = \sqrt{x_0},\ x_2 = \sqrt{x_1},\ \ldots$$

This sequence converges to 1 with rate 1/2. Indeed

$$\sqrt{x} - 1 = \frac{x-1}{\sqrt{x}+1} \sim \frac{x-1}{2} \text{ for } x \sim 1;\ x > 1 \Rightarrow \sqrt{x} - 1 < \frac{x-1}{2};$$

$$x < 1 \Rightarrow y = 1/x > 1 \Rightarrow \sqrt{y} - 1 < \frac{y-1}{2}$$

This algorithm is an important fragment of other algorithms.

$$\underline{\text{if}}\ x > 0 \Rightarrow$$
$$\underline{\text{do}}\ x \neq 1 \Rightarrow x \leftarrow \sqrt{x}\ \underline{\text{od}}$$
$$\underline{\text{fi}}$$

Fig. 4

Example #3. Fast powering with natural exponent

I want to compute a^b for real a and integer $b \geq 0$. First I replace the constants a,b by variables and then I drive the variables toward my goal. I set initially $z,x,y \leftarrow 1,a,b$. Then

(1) $1.a^b = z.x^y$

The idea is to drive y to zero, while keeping (1) invariant. This can be done by the two transformations

(2) $z.x^y = z(x.x)^{y/2} = (z.x)x^{y-1}$.

The first reduces very quickly, but can be done only for even y. The second reduces y slowly, but works also for odd y. Thus we get the program in Fig. 5. If you see this program for the first time, and without comment, you may not understand it. But the addition of the single comment $P : y \geq 0$ and $z.x^y = a^b$ is enough to understand the program.

$$z, x, y \leftarrow 1, a, b$$
$$\underline{\text{do}}\ y \neq 0\ \underline{\text{and}}\ x \neq 1$$
$$\quad \underline{\text{do}}\ 2|y \Rightarrow x, y \leftarrow x^2, y/2\ \underline{\text{od}}$$
$$\quad z, y \leftarrow z \cdot x, y - 1$$
$$\underline{\text{od}}$$

Fig. 5

For it is easy to verify

(1) that P is true initially

(2) that (P and $y = 0$) implies $z = a^b$

(3) that the execution of the program leaves P true

(4) that the loop halts, since each execution of the loop body decreases y by at least once.

This program is of great importance in number theory and its applications.

Example #4. Powers with real exponent

I want to compute a^b with $a > 0$, $0 < b < 1$, both real. This time I have

$$1.a^b = z \cdot x^y = z(\sqrt{x})^{2y} = (z \cdot x)x^{y-1}$$

In the preceding example I have driven y to zero. This time I will drive x to 1 by repeatedly taking square roots and keeping $y < 1$, as shown in Fig. 6. There is a striking similarity between these two programs.

$$z, x, y \leftarrow 1, a, b$$
$$\underline{\text{do}}\ y \neq 0\ \underline{\text{and}}\ x \neq 1$$
$$\quad \underline{\text{do}}\ y < 1 \Rightarrow x, y \leftarrow \sqrt{x}, 2y\ \underline{\text{od}}$$
$$\quad z, y \leftarrow z \cdot x, y - 1$$
$$\underline{\text{od}}$$

Fig. 6

Example #5. Inverse trigonometric functions

I want to transform an isosceles triangle with given sides (Fig. 7) into a circular sector (Fig. 8) by keeping α

and c invariant. My aim is to find r. Then I have α in radians and all inverse trigonometric functions, since

$\alpha = c/r = \cos^{-1} a/b = \sin^{-1} c/b = \tan^{-1} c/a.$

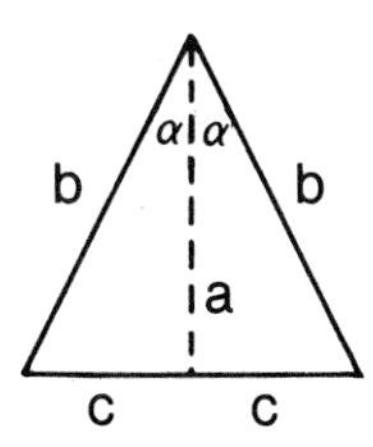

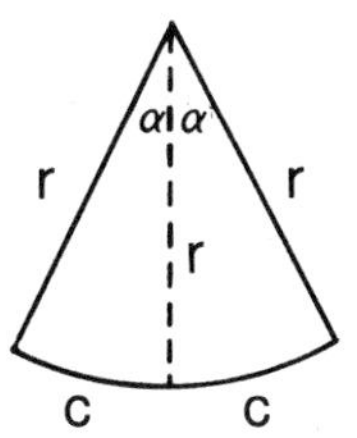

Fig. 7 & Fig. 8

Fig. 9 shows one step which transforms triangle ABC into two triangles A'B'C' and B'D'C'. We get at once

$a' = (a+b)/2, \; b' = \sqrt{a'b}$

and the algorithm in Fig. 10 with input a,b and output α in radians. Initially we have $x < r < y$ and at the end $x = y = r$.

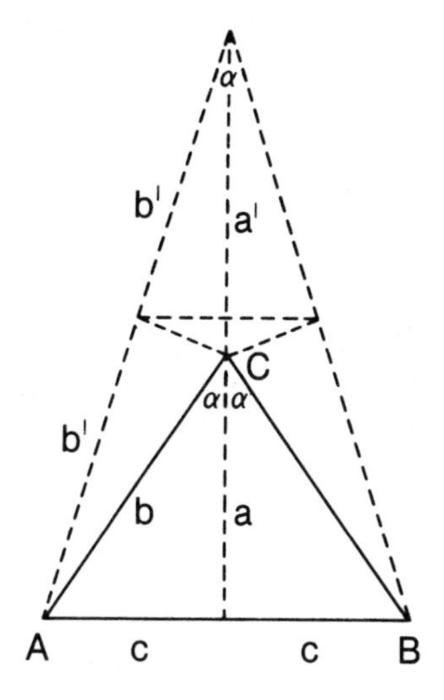

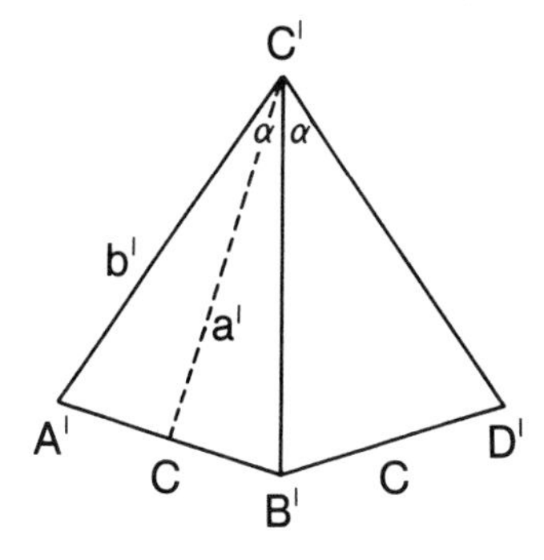

Fig. 9

$$x, y \longleftarrow a, b$$
$$\textbf{do} \;\; x < y \Rightarrow x \leftarrow \frac{x+y}{2}; \; y \leftarrow \sqrt{xy} \;\; \textbf{od}$$
$$\textbf{print} \;\; c/x$$

Fig. 10

How well does this algorithm perform? Fig. 11 shows that $|x' - y'| < |x - y| / 4$.

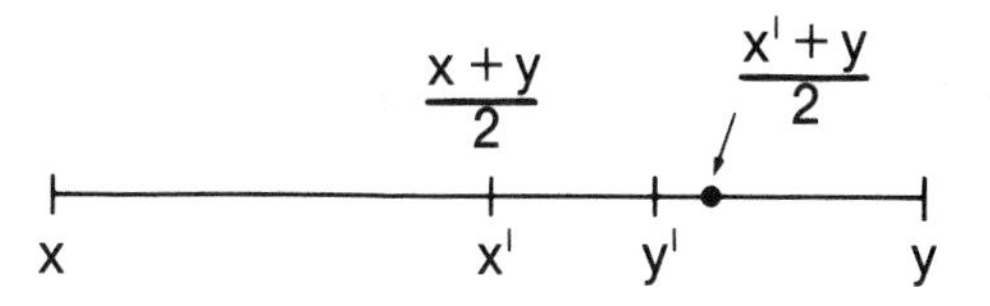

Fig. 11

At each step the gap of uncertainty is reduced at least 4 times. When x - y becomes small we have

$(x + y)/2 \sim \sqrt{xy}$ and $|x' - y'| \sim |x - y| /4$. Thus the convergence rate is 1/4 and we gain two bits per step.

With input $a = 1$, $b = \sqrt{2}$ we get $c = 1, \alpha = 1/x = \tan^{-1} 1 = \pi/4$. By printing $4/x$ intead, we get π.

Example #6. The function ln

The trapezoids in Figs. 12 and 13 have the same area

(1) $s(x) = \frac{1}{2}(x - \frac{1}{x})$

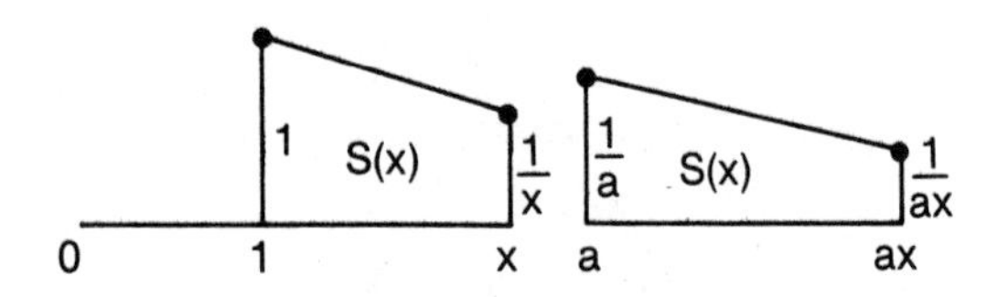

Fig. 12 & Fig. 13

Hence both trapezoids in Fig. 14 have the same area $s(\sqrt{x})$ and their sum is

(2) $s'(x) = 2\ (\sqrt{x})$.

For later we introduce

(3) $c(x) = \frac{1}{2}(x + \frac{1}{x}) \geq 1$.

Then we have

(4) $s(x) = 2s(\sqrt{x})c(\sqrt{x})$,

$$c(\sqrt{x}) = \sqrt{\frac{1 + c(x)}{2}} .$$

(5) $s'(x) = \frac{s(x)}{c(\sqrt{x})}$.

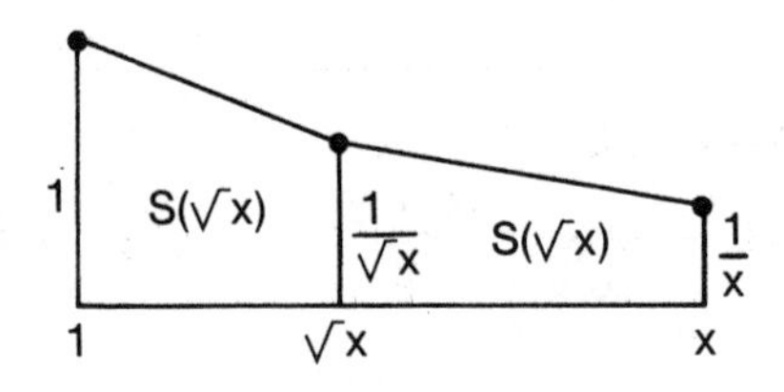

Fig. 14

I want to transform s(x) in Fig. 12 into ln(x) in Fig. 15. To design one step we apply the divide-and-conquer paradigm. A basic guide to good algorithm design is to maintain balance. Hence we split the area of the trapezoid into two parts of equal area. (it would be a disaster to use equal height!) The result is Fig. 14. From (1) to (5) we get for ln(x) the algorithm in Fig. 16.

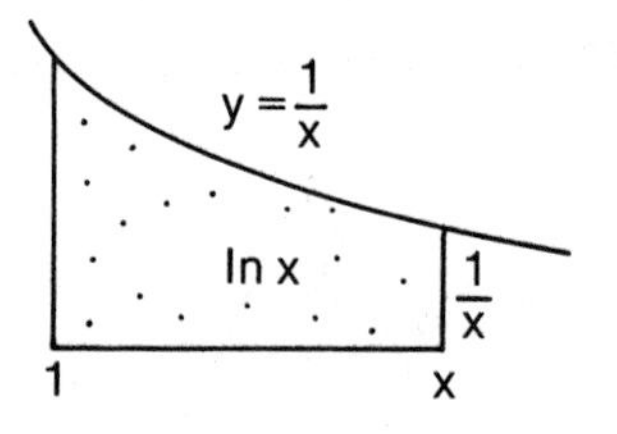

Fig. 15

```
s,c ← (x − 1/x)/2, (x + 1/x)/2
do   c > 1 ⇒ c ←√(1 + c/2; s ← s/c   od
print   s
```

Fig. 16

A look at Fig. 17 shows that for $c' = \sqrt{(1 + c)/2}$ we have $o < c' - 1 < (c - 1)/4$. For $c \sim 1$ we have $c' - 1 \sim (c - 1)/4$. Thus the convergence rate of the algorithm is again 1/4.

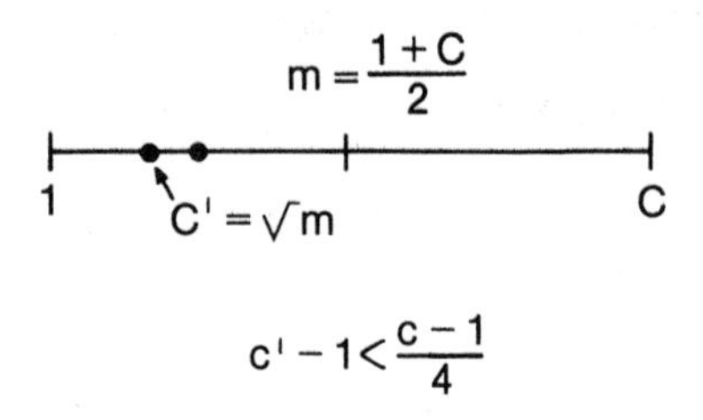

Fig. 17

Example #7. The functions of cos and exp. Avoiding rounding errors

If you try to compute a transcendental function by an obvious algorithm, you will be punished by severe rounding errors. In the two preceding programs we avoided this by a sophisticated approach. But let us be naive for a moment and try to design a cosine program. We use two natural ideas

a) If I know $c(x) = \cos(x)$ for arc (x), then I can easily compute $c'(x) = c(2x) = \cos 2x$ for arc 2x. Indeed, Fig. 18 shows that $4c^2 = 2(1 + c')$, $c' = 2c^2 - 1$ or

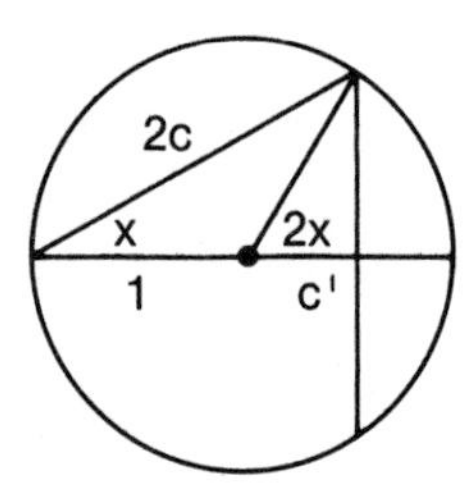

Fig. 18

(1) $c(2x) = 2c^2(x) - 1$ (duplication formula for $c(x) = \cos(x)$)

b) For sufficiently small x we know c(x) to any desired accuracy. Indeed, we get from Fig. 19 $x^2 \sim 2(1 - x)$, or

(2) $c(x) \sim 1 - x^2/2$.

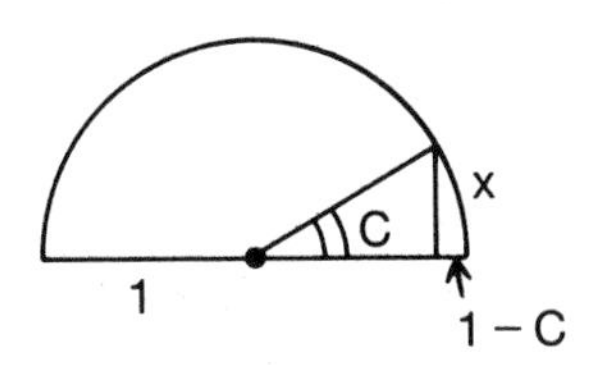

Fig. 19

To find c(x) we start with $x \leftarrow x/2^n$, compute its cosine by (2) and then apply $c \leftarrow 2c^2 - 1$, n times. If your calculator works internally with 2s significant digits, then you will get at most s correct digits. We can easily avoid the severe rounding errors in (2) by a standard trick. We set

$c(x) = 1 + f(x)$.

Then we have from (1)

$f(2x) = 2f(x)\,(2 + f(x))$

and for small x

$f(x) \simeq - x^2/2$.

Thus we get the cosine-program in Fig. 20.

```
x ← x/2ⁿ; f ← −x²/2
do n > 0 ⇒ f ← 2f(2 + f); n ← n − 1 od
print f + 1
```

Fig. 20

Similarly we may compute e^x from its definition in Fig. 21.

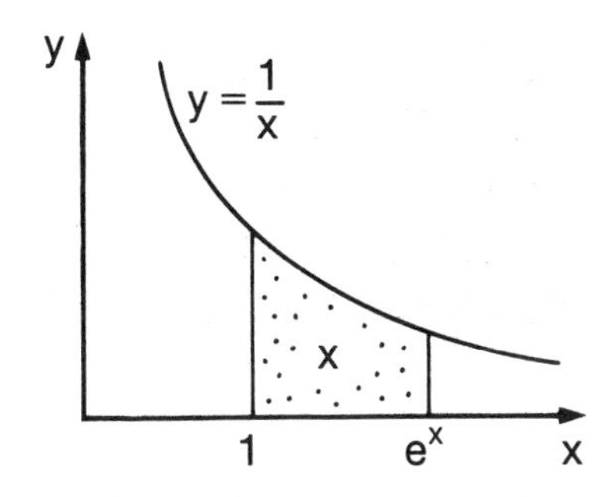

Fig. 21

With $g(x) = e^x - 1$ we get the duplication formula

(3) $g(2x) = g(x)(2 + g(x))$.

For small x we approximate g with the midpoint rule in Fig. 22, which yields $2g/(g + 2) \sim x$ or

(4) $g(x) \sim (2x)/(2 - x)$.

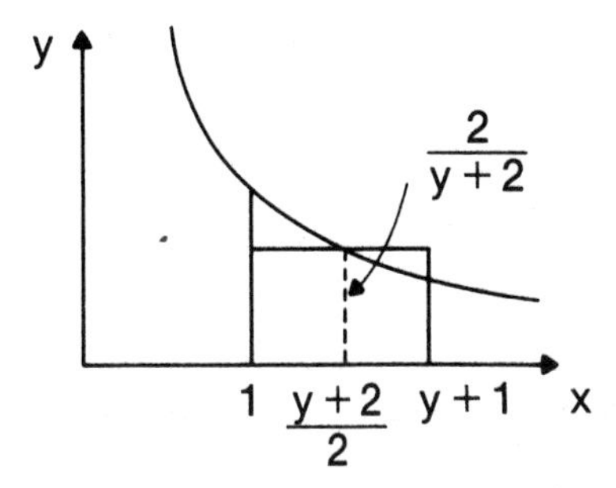

Fig. 22

The program in Fig. 23 is based on (3) and (4).

```
x ← x/2ⁿ; g ← 2 • x/(2 − x)
do n > 0 ⇒ g, n ← g(2 + g), n − 1 od
print g + 1
```

Fig. 23

In Figs. 20 and 23, n = 16 suffices for 10-digit accuracy. Indeed, the programs for $\cos^{-1}x$ and $\ln(x)$ in Figs. 10 and 16 have convergence rates 1/4. Hence 16 steps will reduce the initial error by the factor 4^{-16}.

Solving Equations from an Algorithmic Standpoint

We will write equations in standard form

(1) $x = f(x)$.

A solution of (1) will be called a fixed point of f. Thus in Fig. 24 s is a fixed point of f, while a is not.

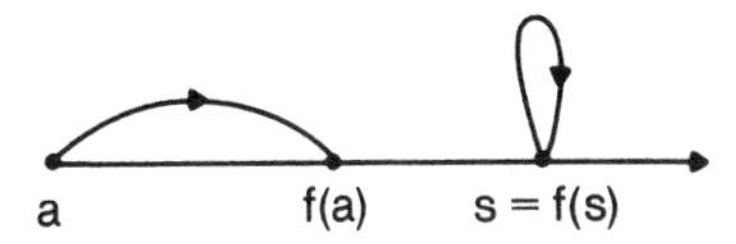

Fig. 24

Solving equations has both an algebraic and an algorithmic aspect. The algebraic aspect consists in producing new names for solutions (pseudo solutions) by formal manipulation. Thus, the equations

$17x = 1;\ y^2 - 10 = 0;\ z^2 - 4z - 1 = 0$

have the pseudo solutions

$x = 1/17;\ y = \pm\sqrt{10};\ z = 2 \pm \sqrt{5}$.

The algebraic approach is confined to linear and quadratic equations.

The algorithmic approach is powerful, instructive and useful. An unknown real number x is an infinite decimal, its digits are hidden behind a curtain. Your job is to find an algorithm which rolls back the curtain, revealing more and more of its digits. The more digits you get per step, the better is the algorithm.

We take a look at the linear equation from an algorithmic point of view:

(2) $x = ax + b,\ a \neq 1.$

Its algebraic solution is

(3) $x = b/(1 - a).$

But this is not an algorithmic solution. In algorithmics you do not study equations, you design and analyze processes. Hence we change (2) into the process in Fig. 25, which generates a sequence of numbers. If it ever stops then we have found a fixed point of f defined by $f(x) = ax + b$.

```
y, x ← 0, b
do y ≠ x ⇒ y, x ← x, ax + b od
```

Fig. 25

But does it ever stop? Fig. 26 shows that

$x - s = (ay + b) - (as + b) = a(y - s),\ |x - s| = |a| \cdot |y - s|.$

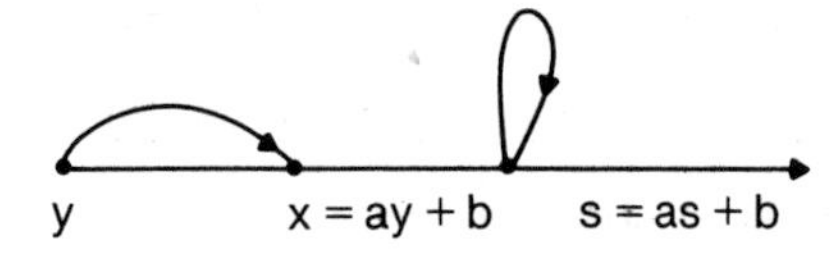

Fig. 26

That is, in one step the distance to the fixed point s shrinks by the factor $|a|$ iff $|a| < 1$. Thus the standard form of a linear equation is

(4) $x = ax + b,\ |a| < 1.$

Fig. 25 gives linear convergence with rate $|a|$. Consider the example

(5) $x = 1/17.$

Algebraically this is the solution. To get the algorithmic solution, you must transform (5) into (4) with a small $|a|$ (the smaller the better!):

$x = 1/17 \Leftrightarrow 17x = 1 \Leftrightarrow 51x = 3 \Leftrightarrow 50x = 3 - x \Leftrightarrow x = 0.06 - 0.02.x.$

Fig. 27 shows the corresponding algorithm with printout.

```
y, x ← 0, 0.06
do x ≠ y ⇒ y, x ← x, 0.02(3 − x) od
```

$x_0 = .06$
$x_1 = .0588$
$x_2 = .058824$
$x_3 = .05882352$
$x_4 = .0588235296$
$x_5 = x_6 = .0588235294$

Fig. 27

Fast Reciprocation of Integers

You should never be satisfied with linear convergence. So we try to design systematically a superfast algorithm for $1/a$. Start with an estimate x with relative error y, $|y| < 1$. Then initially

$x = (1 - y)/a,\ y = 1 - ax,\ |y| < 1$, or

(6) $1/a = x/(1 - y).$

I want to keep (6) invariant while driving y quickly to zero, for instance by repeated squaring. This is easy to achieve, since
$1/a = x/(1 - y) = (x(1 + y))/(1 - y^2) = x'/(1 - y').$
Thus we get the algorithm in Fig. 28.

```
x, y ← x0, 1 − ax0
do 1 + y > 1 ⇒ x, y ← x + xy, y² od
```

Fig. 28

With $a = 17$, $x_0 = 0.06$, $y_0 = -0.02$ we get the printout in Fig. 29.

$x_0 = .06,\ y_0 = \text{-2E} - 2$
$x_1 = .0588,\ y_1 = 4\text{E} - 4$
$x_2 = .0588235\underline{2},\ y_2 = 1.6\text{E} - 7$
$x_3 = .05882352941176\underline{32},\ y_3 = 2.56\text{E} - 14$

Fig. 29

The underlined digits are not correct. If we start now with $x_0 = 0.0588235294$, then one step gives 20 correct decimals, $x_1 = 0.0588235294\ 1176470588$.
Let us now look at quadratic equations. Algebraically the equation

(7) $x^2 - 10 = 0$
is simple and the equation

(8) $x^2 - 4x - 1 = 0$

is complicated. By "completing the square" it is brought into the shape (7). From an algorithmic standpoint it is just the opposite: (8) is simple and (7) must be transformed into shape (8). We first solve (8) by bringing it into the form

(9) $x = 4 + 1/x$.

This is the algorithmic standard form of a quadratic equation. From (9) we switch to the sequence

(10) $x_1 = 4,\ x_{n+1} = 4 + 1/x_n$.

The pocket calculator gives the results in Fig. 30.

$x_1 = 4$

$x_2 = 4.25$

$x_3 = 4.235294118$

$x_4 = 4.236111111$

$x_5 = 4.236068111$

$x_6 = 4.236065574$

$x_7 = 4.236067970$

$x_8 = 4.236067978$

$x_9 = 4.236067977$

$x_{10} = x_8,\ x_{11} = x_9,\ x_{12} = x_{11}$

Fig. 30

The fixed point s lies between any two successive terms of the sequence since

$x_n > s \Leftrightarrow x_{n+1} < s,\ x_n < s \Leftrightarrow x_{n+1} > s.$

$|x_{n+1} - x_n|$ is the n-th interval of uncertaintly about the location of s. How fast does it shrink? We get

$$q_n = \left|\frac{x_{n+1} - x_n}{x_n - x_{n-1}}\right| = \frac{1}{(x_{n-1})(x_n)} \sim \frac{1}{18}$$

where q_n is almost constant. Thus the convergence is linear with rate ~ 1/18.

In solving (7) the transformation $x = 10/x$ does not help, since a sequence defined by $x_{n+1} = 10/x_n$ has period 2. By putting $x = z - 3$ we get

(11) $z = 6 + 1/z$

and the sequence

$z_1 = 6,\ z_{n+1} = 6 + 1/z_n$.

Fig. 31 shows the result. This time we have

$$q_n = \left|\frac{z_{n+1} - z_n}{z_n - z_{n+1}}\right| = \frac{1}{(z_{n+1})(z_n)} \sim \frac{1}{38}.$$

The algorithmic aspect of linear and quadratic equations is thoroughly explored in (5).

$z_1 = 6$

$z_2 = 6.166666667$

$z_3 = 6.162162162$

$z_4 = 6.162280702$

$z_5 = 6.162277580$

$z_6 = 6.162277662$

$z_7 = 6.162277660$

$z_8 = z_7$

$x = z - 3 = 3.162277660$

Fig. 31

Extrapolation to the Limit

Numerical mathematics is very much an experimental science. By numerical experimentation we can discover quite general laws about linear convergence rates. Some of these we can make plausible, and we can use them to double the convergence rate, thus cutting the computational work in half.

a) Rectangular approximation

Suppose you want to find the area a under the curve f in Fig. 32. Let a_n be the lower (or upper) approximation to a by n rectangles. Then

(1) $a_n = a + e_n$

where e_n is the n-th error. Cut each subdivision in half. Then

(2) $a_{2n} = a + e_{2n}$.

For large n there is a simple relation between e_n and e_{2n}. If you take a sufficiently small peice of a decent curve, it will look more and more like a straight line. Fig. 33 shows that for straight lines, $e_{2n} = e_n/2$ exactly. For any decent curve we have

(3) $e_{2n} \sim e_n/2$.

From (1) to (3) we get

(4) $a \sim 2a_{2n} - a_n$.

This extrapolation step to the limit is exact for straight lines and doubles the convergence rate for other curves. In addition, (3) shows that the rectangular approximation converges with linear rate 1/2.

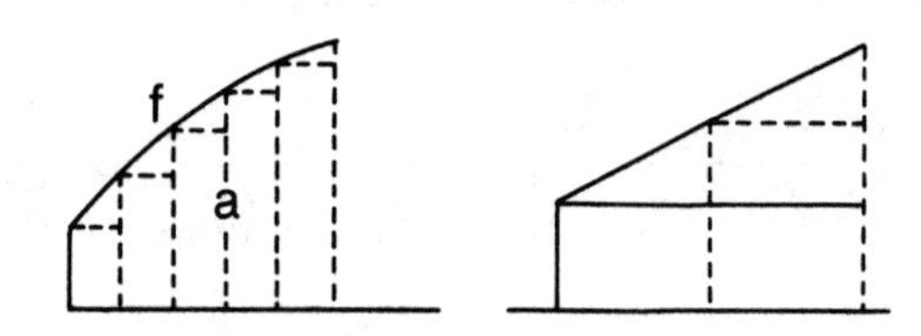

Fig. 32 & Fig. 33

b) Trapezoidal and midpoint approximation

If you take a sufficiently small piece of any decent curve, then it looks more and more like a quadratic parabola. For quadratic parabolas Archimedes proved some remarkable theorems. In Fig. 34 let P be the area of the parabolic segment cut off by the chord AB, and let Δ_1 be the area of the triangle ABC. Then

(5) $\Delta_1 = 3/4P$.

Indeed, a short computation gives, with $h = b - a$

$g = ((b-a)/2)^2 = (h/2)^2$.

Hence

$g' = g'' = (h/4)^2 = g/4$

and

Δ_2 = area of triangle ACD + CBE = $\Delta_1/4$.

By repeating this step indefinitely we get

$$P = \Delta_1 + \Delta_1/4 + \Delta_1/16 + \ldots = \tfrac{4}{3}\Delta_1$$

which is (5). Now let a_n be the approximation of an area $\underline{a}$ by $\underline{n}$ trapezoids. Then

(6) $a_n = a + e_n$, $e_n \sim \Sigma$ (+/- parabolic segments).

If you double the number of trapezoids, then

(7) $a_{2n} = a + e_{2n}$.

Fig. 34 and (5) show that for a quadratic parabola $e_{2n} = e_n/4$ exactly. For other decent curves

(8) $e_{2n} \sim e_n/4$.

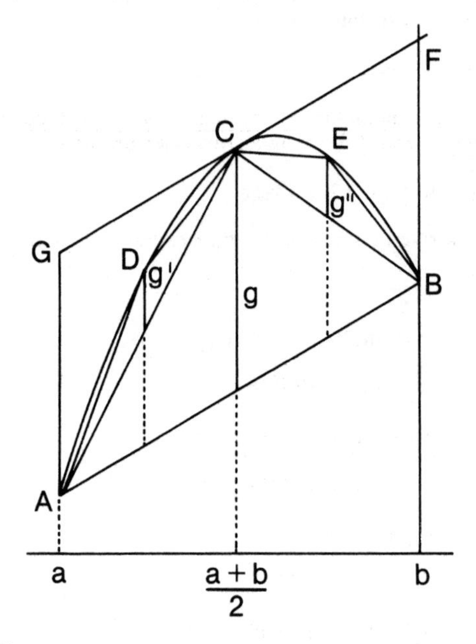

Fig. 34

From (6) to (8) we get

(9)

$$a \sim \frac{4a_{2n} - a_n}{3}$$

For quadratic parabolas (and cubic curves) (9) is exact. For other curves we get a sequence $b_n = (4a_{2n} - a_n)/3$, which converges to a with linear rate 1/16, while (8) shows that the trapezoidal approximation converges with linear rate 1/4. Because of (5) the parabola divides the area of the parallelogram ABFG in the ratio 2:1. That is, the error of the midpoint approximation by the tangent GF is half as large as that of the trapezoid approximation by the chord AB.

Let us take an example. From Fig. 14 we see that

(10)

$$s_n = 2^{n-1}\left(\sqrt[2^n]{x} - 1/\sqrt[2^n]{x}\right)$$

is the approximation of $\ln(x)$ by 2^n trapezoids of equal area. My calculator gives for $x = 2$

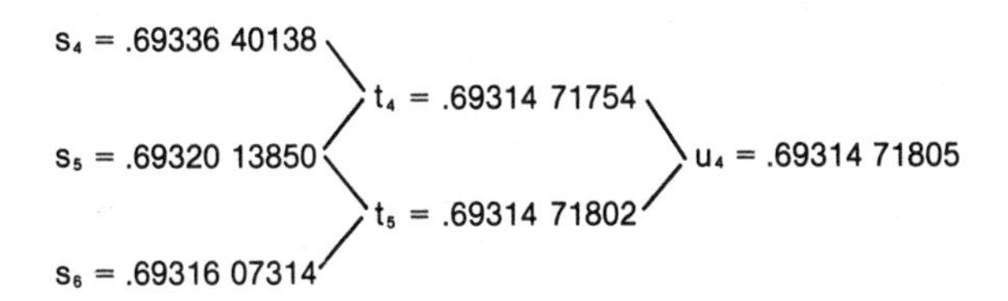

The s_i in the first column converge at the rate 1/4. The $t_i = (4s_{i+1} - s_i)/3$ in the second colum converge with rate 1/16. Using this for one more extrapolation we finally get

$$u_4 = (16t_5 - t_4)/15 = .69314\ 71805$$

The correct value to 11 decimals is $\ln(2) = .69314\ 718056$. On the other hand it is not possible to get a precise value for $\ln(x)$ from (10) becuase of subtractive cancellation for large values of n.

Second Lecture: Seminumerical Algorithms

1. Algorithmic Number Theory

Introduction

Some number theory is indispensible in algorithmics. But all you need can be discovered by analyzing an algorithm from grade school. Gauss, as a boy, developed number theory by studying periodic decimals. He computed a table of reciprocals 1/m out of m = 1000. By observing the behavior of the long division algorithm he got directly to the important concepts and results. This is the algorithmic approach. We will follow Gauss, but with a different motivation. We try to find a practical solution to an insoluble problem:

Design an algorithm which generates a purely random sequence

You need random sequences in probability and statistics, and in cryptography for protecting private data.

For large n a purely random sequence of length n is such a complicated object that you need a very long program to describe it. In fact, the description will be about as long as the sequence itself. We will use short programs to produce very long pseudo-random sequences with hidden regularities. To analyze these sequences empirically you must design algorithms for data analysis, pattern matching, loop detection etc. These are the first steps into combinatorial computing.

The long division algorithm

We look at some computational results:

$1/13 = .076\ 923$, $1/17 = .05882352\ 94117647$

$1/91 = .010\ 987$, $1/56 = .017\ 857\ 142$

How do you compute $1/m$ in base 10 (base $a \geq 2$)? What is one step of the procedure? Starting with remainder $r = 1$ you first multiply r by 10 (base a). Then you divide by m:

$$10r = m.q + r',\ 0 \leq r' < m$$

getting the next digit q and the next remainder r'. Here it is quite natural to introduce

$$x \bmod y = \text{remainder on division of } x \text{ by } y$$

and modular arithmetic. Then the sequence of remainders is $1, 10, 10^2, \ldots \pmod m$, or, in base a: $1, a, a^2, \ldots \pmod m$. These are the iterates of the function f defined by $f(x) \equiv ax \pmod m$. If you iterate any function mod m, you either get a pure cycle, or a tail followed by a cycle, as shown in Fig. 1a - d.

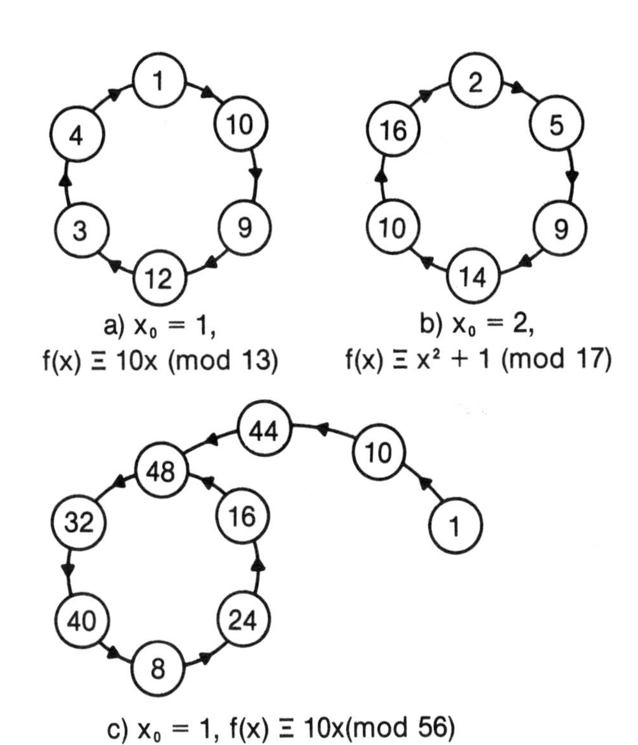

a) $x_0 = 1$, $f(x) \equiv 10x \pmod{13}$

b) $x_0 = 2$, $f(x) \equiv x^2 + 1 \pmod{17}$

c) $x_0 = 1$, $f(x) \equiv 10x \pmod{56}$

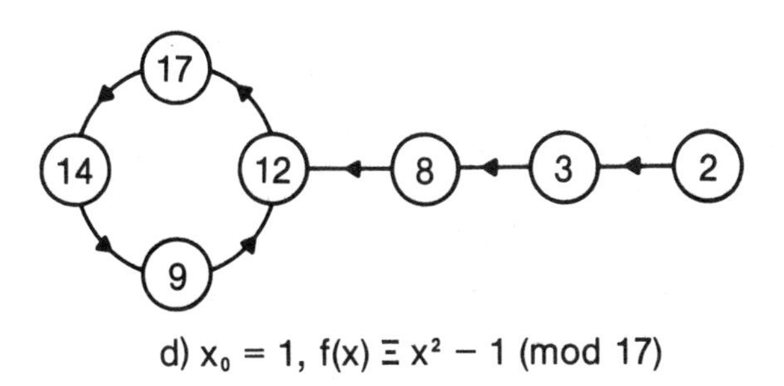

d) $x_0 = 1$, $f(x) \equiv x^2 - 1 \pmod{17}$

Fig. 1

Coin tosses for poor people

If you compute r/m in base a then you get a sequence D of digits from (0, 1, . . . , a-1) and a sequence R of remainders from (0, 1, . . . , m-1). First take a prime p, choose r from (1,2, . . . , p-1) "at random" and compute r/p in base 2. With r = 6, p = 13 one period of r/p is

(D) 011101 100010 = AB with A = 011101, B = 100010 and the remainder sequence is

(R) 6, 12, 11, 9, 5, 10, 7, 1, 2, 4, 8, 3 (see Fig. 2).

Here randomness seems to pop up in many places.

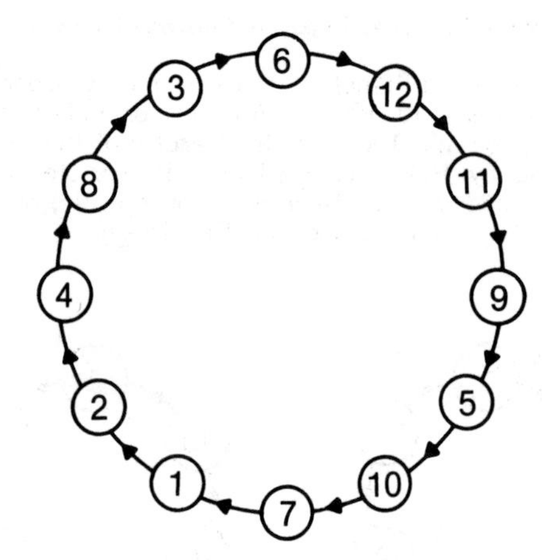

$x_0 = 6,\ f(x) \equiv 2x \pmod{13}$

Fig. 2

(i) D looks like a sequence of 12 coin tosses. A closer look reveals a hidden regularity: B = A = complement of A.

(ii) R looks like a "random permutation". Again there is a hidden regularity. We have $x + y \equiv 0 \pmod{13}$ for any two antipodes in Fig. 2. This implies B = A in D.

(iii) R mod 2 is 001110110001, a cyclic shift of D.

(iv) In R read 1 - 6 as "0" and 7 - 12 as "1". You get D again.

(v) R mod 6 is 0, 0, 5, 3, 5, 4, 1, 1, 2, 4, 2, 3 (rolls of a 6-die).

(vi) R mod 3 is 0, 0, 2, 0, 1, 1, 1, 2, 1, 2, 0 (rolls of a 3-die).

(vii) Apply x<-- (x/2) to R: 3, 6, 6, 5, 3, 5, 4, 1, 1, 2, 4, 2 (rolls of a 6-die).

These results are truly remarkable. Each time we have equal frequency of digits. A bigger example like 28/101 can be done easily by hand. We get the remainders by repeatedly doubling mod 101:

(R') 28, 56, 11, 22, 44, 88, 75, 49, 98, 77, 53, 05, 10, 20, . . . , 73 (#51), . . .

Add the 1st and the 51st remainders: $28 + 73 \equiv 0 \pmod{101}$. Since $x + y \equiv 0 \pmod{101}$ implies $2x + 2y \equiv 0 \pmod{101}$, the second half of R' is complementary to the first half. By replacing R' in each $r \leq 50$ by "0" and each $r > 50$ by "1" we get for one period of 28/101:

(D') 0100011011 1110000110 0101011000 1011011001 1111101011 1011100100 0001111001 1010100111 0100100110 0000010100

The second half of D' is complementary to the first half since $r + \bar{r} = 101$ implies $r \leq 50 \Rightarrow \bar{r} > 50$. So we would use only 50 consecutive digits in D'. These "coin tosses" seem to be of good quality, and they cost merely 1 1/2 additions per bit.

For applications we need long periods. These can be achieved by large primes, or by combining sequences with short periods by bitwise addition mod 2. Thus 28/101 and 40/103 have periods 100 and 51 respectively. Their bitwise sum has period 5100 = LCM (100,51).

<u>Study of cycle length</u>

Let $Z_m = (0, 1, \ldots, m-1)$ and $a \in Z_m$. For $(a,m) = 1$ the linear function f defined by $f(x) \equiv ax \pmod{m}$ is a <u>permutation</u>. Indeed

$ax_1 \equiv ax_2 \pmod{m} \Rightarrow m \mid a(x_1 - x_2) \Rightarrow m \mid x_1 - x_2 \Rightarrow x_1 = x_2$.

So iterates of f are pure cycles, as shown in Fig. 3.

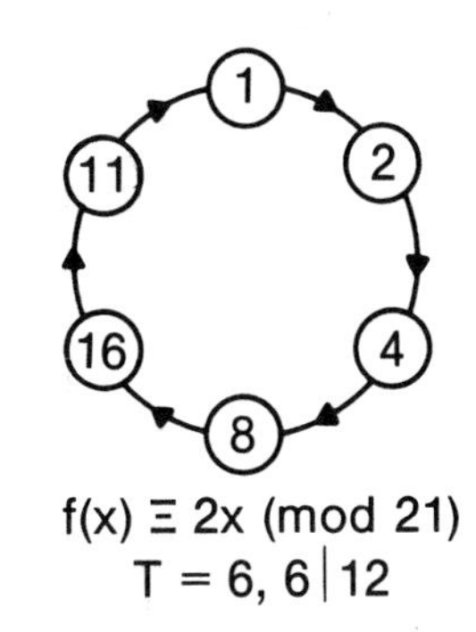

$f(x) \equiv 2x \pmod{21}$
$T = 6,\ 6 \mid 12$

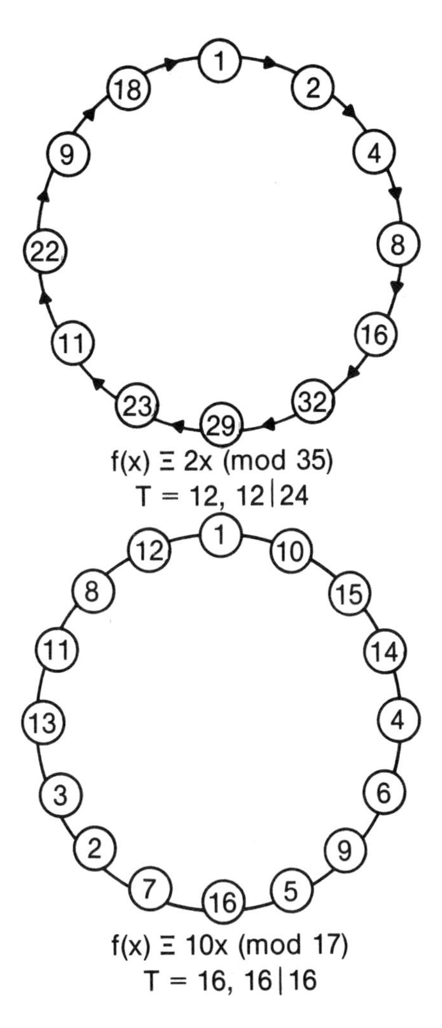

$f(x) \equiv 2x \pmod{35}$
$T = 12,\ 12 \mid 24$

$f(x) \equiv 10x \pmod{17}$
$T = 16,\ 16 \mid 16$

Fig. 3

Take some $\underline{a}$ and find its cycle length T for various moduli m prime to $\underline{a}$. Gauss quickly discovered the following facts: For m = p, a prime, $T \mid p - 1$, or equivalently, $a^{p-1} \equiv 1 \pmod{p}$. For m = pq, a product of distinct primes, $T \mid (p - 1)(q - 1)$, i.e. $a^{(p-1)(q-1)} \equiv 1 \pmod{pq}$. This is quite easy to prove, but we can also check it each time we need it.

For each prime p there are multipliers $\underline{a}$ with period T = p - 1. This is not easy to prove, but we can always quickly find such maximum period multipliers by computer search. The search is easiest for primes p = 2p' + 1 with p' also a prime. For such a prime the possible periods are 1, 2, p', 2p'. There are plenty of these primes, for example: 2063, 10463, 200087, 1000667, 2040287. We can check with a pocket calculator that a = 10 has maximum period 2062 (mod 2063). By repeated squaring (mod 2063) we get the following table

n	1	2	4	8	16	32	64	128	256	512	1024	7	1031
10^n	10	100	1743	201	1204	1390	1132	301	1892	359	975	639	2062≡−1

Thus 10 has period T = 2062 (mod 2063) and 1/2063 has period 2062. Fig. 4 shows the first half of the period. The beginning of the second half is underlined. These are 1031 pseudo-random decimal digits. Fig. 5 shows the frequency h_n of the digits in the full period and in its first half. The digits in the full period are too evenly distributed to be qualified as "random".

It is easy to check that 1/20663 has period 20662. Fig. 6 shows the frequency of digit n in a full period and in its first half.

.00048478097430925836160930683470673776054283869122
63693650024236548715462918080465341735336888027144
93456131846825012118274357731459040232670867668444
01357246728065923412506059137178865729520116335433
83422200678623364032961706253029568589432864760058
16771691711100339311682018480858126514784294716432
38002908385845855550169658841008240426563257392147
35821619001454192922927775084827920504100213281628
69607367910809500727096461463887942413960252060106
64081434003683955404750363548230731943771206580126
03005332040717401841977702375181774115355971685603
49006301502666020358700920988851187590887057882985
94280174503150751333010179350460494425593795443528
84149297140087251575375666605089675230247212796897
72176442074648570043625787687833252544837615123606
39844836068221037324285021812895843915625272418307
56180319922443044110518662142510908446921958313136
20940378090159961221522055259331671255453223460979
15656810470189045079980610761027629665535627726611
73048957828405235094522539990305380513814832767813
8633058652447891420261754726126<u>99951526902569</u> ....

Fig. 4

digit n	0	1	2	3	4	5	6	7	8	9
h_n in period	206	206	206	207	206	206	207	206	206	206
h_n in ½ period	130	105	112	101	101	105	106	94	101	76

Fig. 5

digit n	0	1	2	3	4	5	6	7	8	9
h_n(period)	2066	2066	2066	2067	2066	2066	2067	2066	2066	2066
h_n(½ period)	1044	1037	1029	1041	1019	1047	1026	1037	1029	1022

Fig. 6

<u>Coin tosses for poor but sophisticated people</u>

Starting with a "seed" $r_0 \in (1, \ldots, p - 1)$ and a multiplier $\underline{a}$ with period p - 1 we get a permutation of $(1, \ldots, p-1)$ by means of the recurrence $r_{n+1} \equiv ar_n \pmod{p}$. Uniform random numbers are obtained by $u_i = r_i/p$. This is the usual random number generator of the computer. It yields satisfactory results for big primes and

$$\sqrt{p} < r_0,\ a < p - \sqrt{p},\ a > p/100.$$

We get higher quality by using more than one preceding term. Then even a small modulus like 2 or 5 will produce long periods. What is even more important, these sequences can be generated by hand with high speed.

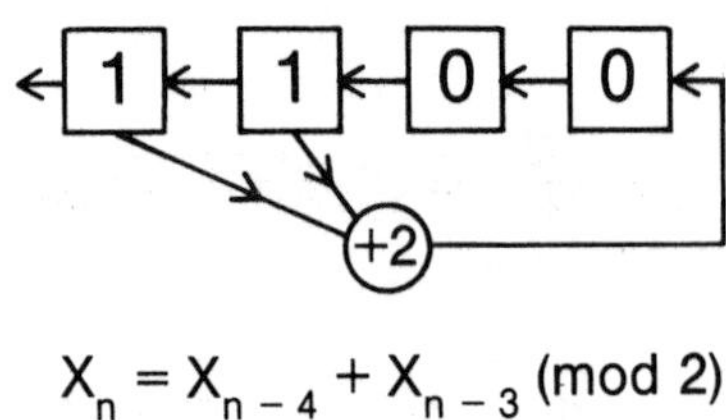

$$X_n = X_{n-4} + X_{n-3} \pmod{2}$$

Fig. 7

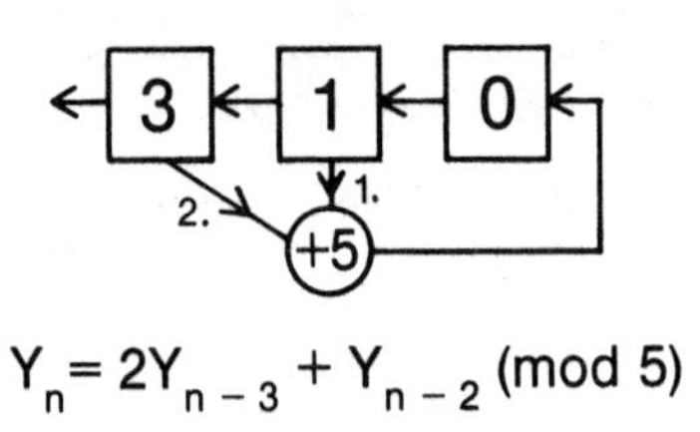

$$Y_n = 2Y_{n-3} + Y_{n-2} \pmod{5}$$

Fig. 8

The two linear shifts registers (LSR) in Figs. 7 and 8 generate streams of binary and quinary digits, respectively. Initially the cells are loaded with a nonzero vector. The generated sequences X and Y can be extended forward and backward in a unique way. Hence they are purely periodic and the zero vector will never come up. The maximum possible periods $T_1 = s^4 - 1$ and $T_2 = 5^3 - 1$ are attained in this case since

X = 11000 10011 01011 <u>1100</u> . . . ,

Y = 3102221103 2332433142 1002024230 2120444220 1411431123 4200404341 0424033344 0232231224 1340030313 2034301113 3041441244 3213001012 1401 <u>310</u> . . .

We use X to transform Y into a sequence of decimal digits with period $T = T_1.T_2 - 1860$ by means of the following algorithm:

If $X_n = 0$, leave Y_n unchanged.

If $X_n = 1$, set $Y_n \leftarrow Y_n + 5$.

This is easy to do by hand. We get the sequence

Z = 8602271158 2837988142 6007529285 7620494275 1916986123 9205909396 5924083399 0737786224 6345535368 7534351168 3546996244 8218506067 . . . .

The digit frequencies of these sequences are

X:

digit n	0	1
h_n	7	8

Y:

digit n	0	1	2	3	4
h_n	24	25	25	25	25

Z:

n	0	1	2	3	4	5	6	7	8	9
h_n	168	175	175	175	175	192	200	200	200	200

We get rolls of a die by means of the LSR's in Figs. 9 and 10. They generate sequences with maximum periods $T_1 = 2^4 - 1$, $T_2 = 3^5 - 1$. The combined sequence has period $T = T_1.T_2 = 3630$. The combination rule is:

<u>if</u> $X_n = 0 \Rightarrow Y_n \leftarrow Y_n + 1$ [] $X_n = 1 \Rightarrow Y_n \leftarrow Y_n +$ <u>fi</u>.

We get the sequence

6154552324 1364252455 4321621663 4146441324 1366161455 4333511661 1166161565 5223411551 4246453325 2164163546

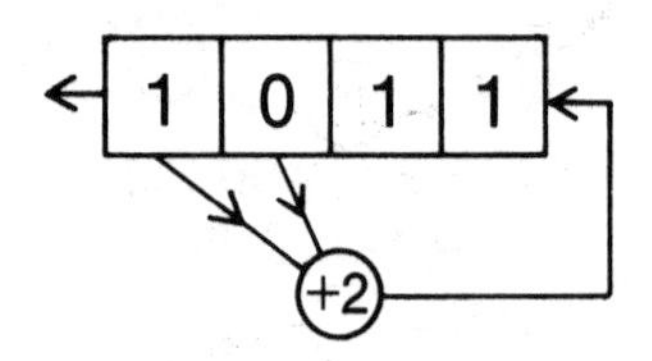

$$X_n = X_{n-4} + X_{n-3} \pmod{2}$$

Fig. 9

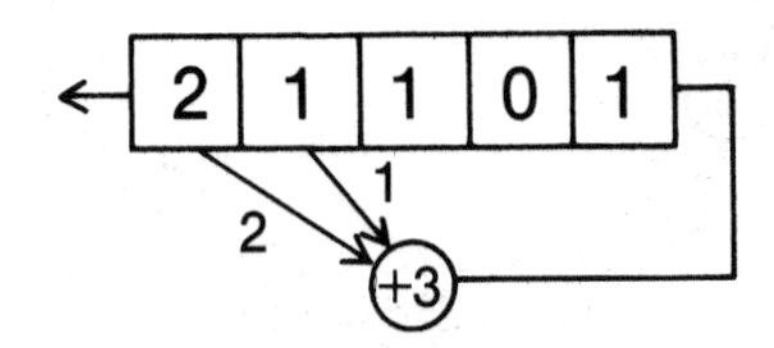

$$Y_n = 2Y_{n-5} + Y_{n-4} \pmod{3}$$

Fig. 10

The registers in Figs. 11 and 12 generate sequences with periods $T_1 = 2^5 - 1$, $T_2 = 5^5 - 1$. By combining them we get a decimal sequence with period $T = T_1 \cdot T_2 = 96\,844 \sim 100\,000$.

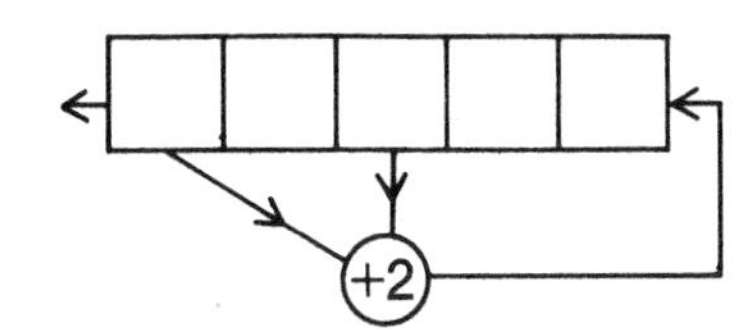

$$X_n = X_{n-3} + X_{n-5} \pmod 2$$

Fig. 11

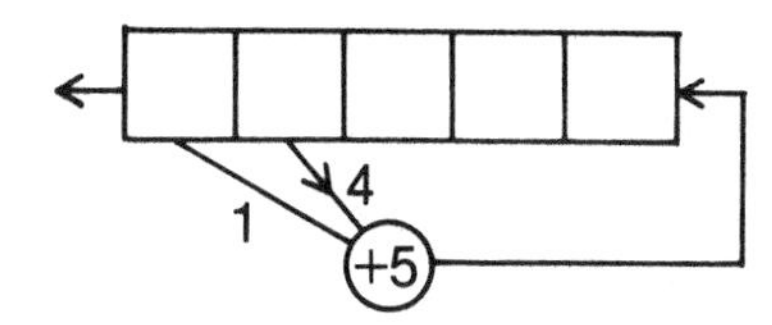

$$Y_n = 4Y_{n-4} + Y_{n-5} \pmod 5$$

Fig. 12

We can easily do better. The recurrences

$X_n = X_{n-14} + X_{n-17} \pmod 2$,

$Y_n = 4Y_{n-1} + 3Y_{n-7} \pmod 5$

generate sequences with periods

$2^{17} - 1$ and $5^7 - 1$, respectively. Their combination is a decimal sequence with period

$(2^{17} - 1)(5^7 - 1) = 131071.78124 > 10^{10}$. Experience shows that the quality of these sequences improves dramatically as the order of the defining recurrences increases. The sequences are efficiently generated by circular arrays.

Important number theoretic algorithms

In designing number theoretic algorithms, invariance plays a decisive role. We list some algorithms, which are important for algorithmics.

a) How to find a mod b

Dijkstra (2) develops in great detail a fast algorithm for this basic operation.

b) How to find $a^n \pmod m$ for fixed a, m

Let s = size of n = number of digits in n. T(s) = computation time as a function of s.

An algorithm for a problem of size s is calle bad if T(s) grows exponentially with s, and it is called good if T(s) is a polynomial in s. Repeated multiplication is a bad algorithm for $a^n \pmod m$. The fast powering procedure of Fig. 5 in Lecture 1 is a good algorithm. To find $1981^{1000000} \pmod m$ it requires merely twice as much time as to find $1981^{1000} \pmod m$.

c) How to find gcd(a,b)

Let $a \geq b > 0$. The move $a,b \leftarrow a - b, b$ leaves the gcd of the pair invariant. Hence also its q-times iteration $a,b \leftarrow a - qb, b$, where q is the maximum value such that $a - qb \geq 0$. Thus we get the basic step $a, b \leftarrow b, a \bmod b$ of Euclid's Algorithm:

```
x,y <- a,b

do y > 0 => x,y <- y,x mod y od

print x
```

Euclid's algorithm is a good algorithm, while finding gcd(a,b) by factoring a and b is a bad algorithm.

d) How to solve ax + by = gcd(a,b) in integers x,y

(Extended Euclidean algorithm). We start with two lines

(1) a 1 0 (i.e. $a = 1.a + 0.b$),

(2) b 0 1 (i.e. $b = 0.a + 1.b$).

Let $q = \lfloor a/b \rfloor$. By subtracting from row (1) q-times row (2) we get

(3) r u v (i.e. $r = a.u + b.v$).

Now cross out the first row and repeat the step with the remaining rows. The last line before r becomes zero is

(4) $\gcd(a,b) = a.u + b.v$.

This follows by invariance. We start with two linear combinations of a,b. The basic step produces again a linear combination of a,b.

e) How to solve $ax \equiv 1 \pmod m$

By means of the extended Euclidean algorithm we solve $ax = my = 1$.

f) How to tell is m is prime

There is a test of Rabin, which decides in random polynomial time if m is prime. Suppose you have a huge, say 500-digit number m. Apply Rabin's test once. If it tells you that m is composite, then m is indeed composite. If it tells you m is prime, then m is prime with error probability $< 1/4$. In this case repeat Rabin's test n times. If each time it tells you m is prime, then m is prime with error probability $< 4^{-n}$.

g) How to factor m

No polynomial time factoring algorithm (in the size of m) is known. The most powerful algorithm is due to Morrison and Brillhart. It is quite complicated.

h) How to solve $x^2 \equiv a \pmod m$

If m is prime I can find x in polynomial time. If m is composite, finding square roots is as difficult as factoring. If I can solve $x^2 \equiv a \pmod m$ efficiently, than I can factor m efficiently. This was shown by Rabin.

The most interesting application of these algorithms is to Public Key Cryptography. We sketch the most efficient encryption and decryption algorithm due to Rabin.

With Rabin's test I find two large primes p,q, $p \equiv q \equiv 3 \pmod 4$, say of about 100 digits each, and I keep them secret. Then I compute m = p.q and publish m. If someone wants to transmit to me a secret message M, then he codes it as a number x < m. If M is too long, he splits it into several parts. Then he computes $x^2 \equiv a \pmod m$ and sends me a. If an eavesdropper intercepts a he is stuck with the formidable task of factoring a 200-digit number m. I know p and q. So I can solve $x^2 \equiv a \pmod p$ and $y^2 \equiv a \pmod q$. Indeed, by Fermat's theorem $x = \sqrt{a} = a^{(p+1)/4} \pmod p$, $y = \sqrt{a} = a^{(q+1)/4} \pmod q$ are solutions. Then I combine these solutions into a solution of $x^2 \equiv a \pmod m$ by means of the Chinese Remainder Theorem. One of the four solutions is the message x.

See references (16) - (20) for this section.

II. Discrete Probability: An Algorithmic Approach

The algorithmic approach radically simplifies discrete probability and makes it almost self evident. No special concepts or any extensive theory are needed. Any problem is practically solved as soon as it is represented by a probability graph.

Discrete probability studies random walks on graphs. Equivalently, it studies the flow of mass 1 through a probability graph, respecting the transition of probabilities. The size of the graph is a good measure of the complexity of the problem to be solved. Each symmetry of the graph can be used to reduce its size, and hence to reduce the complexity of the problem.

Suppose the graph of a problem is so small that it can be represented on paper or inside a computer. If the probabilities are simple rational numbers, we can use the "probabilistic abacus", i.e. we simulate the flow of mass 1 by moving chips on the graph. This applies to most textbook problems. A short description of the probabalistic abacus can be found in Springer's Mathematics Calendar 1981. For more details see (7) and (8). If the probabilities are not so simple, then the mass 1 should be pumped through the graph by means of the computer.

But suppose the graph is so huge that we cannot store it in a computer. It may well be bigger than the universe. Then we can find approximate answers by generating many random walks with the random number generator of the computer.

Below we give two examples illustrating these possibilities.

Example #1. Another poor man's generator of coin tosses

Let 1 < a < b. It is easy to see that the algorithm in Fig. 13 computes the binary digits of $\log_b a$. Usually it does not stop.

```
do a ≠ b ⟹
   a ← a   a
   if a < b ⟹ print 0
   □ a > b ⟹ print 1; a ← a/b
   fi
od
```

Fig. 13

My pocket calculator gives for a = 2 and b = 10:

log 2 = .0100110100 0100000100 1101010000 1001111101 1110001111 0000000011 1010110010 0111110010 0011000001 0111011100 0101110001 1101101001.

Only the first 36 bits are correct. The remaining bits are corrupted by rounding errors.

Is this algorithm a good generator of coin tosses? Obvious frequency counts are satisfactory. There are 63 "zeros" and 57 "ones", and there are 56 runs. These numbers lie within δ-limits of their expectations. But there is a run of 8 zeros. Is this suspicious?

To answer this question we formulate and solve two probabilistic problems: A fair coin is tossed repeatedly. a) What is thé expected waiting time for an 8-run of zeros or ones? b) What is the probability of an 8-run in 120 flips of a coin?

Fig. 14 shows a graph for these problems. The states of 0 to 8 are the run lengths. I am in state i if I have collected a run of length i. At the next step I will have run length i + 1 or 1, each with a probability of 1/2 ($i \geq 1$). All branches except 01 have probability of 1/2 (not shown in Fig. 14).

Now problem a) is easy. In one minute I get with chips the expected waiting time 255 for an 8-run.

Problem b) is computationally more difficult. Here one should use the computer or a programmable pocket calculator. Initially we place mass q at state 0 and clear all other states. Then we pump this mass through the graph in a sequence of steps. At the first step all the mass goes to state 1. At each step after the first the mass at each state is moved to the two neighboring states, each neighbor getting the same amount. After 120 steps we stop. The mass in state 8 will be m(8) = 0.3678. This is the probability of at least one 8-run in 120 steps.

Suppose we initially set e ← 0. After each pumping step we set e ← e + 1 - m(8), until m(8) = 1. Then e will be the expected waiting time for an 8-run! This method works always, whereas the abacus works only with simple probabilities. More details about this method can be found in (3).

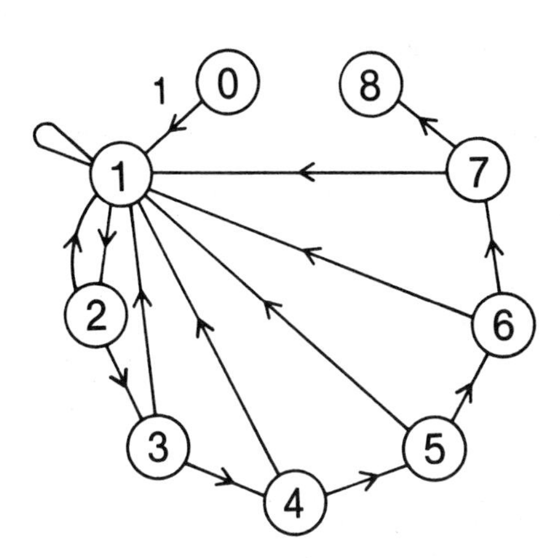

Fig. 14

Example #2. Knuth's method of estimating the size of a tree

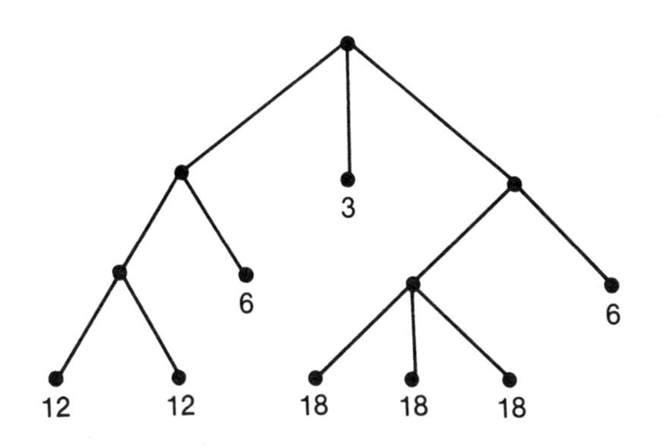

Fig. 15

Fig. 15 shows the tree of all possible evolutions of some process. Walk at random through the tree. That is, when at node i, choose each of the X_i branches leaving that node with the same probability $1/X_i$. The product $X = X_1 \cdot X_2 \cdot X_3 \ldots$ along your path is your gain. What is your expected gain E(X)? It is almost immediate that E(X) = # of paths through a tree. So each walk gives you an estimate of the size of the tree. In this way we can find rough estimates for intractable counting problems. We apply the method to two famous problems in theoretical physics.

a) A dimer is a 2-atom molecule, i.e. the configuration •——• . To find the physical properties of a system of dimers one needs the numer of possible coverings of a graph with dimers.

Take the 8 x 8 square lattice. In how many ways can you cover it with dimers?

Equivalently: In how many ways can you cover an 8 x 8 chessboard with 2 x 1 dominoes?

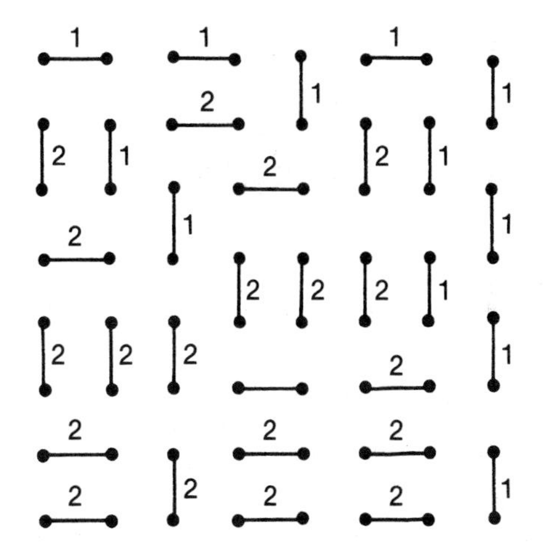

Fig. 16 . 2^{20}

We generate some coverings with a coin. In the first row of Fig. 16 I went from left to right. For each point I assigned its neighbor to the East or North of its mate. If there were two possibilities the coin decided the mate. Similarly I proceeded with rows 2 to 8. The product 2^{20} of the number of alternatives at the 32 stages is our first estimate for the number of coverings. Fig. 17 shows 14 additional estimates. The arithmetic mean of these 15 numbers is about 8.9×10^6. This is an even better estimate for the number of coverings.

2^{21} 2^{24} 2^{21} 2^{17}

2^{22} 2^{24} 2^{19} 2^{21}

2^{22} 2^{13} 2^{22} 2^{18}

2^{26} 2^{22}

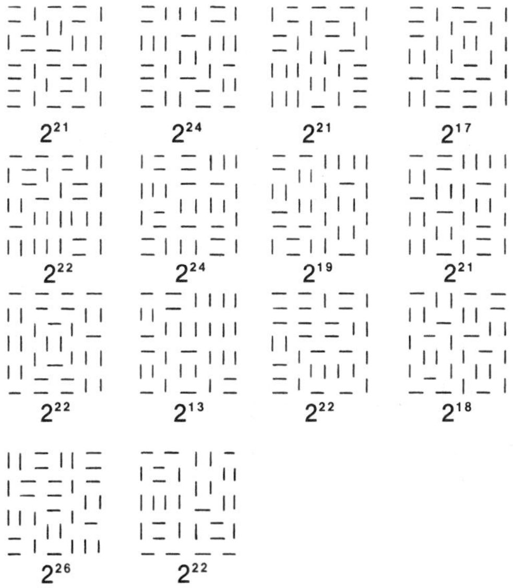

Fig. 17

b) Self avoiding random walks (D.E. Knuth, Science, Vol. 194, pp. 1235-1242)

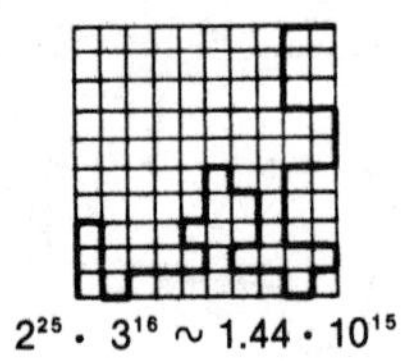

Fig. 18

Fig. 18 shows a 10 x 10 road system. There are

$\binom{20}{10}$ = 184756 shortest lattice paths from s to t. Physicists and chemists would like to know the number of self avoiding paths from s to t, i.e. paths which do not cross or touch themselves. In spite of immense effort no solution is in sight. It may well be that this problem is intractable and we will never find the answer.

To estimate this number we have generated 10 self avoiding paths with a die and for each we found the the product of available choices along the path (Fig. 19). Each of these products is an estimate of the unknown number and their arithemtic mean $x \sim 0.68.10^{23}$ is an even better estimate. Because of the huge fluctuation x is not a reliable estimate. Knuth gets with several thousand paths the estimate $(1.6 \pm 0.3).10^{24}$.

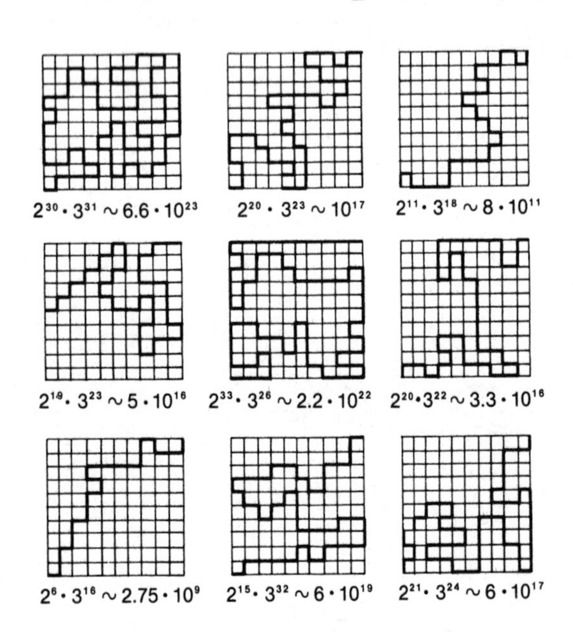

Fig. 19

The approach to probability via graphs is explained in great detail in (6). Short summaries are given in (7) and (13).

III. Statistics and the Computer

In statistics you collect, describe and interpret data generated by some unknown random process. In all three aspects the computer can play an important role. Consider data interpretation. There are computationally simple problems, but the student has trouble to understand the meaning of the result. Here simulation can help. On the other hand there are intractable problems. A substantial theoretical background may be required to understand some approximate solution. With a computer we can always use simulation and quite often exhaustive search. Both procedures do not require any statistical background. The simulation program is usually very simple. Mostly the exhaustive search program must be quite sophisticated to be efficient. We give two typical examples.

Example #1. The taxi-problem. Confidence intervals

In Frankfurt taxis are numbered 1, 2, . . ., T. I do not know the parameter T. So I observe four taxis with numbers 355, 512, 987, 1200 = M (maximum observed number). Obviously $T \geq M$. I want a reasonable upper bound for T.

Suppose I observe n taxis with numbers $N_1, N_2, \ldots, N_n$ and let $M = \max N_i$. Then from Fig. 20 we get $P_n = P(\text{all } N_i \leq M) = (M / T)^n$.

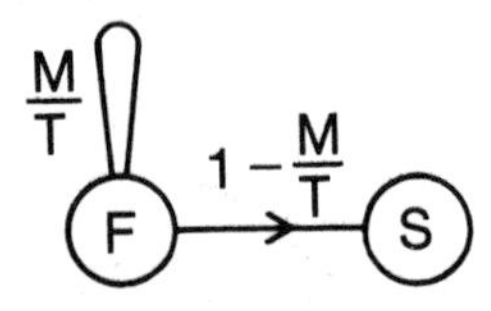

Fig. 20

We consider an estimate for T as "reasonable" if p_n is not "surprisingly low", say $p_n < 5\%$. Then we have

$(M/T)^n \geq 1/20$ or $T \leq M \cdot \sqrt[n]{20}$.

We call

$M \leq T \leq M \cdot \sqrt[n]{20}$

a 95% confidence interval for the unknown parameter T. For n = 4, M = 1200 we get $1200 \leq T \leq 2538$.

These computations are quite trivial, but few students can correctly interpret the meaning of this statement. Here it is very helpful to repeat many times the following computer simulation:

a) Choose T at random for some interval, say between 1000 and 2000.

b) Choose at random n numbers from 1 to T and call their maximum M.

c) Print the statement

$M \leq T \leq M \cdot \sqrt[n]{20}$.

About 19 out of 20 of these statements will be correct.

Example #2. Effect of marijuana on naive subjects (Science, v. 162, pp. 1234-1242)

subject	ordinary cigarette score X	marijuana cigarette score Y	difference D = X−Y	sign of D
1	−3	+5	+8	+
2	+10	−17	−27	−
3	−3	−7	−4	−
4	+3	−3	−6	−
5	+4	−7	−11	−
6	−3	−9	−6	−
7	+2	−6	−8	−
8	−1	+1	+2	+
9	−1	−3	−2	−
	ΣX = 8	ΣY = −46	ΣY − ΣX = −54	

Table 21

Table 21 shows the change in intellectual performance in 9 subjects after smoking an ordinary cigarette and a marijuana cigarette. The scores X,Y are measured from the subject's base level 0.

Consider the hypothesis H: There is no difference between the two types of cigarettes, versus the (one-sided) alternative A: Marijuana lowers the average performance.

Let M be the number of minuses in the last column. If H is true, we have

$$P(M \geq 7) = [\binom{9}{2} + \binom{9}{1} + \binom{9}{0}]2^{-9} = \frac{23}{256} \sim 9\%$$

This is too large to reject H. But we did not take into account the size of the difference D. This can only be done with a substantial theoretical background. With the computer we have two good alternatives.

Simulation. Put all 18 scores into a box, draw 1000 random partitions into two 9-subsets X,Y and each time compute

$$|\Sigma Y - \Sigma X|$$

Count how often this is ≥ 54. In 10 repetitions I observed the counts 36, 31, 27, 20, 27, 28, 27, 30, 30, 29 (out of 1000). So

$$P(|\Sigma Y - \Sigma X| \geq 54) \approx 2.9\%$$

Since we are testing one-sidedly we have

$$P(\Sigma Y - \Sigma X \leq -54) \approx 1.5\%$$

This is a highly significant result and we may reject H in favor of A.

Exhaustive search.

a) Combine all 18 numbers into a set X = (-3, 10, . . . , -3) and consider all the $\binom{18}{9}$ = 48 620 ways of partitioning X into sets X,Y each having 9 members.

b) For each such partition compute the difference $\Sigma Y - \Sigma X$. there are 48 620 such differences, one of which is the observed -54.

c) Count the number L of differences for which $\Sigma Y - \Sigma X \leq -54$ and compute $\alpha = L/48\,620$. If $\alpha \leq 5\%$, decide that the set Y is really smaller than the set X.

The program for exhaustive search is quite complicated. We must generate all permutations of the word 000 000 000 111 111 111 by means of some permutation generation program. Such a program is quite sophisticated and time consuming.

References

1. Aho, Hopcroft, Ullman, The Design and Analysis of Algorithms. Addison-Wesley 1974.
2. E.W. Dijkstra, A Discipline of Programming. Prentice-Hall 1976.
3. A. Engel, Elementarmathematik vom algorithmischen Standpunkt. Ernst Klett 1977.
4. A. Engel, The role of algorithms and computers in teaching mathematics at school. In "New Trends in Mathematics Teaching IV". UNESCO, Paris 1979.
5. A. Engel, Algorithmen fur den Taschenrechner. In "Der Mathematikunterricht", 1979, Heft 6, pp. 52-77.
6. A. Engel, Wahrscheinlichkeit und Statistik, Bd 2, Ernst Klett 1976.
7. A. Engel, The Probabilistic Abacus. Educ. Stud. in Math. 6 (1975), 1-2.
8. A. Engel, Why Does the Probabilistic Abacus Work? Educ. Stud. in Math. 7 (1976), 59-69.
9. D.E. Knuth, The Art of Computer Programming, vol. 1-3. Addison-Wesley.
10. L. Kronsjo, Algorithms: Their Complexity and Efficiency. Wiley 1979.
11. Nijenhuis & Wilf, Combinatorial Algorithms. Academic Press 1978.
12. Reingold, Neivergelt & Deo, Combinatorial Algorithms. Prentice-Hall 19 .
13. Springer's Mathematics Calendar, August 1981.
14. N. Wirth, Systematic Programming. Prentice-Hall 1973.
15. N. Wirth, Algorithms + Data Structures = Programs. Prentice-Hall 1976.

16. D.E. Knuth, Fundamental Algorithms, vol. 2, Second Edition 1981. This is one of the most up to date single references on seminumerical algorithms and public key cryptography.

17. M.O. Rabin, Digitalized signatures and public key functions as intractable as factorization. MIT/LCS/TR-212, Technical Memo MIT (197).

18. A. Engel, Datenschutz durch Chiffrieren. Der Mathematikunterricht, 1979, Heft 6, pp. 30-50. (Rabin's test explained.)

19. M.E. Hellman, The Mathematics of Public Key Cryptography. Scientific American, August 1979.

20. M.O. Rabin, Probabilistic Algorithms. In "Algorithms and Complexity, New Directions and Recent Results", J.F. Traub (ed.) 1976.

10.4 OPERATIONS RESEARCH

OPERATIONS RESEARCH: A RICH SOURCE OF MATERIALS TO REVITALIZE SCHOOL LEVEL MATHEMATICS

W.F. Lucas
Cornell University

1. Introduction. An enormous amount of new mathematics has been introduced over the past forty years. This includes whole new subjects and entirely different mathematical techniques which are proving to be as fundamental and important as the traditional and classical mathematical subjects. Many of these developments provide an enormous and rich source of materials which are highly appropriate for introduction into school mathematics at both the primary and secondary levels. These topics are particularly suitable for developing more exciting and relevant courses in which students can become actively involved in a more personally creative and interactive manner. On the other hand, there is ample evidence that these splendid opportunities are not fully appreciated and are being mostly overlooked by mathematics educators. It behooves the mathematical community to become better informed about these newer scientific developments and their appropriateness for inclusion in the curriculum, especially in light of some of the problems and difficulties currently existing in mathematics education at all levels.

Some comments concerning ongoing discoveries and new intuitions, and their relevance to education will be made in section 2. Some questions about how topics are selected for inclusion in the standard curriculum arise in section 3. A description of one of the newer mathematical subjects, operations research, as well as the nature of mathematical modeling will be given in sections 4 and 5, respectively. A discussion of the personal nature of mathematical learning and the consequences which this implies for teaching appears in section 6. A few simple illustrations of the type of materials now abundantly available for enhancing such learning and teaching are mentioned in the final section.

2. Contemporary Developments. New discoveries are occurring so rapidly in the general mathematical sciences, as well as in many other disciplines, that the total quantity of mathematical knowledge available more than doubles about every decade. This is truly an astonishing rate of growth! The resulting implications for mathematical education are likewise no less astonishing.

There is a major misconception that most of this mathematical growth is in areas so specialized, abstruse or advanced that it is not of importance to school level education. Mathematicians are viewed as building tall narrow towers of abstract specialized theories which reach so high above the clouds that they are obscure to almost every one on earth. Whereas the many new applications of the subject are often pictured as the same old traditional building blocks being scattered about haphazardly in so many different directions. While in fact, many of the recent developments are extremely original and fundamental in nature, and many are also well suited for designing more interesting, creative and relevant courses at all levels of education. The current breadth of applications is such that no one person can really appreciate it.

The most revolutionary aspect of many new developments is probably the fact that whole new areas of human endeavor are being scrutinized from a mathematical point of view and that entirely different basic intuitional inputs are significantly influencing ongoing research. A basic reliance upon ideas about numbers,geometry and physical dynamics is often being replaced by quite different quantifiable concepts which often arise in areas such as the social, managerial and decision sciences. The great philosopher Immanuel Kant said -

> Two things fill my mind with never ceasing awe: the starry heavens above me and the ethical law within me.

Mathematics is clearly moving from considerations originally arising from more physical type realities to ones concerned with more human types of interactions as well. I.e., from the general direction of Kant's first awe towards that of his latter one. These advances are surely providing a rejuvenating return to fresh empirical and intuitional sources, and are significantly revitalizing modern mathematics. The implications for education are certainly profound and enormous in scope. They naturally suggest major changes and experimentation in curriculum content, teaching methods, and teacher training. There is now a great variety of interesting topics and approaches becoming available to motivate and stimulate all but the most recalcitrant of students.

Nevertheless, the mathematical community is faced with some difficult problems. The relevance of mathematics (or at least as it is presently being taught) is being seriously questioned, and perhaps for good reasons. The "new math" is being declared a failure and being replaced. The various declines in our measures of mathematical comprehension or skills are frightening in their magnitude. Employers are appalled at the lack of mathematical competency in our graduates. Although society in general should take the blame for some of these developments, much of it appears to be primarily due to past decisions and policies (and inactions) by the mathemtical establishment and its leadership. The prestige of mathematics has fallen in the U.S.A. to a point where it is a source of jokes appearing in cartoon and comic strips in the newspapers. E.g., the Governor Duke in the popular Doonesbury comic, who is serving in an American territory bemoans the sad state of the emergency relief bureaucracy in Washington and says: "They sent me a dozen math teachers!" At the same time, there are "culturally rich" materials becoming rapidly available that could go a long way in helping to overcome many of the shortcomings in current mathematical education in the U.S.A. An enhanced visibility and improved image as an interesting, relevant and dynamic field could surely be achieved. Despite our problems, some splendid opportunities are readily available. A willingness to change along with a better knowledge of the particularly suitable recent directions of mathematics (both pure and applied) will be necessary. It is surely time to stop listening to the so-called "experts" who are heavily responsible for current problems, to quit analyzing, justifying or apologizing for past mistakes, to discard prejudices and bad attitudes, and to get on with the business of teaching modern mathematics in a manner that will appeal to the next generation of students.

3. Priorities. What parts of mathematics, if any, are really the most important? For research, or for education at various levels? Who decides this, and upon what basis? Are the most distinguished researchers in the more established or fashionable areas, or the best extemporaneous speakers or leaders of professional societies, the ones best able and qualified to make such decisions? Who is?

Few will likely argue with the position that the individual research scientist should be the ultimate judge of what he or she should pursue based upon ones own ability, likes, and judgment of what is important. It is not so obvious, how more global research policies and decisions on educational content should be made. Almost all accomplished researchers and teachers do in fact have a rather minimal understanding of most of the broad field of mathematics, and this includes many things that are rather basic and elementary as well. Some distinguished scientists such as Joh Von Neumann have expressed great concern over how our precious human, financial and institutional resources are allocated to different aspects and areas of mathematics. It is also an obvious fact that a very great majority of what we teach below the graduate level is very "old" in light of the growth mentioned above. That is, almost all of currently known mathematics has been discovered after we learned about the topics which are taught. Such a disproportionate amount of "old stuff" in our courses does not match with the fact that mathematics is a live and growing field. Courses in the other natural sciences, as well as those in many other areas, rarely contain so much dated material, even at the elementary levels.

If a highly knowledgeable and unprejudiced source were to design a mathematics curriculum from scratch, and without any hinderance from the practical constraints which would arise in implementing it, then it is reasonably safe to say that the result would differ greatly from the present emphasis. The idea that the vast majority of our students should all see the same topics, whereas hardly anyone sees others of likely equal importance to mathematics and its applications, is silly. Is group theory really any more basic or useful than graph theory is likely to be? Linear algebra is probably the most fundamental undergraduate course, whereas few mathematics majors learn linear programming. Are vector spaces really that much more fundamental than cones or polyhedra? Or have we been influenced by more historical events, such as, the fact that physics has been around for a long time and uses mostly equations while econimics, which frequently uses inequalities, has only recently been mathematized in a really serious manner? Or simply that all college teachers know the former subject and few the latter one? Can any knowledgeable person really beleive that network flow theory with its extensions and generalizations will not develop into as high and rich a subject as ring theory and its decedents. And network flow theory which is only about 20 years old, has already been used in an essential way outside of mathematics by more people than group theory has in nearly 200 years. One looking into the theory of networks should not become turned off by the tedious algorithms present in so much of the literature. And they should recall that most algebraists in the 19th century could actually produce or construct the things they discovered and studied. Also, that Cayley's theorem in group theory almost "killed" that subject, because what could be more dry and tedious than studying permutations. Even the necessity of sacred subjects like calculus is beginning to be questioned by a few, and is being replaced in part by a few others. In light of digitial computers, finite math will surely prove to be as important in a decade or two as continuous math will be. It is naive to think that some modern topics, such as graphs or networks, could not be much more successful as a framework or idiom in which to embed elementary mathematics as the logic and set theory of the "new math". Many of the nonnumeric and less spatial concepts in many recent developments are also very fundamental in nature and should hardly be ignored until studied in college courses. Some currently basic topics will necessarily have to be squeezed out of the training for at least some of the people in order to make room for the new. But compactification also will go on as the ongoing "trivialization of knowledge" takes place.

4. Operations Research. It is rather interesting how infrequently mathematics courses begin with a serious definition of the term "mathematics" itself. This is somewhat understandable, since it is such an extensive field that it would be impossible to do justice to the subject with a rather concise statement of the type mathematicians like. Nevertheless, it would probably be a good exercise for each mathematics teacher to formulate such a definition that they would be willing to state and defend in a public manner. In lieu of a definition some mathematical scientists or student advisors will attempt to provide a rough outline of the subject. Such attempts often begin with some of the major divisions, such as pure and applied, discrete and continuous, stochastic and deterministic, etc. It is most unfortunate in the U.S.A. how seriously this first division is taken and used to support outdated and harmful arguments on course or degree content. There is a tendency to view utility and beauty as inherently in conflict, or to hold that marketable skills work to undermine a liberal educatin. It is even becoming fashionable to distinguish more finely between pure, applicable and applied mathematics. Perhaps we should extend this into two dimensions and talk about pure pure, pure applied, applied pure, and applied applied mathematics. The author, when advising students, divides applied mathematics into four major areas: (1) the classical applied mathematics growing mostly out of calculus and useful in the physical sciences, (2) applied probability and statistics, (3) the realm of computers and computer science, and a less easy to describe

collection of additional theory and methods which will be called (4) operations research. Admittedly many things fall between the cracks of such a crude partition, but this is a reasonable first list of what type of degree programs, jobs, and rich sources for mathematical education are available. Only ideas related to this fourth area will be discussed further in this paper.

Operations Research (O.R.) grew out of World War II, in particular with problems and support from the U.S. and British navies. There were some notable individuals who made important contributions prior to the War. However, the large scale logistics, supply, allocations and operational problems which arose in the military, and their study by scientists from many different disciplines created the techniques and approaches which grew into modern O.R. O.R. is the study of complex systems, structures and institutions by means of the scientific or applied mathematical method with a mind towards operating such systems more efficiently within various constraints, such as, scarce resources. The results of such studies become one of several inputs into making decisions, often in fields in which decisions were previously made on the basis of less quantitative approaches, such as experience, tradition, hunches, or suchlike. There is frequently a major concern with "people" as well as "things", and the man-system interface in a complex social or public activity. The problems studied involve supply and demand, inventories and queues, resources, forecasting and planning, allocations, reliability, scheduling and maintenance, marketing and budgeting, services, technology and productivity,design, utility and measurement, and many other aspects. The approach is interdisciplinary in nature, and uses common sense, data, and substantial empericism in addition to new, as well as, repackaged traditional mathematical methodoligies. Optimization, understanding complicated interrelationships, and clarifying trade-offs are common goals. O.R. has connections with many other fields such as industrial engineering, systems analysis, management science, logistics, econometrics and decision theory. Above all, O.R. is concerned with problem solving, as are many other subjects. However, the types of problems studied and the tools or techniques used to solve them is what places them in O.R. All of the skills and approaches used in general mathematical modeling (to be discussed in the next section) are essential in O.R. Interesting articles by Daniel Wagner and Gordon Raisbeck about O.R. appear in The American Mathematical Monthly, as well as in the booklet Careers in O.R. put out by the Operations Research Society of America.

The primary concern in this paper is with what O.R. has to offer to education, especially to mathematical education at the school level. It offers a new set of highly motivating, more realistic, and challenging problems which can be adjusted to nearly all grade levels. These allow the students to experience and appreciate all aspects of the mathematical modeling process at much lower levels. To model in the directions of the physical, biological, and much of the engineering sciences normally requires substantial prerequisites in these sciences as well as in mathematics, whereas, many O.R. sorts of problems are often so common and intuitive that fewer scientific or mathematical tools are needed. So that real mathematical creation can begin at a much younger age. It does not follow that problem solving per se is easier in O.R. than other applications, nor that O.R. is a direction in which to direct one's less talented students. Nevertheless, many students are likely to be "turned on" by such intrinsically exciting problems who are not "moved" by ones originating in physics or geometry.

Of course, such benefits as new intuitional and new subject matter input for mathematical research as well as increased job opportunities at the college level are an added bonus for those exposed to contemporary applications such as O.R.

5. Mathematical Modeling. A main ingredient in O.R., as well as all of applied mathematics, is an activity which will be referred to as mathematical modeling. It is the primary methodology of applied mathematics, and more generally it is the general scientific method or the quantitative approach to knowledge. It is a fundamental and essential intellectual activity of those who believe in the rational life based upon knowledge, observation, quantification and analysis. It involves much more than pure reason on the one hand, and it contrasts with argumentation and decision-making based on more emotional or purely intuitive grounds on the other. It is a methodology for problem solving in general, and it is used in many diverse fields by abstract thinking people. It is unlikely to decrease in importance, at least until all of our problems are solved.

The activity of mathematical modeling can be represented to a certain degree of approximation by the enclosed figure. The problem solver begins

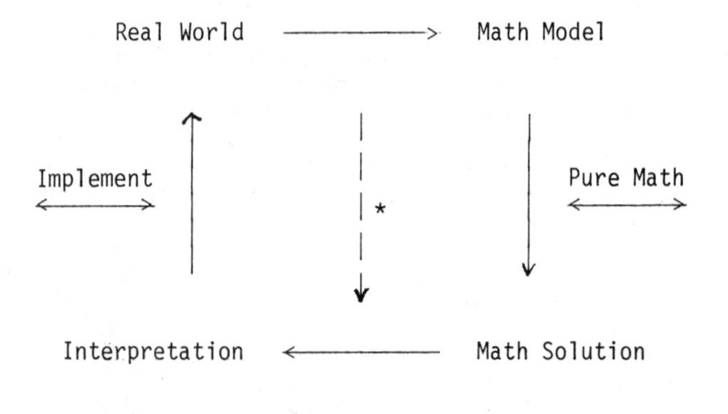

Figure 1

with a problem which most frequently does in fact arise rather directly out of the "real world". One builds an abstract model for purposes of analysis which often takes a mathematical form. The model is in some sense solved first in this abstract setting, but then must be interpreted back in its original context. The analytical conclusions must then be compared with reality. If they fall short of matching the real situation, then a modification or revision of the model may be called for, and one procedes around this cycle again. One often procedes back and forth within a cycle, and iterates around this figure many times before arriving at a satisfactory representation of the real world. Often the abstract model will merely verify from another perspective things that were already known. This is usually not bad or wasted effort as is often implied. It is usually beneficial to have multiple ways to explain things, and reproducing known facts is also a partial verification of the validity of the model being used.

Nevertheless, the real gain from the modeling process occurs when the abstract approach suggests some new knowledge, and then this is actually verified by observation or experiment. The creation of new insights via this indirect route is what theoretical science and applied mathematics is all about.

Discussions of this modeling paradigm are often dressed up in much more flowery language. If one figure is rotated 90°, then one can speak of rising up from the real world into the heavenly cosmos of abstraction, or floating or gliding through the ethereal bliss of mathematical deduction, and of finally plunging down again to reality. If the figure is rotated a negative 90°, then one can think of first diving down into the dark underworld of mathematical symbolism and deep abstraction, of swimming in the murky and turbulent labyrinth of abstract logic, and of ultimately finding or clawing one's way back up to the more concrete things which are real. With the aid of technological advances, such as the computer, one can frequently "short circuit" the network in the figure. Using methods such as simulation one can often build a simpler conceptual model which, instead, involves high amounts of computations of a more routine nature which can be accomplished by the computer. This short cut is indicated by the downward broken arrow indicated by an * in the figure. The activities on the right side of the center square in our figure provide ideas for pure mathematics and vice versa. Off to the left is the very practical and political problem of getting new beneficial ideas or discoveries implemented in the scientific or wider society. The activities concerned with the four solid arrows in the figure will be discussed in turn in what follows. Many writers add other nodes and arrows to describe the creative or modeling process than done here, but these four will suffice for the current discussion.

First consider the downward pointing solid arrow on the right side of the figure, i.e., the one between math model and math solution. This is the activity of finding solutions to well formulated mathematical problems. It is usually the most logical, well defined and straight foreward part of modeling, although not necessarily the easiest. It is often the most immediately pleasing, elegant, perhaps more seductive, and aesthetically beautiful intellectual endeavour, and has a strong deductive element to it. This one "side" of the "modeling square" is probably the one covered best in mathematical education. Unfortunately, this aspect of problem solving is done almost to the exclusion of the other three sides of the square. Whereas these other sides often involve much more creativity, originality, interaction with other disciplines, and communication skills. The typical mathematics (or even physics) lecture in college rarely contains more than a few minutes at the start or at the end that is not devoted to obtaining math solutions to well defined problems already in precise math form. Great benefits could accrue to mathematical education if this unreasonable emphasis could be altered.

The bottom arrow in the figure is concerned with translating and explaining a purely mathematical result in terms of the original real world setting. This involves the need to communicate clearly and in a lucid manner, and this, according to many employers, is the most serious shortcoming in graduates. This aspect of a mathematical training can hardly be left solely to courses in other sciences or to job experience after graduation. More of this aspect in courses would surely demonstrate the relevance of mathematics in a very meaningful way. In describing the meaning of a mathematical solution one must take great care to be complete and honest. It is dangerous to quickly discard some of the math solutions to a physical problem as extraneous or having no physical meaning; there have been too many historical incidences to show that this was not the case. Likewise, one should not select out just the one perceived answer which the "boss" wants to support his or her position. A politician or decisionmaker does not want just one optimal solution, but desires to know a variety of "good" ones and the range of reasonable options available from which to select. And not only the proverbial <u>three</u> options normally presented: one too extreme to the "right" (or overreaction), and one too far to the "left" (or inaction), and a dominant one so prominently placed in the "center of reasonability". The analyst in industry or government can often be the scapegoat. When plans succeed, bosses credit themselves and forget where they got their ideas, whereas if they fail, they blame their advisors (including the analysis group) and never forget where the suggestions came from. However, bosses typically do not act on quantitative recommendations unless they are communicated in a manner which is somewhat understandable to them. This can often be a difficult task because of the technical nature of the formulation and solution, and also because great quantities of data and computation may need to be compressed to a manageable size for the layman to understand in a relatively limited time. Nevertheless, mathematical education must give some attention to this aspect of modeling if it expects any respect of visibility in the non-academic world.

A major step in real world modeling is to validate and verify solutions against the original phenomena being studied. This involves testing, verifying, checking, evaluating and comparisons. Clearly, modern math education rarely "dirties" itself by getting involved on this side of the modeling square. Except perhaps for an occasional "eyeballing" of an answer from this point of view, or the infrequent project arising in the math or statistics clinic, or suchlike. By omitting this activity mathematical education misses an opportunity to interact with real-world decision making, judgmental inputs, the limitations of its tools, and the more human aspects of science.

Finally, consider the top arrow in the figure which represents the heart of the modeling activity. The formulation or construction of an abstract model from a real situation is the really creative activity and most important contribution of theoretical science. Building models involves translation into mathematics, maintaining the essential ingredients while filtering out the great amount of excess baggage, and arriving at realistic and manageable intellectual prototypes. The fabricator of mental models must perceive and observe keenly, recognize and visualize clearly, discern laws and identify properties precisely, conceptualize and idealize with accuracy, and lay bare the essential ingredients. One must use induction, become an absorbing receptor of data, become actively involved, and encounter problems with a challenging and confronting attitude. This initial part of modeling is clearly the most essential and valuable part of the whole process. It is also the most difficult part. Eddington said: "I regard the introductory part of a theory as the more difficult, because we have to use our brains all of the time. Afterwards we can use mathematics." Or as E.E. Cummings says in <u>A Poet's Advise</u> (1955): "A poet is someone who feels and expresses his feelings in words. This may sound easy. It isn't." This quote would remain valid if one replaced "poet" by "applied mathematician", "feels" by "observes", and "words" by "mathematical relationships". Unfortunately, mathematical education does a very poor job in developing model builders. Whereas this is the essence and most exciting part of intellectual creativity. To shortchange students in this way is to cheat them out of the potentially most rewarding and stimulating part of their education. It is not something to be left entirely to on-the-job experience, since few graduates will be given the opportunity to concentrate on this activity unless they quickly show some attribute for it. Clearly, the most glaring error and most gapping shortcoming in mathematical education today is the essential neglect of providing experience for, and of building confidence in, students in this model building activity. This is true for pure as well as applied math and for elementary as well as advanced levels. Although specialized modeling courses are becoming popular at some colleges, this approach (like history) should permeate throughout most courses in the mathematics curriculum.

One main point of ths paper is to make a plea for putting more educational emphasis on the full cycle of the modeling process, especially upon the model formulation link in this cycle, and to assert that many exciting materials developed in recent years are amply suited for accomplishing this and at educational levels much lower than those previously realized using most scientific areas popular in the past. In particular, a great many problems arising in O.R. and finite math have quite realistic and elementary forms which are so natural and obvious that they can be modeled with essentially no scientific prerequisites, including few mathematical ones. Nevertheless, they require ingenuity and creativity to solve, can provide enormous intellectual satisfaction, and indicate the great power of mathematics in aiding to understand real-world phenomna. In short there is ample material available that would enhance motivation and creativity, provide ample quantities of fun, demonstrate the relevance of mathematics, and at the same time develop the essential skills considered important in mathematics. The mathematical community appears to be "missing the boat" to achieving a modern contemporary curriculum, under the false notion that anything really important or fundamental must have been discovered and developed many years ago. This treating of mathematics as one of the classics of antiquity (or as out of the neneteenth century even) is very unfair to students and the subject.

Mathematics and many other disciplines surely need young scientists who can do all aspects of the modeling process well, not just ones good at providing math solutions to well defined math problems. In selecting students for advanced study in many fields, such as, engineering and medicine, math skills are often given enormous weight. However, it is becoming much better known how narrow modern mathematical training has become. It may soon follow that tests of mathematical knowledge will become less important in selecting students for these other areas, which may not be a very bad thing in itself. However, it may also lead to mathematics being viewed as less important as a part of general education, and this could have a serious impact on the mathematical community, its members, and society as a whole.

6. Need for More Interactive Teaching. The learning and doing of mathematics is normally recognized as a very personal and individualized activity. One normally learns this subject by "doing it" and not merely by watching it being done. Mathematics is a participatory game and not a sport for mentally sedentary spectators. The private and internal nature of learning mathematics is evident at all levels, from understanding elementary concepts through original research at the frontiers of knowledge. The great majority of publications in the subject are by sole authors, contrary to what one finds in many other fields. Large cooperative projects in mathematics are less common and rarely produce specific goals or priorities set long in advance, although they may "set the stage" for many individuals in the group to make important contributions. Close and open association and extensive communication plays an important role in mathematical discoveries and advances. Nevertheless, each mathematical reseracher is both king or queen and laborer, and must design, build and toil by himself or herself in a mostly independent fashion. Success in learning and research is most likely to follow individual selection of problems on the basis of fun, challenge and personal need or satisfaction.

It is rather obvious that most mathematics classes in the U.S.A. are not very effective vehicles for learning. Large lectures are usually the most ineffective format, and anyone who doubts this should dare to design a testing mechanism that separates, to the extent possible, lecture learning from other types. Anyone who attends presentations to large groups in classrooms or at professional meetings is usually appalled at the marked contrast of the quality of exposition between mathematics and other scientific disciplines. Mathematics students must really love their chosen field to be willing to so persistently endure so much poor teaching! Even the few supposedly good math lecturers are normally imparting mostly motivation, and distressingly little in terms of understanding. Passivity on the students part is often taken as satisfaction, and protext to horrendous teaching as antiintellectualism. It is likewise well-known that math students rarely read much of their subject before attaining a relatively advanced level. It is quite obvious that much is to be gained if more individual discovery and hands-on experience could be incorporated into mathematical education by means of inherently interesting, worthwhile and suitable materials.

The great need for more personally involving and interactive teaching techniques is all the more necessary in light of the many glamorous and seductive aspects of modern life in the U.S.A. which are competing for the students' attention. These range from the nonprerequisite material one can passively absorb via television, through a long list of other activities available to children and young adults. In comparison, abstract mathematics can appear quite dull, rather pointless, and devoid of real human interest. In an age of instant gratification, "show me" attitudes, permissivism, and so many people telling one how he or she should feel, along with behavioral objectives, standardized tests, and the "new levelers" in education, it will become increasingly difficult to persuade students to keep their noses to the grindstone called mathematics. More and more people will likely "vote with their feet" and march off in other directions. Mathematics will surely lose out unless, of course, mathematical educators rise to the challenge and confront the competition with a reasonable counterattack.

American educators can learn a great deal about the likely success or failure of alternate educational approaches in mathematics by investigating methods used in other parts of the world. It seems most unlikely that the system nor the public in the U.S.A. is up to providing the extensive and very deep grounding in basic fundamentals of science and mathematics whose mastery is necessary for most advancements as in the Soviet Union, or the long school day and large number of required courses so common in many European countries. On the other hand, the shorter but extremely intensive and interactive nature of the mathematics classes in China (P.R.C.), the emphasis on collecting real data at elementary levels in Japan, the early exposure to finite math techniques and the resulting research done by teenagers in Hungary may well provide successful models which could be initiated into more courses in the U.S.A. As Israel moves from covering the traditional twelve years of school mathematics in ten years (rater than the current eleven for higher tracked students) they will likely incorporate some of the newly available materials rather than just move more traditional college courses down into the high schools.

An approach to mathematical education which would counter current negative trends is to involve students directly in a more creative and interactive mode. To do more openly what really takes place as people learn, understand or discover mathematics for themselves. To let students invent their own models, formulas or algorithms rather than using existing ones. To learn by trial and error, and stepwise converging upon the (or one of the) correct answers; rather than being given the result as dogma or as a straight foreward exercise analogous to ones repeatedly done by the teacher. This should include a great deal more questioning and interaction involving students as well as teachers in the classroom. Students should also work towards solutions orally, and not merely report on previously solved problems. All suggestins should receive some attention and wrong answers or approaches should be argued away (by other students preferably) and not abruptly dismissed. Open ended problems whose solutions extend over weeks should be included. If the material is selected so as to be interesting enough, then problems of maintaining active participation should be minimal. However, other methods can be employed if the level of involvement wanes. As one colleague suggested: When the class becomes too quite and noninventive, then it is time for a test. In order to achieve such interaction and excitement in the classroom one needs an appropriate approach and suitable material.

At the college and professional masters level one can make use of a great variety of problem solving and modeling type courses. Formats involving miniprojects, semester long projects, team projects, and design experiments so common in engineering education can be undertaken. The R.L. Moore approach or other discovery type methods can be employed using pure or applied mathematical materials. Various types of simulation or gaming techniques can be introduced for the purpose of learning mathematics. One such famous course by Howard Raiffa in the Business School at Harvard University received some "bad press" in early 1979 for teaching students how to lie (i.e., strategic misrepresentation). Internships in nonacademic settings, cooperative arrangements with other academic areas or industry or other sorts of exchange programs should become more popular and widespread. More oral and written reports should be included in mathematics programs. Math and stat clinics should become more common. Although many of these approaches are more appropriate for college level teaching, rather obvious variants of some of them should prove successful at more elementary levels. This includes the problem solving and modeling approaches in particular. The main ingredient necessary for success in achieving a really creative experience for students is to have a reservior of suitable materials that are either interesting or perceived as useful to the students.

7. New Materials. The main theme of this paper is that there now exists an unbelievable variety of enjoyable and practical materials which are readily accessible and most appropriate for introducing more lively, interactive and discovery-type courses at all educational levels. This should not seem surprising if one really contemplates the implications of a subject's store of knowledge and number of applications doubling every decade. Furthermore, many of the advances being made in the traditional areas of arithmetic, algebra, classical geometry, and calculus and its decedents are much less suitable for elementary levels than so many of the other more recently created subjects. One can so easily expand beyond merely using math tools to solve purely math problems arising within mathematics itself. One can develop mathematical skills and insights while simultaneously seeing how this subject penetrates so many aspects of life. One can appreciate the sheer beauty of mathematics and in addition realize what a critical role it does and will play in his or her future life.

The field of operations research literally includes hundreds of highly motivating, practical and realistic types of models and problem areas suitable for study at all levels. The decision sciences are filled with mathematical models of every day human activity. The behavioral and social sciences also routinely use math models, and many of an elementary nature to investigate activities of a veryhuman and societal nature. Finite mathematics now has an enormous number of exciting things to offer all comers. Graphs and networks can serve as a framework to embed elementary mathematics, and is as basic and fundamental as currently taught topics. Systems analysis provides a fundamental mathematical way of thinking. Interesting conbinatorical games for grade school children can be formulated from finite geometry, combinatorical optimization, and basic combinatorial lemmas related to fixed point theorems (e.g., Sperner's lemma or Tucker's lemma). Actual proofs of such lemma can now be presented in beginning college courses. Mathematical models concerning fair division, equity, utility, bargaining, and many others are keenly interesting topics for most individuals. The axiomatic approach to the apportionment problem taken by M.L. Balinski and H.P. Young over the past decade is a beautiful example of axiomatic modeling which can be appreciated at college or pre-college levels (with different emphasis and expectations). Similar remarks apply to aspects of social choice theory, and many other sorts of common voting or decision making schemas frequently arising in everyday life. In short, volumes upon volumes of mathematics has been created in the last few decades which could greatly enrich and popularize mathematical education at all levels. The slowness of implementing changes in education surely borders on the inconceivable, and some responsible people should be called to task because of it. The difficulty seems to have two root causes. First, there is a gross ignorance by most mathematicans and educators of just what most contemporary developments of mathematics and its applications consist of. This is caused mainly by a college mathematics curriculum which contains very little that has been developed within the last couple of "half lives" in terms of quantity of mathematical discoveries and its uses. More wide spread knowledge of the many new directions of mathematics is badly needed. Continued exclusion by existing groups of people knowledgeable in these areas will only prolong this serious problem. Secondly, the main cause of the problem of lack of current knowledge seems to be bad attitudes. The entrenched belief that the older traditional materials are somehow more basic, more important, more essential to living, culturally richer and intellectually better. The unwillingness of many to accept that the great majority of all mathematics, i.e., that invented in the past forty years, may contain concepts as fundamental or essential as the older "core" of the subject. Surely, the mathematical community owes it to their students and the public to take more advantage of the opportunities provided by contemporary developments in the subject, and the resulting enormous supply of rich materials now available.

10.5 MAXIMA AND MINIMA WITHOUT CALCULUS

MAXIMA AND MINIMA WITHOUT THE CALCULUS

A.J. Lohwater
Case Western Reserve University
Cleveland, Ohio

Summary

We present a sketch of some of the elementary methods which we have used in various of our calculus classes to determine maxima and minima for functions of one and several variables. We have found that the methods, whenever applicable, have had an appeal to the students far beyond what we had expected, particularly when the students were able to avoid long and tedious calculations. In this (necessarily abbreviated) paper we restrict ourselves to the inequality between the arithmetic and geometric means, Schwarz's inequality, Holder's inequality, and Jensen's inequality, and we indicate the types of problems that can be solved by the use of these inequalities. An extended version of these ideas, along with other material, will appear soon in the form of a little book, Elementary Inequalities and Some of their Applications.

I. Arithmetic and Geometric Means. The arithmetic means of n numbers $a_1, \ldots, a_n$ is defined to be

(1.1)
$$\frac{a_1 + \ldots + a_n}{n} = \frac{1}{n}\sum_{i=1}^{n} a_i$$

while the geometric mean of n positive numbers is defined to be

(1.2)
$$\sqrt[n]{a_1 \ldots a_n} = \left(\prod_{i=1}^{n} a_i\right)^{1/n}$$

The fundamental inequality between these means is the following:

(1.3)
$$\sqrt[n]{a_1 \ldots a_n} \le \frac{a_1 + \ldots + a_n}{n}$$

with equality in (1.3) only if the positive numbers $a_1, \ldots, a_n$ are all equal. We shall not prove (1.3) here, but we have found that the most satisfactory proof is the proof by induction. (We must remark at the outset that, despite the well-known fact that all the inequalities that we present follow from Jensen's inequality, in Section 4, we have found it far more satisfactory to derive our inequalities by methods which are familiar to the students; in this way, the students feel that they "have known these results all the time". Furthermore, it is a pedagogical mistake to fail to emphasize the interplay among the elementary inequalities.)

Now if, in (1.3), the sum of the a_i is constant, say K, then

(1.4) $\quad a_1 \ldots a_n \le (K/n)^n,$

so that the product in (1.4) can never exceed the constant $(K/n)^n$. However, the product in (1.4) can be equal to the constant $(K/n)^n$ whenever $a_1 = \ldots = a_n$, and this is what we mean by a maximum. We state this principle as

I. The product of n positive numbers whose sum is constant will be a maximum whenever all the factors of the product are equal.

We illustrate Principle I by means of some examples.

Example 1.1. Find the rectangle, open on one side, of maximum area if the sum of the other three sides is 400. If x be the length of each of two equal sides and 400 - 2x the length of the third side, then the area to be maximized is given by
$A = x(400 - 2x) = 1/2(2x)(400 - 2x)$. If $a_1 = 2x$ and $a_2 = 400 - 2x$, then the sum $a_1 + a_2 = 400$, a constant. Hence the product a_1a_2, and therefore A, will be a maximum whenever $a_1 = a_2$ or $2x = 400 - 2x$, or $x = 100$.

Two remarks are important here: first, we must ensure that the sum of the factors is constant, and this point must be emphasized to the students; second, there is a simpler solution available if we notice that the graph of the function $A = x(400 - 2x)$ is a parabola and that the maximum occurs at the vertex. The use of the quadratic function will be mentioned in the next section.

Example 1.2. Find the rectangle of largest area that can be inscribed in the semicircle $y = \sqrt{a^2 - x^2}$ if the base lies along the x-axis. The area is given by

$A = 2x\sqrt{a^2 - x^2}$. We note that that value of x which maximizes A also maximizes $A^2 = 4x^2(a^2 - x^2)$. If we set $a_1 = x^2$ and $a_2 = a^2 - x^2$, we see that $a_1 + a_2 = a^2$, a constant, so that a_1a_2, and hence A, is maximized whenever $x^2 = a^2 - x^2$, or $x = a/\sqrt{2}$. Note that if we set $z = x^2$, we have $A^2 = 4z(a^2 - z)$, so that the method of the quadratic function may be used; this is always the case whenever $n = 2$.

Example 1.3. Maximize $f = xy^2z^3$ for positive x,y,z under the constraint $x^2 + y^2 + z^2 = 6$. In order to use our Principle, we must ensure that the sum of the factors is constant. We have that $f^2 = 108x^2(y^2/2)(y^2/2)(z^2/3)(z^2/3)(z^2/3)$, the product of 108 and six factors $a_1, \ldots, a_6$. Now $a_1 + a_2 + \ldots + a_6 = 6$, which is constant, so that the product $x^2y^4z^6$, and hence xy^2z^3, is a maximum whenever $x^2 = y^2/2 = z^2/3$. Substituting these conditions into the constraint, we obtain $6x^2 = 6$ or $x = 1$, $y = \sqrt{2}$, $z = \sqrt{3}$, which give us the maximum value $6\sqrt{3}$ for f. Note that we are able to avoid the use of Lagrange multipliers, and that we do not need a second-derivative test to distinguish a maximum from a minimum, since Principle I guarantees us a maximum.

There is another principle, the dual of Principle I, which we state as

II. <u>The sum of n positive numbers whose product is constant will be a minimum whenever all the terms in the sum are equal.</u>

<u>Example 1.4</u>. Show that $f = 5\tan^6 3x + \cot^6 3x$ has a minimum, and find its value. If we consider f as $a_1 + a_2$, then $a_1a_2 = 25$, a constant. Hence f has a minimum whenever $a_1 = a_2$ or $\tan^{12} 3x = 1$, or $\tan 3x = 1$, so that $x = 15^0, 75^0$, etc. The minimum value of f is then 10. We remark that, as usual, there are simpler methods of minimizing f, one of the most useful being the following.

Suppose that α is positive. Then $\alpha + 1/\alpha \geq 2$, with equality (the condition for a minimum) for $\alpha = 1$. We may use Principle II for this result, but it is simpler to note that $(\sqrt{\alpha} - 1/\sqrt{\alpha})^2$ 0 with equality for $\alpha = 1$. If we discard the factor 5 in the function of Example 1.4, then that problem is even more transparent.

<u>Example 1.5</u>. Show that the function $f = 6 + |x| + 1/(1 + |x|)$ has a minimum and find its value. If we set $\alpha = 1 + |x|$, then $f = 5 + \alpha + 1/\alpha$, which is not less than $5 + 2 = 7$ and achieves the minimum 7 whenever $\alpha = 1$ or $x = 0$. It is important to note that f fails to have a derivative at precisely the critical point of the problem.

2. <u>The Quadratic Function and Schwarz's Inequality</u>.

We have remarked in the preceding section that finding the vertex of the parabola represented by a quadratic function is equivalent to locating the maximum or minimum point on the parabola. Thus, for the function $f = 3x^2 - 12x + 17 = 3(x - 2)^2 + 5$, we have a minumum of 5 whenever $x = 2$. If we alter our reasoning a bit, we notice that f is the sum of a constant, 5, and a non-negative term, $3(x - 2)^2$. Thus f will be a minimum whenever we add the least possible amount to 5, in this case 0 when $x = 2$. Thus, this point of view allows us to formulate the following principle, which no longer depends on the properties of a parabola:

III. <u>Let A be a constant, and let F be a non-negative function. Then the function A + F will be a minimum whenever F is a minimum, and the function A - F will be a maximum whenever F is a minimum.</u>

Example 2.1. Examine the function $f = x^4 - 8x^2 + y^2 - 4y + 25$ for maxima and minima. If we factor, we obtain $f = 5 + (x^2 - 4)^2 + (y - 2)^2$, which is of the form A + F with A = 5 and $F = (x^2 - 4)^2 + (y - 2)^2$. Now F is non-negative and has a minimum 0 whenever $x = \pm 2$ and $y = 2$. Thus f has a minimum 5 at the points (2,2) and (-2,2). The second derivative test is still tractable, but it is clear that the addition of more independent variables confounds the problem unnecessarily.

Example 2.2. Determine whether the function $f = (x + 3)(x + 5)/(x + 6)$ has any maxima or minima. If we set $x + 6 = z$, then $x = z - 6$, so that f becomes $(z - 3)(z - 1)/z$, or

$$f = \frac{z^2 - 4z + 3}{z} = z + \frac{3}{z} - 4 = (\sqrt{z} - \frac{\sqrt{3}}{\sqrt{z}})^2 + 2\sqrt{3} - 4$$

This is of the form A + F with F non-negative. Hence we have a minimum, namely $2\sqrt{3} - 4$, whenever $z = \sqrt{3}$, or $x = \sqrt{3} - 6$.

Let us return to the quadratic function $y = ax^2 + bx + c$. Suppose that we are given two sets of real numbers $a_1, \ldots, a_n$ and $b_1, \ldots, b_n$; here we may lift the restriction that the numbers a_k, b_k be positive. For any value of the real number x, the expression

$$y = (a_1x - b_1)^2 + \ldots + (a_nx - b_n)^2$$

cannot be negative, and can be zero only when each item is zero, i.e., when

(2.1)

$$\frac{b_1}{a_1} = \frac{b_2}{a_2} = \cdots = \frac{b_n}{a_n}$$

The fact that the discriminant cannot be positive gives us the inequality of Cauchy and Schwarz,

(2.2)

$$(\sum_{k=1}^{n} a_kb_k)^2 \leq (\sum_{k=1}^{n} a_k^2)(\sum_{k=1}^{n} b_k^2)$$

the proof of which we shall not discuss further. Equality in (2.2) with one side of (2.2) constant leads to maxima-minima problems.

Example 2.3. Find the minimum value of $f = (16 + x^2 + y^4)(1/16 + 1/x^2 + 1/y^4)$. This expression is essentially the right hand side of (2.2) with $a_1 = 4$, $a_2 = x$, $a_3 = y^2$, etc., so that

$$(16 + x^2 + y^4)(1/16 + 1/x^2 + 1/y^4) \geq (1 + 1 + 1)^2 = 9,$$

with equality (minimum) if

$$\frac{4}{1/4} = \frac{x}{1/x} = \frac{y^2}{1/y^2} \text{ or } x = 4,\ y = 2$$

Example 2.4. Find the minimum value of
$f = A\csc^2 x + B\sec^2 x$ if $A>0$, $B>0$, $0<x<\pi/2$. We have

$$f = (A\csc^2 x + B\sec^2 x)(\sin^2 x + \cos^2 x)$$

$$\geq (\sqrt{A}\csc x \sin x + \sqrt{B}\sec x \cos x)^2 = (\sqrt{A} + \sqrt{B})^2$$

with equality (minimum) if and only if

$$\frac{\sqrt{A}\csc x}{\sin x} = \frac{\sqrt{B}\sec x}{\cos x},\ \text{or } \tan^2 x = \sqrt{\frac{A}{B}},\ \text{or } x = \tan^{-1.4}\sqrt{\frac{A}{B}}$$

At this stage in our courses we re-introduce a geometric insight into the inequality (2.2), namely, the notion of the inner product or scalar product. If we denote by $\underset{\sim}{a}$ the n-tuple $(a_1, \ldots, a_n)$ and by $\underset{\sim}{b}$ the n-tuple $(b_1, \ldots, b_n)$, then (2.2) is equivalent to

(2.3)

$$(\underset{\sim}{a} \cdot \underset{\sim}{b})^2 \leq (\underset{\sim}{a} \cdot \underset{\sim}{a})(\underset{\sim}{b} \cdot \underset{\sim}{b}) = ||\underset{\sim}{a}||^2\ ||\underset{\sim}{b}||^2$$

and the condition of equality (2.1) becomes simply

(2.4)

$$\underset{\sim}{a} = k\underset{\sim}{b}$$

for some real number, or scalar, k.

As an example of the kind of maximum-minimum problem which can be based on (2.3) we have

Example 2.5. Find the maximum and minimum of $f = 3x + 4y + 12z$ subject to the constraint $x^2 + y^2 + z^2 = 1$. We note that $f = \underset{\sim}{a} \cdot \underset{\sim}{X}$, where $\underset{\sim}{a} = (3,4,12)$ and $\underset{\sim}{X} = (x,y,z)$; we note too, that $\underset{\sim}{X}$ is a unit vector.
Since

$$f = \underset{\sim}{a} \cdot \underset{\sim}{X} = ||\underset{\sim}{a}||\ ||\underset{\sim}{X}|| \cos\theta$$

where θ is the angle between $\underset{\sim}{a}$ and $\underset{\sim}{X}$, it is clear that f is a maximum or minimum whenever $\underset{\sim}{a}$ and $\underset{\sim}{x}$ are parallel and have, respectively, the same or opposite directions. Thus the maximum 13 occurs whenever $\underset{\sim}{X} = (3/13, 4/13, 12/13)$ and the minimum -13 occurs whenever $\underset{\sim}{X} = (-3/13, -4/13, -12/13)$. Problems involving scalar products may be handled by a direct appeal to (2.2), of course, but both science and mathematics students find the geometric intuition of (2.3) rather attractive. For example, we may write

$$f = |3x + 4y + 12z| \leq \sqrt{3^2 + 4^2 + 12^2}\sqrt{x^2 + y^2 + z^2} = 13$$

with equality, according to (2.1), whenever

$$x/3 = y/4 = z/12;$$

the exact values of x,y,z may now be found from the condition $x^2 + y^2 + z^2 = 1$.

There are some elementary applications of the integral form of (2.2) which have both appeal and motivation for students in an elementary calculus course. If f(x) and g(x) are continuous on $x_1 \leq x \leq x_2$, then, of course, we have

(2.5)

$$\left(\int_{x_1}^{x_2} f(x)g(x)dx\right)^2 \leq \left(\int_{x_1}^{x_2} f^2 dx\right)\left(\int_{x_1}^{x_2} g^2 dx\right)$$

with equality whenever

(2.6) $\qquad f(x) = kg(x)$

for some real number k. We shall not cite any problems involving (2.5) - (2.6) similar to those given above, but rather we shall give an application that can simplify a classroom demonstration which, in most textbooks, is rather tedious. The natural logarithm, log x, is defined as

$$\int_1^x dt/t$$

and it is easy to show the existence of a number e such that log e = 1, along with the elementary properties of log x. What is tedious is the estimation of the size of e; this estimate is usually shown by subdividing the interval to the right of x = 1 and estimating a Riemann sum $\Sigma\, 1/t_k\ \Delta t_k$. Consider instead the application of 2.5 to

$$(x-1)^2 = \left(\int_1^x dt\right)^2 = \left(\int_1^x \frac{1}{\sqrt{t}}\sqrt{t}\,dt\right) < \left(\int_1^x \frac{dt}{t}\right)\left(\int_1^x t\,dt\right)$$

$$= (\log x)\ \tfrac{1}{2}(x^2 - 1)$$

whence

(2.7) $\qquad 2(x - 1)/(x + 1) \quad \log x.$

If we set x = e in (2.7), we have that e < 3; if we set $x = e^{1/3}$, we have $e < (7/5)^3 = 2.744$; while if we set $x = e^{1/32}$, we have $e < (65/63)^{32} = 2.7185$. For a lower bound for e we consider

$$X^2 = \left(\int_0^x e^{t/2}e^{-t/2}dt\right)^2 < \left(\int_0^x e^t dt\right)\left(\int_0^x e^{-t}dt\right)$$

$$= (e^x - 1)(-e^{-x} + 1) = e^x + \frac{1}{e^x} - 2$$

so that

(2.8)

$$e^x > \frac{x^2 + 2 + x\sqrt{x^2 + 4}}{2}$$

If we set $x = 1$ in (2.8), we have $e > (3 + \sqrt{5})/2 \approx 2.62$; while if we set $x = 1/3$ in (2.8), we obtain $e > 2.705$; if we set $x = 1/10$ in (2.8) we obtain $e > 2.717$.

3. Hölder's Inequality

Suppose that x and y are positive numbers and that p and q are positive numbers such that

(3.1) $\qquad 1/p + 1/q = 1.$

The fundamental result upon which most proofs of Hölder's inequality are based is the inequality

(3.2) $\qquad x^p/p + y^q/q \geq xy,$

with equality only when

(3.3) $\qquad x^p = y^q.$

We shall show that, in the case where p and q in (3.1) are rational, the result (3.2) is a direct consequence of the inequality between the arithmetic and geometric means cited in (1.3); in the case that p and q in (3.1) are not rational, it is necessary to make approximations to x^p and y^q with rational p and q and then to apply our result for the rational case to obtain (3.2) for all cases. (We remark that the case of equality (3.3) in the case of irrational p and q involves niceties which we omit in the elementary courses, for the cases which are of immediate applicability usually involve only rational p and q.)

Now if p and q are rational, say $p = r/s$ and $q = r/(r-s)$, then the expression $x^p/p + y^q/q$ may be written

$$\frac{sx^p}{r} + \frac{(r-s)y^q}{r} = \frac{sx^p + (r-s)y^q}{r}$$

which is an arithmetic mean in which the term x^p is added s times and the term y^q is added r - s times. By (1.3), we have, with $n = r$,

$$\frac{x^p}{p} + \frac{y^q}{q} \geq (x^{ps}y^{q(r-s)})^{1/r} = x^{p\frac{s}{r}}y^{q\frac{(r-s)}{r}} = xy$$

with equality only if all terms are equal, that is, only if $x^p = y^q$.

Hölder's inequality, which follows easily from (3.2) under the conditions stated at the beginning of this section, asserts that, for two sets of positive numbers $a_1, \ldots, a_n$ and $b_1, \ldots, b_n$,

(3.4)

$$\sum_{k=1}^n a_k b_k \leq \left(\sum_{k=1}^n a_k^p\right)^{1/p} \left(\sum_{k=1}^n b_k^q\right)^{1/q}$$

with equality only if

(3.5)

$$\frac{a_1^p}{b_1^q} = \frac{a_2^p}{b_2^q} = \ldots = \frac{a_n^p}{b_n^q}$$

Since the proof of (3.4) is almost trivial at this point, let us complete it so that we may include some remarks about our experience in the classroom. If we set $A^p = \Sigma a_k^p$ and $B^q = \Sigma b_k^q$, and write $c_k = a_k/A$, $d_k = b_k/B$, we have, by (3.2),

(3.6)

$$\frac{c_k^p}{p} + \frac{d_k^q}{q} \geq c_k d_k, \; k = 1, 2, \ldots, n$$

If we sum these inequalities over k, we have

$$\frac{\Sigma c_k^p}{p} + \frac{\Sigma d_k^q}{q} \geq \Sigma c_k d_k$$

or, by re-introducing the a_k and b_k,

$$\frac{\Sigma a_k^p}{A^p p} + \frac{\Sigma b_k^q}{B^q q} = \frac{1}{p} + \frac{1}{q} = 1 \geq \frac{\Sigma a_k b_k}{AB}$$

or

$$\sum_{k=1}^n a_k b_k \leq AB$$

which is (3.4). We remark that when $p = q = 2$, we obtain (2.2), but, in this case, (3.2) may be obtained more readily observing that
$(x - y)^2 \geq 0$ or $x^2/2 + y^2/2 \geq xy$, with equality when $x = y$.

We remark that (3.2) is easily extended, if we use (1.3), to the case of positive x,y,z and positive p,q,r with

(3.7) $$1/p + 1/q + 1/r = 1$$

to the fundamental inequality

(3.8) $$x^p/p + y^q/q + z^r/r \geq xyz,$$

from which we have

(3.9)

$$\sum_{k=1}^{n} a_k b_k c_k \leq (\sum_{k=1}^{n} a_k^p)^{1/p} (\sum_{k=1}^{n} b_k^q)^{1/q} (\sum_{k=1}^{n} c_k^r)^{1/r}$$

with equality only if

(3.10)

$$a_1^p : b_1^q : c_1^r = \cdots = a_n^p : b_n^q : c_n^r$$

Example 3.1. Find the largest value of
$f = 5xy + 12yz + 84xy$ subject to the constraint
$x^4 + y^4 + z^4 = 1$. If we use (3.9) with $q = r = 4$, $p = 2$, we have

$$f \leq (5^2+12^2+84^2)^{1/2}(x^4+y^4+x^4)^{1/4}(y^4+z^4+x^4)^{1/4} = 85$$

THe values of x,y,z may be found from (3.10) together with the constraint.

What is emphasized in the classroom up to this point is the pervasiveness of the inequality (1.3) between the arithmetic and geometric means. We have found that the students are attracted to methods which can save them extensive tedious manipulations, even if the price that they pay is that of having to think more carefully about each problem posed to them.

4. Jensen's Inequality

Jensen's inequality is a powerful tool in the event we have a convex function f(x), i.e., a function f(x) such that

(4.1)

$$f(\frac{x_1 + x_2}{2}) \leq \frac{f(x_1) + (x_2)}{2}$$

for any x_1, x_2 on some interval $a \leq x \leq b$. We restrict ourselves for simplicity to the case of a twice-differentiable function, so that a sufficient condition for (4.1) is that $f''(x) \geq 0$ over (a,b). It is shown, usually by induction, that for $x_1, x_2, \ldots, x_n$ in (a,b),

(4.2)

$$f(\frac{1}{n}\sum_{k=1}^{n} x_k) \leq \frac{1}{n}\sum_{k=1}^{n} f(x_k)$$

or, if $\alpha_1, \ldots, \alpha_n$ are positive number such that $\Sigma\alpha_k = 1$,

(4.3)

$$f[\sum_{k=1}^{n} \alpha_k x_k] \leq \sum_{k=1}^{n} \alpha_k f(x_k)$$

with equality only if $x_1 = \ldots = x_k$ or if f(x) is linear between the smallest and largest values of the x_k.

If f(x) is concave, then -f(x) is convex, so that for a concave function the inequalities (4.1), (4.2), and (4.3) are reversed. A sufficient condition that f(x) be concave on some interval is that f"(x) 0 on that interval.

Example 4.1. The function $f(x) = \log x$ is concave on $0 < x < \infty$, for $f''(x) = -1/x^2 \leq 0$ on that interval. If we use (4.2) with the inequality reversed (we have a concave function) for any n positive numbers $x_1, \ldots, x_n$, we have

$$\log(\frac{1}{n}\Sigma x_k) \geq \frac{1}{n}\Sigma \log x_k = \frac{1}{n}\log(x_1 \ldots x_n) =$$

$$\log (x_1 \ldots x_n)^{1/n}$$

or

$$\frac{x_1 + \ldots + x_n}{n} \geq \sqrt[n]{x_1 \ldots x_n}$$

which is (1.3). Since the function is linear nowhere on $0 < x < \infty$, the equality sign can hold if $x_1 = \ldots = x_n$.

If we apply (4.3) with a set of positive α_k with $\Sigma\alpha_k = 1$, we obtain the result

(4.4)

$$\sum_{k=1}^{n} \alpha_k x_k \geq x_1^{\alpha_1} \ldots x_n^{\alpha_n}$$

a very powerful generalization of (1.3), which contains (3.20 and (3.8).

Example 4.2. Maximize $x + y + z$ subject to the constraint $x^p + y^p + z^p = 1$, where $p > 1$. The function $f = x^p$ is convex for $p > 1$, so that, by (4.2), we have

$$\left(\frac{x+y+z}{3}\right)^p \leq 1/3(x^p + y^p + z^p) = 1/3$$

Hence $x + y + z \leq 3/\sqrt[p]{3}$ with equality (maximum) whenever $x = y = z = 1/\sqrt[p]{3}$. (We remark that we have shown an elementary form of Chebyshev's inequality, which can also be proved by less sophisticated methods.)

Example 4.3. If A,B,C are the angles of a triangle, show that sin A + sin B + sin C is a maximum for the equilateral triangle. The function f = sin x is concave over the interval $0 \leq x \leq \pi$, so that, by (4.2) with the inequality reversed, we have

$$\sin((A + B + C)/3) \geq 1/3(\sin A + \sin B + \sin C),$$

or

$$3 \sin\frac{\pi}{3} = \frac{3\sqrt{3}}{2} \geq \sin A + \sin B + \sin C$$

with equality (and hence a maximum) only when A = B = C.

Conclusions

The fact that we introduce elementary methods of solving certain maximum and minimum problems does not mean that we slight the calculus in any way. The calculus is a powerful branch of mathematics, and we show its power in this area by problems which are worthy of its application. The use of inequalities, and of the methods described by Professor Niven in the preceding lecture, shows the student that there are simpler and more elegant methods which are more appropriate to some of the more common problems which he is likely to encounter in his work in other fields. We find, too, that these methods depend, for their success, on a more complete mastery of the principles involved than do the methods of calculus; we find that the students seem to understand better both the power and the limitations of the calculus in connection with this topic.

MAXIMA AND MINIMA WITHOUT CALCULUS

Ivan Niven
University of Oregon
Portland, Oregon

The basic thesis of this paper is that significant questions of an applied type can be introduced into the curriculum much earlier than at present. This can be done by bringing in many important problems in maxima and minima which can be solved by pre-calculus methods. In this way, for example, a teacher of elementary algebra has an answer to the perennial question of the reluctant student, "What is the good of this?"

Although many extremal problems can be solved most readily by means of calculus, many questions are better approached by other methods. For example, a standard problem of calculus books is this: among all rectangles of a given perimeter, which has the largest area? But calculus books are silent on the obvious generalization: among all quadrilaterals of a given perimeter, which has the largest area? Another standard problem in calculus books is the following: among all rectangles that can be inscribed in a given ellipse, with sides parallel to the axes of the ellipse, which has the largest area? Again, it may be noted that calculus books do not ask for the quadrilateral of the largest area than can be inscribed in an ellipse. These two problems will be discussed in further detail later in this paper, along with other examples. Our main purpose is to outline a variety of methods that can be used to solve important extremal problems, in cases where calculus is an awkward instrument.

In another session at this International Congress, a panel is discussing these questions:

> Should calculus keep its unique position as the core of post-secondary and undergraduate mathematics? If not calculus, what should replace it?

Although this paper does not deal directly with these matters, our observations are related. We do offer some alternative approaches to extremal problems that might well be incorporated into the undergraduate curriculum in one form or other. Many an undergraduate student of mathematics, if confronted by a problem of maxima or minima, immediately seeks some function to be differentiated. In the absence of other methods, what else can the student do?

To illustrate this, consider the classical problem of Heron, to find the point P on a given straight line so as to minimize the sum of the distances from P to two fixed points lying on one side of the line. (We note in passing that the corresponding problem in 3-space is rarely discussed, where the two given points do not lie in the same plane as the given line.) If we apply calculus to this problem, we might take the given line as the x-axis, and then assign some specific coordinates such as (0,a) and (b,c) to the two fixed points. Assigning the coordinates (x,0) to the point P the sum of the distances from P to the fixed points is then easily formulated as a function of x, ready for the differentiation process. The trouble with this method is that the function involves the sum of two square roots, with the consequence that the algebraic details of the solution by calculus are messy. Heron's solution is much more elegant, by an application of the reflection principle, or the symmetry principle. That is, if A and B are the two fixed points, we take the mirror image, say A', of the point A in the given line, and it is then easy to prove that the optimal position for P is the intersection point of the given line and the line A'B.

The reflection principle is a technique that can be used in a variety of ways. We cite one further example here. Suppose that we have established that the square gives the maximum area among rectangles of a given perimeter. Then consider the question of the maximum area that can be enclosed by a farmer using a given length of fencing (say b units of length) for three sides of a rectangle, part of the side of a barn being used as the fourth side of the rectangle. Taking the mirror image of the rectangle, with the side of the barn serving as the mirror, we observe that the rectangle together with its mirror image should form a square for maximum area. Hence the fencing should be arranged so as to form half a square, giving a rectangle with sides $b/4$, $b/2$ and $b/4$.

Quadratic Functions

Many a problem in calculus books amounts to nothing more than finding the maximum, or the minimum, of a quadrataic function of a single variable. For example, to find the rectangle of largest area among those of a given perimeter, c, we seek the maximum of $1/2cx - x^2$, where x is the length of one side of such a rectangle. Calculus is not needed here, because the problem can be solved by completing the square. For example, the maximum value of $10x - x^2$ occurs if and only if $x = 5$, because the expression can be written in the form $25 - (x - 5)^2$. Similarly, the least value of $x^2 + 8x + 25$ over all real numbers x occurs if and only if $x = -4$, by reformulating the function as $9 + (x + 4)^2$.

A Fundamental Inequality

Finding extremal values of quadratic functions by completing the square is a device with many applications. Using this method, for example, it is easy to prove that among all points P lying inside a triangle, the one such that the sum of the squares of the distances from P to the three vertices is a maximum, is the centroid of the triangle. This can be done quickly by assigning the coordinates (x,y) to the point P, and some constant coordinates to the vertices, and then using the distance formula from coordinate geometry.

However, this technique of completing the square can be superseded by a much more elegant method, namely the application of the inequality of the arithmetic-geometry means. As is well-known, this inequality asserts that the arithmetic mean of two or more non-negative real numbers exceeds the geometric mean unless the numbers are all equal. For two numbers u and v this inequality can be written in the form $uv \leq (u + v)^2/4$, with equality if and only if $u = v$. Consider again the simple question of finding the maximum value of $10x - x^2$ over all real numbers x. Taking u as x, and v as $10 - x$, the above inequality becomes $x(10 - x) \leq 10^2/4$, or the maximum value of $x(10 - x)$ is 25, and this is achieved by writing $u = v$, that is, $x = 10 - x$.

The principle involved here can be extended to a much wider class of problems. Consider for example, the question of finding the maximum value of $500\,x^2y^3(12 - 4x - 15y)$ over positive numbers x and y. (The factor "500" has been included here to simplify the arithmetic of the problem.) The inequality of the arithmetic-geometric means for six non-negative numbers u_1, $u_2, \ldots, u_6$ can be written in the form

$$u_1\, u_2\, u_3\, u_4\, u_5\, u_6 \leq [(u_1 + u_2 + \ldots + u_6)/6]^6,$$

with equality if and only if the u's are all equal. We write

$$500x^2y^3(12\text{-}4x\text{-}15y) = (2x)(2x)(5y)(5y)(5y)(12\text{-}4x\text{-}15y),$$

setting up the product here so that the six factors have a constant sum, namely 12. Applying the inequality of the arithmetic-geometric means with

$$u_1 = u_2 = 2x,\ u_3 = u_4 = u_5 = 5y,\ u_6 = 12\text{-}4x\text{-}15y,$$

we find that

$$500x^2y^3(12\text{-}4x\text{-}15y) \leq 2^6.$$

Hence the maximum value we seek is 2^6, or 64, attained by taking $2x = 5y = 12\text{-}4x\text{-}15y = 2$.

At first glance it might appear that here we have employed a very special function to which our neat little device can be applied. However, a wider class of functions can be treated than might be expected. In a forthcoming book to be published by the Mathematical Association of America, the author shows how the method can be adapted extensively.

Geometric Techniques

Next we return to the two problems set forth at the outset. Among all quadrilaterals of a given perimeter, which has the largest area? One way to show that the square is the answer is to use the following lemma. Among all triangles of a given perimeter p and one side of given length, say b, the isosceles triangle with sides b, $p\text{-}b/2$ and $p\text{-}b/2$ has the largest area. Assuming this lemma for a moment, the quadrilateral problem is readily solved. Let $PQRS$ be any quadrilateral, not a square. We sketch a proof that the square of the same perimeter has a larger area. (We may presume that $PQRS$ is a convex quadrilateral, because it is easy to see that any non-convex quadrilateral can be replaced, by using the mirror image of the two "interior" sides, by a quadrilateral of equal perimeter with larger area.) Using the triangle lemma above, we replace the triangle PQS by the isosceles triangle $P'QS$ on the base QS with $P'Q = P'S$ and having the same perimeter as triangle PQS. Similarly, replace the triangle QRS by the isosceles triangle $QR'S$. Then with the line segment $P'R$ as a base, replace the triangles $QP'R'$ and $P'R'S$ by the isosceles triangles $Q'P'R'$ and $P'R'S'$. The quadrilateral $P'Q'R'S'$ is a rhombus, with larger area than $PQRS$ (unless $PQRS$ is itself a rhombus) and the same perimeter. The rhombus $P'Q'R'S'$ has smaller area than the square of the same perimeter.

A much simpler proof can be given, if we assume that a quadrilateral of maximum area exists. (Assuming the existence of a solution is not unreasonable in beginning courses.) For by the geometric lemma about the triangle given above, the quadrilateral of largest area must have all pairs of adjacent sides equal, hence all sides equal. Again it is easy to see that the square is larger than any rhombus of the same perimeter.

In the above arguments we have assumed the geometric lemma that among triangles having a fixed side of length b and a fixed perimeter, the isosceles triangle with the fixed side as base has largest area. This can be proved by geometry alone, or by a combination of geometry and algebra. We now give a solution of the second type. The square of the area of any triangle with sides x, y and b is $s(s-x)(s-y)(s-b)$, where s is the semi-perimeter. Since s and $x\text{-}b$ are constants we want to maximize $(s-x)(s-y)$. But $y = 2s - x - b$, so the problem is to maximize $(s-x)(x + b - s)$. This is a quadratic function of x, for which the maximum occurs if and only if $x = (2s - b)/2$, giving the isosceles triangle.

Geometric Transformations

Consider now the other problem mentioned at the start of this paper: among all quadrilaterals inscribed in a given ellipse, which has the maximum area? An easy way to solve this problem is by use of a simple geometric transformation, as follows. Consider any ellipse $x^2/a^2 + y^2/b^2 = 1$. Applying the transformation $x = au$, $y = bv$, we note that the ellipse is mapped onto the circle $u^2 + v^2 = 1$. It is easy to verify, furthermore, that straight lines in the xy plane are mapped onto straight lines in the uv plane. Any triangle T_1 in the xy plane is mapped onto a triangle T_2 in the uv plane, with the property that the area of T_1 is ab times the area of T_2.

Any quadrilaterial Q_1 inscribed in the ellipse is mapped onto a quadrilateral Q_2 inscribed in the circle, again with the property that the area of Q_1 is ab times the area of Q_2. It follows that to find a quadrilateral Q_1 of largest area inscribed in the ellipse, we look for the quadrilateral Q_2 of largest area inscribed in the circle $u^2 + v^2 = 1$. Presuming for a moment that we know that the largest quadrilateral inscribed in a circle is a square, we see that the solution of the problem is obtained by starting with any such square, and then using the mapping $x = au$, $y = bv$ to locate the corresponding quadrilateral in the ellipse. Thus there are infinitely many quadrilaterals of maximum area inscribed in the given ellipse. One of these largest quadrilaterals is a rectangle, with sides parallel to the axes of the ellipse; the others are parallelograms. One of these parallelograms has vertices at the ends of the axes of the ellipse.

One interesting feature of this problem is the infinitude of answers, with no unique largest quadrilateral.

Convexity

In the above argument we presumed that the quadrilateral of largest area that can be inscribed in a circle is the square. One way to prove this is by Jensen's theorem on convex functios. Given any quadrilateral inscribed in a circle, say in a unit circle for convenience, let α_1, α_2, α_3 and α_4 denote the angles subtended at the center by the four sides. The area of such a quadrilateral is one half of $\sin \alpha_1 + \sin \alpha_2 + \sin \alpha_3 + \sin \alpha_4$. Jensen's inequality states (in this case) that this trigonometric sum is at most $4 \sin \alpha$, where α is the arithmetic mean $(\alpha_1 + \alpha_2 + \alpha_3 + \alpha_4)/4$. Clearly $\alpha = 90^\circ$, so the largest quadrilateral occurs in case the angles subtended at the center are right angles. A very slight adaptation of this argument can be used to show that the quadrilateral of largest perimeter inscribed in a circle is also a square.

The form of Jensen's theorem applied here is readily proved without any use of calculus. All that is needed is the simple result $\sin \alpha + \sin \beta \leq 2 \sin (\alpha + \beta)/2$ for positive α and β not exceeding 180°, coupled with an argument by the special induction method used by Cauchy to prove the inequality of the arithmetic-geometric means.

In summary, we have outlined several methods available for attacking problems in maxima and minima without using calculus. There are many ad hoc methods in geometry that we have not mentioned. Although some of these geometric methods are ingenious devices that solve only one problem, patterns can also be observed here. We offer one simple example of this, namely the use of an auxillary circle, or a rotation of part of the diagram, to reveal a way to solve the problem.

10.6 EXPLORATORY DATA ANALYSIS

SYMPOSIUM ON EXPLORATORY DATA ANALYSIS

R. Gnanadesikan
J.R. Kettenring
P.A. Tukey
Bell Laboratories
Murray, Hill, New Jersey, U.S.A.

A.F. Siegel
Princeton University
Princeton, New Jersey, U.S.A.

1. Overview (by R.G.)

To introduce the topic of this symposium and set the stage for the three speakers, Exhibit 1.1 contrasts certain aspects of what we know as mathematical statistics and the more recent developments in a field that has come to be known ad data analysis.

Exhibit 1.1

Contrasts between Mathematical Statistics and Data Analysis

	"Classical" Mathematical Statistics	"Modern" Data Analysis
CONCERNS	Inference Estimation in tightly specified models Prespecified hypotheses & significance levels	Insight Summarization & exposure Data-guided models & assessments
TYPICAL DESIDERATA	Optimality Admissibility Uniformly most powerful	Versatility Resistant/Robust Sensitivity & stability
CHARACTERISTICS	Single analysis Heavily numerical	Multiple analyses Heavily graphical
TECHNOLOGY	Hand calculator	Modern computing

For instance, mathematical statistics is primarily concerned with formal statistical inference including estimation of parameters in very tightly specified models and tests of prespecified hypothesis at prechosen levels of significance - - peeking at the data is a forbidden sin! In contrast, data analysis aims at gaining insights into the data, which means both summarizing the informational content of a body of data and exposing unanticipated phenomena. Such a discovery process has to be closely tied to the data. The models, as well as aspects of their assessment, are essentially evolved out of the data, so that looking at the data is fine and not forbidden!

With respect to desiderata, mathematical statistics has focussed on optimality under often unverified assumptions; descriptions such as "admissable" and "uniformly most powerful" are central in mathematical statistics. On the other hand, in the pragmatic environment of data analysis, versatility in the sense of usefulness for a variety of purposes is key; and one seeks techniques that are resistant to outliers and robust in the fact of deviations from assumptions. The sensitivity and stability of procedures are natural concerns too.

With an optimal procedure in hand, mathematical statistics encourages a single analysis, while a hallmark of the data analytic approach is to subject a given body of data to several different analyses and then to sift through the results learning from both the similarities and the differences among them. Classical statistics has a heavy numerical orientation, while modern data analysis relies heavily on graphical analyses as this symposium will show. Finally, the technology of mathematical statistics was by and large the hand calculator. For data analysis today the technology is varied, ranging from hand-held electronic calculators through personal and mini-computers to large computers with sophisticated hardware and software features. The methods of data analysis should adapt to this range of capabilities, and the volume of data as well as the complexity of an analysis will often dictate the choice of a computing environment.

Under the umbrella of data analysis, John Tukey has tried to distinguish two broad types - - exploratory data analysis, wherein one is merely exploring for insights, and confirmatory data analysis, wherein one is seeking insome sense, to weight the strength of the evidence for the insights gained. While the emphasis in this symposium is on exploratory data analysis, this dichotomy is more a philosophical convenience than anything else, and a good data analyst practices both types.

The three lectures of this symposium will be self-contained but the talks together constitute a whole. Some common themes are: firstly, an emphasis on data-driven analyses rather than on model-driven ones, thus conveying the essential feature that data analysis is far more like empirical science than it is like mathematics; secondly, a shared concern for real-world phenomena, such as "outliers" or extremely deviant observations; lastly, the serendipity often involved in data analysis - - you will notice not only examples of insights into questions formulated prior to the analysis of the data but also instances of detecting unanticipated phenomena in the data.

All three lectures address these questions: 1) What is the particular methodology all about? 2) Why bother, or how is it useful? and 3) How might one teach it? An annotated list of references for help in purusing further details is included.

2. Graphical Methods (by P.A.T.)

This lecture will present some basic but fairly simple graphical methods of exploratory data analysis - plots of raw or derived data (which usually means separate points, not curves) which can help us discover things about the data.

Why are plots useful? Human beings perceive visual patterns more readily than patterns in collections of numbers. This is especially important in exploratory data analysis because pictures dramatically revealthings that we did not expect to find in the data. By contrast, numerical summaries tend to cover up any structure that they were not specifically designed to handle.

Exhibit 2.1 is a set of data on 47 French-speaking Swiss provinces showing the percent of the population that were Catholic in 1888. For summarizing how these spread out over the possible range (0-100), if we were to use the mean and standard deviation we would have missed a very important fact! Instead we can write down all the possible tens digits in a column, and, in rows alongside them, write down the ones digit (after rounding) for each observation. Exhibit 2.2 shows such a "stem-and-leaf" display (leading digits are stems, others are leaves). We see not that the distribution of these data is bimodal.

Exhibit 2.1

Stem-and-Leaf Display

Percent Catholic in 47 French Speaking Swiss Provinces in 1888

Data: 10.0 84.8 93.4 33.8 5.2 90.6 92.9 . . . .
(Complete data in Mosteller and Tukey, 1977)

```
 0 | 5924332354425386659
 1 | 0258417
 2 | 4
 3 | 4
 4 | 2
 5 | 08
 6 |
 7 |
 8 | 5
 9 | 313781998997
10 | 000
```

stem (leading digit(s)) — leaf (last digit)

The stem-and-leaf is like a histogram turned on its side, but more informative since the original (rounded) data values can be read off.

The stem-and-leaf lets us examine in detail how a collection of numbers spreads out, but there are times when we want a quick overall impression without the fine structure. For this purpose we can turn to the box plot. Exhibit 2.2 shows a set of artificial data plotted as x's along a line. The idea of the box plot is to summarize the location of a majority of the data by drawing a box to cover the central 50% of the points (extending from the first to the third quartile), and "whiskers" stretching to the extremes. We can see from the position of the median (shown as a line across the box) and from the length of the whiskers that these particular data are skewed with a longer upper tail. (Obviously, a box plot of bimodal data like the percent Catholic would not reveal the bimodality.) Incidentally, the 5 numbers that we need from the data can all be obtained from an initial stem-and-leaf display.

Exhibit 2.2
Box Plot

Data: .27 1.16 2.95 1.41 . . . .

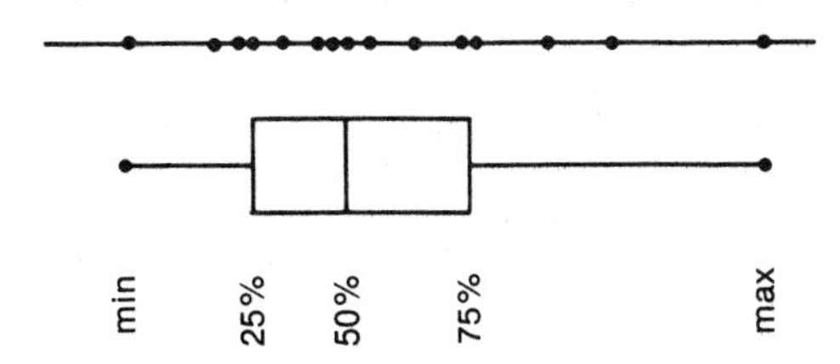

Why should we want to see that much less detail? One reason might be to compare several sets of data. This was necessary, for example, in a recent work-time study of telephone cable splicing based on data from a survey. At one point, the 1042 jobs observed in the survey were separated into 5 (partially overlapping) subsets. Then a box plot was made for the recorded work times for the jobs in each subset, as shown in Exhibit 2.3. (The plotting range was restricted so that the details of the boxes could be seen.) The one on the left is for all the jobs together. The distributions are all very skewed - - in each category the longest jobs take much longer than a "typical" job. Two additional features are the varying widths of the boxes reflecting the sample sizes and the notches to help make visual comparison of medians - - if two notches overlap, the corresponding medians are not judged to be statistically significantly different at about the 5% level. The so-called secondary jobs are clearly much shorter than the primary jobs, as expected from knowledge of what those categories mean, but it was a surprise that the incomplete jobs took longer than the complete jobs. Since each incomplete job is really a complete job that was only partially observed due to the way the survey was organized, we expected them to appear shorter. This paradox is due to a phenomenon called length-biased sampling.

Exhibit 2.3

Box Plots of Splicing Survey Data

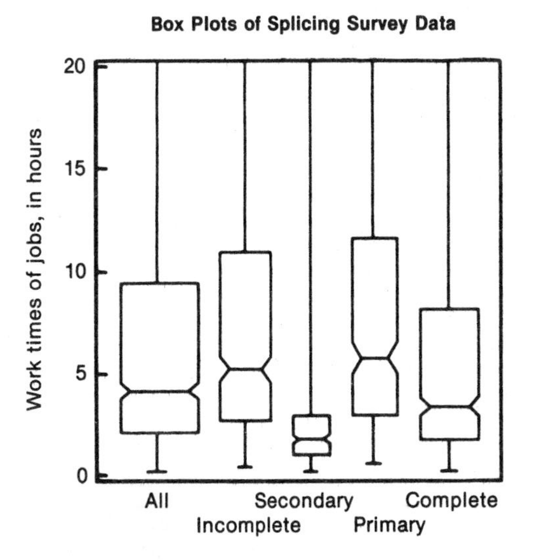

Suppose now that we seek a more detailed comparison of the distributions of two sets of observations. A plot designed for this purpose is the empirical quantile-quantile or EQQ plot. To construct it we can, conceptually at least, first draw the empirical cumulative distribution functions (ECDFs) for the two data sets, as in Exhibit 2.4. The ECDF can be defined as a staircase with a step of height 1/n over each of the observations, where n is the number of data points. We prefer to plot an asterisk in the middle of each vertical line and omit the steps. Now we ask, "How well do the two ECDF's track each other?"

Exhibit 2.4

Construction of an Empirical QQ Plot

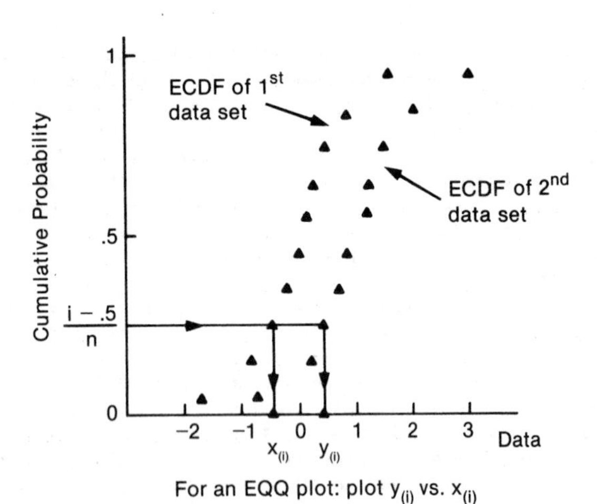

For an EQQ plot: plot $y_{(i)}$ vs. $x_{(i)}$

To answer this question, we can draw horizontal lines and note the positions where they hit the two sets of asterisks. (We'll only do it at the n asterisks.) This gives us pairs of values $x_{(i)}$ and $y_{(i)}$ which are the i-th ordered observations in the two data sets or the percentiles or quantiles of the two sets, and we will use them as abscissa and ordinate values for points in the EQQ plot of Exhibit 2.5. If the two sample sizes are unequal an EQQ plot can still be made; interpolation would be used for one set of quantiles.

Exhibit 2.5

Empirical QQ Plot

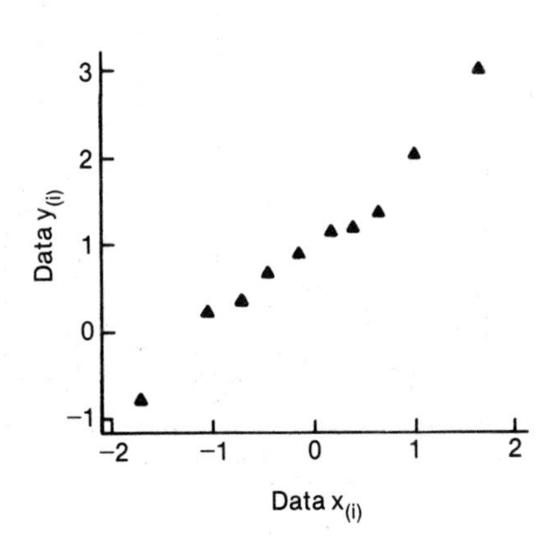

What should we look for in an EQQ plot? First, what should we see if the two distributions are alike? When the two ECDF's track each other closely, the $x_{(i)}$'s nearly equal the $y_{(i)}$'s, and the points in the EQQ plot fall near the line $y = x$. Of course the randomness in real data causes some fluctuations, but the basis for our judgement of goodness of fit is a straight line.

Next, suppose the distributions are basically the same, but with different locations or means. To change the location is like adding a constant to every observation, but the observations (albeit sorted) are being used as x- or y-coordinates in the EQQ plot, so the result is to shift the plot vertically or horizontally, but not to change it otherwise. Similarly, if the standard deviations or scales were different, that would be like multiplying each observation by a constant, which would stretch or compress the EQQ plot vertically or horizontally, but the configuration would still be just as straight or crooked as before.

So it is the straightness of the configuration that tells whether the distributional shapes are similar. (Which straight line depends on the means and spreads of the data sets.) Large or systematic departures from a line indicate lack of fit. For instance, an outlier or two in one data set will appear as points that have drifted away from the rest, but the bulk of the configuration remains straight. Or if one distribution is symmetrical and the other is skewed, then the EQQ plot will be curved. The point of all this is not only can we judge whether the distributions seem to fit each other, but also if they don't fit well we obtain valuable diagnostic information to help us understand how the fit fails.

Exhibit 2.6 is an EQQ plot from a study of air quality in the New York area. The variable involved is the daily average sunlight in Manhattan for several months, and we are comparing the collection of observations for Sundays against those for weekdays using an EQQ plot. The line on the plot is $y = x$, and if the points lie close to it we would conclude that there is no difference between Sundays and weekdays with regard to sunlight.

In fact, the points in the extreme tails are on the line, so we conclude that the very sunniest and the most overcast days have about the same levels of sunlight on Sundays and weekdays. However, the points in the center fall above the line indicating that for the majority of typical days in the centers of the distributions - - the ones that are partially overcast or hazy - - Sundays are indeed sunnier (and so the name Sunday is deserved!)

Exhibit 2.6

Empirical QQ Plot for Comparing
Sunlight on Sundays vs. Weekdays
in Manhattan (May-Aug, 1972 and 73)

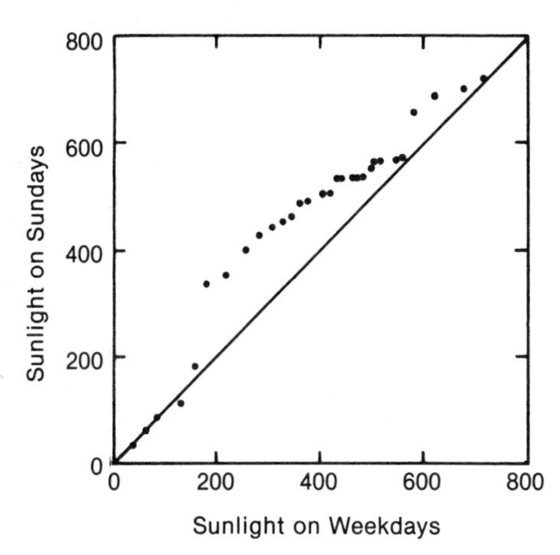

Exhibit 2.7

Cable-Splicing Survey
Scatter Plot of 842 Points (Jobs)

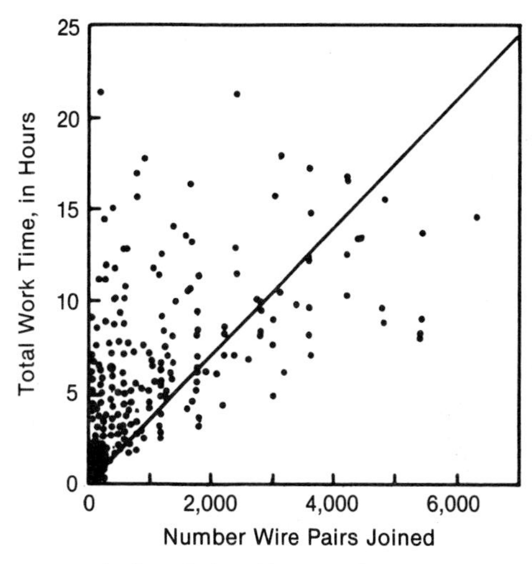

Is the relationship linear? Which line?

We note in passing that the EQQ plot can be modified to deal with another common situation, that of comparing one set of data to a proposed theoretical distribution (for instance, the normal or Gaussian distribution). We only need to replace one ECDF by the cumulative distribution function (CDF) of the proposed distribution, which will be a curve instead of discrete points, and proceed as before. We call the result simply a Q-Q plot.

So far we have only been looking at one variable at a time - - either a single set of numbers or several sets collected under different circumstances. There are many important situations in which we observe two (or more) variables at once and we want to study the possible dependence of one on the others. The simplest case is when the data are a set of ordered pairs of observations (x_i, y_i). The most basic graphical display is a simple Cartesian plot of the separate plots, i.e., a scatter plot. Nearly always with bivariate data it is worth making a scatter plot, even when we have a strong advance belief concerning the nature of the relationship, since often there are surprises in store.

Returning to the cable splicing study, Exhibit 2.7 is the scatter plot of the number of wire pairs that had to be spliced vs. the number of hours it took to do the splicing work for each of the 842 complete jobs. Since splicing is done manually, one might think that to splice twice as many pairs would take twice as long, so one might expect to see a linear relationship between these two variables with zero intercept. There is a relationship, but the points are scattered so widely that it is unclear what shape the relationship takes. A line fitted by ordinary least squares is also shown, but it is hard to lend it much credence.

One way to alleviate this problem is to compute and plot a "smoothed" version of the data. This is done in Exhibit 2.8 by taking groups of 15 points moving across the plot (letting consecutive groups overlap by 10 points), computing the median of the x values and the median of the y values for each group, and plotting them. (The median is robust at outliers like the off-scale point in Exhibit 2.7. See Section 3.) It's now doubtful whether any single straight line can fit this data well. A curve bending down at the right might do, or perhaps two straight lines (both of which, interestingly, point approximately toward the origin). We note that a lower slope here indicates work done more efficiently since it corresponds to less time spent on each 1000 wire pairs spliced, so it looks as if larger jobs are being done more efficiently.

Exhibit 2.8

Cable-Splicing Survey
Medians of 15-point Groups, Overlapped by 10

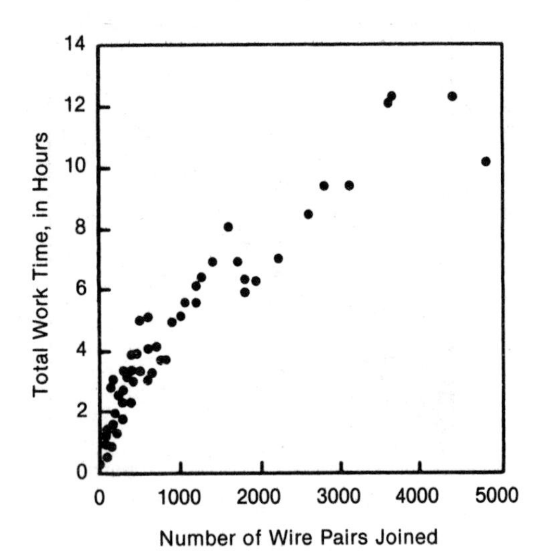

One straight line? A curve? Two lines?

If we look closely at this plot we can see some other interesting things. For instance, there is a curious down shift and change in slope occurring at about 6-9 hours, which represents approximately one day's work by one splicer. All the larger jobs would have taken two or more days, or (as the people who are provided that data later confirmed) some of them were jobs done by two or more splicers in one day!

In conclusion, how does all this relate to teaching? First, exploratory data analysis - - and graphical methods in particular - - can put the discovery stage back into statistics - - the "Aha!" - - making the subject considerably more fun for students than grinding through formulas and writing down estimates and significance levels.

Second, since exploratory data analysis naturally precedes confirmatory statistics in applications, it should be taught either before, or at least along with "classical" statistics. It is important for students to get a feel for working with and manipulating data before studying the more theoretical aspects of statistics - - in the same way that it is important to study arithmetic before abstract algebra.

Finally, since the mathematics used in exploratory data analysis is not advanced, the subject can be taught to people with less mathematical background than classical statistics requires. This is especially true for graphical methods. (In modern math programs, school children have no trouble making and using graphs.) In primary or secondary schools the subject can be taught in earlier grades, at the post-secondary level it can be taught to non-math majors, and beyond the university it can be taught to people in business, management, etc.

3. Robust Methods (by A.F.S.)

Unlike the usual statistical methods whose results tend to be distorted by outliers, robust methods provide protection against such observations and can easily be incorporated into the teaching of statistics at all levels. Because many of the basic concepts are simple, robustness can and should be discussed when the student is being introduced to statistical ideas.

Robust techniques should complement, not replace, standard statistical tools such as means, variances, and least squares estimates which are based on assumptions such as the normal distribution. In fact, many statisticians now recommend that a robust analysis be used routinely to help assess the validity of a more classical analysis, because hidden structure or problems with the data are often brought to light. If the classical and robust analyses approximately agree, this can be taken as a confirmation of the classical results by a secondary analysis. But when they disagree, there is work to be done because either errors in the data need to be corrected, or else unexpected structure remains to be discovered and explained.

The need for statistical robustness can be seen even in the basic problem of finding an "average" value to summarize a list of numbers. For example, to summarize the five numbers (7, 8, 6, 4, 100), the arithmetic mean is (7 + 8 + 6 + 4 + 100)/5 = 25, which is not typical of most (or even part) of the data! For some real-life problems, 25 would be the proper summary; but it is often better to summarize the reasonable portion of the data (7, 8, 6, and 4) and to study exceptional values (like 100) separately to decide, for example, whether they are interesting special cases for further study or simply in error.

The median is a robust measure of an average; it is that number which has half of the numbers smaller and half larger than itself. For this data set, it is the middle value of {4, 6, 7, 8, 100} = 7, which is a typical value.

Robustness, formally, is protection against unusual data and violated assumptions. A few atypical or "bad" observations can ruin an ordinary analysis but will have only a very limited effect on a robust analysis. Using robust methods is analogous to taking out an insurance policy: the premium is paid as a decrease in efficiency of the estimate when the data are well behaved. In real data, errors are often present, and this "insurance" can be vital. Robust methods also aid the detection of outliers and this can be very useful.

The teaching of robustness can proceed at many levels: simple or complex, pencil or computer, in-class or independent project. It can be taught separately as a section by itself, but is also easily integrated with other statistical topics. For example, after teaching a new standard procedure, some time can be spent discussing ways of "robustifying" that method. The use of examples is crucial, of course, to teaching any statistical ideas and maintaining student interest; pictures and graphic displays (see Section 2) should be used frequently.

To illustrate some robust methods for estimation of location (i.e. average or mean), consider the attention spans of 10 hypothetical students:

5, 18, 15, 2, 8, 55, 11, 3, 8, 8 minutes.

The arithmetic mean (which is not robust) is 13.4 minutes. The 10% trimmed mean (which is robust) is formed by (1) ordering the data from smallest to largest, (2) trimming away or removing 10% of the data from each side, and (3) taking the arithmetic mean of what remains:

1. order: 2, 3, 5, 8, 9, 11, 15, 18, 55

2. trim 2 and 55

3. 10% trimmed mean = (3 + 5 + . . . + 18)/8 = 9.6 minutes.

The median (which is very robust against outliers) is (8 + 9)/2 = 8.5 minutes. These estimators (mean, trimmed mean, median) are all examples of a rich family of location estimators called L-estimates, which are linear combinations of order statistics (or sorted data values).

Another useful class that also includes robust numbers is the M-estimates, which are generalizations of least squares and maximum liklihood procedures. The arithmetic mean is also an M-estimate, for it is the value θ which minimizes

$$\sum_{i=1}^{n} (x_i - \theta)^2$$

the sum of squared deviations of the data values from θ.

In place of squaring, M-estimate allow other functions ρ that can be less sensitive to outliers. This is done by differentiating

$$\sum_{i=1}^{n} \rho(x_i - \theta)$$

and solving

$$\sum_{i=1}^{n} \psi(x_i - \hat{\theta}) = 0$$

where ψ = (constant) $(d\rho/d\theta)$. Different choices of ρ lead to M-estimates with different properties. Some examples are given in Exhibit 3.1.

Exhibit 3.1

Examples of M-estimates

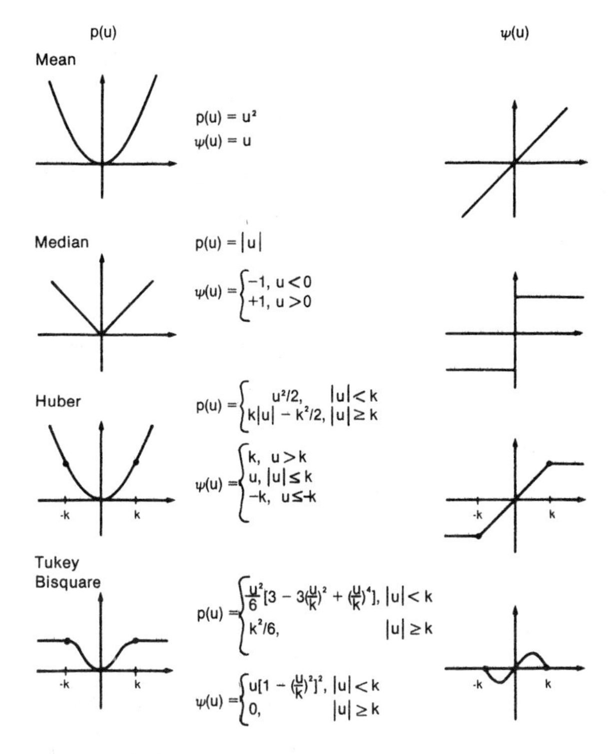

The median, which is an M-estimate with $\rho(u) = |u|$, is extremely resistant to bad data but suffers from "granularity", a lack of responsiveness to data near the central value. The Huber choice for ρ corrects this problem: near zero, it is like the ρ for the mean, allowing data near the average to "fine-tune" the estimate, but it maintains resistance to bad data by behaving like the median away from the middle. Tukey's bisquare also combines efficiency and robustness, but has a ψ that "redescends" to zero. In effect, this means that data that are very far from the middle will have no effect on the estimate.

For easy pencil-and-paper calculation, L-estimates are preferable because the minimization step for M-estimates (other than the mean and median) requires iterations and is best attempted with a pocket calculator or computer.

The proportion of bad data that an estimate can tolerate and still return a sensible answer is its breakdown value. The mean has a breakdown value of zero because, by changing the value of even a single number, the mean can be forced to assume any value, as in Exhibit 3.2a. The median has a breakdown value of 50% because almost half of the data must be changed before the median breaks down completely, as illustrated in Exhibit 3.2b. Note that extreme observations do have an effect, albeit small, upon the median. (Compare the second and third parts of Exhibit 3.2b.) Also note that when 3 of the 5 points (i.e. more than 50%) are moved, the median breaks down, as shown at the bottom of Exhibit 3.2b. The breakdown value for a trimmed mean lies in between those for the mean and median; for example, the 10% trimmed mean has a breakdown value of 10%.

Exhibit 3.2a

Breakdown of the Mean

Exhibit 3.2b

Breakdown of the Median

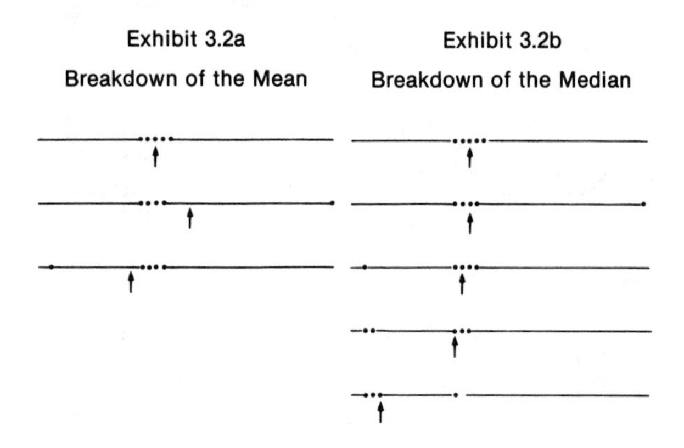

One measure of robustness of an estimate, $\hat{\theta}$, considered as a function of the sample points, is provided by measuring the effect of adding a new point x to a sample $x_1, \ldots, x_n$. The influence function at the value x is defined to be

$$I_+(x,\hat{\theta}) = (n+1)\{\hat{\theta}(x_1,\ldots,x_n,x)-\hat{\theta}(x_1,\ldots,x_n)\} .$$

For example, if $\hat{\theta}$ is the mean, $\hat{\theta} = (\Sigma\ x_i)/n$, we can calculate $I_+(x, \bar{x}) = x - \bar{x}$, which increases linearly with x. This shows that the mean has an unbounded influence function, and is therefore not robust, because there is no limit to the effect of a single new point. For M-estimates, I_+ is essentially proportional to the function ψ.

For estimating scale or spread of a data set, there are several alternatives which robustify the sample standard deviation,

$$SD = \left(\frac{(x_1 - \bar{x})^2 + (x_2 - \bar{x})^2 + \ldots + (x_n - \bar{x})^2}{n - 1} \right)^{1/2}$$

One possibility is the median absolute deviation from the median, or MAD, which is obtained from:

$$MAD = Median(|x_1-m|, |x_2-m|, \ldots ,|x_n-m|) ,$$

were $m = Median\ (x_1, \ldots, x_n)$. If the MAD is to be used as a "standard deviation", then it is customary to divide it by 0.675 to make it more comparable to the SD. (This is the asymptotic factor needed to make the MAD an unbiased estimator of the standard deviation for a normal distribution.)

For example, an initial data set (7, 8, 6, 4, 100) has SD = 42 but it has

$$MAD= Median\{|7-7|, |8-7|, |6-7|, |4-7|, |100-7|\}$$

$$= Median\{0,1,1,3,93\}$$

$$= 1$$

The large value of the SD is due to the fact that 100 is very far from most of the data. However, MAD/0.675 is 1.5 which is much smaller than 42 because the single large number is not allowed to dominate.

Anotherrobust scale estimate is the interquartile range, or IQR, which is simply the upper quartile minus the lower quartile or the length of the box in a box plot (see Section 2). To convert this estimate approximately, as for the MAD, one would divide it by 1.35.

Simple linear regression, i.e. fitting a straight line to points in two dimensions, can also be robustified, for example with M-estimation techniques. However, even M-estimates can break down in a situation such as that in Exhibit 3.3. Assuming a line is appropriate, which one do we want? The answer is "both lines". If the high-leverage point (the right-most one) is in error, we would prefer a robustly fitted line, such as the so-called repeated medians line. But the least-squares line is preferable when that outlying point is correct, because in this case that single point provides nearly all the information about the slope!

Exhibit 3.3

Leverage in Regression

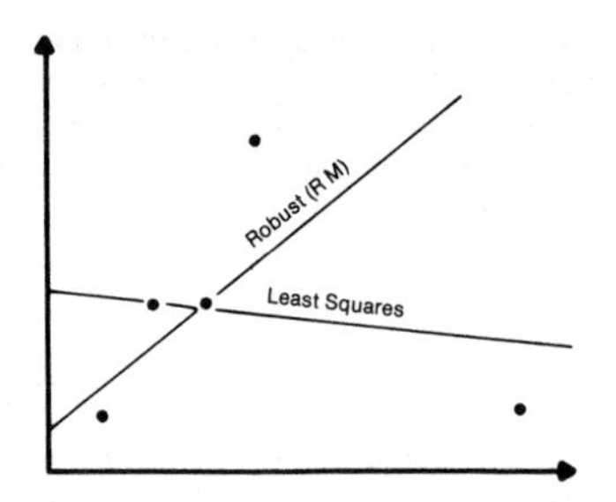

A final example of the usefulness of robust methods is the fitting of two related shapes. Consider a square with one corner point distorted (shown as a dotted shape in Exhibit 3.4) fitted to a perfect square (the solid shape) by allowing rotation, translation, and magnification. Exhibit 3.4 depicts both the least squares fit and the robust fit by repeated medians. Because the robust method "fits what fits well" it indicates clearly that the dotted shape is identical to a square except at one point. The least squares fit, by compromising and trying to get every point to fit reasonably well, makes this sort of inference much more difficult. Practical application of this type of shape-fitting has also been carried out for fossils and human skulls.

Exhibit 3.4

Fitting of Two Related Shapes

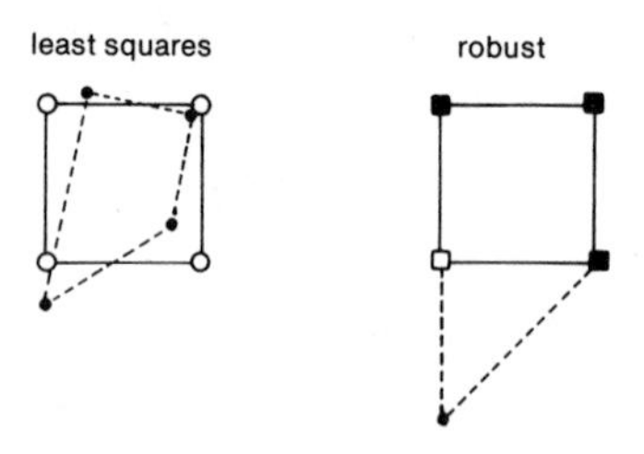

Robust methods are available in a variety of areas, including correlation, time series, and two-way analysis, in addition to the location, scale, and regression problems discussed here. More information may be found in the references.

4. Cluster Analysis (by J.R.K.)

The general problem of partitioning data in statistically sensible ways is what cluster analysis is all about. As a quick example, suppose the data consist of ten medical measurements on 100 patients. We wish to partition the patients into groups so that patients within a group have similar medical profiles. Because the data are ten dimensional, this can be a very hard task. Special methods may be needed to group the data effectively.

To help us get in the right spirit for clustering, consider the three examples shown in Exhibits 4.1 to 4.3, and try to cluster or group the data by eye as you think appropriate. In the first example, it appears that there are three well-separated and reasonably circular clusters. In the second example, there is one very distinct linear cluster and two roughly circular ones with a fuzzy boundary between them. The third example has two distinct elliptically-shaped clusters. An "eyeball solution" would probably be preferable in this case to one from a standard clustering algorithm, which would likely miss the obvious structure.
An education example will be used to show how clustering methods can help put together a curriculum of courses. The potential students in this case work for the Bell Telephone System. They are known as dial administrators and perform a wide variety of jobs dealing with central office switching equipment. The basic objective of the study was to identify the training needs of the workers and to compare these need with the contents of the courses in the existing training curriculum.

Exhibit 4.1
Example 1

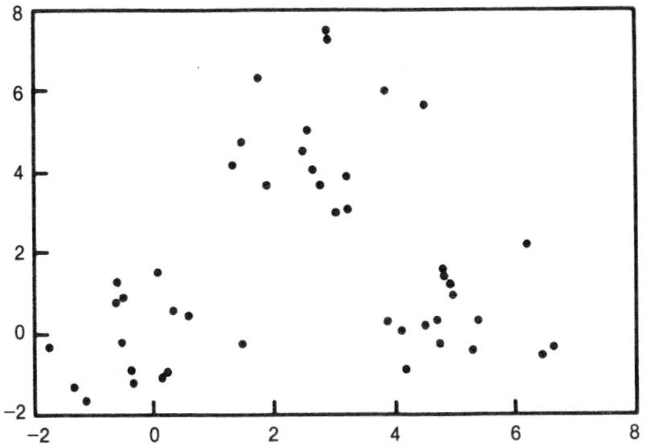

Exhibit 4.2
Example 2

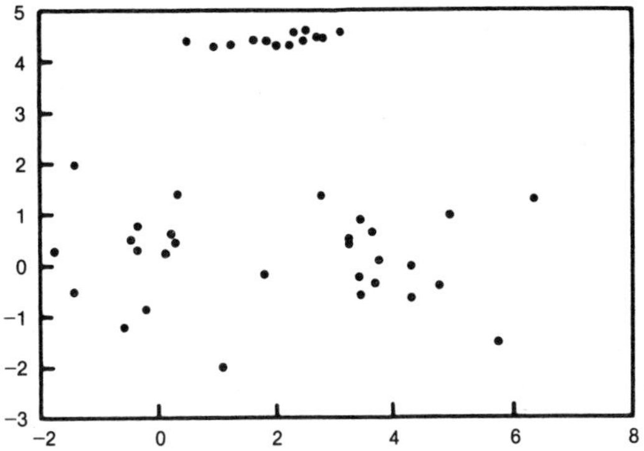

Exhibit 4.3
Example 3

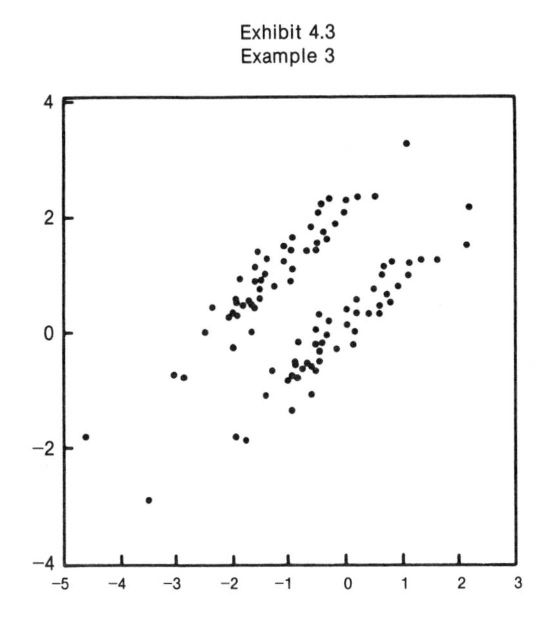

In a survey, a sample of 452 workers each was asked whether or not they wanted training in each of 169 components of the job identified in an initial phase of the study. The survey resulted in a matrix of data with dimensions 169, the number of variables or job components, by 452, the number of workers sampled.

Each cell contains a "0", if the worker did not want training or "1", if the worker did.

Now this is a large matrix to digest simply by staring at it. Clustering methods were used in two ways to try to simplify things. First, a cluster analysis was performed on the rows of the matrix in order to group together similar tasks. Second, a cluster analysis was made of the columns of the matrix to group together those workers who had common training needs.

Exhibit 4.4 gives a summary of the results of the clustering of the 169 job components. Basically, the boxes represent the clusters and these are combined together in a hierarchical or tree-like fashion. The hierarchical structure is a consequence of the algorithm which was used. Hierarchical clustering is often done more for convenience than belief in a hierarchical structure. In fact, the usual prarctice is to cut the three to achieve a partitioning of the objects into groups and then to forget about the hierarchical structure. But in an educational setting where it is natural to combine topics into instructional units, instructional units into courses, and courses into curriculum, a hierarchical structure may be quite sensible.

Look now at the clusters labelled IIA, IIB1, and IIB2. These join together into cluster II farther down the tree. All of the items in cluster II deal with data validation, which is a major part of the dial administrator's job. Cluster IIA has 17 items (or tasks) in it, IIB1 has three, and IIB2 has five.

Each of these clusters was checked against the contents of the courses being offered on dial administration. The items in cluster IIB2 were contained in one of the courses, but neither of the other clusters was covered by any part of the curriculum. The data also showed a high-level of interest in training for both of these clusters. Partly as a result of this study, the curriculum was modified to include these topics.

For many clustering problems, the starting point is a matrix $\underline{Y}$ of measurements on p variables for n objects. The variables can be either discrete as in the education example or continuous. The goal is to cluster the objects. However, the roles of the rows and the columns can be interchanged, as in the education example where both the job components and the workers were clustered but for different purposes.

Exhibit 4.4

Hierarchical Cluster Analysis of Job Components

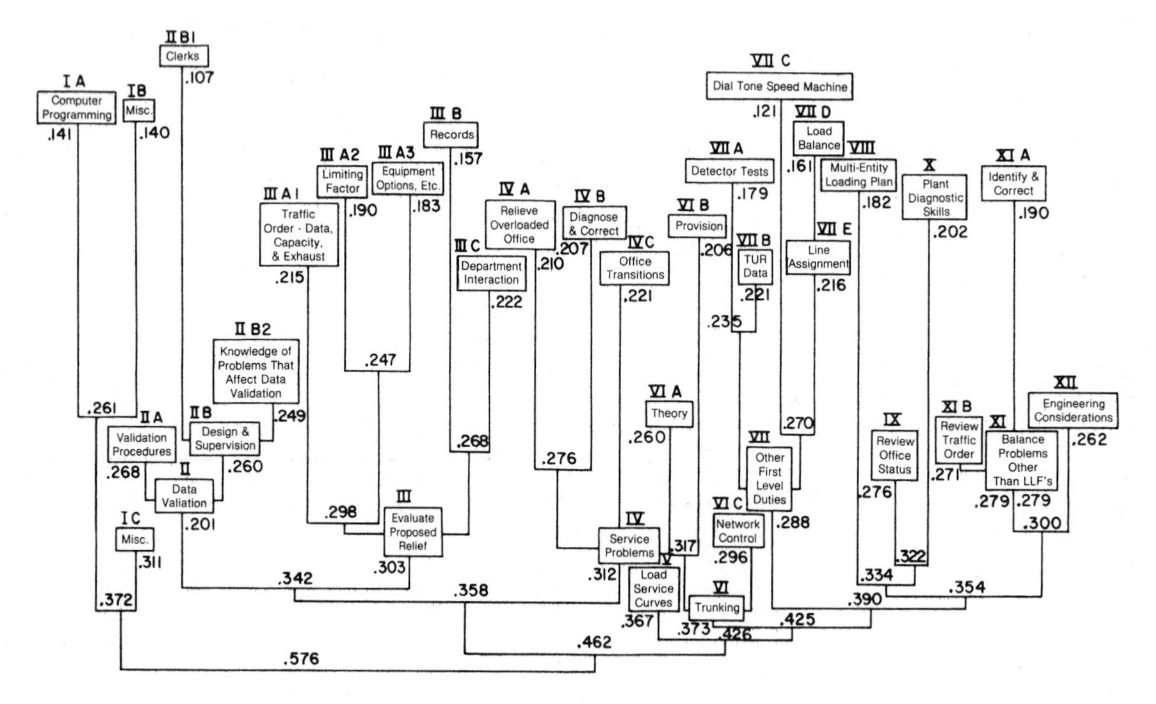

Generally speaking, one can distinguish two different approaches to clustering the n objects using the columns $\underline{y}_1, \ldots, \underline{y}_n$ of $\underline{Y}$: informal clustering by eye (based on graphical displays of $\underline{y}_1, \ldots, \underline{y}_n$) and formal clustering by algorithm (based on numerical manipulations of $\underline{y}_1, \ldots, \underline{y}_n$).

The informal procedures can work on small problems but they have a number of limitations for large ones. None of them would have helped much on the training problem, mainly because it has too many variables and objects.

The ideal approach to clustering would be to consider all possible partitions of objects into groups and then to pick the "best" partition. Computationally, this is very impractical! For this reason, a large number of more or less automatic algorithms have been devised for obtaining reasonably good clusters.

First, there are the nonhierarchical algorithms which shuffle the objects back and forth between tentative clusters until convergence is (hopefully) achieved. The user must prespecify values for certain parameters which govern how these algorithms work. This can involve a lot of guess work and trial and error. In short, these procedures can be very effective but also they can be frustrating and expensive.

The other algorithms are the hierarchical ones. The focus here will be on these - - not because they are superior to the other type but because they are simpler and probably more widely used.

With this background, it is now appropriate to define the three main stages of a cluster analysis: (1) the input stage where the variables and objects are decided upon and the data are put into the "right form" for clustering, (2) the algorithm stage where methods for finding clusters are applied to the data from the input stage, and (3) the output stage where the proposed clusters are studied to see if they make sense.

The input to a hierarchical clustering program is usually a measure of distance or dissimilarity between each pair of objects. For example, in the training problem, so-called Hamming distance was used. This is just the square root of the fraction of pairs of responses which are (0,1) or 1,0). Ignoring the denominator, this is also the Euclidean distance between the vectors of responses for any two questions. This seems very simple and straightforward but there are other ways of doing it (e.g., asymmetric treatment of 0's and 1's), and one wants to get the best possible measure since it will affect the clusters which are obtained.

For continuous variables, it is common to use Euclidean distance. This is most appropriate when the clusters are spherical in shape; that is, when the within-cluster variances of the variables are the same and the variables are uncorrelated. If there are good reasons for using variables on different scales, then Euclidean distance may still be appropriate even if the variances are unequal. However, an arbitrary choice of scales could place too much emphasis on some variables and not enough on others. To avoid this problem, one can use a weighted Euclidean or ellipsoidal metric which adjusts for differences in the within-cluster variances. In addition, the variables are usually correlated. In this case it would be better to use a generalized weighted Euclidean distance which simultaneously adjusts for unequal variances and nonzero correlations.

Unfortunately, the real-life situation is never so simple! For one thing, the within-cluster variability of the clusters is unknown since the clusters are unknown. Even if the clusters were known, they might have different shapes, which makes the problem still messier. These are some of the real difficulties in cluster analysis - - compromises are unavoidable.

Only a few of the basic types of distances which are commonly used have been mentioned. Sometimes it is a good idea to try more than one to see if the clustering results change.

Next a simple class of hierarchical clustering algorithms will be introduced with the help of another example. It involves 21 operating companies or subsidiaries of the Bell Telephone System - - these are the objects. The variables are financial measures including debt ratio , rate of return, etc. for each company. The purpose of the analysis was to find clusters of the telephone companies with similar financial characteristics.

The distance between each pair of the companies was computed using a type of weighted Euclidean distance. These distances formed the input for the algorithm.

To start, it is convenient to think of the companies as 21 individual clusters, labelled as A through U in Exhibit 4.5. At the first step of the hierarchical cluster analysis, the closest pair is identified and merged into a single cluster. Now there are 20 clusters. Next the closest of the 20 is joined, and so on, down to the point where all companies have merged together.

Exhibit 4.5

Hierarchical Tree

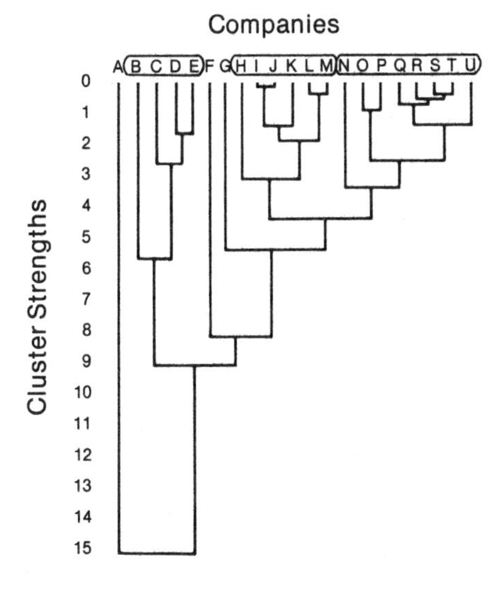

This raises an important question: how does one measure distance between clusters containing more than one object? This can be done in several possible ways, such as using the minimum pairwise distance, the maximum pairwise distance, or the average of the pairwise distances. Each rule gives rise to a different clustering algorithm.

A few properties of these procedures should be noted. If the monotone increasing transformation is made of the distances, the tree structure for the minimum-distance and maximum-distance methods will not change - - that is, the same objects will be joined at each step of the algorithm. This is important for situations where one only wants to rely on rank-order information. The minimum-distance method usually produces a smaller number of major branches or clusters in the hierarchical tree than the maximum-distance method. The minimum-distance method also tends to produce chain-like clusters which wind through the space of objects in contrast to the maximum-distance method which is biased towards compact, spherical clusters. The average-distance method is not invariant under monotone increasing transformations of the distances. It tends to give clusters which are intermediate in shape and number between those from the other two methods.

For the telephone company data, the maximum-distance method was used and Exhibit 4.5 shows the tree which resulted. The vertical dimension gives the "distance" between two groups of objects which were merged. For example, I and J joined at a distance less than one; and H combined with (IJKLM) at a distance near three.

Each node of the tree represents a potentially meaningful cluster. One can decide on a set of disjoint clusters by cutting the tree in what seems to be an appropriate way. In this case, it was cut to give three main clusters, (BCDE), (HIJKLM), and (NOPQRSTU), and three singleton clusters, A, F, and G.

In making this partition, decisions have been made about which parts of the tree are worth thinking of as clusters and which are not. This is not always easy to do in a sensible way. In fact, the main problem is that a complete tree is always constructed by the procedure, even if there are no real clusters at all in the data.

This brings us to the output stage. The idea here is to look hard at the proposed clusters to see if they are statistically separated. There are many informal checks which one could make. Two specific types are distance plots and projection plots.

Data from another Bell System problem will be used to illustrate a simple projection plot. The operating companies of the Bell System were separated into 48 regional entities or regions and the goal was to find clusters of the entities operating in similar environments based on business environment variables. Fifteen clusters were found and the question is whether any of these are really distinctive.

The idea for the projection plot is to find the directions in the space of variables which do the best job of separating the clusters. Mathematically, this can be done by constructing a between-groups covariance matrix $\underline{B}$, finding its first two eigenvectors, and then projecting the data onto the plane defined by these vectors. (Other matrices can be used to achieve different projections.)

Exhibit 4.6 shows an example of such a projection plot for the 48 entities. Entities which are far away from the plane of the picture are shown with smaller character sizes.

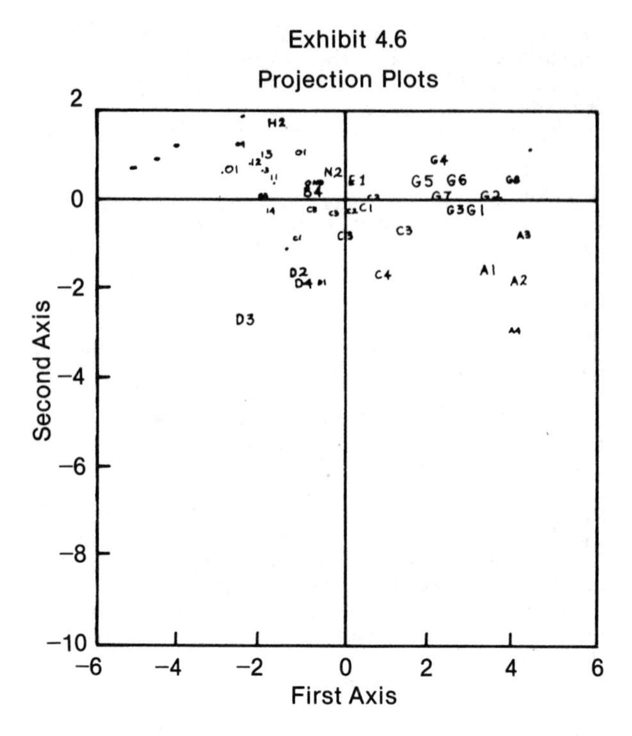

The main feature of the plot is the singleton cluster F1 in the lower left corner. Cluster A, which has four entitites in it, can be seen on the right. Actually, it looks as if many of the clusters are not well separated. To verify this, a next step might be to omit the outlier F1 and to redo the analysis. This process illustrates the iterative nature of data analysis.

How can the subject of cluster analysis be integrated into the classroom? As soon as a child is old enough to use scatter plots, meaningful discussions can begin. Primary students in the fifth or sixth grade could deal with data of the type shown in the top half of Exhibit 4.7, where the problem is to compare air pollution levels in different geographical regions. As a simple ad hoc procedure for analyzing this data, students could be asked to sort the rows into homogeneous groups, as shown in the bottom half of Exhibit 4.7.

Exhibit 4.7

Air Quality 1971
(X - Highly Polluted)

Region	Particulate Matter	Sulphur Dioxide	Nitrogen Dioxide	Carbon Monoxide	Hydro-Carbons
Cheyenne					
Chicago	X	X	X	X	X
Cleveland	X	X	X		X
Denver	X			X	X
Houston	X	X	X		X
Kansas City	X			X	X
New York	X	X	X	X	X
San Diego			X	X	X
San Francisco			X	X	X
St. Louis	X	X	X	X	X
Chicago	X	X	X	X	X
New York	X	X	X	X	X
St. Louis	X	X	X	X	X
Cleveland	X	X	X		X
Houston	X	X	X		X
Denver	X			X	X
Kansas City	X			X	X
San Diego			X	X	X
San Francisco			X	X	X
Cheyenne					

Secondary school students ought to be able to carry out simple algorithms such as the hierarchical ones which were described. Some may be able to suggest their own algorithms.

Computers ought to be used for large problems but small ones can be done without them. For example, a hierarchical cluster analysis of ten objects can be done by hand.

At the post-secondary level, students should be able to appreciate some of the more subtle distinctions between different distance metrics, different algorithms, etc. They ought to be able to analyze real data on the computer. By the way, cluster analysis packages are available today from a variety of source.

The main reason that cluster analysis was selected as one of the topics for this symposium is because of its usefulness in real problems. It identifies clusters whose objects are similar within but different across clusters. Outliers show as singleton clusters. By switching roles, clusters of variables can also be found. This is an important idea because it provides a way for reducing the dimensionality of large problems by replacing a cluster of variables by either a representative or an average variable. Even when there are no obvious clusters, these procedures can still help to organize the data into manageable groups.

5. Summary (by J.R.K.)

Exploratory data analysis can be described as numerical detective work which tries to extract what is useful and interesting from a body of data. This symposium has described three important areas of data analysis: graphical methods, robust methods, and cluster analysis. A graphical display in data analysis is like a photograph in everyday life: it portrays the data in a form that is easy to grasp and remember. Robust methods are like insurance policies: they protect against the potentially catastrophic effects that small amounts of bad date can have on statistical analyses. clustering methods are like filing systems: they help us to organize data into groups which are simpler to think about and work with.

While the three topics are different in many respects, there are also significant connections, some of which are indicated here.

Cluster analyses can lead to effective displayes as in the projection plots.

Graphical displays can be used to suggest clusters.

Robust methods can improve cluster analyses by helping to determine a sensible metric for the data.

Cluster analyses can help robust analyses by locating homogeneous groups of objects and identifying outliers.

Robust methods can improve graphical displays by suggesting what is worth plotting.

Graphical displays can provide robust summaries which can be read directly from the plot.

These methods are a few of the basic tools for exploratory data analysis. We hope this symposium has provided a feel for their importance, the motivation to learn more about them, and the stimulus to include them in courses.

Annotated References

(#) refers to sections of this paper.

Books

Andrews, D.F., Bickel, P.J., Hampel, F.R., Huber, P.J., Rogers, W.H. and Tukey, J.W. (1972). Robust Estimates of Location-Survey and Advances. Princeton University Press, Princeton, New Jersey. (Extensive Monte Carlo study under a variety of sampling conditions.) (3)

Bickel, P.J. and Doksum, K.A. (1977). Mathematical Statistics: Basic Ideas and Selected Topics. Holden-Day, San Francisco. (Brief introduction to robustness in Section 9.5.) (3)

Daniel, C. and Wood, F.S. (1971). Fitting Equations to Data. Wiley, New York. (Approaches to exploratory analysis and model building presented through the extensive analysis of a limited number of data sets. Includes scatter plots and Q-Q plots.) (2)

Everitt, B. (1974). Cluster Analysis. Wiley, New York. Concise introduction, easy to read.) (4)

Gnanadesikan, R. (1977). Methods for Statistical Data Analysis of Multivariate Observations. Wiley, New York. (Modern treatment of multivariate data analysis; distance metrics; clustering procedures.) (1, 2, 3, 4)

Hartigan, J.A. (1975). Clustering Algorithms. Wiley, New York. (Authoritative work on clustering; examples.) (4)

Huber, P.J. (1977). Robust Statistical Procedures. S.I.A.M., Philadelphia. (Advanced mathematical treatment.) (3)

Launer, R.L. and Wilkinson, G.N., eds. (1979). Robustness in Statistics. Academic Press, New York. (Proceedings of a workshop on robustness.) (3)

McNeil, D.R. (1977). Interactive Data Analysis, a Practical Primer. Wiley, New York. (A good introduction, with tutorial examples. Compact. Includes computer programs in Fortran and APL for stem-and-leaf, box plots, etc. Available in paperback.) (2, 3).

Mosteller, F. and Tukey, J.W. (1977). Data Analysis and Regression, a Second Course in Statistics. Addison-Wesley, Reading, Massachusetts. (A data analyst'sview of regression and other more classical tools. In the spirit of Tukey (1977) but more advanced and specialized. Contains a wealth of data sets for exercises.) (2, 3)

Siegel, A.F. and Launer, R.L., eds. (1981). Advances in Data Analysis. Academic Press, New York. (Conference proceedings. Includes regression by repeated medians.) (2, 3)

Sneath, P.H.A. and Sokal, R.C. (1973). Numerical Taxonomy. Freeman, San Francisco. (Comprehensive treatment of clustering with extensive bibliography.) (4)

Tukey, J.W. (1977). Exploratory Data Analysis. Addison-Wesley, Reading, Massachusetts. (Wide ranging. Unique. Starts with simple concepts and tools. No probability theory or classical statistics required. Extensive examples.) (1, 2, 3)

Velleman, P.F. and Hoaglin, D.C. (1980). Applications, Basics, and Computing of Exploratory Data Analysis. Duxbury Press, North Scituate, Massachusetts. (A companion volume to Tukey, 1977.) (1, 2, 3)

Articles and Papers

Beck, B., Denby, L. and Landwehr, J.M. (1976). Statistics in the elementary school. Communications in Statistics - Theory and Methodology. Vol. A5(10), pp. 883-894. (General comments and perspectives on statistics in the school curriculum.) (1)

Blashfield, R.K. (1976). A consumer report on the versatility and user manuals of cluster analysis software. Proceedings of the Statistical Computing Section of the American Statistical Association, pp. 31-37. (Compares 14 programs using 11 criteria.) (4)

Chen, H.J., Dunn, D.M. and Landwehr, J.M. (1975). Grouping companies based on their operating environment. Proceedings of the Business and Economic Statistics Section of the American Statistical Association, pp. 278-283. (Example of clustering 48 Bell System entities.) (4)

Cleveland, W.S. (1979). Robust locally weigted regression and smoothing scatterplots. Journal of the American Statistical Association, Vol. 74, pp. 829-936. (A new approach to smoothing.) (2, 3)

Cleveland, W.S. and Kleiner, B. (1975). A graphical technique for enhancing scatterplots with moving statistics. Technometrics, Vol. 17, pp. 447-454. (Moving medians and other smoothers with examples from an air pollution study.) (2, 3)

Cohen, A., Gnanadesikan, R., Kettenring, J.R. and Landwehr, J.M. (1977). Methodological developments in some applications of clustering. In Applications of Statistics (P.R. Krishnaiah, ed.). North-Holland, Amsterdam, pp. 141-162. (Four real examples, practical problems.) (4)

Cormack, R.M. (1971). A review of classification. Journal of the Royal Statistical Society, Vol. A134, pp. 321-357. (Important review article on cluster analysis.) (4)

Denby, L. and Landwehr, J.M. (1975). Examining one and two sets of data - Parts I, II and III. Prepared for USMES. Available from authors at Bell Laboratories, Murray Hill, New Jersey. (Material developed for secondary school curriculum. Includes histograms, ECDFs, and Q-Q plots.) (2)

Gnanadesikan, R., Kettenring, J.R. and Landwehr, J.M. (1977). Interpreting and assessing the results of cluster analyses. Bulletin of the International Statistical Institute, Vol. 47 (2), pp. 451-463. (Output stage; distance plots.) (4)

Gnanadesikan, R., Kettenring, J.R. and Landwehr, J.M. (1981). Projection plots for displaying clusters. To appear in Statistics and Probability: Essays in Honor of C.R. Rao (G. Kalliandpur, P.R. Krishnaiah, and J.K. Ghosh, eds.), North-Holland, Amsterdam. (Output stage; projection plots.) (4)

Gnanadesikan, R. and Wilk, M.B. (1969). Data analytic methods in multivariate statistical analysis. In Multivariate Analysis III (P.R. Krishnaiah, ed.), Academic Press, New York, pp. 593-638. (An expository survey). (1, 2, 4)

Kettenring, J.R., Rogers, W.H., Smith, M.E. and Warner, J.L. (1976). Cluster analysis applied to the validation of course objectives. Journal of Educational Statistics, Vol. 1, pp. 39-57. (Edcation example.) (4)

Mallows, C.L. and Tukey, J.W. (1980). An overview of techniques of data analysis, emphasizing its exploratory aspects. To appear in a volume in obervance of the 200th anniversary of the Lisbon Academy of Sciences. (A survey of state-of-the-art isues both of the philosophy and the technology of data analysis.) (1)

McGill, R., Tukey, J.W. and Larsen, W.A. (1978). Variations of box plots. The American Statistician. Vol. 32, pp. 12-16. (Details of box plots - - variable widths, notches, etc.) (2)

Siegel, A.F. and Benson, R.H. (1980). Estimating change in animal morphology. Unpublished. (Application of robust methods in fossils and human skulls.) (3)

Tukey, J.W. and Wilk, M.B. (1966). Data analysis and statistics: An expository overview. AFIPS Conf. Proc., Fall Joint comput. Conf. Vol. 29, pp. 695-709. (An early survey of the nature, concerns and needs of data analysis.) (1)

Wilk, M.B. and Gnanadesikan, R. (1968). Probability plotting methods for the analysis of data. Biometrika, Vol. 55, pp. 1-17. (Detailed discussion of Q-Q plots, ECDFs, etc.) (2)

CHAPTER 11 - **Mathematics Curriculum**

SUCCESSES AND FAILURES OF MATHEMATICS CURRICULA IN THE PAST TWO DECADES

SUCCESSES & FAILURES OF MATHEMATICS CURRICULA IN THE PAST TWO DECADES

H.B. Griffiths
University of Southampton
Southampton, England

How may we distinguish good work from bad? How can we devise critical standards for evaluating Mathematics Education in its environment. I shall confine these questions to the general area of mathematical curricula, by consideration of well-known examples: more recent ones will presumably be reported at this Congress. It is necessary to warn you of my (conscious) biasses: (i) I am a University teacher of Mathematics in England (ii) I believe that what finally matters is not the administration, not the hierarchy of inspectors etc., but the teachers and their pupils.

The words 'success' and 'failure' come easily to Americans, but the first is dangerously inappropriate for our discussion, and I begin by showing why we must abolish its use.

Examples:

(i) 'To put a man on the moon within 10 years'. Aims of this project were achieved, but at the cost of diversion of resources. 'Good in parts' rather than simply a 'success'.

(ii) 'To defeat Cancer within 10 years'. Much money was subscribed (in USA). The primary aim (defeat) was not achieved, though much was learned. Success?

(iii) How can a (Mathematical) curriculum be described.? Of several possible desciptions, which are most useful? Suppose first that we change 'curriculum' to 'Stage Play " Is (was) 'Oklahoma' a greater success than 'Hamlet'? (In terms of employment of actors and cash generated, Yes), Note that we have a well developed language of dramatic criticism, whereas that of curriculum criticism is perhaps in the stage of pre-Greek geometry-even though there are plays of comparable age with mathematics.

These examples show that 'failure' can be a useful descriptor, whereas 'success' is inadequate for a complex organism with a potentially long life. A mathematics curriculum is such a complex organism provided we think of it as more than just a body of content. In any reasonable description of such a curriculum I would expect to see references to the following elements, at least: the pieces of work done by teachers and pupils, the mathematical content, the mathematical activity, styles of teaching, styles of assessment, relative emphases on training versus education, and the setting within the general curriculum. Such a curriculum could easily be described as a 'failure' it it were terminated; but before we can attempt to pass any other kind of critical judgement upon it, we must first be able to describe it in a manageable way without too much distortion. Simply to summarise it as a 'success' would be so superficial as to lead to suspicions of sales-talk. Now consider some different types of curricula.

Some curricula are due to strong individualists, who can control a large part of their environment. (This would be especially true of a Head teacher with a holistic philosophy). Examples:

(i) Thomas Arnold, at Rugby School (1828-42), of whom, at his appointment, it was said 'this man will change the face of English Education'. His general aim was to turn 'wicked' boys into 'Christian Gentlemen' (relative to his own definition of the adjectives). His vehicle for accomplishing his aim was Classical culture, studied through the Latin & Greek languages, but he included some arithmetic and early parts of Euclid. It was important, for him, to relate the historical material with contemporary life. However, his pupils do not seem to have been sure what his aims 'really' were, if we compare the accounts of schools run on Arnoldian lines.

(ii) We have anecdotal evidence (in biographies of pupils) of 'Ordinary' teachers who could start off 'ordinary' pupils. These are often forgotten, and yet are extremely important. This leads to a Research Problem. How can we 'know' of such teachers? What is involved when we observe their classes? (See Yates 26).

In countries like the U.S.A.; the problems of a mobile population and the low esteem of teaching as a job have induced the dream of a superficially simple description of a curriculum by means of an algorithm: the pupils are to pass through stages S1, S2,...,Sj,..., and then a pupil is judged to be in stage Sn at time t; the algorithm moves him into stage sn + 1 by time t + h where h depends on n.

Such a possibility was envisaged in an SMSG paper (21) which noted the number of reports concerning the efficacy of some mode of teaching a piece of mathematics: each report was contradicted by another one, so the SMSG suggested the notion of a teaching algorithm as an 'objective' way by which different methods could be evaluated. Some attempts to create such algorithms have been made for limited topics within mathematics,--units of Computer Assisted Learning, self-paced courses of Keller-type, sets of work-cards, or remedial work, but I know of no complete curriculum to span several years of schooling. In any case, such units usually need direction by a teacher, of whom an important aspect is her personality-she is not working on a Volkswagen production line!

When we attempt to be 'scientific' about education, it is worth-while to compare a statement like 'Arithmetic performance falls off when the classroom teperature exceeds 80° F', with 'Miss Smith is an excellent teacher of Class A but not usually so good with Class B. This year however, little John Doe has acted as a catalyst, and with class B she has been very good.' The former is 'scientific' but relatively trivial, whereas the second has great significance but its meaning is difficult to objectify (which is no reason for ignoring it). I have tried not to caricature the behaviourist viewpoint, but I do not find it congenial to me as a teacher, so I now pass to less 'scientific' but more meaningful material.

Some descriptions of 'Traditional" Curricula were commissioned by ICMI around 1912 (see, e.g. the British reports (3). Such curricula were the basis of mathematics teaching until the 'Modern' movement that began in the 1950's. A valuable set of descriptions of 'Modern' Curricula was gathered by Freudenthal (5) in Educational Studies in Mathematics (1978), covering 17 countries. All the writers pay attention to aspects of the following diagram, which I have elaborated further in (6), (8)

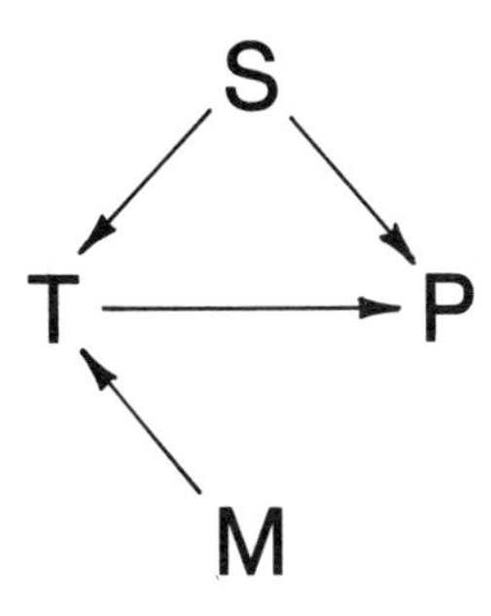

Here T,P,M,S denote respectively teachers, pupils, and mathematics as they appear within each writer's own society S (so there is no misguided attempt to be acultural). The arrows indicate relationships such as the salaries paid by S to T, the communication of mathematics from T to P, explicit aims of S, and the teachers' selection from mathematics of their subject-matter. But here we must think of M as consisting of more than just subject-matter: it must include the mathematical community of the society in question and their links with the international mathematical community. Those links exist through technical literature in space and historical time, and carry attitudes to 'relevance', truth, and authority which may be at variance with those of any individual society S.

The fact that all the writers in question take into account S, T, P and M-even though implicitly and in different degrees-is itself encouraging. We do not always find such sophistication among certain other types of commentator (e.g. when certain indignant theorem-proving mathematicians write complaints to the Press) or even among all curriculum designers. However, in assessing a description of a curriculum, we must remember that the writer's position within the system (as administrator, designer, or teacher) affects his viewpoint and his willingness to admit mistakes or be critical. Also, we must distinguish three versions of a Curriculum as:

intended   implemented   attained

For example, highly academic members of the mathematical community have often discussed 'intentions' merely in terms of content, with slogans involving 'rote-learning' (BAD) and 'understanding' (GOOD), as if there were no such activity as intelligent-memorisation. Perhaps for that reason, several of the writers of these articles do not question the equation 'Modern = Good', and almost all imply criticism of the contribution (or its lack) made by their local Universities.

The general verdict of the articles seems to be that the new curricula have been much better for improving teachers than pupils. Teacher organisations have often been reformed out of their initial complacency, and horizons have been widened; but some teachers have unfortunately been made to feel insecure merely because they taught unfashionable content. A notable article is that from the West Indies, where plans for change concentrated on teachers because there were insufficient funds for a direct approach to the pupils. The first question asked was 'What are the strengths and weaknesses of our teaching force?', and the aim was to improve teacher-handling of existing materials. It so happens that there is an independent report (Cundy, 4) on this work, in much greater detail, written by an outside evaluator who had come with great experience and insight from the English SMP: see Section 9 below.

In some countries, the first requirement for change was to provide schooling itself for all children, preferably in the mother-tongue with mathematics relevant to that country, rather than to what had been inherited from a former colonialist occupation. But, in richer countries with an adequate structure of basic education, the aim was to shift from

'Arithmetic for all' to 'Mathematics for all'.

The best-known model for that was to copy the mathematics of 'elite' schools, an obvious (but crude) technique being to revamp that material for all children, even though many children in the elite schools had fallen by the mathematical wayside. An added complication was injected from the mathematical community M, with the pressure to change the content to 'Modern'-i.e. to introduce sets, logic, probability, matrices, axiomatics.... In so doing this naturally increased the linguistic complexity, and often led to the domination of 'folk' culture by 'high technological' culture. A highly important element of mathematics education was being ignored, which I have discussed elsewhere (Griffiths (7)) and call

'The Problem of the three languages.'

Thus, suppose we decide to teach a piece of Mathematics. At first, it will exist in the Mathematical literature in a 'raw', 'official' version in one of the languages Ω of the community, M. We wish to teach it to pupils whose language Σ has to be inferred and is not rich enough to 'carry' Ω. The problem for a teacher is to infer Σ and construct an language Λ as bridge from Σ to Ω, thus:

$$(\text{pupils}, \Sigma) \xrightarrow{\Lambda} (M, \Omega)$$

'Language' here means vocabulary, grammar, habits of thought, consistency, rationality, questioning processes, - and expectation that questions are worth answering. Some awareness of the scientific tradition is therefore necessary, and this can involve a clash between the long-term values of the general professional/artisan community and the short-term rewards demanded by the pop/entrepreneur attitudes purveyed by the media in many countries. However, the problem is subtler than it appears at first: for example Tanzania had the problem of writing books in Kiswahili that would enrich the mother-tongue with technical words, but the aim was still to generate the activity of getting sums right-whereas with some mathematics, Λ may rarely reach Ω

e.g. Λ = language of Cuisenaire rods, Ω = theory of N;
or Ω = Calculus, Λ = language of engineering students.

(Λ is to reveal the 'spirit' of Ω , omitting some of the rigour of Classical Analysis). Each Λ is chosen, of course, in the hope that its use will make life easier for the students. In view of the complexity of choosing a good Λ, I suggest, then, that thoughtless imposition of an unnecessarily complex and strange mathematical language can create as much resistance as imposing the tongue of a colonising Power.

Outside the USA, public, formal examinations are very important, both to the individual for social advancement, and to society in the selection of elites for certain types of job. Especially, the process can be impartial, and can eliminate nepotism and corruption. But conventional examinations usually corrupt mathematics, via loyalty of Teacher to pupils, or pupils who learn only what is on the syllabus; these practices lead to a ritual of defeating the examiner. Thus, anybody who wishes to describe a curriculum must reckon with the fact that, to most readers outside America;-

No appraisal of a curriculum can be valid unless we know its stability relative to the examining system

Indeed, in the early stages of a curriculum, examinations can provide a minimum indicator of achievement; and they may provide valuable 'activity' for pupils who work at old papers. (Almost all leading European mathematicians were produced by this activity.) But conventional, timed, written, invigilated examinations are easiest to mark when they test low-level skills, and consequently they usually degenerate into stereotyped questions. For example, the Traditional English elite curriculum was designed around 1910, for a 'race of inventors and engineers", but it degenerated via the examination system. Therefore the SMP changed the content in order to force changes in the teaching style. after a dozen years, the conventional mode of examination is now corrupting (as before) the attitudes the SMP had hoped to convey. (see Howson, 9, 10).

Mention of 'skills' raises the following question.
Question What are mathematical skills? How can we assess them (e.g. for purposes of feed-back)?

This question is discussed in McLone (12). New methods of assessment can best be tried out by Unversity mathematicians, who usually have more flexibility and smaller numbers to examine than the public examination systems. Some of us have been trying to do this for some time within a normal British degree structure, through the use of course-work and projects. Where the structure forces the timed or oral examination, a great deal of work needs to be done to analyse the design of questions-for example, what skills are they testing, and how could they be constructed to test desirable skills?

The mathematics curriculum has many contacts with the general curriculum, so it is not surprising that changes to 'Modern' content produced hostility from such users as Science Teachers who (for various reasons) did not appreciate the changes. Conversely, mathematics often did not benefit from the modernisation of Science curricula: two-way co-operation is needed (See Nuffield, 18, Pollak, 20).

If all subjects are 'modernised', primary teachers especially are put under great pressure to increase their knowledge, and to change their teaching styles. So, frequently,

'Johnny can't add' (and perhaps that's true).

- - even though he may be learning skills that the public don't recognise as mathematical (e.g. Visual skills, reading maps and signs, observing number patterns. In the Secondary Sector, the growth of Comprehensive Secondary Schools can end selection at 11+, and allow a richer curriculum at primary level, but perhaps at the cost of more stress on primary teachers. Moreover at secondary level parents may insist on an unsatisfactory 'academic' curriculum because it leads to a prestigious examination (even though their offspring may have no chance of passing it). These parent's worries can be inflamed by the commercial media, and have been used by politicians. This does not happen if the media is controlled by Government, but even with its full backing no mathematics curriculum seems to have fulfilled its declared intentions.

Now, education has always been 'political', because Education partly concerns allocation of scarce resources, and consequent increases in status, privileges and rewards from Society. But recently many countries have seen that, since parents have votes, then mathematics has itself become 'political', because complexity is easy to distort. The old slogans of 'Modern Math' have backfired: simple-minded 'patriots' actually admire rote-learning, and think that 'undertanding' encourages long-haired revolutionaries. In the resulting atmosphere, politicians ride to power with appeals to a nostalgic past and 'Back to Basics', and they can kill off promising curricula (and have done so in several places).

Question: How can a curriculum withstand corruption or abolition by predators outside mathematics, or by incompetence form within? One requirement is clear: design the curriculum to be within the capacity of an average teacher. (This is very dull for 'Star' teachers, but in the past, they have often been able to enhance a dull curriculum). But we immediately have an important research problem:

What is an 'average' teacher?

A prior requirement for designing a good curriculum is that we first make an honest appraisal of reality. On philosophical grounds, we can rarely describe 'reality' accurately but we can at least be honest about our attempts. Now suggestions for change are usually intended to remedy some perceived defect in the existing curriculum. But is the 'defect' really there? If not, the 'change' may quite well impose a curriculum, that was reasonable for a past society, on a modern one in which all kinds of things have changed (including parental support for rigid discipline). For a sensible, work-related version of 'Back to Basic', see Lindsay (11).

Politicians, the Press, and some Academics (when under threat), will deliberately blur or distort even their own limited view of reality. But we ourselves must live with these awkward reminders of a 'Golden Age' of education in the past, - the Age was Golden for a select few. In connexion with the ideal of 'Mathematics for all', we cannot always avoid discussion of what elites should be produced; we may have to accept that raising the general level may mean lowering that of former elites; and those elites (often socially illiterate) may regard the need for change as a criticism of themselves. This brings us back to the Three-language problem again but in a different context.

a. We need an Ω for our own techincal descriptions of curricula-one for a Congress (ICME) comparable with the general core-language of an ICM.

b. The analogue of Σ (pupil-language) is that of the man-in-the-street or parent.

c. Problem: construct Λ to simplify Ω, but honestly.

Examples.

i. Cundy's evaluation (4) of the Carribean Project (see Section 5) is in the form of an Ω, 100 pages long, but it contains simplified 'Conclusions' expressed in a form that administrators can understand.

ii. J. Yates (26) studied just four Mathematical Classrooms. Her description distinguishes different levels of thought by different colours of paper. Still, she includes no Λ which would allow an administrator or Head Teacher to take appropriate action.

There is one candidate that might be developed as a language Ω, namely the 'Q-analysis' of Atkin (1). Although this is expressed within the language of algebraic topology, the methodology has been found acceptable by urban planners (see Atkin, 2 p. 89). It is being further developed in order to describe television programes, by the International Television Flows Project, Geography Department, University of Cambridge, England. Since a programme is a (simplified) analogue of a lesson in a classroom, a lesson might be modelled within the methodology, provided certain improvements can be made.

The age of large-scale new curricula for schools is probably over, and future improvement will probably come from the teachers within an existing system, in a piece-meal way. It is their modes of learning and self-improvement that then become important. In rich countries, initial training is declining in numbers, owing to population trends; but in all systems, In-service training will always be necessary, even though few systems have built in such training as a regular feature of the teaching job.

The early 'handed-down' models of In-service work are not very good, because they have usually consisted of some 'high-up' telling teachers what he thinks they ought to know, rather than what they want to know. One alternative is to follow the good work that has arisen from Teacher Centres, IREM's, IOWO, Entebbe project, etc; unfortunately, such centres are vulnerable to financial cuts and political attack.

A possible stable long-term solution is to establish Teacher Networks of peer-groups, which are not formed via hierarchies, but 'horizontally' though feelings of mutual need, 'on the job'. Enormous quantities of mathematical materials are now in existence from the earlier curriculum reforms, that such Teacher networks might adapt for local use, so provision of content is no great problem, except in applications of mathematics-but this is being overcome through projects such as those of UMAP (22) and Ormell (19). In recent years we have seen also the development in many countries, of journals for teachers which are easier to read than the older, more formal ones. These allow a more rapid circulation and exchange of ideas with immediate classroom relevance, but of course a major problem is to get teachers to read them.

To my mind, an outstanding example of this trend can be seen in the work of the Australian Association of Teachers of Mathematics (AATM). The structure of Australian society has not allowed the introduction of large-scale mathematics curricula, but many small-scale ones have been introduced instead, and these generate considerable teacher-involvement. Biennial Conferences have been organised by the AATM, which are structured to allow working teachers to give talks on their ideas and experiences. Moreover, the speakers must write up their talks, and this discipline itself raises the professional level and rigour of thought. The collected material forms a body of interesting work, for reference and future development (see Williams, 24, 25).

What improvements could we hope for, from such a 'network-curriculum' for teachers? Besides the building of a professional team-spirit, we might expect an increase in mathematics as activity-first in the teacher's own private development, then in the way in which he does mathematics with his pupils. Activity means not only fluency in solving problems set by others who already know the answer (as in Olympiads) but in seeing mathematical situations, asking questions, formulating problems-even learning to design good examination questions. And especially in having the confidence to admit ignorance and to remedy it by using the skills of asking others, and reading books and appropriate teacher's journals.

Such skills will not be valued or cultivated if they are not seen to be valued at the 'top' of the system, i.e. by those who teach in such tertiary instiutions as Universities. If they merely mumble their lectures, so will their students' if they appear to concentrate on mathematics as content, or as examination-fodder, so will their students. They will have nothing seriously to offer if without upgrading their degree of sophistication, they give dogmatic in-service courses to practising teachers. One way of countering this is to make mathematicians in Universities more consious of their role as teachers, to show that hard thinking is needed to teach effectively just as much as to do mathematical research. A start in this direction has been made in Britain with the series of annual Nottingham Conferences, which allow mathematics lecturers from all over Britain to meet each other, with school-teachers and industrialists, in a structure that forces them to write down their thoughts about the teaching-side of their own work. The Nottingham Conference Proceedings (13-17) consist of edited versions of these writings, and the improvement over the years is heartening. Relatively few participants are seriously affected by the process, unfortunately; we do not yet know what leavening influence they have when they return to their own departments.

Whatever the failings of the past era of curriculum-reform, we have nevertheless seen a great widening and improvement of teacher-awareness within mathematics education. The establishment of our International Congresses has also helped not only as an exchange of ideas, but to emphasise that improvement comes form a professional approach ot mathematics education which could not be fostered by the older series of International Congresses of Mathematicians (devoted as they are to fostering mathematical research, with only a sympathetic but largely amateur interest in the teaching of mathematics).

So I believe that this is far too early to be talking of 'successful' curricula, and there is a long hard road ahead before we shall even know what it is reasonable to hope for. Our best improvement to date is, I believe, that facilities now exist for discussion and comparison to take place. But we still have to ensure their survival in a world of economic recession and military rivalries.

References

1. Atkin, R.H. Mathematical Structure in Human Affairs, (1974), Heinemann.

2. Atkin, R.H. A study area in Southend-on-Sea. Urban Structure Research Porject. Research Report III. (1973). Dept. of Math., Univ. of Essex.

3. Board of Education, Special Reports on the teaching of mathematics in the U.K., (1912), H.M.S.O.

4. Cundy, H.M. Carribbean Mathematics Project: an evaluation study, (1976), British Council.

5. Freudenthal, H. (Editor). 'Change in Mathematics Education since the late 1950's-Ideas and Realization,' Educ. Studies in Math. (1978).

6. Griffiths, H.B. 'Mathematics Education To-day', Int. J. Math. Educ. Sci. & Tech. (1975) 6 pp. 3-15.

7. The Structure of Pure Mathematics' (Chapter I of Mathematical Education (1978) Edited by G.Wain. Van Nostrand.

8. Griffiths, H.B. 'Mathematics Education among other contexts for mathematics'. MI Symposium, Helsinki (1978), to appear from Bielefeld.

9. Howson, A.G. 'Article on 'Great Britain' (in 7) pp. 183-224.

10. 'A critical analysis of curriculum development in mathematical education' (Chapter VII of New trends in mathematics teaching, (1979) UNESCO).

11. Lindsay, R. Basic Skills for Mathematics in Engineering, Shell Centre Research Report, Nottingham (1977).

12. McLone, R.R. The Training of Mathematicians (1973) Social Science Research Council.

13. Nottingham Conference Proceedings. Adapting Unviversity Mathematics to Current and Future Educational needs (1975).

14. Unversity Mathematics Curricula and the Future (1976).

15. Teaching methods for Undergraduate Mathematics (1977).

16. Assessment and Service Teaching (1978).

17. Teaching Applications of Mathematics (1979).

(14 - 17 all available from Shell Centre for Math. Education, Unviersity of Nottingham.)

18. Nuffield Foundation. Various Science Projects (1970-) Longman, Penguin.

19. Ormell, C. Mathematics Applicable, (1975) Schools Council Sixth Form Math. Project, Heinemann.

20. Pollak, H.O. 'The Interaction between mathematics and other school subjects'. (Chap. XII of New trends in mathematics teaching (1979) UNESCO).

21. SMSG. Final report.

22. UMSP (Undergraduate Math. Applications Project) Education Development Center, Newton, Mass.

23. Williams, D. (Editor) Learning and Applying Mathematics (1978) AAMT.

24. Mathematics-Theory into Practice (1980) AAMT.

25. Wilson, B.J. Article on 'West Indies' (In 7), pp. 355-379.

26. Yates, J. Four Mathematical Classrooms: an enquiry into teaching method. (1978) REsearch Report, Southampton.

SUCCESSES AND FAILURES OF MATHEMATICS CURRICULA IN THE PAST TWO DECADES: A DEVELOPING SOCIETY VIEWPOINT IN A HOLISTIC FRAMEWORK

Ubiratan D'Ambrosio
Institut fur Didaktik der Mathematic
Unicamp, Brasil

In trying to analyse mathematics curricula in a school system, we have to place ourselves in a global context, adopting a holistic viewpoint to education. This concept of holistic education brings into the discussion in issue all the components involved in the educational processes and the various interrelations between each and every one of these components and the whole. In the special case of developing countries, we see no possibility of a meaningful curricular analysis without placing ourselves in this holistic framework. And the same is true for developed countries as well (see 1, page 196).

In this holistic framework, we see curricula in a three-dimensional representation, with objectives, contents, and methods absolutely unified. It is impossible to look at either one of these components independently from the others. It has been a frequent mistake in developing countries, and the same is true for developed countries, to place all the emphasis for improving mathematics education in contents components. We see, frequently, faculties refusing teachers with advanced degrees because they tend to bring the advanced mathematics they have learnt into the classroom experience. Incredibly enough, several schools opt for lay teachers, in the developing countries, because they are less "pedantic" with respect to their knowledge and hence establish a better learning relationship with students.

Insisting on the point, the emphasis on contents as the focal part of a curriculum has been probably the most damaging aspect of mathematics education. This is particularly true in developing countries, with respect to teacher training. A somewhat subservient attitude towards developed centers places a sort of need of affirmation of capability on what can be measured by universal standards, which is contents, with absolute disregard from what is intimately related to the sociocultural context in which education takes place,in the curriculum, the coordinates objectives and methods. Some teacher training models, putting emphasis on contents even if other training courses on methodology are provided, are the main cause for a noticeable degradation of mathematics in educational systems. We see an increasing recognition of the need for a holistic approach, looking in an integrated and unified way to objectives, contents, and methods, as the major result of these last twenty years dominated by "new" contents. This trend is evident in recent recommendations by UNESCO and OAS, which reflect this global understanding of all the sociocultural implications on the concept of curricula (see 2 and 3). In a visual form I put these ideas in the following way:

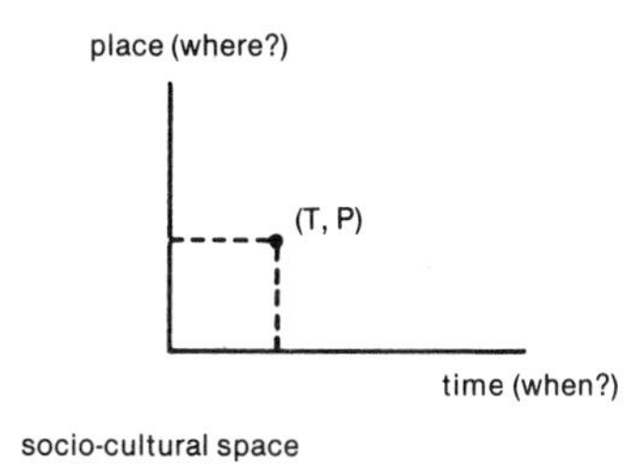

socio-cultural space

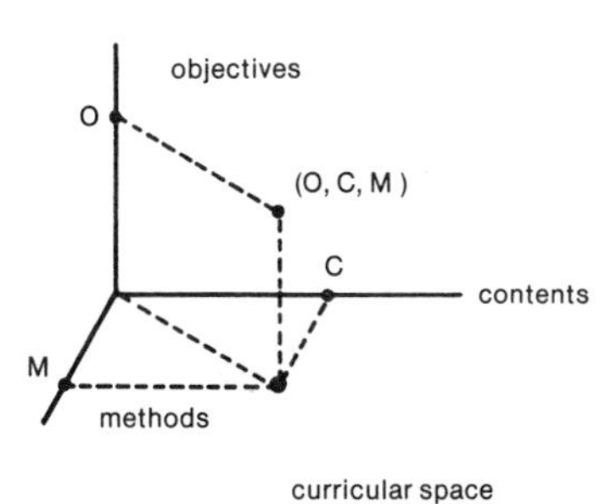

curricular space

It is our function, in a curriculum analysis, to comprehend the correspondence (T,P) ↔ (O.C.M), with the clear understanding that the correspondence is in both ways. In other terms, we must accept the fact that the curriculum as a whole, with its three coordinates of objectives, contents, and methods, interpreted in a unified way, is a function of the socio-cultural instrument, i.e., depending on when?, where? for its definition. But at the same time, we must accept that the curriculum as a whole, again with the three coordinates understood in a total solidarity, will determine the way the point in the socio-cultural and curricular spaces is the basis of our holistic approach to education (see 4).

Notice that we do not see "evaluation" as a coordinate in the curricular space, as some curriculum theories ask for. We see evaluation as an internal component of each and every one of the curricular coordinates (objectives, contents, and methods) and never as a coordinate in itself. Again, there resides probably one of the most damaging aspects of education in developing countries: a look for recognition through external standards of achievement, consistently based on evaluation procedures which lay entirely upon the contents coordinate, independently of the others, and shockingly disregarding global societal and individual implications of evaluation. The search for standards by comparing achievement, through contents based on evaluation procedures, with more developed countries, has caused most of the divorce between educational systems and the societies in which they are immersed. In particular, mathematics has been distinguished in an inglorious way, in this respect. Not only has mathematics education been regarded as the "hard" subject in schooling, accessible only to a very privileged class of individuals, but what is even more damaging, mathematics has been used as a barrier to social access, reinforcing the power structure which prevails in the societies. No other subject in school serves so well this purpose of reinforcement of power structure as does mathematics. And the main tool for this negative aspect of mathematics education is evaluation. All these factors place mathematics as a questionable subject in an emerging or changing educational system, as is the case of most developing countries (see 5, p. 177).

Much of what happened in the last twenty years, since the appearance of the so-called "modern mathematics," was a considerable change in contents, and enormous efforts to inject into the teaching profession more and more contents, often the traditional topics dressed with new names and new symbolism. This made its way to developing countries, carried by the prestige of mathematicians who were overly enthusiastic about this recent challenge: to make accessible to more individuals what usually was restricted to the inner circle of practicioners. This was not accompanied by the absolutely necessary critical analysis of what were the aspirations, the anxieties, and the perspectives of the youth who were beginning to boom into becoming one-third of the entire world population and highly critical of the way previous generations had been conducting world affairs. The prestigious apostles of modern mathematics and their hastly trained followers brought to the developing world seeds of supposedly new contents without knowing the seedbed. They did not understand the dynamics of societies, they did not accept youth as a partner in defining the future of these societies. And, sadly enough, this is not over and the developing countries are still under this lure.

Summing up, mathematics has been viewed uncritically as a subject important in itself, securely placed in the educational system, with emphasis on contents, either became of its usefulness, of because of its beauty, lacking the global understanding of society as a whole and of individuals as components of this society, with very peculiar aspirations and perspectives. How mathematics fits into the very broad texture of society will determine its future in the educational scene, indeed its survival as a school subject.

1. ICMI - New Trends in Mathematics Teaching vol. IV, UNESCO, Paris, 1979.

2. UNESCO - Meeting Experts on "Mathematics Education in view of Societal Needs", Paris, UNESCO House, May 19-23, 1980 (Documents available from UNESCO, Paris).

3. O.A.S Meeting on "Metas Educativas para la Decada de 80", (Educational Goals for the decade of 80"), Panama, 5-9 May 1980 (Documents available from OAS, Washington).

4. Ubiratan D'Ambrosio, La Educacion Cientifica en el Contexto Socio-Cultural (Science Education in the Socio-Cultural Context), Working Paper, Meeting on Integrated Science Education in Latin America, UNESCO, Huaraz, Peru, March 19-28, 1980 (documents available from OREALC-UNESCO, Santiago).

5. Ubiratan D'Ambosio, Secondary Mathematics Education in Brasil, in Comparative Studies of Mathematics Curricula-Change and Stability 1960-1980, Materialen und Studien Band 19, Institut fur Didaktik der Mathematik, Bielefeld, 1980, p.167-178.

SUCCESSES & FAILURES OF MATHEMATICS CURRICULA IN THE PAST TWO DECADES

Stephen S. Willoughby
New York University
New York, New York

In the mid 1950's pre-college mathematics education in the United States generally consisted of mathematical content that would have been familiar to Galileo, taught using the behaviorist psychology of E.L. Thorndike with a small dash of "social applications" (taxes, insurance, mortages, etc.) thrown in for the edification of 12 and 13 year olds. Some mathematics educators of the time thought we should update the content, we should use teaching techniques that emphasized thought more and rote learning less, and perhaps even postpone the social applications until a time when students were morenearly ready to use them. In fact there had been reports, projects and individual efforts to improve the teaching of mathematics throughout the twentieth century and much of the nineteenth.

Several projects to improve the teaching of the mathematics were started in the 1950's, but it was the flight of the first sputnik in October 1957 that gave impetus to the "new math" movement in the form of massive federal grants and an attempt by commercial publishers to jump on the new math band wagon without doing anything radical enough to frighten traditional teachers.

One of the ironies of the new math movement is the fact that while most new math programs emphasized precision of vocabulary and careful use of words, the phrase "new math" itself has remained ambiguous -- meaning very different things to different people. In fact, the phrase "new arithmetic" had been used during the 1920's in the United States to describe content and methods that were almost diametrically opposed to those of the 1950's and '60's.

Although the new math of the 1960's included a wide range of government supported projects and publishing ventures that had varying goals and used disparate methods of achieving those goals, most mathematics programs of the '50's and '60's had the following common characteristics, though the emphasis varied from program to program:

(1) An increased emphasis on logic, structure, proof and rigor. This took many forms. Structure and proof were introduced into algebra through the use of field properties to prove simple theorems. The Euclidean axioms for geometry were modified a la Hilbert or, in some cases, eliminated entirely in favor of an algebraic foundation for geometry. At the elementary school level (ages 5 through 12) arithmentic was taught using proofs and derivations involving the commutative, associative and distributive laws. For esample, one of the better selling mathematics texts of the 1960's shows children how to compute 17 + 6 as follows:

$$
\begin{aligned}
17 + 6 &= (10 + 7) + 6 \\
&= 10 + (7 + 6) \\
&= 10 + 13 \\
&= 10 + (10 + 3) \\
&= (10 + 10) + 3 \\
&= 23
\end{aligned}
$$

In some programs children were expected to give reasons for each of these steps and in some cases the form was provided and children were expected to supply the correct numbers (or numerals-see point 2 below). For example:

$$
\begin{aligned}
17 + 6 &= (10 + \_\_) + \_\_ \\
&= \_\_ + (\_\_ + \_\_) \\
&= 10 + \_\_ \\
&= \_\_ + (10 + \_\_) \\
&= (\_\_ + \_\_) + \_\_ \\
&= \_\_
\end{aligned}
$$

Those of us who had the opportunity to watch children complete procedures like this last one noticed that the children would begin by filling out the top line and maybe even the second line. Shortly, however, they would skip to the bottom line, fill in 23 and work backwards until they had all the blanks filled in. Clearly, such children were not using a deductive approach to discover that 17 + 6= 23 and the formalism was not really helping them understand why we add the way we do.

At the high school level, Euclidean geometry, which had been the epitome of exciting, rational mathematical thought for generations of mathematically inclined students, became a pedantic boring subject in which, after much labor, we could prove that if our axioms were true and three points were all on the same straight line then exactly one of the three points was between the other two. There were some of us who believed that the decreased emphasis on physical models and intuition and the increased emphasis on rigor was leading to intellectual rigor mortis.

(2) A closely allied trend in many mathematics programs of the past two decades was an emphasis on the careful use of language. This includes a distinction between names and the objects for which they stand (numerals and numbers, for example), an avoidance of certain words that seemed ambiguous or gave an impression that seemed misleading (such as cancel, borrow, carry, reduce, etc.) and the use of the language of sets "to unify mathematics." This trend seemed to many of us to be artificial and pedantic at best and either impossible to achieve or thoroughly confusing in many instances. The language of sets and formalism rarely seemed to clarify or simplify a concept and generally seemed to place mathematics in the category of an ancient language--interesting to some, but hardly related to the real world in any meaningful way.

(3) Perhaps the most useful trend during the past two decades has been the introduction of new subject matter into the school curriculum. This has included statistics, probablility, emphasis on functions and graphing, solving of inequalities along with equations, work with calculators and computers, and tidbits from topics such as number theory and topology. Certainly the potential usefulness of such topics in everyday life as well as in various professions justifies their introduction into the school curriculum.

(4) A fourth characteristic of mathematics programs of this country over the past 20 years has been a tendency towards acceleration; that is, learning more mathematics earlier. This is clearly related to the introduction of new material and is, in fact, not a new trend. Since the beginning of formal education in this country there has been a tendency to teach each mathematical topic to children of younger age. With the explosion of mathematical knowledge during the twentieth century there seemed to be no reason to reverse that tendency. However, there is considerable psychological evidence now available that concepts taught too early may be learned in a rote, mechanical, unthinking manner with little understanding and even less ability to apply. Thus, there ought to be some strict limitations on the teaching of subject matter to ever younger children and these limitations ought to be applied long before we reach the obvious extreme of teaching prenatal ergodic theory. Even when teaching an exotic subject to a young child is possible, the question of whether it is desirable should be raised.

(5) A fifth goal of some of the recent mathematics programs has been to integrate various mathematical topics. In this country, as in few others, the subjects of arithmentic, algebra, grometry, etc. were taught in isolation from each other to the extent that they were not even taught in the same year. That is, arithmetic was taught during the first eight years of school, algebra was taught in the ninth and eleventh years with geometry in the tenth. Various subjects, including more algebra, trigonometry, and solid geometry were taught in the twelfth. There had been several abortive attempts to remedy this situation before 1960 but essentially no progress was made until the advent of the new math. While further progress is still needed, the results so far have been quite positive. On the basis of standardized test results and simple observation, seasoned with the experience and prejudice of the observer and interpreter, I believe the successes of the so called new math were largely in the area of improved content and integration of content. If introducing geometry, statistics, algebra and probability into the early grades was acceleration, that too was a positive development. However, the formalism and rigor of the new math discouraged many promising students from continuing their study of mathematics and gave others a totally erroneous impression of what mathematics is. I believe the recently noted drop in problem solving ability reported on National Assesment tests can be more fairly attributed to the new math programs than to the "back-to-basics" movement in light of the ages of those who took the tests and what is known about which kind of books were being used when those children were studying mathematics. Beyond this, the mad rush by many schools to teach calculus in the twelfth year to the exclusion of topics such as statistics, number theory, algebra and even coordinate geometry was, and is in my opinion, a mistake that has prevented many students from appreciating the true usefulness and beauty of mathematics.

As test scores fell and parents saw their children failing to learn simple arithmetic facts, we saw an anti-intellectual and anti-thinking backlash against the new math. Leaders of this movement seemed to be saying "teach our children to do simple arithmetic and maybe a little old fashioned algebra well and forget about all the frills." Of course, this misses the essential point which is that machines can be cheaply and quickly constructed that do simple arithmetic and algebra more efficiently than any human. It is thinking and problem solving that set humans apart from machines. During the 1950's, some space scientist referred to man as the only 150 pound, servo-mechanical unit that can be mass produced by unskilled labor. I believe that people are more than this. People who are properly educated can and will be masters of the machines. Such people wil be the thinkers and problems solvers of the future. Thus, without further evidence I can say that the "back to basics" movement will be a failure in my view to the extent that the authors of such programs forget that thinking is a basic skill--indeed, thinking is THE basic skill.

Recent reports of professional organizations and their developmental projects seem to hold grand promise for the future. The curriculum development projects are designed to develop children's thinking and problem solving abilities by building on the child's experience and previous understandings. They use materials, games, activities, stories, understanding of children's logic, and intensive work with teachers and children to develop so called basic skills along with thinking and problems solving ability and often skills that will be useful in the 21st century. Robert Wirtz, here in California, is responsible for one of these projects. Burt Kaufman is director of another such project that has been supported by federal funds. The third such project is being developed by Carl Bereiter, Peter Hilton, Joseph Rubinstein and me with support from the Open Court Publishing Company. I believe the preliminary results from these projects hold great promise for the future both in terms of what an interested observer sees in classrooms using these materials and in the results children from these projects are producing on standardized test expecially for children form non-middle class backgrounds.

In summary, I believe that mathematics education in the United States is better than it was 20 years ago because of the introduction of new subject matter and the integration of subject matter but worse in so far as we have replaced intuition and thought at the student's level by empty formalism and rote learning. The "back to basics" movement will be a disaster if thinking and problem solving are neglected, but reports of professional organizations and a few developmental projects seem to hold great promise for the future.

11.2 CURRICULUM RECOMMENDATIONS FOR THE 1980'S BY SEVERAL NATIONAL COMMITTEES

PROBLEMS OF CURRICULUM DEVELOPMENT IN SUDAN

Mohammed El Tom
University of Khartoum
Sudan

General Background. Sudan is a sparsley populated (17.8 milions spread over an area of 2.5 million km 2), predominantly rural and culturally diverse (114 distinct languages) country. The rate of illiteracy is 75-80%. It is economically poor (national per capita income fluctuating between 100-250 US $ during the 70's) and is a one party state.

While the country's educational system is hetrogeneous, the overwhelming majority of schools are academic, state owned, financed and run. The educational ladder is 6 (primary), 3 (intermediate), 3 (secondary) and the school age entrance is 7. Teacher education (as of 1973) starts only at the post-secondary stage. Except for some localities in the southern region and higher education institutions, the laguage of instruction is Arabic.

Curriculum Development: 1955-1980. It is not uncommon for the concept of 'curriculum' to equated with that of 'syllabus'. However it is now widely accepted (Griffiths and Howson, 1974) that for meaningful curriculum development it is necessary to view curriculum as encompassing: aims, content, methods and evaluation procedures. It is with this understanding of the curriculum that I now propose to discuss curriculum development in Sudan during the last 25 years.

Organizational aspects. Sudan's educational system is one in which "the Ministry of Education... is responsible for educational planning...and the control of the smallest points of execution of policy." (ILO, 1976).

Thus all aspects of the curriculum for primary and intermediate stages are practically the sole responsibility of the Institute of Education at Bakht Er-Ruda (established in 1934). The Institute has its own schools, teacher training institutes, examination centres and a research centre. The department of mathematics, one of 13, has 5 members of staff all of whom are secondary school teachers.

In contrast to the lower two stages, the institutional implementation of responsibility for curricula at the secondary stage is of recent origin. For six years after independence in 1956, Sudan had the syllabus as well as examination scripts for its school certificate set and marked in England. This tradition was (nominally) discontinued in 1961 and a system of 'panels' was set up for the evaluation (in reality discussion) of the school certificate examination results. It was not until 1967 that a curricula department was established within the Ministry of Education and it included a unit responsible for mathematics having 3 members of staff. The curricula department had been dissolved in 1979 and at the time of writing there is no body responsible for secondary schools curricula.

In-service training is administered by a central unit within the Ministry.

Curriculum reform in the primary and intermediate stages. For nearly 20 years (1955-75) syllabi for the two stages remained stagnant. An opportunity for reform was provided in 1970 when the 4-4-4 ladder in use until then was changed into the present 6-3-3 ladder. However, the classes lost in the process by secondary and intermediate schools kept intact their pre-1970 syllabi.

The first significant reform started in 1974/75 when the Ministry of Education decided to introduce 'modern maths' in all the country's schools. It must be emphasized, however, that rather than being a response to indigenous pressures, the decision was essentially motivated by the desire to emulate other Arab countries.

Experimentation with new materials, prepared by experts from the American University in Beirut, started simultaneously in both stages in 1974/75. The experiments involved 800 primary and 1000 intermediate pupils all of whom belonged to schools in two rather special localities: Bakht Er-Ruda and its immediate neighbourhood and Khartoum.

On the completion in April 1977 of the trial period for intermediate schools, a 2-stage evaluative study (NCER, 1980) was carried out. First, the performance in the 1977 intermediate school certificate of 40% of pupils in experimental classes was compared with that of 6.5% of pupils who followed a 'traditional' syllabus. In the second stage three sets of questionaires were distributed one each to teachers, parents, and headmasters and inspectors. The evaluation study concluded that "Although 'modern maths' is not yet satisfactorily established in the minds of many teachers, parents and citizens in general; and although this evaluation is rather pre-mature, the general attitude of teachers, headmasters, inspetors and parents is largely in support of the continuation in teaching 'modern maths' in intermediate schools." The study further recommends the introduction as of 1980/81 of 'modern maths' in secondary schools.

Curriculum reform in secondary schools. During the first ten years of the period under consideration, the language of instruction in secondary schools was English and the mathematics syllabus was based on the wellknown Durell's series of texts. Shortly after the 1964 mass upheaval, a powerful teachers trade union brought forward the issue of Arabicization of secondary school education and demanded its immediate implementation. Their demand was met in 1965.

As far as mathematics is concerned, Arabicization resulted in an almost word for word translation of the then existing Durell's texts. Soon afterwards it was realized that there was a great deal of overlap (essentially in arithmetic) between the syllabi of the final and first year in intermediate and secondary schools, respectively. However, it took four years for the overlapping material to be removed and the resulting gap was progressively filled by statistics, linear programming and inequalities. The present syllabus is essentially the same one as that of 1970.

In-service training. Originally the intention was to introduce 'modern maths' first in secondary schools. Accordingly, in-service training courses of 2-3 weeks duration, based on UNESCO's Mathematics Project for Arab States, was first started and later (1978) discontinued for secondary school teachers. In-service training of similar duration is still continuing for primary and intermediate teachers and has so far covered about 50% of all teachers.

Major characteristics. Besides its confused nature, curriculum development within the last 25 years has the following characteristics:

i. It dealt almost exclusively with 'content'. In particular, teaching methods and assessment procedures have not been touched by the reforms.

ii. Rather than being the result of indigenous pressures, recent reforms were essentially motivated by the desire to emulate other countries.

iii. Very few individuals have been involved in the production of new materials and the process of change in general. In particular, the role of teachers of mathematics, as a body, in the process of change has been that of spectators.

iv. Absence of serious evaluation studies of developmental work.

Future Trends. Regarding the future, one thing seems to be certain: the Ministry will carry through its policy of introducing a 'modern maths' syllabus in all stages of the system. This has already been achieved in intermediate schools and the process will be completed in 1983 in primary schools. In sofar as one of the outstanding characteristics of Sudan's educational system is the so-called 'Upward Push', the Ministry will be under strong pressure to introduce a 'modern math' syllabus into secondary schools rather soon.

But beyond these changes in syllabi, is Sudan likely to witness in the 80's the emergence of effective curriculum development? The possibility of this happening must be weighed against the dynamic of a number of interrelated problems.

i. Societal pressures. In a society with rampant illiteracy and the majority of whose population view academic certification as the only way out of a miserable existence, any pressures for qualitative reforms in the educational system will be heavily dominated by pressures for its quantitative expansion.

ii. Availability and stability of resources. Without a 'critical mass' of curriculum developers, competent and motivated teachers and a minimum of funds and equipments, curriculum development is, in any meaningful sense, a non-starter. If adequate resources do exist but are unstable, then curriculum development will at best be an erratic process. During the last 7 years not more that 30% of the approved development budget for education has been available for expenditure. In 1977/78 expenditure on books, stationary and printing per pupil averaged US $4.75. Libraries and (even elementary) technological aids are non-existent in almost all schools and teacher training institutes. An (inactive) educational television unit, established in 1963, has just been dissolved. In some provinces, 80% of primary schools are built out of straw. Sudan's educational system has historically suffered from a high degree of mobility amongst its staff and a lack of qualified mathematics teachers. A novel factor which is seriously aggravating the latter problem is a high rate of emigration (to Arab oil producing countries) within the last 5 years.

30% of all Sudanese PH.D's in mathematics have emigrated. The unit responsible for secondary schools mathematics curricula had no less than 7 different directors during its 13 years of existence. In May 1979, 47 secondary school mathematics teachers were recruited; only 5 remained by July 1979. In 1979/80, 32% of secondary school mathematics teachers were expatriates (Egyptian). The National Centre for Educational Research (established in 1976) has only 2 out of its 4 members of staff left. Out of 26 trained tutors in the (unique) Institute of In-Service Training (established in 1972) only 2 remain now.

iii. Independent agencies for curriculum development. Highly centralized educational systems functioning under conditions of underdevelopment (Sudan's case) are characterized by a high degree of inertia. Innovative work can rarely be initiated and/or successfully implemented within such systems. In such cases the role of independent organizations in influencing curriculum development is vitally important. At present there are only two bodies which are qualified to play such a role: the School of Mathematical Sciences, University of Khartoum and the Union of Sudanese Physicists and Mathematicians both of which were established in 1978. While the first enjoys considerable prestige, it suffers from lack of sufficient interest amongst its (highly unstable) staff in school mathematics. The second excludes from its membership both intermediate and primary schools teachers and has been conspicuously inactive since its creation.

To conclude, the nature of curriculum development within the last 25 years and the problems outlined above strongly suggest that curriculum development in Sudan in the 1980's will be as confused, narrowly conceived and 'externally' motivated as it has been in the past-- assuming of course that meanwhile the country undergoes no fundamental political transformation and/or substantial imporvement in its economic fortunes. These I can not predict.

REFERENCES

1. Griffiths, H.B. and A.G. Howson, Mathematics: Society and Curricula, Cambridge Unversity Press, 1974.

2. ILO, Growth, Employment and Equity: A comprehensive strategy for Sudan, International Labour Office, Geneva, 1976.

3. NCED (National Centre for Educational Research), Report on evaluation of modern maths in intermediate schools, Institute of Education at Bakht Er Ruda, 1980.

THE EFFECT OF GOVERNMENT ENQUIRIES INTO THE TEACHING OF MATHEMATICS

W.H. Cockroft
The New University of Ulster
Londenderry, Northern Ireland

1. Introduction

In July 1977 the Parliamentary Expenditure Committee published a Report of its Education, Arts and Home Office Sub-Committee concerned with the attainment of school-leavers. In the report it was stated that "it is clear from the points which were made over and over again by witnesses that there is a large number of questions about the mathematical attainments of childrens which needs much more careful analysis than we have been able to give during our enquiry. These concern the apparent lack of basic computational skills in many children, the increasing mathematical demands made on adults, the lack of qualified maths teachers, the multiplicity of syllabuses for old, new and mixed maths, the lack of communication between further and higher education, employers and schools about each group's needs and viewpoints, the inadequacy of information of job content or test results over a period of time, and the responsibility of teachers of mathematics and other subjects to equip children with the skills of numeracy."

The Report considered that possibly the most important of its recommendations was that the Secretary of State for Education and Science should set up an inquiry into the teaching of mathematics. In its reply, presented to Parliament in March 1978, the Government agreed that the issues listed in the Committee's Report needed thorough examination and announced their decision to "establish an Inquiry to consider the teaching of mathematics in primary and secondary schools in England and Wales, with particular regard to its effectiveness and intelligibility and to the match between the mathematical curriculum and the skills required in further education, employment and adult life generally." Government also committed the Inquiry to examining the suggestion that there should by a full analysis of the mathematical skills needed in employment and the problem of the proliferation of mathematical syllabuses at A level and at 16+.

By the autumn of 1978 the Committee of Inquiry had been formed. Its precise terms of reference are "to consider the teaching of mathematics required in further and higher education, employment and adult life generally, and to make recommendations."

2. The Work of the Committee So Far

The full Committee has met at three-weekly intervals since 25 September 1978, and on occasions has held weekend residential meetings. In addition the Committee has divided into working groups and these meet at least once and often twice between full committee meetings. Since the working groups have, of necessity, overlapping membership, the time given to the Committee by its members has been considerable. Since all but three of the members are in full-time employment, we must be grateful to them for giving so freely of their time.

At an early stage the Committee decided to advertise as widely as possible inviting evidence and informed opinion. Letters were sent out to some 600 associations, colleges, local authorities, employers and other bodies. We have been encouraged by the helpful response to these letters and by the welcome many people have given to the setting-up of the Inquiry. Some 860 pieces of written evidence have been received so far including a significant number from bodies and individuals who were not approached directly. Many are both extensive and carefully argued. In many cases the submissions have been prepared by groups especially set up for the purpose by those we approached. A number include papers and reports prepared before the Committee was set up, showing that consideration was already being given to the mathematical needs of pupils both for employment and for adult life. Industrial and commercial companies have often included examples of entrance tests they are using.

We have recently, through the press, asked for further evidence from employers relating to any noted lack of appropriate mathematical ability in those whom they employ or who apply to them for jobs.

As you can imagine, our reading programme has been formidable. In addition to working in small groups to consider particular aspects of our work, we also divide into small groups with cross-membership to read and summarize the submissions we have received. More recently, some of us have spent whole days re-reading and re-considering the totality of evidence which we have received on particular topics. You will understand our concern to avoid letting slip past us anything of significance, and our need to compare written evidence with the oral evidence which we have been taking during past months.

3. Research Projects

Early in the Committee's existence we became aware of the fact that there were certain areas in which we would require much more detailed information than we were likely to receive in written evidence. We therefore asked the Department of Education and Science to consider commissioning four research projects. All of these are now in the process of producing their reports.

Professor Bailey of the University of Bath assisted by Mr. Fitzgerald of the University of Birmingham (who has been seconded to Bath for the duration of the project) has been directing an inquiry into the mathematics needed by 16 and 17-year old school-leavers entering employment. The firms visited have included large and medium sized companies together with a random sample of small firms. We very much hope that it will be possible to obtain funding for a sequel to this work which will develop applications of school mathematics, found in various types of employment, for use in classrooms. We are hoping that these can be produced, and made available, so as to be of maximum use within our schools.

A further project in connection with the mathematical needs of employment has been directed by Mr. Lindsay at the Shell Centre for Mathematical Education in the Unviersity of Nottingham. This has complemented the Bath project by carrying out in-depth studies into the mathematical needs of clerical and allied workers, starting in the Boots Company in Nottingham and extending to a number of other companies. The grant for this project has also helped to accelerate studies already in progress at the Shell Centre in the fields of engineering, agriculture and health service. Again we hope to see the results of this work disseminated as broadly as possible within our school system.

An interesting outcome of these Industrial and Commercial studies has been the collection of comments we have received from employees, and employers, in retrospect, about the mathematics teaching they were exposed to in school.

The Department has also supported an inquiry in conjunction with the Advisory Council for Adult and Continuing Education into the mathematical needs of adults in daily life. Albeit on a small scale, this has nevertheless proved of great use to us. Interviews and observation have been made attempting to establish not only a typology of the most basic mathematical problems regularly encountered by adults but also the variety of strategies and methods adults use to cope with these problems.

An incidental by-product which we are sure the council will wish to make use of in fulfilling its new remit to consider not only literacy but also numeracy has been an analysis of attitudes towards mathematics and the inhibitions felt by adults in using it in everyday life.

Finally, Dr. A. Bell of the Shell centre for Mathematical Education and Dr. A. Bishop of the University of Cambridge have undertaken for us a critical review of existing research on the teaching and learning of mathematics. Their report has given us an opportunity to have an oversight of both theoretical and practical research relevant to our brief, both here and overseas.

4. Visits made by the Committee

You will not be surprised to know that we have been in no doubt of the need to visit schools of all kinds. In the time available to us we have naturally only been able to undertake a small number of such visits but these have proved invaluable to us and have made us aware of the enormous range of situations to be found in our schools. We have chosen to visit schools on the basis of size, type and geographical situation so as to observe and listen to the teachers and the pupils.

We have also visited a number of industrial and commercial companies chosen to represent a range of different types of employment. This has given us an opportunity to talk to young employees who have joined companies either directly from school or after some kind of further training, as well as to managers, supervisors, foremen and those responsible for training. Just as our visits to schools have enabled us to observe at first hand the work of pupils, so also these industrial visits have provided us with a valuable background against which to consider the reports of our research projects.

Although it has not been part of our task to make a comparative study of mathematical education throughout the world, we have sought information about mathematical education in a number of other countries and we have visited Denmark, Germany and Holland. We have also received helpful information from British teachers who have taught overseas and also from teachers from other countries who have been teaching in England and Wales. We have also visited Scotland and obtained much useful information from the Scottish Education Department.

5. Outside activities relevant to the Committee

The educational world, and in particular that part of it concerned with the needs of the teaching profession, particularly in shortage subjects like mathematics, has not stood still during the life-span of our Inquiry.

There have become available to us a number of documents arising from work that had already been started before the Committee was set up. These include the reports of the National Primary and the National Secondary Surveys, the discussion document produced by Her Majesty's Inspectorate on mathematics for the 5-11-year old age range and the report of the Assessment of Perfomance Unit on the mathematical performance of 11-year olds. In addition to this, the first Assessment of Performance Unit's report on the work of 15-year olds is about to be published. You will also be aware that the Secretary of State has published his discussion document on a Framework for the Curriculum, and of course we have had a contribution to the current debate from the Central Policy Review Staff in its report entitled Education, Training and Industrial Performance.

We have naturally considered all these documents and continue to do so. We have watched their dissemination and are concerned that very often we find teachers in the classroom not aware of their contents or indeed of their existence. We also find a lack of understanding by the general public, reflected unfortunately in the media, of the nature of the nature of the work of the Assessment of Performance Unit. That one should have to explain that when statistically controlled exercises of this kind are carried out, and ratings given between 0 and 100 on this basis, then inevitably the average mark is 50 and half the children obtain a mark below 50 and half above 50, is surely a reflection of the need of the educated public to understand the nature of such statistically controlled survey.

Again in this connection, the lack of appreciation of the percentage of the school population for whom GCE O level is intended-some 25 per cent-and the percentage for whom CSE in intended-going down to the 60th percentile-is a cause for concern. There is a clear need for the media at all levels to eduate the public at large about the nature of our examining procedures and the ways in which our public examinations are designed for particular sections of our school population. For example, it was the announced intention when the CSE Boards were set up, to ensure that a pass at grade 4 in CSE would represent the standard of work achievable by school children in the centre of the ability span of the entire 16-year old group. It is clear that this is not understood by those who publicly talk of the need for all school children to continue to take mathematics "up to O level."

Finally, I cannot close without reminding you of the debate, which continues, concerned with our need to recruit more well-qualified teachers into shortage subjects in our schools. No inquiry of the kind we are committed to can ignore this debate, but whether we shall be able to glean from the evidence given to us anything which has not already been said, remains to be seen. I only hope that as in all matters on which we shall report, we can be as constructive and useful as possible to those at the "chalk face" and to the children in their classes, to whom at the end of the day we owe our duty.

TRENDS IN MATHEMATICS EDUCATION IN CANADA

David F. Robitaille
University of British Columbia
Vancouver, British Columbia

Because of the nature of the educational enterprise in Canada, the task of identifying national trends and emphases is especially difficult. Each of the ten Canadian provinces has complete control over education within its geographical boundaries; there is virtually no federal involvement. Over the years the educational systems of the provinces have evolved along more or less similar lines, but there are notable differences among them on matters affecting both the content and the organization of schooling. The establishment of a secretariat for the Council of Ministers of Education, and its sponsorship of meetings of persons responsible for curriculum matters in each of the provinces should provide better sharing of information and communication on these matters in the future.

Another feature of the educational scene in Canada, and one whose influence must be borne in mind in any discussion of mathematics education in Canada, is the pervasive influence of the United Stes. In most of the provinces, children use texbooks which were written in the U.S. and then adapted for use in Canadian schools. The only nation-wide professional organization for teachers of mathematics in Canada is the National Council of Teachers of Mathematics (NCTM). Attempts to create a separate Canadian association have not been successful. Moreover, there has been a lack of communication and exchange of information among mathematics educators from different provinces.

This situation is beginning to change. Four studies which have been conducted in the past two years have proven to be valuable sources of information on trends and developments in mathematics education in Canada. The first of these was commissioned by the Council of Ministers of Education and consists of an analysis of curriculum guides and related materials published in each province supplemented by information obtained through direct contact with the persons involved. The second, conducted by Professor Shirley McNicol of McGill University, is an analysis of elementary mathematics programmes in Canada. She conducted over 100 interviews with teachers, teacher educators, and representatives from ministries of education regarding current practices and trends in mathematics education. Thirdly, the author of the present paper sent a questionnaire dealing with current trends and projections to a teacher, a teacher educator, and a Ministry of Education representative from each province. Finally, the PRISM project sponsored by the NCTM in the United States had a Canadian component. Although the final report of the study was not available when this report was prepared, the major findings of PRISM-Canada were presented at the general meeting of the NCTM in Seattle in April, 1980.

The present paper is a synopsis of the findings of these four studies. The trends and directions which were identified have been grouped into three categories, as follows:

- trends in the curriculum
- trends in instructional practices
- trends in teacher education

All of the directions and issues discussed here are based on the results of interviews or an analysis of questionnaire responses. The degree to which such information and opinions are a reflection of the true state of affairs is open to some question. Certainly, a number of recent studies in mathematics education have shown that there is often a considerable gap between what people say is or wil be the case and what actually is the case.

Trends in the Curriculum

As of 1977, each of the provinces either had recently adopted a revised mathematics curriculum or was in the process of conducting such a revision. Those new programmes reflect a rather marked shift in emphasis in the school mathematics curriculum away from the New Math. Another way of characterizing this shift would be to say that the programmes now place more emphasis upon mathematics as an applied subject and correspondingly less emphasis upon mathematics as the study of structures.

A common goal of many of the new programmes is expressed in the following statement from the Council of Ministers report to the effect that these changes will "create a curriculum which will provide all students with mathematics competence and confidence." Although one might applaud such a goal as a democratic ideal, there is a danger that the curriculum will become trivialized in seeking to attain it.

Among the specific trends which have been noted in the mathematics curriculum are the following:

1. There is definitely a "back-to-basics" move, but most of the reports include problem-solving and measurement as part of those basics. Although computational skills are to be emphasized, extreme cases will be avoided. Work with 2- and 3-digit numbers will be emphasized.

2. Less emphasis will be placed on terminology and rigour in the 1980's.

3. Work with decimals will be introduced earlier, and fractions will be somewhat de-emphasized. This may be seen as being a result of Canada's adoption of the metric system and of the increasing presence of calculators in the classroom.

4. New programmes will emphasize estimating skills.

5. Consumer mathematics and applications of mathematics will be important areas of the curriculum.

6. In at least half the provinces, there is a continuing trend to integrate mathematics courses at the secondary school level rather than to teach them as separate entities.

7. An introduction to geometric motions (slides, flips, and turns) will be part of most elementary school programmes. No corresponding trend to teach transformational geometry is apparent at the secondary school level.

8. More emphasis will be given to the study of statistics at the secondary school level.

9. Computer literacy will continue to grow in importance although there is some resistance to the idea of introducing new courses in this area into an already crowded curriculum.

Trends in Instructional Practices

In addition to trends in the content of the curriculum, a number of trends in teaching practices, both new and continuing, may be identified.

1. The importance of using manipulative materials to clarify concepts and processes will continue to be stressed at the primary (K-3) level. No such trend is evident at higher levels.

2. There is a tendency to encourage less reliance on a single textbook for a given grade. Instead, the use of several textbooks and resources is being recommended.

3. Although most provinces have yet to state a formal policy regarding the use of calculators, there is a realization that such use will continue to grow. Considerable doubt remains about the use of calculators at the elementary level.

4. There is a trend toward the adoption of mathematics textbooks written and published in Canada. There is no clear indication how, or if, such texts will differ greatly from those published in other countries.

5. The use of computers and the teaching of computer literacy will continue to grow.

6. Testing is regaining prominence. Where there was once fairly strong resistance to testing, particularly standardized test, teachers are now seeking out and using such tests. Several provinces have instituted province-wide assessment programmes more or less along the lines of the National Assessment (NAEP) model in the U.S.A.

Trends in Teacher Education

Trends in teacher education appear to be related to questions of teacher supply and demand. Thus, in those parts of the country where there is an over-supply of teachers, there is a trend to emphasize in-service as opposed to pre-service training. This is less true in other parts of the country where declining enrollments and the availability of teaching positions are not such serious problems. Despite these regional differences, it is possible to identify a number of more or less national trends.

1. There is a trend in teacher education to emphasize methods of teaching mathematics and to place correspondingly less emphasis on mathematics content in the pre-service training of elementary teachers.

2. In some provinces, school-based programmes and internships of various kinds are supplanting the more traditional teacher education courses: at least for elementary teachers. Such programmes seem better designed to produce generalists than subject-matter specialists.

3. One of the reports indicates a continuing decline in enrollment in graduate programs in mathematics education.

4. As noted above, there is evidence of a growing demand for in-service programmes of all kinds, but especially those which will assist teachers in copiing with their day-to-day tasks.

Conclusion

The overall picture that emerges from these four reports is that the 1980's will be a period of modest renewal and retrenchment in mathematics education. Negative reactions to some of the excesses of the New Math era is giving way to a more positive consideration of alternatives. Should these trends continue, the Canadian mathematics curriculum in 1990 will, in all probability, not look very different from the 1980 edition. There is no indication of sweeping change on the horizon at this time.

11.3 CURRICULUM CHANGES DURING THE 1980'S

CURRICULUM CHANGES DURING THE 1980s: ELEMENTARY SCHOOL

Shigeo Katagiri
Yokohama National University
Japan

Introduction

This mini-conference on the curriculum for the 1980s is organized in three sections. The first focuses on needs, problems, and principles that need to be considered in contemplating the future for the elementary school level. The second paper discusses the needs of students in the secondary school who do not aspire to continuing their education beyound the secondary school. The final paper examines issues and problem for the student who intends to matriculate to the university level.

The papers each focus on the need for realistic applications and problem solving that fit the future of the student. Please realize that there is considerable variation from one country to another in terms of the traditions, practices and resources that can be brought to bear on curriculum problems in mathematics. Each paper reflects some of the idiosyncratic features of curriculum and schools in the country of the presenter. The first paper identifies some fundamental purposes of mathematics education that should serve to guide the selection of activities and curricular thrusts at the

elementary school level. The second paper indicates that high expectations are necessary if the mathematics experience is to represent a thoughtful fit between the needs of the country and the needs of workers and citizens in that country. The final paper addresses problems of what secondary school mathematics is needed by students who will matriculate through the university but not in technical and scientific fields and the problems of mathematics' contribution to leadership in a country.
In the elementary school in Japan, students are taught functional and statistical relationships, as well as about numbers and operations, quantities and measurements, and geometrical figures and their properties. This content provides a sound basis for thinking about changes in the elementary school curriculum needed through the decade of the 1980s. Thus, I will expalin it in greater detail.

For numbers and operations, we teach students the concepts of whole number, decimal, and fraction, and their operations. The number line is taught and used from the first grade. For quantities and measurement, we teach the concepts of many kinds of quantities; that is, length, weight, time, area, volume, angle, speed, and so on. Principles of comparison and measurement of each system are taught.

In geometry the concepts and the relations of many plane figures are taught; for instance, children study many kinds of triangles, quadrangles, and polygons. Some solid geometry is taught; for example, the geometric characteristics of cubes, prisms, pyramids, cylinders, and cones. We also teach the expression of the positions of points in the plane and in space in order to develop the concept of space.

Instruction about three types of relationsips is provided. The first type concerns expressing quantitative relationships in formulas and the reading and use of these formulas. The second type is the investigation of some functional relationships and expressing and interpreting those relationships in formulas, graphs, or tables, to study the meaning of ratio, direct and inverse proportions, and the use of them to solve problems. The third area of study is statistical relations, expressing them in graphs or tables, reading and interpreting them, and using the idea of probability in simple cases.

This content will not have to be changed very much; fundamentally, this content should be taught throughout the 1980s.

Given an extensive and careful curriculum of this nature, we can identify several important problem areas for study in order to revise and improve the curriculum The important content and appropriate teaching methods of elementary school mathematics should be clarified in the light of the primary purpose of teaching elementary school mathematics. I think the purpose of teaching elementary school mathematics is this:

> To develop students' abilities and attitudes to mathematize real situations of their own accord and to study mathematical problems independently.

1. To attain this purpose, we should help students acquire the basic knowledge and skills.

Recently, people have indicated that "back to the basics" is the fundamental point for revising the curriculum. Of course, basics are important, but to clarify what is basic is more important and a fundamental first step in our tinking. What is basic must be decided according to its contribution to the primary, central purpose of mathematics in the elementary school. The following is an example of what is basic in computation. In teaching computation with whole numbers, decimals, and fractions, what is important and basic?

Japanese students learn the multiplication and division of decimals in the fifth grade, and the multiplication and division of fractions in the sixth grade. After learning the multiplication and division of decimals, they have to be able to solve independently real problems that require these operations:

Example 1: The price of gasoline went up 92 percent and now the price of 140 liter of gasoline is $94.08. What was the price of 1 liter of gasoline before the increase?

Even if the student can perfom a division like $0.672 \div 1.92$, he will not be able to solve such problems if he cannot decide which operation should be applied.

Performing a given computation correctly is basic, but deciding what operation should be applied is even more basic. The student who cannot decide which operation should be applied can never act independently. Deciding which operation should be used is most essential to being an independent problem solver. After this decision, the ability to perform the computation comes on the scene and may be supplanted by a handcalculator. But the selection of the operation cannot be made by any calculator. We need to have students who are able to select the correct operation or a correct sequence of operations to solve problems. Therefore, more emphasis is needed in teaching the general meaning of each operation, that is, in which cases each operation is to be applied.

As an example, I suggest the meaning of multiplication and division, which I think should be taught in the elementary school mathematics program. The meaning of multiplication of whole numbers should be taught as repeated addition in grade two.

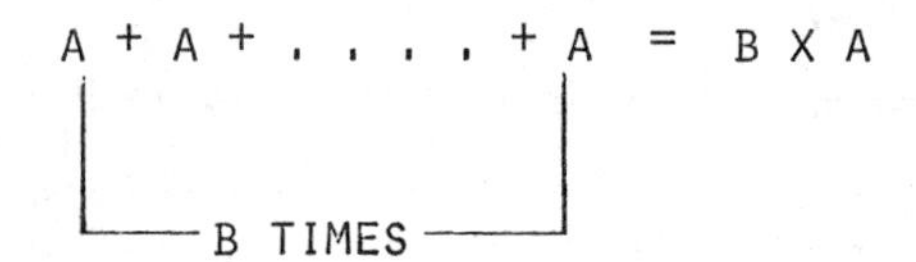

This can serve as a base to establish meaning for division. There are two meanings of the division of whole numbers, $b \div a$ (no remainder). One is the operation used to find out how many quantities with size a are included in the quantity with size b. The other is the operation used to find out what is the size of each part if b is separated into a equal parts. These meanings are applied to problems in which the corresponding number for the multiplier of the divisor is a whole number, even if the corresponding number for the multiplicand or the dividend is a decimal or a fraction. But these cannot be readily applied to the problem if the corresponding number for the multiplier or the divisor is a decimal or a fraction. At this stage, the meaning of division should be extended by using problems like this:

When the multiplier and the diviser are whole numbers, we can recompose the above meanings as follows.

$\underline{b}$ stands for the quantity corresponding to the ratio 1,

$\underline{p}$ stands for the ratio,

$\underline{a}$ stands for the quantity corresponding to the ratio $\underline{p}$.

(See the number lines below).

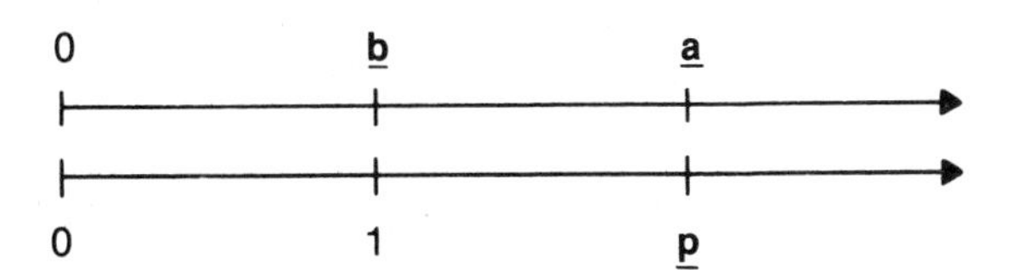

If $\underline{b}$ and $\underline{p}$ are known, then $\underline{a}$ is found by $\underline{b} \times \underline{p}$.

If $\underline{a}$ and $\underline{p}$ are known, then $\underline{b}$ is found by $\underline{a} \div \underline{p}$.

If $\underline{a}$ and $\underline{b}$ are known, then $\underline{p}$ is found by $\underline{a} \div \underline{b}$.

These interpretations can be extended to cases that have, as the corresponding numbers for the multiplier and the divisor, decimals or fractions. For instance, if we look back at example 1, the relationship of the quantities in this problem is shown below.

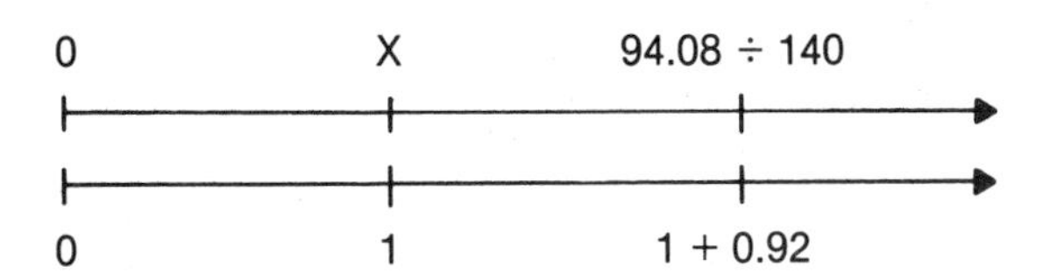

Therefore, we can deduce that $\underline{x} = (94.08 \div 140) \div (1 + 0.92)$.

2. To attain the purpose, we should develop the students' mathematical ways of thinking.

In order to develop the attitudes and abilities for acting independently in problem solving and studying mathematics, we should endeavor to develop not only mathematical knowledges and skills, but also the attitudes and abilities for using the mathematical methods and points of view that are often used to solve probems mathematically.

To develop these attitudes and abilities is the same as the primary purpose of teaching elementary school mathematics that was identified previously. I think these attitudes and abilities are the same as a mathematical way of thinking. An example will clarify what I mean by the mathematical way of thinking.

Example 2: Suppose a teacher gives fourth grade students a problem: "How many squares can you find in this figure? They begin to count the squares at once, but most of them cannot do this correctly.

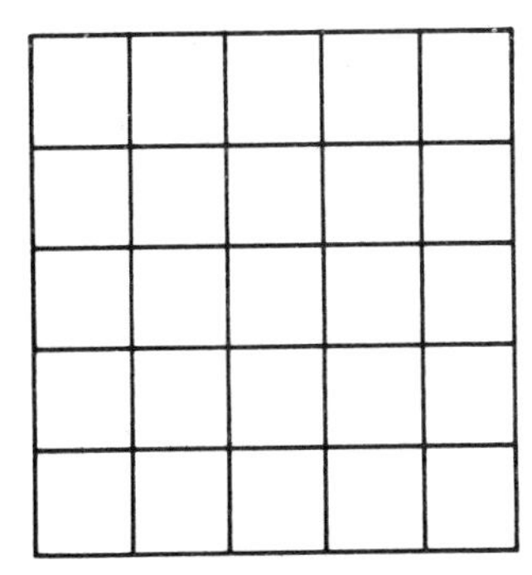

That is, their answers differ from each other. At this stage, the children notice that they must devise a new way of counting and discover the way to classify squares from the point of view of the sizes of the squares. The attitude of trying to find a more elegant way, and thinking of classifying the squares by size, is a step toward a more mathematical way of thinking.

Have them classify the squares by side lengths 1, 2, 3, etc. If the children count and classify them separately, they get the result:

$$25 + 16 + 9 + 4 + 1 = 55$$

Then the teacher can ask a student to explain how to find the answer. If the student counts the squares with side length 2 while tracing their perimeter with a chalk, other students will notice that this way of counting is ambiguous and cumbersome. Then the teacher can help them devise another way of counting and can help them discover it is enough to count the upper left vertices of the squares with side length 2. This discovery has the effect of focussing on another kind of mathematical thinking, functional thinking. The squares with side length 2 correspond one-to-one to the upper left vertices. If and only if we select a lattice point, we can find a square with side length 2 which has the point as its upper left vertex. After this discovery, the teacher can inquire, "If the side length of the original figure is more than five, how can the number of squares be found?"

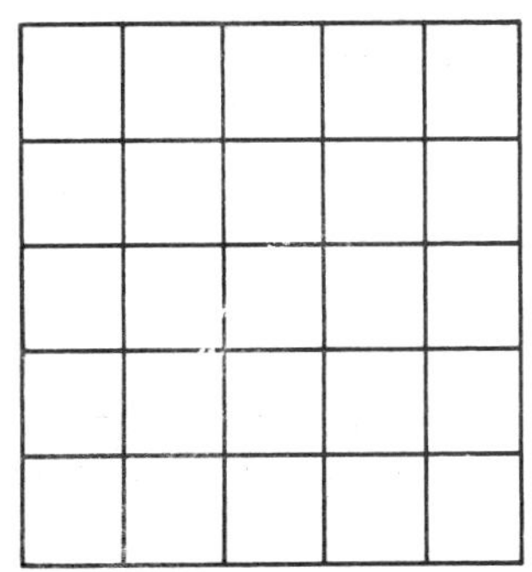

The aim of this question is to let students know the importance in thinking of generalization. By counting upper left vertices of squares for each class of square, they find the answer of the above problem to be:

$5 \times 5 + 4 \times 4 + 3 \times 3 + 2 \times 2 + 1 \times 1$.

They can easily generalize this result. If the side length of the original figure is 6, then the number of squares is: $6 \times 6 + 5 \times 5 + 4 \times 4 + 3 \times 3 + 2 \times 2 + 1 \times 1$.

Moreover, the teacher may ask for other ways of solving the problem. Some students may try to classify the squares into several groups which make up the squares with the same point as their upper left vertex. This is an analogy from the above way of counting. They will find the result:

$5 \times 1 + 4 \times 3 + 3 \times 5 + 2 \times 7 + 1 \times 9$

This example problem exhibits four mathematical ways of thinking that need to be stressed in the future:

(1) classifying,
(2) functional thinking,
(3) generalizing, and
(4) thinking by analogy.

It is possible and necessary to recognize and teach other mathematical ways of thinking. Among these are:

(1) inductive thinking,
(2) deductive thinking,
(3) unifying, and
(4) ideas of sets.

In summary, four points are important for considerations about future curricular needs:

1. We should clarify basic content and select better teaching methods in the light of the primary purpose of teaching mathematics.

2. We should let and help students use mathematical ways of thinking when they solve many kinds of problems by themselves.

3. We should help them identify and summarize what mathematical ways of thinking are used when problems are solved.

4. We should endeavor to find good problems for those purposes. The problems that I think are good should have the following characteristics: (a) allow the students to concentrate on certain mathematical ways of thinking during and after problem solving, (b) require only a few seconds to explain and yet keep a student busy for an hour or more, and (c) make the students feel that they have been involved in a real and natural situation.

COLLEGE AND UNIVERSITY ENTRANCE PROGRAMS AT THE SECONDARY SCHOOL LEVEL

Alan Osborne
Ohio State Unviersity
Columbus, Ohio, U.S.A.

The curriculum in mathematics at the secondary school level has been dominated by university entrance in the majority of the countries of the world. In my opinion, this has been particularly appropriate for countries in which universal education for all youth has not been attained. Such countries must concentrate their efforts and resources on the problems of developing leadership. Given the characteristics of industrial and technological components of modern society that have developed since the turn of the century, no sensible alternative for preparing leaders for that society exists. For countries that have attained universal education at the secondary school level, the economic development and industrial enterprise dictate a continuing need for products of the universities. In either case, the need for knowing mathematics will not decrease but, if anything, will increase. Thus, it is not necessary to urge that more mathematics be included in curricula of the future. Rather, the important issues concern the kind of mathematics and when it should be taught.

Three general problems about the design of university entrance curricula are discussed below before considering specific examples of curricular issues that need attention during the coming decade. The three general problems are:

1. The relation between future leadership responsibilities of students and the mathematical content of the secondary school program.

2. The problem of the pre-university student who will not elect to be a major in science, engineering, or mathematics.

3. The relation between mathematics and the processes of modern science.

It should be noted that these three general problem areas do not necessarily have the same resolution from one country to another. Some of the significant variables are the portion of the citizenry that is educated, the health of the economy, and the academic traditions in the country. That is, the needs of developing countries and developed countries as well as the resources available for education should determine how the problems are best treated. These three general problem areas must be dealt with by a country before it can come to a sensible resolution of many specific curricular issues.

Leadership

Mathematics serves two purposes for the products of universities. One purpose is in terms of functioning within the settings of business technology, science, or education. The other purposes is more personal, representing the "man of the street" level of competence exemplified by decisions concerning best buys, voting, plans for use of materials, and the like. The locus of the second purpose is typically local and immediate. The usual university entrance curriculum in mathematics at the secondary school level gives primary attention to content fitting the first purpose.

Developing and developed countries differ in their needs to feature these two purposes in their curricula. In developed countries, either an individual or an agency can be found to perform the mathematics involved in the second purpose. Although most of us believe that individuals should be able to deal with such personal-use mathematics, the availability of expert advisory service is a characteristic luxury of the developed country. In developing countries, a product of a university cannot expect expert advisory service and must be able to function for himself.

Another aspect of this problem domain is even more critical for developing countries when considering the second, personal-use purpose of attaining knowledge of mathematics. Developing countries have a major need to evolve to a society possessing a sufficient knowledge base about mathematics and its uses to support and sustain the technical expertise that they do have in their scientific-industrial enterprise. An engineer or agricultural expert can contribute more and be more efficient if the workers have the fundamental, basic knowledge of mathematics that allows communication. Small steps in building appreciation of the use and power of mathematics are more readily obtained on a basis of demonstrated applications. But demonstrated applications of mathematics must fit the knowledge of observers if they are to affect appreciation of the power and use of mathematics. Desmond Broomes observes in a recent UNESCO paper that more than half of the children in secondary schools of the world live in communities that are essentially rural. University graduates will in many cases return to essentially rural environments. If they are to function in a sense of helping the society develop a base for support and sustenance of science and industry, they must assume a leadership function. I know of no more effective means than demonstrating the use of mathematics to deal with real, immediate, and local problems. Functioning as a leader in the sense commensurate with university matriculation defines a responsibility to help individuals deal with local problems at this functional level. The mathematics of the commonplace and the immediately real in the sense associated with the second purpose and with the leadership needs of developing countries cannot await the university curriculum for explication. And it is also unlikely content for university level mathematics within typical academic curricula for science, engineering, or business.

Conclusion: Particularly for developing countries, curricular attention must be given to practical problems and local applications of mathematics for the university entrance program in order to develop the leadership potential and power of university graduates.

Programs for Students not Aspiring to Careers in Science, Mathematics, or Engineering

The problem of the curriculum in mathematics for students who are not science, mathematics, or engineering majors assumes different forms in developed and developing countries. For developing countries, failure to encourage majors in mathematics and the sciences establishes boundary conditions or limits for functioning in the modern world. Controlling the reward and support systems (wages, scholarships) for education and within the port-degree industrial world provides a modicum of control for the problem. The problem for the developing country is more one of encouraging the student to aspire to a career that uses mathematics, the sciences, or engineering than one of considering alternatives in designing programs for students aspiring to other fields.

For developed countries, two problems are significant and important. Developed countries have the luxury of encouraging and fostering the arts and other individual endeavours that do no require a high degree of competence in mathematics or science. Many careers that require but a minimum of science or mathematics at the university level can require the making of decisions that demand some technical expertise and understanding or literacy in mathematics. Some of the decisions have a major impact for society. The judge who must make a decision concerning the environmental impact of a particular industry or the legislator who must decide about the merits of nuclear power generation are quite likely to matriculate through programs in universities that require at best a bare minimum of mathematics or science. They are examples of individuals who as adults make decisions requiring considerable mathematical and scientific literacy but who acquire the majority of their mathematical and scientific knowledge in the secondary school. The secondary school program is often their last chance to learn the mathematics they need.

Conclusion: Much more is at stake in the design of university entrance curricula in mathematics than simply the the training of prospective scientists, engineers, and mathematicians.

The second problem in designing of curricula for non-science majors is more subtle and, probably in the long run, of even greater significance and importance than the first. C.P. Snow describes a world of two cultures, one concerned with the arts and humanities and the other concerned with science and technology. Each has many individuals who fail to understand the other culture and its values, who are unable to communicate with the other culture, and who possess a fundamental arrogance and disdain for the other culture and its processes. The society that Snow describes has evolved to an even greater degree of separation, in my opinon, since he wrote the book. Witness the enhanced popularity of pseudo-sciences such as astrology that has been demonstrated during the recent years in developed countries. Issues pertaining to energy, the environment, economy, population, and the like are considered by large segments of the population on a purely emotional basis, with little consideration of appropriate facts or scientific decision making.

Conclusion: Secondary school curricula in mathematics must address the problem of the two cultures on two different levels, one of which builds fundamental scientific and mathematical literacy and the other of which considers what mathematics and science can and cannot do in formulating values and making decisions.

Processes Of Science

The nature of science and the processes of science used currently are of a fundamentally different character than those of even 25 years ago. Three different aspects of being a scientist today are (1) a heavy reliance on the processes of mathematical modeling, (2) an extensive reliance on the computer and electronic data processing, and (3) a number of large, complex problems that require a team effort for solution. Few secondary school mathematics programs respect these characteristics of modern science by providing students with experiences that foster appropriate skills and understandings. In my opinion, most of the skills and understandings cannot await the university level course work to be established.

Conclusion: The secondary school curriculum in mathematics needs to be made consistent with the techniques and processes employed in modern science and technology. This require that some individuals operate in curricular decision making and implemetation who possess an understanding of both mathematics and science.

Specific Curricular Problems

The three general problem areas identified above provide a basis for considering a wide variety of specific curricular problems and issues in the mathematics programs at the secondary school level for university entrance. The conclusions help to identify what is important in attaining a conclusion that fits the needs of the future. Problems and issues are identified below that deserve priority consideration by mathematics educators.

1. Finding "real" practical problems. All too frequently, practical problems found in textbooks are concerned with a routine, algorithmic-like application of previously taught mathematics. The practicality appears to be identified with the use of the mathematics rather than situational practicality of the local society and the environmental setting. Affecting the leadership potential described in the first general problem, students are particularly hampered in discovering problems within situations and building the associated skills. A particularly invidious, related aspect of this problem is the fact that many university entrance and placement tests have content that is tied to the classical applications of mathematics featured in current curricula (thereby serving to preserve the status quo in curricular content). Problem solving is perceived as an opportunity for the practice of mathematical principles through application of rules rather than as an opportunity to discover mathematics in situations. Mathematics for the future can attain much more power for society and for the individual if secondary school curricula attend to the responsibility of helping students see mathematics in the commonplace and the practical. The curricular themes for the elementary program described by Professor Katagiri would help meet this need.

2. Building skills in the processes of estimation and approximate computation. Particularly for students who will need mathematical skills as part of their repertoire of leadership abilities, an understanding and demonstrable skills in this area will enhance communication of mathematical usage by example. Non-paper-and-pencil mathematics is part of this repertoire.

3. Building communication skills in mathematics. Part of the leadership problem as defined above is being able to communicate mathematics and its uses effectively in both a written and spoken format. I judge that very few curricula any place in the world address this problem directly. It is particularly important for those in leadership positions to acquire this skill.

4. The problem of statistics, probability, and the understanding, interpretation, and use of data. In the United States, statistics and probability represent perennial losst causes of mathematics educators in curriculum development and change. Every set of curriculum recommendations since the late nineteenth century has recommended additional attention to these curricular area; no program of implementation has enjoyed widespread acceptance. Indications are that this is the case throughout much of the world. Yet in terms of modern, functioning citizenship, this area of curriculum is a basic skill that needs considerably more emphasis during the coming decade. Many of the critical areas of political decision making require an understanding of this domain of knowledge. The content is important for leadership in both developed and developing countries whatever the career aspirations of the students. Further, the technical knowledge of statistics required in most fields of modern science is large. How can we incorporate emphasis on these topics in the curricululm when there is so little evidence of commitment at the school level? Can we produce effective instructional materials that take the potential college or university student beyond lower level but fundamental understandings of measures of central tendency?

5. The problem of mathematics content selection for the university bound student who does not aspire to a career in the sciences or technological fields is a matter of determining what constitutes mathematical literacy. I would nominate for consideration topics that communicate the nature of problem solving, what mathematics can and cannot do, understanding the computer and its limitations, statistics, the role of deduction and model building in science and applications as well as functional, personal competency in dealing with the mathematics involved in real life. This list is similar to what many thoughtful individuals would generate. However, such lists seldom are in agreement about the depth of knowledge or the degree of competence in associated skills required and, in particular, do not consider the problem of university entrance testing or placement testing. If it is not tested, it typically will not be taught. The topics can be treated via a variety of mathematical contents. Which are most appropriate? What is the role of the content setting in addressing the attitudinal issues and problems associated with the two-culture syndrome?

6. The extent to which computer literacy is expected vis-a-vis skills and understandings in using the computer is an issue that needs to be addressed in both developing and developed countries. However, the problem is different for a country with limited resources to invest in computers than it is for countries with more resources. For the developed country, everyone should have experience with using computers and should acquire fundamental literacy given the rapid, extensive expansion of use of the computer in all aspects of society. To fit the secondary school student for future university life and for the years beyond the university, the real-time encounter with computers is a necessity in my opinion. How best to accomplish this is the problem rather than whether to accomplish it.

For the country with limited resources, the question is quite different. For leaders participating in the international world of business, politics, and science, the country has a responsibility to assure that their experiences at the secondary school and university levels provide experience with computing devices. It is in the country's interest as it looks to the future to design experiences for the potential leaders that will fit them for working with individuals from developed countries. The value of preparing students for a future of understanding and communicating with individuals from a society that make extensive use of the computer must be weighed against the probable limited use of computers within the country. Educational policy making relative to providing computer experience and instruction must be considered within the larger context of the extent of use of the computer foreseen for the future.

7. The role of mathematics in helping develop scientists and technicians is important. Applications of mathematics within the mathematics curriculum should

contribute to the student acquiring judgement about and knowledge of science and its processes. The design of curricula in mathematics has not in the past shown sufficient cognizance of this important role of mathematics. Rather than being constructive and oriented to finding problems in situations, most curricula have involved students in the routine, algorithmic fitting of previously learned mathematical skills to situations. Although the relatively standard, commonly encountered scientific situations, such as distance-rate-time problems, need considerable attention in the curriculum, the current emphases do not produce learning that fits how scientists and engineers must operate in many situations. Another failure in fitting current application experiences in the mathematics curriculum to current scientific practices is learning to use resources external to the student's own thought processes and the textbook; do we need to teach the student how to use and find references as sources of information and processes? Another problem in the mathematical education of potential scientists and engineers is the limitation of application problems to only a few domains of science for the selection of mathematical problems. Seldom are problem settings in psychology, sociology, agriculture, and fields other than the traditional physical and life sciences used as a context for teaching mathematics.

8. University entrance curricula in mathematics have a long tradition of being concerned primarily with helping prepare students for their future work in calculus. Textbooks and entrance and placement tests respect this tradition, for in the past it has been the most important set of skills and understandings needed to assure success. However, both science and mathematics are characterized by making extensive use of other domains of mathematics. Can we project that by the end of the decade of the 19080s that other types of mathematics will be needed for university entrance?

9. Much of the mathematics used by university graduates, particularly if they are not entering technical or scientific professions, is taught and learned during the secondary school years. Often this mathematics is not used during the period of university enrollment. With lack of use, forgetting is a major problem. Can we design instructional and curricular practices to enhance memory over extended periods of little use or practice? Interestingly, although memory is a basic factor affecting cognition, researchers in mathematics education have ignored it. We need more research in this fundamental cognitive function to design curricula for secondary school mathematics.

SOME REMARKS WITH RESPECT TO THE CURRICULA OF SECONDARY SCHOOLS, ESPECIALLY EMPHASINZING NON-COLLEGE-AIMING YOUTH

Hans-Christian Reichel
Department of Mathematics, Unveristy of Vienna
Austria

This mini-conference is concerned with the question "What should be the primary thrusts of mathematics curricula in the 1980's?" and my paper—as a part of it--will concentrate on secondary school curricula with special emphasis on those students who are not aspiring to careers in science or mathematics and those who will not attend university at all. Hereby, the word "curriculum"-- as I understand it--means more than only "syllabus"; it also includes aims, contents, methods, and assessment procedures. The term "secondary school" will mean all types of school for pupils between 11 and 18 (or19) years except special types of trade schools providing vocational education and designated for "producing" specialists of certain professions ("Berufsschulen"). Although we shall refer to a wide variety of secondary schools, the original background of the lecture is those which are assembled by the term "Allgemeinbildende Hohere Schulen" in Austria. (More details later on.)

By the general theme, it is the task of this mini-conference to look into the future, to identify what we might call trends, and perhaps to stimulate some features of such trends. Any such attempt is bound to have subjective elements; I therefore it is necessary to compare the many other papers of ICME IV dedicated to a similar theme. (Compare expecially the paper of Alan Osborne or NCA).
In many cases, the trends will be more wishful thinking than actuality," says G.A. Howson in (HO). And Dr. Faustus in Goethe's drama: "Was Ihr den Geist de Zeiten heibt, das ist im Grund der Herren eigener Geist, in dem die Zeiten sich bespigeln." (I.Teil, I.Augzug).Many people who have not had a special education in science or mathematics show a more or less unreflected appreciation of mathematics on the one hand, and a far-reaching ignorance on the other. In general, people consider mathematics a thing for specialists only. We surely agree that mathematics curricula of the future should try to improve this situation.

With the exception of some kinds of trade schools providing a special vocational education, education in the various types of secondary schools must prepare students as well for university entry as for the so-called professional life. However, during the last decades, curriculum development seems to have concentrated mainly on internal scientific features of mathematics; it was guided by the development of mathematics itself and--more or less-- neglected the role mathematics can play when educating students who do not aim at a study of mathematics or science later on. With respect to this large group of students, curriculum developemnt in mathematics should be governed by three items (and, I am sure, considering these points will also improve mathematics education for those who will specialize in mathematics or science after school):

1. Exploring and focussing on needs and requirements of "professional life"

2. Exploring and focussing on general education goals which a training of mathematical abilities might help to realize

3. Considering that, for the group of students in question, our mathematics courses will effect the remaining impression of what mathematics is, and as is well known, the latter very often deviates from what we want to effect.

The importance of the last point, expecially for the non-college aiming youth, is often neglected, but let me start with the first point. Although this is a very old guideline of any curriculum, much has to be done in the 1980's: empirical investigations are necessary, and these must be carried out locally. The results may be different for industrialized nations and for developing countries, and there will even be differences between countries whose economy concentrates on small companies producing specialized products and those who concentrate on big and world-wide industries.

Although most of the personnel managers responsible for employing young people do not really care about the mathematics syllabi of secondary schools, it is clear that certain mathematical abilities and knowledges are desired, and these are likely to differ from the contents esteemed one or two decades ago. To give an example let me cite a study project started in 1976 in Austria (DP) which is to investigate--roughly speaking--the relations between mathematics curricula, their effects, and the demands of professional life. Its final results will be published next year. One of the preliminary results seems to be the discovery (or confirmation) of gaps and differences between what is effected by presently established mathematics education (curriculum reforms of the last decades included) and that which managers claim to be necessary today. But anyway, before delineating concrete items, I also would like to warn of the disadvantages of a philosophy which connects curricula too strongly with present demands of industry and other business institutions. Firstly, because concrete employment conditions may change very repidly according to technical development, and secondly, since real progress always comes from the converse request: industry and business institutions should provide our youth with chances and possibilities for applying their (theoretical) knowledge and skills as freely as possible.

But nevertheless, we can delineate some mathematical contents and general abilities which definitely deserve special emphasis by future curricula. Besides, one should think that for the group of students in question training of mathematical and general abilities connected with mathematical activities might even be more important than knowledge of concrete mathematical concepts. And indeed it was anothe result of the above-mentioned Austrian study project that employers primarily demand such general abilities.

A few examples will also show what I mean by item (2) above:

a. Being able to understand (and produce) diagrams, flowcharts, and graphical representations of dependencies and issues

b. Being able to understand (and produce) statistically elaborated material

c. Being able to explain matters, issues, problems, and dependencies shortly and clearly

d. Being able to recognize details

e. Being able to disintegrate complex and complicated matters

f. Being able to see connections, relations, and correlations between seemingly independent facts and to integrate them

Future curricula should paradigmatically show teachers how to realize such goals. But now let me give some headlines concerning concrete topics of mathematics curricula which will (and should) be focussed on during the years to come. (It should be noted that many papers of ICME IV are devoted to special problems and details of exactly these topics.)

Statistics and probability theory:

Both fields will be of major importance for future curricula. Following the opinion of may scientists, they should not be added as a new block, but should be incorporated at any level of mathematics education and taught as a way of looking at the real world. "The whole world is a probabilistic process," as A. Engel claims in (EN). And with regard to the group of students in question I would add: this is the way of teaching mathematics at all. Mathematics is a tool and an additional language often better suited for describing situations and posing and solving problems--and not only technical ones, as we know today. Besides, I would like to correct Engel's formulation and say: what mathematics curricula should reflect is that the whole world could be conceived as a probabilistic process. (We should take care of presenting mathematics as a new ideology, as one and only one type of spectacles to watch the world through. On the contrary, mathematics education should try to effect in pupils a kind of immunity from all kinds of ideology and pseudosciences as they grow up nowadays. If well taught, mathematics could contribute to making students open minded and resistant against superstitions of the modern world.

To sum up: Statistics and probability should start when pupils are 10 or 11 years old and should be present at any further year and level. (To cite but one school book which realizes this program: (MH), vol. 5-10.)

It is not the goal of this lecture to go into further details, but there are many well-elaborated suggestions prepared all over the world, e.g., from SMP in Britain (SMP) or - - even more detailed - - a series of books edited by the American Statistical Association and the National Council of Teachers of Mathematics and called "Statistics by Examples" (SBE).

Let me add two general remarks: in my opinion, combinatorics is not the best way to begin probability with. It fixes pupils too much to computational procedures and to only one type of exercise. Moreover, it is difficult to switch seamlessly to statistical reasoning. Easier to start with--especially for younger students--is, for example to work with statistical tables (population statistics, sport events, and so on) and to give pupils a feeling for the probability of an event via percentage computing before they are provided with models of probability functions. In any case (perhaps in the last year of secondary school), students should be confronted with the axiomatic way of defining probability, even--and, I should say, especially--those who will not enter the study of mathematics or science after school. It is true that in connection with structural mathematics and the so-called "new math" axiomatics as a subject of mathematics teaching was strongly overestimated, especially in algebra. On the other side, the axiomatic way of defining the probability concept could affect so many general teaching aims that we should not miss this chance if possible. The probability concept is a prominent item to show students how concepts which have a more or less undefined and vague meaning in everyday language could be handled in mathematics; how such concepts could be "exactified" by proceeding stepwise to "higher" levels of precision; what we could gain by this process and variable (models, computability); what prize we have to pay for that process (e.g., non-measurable sets--the existence of events which cannot be assigned any probability in models with an infinite set of possible outcomes seemingly in contradiction to our "everyday concept" of probability--there are easy examples to show that); and what we may mean by rigour, exactness, and generality in mathematics. There are also many other topics which could realize similar teaching aims, for example: stepwise exactification of "area" or "continuity" by both of which we could show how intuitive "everyday concepts" could be made precise and which role axiomatics can play (REI,2). But surely, both concepts are less suited for non-college-aiming students, and, in my opinion, the same remarks apply to the concept of "number" and all algebraic parts of the syllabus where "new math" had suggested teaching axiomatic aspects. (See also FR , FI .)

Pocket calculators and computer activities

The question whether and from what age on pocket calculators should be used is almost "historical." No question, they are a reality for every pupil, and--becoming cheaper and cheaper--they will become a matter of course like, say, typewriters and television sets. Calculators are used not only as a means for computing; they will--and should--also influence methods of teaching theoretical subjects; e.g., the concept of limit integral (starting with numerical approximations, for example), special functions, and others. The use of calculators enriches mathematics courses not only by new aspects of teaching but also by fun. I do not know anybody who would not like to work or play with calculators. We should make use of this chance at every level of secondary school, although we must admit that the whole problem has not yet been solved completely by educators (compare the many papers dedicated to this topic at ICME IV). Calculators, as a tool should not become a central subject. Here and there we already find teaching frames and exercises specially invented for training the sophisticated handling of calculators and unrealistic pseudo-applications. I partly fear the danger of wrong developments as exemplified formerly by parts of calculus and - - even more - - in algebra, where finally lots of boring procedures were taught: transforming and resolving special types of equations and practicing unimportant routines. A similar warning applies to computer activities, which are one of the most exciting topics of future mathematics curricula. The group of students in question should learn to use computers as a tool in solving mathematically formulated problems. Therefore, computer activities should begin very early. If the teacher had emphasized algorithmic thinking from the very beginning, school work with computers could begin when students are 13 or 14 years old. In Austria several tuition models are being tested at present (MU), (FK), and others. In my opinion, the question as to what programming language should be taught - - if terminals are used - - and generally what hardware should be used is not of primary importance. There are equally good results with BASIC and even COBOL and FORTRAN. In some countries there is a trend towards special computer languages for schools, like ELAN in Germany, ECOL in the Netherlands, PASCAL in Switzerland and (partly) in Austria, and some others, too. Such trends need not, but might, hide dangers of the kind indicated above. Clearly, we should not join mathematics curricula too strongly with the present state of hardware and software. Computer activities in mathematics classes rather should enable students to quickly adapt themselves to new situations. Therefore, instead of teaching one language up to a comparatively high level and proceeding stepwise to more and more complicated problems, it might be more useful to solve rather easy problems in different settings and languages. With the help of the teacher, students should learn to use computer manuals and constructions self-reliantly. In Vienna, we had good success, for example, in dividing a class into two or three groups which had to solve the same problems by using different hardware and languages and afterwards had to discuss their procedures and to discover differences (MU). I think that curricula should observe that - - especially in smaller countries - - only a few computer specialists are needed, in contrast to professionals of any kind who - - additionally - - can employ computers and who can understand and get the full value from the help provided by computer specialists. This should be observed by curricula. "Being able to assess and evaluate computer material for given problems" could be the pretentious motto of education and, in my opinion, mathematics courses could be a "natural" place to learn it. In this respect, it may be more essential to become acquainted with a few but various kinds of desk-top calculators and peripheral devices. In the 1980s almost every secondary school in the USA, Europe, and Japan will have access to computer or to desk-top calculators, but up to now, appropriate curriculum material has had to be developed by schools and teachers themselves, and many of them had to "reinvent the wheel." We need curriculum material which is independent of hardware but which could be adapted easily by the teacher. National curriculum activities in this direction should be established as soon as possible.

It should be noted that some countries - - for example, the Federal Republic of Germany - - strongly tend to establish computer science (informatics) as a new school subject. Others - - e.g., France (Mercouroff plan) - - have the opposite tendency: they tend to incorporate computer activities into the usual school subjects. A good reference about the present state is (EN_2).

The problems we are confronted with could be subsumed by the headline: Computer literacy versus skill. If I am asked whether we should tailor the contents of mathematics teaching specifically to the computer, I tend to answer "use computers as objects to think with" (as S. Papert once literally said). That does not mean that computers should not have any effects on teaching mathematics. One consequence, for example, is to emphasize the algorithmic mode of thinking at every level. (Although I would not follow A. Engel completely when he damands in (EN_2) to pervade the whole mathematics curriculum by algorithms, to replace - - so to speak - - "functional thinking" by "algorithmic thinking." Compare our point 3 above.

Related to calculators and algorithms is another subject which should play a stronger role in future curricula:

Numerical methods and analysis

Most of those equation (integrals, and so on) which arise in realistic problems cannot be resolved by using special formulae. So, either the problem has to be posed artificially, or we have to use approximation methods. Now the latter is possible by using calculators and computers in school. Teaching such methods could replace boring procedures for resolving special types of higher polynomial, exponential, goniometric, and other transcendental equations (DI). A new field could be numerical methods with systems of linear equations as used for example in Computer Topography, one of the most surprising, new, and powerful medical applications of theoretically rather simple mathematics (as far as it concerns CT of the "first generation"; (SSW), (HE). To obtain pictures of cross-sections of, say, the brain, a system of about ten thousand linear equations has to be solved. The so-called "algebraic reconstruction technique" (ART) used hereby could easily and truly be explained with systems of two or three linear equations.

This remark brings me to a topic which generally is acknowledged to be the most comprehensive topic, especially important for the group of students in question:

Interactions between mathematics and other school subjects; problem solving; applications of mathematics

Concerning the latter, I do not have to say much, since we have many recent contributions from which I would like to indicate the article by H. O. Pollak (PO), published by UNESCO in 1979, and the literature list. So far, we have a lot of material for teaching applications in the recent literature all over the world (BE), but I would like to stimulate discussion here about efficient methods of teaching. So - - amongst other things - - we have to discuss progress probably effected by reforming not only the content but also the organization of tuition. One suggestion could be to reorganize the last year of secondary school and hence also the final exams. I could imagine that the syllabus of the last year would not contain many (or even any) new concepts, but rather would contain new relations, connections, and combinations between subjects taught systematically in earlier years, which are now reviewed along new (and mainly problem-oriented) lines of sight. In our suggestion, the mathematics syllabus of the last year would then be split up into four or five compact minicourses, each of which would be devoted to a special (and mainly problem-oriented) theme which - - as I said before - - should not contain new concepts but combine by a new aspect various concepts taught before. I shall give examples below. Let me justify my suggestion first: Usually the order of topics in a syllabus is established and justified by their place in the scientific system of mathematics. Up to now, applications are included occasionally when they can emphasize the importance of the topic in question. By our suggestion, those applications will now appear gathered by a problem-oriented aspect. (Before giving examples, let me say that this tuition problem could be combined effectively with a rather formal reorganization of tuition in the last year of secondary school described as follows: In Austria - - as well as in many other countries - - students have to attend mathematics courses in every year of secondary school and can not choose special topics. During their last year they will have 110 to 150 one-hour-lectures in mathematics. Now, we should discuss whether it could be effective to divide up those lectures into, say, four or five intensive periods of four weeks each. We would then have periods with six hours of mathematics a week and others which are free of mathematics. Each of these "math periods" or, better, "minicourses" would then be devoted to a fixed theme as indicated below. Clearly, the other school subjects must then be organized similarly.

Let me now suggest some examples suited for students who already have had seven or eight years of mathematics in secondary schools. Similar themes will have to be found for other types of schools.

By the way, these themes would also be suitable for teacher training courses or college lectures when educating teachers. At the University of Vienna we have some experience with courses labelled "school mathematics" which propsective teachers have to pass. here, I have already tested some of the following themes:

1. Optimization techniques (from calculus, linear algebra, and probability theory; further techniques)
2. Decision techniques (probability, calculus, mathematics of finance - - computing interests and annuitities)
3. Axiomatic methods (algebra, numbers, vectors, geometry, probability, area)
4. Proving methods (various paradigms of proofs, mathematical arguments, differing examples of complete induction)
5. Defining mathematical concepts and discovering various levels of exactness and rigour (limits, probability, continuity, geometry, natural numbers; from intuitive everday language to axiomatics)
6. Various implications of geometry
7. Numerical and approximation methods (algebra, polynomials, integrals, series)
8. Various paradigms of algorithms
9. Various methods and applications in economy, biology, and elsewhere (linear algebra, probability, geometry, statistics)
10. Various interpretations and applications of derivatives (here one must observe that curricula of secondary schools in Europe generally contain calculus at least during the last two years - - the students are 17 or 18 years old)
11. Various situations leading to $y' = cy$, $y' = c/y$, $y' = y$ (simple difference and differential equations)
12. Various interpretations and applications of integral (e.g., as a mean , area, expectation, "infinite sum" - - as used in physics and so on)
13. Various applications and interpretations of vectors (geometry, $\mathbb{R}^n$, systems of linear equations, polynomials, physics, economics)
14. Comparing different desk-top calculators and deciding which one fits for which kind of mathematics problem
15. Interpreting and drawing graphs of functions and problems which can be solved graphically (equalities, inequalities)
16. Problems which lead to matrices, use of matirces ((FL), (RA))
17. Important kinds of mathematical models (functions, equations, geometric models, probabilistic models)
18. Applications of the concept of limit (decimal numbers, series, derivatives, continuity, integral, numerical methods)

19. Various aspects of the concept of infinity

20. Historical aspects of various mathematical problems

21. Computing geometrical quantities

22. Describing discrete processes by continous models (functions) and converse (applications of calculus, distribution functions, models of growth (e.g., "exponential growth"), population statistics, simple difference and differential equations)

I am sure we can find many more themes, which could be offered alternately and for choice. Always the idea is to gather various methods which could contribute to the same kind of problem or theoretical aspect (e.g., themes 3 to 5) and, conversely, to gather various types of problems which could be handled by the same mathematical concept (e.g., theme 22). Teaching applications means teaching three ways of thinking on which future curricula should focus:

1. to find applications of a given concept (method)

2. to discover concepts (methods) applicable to a given problem

3. to identify problems and to find correct, well-posed, and clear formulations of these problems

Compact minicourses as described above may contribute to this concern, especially for those students who will not aim at the further study of science or mathematics.

A last remark: Reorganizaing, say, the last year of secondary school could be effective only if combined with reorganizing the final exams, respectively. When constructing new curricula, we must not forget the role exams play for pupils and students. To be realistic, a main motive and purpose of learning still is the wish to pass the exams successfully. Therefore, any change of curricula should include the problem of exams; even more it should start with it. This view of the problem still needs research, although we already have had some important papers and books recently (e.g., (BA), (HO)). But this, of course, should be the theme of another paper.

References

(BA) L. Bauer: Mathematische Fahigkeiten; UTB No. 835, Schoneingh, Paderborn 1978.

(BE) M.S. Bell (ed.): A preliminary survey of materials available worldwide for the teaching of applications at the school level; paper prepared for ICME IV, Univ. Chicago, 1980.

(DI) C. Dixon: Numerical Analysis (written for sixth-year school pupils); Black/Chambers, Glasgow/Edinburgh, 1977.

(DP) W. Dorfler und W. Peschek: Die Situation des Mathematikunterrichtes an den Hoheren Schulen; Festband "10 Jahre Univ. F. Bildungswiss. Klagenfurt 1980."

(EN_1) A. Engel: The relevance of modern fields of applied mathematics for mathematics education; Ed. Stud. in Math. 2 (1969), 257-269.

(EN_2) A. Engel: The role of algorithms and computers in teaching mathematics at school; New Trends in Math. Teaching, vol. 4, UNESCO 1979, 249-277.

(FI) R. Fischer: Erste Einfuhrung in die Wahrscheinlichkeitsrechnung in der Schule; Austrian Math. Society, Didaktik-Reihe, vol. 4, 1980.

(FK) E. Fian, H. Keilbauer et al: EDV fur allgemeinbildende Hohere Schulen (Electronic Data Processing for Secondary Schools), Verlag Manz, Vienna 1979.

(FL) T.J. Fletcher: Linear Algebra through its applications; Van Nostrand und Reinhold, London 1973.

(FR) H. Freudenthal: Der Wahrscheinlichkeitsbegriff als Angewandte Mathematik; Echternach Symp. 1973, 15-27.

(GKL) K. Graf, K. Keil, H. Lothe, B. Winkelmann: Computer use in mathematics education; IDM-paper for ICME IV, August 1980.

(HE) J. Hejtmanek: Computer Tomographie; Wiss. aktuell 2 (1980), 62-65.

(HO) A.G. Howson: A critical analysis of curriculum development in mathematical education; New Trends in Math. Teaching, vol. 4, UNESCO 1979.

(MH) "Mathematik heute", ed. by H. Athen, H. Griesel et al., Schroedel-Schoningh, Hannover, Fed. Rep. Germ., 1979.

(NCA) An Agenda for Action, Recommendations for School Mathematics of the 1980s, Nat. Council of Teachers of Math., Reston, VA 1980.

(PI) J. Pieper: Was heisst akademisch? Zwei Vershcuhe uber die Chancen der Universitat heute; Kosel-Verlag, Munchen (Munich), 2. erw. Auflage, 1964.

(PO) H.O. Pollak: The interaction between mathematics and other school subjects; New Trends in Math. Teaching, vol. 4, UNESCO 1979, 232-248.

(RA) C. Rorres and H. Anton: Applications of Linear Algebra, 2nd ed., J. Wiley, New York 1979.

(RE_1) H.-Ch. Reichel: Zur Didaktik der Intergralrechnung fur Hohere Schulen; Didaktik der Math. 3 (1974), 167-188.

(RE_2) H.-Ch. Reichel: Zum Skalarprodukt im Unterricht an der Sekundarstufe, eine didaktische Analyse; Didaktik der Math. 8 (1980), 102-132.

(SBE) "Statistics by Examples," vol. 1 - 8, ed. by F. Mosteller, W. Kruskal, R. Link, R. Pieters, G. Rising, M. Zelinka, S. Weisberg, Addison Wesley 1973.

(SSW) K. T. SMith, D.S. Solmon, S.L. Wagner: Practical and mathematical aspects of the problem of reconsructing objects from radiographs, Bull. AMS 83 (1977).

(SMP) The School Math. Project (Further Math. Series 5: Statistics and Probability) Cambridge 1971.

11.4 THE CHANGING CURRICULUM - - AN INTERNATIONAL PERSPECTIVE

THE CHANGING CURRICULUM - - AN INTERNATIONAL PERSPECTIVE: CASE STUDIES

E.E. Oldham
Trinity College
Dublin, Ireland

Introduction

One feature of the Osnabruck conference (Comparative Studies of Mathematics Curricula - - Change and Stability 1960-1980: A conference jointly organised by the Institute for the Didactics of Mathematics and by the International Mathematics Committe of the Second International Mathematics Study of the International Association for the Evaluation of Educational Achievement. The conference was held at Haus Ohrbeck, near Osnabruck, Germany, in January, 1980) was the attention given to case studies which aimed to look more closely at the processes and products of change in three specific areas of the curriculum. (These case studies, together with some papers offering general discussion, are available in the Proceedings of the Osnaburck conference (Steiner, 1980).) These areas were geometry, algebra, and statistics, and they were chosen for various reasons:

1. Geometry, because of the great contrasts between curricula in different countries, and indeed the controversy surrounding the topic since the Royamount seminar
2. Algebra, because it has been the archtypal "modern" area in the Bourbakiste sense, and also one where much development started a school level
3. Statistics, because it is modern in a different sens - - a growing area in school curricula, and one that is likely to develop further.

In this paper, three issues will be addressed. The way in which a case study actually operated at Osnaburck will be considered; some impression of the results will be given; and there will be a discussion of the power and limitations of the case study method, as it appeared at Osnabruck and as it may be developed in the future. These issues are presented in sections 1 - 3, respectively.

Mechanism of the Case Studies

The studies were organised as follows. For each of the three content areas, a few countries were selected: countries, essentially, that provided contrasting models of development, or had contrasting curricula. (It must be emphasised that the selection was based on diversity, and not on any statistically balanced representation of the participating countries. This naturally affects the way in which the results can be interpreted. The algebra case study can serve as an example. Here, the United States was chosen because it had made the transition to modern algebra. Ireland was picked for two reasons: because the modernisation of the algebra curriculum had gone to considerable lengths, and also because there is centralised control of syllabuses, and so the change in the (intended) curriculum was uniform throughout the country. (The customary distinction is made between the intended curriculum (that set forth in syllabuses, official programmes, and so forth) and the implemented curriculum (that actually taught in the classroom. See Travers 1980.) Germany and Japan were selected to complement these in various ways. These four countries were the scheduled participants; however, other countries which had had interesting experiences were invited to contribute also, and the study was enriched by a paper from England and Wales. (England and Wales share an education system, and thus the one paper applied to both countries.)

The guideline that was given to authors was to look at the intended curriculum and its implementation, and hence consider questions such as the following: What forces produced change? Where was the decision-making power? How were ideas disseminated? And how were they received? These questions had differing relevance in the different cases, and the speakers for each study met before their presentation to note the areas of particular interest for their countries. These aspects were then emphasised in the actual presentations, and in some cases have been further developed in the papers that appear in the Osnabruck Proceedings.

Results of the Studies

It is not easy to summarise the presentations, or to pick out trends in the three studies. As was pointed out above, the countries were chosen essentially for their diversity, and so any synopsis can deal only with the broadest of issues. It is tempting, and generally true, to say that: (1) geometry in many countries has been transformed; (2) there are inequalities in the attitudes to algebra; and (3) it is now a la mode to teach statistics. However, something a little more serious is needed. In the space available, only one case study - - that of algebra - - can be considered in any depth; just the briefest of comments will be made on the other two.

Algebra

The algebra study is the one which lends itself to the most interesting attempt at synthesis. The discussion from the United States focussed on the changes in really elementary algebra, such as the way in which variables were introduced. The actual change in content was quite small, but it generated an apparently disproportionate amount of fear among the teachers. However, some degree of acceptance has now emerged. This seems to be partly because to young teachers the work is not "new," but is something that they themselves learnt at school, and perhaps partly because of a natural tendency to accept the status quo (Hirstein et al., 1980).

In Ireland, by contrast, the equivalent changes seem to have been seen only as fairly trivial alterations in terminology--changes which could, so to speak, be taken or left alone--and teachers' attention focussed on other aspects of the new courses. Insofar as algebra provoked comment at all, it was the study of algebraic structures, such as groups, which caused debate (Oldham, 1980). The algebraic structures seem to have produced insecurity also in England and Wales. Teachers were reported as not knowing "what they were for": where they led, and what role they were meant to play in the whole course (Fletcher, 1980).

It is natural to ask if there are reasons for these different reactions on the two sides of the Atlantic. Perhaps it somehow reflects variations in the structure of mathematics courses and the level at which algebra tends to be introduced. (In Britain and Ireland it is taught, alongside other mathematical topics, to all grades from the seventh or eighth upwards). Perhaps there is some more profound reason. For example, did the American teachers realise, and become overawed by the fact, that the apparently small changes represented a considerable change in philosophy? Were the English, Welsh and Irish teachers less aware of the background to the "new mathematics" movements, and thus were they simultaneously less worried by the new language and more concerned at the apparent lack of raison d'etre for the changes? The case studies have opened up the question, but we must look further for definitive answers. Some clues as to how this might be done are suggested in S3.

The acceptance of new approaches, mentioned in connection with the United States, has not always been apparent. Here, the Irish and German experiences are similar: the pendulum swung considerably, and now has to some extent swung back. In Ireland, there has been a retrenchment in the algebraic content of the courses (Oldham, 1980, pp. 408-409); in Germany, a very theoretical approach to elementary equations was attempted, but now appears to be losing favour (Volbrath, 1980). It will be interesting to see what happens in the next few years.

Geometry and Statistic

A few brief comments can be made on the other two studies. As regards geometry, it became clear (and has been equally clear at this conference) that, in the years since the Royaumont seminar (OECD, 1961), no common approach has emerged. The redevelopment of geometry courses—which in some countries, such as England and Wales (Howson, 1980) and especially Germany, began considerably before the Royaumont debate-is still continuing. If there is such a thing as definitive shape for the courses of the future, it has yet to be found.

In statistics, or stochastics-the term used in the Hungarian paper to embrace both statistics and probability (Nemetz, 1980)--the situation appears rather more uniformly cheerful. For one thing, all the papers reported substantial development in the statistics or stochastics syllabuses in their countries. (It must be borne in mind, however, that countries chosen for the study were ones which had such developments to report, so too much significance should not be attributed to this trend.) Also, general discussion at Osnabruck suggested that elementary, descriptive statistics can be one of the easier subjects to teach, probably because of its relatively concrete and practical nature. In this context, it is worth noting that statistics is a topic which has come to school mathematics courses "from the outside:" from the need for statistical literacy in other subjects and in the world at large, rather than from purely mathematical sources. Possibly connected with this, however, is a tendency to place statistics outside mathematics, either as part of courses in other subjects, or as a subject in its own right (Barnett, 1980; Ohuche, 1980). Here, again, it will be interesting to monitor this development over the new few years.

S3 Power and Limitations of the Case Study Method

It remains to consider the power and limitation the case study method as it was used at Osnabruck: to examine what has been achieved by the studies, and the extent to which they can be developed further as tools of curriculum analysis.

On the credit side, the studies have porvided valuable insights into the evolution of algebra, geometry, and statistics courses in the various countries, and they have raised what may be even more valuable questions. On the debit side, they may be somewhat too diverse in style to allow really useful analyses or syntheses to be made. It is here that there is room for developemnts. Suggestions have already emerged as to how this could be accomplished, and two of them will now be discussed.

Westbury (1980) points out that, as authors were specifically asked to examine stability and change in curricula (intended and implemented), the discussion was to some extent removed from its social context; and, since the last two decades have been a time of growing participation and changing school structure in many countries, perhaps any study that ignores these aspects is essentially incomplete. In fact, some of the case study authors did address these topics; but in future studies, perhaps an examination of the social context could be built in to the specifications in a more uniform way.

Another very interesting suggestion is made by Howson. (1980, pp. 502-508). He points out that, since the various authors wrote of their own countries, they may have been unaware of some of the aspects of their curricula or education systems that are remarkable from an international point of view: they may, infact, have been prisoners of their own culture, bound by that culture's hidden axioms. It is possible also that some of the distinctions in emphasis that the case studies seemed to reveal were a reflection of the varying interests of the writers, and not of genuine differences in the processes and products of change in the countries concerned. Thus, teacher educators might tend to focus on the support given to teachers during times of change, mathematicians might emphasise syllabus content, and so forth. Perhaps a collaborative structure, both within and across countires, countries, could be used in preparing future case studies, to bring some kind of unity and objectively into the perspectives without losing the benefits of local knowledge.

In these ways, the case study method, as envisaged at Osnabruck, might be developed into an increasingly useful tool for curriculum analysis. Along with other methods of examining and understanding curriculum change, it may perhaps help us to provide better mathematics education in the future.

References

Organisation for Economic Cooperation and Development. New Thinking in School Mathematics. Paris: OECD, 1961.

Steiner, H.G. (Ed.). Comparative Studies of Mathematics Curricula--Change and Stability 1960-1980 (Materialien and Studien, Band 19).

Bielefeld: Institut fur Didaktik der Mathematic der Unversitat Bielefeld, 1980.

Barnett, V. "Teaching Statistics in Schools in England and Wales," pp.444-463.

Fletcher, T.J. "Algebra in English Secondary Schools: Change over Twenty Years," 354-369.

Hirstein, J.J., Weinzweig, A.I., Fey, J.T., & Travers, K.J. "Elementary Algebra in the United States: 1955-1980," pp.370-394, especially pp. 390-391.

Howson, A.G. "Geometry in Great Britain in Recent Years," pp.304-325.

Howson, A.G. "Some Remarks on Case Studies," pp. 502-508.

Nemetz, T. "Pre-University Stochastical Education in Hungary," p.478.

Ohuche, R.O. "Developments in the Teaching of Statistics and Probability since 1960: The Experience of Nigeria," pp.486-501.

Oldham, E.E. "Case Studies in Algebra Education: Ireland," pp.395-425.

Sawada, T. "Elementary Algebra in Lower Secondary School in Japan," pp.432-433.

Travers, K.J. "The Second International Mathematics Study: An Overview," pp.188-189.

Vollrath, H.J. "A Case Study in the Development of Algebra Teaching in the FRG," pp.435-443, especially pp.440-441.

Westbury, I. "Case Studies in curriculum Change: Introduction," pp.270-276.

11.5 MODELS OF CURRICULUM DEVELOPMENT

AMI'S REFORMATION OF MATHEMATICAL EDUCATION

Tashio Miyamoto and Ko Gimbayashi
Tokyo, Japan

We would like to report on the results obtained by the Association of Mathematical Instruction (AMI) of Japan, which is organized by teachers, researchers, students and parents who are interested in mathematical education.

I. Character of AMI

In Japan, almost all aspects of education are fully controlled by government, so in order to continue the study and practice of education independently, we cannot but organize civil education associations. AMI is one of them, and has been active and obtained good results for the last 30 years. AMI now has about 2,000 members and issues a monthly which has a circulation of about 10,000 copies.

II. The problem of Curriculum, and Modernization- - It's Necessity and Its Faults - -

The demand for the modernization has arisen from wide penetration of mathematical methods to various sciences and many fields in our society. So it is natural that movements for the reformation of mathematical education have arisen all over the world.

But too rapid modernization has not succeeded and, produced many pupils who dislike mathematics and so many dropouts.

As the causes of the failure have been already discussed, we now wish to mention only following one point:

Almost all modernization was movement from the top down in three-fold sense. Namely, educational reforms were introduced from higher education down to secondary and elementary ones, and naturally concerned to introduce formal and abstract materials of modern mathematics regardless of interests of children. Furthermore they were carried out with the top pupils in mind and little effort was made to raise the ability of the pupils as a whole.

III. The Modernization of AMI

If the analysis just given is assumed to be correct, then the standpoint for a reorientation of the modernization program is also self-evident and that of AMI is just so.

1) the first objective should be to thoroughly analyse and reorganize the material that has been used up to now. Secondly, rather than trying to teach everything at once and hasten the process as much as possible, a plan should be devised whereby the most important material is chosen and sufficient amount of time is spent to have its meaning understood and to have pupils enjoy mastering mathematics.

2) As for reorgnization of material, the idea and method of modern mathematics should be applied to construct the curriculum. In other words, mathematical education itself ought to be considered as one of mathematical sciences.

3) Instead of concentrating only on the top pupils, we should strive to teach high-quality mathematics to all children. Furthermore, material should be taught in accordance to children's mentality, and in the way that pleases them and stimulates their curiosities.

IV. Theory of Quantity

It is natural that in the realm of mathematics proper there should be many different ways of defining or constructing the number system and the rules of operations. But in elementary instruction of mathematics no one will deny that the concept of number should be derived from its connection with the real world. Regarding the method of its derivation, however, there are several different viewpoints possible.

One is so-called "the principle of counting" or "Kroneckerism", which starts from counting and derives other operations from them rather formally according to following diagram:

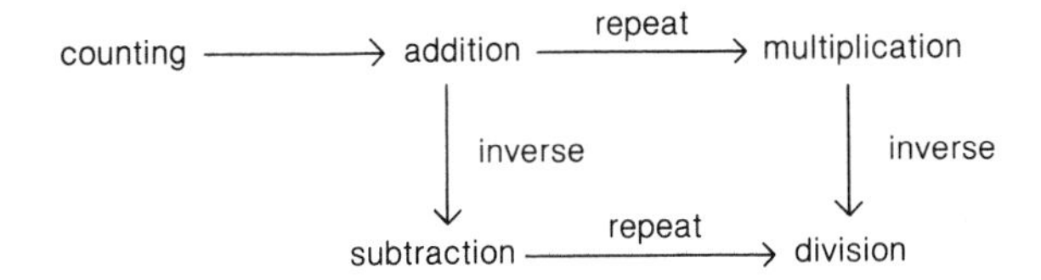

Such a formal method of teaching will not help pupils to grasp the meaning of advanced number systems and operations.

A second, opposite viewpoint is that of the activity curriculum (or core curriculum) which sticks to the real world and unsystematically borrows a variety of quantities as teaching materials. But such a pragmatic method of teaching was not suited to make pupils understand a systematic subject like mathematics.

Needless to say, we attach much important to the connection with the real world, but not exclusively. Among a great many kinds of practical quantities in the world, we must pick them up systematically and give them to pupils in suitable order. So we much have some principle that allows to classify those existent quantities.

Firstly, we can distinguish two kinds of quantities, discrete and continuous:

quantity { discrete quantity------whole number
continuous quantity-----real number }

Since a discrete quantity has a definite natural unit, we can easily derive natural (or whole) numbers from it.

But there is no proper unit for a continuous quantity divisible infinitely, so it is necessary to prepare some unit artificially. We think it more important for pupils to understand this process of preparation than give some ready-made tools of measurement. After a unit is settled, real number is derived by measurement from a continuous quantity. We may well consider decimals and fractions to be the approximate representations of real numbers at the stage of elementary instruction.

Further, continuous quantities should be classified into extensive one representing quantitative magnitude or extent and intensive one corresponding to qualitative intensity:

continuous quantity { extensive quantity
intensive quantity }

While extensive quantities are added together when they are unified, intensive ones are rather averaged when unified. Also while we can represent extensive quantities by numbers by defining unit, intensive quantities are represented by numbers only by division between two kinds of extensive quantities as in the following formulae:

distance (m) / time (s) = velocity (m/s),
weight (kg) / volume (m^3) = density (kg/m^3)

Thus we have such correspondent relation as:

extensive quantity-----addition and subtraction
intensive quantity-----multiplication and division

which suggests that it is desirable to give instruction in addition and subtraction prior to multiplication and division.

V. Suido-Method

In order to put into practice the above described mathematical curriculum in all its aspects, we had to organize the system of arithmetical calculations and devised an efficient method to instruct it.

First, we had to focus our attention on the calculation by writing, for mental calculation is restricted to the extent of smaller numbers.

The calculation by writing is performed on the basis of the principle of location which is the most remarkable characteristic of Indo-Arabic notation. To teach this principle successfully, we use small squares called "tiles" to represent numbers. Namely, ten such tiles tied vertically make one bar representing the number 10, and ten such bars tied horizontally make one larger square sheet representing the number 100. Pupils can easily cut the tiles out of cardboard by themselves and, by manipulating them, can fix the images of number in their brain.

By using such tiles we can easily teach pupils the principle of all arithmetic operations, namely, that all calculations consist of the smallest ones between 1-place numbers which are called "elementary processes". So, if pupils become skilled in elementary processes, they can proceed to the calculations between multi-place numbers which are called "compound processes".

For example, we explain these methods with respect to addition of 2-place numbers.

First, elementary processes are classified according to:

1. whether carrying up takes place or not, and
2. whether they contain 0 or not.

$$\begin{array}{r}2\\+2\end{array} \nearrow \begin{array}{r}2\\+0\end{array} \searrow \begin{array}{r}0\\+0\end{array}, \quad \begin{array}{r}9\\+9\end{array} \longrightarrow \begin{array}{r}9\\+1\end{array}$$
$$\begin{array}{r}2\\+2\end{array} \searrow \begin{array}{r}0\\+2\end{array} \nearrow \begin{array}{r}0\\+0\end{array}$$

(by symbolic notation of "2-9 classification")

Arrows mean the direction of specialization, and we think that more special type is more difficult for children. At first sight, this point of view might be seen to be contrary to common sense, but our accumulated experiences proved it legitimate.

Addition of 2-place numbers can be classified and organized in similar way:

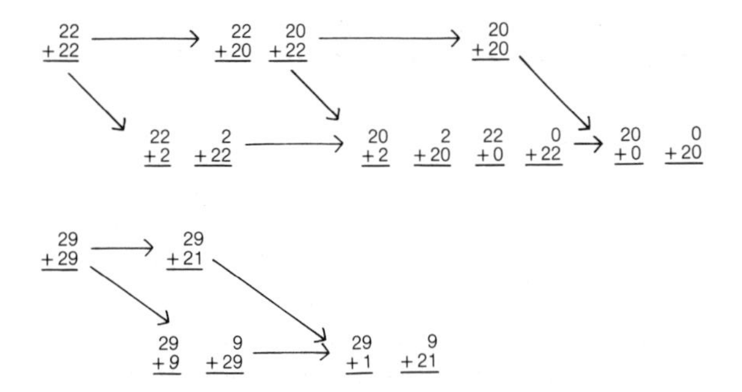

As the shape of this diagram resembles that of the network of city water supply, it was nick-named "Suido Method" (Water Supply Method).

We have roughly introduced here the main results gained by AMI in the field of elementary instruction of mathematics. Our Theory of Quantity is an effort to establish systematic connection between mathematics and the outer real world (semantics). On the other hand Our "Suido Method" concerns to organize procedures that children take, namely algorithm (syntactics).

In addition, we have been successful in producing enjoyable lessons by introducing some games and suitable teaching tools. Our "Enjoyable lessons" is a trial to stress on the subjective side of mathematical learning (pragmatics).

We should like you to see the bibliography for further details.

CURRICULUM DEVELOPMENT-A LARGE SCALE PERSPECTIVE

James M. Moser
University of Wisconsin
Madison, Wisconsin USA

0. This presentation deals with a discussion of a large scale curriculum development project that took place in the United States during the period 1967 to 1976. The end product of this effort is Developing Mathematical Processes (DMP), a complete program of instruction in mathematics for Kindergarten through grade six (ages 5-12). Over $3 million was spent in developing DMP mostly provided by agencies of the US Federal Government. The work took place at the Unviersity of Wisconsin Research and Development Center.

The following sections examine the major components of this curriculum development. First, we shall consider the underlying theoretical notions that that preceded the actual development of materials. Second, a brief description of the actual production procedures will be given. In the third section, the evaluation steps that were undertaken during the period of development and implementation will be considered. In the final section, a historical perspective, taken four years after final development wil be presented.

1. At the outset, the decision was made to carry out the curriculum development effort using a model of continuous but evolutionary progress. DMP was developed on a year-round basis by a staff of full time specialists whose only task was to develop and evaluate instructional materials. Most of the writing was done by persons who were recruited from elementary classrooms. The direction and conceptualization of the program was carried out by professional mathematics educators and mathematicians with advisory input received from other mathematicians, educators, and psychologists. It was a small core of three or four person who developed the philisolophical and theoretical basis upon which DMP was built. There were three areas of consideration: mathematical, psychological and pedagogical.

1.1 The desire was to create a program of well rounded mathematics, including not only arithmetic, but geometry, measurement, elements of probability and statistics, and some simple logic as well. The overriding theme was to approach the subject through problem solving. Our aim was to get children to see mathematics as a useful subject, relevant to their needs. The day process in the mathematization of real world problems was representation. Where number was concerned, we used the process of measurement to assign a number to sets, objects, and events. Not only was the discrete attributes of length, weight, area, volume and capacity, duration, temperature, and likelihood of occurrence.

1.2 Our plans for curriculum development were influenced by the writings of persons such as Brownell, Piaget, Lovell, and Bruner. Mathematical concepts should be rooted in concrete experiences. The plan, then, was to produce materials that would enable children to be active learners and to learn through manipulation of a variety of concrete materials. This notions was to be carried through all grades, with materials increasing in level of sophistication and use as the children became older and more mathematically experienced. Another psychological consideration in the planning of DMP was the provision for motivation. This was effected by posing real, interesting problems, by having a variety of materials and instructional settings, and by paying attention to needs of individuals.

1.3 Pedagogic consideration considerations centered around concerns that materials and activities be planned with both the teacher and the child in mind. Classroom management, especially with a variety of instructional settings and classroom groupings, was factor in the design of materials. For the child, there was need for a management system to keep track of progress towards instructional objectives.

2.1 The initial phase of production of materials involved a two-part analysis of each major curriculum unit. The first was a task analysis of the specific mathematical behaviors to be learned, using techniques suggested by Gagne and Resnick. Hierarchies of behaviors, dependency relationships, and the establishment of prerequisite behavioral objective were the major components of this analysis. The second type of analysis was instructional in which attention was given to concerns such as types of physical materials, patterns of instructional grouping (e.g., large groups, experiments, games, individual seat work, learning stations), sequencing of attributes to be emphasized (e.g., introduce length before area), and interweaving of strands of mathematical content.

2.2 The second and most time-consuming phase of production was the actual writing of materials. A staff of full-time writers was used. Initial writing of a unit of instruction was carried out by a team of two to four persons. Guidelines established by the earlier analyses were followed. Written units were then tried out in local classrooms by teachers who were employees of the school district and not of the DMP staff. Writers who prepared the materials also served as observers of the classroom try-out, being in the school two or three days a week. Results of the observations led to subsequent revision and additional try-out as required. As each new instructional unit was to be written the composition of the writing teams was changed so that all writers had a chance to work with all other writers at some point in time.

The final cycle of this production phase was carried out in cooperation with a commercial publisher, Rand McNally & Co. Editors, artists, materials developers and production staff cooperated with DMP staff personnel in the creation of the final commercial materials which were made available for purchase to American schools in 1974, 1975, and 1976.

3.1 Evaluation of the curriculum materials were made at various stages during and after the developmental process. There were two categories of evaluations, prior to and subsequent to commercial implementation. Pre-commercial evaluation was made of the design of DMP. An advisory committee examined the underlying positions taken and the various initial analyses made. A formative evaluation was carried out as the materials were being written. In addition to the classroom observations, small and large scale field tests in schools geographically removed from the developers were also held.

3.2 Post-commercial evaluation took two forms. The first type were evaluations carried out by Wisconsin Research and Development Center evaluators. Systematic selection of evaluation sites was made to conduct a series of tests, using standardized test batteries. Other internal evaluations were user data studies where teachers using DMP were interviewed for reactions and suggestions for improvement of the program, content reviews carried out by invited panels of experts from the field of mathematics and mathematics education, and an assessment of the effects of DMP as part of a larger evaluation of IGE (Individually Guided Education) schools. External evaluations initiated by agencies other than the developers also took place. Some were carried out by individual school districts as part of their own on going evaluation efforts; others were reviews by experts of the philosophy, content, and pedagogy of the program.

4. One needs to know something about the American system of education in order to be able to interpret the dissemination and implementation aspect of the development of DMP. American education is highly diversified and decentralized. The Federal Government may support education financially in a variety of ways, as it did, for example, in the financing of development of DMP, but it does not control it. The different states have some regulations, particularly in things such as teacher certification, and setting of minimum standards of performance. But basically, control lies in the hands of local school district officials, especially when it comes to matters of choice of programs to use in classrooms. In fact, in many larger urban areas where teachers' unions are strong, decisions are made at the individual school, and perhaps even individual classroom level. Looking specifically at the choice of whether or not to adopt DMP as a program of instruction, and then how to implement it once adopted, it was strictly a local matter.

The commercial publishers made efforts to market the program. A model of teacher in-service was chosen. DMP staff members trained a large cadre of intermediate "change-agents" and consultants who were to be knowledgeable about DMP. These were university and college mathematics educators, school district subject-matter specialists, and key teachers within schools. These persons, in turn, carried out in-service work at the local school level. Through the help of the publisher, a number of DMP Dissemination and Implementation Centers were established on campuses around the country.

5. As a person who was most heavily involved in all aspects of conceptualization, production, evaluation, and dissemination of a major curriculum development effort, the writer welcomes the opportunity to look back on a completed task and to engage in second thoughts. The frenetic pace of the developmental years and the attention to the myriad of details made it figuratively impossible "to see the forest for the trees." Very briefly, what would be done differently if it were to be undertaken now? What things would stay the same?

First, our philosophic orientation towards genuine problem solving at all age levels and within all types of mathematical content was right on target. Given the current interest in problem solving in the U.S., we were at least ten years ahead of our time. However, our subsequent analyses at the mathematical and instructional levels were remarkably naive and inappropirate. Current research interests and evidence suggest that analysis of tasks from the adult or mathematical perspective is incorrect if one wants to produce appropriate instructional materials. Rather, the demands upon a child's memory and information processing capacity should be analyzed and tasks set out accordingly. Also the concern for great vareity and diversity of instructional materials, for individualization, and for careful sequencing of content seems less important now than it did then. Some current research from the University of Missouri seems to suggest that carefully structured, large group activities were simple, straightforward directions on how to accomplish certain mathematical tasks may be more appropriate. This is not to suggest dill, rigid teaching or instructional materials on limited mathematical topics. But the tremendous diversity currently embodied in DMP is probably not the answer either.

The mechanics of production was essentially sound. A critical mass of full-time, expert, professional curriculum developers drawn from the ranks of outstanding classroom teachers is important. Persons whose primary concern is the creation and validation of sound instructional materials are necessary for the development of quality curricula. By the same token, the staff should not be too large and working on too many different components at the same time. This can lead to fractionization and discontinuity of the final end-product.

Finally, I wish to address a most fundamental issue. The ultimate evaluation of a curriculum product is whether it is used by teachers, and if it is used, is it being used the way the developers intended it to be used?

Qualitatively, a product can have rave reviews by many experts and can, in fact, produce change for the better in those who are using the product. But there is the quantitative dimension that really counts. Does a curriculum product have a substantial impact upon the way mathematics is taught and learned by a large segment of society?

The developers of DMP wanted to improve American mathematical education at the elementary school level. To do this, a curriculum was developed which suggested some rather dramatic changes in content structure and sequencing, in classroom organizational patterns, and in the nature and type of student instructional materials. And here is the very crux of the matter. We, the developers of DMP, failed to take into account the sociological and psychological phenomenon of change itself. We actually were asking teachers and school district officials to change the way they thought about teaching and about the nature of mathematics, and then to change the way they behaved in a class. Nature's law of inertia was stronger than we were! We lacked the coercive power of moral or rational persuasion, or of a strong central governmental structure which can require change.

The lesson for future curriculum development seems clear. If you are not going to change anything, then why bother? If you plan fundamental changes in mathematical instruction, be aware of the human factor of the teacher. Careful analysis and well planned production efforts are necessary but not sufficient to achieve your fundamental efforts at reform.

11.6 MATHEMATICS FOR SECONDARY SCHOOL STUDENTS

MATHEMATICS FOR SECONDARY SCHOOL STUDENTS

Harold C. Trimble
Ohio State University
Columbus, Ohio, U.S.A.

Each of you has received or will receive a copy of Recommendations for School Mathematics of the 1980s. The subtitle An Agenda for Action is to be taken seriously. To recommend that "problem solving be the focus of school mathematics in the 1980s" is neither controversial nor productive unless one acts to interpret its meaning.

I've been thinking about four groups of secondary school students, namely, ones for whom:

A. high school is terminal.

B. post-secondary vocational education is anticipated.

C. college education is expected in the life sciences or the social sciences.

D. college education is sought in fields based upon the physical sciences or mathematics.

The questions to be addressed are:

A. What problem solving skills does each group need?

B. Do present curricula provide these skills?

C. What changes are desirable? feasible?

Think with me, first, about students for whom high school is terminal. In the United States, only a few of these students drop out before graduation. But many more graduate without achieving the mathematical skills one would like to expect in grade six. This fact has led us to "remedial" mathematics, to reteaching such topics as common and decimal fractions in the senior high school. National longitudinal studies have shown that, at least temporarily, such reteaching produces some improvement in computational skills; but that selecting which skills to apply in a problem situation remains a mystery to most of these boys and girls. I doubt the usefulness of skills from arithmetic, geometry or algebra that are studied apart from problem contexts.

In a demonstration lesson I have used a loop of rope of length 20. Any unit of measure is fine. We make rectangles and record lengths, widths and area such as:

Perimeter of rectangle 20

length	width	area
9	1	9
8	2	16
7	3	21
6	4	24
5	5	25

We try lengths like 7 1/2 or 6.9. In this context is it possible to "teach":

a. Among rectangles of perimeter 20 the square has greatest area.

b. In building a house you get maximal volume for a given price when the house is square.

c. The graph of the $\{$length, area$\}$ is a parabola.

d. The sense of the problem limits the domain of numbers usable as lengths. But fractions, and even numbers like $\sqrt{3}$ or π , are permissible choices.

And many other things.

More important, for our present purposes, is the use of numbers in contexts sensible to these students. The problem-solving-as-a-focus recommendation means tome that teachers should give much more instructional time to such contexts; that such geometric language as "perimeter" will be more apt to be assimilated; that such problem-solving skills as acting out the situation can be practiced; that such productive questions as "What about regular polygons other than the square? pentagons, hexagons, etc.?" can be asked; that sequences of questions leading to such generalizations as "The circle gives greatest area for a given perimeter" are experienced by the students.

In the United States hand-held calculators are widely available. The shift from teaching algorithms for paper and pencil calculation to teaching estimation and strategies for problem solving is now feasible. I believe that this is the essence of the problem-solving recommendation as it applies to students for whom high school education is terminal. They should be taught mathematical knowledge, skill and appreciation in problem contexts. Time should be provided in mathematics classes to explore; to try several approaches to a problem.

How do students who anticipate post-secondary vocational education differ from high-school terminal students? I think they have greater need for "learning to learn". They need higher level reading skills that include reading in mathematics. They must not give up on a reading task when symbolism strange to them is explained and used. For example, a summation sign frightens many of our present-day students. What sorts of experiences in mathematics will help these students?

Again I would begin with a "sensible" problem. For example, the teacher brings to class the report on yesterday's weather. "It says here we had 0.24 inches of precipitation. How would you measure that?" Here in the United States we still use inches. It is my experience that very few people can visualize 0.01 inch.

Some students may need to begin by setting a cylindrical container near a lawn sprinkler to try to measure the depth of the water after 30 minutes of this artificial rain. Surely they will find it impossible to mesure the nearest 0.01 inch. This context makes it natural to:

a. Read about measuring rainfall.

b. Ask someone working in meterology.

c. Explore, under teacher guidance, using a funnel of diameter 5 inches to feed into a cylinder of diameter $1/2 \sqrt{10}$ inches to stretch one inch of rain into a column 10 inches long.

In pursuit of a "solution" to the how-to-measure-rainfall problem the student may be led to learning about proportions, formulas, variation as the square, length, area, volume, etc. The teacher directs assignments to focus the learning and provide practice as needed. But, beyond the learning of special mathematical content, problem solving procedures are highlighted. In particular, the collection and interpretation of data call for reading and for asking questions. The teacher helps. But, increasingly, students are expected todig out information as needed. The students are expected to "learn to learn."

Students who plan college work in the life sciences, the social sciences or in fields that apply knowledge and concepts from these disciplines need basic mathematical skills <u>and</u> learning skills. Beyond these needs that they share with the two previous groups, they need data management skills. They need elementary statistics as it applies to organizing data. They need probability theory as it applies to interpreting data.

Here, in the United States, college preparatory students are given mathematics intended to prepare them for calculus. There are exceptions, of course. Some high schools do offer courses in probability and/or statistics. Most schools include some work with averages and with theoretical and empirical probability. In my experience this work centers upon vocabulary development, often without much meaning to students. Such terms as "equally likely" or "sample space" describe important ideas. To learn the terms without the ideas is, in my mind, not very useful.

Once again it seems to be that meeting ideas in "sensible" problem contexts will help. Professionals in mathematics education must, as the recommendations suggest, take a close look at the mathematics appropriate for students in the life sciences and in the social sciences. What ideas from statistics, probability, linear algebra, computer science, etc. do these students need? What problem contexts will be most productive in leading students toward these ideas? What applications of these ideas will be most effective as aids to transferring them to the solution of questions as yet unasked? I believe you will agree with me that we professionals do have "An Agenda for Action."

Students planning college work in fields based upon the physical sciences or mathematics need most of what previously-described groups need. In addition, they need much more than a speaking acquaintance with the elementary functions. I went to college believing that logarithms were things you need to solve oblique triangles. It was a great surprise to meet them in chemistry as useful for describing hydrogen-ion concentration; to meet them much later in scales of loudness of sound and in describing near-exponential growth of money, or population or whatnot.

Again, I believe we must learn to help students find the need for a class of functions in a "sensible" problem context. Very young children are, in my experience, fascinated by the question "How many times can you fold a sheet of paper?" The function

Number of folds	0	1	2	3	--	100
Sheets thick	1	2	4	8	--	2^{100}

is easy to develop, beginning with actual paper folding and making five or six folds. The continuation, to evaluate 2^{100} as approximately

$$1.26765 \times 10^{30}$$

by hand-held calculator, or logarithms or as $2^{10} \times 2^{20} \times 2^{40} \times 2^{20} \times 2^{10}$ suitably rounded as one goes along, produces a number beyond my powers to comprehend. The exercise of calculating the thickness in inches, in miles, in astronomical units and in light years is quickly done with simple measurements, a hand-held calculator and tables from astronony texts. The figure 10^{10} light years may mean something when compared with the distance to the spiral galaxies nearest us that are within 2×10^6 light years from the sun. This is a very thick pile of paper. You can't fold even a very large sheet of paper 100 times! Lengths of sides of the smaller and smaller rectangles produced by the folding are also interesting. They quickly become smaller than the filaments of the higher bacteria. You can't fold even a very, very large sheet of paper 100 times!

No example taken alone suffices to communicate what I mean by achieving "more than a speaking acquaintance with the elementary functions." That is why the Recommendations are readily An Agenda for Action. If a student meets the exponential function first in paper folding he must, of course,return to it in other contexts. Perhaps a study of chain letters would be of interest as a second contact. Later he might meet $A = (1.05)^n$ as the amount of one dollar at 5% compounded each period for n periods; at a stage where limits are to be encountered he might consider

$$\lim_{n \to \infty} (1 + \frac{.05}{m})^{mn} = e^{.05n}$$

as this arises in the continuous conversion of interest to principal. If students are to encounter exponential functions in such contexts, mathematics educators do have an "agenda". They must provide materials and help teachers learn to use them. To call such student activities "applications" may miss the main issue. For a teacher or text to point to a use of exponential functions is helpful. I think this is the common interpretation given to teaching applications. For a student to find a need for exponential functions in sensible problem contexts does more than this. It gives him or her what I've been calling a problem solving experience.

I have tried to speak about the needs of four groups of students: high school terminal; post-secondary vocational; college non-physical science; and physical science. Finally, what about the much less numerous, yet very important, group of students for whom mathematics as such is central? What combination of "modern rigour" and insight development through problem solving experiences is right for them?

Perhaps "teaching" that sets contexts for problem solving has been the missing link. You have had, and I have had, students who "knew" a great deal of mathematics. They were much better "educated" than others. Many of these students never found problems for themselves; many who were give productive questions to explore failed to formulate problems growing out of these questions. Perhaps their teachers taught them solutions for well-formulated problems without giving them problem-solving experiences. If so, we in mathematics education surely have an agenda in teacher education.

As the scope of mathematics expands, questions about what mathematics to include in the curriculum gets harder to answer. Specialists in number theory will give different answers than will specialists in topology. I hope the battle for time in the curriculum will continue. As the Recommendations point out, there is, in the United States, much deadwood in the contemporary offerings in secondary mathematics. The likely usefulness of computer sciences, for example using hand-held calculators to ordinate generating machines in studying relations and functions, must be explored. Linear algebra may displace some portions of analysis. Probability and statistics seem more relevant to the needs of students of the biological sciences than the study of quadratic equations. So there are urgent questions about the topics to include in the curricula of the several groups of students identified earlier.

The urgency of curriculum reform may, I fear make it even order to tackle the teaching of problem solving. Gains in problem solving power tend to come slowly; and, they are hard to measure. Surely the need to be "better educated" will make it more difficult to afford the leisureliness required for exploring problem-solving contexts. At worst, instructional time in mathematics might shift to drill and practice on such algorithms as using a hand-held calculator to find the harmonic mean of n numbers. At best, students will learn how harmonic means relate to focal lengths of compound lenses, to radius of action calculations, to resistances in parallel wiring, etc., and will be skillful in using the manual for their calculator. As a result they will be better problem solvers; more ready to tackle problems as yet unrecognized.

In closing I must apologize. What I have said has been said before. It is, under a kindest of interpretations, nomore than a few corollaries to theorems implicit in the sayings of men like Socrates or Felix Klein or George Polya. I did think it deserved saying again. I still think so. I hope you will accept some of what I have said as being relevant in your own work; as an Agenda for Action.

11.7 WHAT SHOULD BE DROPPED FROM THE SECONDARY SCHOOL MATHEMATICS CURRICULUM TO MAKE ROOM FOR NEW TOPICS?

WHAT SHOULD BE DROPPED FROM THE SECONDARY SCHOOL MATHEMATICS CURRICULUM - IN TAIWAN AND ELSEWHERE - TO ENHANCE THE PROGRAMS?

Ping-Tung Chang
Normal University
Tapai, Taiwan, Republic of China
Augusta College
August, Georgia, U.S.A.

The history of mathematics indicates clearly that mathematics grew out of experiences in the real world. All fields of knowledge from the economic, scientific and social world are dependent on mathematics for solving problems. Of course, the mathematics teachers in Taiwan have shown the students the marvelous achievement of our forefathers, such as determining how to count objects, add fractions, solve equations, draw maps and geometric figures, measure forces, times, and energy, determine sizes and properties of objects, and compute taxes and earnings. It seems that it is reasonable that the students can learn mathematics in a similar way, largely on the basis of dealing with and experiencing real world problems. The method of working only with pencil and paper, unfortunately, denies students the opportunity to deal with actual problems from the environment in which they live. Due to the separation of real life and mathematics, most of our students, especially those from elementary and junior high schools - - like their counterparts in the United States - - have decreasing motivation and a poor attitude toward mathematics. Even at the secondary level it is almost impossible for students to see that mathematics will improve their opportunities for meaningful problem solving if they are only engaged in a struggle to memorize formulas, prove theorems in geometry, and to solve, say, mixture problems. The only reason they study mathematics is because mathematics is a required course. Due to the intense competition of the entrance examinatins of senior high schools and colleges in Taiwan, students must achieve high mathematics test scores in order to be admitted to the school of their choice. Because of such pressures, it is not surprising that students fail to understand that mathematics has provided both the tools for discovering new scientific principles as well as those principles dealing with almost every occupational and daily life activity.

Mathematics teachers in Taiwan have the responsbility of motivating the students not only to develop mathematics skills but also to enjoy mathematics and learn how to use it to solve real-world problems. To insure the transfer of significant mathematical ideas and skills, it is clear that they must delete certain topics which have, for many years, been an integral part of most mathematics programs. In a Taiwan eighth grade class, it is not uncommon for students to be asked to factor such expressions as $x^3 + 2x^2 - 5x - 6$. Can such a skill really have a meaning for a fourteen-year old? Later, these students will devote much time to rationalizing the denominator of such expressions as

$$\frac{\sqrt{x}}{\sqrt{2} - \sqrt{x}}$$

or finding the least common denominator, so that they can carry out addition of fractions.

$$\frac{x + 1}{4x^2 - x - 5} + \frac{4}{x^2 + 5x - 6}$$

Usiskin strongly recommends that some topics, such as trinomial factoring in the first-year algebra curriculum be deleted because they do not have enough applications in later mathematics and, additionally, not enough real-world applications to warrant their inclusion (Usiskin, 1980). Of course, we have to be cautious with the deletion of any topic. One must make certain that this reduction will not interfere with the learner's training and his occupational needs. Certainly, with regard to the secondary school mathematics curriculum, adjustments will be made to reduce certain topics, such as the applications regarding insurance, mixture problems, logarithms, and factorization involving multiple variables, and much thought must go into determining the role of proof in geometry.

Much research and study will have to be undertaken to determine which topics should be included in order to obtain both the goals of general education and the aims of occupational necessity. More difficult will be determining the appropriate processes for presenting these topics so that students experience "doing" mathematics.

Teaching mathematics must catch up with the technological advances available today. For example, time spent acquiring computational skills should require much less classroom time than is now expended if one employs hand-held calculators. It is exciting to discover in the classroom that the calculator can be used to help students learn mathematics. Of course, the calculator doesn't have to be a device that merely provides answers to computational exercises (Beardslee, 1978), but the amount of time spent in drill problems of arithmetic and algebra can be reduced drastically, providing more time for teachers to introduce new areas such as problem-solving and estimation. In connection with social and commercial problems, the hand-held calculator allows students to deal with realistic sets of numbers. Major emphasis must be given to finding creative uses of calculators in classrooms. Perhaps the advent of the micro-computer poses the most exciting opportunity for altering school programs (Downes 1979).

In regard to teacher training, such programs should reflect a new attitude of concern for showing the relevance of mathematics if one is to obtain a new generation of enlightened and technically oriented students (Cicero 1979). One way to achieve this goal would be to introduce more finite mathematics, including introductory graph theory, linear programming, elementary statistics and computer programming to replace the classical topics which do not relate directly to the solution of real-world problems.

It is clear that the computer has changed our society, just as the Industrial Revolution changed history centuries ago. Automation controlled by the computer is creating an economic upheaval in industry (Johnson & Rising, 1972). It is reasonable to suggest that a similar upheaval will occur in education. In particular, the micro-computer will challenge designers of school programs to take advantage of their simulation and graphics capability as well as their "word processing" abilities. Creating appropriate software may well occupy a massive amount of time for educators during the next decade (Downes, 1979).

Problem solving should be at the heart of every mathematics course at the secondary level. Perhaps the teacher should introduce a problem whose solution uses the material to be taught next, and then show the class how the new knowledge expands their horizons and increases their value in a technological society. This form of reinforcement will benefit the student and establish a closer relationship with the teacher. Of course, teachers will have to learn how to teach the processes of problem solving. Thus in order to increase their effectiveness as problem solvers, teachers will have to work a great deal on their own. At this time, the textbooks and curriculum are still very classical and constitute a stumbling block rather than an aid. It would appear that teacher training programs are the most efficient route to success if such an effort is to be tried and to succeed.

The author wishes to thank Dr. John P. Downes of Georgia State University, Atlanta, Georgia. Dr. Joseph E. Cicero, of the University of South Carolina, Conway, South Carolina, for their assistance in preparation of this paper. Appreciation is also expressed to Drs. Bill Bompart, John W. Presley, and Betty House, all of Augusta College, for their critical reviews of the manuscript.

This paper was partially supported by a travel research grant from the Ministry of Education, Republic of China.

References

Beardslee, Edward C. "Teaching Computational Skills with a Calculator." Developing Computational Skills, 1978 Yearbook of the National Council of Teachers of Mathematics, edited by Marilyn N. Suydam, pp. 226-241. Reston, Virginia: NCTM, 1978.

Cicero, Joseph E., Moderator of Panel discussion of "What Math for the 1980's", "Mathematics for the 1980's" Conference, Augusta College, Augusta, Georgia, 1979.

Downes, John P., "Mathematical Issues for the 1980's" Speech given at "Mathematics for the 1980's" Conference, Augusta College, August, Georgia, 1979.

Johnson, Donovan A., and Gerald R. Rising. "Guidelines for Teaching Mathematics", 2nd Edition, Belmont, California: Wadsworth Publishing Co., Inc., 1972, pp. 418-435.

Usiskin, Zalman, "What Should not be in the Algebra and Geometry Curricula of Average College-Bound Students?" The Mathematics Teacher 73 (September 1980): pp. 413-424.

WHAT SHOULD BE DROPPED FROM THE SECONDARY SCHOOL MATHEMATICS CURRICULUM

Zalman Usiskin
University of Chicago
Chicago, Illinois, U.S.A.

We all recognize that the thirty-five years since the end of World War II have been ones of great growth in the mathematical sciences. Algebra, topology, combinatorics, graph theory, numerical analysis, etc., have extended the bounds of pure mathematics; statistics, operations research, optimization theory, mathematical biology, etc., have expanded the range of applied mathematics; and computers and calculators have affected not only the content but also the processes by which both the scientifically educated and the consumer do and think about mathematics.

The school curriculum cannot keep ahead of these developments; it must play catch-up. So it is natural that curricula throughout the world do not reflect many of these recent developments. As a result, we have pleas or pressure from the scientific community, business and industry, or governmental agencies to keep up with the times and modify the curriculum accordingly. These agencies usually call for increased attention to new and important developments in pure or applied mathematics that are unrepresented or underrepresented in the curriculum.

The typical school, however, allocates only a prespecified amount of time to the teaching of mathematics in a given student's program. At the secondary level, this ranges worldwide from 30 minutes or less 5 days a week until the age of 16 to as much as 3 hours a day 6 days a week in some technical curricula. This time allotment is traditionally rather fixed from year to year, so that even those who wish to teach something new are faced with the problem of not having the time to teach that unless some other topic, now in the curriculum, is desired.

Sometimes, to fit in a new topic, there is an appeal to efficiency, "Fit the new topic by taking a little of the old." This works in a few situations but generally runs counter to evidence from research, which suggests that both instructional time and time-on-task are on the whole positively correlated with achievement in a given area.

Thus we have a critical curricular dilemma: time devoted to a given topic is a critical variable in school learning while total time available is a fixed constant.

Because a recent article dealing with specifics of the situation in the United States is available, I have chosen to take a broader look at this problem and will focus on individual topics only to illustrate more general considerations.

We recognize that it is very difficult to get a new topic into the curriculum: decisions regarding the selection of material from that topic must be made; materials must be written; teachers must be retrained; and the education community must be convinced that the rationale for the new topic are valid. Since it probably has always been difficult to get something into the curriculum, every topic must have entered the curriculum with what were, for that time, compelling reasons for its inclusion. Thus, to take something out, one must decide either that (1) the situation has changed so that the compelling reasons for inclusion no longer apply, or (2) the reasons for another concept are more compelling. In either case, for any given topic under consideration, it is helpful to know what the original reasons were for inserting the concept into the curriculum and the reasons given today for continuing to teach the concept. These reasons seem to fall into four categories:

Category A: Importance Outside School:

1. The concept is important for a consumer to know - - i.e., it affects everyday life.

2. The concept is important for an educated person to know in order to make decisions or select decision-makers.

3. The concept is useful for getting a job or useful on the job.

Category B: Importance to Student's Later Life in School:

4. The concept (may not be important in its own right but) builds a foundation for other important concepts.

5. The concept is found on college or university entrance tests.

6. The concept appears in a later mathematics course; teaching it earlier will help to accelarate a student or help the student in that later course.

7. The concept will help in a nonmathematics course that applies mathematics.

Category C: General Transfer:

8. The concept helps build important general cognitive traits such as perception, critical thinking, precision, or memory.

9. The concept promotes desirable moral values or disciplines such as neatness, tolerance, uniformity, hard work, or avoiding error.

Category D: Motivation:

10. The concept is fun and encourages students to want to learn more.

11. The concept is easy to learn and helps build confidence.

12. The teacher enjoys teaching the concept.

No small list could include all reasons that are given for including concepts. Missing from the above list are those reasons that might be classified as cultural; some things are said to be taught simply because to learn them is part of the culture.

Sometimes reasons that are given reflect false tautologies. For example: (1) Hard work is good. To learn concept X requires hard work. Therefore concept X is good. (2) A general area (such as algebra) is important. Therefore any specifics falling under that area (such as trinomial factoring) are important. (3) It's good for the students who are best. Therefore it's best for all students. Under scrutiny, these false tautologies are seen often to fall under the notions of general transfer mentioned above (category C).

Sometimes reasons that are given beg the question entirely. For example: (1) It's in the book. The people who wrote the book know more than I do. I should follow them. (2) It's in the syllabus. The people who wrote the syllabus know more . . . (3) I had it when I was in school. If it was good enough for me, then . . .

The reasons given by teachers are often different than the reasons the original curriculum designers had in mind. Teachers' reasons tend more often to emphasize categories B, C, and D, while curriculum designers go for category A.

Taking the categories in reverse order from that given, let us begin with category D, motivation. There are two types of motivation, often confused. The motivation for selection of content comes from other categories and ought to be a necessary condition for a topic to appear in the curriculum; on the other hand, the motivation within a presentation should utilize the other motivations and be present whenever possible, at the least taking advantage of the intrinsic importance of the concept in the first place. That is, in a presentation, if content cannot be motivated, this is a sign that the concept should not be in the curriculum. Thus, lack of motivation is a criterion against inclusion of a topic in the curriculum, but the presence of motivation is not sufficient for inclusion.

Consider category C. The research has long showed that measurable general transfer does not occur from studying particular mathematics (e.g., studying proof in geometry does not improve critical thinking ability). Yet D'Ambrosio in Brazil and Stake et al. in the United States have independently noted how often teachers appeal to moral reasons or general faculties when explaining why they teach what they teach. And the public often appeals to general transfer rationale when voicing opinions regarding the mathematics curriculum. For instance, arithmetic in the elementary school in today's calculator age is often defended because it teaches children orderly thinking, neatness, and precision. The calculator is viewed as taking away what was never there. In general, these transfer rationale seem to be offered when no other rationale can be found.

By far the most common rationales offered by mathematics teachers come under category B, the category of importance to a student's later life in schools. Properties of numbers of arithmetic are taught because they are useful in early algebra; later algebra is important for calculus; and so on. In the United States it is commonly felt that the college entrance examinations known as the SAT's (Scholastic Aptitude Tests) are based upon the secondary school curriculum despite reports by the test creators that the tests are designed to be relatively curriculum free. Importance to a student's later life in school is important if a significant segment of the student population will have such a later life, but seems to be a vestige from the days when schooling was for the elite only.

In an earlier age, the rationales that remain - - importance to the student as consumer, enlightened citizen, or worker - - would not have been enough to insure mathematics other than arithmetic its importance in the curriculum. Frankly, the typical consumer encountered only arithmetic and the simplest geometry of measure, the enlightened citizen needed to know little more, and few workers used mathematics above trigonometry on their jobs. Today, we know the situation to be vastly different.

This entire discussion quite obviously oversimplifies the content deletion problem. Sometimes deletion is not "all or none" but a matter of degree. For example, how much rigor is needed to satisfactorily teach that notion? Sometimes the real world or job situation is in the process of change so that both old and new ideas are needed. For example, iterative solutions to equations are used even with quadratics in some computers, but so also is the quadratic formula. And sometimes a little of a topic may be worse than none. For example, sets or groups or vectors isolated from other content may have so little payoff that either deletion or increased attention might be a better solution. All this simply tells us that the question of deletion is by no means trivial and deserves strong attention.

References

Ubiritan D'Ambrosio, from a talk given at ICME IV.

Robert Stake, Jack Easley, et al. Case Studies in Science Education, Volume II: Design, Overview and General Findings. Urbana, IL: Center for Instructional Research and Curriculum Evaluation, 1978.

Zamlan Usiskin, "What Should Not be in the Algebra and Geometry Curricula of Average College-Bound Students?" The Mathematics Teacher, 73:413-424 (September, 1980).

II.8 ALTERNATIVE APPROACHES TO THE TEACHING OF ALGEBRA IN THE SECONDARY SCHOOL

INTRODUCING ALGEBRA

Harry S. J. Instone
Leamington Spa, England

The School Mathematics Project has a number of teachers involved in the writing of a new mathematics course for children of 11 to 16 years of age. I am one of those involved in writing the algebra part of the course, and I want to say something about the approach we have adopted.
The course is designed for pupils of all abilities, and this should be borne in mind when considering the approach to algebra which I shall outline. I do not wish to define "algebra" closely—let it suffice to say that by algebra we mean the use of letters or symbols to stand for numbers.

There are many ways in which a letter may "stand for" a number, for example:

$$x + 5 = 8$$

$\underline{x}$ is a particular unknown number

$$x + 7 = 7$$

$\underline{x}$ and $\underline{y}$ are variables, they may take any value rom a set of values

$$x + y = y + x$$

an identity, $\underline{x}$, $\underline{y}$, still variables

In using "letters" we have chosen to concentrate on the second of these uses—letters as variables. We have done this in the belief that this is the basis for any real understanding of algebra. It is easy to use "$\underline{x}$" in the first manner, but this can lead to real difficulty later on (the school leaver who asked, "Please, but what was $\underline{x}$, Miss?").

The question is, then, how to most usefully introduce the idea of a variable. We wanted to base the work on the child's own experience and, if possible, "discoveries," but it is difficult to find a "concrete" approach to what is essentially an abstract subject. However, we felt this was imperative if many children were not to get lost immediately. What was needed was a situation with two sets of related numbers. Our first thought was to use chains of squares made from matches and to relate the number of squares with the number of matches. The difficulty with this approach is that the number of squares may appear to have a different status to the number of matches, and that the numbers of squares builds up naturally one at a time. Why this is a disadvantage we shall see shortly.

After a number of other thoughts, we decided to use numbers arising from patterns of red and white tiles. The pupil is asked to put out various numbers of red tiles, then to place white tiles above the red row and then white tiles at the end (Figure 1). The pupil then builds up a table of the numbers of red and white tiles used.

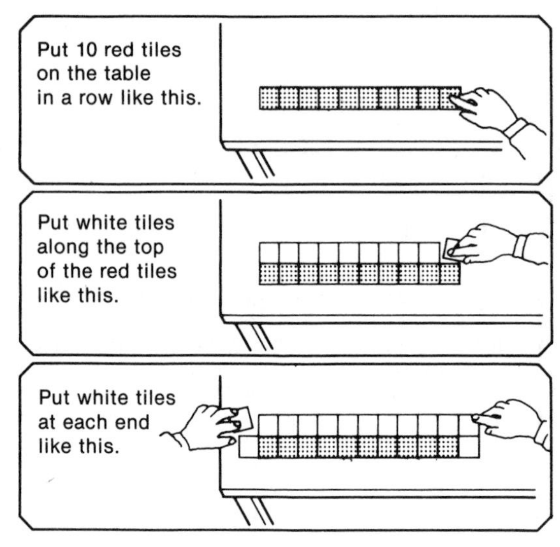

A1 Now you have made a bridge of white tiles round the red tiles.

How many white tiles are there in the bridge?

Figure 1

After using the tiles the pupil is asked to complete the table for 40 and 100 red tiles without actually using tiles. The problem that now arises is how the pupil is to write down the relationship between the numbers of red and white tiles. He or she is first asked simply to write down the rule that connects the number of white tiles with the number of red tiles. However, earlier work should have prepared the pupil for this.

Work on "number machines," with plenty of practice on spotting rules will have been done earlier in the course. Here the emphasis has been on the fact that given one pair of numbers there are many rules that fit the pair, but that given a whole table of pairs there is essentially only one rule that fits (Figure 2).

There are lots of rules that fit this table.

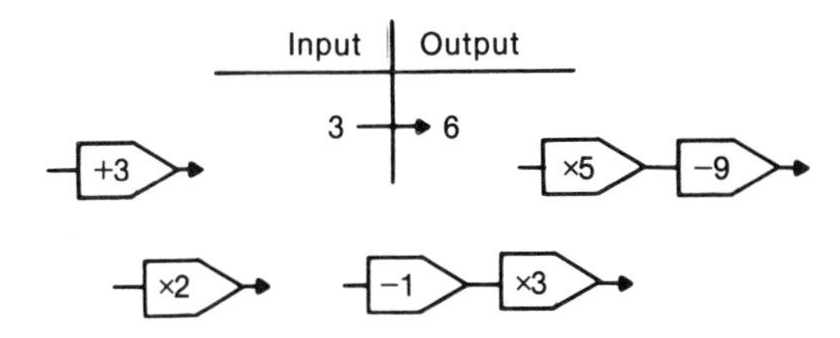

But only one fits this table. Which is it?

Input	Output
3	6
4	8
10	20
31	62

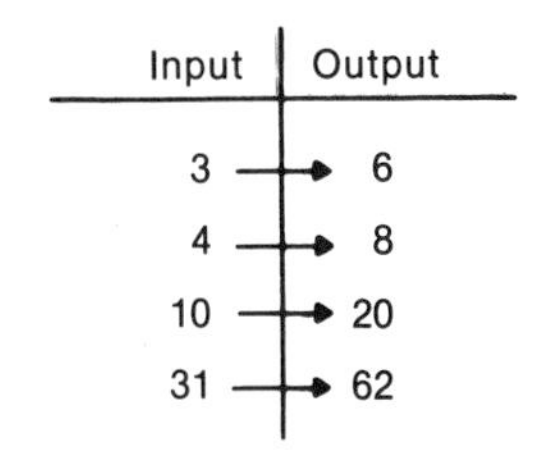

Figure 2

And here I can go back to one of the reasons we abandoned squares and matches. In a table where the inputs go up by ones the eye is naturally drawn to relationships which appear down the table; in Figure 3, "add 3." But we wish to emphasize relationships across the table, and it is hence best not to have ordered inputs (Figure 4).

Input	Output
1	4
2	7
3	10
4	13
5	16

Figure 3

In	Out
10	14
6	10
15	19
40	44
100	104

Figure 4

The pupils, then, are encouraged to find rules in the form

In

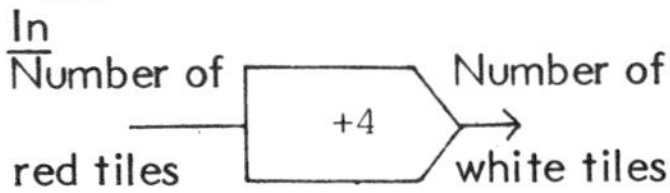

They are asked to work through several examples of patterns with tiles which lead to a "rule." A natural need arises to shorten "number of red tiles," and at this stage, and in later questions, the pupils are encouraged to use:

r to stand for "the number of red tiles"
and w to stand for "the number of white tiles"

There are other situations where the "machine" approach is not the most fruitful, for example, relationships of the type $x + y = 6$. A separate booklet called "Dice Games" introduces relationships of this type (Figure 5).

This time you win a star when

the number on the red dice + the number on the black dice = **6**

We write this winning rule in a short way, like this.

r + b = **6**

r means the number on the red dice.
b means the number on the black dice.

So you win a star when r and b add up to 6.

Here is a winning throw.

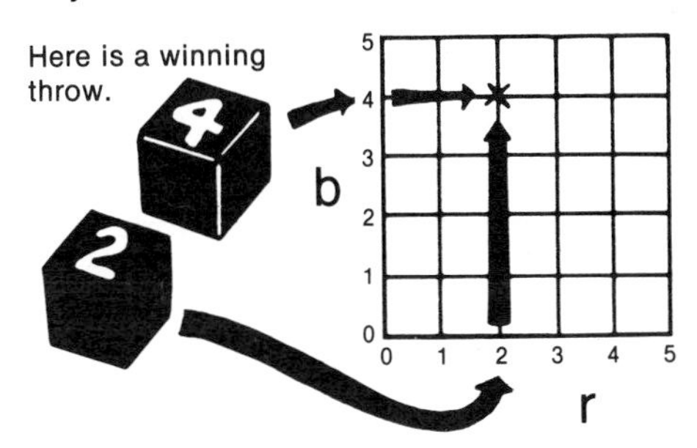

Figure 5

In this booklet r is made to stand for "the number on the red dice"--we apologise for the (already established?) curruption of the singular "die"!

I have concentrated here on the second use of letters--as variables--but we did not wish to ignore the first use of letters as particular unknown numbers, in other words, in equation solving. We decided to adopt a "balancing" approach to this aspect of equation solving (Figure 6)--taking, we hope, a reasonably "practical" approach.

A5 All the scales balance in these pictures.
Find the weight of each animal in kilograms.

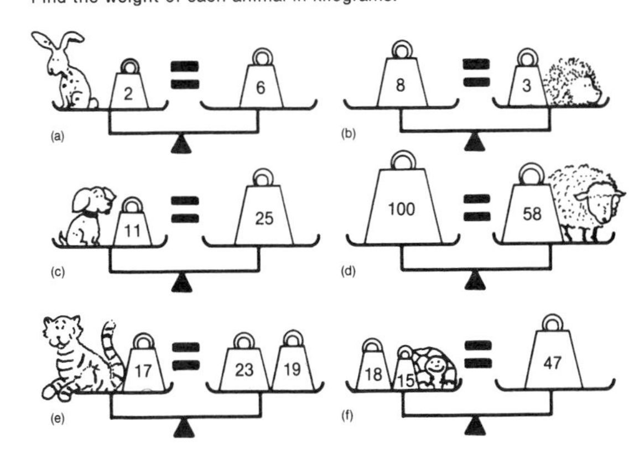

Figure 6

Again, this approach leads to a natural contracting which we feel will appear useful to the pupils (Figure 7).

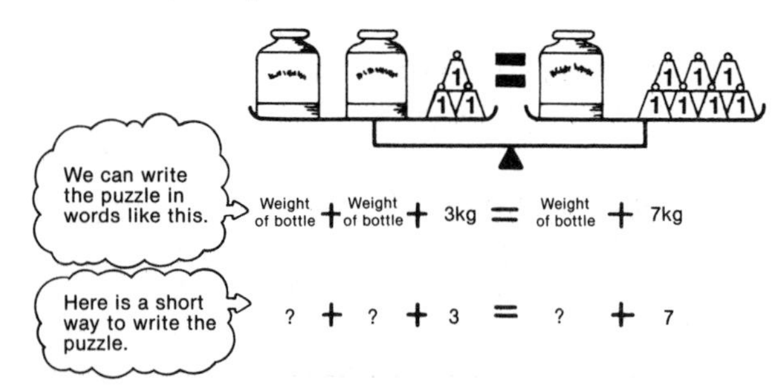

We write ? instead of 'weight of bottle'. ? stands for the weight of a bottle in kilograms. Each bottle weighs the same. So each ? stands for the same number.

D1 What is the weight of a bottle?

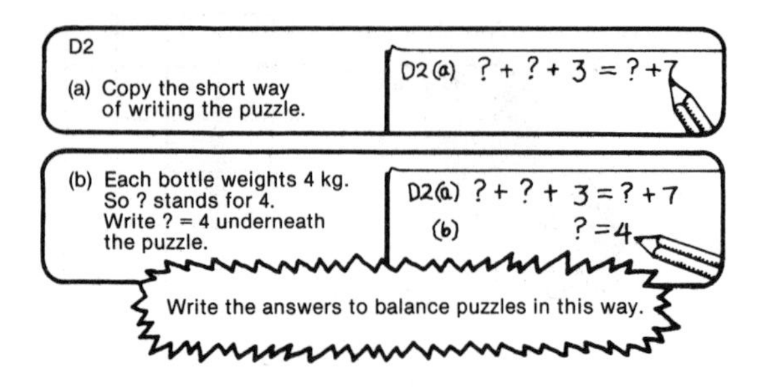

Figure 7

Our decision to use ? and not a letter for fixed numbers is deliberate. We hope thus to avoid the early difficulty for pupils of not knowing when a letter is "fixed" and when not. We also hope to avoid another tendency children have--that if r stands for "the number of red tiles" then 6r stands for "6 red tiles." We have avoided shortening r x 6 to 6r for a considerable time in the early part of the course, and when we do introduce this symbolism we emphasis that r always stands for a number.

It will be seen that we have introduced two ostensibly separate strands to early algebra: firstly, a letter standing for a variable, and secondly, a symbol (?) standing for a particular unknown number. Our actual writing of the course has not yet proceeded to the point where these two strands interweave, but I have outlined in Figure 8 the approach to be taken.

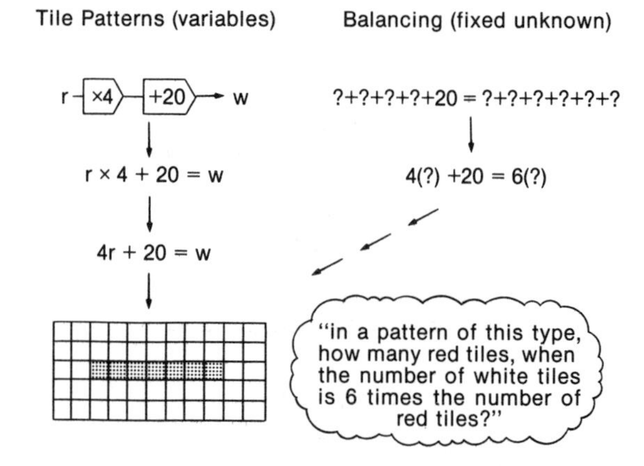

Suppose ? red tiles, so 6(?) white tiles.

In the formula 4r + 20 = w, replace r by ? and w by 6(?)

so 4(?) + 20 = 6(?) — now solve like a balance puzzle

Only much later does pupil use w = 6r

so 4r + 20 = 6r.

Figure 8

11.9 HOW CAN YOU USE HISTORY OF MATHEMATICS IN TEACHING MATHEMATICS IN PRIMARY AND SECONDARY SCHOOLS?

USE OF THE HISTORY OF MATHEMATICS IN THE MATHEMATICS CURRICULUM

Casey W. Humphreys
Minneapolis Community College
Minneapolis, Minnesota

The history of mathematics provides a wealth of material which can enrich and strengthen the total K-12 mathematics curriculum. Moreover, historically based enrichment and extension learning activities have been found particularly appropriate for use with gifted students. Your handout (attached) includes examples of the kinds of historically based materials which have been used successfully with gifted students. Also included in the handout is the outline of a history of mathematics course offered at the Governor's School for the Gifted of the Commonwealth (state) of Virginia. But before we look at this specific example of the use of the history of mathematics, let us briefly consider the need to increase the use of the history of mathematics in teaching and learning mathematics at every instructional level.

Too often our students see mathematics as a mystical system that appears ready-made in their textbooks. Many seem convinced that much of mathematics has little relationship to the real world and no relationship to the other subjects studied in school. In fact, relationships between topics within mathematics frequently are not recognized. Geometry,algebra, trigonometry, probability, etc., all seem to be separate strands in a tangled maze which is impossible to unravel. It is a tragedy that so many students view mathematics as a confusing and static body of knowledge. Failure to recognize the dynamic growing nature of mathematics prevents them from experiencing the thrill of discovery and invention.

Perhaps there is no subject that suffers more than mathematics by being disassociated from its history; for an appreciation of the very nature of mathematics is impossible without some acquaintence with its history. Because mathematics is the study of ideas, understanding can be greatly facilitated by an analysis of the origin of these ideas. Whitehead saw mathematics as the most original creation of the human spirit. And, Jacob Bronowski considered mathematics the most elaborate and sophisticated of the sciences . . ." a ladder for mystical as well as rational thought in the intellectual ascent of man (Bronowski, Jacob, The Ascent of Man, Little Brown and Company, Boston, Mass., 1973). Yet how many of our students would find these views strange and surprising - - if not just plain weird?

Ideally the history of mathematics should be incorporated into, even imbedded in, the total K-12 instructional program, but because the curriculum is frequently determined by the textbook being used, much of the history of mathematics traditionally is relegated to footnotes in the textbooks which students and teachers often ignore. Textbook publishers should be encouraged to incorporate history into the main body of the text. But, more important than convincing publishers of the value of incorporating history into their texts, is the need to convince teachers of the great value of using mathematics history in the instructional program.

Unfortunately, many mathematics teachers are not familiar with it shistory. Frequently they are surprised as students are to learn just how ancient some of "modern" math is. Teacher preparation institutions generally overlook the need to provide teachers with even a minimal background in the history of mathematics. Indeed, it is often difficult to find a single elective course at many of these institutions which deals with the history of mathematics. College and university mathematics education departments have sadly neglected to recognize the instructional importance of the history of mathematics.

Fortunately for all of us, Howard Eves' classis work An Introduction to the History of Mathematics is available. Every teacher, whether teaching elementary school arithmetic or college level calculus, can find appropriate and useful instructional material in this unequaled resource book. It contains many problems which can easily be adapted into the regular instructional program. For example, it is common to hear beginning algebra students express hatred for "word problems". Including the following problems found in Eves' book in a home work assignment won't make students automatically love word problems but they will provide students with a different perspective toward them.

"Work", "cistern", and "mixture" problems were not invented to frustrate you. They have been around for a long long time. See if you can work these three problems from the "Greek Anthology" dated around 500 AD. (There is reason to believe they are of considerably more ancient origin.) - - Good Luck!

A. Brickmaker, I am in a hurry to erect this house. Today is cloudless, and I do not require many more bricks, for I have all I want but three hundred. Thou alone in one day couldst make as many, but thy son left off working when he had finished two hundred, and thy son-in-law when he had made two hundred and fifty. Working all together, in how many days can you make these?

B. I am a brazen lion, my spouts are my two eyes, my mouth, and the flat of my right foot. My right eye fills a jar in two days (1 day = 12 hours) my left eye in three, and my foot in four. My mouth is capable of filling it in six hours. Tell me how long all four together will take to fill it?

C. Make a crown of gold, copper, tin and iron weighing 60 minae: gold and copper shall be two thirds of it, gold and tin three fourths of it, and gold and iron three fifths of it. Find the weights of gold, copper, tin and iron required.

In addition to providing students with a bit of historical perspective, problems like these are fun for the teacher. I especially enjoy the brazen lion. Can't you just see all the water spouting out? The lion is so much more interesting than the well known A's, B's and C's.

Another useful resource book is the NCTM 31st Yearbook: Historical Topics for the Mathematics Classroom. It also contains some examples of topics and problems which can be adapted for use in mathematics classrooms. As is the case with Eves' book, much of the material in the yearbook is particularly appropriate for use with mathematically talented students. Both books have proven to be invaluable sources of meaningful extension and enrichment learning materials for developing programs designed to meet the special needs of the gifted.

Until recently these special needs have frequently been overlooked in the rush to provide services for students at the lower end of the learning scale. The gifted have accurately been described as the most neglected group of students in our schools. Ironically, it is due to a federal regulation, Public Law 941 enacted in 1976, which provides funding for special programs to meet the needs of the mentally, physically and emotionally handicapped students, that the educational needs of gifted and talented students are now receiving national attention. The guidelines established by PL 941 for identifying students with special needs have been interpreted to include the gifted and talented who are considered to be handicapped by the restrictions of the regular classroom instructional program. Prior to passage of PL 941 special programs for the gifted, when they existed, were developed at the local level and in a few instances, at the state level.

One of the first efforts to provide for the needs of very bright students was a program designed by Isabelle P. Rucker for the Commonwealth of Vriginia. The program, named the Governor's School for the Gifted, began in the summer of 1972 and has just completed its eighth successful year. The school consists of a four week total immersion enrichment program for very bright rising high school seniors. It has been my joy to teach the mathematics course at the school each year since its beginning. After trying several courses, (number theory, probability and statistics, matrix algebra) in the first years, it became evident that the course which students enjoyed most and also considered the most valuable was the course, "The History of Mathematics from a Mathematician's Point of View". Much of the content of this course is adapted from the first half of Howard Eves' book. The 36 hours of instructional time allow for a chronological progression from the concept of number up to the time of Newton's "fluxions". An outline of the course and examples from the problem sets which accompany each lecture/discussion session are in your handout.

As you can see, the problems require no more than basic algebra or geometry for solution. Problems I, IV, V and X have been used successfully with junior high school students. The Towar of Brahma problem (V) has been used to introduce exponents to third grade students. Durer's magic square (X) has been used successfully with students and their teachers from the second grade level through high school. Many high school geometry teachers are not familiar with the Franciscan Cowl (VIII) proof of the Pythagorean Theorem found in Euclid's Elements. It provides a meaningful introduction for inservice programs with these teachers. It fact, all of the example problems have at one time or another, been used successfully as part of teacher inservice programs. The history of math course as outlined could be easily adapted to a semester schedule for a teacher's inservice course as well as provide the basis for a course for secondary school students and undergraduate mathematics and/or teacher education students.

All of us are concerned about mathematics education, but we still need to pause from time to time to reflect on the responsibilities associated with our commitment to offer students a full and comprehensive mathematics education. To continue to ignore the instructional treasures found in the history of mathematics is both irresponsible and foolish. We know that students who perceive the structure of mathematical ideas become more effective learners and users of mathematics. Clearly by incorporating the history of mathematics into the mathematics curriculum, we can help our students develop an appreciation for, and insights into, the logical structures and patterns inherent in mathematics. Additionally, the relationships between branches of mathematics and the connections between mathematics and other disciplines become self-evident through familiarity with the history of mathematics.

Mathematics as viewed through its history is human. It grows and evolves as one idea leads to another. Men and women, whether in China, the valleys of the Indus, the windswept slopes of Mount Olympus, or in a challenging classroom invent mathematics . . . sometimes in linear progression, sometimes in complete independence and sometimes out of great bursts of vision. When students and teachers recognize that the excitement of mathematics originates in the mind, the teaching and learning of contemporary mathematics is lifted to a far more dynamic plane. Familiarity with the history of mathematics fosters an attitude which encourages students to use their intuition and imagination, to ask the question, "What would happen if - - ?". As the dynamic and exciting aspects of mathematics become evident to students, they recognize that they too have the opportunity to experience the thrill of disccovery and invention. They begin to anticipate making their own personal contribution to the ever growing and beautiful world of mathematics. Let us give them that chance! Let us use our own natural resource, our own renewable energy source, let us use the history of mathematics.

THE HISTORY OF MATHEMATICS AS A PEDAGOGICAL TOOL

Bruce E. Meserve
University of Vermont
Burlington, Vermont

The history of mathematics is useful to me primarily as an aid to understanding topics that are already in the curriculum. Mathematics evolved from techniques for solving practical problems. The historical evolution of the mathematical sciences is an outstanding source of material for helping students understand mathematics and its applications.

How would you construct a square with twice the area of a given square? Is there a measure of a part of the given square that can be used to construct the desired square? The early Egyptians used a diagonal of the given square as a side of the desired square. We can use paper folding to demonstrate the correctness of this procedure. We can prove the correctness of the procedure using the Pythagorean theorem and in other ways. Does this imply that the early Egyptians knew the Pythagorean theorem a thousand years before the time of Pythagoras? Since early procedures were based upon trial and error, probably this procedure was simply a forerunner of the Pythagorean theorem for isosceles right triangles.

Approximations for areas may be used to emphasize the role of estimation. The early Egyptians sometimes used one-half the product of the length of two sides of a triangle as an approximation for the area of the triangle. A useful approximation was obtained for some isosceles triangles. Our students can determine the condition under which the exact area is obtained. Some students can explain how this procedure might have been extended to obtain the product of the averages of the lengths of the pairs of opposite sides for the area of a quadrilateral. A record of this procedure for finding the area of a field was left on a wall of a temple in northern Egypt.

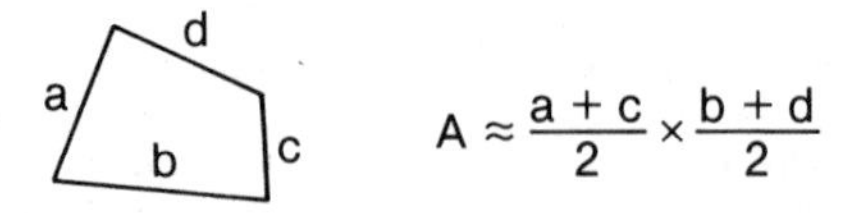

$$A \approx \frac{a+c}{2} \times \frac{b+d}{2}$$

The early Pythagoreans believed that everything could be explained in terms of numbers. All numbers were counting numbers 2500 years ago. Thus any number could be represented by a set of stones or other elements. The square of a number could be represented by a square array of stones. As the arrays for squares of successive numbers are considered, the increments are successive odd numbers.

x x x x
x x x o o o x
x x o o x o o o x
x o x o o x o o o x

1 $1^2 + 3$ $2^2 + 5$ $3^2 + 7$ ...

Starting with the square for any number n, the square for n + 1 may be obtained by adding n elements at the side and n + 1 elements at the top.

$$(n + 1)^2 = n^2 + n + (n + 1) = n^2 + 2n + 1$$

This type of geometric visualization enables students to "see" the arithmetic patterns.

Another use of the history of mathematics as a pedagogical tool may be illustrated by a consideration of the process of completing the square in an algebra class. In my classes I use the geometric algegra of the classical Greeks of the fourth century B.C. Numbers are represented by lengths of line segments. Products of two numbers are represented by areas of rectangular regions. For example, x^2 is represented by a square of side x, 6x is represented by a rectangle 6 by x, and this rectangle may be bisected into two rectangles each 3 by x. Then, as in the extension of the square for n^2 to obtain the square for $(n + 1)^2$, we use the square of side x and the two rectangles 3 by x as parts of a square of side x + 3.

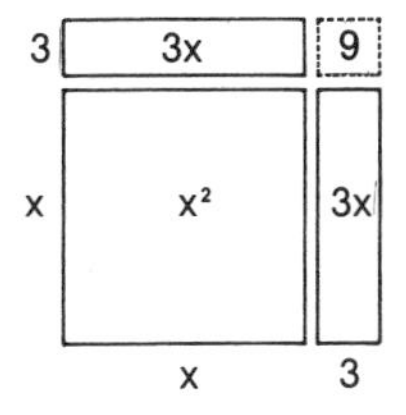

To complete the square of this side x + 3 we need a square of side 3.

$$(x + 3)^2 = x^2 + 6x + 9$$

The geometric picture explains the origin of the term "complete the square" and vividly illustrates why half of the coefficient of x is used. Half of the rectangle is placed on each of two adjacent sides of the square of side x.

The use of geometry to clarify algebraic procedures is not a recent fad. Nearly 900 years ago Omar Khayam wrote:

> Whoever thinks algebra is a trick of obtaining unknowns has thought in vain. No attention should be paid to the fact that algebra and geometry are different in appearance. Algebraic statements are geometric facts which are proved.

Three hundred and fifty years ago Rene Descartes felt that geometry relied too heavily upon diagrams and described algebra as "a confused and obscure art that embarrasses the mind." Descartes considered algebraic procedures necessary to free geometry from its dependence upon figures. He also considered geometric figures necessary to give meaning to the operations of algebra. For Descartes neither geometry nor algebra was by itself sufficient for a satisfactory proof.

Interrelations among the branches of mathematics are now being recognized as important. The history of mathematics provides excellent opportunities for emphasizing such interrelations. Consider Proposition 4 in Book II of Euclid's Elements.

> If a line segment is cut at random, the area of the square on the whole is equal to the sum of the areas of the squares on the parts and twice the area of the rectangle contained by the parts.

The geometric representation helps students to remember to include the term 2ab in the algebraic expansion for $(a + b)^2$. The rephrasing of statements from Euclid and others into contemporary vocabulary, and especially contemporary symbolic notation, is an excellent experience for students. Many statements that appear to be geometric have familiar algebraic representations, often two or more different algebraic representations, depending upon the assignment of the letters. The consideration of both algebraic and geometric representation provides a better understanding of the underlying mathematical concept that can be obtained from either the algegra or the geometry alone.

There is a biological law which says ontogany recapitulates philogemy. It means that the development of the individual repeats the principal stages of the development of the group. As teachers of mathematics we can improve our effectiveness by recognizing that, to some extent, each child's learning of a mathematical concept includes the principal stages of the historical development of the concept. The following seven stages are suggested as significant in the historical development of most mathematical concepts.

1. Experiences - unstructured experiences before the individual's full awareness of them.

2. Observations - as the individual becomes aware of experiences, seeks to understand them, and looks for patterns or structure in them.

3. Experimentation - purposeful structured experiences, trial and error experiments, failures, successes.

4. Inspiration - conjectures of patterns, recognition of patterns.

5. Informal applications - what would happen if . . . ?, testing of conjectures.

6. Formalization - postulates (assumptions), identification of structure, proofs of conjectures.

7. Applications and interpretations.

The concepts of number, function, polygon, parallellism, and variable are only a few of those that can be understood much more easily and fully if the stages of their development are considered. The particular details of the development to be used depend upon the mathematical maturity and interests of the students and the teacher. For any given mathematical concept several stages may be occurring at the same time. For example, in the development of the concept of number, early experiences and observations undoubtedly included

a. the distinction between one and more than one,

b. the recognition of one, two, many, and

c. comparisons of sets of objects.

Introductions of number words, number pictures, and number symbols were forms of experimentation that continued through later stages. Recognition of one-to-one correspondence was an inspiration that provided the basis for tallying and early systems of numeration. Operations with numerals were performed informally for many centuries. Mysticism and astrology were informal applications. Expansion of the number system to include zero, negative numbers, complex numbers, and so forth, proceeded by experimentation. For example, note the approach used in 1545 by Girolamo Cardano in his Ars Magna. As he considered the problem - find two numbers whose sum is 10 and whose product is 40 he observed that: This is obviously impossible ... nevertheless, let us operate . . . . Then, using traditional methods, answers of the form 5 + - 15 and 5 - - 15 were found as new "sophisticated numbers" that satisfy the given conditions. Such symbolic manipulations serve as a form of experimentation before a formal structure has been introduced.

Historical examples are particularly useful in illuminating the need for caution in symbol manipulation. Consider, for example, the infinite series

$$1 - 1 + 1 - 1 + 1 - 1 + \ldots$$

Sums of 0, 1, -1, 2, and so forth may be obtained by associating the terms in various ways. Leibniz, one of the developers of calculus, considered the appropriateness of 1/2, the average of 0 and 1, as the value for this infinite series. Students who develop conjectures for several such sums recognize the need for the guidance of rules in working with infinite series. The history of mathematics is a bountiful source of such examples showing ways in which previous generations have experimented and discovered the need for formal mathematical structures.

Finally, there are many human interest situations in the history of mathematics, both successes and failures. For example, Kepler's work on volumes evolved from his efforts to identify volumes of wine barrels. Such situations may be used in the presentation of mathematics as a significant part of our culture.

THE MATHEMATICS CURRICULUM AND THE HISTORY OF MATHEMATICS

Leo Rogers
Roehampton Institute
London, England

The History of Mathematics, like mathematics itself, means different things to different people. This statement is perhaps trivially obvious, but needs constantly to be underlined in our discussions. Since 1972, when the history of mathematics first appeared on an ICME programme, there has been considerable growth in the literature concerning the teaching of topics, and the approaches to mathematics using its history. Together with this, the circle of interested individuals expands through national and international organisations. Slowly a common language is being developed, and common understandings achieved, but with all such dialogues, the process is continuous, and there is still a long way to go. Nevertheless, a number of distinctions are being clarified.

The 'History of Mathematics' is dominated by the classical view of mathematics and university-based research that has developed in Europe and North America. (One might note this as one of the results of social and economic development - at least as much as the influence of Euclid in our mathematical culture.) It is our duty to encourage our colleagues in the rest of the world to investigate their own national mathematical history and its wider dissemination. Some cultures, notably Arabia, India, China and Japan have already unearthed a wealth of material in the history of their mathematics, relevant to teaching at many levels.

The distinction between teaching history of mathematics and using the history of mathematics in our teaching is now more clear, and while we may not generally think that a course in the history of mathematics is appropriate in school, (thus making it yet another subject to be timetabled and examined), we do agree that using historical material to motivate interests, develop positive attitudes, and encourage appreciation of the nature and role of mathematics in our culture, can only be an advantage. We have now a fairly comprehensive list of objectives for the teaching and use of the history of mathematics, and a range of media and methods of presentation. Film, video and audio tapes, slides, biographies and portraits as well as some good texts are available to assist the teacher,and we have the accounts of teachers' experiences with these, and of story telling and even acting out playlets to encourage pupils' appreciation and enjoyment.

Many of the reasons we use the history of mathematics concern the support it gives to the 'standard' curriculum, which in most countries has been established by a consensus between the requirements of society and those of the universities. An aim such as "to give the pupil basic competence in mathematical skills" is accepted as praisworthy by administrators, but as teachers we have to ask: what mathematics?, what skills?, to who?, what for?, and when?, and each of these questions must address themselves to the history of mathematics as well. What is appropriate for my class may not be so for yours, but my account of my experience might be taken by you, and adapted to your needs, your resources, your capabilities, and those of your pupils.

'Standard' history of mathematics supports the 'standard' curriculum in the sense that it demonstrates how modern mathematics has its roots in the past, and underlies the improvements in mathematical rigor. One can find implications, if not direct statements that, for example, the problem of incommensurability in Greek mathematics is a contribution to the theory of Real Numbers, or Leibniz' use of infinitesimals is reinstated as a contribution of Non-Standard Analysis. While we cannot help looking at past mathematics with present knowledge, we must be very careful how we interpret what we see. Greek culture and that of the seventeenth century are far removed from us now, even though they may be common parts of our heritage. On the other hand, in choosing to investigate, say, Egyptian fractions, we must not treat them as mere curiosities or examples of a failure to appreciate the Rationals, but as a valuable contribution to mathematical thought in a particular social-cultural setting.

How did our standard curriculum come about? Who made the choices that determine its content, its quality, its style and quantity? Is our standard mathematics the only respectable mathematics or is there another mathematics relevant to our teaching that we need to become aware of? This is certainly true, and has been emphasized by recent publications of a wealth of arithmetical processes from India and Arabia, by the investigation of non-university or 'popular' mathematics in England, and by a number of methodological papers inspired by the work in Imre Lakatos.

We know that different ages, cultures and individuals have regarded mathematics in different ways. This disparity is not necessarily a disadvantage, for in the the very recognition of variety, we have a powerful tool for the widening of people's appreciation of mathematics. The communication of the ways in which this variety has evolved and relates to the needs of individuals and societies is a factor generally omitted from mathematics teaching. In fact, mathematics teaching has traditionally been more concerned with the communication of mathematics as a set of tools for the improvement of the intellect and the appliation to the physical world. The omitted part of mathematics, the dialectic concerned with the communication of its variety, is largely a historical-evolutionary study, and as such has not yet found a place in the mathematics curriculum which deals with 'facts', or the history curriculum which continues to ignore a significant part of our scientific and mathematical culture.

In a way, mathematics is itself to blame for its own failure to communicate a large part of its essential nature, for the main activity of the mathematics communicator is the consolidation and systematisation of existing structures, and this deletes the original activity of the mathematics creator from the record. Briefly, mathematics is a subject which defines away its past, and mathematics creators and communicators unwittingly subscribe to the continuation of this situation.

The further we go into modern, abstract mathematics, the less accessible are the records and writings of the individuals who made their contributions in the past. It is often argued that these contributions have no relevance for teaching - for retracing past mathematics only takes up time which should be better spent learning 'real' mathematics. But we know that a large number of children and students are unable to related to this so-called 'real' mathematics, it becomes inaccessible to them. We know that changes in attitude (and consequently achievement) in both pupils and teachers can be brought about by a deliberate attempt to introduce the historical-evolutionary dialectic hitherto omitted from mathematics teaching. Appreciation of the history of mathematics - the evolution of mathematics in our culture, and the development of mathematical concepts - show this dialectic at work, and can provide a means for teachers at all levels to develop attitudes towards mathematics which will help to make it more readily accessible to themselves and their students.

When we look at the evolution of mathematical curricula, we see the influence of the body of mathematical knowledge, and also the philosophical, social and economic influences that shape the demands and expectations that society makes of the educational system. Influences on the curriculum may be from mathematicians who have particular attitudes, preferences and beliefs about the relative importance of particular content, and whose views are carried more by virtue of their position than any real knowledge of particular teaching situations. Few teachers are able to challenge such an authoritarian viewpoint successfully.

There are a number of other influences over which a teacher may have little or no control. Economic demands are shaped by the society we live in and tend to be practical, often responding to the latest developments. Some sectors of society attempt to preserve the social structure in which they succeeded (or which they accept) while others attempt to break away from established norms to cater for minority groups, and influence teaching styles and content by pressure on teachers.

Political factors arise as a result of these decisions affecting the funding of education and its administration which in turn influence the development of subject areas and individual opportunities. The major original political influence was the institutionalisation of mathematics teaching in the nineteenth century, and in our different countries we have a variety of forms which give different degrees of autonomy at different levels.

We are concerned, in our teaching, with the formation of mathematical ideas in our pupils who are both learners and creators (or re-creators) of mathematics. Courses for teachers include some study of concept-formation, but the main emphasis lies in the area of psychology and not in epistemology, or a study of the origins of mathematics. The history of mathematics provides considerable data for investigating both the evolution of mathematical forms, and the inception of ideas in individuals.

A number of questions need continued discussion: Does a knowledge of the history of mathematics suggest any general activities for the classroom which are likely to encourage fruitful mathematical experience? Does the history of mathematics help us to recognise the mental constructs or operations that constitute mathematical experience, and are we thereby able to assist our pupils to appreciate the nature of mathematics without necessarily requiring of them a great deal of technical expertise? Is it possible to convey the spirit of some of the developments in mathematics in a satisfactory way to those who are not mathematics specialists? It is legitimate to tell historical stories to children, if this can be done in an interesting and honest way? How has the mathematics curriculum evolved, and what can we learn from a study of the internal and external influences that have caused this evolution? Is it possible, for example, to isolate any fundamental factors which can be of use to teachers in planning curricula? What data can the investigation of history provide for the study of the origins of mathematics?

Many of the aims we have in our teaching of mathematics have no necessary reliance on a teacher's knowledge of history, but it does seem that much the easiest and most fruitful way of presenting a large part of mathematics is either against a background of history, or in the context of a historically based philosophy of mathematics. Teaching mathematics with an historical perspective is not merely attempting to imitate in teaching one's interpretation of a sequence of historical discovery, neither is it assuming that a growing child passes through the same mathematical experiences as our ancestors; it is the constant awareness of the continuing and dynamic dialectic by which mathematics evolves, and its application whenever we have an opportunity.

THE USE OF HISTORY OF MATHEMATICS IN THE TEACHING OF MATHEMATICS IN SECONDARY EDUCATION

Maassouma M. Kazim
Ain Shams University
Cairo Egypt

No one denies that mathematics is the mirror of civilization, yet many teachers teach mathematics as if it is only a collection of definitions, theorems, rules and problems to be solved. Thus, many students feel that they study mathematics, only, because it is a part of the curriculum and because they have to pass exams in that field. The speaker feels that teachers take care only of the cognitive and psychomotor aspects of mathematics and neglect the affective domain. As a result, many students usually are not satisfied with school teaching, so they turn to special tutors to teach them in groups or individually, so that they can achieve as much as they can of the facts and skills and to do as many drills as possible to pass their exams.

To take care of the affective aspect of mathematics, it is the speaker's belief, that the use of the history of mathematics in teaching may be of some help. She thinks that using the history of mathematics may motivate students and, in addition, it may develop positive attitudes towards mathematics and its values in many situations.

Before going into detail in this paper, we emphasize the fact that a study of the history of mathematics should be included in the program of preparation of the mathematics teacher. A mathematics teacher should be cultured and broad minded. Moreover, he should have a good knowledge about the application of mathematics in different areas. It is important to improve teacher preparation programs, if we want to improve school programs. Therefore, it is the speaker's belief that a course in the history of mathematics should be required in the mathematics teacher's program.

Although, many mathematicians and mathematics educators emphasize the importance of the history of mathematics, not many have discussed the objectives of using the history of mathematics in teaching secondary school mathematics, nor the possible methods which can be followed in teaching.

It is the purpose of this presentation to suggest:

1. Some objectives of using the history of mathematics in teaching Secondary School mathematics.

2. Some methods of using historical material in the classroom.

Objectives of Using the History of Mathematics

1. The students would know something about the economy and power resulting from the discovery and/or development of symboic techniques. Example: Algebraic symbolism simplified the solution of the quadratic equation which was very difficult by the Euclidean method and by Al-Khowarizmi who used words like root and wealth which we denote now by x and x 2.

2. The students would better understand the character of mathematics as a subject which always tries to generalize and specialize. Examples: (1) The theorem that the sum of the angles of a triangle equal 180° is a generalization for all triangles on plane surfaces. (2) The Pythagorean theorem is a special case for the general law $a^2 = b^2 + c^2 \pm 2\,bc \cos A$ (where a, b, c are sides of triangle and A is the angle opposite to side a).

3. The students would know the story of the development of some of the more common mathematical concepts. Examples: The story of the development of the idea of place value, the idea of standard measure, sets by Cantor, and group theory by Abel and Galois just to name a few.

4. The students would know that mathematical truth is independent of place, personality and human authority. Examples: The value of pi or the idea that theorems are mathematically true regardless of the empirical truth of the basic postulates (non-Euclidean geometry).

5. The students would understand that some mathematical solutions are direct responses to a purely intellectual challenge. Examples: (1) Attempts to prove that roots of polynomial can be expressed in terms of its coefficients, (2) Fermat's Last Theorem is still an intellectual challenge.

6. The students would see how mathematical ideas have been developed through the years and how mathematical concepts were developed slowly, hence, they often are learned slowly. Examples: (1) How numbers have been developed, from natural numbers to the knowledge of fractions, decimals and complex numbers. (2) How computing was difficult in ancient Egypt; their knowledge of fractions had not been developed at that time, and so they represented their fractions by the sum of fractions with unit numerators. (3) The slow development of geometry from one geometry to several geometries. (4) Solution of equations by the Babylonians, the Greeks and the Arabs.

7. The students would encounter a few instances where problems which were once considered to be of a purely theoretical interest now have sound practical applications. Examples: (1) Non-Euclidean geometries were once of theoretical interest only, yet they were useful in reaching the moon. (2) Mathematical structures, though abstract, yet they have very important applications, in many areas, such as in chemistry.

8. The students would understand the place of mathematics in the history of civilization and human culture; they would know that there is no civilization without mathematics, by realizing the positive relationship between progress in mathematical knowledge and progress in other aspects of science. Examples: (1) Comparing the knowledge of

mathematics in Ancient Egypt, Babylonia and Greece to the knowledge of mathematics in some other countries that had not developed a number system. (2) Mathematics in Egypt as an empirical science and its place in logical thinking in Greece. (3) Comparing the mathematical knowledge in those countries which have reached the moon to mathematics in the developing countries.

9. The students would know that mathematics has influenced thought in many areas, such as art and philosophy. Examples: (1) Pythagoras discovering some laws in sound. (2) Pythagoras' philosophy was founded on the concept of numbers. (3) The Newtonian influence on philosophy, literature and aesthetics. (4) The philosophical views of Descartes, and Kant's idea about space and time.

10. The students would realize the necessity of mathematics to all fields of science, and how mathematics has given maturity to sciences. Examples: (1) The value of quantitative estimation. (2) Boolean algebra and mathematical logic. (3) Isomorphism in Chemistry.

11. The students would understand that human beings have needed mathematics since time immemorial. Example: History shows that the most primitive peoples know the rudiments of counting, measuring and telling time.

12. The students would develop a growing appreciation of the record of mathematics as a servant of mankind. Example: (1) The story of the development of standard units to facilitate communication and trade. (2) The development of instruments based on mathematical principles for the benefit of man. (3) Studying mathematical models and modern mathematical applications in economics, politics, planning, etc.

13. The students would understand that some mathematics are a direct response to the needs of man. Examples: The rope stretchers in Egypt in drawing right angles in dividing land after the flooding of the Nile. (2) The need for counting and measuring. (3) The need for the concept of set and mathematical structure.

14. The students would understand that some mathematical discoveries are direct responses to the spirit of the time. Examples: (1) The logical character of Euclidean Geometry. (2) The discovery of logarithms in Switzerland by Burgi and in Scotland by Napier. (3) The non-Euclidean geometry which was developed by different mathematicians. (4) The discovery of calculus by Newton in England, and at the same time, by Leibniz in Germany.

15. The study of the history of mathematics would motivate some students to study mathematics, because the history of mathematics provides a large and interesting field of reading and learning.

15. The gifted students would be encouraged to work hard and set high ideals for themselves, through reading and learning something about the lives of famous mathematicians who shared in creating new concepts and changing some common notions. Examples: Reading about: (1) Pythagoras and his philosophy, his followers, and their secret society. (2) Something about Newton's childhood, and the circumstances of discovering the law of gravity. (3) Gauss proving the fundamental theorem of algebra at a very young age. (4) Cantor and the development of the concept and theory of set.

17. Some of the students would enjoy the historical stories about mathematics and mathematicians. Examples: (1) The story of Archimedes and his cry of "Eureka". (2) Galileo and the swinging lamp. (3) Euler and his reply to Diderot about the existence of God. (4) Poncelot's studies in a Russian prison. (5) The story of Newton's discovery of the law of gravity.

18. The students would realize that validity rather than truth characterizes mathematics. Example: The invention of non-Euclidean geometry and the development of mathematical system.

19. The students would realize that great people sometimes erred and revealed faults. Example: Abel thought that he had solved the polynomial equation of degree "n", then he realized his fault and admitted it.

20. The students might have an idea about teaching methods others have used.

21. The study of the history of mathematics in some of the ancient countries develops an appreciation of the respect for those countries that shared in constructing our mathematics. Example: Mathematics in Egypt, Babylonia and Greece.

<u>Methods of Using the History of Mathematics in Teaching:</u>

1. The teacher asks the students to read historical materials which is put down in separate notes, or chapters in the textbooks.

   a. prior to the study of the topic to which it is related,

and, or

   b. after the study of the topic to which it is related.

2. The teacher gives the students names of books about the history of mathematics and suggests some topics to be read in these books; reading has to be followed by class discussions or discussions in a mathematics club.

3. The teacher encourages the students to apply both the ancient and modern methods, to understand the basic ideas, and to know the value of modern discoveries.

4. The teacher mentions some historical material related to what is being taught at that time.

5. The teacher gives the students an historical introduction before beginning a course in mathematics.

6. The teacher shows the students some films about the development of mathematics and about the famous mathematicians.

7. The teacher asks the students to make a report on some topics in the history of mathematics.

8. The teacher asks the students to make a scrap book of historical material.

9. The teacher and students work on a project in which they employ the history of mathematics. Example: A project on the development of measurements beginning with the use of natural units (the limbs, span and foot) as did primitive man. The students would realize that these measurements differ from one student to another; they may choose the limbs of one of them as standard units - so they approach the idea of standard units in their use and development.

10. The school offers a course in the history of mathematics separate from other mathematics courses, to show the development of civilization in different countries, in order not to limit ourselves to political history, which is concerned with wars and the violation of peace and civilization.

C H A P T E R 12 - The Begle Memorial Series on Research in Mathematics Education

12.1 CRITICAL VARIABLES IN MATHEMATICS EDUCATION

STUDENT APTITUDE IN MATHEMATICS LEARNING

Richard E. Snow
Stanford University
Stanford, California, U.S.A.

This paper is one of a series of presentations at this Congress commemorating the work of Ed Begle. Begle's (1979) last work, "Critical Variables in Mathematics Education", served as a basis for these presentations. In particular, I will address "The use of critical variables to organize research, problems of synthesizing research, and the kind of empirical research that would be most useful to mathematics education."

My paper has three parts. First, there are a few words about Ed Begle - - my colleage at Stanford from 1966 until his death in 1978. Second, I will discuss some aspects of the present state of research in mathematics education, as depicted in part by Begle's book, using it as a starting point, but adding in my own views as well. Some examples of recent studies are given to show some complexities to be faced and some promising directions for further work. Finally, a brief summary of some problems and prospects for research in mathematics education is given.

I was introduced to Ed Begle and the SMSG-NLSMA project upon my arrival at Stanford, and for the ensuing 12 years Ed and I had periodic interactions about the problems and prospects of research on mathematics instruction, and particularly about the uses of psychology in that endeavor. I am a psychologist, not a mathematics educator, so we worked in somewhat different areas within the School of Education. But we shared teaching and advising duties for many doctoral students in both mathematics education and educational psychology. He would send students to me to get the psychological side of research experience. I would send students to him to get them to think about the real research problems of a curriculum field. Particularly in his last years, he was concerned that his students be connected with other ongoing research projects. Ed was a committed user but also an astute critic of psychology in education. He had a sensitive appreciation for the strengths and the weaknesses of psychological research in relation to work in the curriculum fields. His library showed it and his book shows it as well. Unlike many educators, and also many psychologists, he believed as I do that the relation between psychological research and mathematics education is a two-way street; technical and conceptual enrichment run both ways. As a result, his students I think have been better prepared for empirical research in mathematics education than those of any other curriculum field, at least in my experience. And some of my students know mathematics education far better than does the typical educational psychologist, thanks to Ed Begle.

The Begle book is indeed a map of the terrain, as the editor's preface suggests. It is a topographic surface or outline map identifying the many mountains that exist in different regions of the literature, showing their respective elevations in terms of the amount of attention that has been paid to each. To be sure, Begle added his own views, clearly identified as such, about which mountains deserved more attention in future research and which deserved less. One can easily disagree with these relative emphases, but the book will still remain useful, for it provides two kinds of starting points for further work. One can choose individual variables identified by Begle to be given attention in more intensive and integrative literature reviews and in more analytic empirical research. Or, one can begin to investigate the complex interrelationship among sets of variables chosen from different Begle categories. One can also do both of these things, of course, and in any case one is free to follow Begle's advice in making these choices or not.

Figure 1 gives a rough schematic array of Begle's categories of critical variables to help indicate how these two kinds of further research might be derived and organized. In the figure, Begle's categories of critical variables are shown as classroom, school, and social environment variables, above the arrow line, and as student variables, teacher variables, curriculum and instruction variables, test and test use variables, and the goals of instruction, below the arrow line. The sequence of symbols and arrows simply indicates that students come into instruction exhibiting variation in a number of prior characteristics. Many of these are outcome (O) variables from previous instruction and experience which serve as aptitudes (A) for learning from new instruction. Student variables such as prior knowledge of mathematical concepts, reasoning abiities, mathematics anxiety, etc. are all aptitudes for new mathematical learning if they indicate individual differences in readiness to profit from the new instruction. Students then experience an instructional task or treatment (T) which represents a complex mixture of teacher variables, curriculum and instructional variables, and test and evaluation variables, to reach outcome (O) of instruction which display educational goals and again serve as student aptitude (A) variables for further instruction. All this takes place in particular educational environments. My research concentrates on student variables and goals, but some examples will be given that show connections between these categories and other regions of the educational complex.

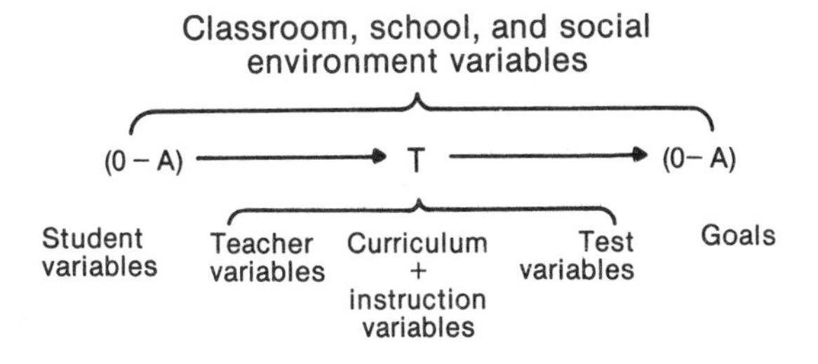

A schematic organization of Begle's (1979) categories of critical variables.

Figure 1

As an example of research in which one or more variables from a single category are chosen for more penetrating review and empirical analysis, consider Begle's category of student variables. He lists intelligence, logical thinking, and mathematical ability as three such variables, noting several important but as yet unanswered questions regarding each. New research might start by reviewing the literature on these variables in relation to mathematics learning and then progress to detailed analyses of the aptitude constructs involved. Some research conducted since the close of Begle's review has done just this. For example, work by Sternberg (1977, 1979, with Guyote & Turner, 1980) Pellegrino and Glaser (1979, 1980), and some done in my own laboratory (Snow, 1978, 1980) has sought to identify the component information processing skills and strategies involved in performance on the kinds of reasoning tasks found in tests of intelligence and logical thinking. The aim is not only to identify component skills but to show how these components are organized in task performance - - to build in effect a cognitive process theory of intelligence. In this work, for example, it has been shown that individual students differ not only on the many component skills involved in intellectual performance but also in their strategies for assembling the components of performance and their flexibility and adaptiveness in shifting strategies according to task demands. It has also been shown that these strategies can be changed to some degree through special training. This work would ultimately feed into an improved understanding of ability organization in general.

Another angle in this work comes from the decades of factor analytic research on cognitive abilities that has produced the now-familiar hierarchical model of ability organization (e.g., see Snow, 1978). In this model, mathematical ability is usually located at an intermediate level, connected to more general fluid analytic and crystallized intelligence above it, and more specific numerical concepts and computational skills below it. It is possible that combined information processing and factor analysis, aimed at mathematical ability but also at ability constructs connected to mathematical ability in the hierarchy, can produce an integrated picture of the component organization of higher-order and lower-order mathematical abilities. Perhaps this work could have been done without Begle's book, but with in on hand it is at least clear where one is working in relation to the rest of the field. Further, such an analysis of mathematical and related intellectual abilities should inform work in other regions, such as educational goals. The identification of higher-order cognitive processes in mathematical ability would help advance educational goal specification in a direction called for explicitly by Begle. To set goals in terms of the development of higher-order thinking and problem-solving abilities, one must have componential information processing models of such abilities. Cognitive psychology has made substantial progress in this direction in recent years, a point to be returned to in a later example. It is work reemphasizing that the hierarchical view of ability organization suggests that performance on quantitative ability tests seems to involve both crystallized (Gc) and fluid analytic (Gf) intelligence. That is, there is both a conceptual knowledge and a procedural knowledge or analytic reasoning aspect to mathematics. This nothing new, but an important implication is that both kinds of performance should be measured as both aptitude and outcome in research on mathematics instruction.

Two other examples show why this is important, and also exemplify the other line of research - - that of investigating relationships among critical variables chosen from different categories. One example comes from a dissertation study by Sharps (1973; see Snow, 1980), which was a field evaluation of Individually Prescribed Instruction (IPI) in fifth-grade reading and mathematics over the whole year. Sharps measured Gc and Gf ability separately as student aptitude variables and related them to reading and mathematics achievement outcomes under IPI and also under conventional treatment conditions. The contrast of IPI and conventional teaching involves a complex of many critical variables from four of Begle's categories. IPI is a system of individually paced instruction relying on specific pretests, geared to carefully specified objectives and sequenced content, with frequent checkpoints as guides and feedback on learner progress, plus mastery tests for each unit. As such it combines many features of the kinds of instructional conditions that have been found to help lower ability students in the past. It is "directed" learning to a far greater extent than is typical of conventional teaching and thus may remove some of the strategic information processing burdens that conventional instruction places on learners.

Sharps found that Gf aptitude showed low positive relation to outcome under both kinds of instruction, while Gc gave striking aptitude-treatment interactions (ATI) for both reading and mathematics outcomes. With conventional teaching, prior standing on crystallized ability correlated strongly with outcome. With IPI this relation was substantially reduced. In other words, the ATI pattern suggested that students high on Gc learned more under conventional conditions - - they were somehow hampered or held back by IPI - - while students low on Gc learned more with IPI - - they were not well served by conventional teaching. This result has since been replicated in another evaluation of IPI in sixth-grade mathematics reported by Crist-Whitzel and Hawley-Winne (1976). It is a good example of the kind of ATI patterns found in many other instrumental settings (see Cronbach & Snow, 1977, and Snow, 1977, for reviews of this literature).

Since reading and mathematics achievement are considered constituents of crystallized intelligence, one can also take these results to imply that conventional teaching develops higher Gc in those students already relatively proficient, while IPI develops higher Gc in those students previously less proficient. Another example suggests further that the contrast of Gc and Gf as outcomes of instruction may be quite important.

The massive Follow Through Evaluation conducted through the 1970's compared instructional treatments that were again complex mixtures of Begle's critical variables. The results suggested that different mixtures affect the development of Gc and Gf differently. The statistical analysis contrasting nine different Follow Through treatments was quite complex (Kennedy, 1978), but results for the five outcome variables used in the evaluation can be condensed to form two dimensions, one for Gc based on reading and mathematics tests and one for Gf based on the Raven Matrices test (see Snow & Yalow, in press). It is then seen that the Follow Through treatments that concentrated on drill and direct instruction on mathematics and reading skills produced the most improvement in these skills, but may as a result have hampered the development of fluid ability for their students. The two treatments that gave the highest Gc scores gave the lowest Gf scores. Treatments yielding higher Gf results emphasized variety and novel problem-solving, independent activity, and games, rather than reading and mathematics drills alone.

These examples taken together show that student aptitudes interact treatment variables, that treatment variables interact with outcomes, and further than the treatments involved in each case include several of Begle's critical variables. In fact, his lists of critical variables can be used to help describe the dimensions along which instructional treatments differ.

Two examples can now be added. One connects student aptitude with a critical environment variable. The other shows the potential of information processing analysis for connecting, A, T, and O variables for potential theory. Both deal with mathematical problem solving.

Webb (1977) reported a study showing the complicated effects of ability grouping. High school students were given instruction and then asked to solve mathematical problems, involving probability, polygonal numbers, and negative number bases. Students worked both individually and with a small group. For group work, they were assigned to either a mixed ability group or to a uniform hgh, medium, or low ability group. It was found that advantages in learning either individually or in groups varied as a function of group composition as well as individual student ability. High ability students did equally well as individuals in mixed ability groups, but not as well in uniform (i.e., high) ability groups. Medium ability students did best in uniform (i.e., medium) ability groups and worst in mixed ability groups. Low ability students did best in mixed ability groups, and worst in uniform (i.e., low) ability groups. In other words, heterogeneous ability grouping appeared better for high and low students, while homogeneous grouping appeared better for middle ability students. Further, observations of group discussion indicated that high ability students in mixed ability groups tended to act as teachers, offering explanations for lower ability students. Low ability students in mixed ability groups were provided with these explanations, and benefitted from them. In uniform ability groups, those medium ability students who offered explanations showed excellent performance; those who did not participate in this way performed worse than in individual learning. Thus, Webb concluded that active group members did better than in individual learning. The tendency to participate was a function of both individual ability and group composition. To predict learning outcome for an individual, then, one has to know that person's ability level, the mix of ability in that person's group, and the role the person takes in group interaction.

Such phenomena complicate immensely the problems of research in mathematics education, for they show the importance of social contexts and social roles in determining the properties of interrelationships among critical variables. Virtually all research on ability and learning conducted in laboratories has assumed that an individual psychological level of analysis would be sufficient.

A final example comes from research by Greeno (1978, 1980) and Anderson (Anderson, Kline & Beasley, 1980), using computer simulation models of problem solving. Greeno used periodic intensive interviews with high school students to build a representation of geometry knowledge as it is acquired, organized, stored, and used in the learner's cognitive system over a year's course work. In his model, knowledge is represented in the form of labeled networks (see also Norman & Rumelhart, 1975), processing is carried out in the form of production systems (see also Newell & Simon, 1972; Anderson, 1976), and the whole construction is expressed as a computer program. Anderson's simulation program is, in effect, a general abstract theory of learning, but again propositional networks represent factual-conceptual knowledge and production systems represent procedural knowledge. The model is used to show how generalizable ability might develop through exercise in performing example problems.

Both Greeno and Anderson have gone on to the important step of applying their theories to the analysis of learning and instruction in high school geometry. It is then seen that conventional texts and classroom lessons are often incomplete with respect to some features that the theories suggest are crucial for learning. Anderson notes that important background features of geometry problems are often left implicit in presented diagrams, and that example exercises are often insufficient to provide students with the critical juxtapositions that build up the cognitive operations his model needs for learning and transfer. Both Anderson and Greeno emphasize the critical importance of strategic planning, i.e., procedural knowledge. Greeno (1978, p. 72 emphasis added) sums up his analysis with the observation that:

> "The main implication of the analysis regarding instruction is in regard to strategic knowledge. Knowledge for pattern recognition and propositions to be used in making inferences are taught explicitly as part of the content of the geometry course. Strategic knowledge for setting goals and choosing plans is not a part of the explicit content of the course, although it seems likely that many students acquire strategic knowledge by induction from example problems that present strategic principles implicitly."

This prompts the hypothesis that some students learn the implicit yet essential strategic planning knowledge by induction from examples, some do not, and the difference would be predictable from intelligence tests. Learning from instruction that is incomplete in this way requires discovery learning whether it is called "learning by discovery" or not, and the learning ability involved is clearly what Spearman (in his 1923 definition of intelligence) meant by "education of relations": given two or more examples, produce the rule that connects them. To the extent that the examples given are also insufficient, then the ability involved is also one of "education of correlates", Spearman's term for producing connected examples given one example and a rule.

The point is, in short, that there are now detailed cognitive process models that seem able to make connections across aspects of student aptitude variables, treatment variables, and outcome variables in a rigorous way. It does not matter that Begle, in his section of problem solving, feared that interest in computer simulation of cognitive processing was dissipating. In fact, it was not. The important point is that he identified the area in 1976, before some of the most educationally relevant work had come out.

In summary, consider the following observations for further research:

1. Begle's critical variables, taken individually or in sets, focus research attention and can be used to derive more detailed descriptions of complex educational events and settings than have been available heretofore.

2. There are, indeed, critical variables to be analyzed, but they do not exist in isolation. Rather, they interact, and the complex of interactions may vary from setting to setting and locale to locale. Simple generalizations about these variables are not likely to be forthcoming.

3. Thus, the emphasis in future research should be on rich description, multivariate measurement, and evaluation of complex interactions in context, as well as on the production of theoretical models of interactions that serve to enrich these descriptive, multivariate evaluations.

4. Literature reviewing, as well as empirical research, should be conducted with this aim in view. One cannot simply count up studies that show one or another effect, as one would count up votes, to reach generalizable scientific laws. Continuous reevaluation is needed. The most important tools in both activities will be theoretical concepts that combine psychological and mathematical educational insights, and sophisticated quantitative methods, aimed at attaining richer description of acknowledgeable complex, dynamic, local phenomena.

The five years since the last congress have seen substantial advance in this direction, and I look forward to the next five years with great anticipation.

References

Anderson, J.R. Language, memory, and thought. Hillsdale, N.J.: Erlbaum, 1976.

Anderson, J.R., Kline, P.J., & Beasley, C.M., Jr. Complex learning processes. In Snow, R.E., Federico, P-A, & Montague, W.E. (Eds.) Aptitude, learning, and instruction Vol. 2: Cognitive process analyses of learning and problem solving. Hillsdale, N.J.: Erlbaum, 1980.

Begle, E.G. Critical variables in mathematics education. Washington, D.C.: Mathematical Association of America and the National Council of Teachers of Mathematics, 1979.

Crist-Whitzel, J.L., & Hawley-Winne, B.J. Individual differences and mathematics achievement: An investigation of aptitude-treatment interactions in an evaluation of three instructional approaches. Paper presented at the meeting of the American Educational Research Association, San Francisco, 1976.

Cronbach, L.J., & Snow, R.E. Aptitudes and instructional methods. N.Y.: Irvington, 1977.

Greeno, J.G. A study in problem solving. In Glaser, R. (Ed.) Advances in instructional psychology. Vol I: Hillsdale, N.J.: Erlbaum, 1978.

Greeno, J.G. Some examples of cognitive task analysis with instructional implications. In Snow, R.E., Federico, P-A., & Montague, W.E. (Eds.) Aptitudes, learning, and instruction Vol 2: Cognitive process analyses of learning and problem solving. Hillsdale, N.J.: Erlbaum, 1980.

Kennedy, M.M. Findings from the Follow Through Planned Variation Study. Educational Researcher 1978, 7(6), 3-11.

Newell, A. & Simon, H.A. Human problem solving. Englewood Cliffs, N.J.: Prentice-Hall, 1972.

Norman, D.A. & Rumelhart, D.E. Explorations in cognition. San Francisco: Freeman, 1975.

Pellegrino, J.W. & Glaser, R. Cognitive correlates and components in the analysis of individual differences. Intelligence, 1979, 3, 187-214.

Pelligrino, J.W. & Glaser, R. Components of inductive reasoning. In Snow, R.E., Federico, P-A., & Montague, W.E. (Eds.) Aptitude, learning, and instruction Vol I: Cognitive process analyses of aptitude. Hillsdale, N.J.: Erlbaum, 1980.

Sharps, R. A study of interactions between fluid and crystallized abilities of two methods of teaching reading and arithmetic. Unpublished doctoral dissertation. Pennsylvania State University, 1973.

Snow, R.E. Research on aptitude for learning: A progress report. In Shulman, L.S. (Ed.) Review of research in education. Vol 4. Itasca, Ill.: Peacock, 1977.

Snow, R.E. Theory and method for research on aptitude processes: A prospectus. Intelligence, 1978, 2, 225-278.

Snow, R.E., & Yalow, E. Intelligence and education. In Sternberg, R.J. (Ed.) Handbook of Human Intelligence. N.Y.: Cambridge University Press, in press.

Spearman, C. The nature of "intelligence" and the principles of cognition. London: MacMillan, 1923.

Sternberg, R.J. Intelligence, information processing, and analogical reasoning: The componential analysis of human abilities. Hillsdale, N.J.: Erlbaum, 1977.

Sternberg, R.J. The nature of mental abilities. American Psychologist. 1979, 34, 214-230.

Sternberg, R.J., Guyote, M.J., & Turner, M.E. Deductive reasoning. In Snow, R.E., Federico, P-A., & Mongague, W.E. (Eds.) Aptitude, learning, and instruction Vol. I: Cognitive process analyses of aptitude.

Webb, N.M. Learning in individual and small group settings. Technical Report No. 7. Aptitude Research Project, School of Education, Stanford University, 1977.

EDUCATIONAL PRODUCTIVITY: THEORY, EVIDENCE, AND PROSPECTS

Herbert J. Walberg
University of Illinois
Chicago, Illinois

Productivity Research

Applied research in medicine, agriculture, and engineering provides useful precedents for productivity research in education. Research in these practical fields drew from theory and research in the natural sciences to formulate and validate policies that have increased productivity by many magnitudes. After estimates of causal relations are made, such policies can be specified within an explicit, objective, rational framework; and the feasible treatments and conditions that maximize output can be recommended. Or, if the costs of the treatments and the value of outputs are known, the treatment that produces the greatest added value can be identified. Should such an approach be taken in education?

To answer this question, two areas of educational inquiry may be distinguished. The first is normative and chiefly concerns the ethics of educational means as well as the values of goals. The second area of inquiry is the scientific description or measurement of causal relations of means and goals. Given the costs in money or resouces as well as the value added by different admissible means, the most productive program to obtain the goals might be specified. Micro-economics and linear programming are routinely applied in agriculture and industry and increasingly in public affairs to combine normative and descriptive results in the formulation of prescriptive policy.

Such an approach to educational decision making has been difficult to apply for several reasons. There are disagreements about the value of goals and admissability of means; and some goals and means cannot be well measured, and their causal relations are uncertain. Moreover, the costs or resources necessary for many educational means are poorly estimated. Nevertheless, in an era of increased concern about the productivity of public institutions, educators must face these problems and attempt to give a reasoned account of the decision-making process of allocating scarce resources and efforts.

Although questions of productivity may sound simplistic and inhumane, increased producitivity can lead not only to lower costs but greater learning and savings of time, an irreplaceable value in human life. Even small percentage increases in productivity, moreover, would lead to immense savings in the precious resources of both educators' and students' efforts that involve nearly all the people at some time during their lives and perhaps a quarter or a third in formal programs at any given time. More could be said, of course, about education as a consumatory value in its own right, as a means to further education, and as a capital investment producing continuing benefits to the individual and society.

Production Theory

A psychological model of educational productivity is based on an economic theory of national, industrial, and agricultural productivity with origins in Austria, England and Sweden. Cobb and Douglas in the United States were the first to put the productivity theory into the form of an explicit mathematical equation and to carry out extensive empirical tests in 1928. The concrete example of the nineteenth century farm which inspired the theory is an intuitive starting point for understanding its key features. Adding more farm labor, land, or plows and other equipment increases grain yield, and each factor is necessary but insufficient by itself for production. Each factor, moreover, can substitute or trade-off for another but at diminishing rates of return; for example, it will eventually become less and less efficient to continue adding labor to a fixed amount of land and capital equipment. In modern general form, such trade-offs are represented in the Cobb-Douglas equation $O = aK^bL^c$ where **O** is output; land is subsumed under capital **K**; **a** is a constant; and **b** and **c** are factor coefficients for capital and labor. This simple equation usually accounts extremely well for economic processes that are known to be highly complex. Samuelson showed "how we can sometimes predict exactly how certain quite complicated heterogeneous capital models will behave by treating them as if they had come from a simple generating production function." Solow, in a discussion of economic growth, commented: "The art of successful theorizing is to make the inevitable simplifying assumptions in such a way that the final results are not very sensitive." The Cobb-Douglas equation has not only led to a great deal of academic research but also important policy and practical analyses such as the comparison of output per hour of labor or per dollar of capital invested across competing technologies.

A clear exposition of the equation, its parsimony and intuitive appeal, and Douglas's replication of its coefficients in various industries, countries, and time periods led to its widespread use. Other more complicated production equations by the Nobel-prize winners Arrow and Leontief, and others may eventually prove superior but they presently lack the long record of successful empirical verification of the Cobb-Douglas equation and are beyond the scope of the present treatment.

An analogous two-factor production function in education could be based on Aristotle's view of learning as determined by aptitude and experience; but these factors must be divided into smaller parts to reflect specialized empirical research in contemporary education and psychology. Statistical analysis of experimental effects and correlations in extensive empirical literature of the last decade in science and other subjects as well as a collection of 35 other quantitative research syntheses by 80 investigators reveal considerable evidence for the positive relations of learning with seven factors identified earlier - - student age, ability, and motivation; the quality and quantity of instruction (including self-instruction); and the social-psychological environment or morale of the class and home (and possibly two others that emerged in the course of the syntheses - - the peer-group environment and exposure to mass media such as television). By analogy to the Cobb-Douglas economic production equation, the psychological production equation relating estimated learning to the educational and psychological constructs

$$\text{Learning} = a(\text{Age})^b(\text{Abl})^c(\text{Mot})^d(\text{Qul})^e(\text{Qun})^f(\text{Cls})^g(\text{Hom})^h$$

suggests the following hypotheses and related points:

1. Increasing any factor increases learning; coefficients **b** through **h** are positive.

2. Increasing any factor while holding the others fixed produces diminishing marginal returns; **b** through **h** are each less than one.

3. A direct extension of the economic production function is that any factor equal to zero results in zero learning. In principle, any factor such as motivation or amount of instruction at the zero point would result in zero learning. Unlike capital and labor, however, the educational production factors may not have validly measurable zero points. Thus, it is more reasonable to hypothesize that when any factor is near minimum, it is unlikely that learning will be great, unless perhaps the other factors are near their maximum levels.

4. The factors substitute or trade off for one another but at diminishing rates of return.

5. The coefficients **b** through **h** are estimates of the percentage increase in learning associated, or, of causality is imputed, determined by a one percent increase in the corresponding factor. It would be possible to plot a profile of the factors for diagnosis and forecasting of learning production for each student (or aggregates such as classes or districts). Those achieving productively will have high and flat profiles. Unproductive students will have low, uneven profiles; and raising the lowest factors with the highest coefficients would produce the greatest returns (although it may be costly, difficult, and beyond the control of educators.)

The production function subsumes several psychological models of educational processes such as mastery learning and aptitude-treatment interactions, and explains a number of important educational phenomena such as the diminishing returns of learning to quantity of instruction; minimal learning when motivation or any other factor is minimal; the academically rich getting richer; the negative evaluations of compensatory programs; and the lack of linear substitution or trade-off of such factors as ability, motivation, and quantity of instruction. Should the model continue to survive reasonable empirical probes, it would make more confident our understanding of school learning and point to the most potent, least difficult, or, if prices of the factors can be estimated, cheapest means of maximizing learning. Combined with modern econometric analysis, moreover, it can serve as a comprehensive starting basis for observational research and educational policy.

Research Evidence

Even though educational research can be criticized for a variety of substantive, methodological, and practical reasons, much can be learned by systematic, quantitative analysis and synthesis of its available results before planning and conducting comprehensive and expensive field trials of the productivity theory, which require simultaneous, multivariate analysis of the complete set of productivity factors. This section first briefly discusses the primary methods of research synthesis and then summarizes results related to the nine productivity factors.

Methods of Research Synthesis

A recent special issue of the Journal Evaluation in Education: International Progress contains six papers on methods of criticism of research synthesis and 35 papers exemplifying substantive syntheses written by some 80 authors. The major centers of research synthesis and points of view are represented, together with extensive references and summaries of work completed and in progress. There is room here only for an explanation of the primary methods and several critical points.

The two primary methods of research synthesis are exemplified in data from Frederick and Walberg's, 1980, review. The first method, called the "vote count" or "box score" is simply the percentage of all studies that are positive. Thirty-one of 34 studies of time and learning measures, or 91 percent, showed a positive correlation.

The second method estimates the magnitude rather than the consistency of relations. Thus, the correlations between time and learning range from .13 to .71 with a median of about .40. Partial correlations that statistically control for the influence of ability, social class, or other variables are somewhat lower, with a median of about .35 on average. When the study results are expressed as means and standard deviations of groups (or functions of them), it is possible to calculate the ranking of a typical student in one group, say, an experimental group, in the distribution of students in another group, say, the untreated control group. Thus, for example, Lisakowski found, in a synthesis of 39 studies of the effects of six types of positive reinforcement or rewards such as praise and tokens on classroom learning, that the average experimentally-reinforced student outranks about 90 percent of the students in control groups.

Two difficult but not insuperable problems - - generality and causality - - confront research synthesis, especially those trying to formulate a research basis for educational policy. To a large extent, these problems are inherent in the original research itself, rather than in the methods of research synthesis in their more advanced applications.

The generality problem may be divided into questions of extrapolation and interpolation: Do the results synthesized generalize to other populations and conditions, in particular to those that have been unstudied or for whom the results are unpublished? And, do the results generalize across the populations and conditions for which results are available? Extrapolation threatens the validity of synthesis of published work because journal editors may favor the publication of positive, significant studies. Smith estimates that effects in unpublished work, mainly in doctoral dissertations, average about a third smaller than those in published studies. A solution is to obtain, at considerable cost and time, the unpublished work; but it can be argued that work found unpromising in yielding positive effects is left undone and thus the selection would still be positively biased.

Rosenthal, on the other hand, shows that, given the great statistical significance of collections of published studies, the probability of null effects in most cases is minimal. Furthermore, the lack of perfect reliability of educational measures and the lack of perfect curricular validity (correspondence between what is taught and what is tested) both diminish the estimates of relation between educational means and ends. It seems likely that less than optimal reliability and validity which lead to under-estimates of effects more than compensate for publication bias which may lead to over-estimates.

The interpolation problem can be readily solved by additional calculations. The most obvious questions in research synthesis concern the overall percentage of positive results and their average magnitude. But the next questions should concern the consistency of the results across student characteristics, educational conditions, subject matters, study outcomes, and the validity of each study. These questions can be answered by calculating separate results for each of the various categories of effects. Frederick and Walberg's data for example, shows that the consistency and magnitude of the time-learning relation are roughly constant across studies in which the measures ranged from years of schooling to the minutes of class time that students engaged in their lessons.

Most research syntheses yield results that are robust and roughly consistent across such categories. Such robustness is scientifically valuable because it indicates parsimonious, generalizable relations. Robust findings can also be applied more confidently and efficiently by educators because they suggest policies that are uniformly effective and relatively easier to implement than complicated, expensive procedures tailor-made, on untested assumptions, to subject matters, grade levels, sub-groups of students, and the like.

Evidence on Productivity Factors

As principal investigator, I have received about $1,000,000 in support of educational-productivity research, which started with research synthesis, led on to analyses of the vast data collected by the National Assessment of Educational Progress, and, finally, to joint research with Professor Hiroshi Azuma of the University of Tokyo on comparing Japan's highly productive schools with our own. The synthesis effort involved a dozen colleagues and students in the search, recording, tabulation, and analysis of about 4,000 effects and correlations in published and unpublished research. This work must be presented abstractly here and the reader is referred to several extensive reports for details.

We have compiled summary ratings on the research, its causal basis, and priorities for implementation for three sets of nine productivity factors. Our findings show that a large number of results are available on the correlations of the quality of instruction, home environment, and ability with learning outcomes; the other factors have been investigated less often. The correlations with learning are highest for ability measures such as IQ and prior achievement and for the educationally-stimulating, social-psychological morale or climate of the home and classroom. The amount and quality of instruction correlates moderately with learning. Less exposure to popular television programming and similar media correlates moderately with school learning. Although a considerable amount of evidence on the magnitude of correlations of the factors with learning has accumulated, much work remains to be done to achieve consensus on theoretical definitions of the factors and inexpensive, practical, and valid measures of several of them.

The causal direction of the correlations, moreover, are open to question. It seems plausible enough, that the amount and quality of instruction and the student's ability and motivation are direct causes of learning; but it is also reasonable to think that more learning leads to greater motivation and that motivated students can stimulate the teacher. It is unclear, moreover, whether the social-psychological factors influence affect-learning directly or facilitate it indirectly by increasing motivation and the amount and quality of instruction, including self instruction. Statistical controls for ability, social class, and related variables have reduced the causal uncertainty with respect to amount and quality of instruction and the class morale. But experimental controls with strict random assignment of students to educational conditions are the most convincing. More experiments on amount of instruction, for example, would allow us to assess more confidently the size of the direct, one-way effect of instructional time on learning rather than examining correlations between the two that are possibly inflated with reverse causation. Since a large number of experiments have been conducted on the quality of instruction and clasroom morale, one can be assured of the causal basis of these factors.

Prospects

Educational researchers must retain some skepticism about both the productivity theory and the evidence, just as medical researchers must continue to retain some doubt about the cigarette-lung cancer connection, notwithstanding statistical controls for social class and area population as well as experiments with animals. Thus more and better research is in order, particular that which includes all nine productivity factors - - quantity, correlation, measures, theory, plausible, statistical, experimental, research, and policy.

Decision-makers working in schools and colleges, however, cannot wait for definitive results and must face the challenge of increasing educational productivity as a day-to-day effort. On the basis of the evidence, the statistical and experimental controls, the plausibility, it seems to me that improvements in the amount and quality of instruction, the educationally-stimulating qualities of the social-psychological environments of the classes and homes, and exposure to mass media are likely to increase both the effectiveness and productivity of learning. It is possible, for example, that doubling the time students actually concentrate on instruction and study might nearly double learning and that improvements in instruction and morale, discussed in the references cited, might re-double it.

It appears that other nations have concentrated their resources and energies on educational effectiveness and productivity and may be pulling far ahead of the United States in certain important subjects. Izaak Wirzup, Professor of Mathematics at the University of Chicago, for example, points out that "recent Soviet educational mobilization, although not as spectacular as the launching of the first Sputnik, poses a formidable challenge to the national security of the U.S., one that will be much more difficult to meet." In extensive analyses of Soviet education, Wirzup found that:

1. 98 percent of the Soviet students in the U.S.S.R. completed secondary school compared with 75 percent in the U.S. in 1978;

2. compulsory mathematics courses for Soviet students cover the equivalent of 13 years of U.S. schooling, and five million graduates of Soviet

secondary schools had studied calculus for only one year; and

3. compulory science education in Soviet schools is equally extensive and intensive. Wirzup concluded that the U.S. is still superior in many areas of science and technology; but he warned that the Soviet investment of human resources and time to educate the masses and the elite in science will continue to challenge us.

In cases where comparative test performances are available, it also appears that our allies have been making outstanding progress in the teaching of science and mathematics, partly attributable to larger time allocations to study. Our recent grant from the National Science Foundation for the comparative study of elementary school science in Japan and the United States will allow specific estimates of class time, quality of instruction, home work, and other factors. It already seems clear from preliminary observations that Japanese students spend more time studying. They attend school from about 8 to 3 or 4 on weekdays, and 9 to 12 on Saturdays, do homework for an hour or two and often more each day in the later grades of elementary school, and attend special after-school tutoring programs in academic and related subjects. They also appear to concentrate intensively on their lessons during these allocated hours; and Japanese parents and educators also seem to provide highly constructive psychological environments in their homes and schools.

Conclusion

Agriculture, industry, and medicine made great strides in improving welfare as doubts arose about traditional, natural, and mystical practices; as the measurement of results intensified; as experimental findings were replicated, accumulated, and synthesized; and as their theoretical and practical implications were forcefully implemented. Education is no less open to humanistic and scientific inquiry and no lower in priority when about half the workers in modern nations are in knowledge industries. Although we need more and better educational research, it now points the way more definitively than ever before toward improvements that seem likely to increase educational effectiveness and productivity.

12.2 SOME CRITICAL VARIABLES REVISITED

BEGLE MEMORIAL SERIES ON "CRITICAL VARIABLES IN MATHEMATICS EDUCATION": CURRICULUM VARIABLES

Christine Keitel-Kreidt
Institut fur Didaktik der Mathematik
Bielefeld, Federal Republic of Germany

It is a great honour for me to be invited to participate in this Begle Memorial series, which by its designation is raised above rote professional communication. However, it is a delicate thing, too, for somebody who comes from far away, both locally and mentally - - from the European traditions.

The portrait of Edward Begle that the editors draw in their preface to the "Critical Variables in Mathematics Education" is very impressive. In a circle of scholars, among whom esteem and gratitude for this fascinating, charismatic personality still are alive, criticism may well look irrelevant or carping. However, as I understand it, this book provokes and encourages discussion at the same time: not only would the author probably not have disliked controversial discussion, furthermore, the book is thoroughly controversial in itself. For instance, take the contrast between the quotations from Begle's Lyon address and his final conclusions of this survey. In 1969 he firmly pointed out the way research in mathematics education should take. The outcomes, however, of progressing on this path for about a decade seem to have deeply disappointed him. There are lots of issues between the covers of this book, and the more we realize them, the more we become aware that Ed Begle did not leave us a heritage we simply can enjoy without working on it ourselves. It is not very likely, but the idea nevertheless is attractive that he did that deliberately. Anyway, in his last book his concern obviously was not so much a documentation of what has been accomplished in the field, but to narrow down the blank spaces on the map accordingly, he was less interested in giving answers than in putting questions. Some of them may have emerged just from the work on this book, and my impression is that the gloomy mood of its end may partly be due to a notion that further inquiry would be beyond the reach of his era. Now, if we look at the book this way, a critical discussion is not inappropriate, but carrying on his work.

The subject of my review is Begle's survey of "Curriculum Variables". This is an extended section, the contours of which are somewhat blurry. Although Begle specifies his understanding of "Curriculum" as pertaining to the content of the instructional program and to its mathematical objects, he does not stick very closely to ths restricted definition. Several of the 21 variables listed here could well be placed in other chapters. The variables themselves differ considerably in complexity and comprehensiveness. If a strick use of variables requires their independence from one another, one may find it difficult to understand how Begle determined and classified his subjects. As I see it, he may have intended to convey a picture of his findings as undistorted as possible and gave it an organisational fabric as neutral as possible. He just grouped specific studies with similar ones and placed them under a common heading. The red thread of the book are the views and assessments of a great math educator. Examining the book means to look at the historical position of its author. If, however, our interest is the subject of the book itself, we shall have to trace some structures in the multifarious material displayed before or eyes. I think the intention of this Memorial Series is to do a bit of both, and so I propose to give some reflexions on two topics: first, how do we perceive the landscape Begle depicted from our standpoint, after the work has been going on for some more years? And secondly, what can we suggest on the relation of empirical research in math education and theory; that will refer to the problems which Begle was mostly worried about: the deficit of a persistent theory in empirical research. It may also cast some light on Begle's own conception.

If we examine the 21 curriculum variables with respect to meaningful relations between them, it is obvious that most of them are closely related to several others. It may help to classify the matter if we subsume these related variables under a few headings as is shown in figure I. (Figure omitted. ed.)

These groups include all of Begle's curriculum variables but one. The last variable ("verbal variables") refers to an unspecified number of related variables for which only different directions of research are reported, which can be subsumed under other headings as well. Topics like "the effects of verbalizing discovered math objects" or "programs to develop math vocabulary" may be alloted to our heading "Advance organiser"; "the effects of achievement of the reading level of textbooks", and "the effects of math achievement of remedial reading programs" to our heading "Textbooks". Now let us have a look at our grouping.

The headings are:

1. Modern Math
2. Meaningful instruction
3. Learning sequence
4. Advance organizers
5. Textbooks

I think this grouping allows us to recognize a structure of levels which fairly well represent the fields in which research in math education has been carried out:

Number I - Modern Math refers to the level of subject analysis and content-oriented research in its proper sense.

Numbers 2 + 3 - Meaningful Instruction and Learning Sequence refer to the level of work on curriculum organisation and instructional programs.

Numbers 4 + 5 - Advance Organizers and Textbooks both refer to the level of instructional instruments and related studies.

Concern for the content level of math education marked the beginning of the reform era in the early 1950s. It seemed self-evident that the enormous gap between the subjects of University Mathematics and the content of School Math and failing subject-orientation of math education were tobe blamed for low achievement. The peak of attention to this field was in the New Math era. Fortunately, the content level of math education seemed to be the least problematic one, in particular because of the systematic lucidity of the discipline itself. Accordingly, research interest later decreased and shifted to the levels of curriculum organization and the instructional tools.

Obviously, studies on the level of curriculum organization and instructional programs had their great time when large curriculum projects came to be initiated everywhere. It was a period of enthusiastic reform activities and readiness for change. Nearly everything could be moved, at least for experiment. A gigantic amount of math educators' work was absorbed by curriculum development and referring evaluation. We know that the chances this period offered are gone and will not return easily. Activities on this level have diminished - educators had to get settled within the more or less imperfect structure they so far had been able to establish.

So its quite natural that research work in the last years more and more concentrated upon investigations into the instruments of teaching and learning. A demand for empirical knowledge on the field resulted from the recognition of the first New Math projects that a merely content-determined curriculum would not ensure efficiency of math education. Research interest was substantiated by the availability of new psychological learning theories: the progress of the behavioral psychology on the one side and the cognition theory following Piaget et al. on the other side. In addition, the realization of the cognitive approach made it necessary to enlarge the store of instructional material, and new technologies offered new instruments such as computer, hand-calculator, video-tape, etc. At first investigations were mainly preoccupied with the material side of these new instruments and with more practical questions: they were to be created and tested, they had to be given a place in the instructional process, and their usefulness was to be explored. After all, however, the textbook remained the by far most important tool for the development of knowledge. But nevertheless, all that helped to make educators in a new and more comprehensive way aware of the significance of the tools of instruction. In the very last years, a considerable amount of research work has been invested in that field, and a better understanding of it took shape, which in turn prompted a new research interest.

It was better recognized that the textbook - which may stand here for any kind of instruments - and the instructional process are neither identical nor homogeneous in their fabric. Instead, the text is an impartial complement to the instructional process and a lasting counterpart to the process-biased classes. Therefore it is particularly apt to develop independent working. However - and that also is more widely being recognized - there are no properties of the textbook which work automatically. Many efforts have been made to construct and test texts with regard to content, but rather little attention so far has been paid to what it means to use a text to develop knowledge. A text is not the mathematical object, but its representative substitute. Accordingly, a mathematical concept is not only within a text, but behind it. The textbook is the catalyzer between the concept and the students' apperceptive activity, and an object of the activity at the same time. This activity must carefully be trained, and we know relatively little about this process. If we are looking for trends, I should say the growing interest in the field is a trend. And we may state that it is a satisfying development not only for filling a considerable need. It also means a return of research to the normal classes, from high-aiming progressive programs to every-day instruction. And it means changes on the most reasonable level of implementation - in the teachers' minds; and it means good chances for successful implementation, for teachers' and teacher-students' interest in this kind of help may rightly be assumed. (1)

For the development of theoretical structures, which might underpin empirical research in math education and would give it coherence and direction, the sketch we can draw is less favourable. In fact, for the deficiency Begle deplored, no remedy is in view, and Begle's concern about it is as well-justified today as it was some years ago. On the contrary, new masses of studies have been added to the accumulation Begle reviewed without any visible progress. And research turning to smaller studies in a more restricted field makes it less likely that a grand design will come out of that. My opinion is that it is time now to abandon the idea that concepts and theoretical structures of math education as a new science would arise from carrying on research in that way. The question just is: why? If we could prove evidence by arguments, that probably would at the same time indicate a way for progress.

I would emphasize that I do not at all intend to deny the necessity and value of empirical research. Begle is right in saying: "To slight either the empirical observations or the theory building would be folly" (Begle's Lyon address). However, one should remember that Begle's objective to "follow the procedures used by or colleagues in physics, chemistry, biology, etc." (ibidem) has limitations and raises some methodological problems in the field of math education.

Doubts may be urged about the question to what extent far-reaching yet reliable conclusions can be drawn from individual studies. In his review on the book in the Mathematical Gazette (1980), Howson refers to the fact that the studies compiled by Begle are very different in dimension and authority. In addition we must ask: if no common objectives, premises, proceedings, or criteria of assessment exist, how can we just put a lot of such studies together in order to derive a sort of majority decision about an issue? We know very well that Galileo Galilei was right, although he had an army of scholars against him. Furthermore, if we know that empirical studies come into being rather randomly, can we deduce just from the fact that they are so numerous that they cover the whole field of research? In the first part we saw that empirical research focused on different subjects at different times. The direction of research interest depends strongly on outside developments. If the direction of research followed in the past obviously did not lead to theoretical structures of a new science, should we not assume that something happened not to be given sufficient attention?

Let us remember the three levels of empirical research we discovered in the first part. Obviously, content, curriculum organization and instructional instruments do not add up to the whole of curriculum problems. They are but one part, for they concern its concentration and realization. They do not pertain to decisions about what is to be realized, namely, the goals of the curriculum.

Clearly the level of goal setting is basic to all decisions about curriculum development, and is not restricted to that: not only is the shape of the syllabus deduced from goals, goals deeply influence the teachers' work and the students' activities as well. Indeed, goals affect the whole of math education on all levels from a general design of the curriculum up to very specific questions concerning the refinement of the instructional process.

A theory of math education (of the curriculum) cannot ignore the heavy impact of the goals. On the contrary, if such a theory is to explain something, if it is to provide criteria for rational decisions and their legitimation, it inevitably has to go back to the goals of math education. A theory essentially, though not exclusively, is a theory of goals, of their determination as well as their effects on the math curriculum.

Then, how did research in math education care for elucidating the level of goal setting and the processes of transforming them into instructional reality? We know that rather little has happened in this field. If we take Begle as representative, there is a striking disproportion between an enormous expenditure of work devoted to all kinds of variables on one side, and very scarce information about goals on the other. In fact, inquiry seems to have been paralyzed by the recognition "that the number of different kinds of interest which insist on being involved in setting goals for Math Education is so large that unanimous agreement on any set of clearly stated objectives is not to be expected" (Begle 1979, p. 13).

In my opinion the crucial point now is what consequences we draw from this fact. Ed Begle contents himself with a pragmatical response to the dilimma: "That each test be broad and wide-ranging enough to include a sufficient sample of the objectives of each interested party" (ibidem, p. 13). This may be a practical solution for the moment. But what happens in that black box of testmaking?

On what do testmakers base their decisions? How can they identify specific goals of particular "interested parties" if these parties often cannot explicate their goals themselves? And, first of all, what do they understand by goals? As a matter of fact, goals may be set on very different levels of generality, in that case, they have little or nothing to do with what really is happening in the curriculum. We all know that following current trends in curriculum revision, project materials easily could change the goals they postulated in their preambula without changing anything else substantially in the materials (2).

On the other hand, it is obvious that goals will not come out from starting with the smallest but concrete entities of earning objectives and from trying to buld up a structure of them. Proceeding step by step leads anywhere, but guidance has to be a premise, not a result. In short, not setting goals is the problem; setting meaningful and justified goals for the math crriculum is necessary.

Theoretical consideration has to be devoted to the connexion of interests, goals, objectives, and results. Presumably, goal setting has to be imagined as a process of condensation, starting from very concrete needs in very different parts of reality. Goals (similar to a liquor) have to conserve the essence of its particular origins and yet, on an intermediate level of abstractness, come to generic characteristics of a wider range of applicability. This does not work without theoretical interpretations.

Then, in turn, these meaningful goals will have to be retransformed into concreteness of yet another kind: the concreteness of classroom materials. The meaningfulness will prevent this process of transformation of being as arbitrary as at present it normally is. That, however, will reveal another problem much clearer than it is recognized today: that a good many goals cannot just be brought together as is done in the plurality model of testing, for they are contradictory to one another. As a consequence, decisions on goal setting will have to be substantiated, much better, and here then a vast field will become visible for empirical research (figure 2).

What is to be desired and why? What is possible and what not? What is really needed, needed by demands of society, by the claim of the individual for autonomy and competency, and needed for the progress of scientific knowledge? (That is not a question of asking somebody what he needs!) (3).

The very few empirical works which studied related questions clearly reveals that here Begle's statement is particularly true: that no progress is to come about "until we abandon our reliance on philosophical discussions based on dubious assumptions . . . ." That, if fact, is on what goals and their meaning for the math curriculum today are based. Begle himself did not apply his statement to extending empirical research in this direction. Maybe, he was too deeply involved in those decades of reform activities which we already perceive as a historical period, historical in meaning both limitations and persistent validity. So in my opinion Begle's work as a math educator is of historical significance in that, on the one side, we recognize its limitations, but on the other side enjoy its comprehensiveness which opens the view on the work to be done.

1. Keitel/Otte/Seeger: Text-Wissen-Tatigkeit. Scriptor. Konigstein 1980.

2. Howson/Keitel/Kilpatrick: Curriculum Development. Cambridge University Press. 1980.

3. Damerow/Elwitz/Keitel/Zimmer: Elementarmathematik: Lernen fur die Praxis? Ein exemplarischer Versuch zur Bestimmung fachuberschreitender Curriculumziele. Klett, Stuttgart 1974.

A version of this paper has appeared in Educational Studies in Mathematics, Vol. 13, No. 3, August 1982, pp. 257-263, Copyright 1982 by D. Reidel Publishing Co., Dordrecht, Holland.

THE VARIABLE OF DRILL REVISITED

Donald J. Dessart
The University of Tennesses
Knoxville, USA

The purpose of this paper is to consider the variable of "drill" as discussed in Begle's Critical Variables in Mathematics Education as a point of departure in order to: (1) briefly survey the research that has been completed since Begle's survey and (2) to suggest directions for future research.

The survey of American research was aided immensely by the annual reviews completed by Suydam and Weaver (1977, 1978, 1979). In searching for studies completed outside of the United States, written correspondence was condicted with Heinrich Bauersfeld and Jens-Holger Lorenz, University of Bielefeld; Alan Bell, University of Nottingham: Geoffrey Howson, University of Southampton; Hendrik Radatz, University of Gottingen; and Bevan Werry of Wellington, New Zealand. Based on this correspondence, it was concluded that empirical research on drill is largely an American enterprise, although the use of drill in instruction has been employed throughout the world and notably in Japan.

Research on the Usage of Drill

Begle (1979, pp. 64-65) noted that during the 1920's and 1930's there was an abundance of studies comparing the effects of the usage and the non-usage of drill in mathematical instruction. The wave of studies began to subside in the 1940's when it became obvious from research that meaningful instruction should always precede drill in order for drill to be effective.

Of seven recent studies related to the comparisons of instruction with and without drill, two found strengths for computer-assisted drill (Cranford, 1977; Palmer, 1973); two favored the use of conventional drill (Davidson, 1975; Parrish, 1976_; and three found no significant differences in favor of drill (Bird, 1978; Cummins, 1974; Starr, 1977).

The Cranford study conducted at the fifth- and sixth-grade levels over a semester found that students achieved at a faster rate in computations and applications in teacher-directed classes augmeted by computer-assisted drill and practice than in conventional classes. Palmer found similar results for elementary school children, but he noted that children in conventional classes had greater gains in mathematical reasoning.

Parrish investigated the effectiveness of drill homework at the ninth-grade level, and Davidson studied the effects of drill on addition-subtraction algorithmic learning with over 1000 students in grades one through nine. Parrish found significant differences in favor of drill homework usage, whereas Davidson concluded that maximum gains were made at the third- and fourth-grade levels when drill procedures (using an overhead projector) were utilized.

Massed Versus Distributed Drill and Practice

Six studies were concerned with investigating the effectiveness of various degrees of massed versus distributed drill and practice. Four of these investigations found significant differences in favor of various forms of distributed over massed practice (Butcher, 1976; Horwitz, 1975; Pence and Begle, 1974; Taylor, 1975); only two studies revealed no significant differences between the two kinds of drill (Begle, 1976; Weaver, 1976).

Butcher compared the effects of distributed and massed problem assignments in the homework of ninth-grade algebra students. Horwitz studied the effects of imemdiate and delayed practice (five days after learning) on the retention of rules involving operations with exponential notation on the 21st day after the learning of the rules. Butcher concluded that homework spread over three assignments was more effective than the same homework in one assignment. Horwitz found that delayed practice is more effective than immediate practice on delayed retention.

Pence and Begle investigated the effects of varying the number of examples and practice problems on the achievement of 7th graders in a unit on probability; they concluded that, for the low ability group, many practice problems produced significant results. Taylor found that varying the amount of practice until students reached criterion levels was more efficient than fixed amounts of practice.

Modifications of the Drill Process

Accepting the drill process as a useful instructional method, a number of researchers concentrated upon testing various modifications designed to enhance the drill experience:

Deines (1974) conducted a study on the effects of pacing in CAI drill and practice, in which (1) students were identified whose performance indicated inappropriate self-pacing, (2) optimum pacing was then determined for them, and (3) external pacing assistance was provided. It was concluded that for the improvement of skills, response accuracy measurement alone does not provide as valid a basis as when time is also measured.

Suppes et al. (1973) investigated five levels of intensity (10, 30, 70, 100, and 130 CAI drill sessions) with random groups of hearing-impaired children. All sessions of six to ten minutes each were equally distributed throughout the experimental period. The largest number of sessions resulted in significantly greater improvement in grade-placement testing than the smallest number of sessions.

Malone (1979) tested ten models for predicing a student's final grade placement from the amount of time spent on a computer-assisted drill and practice program. They found that the best predictions were provided by a piecewise power function model in which grade placement increases with the square root of CAI time.

Hohlfeld (1974) compared the effectiveness of an electronic calculator programmed for immediate feedback of basic multiplication facts with paper-and-pencil experiences without immediate feedback. He found that students who had reached the fifth grade but scored low enough on testing to warrant practice succeeded significantly better with the calculator practice on acquisition and short-term retention measures.

Howell (1978) found that peers following a well-designed flashcard procedure with fellow students could provide effective drill on multiplication facts with students in grades 5 through 8, where maintenance of instructional control is desired.

McClung (1977) stratified sixth-grade children into analytic and global groups by use of the Group Embedded Figures Test and randomly assigned the strata to two treatments (mental or written practice). He concluded that the children, regardless of cognitive style, can do equally well in work with equivalent fractions.

Bright, Harvey, and Wheeler (1979) investigated the use of games for retraining of skills with basic multiplication facts. The study, conducted during the first ten days of the school year withfourth-, fifth-, and sixth-grade children, revealed that the use of games requiring basic computations for 15 minutes a day were effective for reviewing skills.

Katz (1974) arranged items in algebra exercise sets in two ways: (1) unordered and (2) arranged by item analysis in ascending order of difficulty. The lower-ability students achieved significantly greater in achievement and retention using the ordered exercise sets.

Recommendations for Future Research

The following recommendations are offered for future research on drill and practice:

1. Since past research studies have clearly demonstrated that drill has a role in instruction as long as it is preceded by meaningful instruction (Begle, 1979, pp. 64-65), further comparative studies of instruction with and without drill are unwarranted.

2. An international survey on the role and use of drill in mathematical instruction would be most useful. For example, are there significant differences between the use of drill in Japan and England or Canada?

3. Further research on the following aspects of drill appear useful.

a) The most effective distribution of drill before and after criterion levels have been attained by students.

b) The most effective pacing and timing of drill sessions, particularly for slower students.

c) The effectiveness of various arrangements and orderings of the items of a drill experience.

d) The most useful patterns of drill experiences for students of various cognitive and learning styles.

4. More clinical studies of the acquisition of skills investigating the stages of preparation, active learning, practice, and retention as suggested by Suydam and Dessart (1980, pp. 230-235).

References

Begle, E.G. Critical Variables in Mathematics Education: Findings from a Survey of the Empirical Literature. Washington: Mathematical Association of America and National Council of Teachers of Mathematics, 1979.

Begle, E.G. The Effects of Varying the Number of Practice Problems, Number of Examples, and Location of the Practice Problems in Elementary School Geometry. Palo Alto, CA: Stanford University, 1976. (ERIC Document Reproduction Service No. ED 142 408)

Bird, B.A. Effect of Systematic Drill System on Computational Ability of Primary Children (Doctoral Dissertation, Brigham Young University, 1977.) Dissertation Abstracts International, 1978, 34, 1317A. (University Microfilms No. 78-16188)

Bright, G.W.; Harvey, J.G.; and Wheeler, M.M. Using Games to Retrain Skills with Basic Multiplicatin Facts. Journal for Research in Mathematics Education, 9 (2), 103-110.

Butcher, J.E. Comparison of the Effects of Distributed and Massed Problem Assignments on the Homework of Ninth-Grade Algebra Students. (Doctoral Dissertation, Rutgers University, 1975.) Dissertation Abstracts International, 1976, 36, 6586A-6588A. (University Microfilms No. 76-8683)

Cranford, H.R. A Study of the Effects of Computer-Assisted Instruction in Mathematics on the Achievement and Attitude of Pupils in Grades Five and Six in a Rural Setting. (Doctoral Dissertation, University of Southern Mississippi, 1976.) Dissertation Abstracts International, 1977, 37A, 5660A. (University Microfilms No. 77-5932)

Cummins, J.K. The Arithmetic Achievement of Sixth Grade Pupils and the Effect of Short Term, Well Designed Practice on Their Computational Abilities. (Doctoral Dissertation, University of California, Los Angeles, 1974.) Dissertation Abstracts International, 1975, 35, 4031B-4032B. (University Microfilms No. 75-2221)

Davidson, T.E. The Effects of Drill on Addition, Subtraction Learning with Implication of Piagetian Reversibility. (Doctoral Dissertation, Utah State University, 1975.) Dissertation Abstracts International, 1975, 36, 102A, (University Microfilms No. 75-14427)

Dienes, A.B. The Time Factor in Computer-Assisted Instruction. (Doctoral Dissertation, University of Toronto, 1972.) Dissertation Abstracts International, 1974, 34, 4981A. (National Library of Canada at Ottawa)

Hohlfeld, J.F. Effectiveness of an Immediate Feedback Device for Learning Basic Multiplication Facts (Doctoral Dissertation, Indiana University, 1973.) Dissertation Abstracts International, 1974, 34, 4563A. (University Microfilms No. 74-2670)

Horwitz, S. Effects of Amount of Immediate and of Delayed Practice on Retention of Mathematical Rules. Tallahassee, Florida: Florida State University, 1975. (ERIC Document Reproduction Service No. ED 120 010)

Howell, K.W. Using Peers in Drill-Type Instruction. Journal of Experimental Education, 46 (3), 52-56.

Katz, W.H. Effects of Item Placement in Exercise Sets on Achievement in Elementary Algebra. (Doctoral Dissertation, The University of Connecticut, 1974.) Dissertation Abstracts International, 1974, 34, 3691A-3692A. (University Microfilms No. 74-9205)

Malone, T.W. and Others. Projecting Student Trajectories in a Computer-Assisted Instruction Curriculum. Journal of Educational Psychology, 979, 71 (1), 75-84.

McClung, C.J. The Effects of Cognitive Style on Type of Practice. (Doctoral Dissertation, University of Southern California, 1976.) Dissertation Abstracts International, 1977, 37, 5706A.

Palmer, H. Three Evaluation Reports of Computer-Assisted Instruction in Drill-and-Practice Mathematics. Los Angeles, California: Los Angeles County Schools, 1973. (ERIC Document Reproduction Service No. ED 087 422)

Parrish, D.C. An Investigation of the Effects of Required Drill Homework Versus No Homework on Attitudes Toward and Achievement in Mathematics. (Doctoral Dissertation, University of Houston, 1976.) Dissertation Abstracts International, 1976, 37, 2040A. (University Microfilms No. 76-23369)

Pence, B. and Begle, E.G. Effects of Varying the Number of Examples and Practice Problems. SMESG Working Paper No. 7. Palo Alto, California: Stanford University, 1974. (ERIC Document Reproduction Service No. ED 142 405)

Starr, R.J. Modern Math Plus Computational Drills: Affective and Cognitive Results. School Science and Mathematics, 1977, 77, 601-604.

Suppes, P. and Others. Evaluation of Computer-Assisted Instruction in Elementary Mathematics for Hearing-Imparied Students. Palo Alto, California: Stanford University, 1973. (ERIC Document Reproduction Service No. ED 084 722)

Suydam, M. and Dessart, D.J. Skill Learning. In R.J. Shumway (Ed.) Research in Mathematics Education. Reston, VA: National Council of Teachers of Mathematics, 1980.

Taylor, S.C. The Effects of Mastery, Adaptive Mastery, and Non-Mastery Models on the Learning of a Mathematical Task. Tallahassee, Florida: Control Data Corporation, 1975. (ERIC Document Reproduction Service No. ED106 145)

Weaver, J.R. The Relative Effects of Massed Versus Distributed Practice Upon the Learning and Retention of Eighth Grade Mathematics. (Doctoral Dissertation, The University of Oklahoma, 1976.) Dissertation Abstracts International, 1976, 37, 2698A. (University Microfilms No. 76-24394)

STUDENT COGNITIVE VARIABLES AND INTERACTION WITH INSTRUCTIONAL TREATMENTS

L. Ray Carry
The University of Texas at Austin
USA

Begle mentioned Aptitude-Treatment Interaction (ATI) Research in his chapter on Curriculum Variables. He observed that . . . "some of the studies contradict the results of others." (p 62) and went on to say "it is too early, however, to give up this topic." (p 62). When he turned to student cognitive variables he had found insufficient research to discuss the mediating effect of instructional treatment on the aptitude-achievement relationships. His examples make clear his concern with such cognitive factors as spatial visualization and field independence. Considerable research has continued with these variables, e.g., McLeod and Briggs (1980) and DuRapau (1979).

The ultimate toal of ATI research is to adapt instruction to indivudal learner characteristics. The theory has been fully elaborated by Cronbach and Snow (1977). Briefly stated the researcher's task is to identify or develop instructional treatments that yield different linear relationships between aptitude and achievement. If one treatment significnatly favors lows and the other signficantly favors highs on the aptitude, then the interaction is called disordinal. Results are described as more steep or less steep slopes of regression lines or surfaces. Significant differences in slopes represent interactions. The decision rule for describing an interaction as disordinal has been debated, but if the regression lines or surfaces intersect near the center of the distribution I have chosen to consider the interaction disordinal.

My concern with ATI in mathematics began with my dissertation study (Carry, 1967). I desgined two treatments for quadratic inequalities. One treatment (graphical) was based on interpreting the graph of the ralated parabola; the other treatment (analytic) was based on algebraic properties of signed factors. I predicted that the treatments would interact with spatial visualization, the graphical treatment yielding the steeper slope. Without any substantive rationale other than pure intuition, I also measured general reasoning with the E.T.S. Necessary Arithmetic Operations (NAO) test and predicted that the analytic treatment would produce the steeper slope on NAO. A significant disordinal interaction was found in predicting transfer scores, but it was NAO that yielded the steeper slope under the graphical treatment. Since the result was counter to rationale, I searched for explanations and concluded the results were spurious because of low reliability on the transfer test.

Two studies followed using the same mathematical topic after improving the treatments and the testing instrument. The fist (Webb and Carry, 1975) found no significant interactions. The second (Eastman and Carry, 1975) yielded results that we described as confirming the original hypothesis "that spatial visualization will predict success in a graphical treatment and that general reasoning will predict success in an analytical treatment" (p. 148).

Salhab (1973) changed the mathematical topic to absolute value equations, created analogous treatments, but chose to describe the treatment distinctions as geometric vs algegraic. His results were consistent with Eastman and Carry.

A very confusing picture was emerging. In four successive studies where a treatment distinction had been built around the use of algebraic-analytic techniques vs. graphical-geometric techniques, a significant disordinal interaction had been found three times. In each case the formulation of theory focused on spatial visualization as the salient aptitude, but the non-spatial NAO test had been the stronger measure in the interaction finding. In the last two of these studies, the measure used for spatial visualization was the DAT abstract reasoning test. This entirely figural test nevertheless involves inferential reasoning to develop recognition of complex figural patterns. Furthermore, when analyzed in terms of single aptitudes the interactions were in the same direction for NAO and DAT scores. The most confusing results of all were that Carry's finding showed the graphical treatment producing the steeper slope on NAO, Salhab's finding showed the geometric treatment producing the steeper slope on NAO, but Eastman and Carry found that their analytic treatment produced the steeper slope on NAO.

Although no generalizable statement was possible at this point, some conclusions were warranted. First, my disseration results combined with the findings reported in Webb and Carry (1975) strongly suggested that immediate recall learning is not a useful outcome variable in studies of this sort. Second, the four studies combined seemed to show that characterizing treatment distinctions in terms of gross characteristics did not aid in describing the nature of the interactions found. These conclusions were supported by Cronbach and Snow (1977, pp. 510, 514).

We then turned more seriously to Cronbach and Snow's (1969) earlier report on ATI with particular attention to the statement:

> One who finds a high-slope and a low-slope treatment, considering only general ability or some segment thereof as an aptitude, is in a position to capitalize on ATI in instruction. (p. 189)

Since many psychologists consider both the NAO test and the DAT abstract reasoning test as segments of general ability, our findings fit that categroy. The idea that relating graphs and geometric concepts to algebraic entities would strengthen the correlation between spatial tests and achievement still seemed logical enough, but our results didn't provide support. On the other hand, our results with respect to NAO reinforced Cronbach and Snow's emphasis on general abiity. The pertinent issue at this point was the description of the treatment characteristics responsible for the interaction with NAO. That description eluded us, but our recognition that gross descriptions of treatment characteristics was non-productive suggested that considering the NAO test as a measure of General Reasoning Ability might also be simplistic. More specifically, the NAO test measures the efficiency with which a subject can make appropriate choice of arithmetic operations necessary to solve exercises in English words, in other words the ability to relate properties of arithmetic operations to their analogues in plain language.

Skemp's (1978) excellent article aided us in characterizing what we now consider the salient treatment distinction to produce ATI with NAO. Instruction designed to produce efficient immediate recall learning involving an emphasis on drill and practice, is in Skemp's sense of term "instrumental". He says, "If what is wanted is a page of right answers, instrumental mathematics can provide this more quickly, and easily" (p. 12). On the other hand, he argues that relational mathematics is ". . . more adaptable to new tasks" and ". . . easier to remember" (p. 12). He went on to clarify that by "easier to remember" he meant in relation to transfer tasks and to generalizing interrelated ideas.

As we reflected on the treatment contrasts from each of the three studies with significant ATI, we found that the steeper slope case was associated with the more relational treatment. We have conducted two additional studies since then reporting a similar significant ATI. DuRapau (1979) contrasted a transformational and non-transformational treatment of point and line symmetry and found the transformational treatment yielded the steeper slope on NAO predicting achievement. Hickey (1980) conducted an ATI study using standard course content for an entire semester in a freshman college mathematics course. She found essentially the same disordinal interaction with NAO and a mid-term test. The interaction failed to reach significance on the final examination, but the last half of her course was heavily concentrated on matrix computations, a topic which demanded a somewhat instrumental treatment for both groups. Both DuRapau's and Hickey's results occurred in the direction predicted.

In a totally unrelated investigation, McLeod and Briggs (1980) reported a signficant ATI for the NAO test with the topic of equivalence relations. They interpreted their findings as similar to Eastman and Carry (1975) but inconsistent with Cronbach and Snow's (1977) position relative to general ability. Their treatment distinctions were inductive vs. deductive. Having examined their treatments from the point of view of the relational vs. instrumental distinction, I find all the results consistent.

The studies mentioned show a consistent pattern of ATI with NAO as the aptitude measure. Skemp's instrumental vs relational distinction aids in characterizing treatment distinctions. But the underlying question of why the interaction occurs needs much study. Larkin et al. (1980) described a contrast in problem solving by experts and novices that I feel is related. Their position is that, for the novice problem solver, solving behavior seems to proceed from a loosely organized memory of salient facts with success depending largely on a trial-and-error testing of memorized chunks of information. By contrast the expert exhibits much less trial-and-error behavior. The expert memory structures are "not merely an unindexed compendium of facts, however. Instead large numbers of patterns serve as an index to guide the expert in a fraction of a second to relevant parts of the knowledge store" (p. 1342). In the case of problem solving in chess, they characterize the expert as having an "indexed" memory

> organized as a large set of productions consisting of a condition and an action. Whenever the stimulus to which a person is attending satisfies the conditions of one of his productions . . . the action is immediately evoked . . . (p. 1336-7).

Skemp emphasizes interrelatedness of concepts in his description of relational understanding. He does not address the organization of memory but the notion of hierarchical organization of memory is consistent with his view. Larkin's description of the novice solver's memory organization is certainly consistent with Skemp's position on instrumental understanding.

In my view, relational treatments are those which motivate the learner to store the concepts and processes taught into a memory structure hierarchially organized around similarities and dissimilarities and indexed by general characteristics. Instrumental treatments, on the other hand, motivate the learner to memorize chains of specifics and are indexed by specific recall cues.

We feel reasonably confident now in our subjective ability toapply the instrumental-relational labels to treatments that will interact with the NAO test, although the treatment distinction still lacks operational specificity. We also feel confident that the outcome measures most sensitive to this ATI effect are at the higher cognitive levels, particularly application and analysis levels from the NLSMA model (Romberg and Wilson, 1969). The remaining research challenge is to enhance the specificity of the treatment distinction and to explain why the interaction occurs. My conjecture as to the latter question is that the cognitive processes that facilitate higher scores on the NAO test are similar to those elicited by relational treatments and derive from hierarchially organized memory structures as described by Larkin et al. (1980).

I hope the above discussion serves to reinforce Cronbach and Snow's (1977, pp. 492-494) discussion of the complexity of the ATI problem and of the importance of a persistent effort not only to search for ATI but to seek deep understanding of treatment characteristics that produce them. I believe that the greatest payoff from this research effort will come from those understandings and their long-range implications for managing school learning.

<u>References</u>

Begle, E.G. <u>Critical Variables in Mathematics Education</u>. Washington: Mathematics Association of American and National Council of Teachers of Mathematics, 1979.

Carry, L.R. <u>Interaction of Visualization and General Reasoning Abilities with Instructional Treatment in Algebra</u>. Doctoral Dissertaion, Stanford University, 1967.

Cronbach, L.J. and Snow, R.E. <u>Individual Differences in Learning Ability as a Function of Instructional Variables</u>. Contract No. OCE 4-6-061269-1217 USOE, Stanford Unviersity, 1969.

Cronbach, L.J. and Snow, R.E. <u>Aptitudes and Instructional Methods</u>. New York: Irvington, 1977.

DuRapau, V.J. <u>Interaction of General Reasoning Ability and Gestalt and Analytic Strategies of Processing Spatial Tasks with Transformational and Non-Transformational Treatments in Secondary School Geometry</u>. Doctoral Dissertation, The University of Texas at Austin, 1979.

Eastman, P.M. and Carry, L.R. Interaction of Spatial Visualization and General Reasoning Abilities with Instructional Treatment in Quadratic Inequalities: A Further Investigation. <u>Journal for Reserach in Mathematics Education</u>, 1975, <u>6</u>, 142-149.

Hickey, P.A. <u>A Long Range Test of the Aptitude Treatment Interaction Hypothesis in College Level Mathematics</u>. Doctoral Dissertation, The University of Texas at Austin, 1980.

Larkin J. et al. Expert and Novice Performance in Solving Physics Problems. <u>Science</u>, 1980, <u>208</u>, 1335-1342.

McLeod, D.B. and Briggs, J.T. Interactions of Field Independence and General Reasoning with Inductive Instruction in Mathemaatics. <u>Journal for Research in Mathematics Education</u>, 1980, <u>11</u>, 94-103.

Romberg, T.A. and Wilson, J.W. <u>The Development of Tests</u>. NLSMA Report No. 7 (Wilson, Cahen, Begle, Eds.) Stanford: School Mathematics Study Group, 1969.

Salhab, M.T. <u>The Interaction Between Selected Cognitive Abilities and Instructional Treatments on Absolute Value Equations</u>. Doctoral Dissertation, The University of Texas at Austin, 1973.

Skemp, R.R. Relational Understanding and Instrumnental Understanding. <u>Arithmetic Teacher</u>, 1980, <u>26</u>, 9-15.

Webb, L.F. and Carry, L.R. Interaction of Spatial Visualization and General Reasoning Abilities with Instructinal Treatment: A Follow-Up Study. <u>Journal for Research in Mathematics Education</u>, 1975, <u>6</u>, 132-140.

SOME CRITICAL VARIABLES REVISITED

Jens Holger Lorenz
University of Bielefeld
Bielefeld, Federal Republic of Germany

The present article deals with those student variables which Begle called "affective", namely anxiety, mathematics attitudes, motivation, personality, school attitudes, self concept, and test anxiety, and their impact on mathematics learning and achievement.

Summing up the scientific knowledge till 1976, Begle concluded that the "survey of student variables provokes mixed reactions". Though quite a remarkable amount of empirical research with regard to these variables had been conducted, the results were by no means conclusive. In particular, the correlations between some affective variables and mathematics achievement ranged from "considerable" to "none" and the quality of the relationship, the problem of "which causes which", remained unclear.

Before taking a deeper look at each variable and the research since 1976, some more general remarks are wanted. The line of research since and probably because of Begle's book has changed in several respects.

a. Several studies and projects have been conducted to develop and evaluate intervention programs with the intention of increasing self concept, attitudes, and motivation of decreasing anxiety. The starting point for these studies is not necessarily the same, as some argue for a positive change of the affective variable in question because of its causal relationship with mathematics achievement, while others follow Begle's argument, saying that a reduction in anxiety or a positive attitude towards mathematics is in itself intrinsically desirable, even if it does not correlate with mathematics development.

b. The rationale of the studies has shifted in so far as the correlation between one affective variable and mathematics achievement is no longer the main point of interest. Rather, the main point is the relation between the affective variables, between these variables and modes of instruction, and finally between the variables and stages of the learning process.

c. The amount of empirical research seems to have decreased in favour of more theoretical endeavours reconceptualizing the underlying constructs in a mathematics specific way.

The following will not summarize those studies which are "follow ups" confirming or disconfirming previous research already cited by Begle, but is restricted to what seems to be promising lines for future research. Begle's categorization of the affective variables is adopted for this purpose.

Anxiety and Test Anxiety. Correlational studies on anxiety and mathematics achievement seem to be infertile and questionable as research in general psychology has shown that the relationship is not linear but rather of an inverted U-shape. Though these findings are not limited to mathematics learning but apply to all subjects and cognitive tasks, they were mainly established and confirmed in mathematics.

Mathematics differs from other subjects as students' anxiety level is higher than elsewhere. Typical differences in anxiety between the sexes favouring boys are reported in several studies, a problem which led the NSF to support research on its causes and possible remedial strategies. These projects are still being conducted and only partially published.

Begle's categorization of anxiety and test anxiety is equivalent to the state- and trait-anxiety distinction. Analytic research on the differential effects of these anxieties showed that trait-anxiety only indirectly influences mathematics achievement via state-anxiety, the latter having direct impact on the learning and problem-solving process. State anxiety exceeding a critical value prevents hypothesis generation in problem-solving tasks. Unfortunately, for empirical research the effect of anxiety on achievement may be reduced by the greater effort which is undertaken by high-anxious students, something that can hardly be held constant or partialled out.

State anxiety seems to be directly related to another affective variable, namely the student's self concept of his ability in mathematics. High self esteem of one's own task-specific ability reduces anxiety and vice versa, the (negative) correlation being quite considerable. A change towards a cognitive point of view in future research on affective variables may lead to the conclusion that self concept is not only an indicator of anxiety but the prior construct.

State anxiety and motivation relate in a similar way, with high anxiety reducing the motivation to undergo mathematical tasks. The same holds true for the anxiety-attitude relationship.

Some research studies considered the effects of teachers on anxiety and mathematics achievement. It is obvious and well-known that teachers differ in the amount of anxiety their instructional style produces within students. But as shown in these works, teachers even differ in the way student anxiety affects mathematics achievement; that is, in one class the negative effect of anxiety on test results is high whereas the influence of intelligence and creativity is low; in another class the effects are the other way round. It seems to be fertile to study the teacher behaviour leading to high/low anxiety impacts on mathematics achievement, last but not least because of its practical value. It could be unnecessary to let students undergo anxiety-reducing programs such as systematic desensitization; instead, train teachers in behaviours located by empirical evidence.

The influence of anxiety on achievement diminishes as students progress in a content area. If the instructional unit is presented in a hierarchical order (which it normally is), the high influence of anxiety on learning decreases whereas the influence of intelligence, being low at the beginning, increases. It is not clear yet if this result obtained in elementary arithmetic is generalizable over age and content.

School Attitudes and Mathematics Attitude. What has been said on anxiety holds true for attitudes as well. School attitudes are too global a construct to be directly attached or influential to any specific behaviour like mathematics achievement. School attitudes are built up by attitudes towards the several subjects and are more or less their average. The relationship between school attitudes and mathematics thus is not overwhelming and the correlation with mathematics achievement cannot be more than moderate.

Mathematics attitude, as Begle mentioned, improves slightly from fourth to sixth grade and declines afterwards. The inverse linkage with anxiety toward mathematics might explain this and could be due to sex specificities.

The causal relationship between attitudes and achievement has hardly been investigated since 1976, partly due to some theoretical orientation.

Self Concept. Theoretical assumptions confirmed by a considerable amount of data showed that the self concept as a well-defined construct does not exist, but should be split into several parts each related to task-specific activities. Not only could a student assess his mathematics ability differently from his English or science ability, but he could even be better in arithmetic than in geometry. To evaluate the relationship between task- or content-specific self concept and achievement, self-concept measures have been constructed.

Research findings show that mathematics self concept is built up in an analogous way as concept formation: success and failure lead to tentative hypotheses about one's own ability to cope with difficult tasks, these hypotheses are tested in future activities, and they become stable over time. Several points should be mentioned:

a. The student's assessment of his mathematics ability depends on the achievement of his reference group e.g., his class. Quite different achievement outcomes in an elementary school and a special school could correspond to the same self concept, as the levels calling a result success or failure are dissimilar. This could partly explain the relatively moderate correlations between self concept and achievement in field studies.

b. Data from studies in general psychology confirm the close relationship between self concept and persistence at difficult tasks. These findings, though not astonishing, are important as they relate cognitive with behavioral phenomena and elucidate the mechanisms between these two. High self esteem leads students to invest more effort when failing at a mathematics problem, because additional effort seems reasonable and promising to cope with the task. Low self concept, on the other hand, leads to a stop in persistence because success seems to be out of reach even with additional exertion. The construct that links self concept and persistence is effort calculation.

c. Anxiety and motivation as well as attitudes are related to self concept, though in different ways. Attitudes are effective correlates of students' cognitive representations of being successful in certain content areas. Anxiety can be conceptualized as fear of failing at a task (content, subject) which implies a cognitive calculation of perceived ability and assumed (subjective) task difficulty. If a student assesses his mathematics ability to be low, his fear of failing raises. The same applies to motivation, regarded as hope for success. Anxiety and motivation both influence students' willingness to persist at difficult tasks.

d. Persistence as a behavioral construct directly relates to mathematics achievement.

The model outlined in the foregoing discussion might establish a theoretical frame to explain the mechanisms between the effective variables discussed by Begle (see Figure 1). A lot of crucial points still remain for future research, which has to be rather subtle to enlighten the differential relationships between these variables and their interaction with mathematics achievement. A simple causality seems unrealistic in light of the data and does not meet the complexity of classroom interaction and affective and cognitive student variables. A more systematic approach including teacher behaviour and content area is needed.

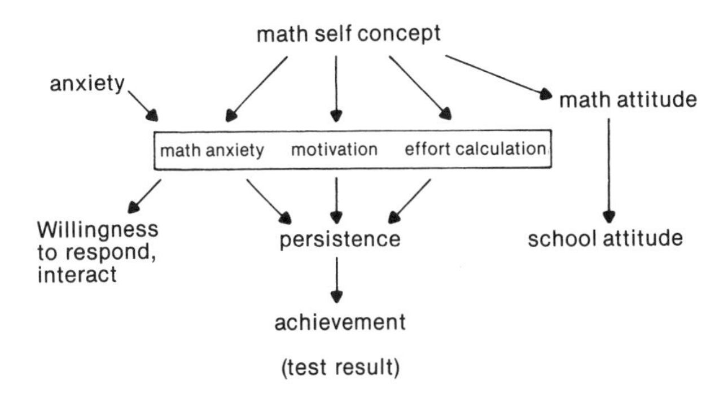

Figure 1. Process model of affective student variables.

One of the crucial points in future research will be the development of alternative methods and techniques to illuminate the process character of mathematics learning and teaching, only partially studied by research during the past four years.

The need for content specific instruments measuring the affective variables discussed in this paper might show (and some research evidence already did) that these variables are not constant over time, not even from one day to the next. These findings contrast the researcher's need for reliability and they limit the results in their external validity to specific content, classes, and even situations. Still, they may increase the practical value of research in mathematics education.

SOME CRITICAL VARIABLES REVISITED PROBLEM SOLVING STRATEGIES

Nicholas A. Branca
San Diego State University
San Diego, California

It gives me great pleasure to be a part of this memorial series honoring Edward G. Begle and his work in the field of research in mathematics education. I was his colleague for three years at Stanford University and was greatly influenced by his work. Professor Begle was strongly committed to empirical research and he, and his students, continually surveyed and criticized the literature resulting in, among other things, the manuscript "Critical Variables in Mathematics Education." I shall focus on strategies for problem solving; I will relate what Begle said regarding strategies, briefly review what has happened since the Begle review, and give some comments on the future in this area. I am presently working with fifth and sixth grade students to see what processes they use, are aware of using, and can be taught to use in the area of problem solving. I am planning to follow these same students for up to three years in a longitudinal study of their development. My work stems from the works of George Polya on problem solving and heuristics (Pflya, 1957) and V.A Krutetskii on mathematical abilities (Krutetskii, 1976). To me a problem (as contrasted with an exercise) is a situation for which no immediate algorithm or solution path comes to mind. There are differences of opinion on what the definition of problem solving is and consequently there are differences in how one believes problem-solving skills should be taught.

I have also worked in the area of strategies, replicating and extending a study by Zoltan Dienes and Malcom Jeeves in which strategies of students attempting to make sense out of seemingly chaotic stimuli were investigated (Dienes & Jeeves, 1965). In my view strategies are planning processes which students are aware of using, and use consistently. As in the case of problem solving, however, there are differences in definitions of strategies. The concepts of problem, problem-solving, and strategy mean different things to different people.

I should also admit that I come to this task with much apprehension. Begle's book has been described by some as an overview of the terrain of research in mathematic education. In reviewing research on problem-solving strategies, I felt as if the terrain became a swamp. Begle's book was divided into chapters, five of which were disjoint, surveying clearly identifiable critical variables. The chapter on problem-solving, however, was treated differently, illustrating the lack of an underlying structure or theory. Rather than summarizing what Begle had to say about strategies in that chapter, it can be reproduced verbatum.

Strategies

A substantial amount of effort has gone into attempts to find out what strategies students use in attempting to solve mathematical problems. The problems investigated have ranged from simple mathematical arithmetic ones suitable for primary students to complex algebraic and geometric problems meant for senior high school students.

Some of the most interesting of these studies have been computer simulations of putative problem-solving strategies. Further efforts along these lines would be welcome.

No clear-cut directions for mathematics education are provided by the findings of these studies. In fact, there are enough indications that problem-solving strategies are both problem- and student-specific often enough to suggest that hopes of finding one (or few) strategies which should be taught to all (or most) students are far too simplistic.

The strategies that Begle talks about deal with problems in arithmetic, algebra, and geometry. Begle was very pragmatic and his sense of problem is closely related to the teaching of mathematics as it occurs in the schools and is best exemplified by textbook problems. Nonroutine or puzzle-type problems are not dealt with. Thus, the sense of problem and strategy that I have is not the same as the sense indicated by Begle's survey. Since the survey was conducted, there have been a large number of studies that focused on strategies or considered strategies in a broader content. I have included a selected list of references which will give you some indication of the extent and breadth of studies and will provide a starting point for those of you who are interested in following up this topic. In general, the studies echo Begle's remark that strategies are problem specific. Since there is not time to summarize each of these studies in detail, I would like to illustrate the differences among the studies by focusing on two of the most recent.

The first is a study in which the effects of different strategies on the problem-solving performance of junior high school algebra students were investigated (Malin, 1979). Students were placed into three groups with each group receiving a different set of strategy instructions to solve typical algebra word problems of the following type:

Amy works 5 hours in the morning and earns $18. She earns $10 in the afternoon when she is paid at a different rate. Her average rate of pay is $3.50 per hour. How many hours does she work in the afternoon.

The problems can be solved using equations that relate the givens and unknowns according to the basic pay rate formula.

During a one-hour instructional treatment the students were asked to memorize the five equations and then taught to use them either using a "working forward" strategy, in which they would combine the givens to derive a new variable until the unknown was found; a "working backward" strategy, in which they would continue finding what variables they needed to use to derive the unknown; and a mixed strategy, in which they were taught to work backwards for the first step and then to solve by working forward. Malin used the time it took to complete a problem as a measure of the efficiency of the strategy used and concluded that there was a strong relationship between that and possible blind alley steps. The working backward strategy was found to be the most efficient, followed by the working forward strategy.

In this study the researcher's notion of problem, problem solving, and strategy are worth nothing. A problem is a specific type of algebra word problem for which standard algorithms exist. A strategy is a particular algorithm used, with no choice involved in its selection, and problem solving is applying the algorithm until a solution is found in a minimum amount of time. One can seriously question how generalizable the results are to problem solving in general and even to other types of word problems. The study is very limited in its application and leads to few if any other areas of investigation.

The second study is an exploratory study on the use of heuristics in problem solving. It was designed by Kantowski to search for regularities in processes used during the solution of nonroutine mathematics problems that would suggest hypotheses for further experimental as well as clinical studies of problem-solving processes, and for suggestions that could be tried in the development of instructional strategies and materials. The heuristic processes focused on were planning, memory for related problems, and looking back. A problem was defined for an individual as a situation for which no algorithm is known. A problem is considered subjective in nature depending on its particular relationsip with the problem solver. Examples of problems used are:

1. If there are 20 people in a gathering and each person shakes hands with all the others, how many handshakes were there?

2. What are the last 5 digits of the expansion in base 10 of 20!?

3. AHAHA
 +TEHE
 TEHAW

Problem solving is seen as a set of behaviors in which an individual combines the information in a way that enables him or her to reach the desired outcome - the solution to the problem. Kantowski considers this somewhat nebulous and states "a definition that satisfies all schools of thought has not been formalized and perhaps never will be."

The study used the methodology of the teaching experiment. Seventeen average and above average students in grades 9 through 12 were grouped together into a class and given a preliminary problem-solving session in which they were asked to solve four problems while "thinking aloud." This was followed by an instructional phase consisting of 22 sessions designed to introduce Polya's four steps in the solution of a problem, with emphasis on the problem-solving heuristics listed above. Three months after the instructional phase, a posttest of four problems was administered. Again students were asked to think aloud and the verbal protocols were recorded.

Problem-solving protocols were coded for heuristic use and analyzed. Because of the nature of the study, results are not generalizable. However, a large number of observations were made in order to give direction for future studies. They included observations on how students solved problems, what aspects of problems they focused on, and what techniques influenced their approaches. The study makes a number of recommendations for future studies, indicating what is needed in the way of research and instructional material. In contrast to the study by Malin, this study is applicable to many other areas of investigation.

There is an abundance of work being carried out in this area. However, because of the lack of agreement on the meaning of problems, problem solving or strategy, communication is difficult, if not impossible. What is needed is better communication among both researchers and practitioners. This can be accomplished by researchers being clear and explicit on what their definitions are. I'm not calling for agreement among researchers. At this point in time we should all pursue our avenues of interest as we view them according to our biases. I do not believe we are at the point where we can synthesize results, but we need to be able to understand what each of us has done and is doing. In the future perhaps a theory of problem solving will emerge that can account for our differences and similarities. Now is not the time to try to speak the same language or arrive at common definitions and methodologies. Attempts at this will take away from the task at hand, which is to gather systematic information.

I think Begle would approve of this. One of the last projects I worked on with him was on a statement of goals and objectives for mathematics education. He delayed this task until the early 70's, despite pressure from some early on in his curriculum development work to come with common, agreed-upon goals and objectives. He felt, twenty years ago, that there was much work to be done in curriculum development and time would be wasted and ill-spent on theory formalization. I think this is where we are today in problem-solving. There is a task at hand and much to be done. I encourage those of you engaged in research on problem solving and strategies to continue so in the spirit of Edward G. Begle.

Selected References

Begle, E.G. Critical Variables in Mathematics Education. Washington, D.C.: Mathematical Association of America and National Council of Teachers of Mathematics, 1979.

Branca, N.A. & Kilpatrick, J. The consistency of strategies in the learning of mathematical structures. Journal of Research in Mathematics Education, 1972, 3, 132-40.

Brown, J.S., Collins, A., & Harris, G. Artificial intelligence and learning strategies. In H.F. O'Neil (Ed.), Learning strategies. New York: Academic Press, 1978.

Days, H.C. Classifying algebra problems according to the complexity of their mathematical representation. Inn G.A. Goldin & C.E. McClintock (Eds.), Task variables in mathematical problem solving. Columbus, Ohio: ERIC, 1980.

Dienes, Z.P., & Jeeves, J.A. Thinking in structures. London: Hutchinson Educational, 1965.

Dienes, Z.P., & Jeeves, M.A. The effects of structural relations on transfer. London: Hutchinson Educational, 1970.

Goldin, G.A. Structure variables in problem-solving. In G.A. Goldin & C.E. McClintock (Eds.), Task variables in mathematical problem solving. Columbus, Ohio: ERIC, 1980.

Kantowski, M.G. The use of heuristics in problem-solving: An exploratory study. (NSF Technical Report SED 77 18543). Gainesville, Florida: University of Florida, 1979.

Klahr, D. Goal formation, planning, and learning by pre-school problem solvers or "My socks are in the dryer". In R.S. Siegler (Ed.), Children's thinking: What develops? Hillsdale, N.J.: Lawrence Erlbaum, 1978.

Kruteskii, V.A. The psychology of mathematical abilities in school children, J. Kilpatrick & I. Wirzup (Eds.). Chicago: University of Chicago Press, 1976.

Kuhm, G. The classification of problem-solving research variables. In G.A. Goldin & C.E.

McClintock (Eds.), Task variables in mathematical problem solving. Columbus, Ohio: ERIC 1980.

Landa, L.N. The ability to think - - how can it be taught? Soviet Education, 1976, 5, 4-66.

Lester, F.K. Issues in mathematical problem-solving research. A paper presented at a Research Pression of the Annual Meeting of the National Council of Teachers of Mathematics, Seattle, WA, April 1980.

Luger, G.F. Applications of problem structure. In G.A. Goldin & C.E. McClintock (Eds.), Task variables in mathematical problem solving. Columbus, Ohio: ERIC, 1980.

Malin, J.T. Strategies in mathematical problem solving. Journal of Educational Research, 1979, 73, 101-08.

McClintock, C.E. Heuristic processes as task variables. In G.A. Goldin & C.E. McClintock (Eds.), Task variables in mathematical problem solving. Columbus, Ohio: ERIC, 1980.

Pereira-Mendoza, L. Heuristic strategies utilized by high school students. The Alberta Journal of Educational Research, 1979, 25, 213-20.

Polya, G. How to solve it. (2nd ed.). Garden City, N.Y.: Doubleday, 1957.

Rigney, J.W. Learning strategies: A theoretical perspective. In H.F. O'Neil (Ed.), Learning strategies. New York: Academic Press, 1978.

Schoenfeld, A.H. Heuristic behavior variables in instruction. In G.A. Goldin & C.E. McClintock (Eds.), Task variables in mathematical problem solving. Columbus, Ohio: ERIC, 1980.

Snow, R.E. Aptitude processes. In Snow, Federico, & Montague (Eds.), Aptitude, learning & instruction: V & I Cognitive process analysis of aptitude. Hillsdale, N.J.: Lawrence Erlbaum, 1980.

Waters, W. Concept acquisition tasks. In G.A. Goldin & C.E. McClintock (Eds.), Task variables in mathematical problem solving. Columbus, Ohio: ERIC, 1980.

Webb, N. A review of the literature related to problem-solving tasks and problem-solving strategies used by students in grades four, five and six. (Technical Report). Bloomington, Indiana: Mathematics Education Development Center, 1977.

Wittman, E. Matrix strategies in heuristics. International Journal of Mathematical Education in Science and Technology. 1975, 6, 187-98.

12.3 SOME NEW DIRECTIONS FOR RESEARCH IN MATHEMATICS EDUCATION

RECENT RESEARCH IN MEMORY AND COGNITION

Richard E. Mayer
University of California
Santa Barbara, California

My assignment for this presentation is to review recent research on human memory and cognition with particular emphasis on research that might have some promise for education in mathematics. Trying to demonstrate the relevance of memory research for real-world problems like mathematics learning is a rather humbling experience. However, recent developments in the psychology of memory are noteworthy for two reasons: (1) psychologists have developed analytic tools for describing mental processes, structures, and knowledge that may be relevant to performance in mathematics, and (2) psychologists have begun to apply these tools to real-world problems including mathematics.

Let's begin with an example of what has been called "one of those 20th century fables", an algebra story problem:

> A sleek new blue motorboat traveled downstream in 120 minutes with a 8 km/h current. The return upstream trip against the same current took 3 hours. Find the speed of the motorboat in still water.

Take a minute and try to solve this problem. As you solve it, write down each step you take, and write down each piece of your existing knowledge you needed in order to solve the problem. Certainly, this problem is not meant to represent all the many facets of mathematics learning, but it does serve as an example of a typical problem present in secondary school algebra textbooks. The goals of this paper is to determine whether there is any useful information from research on human memory and cognition that is relevant to understanding how a student learns and solves problems like these.

The major contribution of modern research and theory in human memory and cognition to date concerns techniques for analyzing human mental life. The most relevant analytic techniques with respect to mathematics education are:

1. Techniques for analyzing the architecture of the human memory system, i.e., what are the characteristics of the basic memory stores and processes that are available to the student? Subsequent sections refer to these techniques as the information processing model.

2. Techniques for analyzing acquired knowledge, i.e., what is the content and structure of knowledge that a student brings to a problem? Subsequent sections refer to comprehension models (based on linguistic and factual knowledge), schema models (based on knowledge of problem forms), process models (based on knowledge of algorithms), and strategy models (based on knowledge of heuristics).

Information Processing Model

What are the memory stores and processes that people use when solving the motorboat problem? Although theorists disagree on many details, typical information processing models analyze human memory into separate stores, such as sensory memory, short-term memory, working memory, and long-term memory, and into many control processes such as attention, rehearsal, search of long-term memory, etc. (see Klatzky, 1980; Loftus & Loftus, 1976).

Analysis of individual differences in mathematics ability. Hunt (1978; Hunt, Lunneborg & Lewis, 1975) has used the information processing model in order to analyze differences between students who score high vs. low on standardized tests of verbal ability. For example, high verbal students are faster than low verbal students on retrieving a letter's name from long-term memory, high verbal students can hold more letters temporarily in working memory than low verbal students, and high verbal students can make a mental decision about whether two letters match faster than low verbal students. However, low verbal students are no slower on reaction time tasks per se, and the differences that are obtained are very small (e.g., 20 to 40 msec per letter). Although, the differences in processing are small, they add up to a considerable effect when you note that they must be performed thousands of times during the course of reading. Thus, Hunt et al. were able to characterize differences in verbal ability in terms of differences in the operating characteristics of the human information processing system.

Analysis of individual differences in mathematical ability. Individual differences in mathematical problem solving may be due to specific acquired knowledge (Greeno, 1980; Simon, 1980), as discussed in the subsequent four sections. In addition, there may be individual differences in the characteristics of the information processing system that are particularly important for mathematics:

1. Holding capacity of working memory. People may differ with respect to how much information they can handle at one time. For example, the motorboat problem requires that you hold the following facts: rate of current is 8 km/h; time downstream is 120 minutes; time upstream is 3 hours; speed in still water is unknown; distance upstream equals distance downstream. Case (1978) has shown that young children may not be able to handle more than two or three items at a time, and has provided instructional procedures that do not overload working memory. In a recent experiment, we (Mayer, Larkin & Kadane, 1980) found some evidence that writing in equation form (rather than words) may help reduce load on working memory. For example, problems were presented in word form such as, "Find a number such that 11 less than three times the number is the same as if 8 more than 3 times the number was divided by 2," or problems were presented in equation form such as, $(3X + 8)/2 = 3X - 11$. Although subjects were able to solve both kinds of problems, the pattern of response times suggested that subjects given equations were able to use a planning strategy - - looking a few steps ahead - - while subjects given word problems used a different strategy that required much less memory and no planning.

2. Type of code used in working memory. Subjects may differ with respect to the mode of representation in working memory such as visual vs. verbal vs. equation representation. For example, Hayes (1978) recently interviewed subjects as they solved simple arithmetic and algebra problems. Subjects differed greatly in their use of imagery; some reported heavy reliance on imagery, including "counting points" in images for digits, while others rarely or never used imagery. In addition, some problems elicited more imagery than others; for example, few people reported imagery in solving $5 + 7 = A$ but most subjects reported that they visually moved symbols in their working memory to solve $K + 13 = 8$. Individual differences in imagery are relevant to algebra story problems such as the motorboat problem, as well. In a recent study, we (Mayer & Bromage, 1980) asked subjects to either draw a picture, write an equation, or write in simple English as a translation for various story problems. Some problems were almost undrawable, such as, "Laura is 3 times as old as Maria was when Laura was as old as Maria is now. In 2 years Laura will be twice as old as Maria was 2 years ago. Find their present ages." Other problems evoked consistent pictorial representation, such as, "The area occupied by an unframed rectangular picture is 64 square inches less than the area occupied by the picture mounted in a frame 2 inches wide. What are the dimensions of the picture if it is 4 inches longer that it is wide?

3. Speed of mental operations is working memory. People may also differ with respect to how fast they can carry out a single mental operation in working memory. For example, Groen & Parkman (1972) and Resnick (1976) have developed models of simple addition and subtraction, and subjects may differ in terms of how long it takes to perform one step. Recently, we measured response times for simple algebraic operations such as moving a number from one side to another; this was accomplished, for example, by comparing time to solve $3X - 8 = 22$, to time to solve $3X = 30$. The difference between these two problems gives an estimate of the time to move a number; there were large individual differences, and time to move was faster for equation format as compared to word format.

4. Speed of search for target in long-term memory. Suppose you learned some formulas like: driving time = arrival time - leaving time, distance = speed x driving time, Distance = gas mileage x gas used, speed = wheel size x wheel speed. Then suppose I ask, arrival time = 6:00, leaving time = 4:00, average speed = 25, find distance. You have to search your memory for the required equation, determine a value for a variable in the equation, and search for another equation, and so on. People differ with respect to how fast they can search for and find target information in long-term memory. In our research (Mayer & Greeno, 1975; Mayer, 1978), search time varies greatly from person to person; also, search time for a new equation is faster when the material is meaningful (as above) than when it is a set of corresponding nonsense equations. Thus, equation format seems to slow down the speed of search in long-term memory, but is more efficient for temporarily holding information in working memory.

5. Selective attention. When you read a problem like the motorboat problem you need to key in on crucial facts such as time to go upstream is 3 hours, but you can ignore irrelevant information such as the type or color of the boat. Recent research by Robinson & Hayes (1978) shows that subjects are quite able to distinguish between what is "important" and what is "garbage" in an algebra story problem, although there

are certainly individual differences in selective attention.

6. Pattern matching. When you read a problem like the motorboat problem, you may look at a few critical features and say, "That's a current problem." You match some features of the problems to a set of features stored in memory. Similarly, in solving an algebra equation such as (8 + 3X)/2 = 3X - 11, you may note that one side is divided by 2. This pattern may alert you to the necessity to multiply both sides by 2. Since pattern matching is an important component of problem solving, individual differences in the speed and efficiency of the matching process could influence performance.

Comprehension Models

According to most descriptions of mathematical problem solving, the first step is to translate the words of the problems into an internal representation such as going from the words of an algebra story problem to an equation. For example, in order to solve the motorboat problem you need at least two kinds of acquired knowledge - linguistic knowledge, such as "motorboat" is a noun, "travel" is a verb; and factual knowledge, such as "120 minutes equals 2 hours" or "rivers have currents that run from upstream to downstream."

Role of linguistic and factual knowledge. Bobrow (1968) developed a computer program called STUDENT which solves simple problems such as:

> If the number of customers Tom gets is twice the square of 20 percent of the number of advertisements he runs, and the number of advertisements he runs is 45,, what is the number of customers Tom gets?

The translation phase of the program involves steps such as: (1) Copy the problem word for word. (2) Substitute words like "two times" for twice. (3) Locate each word or phrase that describes a variable, such as "the number of customers Tom gets" and note if two or more phrases refer to the same variable. (4) Break the problem into simple sentences. (5) Translate each simple sentence into variables, numbers, and operators, such as, (NUMBER OF CUSTOMERS TOM GETS) = 2 X (.20 (NUMBER OF ADVERTISEMENTS))^2, (NUMBER OF ADVERTISEMENTS) = 45, (NUMBER OF CUSTOMERS TOM GETS) = (X).

As you can see, STUDENT performs a very literal translation of words into equations. To do even this, however, requires that STUDENT have knowledge of the English language such as the ability to distinguish between operators and variables, and some factual knowledge such as knowing that a dime equals 10 cents or there are seven days in a week. More recently, Hayes & Simon (1974) have developed a program called UNDERSTAND that translates problems into an internal representation. In a recent study, we (Johnson, Ryan, Cook & Mayer, 1980) gave students a 30-minute lesson on how to translate an algebra story problem from words to equations using a modified version of Bobrow's procedure. Results indicated that instruction on translation improved the performance of low-ability subjects on tests of writing equations. Further work is needed to determine to what extent deficiencies in linguistics and factual knowledge influence students' problem-solving performance, and to determine means of diagnosing and remediating the lack of knowledge.

Schema Models

Let's return for a moment to the motorboat problem. Is there anything else you need to know beyond linguistic and factual knowledge? One basic idea you need to know can be expressed as "distance = rate x time". Further, the specific form of the motorboat problem is: (rate of powerboat + rate of current) x (time downstream) = (rate of powerboat - rate of current) x (time downstream). This equation represents the structure of the problem, and helps the student know what to look for; we will refer to the student's knowledge of the form of the problem as a "schema."

Understanding. You may have noticed that STUDENT does not really "understand" what it is doing, and does not care whether the variables are related to one another in a logical way. Is this the way humans solve problems? Paige & Simon (1966) gave students problems such as:

> The number of quarters a man has is seven times the number of dimes he has. The value of the dimes exceeds the value of quarters by two dollars and fifty cents. How many has he of each coin?

Some subjects behaved like STUDENT by generating translations of the sentences into equations. Other subjects recognized that something was wrong in this problem and corrected it by assuming that the second sentence said, "The value of the quarters exceeds the value of dimes by $2.50." Finally, some students looked at the problem and said, "This is impossible." Thus, while some students may use a literal translation, there is evidence that some students try to "understand" the problem.

How can people be encouraged successfully to "understand" a story problem? Paige & Simon asked subjects to draw pictures to represent each problem. when subjects drew integrated pictures, containing all the information in one diagram, they were much more likely to arrive at the correct answer. When students produced a series of sentence-by-sentence translations, they were more easily led astray. Thus, in addition to linguistic and factual knowledge, the student needs knowledge about how to put the variables together in a coherent way.

Roles of schemas. A further breakthrough concerning how people understand story problems like the motorboat problem comes from the work of Hinsley, Hayes & Simon (1977). Subjects were given a series of algebra problems from standard textbooks and were asked to arrange them into categories. Subjects were quite able to perform this task with much agreement, yielding 18 different categories such as river current (the category for the motorboat problem), DRT, work, triangle, interest, etc. Hinsley, Hayes & Simon found that subjects were able to categorize problems almost immediately. After hearing the first few words of a problem such as "A river steamer travels 36 miles downstream . . . ", a subject could say, "Hey, that's a river current problem." Hayes, Waterman & Robinson (1977) and Robinson & Hayes (2978) found that subjects use their schemas to make highly accurate judgements concerning what is important in a problem and what is not.

Many of the problems people have in solving story problems can come from using the wrong schema. For example, Hinsley, Hayes & Simon presented subjects with a problem that could be interpreted as either a triangle problem or a distance-rate-time problem. Subjects who opted for one interpretation paid attention to different information than subjects who opted for the other interpretation. Early work by Luchins (1942) demonstrated that shifting from problems that require one schema to problems that require another can cause "einstellung" (or "problem solving set"). For example, Loftus & Suppes (1972) found that a word problem was much more difficult to solve if it was a different type of problem from ones preceding it.

Greeno and his colleageus (Riley & Greeno, 1978; Heller & Greeno, 1978) have located schemas for children's word problems such as "cause/change" (Joe has 3 marbles. Tom gives him 5 more marbles. How many do they have together?), and "compare" (Joe has 3 marbles. Tom has 5 more marbles than Joe. How many marbles does Tom have?) Development of schemas seems to be in the order above; for example, second grders perform find on cause/change problems but poorly on compare problems. When asked to repeat compare problems, one-third of the children said, "Joe has 3 marbles. Tom has 5 marbles? How many marbles does Tom have?". Failure to solve word problems may, thus, be due to lack of appropriate schema rather than poor arithmetic or logical skills.

Can anything be done to make subjects more readily "understand" problems, i.e., help students find appropriate schema for problems? Our work on solving algebra equations shows that subjects are much faster at making appropriate deductions when the material is familiar (Mayer & Greeno, 1972; Mayer, 1975). Further, when instruction emphasizes familiar experiences, subjects re better able to recognize unanswerable problems and engage in transfer; for example, in teaching students to solve binomial probability problems, instruction could emphasize previous experience with batting averages, rainy days, and seating r people at n spaces at a dinner table (Mayer & Greeno, 1972; Mayer, 1975). Indeed, further research is needed to determine how to teach problem solvers to effectively use the schemas they have.

In order to gain a broader perspective on the nature of schemas for algebra story problems, I recently surveyed the exercise problems of major algebra textbooks used in California secondary schools (Mayer, 1980). Of the approximately 4000 problems collected, 25 general "families" of problems were located: motion, current, age, coin, work, part, dry mixture, wet mixture, percent, ratio, unit, cost, markup/discount/profit, interest, direct variation, inverse variation, digit, rectangle, circle, triangle, series, consecutive interger, physics, probability, arithmetic, and word (no story). Each type of problem has its own familiar plot line, but there is a major distinction between problems that require use of a formula (such as distance-rate-time for the motorboat problem) and problems that do not (such as the advertising problem solved by Bobrow's STUDENT). Also, for any major family, there are many distinct variants - - there were 14 different basic forms located for current problems, with observed frequencies of from 3 to 14. As an example, some of the common "motion" problems were: simple distance-rate-time, vehicles approaching from opposite directions, vehicles starting from the same point and moving out in opposite directions, one vehicle overtakes another, one vehicle makes a round trip, speed changes during the trip, two vehicles take the same amount of time to travel, two vehicles cover the same distance, etc. The procedure used for describing the format of any particular problem is to list the key information; there were four major types of statements: (1) assignment of a value to a variable, e.g., the time to travel downstream is 120 minutes, (2) designation of a relation between two variables, e.g., Laura's age is twice that of Anne, (3) assignment of an unknown to a variable, e.g., what is the speed of the boat in still water, (4) statement of fact, e.g., the cars took the same road. In a series of recall studies (Mayer & Bromage, 1980) it is clear that subjects tend to focus on these types of propositions when asked to remember a problem.

Process Models

Let's consider the motorboat problem again. As we have seen, in order to translate and represent the problem, the student needs appropriate linguistic, factual and schema knowledge. In order to solve the problem the student needs to know: (1) the rules of arithmetic, and (2) the rules of algebra.

Role of arithmetic algorithms. Groen & Parkman (1972) and Resnick (1976) have provided process models to represent the algorithms that children have for simple addition and subtraction. These models can be represented as flow charts, and can be fit to the reaction time data of children. One particularly interesting aspect of the process model work in arithmetic is that children tend to develop more sophisticated (e.g., larger) models as they get older. More recently, Brown & Burton (1978) have been able to model the procedural bugs students have for three digit subtraction problems. Bugs including borrowing from zero 103 - 34 = 158), subtracting the smaller from the larger number (258 - 118 = 145), and ignoring a zero (203 - 192 = 191). This work builds on previous analyses of error patterns, and allows for a precise description of the child's algorithm for subtraction.

Role of algebraic algorithms. Mayer, Larkin & Kadan (1980) have described a model for simple algebraic operations such as moving a variable from one side to the other. The process involves creating nodes, deleting nodes, and forming links among nodes. Simon (1980) has pointed out that algebra textbooks tend to emphasize algebraic algorithms (such as adding equal quantities to both sides of an equation) but fail to emphasize the conditions under which an algorithm should be applied.

Strategy Models

So far we have listed the types of knowledge needed to translate the motorboat problem (i.e., linguistic, factual, schema) and to solve the motorboat problem (i.e., algorithms). In addition, this section explores one final type of knowledge needed to control and use the knowledge at the right time - - strategic knowledge. For example, Polya (1968) has emphasized general strategies such as working backwards or working forwards to solve mathematical problems, and more recently Wickelgren (1974) has offered some general problem-solving strategies, some especially relevant to mathematics. Attempts to teach these strategies have met with some limited success (Lockhead & Clement, 1979).

Means-ends analysis for algebra equations. Newell & Simon (1972) have provided a technique for representing problems as a problem space. The problem space begins with a concrete description of the given (initial) state of the problem, the goal state, and all intervening states that can be generated by applying allowable operators. For example, the problem (8 + 3X) = 2 * (3X - 11) has the equation as its given state, X = some number as its goal state, and intermediate series like 8 + 3X = 6X - 22, etc. Newell & Simon offer a powerful strategy called "means-ends analysis" for guiding the problem-solving process. The procedure may be represented as a production system, a list of condition action pairs. For example, typical productions could be: "If there is an X term on both sides, move the one on the right to the left side of the equation". Problem solving involves moving through the problems space by executing the relevant productions in the production system (see Mayer, Larkin & Kadone, 1980, for an example). Recent work by Bundy (1965), Matz (1979), Davis & McKnight (1979), and Carry, Lewis & Bernard (1980) is directed offering precise models of the strategic knowledge required for solving algebra equations. Further work is needed to provide models that are closer to the real-world performance of school children.

Finally, recent work by Larkin (1979) and by Simon & Simon (1978) has compared the strategic knowledge of experts and novices concerning how to solve physics problems. Experts tend to rely on better organized production systems with more actions chunked into each production. Similarly, recent work by Mayer & Bayman (1980) concerning students' knowledge of how to use electronic calculators showed that experts relied on more sophisticated strategies than novices Further work should focus on the optimistic implication that expertness involves the acquisition of great amounts of knowledge rather than special mental abilities.

Conclusion

To understand mathematics learning and problem solving, you must decide the hardware (e.g., the information processing model) and software (e.g., comprehension, schema, process, strategy models) that a student brings to a task. There are promising signs that we will continue to see what Larkin (1979) calls a "fruitful interaction" between the needs of the mathematics classroom and the development of analytic theories in cognitive psychology.

References

Bobrow, D.G. Natural Language Input for a Computer Problem-Solving System. In M. Minsky (Ed.), Semantic information processing. Cambridge, Mass.: MIT Press, 1968.

Brown, J.S. & Burton, R.R. Diagnostic models for procedural bugs in basic mathematical skills. Cognitive Science, 1978, 2, 155-192.

Bundy, A. Analyzing mathematical proofs. Edinburgh: University of Edinburgh, Department of Artificial Intelligence, Research Report No. 2, 1975.

Case, R. Implications of developmental psychology for the design of effective instruction. In A.M. Lesgold, J.W. Pellegrino, S.D. Fokkema & R. Glaser (Eds.) Cognitive Psychology and Instruction. New York: Plenum, 1978.

Carry, L.R., Lewis, C. & Bernard, J.E. Psychology and equation solving: An information processing study. Austin: University of Texas, Department of Curriculum and Instruction, NSF Final Report 78-22293, 1980.

Davis, R.B. & McKnight, C.C. Modeling and processes of mathematical thinking. The Journal of Children's Mathematical Behavior, 1979, 2, 91-113.

Greeno, J.G. Trends in the theory of knowledge for problem solving. In D.T. Tuma & F. Reif (Eds.), Problem solving and education: Issues in teaching and research. Hillsdale, N.J.: Erlbaum, 1980.

Groen, G.J. & Parkman, J.M. A chronometric analysis of simple addition. Psychological Review, 1972, 79, 329-343.

Hayes, J.R. & Simon, H.A. Understanding written instructions. In L.W. Gregg (Ed.), Knowledge and cognition. Hillsdale, N.J.: Erlbaum, 1974.

Hayes, J.R., Waterman, D.A. & Robinson, C.S. Identifying relevant aspects of a problem text. Cognitive Science, 1977, 1, 297-313.

Heiler, J. & Greeno, J.G. Semantic processing in arithmetic word problem solving. Paper presented at the Midwestern Psychological Association, 1978.

Hinsley, D., Hayes, J.R. & Simon, H.A. From words in equations. In P. Carpenter & M. Just (Eds.), Cognitive processes in comprehension. Hillsdale, N.J.: Erlbaum, 1977.

Hunt, E. Mechanics of verbal ability. Psychological Review, 1978, 85, 109-130.

Hunt, E., Lunneborg, C. & Lewis, J. What does it mean to be high verbal? Cognitive Psychology, 1975, 7, 194-277.

Johnson, J., Ryan, K., Cook, L. & Mayer, R.E. From words to equations: An instructional study. Santa Barbara: Department of Psychology, Series on Learning and Cognition, Report No. 80-4, 1980.

Klatzky, R. Human memory: Second Edition. San Francisco: Freeman, 1980.

Larkin, J.H. Information processing models and science instruction. In J. Lochhead & J. Clement (Eds.), Cognitive process instruction. Philadelphia: Franklin Institute Press, 1979.

Larkin, J.H. Teaching problem solving in physics: The psychological laboratory and the practical classroom. In D.T. Tuma and F. Reif (Eds.), Problem solving and education: Issues in teaching and research. Hillsdale, N.J.: Erlbaum, 1980.

Lochhead, J. & Clement, J. Cognitive process instruction. Philadelphia: Franklin Institute Press, 1979.

Loftus, G.R. & Loftus, E.F. Human memory: The processing of information. Hillsdale, N.J.: Erlbaum, 1976.

Loftus, E.F. & Suppes, P. Structural viables that determine problem-solving difficulty in computer-assisted instruction. Journal of Educational Psychology, 1972, 63, 531-542.

Luchins, A.S. Mechanization in problem solving. Psychological Monographs, 1942, 54:6, Whole No. 248.

Matz, M. Towards a process model for high school algebra errors. Paper presented at Conference on Cognitive Processes in Algebra, University of Pittsburgh, 1979.

Mayer, R.E. Information processing variables in learning to solve problems. Review of Educational Research, 1975, 45, 525-541.

Mayer, R.E. Effects of meaningfulness on the representation of knowledge and the process of inference for mathematical problem solving. In R. Revlin & R.E. Mayer (Eds.), Human Reasoning, Washington: Winston-Wiley, 1978.

Mayer, R.E. Schemas for algebra story problems. Santa Barbara: Department of Psychology, Series in Learning & Cognition, Report No. 80-3, 1980.

Mayer, R.E. & Bayman, P. Analysis of students' intuitions about the operation of electronic calculators. Santa Barbara: Department of Psychology, Series in Learning & Cogntion, Report No. 80-4, 1980.

Mayer, R.E. & Bromage, B. Recall of algebra story problems. Santa Barbara: Department of Psychology, Series in Learning and Cognition. Report No. 80-5, 1980.

Mayer, R.E. & Greeno, J.G. Structural differences between learning outcomes produced by different instructional methods. Journal of Educational Psychology, 1972, 63, 165-173.

Mayer, R.E. & Greeno, J.G. Effects of meaningfulness and organization on problem solving and computability judgments. Memory & Cognition, 1975, 3, 356-362.

Mayer, R.E. & Larkin, J.H. & Kadane, J. Analysis of the skill of solving equations. Santa Barbara: Department of Psychology, Series in Learning & Cognition, Report No. 80-2, 1980.

Newell, A. & Simon, H.A. Human problem solving. Englewood Cliffs, N.J.: Prentice-Hall, 1972.

Paige, J.M. & Simon, H.A. Cognitive process in solving algebra word problems. In B. Kleinmuntz (ed.), Problem solving: Research, method and theory. New York: Wiley, 1966.

Polya, G. Mathematical discovery. New York: Wiley, 1968.

Resnick, L.B. Task analysis in instructional design: Some cases from mathematics. In D. Klahr (Ed.), Cognition and instruction. Hillsdale, N.J.: Erlbaum, 1976.

Riley, M.S. & Greeno, J.G. Importance of semantic structure in the difficulty of arithmetic word problems. Paper presented at the Midwestern Psychological Association, 1978.

Robinson, C.S. & Hayes, J.R. Making inferences about relevance in understanding problems. In R. Revlin & R.E. Mayer (Eds.), Human reasoning. Washington: Winston/Wiley, 1978.

Simon, H.A. Problem solving and education. In D.T. Tuma & F. Reif (Eds.), Problem solving and education: Issues in teaching and research. Hillsdale, N.J.: Erlbaum, 1980.

Simon, D.P. & Simon, H.A. Individual differences in solving physics problems. In R. Siegler (Ed.), Cihldren's thinking: What develops? Hillsdale, N.J.: Erlbaum, 1978.

Wickelgren, W. How to solve problems. San Francisco: Freeman, 1974.

A MATHEMATICS EDUCATOR VIEWS RESEARCH ON HUMAN INFORMATION PROCESSING AND MEMORY: ON A CLEAR DAY I CAN SEE IMPLICATIONS

Edward A. Silver
San Diego State University
San Diego, California, U.S.A.

In what follows I will provide a mathematics educator's view of recent research of information processing and memory, assessing the implications of this research for mathematics education and identifying some productive directions for future research. Since it is impossible to be comprehensive in the allotted time and space, I will focus on highlights and sketch the implications and new directions I see in broad strokes rather than elaborate detail.

I. How Can Research on Memory Influence Mathematics Education?

Some ways in which research on human memory can influence the enterprise of mathematics education will be outlined and six types of influence, or modes of implication, will be given, with one or two brief examples for each.

Osmosis. Perhaps the best model to characterize the potential influence of memory research on mathematics education is the diffusion model. As Simon suggested in his recent address to the American Educational Association, it is probably unlikely that modern cognitive science research will produce many directly relevant and usable products for educational practitioners; rather, its influence will be more subtle. Similarly, modern research on human memory is likely to have its greatest influence on mathematics education in the subtle ways in which it structures our thinking about tasks and learners.

Example. Perhaps the most direct way in which memory research may influence mathematics education is by identifying particular concepts or theoretical constructs for consideration. For example, the currently popular notion of the memory schema (Bobrow & Norman, 1975; Rumelhart & Ortony, 1977) is one that may prove useful to mathematics educators in understanding the mechanisms of skilled problem solving in certain domains, such as algebra word problems (Hinsley, Hayes, & Simon, 1977). Not only does the notion of schema provide a new descriptor to apply to commonly observed phenomena in mathematical problem solving but it also provides a link to a considerable body of related research on prose text comprehension.

Another useful construct suggested by the memory literature is imagery. The classic writings of Hadamard (1945), Poincare (1913), and other commentators on mathematical thinking have suggested the potential importance of imagery in mathematical activity, and evidence exists that some successful students spontaneously use images to organize their knowledge;yet, imagery has not been carefully studied with respect to mathematical performace. We know that imagery can be useful and important in mental calculation (e.g., Paivio, 1971), solution of linear ordering syllogisms (e.g., Potts, 1972), and complex problem solving (e.g., Larkin, 1977; Simon & Simon, 1978). In general, the role of imagery in mathematical learning and performance presents itself as a rich, possibly fruitful, area for further research.

Reduction. Another type of influence that memory research can have on mathematics education is exemplified by the work of Hunt and his colleagues (e.g., Hunt, Lunneborg, & Lewis, 1975) in analyzing verbal ability. They have found a relationship between individual differences on a standard test of verbal ability and differences in performance on several basic information-processing tasks, such as recognizing that two letters are identical or that they name the same letter or different letters. In particular, Hunt and his colleagues found that subjects with high verbal ability were substantially better at handling information in short-term memory than their low verbal counterparts. Since the particular processes studied by Hunt are not of great importance in performing the test of verbal ability (Carroll, 1976), Hunt's results may inform us about the way in which differences in verbal ability occur accumulatively over time. The reduction of verbal ability to millisecond differences in performance on basic information-processing tasks suggests the fundamental importance of elementary processes that are often ignored in mathematics education research.

A different example of this mode of implication is found in the analysis of chess ability by Simon and his colleagues (e.g., Simon & Gilmartin, 1973). They have proposed that a chess master needs to acquire about 50,000 "chunks" of information in order to be able to recognize commonly encountered configurations in a chess game. Simon (1980) suggested that this estimate of 50,000 "chunks" is reasonable for many areas of expertise. Although there is some lack of clarity on the term "chunk" in mathematical settings, Simon's analysis demonstrates the fundamental importance of basic knowledge acquisition processes.

Complication. Not all influences of memory research in mathematics education may be characterized as reducing complexity. In fact, another very important type of influence comes from work suggesting the complexity of task performance that we may currently regard as rather elementary. For example, a popular and almost universal distinction is made in the mathematics education problem-solving literature between a problem and an exercise, the former being complex (path to the goal unknown) and the latter being rather elementary (applying a known alogorithm). Nevertheless, Greeno (1980) has called for abandoning the distinction. Among the several bases for his call is the observation from recent research on human information processing that, when one carefully examines the performance of apparently "routine" exercises, one observes the essential characteristics of the problem-solving process. Whether or not one agrees with Greeno's analysis, it should influence us to consider more seriously the complexity of routine task performance.

Another example of this type of influence is found in the work of Brown and his colleagues (e.g., Brown & Burton, 1978) in analyzing algorithmic performance. For example, the well-known Brown and Burton analysis of the decomposition subtraction algorithm has suggested the cognitive complexity of this routine procedure.

Specification. One of the most distinctive features of research on human information processing is the detailed specification of task characteristics and performance routines. Of course, task analysis is not new to mathematics educators, but task analysis on the basis of cognitive processes and components offers a new perspective. For example, Sternberg's (1977) componential analysis of analogical reasoning abilities may suggest a powerful technique that is adaptable to mathematical task performance in certain restricted domains, such as equation solving. Furthermore, Carroll's (1976) analysis of tests of mental abilities with respect to their demands on short-term, working, and long-term memories may offer a useful technique for analyzing tests of mathematical abilities. In general, returning to the influence of osmosis, the kinds of detailed task analyses that are characteristic of information-processing research should serve to remind us that we are better able to understand the performance data we obtain when we understand completely the knowledge base and processes needed to perform our tasks. A recent analysis of task variables in mathematical problem solving (Goldin & McClintock, 1979) indicates the sort of productive influence that information-processing specificity can have on mathematics education.

Implication. The last type of influence that I will discuss is fairly obvious: implication by implication. Here I refer to those suggestions found in the memory literature that have direct application to situations of interest to mathematics educators. For example, mathematics educators who study problem solving often use verbal "think aloud" protocols as a trace of the underlying thought processes and problem-solving mechanisms. Ericsson and Simonn (1980) have recently analyzed verbal self-report data from the perspective of memory architecture. Their analysis specifies under what conditions one might expect a subject to be able to report accurately information about internal mental processes as a part of the verbal protocol. Their analysis, based on the notion of information availability in working memory, should have direct and important implications for researchers interested in mathematical problem solving.

Another example of this type of influence is the work of Case and his colleagues (e.g., Case, 1975; 1978) on working memory and instructional task requirements. In particular, Case has argued that the well-known memory limitations of young children should constrain instructional tasks and that instruction can be developed to minimize the effects of memory limitations. His practical implications for instruction, drawn from extensive research on working memory, should be of interest to developers of mathematics curriculum at the early elementary school levels.

2. Some New Directions for Research in Mathematics Education.

I hope that the first section of this paper has, perhaps by osmosis, identified some issues, techniques, and directions for future research. In this section I will outline three specific memory-related areas that I see as important for your consideration as future research directions.

Representational Linkages. Like love, mathematical knowledge is a many-splendored thing. For example, it can be represented in different forms: propositional, procedural, episodic, and imaginal (Gagne & White, 1978). Although some attention has been paid to the nature of propositional and procedural knowledge as it exists in the minds of learners (Shavelson & Porton, 1979), far less is known about the critical linkages that must exist between classes of representations in order to account for skilled mathematical performance. Since the increased use of pre-, co-, and post-instructional games and the continued use of manipulative materials in mathematics instruction heighted the likelihood that episodic and imaginal representations will be formed as well as propositional and procedural representations, we need to study carefully the mechanisms that provide for strong links between representational modes. It might be particularly fruitful to examine the role of imagery as a mediational link between episodic, procedural, and propositional representation.

Schemata. Davis and his colleagues (Davis, Jockusch, & McKnight, 1978) have demonstrated that schemata, or "frames", can be usefully applied to explain many aspects of algebraic task performance. As yet, however, we have few instantiations of mathematical schemata in other domains. Among those areas that might be particularly promising are integration and differentiation in calculus, theorem proving in geometry, mental calculation, and theorem proving in abstract algebra. Nevertheless, we need to go beyond mere instantiations of schemata, merely observing that persons tend to behave in stereotypic ways when faced with highly standardized tasks. We need to know more about how schemata develop and how to encourage their development in positive ways.

Careful observation of individual differences in schema content and/or schema utilization should lead to valuable extensions of current theory. It is well known that all students exposed to the same instructions on solving algebra problems do not acquire the same schemata; far less is known about why. Krutetskii's (1976) observation of differences between high ability students and low ability students in their tendency to notice and to recall information about a problem's structure and some of my own subsequent research on students' perceptions of and recall of problem structure (Silver, 1979; in press) suggest that it may be productive to explore the differences between the schemata of high ability and low ability students. In addition, it might be productive to study schema differences among students at the same ability level.

More attention should be given to the mechanisms of schema formation and adaptation. The processes of assimilation and accomodation suggested by Piaget (Piaget & Inhelder, 1973) would appear to be the appropriate analogues. In exploring these educationally relevant issues we need to proceed with great care, especially since research by a number of Soviet psychologists (e.g. Kalmykova, 1947/1969) reminds us that the formation of certain types of problem schemata may not always be a desirable goal of our teaching.

Metacognition. Metacognition refers to one's awareness of one's own cognitive processes and products. There has recently been accumulated, largely by developmental psychologists, a significant body of research on some aspects of metacognition, such as metamemory (e.g., Brown, 1978; Flavell & Wellman, 1977). Among the more provocative suggestions from this literature is Flavell's (1976) observed link between metacognition and planning processes in problem solving. This observation is generally consistent both with the popular mathematics education folklore that good problem solvers know more about "what they are doing" when they solve problems and with Gurova's (1959/1969) suggestion that problem-solving ability was linked to awareness of one's mental processes while solving problems. Although I have reservations about Gurova's reported observations, perhaps due to the sketchiness of her report, it would appear that the metacognitive aspects of mathematical problem solving are important and worthy of further systematic study. For example, it may be useful to recast Polya's (1957) heuristics in the light of metacognition. In particular, since the heuristics are stated as external metacognitive prompts, they might be studied more productively as they relate to the attainment of expertise rather than its manifestation. Autocriticism is another metacognitive aspect of problem solving that deserves a closer look by mathematics educators.

3. Concluding Remarks.

This paper has presented a sanguine view of ways in which memory research can influence mathematics education. On days that are less clear than the one mentioned in my title, I worry about how far memory research can be applied to mathematics education. In particular, despite the extensive research on human memory, I wonder how much closer we are to answering a fundamental question of interest to most mathematics educators: How is is that, when faced with a situation different from the one in which needed knowledge was acquired, one recognizes the need for that knowledge and retrieves it? I hope that this paper will contribute to continued dialogue among mathematics educators and between psychologists and mathematics educators on this question and on other issues of mutual concern.

References

Bobrow, D.G. & Norman, D.A. Some principles of memory schemata. In D.G. Bobrow & A Collins (Eds.), Representation and understanding. New York: Academic Press, 1975.

Brown, A.L. Knowing when, where, and how to remember: A problem of metacognition. In R. Glaser (Ed.), Advances in instructional psychology (Vol. 1). Hillsdale, N.J.: Erlbaum, 1978.

Brown, J.S. & Burton, R.R. Diagnostic models for procedural bugs in mathematical skills. Cognitive Science, 1978, 2, 155-192.

Carroll, J.B. Psychometric tests as cognitive tasks: "A new structure of intellect." In L.B. Resnick (Ed.), The nature of intelligence. Hillsdale, N.J.: Erlbaum, 1976.

Case, R. Gearing the demands of instructions to the developmental capacities of the learner. Review of Educational Research, 1975, 45, 59-87.

Case, R. Intellectual development from birth to adulthood: A new-Piagetian interpretation. In R.S. Siegler (Ed.), Children's thinking: What develops? Hillsdale, N.J.: Erlbaum, 1978.

Davis, R.B., Jockusch, E., & McNight, C. Cognitive processes in learning algebra. The Journal of Children's Mathematical Behavior, 1978, 2(1), 10-320.

Ericsson, K.A., & Simon, H.A. Verbal reports as data. Psychological Review, 1980, 87 (3), 215-251.

Flavell, J.H. Metacognitive aspects of problem solving. In L.B. Resnick (Ed.), The nature of intelligence. Hillsdale, N.J.: Erlbaum, 1976.

Flavell, J.H. & Wellman, H.M. Metamemory. In R.V. Kail & J.W. Hagen (Eds.), Perspectives on the development of memory and cognition. Hillsdale, N.J.: Erlbaum, 1977.

Gagne, R.M. & White, R.T. Memory structures and learning outcomes. Review of Educational Research, 1978, 48 (2), 187-222.

Goldin, G.A. & McClintock, C.E. Task variables in mathematical problem solving. Columbus, OH: ERIC/SMEAC, 1979.

Greeno, J.G. Trends in the theory of knowledge for problem solving. In D.T. TUma & F. Reif (Eds.), Problem solving and education. Hillsdale, N.J.: Erlbaum, 1980.

Gurova, L.L. Schoolchildren's awareness of their own mental operations in solving arithmetic problems (Originally published in 1959). In J. Kilpatrick & I. Wirszup (Eds.) Soviet studies in the psychology of learning and teaching mathematics (Vol. 3). Stanford: SMSG, 1969.

Hadamard, J. The psychology of invention in the mathematical field. Princeton, N.J.: Princeton University Press, 1945.

Hinsley, D.A., Hayes, J.R., & Simon, H.A. From words to equations - meaning and representations in algebra word problems. In M. Just & P. Carpenter (Eds.), Cognitive processes in comprehension. Hillsdale, N.J.: Erlbaum, 1977.

Hunt, E., Lunneborg, C., & Lewis, J. What does it mean to be high verbal? Cognitive Psychology, 1975, 7, 197-227.

Kalmykova, Z.I. Psychological analysis of the formation of a concept of a problem type (Originally published in 1947.). In J. Kilpatrick & I Wirszup (Eds.). Soviet studies in the psychology of learning and teaching mathematics (Vol. 6). Stanford: SMSG, 1969.

Krutetskii, V.A. The psychology of mathematical abilities in schoolchildren, J. Kilpatrick and I. Wirszup (Eds.). Chicago: University of Chicago Press, 1976.

Larkin, J.I. Skilled problem solving in physics: A hierarchical planning model. Unpublished manuscript, University of California at Berkeley, September 1977.

Paivio, A. Imagery and verbal processes. New York: Holt, Rinehart and Winston, 1971.

Piaget, J. & Inhelder, B. Memory and Intelligence. New York: Basic Books, 1973.

Poincare, H. Mathematical creation. In The Foundations of Science. (Translated by G.H. Halstead.) New York: Science Press, 1913.

Polya, G. How to solve it (2nd ed.). Garden City, N.Y.: Doubleday, 1957.

Potts, G.R. Information processing strategies used in the encoding of linear orderings. Journal of Verbal Learning and Verbal Behavior, 1972 11, 727-740.

Rumelhart, D.E. & Ortony, A. The representation of knowledge in memory. In R.C. Anderson, R.J. Spiro, & W.E. Montague (Eds.), Schooling and the acquisition of knowledge. Hillsdale, N.J.: Erlbaum, 1977.

Shavelson, R.J. & Porton, V.M. An information processing approach to research on mathematics learning and problem solving. Paper presented at the Conference on Modeling Mathematical Cognitive Development, Athens, GA., May, 1979.

Silver, E.A. Recall of mathematical problem information: Solving related problems. Journal for Research in Mathematics Education, in press.

Silver, E.A. Student perceptions of relatedness among mathematical verbal problems. Journal for Research in Mathematics Education, 1979, 10, 195-210.

Simon, H.A. Problem solving and education. In D.T. Tuma & F. Reif (Eds.), Problem solving and education. Hillsdale, N.J.: Erlbaum, 1980.

Simon, H.A. & Gilmartin, K. A simulation of memory for chess positions. Cognitive Psychology, 1973, 5, 29-46.

Simon, D.P. & Simon, H.A. Individual differences in solving physics problems. In R. Siegler (Ed.), Children's thinking: What develops? Hillsdale, N.J.: Erlbaum, 1978.

Sternberg, R.J. Intelligence, information processing, and analogical reasoning: The componential analysis of human abilities. Hillsdale, N.J.: Erlbaum, 1976.

DIRECTION NOUVELLES DES RECHERCHES DANS LA PEDAGOGIE DES MATHEMATIQUES CONCEPTUALISEE D'APRES LES SCIENCES COGNITIVES

Robert B. Davis
Laboratoire de Recherche Pedagogique
L'Universite d'Illinois a Champaign-Urbana

Pour emprunter la definition de Kuhn, un nouveau paradigme a fait son apparition dans la pedagogie des mathematiques et est presente, entre autres, dans les travaux de Papert, Minsky, Schank, Winograd, Winston, Wilks, Sussman, Matz, Rissland, Simon, Larkin, Lenat, Karplus, McDermott, Lement, et Lochhead. Le but de cette notice est d'attirer l'attention sur ce nouveau paradigme qui, je le crois, est en passe de devenir l'un des plus importants courants de la recherche en pedagogie des mathematiques des annees 80 et 90 - peut-etre le courant principal.

Les caracteristiques de ce nouveau paradigme. Bien qu'il y ait evidemment des variantes, il y a probablement quatre caracteristiques qui definissent le nouveau paradigme. 1) Le plus souvent, on en obtient les donnees durant une "task based interview" (une interview basee sur la resolution de quelque probleme mathematique). L'etudiant s'applique a quelque tache mathematique tandis que le chercheur l'observe et/ou l'interroge. Cependant d'autres donnees sont parfois utilisees, y compris les reponses de l'eleve a des problemes presentes par l'ordinateur. ii) Typiquement, le point de concentration de notre recherche est plutot de dimensions restreintes; il se resume souvent en une seule reponse de l'eleve. Par exemple, l'objectif scientifique immediat est peut-etre d'expliquer pourqoui X** vient juste de multiplier 4 par 4 pour obtenir 8. (Bien sur, l'objectif a long terme est d'expliquer comment les etudiants dans leur ensemble pensent aux mathemtiques dans une grande variete de situations differentes.) iii) Les donnees sont interpretees en fonction d'une theorie du traitement de l'information qui a ete posee en postulat. (La recherche pedagogique semble trop souvent s'attendre a ce qu'une theorie jaillisse spontanement d'une foule de donnees, comme d'antan l'on supposait que des matieres fecales naissaient spontanement les mouches. En fait, de la geometrie Euclidienne a la physique de Newton, en passant par la mecanique quantique moderne, la theorie a du etre postulee.) iv) Les concepts principaux sont empruntes de, ou du moins inspires par, les resultats des travaux sur l'intelligence artificielle. (A vrai dire, s'il y a un seul ouvrage fecond dans ce domaine, c'est certainement celui de Minsky publie en 1975.)

Les antecedents. Bien que le "nouveau paradigme", considere dans son ensemble, indique une nouvelle direction de la recherche pedagogique des mathematiques, il a une mine d'antecedents, par exemple: i) En ce qui concerne les "task based interviews" il y a les travaux fameux de Piaget, et plus recemment ceux de Ginsburg, Erlwanger, Easley, Clement, etc. ii) En ce qui concerne la concentration intentionnelle de notre champ d'analyse, cf. l'oeuvre "task analysis" (c'est a dire, la separation des taches necessaires pour resoudre un probleme en ses plus petits mais toujours identifiables components) de Gagne, ainsi que les travaux sur les "behavioral objectives" (les actions qui peuvent etre observees et mesurees de dehors dont l'execution sert de critere a l'accomplissement de quelque objectif pedagogique), ou le contenue est minutieusement decoupe. (Le point de vue de l'intelligence artificielle est cependant different de celui des "behavioral objectives", a savoir que les decoupages des "behavioral objectives" ne sont pas, d'ordinaire, percus clairement en relations avec un context plus general - c'est parmi les principales critiques que l'on adresse a la methode des "behavioral objectives". Au contraire, les pieces de l'intelligence artificielle sont considerees comme les maillons de la chaine hypothetique des evenements durant le traitement de l'information.) iii) L'essort dans cette nouvelle direction a ete donne par les travaux de Kilpatrick, Begle, Wilson et Wirszup qui a rendu accessible aux lecteurs de langue anglaise des traductions d'ouvrages sovietiques, specialement Kretetskii (1976). (Cf. egalement Davis, Romberg, Rachlin et Kantowski, 1979.) iv) A propos de l'important sujet de "conceptualizations", une litterature abondante en a prepare le terrain, y compris les "frames" de Minsky mentionnes ci-dessus, et egalement Newell et Simon, 1961, Newell et Simon, 1973, Minsky et Papert, 1972, Winograd, 1971, Bobrow et Collins, 1975, et Schank, 1977, plus les travaux de Norman, de Rumelhart, d'Ortony, etc. Il faut aussi se referer a Davis, Jockusch, et McKnight, 1978. Pour de bons ouvrages, cependant moins techniques, voir Boden, 1977 et McCorduck, 1979.

Les Concepts. Parmi les concepts principaux de ce nouveau paradigme, dont l'approche s'inspire de celle du traitement de l'information, on trouve: "VMS sequences", "integrated sequences", "operators on sequences", "frames", "meta language" (ou "planning language") et l'heuristique. Voici une breve definition de quelques-uns de ces termes:

"VMS sequence". Une "visually moderated sequence" est un algorithme dont on connauit differents morceaux, chacun d'entre eux recouvrable par un signal visuel approprie. Une longue division est un exemple typique pour l'eleve qui manque de maturite mathematique. Une entree visuelle declenchera (on l'espere) le recouvrement du prochain morceau de l'algorithme; l'effectualization de cette etape va produire une nouvelle entree visuelle qui va a son tour (on l'espere) declencher le recouvrement du morceau suivant, etc. (Cf. Davis, Jockusch, et McKnight, 1978.)

"Integrated sequence". Avec suffisamment d'experience, un "VMS" peut devenir independante des signaux visuels pour declencher le recouvrement du morceau suivant (bien qu'il faille parfois noter sur papier une partie des calculs numeriques); lorsque ceci a lieu, la "sequence" devient "integrated sequence".

"Operators on sequence". Il est necessaire de postuler un nombre de differents types d'"operateurs" qui operent sur des series ou sur des algorithmes. Par exemple les operateurs "look ahead" (regardant a l'avant) arrivent a etre capables de "predire" les etapes suivantes de l'algorithme et de se preparer pour elles. Les operateurs "back off" (a reculons) proposes par Martin (Minsky, 1975) comparent les details des algorithmes avec les declarations des objectifs de type "top down" (haut en bas).

"Rules that make rules" (les regles qui font les regles) posees en hypothese par Matz, seront discutees plus loin en detail. Les "frames", proposes par Minsky, sont des structures formelles pour representer de l'information qui possedent un type particulier (cf. Minsky, 1975, et Davis et McKnight, 1979).

Quelques resultats typiques. Le phenomene "Friend". Friend (1979) rapporte que de ces trois problemes A) 121 + 19 + 23, B) 121 + 119 + 23, et C) 121 + 69 + 23, c'est etonnamment le probleme A qui est le plus difficile. Cela va a l'encontre de l'intuition dans le sens ou le probleme B se complique d'une addition supplementaire dans la colonne des cents, et que le probleme C inclus une retenue supplementaire. Ainsi, on pourrait s'attendre a ce que le probleme A soit le plus facile des trois. Mais si l'on pense en termes de "frames" chaque probleme signale le recouvrement d'un "binary addition frame" (Davis et McKnight, 1979), et chaque "binary addition frame" assume la commande du traitement et exige deux entrees. Dans le probleme A, il n'y a pas de maniere correcte de satisfaire cette exigence et pour cela, d'apres une loi postulee par John Seely Brown, on a recours a une maniere incorrecte. Cette theorie explique parfaitement le type le plus courant des reponses incorrectes donnees pour le probleme A.

"Laws That Make Laws" (les lois qui font les lois.) Matz (1979) postule deux niveaux de "regles": un niveau "a la surface" des regles que nous reconnaissons d'ordinaire (par exemple la loi distributive), et un niveau plus profond qui cree les regles a la surface. (Brown utilise le meme postulat.) Il est bien connu que ce type d'erreur-ci est extremement persistant: Pour resoudre $(x - 2)(x + 3) = 24$, remplacer par 2 equations: $x - 2 = 24$, $x + 3 = 24$ d'ou $x = 26$ ou $x = 21$. Cette erreur est comme le chiendent, difficile a eradiquer car, comme le chiendent elle se regenere. Une regle du niveau profond generalise des nombres concrets aux variables. Sans cette regle du niveau profond, nous ne pourrions pas apprendre l'arithmetique. En effet, de $(x - 5)(x + 6) = 0 \Rightarrow x = 5$ ou $x = {}^{-}6$ nous devrions generaliser que $(x - A)(x + B) = 0 \Rightarrow x = A$ ou $x = {}^{-}B$. C'est a quoi sert la regle du niveau profond. Cependant on ne doit PAS generaliser de la meme maniere avec le zero. Mais, une fois de plus, suivant une loi bien connue, la loi du niveau profond a tendance a trop generaliser, en deduisant que $(x - 5)(x + 6) = 0$ generalise que $(x - A)(x + B) = C \Rightarrow x - A = C \vee x + B = C$.

Plus recemment, Matz (1980) rapporte des erreurs provenant de l'accolement de descripteurs trop succincts a certaines situations et a certains algorithmes.

"The paradigm teaching strategy" (la strategie de l'enseignment par paradigme). Basee sur l'intuition et le jugement de professeurs, une strategie pour enseigner un concept nouveau evolue, comme celle qui concerne l'introduction des nombres negatifs. La strategie est d'abord d'attirer l'attention sur une "histoire" choisie avec soin, histoire dont la structure est a) auto-evidente, b) isomorphe a la structure mathematique que l'on desire. On dit que cela cree chez l'eleve l'"assimilation paradigm" (cf. Davis, 1967). Cette strategie pedagogique est nee de l'intuition d'enseignants mais son fonctionnement a ete hypothese independamment par Minsky (1975): "Un frame general utilisera souvent un des cas specifiques au-dessous de lui comme son modele; le frame 'mammifere' peut simplement utiliser le frame 'chien' ou le frame 'vache' comme son modele, plutot que d'essayer de trouver quelque modele schematique d'un mammifere ideal et nonspecifique" (Minsky, 1975, p. 266).

Cette approche basee sur l'intelligence artificielle et les sciences cognitives se developpe rapidement, a tel point qu'elle peut explique une vaste et impressionnante gamme de comportements mathematiques. Elle va surement devenir un courant dominant de la recherche dans la pedagogie des mathematiques.

Cette notice a ete redigee en vue d'une presentation au Symposion Begle pour le Quatrieme Congres International de la Pedagogie des Mathematiques a Berkeley en Californie, le 13 aout 1980. Les recherches et les etudes de l'auteur mentionnees dans cet article ont ete subventionnees par la National Science Foundation, fonds GS-6331, NSF PES 741 2567 NSF SED 77-18047 et par ceux de NSF/NIE NSF SED 79-12740. Traduction par Catherine D. et Stephen C. Young.

References

Beth E., & Piaget, J. Mathematical Epistemology and Psychology. Dordrecht, Holland: D. Riedel, 1966.

Bruner, J.A. The Process of Education. Cambridge, Massachusetts: Harvard University Press, 1960.

Lesh, R.A. An interpretation of advanced organizers. Journal for Research in Mathematics Education, 1976, 7 69-74.

Lesh, R. Mathematical learning disabilities: Considerations for identification, diagnosis, and remediation. In R. Lesh, D. Mierkiewicz, & M.C. Kantowski (Eds.), Applied Mathematical Problem Solving. Columbus, Ohio: ERIC/Clearinghouse for Science, Mathematics, and Environmental Education, 1979,

Lesh, R. Applied mathematical problem solving. To appear in Educational Studies in Mathematics.

Lesh, R., Landau, M., & Hamilton, E. Rational number ideas and the role of representational systems. To appear in Proceedings of the Fourth International Conference for the Psychology of Mathematics Education, August, 1980.

WHERE DO WE GO FROM HERE? SOME QUESTIONS IN MATHEMATICS EDUCATION TO BE CONSIDERED IN THE NEXT FOUR YEARS.

Gunnar Gjone
Pedagogisk Seminar
Oslo, Norway

Research in Mathematics Education

The situation today.

The teaching and learning of mathematics can be viewed as a complex human interaction in an institutionalized setting (1). The setting is the school, each school being a separate institution, but at the same time part of a local community and also part of the school system of a larger society. Within this society, specific goals are set for the activity in schools, and these may differ from one society to another. Another factor is essential in our description, the subject matter to be communicated - mathematics.

We may say that the overall structure of mathematics education is similar in various countries, but that there are important differences which we whould be aware of.

What do we know about the situation?

It is impossible to give an extensive list of what we know. However, it seems that we have several forms of knowledge.

Knowledge of how some single variables affect some outcome.

Especially in the United States there exists a vast literature and research on "critical" variables in mathematics education, and also a major work compiling the various results: Begle's book" "Critical Variables in Mathematics Education" (2). Here we can find all kinds of information, e.g.:

> "Sex. A positive main effect indicates that students with female teachers scored higher than students with male teachers."

> " - for no single family variable is there enough information to suggest a strong pattern of influence on mathematics achievement".

Begle's book is an excellent source of this type of information, and also today we find a very similar approach to research in mathematics education in the United States.

However, in recent years there has been increasing criticism voiced against this research paradigm, most notably from European mathematics educators (3). As a conclusion we can say that we have extensive knowledge relating to this model, but that its value is being questioned.

Knowledge of the mathematics classroom.

"The mathematics classroom" is the place where teachers teach, students learn, and they interact. Some people, excellent teachers with long experience, have a profound knowledge of the teaching situation. But, this knowledge is often to a very small degree available for their colleagues, so it stays with the individuals and is not common knowledge.

Considering the learning process we have an abundance of data, and the undisputable fact, the people do learn mathematics. However, many today feel that we don't really know so much about learning processes.

Freudenthal (5) asks this question:

> "I agree it is not easy to observe individual developments, and the laboratory of the psychologist is the least appropriate place to do it. The easiest place is the family and the next easiest the classroom. So if there are so many families and so many classrooms in the world, why do we know so little about learning processes?"

What would be the reason for this state of affairs? Could it be that our data are not relevant, that we have observed the wrong variables or the wrong situation?

Knowledge of the education system.

Mathematics education takes place in an institutionalized setting. What do we know about this setting?

We have an extensive knowledge of the general system, since this has for a long time been an important field in educational theory. However, comparatively few such studies have been carried out for the specific subject of mathematics. One promising area centers around the teacher and his professional life. However, the limitations on this research is that it is often specific knowledge relating only to one country.

Then, trying to characterize our knowledge today, we are faced with a very difficult task. In some areas we have a very large detailed knowledge, in other areas we have come to realize that we really do not know so much, and it seems that we lack a structure of our knowledge. It also seems that most of our knowledge is relevant only to some countries, and it is an important question to which degree we can hope for a general knowledge.

Issues in Research in Mathematics Education Today.

Search for a theoretical base.

Starting around the time of the last ICME in Karlsruhe, 1976, there has been an increasing demand by several mathematics educators for a theoretical base of research or a separate "science" of mathematics education (6). Why a science of mathematical education?

As outlined above there exists a wealth of observations, there have been asked numerous questions - and many of them have been answered. But - are they the right questions? Are the input variables to the process of mathematics education, which have been investigated, really critical? We need some guidance, many will call it a theory, to be able to ask the right questions.

To provide such a theoretical base will be the task of the most competent persons only, who combine theoretical insight with practical experience.

Interdisciplinary approaches.

Today we see that mathematics educators use methods from several other disciplines in their research. Sociological methods have found increasing use in the classroom situation. One speaks about a whole new subdiscipline of classroom research. Another method is to use linguistic methods in analyzing the communication in the classroom. Also there has emerged a special interest in the restricted areas of "the psychology of mathematics education", opposed to the more general learning psychology.

We also need to give the question: "What influences research?" some consideration. The economic situation is one obvious factor; another important factor is what we might call the professional life of the researcher. Is she or he a graduate student, with short-term goals, or a person who can carry out a project lasting an extended period of time?

A Research Agenda for 1984.

What should have been accomplished in four years, to the next ICME? First of all it must be noted that four years constitute a very short period. In many countries compulsory education lasts more than nine years, so our ambitions for 1984 should not be set too high. However, let us discuss a few areas where research has already been initiated and where we can hope for some results by 1984.

1. Historical analysis - a key for understanding the present and a guide for the future.

In many countries the history of mathematics education since the early 1950s has had a decisive influence on the situation today. This has been a period of great actions - and reactions. We have experienced a wide variety of teaching techniques and philosophies: "new math", "programmed instruction", and "back to basics" to mention a few.

In many ways our situation now can be characterized as one of confusion: we are desparately seeking knowledge and there is an intense variety of research being carried out. Looking ahead we can see today the contours of new directions:

> NCTM recommends in "An Agenda for Action" (9) that problem solving be the focus of school mathematics in the 1980s.
>
> There is a strong tendency on the secondary level towards discrete mathematics.
>
> In many countries we find a move towards more applicable mathematics.

These new tendencies seem to relate to the preceding period in a variety of ways. Applicable mathematics can be viewed as a reaction to the pure mathematics in the "modern mathematics" programs, and problem solving is perhaps another try to establish a base for school mathematics itself, when it has become clear that mathematical structure is not suitable. Therefore, if we want to implement our new ideas we need knowledge of the former attempts.

Many teachers and more pupils made experiences in this past period. This is a reservoir of knowledge we should use. It is imperative that these experiences are subject to tinvestigations and careful analysis, not by statistical comparisons, but by using different kinds of modern historical and anthropological methods.

Specific questions could be:

> What happens in a school reform? What is the decision process and who has influence? What are the effects of a reform on daily classroom life and how crucial are teachers for the completion of a reform?
>
> Experiences of content-organizing. What are the benefits and drawbacks of organizing according the mathematical structure?
>
> Experiences with teaching methods. Which elements of programmed instruction can be used beneficially by the students?

2. Life in mathematics classrooms.

Educationalists have for some time studied the classroom. One prominent study is presented by Jackson (8). However, not many studies have been carried out concerning the mathematics classroom. The center of the activity in mathematics education is the mathematics classroom, so this should perhaps be the most important object of study.

How should such a study be carried out? There is no "theory" to guide the observer, but the following words by Freudenthal (5) should be noted:

> "Look and listen with an open mind and have the courage to notice and to report events that most people would consider as too silly to be noticed and to be reported - there might be a minority who can appreciate them, and this minority will be right."

More specifically we should consider carrying out similar studies as Piaget's, but in the reality of the classroom. It has come to be acknowledged that problems and concepts are functions of the situation, instead of being constant and stable as Bauersfeld points out (1). However, one should bear in ming the time spent by Piaget and his collaborators on their tasks, so we should have a low level of ambition in this area for 1984.

3. Information processing society - forward to basics

We have now a glimpse of a future where the key word is information processing. Possibly every bit of information ever visually recorded will be available to everyone with a telephone and a TV-receiver. Some aspects which have been predicted for this future are:

> Increasing automatization, and more spare time for individuals. Opportunity for "life-long education".
>
> Demand of greater flexibility: new jobs will be created, others will disappear. Need of "life-long education".

Our task is to prepare people, through education, for a future society, knowing that our acts today also will help shape that society.

Which consequences have been deduced for education? It seems that the various proposals made to meet this challenge can be classified to belong to one of the following "models":

> The flexible educational system. The educational system must be able to absorb new ideas and results in research very rapidly.
>
> Basic knowledge. The object of schooling is to give everyone a foundation on which to build further education.

For mathematics education the "basic knowledge model" has many advantages. The most important advantage is found within the subject itself: Mathematics is cumulative, i.e., new results build on earlier findings, and one cannot enter the subject at any random point.

One way of looking at the last years in mathematics education is to see it as a search for basic knowledge in mathematics. The "new math" was an attempt to use mathematical structure as basic knowledge. Many also hold this view today, but in a country like Norway it is definitely of the past. Computational skills has been another proposal, which has been forwarded by the name "back to basics". Many mathematics educators have been sceptical to this movement and it has not been internationally acclaimed.

Hans Freudenthal (7) has introduced the concept "forward to basics" and today we have several proposals, as to what this could be. NCTM has in its "An Agenda for Action" proposed problem solving as "a focus of school school mathematics in the 1980s." When Zoltan

P. Dienes proposes (4)

> "... that the main aim of mathematics education, instead of being the knowledge of mathematics, should be the development of certain patterns of thinking, certain types of strategies, that people might develop in the face of new situations which they have never encountered before",

he comes up with four aspects:

> "the aspect of abstraction, generalization, then the non-redundancy of mathematics leading to the problems of encoding and decoding".

Perhaps these more general aspects may serve as "basic knowledge"?

For 1984 we should have arrived at some understanding on how some of the elements proposed as basic knowledge will function in "an information processing society".

In what way can problem solving serve as a foundation on which various forms of mathematical education can build? We need curriculums constructed on this basis.

Computational skills must be seen on a background of "man-machine-interaction". What amountof "paper-pencil" computations will be needed? What amount of mental calculations?

4. Forms of representation of mathematics knowledge. Need for alternative forms of evaluation.

In many countries we find a similar situation: Tests in mathematics performance are written tests, both in schools and in various testing instances. On often-observed fact is also that students who are helpless with pencil and paper function well in practical situations.

This situation might be criticized in several ways:

that the testing performed tends to favour special groups of students

that only some special aspects of what we set out to measure are being measured

that the testing itself has a large effect on the educational process prior to it.

In many countries we also find a scepticism as to what can be measured on conventional tests. One has a feeling that what is sought to give the students from their school mathematics is something which cannot easily be measured. Perhaps a totally new approach to educational measurement is needed?

In our opinion some questions to be considered could be:

What guidelines should be established as to what is a "correct answer" to a problem, knowing that an answer is only "correct" in a given context?

How can a solution to a mathematical problem be represented? Can some concrete actions indicate mastering of problems - and are we willing to accept these in schools?

How can we evaluate the problem-solving process and not only the answer itself?

Conclusion

This list of problems could easily have been extended. It represents only a small selection of possible research topics. The two first areas are considered to be essential - "historical analysis" and "life in mathematics classroom". It is felt that no theory or science of mathematics education can be established without profound results in these areas.

References

1. Bauersfeld, H.: Hidden dimensions in the so-called reality of a mathematics classroom. Educational Studies in Mathematics 11, (1980), pp. 23-41.

2. Begle, E.G.: Critical Variables in Mathematics Education. Mathematical Association of American and National Council of Teachers of Mathematics, Washington, 1979.

3. Bussmann, H., Heymann, H.-W., Lorenz, J.-H., Reiss, V., Scholz, R.W. and Seeger, F.: "Begle, E.G.: Critical Variables in Mathematics Education". Zentralblatt fur Didaktik der Mathematik, Jhrg, 12, Heft 1, 1980.

4. Dienes, Z.P.: Learning mathematics, In Wain, G.T. (Ed.): Mathematical Education, Van Nostrand Reinhold Col, Wokingham, 1978.

5. Freudenthal, H.: Address to the first conference of I.G.P.M.E., at Utrecht 29 August 1977. Educational Studies in Mathematics 9, (1978), pp. 1-5.

6. Freudenthal, H. Weeding and sowing. Preface to A Science of Mathematical Education, D. Reidel, Dordrecht, 1978.

7. Freudenthal, H.: New math or new education. Prospects, Vol. IX, No. 3, (1979), pp. 321-331.

8. Jackson, P.W.: Life in Classrooms. Holt, Rinehart and Winston, New York, 1968.

9. NCTM: An Agenda for Action. National Council of Teachers of Mathematics, Reston, 1980.

CRITICAL VARIABLES IN MATHEMATICS EDUCATION: A STATEMENT ON DIRECTIONS FOR FUTURE RESEARCH

John P. Keeves
Australian Council for Educational Research
Hawthorn, Victoria,Australia

A Tribute

During his lifetime as a mathematician and educator, Edward Griffith Begle contributed a great deal to the cause of research in mathematics education and also to curriculum development. The latter contribution was largely through his association with the School Mathematics Study Groups (SMSG), which had considerable influence on the direction of primary and secondary mathematics education in Australia during the 1960s.

In an article by Blakers (1978), it is noted that changes which have occurred in mathematics education in Australia since the late fifties have been greater than all previous changes in this century. Blakers attributes these changes largely to overseas influences and notes that many Australian mathematics teachers gained new insights into their subject through studying the materials produced by SMSG. Publications such as Background in Mathematics, an Australian guide book for primary teachers, also drew heavily on the ideas developed by this group of educators. This work has proved of great value in the revision of mathematics courses at the elementary school level throughout Australia.

Ed Begle has therefore had an indirect influence on the progress of mathematics education in Australia. He strongly believed in the need for a foundation of firm empirical knowledge about mathematics education and his final publication is a fitting testimony to the man and his principles.

A Comment

The book, Critical Variables - Mathematics Education, provides a comprehensive survey of the evidence available at the end of 1976 about the effects of different variables on student learning of mathematics. It is not a review of research literature, since it gives relatively few details of the sources of findings quoted. However, it gives the reader an idea of the amount of attention that has been paid to a given topic, the author's view of the value of this work, and whether further research is indicated. Begle stressed the need for comprehensive reviews of particilar areas and those interested in conducting such reviews can obtain guidelines and inspiration from this survey. It was Begle's intention to point to the existence of gaps in our knowledge and to emphasize the need for a theoretical framework to direct future research in mathematics education.

While reading the book there are occasions when one may well feel exasperated with Begle's approach, with his abrupt dismissal of some variables and lingering emphasis on others. There is a marked lack of uniformity in the level of detail and extent of discussion devoted to the variables. Begle has clearly worked from primary sources in some instances and secondary sources in others and this has probably affected the quality of his comments.

In the survey he sometimes devoted about the same amount of discussion to two topics based on a comparable number of studies. Evidence in both cases was inconclusive, yet Begle recommended further research on one variable and dismissed the other from future consideration. For example, Begle dismissed the topic of mastery learning by saying that results have been mixed and "it is doubtful that further study at the pre-college level would be useful at present'. He summarized the discussion of another instructional variable, reinforcement, by stating again that results had been mixed but there were enough positive results to indicate that further studies would be worth the effort. It is difficult for the reader to fathom the reasons behind some statements. This would be less of a problem if references were identified, but the readers have no way of knowing which studies were and were not included and therefore cannot verify Begle's conclusions for themselves.

Begle has largely ignored the interaction of the variables he identified. He stated that

> . . . the outcome of teaching does not depend just on the teacher but rather is the result of a complex interaction between the teacher, the students, the subject matter, the instructional materials available, the instructional procedure used, the school and community, and who knows what other vraiables . . . (p. 33)

When interactions are ignored it is hard for the reader mentally to link the factors into a pattern. Even within one broad division of variables there are likely to be interaction effects, such as between ability grouping and class size. Begle made no mention of these interactions, and made no attempt to draw the topics together. It is particularly in an interactional way that the influence of many variables may be asserted. Begle was discouraged by the lack of theory in mathematics education, but by segmenting the research to such an extent and discussing topics in isolation, he has made it difficult for the reader to see the relevance of any guiding theoretical structure.

New Directions for Research

It is now necessary to seek new directions for research. There is clearly a lack of theoretical framework to guide studies into mathematics learning. Begle in his review drew attention to some studies that have focused on Carroll's model of school learning, and suggested that further attempts to fit Carroll's model to actual mathematics learning situations should be encouraged (see Carroll, 1963).

Carroll's model is not derived solely from a psychological perspective involving the study of behaviour in individuals. Theories of learning, thinking, motivation, and cognitive development can be related to the model, but it also permits the study of relationships concerned with the classroom, the school, or the school system. The basic principle of the model is that the learner will succeed in learning a given task to the extent that he spends the amount of time that he needs to learn the task. Carroll proposed five factors which were grouped under two headings: determinants of time needed for learning and determinants of time spent in learning. Under the first heading he considered aptitude, ability to understand instruction and the quality of instruction. These factors interact to determine the optimum time needed by an individual student to master a given learning task. Under the second heading, he considered the time allowed for learning, and perseverance, which is a measure of the time the learner is willing to spend in learning, a motivational factor. The degree of learning is a direct function of the ratio of the time spent to the time needed for the learning task.

Studies have already been carried out that have examined the effects of differences in time available for learning and opportunity to learn on achievement in mathematics. Work that has been undertaken in Australia in connection with the IEA studies in mathematics learning both in 1964 and more recently in 1978 (see Husen, 1967; Keeves, 1976; Rosier, in press) has confirmed that both at the individual and at the system levels the time available for learning is related to the opportunity provided to learn mathematical content and to achievement in learning mathematics. Unfortunately, Begle in his survey has largely ignored the important work carried out by IEA in the cross-national studies of achievement in mathematics and by doing so he has failed to argue for the world wide universality of many of the findings that he reports and discusses.

Carroll, however, distinguishes in his exposition of the model between 'elapsed time' and 'engaged time' during which a student is 'paying attention' and 'trying to learn'. Keeves (1972) and in subsequent reanalyses of the data (Keeves, 1974,; and McGau et al., 1979) have confirmed the importance of 'attentiveness' or 'engaged time' for learning in mathematics, and other studies exist (see Bloom, 1979) that endorse these results.

In the future quest for factors influencing achievement in mathematics, it would seem highly profitable to follow Carroll's proposal for the investigation of variables which determine first, how much time the learner spends actively engaged in learning; and secondly, how much time a learner needs to spend in order to learn the mathematical content under consideration. From this standpoint many areas for research in mathematics may be listed:

1. studies of teaching behaviour that serve to increase the engaged time of the students, such as immediate feedback on work completed;
2. studies of classroom organization that serve to increase the time available for learning, such as the planning of instructional procedures;
3. studies of school organization that will provide the optimum time available for student learning;
4. studies of curriculum planning that will determine the time required to achieve specified learning outcomes;
5. studies of student behaviours that are associated with maximum attentiveness in the classroom;
6. studies of learning facilities available in the home that will increase the opportunity and effectiveness of mathematics learning undertaken outside the classroom;
7. studies of factors influencing student motivation to increase the perserverence of a student towards learning mathematics;
8. studies of testing procedures that will provide reward and reinforcement for student learning in mathematics;
9. studies of the prerequisite knowledge and skills for the learning of new mathematical content in order to determine what time might be needed for learning new content;
10. studies of the structuring of teaching, or the quality of the instruction required so that learning will take place efficiently. It must, however, be recognized that there is rarely a single best sequencing or structure of instruction.

The advantages of Carroll's model are several. First, although the model has been developed in a manner to be consistent with psychological thinking it is not totally dependent on a particular psychological theory. It maintains the strength and flexibility necessary for an attack on educational problems that come from the inclusion of the practical and measurable variable - time.

Secondly, it permits the study of the manner in which variables believed to be related to learning outcomes are interrelated through their efforts on the mediating variables of 'engaged time' and 'time needed for learning'. The survey that Begle has carried out takes each variable in isolation and examines its relationship with achievement, without any consideration of how and why the variable might influence learning. Carroll's model, through its assumptions that time actually spent and time needed for learning intervene between the predictor variables and the criteria, permits a level of explanation that has previously been lacking. Thus the input-output approach that has dominated educational research in the past is largely rejected by Carroll's model, being replaced by causal chains or paths between the antecedents, the processes, engaged time, and the products of learning.

Thirdly, by employing time as the key concept, certain measurement and analytical problems are avoided. Time is a variable that has meaning as a fundamental variable. It is not a derived variable whose meaning and measurement changes at different levels of aggregation of data. This is of particular value in the study of school learning, because frequently the unit of sampling, experimentation or investigation, and analysis cannot be the individual. One of the other meaningful analytical units of educational research - the classroom and teacher, the school or the system - must frequently be employed. However, while interpretation must always be undertaken with care to avoid the fallacies that are common when data are analysed across different levels of aggregation, relationships which involve time have a meaning that can commonly be carried across these different levels.

Fourthly, time is a variable that is quantified and that can be measured with accuracy, thus mathematical formulations and procedures can be employed to investigate hypothesized relationships, and research gains in strength according to the extent to which mathematical procedures can be employed in the analysis of evidence.

Fifthly, educational practice is frequently confronted with variables known to be important but that are not malleable. However, time is a variable that is alterable, and changes in time needed or time allowed are readily recognized. Thus the focus of the research into school learning is on malleable and alterable variables and not on those that cannot be changed. From a consideration of alterable variables the potential gains for educational research and educational practice are many.

Begle has performed a much needed task of surveying the field of research associated with mathematics learning and presenting the findings in a concise publication. Using this report by Begle as a starting point, the next step would appear to be the integration of the evidence available around the model of school learning advanced by Carroll. Mathematics is an appropriate field for an initial attempt to examine Carroll's model, in detail, because of the extensive body of research that has been carried out. In this examination of the research evidence in mathematics learning, strategies advanced by Glass and Smith (1978) of meta-analysis and the estimation of effect size should be employed. This investigation will, I believe, lead to the explication and refinement of the model, in a manner that Carroll did not envisage. Carroll suggested that the exploration of interactions constituted a major task for educational psychology, but it would be more appropriate to tease out the causal interrelationships initially by correlational studies and subsequently, if possible, by experimental studies. Interactions may be important, but the study of interrelationships between main effects is surely a necessary first task. Analytical procedures are now available for not only the testing of causal relationships but also for the estimation of effects under differing circumstances, and this must be part of the research program for the future if our knowledge of mathematics learning is to proceed beyond the relatively meagre and unsatisfactory stage at which it now stands.

Once the conceptual model of school learning in mathematics has been developed from that advanced by Carroll nearly 20 years ago, reseach can be undertaken to explore the causal links that are poorly defined and to strengthen knowledge in areas where there are major deficiencies.

Begle was intrigued by Carroll's model. It is unfortunate that he did not explore fully its potential. However, he did in two brief sentences point to the way ahead.

References

Australian Council for Educational Research. Background in Mathematics: A Guidebook to Elementary Mathematics for Teachers in Primary Schools. Melbourne: Department of Education, Victoria, 1966.

Begle, E.G. Critical Variables in Mathematics Education. Washington, D.C.: Mathematical Association of America, 1978.

Blakers, A.L. Change in Mathematics Education since the late 1950's - Ideas and Realisation. Australia. Educational Studies in Mathematics, 1978, 9, 147-158.

Bloom, B.S. Alterable variables: the new direction in educational research. Edinburgh: SCRE., 1979.

Carroll, J.B. A Model of School Learning. Teachers College Record 1963, 64, 723-733.

Glass, G.V. and Smith, M.L. Meta-analysis of Research on the Relationship of Class size and Achievement. San Francisco: Far West Laboratory for Educational Research and Development, 1978.

Husen, T. (ed.) International Study of Achievement in Mathematics (2 vols.) New York: Wiley, and Stockholm: Almqvist and Wiksell, 1967.

Keeves, J.P. Educational Environment and Student Achievement. Stockholm: Almqvist and Wiksell, and Melbourne: ACER, 1972.

Keeves, J.P. The Performance Cycle: Motivation and Attention as Mediating Variables in School Performance. Melbourne: ACER, 1974.

Keeves, J.P. Curricular Factors Influencing School Learning: Time and Opportunity to Learn. Melbourne: SCER, 1976.

McGaw, B., Keeves, J., Sorbom, D. and Cumming, Joy. The Mediated Influence of Prior Performance on Subsequent Performace: An Analysis of Linear Structural Relationships. (Paper presented at Australian Association for Research in Education Conference, Melbourne 1979).

Rosier, M.J. Changes in Secondary School Mathematics in Australia: 1964 to 1978. Hawthorn, Victoria: ACER (in press).

CONTRIBUTIONS OF RESEARCH TO EVALUATION OF TEACHING

Thomas J. Cooney
University of Georgia
Athens, Georgia, USA

In this paper an attempt will be made to deal with the difficult problem of evaluating teaching and to examine the role research can play in the development of criteria (The term criteria is used throughout this paper to denote particular aspects of instruction worthy of consideration for evaluating teaching. As such it is used in a qualitative sense rather than in a quantitative sense which fosters a dichotomous decision.) for such evaluation. The problem is difficult because no clear quantifiable metric for measuring "good teaching" exists. This renders any attempt at evaluation can be developed.

It seems wise to distinguish between evaluating teaching and evaluating the teacher. Consider an analogy. Teachers strive (or should strive) to differentiate the value of students' performances from the worth of the students themselves. We determine whether their performance is acceptable or not based on such evidence as being able to solve quadratic equations or probe the Pythagearean Theorem. We can conclude that the student is or is not a good equation solver or problem solver quite apart from his worth as a human being. Similarly, any evaluation of teaching should differentiate between the performance of a teacher and the teacher's worth in general. I contend that this differentiation can be facilitated by the development of rather specific criteria for the evaluation process.

The Role of Research in Developing Criteria

One of the problems in identifying useful constructs from research on teaching is that much of that research has not focused on mathematical education concerns. As a result variables must be culled out which appear patent for mathematics education. Some of those variables and their constructs will be discussed here.

Nearly a decade ago, Rosenshine and Furst (1971) identified a number of teacher behavior variables which showed potential for being strong correlates of student achievement. The following five variables were identified as having particular promise: clarity, variability, enthusiasm, task-oriented or business like behavior, and opportunity to learn. The identification of these variables has not gone unchallenged. (See Heath and Nielson, 1974.) Nevertheless, some of these variables have stimulated additional research to the point where the associated constructs are at least tenable for mathematics instruction.

The construct of clarity is one which has been studied rather extensively. Bush, Kennedy and Cruickshank (1977) and Cruickshank, Kennedy, Meyers and Bush (1976) have found clarity to be a significant correlate of achievement. In studying the teaching of mathematics, Thornton (1977), Smith (1977), Good and Grouws (1977) and Evertson, Emmer and Brophy (1980) have also found support for clarity as a correlate to mathematics instruction. Thornton defined clarity in terms of the following six rating scales: takes time when explaining, stresses difficult points, explains new words, demonstrates how to do something, works difficult problems on the board, and gives students an example and lets them try it.

Another potent variable is variability of instructional technique. The importance of Dienes' multiple embodiment principle is but one example of support for this construct. An important thrust of the Second International Mathematics Study is to determine the degree of variability of interpretations teachers use when presenting concepts such as fractions, decimals, ratio, proportions, integers, and various geometric and measurement concepts.

One aspect of a teacher's variability is the extent to which examples and nonexamples are used when concepts are taught. Kolb (1977) has developed a model for predicting the effect of various uses of examples and nonexamples. According to his model, examples and nonexamples of concepts produce more learning than characteristics of concepts when students have little prerequisite knowledge. For students with a higher degree of prerequisite knowledge, characterization moves (as defined by Cooney, Davis & Henderson, 1975) are more effective. The model also predicts that there is a "law of diminishing returns" with exemplification moves for the less advance students. Stiff (1978) has found emperical support for the model. Kolb's research highlights the importance of having a variety of ways of presenting concepts and of utilizing those ways in a manner consistent with students' abilities.

While I contend there are elements of teaching mathematics which differentiate it from the teaching of other subjects, it is undeniable that some general characteristics of teaching are important to consider. For example, Tikunoff, Berliner and Rist (1975) found that variables associated with "those familial interactions in the home which have been attributed traditionally to the successfrul rearing of children" (p. 22) are significantly and positively related to achievement.

Classroom management is another area of "general concern" when evaluating teaching. Kounin (1970) studied a number of variables with respect to classroom management and their relationship to achievement. One of the variables identified was called "withitness." This variable dealt with teachers' ability to communicate that they know what is going on regarding children's behavior and with their ability to attend to two issues simultaneously when two different issues are present. Kounin found withitness to be a strong correlate of achievement as did Brophy and Evertson (1976).

Management concerns also include how efficiently time, a precious and quite finite commodity, is utilized in the classroom. Berliner (1978) defined three time related variables: allocated time, engaged time, and academic learning time. Allocated time refers to the time allocated for instruction in a given content area. Allocated time is an upper bound for engaged time which is essentially the amount of time students are on task. Engaged time is an upper bound for the variable academic leaning time (ALT) which is defined as the time a student is engaged in activities with an error rate of less than 20 percent. Berliner contends that variability of ALT is a significant factor in explaining variability in student achievement. Berliner reported a great deal of variance among teachers in how they allocate time for mathematics instruction, particularly for specific topics such as fractions, measurement, decimals, and geometry. Time and how it is allocated is an important factor to consider when achievement, judgments on how well teachers utilize time seems important in any evaluation process.

The Development of Criteria

In working with teachers, I have found that they usually identify the following general categories of criteria for evaluating instruction: knowledge of mathematics, presentation of mathematics, management skills, and classroom interactions, are not specific to mathematics education. This emphasizes what I believe is the case, viz., that being a good mathematics teacher requires not only being a good "school mathematician" but also requires some general characteristics of handling students.

One of the tasks I had this previous year was to help evaluate a series of videotapes of mathematics teachers at the collegiate level. No student achievement data was available. To provide the requested evaluation, a specific set of criteria was developed. Not every criterion was useful for a specific lesson but as a set they were quite useful. The criteria were developed in part from what evidence exists in the literature on effective teaching and from my own values as to what constitutes "good teaching."

The criteria were placed into two main categories: pedagogical and mathematical. The pedagogical category was further divided into three categories: congnitive, affective, and managerial. These three categories were further subdivided into more specific criteria. The mathematical category was divided into three categories: correctness, rigor, and precision of language. A brief outline is given in Table I.

Criterion IA1 (see Table I) has ample support in the psychological literature. In fact, Ausubel (1968) has made the following assertion:

> If I had to reduce all of educational psychology to just on principle, I would say this: The most important single factor influencing learning is what the learner already knows. Ascertain this and teach him accordingly.

Criterion IA2 reflects a somewhat logical orientation to instruction combined with some sensitivity to research like Kolb's and research on clarity and variability. The use of applications and problem solving (Criterion IA3) have long been advocated. In particular the current recommendations for school mathematics in the 1980's by the National Council of Teachers of Mathematics places a strong emphasis on problem solving. Criteria IA4 and IA5 are obvious reflections of research as are criteria IBI and 2. The four criteria listed under IC (managerial) reflect research on time allocation and Kounin's research on management skills. The mathematical criteria were developed primarily from reflections upon previous attempts to evaluate instruction.

The suggested criteria presented above highlights the need for sensitivity to a variety of aspects of instruction. Some are general in nature. But others are not and focus on instructional factors highly specific to mathematics. Both are needed if evaluation is to have any real validity.

Table I. Possible Criteria for Evaluating Mathematics Teaching

I. Pedagogical Criteria (These focus on various aspects of the instructional act.)

A. Cognitive (How was the content treated?)

1. Emphasis on previous learnings (Relate lesson to previous ones.)
2. Use of examples, nonexamples, and illustrations to present content
3. Emphasis on applications or problem solving
4. Clarity of instruction
5. Student involvement in development of content

B. Affective (What was the nature of interpersonal classroom interactions?)

1. Acceptance of students' feelings and thoughts
2. Enthusiasm of teacher

C. Managerial (How effective was the classroom managed?)

1. Time allocation (Was time used wisely?)
2. Pacing of lesson
3. Awareness of what students were doing
4. Attentiveness (Were students on task?)

II. Mathematical Criteria (These focus more on the mathematics per se rather than on methods of teaching the mathematics.)

A. Correctness (Was the mathematics correct?)

B. Rigor (How rigorous was the mathematics presented? Did the level of rigor seem appropriate?)

C. Precision of language (Did the precision of language seem appropriate? How "heavy" was the symbolism?)

Final Comment

Perhaps the essential ingredient in any evaluation of teaching is the ability of teachers to become reflective about their own behavior - - an ability which can be facilitated with the use of specific criteria. It seems to me that teachers can make greater professional strides if they can analyze their own behavior in some objective way. (I am tempted to say that THE most important criterion IS the teacher's ability to be self-analytical.) I have reasoned (maybe rationalized) that if I cannot help the teacher become reflective, then a list of "do's" and "do not's" will have only a minimal and short term effect at best.

Too, one should realize that for a given classroom the teacher assumedly knows more about the particular situation and what will and will not work than anyone else. Hence, those involved in evaluation should take care not to project the image of the "all knowing" Omnipotent who can utter only infallible truths. Those "truths" may not be so "true" in the realities of classroom life.

References

Ausubel, D.P. Educational psychology: A cognitive view. New York: Holt, Rinehart, and Winston, Inc., 1968.

Berliner, D.C. Allocated time, engaged time, and academic learning time in elementary school mathematics instruction. Paper presented at the 56th Annual Meeting of the National Council of Teacher of Mathematics, San Diego, April 1978.

Brophy, J.E. & Evertson, C.M. Learning from teaching: A developmental perspective. Boston: Allyn & Bacon, 1976.

Bush, A.J., Kennedy, J.J., & Cruickshank, D.R. An empirical investigation of teacher clarity. Journal of Teacher Education. March-April 1977, 28(2), 53-58.

Cooney, T.J., Davis, E.J., & Henderson, K.B. Dynamics of teaching secondary school mathemtics. Boston: Houghton-Mifflin, 1975.

Cruickshank, D., Kennedy, J., Myers, B., & Bush, A. Teacher clarity - - What is it? Paper presented at the Conference on Innovative Practices in Teacher Education, Atlanta, January 1976.

Evertson, C.M., Emmer, E.T. & Brophy, J.E. Predictors of effective teaching in junior high mathematics classrooms. Journal of Research in Mathematics Education, May 1980, 11(3), 167-178.

Good, T. & Grouws, D. Teaching effects: A process-product study in fourth-grade mathematics classrooms. Journal of Teacher Education, 1977, 28, 49-54.

Heath, R.W., & Nielson, M.A. The research basis for performance-based teacher education. Review of Educational Research. Fall 1974, 44(4), 463-484.

Kolb, J.R. A predictive model for teaching strategies research. Part I: Derivation of the model. Athens: The Georgia Center for the Study of Learning and Teaching Mathematics, 1977.

Kounin, J.S. Discipline and group management in classrooms. New York: Holt, Rinehart & Winston, 1970.

Rosenshine, B., & Furst, N. Research in teacher performance criteria in B.O. Smith (Ed.), Symposium on research in teacher education. Englewood Cliffs, N.J.: Prentice-Hall, 1971.

Stiff, L.V. The effects of pure C and E strategies, number of moves, and relevant knowledge on learning a contrived algebraic concept. (Doctoral dissertation, North Carolina State University, 1978). Dissertation Abstracts International. 1978, 39, 2803-A.

Thornton, C.D. An evaluation of the mathematics-methods program involving the study of teaching characteristics and pupil achievement in mathematics. Journal for Research in Mathematics Education. January 1977, 8(1), 17-25.

Tikunoff, W.J., Berliner, D.C., & Rist, R.C. An ethnographic study of the forty classrooms of the beginning teacher evaluation study known sample (Tech. Rep. 75-10-5). San Francisco: Far West Laboratory, 1975.

CHAPTER 13 - Research in Mathematics Education

13.1 THE RELEVANCE OF PHILOSOPHY AND HISTORY OF SCIENCE AND MATHEMATICS FOR MATHEMATICAL EDUCATION

THE RELEVANCE OF PHILOSOPHY AND HISTORY OF SCIENCE AND MATHEMATICS FOR MATHEMATICAL EDUCATION

Hans Niels Jahnke
Institute fur Didaktik der Mathematik
Bielefeld, Federal Republic of Germany

Numbers and Quantitities: Historical, Philosophical and Pedagogical Remarks

1. Introduction

The following is intended as an analysis of a specific conceptual change in mathematics during the early 19th century. From this some pedagogical conclusions are drawn. Analyzing a special example seems to be a more appropriate way of handling the relevance of historical and philosophical considerations for mathematical education than dealing with this question in general terms. My example refers to the fact that at the turn of the 19th century one began to distinguish more systematically between numbers and quantities than had been done before. The program of arithmetizing mathematics arose. Certainly this process had various causes. In the following I concentrate on one of these, namely, the relations of mathematics to experimental sciences.

2. The Reference to Applications in the Self-Concept of Mathematics during the early 19th Century

For the following reflections one has to bear in mind that at the turn of the 19th century important and deep changes had taken place in society as well as in science. With respect to science, I remind you of the fact that for the first time mathematized empirical disciplines outside mechanics arose, as for instance, Fourier's analytical theory of heat, or the theory of electricity.

Against this background, I should like to show that the problem of applying mathematics (1) becomes a crucial point for the methodological and ontological ideas related to mathematics, and (2) leads to developing new foundational concepts: "arithmetization of mathematics", "set theory". My arguments will be based on some manuscripts of the mathematicians Bernard Bolzano (1781-1848) and Carl Friedrich Bauss (1777-1855).

During the second half of the 18th century, there appeared in the mathematical sciences, and in the thinking of their leading representatives, phenomena which have been considered indicative for a "crisis of mathematization" by some historians of science (cf. e.g. Belaval, Stuloff). It is an historical fact that leading scientists and philosophers at that time questioned the role of mathematics for a scientific understanding of reality in the future. It was doubted that mathematics would be able to reproduce the complexity of empirical reality. The conclusion leading mathematicians drew for themselves was that mathematics, as an independent theoretical field, no longer had substantial opportunities to grow. Some mathematicians (e.g. Lagrange and Monge) did not work in mathematics for several years, but rather became involved in research in physics, or in chemistry. They returned to mathematics only after the French Revolution and the foundation of the Ecole Polytechnique, doing mathematics not on a methodologically advanced level. This "crisis of mathematization" was accompanied by a "crisis of the concept of quantity". For the entire 18th century, the concept of quantity had been crucial for characterizing the object field of mathematics. By assuming that the totality of empirical reality was "quantity-like", it was simultaneously assumed that this reality could be subjected to mathematical analysis and treatment.

After this conception underwent a crisis, the concept of quantity was bound to become problematical as well. Various difficulties arose when the attempt was made to define what a quantity is. This concept seemed no longer appropriate to characterize the practice of the mathematical sciences. Hence there were various attempts, towards the end of the 18th century, and at the beginning of the 19th, to develop new definitions of mathematics, e.g. attempts to see mathematics as a pure theory of combinations, or as a science studying the relationships between the whole and its parts.

To characterize the new view which was developed by early 19th century mathematicians we turn to Carl Friedrich Gauss. In his works we find two short manuscripts entitled "Zur Metaphysik der Mathematik" (about 1800), and "Fragen zur Metaphysik der Mathematik" (about 1825). Gauss' dislike for most philosophical schools has led many authors to believe that he considered any treatment of questions of the philosophy of science irrelevant. A lot of documentary evidence, however, shows that the opposite is true and that Gauss did examine the problems of a conception of the philosophy of science appropriate for mathematics in a very profound and intense way. In my opinion, the two manuscripts mentioned, despite their cursory character and apparent simplicity, represent the nucleus of a very far-reaching meta-mathematical conception which helps us to decipher important aspects of the self-understanding of mathematics during the early 19th century. This is why I should like to dwell a bit on these two manuscript.

The question whether an object field can be mathematized or not is the crucial starting point for both manuscripts. The manuscript "Fragen zur Metaphysik der Mathematik" (1925) raises quite explicitly, the question: "Which is the essential prerequisite enabling us to conceive of a combination of concepts as referring to a quantity?" In his other manuscript "Zur Metaphysik der Mathematik" (1800), Gauss starts with the opposition between extensive and intensive quantities, saying that all extensive quantities are the object of mathematics, the intensive ones only insofar as they are dependent on extensive ones. Extensive quantities, for Gauss and his contemporary colleagues, are those which are additive in a fundamental sense, i.e. the real bringing together of which will result, mathematically, in an addition of their magnitudes. Hence, extensive quantities are those the "quantity-likeness" (Grossenartigkeit) of which is evident, whereas the intensive quantities do not lend themselves to direct quantification and mathematization, but require extensive quantities as mediation. This means that raising the question of what is the relationship between extensive and intensive quantities is an object field indeed means to raise, in a very fundamental sense, the question how to mathematize this object field, and what are the conditions and methods of quantification or metrization. With these considerations, Gauss has abandoned the naive empiricism of the 18th century in that he no longer assumes quantification and metrization as given and self-evident prerequisites of mathematical theory, but declares these to be the latter's crucial problem. Thus, for Gauss, the problem of applying mathematics is situated at the centre of his definition of the mathematical ontology.

3. Foundational Reflections in early 19th Century Mathematics

With regard to the foundational concepts produced by mathematics in the 19th century, extremely interesting conclusions were drawn both by Gauss and by Bolzano. In his manuscript "Fragen zur Metaphysik der Mathematik" (1825) mentioned above, Gauss states that the essential prerequisite for the possibility of mathematizing an object field or field of science is that this object field contains relationships between things, and not the things themselves. Gauss concludes: "It will be of prime importance to achieve clarity in the theory of oppositions". He goes on to quote the example of leveling, in which the following oppositions occur: "The position of the bubble in the vial is determined, in the quiescent state, by the vial's geometrical axis, and by a line drawn across the plane determined by the feet". It will be most instructive to dwell on this example, as it contains the fundamental problem of every measurement. The problem is that leveling (i.e. determining the horizontal) requires simultaneously to adjust the instrument used, and to accomplish the operation of leveling. The two processes cannot be carried out one after another and separately, but are fundamentally linked to each other. The "vial's geometrical axis", as measured against the horizontal, will yield the amount of the vial's disadjustment, the "line drawn across the plane determined by the feet" (of the tripod) will yield, also as measured against the horizontal, the deviation from the leveled state. These mutually dependent processes of "adjusting the instrument" and of "measuring" can only be carried out by those aware of the functional relationship existing between those two quantities. In the present case, this relationship is particularly simple in that there is simple addition of the disadjustment and of the deviation from the horizontal. Nevertheless, there remains the fundamental problem that the quantities mentioned cannot be determined, i.e. optimized while neglecting the functional relationship to which they are subject. This means that a theory of relationship is required which does not presuppose the quantity concept. (It would be going beyond the scope of this lecture to state the reasons why it is legitimate to translate "theory of opposition" by "theory of relationships".) Rather, quantities will be obtained only as a result of an actual measuring process which, in its general characteristics, is regulated by an awareness of the functional relationship. Gauss in this instance, and by means of a simple and didactically successful example, reflects the problems arising from the theoretical character of empirical measurements, and the resulting circularity of the relationship between regularity (function) and quantity.

In principle, the problem of the basically circular dependence between function and quantity or between natural law and the parameters which are linked by it is an important feature of the structure of modern scientific theories (compare Sneed 1971, Jahnke 1978). It is frequently discussed by philosophers of science asking for example whether Newton's second law,

force = mass x acceleration,

represents a definition for the concepts of force and mass, or a natural law. In my opinion it is this problem and the combined one of the theoretical character of scientific measurements which has forced the mathematicians of the early 19th century to give up the idea of mathematics being a "theory of quantities". One could no longer maintain an empirical conception of the quantity concept and was led to develop an idea of mathematics as a general theory of relations. Within this framework the function concept, of course, is entitled to a prominent position. It becomes important as empirical research is now examining complex natural phenomena which can no longer be modeled so as to achieve a correspondence between a definitely given parameter, and a definitely given result. This conception which may be characterized as "theory of the one-factor-experiment" is no longer adequate to the situation. Rather, it is a question of various parameters varying against each other. Of course, a changed attitude towards the experiment results from the more complex understanding of the processes. It is no longer the single experiment which will serve to clarify a natural phenomenon. Rather, the transition to sequences and series of experiments, that is, to the establishment of an experimental practice, is the essential methodological characteristic of this process. In my opinion, the emergence of the general function concept at the turn of the 19th century does not just represent the development of a new concept. Rather, the function concept represents a fundamentally new model of scientific generalization.

In the manuscript quoted, Gauss goes on to say - and we are approaching a very crucial issue here - that the points on a line give the general idea of relation and distinction between any two of a set of objects. If one of these points may have a relationship to more than one other, Gauss says, this will create the image of points situated on a plane, connected by lines. For an illustration, he draws a network on the margin emphasizing that everything gets much simpler, if we begin by abstracting from the infiniteness of divisibility, by considering merely discrete quantitites. For him, this is, as was already said, only a matter of relating arbitrary things to each other, i.e. a matter of referring to any field of mathematical application. At the same time, the situatin here is similar to the illustration of the concept of complex number, as it was given by Gauss in his famous second announcement, "Theorie der biquadratischen Reste" (theory of bi-quadratic residues), published in 1931. In a letter to Drobisch, Gauss offered an exlanation for this "illustration", saying that it did not express the essence of complex numbers at all. Rather, this essence was to be conceived of in a more advanced and more general way. Representation in the plane, Gauss says, is but the purest example of their application. For Gauss, therefore, the geometric way of speaking and the geometric illustration have a purely metaphorical or symbolic meaning. His general idea is that points in the plane are symbols or metaphors for the general conception of things which are distinct from one another. If we examine the relations between several points (several objects), we focus on the higher level relationships. It is to be noted that this metaphorical conception of geometry, i.e., understanding geometry no longer merely as a science of the space in which we live, but as a more generalmeans of representing causal or other relationships between objects of the empirical sciences, has always played a part in modern mathematics, and acquired particular importance in the 19th century. Without that, the development of higher-dimensional geometries cannot be rationally understood.

In summary, Gauss' reflections on the problem of applying a mathematical result in a foundational concepts, places the discrete quantities, i.e. the natural numbers, in the centre for the very reason that they represent the most universal language in which the most general concept of the relation between two things can be represented. Geometric representation gives a true mapping of this relation.

There is no time here to quote substantially from Bolzano, but I should like to note the following: like Gauss, Bolzano starts from a relative separation of numbers and quantities in order to make the (natural) numbers the instrument of the universal simulation of quantity relationships. Unlike Gauss, however, Bolzano introduces an extensive and, as subsequent developments have shown, most far-reaching additional set of tools, defining a general concept of relation by means which foreshadowed set theory. He then uses his set theoretical tools to define the concept of quantity. Thus, the concept of relation with Bolzano becomes the nucleus of the concept of quantity.

The entire context of Bolzano's work shows that his ideas on sets provides a means of modeling such a universal concept as relation. In the last instance, this is solely characterized by the two aspects of the difference between things ("elements") and their relationship ("set"). For reasons of brevity a detailed presentation of this idea cannot be given here. I should like to refer you to the pages from 139 to 151 of Bolzano's Grossenlehr, in which Bolzano, just as Gauss, developes a "Theorie des Gegensatzes ohne Grossen" ("theory of opposition without quantities").

In conclusion, we may study, in Gauss' and Bolzano's work, how the development of the most important foundational programs of the 19th century, that is of "arithmetization of mathematics" and of set theory, have a common root. Both programs are most profoundly linked to the problem of applying mathematics within the framework of an emerging experimental practice. As we have shown, geometrical concepts, which, however, should not be empirically misunderstood, play an important part in this connection. Within this development, the concept of relation is the common point of reference of all the modes of representation, of the arithmetical mode, of the set theory mde, and of its geometric representation.

4. Conclusions

In my opinion the following conclusions may be drawn from the above considerations.

a. In order to understand the roots of the foundational programs in mathematics which led to its crisis at the turn of the 19th century one has to go back to its beginnings. The origins of "arithmetization of mathematics" and "set theory" are to be found in the growing problems of applying mathematics in the empirical sciences. If we consider how critically these foundational programs have influenced the didactics of mathematics during the last fifteen years, we may realize that a host of didactical consequences may be derived from such a change in viewpoint.

b. One fundamental misunderstanding concerning "new math" may be immediately pointed at. In primary education set theory has been made the foundation of the number concept starting from the assumption that adding 5 apples to 2 apples amounts to nothing else than joining sets of 5 and 2 elements. Thus in mathematics education some people had the impression of being faced by the unique situation that the scientifically most advanced and most profound concept was at the same time the concept most obvious to the everyday knowledge of the pupil. The above analysis shows that this is totally wrong insofar it fails to see that the basis of set theory is the idea of relation. This idea is not a simple one, but a most complex one, since it is deeply rooted in our whole scientific world-view.

c. To make the idea of relation the nucleus of an understanding of the number concept is the conclusion which, in my opinion, has to be drawn. But to do this it is not at all necessary to start with set theory. In my opinion, it may be better, in a more classical approach to develop the number concept step by step alongside that of quantity; the emphasis being on the problem of measurement. Numbers are introduced by means of the formula

$$x = b/a$$

b being an arbitrary object represented as a quantity, and a being a standard (cf Dawydow 1977, Freudenthal 1974, Otte 1976). Thus, numbers have been made relations of magnitudes and developing the number concept amounts to conceiving of relations, algebraic operations, and measurements, more and more generally according to the extensions of the range of applications.

d. The above mentioned empirical understanding of set theory has led to a very static image of mathematics. As a result of this the ontological ideas of mathematics which have been produced by many authors of "new math" were largely opposed to more dynamical views as they are contained in F. Kleins "Meraner Program", putting the function concept in the focus of the curriculum. Going back in our historical analysis to the begininng of the 19th century we see the connections between the concepts of function and set and the program of arithmetizing mathematics, the kernal of these being a kind of thinking in relations as opposed to a thinking in substances. One may imagine that it will be possible to develop a unified view integrating substantially functional and set theoretical ideas.

References

Belaval, Y.: La crise de al geometrization de l'univers dans la philosophie des Lumieres. In: Revue internationale de Philosophie, VI (1952), 337- 335.

Bolzano, B.: Beytrage zu einer begrundeteren

Darstellung der Matematik, 1810. Reprinted Darmstadt 1974.

Bolzano, B.: Grossenlehre, 1830-1848. Published J.Berg (ed.): B. Bolzano, Einleitung zur Grossenlehre und erste Begriffe der allgemeinen Grossenlehre. Stuttgart 1975.

Cassirer, E.: Das Erkenntnisproblem in der Philosophie und Wissenschaft der neueren Zeit, vol. 2, Darmstadt 1974.

Freudenthal, H.: Soviet research on teaching algebra at the lower grades of the elementary school. In: Educ. Studies in Math. 5 (1974), 391-412.

Gauss, C.F.: Brief an Drobish. In: Werke X/1, 106-107.

Gauss, C.F.: Fragen zur Metaphysik der Mathematik. In: Werke X/1, 396-397.

Gauss, C.F.: Theoria Residuorum Biquadraticorum Commentatio seconda. In: Werke II, 169-178.

Gauss, C.F.: Zur Metaphysik der Mathematik. In: Werke XII, 57-61.

Jahnke, H.N.: Zum Verhaltnis von Wissensentwicklung und Begrundung in der Mathematik - Beweisen als didaktisches Problem. Materialien und Studien des IDM, Vol. 10, Bielefeld 1978.

Otte, M.: Die didaktischen Systeme von V.V. Dawidow/D.B. Elkonin einerseits und L.B. Zankow andererseits. In: Educ. Studies in Math. 6 (1976), 475-497.

Sneed, J.D.: The logical structure of mathematical physics: Dordrecht 1971.

Stuloff, N.N.: Uber den Wissenschaftsbegriff der Mathematik in der ersten Halfte des 19. Jahrhunderts. In: A. Diemer (ed.): Beitrage zur Entwicklung der Wissenschaftstheorie im 19. Jahrundert. Meisenheim 1968.

RESTRICTED PLATONISM AND THE TEACHING OF MATHEMATICS

Rolando Chuaqui
Pontificia Universidad
Catolica de Chile
Santiago, Chile

Two of the main objectives on the teaching of a science should be to show what the science is and how it grows (i.e. how discoveries are made). Thus, the conceptions about the nature of mathematics should influence its teaching. In this paper, I intend to analyze the implications that restricted Platonism, as discussed in Chuaqui 1980, has for teaching. Summarizing, this view of mathematics is as follows. Mathematics deals with abstract real objects: nonspatial and atemporal objects existing independently of the human mind. Some of these objects are related to physical objects, but they are different from them. Mathematicians discover these entitites and their main properties by reflection.

The main reason for admitting mathematical objects existing independently of the human mind is the impossibility of giving a reasonable foundation for Mathematics as it exists today without this assumption. As Godel says:

"It seems to me that the assumption of such objects is quite as legitimate as the assumption of physical bodies and there is quite as much reason to believe in their existence. They are in the same sense necessary to obtain a satisfactory system of mathematics as physical bodies are necessary for a satisfactory theory of our sense perceptions and in both cases it is impossible to interpret the propositions one wants to assert about these entitites as propostions about the "data", i.e., in the latter case the actually occurring sense perceptions". (Godel 1944).

However, it is not possible to accept indiscriminately the existence of abstract objects. (Thus, the name: restricted Platonism.) For instance, the set-theoretical paradoxes show that not all properties have extensions. Thus we have first to postulate the existence of certain objects following basic intuitions. These objects are not directly present in intuition, but are discovered in it by an intellectual process that may be called reflection. Properties of these objects are given as axioms and consequences of these are obtained - there should be no contradiction.

In Chuaqui 1980, I discuss more fully the necessity of mathematical objects for giving an account of Mathematics. I believe that Jahnke's analysis of the distinction between numbers and quantities also provides supporting evidence in favor of this restricted Platonism. Jahnke shows that number and other abstract objects could not be identified with empirical properties of physical objects and, thus, have a different status that these properties.

I shall now schematize (oversimplifying) the main features of Mathematics according to restricted Platonism.

I. Basic intuitions.

Basic mathematical ideas are inspired from many different sources: everyday life, other sciences, philosophy, mathematics itself, etc. For instance, in case of geometry the inspiration is spatial. the objects mentioned in the mathematical theories of geometry are not the physical objects appearing in space, but are abstract objects that approximate the physical ones. Geometrical oints, lines, etc. are not physical points, lines, etc. but only their abstract counterparts. In the case of logic, on the other hand, the source of inspiration is mainly philosophical. The concepts dealt with are those of truth, logical validity, etc. In other cases, the notions are abstracted from mathematics itself. Thus, I believe that the main inspiration for the concept of set is mathematical. Sets of X where X is a well defined collection (numbers or functions, or . . .) had appeared in mathematical practice for a long time when Cantor introduced Set Theory.

Although the consideration of a certain mathematical object may be inspired by empirical facts, they are not themselves empirical. However, it is important that the source of inspiration be made known to the students, so that they see the motivation for the study of a theory and do not get the impression that mathematics is just a game.

An important ability that everybody should learn is how to formulate abstract models for physical situations. In particular, this abstraction process, should be learned. I believe that the only way to learn this is by examples of increasing difficulty. The simplest case is that of problems in which an imaginary but possible situation is presented. The first step, to formulate this problem mathematically is a good example of this abstract modelling. More complicated cases are the formalization of mathematical theories, as in geometry.

When a notion is presented to intuition it may be too vague to formulate mathematically. Thus, the first step is to delimit and exlain this notion sufficiently. Sometimes, a notion that appears unique is split into several by this process. This informal process of analyzing notions, should also be exemplified in teaching.

2. Formalization.

Once the mathematical objects that are to be studied are determined as precisely as possible, some properties of these objects are chosen as axioms. I believe it important for teaching that axioms do not appear arbitrary; they should be justified intuitively. But it should be clear that this justification is not a formal proof.

In some cases, abstract axiomatic systems may be introduced, such as axioms for groups. But, in order that they do not appear as a mere game, they should be exemplified profusely. Thus, it would be also possible to show how second level abstractions (abstractions over mathematical theories) operate and how mathematics unifies different theories into one.

3. Deduction.

New concepts are defined (derived concepts) and new properties (theorems) are discovered (and then deduced) from the fundamental concepts (primitive terms) and properties (axioms).

Although it might be possible to discover new properties directly by reflection and be convinced of their applicability on intuitive grounds, this procedure is risky. It has often happened that our reflections are wrong. Thus, it is necessary to prove the theorems by methods which are beyond doubt. When axiomatic systems are completely formalized (in the sense of a logician), proofs can be done in such simple steps that their validity can be checked in elementary number theory, the safest part of mathematics. But, it is seldom necessary to go so far. What is needed is to perform the proof in systems that are sufficiently safe. As Jahnke 1978 has pointed out, there is a subject-related element in proofs. What is proved for one person may not be proved for another. It all depends on which system is safe for each person. And this, on its turn, depends on how well the person is acquainted with the system. However, proofs, in another sense, are independent of the subject, since it is always possible to find a system which everyone would find safe.

To discover by reflection often is also difficult. In some areas of mathematics, methods of discovery can be codified as algorithms. If we have proved that an algorithm works well, then we can use it without proving each time that the result obtained is correct. In general, an algorithm is similar to a proof - it reduces difficult discoveries to a series of simple processes.

Mathematics is the science where proofs and algorithms are used most extensively. Thus, it is the best place for students to learn these methods. To recognize when a proof is correct is an acquired ability and the only way to acquire it, is to do many proofs. It is very important not to confuse the students by calling heuristic explanations, proofs. Also, some nontrivial proofs should be performed.

The consequences that are obtained from the axioms are important for clarifying and fixing the basic objects and notions. A reciprocal influence between the informal notions and formal systems is thus produced obtaining, in this way, a continuous refinement of our knowledge of basic objects and an improvement of formal systems.

4. Applications.

Once we have proved some theorems or calculated some expressions, these results should be compared with the intuitions from which our theory originated. This comparison serves to improve our understanding of the basic concepts and, in some cases, to correct them.

It might be convenient to look for other interpretations of the primitive terms that make the axioms true and show how the theorems also hold in this case.

5. Growth.

I believe it is very important to present mathematics as a growing science. Growth in mathematics can be classified into two groups: growth in extension and growth in depth. Growth in extension may be realized by the discovery of new theories or the solution of problems in developed theories (in particular, the proving of new theorems). When growth is in depth no new theories are discovered nor old problems solved. In this case, new foundations are given for old theories. For instance, Dedekind and Peano at the end of last century gave an axiomatization for number theory and showed that it is sufficient. This helps to understand the nature of numbers, because it tells us which are the fundamental properties of numbers from which all others are obtained. This type of growth also puts theories on a safer base, since it is easier to justify by reflection a few properties, the axioms, than many.

Both types of growth should be shown to the students, explaining in each case which is the intention involved. thus, it may be useful in some cases to prove from axioms a statement that is intuitively obvious. This would show that the axioms are sufficient for the theory at hand. However, it should be made clear that this proof is not given for verifying the truth of the statement, but only for showing that it can be derived from the axioms.

Finally, I would like to add a few remarks about the cultural importance of the study of the philosophy of mathematics. One of the main purposes of education is to form human beings who understand what they are doing. Thus, if somebody is studying mathematics, he should understand what mathematics is; so some philosophical issues shold be discussed in class. It is especially important that teachers of mathematics understand their discipline. That is why in the formation of teachers, philosophy of mathematics should be stressed.

References

Chuaqui, R., Platonism as philosophical foundation of Mathematics, to appear in the Proceedings of the Third Brazilian Conference on Mathematical Logic, U. de Sao Paulo, 1980.

Goedel, K., Russell's Mathematical Logic, in the Philosophy of Bertrand Russell, P.A. Schillp (ed.), The Library of Living Philosophers, pp. 123-154.

Jahnke, A.N., Zum Verhaltnis von Wissensentwicklung und Begrundung in der Mathematik - Beweisen als didaktisches Problem, Institut fur Didatik der Mathematik der Universitat Bielefeld.

WHAT WE CAN GET FROM THE HISTORY OF ARITHMETIC

Gilles Lachaud
Universite de Nice
France

1. Why History of Mathematics in Mathematical Education?

Most of the time, Mathematics is taught as if it were an eternal truth, without any alteration in past or in future, as if theories were born armed, like the goddess Athena, from the brow of Zeus. It was, I guess, in regard to this kind of attitude that the late Roland Barthes was speaking of the arrogance of Science.

This behaviour in teaching generates, in my view, one of the main types of pedagogical obstructions: the student can't integrate a notion which is presented to him without any reference to his own reality.

In fact, every mathematical concept has its own history, with its birth, its position in the general web of the Mathesis where it is occuring, etc. If we see it from this viewpoint, we are not directly faced with the compulsive side of the mathematical activity, once we have seen how relative and evolving is a mathematical concept.

From several examples, historical or not, we see that if someone is studying a mathematical concept, he is always establishing a connection between the history, or rather, the "story" of the concept and his own "story." (The word "story" is used here in its psychological meaning). This gives rise to a kind of epistemological relation between the student and the concept which would be called analytic rather than genetic.

I should add that the historical approach allows us to take some distance with the concepts studied: we see the rise of the concept, its growth and, in the better cases, its achievement. As Novalis said: An unachieved theory takes us far from reality, while a complete one brings us back to it.

And from this position, it's possible to go closer and closer until we arrive to the coincidence point where a kind of empathy is established between the topic and the mind of the man which is studying it. In this approach, Arithmetic (this word is used here as synonymous with "Number Theory") offers a very distinguished topic to work with; firstly, because of the permanence of the concept of Number through the ages; and secondly, because it is possible to obtain arithmetical results at any level, from the most elementary to the most sophisticated.

2. Experiences and experiments.

The above may appear theoretical. However, I'll give some examples of this approach.
a) An example of teaching: During the lectures on number theory given at the graduate level, we used the following procedure. Starting from a summary of the topics appearing in the history of Arithmetic, especially those studied by the early Greeks, and observing the evolution of these topics up to modern times, it was offered to each student to choose one of the topics involved, and to study it, starting from some original text.
b.) An example of a "new" approach to a concept: this process can also lead to a "new" approach of some contemporary concepts. For instance, I begin by quoting the work of Houzel (2) and Weil (7) on elliptic functions. Here, starting from the work of Abel and Eisenstein and so on, these works give an historically progressive approach to the topic, up to contemporary results which are rather hard to master directly with the current literature.

Also, it is possible to give a reading of the newly discovered work of Diophantus in the framework of elementary algebraic geometry (4). This leads to an example of applications of elementary algebraic geometry to diophantine analysis avoiding the megamachines often needed in these kind of results.

Finally, in some sense, I should add that Ribenboim followed the same approach in his book on Fermat's last Theorem (5), as did Ellison for the Theory of Numbers during the 19th century (1).

3. What's in the core of Arithmetic?

The crib of Number Theory. Arithmetic can be seen as a part of "Pure Mathematics" (i.e. without a priori applications). Therefore if one wants to look at the meaning of the mathematical activity in the human behaviour, Arithmetic seems to be a good example.

In the West, or rather in the Middle-East, the oldest trace of an exposition of arithmetical assertions are to be found in the Rhind papyrus, and later in the Pythagorean school. (We leave aside the Babylonian Mathematics for our purpose, since the mathematical cuneiform texts don't give any theoretical explanations of the computations inscribed, so the interpretation of these texts is multiple-valued.) The "table of Egyptian fractions of 2/n" of the Rhind papyrus shows that ontological considerations were at the bottom of the computation of fractions: the purpose of the table (to reduce fractions like 2/n, because only the unity can be divided), and the use of hieroglyphs testify that the author was, at the background of the theory, merely concerned by philosophical rather than logistic considerations.

Later, we find in the text of Euclid several examples of mathematical assertions which seem to be enunciated as a background for the understanding of the texts of Plato, like the Pythagorean triples, the assertion on perfect numbers, the dialectic of odd and even, and the tentative classification of irrationalities (in book XI, according to the reading of Andre Weil).

On the other side, we see also in the East, namely in the Ta Chuan (a chapter of the I Ching), arithmetical considerations as a support for the study of the laws of Change in the Universe.

From this we deduce that the roots of Arithmetic are to be found in a place where the numbers were seen as meaningful in the realities of perception of the exterior world.

The outbreak of axiomatics. An epistemological rupture appears with Diophantus in the West; his writings are "modern mathematics" in the following. Firstly, like Euclid, he takes care to give the rules of computation (so the assertions given are no longer universal, but are depending on the rules introduced). Secondly, his only goal is to solve equations (when he has given a solution of an equation, he studies another equation). Thirdly, he makes no mention of the unsolved problems of the family that he is studying.

This same line will be followed by the majority of the Arabic mathematicians (with the notable exception of Thabit Ibn Qurra), and later by Fermat and his successors.

"Was sind und was sollen die Zahlen?". We have henceforth to consider two distinct realities of the notion of Number, (By reality I mean the meaning given by a receptor, in some specified situation, to a system of data received by him.) If the second reality (the one given by the system of axioms, those of Peano, say) is inner to the theory where they are introduced, the first one (as illustrated by Plato) is exterior to the axioms introduced; the Theory of Numbers is only a part of "Natural Philosophy". It is in this point of view that C.G. Jung hazarded the conjecture that "it might be possible to take a further step into the realization of the unity of psyche and matter through research into the archetypes of the Natural Numbers (6) and also (3).

Since the two approaches aren't opposed, but complementary, they are in fact, at a conscious or at an unconscious level, constantly intertwining; so a system of education which is taking these two approaches into account cannot be unfruitful.

References

1. Ellison, W.J., Theorie des Nombres, in "Abrege d'Histoire des Mathematiques", J. Dieudonne ed., Paris, Hermann 1978.

2. Houzel, C., Fonctions elliptiques et integrales abeliennes, in "Abrege d'histoire des mathematiques", J. Dieudonne ed., Paris, Hermann 1978.

3. Lachaud, G., Nombres et Archetypes: Du vieux et du neuf, Eleutheria, 2 (1979), p. 133-140.

4. Lachaud, G., Rashed, R., Une lecture de la version arabe de "Arithmetiques" de Diophante, preprint; see also a translation of the work of Diophantus, to appear, Les Belles-Lettres, Paris.

5. Ribenboim, P., 13 Lectures on Fermat's Last Theorem, New York, Springer, 1979.

6. Von Franz, M.L., Number and Time, Evanston, Northewstern University Press, 1974.

7. Weil, A., Eliptic Functions According to Eisenstein and Kronecker, New York, Springer, 1977.

WHY THE HISTORY OF MATHEMATICS SHOULD NOT BE RATED X - THE NEED FOR AN APPROPRIATE EPISTEMOLOGY OF MATHEMATICS FOR MATHEMATICS EDUCATION

David Pimm
Department of Science Education
University of Warwick, England

In 1974, in the journal Science, a historian of science Stephen Brush published an article entitled, "Should the history of science be rated X?" In my talk, I should like to adapt a couple of his points to the situation of mathematics and then proceed to discuss one possibility (out of many) in which the history and philosophy of mathematics can be of use to mathematics education, namely that of informing our understanding of mathematics. As an additional point of reference, let us recall what mathematician Rene Thom said in his address to this congress eight years ago in Exeter. The real problem which confronts mathematics teaching is not that of rigour, but the problem of the development of 'meaning', of the 'existence' of mathematical objects". If my total effect today is to entice you to re-read Thom's address, I shall have spent my time well.

Brush argues that one current viewpoint among many contemporary historians and philosophers of science about science is one of cultural relativism. "I will examine arguments that young and impressionable students at the start of a scientific career should be shielded from the writings of contemporary science historians - namely that these writings do violence to the professional ideal and public image of scientists as rational, open-minded investigators, proceeding methodically, grounded incontrovertibly in the outcome of controlled experiments, and seeking objectively for the "truth". Thus if you, as a teacher responsible for the education of young scientists, wished to promulgate this view of scientific activity, his advice was for you to keep your students away from the proliferating undergraduate courses at American campuses on the history of science.

However, Brush also argues that scepticism about established dogma is supposedly a prized asset in science. Clearly one possible root of such resistance will be a broad knowledge of the history of that discipline. But where is the tradition of criticism and scepticism in mathematics? A recent instance of conflict between two published papers in homotopy theory (involving mathematicians Zahler and Toda) provides one of the few publicised instances of controversy (coincidentally Zahler is also one of the protagonists in the Catastrophe Theory dispute). But Zahler's reaction to the controversy was to leave mathematics for medical school.

Imre Lakatos' work Proofs and Refutations provides one instance of what might be termed mathematical literary criticism (as well as much else besides) and in it he draws attention to precisely the reverse tradition from one of scepticism in mathematics, namely one where mathematical maturity is equated with willingness to suspend disbelief (and questions) until the results are proved (a classic instance is provided by analysis proofs which begin, for example, "Choose $\delta = \varepsilon^2/27\varepsilon + \pi$ "). Larry Copes has also written on this point, namely the antagonistic reversal of the meaningful order in order to produce the customary formal one. This provides a new twist to the doctrine of the ends justifying the means. Perhaps Hilbert was referring to this spirit of pragmatism when he said, "In mathematics, as elsewhere, success is the supreme court to whose decisions everyone submits". However, implicit in that statement is the absolute nature of such decisions. Sociologist David Bloor, in his splended stress on the 'negotiability' of meaning and hence of mathematics itself, has highlighted the social nature of mathematics. The history of the Euler-Descartes conjective can be viewed as an attempt to preserve the insight that V - E + F = 2 for polyhedra and the negotiation of definitions and concepts that mathematicians engaged in to that end.

Proofs and Refutations looks at mathematical methodology, and as examples to show that even this negotiable, consider the alleged computer proof of the 4-colour theorem or the methodological monster - barring exclamation of Gordan when confronted with Hilbert's non-constructive proof of the existence of a finite basis of any system of invariants. He claimed, "Das ist nicht Mathematik - Das ist Theologie". Constructivist Everett Bishop's rejection of Jerome Keisler's approach to calculus via infinitesimals is another contemporary instance of this. The historical development since Gordan has not agreed with him - but who is to say that it may not do so later, in a move of Copernican conservative zeal (for example, rampant constructivism). The potential of future mathematics to rewrite that of the past (and along with it all our judgments on validity which seem so timeless to us) has been little appreciated, I feel. For this reason, among others, I would include the discovery of the legitimacy of infinitesimal arguments within the current canons of mathematical acceptability as one of the most interesting items of twentieth century developments in mathematics. It ranks with the discovery of Non-Euclidean Geometry in the previous century in the attack it presents on absolutism in mathematics.

Lakatos' work on mathematics provides a method of criticism for improving conjectives and proofs, as well as a taxonomy of counter examples, organized according to their function and point of contact with the above. He stresses the essential role of criticism in mathematical progress and it is a sad fact that this is a virtually non-existent area within mathematics education. Are we aiming to transmit mathematical methods or a body of knowledge? Surely it is equally essential to explain and communicate the nature of mathematical discovery.

We are provided with a description of the way he believed mathematics developed and an analysis of the process. It provides a more exciting interpretation of "ontogeny recapitulates phylogeny", a precept often quoted in this area for mathematics education. This is the claim that we should teach according to the dialectic method of proofs and refutations in the classroom, augmenting our insight and enriching our concepts, because this reflects the way mathematics was and is actually done.

What is the current involvement of philosophy of mathematics in the mathematics curriculum? Overtly, virtually nothing. Yet, as Rene Thom so brilliantly proclaimed, "In fact, whether one wishes it or not, all mathematical pedagogy, even if scarcely coherent, rests on a philosophy of mathematics". What then is the "hidden agenda" to be metalearnt about mathematics? Clearly that mathematics is the formal study of abstract systems presented in strictly deductivist style. Attendance at courses called philosophy of mathematics which are often in fact courses in logic (identifying the former with formalism) will accentuate this belief. Mathematician Jean Dieudonne has written, "D'ou la necessite absolue qui s;impose desormais, a tout mathematicien soucieux de probite intellectuelle, de presenter ses raisonnements sous forme axiomatique, c'est-a-dire sous une forme ou les propositions s'enchainent en vertu des seules refles de la logique, en faisant volontairement abstraction de toutes les 'evidences' intuitives que peuvent suggerer a l'esprit les terme qui y figurent." But surely in mathematics education, even if it is so in mathematics (which I doubt), we are not even predominantly interested in the presentation of formally correct arguments, thus reflecting Thom's concern over the prevailing balance of rigour over meaning. His parallel reply to Dieudonne was "Any mathematician endowed with a modicum of intellectual honesty will recognise then in each of his proofs he is capable of giving a meaning to the symbols he uses".

Thus the parallel of learning the scientific method becomes learning the axiomatic method. Are we not, in mathematics, in the same position that science was a couple of generations ago, that is we have an officially enshrined methodology - at least as regards public mathematics. My point is that the majority of mathematics instruction at the university level produces a false picture of mathematics, private mathematics, creative, human mathematics by the constraints and beliefs concerning public mathematics - the formal presentation via the lecture hall, the textbook and the research journal. Even if mathematics had a Bridgman figure arguing that there is no such thing as the mathematical method (as Bridgman did for science in the 1920's), will students absorb the "correct" attitude through their mathematical education? Only if mathematicians behaved in private as they do in public. But do they?

What is the current involvement of history of mathematics in the mathematics curriculum? The most common use of history at the earlier levels is in trying to humanise mathematicians by regaling classes with foibles or often apocryphal tales of their youth, rather than a serious attempt to humanise the subject. We thus see mathematicians with a human face rather than mathematics as a humanity, a product of human vision. Courses on History of Mathematics itself tend to be cultural history first and mathematics second. They are either kept isolated from twentieth century mathematics or are used in a Whig historical manner. This latter tendency is encouraged by the looting of the past, ideas being wrenched from their context and, more devastatingly, translated into the current idiom. An attempt is made, in this way, to force current notions and language back in time, purportedly to explain what certain mathematicians were trying to do, but in fact ensuring the continued currency of these ideas.

What is missing from the formal presentation in the customary Satz-Beweis manner and how can the history and philosophy of mathematics help alleviate the problem? Firstly, one lacks any discussion of the problem background. What questions were the mathematicians involved in trying to solve. How were she and her contemporaries viewing the problem and in what way is this theorem a solution? More importantly, how do the definitions of the terms involved relate to the theorems as stated and proved. This whole murky but fascinating area has been incandescantly lit by Lakatos' work (for instance, in the telling concept of proof-generated definition) which has only recently seen more development or criticism (e.g. by David Bloor and Solomon Feffermann). I said earlier that history and philosophy of mathematics can inform our understanding. What are the sorts of things in mathematics we can understand?

a) definitions: e.g. of angle, group, $a^{-2} = 1/a^2$, differentiability. Examination of alternatives. What makes a definition a good one?

b) Statement of a theorem: particular examples included and excluded. e.g. fundamental theorem of calculus.

c) Proof of a theorem: globally or line-by-line. Abstract a method (e.g. induction, contradiction, compactness, e.g.) What makes them successful? Why them rather than others - standard 'moves'? (Lakatos' Proofs and Refutations argues forcefully for the essential interrelatedness of (b) and (c), with regard to understanding.)

d) a problem: Why is it a problem? Consequences of its solution. Polya's heuristic here.

e) a solution: Why is it a solution? Is it the only one, e.g. differential equations?

e') process of solution: e.g. algorithm. Justification of solution (e.g. Hilbertian pragmatism, "It works").

f) a concept: derivative, exponent, factor, prime, ring. Why is it useful? What is its use?

g) an area of mathematics: e.g. analysis. Characteristic arguments, central concepts, methods of working, phenomena of interest, relations between ideas. Why does it cohere? Does it? (e.g. I think "calculus" fails to cohere).

h) mathematics itself.

With this kind of richness available, illuminated and illustrated by the forces and vagaries of historical development, does not the axiomatic presentation seem somewhat stark and yet pale also?

A new epistemology of mathematics for mathematics education is required which involves as the confluence of mutually-permeable strands of the now-separated disciplines of mathematics, history of mathematics and philosophy of mathematics. Lakatos, paraphrasing Kant, claimed "The history of mathematics, lacking the guidance of philosophy, has become blind, while the philosophy of mathematics, turning its back on the most intriguing phenomena in the history of mathematics has become empty." But what hope has mathematics education without either? My aim is the creation and development of personal knowledge of mathematics rather than the transmission of "absolute" knowledge. The growth of understanding can only be enriched and encouraged by an awareness of the problem sources which renew and maintain our mathematics and allow us to observe the changes in what it has been and to inform our creativity with regard to what it might be. A polished, logical presentation of mathematics (that is, historic, prepared with hindsight) shows none of the difficulties, errors, guesses, stumblings which went into its creation and attainment of its present form. Unifying concepts and proofs can't unify if there has been no awareness of a previous state of disparateness and generalizations have to generalize something. In this light, Sarton's claim that "nothing suffers so much when divorced from its history as does mathematics" can be seen in its full force.

The beauty of the study of history is that it can give a sense of place (and hence, for me, meaning) in the scheme of things rather than merely a set of disembodied concepts, while at the same time providing an elevated point of view from which to survey the current position. A sense of history can provide an awareness that it was not always the way it is today and hence that it might have been otherwise. In this way, history can hopefully convey the notion of a culturally-based and culturally-bound mathematics which is open-ended and changing, a challenge to the more prevalent view of mathematics as a static list of accumulated truths.

An amended version of this paper appears in For the Learning of Mathematics, Vol. 3, No. 1, 1982.

13.2 RESEARCH IN MATHEMATICAL PROBLEM SOLVING

CHARACTERISTICS OF PROBLEM TASKS IN THE STUDY OF MATHEMATICAL PROBLEM SOLVING

Gerald A. Goldin
Northern Illinios University
DeKalb, Illinois

This paper presents some perspectives on the role of task variables in research on the psychology of mathematical problem solving. Hopefully, it will serve as an introduction to the contents of the book I was recently privileged to edit on this topic (Goldin & McClintock, 1979). Where appropriate, the paper refers to chapters contributed by various authors.

1. The Task as a Measuring Instrument in Research

Despite widespread recognition of the importance to scientific inquiry of the replication of empirical findings, sufficient information to repeat problem-solving experiments is rarely published, and few replications are actually performed. Of course, inevitable difficulties exist in defining comparable subject populations and in preparing identical experimental treatments. However, it is my view that the problem tasks themselves are the crucial elements in establishing a standard of replicability in problem-solving research. This perspective has been one of the considerations motivating the examination of task variables.

To explain this opinion let me compare the measurement of problem-solving processes in children or adults with the measurement of a physical property such as the pressure or temperature of a system. Problem tasks function as the measuring instruments used to obtain information about problem solvers. When a subject is presented with a mathematical problem, complicated events occur which can be recorded and classified in various ways. If the experimenter is unaware of important properties of the task itself, or does not describe it sufficiently well for other researchers to construct identical instruments, the observations made will be less valuable. Indeed, evidence abounds that small changes in a problem's syntax, content, or structure can produce large changes in problem difficulty and in the processes used by subjects. That is, problem-solving outcomes do not vary slowly and continuously with the variables describing problem tasks. This situation is much worse for reproducibility of measurement than in the case of a barometer or thermometer, where the reading is most probably a well-behaved function of the instrument variables. Therefore, attention to task variables is all the more important in the measurement of problem solving.

In some published experiments, a single problem or set of problems is presented to a number of subjects and their problem-solving processes are recorded. But without information on how various characteristics of the problem or problems are influencing the observed processes, the findings cannot be interpreted as measurements - - it is not possible to separate knowledge gained about the problem itself from knowledge gained about the problem solvers. In other experiments, performance on different but related problems is examined, If, as is often the case, the tasks vary in many characteristics simultaneously, differences in outcome cannot be attributed to particular variables, and interpretation of the findings is again subject to limitations. In order to make progress in developing instrumentation, it is important to be able to control as many task variables as possible, changing one characteristic at a time to discover the effects of the change.

Just as standard pressures and temperatures are defined with respect to reproducible physical states, problem characteristics must sometimes be defined with respect to "reference populations." This idea is familiar to psychometricians, who carry out item analyses on multiple-choice tests, but seems to be rarely used by those studying problem solving processes. A problem "has" a given syntax for populations of speakers of standard English, French, etc.; likewise, it only "has" a given mathematical structure for populations which use certain mathematical notations and rules of procedure. Nevertheless, problem syntax and structure can be defined independently of individual subjects within such a population. Reference populations may also be required to establish appropriate quantitative scales or classificatory categories for reporting problem-solving outcomes.

The above considerations all suggest the need for an exhaustive understanding of the tasks used in problem-solving experiments before we can hope to study problem-solving processes in a scientific way.

2. The Categorization of Task Variables

This section outlines the proposed classification scheme for task variables. The main headings are syntax variables, content and context variables, structure variables, and heuristic behavior variables. These are described in terms of the methods of task analysis needed to define variables in each category, and the stages in problem solving which the variables are most likely to affect (Kulm, Ch. I).

Syntax variables describe the arrangement of and grammatical relationships among the words and symbols in the problem statement, and are expected mainly to affect "decoding" processes. Included are variables of length, grammatical complexity, numeral and symbol forms, verbal sequence, and variables describing the position and form of the question sentence. Of special interest in this area are efforts to describe grammatical complexity by means of a single quantitative index, such as Yngve's measure of depth or the syntactic complexity coefficient of Botel, Dawkins, and Granowsky. Linear regression models have been used by Jerman, Loftus, Suppes, and others to try to predict problem difficulty using task variables, and syntax variables have figured prominently in these studies (Barnett, Ch. II).

Content variables describe the mathematical information contained in the problem statement, with reference to the meanings of terms and phrases, but without reference to further mathematical processing of this information. Included in this category are classifications of problems by subject area, by conventional "types"or by field of application; the "key words" and technical mathematical vocabulary which a problem may employ; variables describing the elements (given conditions, numerical information hints, goal conditions, conditions implied but not explicitly stated, etc.); and the specialized mathematical equipment which a problem may require. Context variables describe non-mathematical information, including features of the problem embodiement (symbolic,manipulative, etc.), the verbal context or setting, and the information format. Both content and context variables are expected to affect "understanding" and "translation" processes to the greatest extent (Webb, Ch. III).

Structure variables describe mathematical properties of a problem representation, as distinct from the problem statement, and generally require some mathematical analysis of the problem for their definition. State-space analysis is a major tool for defining structure variables, for describing the symmetry and subproblem characteristics of problems, and for describing relationships between problems. Numerical variables describing the problem state-space can include the total number of states, the length of the shortest solution path, the number of blind alleys, the number of possible first moves, the number of goal states, etc. Non-numerical structural characteristics of the state-space include equivalence classes of states under the action of a symmetry group, sub-groups of the symmetry group, forward-backward symmetry, and subspace decompositions. Homomorphisms and isomorphisms can describe injective, surjective, and bijective relationships between state-spaces. Structure variables can also be defined which describe the initial state in a standard representation (such as the number of equations and unknowns in an algebra problem), or which describe the problem with respect to a particular algorithm or strategy (such as the length of a solution path generated by the algorithm, or the number of times a particular operator is called for). Structure variables are expected to affect search processes within a representation and the transfer of learning between related problems (Goldin, Ch. IV).

Finally, heuristic behavior variables include those characteristics of the task which elicit particular heuristic processes from solvers, including processes for understanding the problem, for selecting a representation, for exploiting the representation, and for utilizing alternate representations. Processes which are particularly valuable in the solution of a problem contribute importantly to the description of that problem. Sometimes identifiable cues in the problem statement suggest particular heuristic processes; thus the request for a uniqueness proof may suggest the method of contraposition. Heuristic behavior variables as task variables are expected to have their most important influence on planning processes (McClintock, Ch. V).

3. The Task as an Instrument of Instruction

We have briefly surveyed the major classes of task variables which have been studied. Such a framework, once established, can have important applications to teaching as well as research. Among these are the following: (1) controlling certain task variables and systematically changing others to achieve instructional objectives with respect to problem syntax, content, context, or structure; (2) sequencing problems with respect to measures of structural or algorithmic complexity; (3) construction of test items which are "parallel" on a defined set of task variables; (4) construction of problems having desired kinds of structural relatedness for classroom use; and (5) construction of problems requiring particular problem-solving strategies or heuristic processes. Thus reserach on task variables is likely to contribute in an immediate and practical way to classroom instruction in problem solving (Caldwell, Ch. IX; Luger, Ch. XA; Schoenfeld, Ch. XB).

4. The Description of Problem-Solving Behavior

Several experiments have examined the conseqences of holding particular task variables fixed while others are altered. In problem-solving experiments, as in physical measurements, it is important to know in advance the set of possible outcomes, so that the meaning of a given observation can be understood in relation to other outcomes which might have occurred but did not. Three methods have been employed in addition to traditional methods in which outcomes are described in terms of problems answered correctly or incorrectly. The methods have in common that they do not require advance commitment to a specific model for human problem solving, but the categories of behavior are nevertheless established ahead of time and do not change with each subject studied.

First is the idea of establishing "strategy scores" for the choices made during problem-solving, based on "ideal" behaviors associated with the consistent application of one or more well-defined strategies. This approach has been taken previously by authors such as Bruner, Goodnow and Austin, and Dienes and Jeeves. The scoring system is task-specific, reflecting the degree of correspondence between actual behavior and task-specific strategies.

A second method is the description of behavior by means of paths through state-spaces associated with problem representations. As problems vary in structure, their state-spaces provide a way to describe behavior that reflects the changing structure; patterns in the observed paths generated by subjects can be attributed to structural features of the state-spaces. When two problems are related by means of a state-space homomorphism, one can examine similarities in the state-space paths followed by subjects who solve both problems.

Finally, there is the development of "process-sequence coding", in which a subject's verbal protocol is analyzed into a sequence of finely described process categories ("introduces equation," "draws a diagram," etc.). The most detailed scheme to date involves more than fifty symbols representing process and outcome categories. Here the scoring system is task-independent, but difficulties with inter-scorer reliability remain to be resolved (Goldin & Caldwell, Ch. VI; Waters, Ch. VIIA; Days, Cha. VIIB; Luger, Ch. VIIC; Harik, CH. VIIIA; Lucas, Branca, Goldberg, Kantowski, Kellogg & Smith, Ch. VIIIB).

This brief report mentions but a few of the many issues raised by the explicit study of task variables in problem-solving research.

Reference

G.A. Goldin & C.E. McClintock, eds., Task Variables in Mathematical Problem Solving. Columbus, Ohio: ERIC Clearinghouse for Science, Mathematics and Environmental Education (1979)

TOWARD A TESTABLE THEORY OF PROBLEM SOLVING

Alan H. Schoenfeld
Hamilton College
Clinton, New York

The four main issues addressed in this paper are

1. What, beyond basic subject matter mastery, serves to explain "expert" mathematical problem solving behavior?

2. What traits do students lack, or what inappropriate traits do they have, which prevent them from approaching problems with the flexibility and resourcefulness of experts?

3. Can we teach students to "solve problems like experts" - - and how?

4. What clear, scientific evidence can we offer to support our opinions regarding the first three questions?

The students I refer to will be advanced high school or lower division undergraduate students and the problems are "nonroutine," of the type discussed by Polya in Mathematical Discovery. "Experts" can be defined (for the sake of simplicity, and not uniquely) as college mathematics faculty.

The answers from the mathematics education community to the first three questions would, I suspect, involve the word revived by the Honorary President of this Congress, George Polya: "heuristics." The fourth question is harder. There has been little conclusive evidence to date that heuristics "work" - - in the sense that students can learn to use them, and thereby, improve their problem solving performance. In fact, for the most part, heuristics are ignored, dismissed or disdained outside the math-ed community. Herbert Simon, writing to "christen" the new domain of cognitive science, spoke of "cognitive psychologists, researchers in artificial intelligence, philosophers, linguists, and others who seek to understand the workings of the human mind."[1] Allen Newell, coauthor with Simon of Human Problem Solving, wrote that we are working in the wrong direction: "There is hardly any evidence . . . of a scientific kind that Polya is right. We are just impressed and pleased, that's all."[2] For its own part the math-ed community ignores with equal impugnity the advances made in cognitive science: the 1980 NCTM Yearbook, Problem Solving in School Mathematics, would essentially be unchanged if all the fields listed above by Simon did not exist The result is a loss to both schools. The interplay between then can, and should, be fruitful. I shall discuss here some adaptations of ideas and techniques from cognitive science to examine problem solving via heuristics - - and to provide some of the evidence asked for by Newell.

We shall outline the framework of a theory. In brief, we argue that there are (at least) three components which are essential for competent problem solving performance in any nontrivial domain:

1. An adequate knowledge base, including access to basic facts, relations, and procedures.

2. The mastery of relevant problem solving techniques - - in the case of nonroutine problem solving like that discussed here, the mastery of certain heuristics; and

3. An efficient means of selecting appropriate techniques for application, and in general for using efficiently those resources which the problem solver has at his or her disposal. We shall call this efficiency expert a managerial strategy.

The first observation to make is that the knowledge base is more important than it might at first appear. Of course, the problem solver must have the "basic facts" at his disposal. But recent work in cognitive science stresses the difference between factual knowledge and procedural knowledge. The latter includes a knowledge of the conditions under which a particular procedure may or may not be "legal," to what arguments it applies, and so on. Note that the organization of the knowledge base is important: in addition to "knowing" something (that is, being able todiscuss it when asked about it) one must know when it is relevant to a particular problem. An experiment in the psychology of physics learning[3] shows that the way that the knowledge base is organized (under aritificial circumstances) has a strong effect on one's success in solving problems. A recent result I obtained indicates that experts and novices actually "see" different things in problem structures: the criteria experts use for judging whether two mathematical problems are related are quite different from the novices' criteria.[4]

The experiment was conducted as follows. Each person was given a collection of 32 problems, and asked to sort the problems into anywhere from 1 to 32 piles, each pile containing problems which were "mathematically related, or would be solved the same way." A computer analysis of the cumulative sorting (the HICLUS program), revealed that the (apparent) criteria for sorting were quite different. Novices clustered problems by what we call their "surface structure:" that is, by the objects which the problems deal with. For example, problems dealing with the roots of polynomials would be called "related" by the novices, even if one was most appropriately solved by graphing, another by examining special cases, and a third by contradiction. In contrast, the "expert" clusters were sorted by what we call "deep structue." Problems which one might approach by induction were clustered together, even if one dealt with points and lines, a second with the last digit of a complicated numerical expression, and a third with the coefficients of a polynomial. We are just beginning to deal with the complexity of knowledge structures, and there is much to discover.

The second major component for nonroutine problem solving, as we discussed it above, is the ability to use certain heuristics. Most attempts to document the role of heuristics in problem solving have yielded very equivocal results. This is not surprising if one considers (1) the quite complex and usually underestimated web of skills needed to correctly employ individual heuristics, and (2) our hypothesis that the heuristics alone are not sufficient to guarantee improved problem solving performance. Our treatment will be brief here: we will mention only two studies designed to "tease out" the role of heuristics. See the 1980 NCTM Yearbook and the NCTM's Research in Mathematics Education for extensive discussions of the literature. The first study[5] was designed to see whether students will intuit problem solving heuristics simply by working problems. Under controlled laboratory conditions, two groups of students were trained for identical periods of time on identical problems, with only one group given the explicit heuristics underlying the solutions they were shown. There was a significant difference in performance; the "control" students were not able to use their problem solving experience to solve related problems, while the experimental groups was explicitly using the heuristics. However, the journey from the laboratory to the classroom is a long one.

The second study took place in the classroom. It had two goals: (1) to define some useful and replicable measures, so that other teachers and researchers could replicate the results; (2) to use those measures to verify a substantial improvement in students' problem solving performance. In that study[4], students were both taught heuristics and a managerial strategy (of sorts); thus the effects of heuristics (or the managerial strategy) alone are hard to sort out. But there was much greater heuristic fluency, correlating with dramatically improved problem solving.

Finally, we come to the presence of an efficient "manager" itself. This is also difficult to sort out, for all problem solvers obviously have some managerial abilities. Our first argument is by analogy. In an early study[6], students who had learned the techniques of integration were divided into two groups. One group studied the "usual" way, each working problems for an everage of nine hours. The other group was given a strategy which helped them to select the appropriate techniques. The experimental group averaged seven hours study time, and significantly outperformed the control group Now, the argument to be made here is that, even in a simple domain, students who lack an efficient manager squander some of their resources. In general, where there are many more choices and many more opportunities to go wrong, the absence of an efficient manager can be debilitating - - even if one has the appropriate heuristic abilities.[7]

We are currently developing a scheme for analyzing transcripts of problem solving sessions which focuses on managerial actions. Although we have only preliminary results, we believe that this scheme may allow us to 1) characterize some "expert" managerial actions which account for efficient problem solving, 2) demonstrate the consequences of poor managerial actions in students' problem solving, and 3) correlate improved performance in problem solving with both heuristic and managerial improvement. If all goes well, the synthesis of ideas from the heuristic school with the techniques from cogniive science will help us to better understand, and teach, problem solving.

References

1. Simon, Herbert. Cognitive Science: The newest science of the artificial. CIP Working paper. Department of Psychology, Carnegie-Mellon University. 1979.

2. Newell, Allen. Personal Communication. January 1979.

3. Reif, Fred. Cognitive Mechanisms Facilitating Human Problem Solving in a Realistic Domain: The Example of Physics. Physics Department, University of California, Berkeley, 1979.

4. Schoenfeld, Alan. Measures of students' problem solving performance and of problem solving instruction. Mathematics Department, Hamilton College, 1980.

5. Explicit Heuristic Training as a Variable in Problem Solving Performance. Journal for Research in Mathematics Education, May 1979.

6. Presenting a strategy for indefinite integration. American Mathematical Monthly, October 1978.

7. Teaching Problem Solving Skills. American Mathematical Monthly, in press.

13.3 RESEARCHABLE QUESTIONS ASKED BY TEACHERS

RESEARCHABLE QUESTIONS ASKED BY TEACHERS

Elaine M. Bologna
Summit School
Winston-Salem, North Carolina

Having been a primary classroom teacher for more than twenty-seven years, I feel confident to express some of the frustrations, anxieties, and questions that teachers experience and ask. Having just conducted a four-day problem-solving workshop for primary teachers, I have a fresh recollection of the concerns of these people.

Through the years I have always felt that the beginning school mathematics experiences of young children are the most important ones. They are the foundation for all future mathematics learning; they shape the attitudes students develop toward mathematics; and they can provide enjoyment and excitement when they are fruitful experiences.

There is, however, a negative factor that must be considered, and I am certain you are all aware of it. Most mathematically knowledgeable teachers are not teaching primary grades. On the contrary, a high percentage of math-anxious teachers, as well as those with very meager mathematics background, are working in the primary grades. These teachers are a tremendously conscientious group of educators, but they certainly need help. You, as mathematics educators and researchers, together with responsible people in government, must somehow find a way to reach out to these very typical elementary classroom teachers in the far corners of our land. Remediation is fine, but prevention is better! I've often wondered how many teachers of young children really feel adequately prepared for their tasks and how many of them actually are well prepared to teach mathematics.

So I would like to pose for you some questions teachers have asked over and over again. First:

> How can teachers bridge the gap between the concrete and the abstract?

Most primary teachers agree that young children need materials to manipulate as they learn mathematics, and they know children must progress from the concrete to the abstract. But most such teachers, because they are aware of their own inadequacy, rely solely on their textbooks and are very anxious to have children "reach" the symbolic stage so as to have some tangible evidence of mathematics learning.

The typical primary teacher usually introduces a concept to the entire class with some type of manipulative. After children use the materials for a while (and for many children it is a short while), the teacher then brings all students to the workbook or text, which generally means filling in blanks. How many teachers are aware of their important role in constructing that link between those manipulations and that symbolism? How can teachers tell when little children have related in their minds their previous actions on the materials with the written symbols they now put on paper? This is very critical, yet most teachers rely on paper-and-pencil tests for an answer to this question.

This, of course, poses another, related question:

> How do teachers go about diagnosing students to determine if and when they have bridged the gap between concrete and abstract?

Certainly you must know that the small amount of time alloted for mathematics makes it difficult for teachers to group their students for instruction, and few teachers do so. Therefore, how does the classroom teacher construct a good, workable diagnostic instrument that would make it possible to determine if children are acquiring real understanding? Teachers must know how to be diagnostic, for without knowing when each child can mentally generalize about a mathematics idea, such teachers will continue to have students marching through workbooks and texts like an army of robots.

There is another question teachers have been asking and for which there is need of an answer:

> How does a classroom teacher recognize which physical materials are the most versatile and appealing and at the same time are worthwhile to the development and application of mathematical ideas?

Most teachers have very limited funds (if indeed they have any at all) for purchasing materials. They, therefore, need to spend wisely. But I know many are bewildered by the different types of mathematics aids and actually feel unable to make a proper choice. Certainly they would like materials that serve more than one purpose. But which ones have proven most worthwhile? How helpful it would be to find a good unbiased source of such information. For those teachers with no funds at all, what collectible materials could serve the same purposes? Here again, teachers' lack of understanding puts in jeopardy the proper use of manipulatives. Some real guidance from mathematics educators is needed to help teachers understand the ideas they are trying to develop through action on manipulatives.

My next question is:

> How can researchers make available to teachers the data they have gathered on various topics in mathematics?

So often research literature is written in such techical language and is so lengthy as to discourage even the most ambitious teacher. Tables and graphs tell much, but are you aware how few teachers understand the meaning of those tables and graphs? Teachers will read what they can understand and, particularly, what will not take too much time. I'm not certain this is a good situation, but it is realistic. Elementary teachers are generalists and mostly language arts oriented. Mathematics is a subject they must teach, but too often the textbook is the real teacher. That is why I repeat: mathematics educators and researchers must join hands with responsible people in government to find a way to get understandable research information to the classroom teacher as quickly as possible. These teachers are the people to whom your research is being addressed, but these are the people who are the last to know. I do hope that somehow researchers and classroom teachers will have a "hot-line" between them to impart new ideas and findings from those "who are in the know" to those "who need to find out."

And now if I may be permitted, I would like to submit a question that has been in my mind for quite some time. As I read research articles and various mathematics books I note the many references to the importance of helping children develop their spatial ability. I can readily understand such need in the study of geometry. What I would like to ask is:

> How do spatial skills affect overall mathematics learning, and what are some ways of developing such skills in mathematics?

There does not appear to be a great deal written on developing spatial ability, and if it is indeed an important factor in successful mathematics experiences, then perhaps we need to have this information at the primary level. How we can get this data goes back to my previous question.

Just as teachers must bridge the gap between concrete and abstract, so researchers must bridge the gap between their world and the world of those primary teachers in the far corners of this country. I do hope you can find a way.

RESEARCHABLE QUESTIONS ASKED BY TEACHERS

Sadaharu Fujimori
Keio Senior High School
Yokohama, Japan

It is a great honor to have been favored with the opportunity of taking part in this ICME - IV panel discussion. And I also feel responsibility for the honor of being one of the participants here. Today I will talk about the status quo in Japan first and then about some of the problems which I hope to be solved as soon as possible.

On August 4 and 5 of this year, the 62nd Teacher-of-Mathematics Conference was held under the sponsorship of the Japan Society of Mathematical Education, and about 4000 teachers of mathematics from elementary schools, low and upper secondary schools, and universities attended it. Today in Japan, particularly in the large cities where more than 95 per cent of the lower secondary students go on to the upper secondary schools, about one-third of the upper secondary school students go on to universities and colleges. And there are heated competitions among the students trying to pass the entrance examinations.

The competitions are much more heated than the figure indicates because more and more students want to enter a limited number of well-known colleges or universities. And these colleges and universities often impose incredibly difficult problems upon their applicants. As the result, students are obliged to be capable, and we teachers are also obliged to make them capable, of solving such complicated problems within a limited number of classroom periods. The situation is not so bad in lower secondary schools and colleges, but the worst are the upper secondaries. There the teachers are required to do so much in so small a number of hours. Another reform is coming next year, and it probably will make the upper secondary situation worse than it is. A few drastic reforms have been attempted but, honestly, none can be called a success so far.

Today in Japan people often say there are more slow learners than before, but people also say that is because many more students are now studying at upper secondary schools. These people say Japan is keeping up the pre-war level. And also in my opinion, the level itself has not lowered.

So much for the status quo, and now I am going to talk about some of the problems which, I feel, are important:

1. Selection should be made about what is to be taught at schools in order that the teachers can take time teaching it. The greatest problem here is that it is difficult for the teachers to agree on what is to be selected. This, however, I think, must by all means be accomplished. I am now trying to find the answer.

2. Good mathematical problems instead of bad educational problems are needed. These good problems must be neither too difficult nor too simple. They must help cultivate the students' thinking power. Some of the university entrance examination problems discourage it rather than cultivate it.

3. Unification of mathematical terminology must be made so that students can naturally and smoothly move themselves later into the study of advanced mathematics. Symbols must also be unified.

4. Support should be extended to the particularly excellent students or the fast learners. It could be a good book which students study by themselves with little help from their teachers. Or it could be special courses for advanced learners. Needless to say, this is sure to help enhance the level of mathematical education and also of mathematics itself in Japan.

These four problems are, in my opinion, some of the important points I would like to point out. The final point is about the use of electronic calculators in the classroom.

This amazing machine has been rapidly improved and it is now within easy reach of most learners. Some of the teachers insist it should be used on the grounds that it will save time. Others insist it should not be because it may make learners incapable of doing calculations with pencil and paper. But, frankly speaking, I think it should be used in the classroom because it is hoped that the calculator will help make mathematics an enjoyable subject to learn.

Finally, I would like to express my cordial thanks to the persons who collaborated to realize this panel and for the invitation they kindly sent me. Thank you every so much for your kind attention, ladies and gentlemen.

RESEARCHABLE QUESTIONS ASKED BY TEACHERS

Douglas E. Scott
Amphitheater High School
Tuscon, Arizona

The increasing emphasis on "back to basics" and on required testing mean that we are likely to find an increase in the number of remedial students - - students who are being told that they must be able to do arithmetic or they will not graduate.

Very little research has been done with students of this type, and much of what has been done serves mostly to suggest that some things we take for granted are, reassuringly, true: Richardson (1975) found that instruction in problem-solving skills helps improve socres on a problem-solving test; Thornburg (1975) found that a special class using positive reinforcement and team teaching in English and mathematics lowered the dropout rate; Austin (1976) found that writing comments on homework papers could sometimes (but not always) be a factor in improving student achievement; (Griggs (1976) found that small-group and individual instruction resulted in improved scores.

One thing we do not need is research that tells us that individual (not individualized, but individual) instruction gets good results. In the first place, teachers have known that since at least the time of Socrates, and in the second place, my school district (and perhaps yours too) is prevented by law from spending the kind of money that would be needed to permit individual instruction even if we wanted to.

There has been some helpful research: Wheatley and McHugh (1977) compared speed and accuracy in adding a column of figures either directly or by mentally grouping digits that added to 10. They found that adding straight up (or straight down) the column was faster than the "tens" method and that neither method had an edge in accuracy.

We need more research of this type: research that gives us clues as to the "best" algorithm for a particular kind of problem. Note, though, that the concept of "best" algorithm is not one that is easily defined.

"Best" for one student may not be "best" for another - - if we could have the opportunity to discuss procedures with each student individually,we might be able to identify the algorithm that would in fact be "best" for that student. Dealing as we must with students in classroom groups, however, about the only way to proceed, is to find an algorithm or a procedure that seems to be "best" for a majority of students and to use it for our primary presentation, but then to be ready with 2 or 3 dozen other methods, one of which may be just what the doctor ordered for a specific individual.

We should be aware also that different countries use different algorithms, particularly for multiplication and division, and should try some of those as alternatives. And we should be willing to learn from our students, too - - many times students in "remedial" classes will manage to come up with empirically derived algorithms which are quite satisfactory for the particular kind of problem under discussion. Such empirically derived algorithms may not generalize, and may not have a theoretical foundation, but the work - - and the student knows it!

What about innovative methods? The most widespread innovation currently being investigated is that dealing with the use of electronic media - - calculators and computers - - in the classroom. There has been some research published concerning the effectiveness of these devices, but almost none dealing with their efficiency - - that is, the relationship of cost in time or money to whatever gains may be accomplished - - and this latter is a vital concern in these times of budget restrictions.

One study by Gaslin (1975) showed that low-achieving students in ninth-grade general mathematics classes showed improvement in computation with fractions when they used calculators to convert fractions to decimal equivalents. The same study showed increased student success in semi-novel situations such as estimation, ordering of rational numbers, and simplifying rational expressions involving more than one operation. Students in a study by Coward (1978) showed improvement in arithmetic performance when using a calculator, withless-able students showing proportionally much greater improvement than the more able. Similar results were found by Creswell and Vaughn (1979), while Szetela (1980) and Wheatley et al. (1979) report little or no significant difference between those students who used calculators and those who did not.

Perhaps the question is too broad - - perhaps researchers would do better to investigate questions like this: Are there kinds of students and/or kinds of problems on which calculator usage does make a difference? If so, why those kinds and not others? It may turn out, in fact, that the electronic calculator, despite is glamorous appearance, is no more important from the instructional standpoint than the old-fashioned adding machine: a tool that enables tedious jobs to be done faster, but with no instructional significance.

A study by Robitaille, Sherrill and Kaufman (1977) came up with a result that is at first glance surprising: algebra students who used a computer for part or all of the year did less well on an achievement test than a control (non-computer) group. The explanation, I believe, lies in the fact that the students are writing their own programs to solve equations rather than simply plugging appropriate data into existing programs. This explanation rests on two observations I have made concerning the use of computers by high school students: One, writing computer programs can be a difficult and time-consuming task for the students who do not think (or cannot train themselves to think) in a step-by-step fashion - - programming seems to be very much a left-brain activity. It is easy to imagine students getting bogged down in trying to correct errors in the computer program and thus not spending enough time on actually doing algebra. The second observation is almost the reverse: students who do enjoy programming will, if not stopped, spend all their time at the computer console while neglecting all their other studies.

Another technique that might be called innovative (although it has been used in one form or another for centuries) is the use of peer tutoring - - both in-class and out-of-class. One study by Howell (1978) with middle-school students aged 9-13 showed that students could successfully drill each other on multiplication facts. Can this be used with older students? Has anyone done any research on this? I have been unable to find any.

An interesting study by Flaherty (1975) concerned with students' thought processes asked students to "think out loud" as they solved problems. The study proved inconclusive as far as categorizing the thought processes used, but the investigator found an increase in error rate when students were asked to verbalize their thinking.

This illustrates a major problem related to research on methods in mathematics education: How does one find out what's going on in the student's head? Pencil-and-paper testing may avoid the pitfall that arose in the Flaherty study, but it has its own hazards. Certainly a multiple-choice format, unless the distractors are very carefully selected, is seldom appropriate; on the other hand, poring over pages of detailed calculations consumes a great deal of the investigator's time and even then may not reveal how the subject went about solving the problem. The choice of subject matter is important, too - - there was a period a few years ago when formal logic was the subject of a number of studies in mathematics education, even though almost no one is teaching formal logic in grade school or high school any more - - and certainly not to remedial students.

In that connection, a study by Lloyd (1976) showed that scores on an arithmetic achievement test were positively correlated with the likelihood of remaining in school. It would be too much to claim that failure in arithmetic is the cause of dropping out of school - - but is might well be one factor affecting a student's decision. The whole question of attitude toward mathematics is one that is closely tied to the success or failure of the low-achieving student, and a number of researchers have verified that low self-esteem and low achievement go hand-in-hand, but where is the research that shows the classroom teacher how to break the cycle? Do we concentrate on improving skills in the hope that attitudes and attendance will improve? Or do we concentrate on attitudinal aspects in the hope that the improved attitude will lead to improved skills?

And what about class size? The previously mentioned budget restrictions are such as that unless a teacher can cite research that shows a tremendous and unmistakable gain by using small classes or special equipment, school authorities are going to continue putting 25, 30, or more remedial students in a class. Where is the research that shows the best way to teach multiplication to students in large classes? Techniques that may be appropriate in a one-to-one situation or in small groups may not be the most effective for a class of 30 - - particular for a class of 30 students who are hearing it all for the third or fourth time.

Let me summarize by listing three of the areas I touched on under the general heading of needed research dealing with the remedial student in the secondary school:

1. We need research concerning algorithms: Which of the standard algorithms seem best suited to the remedial student? Are there alternative algorithms that would be even more useful?

2. We need research on the use of computers and hand-held calculators: In what specific ways, and in what remedial situations, are these devices useful? Can students who use calculators then perform pencil-and-paper arithmetic without the calculator later on?

3. We need research on the cause-and-effect relationship, if any, between attitude and achievement for remedial students.

References

Austin, J.D., Do comments on mathematics homework affect student achievement? Journal for Research in Mathematics Education 76, (Feb. 1976), 159-164.

Coward, P.H., Electronic calculators in further education. Mathematics Teacher 82, (Mar. 1978), 26-28.

Creswell, J.L. and Vaughn, L.R., Hand-held calculator curriculum and mathematical achievement and retention. Journal for Research in Mathematics Education 10, (Nov. 1979), 364-367.

Flaherty, K.G., The thinking-aloud technique and problem-solving ability. Journal of Educational Research 68, (Feb. 1975), 223-225.

Gaslin, W.L., A comparison of achievement and attitudes of students using conventional or calculator-based algorithms for operations on positive rational numbers in ninth-grade general mathematics. Journal for Research in Mathematics Education 6, (Mar. 1975), 95-108.

Griggs, S.A., An experimental program for corrective mathematics in schools for the socially maladjusted and emotionally disturbed. School Science and Mathematics 76, (May/June 1976), 377-380.

Howell, K.W., Using peers in drill-type instruction. Journal of Experimental Education 46, (Spring 1978), 52-56.

Lloyd, D.N., Concurrent prediction of dropout and grade of withdrawal. Educational and Psychological Measurement 36 (Winter 1976), 983-991.

Richardson, L.I., The role of strategies for teaching pupils to solve verbal problems. Arithmetic Teacher 22, (May 1975), 414-421.

Robitaille, D.F., Sherrill, J.M. and Kaufman, D.M., The effect of computer utilization on the achievement and attitudes of ninth-grade mathematics students. Journal for Research in Mathematics Education 8, (Jan. 1977), 26-32.

Szetela, W., Calculators and the teaching of ratios in grade 7. Journal for Research in Mathematics Education 11, (Jan. 1980), 67-70.

Thornburg, H.D. Attitudinal differences in holding dropouts. Journal of Educational Research 68, (Jan. 1975), 181-185.

Wheatley, G.H. and McHugh, D.O., A comparison of two methods of column addition for pupils at three grade levels. Journal for Research in Mathematics Education 8, (Nov. 1977), 376-378.

Wheatley, G.H., Shumway, R.J. and others, Calculators in elementary schools. Arithmetic Teacher 27, Sept. 1979), 18-21.

RESEARCHABLE QUESTIONS ASKED BY TEACHERS

Richard J. Shumway
Ohio State University
Columbus, Ohio

I would like to thank Marilyn Suydam for most of the organizational efforts on behalf of this session and am most pleased for the opportunity to act as reactor to the papers of Bologna, Fujimori, and Scott.

I will attempt to wear two hats, one of a teacher and one of a researcher. Bologna raises several issues I see as critical to teachers. She is interested in the connections between concrete and abstract, spatial relations, insightful testing, and communications. Fujimori is most concerned about mathematics content, problem selection, unification of terminology, gifted, and the role of the calculator. Scott raises several issues but summarizes with a focus on remedial students and questions of best algorithm, role of calculators, and attitude.

If you were to corner me on some of these issues, I would probably respond in the following way:

1. "Connection between concrete and abstract"

This issue goes to the very heart of mathematics, modeling, and applications. We think we know that concrete experiences are critical to the development of mathematical ideas and their applications. The question of how people make isomorphisms between reality and mathemtics is not clear and is of great interest to researchers today.

2. "What mathematics should be taught"

Here is an issue for which research may be fairly silent. As a teacher I would say, teach some of what is expected in later courses and some things you like, and make sure it is "good" mathematics. The answers from research are not much different. Survey research can give us guidance about what is being used in society, and we know students learn better what is taught than what is not taught. Basically, curriculum questions are not research questions, but rather questions of intent. What do you want students to learn? Arguable, debatable, but no so researchable.

3. "Remedial students: Best algorithm"

My response here would be the best algorithm is the calculator. The quickest results and the fewest errors occur with calculators. Of course teachers already know this. The real question is probably philosophical. should we let children use calculators? The answer depends on our goals. Research gives us clues, but we have to choose our own goals.

The issue which I see as most critical is communication. Some of the issues raised, I have said, have answers; others, I have said, are not research questions; and others, I have said, are of critical interests. My guess is that if Bologna, Fujimori, Scott, and Shumway were to work together on a research and teaching project for a year, we would restate the above issues in researchable terms with results of high interest to teachers.

I believe the most important question is one of communication and the first step is the consistent, regular interaction of researchers and teachers on research problems. The best way to accomplish valuable research is to do it together, teacher and researcher. I believe teachers and researchers are ready. Let's plan procedures, grants, and school organizations which promote such cooperative efforts.

13.4 ALTERNATIVE METHODOLOGIES FOR RESEARCH IN MATHEMATICS EDUCATION

ALTERNATIVE METHODOLOGIES FOR RESEARCH IN MATHEMATICS EDUCATION CLINICAL INTERVENTION RESEARCH

George Booker
Kelvin Grove College of Advanced Education
Brisbane, Australia

Research in mathematics education has traditionally been concerned with the adequacy of alternative techniques or approaches under rigidly controlled experimental conditions. However, confirmative research of this nature, using procedures evolved from agricultural and medical psychometric models, has had little impact of classroom practices. Nor has it provided insight into individual learning styles and learning differences amongst the children in those classrooms. The main reason for this lack of impact is that assumptions made in the original settings do not necessarily apply in an educational framework. Subjects are not randomly chosen; samples are not representative of the whole population; experiments are not free of bias; there is a significant divergence between the theory and practices of education which tend to make the experimental situation more "ideal" than the typical classroom.

On the other hand, recent concerns with the manner in which an individual learns, or fails to learn, mathematical concepts and skills has focused attention on the clinical investigation as an alternative researchmethod. This method has its roots in the work of J. Piaget and other cognitive psychologists, although studies by Brownell during the 1940's (Weaver, 1976), Weaver (1955), and Lankford (1972) afford examples within mathematics education. It usually takes the form of a dialogue or conversation held in an interview situation between an adult (the researcher) and a child (the subject). The interview is centered on a task or problem which has been carefully chosen to give the child every opportunity to reveal mental mechanisms used in thinking about the task or solving the problem. For mathematics education research is as much concerned with an understanding of the cognitive processes underlying alternative procedures as it is with their specification and organisation.

While confirmative research is designed to asses the truth of provable hypotheses, clinical investigations seem to lie within the domain of generative research, intended to generate hypotheses with a priori probability (Wilson, 1976). This may take the form of status studies where facts and the connections between facts are reported, as in the case of Piagetian studies and clinical studies which simply describe the setting. However, a more fruitful line of inquiry involves what Wilson has termed "Clinical Intervention". Rather than trying simply to discern a child's knowledge, an intermediate factor of instruction is introduced after an initial interview and subsequent performance is related to the initial assessment of the child and analysis of the tasks.

Much of this research has grown out of attempts to diagnose student weaknesses in mathematics and as such has been invaluable in understanding learning (or learned) problems in mathematics. But it has also been of great benefit in clarifying the methodology and criteria for using the interview techniques. For the clinical method has had its critics, chiefly on the grounds of the unreliability of a single or limited number of encounters, and the lack of standardisation of procedures. In drawing up hierarchies and taxonomies of skills to be assessed, those involved in diagnosis of learning problems have provided the structured protocols for both administering tasks and coding and interpreting the results that has allowed the acceptance of the method within the general research framework (Campbell, 1975; Johnson, 1980). Clearly, without such procedures precise replication is not possible, nor can the interpretations of the interviewer be assessed.

Thus clinical intervention research involves extensive systematic description of treatments and recording of apparent effects to generate predictions or expectations about the mathematical behaviour of both individuals and groups. In time this may lead to the generation of theories of mathematical learning, or to the selection and synthesis of aspects of existing theories, but in the first instance it will lead to hypotheses whose generality may be later tested in a controlled experimental setting. Such research may also lead to the creation of techniques and materials the classroom teacher can use to better achieve the objectives of the mathematics program. In this way children may be lead to avoid the pitfalls so often encountered in the past.

Clinical research may also be characterised by its methods, which differ significantly from those of experimental research. The collection of data tends to be qualitative rather than mere recording of test scores and usually takes the form of verbatim reports, detailed descriptions or videotapes of the tasks and protocols used during the interview. These responses may be categorised on a simple yes/no basis, but it is more usual for them to be analysed against a hierarchy of levels of maturity. Further, the subjects are not selected randomly but may be deliberately sought according to predetermined criteria, and one or a few subjects may be interviewed repeatedly over an extended period of time.

An example of the clinical intervention model being utilised in these ways is the work being conducted by the writer and his colleagues at the Learning Assistance Centre at Kelvin Grove College of Advanced Education in Queensland, Australia. In this research, students with specific learning problems in mathematics undergo continual diagnosis and prescription over an extended period of time. The goal of this study is to investigate the effect of various learning sequences on particular problem areas in primary school mathematics. A major focus has been on the area of numeration, since it was felt that formal tests that existed at the time did not provide a thorough diagnosis of the level of understanding children had in numeration. In addition, the items included in most tests labelled numeration did not always test that area. Thus a theoretical background did not seem to exist concerning the whole content of numeration. As children were assessed at the Centre, clinicians tested them for numeration using questions they devised on-the-spot, in the interview session. In all of these cases, the interviewers did not follow any particular sequence, but made decisions regarding the next question based on their personal interpretation of the child's ability. Generally, these questions began to follow a pattern, with certain types of items being asked each time, and in time a structured guide was evolved.

In the first instance, specific problem areas such as the teen numbers, the interpretation of large numbers and the meaning of decimals led to the content sequence: basic numbers (0 - 10); two-digit numbers greater than 20 (20 - 99); the teens (11 - 19); three digit numbers (100 - 999); numbers greater than 999; tenths; hundredths;

As well, children seemed to have a very narrow understanding of numbers, largely in terms of symbols. In this regard, the paradigm introduced by Payne and Rathmell (1975) to relate the concrete, verbal, and symbolic aspects of numbers has proven valuable. It has provided clinicians at the Learning Assistance Centre witha guide for determining children's understanding of number as they might be encountered in the primary school. The paradigm also suggests the cognitive level of functioning that children have reached in respect to the use of concrete referrants, verbal cues, or written symbols. However, there are other understandings expected of children as they relate to further mathematics content areas such as algorithms and estimation. These numerations understandings are:

an understanding of the place value concept associated with numbers

an ability to sequence numbers

an ability to compare numbers

an ability to count in both routine and non-routine situations.

The first product of this research, then, has been the clarification of the concepts of numeration and the delineation of areas of difficulty encountered by large numbers of children (Irons, 1978; 1980; Booker, 1980). The systematic investigation of of instructional procedures designed to overcome these weaknesses, by means of intervention and reassessment over a considerable period of time, has led to the production of materials to assist classroom teachers remedy and avoid difficulties with numeration and computational skills for whole numbers and decimals (Booker, 1979; Booker, Irons, and Jones, 1980). The clinical procedures which have evolved have also been shaped into a systematic diagnostic profile for numeration which thus fills a gap in available tests (Booker, Irons, and Reuille, 1979).

One of the hypotheses that has emerged from this clinical intervention work is that an inability to understand and apply numeration concepts underlies many of the difficulties associated with the computational algorithms for whole numbers and decimals. An experimental study designed to test this hypothesis with addition was carried out among four entire grade 3 classes during 1979 (Irons, 1980). The results suggested that not only was a lack of numeration understanding correlated with lack of success on the addition items, but also that a knowledge of numeration on its own did not necessarily ensure an understanding of numeration concepts within the algorithm. Thus, experimental methods were able to confirm that aspects of mathematical learning disabilities located with selected subjects were also prevalent amongst a general population. Further experimental studies are currently in the planning stages, including the large-scale investigation of proposed changes to the teaching sequence for decimal numeration and operations.

While the initial goal was to help children referred to the Learning Assistance Centre to overcome learning difficulties, the primary aim of the clinical intervention program now is to generate hypotheses on task, method, teacher, and learner variables. Systematic investigation of further aspects of mathematics learning can be made using the complete profile as a guide and a recent taxonomy describing different levels of judgement and reasoning ability (Collis and Biggs, 1979) has proved to be invaluable for classifying children's responses to a broad spectrum of tasks. The writer is currently investigating the learning difficulties encountered amongst high school students to determine whether the problems and solutions formed with younger children carry over to this group, and others at the clinic are investigating children with measurable discrepancies between their reading and mathematical abilities.

While the program undertaken at Kelvin Grove College of Advanced Education probably is the most systematic and extensive example of clinical research methodologies over recent years, and studies of this nature have begun to appear at the annual conferences of the Mathematics Education Research Group in Australia. Error pattern analysis findings have been presented at each annual conference since its founding in 1977 (Newman, 1977; Clements, 1979; Ransley, 1980), and case studies have emerged on the way children deal with algebraic concepts (Firth, 1978) and area and fraction ideas (Low, 1977). A discussion paper on the methodological aspects of clinical research was presented at the 1979 meeting (Irons, 1979), and at the 1980 meeting a Special Interest Group on Clinical Investigations in Mathematics Education emerged along with the presentation of several papers involving clinical research.

However, this Australian experience also provides a cautionary note, for unless the generative functions of this type of research are borne in mind and instrumentation to guide the conduct of the interview is developed, the effort can deteriorate into an exercise of pointing a finger at "unusual" behaviour. Potential researchers also need to obtain experiences to guide them into a difficult new area, and to this end, service programs such as clinical internships and the establishment of courses at both the undergraduate and graduate levels are crucial. However, replication studies are certainly necessary, since our information on how children learn mathematics is meager, but generating further examples of well-known errors or misconceptions is of no value unless underlying reasons are also determined. We need to build up an understanding of how mathematics is understood from a child's point of view, for as Brownell (1944), Buswell (1949) and Ginsburg (1972) have shown, meanings in mathematics are what children think, and do not necessarily coincide with what adults think. Clinical intervention is an alternative research methodology which has shown itself to be valuable to understanding learning problems in mathematics. It is now time to extend its enquiry techniques into new areas, to find out why certain aspects of mathematics do not generate learning difficulties, and to determine how some children avoid the difficulties others fall into.

Bibliography

Brownell, W.A. Rate, Accuracy and Process in Learning. Journal of Educational Psychology 35:321-337; 1944.

Buswell, G.T. Methods of Studying Pupils' Thinking in Arithmetic. Supplementary Educational Monographs, No. 70. Chicago: University of Chicago, 1949.

Booker, G., Irons, C., and Reuille, R. Kelvin Grove Numeration Profile. Kelvin Grove College of Advanced Education, 1979.

Booker, G. & Reuille, R. Teaching the Basic Facts of Arithmetic in the Primary School, Canberra, Curriculum Development Centre, 1980.

Booker, G., Irons, C., and Jones, G. Fostering Arithmetic in the Primary School, Canberra, Curriculum Development Centre, 1980.

Booker, G. Can You Count on Counting? Math Matter, No. 4, 1980.

Campbell, D.T. "Degress of Freedom" and the Case Study. Comparative Political Studies 8: 178-193; 1975.

Collis, K. and Biggs, J. Classroom Examples of Cognitive Development Phenomena: The Solo Taxonomy. University of Tasmania, 1979.

Clements, M.A. Analysing Children's Errors on Written Mathematical Tasks. Educational Studies in Mathematics 11: 1-21; 1980.

Easley, J.A. On Clinical Studies in Mathematics Education. Columbus, Ohio; ERIC/SMEAC, 1977.

Firth, D. Algebra by Rules: Do Pupils Understand the Procedures They Apply? In Conroy, J. (Ed.), Research in Mathematics Education in Australia 1: 119-126; 1978.

Ginsburg, H. The Case of Peter: Introduction and Part I. Journal of Children's Mathematical Behaviour. 1: 60-70; 1972.

Hunting, R. Emerging Methodologies for Understanding Internal Processes Governing Children's Mathematical Behaviour. Paper presented to the 4th Annual Conference of the Mathematics Education Research Group of Australia, 1980.

Irons, C.J. Learning Problems in Mathematics: Their Identification, Correction and Prevention. Kelvin Grove College of Advanced Education, 1978.

Irons, C.J. Research Related to Diagnostic and Remedial Aspects of Mathematics. In Booker, G. (Ed.), Research in Mathematics Education in Australia 2: 162-176; 1979.

Irons, C.J. Diagnosing Numeration Problems in Primary School Mathematics. In Williams, D. (Ed.), Mathematics Theory into Practice. Canberra: Australian Association of Mathematics Teachers Conference, 215-224, 1980.

Irons, C.J. An Investigation of the Role of Numeration in Solving Addition Problems. In Foster, B. (Ed.), Research in Mathematics Education in Australia. Vol. 2, 1980.

Johnson, D.C. Types of Research. In Shumway, R. (Ed.) Research in Mathematics Education. Reston, Virginia: National Council of Teachers of Mathematics, 10-28; 1980.

Lankford, F.G. Some Computational Strategies of Seventh Grade Pupils. Charllottesville, Virginia: University of Virginia, 1972.

Low, B. Measurement of Length, Area, and Volume - Ages 10 through 17. In Clements, M. (Ed.), Research in Mathematics Education in Australia. 1: 199-208, 1979.

Newman, M.A. An Analysis of Sixth-Grade Pupils' Errors on Written Mathematical Tasks. In Clements, M. (Ed.), Research in Mathematics Education in Australia, 1: 239-258; 1979.

Payne, J. and Rathmell, E. Number and Numeration. In Payne, J. (Ed.), Mathematics Learning in Early Childhood. Reston, Virginia: National Council of Teachers of Mathematics, 1975.

Ransley, W. Individual Difficulties Experienced by Children in Mathematics Problem-Solving as Revealed by a Diagnostic Interview Technique. In Foster, B. (Ed.), M.E.R.G.A. Conference Proceedings: 173-186, 1980.

Weaver, J.F. Big Dividends from Little Interviews. Arithmetic Teacher 2: 40-47; 1955.

Weaver, J.F. The Use of "Individual Interviews" in Assessing Pupil's Mathematical Knowledge or Learning: A Chronology of Selected Illustrative References. Madison, Wisconsin: The University of Wisconsin, 1976.

Wilson, J.W. An Epistemological Based System for Classifying Research and the Role of Clinical Intervention Research Within That System. Kent, Ohio: Kent State University, 1976.

THE NEED FOR A COGNITIVE ETHNOGRAPHY OF MATHEMATICS TEACHING

Jack Easley
University of Illinois at Urbana/Champaign
USA

Case studies and ethnography are attracting more attention from educational researchers in recent years, which suggests that there may be a bit of uneasiness about the value of standard psychometric and statistical studies. As far as I know, there is no formula or cookbook of designs into which alternative kinds of studies can be placed, and that raises uneasiness as well. My view is that useful research can't come out of a formula anyway, but requires a great deal of courage. I understand the natural tendency to want to rehearse a new form before trying to put it to work to produce something useful. But I don't know whether or how to help someone do that even if I thought this audience wanted that. So I'm putting aside forms of research in favor of discussing how something useful for mathematics education might emerge from some forms of ethnographic approach, why human cognition needs to be represented in the result, and what advantages that confers as well as what problems it raises.

I follow Thomas Kuhn (1961) when he concluded from a review of the role of measurement in modern physics that qualitative theory almost always preceded and guided developments in physical measurement. This means that scientific laws weren't discovered in the results of measurement, but were postulated first and became a reason for measurement. By qualitative theory in mathematics education, I refer to explanations of social and mathematical cognition. Once we understand how teachers' and pupils' ideas control their perceptions and their actions, we may know what it would be worthwhile to measure and what it would be worthwhile to sample from what population. Consequently, I see little use for any research that does not aim at the discovery and description of cognitive systems tht underlie teachers' and pupils' classroom behavior.

Two Cognitive Social Systems.

Some of the Case Studies in Science Education (Stake and Easley, 1978) were productive in this regard, as I have attempted to point out elsewhere (Easley, 1978, 1979). Socialization processes, for example, interact with math lessons in a way that limits inquiry teaching and the help offered teachers by university professors and curriculum supervisors. Another example is the way society pressure to guarantee minorities and girls a minimum level of competence - - because primary school teachers tend to see this as requiring practice of standard algorithms (Easley et al., 1980) - - reduces their freedom to explore algorithmic ideas of their own considerably below that allowed independent-minded white males by default. The result is a perpetuation of the underrepresentation of minorities and women in advanced mathematical programs and careers which reward the high level of self confidence in tackling new problems many white males manage to develop.

Such qualitative mechanisms or theories need appropriate criticism by counter-examples and analysis, if one follows the methodology of research programs of Imre Lakatos (1978). This philosophical answer to the debate between Kuhn and Popper (Lakatos and Musgrave, 1970) generalizes from his famous case history of the Eulerian research program in topology (1977) to scientific research of any kind, and it should be particularly likely, because of its origins, to find followers in mathematics education research. For example, having conjectured certain cognitive and social mechanisms, it is now very important to search for counterexamples. Once we locate a teacher who socializes children into discovery mathematics lessons, or find a minority or female pupil who enjoys the freedom to explore nonstandard algorithms without being pressured into using only standard algorithms, the conditions need to be closely examined to find out why the conjectured mechanism is not working. Such counterexamples can lead us to a much more accurate definition of the mechanisms involved. For example, they may involve a certain kind or level of math anxiety on the part of the teacher, or the absence of a support group or resource person for the teacher.

I have recently been providing resource persons to work with primary school math teachers to see whether these and other social mechanisms are removed or modified. The resource persons become participant-observers and the reports they write are presented to the teachers they work with. Here is one of the advantages of the cognitive ethnographic approach over usual psychometric and statistical approaches. The kind of qualitative theory that guides this research is not just conjectured quantitative relations to be checked statistically according to whether they can be supported by the current paradigm. These conjectured ways of thinking, feeling, and seeing one's working environment can be checked directly with the teachers and pupils who are the subjects of the research. A high correlation between IQ and SAT scores can't be checked by asking the students how they feel about it. However, the checking that is possible here is not only for additional validation, but for utility as well. If a teacher or a student finds it personally useful to think about social mechanisms of their own mind, then there is a broader practical advantage to the research than theoretical inquiries destined for implementation through teacher education programs.

The Counting Trap

Primary school teachers report they have often worried about children becoming dependent on counting and not remembering the addition table, and wondered why place value operations are so difficult for some children to learn. A few teachers have even noticed a conceptual conflict between counting algorithms and place value operations. We have been introducing teachers, who ask for help in teaching the so-called "renaming" algorithms, to ways of avoiding counting dependence, e.g., decomposition and recomposition based on subitized facts from domino patterns, Cuisenaire rods, and other concrete materials. It seems clear from research on children's counting schemes (Gelman and Galistell, 1978; Stake, 1980) that they are deep-seated and relatively inflexible, with strong social support at home and in schools. However, multiplication requires a triple application of cardinal numbers to a collection of objects, violating the correspondence schema used in counting (three two's equals six). Visual patterns and decompositions seem to involve other schema with no contradiction, so there is a better overall payoff in using them for addition and subtraction. Place value, although usually taught before multiplication, involves the ten's, hundred's, etc., times tables, and, therefore, the conception of triple application of numbers. Once some of us saw this we found ways of teaching children with problems in learning place value. We now need to relate this cognitive mechanism to the child-based ideology of teachers which supports counting algorithms.

Once a social cognitive mechanism has been defined which is responsible for certain adverse or beneficial kinds of mathematics learning situations, subjects themselves can also begin to learn to control these mechanisms. One result may be a declining frequency of the occurrence of the mechanism, this kind of theory building is quite different from trait theory. A teacher who is math anxious in a given context may not be at all in another. Moreover, the controlling situational variables are qualities perceived by the person, not absolute, objective qualities. This is both a challenge to empathetic theory development and a way to get rid of a lot of "noise" or "error" in the relational models.

Researchers who are looking for policies that will select persons for certain situations (e.g., teachers and pupils for class assignments, for promotions, etc.) according to personal traits may be disappointed in a theory that selects ways of thinking or ideas as helpful or harmful in the educational development of children. Personnel decisions are now so much in the public domain, and competitive in terms of public criteria of qualification, there is no place left for a decision in terms of whether or not the candidate for placement is interested in certain ideas about classroom work. Objective qualifications take over public administration, replacing much of the function of context design. Consequently, it is important that research into the social and cognitive mechanisms of mathematics education be addressed primarily to educational administrators. They need to find research reports that are so readible and enlightening regarding the possibilities for improving contexts of teaching mathematics that they will use all the time they can spare to promote better contexts through sympathetic listening and reflecting back the ideas they hear, injecting a bit of encouragement here and there and a new idea when it fits the problem raised. Some of the connections with other traditions and this view of research in mathematics are further developed in Easley (1977, 1980).

References

Easley, J.A., Jr. (1977). On clinical studies is mathematics education. Mathematics Education Information Report, ERIC Clearinghouse for Science, Mathematics and Environmental Education, Columbus, Ohio.

Easley, J.A., Jr. (1978). Toward acculturation between traditional, creative, and technological approaches to mathematics education. Osnabrucker Schriften zur Mathematik (Reihe D Mathematisch-didaktische Manuskripte, l: 120-143. Proceedings of the Second International Conference for the Psychology of Mathematics Education, edited by E. Cohors-Fresenborg and I. Wachsmuth.

Easley, Jack (1979). A portrayal of traditional teachers of mathematics in American Schools. Critical Reviews in Mathematics Education. Materialien und Studien (IDM) 9:84-108.

Easley, Jack (1980). Alternative research metaphors and the social context of mathematics teaching and learning. For the Learning of Mathematics 1:32-40 (Concordia University, Montreal, July).

Easley, Jack, et al. (9180) Pedagogical Dialogs in PrimarySchool Mathematics, Bureau of Educational Research, University of Illinois at Urbana/Champaign.

Gelman, Rochel and Gallistel, C.R. (1978) The Child's Understanding of Number. Cambridge, Massachusetts and London: Harvard University Press.

Kuhn, T.S. (1961). The function of measurement in modern physical science. In H. Woolfe (Ed.) Quantification: A History of the Meaning of Measurement in the Natural and Social Sciences. Indianapolis: Bobbs-Merrill.

Lakatos, I. (1976). Proofs and Refutations: The Logic of Mathematical Discovery. Cambridge, England: Cambridge University Press.

Lakatos, I. (1978). The Methodology of Scientific Research Programmes. Philosophical Papers, Volume I (Ed. by John Worral and G. Currie). Cambridge, England: Cambridge University Press.

Lakatos, I. and Musgrave, A. (1970). Criticism and the Growth of Knowledge. Cambridge, England: Cambridge University Press.

Stake, Bernadine (1980). Clinical studies of counting problems with primary school children. Doctoral Dissertation, University of Illinois at Urbana-Champaign.

Stake, R.E., Easley, J.A., Jr. and collaborators (1978). Case Studies in Science Education. Washington: The Government Printing Office (Stock Nos. 038-00-00377-1), -00376-3, and -00383-6).

VARIATIONS DE QUESTIONS QUESTIONNAIRES A MODALITES

Francois Pluvinage
Universite Louis Pasteur
Strasbourg, France

1. Introduction

1.1. Depuis un certain nombre d'annees, nous sommes de ceux que les problemes poses par la didactique experimentale des mathematiques ont conduits a utiliser des dispositifs ou des outil d'observation appropries. La didactique experimentale des mathematiques poursuit essentiellement un objectif de connaissance: l'apprentissage mathematique lors d'un enseignement souleve des questions specifiques, est tributaire de variables qu'il s'agit de determiner, avant d'en determiner les principes de fonctionnement. On pourra se reporter a des textes de G. Brousseau et de G. Glaeser pour des descriptions plus precises du champ de la didactique des mathematiques.

1.2. Plus particulierement ici, nous nous interesserons a un aspect de la connaissance des conditions qui amenent l'enseignement mathematique a des effets d'apprentissage, c'est-a-dire des modifications stables de conduites ou de comportements: il s'agit du point de vue de l'etudiant, de l'individu qui apprend.

1.3. Toute etude bien determinee oblige a ne pas se limiter au point de vue precedent. Insistons, sans preciser, sur l'importance des analyses de contenus, notamment en termes de traitements d'informations, et sur l'utilite de la reference a des classifications de type taxinomique (comme celle de la NLSMA) ou factoriel (comme le modele de Guilford).

1.4. Une comparaison avec des apprentissages dans le domaine sportif peut aider a comprendre certaines de nos preoccupations: pour apprendre a bien executer certains enchainements qui exigent adaptation de la perception et coordination, les sportifs actuels non seulement se font observer par un entraineur, mais s'observent eux-memes grace aux techniques audio-visuelles.

1.5. Cet interet de l'observation se retrouve dans le succes des travaux de Piaget. Mais ceux-ci ne sont souvent pris en compte qu'a l'interieur d'une demarche purement skinnerienne de renforcements immediats, sans que l'interet theorique, en didactique des mathematiques, de cette juxtaposition soit justifier. Or, me semble-t-il, il est possible de risquer une explication, en se referant a un modele de l'apprentissage individuel.

2. Un Schema D'Apprentissage

2.1. La these que j'avance est la presence de la structure suivante chez un individu qui apprend, des que l'on s'ecarte des apprentissages purement repetitifs (la memorisation d'information).

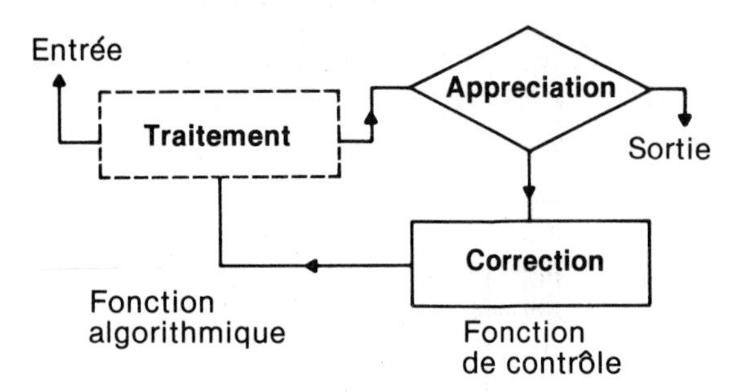

2.2. Le fonctionnement de la structure indiquee peut etre ainsi decrit: implique dans situation-probleme, un individu peut faire appel a des traitements deja en place. Si ceux-ci ne le conduisent pas a un resultat qui lui parait satisfaisant, la fonction de controle pourra declencher une modification des traitements. Les donnees seront alors soumises aux traitements corriges, et le resultat sera a nouveau apprecie. Et ainsi de suite jusqu'a satisfaction suffisante.

2.3. Un apprentissage determine se caracterise par une stabilisation des traitements. La partie "correction" de la fonction de controle n'est plus guere sollicitee, et la case "traitement" peut etre entouree d'un trait continu symbolisant une acquisition, une mise en place.

2.4. L'observation de changements spontanes de procedures est en faveur de notre schema. Un tel exemple est rapporte par H. Ter Heege dans la revue Educational Studies in Mathematics. Le diplome de A. Bodin (IREM de Besancon) donne le compte-rendu d'une tres belle observation de ce genre, mais ou une reponse fausse est donnee alors que la reponse correcte a ete obtenue auparavant.

2.5. L'etude de diverses pratiques pedagogiques met en evidence une action frequente sur les traitements (pedagogie de l'exposition, hierarchie de J. S. Bruner, ...) ainsi que sur l'appreciation (procedes de verification: preuve par 9 ou verifications de dimension; technique des renforcements immediats de Skinner). Beaucoup moins exploitee a ete l'action sur la correction: dans l'enseignement traditionnel, il existe la correction detaille de copies qui peut se rapprocher d'une telle action. Recemment, les "dialectiques" (de l'action, de la formulation, de la validation) proposees par G. Brousseau concernent la totalite de la fonction de controle: les criteres de satisfaction variant d'une dialectique a l'autre, la correction agira differemment a differentes phases de l'enseignement. (Note: ces dialectiques ont aussi un aspect epistemologique.)

2.6. Une consequence, pour l'observation d'activites didactiques, d'un tel fonctionnement individuel est la necessite de prendre en compte la tendance a optimiser qui a abouti a une production mathematique. Par exemple, cette optimisation est mal definie par une consigne comme "representer une situation": sans indication supplementaire, l'objet "ideal" vise par certains eleves sera aussi bien une caricature, une image de bande dessinee ou un tableau qu'un schema mathematique (de quoi?) par exemple.

3. Questions et Reponses

3.1. On l'a vu notre conception d'une interrogation est dynamique, comme celle de R. Thom: fournir une reponse, c'est donner l'image d'un etat d'equilibre atteint apres l'excitation causee par la reponse.

3.2. Une conception statique peut suffire pour l'observation d'un apprentissage mene a terme, puisqu'il y a alors stabilite de traitements. Mais cette conception ne permet pas de savoir si, precisement, l'apprentissage est acheve et ne peut renseigner sur les domaines de stabilite des traitements.

3.3. Pour nous, l'interpretation d'une reponse demande que l'on ait une idee des variations de la question posee qui sont sans effet sur la reponse, et au contraire de celles qui peuvent changer la reponse.

3.4. Notre objectif conduit a ecarter d'emblee une demarche courante dans une perspective d'evaluation: les reponses des eleves sont reparties en seulement deux classes, reussites et echecs, et interpretees isolement. Il est bien evident qu'une enquete qui se donnerait seulement cette possibilite d'exploitation ne pourrait permettre de reperer des comportements. (Citation d'un passage de "Demarches de reponses", de R. Duval et moi-meme.)

4. Un Microscope a Reponses

4.1. L'Analyse Factorielle des Correspondances (en abregee AFC), mise au point par le laboratoire de J. P. Benzecri, est une methode de traitement par ordinateur des croisements de reponses. Il s'agit d'une methode du meme type que l'Analyse en Composantes Principales (ACP): un tableau de donnees est transforme en un nuage de points (ensemble de points munis d'une masse), dont l'analyse fournit les axes principaux d'inertie successifs. Mais la metrique qui fournit les distances entre points est particuliere en AFC: elle est nommee metrique du χ^2 (chi deux) et permet de reperer en quoi un tableau determine differe d'un tableau de memes marges mais rempli au hasard (hypothese nulle d'independance statistique).

4.2. Pour certains tableaux donnes, appeles disjonctifs totaux, le nuage de points associe est simple a determiner. Un tableau I X J = (k(i,j)), constitue de 0 et de 1 est dit code en disjonctif total si l'ensemble J des modalites de reponses se partage en questions, de telle sorte que pour tout individu i et toute question q on ait $\sum_{j \in q} k(i,j) = 1$. Autrement dit, chaque individu a exactement un 1 dans l'une des colonnes de q, et des 0 dans les autres colonnes. (Par exemple, si le codage ne distingue que reussite et echec a une question, les "notes" possibles sur cette question seront (1,0) pour reussite, (0,1) pour echec.) Soit I X J un tel tableau disjonctif total, avec n = card I. Un nuage de points associe dans l'AFC est, a une homothetie pres (sans importance pour les resultats), le nuage des vecteurs colonnes du tableau dans R^n euclidien, le j-eme vecteur colonne etant muni de la masse: $(\sum_{i \in I} k(i,j))/np$ ou p est le nombre de questions.

4.3. Sur un tableau I X J de questions codees seulement en reussite et echec, l'analyse met en evidence des phenomenes deja interessants: le nuage projete sur le plan des deux premiers axes centraux d'inertie peut avoir une apparence generale d'haltere, indiquant une discrimination generale sur reussites et echecs, ou de papillon, indiquant l'existence de deux niveaux de reussite nets, ou d'autres apparences issues de hierarchies a la Guttman.

4.4. Mais nous avons ete amenes a adapter cette methode d'analyse, pour visualiser les effets de variations de questions. C'est un peu ainsi que des dispositifs appropries font passer de la photographie au cinema, ou, plus modestement a la chronophotographie.

4.5. Une technique que nous avons mise au point consiste a farbriquer plusieurs variantes, en general au nombre de trois pour des raisons techniques, d'un meme questionnaire. Nous parlerons des modalites A, B, C du questionnaire. Dans chaque etablissement scolaire interroge, nous distribuons de facon aleatoire un meme nombre de questionnaires A, de questionnaires B et de questionnaires C. Ainsi la population est partagee en trois tiers a priori homogenes.

4.6. Ceci permet un controle des variables d'enonce qui serait souvent impossible avec un questionnaire unique, car la memoire des sujets interroges changerait la signification des reponses.

4.7. Plusieurs techniques d'utilisation de l'AFC sont alors possibles. Nous en avons essentiellement utilisees trois, que voici.

- La technique des variables supplementaires: l'analyse peut prendre en compte certaines donnees (les variables principales, et placer d'autres donnees, non prises en compte pour la determination des axes principaux d'inertie, par rapport aux axes obtenus. Ces dernieres sont nommees variables supplementaires. Dans notre cas, on peut traiter un questionnaire comme unique et placer les modalites en variables supplementaires (etude a paraitre dans le tomme II de la brochure A.P.M.E.P.: Analyse de Donnees, Lyon (France) 1980).
- La technique de recours a la modalite "Question-non-posee". Cette technique a ete exposee dans ma these de doctorat.
- La technique de projection, qui est une combination des deux idees precedentes: utilisation de variables supplementaires et incorporation dans l'analyse de donnees fabriquees par l'experimentateur, a cote des donnees provenant des reponses. Bien sur, des precautions soigneuses doivent eviter les artefacts.

5. Conclusion: Travaux et Perspectives

5.1. Les analyses decrites permettent d'une part de situer des reponses variees par rapport a la reussite d'ensemble a un questionnaire, d'autre part de mettre geometriquement en evidence des variations sous forme de deplacements.

5.2. Parmi les phenomenes les plus spectaculaires ainsi observes, citons: les phenomenes de sous-comprehension (reussite par traitements partiels par rapport aux traitements d'information associes a un concept mathematique), l'inversion de reussite (l'enrichissement du a l'apprentissage fait passer par une phase d'echec a des questions initialement reussies), le caractere tres significatif de retours en arriere spontanes apres une contradiction, le role particulier des enonces ecrits et la mise en evidence de consequences de conduites d'optimisation.

5.3. De nombreuses observations recentes sont issues d'idees de controle des enonces exploitees selon d'autres methodes. Citons les travaux de G. Noelting, de G. Vergnaud et al., de B. Dumont et de J. P. Fischer, mais il y en a beaucoup d'autres. Les analyses d'erreurs, comme celle de H. Radatz mettent egalement en evidence des preoccupations voisines.

5.4. Nos etudes ont attire notre attention sur des phenomenes que nous n'avions pas l'idee a priori d'examiner. Ainsi les phenomenes lies a la contradiction et a la negation, qui sont la cle des dialectiques.

5.5. Les perspectives d'etude apparaissent comme multiples et destinees a une diffusion ample, mais la bureaucratie vise desormais a nous mettre des batons dans les roues, par exemple en supprimant tous les D.E.A. de didactique des mathematiques en France. Peut etre les resultats de l'enseignement mathematique n'apparaissaient-ils pas comme suffisants dans nos etudes ... ou vice versa.

References

J. P. Benzecri et al., Pratique de l'Analyse de Donnees (2 tomes), Dunod, Paris 1980.

A. Bodin, Diplome de DEA de Didactique des Mathematiques, IREM de Besancon (France) 1980.

G. Brousseau, L'observation des activites didactiques - Revue Francaise de Pedagogie, 45, 1978.

B. Dumont, ... Sur l'implication, These de 3e cycle, Universite de Paris VII, 1980.

R. Duval et F. Pluvinage, Demarches de reponses en Mathematiques, Educational Studies in Mathematics 8, 1977.

J. P. Fischer, La perception des problemes soustractifs... These de 3e cycle, Universite de Nancy I, 1980.

G. Glaeser, La didactique experimentale des mathematiques, Cours Universite Louis Pasteur, Strasbourg, 1980.

F. Hitt, Retours en arriere... These de 3e cycle, ULP Strasbourg (France) 1978.

G. Noelting ... Proportional reasoning ..., Educational Studies in Mathematics (11) 1980.

H. Radatz, Error analysis ..., Journal for Research in Mathematics Education, Vol. 10(3) 1979.

H. Ter Heege, ... Multiplication, Educational Studies in Mathematics (9) 1978.

G. Vergnaud et al., ... Un sondage, Bulletin A.P.M.E.P. (France) No 313, 1978.

METHODOLOLOGICAL PROBLEMS OF OBJECT-ADEQUATE MODELLING AND CONCEPTUALIZATION OF TEACHING, LEARNING, AND THINKING PROCESSES RELATED TO MATHEMATICS

R.W. Scholz
Institute fuer Didaktik der Mathematik
Bielefeld, Federal Republic of Germany

1. Introduction

By the term methodology we mean the theory of the methods which are applied to generate knowledge of a specific object. By methods we mean procedures for a systematic investigation of the object, based on adequate modeling. Hence, problems concerning adequate modelling or alternate methods cannot be isolated from problems concerning the conceptualization of the object itself. The possibilities of acquiring knowledge are considerably determined by the previous understanding of the object.

This is why we begin with some basic remarks on the object of our research: on abilities to be developed, learning processes, and teacher activities in mathematics education. After these conceptual remarks, we are going to discuss methodological consequences which will be illustrated by the description of some of our research projects.

2. On the Objects of Research in Mathematics Education

Research approaches on mathematical abilities can be classified along two lines. They may be based on a static or on a dynamic construct of ability, and they may differ according to whether abilities are derived from the structure of mathematical knowledge, or whether general abilities of thinking and activity are taken as a starting point.

We conceive of the genesis and application of mathematical abilities as a process, i.e., as a sequence of empirically distinguishable stages. In doing so, we are taking a position against product-oriented approaches, particularly against those which start from the structure of mathematical knowledge and attempt to determine mathematical ability factors by means of multivariate test analysis. Even product-oriented approaches which try to determine mathematical abilities as a combination of general ability factors or intelligence factors obtained from current test batteries will contribute little to an understanding of mathematical abilities (see Begle, 1979, p. 91).

It is important to understand mathematical abilities not only on the basis of the systematics of mathematics as a discipline. From an educational perspective, allowance must be made for the context of use and application of mathematics from the very beginning. Even if there are up to now hardly any insights about the "meaning" of mathematical concepts in application contexts (even in school tests) and the genesis of this meaning, we consider working on this problem to be impossible without taking into consideration the individual learning history.

The trend observed in the seventies to increase research efforts in teaching takes into account that the teacher is responsible for the organization of learning processes in the classroom, and that didactical measures and innovations are always introduced via the teacher. The teacher's "knowledge" is important here under the following aspects: It mediates between mathematical science and the school curriculum, and it provides the basis for the teacher's teaching behavior. Activity-guiding knowledge is constituted both of scientific theories and of everyday experience. It will often appear in an idiosyncratic shape and is only partly accessible to conscious explication. As a part of the socially existent stock of knowledge, however, it is accessible to empirical and conceptual analysis.

The understanding of teacher activity - with regard to teacher education, too - will require conceptualization of its cognitive foundations. We shall term this research approach, according to the tradition of cognitive psychology, as that of studying the teacher as an expert. Besides teachers, other people who apply mathematical knowledge (in various vocations) may be studied as experts, as well.

3. Methodological Consequences for Studying Teaching, Learning, and Thinking Processes in Mathematics Education.

Knowledge about abilities (qualifications) and learning (acquiring qualifications) can be distinguished according to (1) the object of qualification, (2) physical operations, and (3) the organization of learning processes (cf. Zeiher and Zeiher, 1977). Existing theoretical approaches are deficient, first, because they are almost exclusively static, and second, because they have hitherto insufficiently treated the problem of relations between these three structurally different areas.

Case studies allow us to examine the relations between these areas without having to draw on a priori formulated, more comprehensive models. The holistic approach by means of case studies permits us to analyze characteristics and connections inherent to the case. Case studies thus maintain the system character and the specific dynamics of the processes concerned. Teaching is a social system which is constituted by the interaction between the participants and thus characterized by idiosyncratic components. Maintaining this specificity is important for re-applying the knowledge obtained to the system studied.

Thus, it is important, to carry out naturalistic studies which lead to statements about the object of research, e.g., teacher activity - in its natural context. The study of actual behavior in everyday situations will also provide the opportunity of elucidating the reference of mathematical knowledge to more general thinking and acting abilities, which will be relevant for the determination of desirable mathematical abilities.

Naturalistic case studies offer a suitable approach for the study of experts. Implementation of such studies of experts requires a special kind of co-operation between the researcher and the subject, with the latter as an active partner. The special role of interpretative methods in the evaluation of case studies and similar studies does not imply, of course, incompatibility with quantitative analyses, e.g., within content analyses.

4. Description of Individual Studies

In order to illustrate the above considerations concerning the relationship between object and method by means of concrete examples, we shall briefly describe some of our research projects.

H. Bussman (1980) has examined, by a case study, the genesis of mathematical abilities in students. Using various illustrations, crucial mathematical concepts (e.g., Pythagoras' theorem) were developed with a small group of students. The teacher-student dialogues, led by the teacher with empathy, permitted identification of different levels of concept development which must be passed before the student will be able to grasp the concepts concerned in a mathematically precise fashion.

By means of case studies, R.W. Scholz (1980) has studied the activities of professional experts (branch managers of a bank, neurologists). These studies have shown that "stochastical thinking" in the sense of obtaining and processing information of random character, and decision-making under conditions of uncertainty, are important for professional activities. Explicit application of general mathematics and of methods taken from probability, calculus, and statistics plays an almost negligible part, while implicit and non-formalized application of principles and methods of stochastics will frequently be crucial. Moreover, the case studies show that coping with stochastic phenomena is of process character. Receiving information, cognitive structuring, processing, and decision-making resulting from such processing are linked to each other. Insights gained from case studies can be used as a component for determining goals and objectives of mathematics education.

H.W. Heymann (1980) has studied implicit teaching theories. Mathematics teachers are confronted with their own classroom activities recorded on videotape and questioned in open interviews as to the background (considerations, concepts, experience rules, general evaluations) of their actual behavior. The emphasis lies on their behavior when confronted with learning difficulties (students errors, refusals to answer questions, etc.), as well as when introducing mathematical concepts. Teachers' values and value conflicts, their knowledge gained from experience, and preferences for certain activity patterns are elaborated on in an interpretative way and realted to each other on the basis of the interviews. In this way, sections of individual implicit theories are reconstructed.

V. Reiss conceptualizes teaching as the guiding of classroom activities. Classroom activities function as an intermediate variable between the behavior of the teacher and the students' learning processes. They are organized both sequentially and hierarchically. Teachers employ a wide variety of techniques to minimize irrelevant activities, and to maximize students' active learninng time. The study analyzes videotapes of mathematics lessons in primary school, which have been recorded under conditions which were held as "naturalistic" as possible. It aims at identifying effective and less effective techniques of guiding classroom activities. The process of hypotheses formation within the study can be described as alternating between an elaboration of existing theories and a phenomenological analysis of the videomaterial. Theoretically, the study draws on ideas advanced by Kounin (1970) on signal systems in classroom management processes, and on pragma-linguistic concepts of discourse analysis of classroom language (Sinclair and Coulthard 1977).

R. Bromme (1980) has studied the thinking of teachers during lesson planning. The mathematics teacher is conceived of as an expert for this activity. The study examines which theories of cognitive psychology are adequate for the description of the cognitive processes under such complex and unstructured requirements. It draws on problem-solving theories and on theories about the growth of structured knowledge in text comprehension. The latter contain elaborate models of knowledge representation and can be fruitfully used for the analysis of thinking aloud protocols. Lesson planning of mathematics teachers was recorded by means of the thinking aloud technique. Evaluations by categories of content analysis and cognitive psychology have shown that planning is predominantly a selection of mathematics tasks, and an anticipation of the work done on them by students. This focus on tasks nevertheless permits the teacher to make allowance for other problems of teaching - especially to anticipate students' activities.

References

Begle, E.G. (1979): Critical Variables in Mathematics Education. Washington.

Bromme, R. (1980): Die alltaegliche Unterrichtsvorbereitung von Mathematiklehrern. Zu einegen Methoden und Ergebnissen einer Untersuchung des Denkprozesses. In: Unterrichtswissenschaft, 1980, 2.

Bussmann, H. (1980): Erkenntnis und Unterricht: Zur Genese mathematischer Faehigkeiten. In: Unterrichtswissenschaft, 8 (in press).

Heymann, H.W. (1980): Subjektive Hintergruende unterrichtlichen Handelns bei Mathematiklehrern. In: Beitraege zum Mathematikunterricht 1980, Hanover, pp. 132-136.

Kounin, J.A. (1970): Discipline and Group Management in Classrooms. New York.

Scholz, R.W. (1980): Berufliche Fallstudien zum stochastischen Denken. In: Beitraege zum Mathematikunterricht 1980, Hanover, pp. 298-302.

Sinclair, J.M. & Coulthard, M. (1975): Towards an Analysis of Discourse. London.

Zeiher, H. & Zeiher, H.J. (1970): Ueberlegungen zur Schulforschung Teil II. In: Roeder, P.M. et al.: Ueberlegungen zur Schulforschung. Stuttgart, pp. 125-146.

THE TEACHING EXPERIMENT METHODOLOGY IN A CONSTRUCTIVIST RESEARCH PROGRAM

Leslie P. Steffe
University of Georgia
Athens, Georgian, U.S.A.

A constructivist research program in mathematics education has, as its central problem, the explanation of the process of construction of mathematical objects as it actually occurs in children. Mathematical objects, as constructed by children, are structural in nature in a manner not unlike Piaget's view of operations (Piaget, 1970, p. 22). The teaching experiment, as a methodology, is crucial to the Constructivist program of research especially because operational structures are never acquired in one piece, but must be built up by the child in the course of his or her own experience. In general, methodology refers to a set of techniques by which a researcher constructs theory (Skemp, 1979, p. 2). The teaching experiment is a technique which is used to construct explanations of children's constructions. But, as Skemp (1979, p. 27) has observed, "the methodology of the teaching experiment does not apply exclusively to a particular theory, as demonstrated by its having arisen independently in the Soviet and among Constructivists; but rather to a type of theory, or to a general theoretical stance which may predate the advent of a theory" (p. 27). The teaching experiment, as we use it, includes three basic aspects not universally shared by other users. These three aspects are modeling, teaching episodes, and individual interviews.

Aspects of a Teaching Experiment

Modeling. The most important aspect of a teaching experiment is modeling. In some of the literature on the teaching experiment (Menchinskaya, 1969; Kantowski, 1978) modeling has not been included as a basic research paradigm. But in El'Konin's (1967) assessment of Vygotsky's research, the essential function of the teaching experiment was the production of models: "Unfortunately, it is still rare to meet with the interpretation of Vygotsky's research as modeling rather than empirically studying, developmental processes" (p. 36).

The explanations we formulate in our research constitute models. It must be understood that these models are formulated in the context of intensive observation of children's construction of mathematical objects. But while they are based on actual human behavior, they also are based on our interpretation of the meaning that the children attribute to their behavior or, in other words, why they exhibit the observed behavior. The nature of explanation, as we use it, is consistent with the view of Maturana (1978) that "an explanation is always an intended . . . reformulation of a system or phenomena, addressed by one observer to another, who must accept it or reject it by admitting or denying that it is a model . . . " (p. 29).

The model, as we use it, can be understood using the metaphor of a black box as described by von Glasersfeld (1978); the black box "is used for items that one suspects of possessing the wherewithall to perform some functions, but which . . . one cannot dismantle to see what is going on inside" (p. 14). Unlike behaviorists, however, who are not interested in the machinery inside of the box, a constructivist attempts to design models which ideally will, given an "imput", produce the same output as the black box. One never speculates as to the nature of the function of the black box, but only attempts to design models which seem to be a viable explanation of the input-output relation.

Teaching episodes. Another aspect of the teaching experiment is the teaching episode and its interpretation. The participants in a teaching episode are the teacher and the child. The witness of a teaching episode is an adult other than the teacher whose primary purpose is to interpret the communication between the two participants during the course of the episode. The teacher, as a participant, intends (1) to test the limits of a model he has of the child's knowledge with regard to particular content structures and (2) to investigate how various components of that model may change under the pressure of directed interference. In the test of the limits and in the investigation of change of the model, interpretations of the child's behavior and intentions are necessary. The teacher, of course, can interpret the child's behavior only by means of his or her own conceptual structures.

In a teaching episode, no simple cause-effect relationship is assumed between the teacher's acts and what the child does. There is a break at the child's experiential interface as well as at the teacher's, for experience, as experience, is the individual's own. The researcher, then (as teacher), has the tasks of (1) interpreting what he or she sees the child doing in terms of a model and (2) attempting to perform the ultimate act of decentering by conceiving of his or her own actions from the child's perspective. It is therefore critical that another adult who is familiar with both the model and the methodology witness the episode as it takes place, in order to formulate possible alternative interpretations of the child's behavior and to make them known to the teacher. The witness also has the function of suggesting possible alternatives to the teacher's actions.

The course of teaching episodes is determined by the child as well as by the teacher, because both are participants. All tasks the teacher contemplates using in an episode are designed especially for the particular child. The teacher's model of that child and such modifications of it as the teacher intends constitutes the rationale for each task. Both teacher and witness are responsible for devising impromptu tasks during the teaching episode after they have interpreted the child's responses to prior tasks. In fact, these spontaneously (and appropriately) designed tasks represent a major modus operandi of the teaching experiment.

As a matter of course, the teaching episodes are audio-video taped. This record is used for three purposes. The teacher may make public, in the context of the record of the on-going teaching episode, his or her intentions and interpretation of the child's behavior. Second, through alternative interpretation by other adults, the teacher may form a revised mode of the child's knowledge and new hypotheses to be tested. Third, through analysis of each child's behavior, the next teaching episode may be planned.

Individual interviews. Each teaching episode contains all of the elements of individual interviews with the added feature of intentional intervention by the investigators. Intentional intervention, however, radically changes a "task" from an interview to a teaching episode. In an interview (as well as a teaching episode), one must differentiate between the observer's "task" and the child's "task". Neither reflects a reality that is objective and independent - - which is to say that the child's "task" is not necessarily isomorphic to the observer's. The "task", as constructed by the child, can never be experienced directly by the observer! The observer, as a scientist, wants to explain how the child constructs and solves his (the child's) "task". But if there is no sharp differentiation between the observer's "task" and the observer's interpretation of the child's "task", the explanation will contain operations performed by the observer in the act of observation. Maturana (1978) makes this point crystal clear: "we generally take the observer for granted and, by accepting his universality by implication, ascribe many of the invariant features of our description that depend on the standard observer to a reality that is ontologically objective and independent of us" (p. 29). Children's solutions to tasks should be explained in terms of hypothetical constructs that the observer would not normally use to describe his own knowledge of the "tasks". Focusing on the child rather than on the task constitutes a shift in paradigm from the attempt to establish causal object-subject relationships to formulating models of what may go on in the head of the child. The tasks used in the individual interviews are crucial in our research program, but our focus is not the tasks as we conceive them, but rather as they are constructed and solved by the child.

Instruction

Teaching experiments consist of a series of teaching episodes and individual interviews for extended periods of time - - anywhere from six weeks to two years. The type of micro-analysis we carry out would be practically impossible with a large number of subjects given this extended time period. We agree that working largely single-handed and with only three subjects is not an ideal to be emulated; but to work with more than six to eight subjects would be for us an obvious impossibility. One of our most fundamental assumptions is that there is regularity in children's construction of mathematical objects.

It cannot be stressed too much that instruction in the teaching experiment occurs under optimum laboratory-type conditions. The teachers are expertly trained researchers in mathematics education, the number of students is small, and the interaction of the teachers and students is intensive as well as extensive. But laboratory-type conditions are essential if the child's construction of mathematical objects is ever to be explained in any significant way. The researchers must observe children's constructive processes first-hand if they are to gain the experiential base so crucial in formulating explanations of those constructions. Researchers in mathematics education who do not assume the responsibility of engaging in intensive, and extensive, interaction with students run the great risk of their models explaining only their own mathematical behavior.

Using individual interviews and teaching episodes, it is our intent to formulate a series of models which reflect the development of a child with respect to particular content structures under the influence of instruction. According to El'konin (1967), this was also an essential part of Vygotsky's experimental-genetic method. His description of the process of experimental analysis as allowing "a dissection in abstract form of the very essence of the genetic process of concept formation" (p. 36) agrees with our own understanding of the modeling process. However, a child develops in mathematics only because of the influence of instruction, and if one wants to observe development, one has to observe the child under instruction.

Final Comments

Research in mathematics education has not provided an understanding of the child's mathematics and how that mathematics may be built up, piece-by-piece. It is our contention that this state of affairs is in part the result of researchers using inappropriate methodology in their study of children's mathematical behavior. With few, notable exceptions, researchers simply have not observed students do mathematics for extended periods of time with the intention of expanding their behavior. Essentially, they have abandoned mathematics education as a teaching field and, concomitantly, have created an unnecessarily large chasm between their research and school mathematics.

References

El'konin, D.B. The problem of instruction and development in the works of L.S. Vygotsky. Soviet Psychology, 1967, 5 (3), 34-41.

Kantowski, G. The teaching experiment and Soviet studies of problem solving. In L.L. Hatfield (Ed.), Mathematical problem solving. Columbus, Ohio: ERIC Clearinghouse of Science Mathematics and Environmental Education, 1978, 43-52.

Maturana, H. Biology of Language: The epistemology of reality. In G.A. Miller & E. Lennenberg (Eds.), Psychology and biology of language and thought: Essays in honor of Eric Lennedberg. New York: Academic Press, 1978.

Menchinskaya, N.A. Fifty years of Soviet instructional psychology. In J. Kilpatric & I. Wirszup (Eds.), Soviet Studies in the Learning and Teaching of Mathematics. Chicago: University of Chicago, 1969.

Skemp, R.R. Theories and methodologies. Paper presented at the Wingspread Conference on the Initial Learning of Addition and Subtraction Skills, 1979.

Piaget, J. Genetic epistemology. New York: Columbia University Press, 1970.

von Glasersfeld, E. Radical constructivism and Piaget's concept of knowledge. In F.B. Murray (Ed.) Impact of Piagetian theory. Baltimore: University Park Press, 1978.

A METHODOLOGY FOR RESEARCH INTO THE LEARNING OF MATHEMATICS

Joan Yates
University of Bristol
Bristol, England

Probing the learning process obtaining in a classroom involves probing the interaction between teacher and pupil, pupil and pupil, teacher with himself, and pupil with himself. This interaction can be considered as a circular movement in that it is in a state of continual motion for any one individual.

Who initiates this learning process? As teachers we may like to think that we do. Admittedly we may create a focussing point for the lesson, but the learning process for that focussing point will, I suggest, in many cases have commenced prior to its presentation. So when observing the learning process we are observing a process of which we are privileged to see only a very small part - its beginnings will have had their origin at some point outside our control. Connections made by our centering the class on a particular focus are dependent on an individual's prior experiences - what Gagne terms the schema of an individual. This learning process is a process in operation in both teacher and pupil.

As every class is different, the teacher has the task of accomodating to the pupils of a particular class. Whilst the content to be presented may be the same as that imparted on a previous occasion, the way in which it is done will perforce be different. The teacher is involved in a learning process drawing from past experiences whilst continually adjusting to responses of the particular class of pupils. All are involved in a network of connections that are multidimensional. It is this multidimensionality that has to be probed and any attempt to reduce it to something less is to involve oneself in a similification which is, I suggest, both unjustifiable and an underestimation of the task in hand.

Is it the vastness of this task that has caused some researchers to attempt to reduce this multidimensionality to an analysis of what appears to them as key variables in the learning process? I deliberately use the words 'what appear to them', for the point I wish to highlight here is that they have been involved in a judgment which is subjective: the decsion regarding which key variables they choose to observe. This in turn implies that the researchers' findings are dependent on those original subjective judgments directing their decisions as to what is to be observed. Nor can we ignore the fact that any observation of phenomena during an enquiry is a function of the person observing. This acknowledgement is central to any research but is of particular importance when the object of observation is human behavior. The status we ascribe the results of an enquiry will depend on the probity we accord the researcher and the extent to which his findings resonate with our world as teacher.

At the point we enter this exercise, be it as researcher, teacher involved, or reader of findings, we will be interpreting. Interpreting is an activity in which we are continuously involved. However, I would suggest that the Act of Interpreting is one which can be inextricably caught up with what is observable to us. I do not use the words to us lightly since what we see is inevitably influenced by and made to accord with our experience. All observings are the results of our previous encounters; we observe such and such is a situation according to our schema; it is how we see it. Neither is it a simple one-dimensional observation, several facets of the situation impinge simultaneously - it is multidimensional.

When we try to write or talk about our observings we are immediately confronted with a problem. Both writing and talking is linear so we must record or speak one thought at a time. Thus we can give the appearance of an ordered sequence of events one happening after the other, contrasted with the reality which was a melange of these events.

In the Act of Interpreting we are trying to get behind the observable. As an observer in a classroom one may ask "Why did a teacher do this?" or as a teacher of the class "Why did I do that?" In either case we are searching within ourselves for the answer relating to our experience both in the realms of practice and theory. We consider, we interpret, and the consequences of the latter carry the implication of a personal structure of connections with respect to the situation we have observed. I feel at this juncture I must refer to the metaphysical problem for an observer in interpreting, which is the status he gives to his conceptions of what is 'unseen' when he observes a teacher in action. By 'unseen' I mean the intentions or pressures which lie behind overt observable behaviour. These 'unseens' affect performance in the classroom and their effect can in no way be measured.

Note my question, 'Why did a teacher do this?'. What I am interested in probing is, as I indicated in my opening sentence, the learning process. Since I view the teacher as central to any classroom learning process, any enquiry into mathematical learning taking place in this context must, for me, be teacher-based. I would suggest that this can be done effectively by (a) a sympathetic oberver working with a teacher or (b) a teacher reflecting on his own thinking, actions, decisions in a manner that is systematic for the teacher concerned. I feel it is important to stress the latter point as each teacher's philosophy of learning will perforce dictate his thinking, actions, decision, and process of systematisation.

Quite a lot of research has focused on categorization of behaviour by an external observer (e.g., Flanders' Interaction Analysis Categories), but I suggest that is has failed to probe the learning process as such. It is the necessity for adopting a methodology that will probe the learning process that I now wish to discuss. It is also of paramount importance that the methodology be capable of adaptation as the unfolding of the research reveals the 'real' problems in contrast to the problems which are thought to be there at an earlier stage.

The methodology which I would suggest as being suitable has its roots in the theory of cybernetics. Stafford Beer in his book Platform for Change states,

> To handle any situation with a given amount of variety we must be able to match it - and there are in principle just two approaches to the problem. The first is to increase the variety of the controller until it matches the variety of the situation to be controlled. The second is to take action which will drastically reduce the variety to be controlled.

This statement is applicable to a classroom situation where the given amount of variety is the network of connections that are multidimensional. Of the two approaches, I propose to consider the former, as I feel the latter demonstrates what has all too often happened in educational research in areas where such action has been inappropriate. The controller can be replaced by an observer.

What then is the variety open to an observer? I suggest it is an observer's interpreting of the dynamic situation in which he is involved. The teacher takes responsibility for the network of connections in his classroom and it is his ability in interpreting the variety that is obtaining at any given moment that will govern his control of the classroom situation. This control is transient in the sense that it is in a state of flux, but at each moment in the space-time continuum there must exist using the terminology of cybernetics - homeostasis. For the cybernetician, homeostasis is the capacity of a system to hold its critical variables within physiological limits - note not prescribed limits. This is the task with which the teacher is confronted in a multidimensional situation. Whether control obtains is, I suggest, dependent on the teacher's interpreting. This in turn, is dependent on this awareness, which will be at different levels according to the stimulus to which he is responding and where he is in his own space at that time.

To reflect meaningfully on the learning process of others is, I believe, only possible if one is in close contact with one's own learning process and this is the task of an observer. An observer of the network of connections obtaining in a classroom is all the time interpreting the observable. It is the depth to which he can probe the observable that will give enlightenment to the learning process. The observable may be an action, a feeling, a response of pupil to teacher. Taking the latter case as an example - it is the inspiration that provokes the response that must be probed. Frequently a teacher accepts a pupil's response thinking he knows what has provoked it, sometimes he assumes he knows why - and sometimes the teacher dares not probe the inspiration behind the response. Yet the observable in this situation can be thought of as the tip of the iceberg. It is the inspirations behind a pupil's response or a teacher's performance in the classroom that are, in the end, the effective factors. It is these effective factors that I feel govern the learning process and can only be probed if we accept a theory of interpretation. Such a theory gives an authority to the subjective judgements of the observer. He is no longer constrained by a fear of not being objective, which for those involved in authentic classroom enquiry is, I suggest, experienced at some stage. When this is experienced it is for the person concerned a crux point. He must either acknowledge the impossibility of being objective or remain restricted by inappropriate measuring instruments such as questionnaires, attitude tests, diagnostic tests, etc., which can give a sense of security to the person using them. I would suggest, however, that this security, if the person has the courage to probe, will be found to be divorced from his world of experience. I do not wish to imply that questionnaires, etc., have no place in classroom enquiry. The danger arises when such instruments are accorded a status of measurement which is considered to be devoid of subjectivity. The questionnaire is itself the fruit of the compiler's thoughts, and the person using it, if he is not the compiler, will have superimposed his interpretation on these thoughts. In either case we are in the realm of interpretation.

At the outset of an enquiry, a stance must be taken which acknowledges and recognises the value of subjective judgments. Areas on which to focus in the initial stages must be agreed by both the observer and teacher. These areas may be, for example:

the atmosphere permeating the classroom

the pupil grouping

the way pupils enter the classroom and settle to work

the teacher's functioning during the lesson

the pupils' response(s) to the teacher

In focussing on these areas it is important that both recognise the probity of the other (this relates to my previous references to a sympathetic observer) and be aware of the conditions that commonly cause distortion as delineated by Arne Trankell in Reliability of Evidence:

a. The selective character of percetion, which limits the interpretation of the external signals to that which has foundation in the individual's earlier experience.

b. The logical completion mechanism which often results in a false picture of the series of events.

c. Attitudes, personal wishes and preference which prejudice our interpretation of sense data.

Illumination of the areas under scrutiny is obtainable by:

Discussion between teacher and observer, teacher and pupils, observer and pupils (i.e., if observer decides to be a participant-observer)

Videotape, tapes, pupils' work

Written accounts of pupils

Interviews.

In fact, any instrument can be used that, in the opinion of both teacher and observer, serves as appropriate for probing the area under consideration.

It is important that both teacher and observer write about their feelings, reactions and thinking at different states in the enquiry. This allows the possibility of tracing connections in the act of interpreting in which both have been involved. Connections having been located, they are then open to further probing. The methodology adopted at this stage must again be that which is considered appropriate by both teacher and observer. Thus a filtering process is taking place in which decisions taken are those which, in the judgement of both teacher and observer, illuminate the complexity of interactions in lessons in which both are involved. Thus insight into the learning process of a particular class with a particular teacher emerges through the Act of Interpreting of both teacher and observer. Enlightenment so obtained can enhance the quality of future learning. The place of the observer cannot be ignored as he too will have played his part in the learning process of the particular class. If he had not been there, the learning process would have been different.

Such an enquiry can result in the surfacing of certain findings. These findings, whilst embedded in the philosophy of learning of those actively involved in the enquiry, can be offered to others. There generalisability will be dependent on the credibility they find with others. If the credibility allows others to begin to interpret for themselves, then the findings may begin to be generalisable, but the generalisability, I stress, is dependent on their interpretation by others.

The generalisability that I offer lies in the methodology. Probing the learning process becomes a structuring of what is observable through a process of concentrated focussing with the consequent acceptance of a theory of interpretation.

13.5 ERROR ANALYSES OF CHILDREN'S ARITHMETIC PERFORMANCE

ANALYSE D'ERREURS DANS L'UTILISATION DE LA SUITE DES NOMBRES PAR LES ENFANTS DE LA IERE ANNEE DE L'ENSEIGNEMENT OBLIGATOIRE EN FRANCE OU COURS PREPARATOIRE (ENFANTS DE 6 A 7 ANS)

Annie Bessot
IREM de Grenoble
Grenoble, France

Presentation Global de la Recherche

Mon expose s'inscrit dans le cadre d'une recherche collective (avec C. Comiti et C. Pariselle: la presentation globale de cette recherche est donc commune avec celle de l'expose de C. Commiti sur "The child's concept of number") de didactique des mathematiques, sur le processus d'acquisition de la notion de nombre naturel par le jeune enfant entrant a l'ecole obligatoire: la notion de nombre naturel ne se construit pas d'emblee chez le jeune enfant, ni au cours d'une suite d'activites de complexification croissante. Si l'on suit la theorie operatoire de l'intelligence developper par Piaget (Piaget 1975), on sait que l'enfant, pour apprehender un champ plus large de prioprietes, ne generalise pas simplement l'application des outils cognitifs dont il dispose; il passe au contraire par un processus de reconstruction: ceci l'amene, chaque etape, a se fabriquer une classe d'instruments intellectuels nouveaux qui continuent par la suite, d'une part a se developper en tant que tels, d'autre part a s'integrer et a se ....ionner selon des .....ites diverses.

C'est dans un but de clarification des sensitions dans lesquelles l'eleve construit et s'approprie le concept de nombre naturel (point de vue didactique) que notre recherche a pour objectifs generaux:

- de mettre en evidence les differents modeles implicites et ine..?..lets du nombre fonctionnant chez l'eleve a un moment donne et dans une ,,,certaine numerique donne:

- de mieux comprendre comment les differents modeles qui coexistent, a un moment donne, chez un meme enfant interviennent selon des cacher qu'on lui propose.

Qu'appelons-nous modele implicite du nombre naturel?

Ce sont certains invariants, certaines proprietes de ce concept que l'on peut inferer du comportement d'un enfant dans les situations ou intervient ce concept. La recherche de tels modeles suppose l'hypothese que, quelles que soient les conditions dans lesquelles sont fait les apprentissages scolaires, il existe, a un certain niveau, des regularites qui caracterisent l'appropriation d'une connaissance donnee chez tous les sujets. Les differents modeles qui coexistent, a un moment donne, chez un meme enfant, constituent son "systeme de connaissance" dont nous voulons etudier les differents fonctionnements selon la tache a resoudre.

Methode Generale de la Recherche

La formulation de tels modeles implicites est assujettie a diverses analyses interdependantes:

- Une analyse mathematique du concept de nombre naturel afin de caracteriser les invariants et les proprietes de ce concept;

- Une analyse des taches proposees a l'eleve afin de degager des classes de procedures possibles relativement a la tache;

- Une analyse des comportements du sujet: en particulier en terme de type de procedures et de types d'erreurs;

- Une analyse du systeme d'interactions eleve-tache, sous-systeme du systeme d'interactions plus large eleves-tache-enseignant: notamment une mise en evidence de certaines variables de la situation pouvant modifier le rapport du sujet et la tache en rendant couteuses ou peu fiables certaines procedures (1).

Je m'attacherais tout d'abord a montrer l'importance de l'analyse des erreurs dans notre demarche en precisant notre conception du role de l'erreur dans tout processus d'acquisition.

Conception sur le Role de l'Erreur dans un Processus d'Acquisition

Cette conception depend de notre conception du developpement cognitif. Certaines theories considerent le renforcement externe comme principal mecanisme de ce developpement: de ce point de vue les erreurs sont l'effet de l'ignorance ou de l'inattention et en cela sont a eviter dans tous processus d'apprentissage. Les conceptions de Piaget font jouer au contraire a l'erreur un role constructif vis a vis du developpement cognitif: c'est l'erreur et sa prise de conscience qui, en provoquant un desequilibre du systeme de pensee du sujet, permettent son depassement en un nouvelle "equilibre majorant" (2). Si certaines erreurs sont bien fugaces et isolees, d'autres sont "liees entre elles par une source commune, une maniere de connaitre, une conception caracteristique, coherente, sinon correcte, ancienne et qui a reussi dans tout un domaine" (3). Ces erreurs reproductibles, persistantes sant donc constitutives de la connaissance du sujet. Si nous nous accordons avec cette conception le rapprochement des erreurs, et leur interpretation en liaison avec l'analyse du concept en cause, peuvent etre un facteur priviligie pour identifier les modeles implicites d'un sujet.

Des Questions.

Dans le processus d'acquisition de la notion de nombre naturel un objectif essentiel en mathematiques de la lere annee de l'enseignement obligatoire en France, est l'apprentissage du comptage jusqu'a 99.

En rapport avec cet objectif d'apprentissage et dans le contexte d'une analyse des erreurs dans l'utilisation de la comptine numerique, les questions que je poserai ici sont les suivantes:

1. Quand le comptage est mis en concurrence avec d'autres procedures, telles la correspondance terme a terme, est-il utilise de facon preponderante? Quel est sa fiabilite selon le moment de l'annee ou l'on se situe? (Le domaine numerique est bien

entendu une variable fondamentale pour les reponses a cette question, mais ce n'est pas l'objet de cet exposee. (voir (4))?

2. Quels sont les types d'erreurs les plus courantes et les plus persistantes dans les taches proposees? Comment les interpreter dans le cadre de la conception sur l'erreur presentee?

Ce questionnement se fera relativement au domaine numerique voisin de 30: En effet, a la fin de la lere annec (CP) et a fortiori au cours de la 2eme annee de l'ecole obligatoire (CE I) l'enfant est sense bien maitriser la comptine numerique dans ce domaine et meme au dela. Qu'en est-il?

Pour envisager les reponses a ces questions, nous nous appuierens sur des analyses quantitatives de comportements d'eleves face a certaines taches de nature numerique.

1. Utilisation et fiabilite du comptage dans une tache ou celui-ci est mis en concurrence avec d'autres methodes.

Description de la tache:

L'enfant dispose d'une boite de 45 jetons bleus. Sur la table sont poses en vrac 37 jetons rouges. L'experimentateur demande a l'enfant "met sur la table autant de jetons bleus que de jetons rouges". (Construction d'une collection equipotente a une autre). L'enfant ayant termine, en met en doute son resultat par le consigne "es-tu-sur...?". Puis on sollicite une verification, sauf si l'enfant est sur a juste titre de sa construction (comparaison de deux collections).

Nous avons classe les procedures des enfants en D pour denombrement ou comptage, en C pour correspondance terme a terme, en A pour approximation perceptive.

Nous notons la disparition progressive des procedures d'approximation perceptive. Cette disparition s'accompagne d'une augmentation des procedures de correspondance terme a terme predominantes (dans cette tache) jusqu'a la fin du cours preparatoire; ces dernieres ne s'effacent au profit des procedures de comptage qu'en CE I, avant meme que ce comptage ne soit reellement fiable.

Il serait donc errone de penser qu'a la fin de la lere annee de la scolarite obligatoire, le comptage soit suffisament interiorise par les eleves n'importe quelle situation: Dans une situation ou le comptage est en concurrence avec d'autres methodes, c'est la correspondance terme qui conduit au succes, meme en debut de CE I.

2. Quelques exemples d'erreurs:

a. Dans la connaissance de la comptine orale (sur commande) en juin du cours preparatoire (58 enfants presents).

Les effectifs cumules des enfants ne sachant pas reciter la comptine numerique au dela de N montre que:

-10 enfants (soient plus de I sur 6) n'arrivent pas a reciter la comptine au dela de 29.

-42 enfants (58-16) comptent jusqu'a 60 et au dela. Mais seuls 23 de ces enfants ne font pas d'erreurs au niveau de la recitation, ce qui ne veut pas dire qu'ils n'en feront pas au niveau de l'utilisation.

Les erreurs les plus frequemment rencontrees et les plus persistantes situent au passage des dizaines (5 au passage de 20; 18 au passage de 30, 40, 50; 5 au passage de 60), pouvant donner lieu a des boucles...29 20 21 Etc...

Ces erreurs, par opposition a d'autre types d'erreurs se manifestant dans l'utilisation de la comptine, sont d'ordre lexical. De notre point de vue, elles sont, dans l'effort de memorisation de la comptine...?...d'une tentative de l'enfant a "operer systematiquement, en suivant certaines regles" (5): L'enfant prend en consideration existence de regularites dans l'expression orale de la comptine et vient buter sur les anomalies que sent les noms des dizaines.

b. Dans l'utilisation de la comptine orale pour denombrer les objets d'une collection ou construire une collection de cardinal donne.

Pour Ginsburg (6), cette utilisation, pour se faire sans erreurs, exige au minimum:

1. La memorisation de la suite de nombres dans le domaine numerique considere;

2. La prise en compte de chaque objet de la collection une fois et une seule;

3. L'etablissement d'une correspondance terme a terme entre chaque nom de nombres et chaque objet.

On peut trouver chez cet auteur (6) divers exemples d'erreurs d'enfants relatives au non respect des contraintes 2) et 3), ainsi que des interpretations montrant que ces erreurs sont constitutives du sens de la connaissance de l'enfant.

Je me contenterai de presenter ici un autre type d'erreurs que l'on peut rencontrer en debut de CP mais aussi plus tard en mars de la meme annee scolaire.

Tache: Construction d'une collection de cardinal donne m.

Debut CP: Delphine semble connaitre sans hesitation la comptine jusqu'a 30 environ. Jusqu'alors, elle n'a pas fait d'erreurs repetees de comptage (non respect de 1) 2) ou 3)). Cependant elle a deja attire notre attention par son incomprehension du nombre d'objets d'une collection (18 elements) comme invariant quand on modifie la configuration spatiale de la collection.

On lui demande d'extraire 15 allumettes d'une boite en contenant 32: elle sort toutes les allumettes en les comptant et en trouve 26 (erreurs de comptage). Lorsqu'on repete la consigne, elle nous dit: "je ne peux pas, il n'y en a pas assez." Elle recommence pourtant l'operation precedente en comptant a haute voix; elle depasse 15 sans s'arreter alors qu'elle veut prendre 15 allumettes dans la boite!

Lorsqu'on pose la question pour m = 7, Delphine n'a aucune hesitation.

Mars CP: Dans la tache de construction d'une collection equipotente a une autre decrite plus haut (par. 1-), Carole compte bien 37 jetons bleus: elle compte alors les jetons bleus a toute vitesse, depasse 37 et arrive a 45 (elle les prend tous).

Les erreurs de ces enfants ne semblent pas provenir d'un non respect des exigences 1), 2) ou 3) exprimees par Ginsburg. Construire une collection de cardinal m exige aussi que l'enfant ait suffisamment interiorise la comptine numerique pour pouvoir anticiper sur elle et arreter de prendre des objets au moment voulu: ne pouvant pas prevoir le nombre m dans la suite des nombres, ces enfants ne l'entendent pas passer. Leur connaissance de la comptine est celle d'une chanson globalement apprise.

Conclusion

L'analyse des erreurs dans le cadre de notre recherche ne fait que debuter. Nous avons propose a des eleves de cours preparatoire d'autres taches certaines plus complexes que celles decrites ici; il reste a rapprocher les erreurs d'un mere eleve dans ces taches numeriques et a identifier le systeme de connaissance expliquant ces erreurs.

References

1. Brousseau, G. "L'etude des processus d'apprentissage en situation scolaire." Cahier n°18, IREM de Bordeaux: 1978.

2. Piaget, J. "L'equilibration des structures cognitives." EEG XXIII, P.U.F., Paris: 1975.

3. Brousseau, G. "Les obstacles epistemologiques et les problemes en mathematiques." Compte-rendu de la CIEAEM, Louvain-la-Neuve: aout 1976.

4. Comiti, C., Bessot, A., Pariselle, C. "Analyse de comportements d'eleves du cours preparatoire confrontes a une tache de construction d'un ensemble equipotent a un ensemble donne." Recherches en didactique des mathematiques vol. 12, La pensee sauvage ed., Grenoble: aout 1980.

5. Donaldson, M. "L'erreur et la prise de conscience de l'erreur." Bulletin de psychologie, 327, XXX: 1976-77.

6. Ginsburg, H. "Children's arithmetic: the learning process." Van Nostrand Company, New York: 1977.

AN ANALYSIS OF 'ERROR' IN ERROR ANALYSIS

Leroy G. Callahan
State University of New York
Buffalo, New York

In the United States research on error analysis followed quite naturally from the testing movement, which formed and expanded significantly during the first few decades of the twentieth century. The administration of computational tests to large numbers of children provided opportunities for researchers to examine pupil's performance on each type of example on the tests. Judd (1921) observed, "We are at the opening of a new era in the scientific study of educational proglems" (p/ 655).

The ensuing half-century of error analysis study has been marked by a trend from a simple to a more complex interpretation of 'error' in error analysis research. Not only has the research broadened in scope beyond consideration of simple computational errors in arithmetic, but more significantly there has developed an awareness of the incredible complexities in the interplay between thought and performance in students' responses to meaningful mathematical tasks.

Interpretation of 'error' during the early years of error analysis research was generally restricted to the level of observed behavorial performance. Paper-pencil performance was evoked; students' behaviorial performance was compared with expected conventional performance; an 'error' interpretation was then made on the basis of comparison. The procedure was simple, objective, and reliable.

Seeds of uncertainty and dissatisfaction with the simple procedures for determining 'errors' can be found in the early research, however. Burge (1932) reported that relatively few type errors and questionable habits could be analyzed with certainty from the written responses. Sangren (1923) concluded that there was much need for further study of the psychological reasons underlying error. Judd (1921), discussing a boy's problem with subtraction, observed that there must be an explanation reaching back into the boy's method of subtraction. There was some dissatisfaction with the simple phenomenological assessment of error. Knight and Ford (1931) concluded, "Much of our treatment of error . . . is superficial and unsuccessful because it is based on too little understanding of the psychological nature of the error and the processes leading up to it" (p. 112). In addition, early researchers observed that some student errors had a rational basis and stability similar to correct responses. Reflected in their work was the suggestion that mental states of students should be considered in interpreting 'error'.

The inclusion of both thought structures and processes, along with behavioral performance, in the frame of reference for 'error' interpretations has added a significant increment in complexity to error analysis research. That expanded frame of reference has made error analysis procedures less simple, less objective, and correspondingly less reliable in the measurement sense, while creating a potential for more insightful understanding of students' learning and development vis a vis meaningful mathematics tasks. In this case less may lead to more. Kety (1960) has pointed out that to deny the existence or the importance of mental states merely because they are difficult to measure or because they cannot be directly observed in others is needlessly to restrict the field of the mental sciences and to curtail the opportunities for the discovery of new relationships (p. 1862).

Interpreting a performance of a meaningful mathematics task as being in 'error' has become a more thought-filled proces. The incorporation of mental states into the sphere of influence on performance required the development of more complex assessment procedures since mental states are not accessible by direct observations. In turn the inference of 'error' by an examiner has itself become more error-prone. The complexity of the situational sample space that is operational in interpreting a student's response to a meaningful mathematics task has increased significantly.

Figure 1 presents a situational sample space for interpreting 'error' on a meaningful mathematics task in a situation where a cognitive process or structure is operational. It does not include the option where a student has no structure or process to address the task. Ginsburg (1977) and other investigators have observed that students' errors are typically based on systematic rules and are seldom random happenings. Although the sample space is certainly oversimplified, it does allow for a consideration of some sets of contingencies that lead to 'error' interpretations.

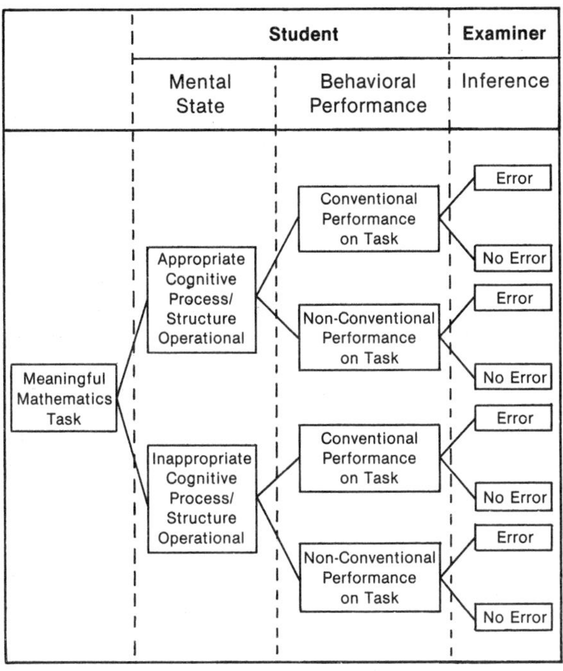

Situational Sample Space for Interpreting 'Error' on a Meaningful Mathematics Task

The sample space suggests the increase in complexity in the interpretation of 'error' even when only the simplest cognitive process or structure is considered. An examiner's inference of 'error' could be based on four different sets of contingencies in the situation. One could come about because of a breakdown in the mechanics of testing; e.g., an errant mark on an answer sheet or an improperly positioned answer key. Another could come about because of a misconception or misunderstanding in the student's mental state, reflected in unacceptable performance on a task. Others could come about through combinations of these contingencies. Hence, there is increasing need to assess sensitively operational process and structures in the mental state, along with performance indicators, so that the nature of the 'error' is better understood. Similar attention should be given to 'no error' inferences.

Certainly the restricted situational sample space is overly simplistic and conveys a very conservative estimate of the uncertainty involved with 'error' interpretations. Though there may be a dominant cognitive process or structure operational in response to a meaningful mathematics task, it is only one of a constellation of interrelated processes and structures that affect thinking about the task. Kent (1978) has written about Jennifer, who was able to respond to tasks such as 1/2 x 1/3 correctly. In the context of the sample space, an appropriate process was operational, there was a conventional performance of the tasks, and the typical examiner would infer 'no error'. She was asked, however, which one of the fractions 1/2, 1/3, or 1/6 was larger while discussing 1/2 x 1/3 = 1/6. She indicated 1/6 was larger because, "you are doing a times sign and that makes things bigger" (p 28). This one misunderstanding (structure) in the constellation of processes and structures that affect thinking about 1/2 x 1/3 suggests the further degree of complexity operational in the interpretation of 'error' on mathematics tasks.

As the incredible complexities involved with 'error' interpretations are recognized, there is the corresponding need for continuously more sensitive assessment procedures in error analysis research to assess structures and processes in the mental state that affect particular performances. Only then can there be an increase in certainty in 'error' or 'no error' inferences. As more sensitive assessment is employed, there will be more opportunity for error analysis research to contribute to the solution of educational problems - - a potential envisioned by Judd over fifty years ago.

Curriculum and instruction programming should benefit from more sensitive error analysis research. Kent (1979), as an example, has pointed out from his work how the limited idea of subtraction as 'take away' may become a limiting stereotype about subtraction that may "ultimately trap the child and create a barrier which he or she cannot destroy" (p. 25). Likewise, Erlwanger (1973) in his study of Benny uncovered learning habits and views about mathematics that could impede his progress in the future. Continued research along this line may contribute to more sensitive programming so that there will be more facilitative flows of instruction with few inhibiting instances.

Remedial work with individual students should also gain from the expanded frame of reference in error analysis research. Lepore (1979) has accurately observed that if a child is working with an inaccurate set of rules (cognitive structures or processes) failure is probable. A good remediation plan requires that the child's operational structures and processes be analyzed. As more enlightened assessment over the broadened frame of reference contributes to better understanding of the interrelations between mental states and performance, more appropriate remediation procedures can be determined.

The testing movement that spawned error analysis research could benefit from more sensitive assessment procedures employed in research. Callahan (1977) has observed that misconceptions in students' mental states may result in their responding reasonably to test tasks, but the tasks measure content different fromthat intended by the test maker. The information-processing model for analyzing error patterns suggested by Radatz (1975) could be particularly useful for insights into improving the validity of educational tests.

For years persons evaluating others' performance have rhetorically inquired, "What in the world were you thinking about when you did that?" As error analysis research extends its frame of reference to the sensitive assessment of the constellation of structures and processes in the mental state that affects performance, we may gain real insights into the question as it pertains to meaningful mathematics tasks.

References

Burge, L.V. Types of errors and questionable habits of work in multiplication. Elementary School Journal, 1932, 33, 185-194.

Callahan, L.G. Test-item tendencies: Curiosity and caution. The Arithmetic Teacher, 1977, 25, 10-13.

Erlwanger, S.H. Benny's conception of rules and answers in IPI mathematics. Journal of Children's Mathematical Behavior, 1973, 1, 7-27.

Ginsburg, H. Children's arithmetic: The learning process. New York: Van Nostrand, 1977.

Judd, C.H. Analysis of learning processes and specific teaching. Elementary School Journal, 1921, 21, 655-664.

Kent, D. Some processes through which mathematics is lost. Educational Research, 1978, 21, 27-35.

Kent, D. More about the processes through which mathematics is lost. Educational Research, 1979, 22 22-31.

Kety, S. A biologist examines the mind and behavior. Science, 1960, 132, 1861-1870.

Kright, F.B. and Ford, E. Temporary lapses in ability and errors in arithmetic. Elementary School Journal, 1931, 11-124.

Lepore, A. A comparison of computational errors between educable mentally handicapped and learning disability children. Focus on Learning Problems in Mathematics, 1979, 1, 12-33.

Myers, G.C. Persistence of errors in arithmetic. Journal of Education Research, 1924, 10, 19-28.

Radatz, H. Error analysis in mathematics education. Journal for Research in Mathematics Educatin, 1979, 10, 163-172.

Sangren, P.V. The Woody-McCall mixed fundamentals test and arithmetical diagnosis. Elementary School Journal, 1923, 23, 206-215.

ERROR ANALYSIS

Roy Hollands
Dundee College of Education
Dundee, Scotland

Background

The data quoted has been extracted fro the presenter's two years of research (1977-1979) into Diagnostic Tests and Remedial Action in Arithmetic for Secondary Pupils ages 12 years.

Objectives were written and submitted to teachers and other educators. The returns were analyzed, modification made, and then items written for each of the objectives. The resulting trials led to further improvements and the main trials were then constructed with four tests, each with 20 items. These tests were on whole numbers, fractions, decimals, and miscellaneous problems.

Over 5000 pupils were tested and the results of the bottom 15% were treated in detail. Computer analysis was used to obtain statistical information and one-to-one interviews played a major role.

The error analysis proved valuable when investigating the work of younger children from 7 1/2 to 12 years old.

Remedial treatment for each error type was designed and tried out in the classroom.

The following table shows an analysis of four questions on subtraction. The notes after the table give the meaning of the data.

Whole Number Subraction (W) S

	8. 14 -8 6	9. 170 -28 142	10. 500 -76 424	11. 413 -28 385	Total
Number right	325 (79.3%)	268 (65.4%)	231 (56.3%)	212 (51.7%)	1036
Number of omissions	4	8	10	23	45
Error Type					
S1 **Number combinations**	18	7	10	47	82
S2 **Wrong operation**	11	9	4	5	29
S3 **Reading or writing digits**	4	5	2	6	17
S5 **Procedure**					
a. Subtracted upper digit from lower digit.	31	32	28	38	129
b. 0 or blank in answer when there is no digit in subtrahend.	7	35	19	19	80
c. 0 in minuend mishandled.	–	27	23	–	50
d. Failed to compensate.	4	14	63	38	115
e. Compensated unnecessarily.	1	4	36	5	46
S6 **Not listed above**	22	20	20	36	98
Most frequent wrong answers	14(15)5a 74(9)5a 22(7)2 4(7)5c 7(7)1	158(28)5a 42(25)5b 150(20)5c 152(12)5d 198(8)2	576(27)5a 434(23)5d 334(18)5e 500(18)5c 334(12)5d 524(12)5d 76(9)5b	395(20)5d 415(16)5a 375(10)1 85(8)5b 425(7)5a	

Notes on the Error Analysis

These notes should be read in conjunction with the above table.

Several errors could be made by a pupil when answering an item and more than one entry would then be recorded.

The four Survey Test questions on subtraction are given on the table together with part of the data. Pupils making one or more errors were required to do the Subtraction Diagnostic Test (33 items).

When place value errors were suspected pupils took all or part of the Place Value Diagnostic Test (87 items).

Number Right

This row shows the number of pupils out of 410 who made each type of error and also expresses that number as a percentage of 410.

Number of Omissions

It was considered important to know how many pupils could not even attempt an item.

Wrong Operation

This error was sometimes due to the pupil not knowing which operation the minus sign indicated, but more often it was due to continuing to add. The pupils had previously been doing addition items.

Reading and Writing Digits

This was not a frequent error but for particular pupils it was important, for carelessness and badly written figures were affecting their work in other sections of the test.

Procedure

		Example
a.	Subtracted upper digit from lower digit. (This was the most common error)	214 - 179 165
b.	0 or blank in answer when there is no digit in subtrahend. (Pupils reasoned that two numbers were needed before there was any need to subtract)	261 - 36 45
c.	0 in minuend mishandled. (Nothing and 4 is to be taken away. I'll still have nothing)	210 - 34 180
d.	Failed to compensate (This was mainly due to forgetting)	782 - 314 478
e.	Compensated unnecessarily (This was mainly done by the weakest pupils who thought you had to in subtraction)	382 - 120 152

Not Listed Above

These included many miscellaneous errors each being made by only a few pupils.

Most Frequent Wrong Answers

14(15)5a The first number is the wrong answer. The number in brackets shows how many pupils made that error. Finally the Error Type is given.

ANALYSIS OF CHILDREN'S ERRORS: A FUNCTION OF OUR ERRORS AS MATHEMATICS EDUCATORS?

Fredricka K. Reisman
University of Georgia
Athens, Georgia, U.S.A.

There are four assumptions that underlie this paper:

1. There are generic factors that influence learning in general and that are related to understanding children's errors in mathematics.

2. It is important to be aware of the psychological nature of the elementary school mathematics curriculum

3. A mismatch between the special educational needs of learners and the psychological characteristics of the mathematics to be learned often underlie childen's errors in mathematics.

4. Teachers must look beyond computational errors and consider issues such as mode of representing mathematical relationships, concepts, and generalizations, sequencing of mathematics topics, and generic influences on learning that are relevant to a specific mathematics topic.

Assumption I: Generic Influences

The term generic is used in the Latin sense of genus, meaning roots or origin. Generic factors are those conditions that affect learning. They are influences of a general nature that modify how a person learns, regardless of their gender or race. Following are some examples of generic influences on learning and their effects and how teachers can address these:

1. Cognitive rates: This means that some are able to grasp new ideas very quickly and others take a long time for understanding to come about.

Teachers can accomodate these differences by presenting the same mathematics topic in a number of lessons for those who learn at a slower rate; those who grasp ideas quickly may need a one-time exposure. For example, a videotape format can accomodate these differences by presenting five-minute segments of the same mathematics topic designed for different rates of learning and different levels of complexity.

2. Complexity. Some can learn complexities; some can understand only the simplest of circumstance.

Ability to deal with complex situations has implications for the design of instructional materials. For example, work sheets, textbook pages, overhead transparencies, board work, and bulletin boards must be kept simple for those who are confused by irrelevant stimuli such as color, embeddedness of ideas, or overload of information.

3. Memory: Some persons can remember ideas easily and need a minimum of repetition; some need a great deal of repitition and practice.

For those who need more repitition than others, a mathematical idea may be presented many times during a single lesson or over a period of time. Following each presentation, the learner may be asked to write each new idea about the lesson topic that he or she learned. Another way of accomodating those who need practice is to present the same idea in a number of different ways using different materials.

4. Incidental learning: Some can learn incidentally, e.g., pick out the important aspect of a situation on their own, while others need a lot of structure to become aware of what it is they must notice.

For example, if a student is to learn simple geometric shapes such as square and circle, the form of objects must be attended to and irrelevant attributes such as color and texture ignored. Therefore, introducing irrelevant attributes such as different colors or textures should be avoided. Thus, using objects that point up only the attribute the student is to notice is an important strategy for those who do not notice the salient aspects of situations. Cueing techniques are helpful. Cues include emphasis of patterns, or chunking information such as when learning one's phone number, social security number, or license plate number.

5. Energy level: Some students may have abundant energy and stick with tasks for long periods of time. Some may fatigue quickly due to a physical condition such as diabetes or epilipsy or because thinking is hard work and uses a lot of their energy.

For these students, the teacher must adjust the amount and pace of instruction.

6. Distractibility: Some thrive in a colorful busy environment, some become overly excited and distracted easily and need a calm sterile environment free of distracting stimuli.

Use of behavior modification techniques including reward and reinforcement of desirable behavior may be incorporated. Providing a structured environment that minimizes distracting stimuli is often effective. Developing clear expectations regarding the student's behavior in various learning environments is helpful. Classroom routine should be predictable and consistent for such students. The use of self-instruction where a student talks his or her way through a task is effective. Eliminating irrelevant or potentially distracting stimuli from instructional materials is appropriate when hyperactivity or distractibility is a factor.

For some students a combination of generic factors may influence learning. For example, a slowly developing cognitive rate, poor memory, inability to engage in abstract reasoning, inability to cope with complexity, and the further compounding effect of a visual and/or auditory impairment will have a greater impact on learning in their combination rather than singularly.

Assumption 2: Psychological Nature of Curriculum

Curriculum is defined here as the body of knowledge of a discipline. The psychological nature of mathematics refers to the type of cognitive processing the learner engages in as he or she constructs mathematics knowledge, and includes the following cognitive products: arbitrary associations, basic relationships, lower-level generalizations, concepts, higher-level relationships, and higher-level generalizations. Mathematics educators should be aware of the psychological nature of a learning task in order to understand the psychological processing involved in that task. This enables educators to develop effective instructional strategies and to organize curriculum in a sequence that takes into account its psychological nature. When the psychological nature of curricula is violated, children's errors occur. These errors are often logical, resulting from inappropriate instructional activities and are usually a function of traditional sequences of curriculum that ignore cognitive prerequisites.

A curriculum analyzed according to the psychological nature of topics has been described as follows (Reisman and Kauffman, 1980; Reisman, in press):

> Arbitrary associations involve associating symbols with their referents. This includes assigning words, numerals, geometric symbols, equivalence signs, and all other mathematical notation to the abstractions which they represent. Examples of basic relationships in mathematics are sequencing and matching elements of a set to elements of another set, equal in number on a one-for-one match. The ability to sequence involves attending in succession and then arranging objects or events into a sequence. Out of the ability to sequence emerges the notion of one-to-one correspondence. One-to-one correspondence is a most basic relationship to mathematics. Unless a child understands the one-to-one relationship he or she does not comprehend identity nor the relation of equality. One-to-one correspondence underlies the following mathematics ideas: many-to-one, equality, and greater than-less than. When the one-to-one relationship is generalized to more than two sets, the generalization of equivalent sets is developed. This is an example of a lower-level generalization. Next the child may state that several sets have the same number of elements. The child who can identify the cardinal number property of equivalent sets is displaying a specific concept such as "threeness". Examples of higher-order relationships are binary operations including addition, multiplication, subtraction and division, and unary operations such as squaring or cubing a number and finding the root of a number. When a child applies an arithmetic operation to numbers, he or she is generalizing at the rule or principle level. For example, putting two concepts together into some sort of relationship as in "two plus three equals five" ($2 + 3 = 5$).

Various taxonomies and hierarchies have been developed that account for the psychological nature of curriculum (Bloom et al., 1956; Avital and Shettleworth, 1968; NLSMA, 1968; Brownell, 1950; Gagne, 1965; Vygotsky, 1962). The components of these hierarchies are referred to as "types of learning", "levels of learning", "learning products", "cognitive processes", etc. The major modifications in the analysis presented here include: (a) recognition that specific concepts emerge from generalizatins (Vygotsky, 1962), and (b) a distinction is made between simple versus complex relationships and generalizations.

In order for mathematics educators to organize and present learning tasks in a learnable sequence, they should know the psychological processing involved in learning and understanding components of a curriculum. However, it is not enough to know the psychological nature of a curriculum; those who are involved in creating and presenting a curriculum must be aware that they know the nature of the thinking needed to learn the various tasks and consciously use this knowledge to facilitate students' learning.

There is a distinction between doing and knowing what you are doing. Often neither the learner nor the teacher is aware of the psychological processing that is inherent in bits of knowledge to be learned. They do not realize that certain basic relationships precede the construction of a particular concept; that some complex relationships arise from previously learned concepts; that rules are made up of relationships among concepts, etc. However, unless such awareness occurs, gaps in prerequisite components of a curriculum will continue to develop, topics inappropriate for some learners will continue to be presented, and instructional techniques and methodology that are ineffective in regard to specific topics will continue to be applied. This, I propose, is what analysis of errors should be about.

Assumption 3: Match Learner Needs and Curriculum Needs

Topics in mathematics must be analyzed to determine what a student needs to know in order to understand a particular bit of curriculum.

Prerequisite knowledge is identified through task analyses, that is, identifying what a student needs to know or do in order to learn a particular topic or perform a particular task. However, a task analysis strategy must go beyond identifying prerequisite objectives within a discipline. The psychological nature of components of a curriculum must also be considered so that both learner and teacher are aware of the cognitive demands of learning tasks. Students often are not only unaware of the cognitive processing demands of a learning task, they also may be unaware that they are not understanding something. Markman (1979) found that "third through sixth graders do not spontaneously carry out these processes that they are capable of carrying out" in tasks involved in assessing their own comprehension failure. Markman (1977) also found that children in first through third grades who were asked to evaluate instructions that contained glaring omissions seemed satisfied with the instructions. It was not until third grade that children noticed the incompleteness. In clinical interviews I have found that mathematics students of all ages (K through graduate school) are often not aware of what it is they do not understand.

When students do become aware of comprehension failure, they begin to monitor their own cognitive processes while making use of the psychological nature of the curriculum (e.g., Do I need to generalize or form a relationship? What is the relationship? I missed that before. I need to pay attention to the important stuff and not let my mind wander. I need to draw this in order to see the relationship.)

It is helpful when the teacher systematically prods the learner to monitor his or her thought processes. The following probe technique, originally used with children (Markman, 1979), is adapted here for use in teacher training:

Situation: A student teacher is guided by her college supervisor to explain the psychological processes involved in understanding the addition operation.

Probe 1. What are you doing as you decide what the psychological nature of this topic is?

Probe 2. Do you have any questions?

Probe 3. Did I tell you everything you need to know or can you tell me what information is missing?

Probe 4. Can you list the components of the developmental curriculum? At this point the student's spontaneous recall was noted. If the student omitted some of the components some general questions were asked: What are the three broad categories other than arbitrary associations? Which two have simple and complex phases?

Probe 5. Do you realize what part(s) you did not mention?

Probe 6. Does it make sense?

Probe 7. Does addition transform numbers?

Probe 8. Do concepts transform numbers? Explain your answer. How does the addition relate numbers?

Probe 9. What happens when you add numbers? What is the psychological nature of the result of adding numbers?

It is suggested that guided curricular analysis such as this is useful as an extension of task analysis procedures. Traditional task analysis focuses on content objectives and does not involve consideration of a task's cognitive nature. The strategy described in the student teacher probe subsumes the notion of task analysis but also incorporates the psychological processing necessary for task execution.

The mathematics teacher must also be aware that the psychological nature of a particular topic may change as a result of students' learning experiences with specific topics. For example, a basic addition fact such as $3 + 2 = 5$ is comprised of concepts (threeness, twoness, fiveness) and higher-level relationships (addition, equals), thus classifying it as a higher-level generalization. However, after the student has learned $3 + 2 = 5$ as a higher-level relationship between concepts, and after a period of practice occurs, the basic fact begins to be processed as automatically as an arbitrary association.

Thus, what in early learning demanded complex psychological processing eventually takes on a simpler nature in terms of cognitive demands. The trouble with initial use of flash cards as a teaching aid becomes apparent when a teacher fails to distinguish a basic fact as a higher level generalization or as a rote arbitrary association. There is a difference between teaching a complex generalization as a short-cut arbitrary association, and allowing a student to construct the complex generalization, understand its relationships and concepts, and then practice it until it becomes an automatic response. Thus, being aware of the psychological nature of curriculum serves as a structure for the organization and effective presentation of learning tasks.

Assumption 4: Look Beyond Computational Errors

As a result of analyzing common errors "trouble spots" in the elementary school mathematics curriculum become apparent. These are summarized in Reisman and Kauffman (1980, Appendix E).

In summary, children's errors are often a function of their teacher's errors, and it appears that their teacher's errors may be a function of our errors as mathematics educators. I, as a mathematics educator, find it hard to believe that I taught time to the minute last instead of first since children count by "ones" first; that I taught place value before multiplication since place value is a product; that I ever accepted the notion of reversibility of thought (there is no such thing); and that I introduced use of a number line, when values increase from right-to-left, at the same time as place value, where values increase from left-to-right. The comfort of the "usual" - - the traditional - - in mathematics instruction must be re-examined as an integral part of analyzing children's errors in mathematics.

References

Avital, Schmuel M. and Shettleworth, Sara J. Objectives for Mathematics Learning - Some Ideas for the Teacher, Ontario Institute for Studies in Education Bulletin No. 3, Toronto, 1968.

Bloom, Benjamin S., Englehart, Max D., Furst, Edward J., Hill, Walker H., and Drathwol, David R. Taxonomy of Educational Objectives Handbook I: Cognitive Domain. New York: David McKay, 1956.

Brownell, William H. and Hendrickson, Gordon How Children Learn Information, Concepts, and Generalizations. The Forth-Ninth Yearbook of the National Society for the Study of Education, Part I. Chicago: University of Chicago Press, 1950.

Gagne, Robert M. The Conditions of Learning. New York: Holt, Rinehart and Winston, 1965.

Markman, Ellen M. Realizing That You Don't Understand: A Preliminary Investigation. Child Development, 1977, 48, 986-992.

Markman, Ellen M. Realizing That You Don't Understand: Elementary School Children's Awareness of Inconsistencies. Child Development, Vol. 50.

Reisman, F. Strategies for Alleviating Mathematics Disorders. In Handbook for School Psychologists (Edited by Cecil Reynolds and Terry Gutkin). New York: John Wiley Interscience, in press.

Reisman, F.K. and Kauffman, S.H. Teaching Mathematics to Children with Special Needs. Columbus, Ohio: Charles E. Merrill, 1980.

Wilson, J.W., Cahen, L.S., and Begle, E.G. (Eds.). Description and Statistical Properties of X-Z Population Scales. NLSMA Report Nos. 4-6. Stanford, California: School Mathematics Study Group, 1968.

Vygotsky, L.S. Thought and Language. Cambridge, Massachusetts: MIT Press, 1982.

13.6 COMPARATIVE STUDY OF THE DEVELOPMENT OF MATHEMATICAL EDUCATION AS A PROFESSIONAL DISCIPLINE IN DIFFERENT COUNTRIES

COMPARATIVE STUDY OF THE DEVELOPMENT OF MATHEMATICS EDUCATION AS A PROFESSIONAL DISCIPLINE IN DIFFERENT COUNTRIES

GENERAL TREND REPORT

Gert Schubring
Institut fur Didaktik der Mathematik
Bielefeld, Federal Repuplic of Germany

During the last two decades, mathematics education has been subject to a very rapid growth. At the same time, the emphasis, the "paradigms" have shifted. A first stage of curriculum reforms was characterized by an almost exclusive orientation towards content, and towards teaching the content directly to the student, bypassing the teacher. This was the stage of the so-called "teacher-proof curricula". The insufficient reform result of this stage has revealed the systematic importance of the teacher's role. Increased demands to improve teacher education were almost never addressed to those persons who actually educate teachers. Generally, it is not seen that the professional group of teacher educators is not a constant, but a variable, the quality and conceptions of which crucially determine the result of teacher education.

This tendency to neglect the professional group of teacher educators is obvious in practically all of the publications and research on teacher education. It seems appropriate, on this fourth conference, to begin thinking about the discipline which has the theory and practice of mathematics teaching as its research field, and teacher education as its decisive field of application.

An extensive international evaluation has been organized for this panel on the basis of a questionnaire with four main dimensions: Teaching/Research/History/Professional Conditions of mathematics education. Reports have been submitted not only by the countries represented on the panel, the USA, the Federal Republic of Germany, France, Great Britain, Poland, Tunisia, but also by the following countries: Argentinia (L. Santalo), Egypt (W. Ebeid), Hungary (J. Suranyi), India (D.K. Sinha), Japan (I. Hirabayashi), Kuwait (M.W. Al-Dharir/Adel Yaseen), Mocambique (P. Gerdes), Tanzania (A. Mwambogela), Thailand (S. Chantrasomsak). There was some additional material from other countries. I shall first give an overview of some main problems and tendencies, which will be represented in more detail by the panelists from the perspective of experience in their respective countries.

The task was not only to evaluate the development of mathematics education in such countries which could be supposed to have attained a high level, but also in countries which have not yet progressed so far. Many states which have recently become independent are faced with the task of having to create an educational system of their own with very limited financial resources and a small staff of not yet fully educated teachers. They are thus facing about the same situation several European countries were confronted with in about 1800, when they were creating, for the first time, a national, public, and general educational system.

In view of the urgent problems in the countries of the Third World, it seems important to learn from this historical development, and to apply this knowledge in the development of conceptions for the organization of teacher education. Vice versa, this situation of development should be a good opportunity for the developed countries to re-examine their own prevailing conceptions.

There is a precedent for the work carried out by this panel: in 1914, the ICMI had developed a questionnaire for an international evaluation of the mathematical and practical training of the mathematics teachers. Like the entire work of the contemporary ICMI, this questionnaire was confined to secondary education. It contained several items referring to the didactics of mathematics. Due to World War I, evaluation of this international comparison could not be done until the 1932 mathematicians' congress. The results showed a feeble development of the discipline: lectures on mathematics education at universities, e.g., were given in seven countries only. Accordingly, Gino Loria, who reported to the congress, did not confer an important status on mathematics education, and considered it best to provide professional training, separated from scientific education, only during practical training in school (1).

This relationship between scientific education and professional training for teachers has led us up to a question which is crucial for constituting didactics of mathematics, or, in Anglo-American terms, mathematics education, as a discipline of its own. First, however, I must give some explanations concerning the terms of discipline. The panel's title contains the terms of profession and discipline, and this is no accident, as the two are not identical.

The concept of profession has been developed by vocational sociology, referred to the traditional professions of doctor, clergyman, lawyer. Typically, a profession is characterized by the following features; 1) Highly specialized knowledge, 2) Corporate character, 3) Certain degree of self-determination and autonomy. 4) Most important is the existence of (individual) clients and therefore the emphasis on application of knowledge.

As sociology of science states an absence of analogous groups of individual clients for scientific activity, and has considered the latter's focus to be the development rather than the application of knowledge, it has bypassed the concept of profession, and has gone on to develop the concept of discipline.

The term of discipline now means the forms of social institutionalization of processes of cognitive differentiation in science. The essential characteristics of a scientific discipline have been comprehensively listed: 1. A "scientific community"; 2. a stock of theoretical knowledge represented in textbooks, i.e. characterized by codification, acceptance by consent, and basic teachability; 3. a plurality of problematical questions at any time; 4. a "set" of research methods, and paradignmatic problem solutions; 5. a discipline-specific career pattern and institutionalized socialization processes which serve to select and educate candidates according to the prevailing paradigms (2). But, since mathematics education is an essentially applied discipline and has got clients, one should look for connections between "discipline" and "profession". It makes sense to distinguish between the two terms, as will be shown by the French example: in France there is a discipline of didactics of mathematics in the sense of a scientific community. There are, however, no clients, and insofar there is no application of knowledge, and hence no professionalization. At the same time, however, this French example shows the necessary innerconnection between discipline and profession, for the discipline will not be able to develop without clients and without application in the long run. Indeed, historical evidence shows, that it was the dualistic character of the teacher-researcher role of scientists, which made possible the institutionalization and professionalization of scientific activity (3).

It seems that there are two main dimensions for the development of the discipline of mathematics education, which are essentially specific in each nation: 1) the conception concerning the relationship between theory and practice in teacher training; 2) the conception concerning the relationship between the functions of education, and the functions of instruction aimed within the education process. While the first dimension essentially determines the emergence and growth of a discipline, the second determines its forms of institutionalizations.

The first dimension is expressed in the role assigned to professional training within teacher training. This concerns the question whether training for professional activity is considered also as requiring a scientific educatipn or whether it is merely treated as a problem of practice, and provided after graduation during traineeships in schools. Professionalization of teacher education indeed means integration of the professional, vocational part of training into higher education.

It is remarkable that the onset of the first stage of the discipline's development is about 1900 in several countries, together with the introduction, at the universities, of lectures on mathematics education for teacher training, as supplements to subject-specific lectures (in the U.S.A., Great Britain, Germany, Belgium). At that time, the first chairs for this discipline were established. The fact that mathematics education became recognized as an independent scientific field is also expressed by the first doctoral theses in mathematics education published at that time. Nevertheless, mathematics education still remained a marginal discipline, which was mainly misconceived as being a craft, a "trade".

Development continued only after a fundamental change in the conceptions concerning teacher education had taken place. In the sixties and seventies, a professionalization of teacher education was introduced in a number of countries which served to supplement the subject-specific scientific education by a scientific education referring to professional practice. The supplement was, in general, an educational science part. Due to the fact that studies, before, had frequently been entirely separated as to subject-specific science and educational science, courses on the didactics of mathematics attained a new integrative importance. The new scientific requirements also served to raise the level of course subject matter. to dissociate it from mere question of methods and practice, and to orient it towards combining theory and practice.

The concrete shape taken by the institutionalization of subject-specific didactics will always depend on the shape of the relationship between education and instruction which prevails in a nation. Traditionally, this relationship is one of separation: elementary school is assigned the task of education, of general personality formation, especially in those countries where elementary school is the school for the lower classes of society. The task of instruction, of disseminating knowledge, is given to secondary education in a one-sided manner. This is possible particularly where secondary education is reserved for the upper social strata, the specific socialization is no longer necessary because of the selection already effected. The split in the education/instruction relationship according toschool types, however, is not specific for nations having a socially selective school system. Gnedenko reports an analogous split between elementary and secondary schools in the USSR (4). In both of these one-sided forms, the didactics of mathemtaics has no important part in teacher training. This can be shown in international development: the curriculum reform first lead to a persuasive shift towards instruction, especially towards contents, i.e. there was, for the first time, an orientation towards contents in elementary school. The didactics of the subject, however, were not yet assigned an essential role during this stage of "teacher-proof curricula", as the subject matter to be taught to the students was immediately derived from science. Only during the subsequent stage, in which the crucial role of the teacher emerged, did it become obvious that the teacher will need subject-specific didactics for his professional knowledge. It is remarkable that pressure in favor of developing and promoting subject-specific didactics was not brought to bear by the discipline itself, but by the teachers or trade unions (Poland, France, Federal Republic of Germany).

On the whole, there is now a tendency in many states to develop a unified discipline, and to overcome the split according to school types in teacher education. This is especially true where the discipline's clients are not only teachers, and where there are other applications for this knowledge as well (e.g. Poland: in medicine, computer science).

Generally, it may be said that the concrete shape taken by institutionalization is in some way of secondary importance as compared to the fundamental attitude taken by a state with regard to education and the system of knowledge. An instructive example for that is Mozambique. The educational system there is of prime importance for overcoming the heritage of the colonial system, which had been particularly inclined to promote magical elements. The basis ofthe present educational system is orientiation towards scientific knowledge, and continuous raising of the level of knowledge for all. Mathematics is assigned a crucial role in overcoming prejudice, tribal dogma and mysticism. Such an impressive support of the educational system is in stark contrast to the situation in some developed countries, where teacher and university teacher posts are deleted, institutions of higher education closed down, and courses of studies abolished for reasons of a short-sighted policy intending to economize.

To sum up: It may be gained from the reports that mathematics education is increasingly becoming a university discipline, not only in the traditionally highly developed Western countries, but also in Japan. This emergence of mathematics education is evident in the increase and diversification of research, beyond doctoral theses, which are the first step to extend research. It is also evident in the establishment of a career pattern which corresponds to the academic careers in the established disciplines. This is true for junior scientists who, as scientists specifically educated for this career, are increasingly joining the ranks of the self-taught or otherwise scientifically trained senior scientists. Projects on the subject matter of mathematics education courses, such as exist in England, the German Federal Republick, and on an international level (BACOMET), show that the discipline increasingly reflects its own activity. Communications are enhanced by conferences and specialized periodicals. Textbooks are doubtlessly a bottleneck: although there are quite a number of textbooks on some subjects, there are no comprehensive ones.

The training of a prospective mathematics educator will require a long time, since generally it is not his first scientific education, but will, as an interdisciplinary education, continue a scientific education, say, in mathematics, psychology, or pedogogy. As such a special qualificatin will always require some break with the discipline of origin, it will not only make sense to establish departments of its own,which will provide a sufficiently institutional basis for interdisciplinary research and teaching activity, but the stronger reference to applications as compared to traditional disciplines can be more easily organized in this way as well.

Through a participation of mathematics educators in in-service-training of teachers would strengthen the relation of the discipline to its applications, there are political impediments in some states against raising the level of in-service-training (France, Tunisia).

States offering very little opportunities as yet of qualification in mathematical education are employing scientists who have obtained their degrees and qualifications elsewhere (Tunisia, Egypt, Kuwait). These scientists are acting there as a "critical mass" in order to promote the national development, and to train scientists for other states (Egypt).

It is remarkable that it is the countries of the Third World which demand that mathematics education be further professionalized (Egypt, Argentina, Tanzania). They request the establishment of a professional body of mathematics education, the enhancement of research, the establishment of special doctor's degrees, the establishment of special departments for mathematics education, and an international declaration of the existence of "mathematics education as a specialized and a separate discipline".

Notes

1. The questionnaire is published in L'Enseignement Mathematique, 1915, 9 - 14, the evaluation in: L'Enseignement Mathematique, 1933.

2. R. Stichweh, Differenzierung der Wissenschaft in: Zeitschrift fur Soziologie, 8 (1979)

3. see: R.S. Turner: The Prussian Universities and the Research Imperative, 1806-1948. Ph.D. thesis, Princetown, 1973.

4. see: F. Swetz (ed.), Socialist Mathematics Education. Southhampton/USA, p. 86.

COMPARATIVE STUDY OF THE DEVELOPMENT OF MATHEMATICS EDUCATION AS A PROFESSIONAL DISCIPLINE IN DIFFERENT COUNTRIES: TUNISIA

Mahdi Abdeljaouad
Ecole Normale Superieure
Tunis, Tunisia

La democratisation de l'enseignement et l'explosion demographique incessante ont entraine la creation de milliers de classes nouvelles et ont amene le Ministere de l'Education Nationale a recruter des centaines de nouveaux professeurs.

La volonte d'independance vis a vis de l'ancienne metropole a amene les autorites a remplacer les cooperants francais par des enseignants nationaux. Cette politique - juste dans son principe - s'est averee source de nombreux problemes nouveaux en ce qui concerne les enseignants de mathematiques.

Le Ministere de l'Education Nationale employait en 1978-79 1233 professeurs de mathematiques dont 213 etaient des cooperants etrangers. La gravite de la situation apparait lorsque l'on sait que les besoins dans les lycees en professeurs de mathematiques sont largement insatisfaits et que l'Universite n'arrive pas a repondre a ces besoins.

La Faculte des Sciences de Tunis et l'Ecole Normale Superieure delivrent la maitrise en Mathematiques. Bien qu'essentiellement orientee vers la formation des professeurs de Lycee, cette maitrise est consideree par les etudiants comme extremement difficile. Par ailleurs, le professeur de Lycee est mal renumere, c'est pourquoi, la plupart des diplomes de la Faculte des Sciences de Tunis preferent continuer leurs etudes vers un 3e cycle (soit de mathematiques pour acceder a un poste d'enseignant a l'Universite, soit vers des ecoles d'ingenieurs).

Les eleves orientes vers l'Ecole Normale Superieure doivent, par contrat, enseigner dans un Lycee a l'issue des quatre annees d'etudes superieures. Lors de leur sejour a l'Ecole, les eleves recoivent une bourse preferentielle et un enseignement specifique qui les motivent a leur future carriere d'enseignants.

La penurie en enseignants de mathematiques et la formation scientifique insuffisante de plus de 50% de ceux qui exercent actuellement, sont des problemes si aigus qu'ils placent au dernier rang des preoccupations des responsables la recherche pedagogique et en particulier la recherche en didactique de la matiere.

L'etat d'esprit dominant a l'Universite considere que la meilleure formation que doit recevoir le futur professeur de Lycee est celle qui lui permettra de devenir un bon chercheur en mathematiques pures. Cette conception est bien illustree par la maitrise de mathematiques pures offerte par la Faculte des Sciences de Tunis. Elle ne possede qu'un seul profil, essentiellement base sur les mathematiques theoriques et abstraites et n'est pas axee sur la formation professionelle du futur professeur.

A l'Ecole Normale Superieure (E.N.S.), l'accent est mis plus explicitement sur la formation du futur professeur.

L'enseignant de didactique est avant tout un universitaire diplome de mathematiques qui accepte - au depens de sa carriere - de consacrer son temps a des recherches en didactiques et a assurer un enseignement nouveau. L'enseignement de la didactique est de la competence des individus volontaires ayant eu une experience dans le secondaire ou ayant participe a des activites liees au secondaire. Aucun departement de didactique de mathematiques n'existe actuellement. Les didacticiens des mathematiques se retrouvent au departement des mathematiques. Deux assistants diplomes en didactiques des mathematiques de Paris exercent a l'E.N.S. mais ils doivent completer leur service hebdomadaire d'enseignement en enseignant des mathematiques pures.

En 1979, l'approbation de la didactique comme discipline scientifique et universitaire s'est traduit par la creation du Jury National de Recrutement des Assistants titulaires d'un diplome de "Didactique des disciplines".

L'enseignement des mathematiques prend un essor et une importance considerables, les mathematiques deviennent, en depit de la resistance des enseignants, un instrument de selection et d'orientation, l'enseignement de masse engendre, en particulier pour les professeurs de mathematiques, des difficultes de communication, d'apprentissage et de gestion de la classe, accentuees par les problemes crees par la langue de l'enseignement des mathematiques (en langue nationale, l'arabe au primaire, puis en langue etrangere, le francais au secondaire et au superieur), le recrutement de nombreux enseignants de mathematiques qui ne possedent pas les connaissances de base minimales necessaires a leur enseignement et les missions pedagogiques que doivent accepter les professeurs de mathematiques alors qu'ils y sont peu ou pas prepares, tous ces facteurs amenent l'administration centrale, l'Inspection et les professeurs dans leurs classes a definir des objectifs didactiques et a effectuer des choix didactiques conscients ou inconscients. Le didacticien de la discipline, par ses etudes theoriques, ses recherches appliquees au contexte national et par son ouverture sur les experiences etrangeres, joue un role important et aide l'institution scolaire a definir ses objectivs et ses choix didactiques dans l'interet de l'enseignement.

U.S.A.

Phillip S. Jones
University of Michigan
Ann Arbor, Michigan, U.S.A.

The preparation of teachers in the United States of America, like the schools themselves, varies within the state. In general, today, all elementary and secondary school teachers must have a bachelor's degree and a teacher's certificate. The latter is issued by the state's department of public instruction along with the B.A. requirements - usually at the end of the four post-secondary years of attendance at a college or university.

The professional education requirements include courses in the history and psychology of education, general and special "methods" courses, and "student teaching" - working with classes in the schools under supervision.

The special methods course in the teaching of mathematics is most likely to be taught in an education department or school in a college or university. In some cases the methods course is taught in a mathematics department. In both cases the instructors (professors) will have an advanced degree, usually a doctorate, in either mathematics or education.

Such degrees are awarded by universities on the basis of advanced courses and a research thesis. The most common theses report experimental studies of teaching methods, courses, tests, curricula, often using control groups and statistical techniques. However, theses may be status surveys, curriculum analyses, or historical studies.

The holders of these degrees are most likely to be employed as professors of education or mathematics. However, they may become supervisors, department chairman, or, rarely, teachers in school systems. They may also work as researchers, or as designers or writers of texts, tests, or school materials, usually while also employed as a teacher by a school or university. These people are a major component of the group which is implicitly working to make mathematics education a professional discipline. The other component includes many teachers and supervisors, often with M.A. degrees who plan and administer school mathematics programs and are active in professional groups, especially the National Council of Teachers of Mathematics.

There is little explicit stress on mathematics education as a separate scientific professional discipline. It is more nearly regarded as a yielding of components from the disciplines of professional education and mathematics.

There are a number of current trends in education which may produce a trend toward a separate discipline. These include such new aspects of mathematics as the use of calculators and computers, new stresses on problem solving, applications, and vocational preparation.

However, these forces may be partially off-set by the increasing non-mathematical services and competencies being required of all teachers such as "multi-cultural" training and preparation for the "main-streaming" of handicapped children. The recent emphases in the schools on "back to basics" (as opposed to "modern math") and on career preparation tends to direct interest away from the theoretical matters which often interest professional educators. Further, the growing teachers' unions are attempting to limit the supply of teachers and to stress seniority in their retention. This may make professionalism in mathematics education less important.

The final outcome of these developments may be slower progress than we would wish toward mathematics education as a profession, and may direct that progress toward more concern with didactical matters than with the psychological and philosophical constructs which some see as the core of a profession of mathematics education.

LA RELATION ENTRE DISCIPLINE ET PROFESSIONALISATION: UNE SITUATION PARADOXALE: FRANCE

Janine Rogalski
Centre d'Etude des Processus Congnitifs et du Langage
Paris, France

Comme discipline scientifique, la didactique des mathematiques est active et l'interet des enseignants pour le travail qui s'y elabore est manifeste: neanmoins l'institutionalisation de la didactique comme profession n'a toujours pas eu lieu, et l'evolution actuelle est inquietante.

1. la communaute scientifique des "didacticiens" traduit son existence par une serie de manifestations scientifiques:

- un seminaire national rassemble quatre fois par an les chercheurs en didactique pour confronter l'etat de leurs travaux theoriques et experimentaux;

- la nouvelle revue "recherches en didactique des mathematiques" a publie en 1980 ses deux premier numeros (avec le concours du Centre National de la Recherche Scientifique) (1);

- une premiere ecole d'ete, avec la participation de chercheurs etrangers, a fait travailler un tres grand nombre de chercheurs et d'enseignants sur les divers champs de recherche de la didactique;

- la participation des "didacticiens" aux debats et aux recherches internationales est importante et reguliere.

On peut dire qu'une "masse critique" a ete atteinte dans la recherche en didactique des mathematiques.

2. Le travail conduit dans les IREM, d'experimentation, de reflexion theorique, de recherche en didactique, associe des chercheurs et des enseignants de tous les niveaux de l'enseignement. L'ampleur de ce travail, le tres large echo donne par le Bulletin de l'Association des Professeurs de Mathematique, temoignent de l'interet qu'attachent les "praticiens" a la recherche en didactique.

Dans le meme temps un grand nombre d'universites ont organise des enseignements specifiques pour les futurs enseignants de mathematique; ces cours rencontrent un grand interet de la part des etudiants, qui y fournissent un travail personnel considerable.

3. Les racines historiques recentes de cette situation sont certainement a la fois le role joue par la creation et l'extension des IREM apres 1968, l'importance mise dans les annees 60 sur les contenues de l'enseignement en mathematique et l'existence en France d'une double tradition d'epistemologie des sciences (Bachelard ..) et de psychologie cognitive et d'epistemologie genetique (Piaget..). (La tradition mathematique en France a joue dans le meme sens.) Ces conditions historiques ont donne un caractere particulier a la didactique en France et fournissent d'autre part les bases essentielles pour une professionalisation reussie et pour un developpement positif des relations entre theorie et pratique en didactique.

4. Mais l'histoire du systeme educatif en France, la conjonction de la centralisation des decisions essentielles et du controle au niveau ministeriel national

et de la dispersion des modes de formation des enseignants selon leur ordre d'enseignement (primaire, premier cycle du secondaire, second cycle, enseignement technique...) n'a pas donne les conditions institutionnelles favorables.

De plus, l'evolution actuelle - telle qu'elle se manifeste par les textes sur la formation des maitres, par les decisions ministerielles budgetaires, par les mesures prises a l'egard de la recherche didactique et pedagogique, va a l'encontre de l'insertion de la didactique dans la profession.

Ainsi: les ressources financieres des IREM sont reduites avec les possibilites de participation des enseignants; plus de la moitie des enseignements de didactique ont ete supprimes par le ministre des universites; la part de formation obligatoire en mathematique pour les enseignants du primaire est reduite, aucune place n'est faite a la didactique de la discipline et les formateurs dans les ecoles normales sont professionnellement "interdits de recherche"; a l'INRP (Institut National de Recherche Pedagogique) les recherches en didactiques sont reduites drastiquement dans les nouveaux projets; les contrats de recherche diminuent; la preparation a la recherche par les doctorats de 3° cycle est supprimee partout ou elle existait...

Il est difficile de dire aujourd'hui comment se resoudra la contradiction profonde entre des choix institutionnels, financiers et d'orientation qui relevent de la politique de l'education actuelle en France, et le mouvement des enseignants de mathematiques et des chercheurs en didactique pour le developpement scientifique de la discipline et pour son insertion dans la formation et la pratique des enseignants. Ce qui est certain, c'est que les enseignants et les didacticiens francais ont la volonte commune d'une evolution qui soit positive pour l'enseignement des mathematiques et pour ceux qui ont a le vivre: les enseignants et les eleves.

References

1. Publication des editions de la Maison des Sciences de l'Homme.

2. La tradition mathematique en France a joue dans le meme sens.

WESTERN GERMANY

Gert Schubring
Institut fur Didactik der Mathematik
Bielefeld, Federal Republic of Germany

The first decisive steps for the development and institutionalization of didactics of mathematics at universities were taken by Felix Klein at the beginning of the 20th century. The premise for these developments was the emergence of the professional group of gymnasium teachers as a scholarly oriented community during the 19th century. Klein demanded that "lectures and exercises in the didactics of mathematics and natural sciences" be held at the universities for the training of secondary school teachers. It was in this spirit that the first doctoral theses were completed, that the academic degree of "Habilitation" was awarded for the first time, and that serious research in mathematics education began, as documented by the famous volumes "Abhandlunger uber den mathematischen Unterricht in Deutschland". But the didactics of mathematics remained a marginal discipline at the universities - taught by assistant lecturers as a part-time duty. The professional training of teachers was done in a second stage of practical traineeships in schools, separated from the scientific studies.

A principle change of status for didactics of mathematics, however, was associated with a radical change of the teacher education for elementary schools. After the elementary schools, which had been the schools of the poor before 1918, were promoted to be the first stage of a consecutive educational system in the 1920s, Pedagogical Academies were established as institutions of higher learning in order to train teachers for these schools. The focus of these institutions was the dimension of education: pedagogics was cultivated there as the sole scientific discipline, preparation for the instruction of subject matter was done exclusively by methods courses. The relationship between these two dimensions was reversed when the institutions were re-established as Pedagogical Colleges in Western Germany after World War II. Mainly due to teachers' demands to achieve scientificness for their professional training, didactics and the didactics of subject matter ("Fachdidaktik") succeeded to pedagogics as the scientific focus of these institutions. It was in this spirit that the didactics of mathematics emerged as a scientific discipline, a visible result is e.g. the new name "didactics of mathematics" instead of "methods of mathematics instruction".

The new conceptional basis succeeded in achieving professionalization for teacher education and academic recognition of the Pedagogical Colleges. The educational reforms also changed the character of the university: it had to integrate professional training of teachers into its courses. Didactics of mathematics was assigned a more important role at these institutions, too.

Thus, didactics of mathematics has undergone a quite remarkable process of institutionalization during the last twenty years. It would seem that this process has been somewhat ahead of the real development of the emerging discipline and its scientific community.

Whereas, traditionally, the teaching and research personnel had been self-taught, there has now emerged a career pattern which is formally identical to that of established disciplines. There is a growing number of doctoral theses and "Habilitations" in mathematics education. Two specialized journals exist, as well as a separate professional organization of mathematics educators. But there is a lack of comprehensive textbooks.

ENGLAND AND WALES

Derek Woodrow
Manchester College of Higher Education
Manchester, England

The initial training of teachers in England and Wales allows for two distinct patterns; a first degree in an academic subject followed by a one year professional training or a three or four year concurrent B.Ed. degree incorporating both academic study and professional preparation. Mathematical education as an identifiable discipline (providing a theoretical and structural framework for the training in mathematics teaching methods) emerged initially for the B.Ed. degree but is not sometimes offered as a third year option in mathematics degree courses. In common with most developed countries there is a long tradition of 'methods' courses within teacher training with a record of theses and texts available since the beginning of the century. The development of a body of knowledge and a literature has, however, been most obvious since about 1960. This recent growth in Mathematical Education in England coincides with two separate developments during this period.

Firstly the college training courses changed from two-year courses to three-year courses. this led to a greater academic input and the typical college tutor changed from being largely a 'master of method' to being a mathematician. During the decade of the 60's these teachers' certificate courses developed into full degree courses. These new professional degrees demanded a surer and firmer link between academic and professional concerns and from this arose a stimulus to develop Mathematical Education.

The second factor in the development of Mathematical Education was the changes in the mathematics taught in schools. As in most countries the 1960's saw the dramatic changes in content at secondary school level, and this was balanced in England by equally far-reaching changes in primary school mathematics. These 'modern' syllabuses disturbed the long standing (if, in England, unwritten) agreements on content, method and objectives. There is a saying in England that 'an Englishman's home is his castle' and a school head in England and Wales has almost dictatorial powers is deciding the curriculum within his school. This resulted in a very great variety of schemes, texts, classroom organizations and teaching styles. The mathematics educators were thus faced with an extremely difficult task when asked to teach 'the method' to be used or in teaching students how to use 'the' school text. They had perforce to develop general principles and guidelines rather than instruct in narrow direct techniques (see Mathematics Teacher Education Project). This is re-enforced in England by the necessity to prepare teachers above all to take professional decisions, since each teacher is responsible for their own teaching schemes and strategies within the structures of the main school syllabus. The loose structure of the primary school curriculum presented teachers with a growing awareness of the need for an assured curriculum model in mathematics to guide them. This had led to the inception of courses in Mathematical Education designed to prepare selected primary school teachers with some knowledge against which a school can design its courses (many of these courses are validated by the Mathematical Association).

The development of mathematical education has been largely the concern of three groups of professionals; University Department of Education tutors have generally combined work in mathematical education with research, College Tutors have usually combined Mathematical Education with the teaching of Mathematics, and the Local Authority Advisors have been increasingly influential in bringing the results of mathematical educational discussions to the notice of teachers. Large conurbations and easy travel have enabled a complex and important structure of inservice work to be developed which involves all three of these groups very heavily. During the last decade a small number of research centres in Mathematical Education have been established and these groups of researchers are beginning to make significant contributions. Three centres of mathematical education research is however quite tiny for such a large educational system. There is clearly a need to develop more centres at which information and material is readily available for teachers. The involvement of mathematics teachers in developing Mathematical Education is an area of real importance which has yet to be achieved. There are some promising developments in the growth of cooperative research rather than that by a single worker, this has been helped by the establishment of the 'British Psychology of Learning Mathematics Society'.

POLAND

Waclaw Zawadowski
University of Warsaw
Warsaw, Poland

Mathematics education, or in Polish "dydaktyka matematyki" is now an established part of institutionalized teacher training courses, both pre-service and in-service courses. There is however much felt painful shortage of qualified personnel in this field. There are two main roads for somebody to become an expert in mathematics education in Poland: 1. by systematic study at a university of mathematics education as a discipline together with research work, which is recommended to young people, starting their academic career, and 2. by personal teaching experience and individual systematic studying of literature and taking part in various research projects. This is usually the way of those who started as teachers of mathematics, and it is rather a hard way. There is also a third road by negative selection from other sources, including all types of failures in pure mathematics. In rare cases this is a good source of valuable personnel. Maturation of an expert in mathematics education takes a very long time and requires special abilities besides of mathematics, e.g. poor knowledge of foreign language as English or German or French will do in pure mathematics, but in mathematics education it will not. It should also be recommended that future academic teachers in mathematics education study some additional discipline, besides of mathematics, like psychology, biology, sociology, philosophy or medicine. Special motivation of the persons involved is essential as well as much perserverance. Creation of special departments within all universities with the aim of promoting mathematics education was a very important factor consolidating all persons involved in mathematics education as a profession. Still it is felt that what is being done is too little to satisfy the growing demand for quality of intellectual life of the pressure group of teachers of mathematics in Poland. In many universities mathematical education is an established discipline, but still it might be that showing interests in mathematics education is considered a highbrow activity for established mathematicians, but full time in this field should be taken as a kind of degradation. Sometimes even important achievements in mathematics education as a research field might be hard to discount on the premise of the university. It is sometimes hard to accept for mathematicians that mathematics education as a discipline changes fast, much more like, say, medicine than mathematics.

There is no independent professional organization for mathematics education research workers. They are usually members of the Polish Mathematical Society and have there their special sections. The Society has always been very active on the Polish educational scene in all respects: promoting research in mathematics education, helping establish it as a discipline of its own academic rights, promoting special publications, etc. The Society is open for teachers of mathematics, regardless of age-group of their teaching involvements or interests. In the past the Society had a very essential role contributing to the quality of often hot discussions on educational matters: general social changes in education, mathematics education for all school age groups, academic education of teachers in mathematics, academic education in mathematics of other professional groups, etc. Mathematics education as a discipline in attractive not only for community school systems but also in other places, to help improve mathematical training there, e.g. Medical Academies, Technical Universities, various groups of computer users to help them understand background mathematics etc.

The main task for the future is to enlarge the body of highly qualified research workers as well as professional experts in the field, and to communicate to all who might be concerned how knowledge and skill in mathematics education as a discipline can help them.

13.7 THE DEVELOPMENT OF MATHEMATICAL ABILITIES IN CHILDREN

SOME QUESTIONS ABOUT MATHEMATICAL ABILTIES

Jeremy Kilpatrick
University of Georgia
Athens, Georgia, U.S.A.

In the words of Jonathan Swift: "A fool will ask more questions than the wisest can answer." I have limited my questions to five, so as not to appear to great a fool. And I have not attempted definitive answers, so as not to appear too wise. In fact, I hope to validate my credentials as an academic by showing that the questions, as posed, do not have adequate answers. They are either wrongly posed or meaningless or simply unanswerable. Her are my five "straw questions":

1. What do we mean by ability?
2. What is mathematical about mathematical abilities?
3. What is the structure of mathematical abilties?
4. How can we measure mathematical abilities?
5. Why should we study mathematical abiltiies?

These questions - - poorly formulated as they are - - touch on what I consider the key issues in research on mathematical abilities.

What Do We Mean by Ability?

Haig (1975) has analyzed how the term ability has been used in the literature. Psychologists have tended to fall into two camps in their usage: (1) the positivists, who restrict ability to overt behavior, and (2) the realists, who view ability as a theoretical (unobservable) entity that is inferred under standard conditions from overt behavior. Haig makes a compelling case for the realist view, arguing that ability is a "dispositional concept" and that an explicit definition is impossible. Following Haig, we can use the following working implicit definition: "X has ability A" means "X has a property A that enables X to respond in manner M under conditions C."

The definition does not seem too informative. Another tack is to attempt to distinguish ability from related concepts. Capability and capacity seem to refer to a limit or potential, whereas ability is more present oriented, more dynamic in connotation. An ability is what you have now, it may have changed before; it may change again.

We can contrast ability with skill or habit. Krutetskii (1968) argues that skills and habits are actions within a person's activity, whereas abilities are personal traits that permit the person to perform a given activity rapidly and well. I do not agree with his distinctin, but he does seem to have caught some of the difference: skills and habits are narrower in scope and more automatic in performance than abilities. However, skill carries a connotation of of expertise that ability does not ordinarily convey.

We can contrast ability with aptitude or achievement. The American literature is full of wrangles over this terminology (see, for example, Green, 1974; Anastasi, 1980). Aptitude has historically carried the connotation of innate capacity to perform in some arena, independent of learning, but its meaning seems to be shifting to ability to learn. Achievement is more dependent on learning; it looks back at past accomplishment, whereas aptitude looks forward to future attainment. And ability sort of floats in between.

I have used ability in the singular, but one should refer to mathematical abilities rather than mathematical ability, if possible, to emphasize that proficiency in mathematics can be exhibited in a variety of ways. There are many facets of mathematics, however you define it, and no one is equally good or equally poor at learning or doing all of them. Furthermore, as Alan Bishop has pointed out, mathematical abilities somehow connote things that teachers might influence, whereas mathematical ability seems more fixed and permanent.

It is sometimes useful, I think to distinguish between an ability to do something in mathematics and an ability to learn something in mathematics. Ability to do mathematics is essentially a power concept; ability to learn mathematics is essentially a rate concept.

In sum: ability is a concept that we cannot adequately define, but that seems to be useful in trying to understand why some people can do things that others cannot and why a given person can do one thing and not another. Ability looks both backward and forward. It looks backward because what one can do now is a function of what one has done in the past. It looks forward because to say that someone has a particular ability right here, right now, is essentially to make a projection into the future. There is a probabilistic flavor to the concept, and we need to think of it as having both static and dynamic qualities.

My long response to the first question has not been an answer. The question does not have an answer. It is, however, an important question to ask.

What is Mathematical About Mathematical Abilities?

If one examines carefully the list of mathematical abilities put forth by Krutetskii (1968), one may be led to ask the above question. For example, Krutetskii lists "the ability for rapid and broad generalization of mathematical objects, relations, and operations" (p. 350), and one can ask whether this is not simply the ability to generalize that is being applied to mathematical material. Krutetskii does not think so, and I agree, but this seems to be an issue that is a matter of one's theoretical viewpoint and that cannot be settled by empirical research.

The investigation of mathematical abilities seems to depend on having a task or a set of tasks to give people to do. We cannot decide what is mathematical about mathematical abilities unless we can decide what is mathematical about a mathematical task. Wheeler (1980) points out that, although we expect almost everyone to learn to speak his or her native language, we do not have the same sort of expectation with respect to the learning of mathematics - - primarily because we have, until recently, had a rather narrow view of what mathematics is.

We need to enlarge our definition of mathematics so as to incorporate into our concept of mathematical abilities the inventive powers children have developed before anyone starts to teach them mathematics. Our attention should not be so heavily focused on the mathematics of the tasks we give to people to do. It should be focused on the mathematics of the situations in which they find themselves and of their responses to those situations.

What is the Structure of Mathematical Abilities?

This question is wrongly posed. Mathematical abilities do not have a unique structure. The work of Krutetskii, like that of the factor analytic researchers before him, shows that the structure of mathematical abilities one obtains through empirical means is nothing more than the structure of the mathematical tasks one has used in the investigation. As Wesman (1968) observed in talking about intelligence, "Such structure as we perceive is structure which we have imposed" (p. 273). It is a fiction - - although probably a useful fiction - - to ask about the structure of mathematical abilities. We should not delude ourselves about the answers we will find.

How Can We Measure Mathematical Abilities?

To measure something, we need to know where to look for it. Mathematical abilities are not objects or entities; they are attributes. They are inferred from behavior. We tend to think of them as being located somewhere inside a person, but it would be more reasonable to think of them as social constructions that appear in the behavior negotiated in a situation between someone whose abilities are being assessed and someone else - - a teacher, a tester, an oberver - - who is making the assessment. Attending to the social dimension of the measurement of mathematical abilities helps us see how narrow our assessment practices have been. Krutetskii's scolding of Western researchers for their dependence on tests as measures of abilities is richly deserved.

We need to extend the measurement of mathematical abilities in two directions. The first, and the more difficult, is to move from tasks to situations, from short bits of behavior to extended samples of behavior over time, and from obtrusive intervention to unobtrusive observation. The second direction is to analyze more thoroughly the behavior we obtain through the tasks we are using now. Information-processing models of cognition hold more promise in helping to characterize the cognitive processes used by capable students. We need to become better acquainted with this work - - overcoming our perhaps natural aversion to the repetitive tasks and response-latency measures that are so commonly used. Although we should be wary of the cognitive psychologists' assumption that a complex task can always be decomposed into a set of simle tasks, we should also recognize that this assumption is not unreasonable as a first approximation.

How can we measure mathematical abilities? There is no good answer except to say that we do not know because we have taken such a limited view of the measurement process.

Why Should We Study Mathematical Abilities?

This should have been the first question, but I put it last because it is so hard for me to knock down. David Wheeler has forced me to think about it because he has asserted, in effect, that knowledge of one's pupils' mathematical abilities is essentially useless knowledge for a teacher to have. The teacher must accept pupils as they are, whatever their so-called abilities as seen through the medium of some test or some other person. I do not agree that it is useless to know one's pupils' abilities, but I am far from understanding how to help teachers see their pupils' abilities in a clearer light.

We study abilities primarily because we are interested in improving them. Mathematics educators should become more informed about mathematical abilities because we are more hampered in our work than we realize by unexamined preconceptions about our pupils' mental life.

The last few years have witnessed a boom in programs that claim to teach people how to think. The effectiveness of these programs is less important than their effect on our beliefs about the modifiability of reasoning abilities. People used to think that the ability to discern musical pitch was fixed until programs were developed to improve that ability (Faris, 1961), and a similar shift in perceptions seems to be occurring in the realm of spatial abilities (Krutetskii, 1973; Lean, 1980). One reason for studying mathematical abilities is to change people's views as to what they are and to get people thinking about how to improve them.

But that is only one reason, and the question remains without an adequate answer. Very likely there is none.

I would be happy if this list of sensible questions and my infuriating refusal to treat them sensibly has provided a spark to ignite some thoughts about mathematical abilities. As the American humorist James Thurber once said, "It is better to know some of the questions than all of the answers."

(This paper is based upon work supported by the National Science Foundation under Grant No. SED 77-17946. Any opinions, findings, and conclusions expressed in the paper are those of the author and do not necessarily reflect the views of the National Science Foundation.)

References

Anastasi, A. Harrassing a dead horse: Review of D.R. Green (Ed.), The aptitude-achievement distinction: Proceedings of the Second CTB/McGraw-Hill Conference on issues in educational measurement. Review of Education, 1975, 1, 356-362.

Faris, R.E.L. Reflections on the ability dimensions in human society. American Sociological Review, 1961, 26, 835-843.

Green, D.R. (Ed.). The aptitude-achievement distinction. Monterey, CA: CTB/McGraw-Hill, 1974.

Haig, B.D. The logic of ability concepts. Educational Philosophy and Theory, 1975, 7(2 , 47-67.

Krutetskii, V.A. The psychology of mathematical abilities in schoolchildren. Chicago: University of Chicago Press, 1976.

Krutetskii, V.A. The problem of the formation and development of abilities. Soviet Education, 1973, 15(5-6), 127-145.

Lean, G. The training of spatial abilities: A bibliography (Report No. 8). Lae, Papua New Guinea: University of Technology, Mathematics Education Centre, March 1980.

Wesman, A.G. Intelligent testing. American Psychologist, 1968, 23, 267-274.

Wheeler, D. Curriculum reform in the seventies - - Mathematics. Journal of Curriculum Studies, 1970, 2, 144-151.

ACERCA DE LA HABILIDAD MATEMATICA EN LA INFANCIA

Horacio J. A. Rimoldi
CIIPME
Buenos Aires, Argentina

El objetivo del presente trabajo es mostrar los resultados obtenidos experimentalmente en ninos de Jardines de Infantes y Escuelas Primarias en la resolucion de problemas matematicos. Esta presentatcion incluira las siguientes secciones: 1 - Operaciones de union, complemento e interseccion; 2 - Estructuras de igualdad-desigualdad.

1 - Operaciones de Union, Complemento e interseccion

El nino en su diario vivir, por ejemplo al jugar, resuelve problemas que implican estructuras matematicas y logic as definidas de complejidad variable. La cualidad distintiva de estos problemas esta, en buena medida, en el lenguaje o modalidad en que son presentados.

Supondremos que los ninos de temprana edad son capaces de realizar operaciones matematicas de mayor complejidad que lo que es usual aceptar. Lo que de acuerdo a su edad cronologica pueden no ser capaces de realizar es la representacion simbolica de la estructura matematica encuestion y sus correspondientes operaciones.

Nos interesa averiguar la capacidad de los ninos para realizar ciertas operaciones y no la cantidad de mateiral que son capaces de retener (informacion y/o cantidad de material resuelto en un tiemp limite). En otros terminos, nos concierne averiguar la capacidad o potencialidad para comprender estructuras o para resolver operaciones usando un lenguaje que sea apropiado.

Con el objeto de establecer los "umbrales cronologicos" de las operaciones de union, complemento e interseccion se administraron a ninos de Jardnes de Infantes problemas basados en esas operaciones (1). La forma de presentacion fue planeada de manera tal que el nino debia realizar las operaciones indicadas utilizando como variable el tamano, el color y la forma de objetos concretos, cubos y esferas, de color rojo y azul y de tamano grande y pequeno. Cada operacion se estudio con relacion a cada una de esas variables. Por ejemplo, cuando se solicitaba al nino la operacion de union, teniendo en cuenta la variable tamano, el sujeto debia enhebrar un collar con cuentas grandes y pequenas pero todas ellas azules y esfericas. Es decir, color y forma eran constantes. Si la operacion solicitada era la de complemento, y la variable color, el nino debia hacer un collar manteniendo constantes el tamano y la forma, y asi similarmente para otras variables y operaciones. En el caso de complemento debia desechar de la caja que contenia los objetos, aquellos no necesarios, es decir el complemento. En el caso de interseccion todo el material se presentaba dentro de dos cajas que contenian los cubos y bolillas. El sujeto debia seleccionar de las dos cajas, aquellos que obedecian a las instrucciones dadas. Por ejemplo, en la interseccion que usamos la variable color, debian enhebrar un collar con algunas esferas grandes y de color rojo y azul, ignorando los cubos grandes y chicos de ambos colores y las esferas pequenas.

Se estudiaron ciento veinte ninos de ambos sexos entre las edades de tres a cinco anos a razon de cuarneta ninos por edad.

Para nuestros fines se considera respuesta correcta solo aquella en la que el sujeto resuelve satisfactoriamente el problema.

En la Table a se presentan los procentajes de soluciones correctas, (los asteriscos indican proporciones signficativas, al menos al 5%, de acuerdo a los resultados del test Binomial y partiendo de la hipotesis de que las probabilidades de exito y las de fracaso son iguales).

En cuanto a la operacion de union (Ver Figura 1) con la varible tamano, la misma se realiza perfectamente desde los tres anos, mientras que con la variable forma, adquiere el 100% solamente a los cuatro anos. La variable color muestra, como se indica en el grafico, un progresivo incremento entre los tres-cinco anos, alcanzando en esta ultima edad una proporcion del 100%. En resumen, la variable tamano facilita la operacion de union, y esta operacion la realizan ya a los tres anos el 90% de los sujetos.

En cuanto a la operacion de complemento (Ver Figura 2), la misma revela en todas las variables un progresivo aumento desde los tres anos, para llegar al 100% a los cinco anos de edad. Notese que con la variable forma, el incremento mas abrupto ocurre entre los tres y cuatro anos de edad, mientras que en las otras los incrementos son mas regulares. Por otro lado la variable tamano es la que da una mayor proporcion de exitos a los tres anos de edad. Como en el caso de union, la varibles tamano "facilita" la operacion que estamos estudiando. Con respecto a la operacion de interseccion (Ver Grafico No. 3), la msima alcanza aproximadamente un 70% de exitos a los cinco anos de edad con la variable forma, mientras que con la variable tamano os resultados son muy precarios tal como se muestra en la Figura 3. Es de notar que, aunque los problemas de interseccion solo se resuelven en pocos casos, la tendencia a mejorar con el aumento de la edad es similar al de las otras operaciones con la interesante observacion de verse dificultada por la variable tamano. Es un problema a dilicidar si tal efecto se debe a la variable en cuestion o a otros factores inherentes a esa edad cronologica.

Si se define el "umbral de realizacion" a un nivel de 75% de exitos, podemos concluir que la operacion de union con cualquiera de las variables se realiza ya a los tres anos, que complemento se realiza a los cuatro anos en todas las variables, excepto tamano, que alcanza el 75% a los tres anos, y que la operacion de interseccion no alcanza el umbral fijado con ninguna de las variables ni a los tres ni a los cuatro ni a los cinco anos de edad.

Estos resultados, ademas de demostrar la asociacion que existe entre las variables tamano, color y forma y las operaciones de union, complemento e interseccion, sugieren que el mateiral didactico a emplear debe ser cuidadosamente considerado de acuerdo con las edades, desde que, todo parece indicar una interaccion entre edad y variable. De alli, que no sea posible concluir, que un nino en edad preescolar, es incapz de realizar una cierta operacion, de no mediar un analisis exhaustivo en el cual se emplean muchas formas de presentacion del material, pero manteniendo la estructura formal de los problemas, invariable, (isomorfismo estructural).

2 - Estructuras de Igualdad-Desigualdad

Los resultados recien citados sugirieron la conveniencia de explorar otras edades y otros tipos de estrucuras. por diversas razones decidimos investigar aptitud matematica en ninos de escuela primaria, teniendo presente que, a diferencia con el caso anterior, aqui se sumarian no solo los efectos de la maduracion, sino aquellos que se refieren a escolaridad. Por otro lado, como las operaciones de union y complemento se realizaban satisfactoriamente a los cinco anos de edad, una razonable extrapolacion permitia suponer que a los seis o siete anos ya se lograria pasar el umbral con la operacion de interseccion. Por esta razon, decidimos explorar otras estructuras. De alli el material que pasamos a detallar: Se prepararon pruebas con contenido matematico presentadas en un lenguaje cotidiano para expresar conceptos tales como mas, menos, iqual, (2 y 3), etc. Estos problemas eran bien comprendidos por los ninos y requerian para su solucion las operaciones de suma, division, resta y multiplicacion de igualdades y desigualdades.

Por ejemplo, el problema que supone operar con igualdades va presentado de la iguiente manera: hay dos frascos de vidrio con igual cantidad de bolitas de color azul. Un nino agrega igual cantidad de bolitas rojas a ambos frascos. Puedes decirme, si?:

a) En ambos frascos hay diferente cantidad de bolitas azules.
b) En ambos frascos hay diferente cantidad de bolitas rojas.
c) En ambos frascos hay igual cantidad de bolitas azules e igual cantidad de bolitas rojas.

El problema que solicitaba el nino que operara con desigualdades era enunciado como sigue: En una canasta tengo mas manzanas que peras. En un cajon tengo tambien mas manzanas que peras. Vuelco el contenido del cajon en la canasta. Dime, que tengo ahor en la canasta?

a) Menos manzanas que peras
b) Mas manzanas que peras
c) Igual cantidad de manzanas que de peras
d) No se puede saber.

Estas pruebas fueron adminstradas a ciento veinte sujetos de 3°, 5° y 7° grado (2). A titulo informativo, diremos que en este conjunto de problemas habia algunos que requerian que el nino operara con la nocion de combinatoria, simetria e infinito. Las caracteristicas de los nueve problemas van dadas en la Table 2, en donde, ademas se indica la significacion de las diferencias de proporciones para los diferentes problemas y entre los diferentes grados.

Notese que en los problemas de desigualdad (1-4-5) ninguna proporcion es significativemente diferente entre 3°, y 5° grado, no siendo asi entre 3° y 7° y entre 5° y 7°.

En la Figura 4 se muestran las proporciones de respuestas correctas en todos los problemas de igualdad y de desigualdad para 3°, 5° y 7° grado. Notese que los valores para los problemas de desigualdad son mas elevados que para los de igualdad. A titulo informativo, se nota tambien que en el problema que requeria la nocion de infinito (no. 9), las diferencias son significativas entre 3° y 7° y entre 5° y 7° grado, tal como se ve en la table 2, no asi entre 3° y 5° grado. En el problema 2 (resta de una desigualdad y una igualdad), las diferencias son significativas entre 3° y 7°, y entre 5° y 7°, no asi en el problema 3 (suma de una desigualdad con una igualdad). En el problema 6 (division de una igualdad), las diferencias son significativas entre 3° y 5° y 3° y 7° grado. En el problema 7 (combinatoria), no existen diferencias significativas. Y en el problema 8 (simetria), las diferencias son significativas entre 3° y 7° grado.

De estos resultados se podria concluir, que alrededor del 5° grado, ocurre un cambio apreciable en la habilidad de los ninos para manejar los conceptos matematicos involucrados en las pruebas que analizamos.

En un analisis factorial de una bateria de treinta y tres pruebas se encontraron los factores que se designan en la primera columna de la Tabla 3, a saber, razonamiento numerico, perceptivo, computo, cierre, clasificacion verbal, razonamiento general, razonamiento verbal (4). En la segunda columna de esta misma table se indican los tests que mejor representan a esos factores. Las puntuaciones obtenidas en los mismos se correlacionaron con la puntuacion obtenida en el total de problemas de suma y resta de desigualdades (3 en total), recientemente descriptos. Estas correlaciones aparecenen la ultima columna de la Tabla 3 (4) (5). Es importante hacer notar, que la muestra de sujetos a los que se laes administro la bateria de treinta y tres pruebas y que fue posteriormente factorizada, es diferente de aquella a la que se administraron las pruebas de igualdad-desigualdad (3).

Los valores de correlacion mas altos se obtuvieron con los factores de razonamiento numerico y razonamiento general. Es clara la ausencia de significacion en la correlacion con el factor perceptivo. Esto, no permite sugerir que estas pruebas de desigualdad, comparten parte de su variancia con los factores de razonamiento general y numerico que se suponen son componentes importantes para definir habilidad matematica.

Con el objeto de explorar la evolucion de las operaciones de suma y resta de igualdades y desigualdades y su relacion con el lenguaje empleado, se preparo una prueba que consta de 16 articulos. Para resolverlos, el sujeto debe operar con igualdades y desigualdades presentadas en los lenguajes concreto, numerico y abstracto. El primero va ejemplificado en la forma ya descripta. El numerico requiere operar con cantidades, y el abstracto con simbolos desprovistos de significado.

La influencia que tienen estos tres lenguajes en la solucionde estos problemas va demostrad graficamente en las Figuras 5, 6, 7, y 8. En las dos primeras se idnican la proporcion de sujetos desde 2° hasta 7° grado que resuelven en forma satisfactoria problemas de adicion y sustraccion respectivamente. Las dos ultimas figuras se refieren a problemas de idualdad y de desigualdad. Mientras que, en estas dos ultimas figuras no se tiene en cuenta si las operaciones son de suma o de resta, en las dos primeras no interesa si la operacion de suma o de resa se efectua manajando la elacion de igualdad o la de desigualdad. Asi pues, la Figura 5 resulta de las puntuaciones de todos los articulos que usan la operacion de suma, la numero 6, de la operacion de resta, la numero 7, del manejo de igualdades, sea la operacion de resta o de suma, y la numero 8 corresponde a la relacion de desigualdad.

En todos los casos y para todos los lenguajes, se nota un incremento acentuado entre 2° y 3° grado, mostrando despues, ya sea un aumento pequeno o estabilizandose. En general las curvas correspondientes al lenguaje abstracto se ubican por debajo de las otras, mientras que el lenguaje numericoo es el que en general, excepto en la Figura 5, sumas, muestra la mayor proporcion de resultados correctos. Un analisis de variacion de tres factores con medidas repetidas en dos de ellas (7) (relaciones y lenguajes) dio para todos los sujetos, el resultado indicado en la Tabla 4. Para los fines de este analisis se clasificaron las puntuaciones en terminos de grados que equivale a edad, en terminos de lenguajes y de relaciones. Notese que en etos dos ultimos factores el mismo sujeto era evaluado repetidas veces en cada lenguaje, a traves de los problemas y en terminos de relaciones igualdad-desigualdad, presentada cada una de ellas en los tres lenguajes: concreto, numerico y abstracto.

Como era de esperar, el factor grado, muestra diferencias significativas, corroborando los resultados de la inspeccion ocular de las figuras 5 a 8, antes descriptos. En cuanto a variabilidad dentro de los sujetos (within subjects), esta comprende la variancia debida al efecto igualdad-desigualdad y la variabilidad correspondiente al efecto lenguajes. Ambos son altamente significativos, asi como la interaccion entre lenguajes y tipo de relacion o estructura del problema. Estos resultados indican que: hay diferencias significativas resultantes de la variable grado o edad, de la varible estructura o igualdad-desigualdad y de la varible lenguaje (concreto, numerico y abstracto). Es altameante sugestivo que las interacciones grados - estructuras y grados - lenguajes <u>no</u> sean significativas, no asi la interaccin lenguajes estructuras cuya significacion alcanza un nivel de mas del 1%. Estos resultados parecen indicar que el cambio en puntuaciones entre grado y grado en los problemas de igualdad y entre los de desigualdad es practicamente identico, o sea, que el cambio en nivel educacional no favorece en mas o en menos la solucion de problemas de igualdad or desigualdad. Un fenomeno similar ocurrre con los lengaujes sugiriendo que no hay asociacion entre nivel educacioneal y facilidad en el manejo de uno u otro de los lenguajes explorados. Sin embargo, la interaccion significativa entre lenguajes y estructuras indica que es <u>importante</u> la manera en que un problema se presenta. En nuestro caso todos los problemas de igualdad y todos los de desigualdad son estructuralmente isomorficos y en consecuencia el resultado obtenido verifica nuestra opinion de que ciertos lenguajes facilitan o dificultan la resolucion de estos problemas y que por ello antes de concluir que un sujeto no es capaz de operar con un sistema relacional dada, es menester explorar muchas formas de presentacion del material. Esto tiene evidentes implicaciones educacionales que no entraremos a analizar.

Dados estos resultados se realizaron seis analisis de variacion de clasificacion doble (8), uno por grado, considerando las variables lenguaje y estructura de los problemas (igualdad-desigualdad). Los resultados se presentan en forma sumaria en la Tabla 5. La unica variable significativa, al menos al 5% en todos los grados, es la que corresponde a lenguaje. Es curioso que la variacion debida a estructuras no sea significativa en 2° grado que se diferencian, al menos con los lenguajes aqui empleados, la aptitud para procesar igualdades de la requerida para procesar desigualdades.

En cuanto a la interaccion lenguaje - estructura, la misma es significativa en todos los grados, excepto en 2° y 6°. Estos resultados confirman nuevamente la importancia de 1) distinguir entre la estructura y el lengauje de un problema y 2) considerar su interaccion. Se hace asi casi perentoria la necesidad de proceder probando multiples presentaciones antes de concluir sobre la ausencia de la habilidad necesaria para procesar una determinada estructura.

En el momento actual se esta explorando mas a fond este problema estudiando el proceso de pensamiento aplicando la tecnica descripta por Rimoldi en 1955 (6).

En el diseno se incluyen cuatro variables fundamentales, a saber: relacion de igualdad y relacion de desigualdad, lenguaje abstracto y lenguaje concreto, contenido cunatitativo y contenido cualitativo y tipo de informacion suministrada (datos de tipo general o datos especificos). Se han construido asi, treinta y dos problemas isomorficos, empleando las estructuras, cada una de ellas realizada en diez y seis problemas. Una estructura corresponde a una doble dicotomia y la otra a una dicotomia mas una cuadritomia. Los resultados de parte de este estudionestan hoy analizandose. Surge de los mismos que la casi totalidad de los alumnos de 3° a 6° grado da la solucion correcta. Sin embargo las tacticas empleadas para llegar a la solucion, difieren notablemente. Esto queda evidenciado por las preguntas que los sujetos hacen para llegar a la solucion. Algunas tacticas llevan directamente a la respuesta final, sin requierir informacion redundante o irrelevante y procediendo logicamente. Estas tacticas reducen toda la incertidumbre de la manera mas economic y eficiente. Pero existen otras tacticas que, a pesar de reducir la incertidumbre en forma total, lo hacen por medio de preguntas redundantes e innecesarieas, acompanadas de preguntas irrelevantes, lo que se traduce en una tactica frondosa, inceirta y titubeante. Es decir, que hay maneras y maneras de reducir la incertidumbre y que hay diferencias marcadas en el estilo seguido para resolver problemas y que todo esto puede o no acompanarse con una solucion correcta. Ademas si el sujeto no ha reducido a cero su incertidumbre, aun asi, da una solucion correcta, la inferencia logica es que ha adivinado.

No interesa pues averiguar, mas alla de las respuestas, estilos cognoscitivos a traves de las tacticas que los sujetos emplean y como estas tacticas cambian de acuerdo con los lenguajes y las estructuras de los problemas. Es obvio que la educacion es en buena medida un proceso en el que alternan preguntas y respuestas y que es importante saber cuales son las preguntas y como deben ser las respuestas para adecuarse al estilo cognoscitivo de cada individuo. El conocimiento de todo aquello que favorezca la realizacion de procesos mas discretos y economicos es tambien, o deberia ser, un objetivo educacional.

Conclusiones

De los resultados obtenidos experimentalmente podemos concluir que:

1. Ya en edades tempranas (preescolares) los ninos son capaces de resolver problemas con estructuras matemticas definidas.

2. Con la iniciacion del aprendizaje formal y de la maduracion se facilita el uso de ciertos lenguajes.

3. El nino esta potencialmente capacitado para resolver algunos problemas matematicos, y efectivamente lo realiza, cuando estos son presentados en un lenguaje accesible a su edad.

4. La ejercitacion y fijacion de los conceptos impartidos debe hacerse de acuerdo al nivel de lenguaje alcanzado.

5. La determinacion de los "umbrales cronologicos" para la ensenanza de las diferentes estructuras matematicas se convierte asi en una tarea de importancia central para la pedagogia.

6. Nuestros resultados paracen indicar que una buena resolucion de problemas matematicos esta positivamente correlacionado con los factores de razonamiento general y numerico.

Bibliografia

1. Figueroa, Nora B.L. De: Evolucion de las Operaciones de Union, Complemento e Interseccin en ninos de tres a cinco anos de edad. Revista Interamericana de Psicologia, 1976, 10(1), pp. 89-97.

2. Figueroa, Nora B.L. de, Rimoldi, Horacio J.A.: Nota Preliminar sobre el Analisis de Estructuras Logico-matematicas en ninos de Escuela Primaria. Buenos Aires, CIIPME, 1979. Puclicacion No. 70.

3. Figueroa, Nora B.L. de, Rimoldi, Horacio J.A.: Sobre Algunos Aspectos de la Habilidad matematica. Buenos Aires, CIIPMF. Publicacion No. 71 (en prensa).

4. Figueroa, Nora B.L. de: Desarrollo intelectual y rendimiento escolar. Acta Psiquiatrica y Psicologia de America Latina, 1978, 24(1), pp. 41-45.

5. Rimoldi, Horacio J.A., Figueroa, Nora B.L. de: Analisis de algunos factores relacionados con Inteligencia en Escuela Primaria. Revista de Psicologia General y Aplicada, 1978, 33(153), pp. 485-595.

6. Rimoldi, Horacio J.A.: A Technique for the Study of Problem Solving. Educational Psychological Measurement, 1955, 15, pp. 450-461.

7. Winer, B.J.: Statistical Principles in Experimental Design. McGraw Hill, 1971.

8. MacNemar, Q.: Psychological Statistics. John Wiley & Sons, Inc., New York, 1969.

DISCUSSION OF A LEARNING ATTAINMENT MODEL

Raymond Sumner
National Foundation for Educational Research
in England and Wales
Slough, England

My basic premise is that a variety of theories about intellectual activity share several concepts. These are relevant to learning only in the context of a curriculum; i.e., the field of knowledge perceived by the learner. These perceptions are conveyed by various modes of communication when the learner receives and produces information. The interaction between the external curriculum and the individual essentially involves his transforming the knowledge from the form in which it is perceived into fresh structures when the knowledge is 'understood'. The input of information not only requires transformation; it has to be expressed as a product. The transition from input to output may involve a change of communication mode.

The framework for a composite mode of attainment is a three-dimensional representation with axes corresponding to the learner's mental attributes (abilities), the forms in which information can be represented (communication modes), and the types of operation that may be developed during learning activity (process ends).

The Abilities are verbal, non-verbal, spatial, mechanical, assembly and motor abilities. The Communication Modes are action, sound, words, numbers, symbols, pictures and objects. Process ends the third dimension encompass cognition/recall, comprehension, application, analysis, selection, synthesis and evaluation.

The importance of the structure of knowledge fields is becoming increasingly recognised, as evidenced by the work of Collis and Biggs (1979) when advancing their SOLO taxonomy as a tool for curriculum analysis, setting learning objectives, and evaluating the quality of a students' attainment. The model has some affinity to that propounded by Guilford (1966) as the Structure of the Intellect Model, but it draws on numerous other sources (e.g., Carroll, 1974; Bruner, 1966; Dressel and Mayhew, 1954; Lunzer, 1968; Ausubel and Robinson, 1969; Pask, 1976; and Skemp, 1976). Guilford embodied three main components. These were Operations (divergent production, convergent production, evaluation, memory, cognition); Contents (figural, symbolic, semantic, behavioural); and Products (units, classes, relations, systems, transformations, implications). Clearly, Contents refers to the vehicle for communications, whilst Product refers to the structural form of the knowledge dealt with by the learner. Meeker (1969) interprets Operations as "processes" when presenting a diagram where it is noted that "1. Learning takes place through cognition and evaluation; 2. Storage of learned material in the memory; 3. Production of learned material may be . . . unchanged = convergent: re-oriented or invented = divergent."

I have preferred to draw on the taxonomy devised to describe activity in the cognitive domain by Bloom et al. (1956). It is used in many settings for curriculum design as well as for laying out test and examination specifications. The rationale takes the line that the individual must be capable of recognising or recalling representations of information as a prerequisite to comprehending its meaning before using it for an application or employing it in the analysis of a problem; the knowledge could then be incorporated in an original production which may, in turn, be evaluated by reference to internal criteria or recognised external standards. I have inserted an extra stage called 'Selection' to signal that critical choices may be entailed when choosing elements for a creative production. There is some conflict in meaning with Guilford's and Bloom's usage; I tend to see analysis as both convergent (defining separate features by applying criteria) and divergent (hunting for further possibilities, testing hypothesized criteria). Conversely, synthesis is complementary in that it can be divergent (trying out a variety of design combinations) and then convergent (settling on one particular solution).

The Abilities appear in my model for several reasons. Firstly, they were established by innumerable theoretical and empirical studies from the time of Spearman and his contemporaries. Secondly, investigations have shown that scholastic attainment is associated with these abilities. Thirdly, there are implications for learning aptitude from which educators might profit either in dealing with individuals or when preparing situations which, hopefully, will cause appropriate learning to occur.

Whereas Guilford's model mentions only products, I see the learner's problems as primarily those of coping with inputs. Hence the organisation of inputs by the teacher (textbook author, etc.) assumes extraordinary importance. The naive learner must perceive a succession of isolated bits of information, sometimes communicated in an unfamiliar mode. Any subsequent differentiation, unless it is arbitrary, must organise the units in the information stream into fresh entities based on some relations principle, independent of the temporal sequence detected in the information as it is received. A further problem for the learner is that representations of information are more often than not themselves abstractions. (E.g., two children share a cake; one has a piece twice the size of the other. They eat all the cake. What fraction of the cake was the smallest piece?) Thus, translation from abstract concepts to concrete representations is sometimes required.

Coping unselectively with fresh information tends to confound the learning activity. Relating data to current objectives (i.e., the process ends) is, therefore, crucial in assisting the learner to choose, devise, or adapt appropriate strategies.

When process ends are well-defined and congruent with the learner's experience, relevant information can be chosen selectively provided it is most overwhelmed by superfluous inputs of 'noise'. The restructuring of information poses the learner with questions as to how to devise larger systems and increase generalisability. Memory services relevant higher-level abilities (e.g., handling complex similies, contrasts, images, associations, deductions, inferences, probabilities, rule devising, rule testing, analogies, series, juxtapositions, movements, and combinations of operations) as well as basic recall and storage functions. The skilled person knows how to develop strategies as well as when to employ those he has added to his repertoire (Polya, 1945).

The various abilities, then, may be thought of as characteristic types of information handling systems. These interact with information perceived serially by the learner when attempting to satisfy the immediate goals implicit in the process ends. Two major production phases are postulated, (1) Retrieval and Assimilation, and (2) Re-structuring and Evaluation. These are described as loose hierarchies in the lists below:

1. Information retrieval and assimilation (Initial transformation)

INPUT STAGE

↓

i. scanning of data presentation
ii. recognition of symbols
iii. selection of specified data
iv. identification of pattern
v. rearrangement of order
vi. relating associated presentations
vii. differentiation of overlapping structures
viii. completion of incomplete patterns
ix. focusing on detail
x. extracting major features
xi. merging related systems
xii. description of the elements of a concept
xiii. the inference of the nature of a concept
xiv. modification of a concept given fresh data
xv. forming new concepts from unfamiliar data

OUTPUT STAGE

2. Restructuring and Evaluation (Re-transformation)

DEPENDENCE

xvi. finding associations between concepts, theories, principles
xvii. extracting main theme and subsidiary theories according to given rules
xviii. devising systems to carry data in sequential form
xix. suggesting rules for appraising information relative to given criteria
xx. proposing attributes inherent within a presentation of data
xxi. attaching incoming data systematically to the components of an existing system
xxii. elaboration of elements, basic concepts, outline systems, etc.
xxiii. devising simulations intended to parallel a given structure or operation
xxiv. transforming heterogeneous sets into homgenous groups
xxv. translating from one code to another
xxvi. locating areas of doubt
xxvii. changing unit bases
xxviii. defining a problem, boundary, intention, goal
xxvix. identifying similar problems, boundaries, intention, solutions
xxx. testing consistency between systems, structures, concepts, goals, etc.
xxxi. devising consensus criteria
xxxii. assessing operational effectiveness
xxxiii. application of compound/serialised tasks
xxxiv. forseeing short- and long-term goals
xxxv. personal method for assessing progress - group vs. individual standard
xxxvi. generation of original structures

SELF-REGULATION

Confirmation that learning is 'on the right lines' is a major factor in continuing with learning activity. Identification of the objectives to be attained and foresight of future goals assists self-regulation on two counts: through (a) awareness of intellectual potential and (b) understanding of the tangible features of the knowledge structure. Personal standards and group norms may be used to mark off progress towards defined goals.

Effective learning will leave the pupil capable of regulating his own learning. In this sense, he will have attained more than knowledge of a discipline. Heuristic methods may have little to commend them, however, if the learner is not capable of meeting the intellectual demands of restructuring inputs, by applying relational operations to the information located in various source.

Kent (1979) quotes Brooks as suggesting how teachers' and pupils' pathways may be modeled. He remarks that "Experienced teachers know that children who have difficulty are often unable or unwilling to share their perceptual space." The obverse case can apply, though, when a teacher is unable to conceive the nature of the pupil's existing knowledge. In such an event, the teacher has neglected to enquire into the elements and relationships which comprise the knowledge field as developed by a naive learner. Equipped with even these insights there is a chance that explanations (examples, parallels, illustrations, logic diagrams, rules, algorithms) might help the learner. In this context, the type of individual assessment procedures described by Foxman et al. (1979) offer more prospects of success than so-called diagnostic tests which merely catalog facets of attainment.

The model has implications for curriculum designers, programme creators, textbook writers, and apparatus constructors. They should focus more on making explicit the relationships which structure knowledge and pay rather more attention to the degree of abstraction introduced when knowledge is represented by various forms of communication. Unlike teachers, they cannot know their learners intimately, so should consider (i) the individuals whose abilities are biased away from verbal contexts towards the spatial, or (ii) supplementing symbolic representations at key points with diagrams, verbal statements, or actions with objects.

Those skeptical of this model might try using it to analyse the introduction of a fresh topic first as (i) a knowledge field organised structurally through the relation of elements (thus identifying separately the elements, relationships, and transformations entailed); (ii) the outputs required to demonstrate properly conceptualised learning; (iii) the process ends entering into attainment of intermediate and final goals; and (iv) the kinds of ability appropriate to the topic. Finally, the chain of events activated during information retrieval and assimilation could be used as a guide to how pupils might proceed from a starting point of no understanding to one of operational competence and then relational understanding. An alternative would be to design and carry out validity studies, using multi-trait, multi-method designs (Campbell and Fiske, 1959).

References

1. Ausubel D.P. and Robinson F.G. (1969) School Learning: An Introduction of Educational Psychology, Holt.

2. Bloom B.S. (Ed.), Englehart M.D., Furst, E.J., Hill W.H. and Kratwohl D.R. (1956) Taxonomy of Educational Objectives: Handbook I: Cognitive Domain.

3. Bruner, J.S., Goodnow J.J. and Austin G.A. (1956) A Study of Thinking, J. Wiley and Sons, New York.

4. Campbell D.T. and Fiske D.W. (1959) 'Convergent and discriminant validation by the multitrait-multimethod matrix' Psychological Bulletin 56, 81-105.

5. Carroll J.B. (1974) Psychometric Tests as Cognitive Tasks: A New Structure of the Intellect Model. Research Bulletin 74-16, E.T.S. Princeton, New Jersey.

6. Collis K.F. and Biggs L.B. (1979) Classroom Examples of Cognitive Development Phenomena: The Solo Taxonomy, University of Tasmania.

7. Dressel P.L. and Mayhew L.G. (1954) General Education: Explorations in Evaluation, American Council on Education, Washington.

8. Kent D (1979) 'More about the processes through which Mathematics is lost', Educational Research, 22.1.

9. Lunzer A.E. (1968) 'Formal Reasoning' in Development of Human Learning (Ed. Lunzer A.E. and Morris J.G.), Stapless Press, London.

10. Pask G (1976) 'Conversational Techniques in the Study and Practice of Education', British Journal of Educational Psychology, 46, 12-25.

11. Polya G (1945) How to Solve It, Princeton University Press, Princeton, N.J.

12. Skemp R (1976) 'Relational Understanding and Instrumental Understanding' Mathematics Teaching, No. 77, December issue.

13. Sumner R and Warburton F.W. (1972) 'Achievement in Secondary School: Attitudes, Personallity and School Success', NFER, Slough.

THE STRUCTURE AND DEVELOPMENT OF MATHEMATICAL ABILITIES IN CHILDREN

Ruth Rees
Brunel University
Uxbridge, England

The studies described have been carried out in the Department of Education at Brunel University during this decade under the overall supervision of Professor W.D. Furneaux. They originated in 1970 with the specific concern in the Faculty of Education sector over student performance in basic mathematics. Faculty of Education staff assumed that the concepts and skills involved in 'workshop calculations' had been acquired by their students prior to entry, i.e., at school. The studies thus subsequently spread to secondary schools and later to teachers in training. University engineering undergraduates were also included. This series of studies therefore became a pioneering investigation which attempted in a modest way to quantify the nature of the problem at various educational levels.

In 1977 the Mathematics Education Group was set up at Brunel to intensify the work both in the longer-term psychological studies and in the shorter-term feedback to teachers, and to coordinate the research work with the teacher training activities. The initial studies and the programme of work which the group is now carrying out appear to be of particular relevance in view of the increasing national concern of recent years over the apparent lack of student's computational abilities. The national agitation is part of an international concern over computational standards. This was made explicit at the Third International Congress on Mathematical Education at Karlsruhe in 1976 and again in the proposed programme for the Fourth Congress at University of California in 1980.

Brunel Studies: Philosophy and Methodology

In these studies, essentially the following questions have been asked:

1. Do school pupils and college students experience particular difficulties in learning of mathematics?

2. If so, what is the nature and extent of these difficulties?

3. What are the reasons for these difficulties?

4. What can be done to resolve the difficulties?

It was felt that an attack on these questions needs a combination of approaches. The traditional paper-and-pencil method enables analyses to be carried out on a considerable quantity of test data. The more currently popular method of case studies, on the other hand, gives insight into the perception and thought processes of the individual student. The writer thinks there has been unfortunate and unnecessary controversy over the relative merits of these two approaches. It was decided for the purposes of these studies to use these approaches in a complementary manner.

In the U.K., the Assessment of Performance Unit's first report (1) of mathematics at 11+ suggests that most pupils can do mathematics involving the basic ideas and skills but there is sharp decline in performance as pupil's understanding of mathematical concepts is probed more deeply. Her Majesty's Inspectors of Schools' report (2) states, as one of many observations, that teachers show real concern about the lack of basic skills in arithmetic but efforts to remedy this take the form of more and more sets of questions on fractions and decimals without identifying the basic number deficiencies that are at the heart of the problem (writer's underlining).

The overall aim of the studies is twofold. The first part is to identify the nature of the deficiencies, whilst the second is to move towards the development of a theory for mathematics learning and teaching.

R.R. Skemp in Intelligence, Learning and Action (3) highlights the need for a unified body of knowledge: "Every new teacher has to learn mostly from his own mistakes". What an inappropriate state of affairs for teaching and therefore learning! The late Ed Begle of the U.S.A. in Critical Variables in Mathematics Education (4) also laments the lack of a theory despite the volume of research carried out.

Research Methodology

The philosophy underlying the research method has developed with the studies undertaken. Essentially what has evolved is a complementary combination of psychometric techniques and a new form of the case-study method, i.e., use of the language laboratory. This approach enables simultaneous monitoring of the thinking of several students. Shortly after this method was applied, Krutetskii's detailed longitudinal study (9) was published and it was of great interest to the writer to compare both studies with regard to the insights obtained for mathematically able pupils.

The psychometric and case study approaches of our work were combined as outlined in Figure 1.

Complementary Research Methods

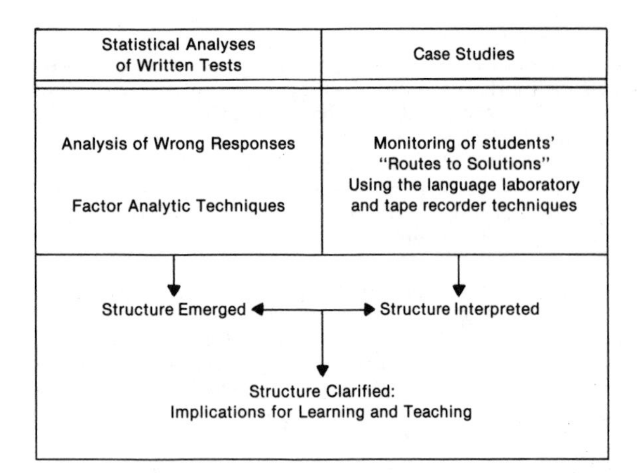

Figure I

The Components of the Psychometric Approach

The Instruments

1. Mathematics Variables. Especially designed sets ('tests') of mathematical tasks in a multiple choice format. These tasks were essentially computational in nature.

2. Psychological Variables.

   (i) The Thurstone Primary Mental Abilities. This is an established battery of verbal, number, reasoning and spatial abilities for the age range covered in the samples.

   (ii) The Nufferno Model of Intelligence. This model has the components of speed (stressed and unstressed), and accuracy (stressed and unstressed) and continuance levels.

3. Eysenk's Personality Variables of extraversion, introversion and psychoticism.

The Sample

1. Continuing Education Students: Craft, technicial and university undergraduates in their first year of study; total sample size, 1800 (approximately).

2. School Pupils: Secondary fourth, fifth, and sixth-form pupils; total sample size, 1800 (approximately), primary-early secondary.
3. Teacher Education Students: Primary School and Secondary School Trainees; total sample size size, 600 (approximately).

The Data Analyses

1. Performance levels on all instruments of various student groups, e.g., mean scores, standard deviations, etc.

2. Item analysis of mathematical instruments for various student groups. These included percentages of students selecting the wrong responses in the multiple-choice format.

3. Factor-analytic techniques applied to all instruments. The original concept in the applicatin of this technique was to insert each mathematical task as a variable in the analysis. This was a point of departure from all previous studies which had entered a total test score as the variable (compare Furneaux and Rees, 5).

Results
Structure and Development

A two-factor structure emerged from the factor analyses. The tasks defining one factor with the measures used of general ability 'g'. The other tasks define a factor, the 'influence'factor, which is relatively independent of 'g', but the strength of the relationship appears to be age dependent. Table I gives values of the ratio of variance accounted for by the 'g' factor to that account for by the 'inference' factor.

	g : I
Further Education Students (17-19 years)	2:3
Secondary Pupils (14-15 years)	2:1
Junior School Pupils (10-11 years)	5:2

Table I

It must be pointed out that the values of this ratio will depend to a certain extent on the composition of the test.

Rescoring the mathematics tests in terms of 'g' and 'inference' scores respectively for secondary school pupils age 15 years gave three categories:

1. Pupils with high 'g' and high 'inference' scores.

2. Pupils with low 'g' and low 'inference' scores.

3. Pupils with high 'g' and low 'inference' scores.

This category would include those pupils whom teachers will recognise as being high achievers in most subjects except mathematics.

There were no pupils with low 'g' scores and high 'inference'. The 'inference' factor may therefore be connected with specific mathematical ability: whether or not the processes involved in this ability are peculiar to mathematics or are more general in nature has yet to be researched.

The characteristics of the two sets of tasks, taking account also of the language laboratory studies, are summarised in Table 2.

General ability mathematical tasks	'Inferential' mathematical tasks
1. Students perform reasonably well provided they **recognize** the problem.	1. Performance is relatively worse for all kinds of students. Solutions have to be **deduced.**
2. Teachers' estimates of students' performance are reasonably accurate.	2. (i) Teachers often do not appreciate the nature of the difficulties and underestimate these difficulties. (i) A learning-teaching cycle appears to be generated by these problems.
3. Students appear to solve algorithmically.	3. Students appear to need more mathematical 'inference' for solution.

Table 2

The second characteristic in both columns refers to the teaching effect: this has been discussed elsewhere (5).

Examples of the two kinds of tasks for 'average' secondary pupils are given in Table 3.

Topic	M_g	M_I
1. $\otimes$ of integers	1. 41×43 is	1. 41×40 is .....
2. $\otimes$ of numbers less than 1	2. 0.4×0.4 is	2. 0.3×0.3 is .....
3. Fractions	3. $\frac{9}{16} + \frac{5}{64}$ is	3. $5\frac{7}{16} - 3\frac{5}{8}$ is .....
4. Equivalent fractions/ratio/ equation/ reciprocal	4. If $\frac{1}{x} = \frac{2}{4}$, x is	4. If $\frac{1}{x} = \frac{3}{4}$, x is .....

Table 3

This table illustrates the subtlety of diagnosing the 'blockages': within the same topic involving apparently similar processes, some tasks may be far more inferential in nature than others.

Implications

To a certain extent there exists a dilemma for teachers. Should we teach to bring about inferential thinking? Different ways of presentation are needed to generate a concept, but for the average and less able student this may result in confusion (8). Much more could be said (e.g., comparison with Skemp's relational and instrumental understanding, Piaget's formal operational stage, our sex-related results, etc.) but time is a constraint. Should our aim as mathematics educators be to teach so that, however demanding it may be for both teachers and learners, M_I ----- M_g?

References

1. Mathematical Development, Assessment of Performance Unit, Primary Survey Report No. 1, Department of Education & Science, HMSO.

2. Aspects of Secondary Education in England. A survey of Her Majesty's Inspectors of Schools, Department of Education & Science, HMSO.

3. Intelligence, Learning, and Action. Richard R. Skemp, 1979, Wiley & Sons.

4. Critical Variables in Mathematics Education: Findings from a Survey of the Empirical Literature, E. Begle, National Council of Teachers of Mathematics, 1979.

5. "Mathematics in Teacher Training Institutions. Some Difficulties Experienced by Teachers in Training", Ruth Rees, I.M.A. Bulletin, December 1976.

6. "The Dimensions of Mathematical Difficulties". A set of problems proving more difficult than teachers expect at all levels in the educational system. W.D. Furneaux and Ruth Ress, Brunel University, Department of Education. Occasional Publications Series No. 1, September 1976.

7. "The Structure of Mathematical Ability". W.D. Furneaux and Ruth Rees. Brit. J. Psychology, November 1978.

8. "The Structure of Mathematical Ability". Proceedings of the Third International of IGPME. James Curnyn, University of Warwick, 1979.

9. The Psychology of Mathematical Abilities in Schoolchildren, V.A. Krutetskii, The University of Chicago Press, 1976.

13.8 THE CHILD'S CONCEPT OF NUMBER

THE DEVELOPMENT OF COUNTING WORDS AND OF THE COUNTING ACT

Karen C. Fuson
Northwestern University
Evanston, Illinois, U.S.A.

This paper briefly summarizes recent research conducted by my colleagues (James W. Hall, Diane Mierkiewicz, and John Richards), students, and myself on the development of counting and on its relationship to other early number concepts. I will focus on three aspects: first the acquisition of the counting word sequence and subsequent elaboration of relations on the elements of that sequence; second, errors children make in counting; and third, developmental changes in the uses of the counting sequence. All of the research was done with American children using the English counting word sequence. Some of the results may not apply to counting with other languages or in other cultures.

The Acquisition of the Counting Word Sequence and Elaboration of Relations on the Elements of that Sequence

Middle-class American children seem to learn very early the basic distinction between words which are counting words and words which are not. Words produced in counting contexts are confined almost exclusively to counting words, even by two-year olds. After differentiating counting words from other words, a child must learn all of the counting words and must learn to produce them in their standard order. Until this learning is accomplished, most children produce counting sequences composed of three separate parts which we have termed the standard portion, the stable non-standard portion, and the spewed-out portion (Fuson & Mierkiewicz, Note I; Fuson & Richards, Notes 2 and 3). The standard portion occurs at the beginning of a child's sequence and consists of an initial group of words from the standard correct sequences. The next, stable non-standard portion is a group of words produced in the same way by the same child over several trials but the group of words either omits some of the correct words or, more rarely, repeats a word or changes the order of some words. The final spewed-out portion has little or no consistency or pattern over repeated productions. Spewed-out portions consist of single unrelated words, runs of two to four words in the correct order with no omissions, and/or chunks of two to four words in the correct order but with omissions. Half of the spewed-out portions contain words from the earlier standard and stable non-standard portions, and half do not. The second stable non-standard portion varies from child to child, while the third spewed-out portion varies within each child over repeated sequence productions.

The counting word sequences produced during the period of acquisition thus seem to involve two somewhat different processes. The first consists of the retrieval from memory of a whole structure composed of the connected "string" of standard words followed by a "string" of stable non-standardly ordered words. Words from the "string" structure are produced one at a time in the same way from trial to trial. The second production process involves the "spew". Separate elements are retrieved from a different non-ordered memory structure and vary considerably over repeated trials by a given child.

The acquisition of the standard sequence of counting words up to 100 begins before or soon after the age of 2 years and ends for most sometime during age 6 (Fuson & Mierkiewicz, Note I). The age of acquisition of the correct sequence is extremely variable, with some three-year-olds producing longer correct standard sequences than some five-year-olds. Most three-and-a-half year olds can produce correct sequences to 10 and are working on the numbers between 10 and 20. Children aged 4 1/2 and older can count to20 and know that the numbers to 100 are in cycles from x-ty-one to x-ty-nine, but they are working to learn the correct order of the decades.

Immediately following the initial acqusition of the counting word sequence, it exists as a single structured whole. The words are stored embedded within the sequence; they exist and can be produced only in a list recitation context, i.e., only one after the other as they are ordered in the sequence and in its entirety. Relations between individual words in the sequence have not yet been established. Then follows is a period of decomposition of this whole list structure into the individual words and groups of contiguous words. The relations derive from the "ordering" of the words in the counting sequence. The relations that become established on individual words are the "immediate successor" and the "immediate predecessor" relations ("seven comes just after six" and "nine comes just before ten"), the general "comes after" and "comes before" relations, and the "between" relation. Open segments of the counting sequence become able to be produced (i.e., children can start counting with any word of the sequence) sometime between 3 1/2 and 5, and closed segments of the counting sequence (counting that starts and stops at specified words) can be produced a bit later than open segments. Ages for learning these relations are not clear, but most three-year-olds have none of these relations while most six-year-olds can produce the "immediate successor" of numbers below twenty without error, the "comes after" relation is farily well-established and the learning of the "between" relation seems to be fairly well underway (see Fuson & Hall, Note 2; Fuson & Richards, Not 3: and Fuson & Richards, Note 4, for more details).

Errors Children Make in Counting Objects

We have undertaken a detailed analysis of the errors children aged 3 to 5 make in counting objects (Fuson & Mierkiewicz, Note I). Some of these errors involve the production of an incorrect sequence of counting words, discussed above. The other type of error made in counting is correspondence errors. The act of counting requires that a one-to-one correspondence be made between a set of objects (which usually exist in space) and a set of counting words (which exist in time). This correspondence is created via an act of indicating, usually a pointing gesture. The pointing gesture creates two correspondences - - one in space between an object and the focus of the point and a second in time between the pointing gesture and the spoken word. The simultaneous co-occurrence of the point-object correspondence in space and the point-word correspondence in time results in an object-word correspondence within the space-time unit created by the pointing gesture.

This analysis generated three types of single object correspondence errors: point-object violations, point-word violations, and violations of both correspondences. Our counting data are too complicated to present in detail here. Some highlights of the findings are: (I) Counting accuracy was generally higher than reported in the literature (e.g., 3 1/2-year-olds counted a row of 5 to 9 blocks with no errors over half the time and 4 1/2-year-olds performed similarly on rows of 15-19 blocks. (2) When children were urged to "try hard" on a counting trial, significant decreases in Skims, Skip Blocks, and Point-Word Other errors resulted. (3) When the number of objects increased, the errors that increased disproportionately were roughly the same ones (Skip Blocks, Skims, and Flurries) that increased with lack of effort, suggesting that the result of more objects to count may be decreased effort, especially by younger children. (4) Proportionately fewer errors occur at the beginning of a row, somewhat more occur at the end, and the highest proportion of errors occurs with objects not at the ends of a row. (5) When objects are maximally disorganized rather than in a row, recount errors increase enormously (from a mean of 0.7 to 17.6 per 100 blocks), as do Skip Block and double Count errors and Point-Word Other.

Uses of the Sequence of Counting Words

One of the currently most intriguing questions about the development of the use of counting is, "How and when do children learn in what situations it is helpful for them to count?" A closely related question is, "When can children use the information gained from counting to facilitate various sorts of quantitative thinking? Many four- and five-year-old children do not count when asked to make equivalence of number judgements in conservation situations. This is in spite of the fact that a recent study we did (Fuson, Secada, Hall, Note 5) indicates that such counting will lead to a judgement of equivalence while a lack of counting will lead to a judgement of non-equivalence. Research by Brainerd and Siegel (Note 6) indicates that the use of counting seems to be unrelated to performance on discrimination tasks requiring a child to choose either a dot pattern with a certain cardinality or the greater of two dot patterns. Gelman and Starkey (Note 7),Gelman and Gallistel (1978), and Brush (1978) likewise report a failure of four-year-old children to ascertain exact amounts added or subtracted when such information could help to avoid errors they otherwise make in judging which of two groups failed to count when asked to make displays containing the same number of objects. All of these studies except the last involve very small numbers (two to four) and are thus well within a fairly accurate counting range of these children. We hope in the coming year to begin to explore the question: Why do children not count in situations in which counting would seem to be helpful to them?

Our chief interest to date in the use of the sequence of counting words has been in its use as a mental representational tool for the solution of addition problems. We have studied the use of the counting sequence in the counting-on procedure for addition problems (Fuson, 1980a). In this procedure, rather than counting words or objects for both addends, the child abbreviates the counting for the first addend and merely produces counting words for the second addend. For example, 6 + 3 would be done as "six, seven, eight, nine". Our analysis of this procedure and some preliminary empirical work with children indicate that the simple statement of the first addend is a very complex cognitive achievement which requires a considerable amount of understanding. The production of this single first addend word (e.g., "six") indicates that the producer has:

1. learned that the single word "six" can be the cardinal value of the count sequence "one, two, three, four, five, six"

2. abbreviated the one to six enumeration of the first addend to the last counting word "six"

3. connected the cardinal meaning of the word "six" given in the addition problem to the counting meaning of the word "six" used in the problem enumeration

4. embedded the abbreviation word "six" in the enumeration of the total addend so that the counting of the second addend can continue on from the first.

Research by Steffe, Spikes, & Hirstein (Note 9) indicate that the ability to count-on may be associated with concrete operational thought, though a study using different measures found little such association (Hiebert, Note 10).

Counting-on can be done in the presence of objects, in which case the objects for the second addend serve to keep track of the enumeration of the second addend. When counting-on is done in the absence of objects and the second addend is large, other methods of keeping track of the second addend must be used. We have explored the use of various methods and have temporarily classified them into the following four types which vary in what is counted for the second addend; external objects present (these objects get counted) mental representations (visual or auditory representations get counted - - patterns are used for larger numbers), simultaneous use of fingers and counting (counting words are conted through the use of finger patterns), and double count word sequences (counting words are counted through the explicit use of a second counting sequence). In the first two types of keeping-track, counting words are used to enumerate objects for the sums; in the last two types, the counting word sequence for the second addend part of the sum is itself counted by a pattern or by other counting words. See Fuson, 1980a, for more details concerning children's use of these types, and see Secada, Notes 11 & 12, for some preliminary work exploring differences in deaf children's keeping-track methods.

Six developments during the period from age 2 to age 8 seem to contribute to the change of the counting word sequence from 1) a sequence not employed in quantitative situations to 2) one used to count objects in various quantitative situations to 3) one which itself is counted such as in counting-on. In Steffe, von Glassersfeld, & Richards' terms, these changes in what is being counted move from 1) to the production of no unit items to 2) the production of perceptual unit items to 3) the production of abstract unit items (Note 13). The six developments we see occurring during these changes are: an increase in the speed of production of the counting word sequence, the derivation of relations on the counting words, the acquisition of sequence meanings for the words, the acquisition of ordinal and cardinal meanings for particular words, the acquisition of symbolic and written forms of the words, and the linking of sequence to cardinal or ordinal meanings through the use of the sequence in counting in cardinal and ordinal contexts. (See Fuson, 1980b, and Fuson & Hall, Note 8, for a more detailed discussion). Future exploration of these developments would seem likely to lead us to a much fuller understanding of the development of early number knowledge and of the growth of the sequence of counting words into a representational tool used in the solution of arithmetic problems.

Reference Notes

1. Fuson, K.C., & Mierkiewicz, D. A detailed analysis of the act of counting. Paper presented at the Annual Meeting of the American Educational Research Association, Boston, April, 1980.

2. Fuson, K.C., & Hall, J.W. The representation of young children's counting words: Evidence for complex chaining. In preparation.

3. Fuson, K.C., & Richards, J. Children's construction of the counting numbers: From a spew to a bidirectional chain. Paper presented at the Annual

Meeting of the American Educational Research Association, Boston, April, 1980.

4. Fuson, K.C., & Richards, J. The acqusition and elaboration of the sequence of counting words. In preparation.

5. Fuson, K.C., Serada, W., & Hall, J.W. Effects of counting and matching on conservation of number. Paper presented at the Annual Meeting of the American Educational Research Association, Boston, April 1980. Later revised as Counting, correspondence, and numerical equivalence. Manuscript under review.

6. Brainerd, C.H., & Siegel, L.S. How do we know that two things have the same number? (Research Bulletin #469). London, Canada: Department of Psychology, November, 1978.

7. Gelman, R., & Starkey, P. Development of addition and subtraction abilities. Paper presented at the Wingspread Conference on The Initial Learning of Addition and Subtraction Skills, Racine, Wisconsin, November 27, 1979.

8. Fuson, K.C., & Hall, J.W. The acquisition, properties, and functions, of the counting word sequence. In H. Ginsburg (Ed.), The Development of Mathematical Thinking. New York: Academic Press, expected 1981.

9. Steffe, L.P., Spikes, W.C., & Hirstein, J.J. Summary of quantitative comparisons and class inclusion as readiness variables for learning first grade arithmetical content. Working paper No. 1, University of Georgia, Center for Research in the learning and teaching of mathematics, 1976.

10. Hiebert, J. The relationship between cognitive development and first-grade children's performance on verbal addition and subtraction problems. Paper presented at the American Educational Research Association, Boxton, 1980.

11. Secada, W. Deaf children's spontaneous use of manual counting to solve a set union (addition) task. In preparation.

12. Secada, W. Deaf children's spontaneous use of manual counting to solve a set decomposition (subtraction) task. In preparation.

13. Steffe, L.P., von Glassersfeld, E., & Richards, J. Children's counting types. Manuscript under review.

References

Brush, L. Preschool children's knowledge of addition and subtraction. Journal for Research in Mathematics Education, 1978, 9(1) 44-54.

Gelman, R., & Gallistel, C.R. The child's understanding of number. Cambridge, Massachusetts: Harvard University Press, 1978.

Fuson, K.C. An analysis of the counting-on solution procedure in addition. In T. Romberg, T. Carpenter, & J. Moser (Eds.), Addition and Subtraction: A Developmental Perspective. In press, (a)

Fuson, K.C. The counting word sequence as a representational tool. Proceedings of the Annual Meeting of the International Group for the Psychology of Mathematics Education, 1980. (b)

Saxe, G.B. A developmental analysis of notational counting. Child Development, 1977, 48, 1512-1520.

SOME NOTES ON ONE-TO-ONE CORRESPONDENCE

Shuntaro Sato
Fukushima University
Fukushima, Japan

1. The Actual State of Affairs in One-to-One Correspondence.

It seems to be easy for youngsters to count before they enter elementary school. But, it's not so easy. One of the reasons depends upon the fact that understanding of one-to-one correspondence operation is not easy. According to Jean Piaget, levels of understanding of one-to-one correspondence operation are divided into the following three stages:

Stage I: No Understanding

For example, here we have 6 bottles and 12 glasses. Then a child is asked, "Put one glass in front of each bottle". Then, the children at stage I put them in the following situation: He takes all 12 glasses and places them near the bottles but in a shorter row.

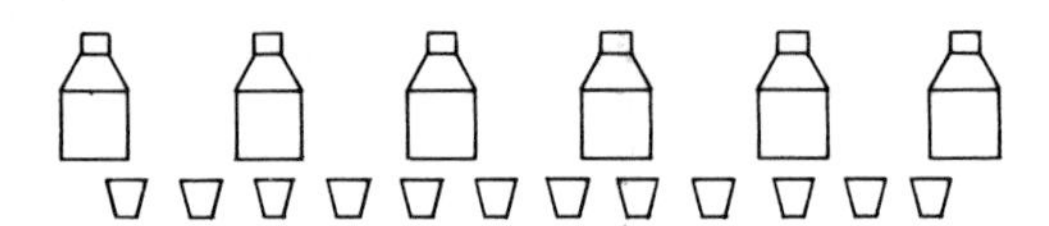

If he is asked which are more, glasses or bottles, he answers that there are more bottles than glasses. If he is asked to put one glass for each bottle, he makes two rows the same length. Now he thinks there are as many glasses as bottles. If the bottles are then spread out more than the glasses, he again thinks there are more bottles.

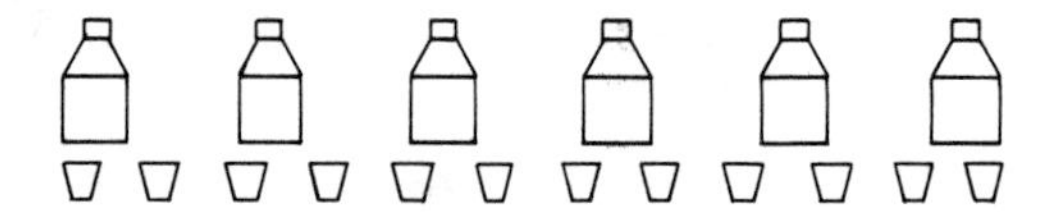

The youngster at this stage can't establish one-to-one correspondence or can't match a glass with each bottle. And the length of the set or the spacial configuration of the set overrules his judgement. So, 6 bottles are more than 12 glasses when the bottles are spread out.

We call this phenomenon a Piagetian phenomenon or the Piagetian phenomena. P.Y. Galperin, a psychologist in the USSR and professor at Moscow University, named it. The Piagetian phenomena means "perception grasps the evaluation of quantity or region based on the dominated property." And, according to Piaget, the cause of the Piagetian phenomena results in the lack of conservation (or invariance) of number or the lack of reversibility.

Stage 2: Partial Understanding

A child at this stage does not fully understand the invariance of number. He is working at a level called "intuitive". Thus he makes the same mistakes as the youngster at stage I. But, by trial-and-error method, he can arrive at the right answer.

Stage 3: Complete Understanding

A youngster at this third stage masters both the idea of one-to-one correspondence and also the idea of everlasting equivalence. His answers are based on the logic of invariance.

According to my investigation, the actual state of affairs in one-to-one correspondence is as follows (1):

age	stage		
4	65	30	5%
5	30	65	5%
6	20	50	30%

2. Can We Promote One-to-One Correspondence?

From now on, it's our problem to extinguish the Piagetian phenomena and to promote the one-to-one correspondence. This problem has much to do with "What is education?"

Let's define "education" from the epistemological point of view: "Education is a social service which accelerates stages of development and gives necessary supports to advance the development in some direction". In this definition, there exists a big hypothesis that stages of development can be accelerated. This is a great big question.

Gesell is a representative figure on classical readiness theory. According to his theory, "In the process of child mental development, there exists the optimum time and before than time we can't teach a child materials effectively but after that time we can teach him them reasonably". That is to say, "Teacher's job is at first to wait for the age of maturity and after that to give a suitable education". As it were, it is "waiting for maturity" education.

Jerome Bruner is against this "waiting for maturity" education (2). Whereas, Piaget says we can't accelerate the stages of development, and he calls it the "American Question" (3).

L.S. Vygotsky in the USSR advocated the zone of proximal development in the 1930's, and voiced opposition to "waiting for maturity" education. He was the first scholar who used this word in the psychological world, and said "education which goes on one step ahead of development has a superiority over others".

By the way, in order to extinguish the Piagetian Phenomena, scholars in the USSR set up first, second and third periods resembling Piaget:

1st Period: Form an ability to use markers in order to build up the conservation concept. We have 2 drawing papers. On one is drawn the picture of 30 fish, and the other is of 30 ships. Of course they drawn in a disordly manner. These drawings are shown to children who can count at most to ten. Children asked, "Which are more, fish or ships?" but they can't move them to count because they are drawn on the papers. So, children must devise how to count the objects.

We prepare cubic building-blocks to count fish, and rectangular parallelepiped building-blocks to count ships. Then we can guide children as follows: When you count fish you put a cubic building-block on each fish, and when you count ships you put a rectangular parallelepiped building-block on each ship. As it were, children are doing thus one-to-one correspondence operation, and thus can form the ability to use markers.

2nd Period: Form an ability to compare two sets by using tools. For example, 2 keys are drawn on a drawing paper. When we want to know which is longer, we can't put one key on the other because they are drawn only. So, at first, children cut out the tape just the same length as one key, and then they put the tape on the other key. Thus, children can compare the length of 2 keys. Anyway, using tools such as tape is an essential ability to compare 2 sets.

3rd Period: Form an ability to use scales. We give the children a drawing of 2 polygonal lines and a short tape. Then children measure the 2 polygonal lines by using the short tape. So, they can get 2 measured values which show how many times the short tape is included respectively. Thus, children can grasp an ability to use scales.

Let us suppose that children have already grasped the abilities mentioned above through the first, the second and third periods. At the first period, children put a cubic building-block on each fish. So we prepared a board on which are cut two rows of rectangular holes so that the children can put only one building-block in each hole. Then, we guide the children to put the building-blocks in the holes; cubic building-blocks in the first row and rectangular parallelepiped building-blocks in the second row. If the first row is shorter than the second row, the children can understand that the number of fish is smaller than the number of ships. If the length of the first row is the same as that of the second row, children can understand both numbers are the same.

Thus, if we devise and design such guidance, we can conquer the difficulties successfully, and we can accelarate the stages of development. The above mentioned is a counterplan in the USSR psychological world, in order to extinguish the Piagetian phenomena.

The Curriculum for Mathematics for Elementary Schools is determined by the Ministry of Education in Japan and is revised almost once very ten years. The courses of study were revised by the Ministry of Education in 1977, and were put into practice on April 1, 1980. For children in 1st grade (6 years old) this objective is included:

To help children be able to express correctly the number and order of objects by using numbers, and through these activities to help them understand the concept of number.

a. To compare the numbers of objects by an operation such as correspondence.

b. To count or express correctly the number and order of objects.

c. To know the size and order of numbers, the make a sequence of them and to express then on a number line.

The above-mentioned are the contents on "numbers and calculations" in Japan. So, we can say the one-to-one correspondence operation is suitable for children's developmental stages.

References

1. One, two, three, . . . Try a little more and you can count number: Shuntaro Sato: Memoir No. 26, Institute for Education Research Center, Fukushima University, 1963, pp. 52-63.

2. The Process of Education: Jerome S. Bruner, Vintage BOoks, 1963, p. 33.

3. How Children Learn Mathematical: Richard W. Copeland, Macmillan, 1974, p. 361.

EVOLUTION DU STATUT ET DU ROLE DU COMPTAGE AU COURS DE LA 1e ANNEE DE L'ENSEIGNEMENT OBLIGATOIRE EN FRANCE (COURS PREPARATOIRE: ELEVES DE 6 A 7 ANS)

Claude Comiti
Universite I de Grenoble
Grenoble, France

Presentation Globale de la recherche

Cette recherche est une recherche Didactique des Mathematiques. Sa problematique est nee du souci de comprendre le processus d'acquisition, par le jeune eleve entrant dans l'ecole obligatoire de la notion de naturel. Or, cette notion ne se construit pas d'emblee, chez le jeune enfant, ni au cours d'une suite d'activites de complexification croissante. Si l'on suit la theorie operatoire de l'intelligence developpe par Piaget, on sait que l'enfant, pour apprehender un champ plus large de proprietes, ne generalise pas simplement l'application des outils cognitifs dont il dispose, mais passe au contraire par un processus de reconstruction: ceci l'amene a chaque etape, a se fabriquer une classe d'outils intellectuels nouveaux qui continuent par la suite, d'une part a se developper en tant que tels, d'autre part a s'integrer et a se coordonner selon des modalites diverses.

C'est dans un but de clarification des conditions dans lesquelles l'eleve construit et s'approprie le concept de naturel que notre recherche a pour principaux objectifs:

- de mettre en evidence les differents modeles implicites et incomplets du nombre fonctionnant chez les eleves a un moment donne et dans un domaine numberique donne;

- de mieux comprendre comment les differents modeles qui coexistent, a un moment donne, chez un meme enfant, interviennent selon la tache qu'on lui propose.

Qu'appelons-nous modele implicite? C'est l'expression de certains invariants du concept de nombre, de certaines proprietes, sans que l'on puisse savoir ce qui, au niveau du sujet observe, est conscient ou pas vis a vis du concept en jeu. La recherche de tels modeles suppose l'hypothese que quelles que soient les conditions dans lesquelles sont faits les apprentissages scolaires, il existe, a un certain niveau, des regularites qui caracterisent l'appropriation d'une connaissance donnee chez tous les sujets, donc ici celle de nombre naturel. Les differents modeles qui coexistent, a un moment donne, chez le meme enfant, consituent son "systeme de connaissance" dont nous voulons etudier les differents fonctionnements selon la tache a resoudre.

Pour conduire cette recherche, nous avons decide de privilegier trois variables qui nous paraissent fondamentales:

- la nature de la tache proposee a l'eleve (variable Tache);

- la grandeur du domaine numerique dans lequel on se situe (variable Taille);

- le moment de l'annee ou l'on se place (variable Moment).

Evolution du statut et du role du comptage, dans une tache ou il y a concurrence entre le comptage et les procedures de correspondance, lorsque l'on fait varier la taille du domaine numerique et/ou le moment du cursus scolaire.

On pourra trouver une analyse detaillee des resultats presentes dans cette communication dans l'article "Analyse de comportement d'eleves de Cours Preparatoire" publie dans la revue "Recherches en didactique des mathematiques" Vol. 1.2. Aout 80 - Editions - La Pensee Sauvage et la Maison des Sciences de l'Homme (France).

Descriptif de la tache

L'enfant dispose d'une boite de n jetons bleus. Sur la table sont poses en vrac m jetons rouge (m n). Sur demande de l'experimentateur, l'enfant etale les jetons rouges sur la table. La consigne est ensuite "maintenant met sur la table autant de jetons bleus que de jetons rouges". Lorsque l'enfant a termine, l'experimentateur lui demande "es-tu sur?". Il exige ensuite de lui une verification, sauf dans le cas ou l'enfant est sur a juste titre de son resultat.

Choix du dispositif experimental

Nous avons choisi un materiel

- deplacable, dans le but de favoriser la multiplicite des procedures permettant de resoudre le probleme pose;

- homogene, de facon a ce que, chaque collection etant constituee d'elements indifferenciables entre eux et donc permutables a l'interieur de la collection; ceci elimine toute intervention, dans une mise en correspondance eventuelle, des qualites ou des aspects propres aux elements.

De plus, nous avons laisse sur la meme table la collection rouge temoin et la boite de jetons bleus, de maniere a ne pas privilegier les procedures de denombrement.

Methodologie

Nous avons retenu la methode de l'entretien individuel de preference a l'observation des comportements d'eleves en classe. En effet, dans une situation de classe, certaines procedures, propres a un eleve donne, risqueraient de ne pas apparaitre du fait de certaines variables difficiles a maitriser (role du maitre, des leaders, effet de contamination...); certains tatonnements pourraient passer inapercus ou meme ne pas pouvoir s'etablir. De ce fait, l'interpretation, au niveau du fonctionnement du sujet, des procedures utilisees serait trop complexe. Au contraire, la situation d'entretien individuel nous permet de saisir, pour chaque enfant, la dynamique des procedures utilisees.

Principales procedures observees chez les eleves

- les procedures de denombrement (que nous noterons D): comptage du nombre m de jetons rouges puis extraction de m jetons bleus de la boite.

- les procedures de correspondance (notees C): le plus souvent correspondance terme a terme, plus rarement copie de configuration spatiale.

- les procedures qui semblent ne faire appel a aucune methode apparente, procedures le plus souvent basees sur une estimation perceptive (nous les noterons A).

Choix des variables taille et moment

Iere experimentation:	m = 15	mars	Cours Preparatoire
2eme experimentation: avec les memes eleves:	m = 37	mars juin	Cours Preparatoire
3eme experimentation:	m = 37	novembre	Cours elementaire Iere annee.

Pourquoi ce choix?

- En mars au C.P., les eleves ont appris, ou sont en train d'apprendre les techniques de denombrement et leurs maitres pensent en general acquis et donc utilisable, la suite des nombres jusqu'a 20. Au voisinage de 15, il y a donc concurrence effective entre les procedures de denombrement et celles de correspondance, ce qui doit permettre d'etudier les methodes les plus utilisees ainsi que leur operationalite.

- Par contre, a ce moment de l'annee les nombres superieurs a 20, et en particulier 37 font partie d'un domaine numerique inconnu de l'eleve de C.P., le choix de 37 doit donc nous permettre de bloquer provisoirement l'apparition des procedures de denombrement et d'etudier les comportements des enfants dans un cas ou ils n'ont pas a leur disposition les outils necessaires pour mener a bien la tache par comptage.

- Mais le programme de Cours Preparatoire comprend l'apprentissage des nombres de I a 99. L'eleve est donc sense bien maitriser la comptine numerique au voisinage de 40, en fin d'annee de C.P., et a fortiori l'annee d'apres, en CEI. L'etude de m = 37 en juin de C.P. et en novembre de CEI nous permettra donc d'etudier l'evolution des procedures mises en oeuvre par les enfants, compte-tenu de leur apprentissage scolaire et de leur developpement.

Comparaison des resultats obtenus au meme moment

Le changement de domaine numerique (passage de m = 15 a m = 37) en mars de C.P.

- fait evidemment chuter brutalement le nombre d'enfants utilisant le comptage au profit essentiellement des procedures d'estimation mais aussi des procedures de correspondance;

- augmente du simple au double le nombre d'enfant remplissant la tache par une estimation basee sur la perception;

- augmente bien entendu les procedures de correspondance terme a terme tout en modifiant la repartition au sein de ces procedures (apparition des pointages simultanes et augmentations des superpositions;

- diminue la confiance dans la fiabilite des methodes utilisees pour les enfants reussissant des la construction (le pourcentage des enfants surs a juste titre passe de 92% a 71%);

- provoque, comme prevu, une baisse brutale de l'operationalite du comptage, ce qui n'a pas de grande repercussion sur le taux global de reussite, etant donne le petit nombre d'eleves ayant recours a une procedure de ce type.

Role de la variable 'Moment' pour la taille m = 37

De mars a juin on constate une nette diminution de nombre d'enfants repondant par estimation perceptive, une forte augmentation du nombre d'enfants remplissant la tache par correspondance terme a terme, une certaine stabilite quant aux procedures de denombrement.

De plus, en juin, les enfants qui ont reussi des la construction sont cette fois tous, surs de leur resultat. Leur confiance en la fiabilite de leur procedure est donc plus forte qu'en mars, confiance sans doute liee a une interiorisation progressive de la correspondance terme a terme. Pour les enfants qui ont echoue a la construction, la consigne de verification les incite comme en mars, a mettre en oeuvre une activite qui constitue un feed-back. Mais ici le feed-back repose sur des procedures de denombrement contrairement a ce qui se passait en mars ou les procedures de correspondance dominaient fortement.

Enfin, en novembre de l'annee d'apres (CEI), les enfants recourent cette fois massivement au comptage dont l'operationalite laisse pourtant beaucoup a desirer dans un domaine numerique pourtant suppose acquis en fin de C.P. et donc a fortiori au CEI. De plus l'eventail des procedures s'est terriblement referme avant meme que le comptage soit reellement efficace: bien que les enfants expriment clairement qu'ils n'ont pas confiance dans leur comptage, ils n'envisagent cependant pas d'utiliser une autre methode.

En resume de mars a novembre, on constate la disparition progressive des comportements par estimation basee sur la perception, disparition s'accompagnant d'une augmentation des procedures de correspondance (juin), qui vont elles-memes s'effacer par la suite (novembre) au profit des procedures de comptage qui deviennent plus operationnelles dans le domaine numerique considere.

Ceci confirme que l'accession aux procedures de comptage, dans un domaine peu familier a l'enfant est precede par la mise en oeuvre de procedure intermediaire de mises en correspondance sur lesquelles elle s'appuie.

L'importance de la variable "moment" est donc due, bien entendu, a la maturation psycho-genetique de l'enfant mais ne saurait etre expliquee par ce seul processus.

L'apprentissage scolaire joue en effet un grand role puisque:

- en mars de C.P., le domaine numerique dans lequel l'enfant est ici place est percu comme "beaucoup, le nombre 37 n'etant pas encore situe dans la suite des nombres connus;
- en fin de C.P., suite a l'apprentissage de la numeration, 37 fait partie des nombres connus de l'enfant encore peu utilises et donc peu familiers;
- ce n'est qu'au CE I, que la plupart des eleves, par l'intermediaire d'activites portant sur la numeration, sur la somme de deux nombres, sur la comparaison de plusieurs nombres, se familiarisent veritablement avec ce domaine numerique qu'ils s'approprient au meme titre que precedemment celui des petits nombres.

Tous ce qui precede montre la necessite pour l'eleve du cours preparatoire, de disposer, pendant la periode ou il s'approprie le comptage, de procedures conceptuellement plus simples de correspondance et en particulier de correspondance terme a terme: ces dernieres jouent en effet le role de procedure intermediaire sur laquelle s'appuie l'interiorisation progressive du comptage tout en relayant ce dernier dans les situation ou il est defaillant (par exemple lorsque la taille du domaine numerique augmente).

PARTITIONING, EQUIVALENCE AND THE CONSTRUCTION OF RATIONAL NUMBER IDEAS

Thomas E. Kieran
University of Alberta
Alberta, Canada

The rational numbers represent a sophisticated system of knowledge for a person to acquire. This is true in conceptual terms, as the rational numbers can be used to model a variety of real-world situations as well as being a prime example of a mathematical field. The symbol system using ordered pairs represents a level of abstraction and the formal system, a/b, can connate other informal pair systems (e.g., a divided by b, a to b, a for b). Also the formal algorithms provide examples of human invention but do not appear natural, particularly addition. Given the abstract and seemingly complex nature of such knowledge, how can it be developed in persons, particularly in school children and young adults?

The philosopher Bateson (1979) makes use of the idea "about" in describing knowledge, particularly symbolic knowledge development. To be useful, a person's knowledge must be "about" something. At a first level, knowledge is derived from perceived patterns in "physical" experience. What is rational number knowledge about (at least ideally)? It has been argued in other papers (e.g., Kieren, 1980; Kieren and Southwell, 1979) that rational number knowledge is about four subsystems or subcontracts - - measures, quotients, ratios, and operators. These four idea sets are derived from and allow a person to apply knowledge to four different kinds of phenomena for which rational numbers can serve as a model. The ordered pair language of rational numbers is based on a fifth construct and kind of experience, the part-whole relationship. This language-generating construct can be related to each of the other four through the identification of a unit under each circumstance. It is not the purpose of this paper to elaborate further this total view of rational numbers as a system of personal knowledge.

Bateson has argued that knowledge is based on patterns that connect disparate situations. In these terms one might say that rational number knowledge is a set of patterns which connect measure with quotient with operators, for example. It is in this sense that rational number knowledge is "about" the four sub-constructs. Hofstadter (1979) has suggested that persons can function mechanically or intelligently in exercising mathematical behaviors. In functioning intelligently a person makes an isomorphism between a symbol/idea and some reality. Thus in rational numbers a person may line "3/4" with dividing three bars of chocolate among four children.

The question is by what means can a person make the patterns or isomorphisms for rational number knowledge? What mental tools does a person use to construct rational number knowledge? These questions represent more general ones about mathematical concept systems in general, in answering such questions one might posit two kinds of mental tools or mechanisms, developmental and constructive. While it is not easy to form a sharp deliniation between them, one might consider the former to be more general and based more on mental maturity, while the latter (and our focus here) are more specific, related to experience, and can be thought of as objects for instruction.

There is much current discussion of such mechanisms with regard to whole number idea development. In this regard one might describe the mechanism of conservation of number, compensation, identity, or reversibility as developmental in character. The prime constructive mechanism is, of course, counting. Other whole number constructive mechanisms are those relating to numeration and related language use.

The mechanisms which relate to fractional or rational number development have not been studied to the same extent as those for whole numbers. Two developmental mechanisms (although obviously related to experience) are conservation of the whole (Piaget, Inhelder, and Szeminska, 1960) and proportional reasoning (Noelting, 1978); Kurtz and Karplus, 1979). In addition, one might think of reversibility (for thinking of two complementary inverses) and simultaneous comparison (for generating non-unit fractional equivalences) as other developmental mechanisms. It is suggested that partitioning and equivalence mechanisms are constructive mechanisms used in developing the five intuitive or informal rational number sub-constructs and more formal rational number knowledge. It is the purpose of the rest of this paper to give a brief description of these two mechanisms in rational number thinking. While a complete theory encompassing the use of these mechanisms has not been developed (these mechanisms are in that sense speculative), the descriptions here are based on interviews with several hundred subjects aged 3 to 16 as well as paper-and-pencil testing of over 500 children and young adults.

Partitioning

Partitioning is defined here as the equidivision of a quantity into a given number of parts. A completed portion (of continuous phenomena or sets of objects) is the basis for the part-whole fractional language.

The act of partitioning is central to the generation and application of rational number knowledge. For example, if one is asked to find 3/4 of 8, one might partition 8 into 4 parts and take 3 parts or 6. This is reflected in the explanation of children and young adults doing such a task "divide by 4 and multiply by 3" as seen by Kieren and Southwell (1979).

There appear to be four aspects of the partitioning activity. First, partitioning is a kind of classification or allocation based on the criterion of "evenness" or "just enough." This particular classification has a social genesis in the act of sharing. This action can pertain to discrete or continuous phenomena and can serve to generate systematic partitioning activities such as dealing out. A third aspect of partitioning relates to language describing the act and results of partitioning. This can be exemplified in the action language of young children separating a quantity into 2 parts and saying "here's a half." It is interesting to note that such English language of young children contains precursors of both the quantitative and relational aspects of rational numbers ("I'm four and a half"; "Take half of it"). A fourth aspect of partitioning is the attachment of parts to measure or number. For example, regardless of how a quantity is partitioned into 3 parts, the size of the part is the same.

Perhaps a couple of examples will further illumine this activity. Children (aged 3 to 8) were individually presented with three trucks and simulations of 15 crates (6 red, 5 blue, 4 green). The trucks had space for exactly 6 crates. The child was asked to place the crates on the trucks so that each truck contained the same amount. One boy (aged 5) lined up the trucks and systematically placed one on each (from left to right) until all were exhausted and said, "There." He knew they were the same but did not have to check their number. The crates were dumped off and the boy asked if there were another way to accomplish the goal. After much thought he said, "Well, you could do it in the other direction", and indicated right to left. A girl, aged 8, performed a systematic one-by-one deal and said, "There's five each." When asked about a different way she quickly started to deal 2 per truck. When she was left with three she looked perplexed and finally placed one on each and said, "Oh well, five again." When asked about another way she tried 3 per truck and was left with 6 crates. At this point she drew out 3 15 and said, "Oh, I guess I'll always get five." These examples illustrate some of the aspects of partitioning. Further, they show that partitioning and size and measure can be quite independent for children.

Expected parttioning behaviors of children can be elaborated in the five sets of behaviors described below. These categories represent a crude description of the growth of partitioning as a constructive mechanism.

A. Separation:

   dividing the given set (or quantity);

   dividing into the appropriate number of sets;

   classification on bases other than number;

   use of division criteria other than "evenness".

B. Evenness:

   some sub-sets even, some not;

   rearrangement to an evenness criteria;

   dividing in two by matching.

C. Algorithmic Partitioning"

   "dealing" behavior by ones;

   "mixed" dealing behaviors.

D. Partitioning and Number:

   relating "evenness" and number or size of the finished partitions;

   realizing that quantity is preserved under partitioning;

   generalized partitioning - describing the process and results of large number partitioning - e.g., into 100 parts.

E. Advanced Partitioning:

   repeated partitioning;

   related number and size of partitions;

   given a partition, change it by adding to the number of partitions or reducing the number;

   given a partition, see another within it (e.g., given a partition into quarters see that a partition into two parts already exists).

Equivalence

The relationship between equivalence and rational number is well known formally (and would seem almost to dominate curriculums on fractions and rational numbers). In an informal sense the realization of equivalence is a basis for concepts of rational or fractional numbers in the various subconstructs. For example, if one considers 2/3 as an operator, a (proportional) equivalence concept underlies a child's seeing that this operator maps 12 to 8, 15 to 10, 30 to 20, etc., and in fact enables a child to generate the pair p — 2/3 — q. Secondly, equivalence arises in the same or identity sense as a child notes that "2 for 3" and "4 for 6" operators do the same thing or are the same or generate the same range element given a domain element. Such examples could be shown in the other subconstructs as well.

In its mature form the child's concept of equivalence is multiplicative in nature and closely related to proportional reasoning (e.g., Karplus and Kurtz, 1979). It also can come under control of formal algorithms with these equivalence algorithms generating other rational number algorithms (e.g., addition).

Yet there are also less formal equivalence notions which are potent for children and young adults developing rational number ideas. Perhaps the first such is an extension of quantitative equivalence. Here a child sees that 1/2 is the same as 2/4 in the sense that "1/4 put with 1/4 makes 1/2." Another such "equivalence" is "3/4 is 1/2 and 1/4." An eight-year-old possessing such "equivalences" was asked to add 3 3/4 and 5 1/2. He said, ""3 and 5 take you up to 8. Then 3/4 and 1/4 take you up to 9 and you have 1/4 left so it's 9 1/4." This example illustrates the early constructive role of quantitative equivalence in rational number thinking.

The proportional side of equivalence can also be seen in various levels of equivalence thinking described below.

Numerator to denomenator "works like" rules.

with the number 2/3 - "If 9 is the denomenator I divide by 3 and multiply by 2 and get 6 for the numerator."

Pair wise equivalence

"I know 6 is paired with 8 so 18 (3 x 6) must go with 3 x 8 = 24."

Equivalence rules

1. concrete: "If I have 3/4 I can get other fractions which are the same by multiplying 3 and 4 by the same number."

2. mathematically formal: "3.4 36/48 because 4 x 36 = 3 x 48."

Class of pairs

"There is an infinite set of pairs which are like 3 for 4."

Equivalence classes

"Equivalence divides the set of fractions into classes each of which is a number."

Finally, equivalence manifests itself in rational or fractional number language use as seen in the statement, "There are many names of 3/4 - - .75, 12/16, 75%, 90 for 120, etc." What is implied here is both the algorithmic ability to convert symbol forms and the knowledge that there are numerous "equivalent" symbol systems for rational numbers. This symbolic equivalence thinking enables a person to apply to rational number concepts in a wide variety of situations.

Summary

Rational number concepts are both extensive and complex in nature. Partitioning and equivalence are seen as two constructive mechanisms which enable a child or young adult to build up such complex ideas. Because such mechanisms can be taught more attention should be given to their place in rational number curriculums. In particular, attention should be paid to early and informal manifestations of these mechanisms. Research is ongoing to further elaborate aspects of these mechanisms described above and to integrate them into a theory of rational number knowledge.

References

Bateson, G. Mind and Nature. New York: Dutton and Co., 1979.

Hofstadter, D. Godel, Escher and Bach. New York: Basic Books, 1979.

Kieren, T. "Rational Numbers: Ideas and Symbols." In Mathematics, M.M. Lindquest (ed.), Chicago: NSSE, 1980 (in press).

Kieren, T. and B. Southwell. "The Development in Children and Adolescents of the Construct of Rational Numbers as Operators." Alberta Journal of Educational Research, 1979, 25 (4), 234-247.

Kurtz, B. and R. Karplus. "Intellectual Development Beyond Elementary School VII: Teaching for Proportional Reasoning." School Science and Mathematics, 1979, 79, 387-398.

Noelting, G. Constructivism as Model for Cognitive Development and (Eventually) Learning. Unpublished paper. Quebec: Universite Laval, 1978.

Piaget, J., B. Inhelder, and A. Szeminska. This Child's Concept of Geometry. New York: Basic Books, 1960.

NUMBER LEARNING AS CONSTRUCTING COHERENT NETWORKS BY USING PIAGET-DERIVED OPERATIVE PRINCIPLES

Gerhard Steiner
University of Basel
Basel, Switzerland

Number learning in grades 1 to 6 usually goes along with performing elementary arithmetic operations such as counting, adding, subtracting, etc. There are two starting points to cope with the problem of number learning. First, a very practical one: numerous complaints were coming from teachers as well as from parents of first-, second-, and third-grade children about their difficulties in number and arithmetic skills, the main problem being the fact that they could not or almost could not refrain from using arithmetic materials or that they fell back into finger counting whenever they got into troubles or rote learned facts were forgotten. Second, a more theoretical one: teachers as well as psychologists interested in teaching and learning processes were asking of what help Piaget's research on number conservation could be for number learning in the school grades 1 to 6. There is no doubt that having acquired number conservation is a prerequisite for number learning, but this ability is not sufficient for later coping with number problems or arithmetic tasks. So the question was raised: What can be derived from Piaget's theory for number learning? What principles should be used by the teachers in planning their teaching as well as in trying to help students with learning difficulties or to introduce arithmetic activities to low-IQ students?

Let us look for a framework within which we can consider number learning in a more comprehensive way. Numbers are concepts, though of a somewhat special kind compared to everyday concepts like "democracy" of "inflation" or whatever moves us. Semantic memory research tells us how concepts can be conceived of as "matters of knowing or knowledge". What we want to do is to see how semantic memory research can be applied to our problem of number learning.

Concepts are parts of larger systems of knowledge, in other words: parts of semantic networks. Concepts like "inflation" include other, more elementary concepts as elements that are connected by a variety of relations; e.g., the concepts of "money", "goods", "services", "the money-goods-circle", etc. Together they constitute the concept of "inflation". Concepts can - on every level - be connected to other concepts so that new, higher concepts are formed like, e.g., "monetary policy" when you connect "inflation" with other particular concepts.

Conceptual elements are connected by "named associations", i.e, by relations of different kinds; e.g., by "belongs to" or "is part of" or quite simply by "is a" or "gives" or "eats" or any verb or other word that connects a grammatical subject to an object. There are several theoretical approaches to what rules determine these connections that I cannot go into now. But two things are important for math learning and should be stressed: First, concepts in a network can be unfolded just as the concept "inflation" can be unfolded into the phrase "the situation where there is too much money circulating compared to the available goods so that the value of one unit of money decreases". Second, a part of a network consisting of several elements of relations between them can be condensed and further used as one element, one concept, that, in turn, can be connected to others by particular relations. Conceptual learning is thus the process of connecting elements into new parts of a network whereby processes of condensing as well as unfolding may take place. The variety of elements in a conceptual network as well as the number and diversity of relations are unlimited. The main characteristic of this process is that it constructive. At this point cognitive learning theory and Piaget's developmental and epistemological theory touch each other.

Let us now have a look at number learning. Numbers, too, are concepts. They are countless, but no so their variety: this one is limted to natural numbers, integers, rational, irrational, real, imaginary, and a few other kinds of numbers. The number of relations that can be established between numbers is equally restricted compared to the infinite number of relations in a linguistic conceptual network. Thus, we can expect the numerical network to be rather simple, easier to glance over, more transparent.

Numbers as conceptual elements can be connected by arithmetic operations into new elements, as can linguistic conceptual elements. They can also be unfolded into constituting elements and particular arithmetic relations. For instance, "8" can be unfolded into several elements and relations that - in each case - constitute the number 8:

8 is 1 more than 7

8 is 1 less than 9

8 is half of 16

8 is the double of 4

8 is a third of 24

in higher grades:

8 is the square root of 64

8 is 2 in the third power

8 is the log of 100 millions

etc.

Theoretically the number of such propositions is infinite!

If you would make a drawing of these propositions that all coincide in the network node "8", you would get a configuration that looks like a conceptual network as postulated by the current semantic memory research. What is constituted by this large number of propositions connecting the interger 8 to many other numerical elements of the conceptual framework is the meaning of 8. Moreover, the network contains, like all conceptual networks, tacit information that can - under certain circumstances - be inferred from what is already given. For example, we can infer tat "if 8 is 1 less than 9 and 9 is one less than 10, the 8 is 2 less than 10"; and "4 is a fourth of 16 if 8 is half of 16, etc.

Number learning is in my opinion constructing numerical networks. What is the starting point of this construction, and what are the crucial learning processes?

Piaget's constructive theory of cognitive development provides us with concepts that prove to be fruitful in number learning. A numerical network can be considered as a Piagetian "structure of a whole" or, as he calls it in French, a "structure d'ensemble". The theoretical coincidence of Piaget's "structure of a whole" and our network-structure is far from perfect. I do not go into this! Networks are more comprehensive and more suitable for explaining open processes like number learning that virtually have no limit. A second concept of Piaget's is more important: what he calls "mise en relation" or "putting into relation"; i.e., putting two (or more) elements (two lengths, two dimensions, two quantities, etc.) into a particular relation instead of "reading off certain features or properties from them" (instead of what he calls "lecture des donnees"). In number learning the very learning process will be the establishing of numerical relations - that are originally action-based - between numerical elements. What are, at least at the beginning, these relations?

First of all, there is $\pm$ 1 relation that is established by the iteration; i.e., the going on forward or backward in steps of 1 within the set of intergers or natural numbers: 8 is 1 more than 7, 7 is 1 more than 6, etc. Iteration (and the respective relations) is not necessarily established by the process of counting. It includes a certain reversibility, namely to realize that if 8 is 1 less than 9, then 9 is 1 more than 8. Iteration allows going into the realm of natural numbers and integers in small steps; it introduces the addition/subtraction aspect into number learning.

A second relation between numbers is that of the double and the half, respectively. As can be seen easily, this, too, includes a certain reversibility: 8 is the double of 4, 4 is the half of 8. Doubling a number (or a certain amount of objects) means taking a certain amount, adding (probably by one-to-one correspondence) once more the same amount, then refraining from focussing the first amount but instead focussing the whole. This is the double of what has been taken at the start. This process not only includes number conservation and one-to-one correspondence but also an elementary understanding of the part/whole relationship. Doubling allows going into the realm of numbers in much larger steps than iteration did. Basically, doubling introduces the multiplicative aspect into number learning.

Now, how are iteration and doubling applied to constructing numerical networks? Already known facts can now be put into relation to as yet unknown or already forgotten facts. For example, the proposition 3 + 3 (known from former learning as the double of 3 equals 6) can be put into relation to 3 + 4. Be sure, this is a very elementary example, but this is exactly what number learning looks like in the beginning. This first step is to get the insight that 3 + 3 is not the same as 3 + 4, the outcome of the latter must be different from the former. For low-IQ children we observed over a six-month period, this could be a very large but important step towards the construction of a coherent numerical network. The second step is to learn that 3 + 4 must be more than 3 + 3 because 4 is more than 3; and the third step, that it is exactly 1 more. Similarly, if you put 5 + 6 or 5 + 7 into relation to the already known doubling of 5 (5 + 5), the problem will be solved on the basis of iteration and doubling. The presupposition certainly being that some doubling is already known such as 2 + 2 up to 9 + 9 or 10 + 10, doublings that form a kind of pillars in the as yet small building of numbers of the young child.

What about Piaget's opposing of the two concepts of "putting into relation" versus "reading off properties"? An elementary example will show what this could mean for number learning at school (Piaget has amply shown what is means for the acquisition of several conservations!). We should keep in mind that most number learning curricula recommend or even prescribe the use of special material(s), e.g., chips or stripes or colored wood sticks or whatever.

Reading-off Situation

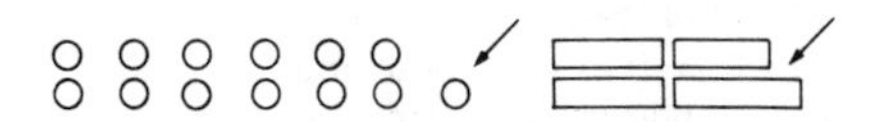

Figure 1. Reading-off situation. Both configurations stand for the comparison, "If I know 3 + 3 = 6 what is 3 + 4?" Chips as well as rods (e.g., Cuisenaire rods) are put in rows, the known above the unknown. (The rows could stand in an upright position as well.) In both cases the empty space or the "one more piece" can be read off (arrow). The child is only weakly supposed to put the two rows into relation.

There is a certain danger that students who get accustomed to using this kind of material never really establish relations between numbers or propositions about numbers, but only read off a certain length of a particular configuration of the material used because the configuration of the material suggests doing so! Reading off certain characteristics (here: numerical characteristics) is exactly what often prevents children from making progress in building up structures; it does not lead to any relation, to any new part of the network. Teachers have to be careful to prepare situations where their students really can establish relations even if they are most simple; they have to prevent a reading-off strategy (including counting the chips that is also a reading-off process!). The easiest way to do so is putting the material apart:

Putting-into-relation Situation

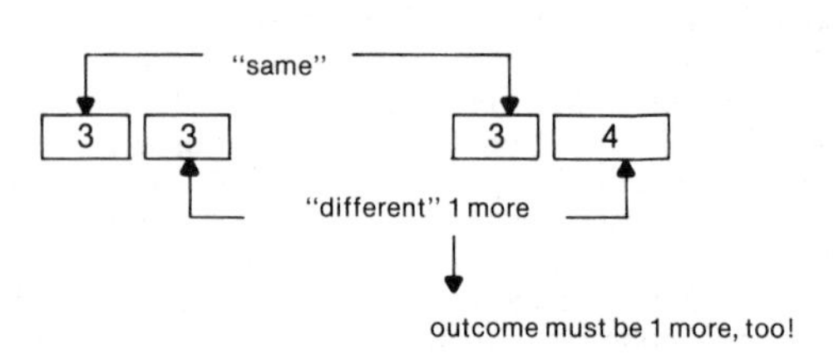

Figure 2. Putting-into-relation situation. The configurations 3 + 3 and 3 + 4 are apart on the child's desk. He/she can point with the first fingers of each hand to what is equal (the equal parts are put into a relation - "equal"): the left hand "3" of both configuration. The he/she can point to the unequal and state: "The one here (right hand "4") is longer. It is 1 longer (or more). So the whole line is 1 longer." The whole configuration inhibits a "reading-off" habit and leads to comparisons; i.e., to putting-into-relation activities.

Constructing a numerical network does not primarily mean to find the correct answers to numerical tasks but to find relations to other numerical tasks or propositions. That is why the teaching and learning process, the program in the classroom, has to look different from other programs: Students are not looking for the outcome of 5 + 5 but for different propositions relating 5 + 5 to 6 + 5 or 5 + 6, to 5 + 4 or 4 + 5 or even to 4 + 6, the last one being an example where a compensation of + 1 ad - 1 relates to the original 5 + 5! Or 10 - 5 will be related to 10 - 4 or 10 - 6, to 9 - 5 or 11 - 5 or to 9 - 4 or 11 - 6! Each time the primary task is to put the first task or proportion into relation to another one and to tell exactly what the relations between the two are and what can be inferred from this relational knowledge.

This is what I can constructing coherent numerical networks by using Piaget's concept of operative principle "putting into relation". This process goes on and on during the school grades and by it propositions such as 7 x 8, the unknown, will be linked to 8 x 8, the perhaps known, and the answer will be constructed from the latter. Known outcomes may serve as reference points that can at any time be put into relation to unknown ones. The learning process is the process of "working through the network" as opposed to a rote learning of many tasks or propositions that stand more or less isolated beside each other. In higher grades this "working through the network" process enables the student to predict, e.g., from knowing that 4704 ÷ 98 or what the task would have to l k like if X ÷ 48 = 99. Should X be larger or smaller? How much? Are there regularities?

And what about all the teachers' and the students' endeavors when one fine day the electronic pocket calculator takes over all number learning activities? The student has not primarily learned to find outcomes - that is what the calculator can do and is made for. He has learned to find relations and by doing so he got accustomed to know that he can always find a solution, that every problem is solvable. He has even learned to find whole sets of solutions - a plurality of solutions - whereby two of them suffice to arrange a self-control over whether he had done a good job or not. This in turn enhances the student's independence from the teacher as well as from the program. Moreover, finding a whole plurality of solutions is a kind of divergent thinking, a kind of creativity. Being able to give reasonable, explainable predictions by putting certain numerical propositions "into relation" is a capability that cannot be replaced by any calculator. Of course, all these activities or capabilities do not appear automatically; after the first construction of the steadily growing numerical network, it has to be worked through as mentioned above. This means: always new or different relations have to be looked for, in several directions until the numerical thinking reaches a relatively high mobility. Then parts of the network are ready to be condensed and used as new, higher elements to be connected to other elements. This is the way number learning goes towards higher kinds of numbers such as rational and other ones.

13.9 RELATION BETWEEN RESEARCH ON MATHEMATICS EDUCATION AND RESEARCH ON SCIENCE EDUCATION. PROBLEMS OF COMMON INTEREST AND FUTURE COOPERATION

COOPERATION BETWEEN TEACHERS OF MATHEMATICS AND OF THE SCIENCES

Charles Taylor
University College
Cardiff, U.K.

You may, like me, become quite bewildered by all the initials and acronyms that float around these days and I would like to begin by explaining some of them. You will be aware that there are about 18 Unions that are concerned with the various scientific disciplines. In particular you know of I.M.U. - the International Mathematics Union. Then there is I.U.P.A.P. for Pure and Applied Physics; I.U.B.S. concerns the Biological Sciences; I.U.P.A.C. looks after Pure and Applied Chemistry; I.U.Cr. covers crystallography and so on. Now many of these unions have their own committees or commissions on education. For example I.M.U. has I.C.M.I. and I.U.P.A.P. has I.C.P.E.

There is also an umbrella organisation which coordinates the work of all the separate unions and this is I.C.S.U. - the International Council on Scientific Unions. But I.C.S.U. itself also has a Committee on the Teaching of Science - I.C.S.U. - C.T.S. - and this consists of a representative of most of the individual unions together with a few other people. Several representatives of C.T.S. are here today: Professor B. Christiansen is the I.C.M.I. representative; Professor A.P. French represents I.C.P.E.; Professor D. Lockard is a biologist on the committee; and I am the current chairman and a crystallographer.

Our terms of reference are to be concerned with science teaching at all levels from primary to post-graduate in all sciences, and in all countries of the world, an incredibly wide brief. Because of limited resources - both of money and of personnel - we can only take on one or two projects at any one time. For obvious reasons we try to choose areas which cut across the individual disciplines and which it would clearly be more difficult for the individual unions to handle. One of four earlier projects led to the establishment of the International Council of Associations for Science Education (I.C.A.S.E.) which links science teacher associations world wide. One of our current projects, which we started about three years ago, concerns cooperation between teachers of mathematics and teachers of other sciences. We felt that there had been a history of conflict and that a valuable role for C.T.S. would be to try to encourage a more cooperative spirit. After considerable planning effort we managed to bring about 60 people together in Bielefeld in September 1978. I.C.M.I., C.T.S., U.N.E.S.C.O. and I.C.P.E. all collaborated and the finance came largely from U.N.E.S.C.O. and from an Industrial Foundation. A full report, prepared by Prof-Dr. H-G Steiner has now been published by the Institute fur Didaktik der Mathematik, Universitat Bielefeld, Universitatsstr.l, 4800 Bielefeld l, F.R.G.

We were particularly anxious to produce some pragmatic solutions - or rather modes of operation that might help and, among other things, have produced a series of booklets which it is hoped will be published by C.T.S. in the autumn of this year. Ordering information will be published in newsletters and journals, or may be obtained from the Secretary of C.T.S. Mr. John L. Lewis, Malvern College, Malvern, Worcester, U.K. or from the I.C.S.U. secretariat at 51 Bd de Montmorency, 75016 Paris, France.

We kept very much in mind not the teacher who is enthusiastic enough to come to a conference such as this, but rather the ordinary class teacher in an average school with large classes and a very full time-table. We felt that, if we could produce some short booklets written in most cases jointly by math and science teachers, we might at least give them a talking point. There are booklets relating Mathematics to Biology, Chemistry, Geography and Physics and also some on mathematical modelling.

As an example, the Physics booklet is called 'Functions and Physics' and uses Hooke's law to illustrate a linear function, acceleration to illustrate a quadratic function and Ohm's law to illustrate a more general function. In each the text deals with the physics, the traditional mathematical approach and the "New Mathematics" approach. The aim throughout is to encourage mutual understanding and cooperation between the teachers.

Throughout the development of the sciences there has been constant interplay with mathematics and we need to reflect this in our teaching.

An example that I used at Bielefeld is taken from my own specialism of crystallography. The phenomenon of X-ray diffraction was discovered experimentally in 1912. A mathematical formulation was produced quickly by von Laue but, though quite complete, was not in a usable form. Sir William and Sir Lawrence Bragg in the course of a year or two brought their pysical intuition to bear on the problem and the led to the now famous Bragg Law. This can now be seen to be merely a restatement of Laue's equations, but it is a form that is of immediate practical use. There were several more cycles of physics-math interaction before the elegant Fourier optics methods could be applied and we now see X-ray crystallography as a branch of imaging theory and indeed many of us now view the Bragg law as slightly misleading. But it is abundantly clear that the physical insight that led to it was essential to the development. The strength of our subject has been the constant interplay and cooperation between physicists, mathmematicians, chemists and many others.

C.T.S. is now anxious to pursue the theme of cooperation through to University level teaching. The problems are more complex: University departments are jealous of their independence and autonomy and, given the present situation with established departments, the problems of encouraging cooperation are formidable. Dr. Niss has told us of his integrated courses in Denmark, but they were made possible because new administrative and course structures were being set up de novo. However, we want very much to use this session at ICME IV to discuss possibiities with you and to hear your suggestions for roles that C.T.S. might adopt.

One positive suggestion already made is that we might be able to sponsor the preparation of banks of problems which are authentic pieces of physics, chemistry, geography, biology, etc. but which are classified and identified in terms of the mathematical techniques used in their solution.

On the general note of cooperation I should like to conclude by reading to you an extract from a letter which was written by Michael Faraday to James Clark-Maxwell on November 13, 1857.

"There is one thing I would be glad to ask you. When a mathematician engaged in investigating physical actions and results has arrived at his own conclusions, may they not be expressed in common language as fully, clearly and definitely as in mathematical formulae? If so, would it not be a great boon to such as we to express them so - translating them out of their hieroglyphics that we also might work upon them by experiment. I think it might be so, because I have always found that you could convey to me a perfectly clear idea of your conclusions, which, though they may give me no full understanding of the steps of your process, gave me the results neither above nor below the truth, and so clear in character that I can think and work from them. If this be possible, would it not be a good thing if mathematicians writing on these subjects, were to give their results in this popular working state as well as in that which is their own and proper to them?"

RELATING THE TEACHING OF PHYSICS AND MATHEMATICS

A.P. French
Massachusetts Institute of Technology
Cambridge, Massachusetts

Introduction

I should like to join my non-mathematics colleagues on ICSU-CTS in expressing appreciation for the opportunity to speak to this group of mathematics teachers about questions of common interest. In doing so I am very conscious of the fact that there are many here who have devoted far more time and effort than I have to these matters.

Although it is well agreed that physics is much closer to mathematics than is any other science, Istill find it a little forbidding to appear as a single physicist before so many mathematicians. I have something of the feeling expressed by Goethe, who once wrote (and perhaps I should apologize to my French friends in quoting it): "The mathematician is a kind of Frenchman. If you say something to him, he immediately translates it into his own language, and it becomes something quite different!"

In thinking about what to say about the relation of the teaching of mathematics and the teaching of physics, it occurred tome that it might be useful (and perhaps this would be the case with respect to other sciences also) to regard the problem as having three aspects:

1. The relation and interaction between mathematics and physics as scholarly disciplines per se.
2. The relation between mathematics and physics teaching.
3. The relation between mathematics teachers and physics teachers.

I. The Relation between Mathematics and Physics

I am sure that there is no need, before this audience, to discuss in any detail the close historical relationship between mathematics and physics. One has only to think, for example, of geometry, and its roots in geodesy and astronomy, or of the role of calculus and differential equations in celestial mechanics, etc., or the origin of Fourier analysis in the problems of heat conduction

Professor J. Provost, in a talk earlier in this congress (Session 268, Mathematics and the Physical Sciences and Engineering), went so far as to say "If most mathematics concepts had not originated in physical situations, one could not understand the applicability of mathematics to physics." I would not go quite that far myself, for it seems to me that there are branches of mathematics that were developed before their relevance to physics was recognized - - for example, the theory of matrices and its later relation to quantum mechanics. But of course it remains a marvelous fact that mathematics does provide the appropriate language throughout physics - - a fact that the great theoretical physicist Eugene Wigner made the subject of his essay entitled "The Unreasonable Effectiveness of Mathematics in the Natural Sciences (Richard Courant Lecture, New York University, 1959, reprinted in E.P. Wigner Symmetries and Reflections, Cambridge, Mass., MIT Press, 1970).

There is nothing further that I would wish to say on this aspect of the relationship.

II. The Relation of Mathematics Teaching and Physics Teaching

This is the main part of my talk, and under this heading I would distinguish three components.

The first, and simplest, is just that of language. The different branches of physics have developed a huge variety of symbols, in contrast to the relatively limited vocabulary needed for basic mathematics instruction, and every physics teacher knows that this can make it very difficult for students to transfer their knowledge of mathematics to the solution of physical problems. It may seem absurd, but can nevertheless happen, that a student who can calculate the area under a given part of the hyperbola xy = constant may be daunted by the task of calculating the work done in the expansion of a gas as described by the equation pV = constant. And I know of a question asked in all sincerity by a freshman student of his university physics teacher: "Are the sines and cosines you use in physics the same as the ones used in mathematics?"

This question of symbolism is perhaps trivial (which does not mean that it is umimportant) but a more profound difficulty of language is, I think, presented by the fact that mathematics is the science of pure numbers, whereas the equations of physics are for the most part in terms of quantities with physical dimensions attached. In my opinion, it is misleading to present students with an equation such as

$$x = t + t^2 + t^3,$$

where x is the distance and t is time, because the relation between x and t necessarily involves coefficients whose values depend on the particular system of units being employed, and I think it is confusing if the presence of these quantities is not made explicit by putting

$$x = at + bt^2 = ct^3.$$

Admittedly, the statement of the connection between x and t is ultimately a connection between their numerical measured values, but it must be borne in mind that the numerical correctness of the relationship will be vitiated by a change of units unless the coefficients a, b, c are modified accordingly.

On the positive side of the "linguistic" aspects of mathematics in physics, I would urge the wonderful power of the graphical representation of the connection between two quantities. To display at one glance the whole dependence of a function upon its independent variable is, I believe, far more illuminating than simply to express the relationship as an equation, and I always advocate the fullest possible use of graphical representations.

Turning now to the question of style, it is of course well known that the average student of physics is interested in results and applications. A concern with mathematical rigor, or a presentation heavy in formalism, is generally repellent to physics students.

Professor F. van der Blij made this point in his stimulating talk at another session (Session 323, Teaching Applications of Mathematics).

A closely related matter is that the mathematical statement of a physical relationship may involve terms of different degrees of importance. The equation is not sacrosanct, and it may be that certain terms can be ignored without significant loss of accuracy - - and perhaps with a great pain in the mathematical tractability of the equation. As a physics teacher, I know that one has to work to impress this on students. They tend to treasure the seeming exactness of a mathematical equation; it is a sort of security belt. Moreover, students tend to think that putting numbers into an equation, to explore the relative importance of different terms, is somehow intellectually rather low-brow - - overlooking the fact that the equation, as a summary of physical experience, is empty until one puts in specific values of the quantities.

One last remark on this matter of style. In teaching the mathematical formulations of physical principles we physics teachers far too often present the theory as if it were an axiomatic, deductive business. By and large, students like this - - it seems clear and comfortably definite and authoritative. But I think it is deceptive, and does not correspond to the way physics has developed. When we do a piece of research, it seldom happens in the organized and orderly way in which it gets described in the published paper. And I suspect that the same must be true in mathematics also. I imagine that mathematicians play and experiment much as physicists do before writing a neatly organized statement of a new theorem. I think that we ought to show more of this real-life aspect to our students, to wean them away from an uncritical acceptance of seemingly exact and external truths.

These last remarks provide a convenient introduction to my third topic concerning mathematics courses for physics students - - namely the question of actual content.

The first year mathematics at university level, calculus dominates the scene. The session "Is Calculus Essential?" (Session 131) at this congress seemed to confirm its position in this regard. But calculus is the language of precisely deterministic physics, and taken by itself it gives a misleading picture of how nature operates. Students ought (as Professor Phillip Morrison at MIT has urged) to be exposed to probability and randomness as well. The need for this in atomic and quantum physics is well known, but what is less often stressed is that the exact predictive power of classical theory is often illusory in practice. Consider one of the basic exemplifications of probability theory - - the tossing of a coin. Although it is in principle a problem in classical dynamics, the phenomenon is in practice so complex that only probabilistic statements about it are, in general, possible. I agree with Phillip Morrison that some of the mathematics of stochastic processes should be a part of introductory mathematics courses for physics students.

Another topic of great relevance is information theory. The use of relevance of this, for example in the microscopic theory of thermodynamics, was discussed by Professor Roman Sexl at the Bielefeld conference in 1978 ("Cooperation between Science Teachers and Mathematics Teachers," ed. H.G. Steiner, publ. by Institut fur Didaktik der Mathematik der Universitat Bielefeld, 1979, pp. 42-49). And, of course, there is the whole enormous impact of computers and pocket calculators, which open the way to handling almost any physical problem, regardless of whether it can be analyzed in terms of clean mathematical functions.

This brings me to my last remarks, as a physics teacher, about mathematics courses for non-mathematicians.

I have long been struck by the fact that the variety of functions and functional relationships that are really essential for the teaching of elementary physics at university level is, in fact, very limited. One can go a very long way with nothing more than the following:

x_n

$\sin x, \cos x$

$e^x, e^{-x}, \log x$

Now it certainly does not require a whole calculus course to teach students to become proficient in handling these few functions. What message does this carry for the mathematics teacher? I would suggest that it should encourage the mathematics teacher not to feel under pressure to make his (or her) mathematics course totally subservient to the perceived needs of physics students, but rather to aim primarily to convey something of the nature of mathematical thought. (As a physicist, teaching many students who are "comsumers" of physics, not future physicists, I feel the same way. There is no general consensus by the consumers as to what specific topics ought to be included in the syllabus.) It seems to be that, in the teaching of both mathematics and the sciences, we tend to accept an excessively narrow view of our role. We concentrate on the direct application and utility of our subjects, and in doing so we do not convey, as teachers in the humanities do, a picture of our subject as an important part of the intellectual heritage of mankind. I believe that this is something to which we should devote very serious attention.

To conclude these remarks about the relation of mathematics teaching to physics, I should like to urge that the mathematics teacher see his or her primary role to be the communication of mathematical ideas and ways of thinking. By all means illustrate and motivate them with the help of physical examples and applications, but do not let the physics (or other science) usurp center stage.

III. The Relation of Mathematics Teachers and Physics Teachers

This third aspect of the question I have tried to discuss is no doubt the most difficult and intractable. It involves not only the complications that are always liable to arise when dealing with human individuals, but also the pressures and constraints resulting from the fact that the teachers are embedded in educational systems and cannot, even if they wish, function as free agents.

Even in the most favorable circumstances the difficulties may be severe. I can cite my own personal experience of discussing with a good mathematical colleague the appropriate design of a calculus course to accompany our introductory physics course at university level. With the best will in the world my colleague could not see his way to fully keying his discussion of mathematical topics to their use in physics. For example, the calculation of work as the integral of a force along a path is a topic that comes early in the treatment of mechanics in our first physics course, whereas the mathematician felt that he could not begin to do justice to the mathematical concepts involved until quite far along in the second course in calculus. The moral to be drawn from this is perhaps that, although a maximum of consultation is desirable, one should not be disturbed if the meshing is less than perfect. The student will see a piece of mathematics presented in two different ways - - first as a blunt tool and then as a delicate instrument. He will benefit from both. We are back, in fact, at some of the considerations of Part II of this talk.

At a deeper level there is the question of how physics teachers and mathematics teachers can or should, in principle, interact in a mutually fruitful way. I could not do better here than to refer again to the talk by Professor Provost in Section 268, and to the extensive discussion of such questions by himself and Dr. H. Naggerl at the Bielefeld conference (see Bielefeld Conference Proceedings, op. cit., pp. 83-106, 121-145, and 230-254).

I was particularly struck by a diagram that Professor Provost presented to indicate one (undesirable) picture of the way in which mathematics and physics teachers seek to cooperate (see figure). The assumption on which it is based is that the realms of physics and mathematics have a meeting ground in the mathematical modeling of physical phenomena.

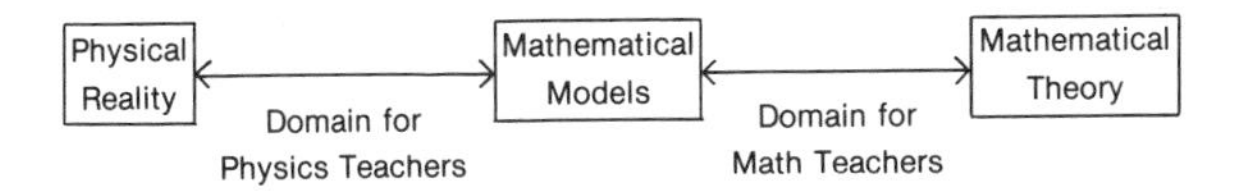

His claim is, however, that this view of the process encourages the development of a situation in which the physics and mathematics teachers accept different areas in which to operate, and that the mathematical model, instead of representing a fusion of their efforts, acts instead as a barrier between them. His view is that there is the possibility of a genuine partnership based upon (a) more specific attention to the historical development of mathematics along with physics, and (b) a greater awareness of how mathematical concepts underlie certain areas of physics but could be made much more explicit - - for example, electrical resistance (possibly non-linear) as representing a functional relationship between voltage and current.

I cannot pretend to have any solutions to offer in this whole question of the relation of mathematics and physics teaching, nor am I under the illusion that anything I have said is novel or unfamiliar. But perhaps this is a good place to stop, with the knowledge that mathematics and science teachers have plenty to discuss with one another in the future, and that our getting together at this Congress is only a beginning.

SOME LESSONS FROM RESEARCH IN SCIENCE EDUCATION

Robert Karplus
University of California
Berkeley, California, U.S.A.

Consider a teacher who writes the following items on the chalkboard (Kurtz, 1976):

X = 4	X = 6
Y = 8	Y = ?

The teacher then challenges the students to propose numerical values of Y that should be associated with X = 6, and to justify their solutions with reference to illustrative examples. Many possibilities exist, of course. Two simple ones are Y = 12 (4 pieces of gum cost 8¢), 6 pieces of gum cost 12¢) and Y = 6 (two children divide the cookies in a box; when one takes 4 the other gets 8, when one takes 6, the other gets 6).

This activity illustrates the fact that unambiguous computation with numerical values can only be carried out when certain other information is provided also: either explicit conditions that determine the mathematical operations to be used or a context that gives the numerical values a significance from which further conditions may be inferred. In scientific applications of mathematics, the context (motion, chemical change, electric circuits, populations growth) provides these conditions. Science teachers have been very concerned with the issue of context-determined mathematical operations. We believe that the heavy early emphasis in mathematics classes on context-free operations and activities (i.e., dealing with pure numbers) results in a severe handicap for many students who experience difficulty in applying their mathematical experiences to problem solving and scientific data analysis.

The sample activity also illustrates the notion of divergent tasks, distinguished from convergent tasks. A convergent task is one that has a single correct answer, as the questions "5 + 2 = ?" or "How many distinct menus can a restaurant assemble if it has two soups, four main dishes, and three desserts?" A divergent task has several correct answers in addition to numerous incorrect answers, as in the sample activity above or the question "? + ? = 7." Both types of task require the respondent to discriminate. Yet focusing on a single answer emphasizes applying a single remembered procedure, while considering various items in a set of answers calls attention to the concept or principle on the basis of which acceptable replies differ from unacceptable ones. Because they require reasoning and stimulate conceptual learning, divergent tasks could be included beneficially in mathematics teaching to a much greater extent than they are at present.

Divergent tasks, properly designed, can make two further contributions to mathematics instruction: first, they can be interpreted by each student at his or her own level of knowledge and reasoning, thereby addressing the problem of individual differences; second, the possible multiple responses allow for wide student participation in a discussion of the task.

The recently published 1980 AETS Yearbook The Psychology of Teaching for Thinking and Creativity (Lawson, 1979) describes these and other aspects of science teaching that have come from research and curriculum development efforts during the last twenty years. Work in mathematics education has contributed to these ideas also, but has suffered from the limitation that the general public thinks of school mathematics as being largely arithmetic. Since the public has had few preconceptions regarding science teaching, experimentation in this field has been more free and diverse.

In this presentation I will mention in somewhat more detail a few of the ideas that have come from my work with the Science Curriculum Improvement Study (SCIS, 1970-74; SCIIS, 1978; Eakin and Karplus (1976) and the Reasoning Workshops (Karplus and others, 1977). Most important of these ideas is the learning cycle (Karplus and Lawson, 1974; Karplus, 1979), an organizing principle for teaching that provides for self-directed activities by students and instructinal guidance by teachers. It is an eclectic approach that includes active learning by students (Piaget, 1964), presentation of advance organizers by teachers (Ausubel, 1963), and practice through applications. Other features I will describe in this article are wait-time, classroom control, and student autonomy.

The learning cycle consists of three phases (exploration, concept introduction, and concept application) rather similar to suggestions made by Edith Biggs at ICME II (Biggs, 1973), but only rarely employed in mathematics teaching. During exploration the students learn through their own actions and reactions to a new situation, new materials, and new ideas, with a minimum of guidance. The exercise described at the beginning of this article was developed as an exploration activity. The new experiences should raise questions the students cannot answer with the accustomed reasoning and familiar knowledge. The students' actions give the teacher valuable diagnostic information because the unstructured situation does not limit their thoughts and allows them to express their preconceptions.

The second phase, concept introduction, builds on exploration and focuses on the introduction of a new concept, principle, or technique - - the principle that the context determines the mathematical operation, for instance, as in my introductory illustration. The concept may be introduced by the teacher, a textbook, a film, or a knowledgeable fellow student, but it is presented within the background furnished by the exploration experiences.

In the last phase of the learning cycle, concept application, the students apply the new concept to additional examples. Problems in which a context for data is presented would be suitable application activities involving constant sum, constant product, constant ratio, and constant difference constraints in a learning cycle begun by my illustrative example. The application phase is necessary to extend the applicability of the new concept or principle. It provides additional time, experiences, and review for students whose understanding grows more slowly than average, or who did not relate the new idea to their experiences. Group discussion of everyday examples, posing of related problems by members of the class, and comparing various interpretations furnish a social setting that helps many students refine their thinking.

A classroom technique developed by Mary Budd Rowe during her work with the Science Curriculum Improvement Study depends on what she called wait-time (Rowe, 1973). Rowe found that the average time allowed by a teacher for a student to answer a question was about one second. When students did not respond quickly enough, most teachers repeated their question, asked another question, answered their own question, or called on another student. Only the fastest students could respond during the short time the teacher waited, and their responses were brief and superficial.

While a very fast pace of questioning may have value for certain special purposes, it is clearly unsuitable for teachers who would like to stimulate thoughtful participation in science or mathematics by their students. In Rowe's invesgiations, teachers were directed to extend their wait-time, the time before they spoke again, to five or ten seconds. The results were very worthwhile. Students expressed their thoughts in whole sentences rather than single words. They tended to speculate about possibilities and implications rather than confining themselves to the narrow aim of the original question. Their self-confidence increased and they supported their answers with evidence. During the seconds of silence after one student's reply, others often described concurring or contradictory ideas and asked questions of their fellow students. Many students who had previously remained silent participated in discussions.

The control a teacher exerts in the classroom through pacing or other techniques can either limit or enhance the students' participation. Though silence is often threatening to the teacher, it creates opportunities for student involvement that are very valuable.

Another aspect of teacher control in the classroom is the distinction between social and intellectual control. The former has to do with the procedures and responsibilities in the classroom, the latter with the content or substance of activities or discussion. Many teachers blur the distinction between these two areas of control and act as though a student's questioning of a conclusion were a disruptive act. If genuine intellectual issues in science or mathematics are to be treated in the classroom, teachers must recognize that they can maintain social control - - an orderly flow of activities and discussion - - while relinquishing intellectual control and encouraging students to challenge one another's ideas and the teacher's, too.

Underlying all of my remarks has been a stress on student participation, involvement, and autonomy. This emphasis reflects my conviction that individuals construct knowledge for themselves, and that learning can take place only if provision for such construction is made. Research in mathematics education should address student autonomy in mathematics classrooms by investigating ways in which it can be enhanced; this should result in more effective learning within the setting of group instruction characteristic of most schools.

In conclusion, I should like to mention that student autonomy has great motivational value as well as contributing the cognitive growth. Doing something because you want to do it, and doing it in the way you want to do it, are powerful incentives. After a class visit during which I administered a group test on proportional reasoning (Karplus, Karplus, Formisano, and Paulsen, 1979), one student wrote on her test paper, "I . . . myself have really enjoyed you here today. And I have really learned how to say and write what I and only I think, with no one else to try to tell me. So I hope you come back again." Unfortunately, I did not have the opportunity to go back. But I do hope that some teachers will allow this student more autonomy than she apparently had experienced.

References

Ausubel, D.P. The Psychology of Meaningful Verbal Learning. New York: Grune and Stratton, 1963.

Biggs, E. Investigations and Problem Solving in Mathematical Education. In Howson, A.G., (Ed.), Proceedings of the Second International Congress on Mathematical Education. London: Cambridge University Press, 1973.

Eakin, J.R. and Karplus, R. SCIS Final Report. Berkeley, CA: Lawrence Hall of Science, 1976.

Karplus, R. Teaching for the Development of Reasoning. In Lawson, 1979, cited below.

Karplus, R., Karplus, E.F., Formisano, M., and Paulsen, A-C. Proportional Reasoning and Control of Variables in Seven Countries." In Lochhead, J. and Clements J. (Eds.) Cognitive Process Instruction. Philadelphia, PA: Franklin Institute Press, 1979.

Lawson, A.E. Ed. The Psychology of Teaching for Creative Thinking and Creativity. 1980 AETS Yearbook. Columbus, OH: ERIC/SMEAC, 1979.

Kurtz, A Study of Teaching for Proportional Reasoning." Doctoral Dissertation, University of California, Berkeley, 1976. Ann Arbor, MI: Xerox University Microfilms, 77-15, 747.

Piaget, J. Cognitive Development in Children: Development and Learning. Journal for Research in Science Teaching. 2, 176-186, 1964.

Rowe, M.B. Teaching Science as a Continuous Inquiry. New York: McGraw-Hill, 1973.

SCIS (Science Curriculum Improvement Study) Chicago: Rand McNally, 1970-74.

SCIIS (Science Curriculum Improvement Study, Rev.) Chicago: Rand McNally, 1978.

CROSS REFERENCES OF ARITHMETICAL TASKS TO PHYSICS AND MATHEMATICS IN THE ELEMENTARY AND SECONDARY SCHOOLS

Gerard Vergnaud
Institut National de Recherche Pedagogique
Paris, France

I am concerned with the psychological problem of "how concepts grow in the student's 'mind'", and it is mainly for psychological reasons that I find it important to coordinate research of maths education and science education. Probably my view of the problem comes also from the fact that the French system of education is especially bad at linking to each other maths and science curriculae. But I will emphasize other reasons that seem to be more meaningful.

Let me recall two Piagetian theses.

> The first one says that concept development consists essentially of building up new invariants, for classes in situations in which transformations occur and modify some aspects of the objects involved.
>
> The second one says that mathematical knowledge relies upon abstraction from the subject's actions, whereas physical knowledge relies upon abstraction from the objects' properties.

I will disucss this last thesis in my conclusion.

Most of my recent research work has been devoted to the study of arithmetical tasks. They are generally viewed as involving pure mathematical concepts: the set of natural numbers and its different extensions (directed numbers, rational and decimal numbers, real numbers . . .)

The main point I want to raise is that solving arithmetical tasks, especially natural language problems (stories), cannot be studied in the light of the number framework alone, but requires other frameworks. It is necessary to know more about the relationships that students take into account. Two concepts appear to be very important, those of time and dimension, but familiar to physicists; and this is the reason why I think that it would be misleading to separate research on maths education and on science education.

Another important thesis is that, in analysing problems, students need to extract relationships and transform or compose them. It has led me to identify what I call "relational invariants" and "relational calculus". By this I mean that it is not sufficient to identify qualitative invariants (like characteristic properties of a class) or quantitative invariants (like those met in conversation tasks); there are also relational invariants, the properties of which are used in a relational calculus. I will give some examples:

Multiplicative Structures

Let us start with an elementary example "A pencil costs 4 francs. Peter buys 5 of them. How much does he have to pay?"

There are four quantities involved: 1 pencil, 5 pencils, 4 francs, ___ francs (unknown)

pencils	francs
1	4
5	___

and the child has to extract from this four-term relationship a three-term relationship. He can extract it in different ways.

- Binary law of composition in N (natural numbers),

$5 \times 4 = \_\_$

This does not make meaningful the fact that you get francs by multiplying pencils and francs.

Unary operation of $\mathbb{N}$ upon $\mathbb{N}$,

first case : $4 \xrightarrow{\times 5} \square$

second case : $5 \xrightarrow{\times 4} \square$

These two cases are not equivalent.

The first case represents a scaler operation, using the idea that the same relationship exists between I pencil and 5 pencils on the one hand and between 4 francs and __ francs on the other hand.

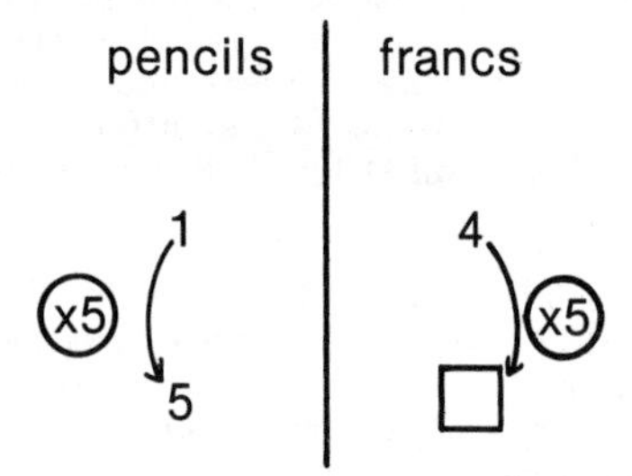

The invarient is scaler. The linear property used is $f(\lambda x) = \lambda f(x)$.

The second case is a functional operation, using the idea that the same relationship exists between I pencil and 4 francs on the one hand, and between 5 pencils and francs on the other hand.

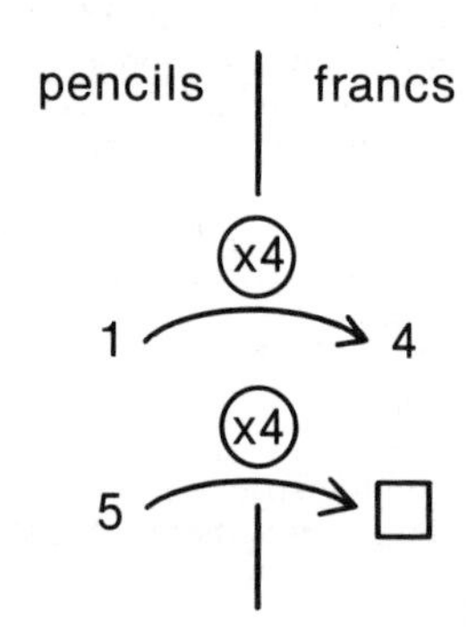

The invariant is a function, and a quotient (or ratio) of dimensions (francs per pencil). The linear property used is $f(x) = a\,x$.

From this elementary example, one can easily see that there are two sorts of divisions:

pencils	francs
1	□
a	b

find the unit value f(1)

pencils	francs
1	a
□	b

find a quantity x; f(x) = b

Thus, there are different procedures to solve classical "rule-of-three" problems:

pencils	francs
a	b
c	□

Several experiments show that scalar operations are more naturally used by 10 to 15 year-old students, even when it is arithmetically easier to use functional operations. Among 25 different procedures observed for "rule-of-three problems", one can see the importance of the dissymmetry above-mentioned, not only among successful procedures but also among failures.

These results will be shown in the miniconference and have been published in references I and 2.

I suggest to use the term "isomorphism of measures" to name the simple proportion structure between two measure-spaces. I have two reasons for this:

the first reason is that students use preferably, and master better, the isomorphic properties of the linear function

$$f(x + x') = f(x) + f(x')$$

$$f(\lambda x) = \lambda f(x)$$

$$f(\lambda x + \lambda' x') = \lambda f(x) + \lambda' f(x')$$

the second reason is that this structure must be distinguished from another structure, the "product of measures", in which one measure is the product of two others: for instance area, volume, cartesian product . . . It is confusing for students not to make distinct analysis for distinct cases of multiplication and division such as

pure numbers n x m = p (Pythagoraus's table)

product z = l x w (rectangle area = length by width)

scalar operator $f(x) = f(1) \times x$

functional operator $f(x) = a\,x$
example d - v t (distance = average velocity by time)

The "product of measures" involved bilinear (or trilinear or n-linear) functions, whereas the "isomorphism of measures" involves linear functions. Some experiments on the concept of volume show that it is a difficult concept up to the age of 15, especially as regards the trilinear aspects, which are essential in the concept. (Results shown in the miniconference.) There are important didactical consequences.

1. The best-used properties of proportion (isomorphic properties) are not often explicited in teaching. It is worth doing it and helping students to do it themselves.

2. For this, the equation symbolism is probably not the most adequate; and other symbolisms such as functional tables, arrow-diagrams and operators, or dimensional considerations can be more efficient, at least at the elementary and secondary levels (up to 15).

3. The link between the physical approach of concepts and the mathematical analysis of them is essential. Density of objects along a line, on a surface, or in a volume, area, velocity, volume, volumic mass can be and should be approached and explained from both the points of view of physics and mathematics. There are difficulties due to the physical aspects (volumic mass and velocity are both quotients of dimensions, but are not mastered in the same way). There are also mathematical difficulties (composition of ratios, multiplication and division of numbers smaller than 1) that are common to different physical concepts and to different structures.

Additive Structures

Mathematicians and maths teachers usually get rid of the concept of time, although situations grasped by students involve the time as an essential component. Additions and subtractions occur very often in situations that are better fitted by the transformational model:

transformation (positive or negative)

$$\underset{\text{Initial State}}{I} \xrightarrow{\quad T \quad} \underset{\text{Final State}}{F}$$

than by the binary composition model (addition of measures):

$$\left.\begin{matrix} M_1 \\ M_2 \end{matrix}\right\} M_3$$

There are two major consequences.

1. The relational calculus needed to solve the problem is distinct from the numerical calculus.

2. Equations in N are not adequate to represent problems in which transformations occur, because the set of possible values of transformations is the set of directed numbers (whole or decimal numbers).

The same inadequacy exists for the representation of static relationships. The situation is still worse when modeling the composition of two transformations, of two relationships, or the operation of a transformation upon a relationship.

These considerations are very important for the study of developmental complexity, and one can show that the relational calculus is a prominant factor of complexity. Let us take three problems as follows:

A. There are 4 boys and 7 girls around the table. How many children are there together?

B. John has just spent 4 francs and has now 7 francs in his pocket. How much did he have before?

C. Robert has played two games of marbles. On the first game he lost 4 marbles. He played the second game. Altogether he has won 7 marbles. What happened in the second game?

Although a simple addition 4 + 7 is needed in all three cases, B is solved one or two years later than A, and C is failed by 75% of 11 year-old students.

So, the concepts of time and transformation play a very important part in the facility rate of different classes of problems. They may make things easier: young children can easily solve addition and subtraction problems in the "find-the-final-state" case;

it may also make things worse: for instance in the "find-the-initial state" case that requires the inversion of the transformation (problem B) or in the case exemplified by problem C (subtraction of opposite-sign transformations).

Conclusion

One of the main cognitive problems for children is to understand the physical world (and also the socio-economical world), to act efficiently upon it and to make reliable predictions upon what is going to happen. Classifying, counting and measuring, operating on classes, on measures, on relationships and transformations are essential activities.

Most of these activities are both mathematical and physical. The concept of number would not even exist if there were no physical quantities to count (discrete) or measure (continuous). Reciprocally, arithmetization of physical concepts is essential to physics. The monotonous character of proportions is acknowledged by young children (6 or 7 year-olds), but it is essential that they go beyond monotony and come to add, subtract, multiply and divide. Volume as occupied by space, or as empty space, is qualitatively grasped by young children, but measuring volume (capacities, for instance) goes one step further, and relating volume to elementary dimensions goes still further; analysing the trilinear character of volume is not a trivial job for 15 year-old students.

Similar problems exist for additive structures, which rely upon the concepts of measure, of transformation, of relationship.

The concept of number appears to be abstracted from all these physical concepts and not, as Piaget says, from the subject's actions. Measures and additive transformations are synthetized in the concept of directed numbers. Similarly the concept of rational number would hardly be constructed with all its properties if it was not referred to the concepts of fractional quantities (that can be added) and fractional operators (that can be multiplied by one another).

The concept of function is undoubtedly mathematical but part of its meaning comes from physical operations like transformations, comparisons and correspondences. So there is no reason to oppose maths and physics, but on the contrary, it is fruitful for research on education and on concept development to keep in mind both references.

This does not mean that maths can be reduced to physics. Once the concept of number is built up, it can be involved in problems which have not much to do with physics. The same is true for other mathematical concepts. But at the primary and secondary schools, it is better not to forget the intricate connection of mathematical and physical concepts.

References

1. Vergnaud, G., Rouchier, A., Ricco, G., Marthe, P., Metregiste, R., Giacobbe, J. - Acquisition des structures multiplicatives dans le permier cycle du second degre. I.R.E.M. d'Orleans 1979.

2. Vergnaud, G. et al. - Didactics and acquisition of "multiplicative structures" in secondary schools. In Archenbold, W.F., Orton, A., Driver, R.H., Wood-Robinson C. (Eds.) Cognitive Development Research in Science and Mathematics. The University of Leeds Printing Service, 1980, 190-201.

3. Vergnaud, G. - ·A classification of cognitive tasks and operations of thought involved in addition and subtraction problems. In "Seminar on the initial learning of addition and subtraction skills" held at Wingspread and organized by Carpenter, T.P., Moser, J.P., Rombert, T.A. University of Wisconsin, 1979.

4. Vergnaud, G. - The acquisition of arithmetical concepts. Educational Studies in Mathematics, 10 (1979), 263-274.

5. Rouchier, A. - Situations et processus didactiques dans l'etude des nombres rationnels. Recherches en didactique de mathematiques, 1 (1980), 115-276.

6. Vergnaud, G., Errecalde, P. - La coordination de l'enseignement des mathematiques entre le cours moyen 2eme annee et la classe de 6eme. Recherches Pedagogiques, 102, 1979.

13.10 CENTRAL RESEARCH INSTITUTES FOR MATHEMATICAL EDUCATION. WHAT CAN THEY CONTROBUTE TO THE DEVELOPMENT OF THE DISCIPLINE AND THE INTERRELATION BETWEEN THEORY AND PRACTICE?

MATHEMATICAL EDUCATION AT THE NATIONAL INSTITUTE OF EDUCATION

Edward Esty
National Institute of Education
Washington, D.C., U.S.A.

In the United States, education is an area that, according to our Constitution, is the responsibility of sate and local government. There is no central direction provided by the Federal government, whether in curriculum, teacher certification, testing, or any other aspect of education. Nonetheless, there is an Education Department at the Federal level, one of several departments within our executive branch. Its major purpose is to provide Federal assistance, as directed by our Congress, to state and local governments via a large number of programs aimed at various audiences. The total budget for the United States Education Department for the next fiscal year will be about $14.2 billion.

One relatively small part of the Education Department is the Office of Educational Research and Improvement; a part of that office is my agency, the National Institute of Education (NIE). NIE was formed as a separate agency in 1972, as the chief agency at the Federal level to support educational research. During its relatively brief life it has undergone several changes in direction, several changes in Director, and several reorganizations. Now, however, it looks as if we have a stable organization, a Director who will be around for a while, and a secure niche within the Education Department. To keepmatters in perspective, I should state that our budget for the next fiscal year will be about $91.2 million, roughly six-tenths of one percent of the total for the Education Department.

NIE is divided into three major program areas. One is called Dissemination and the Improvement of Practice, it operates regional information exchanges, promotes the dissemination of research results, and supports the national educational information clearninghouse - - the ERIC system. The second program area is called Educational Policy and Organization; it supports research on matters like educational finance, school organization, and the effects of legal decisions on education. The third program area, by far the largest, is the Program on Teaching and Learning. It in turn in split into five divisions: (1) Reading and Language Studies, (2) Education in the Home, Community and Work, (3) Teaching and Instruction, (4) Testing, Assessment and Evaluation and (5) Learning and Development. All of these divisions except the first, Reading and Language Studies, conduct some research that is related to mathematics, but most of the mathematics-related work is concenrated in the division called Learning and Development.

It would be impossible to give a complete account of all the sorts of activities we support in manthematical education in the short time that has been allotted to this session, so instead I will try to highlight some of the more important one:

1. The annual Teaching and Learning research grants competition. For several years now, one of the areas within the Teaching and Learning grants competition has been a section on Mathematics Learning. Through this grants competition we support research studies in three broad areas: (a) the development of mathematical concepts in children and adults, (b) analyses of mathematical tasks, and (c) the nature and development of mathematical problem-solving and reasoning abilities. As a result of the FY 1980 competition we will fund fifteen projects, from one to three years in length, spanning a very wide range of topics. Experimental subjects range from pre-schoolers to adults from various cultural and ethnic groups, and the topics from mathematics include arithmetic, geometry, probability, computer programming, and algebra. We feel that it is important to have a consistent competition from year to year, so that researchers will be able to count on a source of support for excellent research ideas. We hope to be able to attract top-flight researchers in more general areas of human learning

into studying some of the fascinating problems connected with mathematical education, and of course we hope to retain the enthusiasm and commitment of talented researchers who are already in the field. Unfortunately, because of severe budget constraints in the next fiscal year, it now looks as if the subtopics of the Teaching and Learning grants competition are going to have to alter starting in FY 81. This means that the Mathematics Learning area will not be active in FY 81, but I think that it will be possible for researchers to find another area within the T&L competition that will accept their applications, perhaps slightly modified.

2. A second mechanism through which we support research studies is the Unsolicited Proposals Program. This program is for people who have ideas that do not fit nicely within any other program in the agency. Proposals in Mathematics Learning would be accepted here during the fiscal year beginning October 1.

3. The NSF-NIE joint program. The National Science Foundation is collaborating with the National Institute of Education in a joint program on the improvement of mathematics education using information technology. During its first stage the program will support the development of prototypes of educational software and courseware, using microcomputers, videodiscs, and similar technological devices. The first deadline was in February and the second is next week.

Nine projects will be supported jointly by the two agencies as a result of the first deadline, and NSF will support an additional ten projects. We think that a similar number will be supported as a result of the second deadline. Part of the total program involves studies of the implementation of microcomputer technology in classrooms, for of course the wonders of microporcessor technology will have little effect on education if the devices are not actually used in classrooms. A later phase of the program will be devoted to a smaller number of larger curriculum development projects.

4. Programs with "labs and centers". NIE supports seventeen educational laboratories and centers throughtout the United States. Some of these, like the Wisconsin Research and Development Center for Individualized Schooling, have mathematics components that engage in research on mathematics learning. One of them, the CEMREL laboratory in St. Louis, has been involved for many years in the development of a new mathematics program for grades K-6. Representatives from the Comprehensive School Mathematics Program are here at Berkeley, and I would urge you to speak with them about this very exciting and innovative program.

5. The IEA mathematics study. We are about toenter into an agreement with the University of Illinois to support the final phases of the Second International Mathematics Study. Our support for this important international study started in 1977, and we have provided assistance to both the International Mathematics Committee and the U.S. National Mathematics Committee since that time. The National Science Foundation has provided some money for the work with the younger population (the thirteen-year-olds), but the bulk of the support has come from NIE. I understand that the IEA group has set up a booth in the exhibit area; I urge you to get the latest information on the Study directly from the people who are conducting it.

Space will not permit me to describe our other activities. Instead, I should mention some of the sorts of things that NIE does not do. First, we conduct very little intramural research; Congress has specified that most of money should go to support others to do research. Second, we support no direct teacher training programs, although we do sponsor research on teaching and instruction. Third, our support of curriculum development is very limited. Five or six years ago we were much more heavily into curriculum development, but our Presidentially appointed policy-making board has declared that we should engage in curriculum development only under very unusual circumstances. The two development projects that I have already mentioned, namely the Comprehensive School Mathematicsl Program and the NSE-NIE prototype projects, are among the very few in the whole agency.

In closing, let me say a word about the future. In my view, mathematics education in this country is going through a very critical period. There is a great amount of concern, at the highest levels, about the mathematical and scientific preparation of our citizenry. All indications are that the next few years will see a considerable increase in the amount of support given to mathematical education, support that is desperately needed. I hope that the National Institute of Education, though it has only modest resources, will play an important role in the strengthening of mathematical education in the years to come.

RESEARCH ACTIVITIES IN A FEW IREM'S*

Georges Glaeser
IREM de Strasboug
Strasbourg, France

I. Science vs. Bureaucracy

We cannot tackle the theme of this panel in terms of administrative structures. A research institute is not just a building, an Executive Board, a director, a secretariat . . . it must only be judged in terms of the scientific work actually accomplished there.

Unfortunately, poor material conditions can dampen the zeal of even the most competent and active research teams. A small dose of administrative infrastructure can thus be useful in reducing minor worries for the scientists. But, we must refuse to pay too high a price for such help. The direction should not be allowed to distract those who are working with its bureaucratic red tape.

The history of the "Instituts de Recherche pour l'Enseignement de Mathematiques" (IREM) is a good example. The first IREM's were created in France 11 years ago. There are not about 30 of them. Initially each Institute was given the some wide range of autogestive power. For the last 3 years, by means of continual financial pressures, the central administration has been trying to elininate them.

The IREM's resist these attacks quite differently and their results differ greatly depending on their objectives, the quality of their recruited teams, and the support (or hostility) of the surrounding mathematical community.

This appears particularly evident to me when I look closely at the library shelves in Strasbourg where photocopies of all the different IREM's publications are kept. Several shelves contain many cartons stuffed with documents; others are practically empty.

The contrast is even more apparent when we consider the diversity of the Institute's activities and the quality of their production. Those which have proven to be the most dynamic are precisely those which have the means and the willpower to resist administrative bother and bureaucratic imperatives, and which do not permit themselves to be sidetracked from their professional priorities.

Moreover, the overall achievement of the IREM's is quite worthy. A dozen IREM's are veritable institutions of research. In a dozen others there are rather isolated teams which have produced work of general interest. All the IREM's have participated in continuing education programs whose results are quire apparent today. Very few spent most of their time in self-management (which unfortunately is not the case with a great many other intitutions subjected to Parkinson's hard law).

II. Continuing Education

A large part of IREM's activities are spent in joint reflection on different aspects of the educational profession. Every two weeks in each IREM, several hundred teachers participate in working groups formed around themes of common interest.

Sometimes, it's only a question of acquiring additional knowledge in certain mathematical areas because certain things were not taught or available during the teacher's initial professional training. But, in most cases, the work of each group is centered on questions concerning student participation and classroom tactics; in brief, all aspects of the teaching profession.

For example, a large number of groups concentrated on how to use computers or audio-visual aids in the classroom. Others preferred to study the problems raised by evaluation and statistical analyses of tests. Numerous teachers wanted more information on psycho-pedagogy and group dynamics applied to teaching situations. A large part of the work was consecrated to heuristic training of teachers who were asked to solve problems themselves, and afterwards to pose the actual problems in their classes and to relate their observations of their students' solving behavior to the group. Recently, there has been a great demand for an introduction to the history of mathematics and for reading the classics as well as for epistemology.

All of these different groups have often produced annual activity reports consisting of documents susceptible of helping other teachers.

III. Innovations

A number of IREM's have also tested new pedagogical methods, original teaching materials, and new curriculum proposals. Several IREM's have collaborated in the conception of new textbooks and have tested them in the classroom, noting the different reactions of the teacher-users.

IREM's have also published collections of activities, exercises, and problems either for the students' use or as teaching documentation for educators. Class observation, either by actual classroom visits or by recordings, have been greatly practiced. Finally, together withthe "Institut National de Recherche Pedagogique", alternatives to officially practiced programs have been attempted on a large scale.

Several IREM's have permanent and easy access to a scholarly establishment where researchers can make continual suggestions. But thanks to the close collaboration which exists between the Instituts and math teachers in general, at all levels, it is easy to have access to wide-ranging experiences.

In about a half-dozen IREM's there has been another kind of innovation: the creation of mini-Olympiads, at local level, called Rallys. Thousands of youngsters who are more than happy to solve extra-curricular math problems are assembled during a summer afternoon.

IV. Didactic research

Recently, several IREM's have been engaged in fundamental research concentrated on the elucidation of the mechanisms of mathematical understanding, assiduously following the rigor of pastorian experimentation by using laboratory techniques. This tendency has given rise to a national seminar which now meets four times a year in Paris, during which time researchers submit their work to the critical examination of other experimental didacts. The first edition of an international journal consecrated to this activity has just been published.

This aspect of research in mathematical education is, in my opinion, the trait most characteristic of our time. At last we're trying to study what actually goes on during the cognitive process, when a student understands or doesn't.

Modern teaching is no longer based only on the contradictory opinion of a few VIP's who claim to know how mathematics should be taught. Today, on the contrary, we constantly submit didactic conjecture to the verdict of experience with a rigor comparable to that so highly acclaimed by our founding fathers in experimental medicine a century ago.

I hope that the next ICME conference will devote more time to the latest results obtained in this field.

V. Research Teams

Ten years of experience have taught us the importance that should be accorded to the recruitment of a highly qualified multidisciplinary research team.

In an IREM University professors meet math teachers from varying levels when they study the possible reactions a child may have before a given activity. A mathematician usch as myself must learn to efface himself before more competent colleagues. And often the kindergarten teacher is the expert we should listen to.

The recruitment of psychologists is essential for the formation of a good team. Besides his specific competence in psychology, he must be able to communicate easily with mathematicians. He must be able to use the appropriate vocabulary while expressing his or her point of view. Generally, it takes several years for a good psychologist to attain this quality. The IREM's have also included physicists, linguists, geographers, informaticians, statisticians, and even actors on their research teams.

*This lecture was presented in English, French, and Spanish. The French and Spanish versions may be obtained from G. Glaeser, Universite Louis Pasteur, IREM de Strasbourg, 10 Rue du General Zimmer, 67084 Strasbourg, France.

SHELL CENTRE FOR MATHEMATICAL EDUCATION

Heini Halberstam
University of Nottingham
Nottingham, England

1. The Shell Centre for Mathematical Education began some 12 years ago, chiefly as an in-service unit, with a grant of $140,000 from the Shell Oil Company; four years later it was incorporated within the University of Nottingham as an inter-disciplinary research group with four permanent members of staff. The Shell Centre of Science Education at Chelsea College, London, has had a similar history.

The Centre still has only four established members (plus secretarial help), but now, through its various research contracts (including renewed sponsorship from Shell), there are also nine research fellows, consultants who visit intermittently, and many academic visitors (including school teachers on secondment) who work in the Centre for substantial periods of time. Last but by n4 means least, other university teachers in Nottingham are engaged in innovative work in mathematical education and see themselves as de facto part-time members of the Centre, as also do colleagues from neighbouring teacher-training institutions and numbers of teachers from the East Midlands area.

Thus the Centre is certainly 'central' in that it spearheads a massive regional involvement. It is well-known in local schools where much (though not all) of its experimental work is done and the Centre's members visit regularly in the course of their work. In my opinion, the strong roots which the Centre now has in the East Midlands region, and the respect it enjoys among local teachers, are the foundations, indeed essential components, of its success.

The Centre is 'central' in a national sense only in so far as its own efforts succeed in commanding national or governmental attention. Recently the University Grants Committee indicated to my university that it viewed the activities of the Shell Centre as important and so (by implication) not to be damaged by economies now being enforced in the higher education sector. In such time, being left alone is a kind of approval; from such and other evidence (e.g., being cited in the course of parliamentary hearings on educational matters) it is apparent that the Centre is now seen as a national resource.

2. We started the Shell Centre to provide the environment and opportunity for innovative work in mathematical education to flourish. We could not see any other reliable way of promoting what is essentially multi-disciplinary work within the resources of the existing department on anything like an adequate scale. Some ten years ago a group of us in the UK, including Sir James Lighthill (a past president of ICMI), attempted to raise funds from British industry to start similar centres in other universities and so to provide teacher support for the whole country on a regional basis. Of course, we failed (except here and there); but it is still, in my view, the right way to proceed in a densely populated country. A network of modestly sized, regional, university-based centres represents a structure that is less open to the vagaries of governmental funding than isolated larger institutions might be; that is more likely to reflect the true complexities of teaching a difficult subject than 'ex cathedra' pronouncements from a central national institute; that has the possibility of sustained contact with associated disciplines at a high level of competence; and, above all, a structure that is intimately bound up with the local communities of serving teachers.

I stress these points because I should not wish to associate myself with a belief in the absolute necessity for research institutes as such in any or every area of scholarship.

3. The multi-disciplinary aspect of mathematical education requires special emphasis. First of all, our kind of Institute must have good and sympathetic relations with professional mathematicians of university calibre. It has come into being precisely because mathematics is, uniquely, both useful and difficult. Given that the mathematicalsciences are expanding at a prodigious rate in content and diversity, an Institute divorced, or even distanced, from mathematicians would be courting disaster. (Actually, our Institute should be able to show mathematicians that it has valuable expertise to offer them, too: there are few university departments that could not profit from reflecting, under guidance, on the problems of learning mathematics. In the UK we now have an annual conference on this and related themes organised from the Shell Centre in Nottingham.)

But mathematics is not the only discipline with which our Institute must maintain close connections. Right now we are witnessing the reinstatement of applicability as a legitimate driving force in the creation of good mathematics, and therefore it is important to study the ways in which mathematical ideas feature in the development of other subjects - in genetics (and biology generally), linguistics, geography, and economics, in the preparatin for engineering and medical professions, etc. Again, the Institute must have direct access to where such development is taking place, and to steer knowledge of it into school classrooms.

I have heard it said (perhaps in different words) that educational psychology is no more like psychology than a sea-horse is like a horse. Now there can be no question but that the psychology of learning mathematics is the great challenge to research in mathematical education that every Institute of our kind must accept; and to escape the label of 'mere' educational research, an Institute needs towork in close touch with professional psychologists and others who study the functioning of intelligence, and at the very least to make its researches available to their appraisal.

4. I suppose society has always regarded education as too expensive; but the cost of universal education in our own day is deemed to have reached frightening magnitude and politicians see it as part of their responsibility to demand returns on 'investment' in education in terms of mastery of useful knowledge. Mathematical skills are very much in the firing line. I am sure an Institute of Mathematical Education cannot keep away from the firing, but on the contrary must endeavour with its expertise to help to form public opinion and also be provide specialist services. For example, two years ago the then British Government set up the Cockroft Inquiry into the Teaching of Mathematics in Primary and Secondary Schools; and among a number of specialist reports it commissioned, two are being prepared from the Nottingham Shell Centre, one a critical review of current research on mathematics teaching and the other an investigation into the mathematical needs of school-leavers in their new employments. At about the same time, a mathematician colleague of mine prepared at short notice, in association with an educational psychologist, what has proved to be rather a successful 13-part television programme - 'Make It Count' - to tackle the problem of adult innumeracy. A companion book had also to be produced (for those who were not simultaneously illiterate!). This kind of responsiveness to pressures of the moment should not be seen as an undesirable distraction, but as part of our Institute's work. Education is of the market place, and that is where some of its battles have to be fought.

5. So far I have tended to stress the irreducible academic milieu and the inescapable 'service' aspects of a successful Institute. Of course, a research institute must involve itself also in research! Research is nowadays a much abused word and so especially say all mathematicians who cultivate an unusually exalted view of the nature of research; and so it must be stressed that much of research in mathematical education is, inevitably, of the fact-finding kind or experimental in character. A quick look through the Shell Centre record shows the following activities listed as research - and I should add that several have involved successful thesis supervision:

Developmental research:

a. A project (funded by the Social Science Research Council) looking into the acquisition of strategies of learning mathematics, such as the art of generalising, of symbolising and the ability to explain and justify the outcomes of investigations.

b. Understanding of graphs - found to present serious difficulties to most pupils.

c. Model-building - analysis of skills required in, and development of tests to identify, modelling ability.

d. Collecting examples of everyday life applications of mathematics.

e. Studies in computer assisted learning.

f. Evaluation of a children's TV mathematics program, accompanied by a large scale feed-back study to improve the design of future programs.

Basic research:

g. SCAN (supported by the Science Research Council) - a systematic classroom analysis notation to furnish an accurate record of a lesson as a basis for subsequent discussion between pupil-teacher and tutor.

h. A detailed study of mathematics required in nursing.

i. A detailed study of mathematics required in engineering.

j. Relationship between high school and university performance.

k. Aspects of teacher development.

l. Computers in the classroom.

m. Development of tests of arithmetical skills of school-leavers that will be acceptable to employers.

6. Curriculum studies, leading to periodic reforms, are certainly a major concern of any Research Institute of Mathematical education. They involve the formulation of objectives, trials of viability and plans for sustained implementation. At the discussion stage of objective, it obviously is important to be in close touch with mathematicians and the users of mathematics; whereas tests of viability must involve cooperation with, and sympathetic briefing of, teachers. It is so important to ensure that any appropriate reforms can be implemented without dilution or corruption (as undoubtedly happened with the grand aspirations of the UNESCO Reports of the '50's). 'Reform in haste, repent at leisure!' should be inscribed over the portals of our Institutes! The safest control on hasty change would come automatically from a well-informed teaching profession; we in Nottingham have long maintained in the School of Education a Mathematics Advisory Unit, consisting of teachers' representatives, local authority subject advisers and university personnel, to discuss current problems and initiate cooperative activities. Also, we have since the mid-sixties an effective apparatus for provision of continuing education (in-service training) of teachers, based on the premise that such education is best carried out in the teachers' own schools and, realistically, channelled through heads of departments or specially trained teachers who are able to act as mathematical consultants. None of this would be possible without good relations between the School of Education and the Shell Centre with local authorities and local teachers - a point I have stressed already, but one that is so important that it will serve also to conclude my contribution.

ROLE AND ACTIVITIES IN MATHEMATICAL EDUCATION OF NIER IN JAPAN

Yoshihiko Hashimoto
National Institute for Educational Research
Tokyo, Japan

1. The role of NIER:

There are about 500 educational institutes and educational centers supported by local governments (Tokyo Metropolitan, Osaka Fu, Kyota Fu, Hokkaido, and 43 prefectures and some of cities and towns.) About 200 of them including NIER make up the National Federation of Educational Research Institutes. NIER coordinates its activities through promoting cooperative research projects, such as educational achievement and career guidance, and serving as coordinating centers.

NIER is the institution supported by the government of Japan which performs practical and fundamental research concerning education. While other institutions emphasize in-service education for teachers, our Institute does not carry out any in-service or preservice education.

The section of mathematics education belongs to the Science Education Research Center which is one of eight departments in our Institute. There are four on the staff, of whom three are researchers and one an assistant, and we are engaged mainly in researches on mathematics curriculum and teaching method.

Characteristics of our work may be summarized as:

A. Engaging in an international work (i.e., Second Study of Mathemtics by International Association for the Evaluation of Educational Achievement. Field works of this study in the national level are promoted with the cooperation of some of the institutions mentioned earlier in this paper.

B. Engaging in long-term developmental researches all over Japan.

C. Maintaining a close relationship with the Japan Society of Mathematical Education (JSME) and helping its activities, because the facilities we use were built as one of the memorial undertakings at the 50th anniversary of the JSME in 1970. For example, the U.S. - Japan Seminar on Mathematics Education in 1971 and the ICME - JSME Regional Conference in 1974 were held at our Institute.

2. Theory and Practice in our Researches:

The First Project (1971 - 1977):

In order to evaluate "higher objectives" of mathematics education, we have developed a method using open-ended problems which allow several different solutions according to how students view the problem situation. From the beginning we made a team consisting of members of our Institute, several university teachers, and school teachers from elementary school to senior high school.

Summary of the six years study is as follows:

April 1971 - March 1972 - 1973, developing open-ended problems, implementation in three districts.

1973 - 1974, classroom teaching using open-ended problems in three districts and administration of attitude scale to measure changes in student's attitudes.

1974 - 1975, 1975 - 1976, classroom teaching using open-ended problems in six districts.

1976 - 1977, summary of research.

Findings from classroom teaching using open-ended problems were presented at the Third ICME (Karlsruhe, 1976) with the title, "Developing study on a method of evaluating student's achievement in higher objectives of mathematics education", and included in its proceedings.

As it is very difficult to use a theoretical sampling method strictly in an education research because of administrative and ethical reasons, case study methods become indispensible in order to validate a theory.

The results of these researches were presented at the JSME annual meeting and other regional meetings and published in Journals of JSME.

In addition to this, papers were presented in the commercial journals for education.

Results are summarized in the following books.

S. Shimada (ed.): On lessons using open-ended problems in mathematics teaching. Mizuumi Shobo Co., 1977.

T. Sawada, Y. Sugiyama (ed.): Evaluation in mathematics education. Daiichi Hoki Co., 1978.

Both are written in Japanese. I am sure it is important to announce the results of research as mentioned here in order to disseminate what we have done.
The Second Project (1978 -):

The research has been carried out from 1978 in the following way.
Mathematics teaching by developmental treatment and its evaluation.

April 1978 - March 1979 (grant: 1.5 million Yen)

1979 - 1980 (2.4 million Yen)

1980 - 1981 (2.5 million Yen).

At present, two university professors in mathematic education (Yamagata and Fukuoka) and about thirty teachers are cooperating on this project. This project is the succession of the first project in the sense of evaluation of "higher objectives" of mathematics education.

One of the findings in the last year is as follows:

We found from case studies that the following method is effective on mathematics teaching by developmental treatment.

1. given a problem,
2. to solve the given problem,
3. to discuss the method and the solution,

4. to derive some new problems by changing parts of problems and to propose them to a whole class (the idea of generalization and analysis are required,

5. to discuss some of new problems.

6. to solve a common problem, and

7. to solve their own problems.

These projects have been supported by the grant-in-aid for scientific research from the Japanese Ministry of Education. It seems difficult to continue this study only with our limited usual budget assigned.

Conclusion

In Japan, the contents of teaching are regulated by the Course of Study published by the government, and so many teachers depend on textbooks only in their teaching and generally they are reluctant to try new things which are not written in the textbook. Therefore, it would take much time to disseminate a new teaching method. And for the wide dissemination, I think it is important that the new method allows a little deviation from a standard one. The progress of research up to now seems encouraging to this aspect.

In my experience over nine years, the fundamental research in mathematical education should be studied in close cooperation with classroom teachers to provide case studies which should be combined with statistical method in order to set up a theory applicable to practice.

For example, before the outset of the first project, we thought it would be easy for classroom teachers to practice the lessons using open-ended problems. However, when they were tried out in classrooms we found some cases where students' responses were not as we anticipated. We had several discussion meetings on ths point, with teachers and observers of their teaching, and found most of these cases were due to the type of problems and not to the individual teachers. At the next time of teaching, teachers modified the original problem and gained much better responses. This resulted in some modification in our original theory.

RESEARCH IN MATHEMATICAL LEARNING AND TEACHING AT THE UNIVERSITY OF WISCONSIN RESEARCH AND DEVELOPMENT CENTER

Thomas A. Romberg
University of Wisconsin
Madison, Wisconsin

The Wisconsin Research and Development Center is a non-teaching department of the School of Education at the University of Wisconsin in Madison, It was organized in 1963, has received funds from several sources, and now is funded by the National Institute of Education. Although a department of the University, the Director and all of the principal investigators hold academic appointments in other departments at the University. The work of the WRDC is defined by the principal investigators and the scope of work and budget is negotiated with NIE.

Since the Center's inception, mathematics has been a major part of its work. The first work was directed by Henry Van Engen. In 1966 I joined the R & D Center and started the Analysis of Mathematics Instruction Project with a developmental psychologist, Harold Fletcher, and a mathematician, John Harvey. Ten years later I started the Integrated Studies in Mathematics Work Group with Tom Carpenter and James Moser. In addition, mathematics has been included in several other projects in the center.

In mathematics, the products of these efforts have been two major curricula - - Patterns in Arithmetic (Van Engen & Romberg, 1970, 1971) and Developing Mathematical Processes (Romberg, Harvey, Moser & Montgomery, 1974, 1975, 1976), a book on problem solving (Harvey & Romberg, 1980), a forthcoming book on initial learning of addition and subtraction, and a large number of research reports and papers.

While mathematics has been an integral part of the Center, WDRC is not a Center for the study of just mathematical learning. The focus of the Center is on general questions of learning, teaching, and schooling captured in the label "Individualized Schooling." Our present approach to learning, which has evolved over the life of the Center, has been operationalized in terms of both skill variabilitity from one identifiable subgroup of students to the next (group differences). Differentiating cognitive process variables is based on adaptations of components of a general model of intelligence, whichincludes knowledge, memory capacity, schemes, strategies, and metacognition.

Differentiating group membership variables include ethnic and sociocultural factors, as well as age and gender. Sutdetn diversity is addressed in our research via both capitalization and compensation approaches (Cronbach & Snow, 1977). With a capitalization approach, one devises instructional materials and strageties that are well-suited to students' "strengths." In contrast, with a compensation approach one devises instructional materials and strategies that will hopefully overcome students' "weaknesses."

Our approach to teaching reflects our experience with Individually Guided Education (IGE) (Klausmeier, 1977) and WDRC's attempt to reform elementary schools in America. We now view teacher actions in relationship to what pupils do in classrooms and in turn how knowledge is acquired in classroom settings. We are particularly interested in teacher planning and decision-making within the context of diverse mathematics and reading instruction

We view schools as institutionalized settings which contain underlying rules and procedures by which schooling is given coherence and meaning. These rules are transferred through regularized patterns of behavior, specific vocabularies (a child in school is a "learner," his learning is "achievement") and in particular roles (one is a teacher, pupil or administrator in the school). The importance of institutional patterns is that the social structuring of experiences channels both the thought and action of the participants, giving definition and meaning to school learning and pedagogical practices.

The current work of the WRDC mathematics project is studying the processes that children use tosolve verbal addition and subtraction problems and we are attempting to identify how these processes evolve as the children mature and as they are taught. We believe that this investigation will also help us to understand how youngsters acquire basic addition and subtraction concepts and skills and perhaps improve instruction.

Strengths and Weaknesses of Research Centers

There are two related basic strengths of central research institutes. First, in a center it is possible to do long-term, large-scale research. Such effort is necessary for significant chains-of-inquiry on important problems to be carried out. To do this type of research takes money, time, talented people, and freedom to inquire. University scholars burdened by the usual teaching loads, working with graduate students and having the usual university commitee work, would find it impossible to organize, direct, and carry out such chains-of-inquiry.

Second, a center provides one with an opportunity for collaboration. Over the years I have worked with mathematicians, statisticians, psychologists, linguists, sociologists, curriculum theorists and lots of classroom teachers in planning and conducting studies. One must tackle important problems with research teams. It would be difficult to do without having a central institute.

To illustrate, let me contrast "today's" central research paradigm with "yesterday's" methodology. Twenty years ago the dominamt approach to educational research involved small scale psychometric studies, the outcome to be a journal article. The typical researcher got permission from a school to gather some data, spent a short period of time collecting data from teachers and/or students, and, to give it a semblance of scholarship, drew heavily upon statistics to draw inferences. Today, our methodology involves an attempt to examine the operating processing characteristics of children and/or teachers in schools. This may include clinical interviews, computer simulation of behavior, mathematical models, time-on-task observations, structural modeling, time series analysis, videotaping of classrooms and/or individual students, complex task presentations with multiple protocol analysis . . . The output of such research is not specific journal articles but monographs or books with detailed analysis of how children think and/or how classes operate. The former approach took little resources and yielded a few snapshots of events in isolation. Today's approach takes lots of resources but yields much clearer pictures of how learning takes place and how classes and schools operate.

My claim is that only central research institutes have the resources and necessary collaboration for this kind of research. An individual scholar with a graduate student cannot do this kind of research. It involves observers and interviewers, and includes examining complex data from several perspectives.

There are three weaknesses of research centers which I should mention. First, there are not many people trained to do systematic long-term research. Too many persons in mathematics education are only trained to do small-scale "journal article" research. Many centers, ours included, often have had only a collection of "journal article projects." Training for this type of work needs to include courses in mathematics, psychology, and curriculum theory. In addition, students need to be involved in such research programs in their training.

Second, there is a potential for research done at centers to become conservative. The field of educational research is growing and changing very quickly. Research centers need fresh ideas and fresh bodies trained, up-to-date in current techniques. Periodically, the Wisconsin R & D Center has had to recruit new principal investigators and terminate others. Thomas Kuhn (1962) described the problems associated with normal science in an era of scientific revolution. Since I believe we are in a period of scientific revolution, there is a continual need to challenge how research is carried out.

Third, centers are politically vulnerable (in fact all of educational R & D is politically vulnerable). Legislators and bureaucrats are not in a position to question cancer researchers, mathematicians, military design specialists, etc., because they recognize the technical capabilities associated with doing significant work in those areas. But in education everyone, especially politicians, think they are experts. Also, I would note that education is central to the political structure of the United States (the basic school districts are governed by elected bodies). This centrality is also reflected in the ever-changing organization and goals of the funding agencies (usually with every election or every change of political appointees). Thus, as one administration changes to another, emphases change. New appointees place different priorities on basic research, psychological research, sociological studies, work on teaching, development of curriculum materials, the handicapped student, the gifted student, etc. The priorities become political vehicles for the maintenance of a bureaucracy.

Add to this the burden we face in the United States of the negotiation process with a funding agent which takes an inordinate amount of time and resources. I do not deny the importance of these negotiations, but the negotiations become difficult because of the political vulnerability of education. Desire for particular agendas by NIE or NSF shift over time. This quite often results in lack of dissemination or implementation of completed work. For example, NSF funded the National Longitudinal Study of Mathematical Abilities, but by the time the data was collected (5 years later) there was little interest in reporting the results. My argument is that political considerations often mitigate against the long-term reserach that centers do best.

In conclusion, central research institutions are important because long-term collaboration chains-of-inquiry are important.

References

Cronbacl, L.J. & Snow, R.E. Apptitudes and instructional methods. NY: Irvington, 1977.

Harvey, J.G. & Romberg, T.A. (Eds.) Studies in Mathematical Problem Solving. Madison, WI: University of Wisconsin R & D Center for Individualized Schooling, 1980.

Klausmeier, H.J. Origin and Overview of IGE. In Klausmeier, H.J., Rossmiller, R.A., Saily, M. (Eds.) Individually Guided Elementary Education: Concepts and Practices. NY: Academic Press, 1977.

Kuhn, Thomas, S. The Structure of Scientific Revolutions. International Encyclopedia of Unified

Science Vol. I, No. 2, The University of Chicago Press, 1970.

Romberg, T.A., Harvey, J.G., Moser, J.M., and Montgomery, M.E. Developing Mathematical Processes. Chicago: Rand McNally, 1974, 75, 76.

Van Engen, H., & Romberg, T.A. Patterns in Arithmetic Revised Teacher's Manual. Madison: Wisconsin Research and Development Center for Cognitive Learning, 1970, 71.

THE INSTITUT FUER DIDAKTIK DER MATHEMATIK DER UNIVERSITAT BIELEFELD

Christine Keitel

The Institut fuer Didaktik der Mathematik
der Universitat
Bielefeld, Fedaral Republic of Germany

The Institut fuer Didaktik der Mathematik (IDM) in Bielefeld was founded in 1973 by the Volkswagen-foundation, with financial support until 1976 as a central, supra-regional, national research institute for mathematics education.

Now it is in the University of Bielefeld, but relatively autonomous in its research program and research organization; i.e., independent from usual university activities as teaching and researching in the math or educational departments.

To understand the role of the IDM in the Federal Republic of Germany in general (i.e., among other research activities or universities or educational institutions), some historical remarks seem appropriate which may give some reasons and explanations for the foundation of the IDM and its program.

Until about 1967 there were only isolated, uncoordinated, activities concerned with math education and related fields in Western Germany. There were two strictly separated branches following the tradition of math education in relation to school types:

Method-oriented activities in a restricted area of topics in elementary arithmetic and some daily life applications in math education for elementary and Real schul (= modern school) teachers on one side.

Topic-oriented activities, discussions, and considerations in math education for teachers of "Gymnasium" on the other side.

Claims for reform and changes in the school system have been one cause for the beginning of coordination and cooperation. Common theoretical reflexions in math education started about 1967 with some national workshops to bring together these two branches of orientation. These approaches were disturbed and nearly stopped with the decision of the "kultusministerkonferenz' in 1968 to suddenly change all syllabi and thereby the textbooks claiming quick answers and practical offers for the reform movement. This bureaucratic model of innovation tried to change the school curriculum and teaching in the classroom by changing syllabi and textbooks in a situation in which there were no fundamental research results nor curriculum developments which could support or transport the change into an unprepared teacher community. The failures of the reforms (which are described by Damerow in his contribution to ICME IV) were caused by the lack of basic developmental work or adaptational work of outside developments, and the gaps between the FRG situation and international curriculum developments.

This was the situation when the IDM was started. It seemed to be necessary to institutionalize research and developmental work in the area of math education. The IDM is one of the last reform institutes. Its general program can be summarized as:

1. to study and analyse international development in math education, to look for successful innovation strategies or curriculum development, and to analyse international research progress;

2. to build up international communication and cooperation and to organize international workshops and common publications in the FRG;

3. to document research results and research directions and to disseminate scientific knowledge to teachers, administrators, and researchers;

4. to develop fundamental research and interdisciplinary work in a yet undefined new science of math education;

5. to develop models of cooperation with practice and administration, especially to develop networks of cooperation with the teachers.

For all these tasks the IDM has permanent staff and additional project groups. The tasks are the organizers for the teams and working groups.

In the IDM we have individual and group work which differ in levels of school ages and special tasks, but there exist several kinds of cooperation between the groups and the individuals. The main work of the institute is done in four groups, which is described below. Besides, there is some individual work on development of curricular units on secondary level with problem orientation and use of studies in the history of mathematics (Stowasser); and on research on development of students' mathematical abilities, strategies, and skill in problem solving (Bussmann). A working group across the main groups is concerned with analysing and discussing methodological problems and methodological alternatives in research in mathematics education (cf. the contribution of Scholz to ICME IV.)

The main groups of the IDM are interdisciplinary teams which include mathematics, history, psychology, education, and sociology competences. There are four groups:

1. Documentation and information (leadership Gert Schubring). This group is concerned with the development of a thesaurus for math education and theoretical frames for annotated bibliographes and trend reports on one side, the evaluation and dissemination of the IDM materials on the other side. There are some projects including one on problem-oriented documentation especially concerned with documentation of curriculum material of the last 30 years (it will end in 1981).

2. Special aspects of the interrelation between teaching and learning of math (leadership Prof. Bauersfeld). This group is concerned with analyses of communication in the mathematics classroom and their implications for initial and in-service teacher training. Special projects are related to the implicit theories of teachers, to lingustic and para-linguistic aspects of classroom management; and to theoretical studies about alternative models (methodology) for research on the teaching-learning processes.

3. Problems of professional training and life of math teachers (leadership Prof. Otte). The group tries to define basic components of teaching in mathematics which include basic concepts and methods of math didactics as a science on one side, and theories of teaching (as contrary to learning theories) and concepts of teachers' activity and practical knowledge on the other side, to combine these two theoretical approaches to relate theory into practice and to integrate subject orientation and models of teaching, the two basic concepts of professional knowledge of math teachers. There are special projects, one concerned with the history of mathematics and math education in the 19th century; and a practice-oriented project which develops materials for teachers' in-service and pre-service education and tries models of cooperation with practitioners. The main goal of this project is to integrate pedagogical and mathematical training into the practical knowledge of the teachers on one side and to integrate the different demands of school types in the education of math teachers on the other side, (i.e., to integrate method-oriented education for the elementary and modern school teachers with the topic-subject-oriented education of the teachers for the gynmasium). Some materials of this project are concerned with preparing daily lessons as part of the teacher's activity, the problem of the textbooks, and the problem of proof in mathematics education. Materials are developed together with practitioners and are the result of the cooperation with the practice. They include case studies written by teacher-trainers about practical problems of the theoretical concepts pointed out in the materials and experiences in the classrooms. The main task of the project is to stabilize these cooperations and to keep good connections to school practice and practical teacher training institutions.

4. Mathematics education for the 15 to 20 year-olds (leadership Prof. Steiner). The group deals with the problems of mathematics training mainly for grades 11 - 13, including academic, university-bound courses (Gymnasium), vocational education, and the training of related teachers as well. To cope with institutional, mathematical, pedagogical, and social aspects of the teaching-learning process of the adolescents, specialists for subject matters, philosophy of mathematics and its applications, school systems, social learning, social psychology, and empirical research closely work together. A rather comprehensive working philosophy on educational goals, possibilities, constraints, and deficits of mathematics education of this age group could evolve, which has to be developed further through deeper studies on the impact of calculators and computers, the mathematicalneeds of various trades, the nature of probabilistic thinking, and real outcomes and processes of mathematics teaching and learning, especially in the kinds and effects of differentation made by administration, teachers, and - typically of this age group - choices of the students.

There are two series of publications, the "Schriftenreiche des IDM" and "Materialien und Studien" which include reports on research and results of IDM group work on international or national work shops organized by the IDM.

ORGANIZERS COMMENTS

B. Winklemann
The Institut fuer Didaktik der Mathematik
der Universitat
Bielefeld, Federal Republic of Germany

Introduction

The label "Central Research Institute for Mathematical Education" does not denote a specific, well-defined species in the genus of institutions of the educational scene. Instead, it labels a set of rather different individuals, as you will see in the presentations of some of the institutes themselves. These individuals have some genetic and functional properties on common, by which they may be loosely separated from other, more normal institutions of mathematical education. These properties may be hinted at in going through the different terms in the qualifying label "Central Research Institute for Mathematical Education".

The term "central" includes the notions of supra-regionality and importance, magnitude. The responsibilities of these institutes are not restricted to local tasks within one university or to regional ones, say to school-counseling or inservice teacher-training; their responsibility or center of research is mathematical education, may it be national or international. This does not exclude, however, regional studies and local exploratory activities. In order to make this national responsibility more than a mere claim, such institutes must have a certain mignitude in terms of scientific potential, and there seems to be a certain critical mass of manpower below which the effectiveness of such wide-spread work drastically diminishes. On the other hand, that these institutes are central does not necessarily mean that they have some executive power or that they are set in and/or controlled by the government. In fact, some of these institutes have been founded by more private organisations, such as Volkswagen Foundation or Shell Company.

The label "research" in our theme excludes some central curriculum projects such as CSMP in the U.S. or SMP in England, the main task of which is in other fields than in research. The term "research" here is by no means confined to empirical research, but includes theoretical studies in the mathematical needs of various populations, the history of education, the philosophy of mathematics, the nature of knowledge, the psychology of learning, and the social behaviour of mathematics teachers as well, to name only few. In addition to such empirical and theoretical research, most of our institutes devote much time to some other work, such as teaching, teacher training, curriculum development, and implementations.

The term "institute" (as opposed, e.g., to "projects", etc.) qualifies our institutions as organisations which are planned for a longer period of time, so that they have the possibility of pursuing rather complex goals in a stepwise, long-term manner, of accumulating knowledge and experience, and of qualifying their own scientific juniors.

In opposition to the first three terms, the term "for Mathematical Education" is rather less exclusive. In this regard I am of the opinion that the central institutes should deliberately play a role which is complementary and supplementary to work normally done and achieved in other institutions which contribute to mathematical education. As you all know and is actually seen in this congress, there is a large span of activities in the field of Mathematics Education, reaching from fundamental research over developing teaching models, teacher training, and textbook writing to the teaching of mathematics in the classroom itself. Many of these tasks can best be accomplished on a local or regional level, others are traditionally done in teacher education institutions, but some can hardly be undertaken in small institutions devoted to narrowly defined activities. Such otherwise seldom systematically pursued tasks are especially those, which are rather complex, interdisciplinary, and comprehensive, or which are less attractive since they yield services which are needed but not directly paid for in money or reputation. These tasks, which to a certain extent may be different according to different national traditions, are natural task-fields of Central Research Institutes for Mathematical Education.

To be more concrete and as a result of the forgoing defining remarks, let me list some features which I believe are characteristic of central institutions and which make them different from curriculum projects or usual mathematics education departments at university schools of education or within mathematical facilities:

a. Their task is to reflect mathematics education in its totality; i.e., with respect to

- the societal and organizational structure and conditions, e.g., the role and structure of the different school types, the status of teachers and subjects, and societal pressures on the contents of mathematics teaching;

- the interdisciplinary nature of the research problem which in the long run can only be tackled by teams in whom the different disciplines involved are represented;

- the complex interrelation between theory and practice, which in mathematical education comes in at different levels, e.g., in the relation between the mathematical knowledge the pupil has learned in school and its intended applications in later life, or the relation between the mathematical and educational knowledge and skills of the teacher and his ability to teach, or the theories of parts of mathematical education and their relation to the implemented educational programs.

b. They should have a linkage function between the various components and groups involved in the field: administration, curriculum groups, research groups, teacher training institutions, teachers, schools, professional organizations, etc. This is, of course, only possible if there are people in such institutes who know the problems and constraints of these groups and are able to communicate on related, different levels. It is necessary to establish contacts to those various groups and to develop them further, which by itself is a quite demanding task that cannot be fulfilled by smaller groups who have other tasks as well.

c. They should contribute to more continuity in research and developmental efforts in mathematical education through basic research and long-term projects. Curriculum projects normally have a very limited existence with rather specialized tasks. Research done at university departments at the Ph.D. level predominantly consists of isolated and single-event activities which do not accumulate.

d. They should help to coordinate, evaluate, and synthesize diverse activities in research and development done elsewhere, collect, digest, and disseminate information. These can be part of the non-attractive service tasks mentioned above, and it normally takes a longer period of time till the value of activities like these are assessed by the majority of mathematics educators; besides the needs of task c), this is one of the main reasons why institutes of this type should consist mainly of members who are permanent at the institute and not just hired and fired.

e. They should develop models for research and developmental activities which set standards for good work in ths field and which may be imitated by others.

Sometimes there may be conflicts between tasks of this type and other demanding tasks, which by their very nature can also be fulfilled by others, but which can best be done by the members ofsuch an institute, since the agglomeration of competence in various fields within such an institute allows a better or faster fulfillment of the work to be done. This may happen, for example, in the writing of an urgently needed textbook or in curriculum development with respect to the hand-held calculator. I think one topic in our panel discussion should be which strategies could be justified with respect to such conflicts between long-term characterstic tasks and actual urgent needs.

13.11 THE FUNCTIONING OF INTELLIGENCE AND THE UNDERSTANDING OF MATHEMATICS

ON INTUITION

Richard Lesh
Northwestern University
Evanston, Illinois, U.S.A.

Questions about the nature of intuition center around three general issues: (a) What is it? (b) What does it affect? (c) How can we help students acquire, develop, or cultivate it--perhaps through role modeling or by teaching certain heuristic procedures (e.g., the use of analogy or symmetry, the examination of limiting conditions, the visualization/concretization/particularization of problem situations, etc.)?

Fischbein's research, his preceding presentation, and several of his recent papers have focused on a series of prior questions leading to the above issues. Can we identify learning or problem-solving episodes which are more intuitive than others? Can we identify characteristics of intuitive thinking? What influences do intuitions have on other types of learning and problem solving? What factors seem to effect intuitive thinking?

This paper will give a brief interpretation of what intuitions are, together with some implications for ways that they might be cultivated. My central claim will be that intuitions are closely associated with the development of cognitive structures, and the problem of facilitating the use of intuitions is essentially the problem of understanding the functioning of these structures or cognitive models. The example below is prototypical of one type of intuition, and it illustrates the close relationship between intuitions and structures.

> In normal teaching situations, the teacher typically confronts a welter of information, all of which is relevant to instructional decision making, from which a small number of cues must be selected, weighed, and organized in a way that is most useful and which minimizes the chances of errors resulting from information that had to be neglected because it could not be processed analytically. Intuition consists of using a limited set of cues, because the teacher knows what things are structurally related to what other things. Intuition allows the teacher to grasp the significant features of a situation without explicit reliance on analytic procedures, because s/he has a model (i.e., a cognitive structure) which allows information to be selected, organized, and interpreted.

The above example illustrates a situation in which a well organized structure (or model) is available, but the amount of information needed to deal with the situation in an analytic fashion exceeds the processing capabilities of the individual. A second category of intuitions arise because a well-organized structure may not be available. Both categories result from the fact that a model is not available which "fits" the situation. In the first category, the lack of fitness derives from the inherent complexity of the situation; whereas, in the second category, the lack of fitness derives from the fact that the model itself is insufficiently organized. In either category, the lack of fitness between model and environment lead to two cognitive characteristics which influence the accuracy of intuitions: (1) centering - - i.e., the individual will not "read out" all of the information that is available; he will focus on only the most obvious features of the situation and will fail to notice less obvious features; (2) egocentrism - - i.e., the individual will "read in" meaning and information because of his own preconceived biases; he will distort the situation to fit his own understanding even when his own ideas do not correspond to objective reality.

To categorize some other types of intuitions, it will be helpful to identify three of the most important types of structures that occur in mathematics learning. The are:

a. "within idea" structures--coordinated systems of relations,operations, or transformations that a child must use in order to make judgments concerning a given mathematical concept. It is the nature of mathematical concepts that they must have these structures, and it is this structural aspect that distinguishes mathematical ideas from other classes of ideas. A child who has not yet coordinated a system that is related to a given concept is called "preoperational" with respect to that concept (Beth & Piaget, 1966).

b. "between concept" structures--which arise because the meaning of any given mathematical idea derives from its relationship to other ideas or from the system in which it is imbedded (Lesh, 1976).

c. "between mode" structures--related to the use of various mathematical representational systems together with organized systems of translation processes linking one representational system to another (Lesh, 1979).

Intuitions which are related to the above types of structures include the following:

a. "within concept" structure intuitions: At the beginning of Piaget's period of concrete operational thought, a child begins to treat properties like transitivity and invariance (i.e., conservation) as intuitive and obvious; whereas, earlier, these properties were denied. Why did this self-evidence and obviousness not impose itself earlier? The answer is that transitivity is a property of a system of relations, that conservation is invariance under a system of operations, and that these two types of systems must achieve a sufficient degree of coordination before they are treated as whole structures--having properties that apply to the structure as a whole and are not simply derived from the constituent parts of the structure.

b. "between concept" structure intuitions: A clever fifth grader, who realizes that multiplication is sometimes like repeated addition, may have some good intuitions about sensible estimates for certain whole number multiplication problems. On the other hand, these same intuitions may make it difficult for the child to accept with understanding the consequences of multiplying $1/2 \times 1/3$; the product is smaller than either of the factors, and the operation cannot be sensibly interpreted as repeated addition.

c. "between mode" structure intuitions: A fifth grader may have certain understandings about rational numbers that are associated with each of the following representational modes:

1. written symbols e.g., it may seem intuitive that $a/b \times c/d = (a \times c)/(b \times d)$
2. spoken language (e.g., it may seem intuitive that 3 eighths plus 4 eighths equals 7 eighths),
3. manipulative materials (e.g., experiences with paper folding may furnish intuitions about multiplication of fractions), and
4. static pictures (e.g., a straight line on a coordinate graph may furnish intuitions about equivalence of fractions).

The child may be able to solve problems that are presented in one mode by using intuitions derived from another mode. Each mode has certain assets and liabilities which emphasize different structural aspects of the problem. Beyond these types of intuitions, however, another level of intuitions is needed to tell a child which mode will be best for dealing with a given problem—or, perhaps, which concrete model, picture, verbal statement, or written sentence will best fit a given problem. For example, Cuisenaire rods may fit one problem, poker chips may fit another, and paper discs may fit a third. Knowing how various concrete models are related to one another, and which ones fit particular kinds of situations, is a particular type of "within mode/between concept" intuition.

Whether one is dealing with "within concept" structures, "between concept" structures, "between mode" structures, or even "between topic" structures relating mathematics to physics or other content areas, the following two properties of cognitive structures have important consequences for related intuitions:

a. A whole structure is more than the sum of its parts. That is, the whole has certain properties which are not derived from its constituent elements, and the parts derive a portion of their meaning from the whole. When a structure is not sufficiently coordinated, or when it is not sufficiently complex to fit the situation it is intended to model, two distinct types of intuitions occur--both of which have to do with the lack of coordination of parts and wholes. The first has to do with "seeing the forest but neglecting the trees" (i.e., focusing on properties of the whole but neglecting constituent parts), and the second has to do with "seeing the trees but not the forest." In either case, the resulting intuitions may be either correct or erroneous. But, when a structure is well-coordinated and fits the event, the whole and the parts are in balance, and the interpretation of the event is the same regardless of whether the whole or the parts are temporarily emphasized. Therefore, the event is correctly and unambigously interpreted (or understood), and intuitions are not needed. They are replaced by known facts.

   In Figure 1, if set A includes all of the events which are interpreted as fitting a given cognitive model when the whole structure is emphasized, and if set B includes all of the events which are interpreted as fitting the model when the parts of the structure are emphasized, then the intersection of these sets represents the events which are "understood" by the model (and for which intuitions are not needed), and the shaded portions correspond to two distinct types of intuitions--depending on whether the whole structure or its constituent parts are emphasized. Examples of these two types of intuitions will be given later.

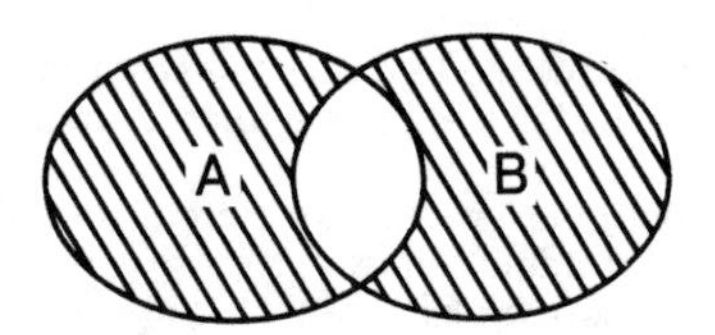

Figure 1

b. Structures must be constructed (i.e., coordinated) before they can be analyzed, and it is during this pre-analytic phase that intuitions are most needed. Bruner (1960) wrote:

> Analytic thinking characteristically proceeds a step at a time. Steps are explicit and usually can be adequately reported by the thinker to another individual. Such thinking proceeds with relatively full awareness of the information and operations involved. It may involve careful and deductive reasoning. (p. 57)
>
> In contrast to analytic thinking, intuitive thinking characteristically does not advance in careful, well-defined steps. Indeed, it tends to involve maneuvers based seemingly on an implicit perception of the total problem. The thinker arrives at an answer, which may be right or wrong, with little if any awareness of the process by which he reached it. He rarely can provide an adequate account of how he obtained his answer, and he may be unaware of just what aspects of the problem situation he was responding to. (p. 58)

It may be that whenever a structure (i.e., a well-coordinated system or model) must be learned, intuitive use precedes formal awareness. For example, children commonly use perfectly correct systems of grammar long before they are explicitly aware of these rules, and average adults can use systems of algebraic or topological relations, even though they may never become consciously aware of the mathematical structures that underlie their reasoning. Through intuitive use, an entire system can be gradually coordinated until individual elements within the system cease to be considered one at a time in isolation and begin to take on new significance by being treated as part of a whole system of ideas. As a result of this reorganization, new self-evidence appears with respect to properties that depend on the existence of the structured whole. Some models or structures, such as those used in teacher decision-making, never become completely analyzed and formalized. Yet, intuitions related to them can be refined and improved. During the early stages in the development of overt or cognitive systems (or structures), it may be confusing to force a novice to be come explicitly aware of components of the system s/he is attempting to coordinate. The situation is similar to the folktale about the centipede who became paralyzed when asked to explain the order in which he moved his legs. Paralysis through analysis is a common phenomenon in the development of either cognitive structures or complex systems of actions.

Cognitive structures, like other adaptive systems, develop into more complex or higher order structures through two basic processes: (a) integration - - i.e., forming relationships among lower order structures, or performing operations on lower-order structures, and (b) differentiation - - by which a single system becomes sufficiently coordinated to detect new subtleties in the environment and in its own functions. Consequently, a crude structure may become subdivided into two more refined structures.

The process of coordinating, integrating, and differentiating cognitive structures is similar to the way people learn to coordinate, integrate, and differentiate overt actions needed to ride bicycles or hit tennis balls, etc. The learner begins in situations in which the complexity of the system and the degree of coordination are minimal (e.g., all of the tennis balls come waist high, on the forehand side, just within arm's reach), and gradually moves to situations which require more complex and well-coordinated systems. New, more complex systems, such as those involved in learning to serve, are formed by: (a) integrating lower-order systems like throwing the ball up properly, hitting it properly, and following through with the serving motion, or (b) differentiating previously learned systems, such as when a serve is varied slightly to produce top spin or back spin.

Two basic processes, integration and differentiation, by means of which structures grow, are closely related to the two types of intuitions depicted in Figure 1. In athletics, these two types of intuitions are often referred to as "instincts" which are developed and refined in one sport and which transfer to another sport or to new and progressively more complex situations within the given sport. For example:

a. Type A intuitions, in which a whole system is emphasized: Serving a tennis ball is globally similar to throwing a baseball. So, an accomplished baseball pitcher can use his pitching instincts (i.e., intuitions) by modifying the overall system of actions used in pitching to fit tennis serves.

b. Type B intuitions, in which parts of the system are salient: In a new situation, an individual action may serve as the cue for a larger system of actions. For example, in tennis, an event may require a defensive action which triggers a more complex system which also includes an offensive response.

In teacher decision-making, type B intuitions may be needed in situations in which decisions must be based on inadequate information. In such situations, a model is selected which not only accounts for the given situation but which also "fills in the gaps" created by missing information and which suggests additional information which might be obtainable.

In mathematics, the "between concept" intuitions that were given earlier in this paper were examples of type B intuitions. They each involved: (a) selecting a model which accounted for some aspect of the initial problem; (b) generalizing, by treating this model as a subcomponent of a larger system; and (c) attempting to apply the larger system to the problem.

For either type A or type B intuitions, the goal is not to develop intuitions (or instincts). Rather, it is to coordinate progressively more complex and sophisticated systems, cognitive structures, or models -- and to have related intuitions facilitate, rather than hinder their development and usefulness. However, as progressively more sophisticated intuitions evolve, they automatically are accompanied by progressively more sophisticated intuitions whose use should be encouraged.

If cultivating sophisticated intuitions becomes a goal of instruction, it should be emphasized that research evidence does not support the existence of general intuitive thinkers. Intuition has proved to be domain-specific, and even within a given content area, profiles of intuition vary depending on the complexity of the problems - - or the complexity of the cognitive structures needed. Therefore, research and instruction about mathematical intuitions should be carried on in conjunction with research and instruction concerning the cognitive structures and conceptual models children develop or acquire during mathematics instruction.

At Northwestern University a current research project on applied problem solving (Lesh, 1980) is attempting to identify what it is, beyond having a mathematical idea, that allows an average ability student to use it in everyday situations. A major hypothesis underlying this project is that useful conceptual models involve well-coordinated systems of ideas and processes, and that intuitions play a key role in the acquisitionn and use of these models. A second project (Lesh, Landau & Hamilton, 1980) is attempting to trace the development of the conceptual models children use concerning one particularly rich mathematical idea (i.e., rational number), and to investigate the role that various representational systems play in this development. Again, intuitions, as they are described in this paper, play an important role in both the development and use of these ideas.

THE FUNCTIONING OF INTELLIGENCE AND THE UNDERSTANDING OF MATHEMATICS

Richard R. Skemp
University of Warwick
Warwick, England

It is my great pleasure to have this opportunity to share with you some of my recent thinking on a subject at which I've been working for quite a time: the understanding of mathematics. Ten years ago I wrote (1):

> "What is understanding, and by what means can we help to bring it about? We certainly think we know whether we understand something or not; and most of us have a fairly deep-rooted belief that it matters. But just what happens when we understand, that does not happen when we don't, most of us have no idea."

And I went on to suggest that until we did, we would not be in a very good position to try to bring about understanding in others. I wrote from experience: for it was as a teacher of mathematics in school that my interest in these problems first arose.

Later in the same publication I offered an answer to this question, "What happens when we understand what does not happen when we don't?" There's a follow-up question of a more practical kind, which I didn't tacke at the time: "What can we do when we understand that we can't do when we don't?" Since I fully agree with Dewey, who wrote (2) "Theory is in the end . . . the most practical of all things", I would now like to offer three answers to the second question, in the particular context of mathematics. In the process it will also become clearer why mathematics, one of the most abstract of theories, is also one of the most practical.

To give an idea where this is leading, I'll say in advance what these answers are. By understanding, we are better able to do three things:

a. to achieve our goals;

b. to cooperate with our fellow-beings;

c. to create.

I came to these answers by way of some work on an apparently different topic, the development of a new model of intelligence. Why I should have embarked on such a demanding and ambitious attempts needs a sentence or two.

All the time I was working on the psychology of learning mathematics. I was also working on the psychology of intelligent learning. Initially I didn't realise this. But gradually I came to regard mathematics as a particularly clear and concentrated example of the activity of human intelligence. From this came a desire to generalise my theory of the learning of mathematics into a theory of intelligent learning which would be applicable to all subjects, and of teaching which would help this to take place. For it became ever more clear that mathematics was not the only subject which was badly taught and ill-understood. It just showed up more clearly in mathematics.

This ambition became intensified in 1973, when I moved from Manchester University to Warwick, and from a Psychology to an Education Department. Over the next five years, I was working on this during all the time I had left from being chairman of a department. The outcome, only partly foreseen, was nothing less than a new model for intelligence itself (3).

Perhaps this description will seem less audacious if I emphasise that I don't see this as a finished job, but rather as a proposal for a new direction to our thinking about intelligence. Earlier models based on 'I.Q.' and its measurement have been with us now for about 70 eras; and while they have helped us to be quite good at measuring intelligence, they don't tell us much about how it functions, why it is a good thing to have, and how tohelp learners to make the best use of the intelligence they possess. Until we begin thinking in these directions, some of the most important questions about intelligence will remain not only unanswered, but barely even asked.

One of the payoffs of this work has been seeing mathematics in a new perspective, as an important special case of this new model of intelligence. Since few of you are familiar with it, I need to give an outline. I'll do this as briefly as I can, which - as you'll appreciate - will involve much condensation, and also a number of omissions.

The new model is a synthesis of ideas taken from a variety of sources, many of which you will recognise; and some new ones of my own. It takes as starting point the simple, common-sense, everyday observation that much human activity is goal-directed. This implies that if we want adequately to understand what people are doing, we need to go beyond the outward and easily observable aspect of their actions, and ask ourselves what is their goal. To say that Richard Skemp is riding his bicycle may be perfectly true, but it is only part of what matters. I may be riding to my office; or I may be enjoying physical recreation with my son; or I may be testing the adjustment of my three-speed gear.

Moreover, taking the first goal state, this may underlie actions which look quite different. For example, instead of bicycling I might be standing quite still by the side of the road, waiting for a bus; or I might be walking (which in this case would also be along a different path); or I might be driving my car (in which case I would start off in a totally different direction). To limit a description of what was happening to the observable behaviours, superficially very different, would be to miss what they had in common, namely the goal state.

I am not claiming that all our behaviour is goal-directed; only that some, perhaps much of it, is; and that where this is the case, to ignore it is to ignore one of its most important feature.

If this assumption is true, it leads us to another question. If we can identify a goal underlying certain observable activity, it is reasonable to ask what is the significance of that goal rather than some other. This question is answered, at some length, in the full description of the model. But there's an old Chinese saying (which I heard from a Russian, at a conference in the U.S.A.) "To tell all is to tell nothing". Repeatedly, I have had to leave things out of this talk which I would much have liked to include, or we would not have had time to look at the application of the model to the learning of mathematics. You will understand, I am sure, that all I can offer in the middle part of a single talk is no more than an outline sketch of the model, concentrating on particular aspects so that we can examine the consequences of these.

The starting point of this model is the assumption which I have already proposed: that (for the most part) our actions are not random, but systematically directed towards bringing about goal states. Now, the situations in which we find ourselves doing this are not always the same. Nevertheless, we are able to achieve our goals by varying our actions appropriately. A system for doing this I call a director system, and since the idea is borrowed from cybernetics, the ways in which a director system works will probably be somewhat familiar. But I need to recap them briefly for the sake of what comes after.

Its essence is a comparison between the present state of some operand and its goal state, combined with a plan of action directed always so as to reduce this difference until the present state coincides with the goal state. This means that both the present state and the goal state have to be represented within the director system in some way: otherwise, how can they be compared?

So, in slightly greater detail: We need a sensor, which takes in information about the present state of the operand.

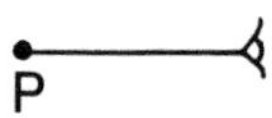

We need a representation of the goal state. And a comparator, which compares these.

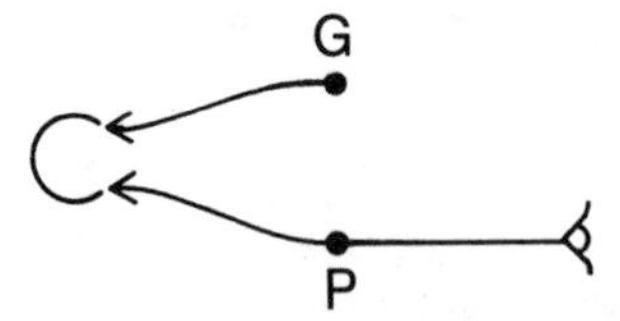

While changing from its present state to the goal state, the operand passes through a number of intermediate states. Each of these must be represented within the director system in order for the process of comparison to continue, without which the director system cannot continue successfully to direct action.

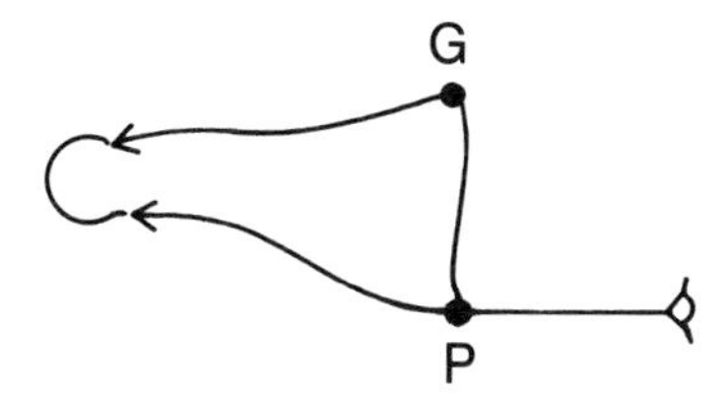

This is equivalent to saying that not only the initial state and the goal state, but also a path from the present state to the goal state, must be represented within the system.

In the lower animals, many of these director systems are innate. They form part of the survival equipment of the species. But there is an upper limit to what can be transmitted genetically; and there are other disadvantages, such as slowness to adapt to environmental changes. So it is not surprising that some species have evolved the ability to set up new director systems during the lifetime of an individual, and to improve the ones they have. This is how I now conceptualize learning: as a change in an organism's director system towards a state of better functioning. Other animals can learn, too, of course. But we have so far outdistanced other animals in our ability to learn that I regard our own species an an evolutionary breakthrough. It is not so much that we can do better than other animals at the same kind of learning. At maze learning, for example, rats are better than humans. But we have available a more advanced kind of learning, which is qualitatively different from the kind exemplified by maze learning. It's the ability to learn in this more advanced kind of way that I would now call intelligence.

Intelligence is a kind of learning which results in the ability to achieve goal states in a wide variety of conditions, and by a wide variety of paths. And if we've been able to agree that action (much of it) is not random but goal-directed, you may be willing to take with me the further step of conceiving learning as goal-directed also or at least, some kinds of learning.

The new model uses the concept of a director system at two levels. Delta-one is a director system whose operands are physical objects in the environment.

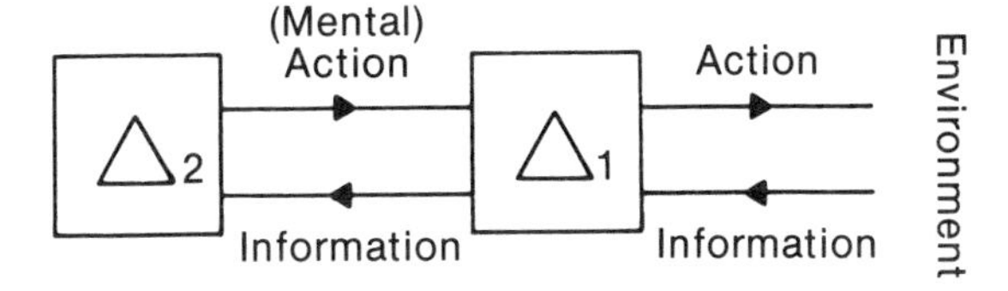

Delta-two is a second order director system which has delta-one as its operand. Its function is to take delta-one to states in which delta-one can do its job better. It does this in several ways. We have already seen that there has to be, represented within delta-one, a path from present state to goal state.

A simple path of this kind may be innate, or it may be learnt. Either way, it is closely tied to actions; and a particular sequence of actions at that. There's not much adaptability; and if one gets off the path, one doesn't know what to do to get back on. So it's much better to have a cognitive map, within which can be represented a variety of present states and goal states.

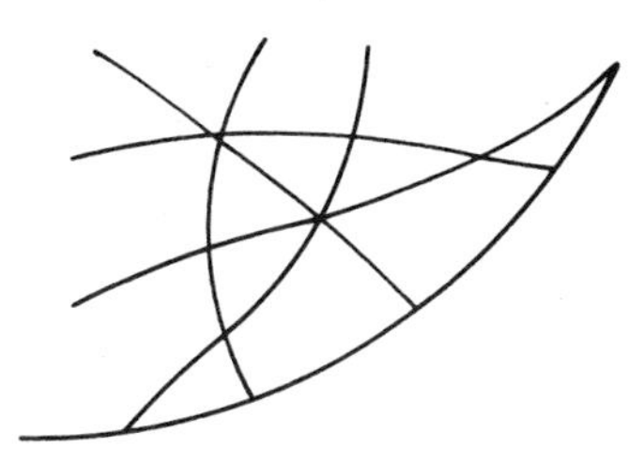

and from which a great number of possible paths can be derived, from any initial state to any goal state, provided that both can be represented within the same cognitive map.

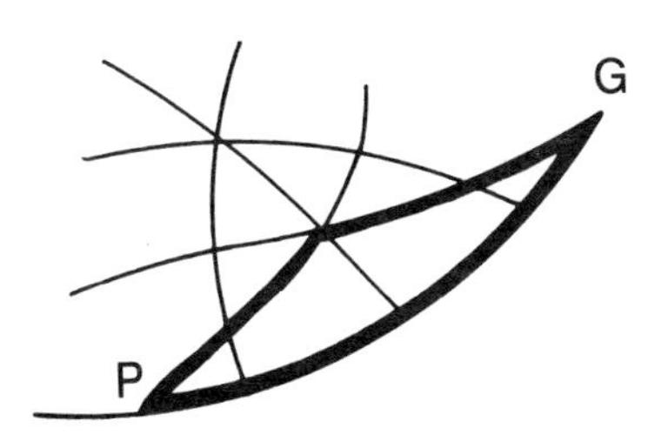

This is a much more effective and economical way of storing knowledge, since it gives the potential for constructing a very large number of plans, which to remember separately would impose an almost impossible burden on the memory.

It's learning of this kind with which intelligence is particularly concerned; and through its function is one step removed from physical actions, it makes these actions more likely to succeed because they are better adapted for each particular task.

Intelligence thus contributes to adaptability in two ways:

1. By the construction of these cognitive maps - not just one, but a large number, for different kinds of jobs that delta-one does.

2. By deriving these particular plans appropriate to different initial states and goal states. These plans can then form the basis of goal-directed action as already outlined.

Next, I would like to look at the nature of these cognitive maps in greater detail. We can think of them as mental models of certain features of the outside world, without which goal-directed activities cannot take place.

Now, as Heraclitis has told us, "We cannot step twice into the same river". The present experiences from which (sometimes) we learn become part of our past, and will never again be encountered in exactly the same form. But the situations in which we need to apply what we have learned lie in the future as it becomes present, or as by anticipation we bring it to our present thinking. It follows that if our mental models are to be of any use to us, they must represent, not singletons from among the infinite variety of actual events, but common properties of past experiences which we are able to recognise on future occasions. A mental representation of these common properties is how, for many years now, I have described a concept; and for this process of concept formation I use the term 'abstraction'. Concepts represent, not isolated experiences, bu regularities, and the organising of them into conceptual structures which are themselves orderly. These conceptual structures, or schemas, are like cognitive <u>atlases</u>, in which a large dot representing (say) San Francisco on a map of America can itself be expanded into another map, a street map of the city, in which our hotel now appears as a dot. But we could expand this dot also, if we liked, into a three-dimensional plan of the hotel. So a schema is a cognitive map in which each point has <u>interiority</u>. Schemas of this kind provide our delta-one systems with mental models which are useable for many different situations, and from which a variety of plans can be derived as required. They are essential for the functioning of delta-one.

<u>Constructing</u> these schemas, and deriving from them plans for use by delta-one, is a job of delta-two. Where does <u>understanding</u> fit into this model?

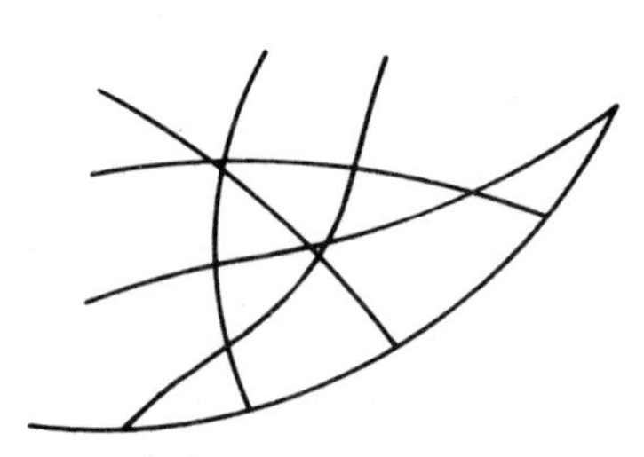

The diagram above can represent ideas at three levels of abstraction. At the concrete level, it can represent what it looks like - a road map. At a more abstract level, it can represent a cognitive map, using the idea of a map now as a useful metaphor. And it can also represent a schema, which includes the earlier two as particular cases. We can work at whichever level we find convenient, and the idea of understanding applies equally to all.

Our schema, as we have seen, are the sources from which delta-two constructs plans for use by delta-one. But suppose now that we encounter something which is unconnected with any of our existing schemas.

P•

Delta-two cannot make any plan which includes this point (meaning, what it represents) either as an initial state, goal state, or on the path between them. If the diagram is thought of as a road map, then we're literally lost at this location; if metaphorically as a cognitive map or schema, we are mentally lost. The metaphore is a very close one: we do not know how to act in order to achieve our goal. And this, in general terms, is our state of mind when confronted with some object, experience, situation, or idea which we do not understand.

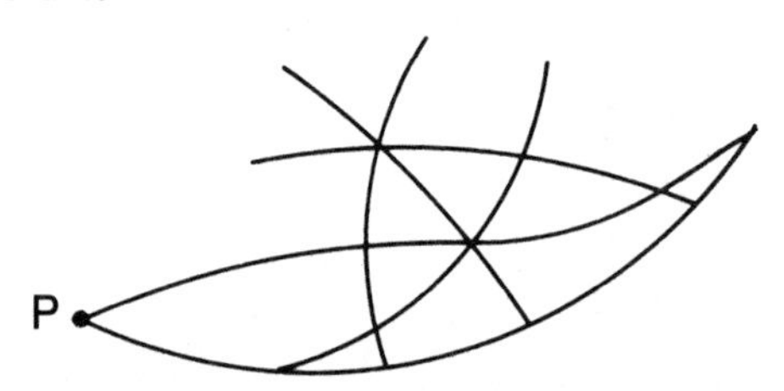

The achievement of understanding makes connections with an existing schema. We now know that in such a state, we could cope. Metaphorically (and in particular cases literally), we know where we are, so we can find a way to where we want to be. This change of mental state gives us a degree of control over the situation which we didn't have before, and is signified emotionally by a change from insecurity to confidence.

This conception of understanding is quite general. Here we are interested in mathematics as a special case, and in the short time remaining I'll outline three answers to the question, "What can we do when we understand mathematics that we can't do when we don't?" They are all examples of goal-directed activity by delta-two.

I. <u>We can make plans for goal-directed activity by delta-one.</u> Mathematical models are a very productive source of plans for the achievement of goals in the physical world. During my flight from London to San Francisco, we were out of sight of land for much of the time, either above cloud or over the North Atlantic. The navigator, by his understanding of the mathematics of navigation, brought us exactly to our destination. Had we been blown off course, or been diverted to a different airport, from the same source he would have made new plans. Both in making and using these plans he relies on sophisticated equipment, largely electronic. These complex electrical activities on which, for the duration of the flight, our lives depend are even more inaccessible to our senses than the landmarks relative to which our aircraft is guided. Understanding of these requires knowledge of electronic theory, and this is also formulated largely in terms of mathematical concepts. Among these concepts are imaginary numbers - - which help to get us to a real destination - - and the trigonometrical functions.

Early navigators used the more primitive method of dead-reckoning, based on the mathematical model $d = ts$. Simple electrical theory uses Ohm's law, of which one form is $E = IR$. Maps, and all scale drawings (including those used for constructing the aircraft), use the model $d = kD$, where d and D are respectively distances in the map or drawing and on the ground or actual object, and k is the scale factor. With different units, these are all uses of the same mathematical model $a = bc$.

Adaptability is a key feature of intelligence, and nowhere does it show better than in mathematics. Foir example, the same mathematical models - trigonometrical functions - are used in both electronics and navigation. They help us to preduct and control the movements of objects as small as electrons and as large as a Boeing 747. Mathematical models are so adaptable for making such a variety of plans, to achieve so many different goals in our physical environment, that it is hard to see how today's science and technology could exist without them.

2. <u>We can cooperate with our fellow-beings</u> for the achievement of our goals. Another of the questions to which I have addressed the model is "What are the requirements for intelligent cooperation?" One of these requirements (there are others) is complementary plans - plans which fit together to form a unified plan by which the various delta-ones involved can successfully coordinate their activities. Since the successful working of any society or profession depends on many and various people cooperating in many different ways for a great diversity of tasks, any system of thought which can facilitate the production of sets of complementary plans must be of great social and economic value.

Mathematics is the foundation for many such systems. Even at a descriptive level, it is often only by the use of mathematics that we can give descriptions of physical objects and events which are exact enough for everyone's plans to fit together. Manufacturer's couldn't make nuts to fit bolts, tyres to fit wheels, without the measurement function of mathematics. Still less, when describing invisible quantities like the capacity of a condenser, the impedance of a coil. It would be no use saying, "Make me a small condenser, about this size."

Quite a different application of mathematics is the agreed system by which, nationally and internationally, we cooperate in the exchange of goods and services. Money earned in the U.K. can be spend in the U.S.A. by using a conversion factor which is the same for exchanging 6 pounds for dollars as it is for 87 pounds, and the same for you as it is for me. This model is our old friend $a = bc$, this time in the form of $D = rP$. When we convert back from dollars to pounds, r has a different value. This difference corresponds to the banks' profit, which they get in exchange for a necessary service to travellers abroad. The many individual transactions which take place daily all fit together because they are governed by particular plans derived from an internally consistent structure internal consistency being a requirement of any mathematical system.

3. <u>We can create new knowledge</u>. When particular regularities have been perceived within a schema, we are able to extend these to new cases, beyond those from which the regularity was first discovered. Indices are a good example of this. Starting with meanings for a^2, a^3, . . . we find patterns common to particular cases like $a^3 \times a^5 = a^8$, a^7 $a^2 = a^5$. These patterns, in the form of general methods for multiplying and dividing powers of the same base, we then extrapolate. By this means we arrive at meanings for expressions like a^{-3}, a^0, $a^{1/2}$. Further extrapolation leads to meanings for expressions such as e^i . The use of de Moivre's theorem $e^i = \cos + i \sin$ in electronic theory requires extrapolation of the meanings of cos and sin for angles greater than 360°. These extrapolations are normally learnt by students ahead of the applications. The new mathematical knowledge, created as an expression of its own self-fertility, provides a foundation for an extension of our understanding of the physical world. Though applications of this kind are useful and powerful, I myself would see mathematical creativity as something also to be valued in its own right, in the same way as works of art are valued as such without being expected to be useful.

If mathematics provides, as I believe it does, a particularly effective way of using our intelligence, then to improve our teaching of mathematics we need first of all to know how intelligence functions, both in general and in the particular case of mathematics. This is what my talk this morning has been about. The next step is to apply this knowledge to what we do in the classroom, and that is what I'm at work on now. I've been describing the way mathematics is for people like air navigators, scientists, technologists, bankers, industrialists, creative mathematicians. For many children, this is not the way it is. They see it variously as a job to be done, a threat to be averted, a complicated exercise for fulfilling teacher expectations, or (in Erlwanger's memorable words), as "a set of rules for making arcane marks on paper". So, with the help of a group of experienced teachers, I'm in process of developing methods and materials by which we hope that all children will experience mathematics as an activity of their intelligence, (i) which enables them to achieve goals in their physical environment; (ii) as a shared reality which enables them to cooperate with their fellows; and (iii) in which they can experience the mental excitement of creativity. There is a lot of work still to be done, but the preliminary results are encouraging; and the pleasure one sees in children when one succeeds in these teaching aims is heart-warming.

It will take about three more years to finish preparing this material, and piloting it in schools. This will be in good time for the next ICME - I hope we shall meet again then if not sooner.

<u>References</u>

1. Skemp, R.R. (1971). <u>The Psychology of Learning Mathematics</u>, Penguin, Harmondsworth.

2. Dewey, J. (1929). <u>Sources of a Science of Education</u>. Liveright, New York.

3. Skemp, R.R. (1979). <u>Intelligence, Learning, and Action</u>. Wiley, Chichester, New York, Brisbane, Toronto.

RESPONSE TO SKEMP AT ICME IV

Laurie Buxton
Inner London Education Authority
London, England

It is a great privilege to be asked to speak at this international conference, but a special delight that it should be in response to Richard Skemp. I have worked with him now for over a dozen years, in my spare time from running school mathematics in Inner London. (That is part of a town in England.)

Various approaches sprang to my mind. It would be interesting to initiate a discussion on just how far we are in fact goal-seeking organisms (I know Richard is!). Again I might have sought to reinterpret his reality construction in terms of Karl Popper's three worlds. In the event, I decided th sketch in some features of work I have done with him on the cognitive-affective interaction, and an extension of the director system model he has just spoken of.

Richard's goal-seeking model of learning had one point of particular interest to me. The feed-back of success, or, more properly, likelihood of success in reaching a goal, led to emotional responses on the pleasure-unpleasure spectrum, and in a manner similar to Freud's "hedonic tone" depending on satisfaction of basic drives.

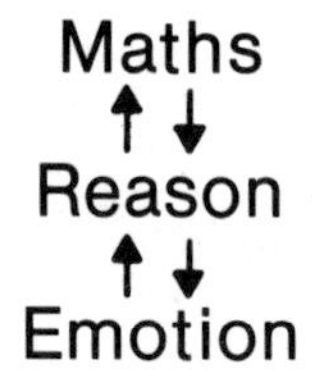

I was also very conscious of the well-known but not in my view deeply studied phenomenon of the mass of the population's deep antipathy to mathematics. The effect of mathematics teaching (note that I do not say "mathematics") has been to reduce them to a state of panic when threatened with it.

So I studied articulate adults who had this particular fear. They were relatively few in number but some were studied in considerable depth. Many things emerged. Some were purely cognitive, but most were in the affective area. We have the interesting paradox that a subject free of emotional content and not value-laden, taught with no consideration of emotional response, should in fact produce the strongest negative reaction in the emotional area of any curriculum area. My subjects had been put under very strong pressures at school - as had many of their fellows - and the significant word that began to emerge, unbidded was: PANIC!

Various elements combined to produce this effect: we shall group them under two main headings, Authority and Time:

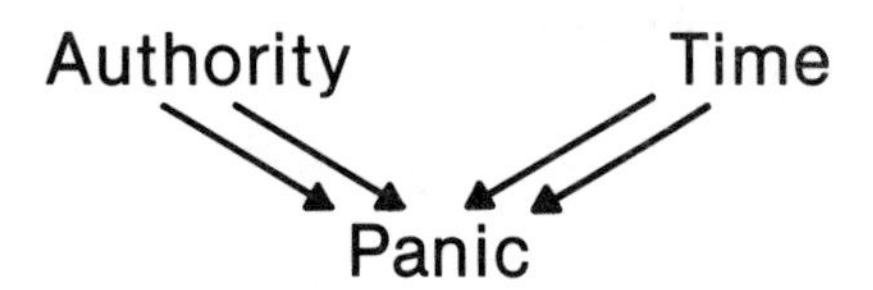

The fieldwork was deeply interesting and more than filled the year's secondment I was given for this work. There is a wealth of material describing people's experiences and leading into strange areas. At times we seemed to be deep into psychotherapy.

For now, however, I would like to develop a model for this single specific reaction - panic.

$$\Delta_1 \longrightarrow \begin{matrix}\text{Env.}\\(W_1)\end{matrix}$$

Our director system, delta-one, deals with our contacts with the environment, "Richard's "actuality", roughly the same as Popper's World One. It is from outside that threats arise; these are from the teachers! Not you or me of course, but no doubt you can think of some. These threats demand response, and delta-one needs a plan.

$$\begin{matrix}\Delta_2 & \xrightarrow{\text{Planning}} & \Delta_1 & \longrightarrow & \begin{matrix}\text{Env.}\\(W_1)\end{matrix}\\ \downarrow & & & & \\ W_2 & & & & \end{matrix}$$

Delta-two deals with out own internal images and models - and it is the plan maker for delta-one. The threat to delta-one produces emotional response, and one function of the emotions is to signal danger.

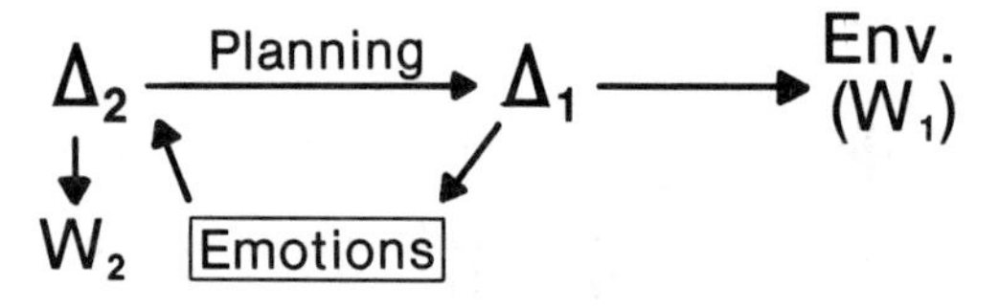

The effect of the emotions is to attract consciousness to the resolution of a threat. If the bringing into action of delta-two produces a plan - well and good. If it either does not have one or cannot provide one quickly enough, the external threat draws the consciousness to delta-one again (and delta-one cannot cope):

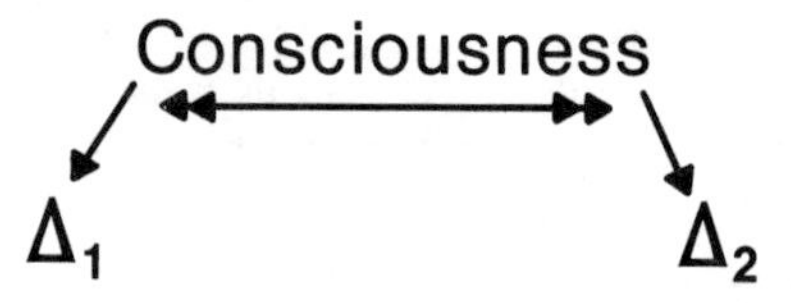

The consciousness now begins to flicker between delta-one and delta-two with a repetition of the neurotic circle.

Sorry, haven't got one! Help-I'm threatened - need a plan.

This produces a state of immobility - we are in panic!

Have we a way out? Perhaps the threat is simply implemented, leaving us humiliated in some way, and even less prepared to engage in mathematics. But at times we do "control ourselves" (and think on those words) and manage to solve our problems. To do this we need to stand aside, observe our state of mind (panic), and somehow operate on the reason-emotion complex to quieten it and break into its neurotic behaviour. We postulate delta-three!

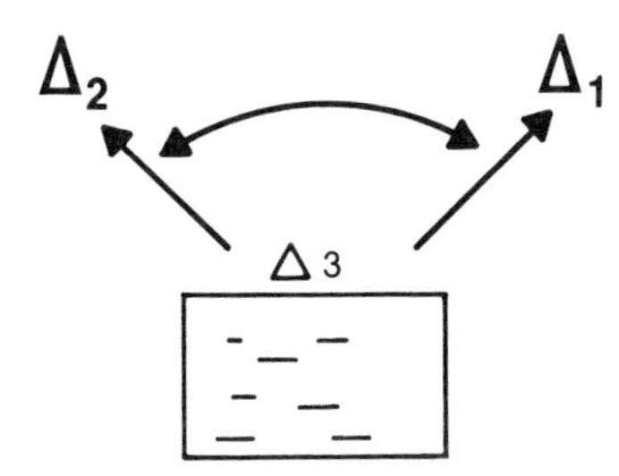

It is the psychologist in the mind that stands aside, observes, and if we can develop and use it, intervenes to prevent reason being shattered by emotion

Let us stop before we enter an infinite regression. There is much more - but please read my book: "Do You Panic About Mathematics?" by Laurie Buxton, to be published in Heinnemann Educational Books about March 1981.

COMMUNICATION REGARDING "THE FUNCTIONING OF INTELLIGENCE AND THE UNDERSTANDING OF MATHEMATICS

Nicolas Herscovics
Concordia University
Montreal, Canada

I quite readily agree with Richard Skemp's statement that the psychometric models of intelligence never told us much about key questions such as the functioning of intelligence and how to use it in order to optimize learning. Perhaps the only good thing one can say about these models is that they are very practical and provide some measure of the difficulty of a task and the pupil's ability to cope with it. Thus I welcome a new model of intelligence which attempts to answer questions we've been struggling with for ages.

Let me start with some general comments by reminding you that all good theories raise some questions than they can answer. I first note that Skemp's model of intelligence is a two-tier cybernetic model which allows for continuous feedback at every level. Such a two-level model is very much in tune with present-day interests, for the second level, delta-two, can be associated with meta-cognitive functions. I am particularly interested in its relevance to education and will address myself to two specific but interdependent questions.

My first question deals with the basic assumption of the model: that for the most part, our actions are not random but goal-directed. I quite agree with this premise when restricted to an individual practicing his free will and I think that a director system model applies. For I can suppose that the individual's actions are within a frame of reference in which he has the freedom to select his choices. I even think this model can apply to education when limited to the case of self-learning. If I decide to study by myself a certain topic, say topology, I have the freedom to choose among many different textbooks and will probably select the least formal, one the one containing many problems and espeically answers which I can use to verify my thinking. But what about the student in our school system? As pointed out in Skemp's paper, many children view mathematics as a "job to be done, a threat to be averted, a complicated exercise for fulfilling teachers expectation". Can we say that such a student has choices? He has only two: to study and hopefully learn or not to study, to simply "turn off." I am not concerned here with the child in elementary school who is still quite maleable and will work to please the teacher or with the university student who has a professional orientation. But what about the student at the secondary level? He is still too young to worry about work. In our day and age, few high school teachers can tell their students, "Learn this today - - trust me, I know what is best for you." The problem with goals at this level is one of motivation. And we have tried many things to motivate the pupil: consumer mathematics, contract learning, mathematics through applications, etc. I am not quite sure that a director system model can apply to a student who has no specific goal in mind, who is in school by obligation. The problem thus is how to motivate the student to learn, how to make learning one of his goals. I will come back to this question for I believe Skemp's work on the understanding of mathematics provides us with an important avenue to explore, that of relational understanding.

The second question which I wish to raise has been implicit in the discussion of the first one. A director system involves a goalstate, a sensor which takes information about the present state of the operand, and a comparator which monitors the intermediate states. Presumably, the sensor and comparator are internal parts of the system and have been programmed to detect the various possible states. does such a model reflect a typical learning situation? I would be tempted to suggest that it may require some modification, for, the learner in the classroom has to cope with material which is new to him and which he is in the process of assimilating. He does not as yet possess the means to judge whether or not he has achieved the desired goal state and is dependent on the teacher or the answer in the textbook to pass a judgment on his state of cognition. Here we find that the sensor and comparator are external to the learner and I suggest that this may be important. In fact, this problem has been tackled by some leading educators who have recognised the learner's need for a broad content, for some global frame of reference. I'm referring here to Ausubel's "advance organizers" and to Diene's "deep end strategy". Nevertheless, the learner will always depend to some extent on external sources to monitor his progress, for we cannot presume that he has in the initial stages of learning a well-defined "cognitive map" or the means to represent his goal state.

Thus it seems to me that Skemp's model is more appropriate to more advanced stages of learning, more sophisticated levels of cognition in which the pupil has at hand a well-defined cognitive map and the means to represent the necessary schemas. I am thinking in particular of problem-solving activities. Much of today's research in this area consists of computer simulation and of coding students' behaviour. Perhaps much could be gained byusing a two-tier model of intelligence to investigate problem solving. It lends itself naturally to the study of strategies used in the solution of problems.

Skemp has raised two distinct questions: "What is understanding?" and "What can we do when we understand?" I will limit myself to a discussion of the first question.

Skemp has defined "understanding" in a highly constructivist manner in terms of connection with one's existing cognition. In fact, he has related the understanding of mathematics to his model of intelligence. In 1976, he contrasted instrumental understanding (rules without reason) and relational understanding (knowing what to do and why). A year later, Byers and Herscovics suggested a four-dimensional model of understanding which included the two modes described by Skemp and two additional modes, intuitive understanding and formal understanding. These last two modes were inspired by Bruner's descriptions of intuitive and analytic thinking and also by the need to distinguish in mathematics between content and form.
In 1979, Skemp published his latest thinking on the subject. He accepts the need to identify a third kind of understanding mathematics which he calls "logical understanding". Connecting his different modes of understanding with his model of intelligence, he associated instrumental understanding with delta-one director systems, relational and logical understanding with delta-two director systems. He does not accept intuition as a form of understanding but prefers to use it in the Brunerian sense as a mode of thinking. He describes two modes of mental activity, the intuitive mode, in which consciousness is centered in delta-one, and the reflective mode, in which consciousness is centered in delta-two. Relating the different kinds of understanding with the different modes of thinking, he suggests the following classification:

MODES OF MENTAL ACTIVITY	KINDS OF UNDERSTANDING		
	INSTR.	REL.	LOG.
INT.	I_1	R_1	L_1
REFL.	I_2	R_2	L_2

The first thing I notice is that in this model, the different kinds of understanding can no longer be associated with only one director system, for if they occur at two different levels of mental activity and each one of these is centered in different director systems, then so is each kind of understanding. My second observation deals with the difficulty of finding examples to fill the matrix. Is it possible to take a given mathematical topic and to describe it in these six ways? Finally, I must admit that, as an educator, I am somewhat loathe to introduce the concept of consciousness in my work, for I just don't know how to handle it in the pedagogically relevant way.

I basically agree with the need to consider thinking processes and their relationsip to different kinds of understandings. But I would suggest that we can use Skemp's basic philosophy to describe this relationship. Whether in a learning or a problem-solving situation, thinking implies revolving ideas in one's mind and this can never be dissociated from one's existing cognition (cognition being interpreted here as organised, structured knowledge). But productive thinking is not a random process; ideas are not revolved in one's mind aimlessly. Productive thinking is a highly goal-oriented process whose objective is to achieve understanding, that is, the perception and discernment of relationships.

This last comment allows me to come back to the problem of motivation. For, if we can accept the perception of relationships as the objective of any productive thinking, then we are talking about achieving relational understanding and I wish to quote from Skemp's last article:

> Whereas the pleasure (if any) resulting from instrumental learning derives mainly from pleasing someone else - teacher, examiner - the delta-two activities involved in relational learning are a source of very personal pleasure. We ourselves know that the achievements of relational understanding can be very pleasurable; but many - perhaps a majority - of pupils unfortunately do not know this because they have never experienced it.

I would suggest that in this quotation lies an important message, especially if we look at past experiences. Whereas the traditional curriculum fostered instrumental learning, the modern programmes were unduly formalistic. Perhaps a relational approach which encourages delta-two activity is worth exploring in the future.

References

Ausenbel, D.P., Facilitating meaningful verbal learning in the classroom, Arithmetic Teacher, 1968, 15

Bruner, J.S., The Process of Education, Cambridge: Harvard University Press, 1960

Byers, V. & Herscovics, N., Understanding school mathematics, Mathematics Teaching, 1977, 81

Dienes, Z.P. & Golding, E.B., Approach to Modern Mathematics, New York: Herder & Herder, 1971

Skemp, R.R., Relational understanding and instrumental understanding, Mathematics Teaching, 1976, 77

Skemp, R.R., Goals of learning and qualities of understanding, Mathematics Teaching, 1979, 88

13.12 THE YOUNG ADOLESCENT'S UNDERSTANDING OF MATHEMATICS

THE MATHEMATICS AND PROBLEM-SOLVING SKILLS ADOLESCENTS SHOULD KNOW FOR APPLICATIONS

Stanley J. Bezuszka, S.J.
Boston College
Chestnut Hill, Massachusetts

A long standing and presumably a continuing concern of mathematics teachers is problem solving, especially in the context of applications. A little historical orientation will place in perspective the specific difficulties connected with applications and problem solving.

In the early stages of the development of mathematics, applications not only enhanced the subject in the eyes of the general public but also helped to expand the subject itself. Very often, new fields of mathematics were developed in the attempt to solve problems previously without a solution. Problem solving and especially applications came to be identified with mathematics - they were the strongest motivations for the learning as well as the teaching of mathematics. Young students of mathematics, during those early periods of mathematical history, were grounded in basic skills for the purpose of problem solving and trained in applications in particular as the latter touched upon the economic and societal needs of the period.

With changing times, other goals and objectives for mathematics, in addition to, or sometimes in place of, applications and problem solving were emphasized by those responsible for the training of the young and the preparation of teachers. A rather popular objective was:

> Students were to absorb from their study of mathematics certain norms, standards, and ideals of the logical rigor, the precision of thought, and the logical structure of a science or body of knowledge.

Mathematics was lauded as an excellent subject for the training of the mind. It was praised for its rigorous methods and its unrelenting thought processes. Mathematics was one of the better instances of systems development based on premises requiring only deductive principles. Some teachers regarded mathematics highly for its disciplinary capabilities and effectiveness. Many a student learned basic facts after school in detention periods where addition, subtraction, multiplication, and division were the punishment.

Imperceptibly, more and more professional mathematicians were irresistably attracted to the kind of mathematics in which assumptions were clearly and explicitly assigned and conclusions were derived with decision and certainty. The problems of nature, on the other hand, had little appeal since the premises of the real world were difficult to discover and define, and the data was frequently enshrouded with varying degrees of probability and plausibility.

As the separation of research from the influence of applications grew wider and wider, adherents joined one or the other of the extremes. There arose a classification of mathematics into "pure" and "applied", causing further ruptures in the unity of mathematics. Departments and divisions within universities were formed to serve both pure and applied mathematics.

Following the period of student unrest and influenced by the reaction to what had been the "New Math", administrators, curriculum developers, and teachers united in the mid-70's under the slogan "A return to the basics". This latest trend was especially welcomed by those teachers who had experienced frustrations with the New Math and who were inept at teaching problem solving. Now operational techniques were again in the ascendency - overriding logic, structure, concepts, problem solving, applications, and the beauty of mathematics. The general public gave its approval to the change because National Assessment scores had underlined the poor performance level of students in the basics of mathematics.

Nevertheless, there were concerned teachers, administrators, and professional mathematical organizations that were disturbed by the narrowness of the objectives being adopted by the proponents of the back-to-basics movement. A counterprogram was an absolute necessity, but these concerned groups well realized that emphasizing once again logic, structure, and concepts of mathematics would only expose them to vehement and articulate condemnations. Perhaps the strongest and most compelling factors motivating a revision of the existing basics curriculum came from the results of the most recent National Assessment, which underscored the low level of student achievement in problem solving. The lack of ability of youth in solving problems has been confirmed by representatives from industry who maintain that those being hired are not sufficiently prepared or capable of solving problems. Under these pressures, a new slogan has emerged for the 80's: Problem solving and applications. The cycle of change has run its course - from applications to logic, to structure, to beauty, to culture, to skills and techniques, and back to applications.

The National Council of Teachers of Mathematics released an Agenda for Action in April 1980 and gave problem solving and applications top priority in its recommendations for the 80's. The Council presented some specific guidelines for the achievement of the problem-solving objective. The charge to teachers and educators is briefly summarized as:

> During the decade of the 1980's, the continuing appearance of new concepts and theories in mathematics, in the applications of mathematics, and in the teaching-learning process will affect both curriculum and instruction in school mathematics. In order to remain professional, teachers must continue to study all three areas. (Agenda for Action, 1980, p. 24)

What then are some of the perennial difficulties of teaching problem solving and applications.

1. Inadequate reading comprehension skills combined with limited vernacular and mathematical vocabularies on the part of the students.

Students generally have restricted vocabularies, inadequate reading ability, and poor comprehension. These, aggravated by extremely short concentration spans, have been mentioned as a constant source of student failure in problem solving.

Apparently, no solution has been found to these problems in the past, and it is highly improbable that mathematics teachers presently have the necessary technical proficiency in the teaching of language, let alone the time or mandate, to cope effectively with these student deficiencies.

Adolescents need: to read problems meaningfully, to analyze data critically, and to think and conceptualize problems quantitatively.

Thus, one of the primary problem-solving skills that the adolescent must acquire is the ability to read, analyze, and conceptualize problems. The adolescent must not only be familiar with the vernacular, but also be proficient with the technical vocabulary of mathematics.

2. Students dislike for problems and applications which are not useful, not interesting, not relevant, not realistic.

Although both reading deficiencies and vocabulary weaknesses are contributing factors to a student's mediocre performance in problem solving, these are not the complete reasons, nor can their elimination serve as a final remedy to the difficulties experienced with problem solving. Another reason for the distaste towards, dislike of, and poor performance in problem-solving situations is psychological, for word problems are often not useful, not very interesting, artificial, and usually irrelevant in the everyday experience of the student.

It is not a simple task to change students' opinions about the relevance, value or interest level of the problems we assign in our mathematics classes. However, accepting that teachers generally have a limited background in the various specialized applied fields and the students themselves have even less familiarity with the language and problems of modern technology, we must for a time make good use of problems that treat everyday familiar subjects.

3. Student mistakes in computation and in algebraic manipulation serve as a pretext to substitute unmotivated and boring drill for problem solving.

Assume that the student has an adequate vernacular and mathematical vocabulary, and that the problems can be read with understanding. Suppose further that the student is not psychologically repelled or unduly bored by the problems and is willing to attempt them. Are these premises sufficient to insure adequate problem-solving ability and the acquisition of skill in applications?

We come to the most common of classroom practices sufficient to kill the enthusiasm of even the best and well disposed students towards problem solving and applications.

A homework or classroom assignment involving problem solving shows that some students have made errors in computation, while others have incorrectly applied algebraic rules. The solution is simple and obvious. The students need more practice in basic arithmetic skills and algebraic manipulations. The end result is clear, too. The slow and average students get bored as well as disgusted with drill that leads nowhere. The bright students look around for a new interest to replace mathematics.

4. Teachers need expertise in problem-solving theory and practive in applications experiences before attempting to teach students.

Teaching effective problem-solving strategies to students is not easy. Teachers need to develop a better understanding of problem solving theory and practice. They should be made aware of and exposed to research efforts into the various aspects of problem solving, to learn from these expriences.

The recommendations that follow expand and supplement some of the points proposed by the National Council of Teachers of Mathematics in its Agenda for Action.

A. Avoid long, cumbersome, tedious computations with pencil and paper. Use hand calculators or computers.

B. Even with calculators, use cross checks, approximations, and estimation to minimize possible errors.

C. The following should receive priority in instruction:

1. Percents

2. Ratio, proportion, and rates

3. Graphical representation and interpretation of collected data

4. Pattern recognition in arithmetic, algebra, and geometry.

Basics in arithmetic should not be limited to mere routine computation in the fundamental operations. Further, the addition of long columns of numbers and the addition and subtraction of numbers with more than two or three digits should be eliminated from the paper-and-pencil activities of students. If practice with large numbers is considered necessary, then a hand calculator should be used. Essentially the same remarks can be applied to multiplication and division.

Hand calculators are not error proof in the hands of a student. Precautions must be taken to minimize mistakes; meaningful instruction in the use of the calculator must be provided. Solving problems even with the aid of a hand calculator will involve utilizing cross-checks, approximations, estimation - - all the usual strategies that were also necessary when the problems were solved by laborious paper-and-pencil techniques. Calculators eliminate the labor of comuting; they are not a substitute for common sense or thinking.

In spite of the fact that amongst parents, educators, and mathematicians there is a lack of understanding and acceptance of exactly what "basic skills" should comprise, there are nevertheless some components of mathematics that are used more frequently, or have wider applicability, than others. These elements should receive priority in teaching whether they are labeled basic or not. We identify a few of these elements:

1. Reasonable practice to insure the understanding of and the ability to apply ratio, proportion, and percent should be given to adolescents. Rates are generally difficult for students. Educators need to spend more time on the presentation and development of this concept, which actually underlies the invention of calculus.

2. Ratio and proportion are more effective when they are taught concretely and provide a nice application of geometry.

3. Students in the early grades should be encouraged to collect data and represent the information graphically.

4. Some say that mathematics is the study of patterns. In any event, the search for patterns is an effective problem-solving technique. Data, whether computed, collected, or presented, provides ample opportunity for students to seek and discover patterns in arithmetic, algebra, and geometry.

Incidentally, because we have access to the hand calculator together with the fact that many problems involving algebra can be solved by computational techniques, we are now in a position to give students a number of problems formerly restricted to more advanced courses. For example, maxima and minima problems abound in business, industry, and our economy. Problems in maxima and minima open up an area rich with applications and are within the reach of students using computational techniques.

Assuming that the students are taught exclusively the "Let x be . . . " approach to solving problems, why do they so often fail to arrive at the correct equation? One contributing factor may be the students' lack of knowledge and facility in the use of some basic dimension theory. This lack accounts for many of the errors students commit in deriving equations and also prevents many students and teachers from pursuing problems in the physical sciences, the social sciences, and economics, and even the standard word problems in most secondary school textbooks. It is a fact that extremely little on dimensions is taught in our elementary and secondary schools. Consequently, a student who derives an equation for a word problem has no way of checking whether or not the equation is dimensionally correct. Of course, dimensional consistency is not by itself a guarantee of the correctness of the equation. However, dimensional inconsistency in the terms of an equation is a positive indication that the equation cannot be correct.

Let me conclude with just one more observation. With regard to applications, more has been established in this century than in the whole recorded past. They have developed so rapidly that we are no longer at all sure of what we know. We can take no comfort in the belief that what appears to be the whole truth today will be the whole truth tomorrow. There will be many obstacles to overcome in the process of developing, formalizing, and actively instituting in the classroom a curriculum whose essence is problem solving and applications. Surely, solutions will not be plentiful to the various problems that will confront the teacher, nor will they in most instances be simple. Therefore, for the interested, dedicated, and concerned teacher who will accept the challenge of promoting problem solving and applications in the 80s, let me recall the spirit of the true teacher that is never beaten or who never yields to obstacles. Let the adage for the 80s be:

> Aut inveniam viam aut faciam.
> I will find a solution or I will create one.

References

Denmark, Tom, & Kepner, Henry S., Jr. "Basic Skills in Mathematics, A Survey" Journal for Research in Mathematics Education, Vol. 11, No. 2, March 1980.

Yeshurun, Shraga The Cognitive Method, National Council of Teachers of Mathematics, 1979.

Shane, Harold G. Curriculum Change Toward the 21st Century, National Education Association, Washington, D.C. 1977.

An Agenda for Action: Recommendations for School Mathematics of the 1980s, The National Council of Teachers of Mathematics, 1980.

Applied Mathematical Problem Solving, edited by Richard Lesh, Diane Mierkiewicz, & Mary Kantowski, ERIC Clearinghouse for Science, Mathematics and Environmental Education, Ohio State University, 1970.

Applications in School Mathematics, National Council of Teachers of Mathematics, 1979 Yearbook, Sidney Sharron, yearbook editor.

The Role of Applications in the Undergraduate Mathematics Curriculum, National Academy of Sciences, Washington, D.C., 1979.

School Science and Mathematics Volume 78, No. 3, March 1978 (the whole issue is on problem solving and research on problem solving).

METHODS CHILDREN USE TO SOLVE MATHEMATICS PROBLEMS

K. Hart
Chelsea College
London University
London, England

The research project 'Concepts in Secondary Mathematics and Science' was financed by the Social Science Research Council for five years (1974-1979) and based at Chelsea College. Its main aim was to give information to teachers and developers of curriculum on levels of understanding in science and mathematics. These levels of understanding were viewed as forming a hierarchy based not exclusively on the logic of the subject, but also on what children appeared to understand and the order in which they understood. The investigation was specifically concerned with the age range 11-16 years. The mathematics team of the project carried out a three-phase investigation of 11 topics which commonly appear in the British secondary school mathematics curriculum. The 11 topics were Number Operations (Whole Numbers), Place Value and Decimals, Fractions, Measurement (Length, Area,

Volume), Positive and Negative Numbers, Ratio and Proportion, Algebra, Graphs, Vectors, Matrices, and Reflections and Rotations.

The first phase of the research was a search of curriculum materials to ascertain the key ideas in these topics (as taught in the schools) and then to write word problems which embodied some of these aspects. The investigation was not designed to test the efficiency of a teaching programme, nor was it carried out immediately after a topic had been taught, and technical terms and computations were kept to a minimum. The word problems in each topic were then used as the basis of an interview schedule with about 30 children (300 children in all) to find the methods, both correct and incorrect, that each child used when solving the problems. The problems were revised on the basis of information obtained from the interviews, rewritten in paper-and-pencil class-test form, pilot tested with whole classes, and finally given to a large representative sample of the English secondary-school population (10,000 children aged 11-16). For each topic in each age range tested, the sample was taken from at least six school (both urban and rural) and was normally distributed with respect to IQ scores. Thus the data gathered were thought to reflect the attainment of a typical year group in a secondary school when all abilities were represented and the children came from a mixed socio-economic background. Details concerning the analysis of the results and the formation of the hierarchies are contained in a research monograph (Hart, 1980) and a book for teachers (CMS team, 1980).

A significant outcome of the research was the identification of certain widespread errors (codes on the written tests). These errors were interpreted in the light of the explanations children gave in the interviews. The SSRC has now financed a further research project at Chelsea called 'Strategies and Errors in Secondary Mathematics', in which these errors and the strategies which lead to them are being investigated in depth. Recently we have interviewed 50 children who committed specific errors on the Ratio test paper and 50 children who were making mistakes on the Algebra paper.

It is overwhelmingly clear that many (over 50 per cent) of our secondary school children find what they are taught in mathematics very difficult to comprehend. In British secondary schools it is probably assumed that 11-12 year olds arrive from their primary education with a working knowledge of the four operations on whole numbers and some experience of fractions, decimals, and geometry. The transition from success with whole numbers and the models on which that knowledge was built, to a working knowledge of fractions and decimals is seen in many textbooks as natural and relatively straightforward. This appears to be far from the case. On the CSMS Number Operations test paper (Brown & Kuchemann, 1976) children aged eleven and twelve were asked to write a story to show the meaning of operations such as 9 divided by 3. Predominantly for division the stories were about sharing sweets between friends. This model for division is perfectly adequate when the elements are whole numbers, but makes no sense when elements are fractions or decimals. The problem 3/5 divided by 2/9, for example, cannot reflect a sharing of objects. One question on the Decimals paper illustrated this: given the opportunity of saying 'there is no answer' to the question of 16 divided by 20, the percentages who replied that no answer existed were 51 per cent aged 12, 47 percent aged 13, 43 percent aged 14, and 23 percent aged 15.

A popular incorrect strategy on the Ratio questions was to employ addition of the difference between two lengths to effect an enlargement. So that given

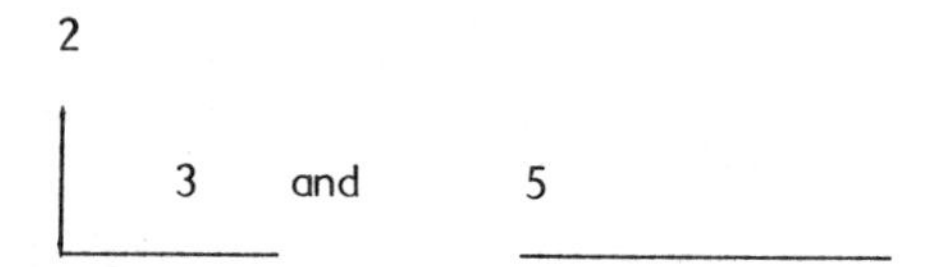

and asked to provide a new upright so that the resultant figure was 'similar' to the original, over forty per cent of the sample (n = 2257) gave the answer 'four'. Karplus (1975) and Piaget and Inhelder (1967) have commented on the use of this strategy. Asked whether there was a way other than adding by which one could get from 3 to 5, the children who added invariably replied that there was not. Similarly, asked whether there was a number one could multiply by nine to obtain six, children said there was no such number. Do children see fractions as numbers, or as labels for pieces of cake, apples, etc? Are we in fact talking the same language as the children we teach?

The children we have interviewed are certainly not using the methods or rules we have taught when they are asked to solve problems out of the immediate context of practising that rule. The easiest questions on all the CSMS topic papers could be solved by counting or addition and knowing the first conventions for labelling fractions or decimals (see, for example, the questions in Figure 1.)

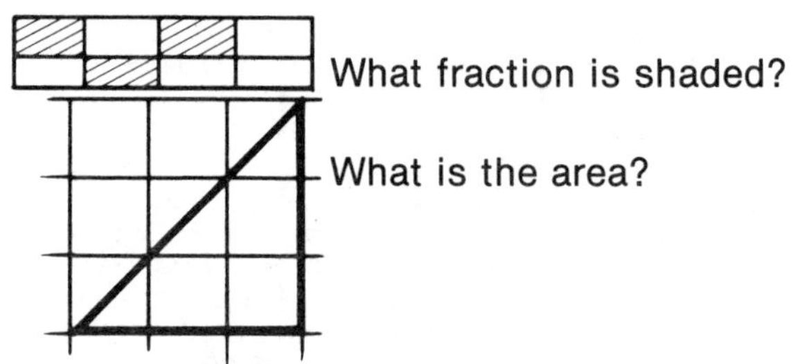

Figure 1

This is not surprising, but what is astonishing is how many children continue to use additive methods long after they have outgrown their usefulness and are very cumbersome. For example, a very popular method of solving the following problem (adapted from Piaget, 1968) was to avoid multiplication by a fraction and build up to an answer using addition:

> Three eels X, Y, Z are fed with fishfingers, the length of the fishfinger depending on the length of the eel.

25 cm Z

15 cm Y

10 cm X

If X has a fishfinger 2 cm long, how long should the fishfinger given to Z be?

A typical reply was:

> "Y gets three. You have to add one because you half it. Z is ten centimetres bigger which we know has two. So Z gets three add two, five

Or sometimes:

> "If Z was double he'd have four centimetres, then he would be 20. But he's 25 so he needs another half of X which is one centimetre. Two add two add one, five".

There were many different additive methods used on the eight questions of this type; only two children of the 28 originally interviewed used the same method throughout, however,

Similarly, to find the volume of a cuboid children often counted the number of small cubes in the top layer and then added that number to itself four times. This method is adequate for examples of the kind in Figure 2, but cannot be easily applied to find the volume of the cuboid in Figure 3. The per cent correct for 12, 13, and 14 year olds, respectively, was 14.2, 18.7, and 27.9.

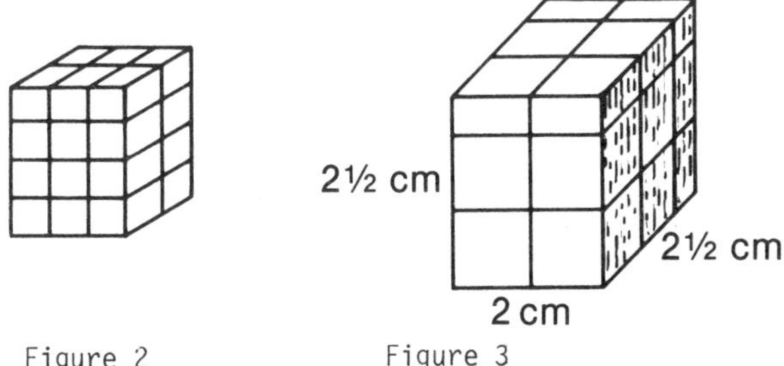

Figure 2 Figure 3

The correct additive method in Ratio is seldom applied when, in order to use it, the process of halving is unsuitable, as in the earlier example of 5 : 3. Here, as we saw, the child again opts for addition but uses it incorrectly. When a child is 'getting the problems right' we often assume that the method being used is the one we have taught as being suitable for that type of problem. Therefore, when the child gives an incorrect answer, we try to remedy the situation by repeating 'our' method; but we are often correcting an error that was never committed because it was not an erroneous use of the suggested rule. The rule, algorithm, or method is often introduced to pupils with very easy illustrations and verified by an appeal to something already known. For example, to inroduce fraction division we might use 4 divided by 1/2 and verify the answer by saying "it is the number of halves which must be added to give 4'. It is little wonder that the child ignores subsequent complications and continues to add a number of halves. In Ratio we tend to introduce a method which can be used for all ratios by using the particular case of a 2 : 1 ratio; but 2 : 1, or doubling, does not need this generalised method, as the alternative method of adding a number to itself is relatively easy.

Many adolescents are still firmly tied within the set of whole numbers and the operation of addition (or counting), and they appear unaware of the nature of fractions and decimals. Their method of solution of problems are often different from those that were taught them. Because they are adolescents, we as teachers are loathe to start again on an 'old' topic with a new introduction. When a topic is introduced to young children, it is usually accomplished by the use of an amount of concrete materials. Using concrete materials with fourteen year olds seems "childish" to both teenager and teacher, and yet it is highly likely to be what is needed. In addition, by the age of fourteen or fifteen the attitude of the adolescent to learning mathematics is far from favourable. The continued statement from adults that the subject is 'useful' is beginning to pall since the child has as yet found very little use for it in his everyday world. The child has often by now discovered that he can adapt what is taught by his teacher, and since he sometimes succeeds by using his own method, then this is better than always failing as the result of using a 'foreign' method which he only partly understands. We must radically rethink our goals and teaching of mathematics, and above all we must talk to children about what they are doing rather than talk at them about what they should do.

References

M. Brown & D.E. Kuchemann. "It it an 'add' Miss?" Mathematics in School, 5, 5, and following issue (part 2), 1976.

CSMS Mathematics Team (Editor, K. Hart). Children's Understanding of Mathematics (11-16): John Murray, London, 1980.

K.M. Hart, Secondary School Children's Understanding of Mathematics (Research Monograph): Mathematics Education, Centre for Science Education, Chelsea College, London University, 1980.

R. Karplus, E. Karplus, M. Formisano, and A.C. Paulsen. Proportional Reasoning and Control of Variables in Seven Countries. Advancing Education Through Science Oriented Programs, Report ID-65, June 1975.

J. Piaget & B. Inhelder. The Child's Conception of Space. Routledge & Kegan Paul, London, 1967.

CHAPTER 14 - Assessment

14.1 ASSESSING PUPIL'S PERFORMANCE IN MATHEMATICS

ON THE DEPENDENCE OF THE INFORMATIVE VALUE OF EDUCATIONAL MEASUREMENT ON THE STRATEGIES USED IN TEST CONSTRUCTION AND EVALUATION

Norbert Knoche
Universitat Essen
Essen, Federal Republic of Germany

Let us begin our considerations with an example: Suppose that among 1000 applicants for a certain training program there are 200 persons who are really suitable for this training, which means that these persons are considered satisfactory before testing. Let us further suppose that only 200 places are available, so the 200 persons have to be selected from the group of applicants. In order to select these persons an aptitude test with validity of r = 0.5 is administered to the group, and the 200 examinees having the highest test scores are selected. Under the (restrictive) assumption that the test variable and the criterion variable have a bivariate normal distribution, we find among the 200 selected persons (only) 88 of the 200 examinees who were judged to be satisfactorily prepared, which means that 112 poorly prepared persons are selected by the test and thereby 112 of the applicants whose performance was satisfactory are rejected.

Does this example give an answer to the question of whether or not the use of tests for measuring pupil's performance can be justified? One naturally can require that the validity of a test should be higher than 0.5, but this demand is illusory in practice. What would have been the effect if one had dropped the test and and selected the 200 persons at random? In this case we would find among the 200 selected persons only 40 of the 200 presons initially judged to be satisfactorily prepared.

Let us discuss this example a little further. With 400 applicants, of which 200 have satisfactory credentials, and a test with validity coefficient of r = 0.5, we get the Table 1 (see Taylor & Russell, 1939).

Table 1.

Number of selected persons	Number of suitable persons among the ones selected. (a) by using the test	(b) by using a random procedure
200	134 ≙ 67%	100 ≙ 50%
100	76 ≙ 76%	50 ≙ 50%
50	41 ≙ 82%	25 ≙ 50%
20	18 ≙ 88%	10 ≙ 50%

With 400 applicants, of which 200 have satisfactory credentials 18 persons of 28 are selected correctly by the test and only 10 by using a random procedure.

Naturally, this example does not give an answer to the question "Testing - - is it helpful or harmful?" but it may be considered as an example of the question raised in this presentation. The magnitude of the validity coefficient, for examle, cannot as such be regarded as a measure of the general validity of a test. The question about the informative value of test scores is, on the one hand, a question about the informative value of the test model, and its characteristic coefficients, used in a certain specified testing situation. On the other hand, the above-mentioned question is a question about the construction of test items with respect to the special evaluation problem for which they are used. The construction of special test items is naturally not independent of the special evaluation problem. The example discussed above showed the dependence of the informative value on the validity coefficient on the selection ratio.

In the following I will not discuss generally the problems of test procedures that arise in the domain of educational measurement. These problems are well known. The complexities of these problems are conditioned by the complexities of the factors involved in educational measurement: the complexity of the learning process, the complexity of the process of teaching, the complexity of the idea of a pupils' performance, the uncertainties in the interpretation of test scores of a special measurement taken in a special situation as an indicator of a person's proficiency, and so on.

These factors make it easy to formulate a catalogue of postulates on the quality of a test that in general cannot be fulfilled even if one restricts one's considerations to special questions. On the other hand, it is evident that anyone who is concerned with issues of evaluation, with problems of selectin or prediction, must endeavor to get out of the stage of the intuitive interpretation of his findings and to reach a quantification and thus, if possible, an objectification of his findings.

We must concede that frequently, or perhaps usually, the procedures of test construction fall short of the standards desired in the catalogues of postulates on the quality of a test as mentioned above. Some reasons are the complexity of the factors involved in educational measurement as mentioned above, but others stem from a relatively unreflective use of statistical procedures for the evaluation of tests and test scores. Some time ago a "coefficient of concordance" was developed for measuring the concordance between observers, when the observations have been dichotomized, as follows: Suppose that n examinees have been judged by k observers. We get Table 2.

Table 2.

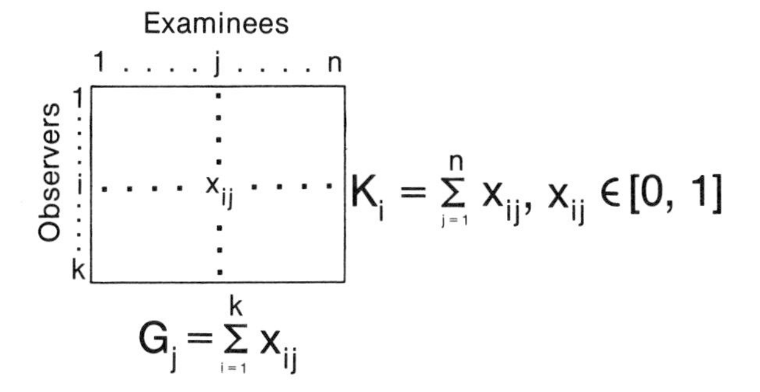

Let

$$S^2 = \frac{1}{n-1} \sum_{j=1}^{n} (G_j - \bar{G})^2$$

be the variance and

$$\bar{G} = \frac{1}{n} \sum_{j=1}^{n} G_j$$

be the mean of G_j. The value of S^2 naturally depends on the distribution of the ones and zeros in the matrix. In addition to the value S^2, therefore, a value S^2_{max} is calculated by arranging the ones and zeros of the given matrix in each row as follows: First, all zeros are listed and then all ones such that the values K_i remain constant in each row. Let G_j be the value corresponding to G_j in the rearranged matrix, then

$$S^2_{max} = \frac{1}{n-1} \sum_{j=1}^{n} (G^x_j - \bar{G})^2$$

Then

$$L = \frac{S^2}{S^2_{max}}$$

is defined as the 'coefficient of concordance.' So L is the quotient of the observed variance and the maximal possible variance, which arises with maximal concordance of the observers under the assumption that each observer will give the same number of ones as he did originally. The author has motivated the definition of S^2_{max} as the measure of maximal concordance as follows: In calculating S^2_{max} the values K_i must remain constant because these values are characteristic of the individual observers. The reason for this assumption, I think, is a purely mathematical one.

1. The L-coefficient is defined analogously to Kendall's coefficient of concordance, which was developed for measuring the concordance of obervers when the observed examinees are ordinally ranked.

2. The assumptions that the values K_i are made in order to use the test criterion of Cochran's test. Yet in Cochran's test k individuals are observed in a repeated measures design in which each subject is tested under n conditions, where the observations are dichotomized. The hypothesis is that there is no effect associated with the various conditions. Under this hypothesis Cochran's coefficient of concordance Q, which in the notation of Table 2 is of the form

$$Q = \frac{(n-1) \sum_{j=1}^{n} (G_j - \bar{G})^2}{\sum_{i=1}^{k} K_i (1 - K_i/n)}$$

follows aproximately a χ^2-distribution with (n-1) df. The author, who developed the L-coefficient, formally replaced "examinees" in Cochran's test by "observers" and "conditions" in Cochran's test by "examinees", and because of the relation Q = const. L the distribution of the L-coefficient is known, too.

Here two statistical procedures are formally carried over to a new situation. The uncertainties of this procedure may be shown in Table 3.

Table 3.

Examinees

	1	2	3	4	5
1	1	0	1	1	1
2	0	1	1	1	1
3	1	1	0	1	1
4	1	1	1	1	0
5	1	1	1	0	1

From this table we get L = 0. Therefore, the hypothesis of minimal concordance cannot be rejected. But we really cannot conclude from this matrix that there is a minimal concordance between the observers. If we substitute the data of Table 3 in Cochran's test, then Q = 0. But this gives a significant statement for the matrix of Table 3.

The quantification of findings and the definition of a coefficient as such does not guarantee a precise measurement. The use of special statistical procedures in the domain of educational measurement will give results of an adequate value only if these procedures are adequate to the situations in which and for which they are used. On the other hand, a well-grounded theory and technique of test construction and evaluation may not be developed without an important contribution by mathematical statistical considerations.

Let me give one more example, this time from the theory of latent trait models. Under the assumption of local independence of items, the fundamental equation of the so-called latent trait models may be written in the form

$$p_I = \int_B [\prod_{i \in I} \varphi_i \;(\xi)]\, g(\xi)\, d\xi$$

where

(1) P_I is the probability of the event

$$A_I = [(\delta_1, \ldots, \delta_N) \in \{0, 1\}^N, \\ \delta_i = 1 \text{ for } i \in I, I \subset \{1, \ldots, N\}]$$

where N is the number of items;

(2)

$$\varphi_i(\xi) = p(A_i/\xi)$$

is the item characteristic curve (ICC); and

(3) $g(\xi)$ is the density function of the latent parameter in the domain B.

The model of latent structure analysis developed by Lazarsfeld is a universal test model in the sense that all response patterns are considered without using scoring formulas. If such a model is to be used in criterion referenced tests, then it is meaningful to require that functions $\varphi(\xi_i)$ be monotonic (if possible continuous) functions. We will not discuss the special forms of the ICC's arising from further assumtions such as that the (possibly weighted) number of positive answers is sufficient statistic for ξ. Let us ask the question under which conditions the data P_I, $I \subset \{1, \ldots, N\}$ are generally traceable to monotonic tracelines, which means that there exists a solution of the system

$$(I)\; p_I = \int_B \prod_{i \in I} \varphi_i(\xi)\, d\xi$$

with $\varphi_i(\xi)$ monotonic functions.

(Without loss of generality, we assume that ξ is measured in an ordinal scale. This assumption is made to reduce the problem of solving the system P_I to the problem of solving the system(I).).

Let N, the number of items, be fixed in the following. Then we have Theorem I. A necessary condition for the existence of strongly monotonic functions

$$\varphi_i(\xi),\; \xi \in [0,1]$$

which solve the system of equation (I), is that the following condition holds for all $I \subset \{1, \ldots, N\}$

$$(II) \sum_{J \supset I} (-1)^{|J-I|} p_J > 0$$

Theorem 2 Let

$$P = \{(p_I)\} \subset \mathbb{R}^{2^N - 1}$$

be the set of all data sets

$$p_I,\; I \subset \{1, \ldots, N\},\; I \neq 0$$

for which there exists trace lines (not necessarily monotonic) $_i(\xi)$, whichsolve the system (I). Let $A \subset P$ be the set for which there exist solutions with monotonic functions $_i(\xi)$. Then

(a) $P \neq \emptyset$,

(b) A is a closed set, and

(b) the interior of P - A is not empty.

These theorems among other things show

1. The requirement for the existence of monotonic ICC's cannot be fulfilled in an open non-empty set of the set of all data for which there exist solutions of (I).

2. The well-known effect that adding only one further item to a given test which in itself follows a special model, which means that it is in reasonable agreement with the data, may prove this model to be false for the enlarged set of data.

Theorem I explains mathematically what is emphasized by Lord (1952), who writes: "In fact at least with psychological models it can be taken for granted that every model is false if only we collect a sufficiently large sample" (a remark based on experience).

The question of whether or not the use of tests can be justified cannot be answered yes or no. There are situations and problems that certainly cannot be solved by mathematical statistical procedures. In each lesson the teacher is concerned with questions of evaluating his pupils where he cannot always use statistical procedures. On the other hand, one must recognize, as has already been mentioned above, that an observer must try to get out of the stage of the intuitive interpretation of his findings.

The more one is concerned with questions of fundamental research, the more problems arise. The question is whether it will be possible to solve or to reduce some of these problems, which are problems from a statistical point of view as well as from a psychological point of view.

References

Lord, F.M. A theory of test scores. Psychometric Monograph, 1952, No. 7.

Taylor, H.C. & Russell, J.T. The relationship of validity coefficients to the practical effectiveness of tests in selection: Discussion and tables. Journal of Applied Psychology, 1939, 23, 565-578.

SLAPONS (THE SCHOOL LEAVER'S ATTAINMENT PROFILE OF NUMERICAL SKILLS) - PROFILES AND TEMPLATES

Robert L. Lindsay
Shell Centre for Mathematical Education
Nottingham University
U.K.

School Examinations at the Minimum School Leaving Age of 16 in (England and Wales)

The General Certificate of Education (GCE) examinations for the top 20% of the school learners are well established and provide certification generally acceptable to employers, but the Certificate of Secondary Education (CSE) examinations for the next lower 40% do not always have the same degree of credibiilty. This is because the CSE is awarded without its certification disclosing: (a) what mode of control was employed i.e., whether the syllabus was designed or the papers marked by an official Examining Board or the user school(s) under moderation by the Board; (b) whether part of the assessment was on classwork and (c) what the content of the syllabus was. These details are not easy for every employer to discover or assess, especially for Mode III and high grades do not necessarily correlate with those mathematical skills considered important by the employer for the jobs usually open to this 40%.

Of the remaining lowest 40%, who are mostly not examined by any public examination board, some pupils had received an almost purely arithmetical syllabus (thus forming a mathematical sub-culture); an increasing number of pupils have become absorbed into the CSE, even to the extent of having a weaker form of examination with a maximum of Grade 3 for the best of this set of candidates; and others have been liberated, hopefully to enjoy a varied and more suitable syllabus. Whatever their fate, it is decided not by any central authority, but by their school.

Employer's Tests of Selection

Many employers ignore the CSE results in August since, as they say, they know more about the selected school leavers after applying a battery of tests at selection interviews in March or April, and after closely observing them at work in July and August, than the schools could ever tell them.

Most of the jobs which are in reach of the lower 60% of the population, and which involve some small but regular use of mathematics, are principally dependent on arithmetic; therefore, most test batteries applied by employers to candidates at interview include a test of the basic numerical skills. Many of the most reputable tests were validated 20 to 25 years ago when our schools spent a greater proportion of time on arithmetic, so that their norms of timing now seem severe.

The general acceptance of candidates with a lower thatn 50% score also leaves open the possibility that they won their marks on the questions least relevant to the job. These tests are so lacking in obvious structure that they neither indicate such an event nor make explicit the potential information they contain.

The contents of these perennial tests are also secret, so that teachers and students are unable to prepare for them as they do for all other examinatins. This militates against both student and employer: some students might fortuitously be in practice because their recent school work was closely related to the test material, while others will not be so well prepared; therefore, the employer may be accepting or rejecting students unfairly. Moreover, the employer would be wise to assess students when in practice, since this simulates the conditions of the working day with the test content (of a truly relevant test) will be continually repeated.

Slapons

The School Leaver's Attainment Profile of Numerical Skills (Slapons) test has been designed for schools to generate, in readiness for the employer's interviews a profile of a student's attainment in 18 distinguishable classes of numerical work.

It is especially useful as a diagnostic test, showing clearly strengths and weaknesses in a visually explicit format.

Employers are able to prepare templates which can be superimposed upon a student's profile, as is shown below, to help decide whom to employ, or what remediation will be necessary of employed.

Mathematics in Education and Industry
SLAPONS 1980
16+ school leavers attainment profile of numerical skills

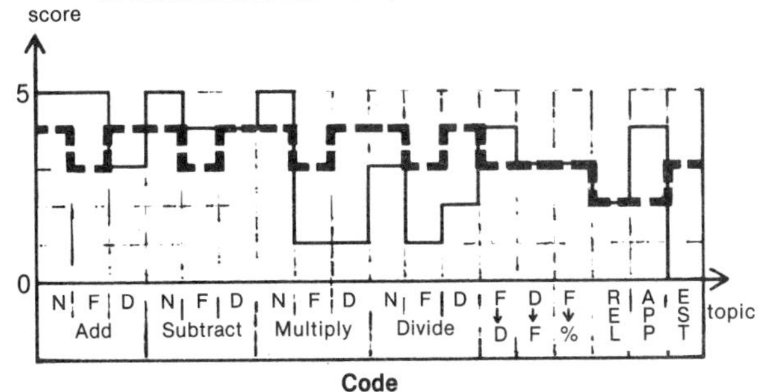

Code

N : Natural (whole) numbers
F : Fractions
D : Decimals
REL : Related thinking
APP : Approximations
EST : Estimations

Student's **profile**

Employer's **template**

Slapons test may also be used in school as a pretest for 15-year-old students to alert them and their teachers to any necessary remediation in the year of run-up to the formal test at 16.

Fresh, equivalent Slapons tests are provided whenever necessary and at frequent intervals in the year; there is no secrecy about the content.

Research

A variety of information can be extracted from both profiles and templates. Locally, employers can circulate their templates to inform schools of their expectations from candidates for employment. Nationally, these expectations can be compared with the distribution of student performances and the general expectations of similar, or different, employers, thereby monitoring the often emotional debate on numeracy between school and employers, and identifying the changing demands of new technology.

Future

The Slapons Test is now entering its fourth year and the increasingly heavy burden of administration will be undertaken by the Royal Society of Arts in September 1981; control of research will remain with the originators.

AN OVERVIEW OF CURRENT ASSESSMENT PRACTICES IN MATHEMATICS IN THE UNITED STATES

Ann F. McAloon
Educational Testing Service
Princeton, New Jersey, USA

Current assessment programs in the United States differ tremendously and include standardized testing, criterion-referenced testing, minimum competency testing, basic skills testing, and diagnostic testing. Some of these are described as state assessment programs. There is no such thing as a national testing program, per se, in this country, although the same achievement tests may be administered in many states. It is impossible, in such a brief paper, to give a full historical development of the different program. Therefore, some brief descriptions will be provided, together with some information about the purposes of, criticisms of, and need for improving the programs. The most highly publicized anti-testing concerns have been, for the most part, socially and politically related. However, criticisms of testing from an educational perspective will be discussed in this paper, with a major focus on mathematics testing as related to mathematics teaching and learning.

Standardized Testing

In the United Stataes, standardized testing has had a phenomenal growth since the 1930's. Over 75% of American school districts administer tests to millions of children in the elementary and secondary schools. In general, the tests are given in order to: provide information that will permit comparisons of learning achieved by students of different teachers, schools, or districts; hold teachers, schools, and school systems accountable; make decisions about individual students such as selection or placement; evaluate educational innovations and experimental projects; and provide instructional guidance to teachers in the classroom. Standardized tests are generally commercially published, contain carefully selected multiple-choice questions, are accompanied by technical manuals, and most importantly, are normed so that scores can be compared.

In recent years, there has been growing criticism of standardized test based upon social, political, and, at times, educational grounds. While there are numberous complaints, perhaps the most publicized are: the tests do not reflect the full range of cultural backgrounds of students and thus lead to decisions that are unfair to minority students; the tests are misused and misinterpreted, resulting in accountability decisions that are not valid; the tests have a limited effect on classroom teaching; and the tests are too narrow in scope to provide a fair evaluation of new approaches to teaching.

Despite concerns such as the above, recent Gallup Polls and other surveys indicate that 75% of classroom teachers and parents consider standardized tests useful for measuring growth of individual students and in detecting system-wide strengths and weaknesses. This general acceptance, however, does not mean that teachers and parents do not recognize the abuses and needs for improvement.

Criterion-Referenced Testing

While criterion-referenced testing (CRT) is still in search of a universally accepted definition, it is used to estimate an individual's performance on the entire universe of items of which the test is a sample or to classify a person as a master or nonmaster of the materials being tested. In contrast to a norm-referenced test (NRT), it is not used to compare an individual's perfrmance with that of others. Rather, for a CRT, a goal or standard is set before the test is administered and the test is desgined to compare an individual's score to that standard so that inferences can be made about what the person did or did not do. Objectives are stated and items are developed to sample the domain of the objectives. Depending upon the specificity of the objectives, a certain number of items are administered to determine whether a student is a master or nonmaster.

Since a NRT is concerned with comparisons of scores of groups of people, it is useful only to the extent that different people get different scores; the more the scores are spread out - - the larger the score variance - - the better the NRT will serve its purpose. However with a CRT, variability is irrelevant. If, for example, a CRT measures telling time on the hour and every student knows how to tell time, there will be no score variance. All items will be answered by all students, and the CRT will have met its purpose by showing what the students are able to do in the absence of any score variance.

A great deal of time is wasted on the argument of whether CRTs are better than NRTs. Each serves a different purpose; each has strengths and weaknesses as well as appropriate and inappropriate uses. The two complement rather than compete with each other.

Minimum Competency Testing

One point of agreement of testing specialists, school administrators, and educational policy makers is that there is no consistent terminology in use for "minimum competency testing." Competencies are used by some to mean skills, by others to mean objectives, and by still others to mean standards. In spite of the difference, it is agreed that minimum competency testing involves the use of objective, criterion-referenced tests; the assessment of basic or real life skills, usually in mathematics, reading, and writing; the requirement of a specified mastery level for determining acceptable performance; and the early introduction of such testing for purposes of identifying students needing remediation.

Mathematics Testing

One of the major stated purposes of most educational tests is to assist in the improvement of instruction. It is an aim that is seldom attained, either in standardized testing, criterion-referenced, or minimum competency testing programs. Many classroom teachers seldom even see the results of standardized tests which are required at periodic points in the students' schooling. While over 35 states have mandated minimum competency testing and have allocated funds for the administration of these tests, to my knowledge only one state has allocated funds to assist in the remediation of students who have been designated as nonmasters. In other cases, funding for such remediation has been left to local districts, already hard pressed for finances for regular education.

With respect to the diagnostic aspects of minimum competency tests, many elementary teachers do not have the necessary mathematics background to provide effective instruction, let alone an interpretation of the diagnostic needs of the students. For example, one of the common errors made by students in computing with fractions is a failure to recognize equivalencies such as 1/2, 2/4, 5/10, etc. Yet when a statement is made that students need more work in recognizing such equivalencies, the usual procedure is to give more of the same problems rather than to analyze the errors and to adjust instruction to correct them.

Relating testing, teaching, and learning presupposes knowledge of the subject matter (in this case, mathematics), knowledge of principles of measurement and evaluation, familiarity with effective teaching practices, and an understanding of how people learn mathematics. Therefore, it seems obvious that the problems associated with mathematics testing are twofold - - those originating from lack of knowledge of measurement and those arising from lack of understanding of the learning of mathematics.

Some of the problems associated with mathematics learning stem from people's attitude toward the subject, their anxiety or fear of it, and their lack of understanding that mathematics is much more than mere computational facility. Too many people think that if people can compute, then they are "mathematical." The National Assessment of Educational Progress (3), among many other programs, indicates that this is not so - - that Americans can compute fairly well, but the cannot apply the mathematics or solve problems. Unfortunately, many minimum competency programs concentrate on objectives that are almost exclusively oriented toward computational skills with little, if any, emphasis on the application of skills and problem solving.

The National Advisory Committee on Mathematical Education (2), has addressed most of these issues in making the following recommendations on evaluation in its 1975 report:

> Given the prominence and value of evaluation in the scheme of education today, more critical attention needs to be given to this area.
>
> Implications:
>
> a. that evaluation instruments be selected after program or individual goals are identified and that they be matched to these goals,
>
> b. that grade-level score reporting of student performance on standardized tests be abandoned,
>
> c. that an intensive efforts be made to develop objective-directed tests to replace the standardized norm-referenced tests now commonly used for student and program evaluation,
>
> d. that sampling techniques be used in program evaluation wherever appropriate, to minimize the over-testing problem,
>
> e. that evaluation results be reported in a multicomponent form corresponding to the multiplicity of goals normally associated with education programs,
>
> f. that extreme care be used in tests construction and administration to minimize potential cultural biases,
>
> g. that evaluators be more sensitive to the effects upon performance of certain factors of testing conditions, e.g., time limitation, over-testing, lack of motivation, unfavorable physical conditions, attitude of test adminstrators and teachers. (pp. 142-143)

Five years later, the National Council of Teachers of Mathematics has reaffirmed many of these recommendations in its An Agenda for Action (4):

> It is imperative that the goals of the mathematics program dictate the nature of evaluations needed to assess program effectiveness, student learning, teacher performance, or the quality of materials. Too often the reverse is true: the tests dictate the programs or assumptions of the evaluation plan are inconsistent with the program's goals.
>
> Recommended Action
>
> 5.1 The evaluation of mathematics learning should include the full range of the program's goals, including skills, problem solving, and problem-solving processses.
>
> 5.2 Parents should be regularly and adequately informed and involved in the evaluation process.

5.3 Teachers should should become knowledgeable about, and proficient in, the use of a wide variety of evaluative techniques.

5.4 The evaluation of mathematics programs should be based on the program's goals, using evaluation strategies consistent with these goals.

5.5 The evaluation of materials for mathematics teaching should be an essential aspect of program planning.

5.6 Mathematics teachers must undergo continuing evaluation as a necessary component in improving mathematics programs.

5.7 Funding agencies should support research and evaluation of the effects of a problem-solving emphasis in the mathematics curriculum. (pp. 14-16)

If the above recommendations are to be carried out in the educational community, it would seem that some additional concerns should be met. Among these are the following:

1. There should be a closer working relationship between measurement and instruction personnel. The task of instructional personnel is to guide the learning of the students, and the task of measurement personnel is to determine the effectiveness of that effort. If testing is to serve instruction, then the testing process must be inextricably linked to instruction.

2. Mathematics curriuclum specialists and competent teachers should become more deeply involved in mathematics testing at all levels. In state programs, they should be involved from the beginning, so that test objectives reflect the total mathematics program.

3. Evaluation in mathematics should be concerned not only with minimal skills usually oriented to content but should also include cognitive competence. Ebel (1) characterizes cognitive competence as a combination of knowledge built from information by thinking and intellectual skills such as observing, classifying, measuring, communicating, predicting, inferring, experimenting, formulating hypotheses, and interpreting data.

4. In setting standards for minimum competency tests, educators should be concerned both with mastery and development and should apply methods which are the least arbitrary as possible. There should be recognition of the diversity of difficulty of many mathematical objectives. It is much easier to add whole numbers, for example, than to add fractions, yet many state programs consider 80% correct as necessary to indicate mastery of both objectives.

5. Concerted and collaborative efforts should be made to assist teachers in developing affective evaluative instruments for the classroom. This has been a neglected area despite the fact that most testing is done at the classroom level. Classroom teachers need help in organizing objectives, in fitting test items to objectives, and in diagnosing and remediating student weaknesses.

6. The rapidly increasing capacity and decreasing cost of electronic information-handling technology make interactive teaching-testing arrangements both feasible and economically attractive. Mathematics educators need to be involved in the use of technology for the development of more programs which are innovative and provide immediate feedback. Some such programs have already been developed, for example, TORQUE (6), PLATO, BUGGY (5), DICOM (7), but more are needed.

References

1. Ebel, Robert L. Practical Problems in Educational Measurement. Lexington, Massachusetts: D.C. Heath and Co., 1980.

2. National Advisory Committe on Mathematical Education. Overview and Analysis of School Mathematics, Grades K-12. Washington, D.C.: Conference Board of the Mathematical Sciences, 1975.

3. National Assessment of Educational Progress. Changes in Mathematical Achievement, 1973-1978. Report No. 09-Ma-01. Denver, Colorado: National Assessment of Educational Progress, 1979.

4. National Council of Teachers of Mathematics. An Agenda for Action: Recommendations for School Mathematics of the 1980s. Reston, Virginia: National Council of Teachers of Mathematics, 1980.

5. National Institute of Education. Testing, Teaching, and Learning: Report of a Conference on Research on Testing. Washington, D.C.: U.S. Government Printing Office, 1979.

6. Project Torque: A New Approach to the Assessment of Children's Mathematical Competence. Newton, Massachusetts: Education Development Center, Inc., 1976.

7. Sherman, Matthew. The DICOM Concept. Princeton: Educational Testing Service, 1976.

Additional References

1. American Association of School Administrators. The Competency Movement: Problems and Solutions. Arlington, Va.: American Association of School Administrators, 1978.

2. Begle, E.G. Critical Variables in Mathematics Education: Findings From a Survey of the Empirical Literature. Washington, D.C.: Mathematical Association of America, 1979.

3. Bloom, Benjamin, Hastings, J. Thomas, and Madaus, George F. Handbook on Formative and Summative Evaluation of Student Learning. New York: McGraw-Hill Book Co., 1971.

4. Braswell, James. "The College Board Scholastic Aptitude Test: An Overview of the Mathematical Portion," Mathematics Teacher, March 1978, vol. 71, pp. 168-80.

5. Bunda, Mary A. and Sanders, James (Eds.). Practices and Problems in Competency-Based

Measurement. National Council on Measurement in Education Monograph, 1979.

6. Committee on Education and Labor, House of Representatives. Needs of Elementary and Secondary Education in the 1980's: A Compendium of Policy Papers. Washington, D.C.: U.S. Government Printing Office, 1980.

7. Committee on Education and Labor, House of Representatives. Oversight Hearing on Mathematics Achievement. Hearings before the Subcommittee on Elementary, Secondary, and Vocational Education. Washington, D.C.: U.S. Government Printing Office, 1979.

8. Ferguson, Richard and Schmeiser, Cynthia. "The Mathematics Usage Test of the ACT Assessment Program: An Overview of Its Purpose, Content, and Use," Mathematics Teacher, March 1978, vo.. 71, pp. 182-191.

9. Glass, Gene V. Standards and Criteria. Kalamazoo, Mich.: Evaluation Center, Western Michigan University, 1977.

10. Herndon, Enid. "NIE's study of minimum competency testing: A process for the clarification of issues." Paper presented at the 10th Annual Conference on Large-Scale Assessment, National Assessment of Educational Progress, Boulder, Colorado, 1980.

11. Jaeger, Richard and Tittle, Carol (Eds.). Minimum Competency Achievement Testing: Motives, Models, Measures, and Consequences. Berkeley, Calif.: McCutchan Publishing Company, 1980.

12. The Mathematical Association of America. Prime 80: Proceedings of a Conference on Prospects in Mathematics Education in the 1980's. Washington, D.C.: Mathematics Association of America, 1978.

13. National Council of Supervisors of Mathematics. Position paper on mathematical skills. Mathematics Teacher, February 1978, vol 71, pp. 147-152.

14. National Education Association. Measurement and Testing: An NEA Perspective. Research Memo. Washington, D.C.: National Education Association, 1980.

15. National Institute of Education. A Study of Minimum Competency Testing Programs: Comprehensive Report. Washington, D.C.: National Institute of Education, 1979.

16. National Institute of Education. Achievement Testing and Basic Skills: Conference Proceedings. The National Conference on Achievement Testing and Basic Skills. Washington, D.C.: National Institute of Education, 1979.

17. National School Boards Association. Standardized Testing: A Research Report. Washington, D.C.: National School Boards Association, 1977.

18. National School Public Relations Association. The Competency Challenge: What Schools are Doing. Arlington, Va.: National Schools Public Relations Association, 1978.

19. National Science Foundation. Report of the 1977 National Survey of Science, Mathematics, and Social Studies Education. Washington, D.C.: U.S. Government Printing Office, 1978.

20. National Science Foundation. The Status of Pre-College Science, Mathematics, and Social Studies Education Practices in the U.S. Schools: An Overview and Summaries of Three Studies. Washington, D.C.: U.S. Government Printing Office, 1978.

21. Orr, David. Measurement in Education in the 1980's--A Federal Perspective. Washington, D.C.: National Center for Education Statistics, 1980.

22. Suydam, Marilyn N. Evaluation in the Mathematics Classroom: From What and Why to How and Where. Columbus: ERIC Information Analysis Center for Science, Mathematics and Environmental Education, 1974.

23. Testing Team Research Program for 1980. Washington, D.C.: National Institute of Education, 1980.

24. Thorndike, Robert (Ed.). Educational Measurement. 2nd edition. Washington, D.C.: American Council on Education, 1971.

25. Thorndike, Robert and Hagen, Elizabeth. Measurement and Evaluation in Psychology and Education. 3rd edition. New York: John Wiley and Sons, Inc., 1969.

26. Department of Health, Education and Welfare. What Do We Know About Standards for Effective Basic Skills Programs? Washington, D.C.: HEW, Office of Education, 1979

14.2 ISSUES, METHODS AND RESULTS OF NATIONAL MATHEMATICS ASSESSMENT

NATIONAL ASSESSMENT OF MATHEMATICS ACHIEVEMENT IN PAPUA NEW GUINEA

R.E. Roberts
Department of Education
Papua, New Guinea

This paper outlines Papua New Guinea's current innovative thinking and action which attempts to tackle the issue of how to achieve a sufficiently high standard of mathematics achievement to meet national development needs.

The Papua New Guinea Context

Papua New Guinea is a newly independent nation of three million people in the southwest corner of the South Pacific. Since it gained independence from Australia in September 1975 its then-expanding education service has continued to grow at a considerable rate coupled with a similarly rapid rate of localisation, particularly at the administrative and primary school levels. A comparatively sophisticated modern sector grows daily, employing advanced technology, e.g. microwave communication systems, and places heavy demands upon the limited numbers of school leavers with a sufficiently high level of mathematical education.

The Education Department seeks ways of improving standards of mathematics achievement. To date the four national examinations in Mathematics, i.e., one at Grade 6, two at Grade 10, and one at Grade 12, serve only to rank students for selection purposes. Very few achievement criteria are required to provide teachers and students with sufficiently high achievement goals. The most serious outcome of low levels of achievement within the school system is the subsequent failure of Grade 10 graduates to meet the demands of further education or technical training courses into which they enter. An extreme example might be the lethal errors which could be made by a trainee nurse with limited knowledge of the metric measurement system, being required to prepare and administer dangerous drugs.

Outlined below are our recent attempts to (a) establish what mathematics is being learned in schools and so to best judge where to place the development emphasis to assist classroom teachers in their day-to-day teaching of the subject and (b) to set realistic achievement goals for teachers and students to work towards.

Primary Mathematics Achievement

In 1977 a small committee of curriculum developers, college lecturers, and researchers was formed to develop a series of achievement tests to assess pupil performance in primary Mathematics. This development coincided with the introduction of a new Dienes-based primary mathematics curriculum, known as Mathematics for Community Schools (MaCS). The PMAT programme was to be in five phases:

Phase 1: The design of a preliminary test consisting of questions ranging the content of the whole course and for all grades.

Phase 2: The trial and analysis of the preliminary test.

Phase 3: The design of an expanded test to be in five parts, one for each of the strands of the course: logic, relations, number, measurement, and geometry. Each of the five tests was to contain questions for Grades 2 to 6.

Phase 4: The trialling and analysis of the strand tests.

Phase 5: The design of an improved batch of strand tests and the setting of grade norms of achievement in each strand of the course. Information gained during the project would enable teachers to judge better what children could be expected to master at each grade level.

To date the PMAT Programme is at Phase 2.

The 50-item preliminary test was structured in such a way that the first ten items were taken from the Grade 2 objectives, the next ten from the Grade 3 objectives, and so on, so that the items 41-50 were testing Grade 6 objectives. Response modes included matching items of the same type or value, circling the answer, colouring the answer, simple computation, and filling-in the answer. The PMAT has been administered late each year for the past three years to an average and representative sample of thirty community schools.

Using the criterion "proportion of their ten questions for which pupils got full marks," i.e. for Grade 6 pupils only questions 41-50 were examined, mastery levels were calculated. Only the Grade 2 pupils were achieving a satisfactory level of mastery of their level of questions. Thereafter, mastery levels dropped rapidly. There had been no significant change in mastery levels during the first three years following the introduction of the MaCS curriculum. It is beyond the scope of this paper to give any detailed analysis of this situation (background can be gained by reading Roberts 1978 and Roberts and Kada 1979). The course content is currently under review as part of the Indigenous Mathematics Project (Lancy 1979 and Souvine 1980) in preparation for the development of a student textbook to replace the present brief teacher's notes (the sole source of information and guidance to being proven to be) falling, this evidence would seem to indicate that standards have not fallen during this period of curriculum change, though they are clearly low in relation to the MaCS syllabus objectives. To date, very little information from the findings of the PMAT programme have been translated into teacher-usable information. An article reporting the findings (in broad terms) of the 1977 PMAT survey appeared in the Department of Education's Gazette in April 1979.

Basic Arithemetical Skills at Post-Secondary Level

During 1979 the Director of the Mathematics Education Centre, Lae, Mr.Allen Edwards, conducted a survey of Grade 10 graduates who had entered post-secondary education and training institutions that year. The survey aimed at identifying "the types of questions which post-secondary entrants could or could not answer correctly" (Edwards 1979:2). A total of 2604 students and trainees, i.e. 42% of all 1978 Grade 10 graduates and 55% of those who entered post-secondary education or training institutions, were tested. The survey covered 88% of all such institutions. The test contained 84items of basic arithmetic ranging from simple additions of whole numbers to comparisons of decimal fractions, time telling and simple measurements.

The mean scores in the 43 institutions ranged from 77.5 in a university to 36.5 in a school of nursing; the median of these mean scores was 63.6. The range of scores on what was essentially a mastery test was very wide. Edwards made a particularly strong comment about the inability of trainee nurses to do simple arithmetic and the inherent dangers when such nurses mixed and administered drugs. Serious deficiencies in the basic arithmetical skills of university students were revealed (Edwards 1979: 12-16). One disturbing example was that amongst 18 second year students tested at the University of Technology, "who were among the cream of the nation's future businessmen, apparently could not divide K75.00 by 100 correctly, when 7 failed to multiply 18.3 by 100" (Edwards 1979: 30-31).

Concerns over low levels of achievement in basic arithmetic amongst entrants to post-secondary education and training institutions in Papua New Guinea has for some time been a matter of serious concern (note Bajpai 1971, Roakeina 1977, for example). The current concern appears widespread amongst educators in the more than sixty post-secondary institutins in Papua New Guinea today. Some of the difficulties faced by post-secondary instructors appears to result from the failure of the present Grade 10 examination system to optimise student ability at that level.

An Alternative Grade 10 Examination Service

An alternative assessment system to the current MYR and MYE examination has been proposed (Britt 1979, Siford 1979). The system, based on one successfully operated in the Nelson and Marlborough Counties in New Zealand during 1975-79 (Nightingale 1979), would relate more closely to the syllabus, testing factual recall, comprehension, application, and some analysis and synthesis in a five-tier testing programme offered by two testing periods. The system is presented schematically in Table 1.

Table 1

A Five Level Grade 10 Criterion-Reference Testing System

Level of Test	Test Emphasis	Approximate grade equivalent and purpose
1	Factual Recall Comprehension	Grade 6 (Basic knowledge/skills)
2	Factual Recall Comprehension Application	Grade 8 (Emphasis practical skills)
3	Comprehension Application Analysis	Grade 10 (Emphasis on technical training)
4	Comprehension Application Analysis	Grade 10 (Emphasis on further education)
5	Application Analysis Synthesis	Grade 10 (Similar to MYRE testing skills not content)

Note: Papers 1 - 4 would be marked at the school level. Paper 5 would be marked at national education headquarters.

Note: Papers 1-4 would be marked at the school level. Paper 5 would be marked at national education headquarters.

There would be five levels of achievement. There would be two testing periods. May and October/November. During the first period (May) four tests would be available - - one for each of the first four levels. Five tests would be available during the second testing period, one for each of the five levels.

Students may take one, or at most two, of the tests at each testing period. Which test or tests should be taken and when, would be mutually agreed by the teacher and the individual students. As a general rule, in the first testing period poor students would take Tests 3 and 4. In the second period, Test 5 may be taken by students who have passed at least Test 3 during the first testing period. The minimum content and skills required and concepts to be understood for each of the first four levels would be carefully specified in terms of syllabus objectives. Students would be required to score at least 75% to demonstrate mastery at that level before being allowed to sit for the test at the next level.

The advantage of this system would be to motivate teachers and students towards the kind of mathematical learning which Britt (1979: 1) describes as primarily "not to perpetuate knowledge or just to extend existing knowledge but rather to foster the creation of new knowledge derived from the solutions of new problems that are being recognised now, or will be in future."

A National System of Criterion-Referenced Tests

A recently completed research study into the possibility of introducing a system of criterion-referenced tests into primary and secondary schools, in a sense, extends the ideas of Britt and Siford to cover assessment in Grades 1 to 10. This viability study was carried out by Gay Townsend of the Education Research Unit, UPNG, at the request of the Department of Education (see Townsend et al. 1980).

The items from which the tests would be constructed would be stored on a computer and drawn upon to provide national examinations as and when required and also tests for school inspectors and curriculum developers. Clearly, a wide ranging and flexible system of assessment, accessed from a computer data base, as suggested in Townsend's draft report, would help to correct many of the faults existing in the present assessment system. Townsend recommended that a system of bi-annual criterion-referenced tests in English and Mathematics be introduced for Grades 4, 5, and 6 and that unit (usually two or three weeks of work) tests in English and Mathematics be developed for Grades 7 - 10.

Conclusion

Assessment of mathematics achievement in Papua New Guinea today does little more than facilitate the selection by rank (as opposed to measuring the achievement of objectives set by the curriculum) of students for higher education or training. Little information on students' actual achievement levels on tests and examinations is returned to curriculum developers, and that which is remains largely untouched for want of time and expertise to analyse and report on it. Educators and training personnel in post-secondary institutions are seriously concerned about the inability of the entrants to their institutions to cope with the basic mathematical content of the courses offered. A system of criterion-referenced tests has been proposed which, it would appear, could provide the objectives, the motivation, and the information presently lacking in the country's national assessment system. It would appear that the Department of Education has not to either make its decision on the priority status of this proposal to develop some other means of assessing the results of classroom instruction in relation to prescribed syllabus objectives.

References

Bajpai, A.C. 1971: A Feasibility Study into the Teaching of Mathematics at the Tertiary Level with Particular Reference to Technical Education in the Territory of New Papua New Guinea, Nuffield, University of Loughborough, England.

Britt, M. 1979: Evaluation of Secondary Mathematics Learning, UNESCO Mathematics Project, Port Moresby.

Edwards, A. 1979: Preliminary Report on the Basic Skills Arithmetic Test Carried Out in Post-Secondary Institutions in Papua New Guinea in 1979, Mathematics Education Centre, Report No. 6, Papua New Guinea University of Technology.

Lancy, D.F. 1979: An Examination and Evaluation System, Education Research 1976-1979: Report and Essays, Department of Education, Port Moresby.

Nightingale, D. 1979: Mastery Levels Assessment: The Nelson-Marlborough Experience 1975-1979, Nelson Community Education Service, Nelson, New Zealand.

Roakeina, G. 1977: Report of the Committee of Enquiry into Standards of High School Students Entering Colleges, Department of Education, Port Moresby.

Roberts, R.E. 1978: Primary Mathematics in Papua New Guinea, Papua New Guinea Journal of Education, 14: 201-216.

Roberts, R.E. and Kada V. 1979: The Primary Mathematics Classroom, Papua New Guinea Journal of Education, 15: 174-201.

Siford, M. 1979: Research Project to Investigate an Alternative Grade 10 Examination System, University of Papua New Guinea.

Townsend, G., Guthrie, G., and O'Driscoll, M. 1980: Proposals for a Criterion-Referenced Measurement System for Community and Provincial High Schools, Report to the Secretary for Education, Port Moresby.

CHAPTER 15 - Competitions

15.1 MATHEMATICAL COMPETITIONS, CONTESTS, OLYMPIADS

STIMULATION OF GIFTED STUDENTS BY CORRESPONDENCE LESSONS

Jan van de Craats
University of Leiden
The Netherlands

Like in many other countries, in The Netherlands we have various means to stimulate general interest in mathematics among young people and to discover extraordinary talents. I mention:

Pythagoras, our mathematics magazine for the youth (5 issues a year, about 15,000 subscribers).

Pythagoras Olympiad, a competition in Pythagoras, started this year (3 problems in each issue).

The Dutch Mathematical Olympiad, a yearly competition, started in 1962. The First Round (10-15 problems, 3 hours working time) is organized in the secondary schools (3000-3500 participants). The best 100 or so of them are invited to take part in the Second Round, consisting of 4 more difficult problems, again with 3 hours time available. The best 10 of the Second Round get prizes.

Although these olympiads certainly are effective in stimulating general interest in mathematics and discovering highly talented young people, they do nothing to develop great talents, once they are discovered. To stimulate these very gifted pupils, and also to enhance our results in the International Mathematical Olympiad, we started in 1976 a program of "correspondence lessons." The best 15 participants of our olympiad are sent small expository papers on subjects like number theory, combinatorics, polynomials and geometry, mixed with lots of exercises. The pupils are requested to work through the paper, try all exercises and send in their elaborations of a small number of problems, together with any questions left on the theory or the problems. We then send them our comments together with the next unit. In this way every interested student gets a kind of individual guidance in accordance with his personal abilities and the time he is willing to spend on it. Preparation for the International Mathematical Olympiad is an important feature of the program, but our main goal is independent of any competition or olympiad. We believe it is most important to enlarge the mathematical knowledge and ability of the students, and to bring them in close contact with elementary branches of mathematics, usually not represented in the school curriculum. It also happens more than once that participants ask questions on mathematical subjects they are interested in, and this offers us the opportunity to help them find their way in a library of a Mathematical Institute.

A serious drawback of competitions like the I.M.O. is that the participants have only a few hours to solve the problems. This condition is alien to "real" mathematical work, and in fact many very good mathematicians simply cannot work this way. The correspondence lessons program also gives the highly gifted, but not so quick-minded, students opportunities to develop their talents.

We can work with our program in this way only with a very limited number of pupils. We hope, however, that the correspondence lessons also may serve wider purposes, such as broadening the horizon of interested school teachers. On an announcement in the teachers' journal, more than 150 teachers were willing to pay a small amount of money to get copies of the lessons.

MATHEMATICAL COMPETITIONS IN NEW ZEALAND

Neville Gale
Christchurch, New Zealand

The "Senior Mathematics Competition" for N.Z. students at Form 6 and 7 level (ages 16, 17+, or the last two years of High School) is held each year and is open to all schools throughout the country with no entry fee. The competition is organised by one of the local Mathematical Associations and began in 1967 as a local competition for schools in the Canterbury area. In 1977 it was opened to all schools in the South Island of New Zealand (New Zealand is made up of two main islands - North Island and South Island). Since 1978 it has been a national competition open to all 400 secondary schools. The competition is in two stages. Stage I consists of a one hour test sat by over 2,000 students in their own schools. The schools mark these and send their best five scripts to their area Competition Secretary (the country is divided into nine areas for this purpose). Each area Secretary organises a team of markers to re-mark these scripts and sends to the Competition Director the best ten scripts from that region. These regional finalists are often recognised by the local association and awarded prizes locally. The area finalists' scripts are marked again and the best twenty students are flown to Christchurch to sit a final paper. The students sit this final paper in the morning and are entertained for the day with prized being awarded at an evening function. The prizes are valued at $2,500 and consist of cash, books and calculators. Very little sponsorship from private firms has been available to cover the costs of this competition.

This year all New Zealand schools were invited to participate in the "Australian Mathematics Competition" described by Mr. O'Halloran. In 1980, 5,700 students from 110 schools have entered the competition at the three levels.

New Zealand has not participated in the Mathematical Olympiads but it is hoped that this will take place at some time in the near future.

In New Zealand there are no National Mathematical Competitions at "Junior" level but local associations organise annual "Mathematical Fairs" for students at Form 1 (12 years), Form 2 (13 years), Form 3 (14 years), and Form 4 (15 years). The idea for this type of contest originated with the Canterbury Mathematical

Association in 1970 so their format is described here. It is repeated with slight variations in other centres. It is called CANTAMATH 80 (depending on the year) and involves several sections with an emphasis in all of getting the young people involved and doing something with a mathematical bias or influence. The sections are - Poster - displaying a poster with mathematical theme. Design - again with a mathematical theme. Models and Projects - both class and individual projects or some mathematical theme (e.g. statistical survey, mathematics in architecture and even song and dance items). Verse - poetry and mathematics, e.g.

Maths is sensical, Maths is fun, Maths can be done By everyone	Maths is forms, Maths is shapes, Maths can be measured With a tape.
Maths is problems, Maths is sums, Maths is learning - Knowing just comes.	Maths is rectangles, Maths is spheres, Maths can even Be done by squares.

(14 years old)

But perhaps the most exciting part of the CANTAMATH type of contest is the Team Competition held during the evenings. Schools enter teams of four students who are required to answer questions as a team. They are given the first question in an envelope and when they think they have the correct answer their "runner" takes it to their particular marker (one marker to two teams). If it is correct, they get the next question. If it is not correct, they must go back until it is correct. They can give up on a question but, of course, this could count against them at the end of the contest. With forty-five teams competing at one time there is a mass of activity with students running everywhere and barracking from the teams' supporters on the sideline. CANTAMATH is held over three days in the Christchurch Town Hall and attracts large numbers of the general public to view the exhibits and to watch the team competitions. The enthusiasm of the competitors (some wearing distinctive "Mathematical" hats), and supporters (parents, teachers, students) has to be seen and heard to be believed. Visitors cannot believe that the enthusiasm is for Mathematics. The questions are lively, topical and interesting mathematical questions and teams develop a skill and technique in handling them. Many school start training their teams months before the actual event. Each contest lasts 30 minutes and the team with the most points wins. If there is a tie, a tie-breaker decides the winner. The audience is kept informed with scores of each team being displayed and some of the questions displayed on large screens. Prizes to the value of more than $3,000 are distributed at each CANTAMATH competition.

A variation of this team competition is also repeated with our Senior Students. Two Form 6 and two Form 7 students form a team and start at one end of the hall. In front of each team is a column of twenty tables with the marker at the other end. Each team works in pairs as they answer each question they move up to the next table. They can call in the other pair of their team to assist them answer a question without penalty but only after they have had one refusal from their marker. If they give up on a question they do not proceed to the next table. In this way the audience can see the overall position of the teams. For this variation the number of teams that can be accomodated is very much reduced so this is run as two heats. The winners of each heat play off in a Mathematical Challenge contest. In this, both teams are given a "starter" question. The first team member to press his buzzer and answer correctly gains ten points for his team and the team then has a chance to gain a further ten points with two bonus questions worth five points each. For the starter questions, no team consultation is allowed but the teams can discuss the bonus questions. This also has proved very successful and holds the interest and attention of the audience as well.

The result of these types of mathematical extravaganzas is a terrific upsurge in interest in mathematics by both students and teachers so much so that we now have English, Science and Language teachers organising similar Subject Fairs.

Maths Camps are held annually and organised by the Wellington Mathematical Association. Each regional Mathematical Association is invited to send selected senior students to attend a four day residential school where the students are conducted on a tour of mathematical exploration in areas they have not been exposed to in their normal classroom courses. These annual schools always have far more students wishing to attend (in their term holidays) than can be accomodated.

Footnote: New Zealand is a small country (about the size of the British Isles) in the South Pacific with a population of 3,000,000 people and 70,000,000 sheep!!!

SOME PREDICTIVE ASPECTS OF THE WILLIAM LOWELL PUTNAM MATHEMATICAL COMPETITION

Jose Gabriel Ipina
Universidad Boliviana de S.F. Xavier
Bolivia and
Laguardia Community College, CUNY
New York, New York, U.S.A.

Purpose

The purpose of this study was to investigate the predictive validity of the rankings and other related variables of the William Lowell Putnam Mathematical Competition in terms of professional and/or academic accomplishments in the mathematical sciences.

Procedures

Besides the rankings of the Putnam Competition - - provided to the general public by the Office of the Director of the Putnam Competition in the form of ordinal categories - - two other related variables were identified, the frequency with which a participant received a public ranking in the Competition and the geographical origin of the contestant.

Eleven criteria of professional and/or academic accomplishments in the mathematical sciences were identified: (1) attainment of the doctoral degree in the mathematical sciences, (2) a quality rating of the graduate program in which the doctoral degree was awarded, (3) the time lapse between the baccalaureate and doctoral degrees, (4) National Science Foundation and/or Woodrow Wilson Fellowships for graduate studies in mathematics, (5) postdoctoral fellowships in mathematics, (6) number of publications reviewed in Mathematical Reviews, (7) number of citatins as listed in the Annual Citation Index of Science Citation Index, (8) a quality ratio of number of citations per number of publications, (9) invitations to deliver addresses at national meetings of the American Mathematical, the Mathematical Association of America, the Society for Industrial and Applied Mathematics, and at the periodic meetings of the International Congress of Mathematicians, (10) prizes and awards conferred by the professional organizations cited above and the International Mathematical Union, (11) professional acquaintance by full professors in the ten most highly rated graduate departments of mathematics (Roose-Andersen rating scale).

The sample of this study consisted of all publicly ranked Putnam contestants from the first competition (1938) through the 38th (1977). The size of the total sample was 7,197 (some contestants entered more than once). The study was subdivided in two parts - - on the grounds of number of ranking categories available for the competition - - the first part (1st through 24th competitions, with only three ranking categories) consisted of an effective sample size of 351. The second part of the study (25th through 38th competitions, with five ranking categories) consisted of an effective sample of 4,774.

The three Putnam-related variables were treated as independent variables and each of the first 10 criteria was treated as a dependent variable. The data collected were first analyzed using multi-dimensional tests of independence, and, to allow for the order of the ranking categories, then through stepwise hierarchical multiple linear regressions with the Putnam variables as predictors and each criterion as dependent variable.

Findings

The tests of independence for Part I of the analysis showed no significant results with the exception of the rankings and the attainment of the doctoral degree in mathematics. The tests of independence for Part II of the analysis showed consistent statistically significant results (with p not more than 0.05) but the "effect size" of association between the independent variables and the criteria was at best of medium size ($W = .30$).

The multiple linear regressions, controlling for a years-since-doctorate variable or a covariate for number of publications and citations, gave at most a squared multiple correlation coefficient, R^2, of 0.07 for the three Putnam variables taken together in the prediction of the quality of graduate program.

The "professional acquaintence" ratings by the professors of mathematics in the ten most highly rated departments of mathematics provided a mean rating of approximately 18 for the Putnam group and 14 for the Non-Putnam sample, however the difference in means was not statistically significant ($p > 0.10$).

SOME FEATURES OF THE MATHEMATICAL CONTESTS HELD IN LUXEMBOURG

Lucien Kieffer
Luxembourg

The Grand-Duchy of Luxembourg (350,000 inhabitants - 1,000 square miles) is too small a country to organize proper mathematical contests.

Lying on the linguistic border between Germany and France all Luxembourgers in everyday life use a German dialect which, however, is no written language. So, from primary school on (ages 6-12) German is the teaching language; from grade 2 on, French is taught and becomes predominant in the post primary "lycees" (grades 7-13). From grade 8 on, English is the third compulsory language.

Besides technical and vocational schools there are 10 Lycees with a total of 10,000 students. In the upper grades there are 4 streams: Languages, Mathematics, Science, and Economics. Mathematics is taught in French, the textbooks are French or Belgian, our curricula very near to the French or Belgian ones. The contests are mostly taken by students of the mathematics stream in grades 12 and 13, a total of 250 students, 60% of whom compete.

American HSME

From 1962 on, this multiple choice contest (30 questions in 90 minutes) has been held in Luxembourg. An article in the Dutch magazine "Euclides" brought me in contact with the MAA-Executive Director, Professor Salkind (Brooklyn) who sent me the question booklet I translated into French. A Government Office provided copies which have been used in Luxembourg and in some French-speaking schools in Belgium.

The Contest is organized in 3 towns. In Luxembourg-City all contestants from 6 Lycees compete in the same room, invigilated by teachers. The marking has been done by myself. Individually, each teacher informs his students of the obtained scores. Only the best results are published.

In the last decade there was a slight ascending trend in the participation. Especially the number of girls increased, up to 20%. Since 1979 the test is taken the same day as in USA and Canada. A report has been sent to the MAA-Committee all the years.

Some Considerations

There is no special training, unless teachers hand their pupils the question sheets used in former years. Our contestants are slightly older (17 to 19 years) than the American competitors, but the curricula differ in many points (no questions about calculus nor linear algebra). Though multiple choice tests are unknown in our schools, our students are eager to compete, for the questions are of increasing difficulty, so that not only bright students, but even "good average" may find some solutions.

A second point is the publicity given to this Contest. In 1966 the American Ambassador inaugurated the Award Ceremony at the Embassy for the winner and the top scorers. This ceremony has been held all these years, and pictures and reports in the newspapers have attracted great attention among students, teachers, and parents. The only prize is the MAA pin for the winner; no books, no calculators, no cash!

In multiple choice tests there must be a penalty for wrong answers. Indeed lucky guesses without reasoning in difficult questions are more profitable than simple omission of such problems. In the last years we have noticed an increasing number of contestants who filled out all the boxes of the answer sheet.

Though the results of only 150 contestants cannot be considered as wholly significant, they were concordant with the Dutch ones in the early 1960's. In the last years our top scorers have confirmed their excellent HSME results in other competitions.

From 1962-80 we had four Honor Roll scorers: in 1972, 1974, 1977, 1980. Our 1980 winner with a score of 112 would have been selected, in the USA, for the USA-Olympiad.

A peculiar feature: in 1970 and 1971 girls had been the winners. In 1980 there were 2 girls among the 6 best top scorers.

Our winners are not always the brightest students of their forms. Therefore some parents, even teachers, whose children are very able, prevent them from competing. They fear the contest mark would not match their brilliant school results. Officially they put forward the stress of the short 90-minutes time. So we are cautious: failures which could discourage participants are never published.

All the winners of our 19 contests remained faithful to mathematics in their future studies.

Other Contests

As our students are fluent in German, almost every year since 1972, one or two top scorers have participated in the German Competition BWM. The first and the second round, in which a set of 4 difficult problems are to be worked at home, can be taken by foreigners, but not the third. Three Luxembourgers, amongst whom is our 1980 HSME winner, have been admitted to the second round.

Since 1977 our students have been participating in the Belgian Mathematical Olympiad. The first round is a multiple choice test (40 questions in 4 hours); the final round consists of 4 hard problems to be solved in 4 hours. In all the 4 years Luxembourgers won second, even first prizes (books, calculators). In 1980, one hundred Luxembourgers competed; among the 40 second round contestants there were 7 Luxembourgers and our HSME winner got a first prize. In the Junior Contest our pupils were not so successful.

International Mathematical Competition, Mersch (Lux.)

As IMO 1980 could not be held in Mongolia, Luxembourg was asked in February 1980 to organize a limited replacement competition for some European countries. For lack of time, lack of funds, lack of enough accomodation in the only boarding school, a competition for 40 students was held in Mersch, a nice tourist resort. GB, NL, B, YU sent teams (8 contestants), Luxembourg a mini-team. Four top scorers of the French "Math. Rallye of Alsace" were observers.

For all points: selection of the problems, marking, and prizes, the IMO regulations were strictly observed. A Dutch student was the individualwinner; the YU team was first before GB. Similar international competitions were held in July in Finland (SF, S, H, GB) and in Poland (P., A).

The present survey on mathematical competitions for students mostly of the upper level of secondary education, may serve as a complement to Prof. Freudenthal's ICMI Report on the same subject Educational Studies in Mathematics 2 (1969) 80-144.

After 1945 the popularity of mathematical contests spread in the European socialist countries which, like Hungary (Eotvos Contest 1894), Romania (Gazeta Mathematica Contest 1902), USSE (Leningrad 1934 and Moscow 1935 Olympiads) had a tradition, but also in Poland, Bulgaria, Czechoslovakia - not to forget USA (Stanford Examination 1946 and New York MAA-Contest 1950). Many local or regional competitions became nationwide when governments or private sponsors provided the necessary funds for the organization.

The unique role of these contests in discovering gifted students was clearly seen by many famous mathematicians, such as G. Polya and G. Szego (USA), Schnirelman, Dynkin, Gel'fand (USSR). Especially Hungary could boast many outstanding mathematicians (Fejer, Haar, Riesz, Konig . . .) and physicists (von Karman, Teller . . .) who, as students, had been winners of the Eotvos Contest.

In the decade after the inauguration of the International Mathematical Olympiad (IMO) in Bucharest (1959) more countries held National Contests: GDR, Jugoslavia, Sweden, Netherlands, Luxebourgh, Italy, GB, Finland, Canada, Israel. In the 1970's the movement attained Austria, Western Germany, Belgium . . . and also Asia, Africa, Latin America and Australia.

In the early contests the number of competitors was small (less than 100), so one round could suffice. For nationwide contests with tens of thousands or even some hundred thousands of participants (in USA and USSR) preliminary screening examinations became necessary. So, most of the competitions are held in 2, even 3-5 rounds. The first is taken at the school level, the top scorers are admitted to the second, and so on. In order not to discourage the participants the questions of the first round are not too difficult, or a choice among the given problems is possible, or students with only 25 to 30% of the maximal are admitted to the next step.

In many countries the first round is a multiple choice test. Best known is the "Annual High School Mathematics Examination" (HSME) inaugurated in USA and Canada (1958), with at present 30 questions to be solved in 90 minutes. GB, L. H, . . . adopted it. The marking, very easy, is done in the schools or, as in Australia and Belgium, by a computerized system.

In other countries a set of 4-10 essay-type problems are sent to the participating schools which pupils, working at home, have to solve individually; the alloted time varies from one week to 2 or 3 months. The first round is held in that way in S, SF, PL, . . . and the same applies to the 2 rounds of the German BWM.

In the last rounds, especially in the final examination with a small number of competitors (less than 200, even in USA and USSR), there are invigilated test papers; 3 to 10 hard problems to be solved in 3 to 5 hours. In IMO's there are two-day examinations with 3 problems to be done in 4 hours, each day. Two papers are also in the final rounds of PL, GDR, USSR and A.

For lack of space no detailed mention can be made about historic, regional or local contests still existing in USSR, USA, H, . . . nor about contests of university level in USSR, H, USA (W.L. Putnam 1938), nor about the many journals with problem corners for the training of the competitors: Kozepiskolai Mat. Lapok (H), Gazeta Mathematica (R), Quant (USSR), Pythagoras (NL), Crux Mathematicorum (Can), Alpha (GDR), Delta (PL), . . . In some countries such as USSR, H, China, . . . a high score in the National Contest replaces university entrance examination. In Eastern countries the training of bright competitors is promoted by mathematical circles, a library of special textbooks with problems, summer courses, even by special boarding schools with continuous competitions.

Evaluation

At the Exeter Congress in 1972 the working group "Mathematical Competitions" with chairman E. Hodi (H) and attended by Prof. G. Polya already gave this topic attention.

At the present mathematical contests are held in more than 50 countries. If their aim is the early discovery and the fostering of the mathematically gifted students, the procedure is different.

In sport the competitive element often leads to greater achievement. Thus, some countries seen in contests a complement to the low standard of their mathematical curricula or to the lack of problem solving in the courses, too much emphasis being given to structures in secondary education.

Good competitions may improve the mathematics teaching by stimulation of the teachers, too. But many teachers donot collaborate as they do not feel at ease when competing students raise questions. Some countries only turn to good average students; the competition must remain a joyful sport and not become a duty of stress. Failure in competition may even lead to disgust with mathematics. Two more critical remarks: Invigilated contests are alien to real mathematical work because of the time stress (NL). Too much publicity, too high prizes in cash are dangerous (S), winners must not become stars (Eastern countries).

For A. Engle (FRG) gifted students are the most neglected group in school. Many topics they could easily learn (number theory, recursions, . . .) are not treated in the courses (poor curricula, no spare time for teachers). Thus training of the contestants is necessary.

For Sam. Greitzer (USA) training is useful, necessary, but not sufficient, for problem solving ability is innate, cannot be taught but only improved if the teacher is himself a skilled problem solver.

As the Hungarian example shows the fostering of the bright students is the basis of progress in science (Austria).

References

1. H. Freudenthal ICMI Report on Mathematical Contests, Educ. Stud. Math. 2, 1969.

2. A. Engel Internationale Mathematik-Olympiade Math. Unterricht 25 -1- 1979 Klett.

3. H. Sewerin Mathematische Schulerwettbewerbe 1979 Manz.

4. Manuscripts of ICME IV, 267, 277.

5. Mathematical Magazines from B, NL, F, FRG, GB, USA, Can.

6. Information received from A, H, S, SF, GB, Can.

MATHEMATICAL COMPETITIONS, CONTESTS AND OLYMPIADS

Murray S. Klamkin
University of Alberta
Edmonton, Alberta, Canada

This session is a two hour panel discussion on mathematical competitions, contests and olympiads. The panelists will cover what is happening is various countries concerning these competitions, what effects these competitions have and what kinds of preparation are most effective for these competitions.

Unfortunately, three of the panel members, Juan Delmasso, Argentina, Jose Ipina, Bolivia and Janos Suranyi, Hungary could not attend. Fortunately, there are still six panelists left. Again, unfortunately, the chosen presider M.K. Singal, India, could not attend. Consequently, I was asked the other day to preside over this session. Since I have to introduce myself, my name is Murray Klamkin and I am presently Chairman of the Department of Mathematics of the University of Alberta and also a member of both the Canadian and U.S.A. Mathematical Olympiad Committees. I will be wearing two hats here since I am also one of the panelists.

To start off this session, I will comment on mathematical competitions in both Canada and the U.S.A. It is well known that the European countries have conducted national mathematical competitions for a long time (also in Africa, China, South America) (1). For example, the Hungarian Eotvos Competition began in 1894. The IMO (International Mathematical Olympiad) was initiated by Romania in 1959. This competition has run consecutively for 21 years. The last one was held in England in 1979. Unfortunately, no country was willing to host the 22nd IMO in 1980 (this involves a lot of work, organization and expense). It is still an open question whether or not the IMO will continue to be a yearly event. There are pressures to change it to a biennial event. Nevertheless, the 1981 IMO is set to be held in the U.S.A.

Canada had its first national olympiad in 1969 (2). There are also individual provincial contests and each province uses these to select its quota of students who will write the Canadian Olympiad. For example, Ontario has four competitions, 1. Junior Mathematics

Contest, 2. Euclid Mathematics Contest, 3. The Gauss Mathematics Contest, and 4. The Descartes Mathematics Competition. These are for different classes of students and are also used in some of the other provinces. Additionally, there are non-quota students who can write the Olympiad. These students must be recommended by their school principals and must have had some good reason for not participating in their provincial competition. There is also a fee of $5.00 for each non-quota student to discourage indiscriminate nominations. So far, Canada has not participated in the IMO due to a lack of financial support. Hopefully, there will be a Canadian team in the 1981 IMO.

The U.S.A. started its national olympiad in 1972 and participated in its first IMO in 1974. The students invited to write the U.S.A. Olympiad are the top 100 or so students from the National Mathematics Contest (3), a contest considerably different from the Olympiad (4). Hopefully, in the near future, there will be a new buffer competition between the two. There are, also very many individual state mathematics competitions. I have given references to a number of these and a bibliography of some 100 papers on competitions in Crux Mathematicorum (5). Many of these papers are good sources for challenging problems.

In my view, these competitions do stimulate the more gifted mathematically inclined students of both countries. To some extent, these competitions help to make up for the poor mathematical programs and low standards found in many of our schools and which pale significantly when compared to those, say in the U.S.S.R. (6).

To help students in the U.S.A. for the IMO, there has been a training session each year since 1972 for the top 8 students of the national olympiad plus 16 others who have done well and who still will be in secondary school the following year. These sessions last about three weeks and are given just prior to the IMO. It is felt that without these training sessions, even if brief, the students who have taken the IMO would not have fared as well as they have.

In order to help challenge and stimulate much greater numbers of students, there is a Problem Corner in the N.C.T.M. publication, The Math Student Journal (5 times a year) which is edited by G. Berzonyi and an Olympiad Corner edited by me in Crux Mathematicorum (10 times a year). The latter column contains olympiad type problems, copies of various national and international olympiads, references, etc. For another source of problems, but not as sophisticated as the olympiad ones, see 1001 Problems in High School Mathematics (7). All this is something the European countries have been doing for a long time.

Finally, one is usually interested in the future of the mathematically gifted students who take these olympiads. Since the U.S.A. Olympiads started in 1972, there is not much we can say here. However, I have noted that when students who have been in our training sessions attend college and write the Putnam Intercollegiate Mathematical Competition, they do very well. Indeed, in each of the last two years, they dominated almost all the top positions. If one looks at the list of winners of the Hungarian Schweitzer Mathematical Contests (8) for the years 1949-1961, one will recognize the names of a number of presently well known mathematicians.

References

1. Freudenthal, Hans (ed.), ICMI Report on Mathematical Contests in Secondary Education I, Educational Studies in Mathematics, 1(1969) 80-114 (contains 105 references).

2. Barbeau, E., Moser, W. (ed.), The First Ten Canadian Mathematical Olympiads (1969-1978), Canadian Mathematical Society.

3. Salkind, C.T., Earl, J.M. (ed.), The MAA Problem Book I, II, III, Mathematical Association of America.

4. Gretizer, S.L., The First U.S.A. Mathematical Olympiad, Amer. Math. Monthly, 80 (1973) 276-281. Also, see references numbered 30-40 on p. 22 of the next reference no. 5).

5. Crux Mathematicorum 5 (1979) 62-69, 220-228.

6. Preliminary Report by Wirszup, I. on the present status of Soviet mathematics and science training at the pre-university level.

7. Barbeau, E., Klamkin, M., Moser, W., 1001 Problems in High School Mathematics I, II, III, IV Canadian Mathematical Society (see 2.).

8. Szasz, G., Geher, L., Kovacs, I., Pinter, L. (ed.) Contests in Higher Mathematics, Akademiai Kiado Budapest 1968.

REASONS FOR THE SUCCESS OF THE AUSTRALIAN MATHEMATICS COMPETITIONS

Peter J. O'Halloran
Canberra College of Advanced Education
Belconnen, Australia

In 1979 the Australian Mathematics Competition was taken by an average of 80 students/school from 60% of Australian high schools. This paper will endeavour to identify some of the reasons for this extraordinarily high participation rate.

A Community Need

During the last few years the community, whether they be employers, parents, teachers or students, has become more aware of the importance of mathematics as a basic core subject in the school curriculum. There has also been considerable publicity in the news media about apparent decline in the general level of numeracy of school leavers. That parents are concerned about numeracy is clearly demonstrated at any parent/teacher night when parents queue up to see the mathematics teachers of their children.

Factors causing ths concern include:

1. a larger percentage of students, including the less able, who are staying at schol longer and entering college or university;

2. a decreasing number of hours spent on quantitative subjects at school due to pressures on students to

do subjects which are often more fashionable, less demanding and perhaps of no great consequence;

3. the disproportional time spent on some aspects of the so-called "Modern Mathematics" at both primary and secondary levels at the expense of obtaining high mastery levels in basic arithmetical and algebraic skills. The learning of Euclidean geometry is also a victim of this modern mathematics;

4. many students are being taught by teachers who are insufficiently trained in mathematics. Many school administrators mistakenly have the view that anyone can teach mathematics. I believe that as a result of this administrative convenience many students are "turned off" or develop a fear of mathematics because of these teachers who themselves have a fear of mathematics.

Basic Aim of the AMC

The Competition is aimed at good average and competent students of mathematics who are in any of the last six years of school. It is not designed specifically for students who have high innate ability in the subject, although such students could be expected to do well in the Competition.

One of the main objectives of the competition is that all entrants will gain a sense of achievement from attempting the papers. These papers are designed to consiste of well graded questions ranging from straight-forward to the challenging type.

Modus Operandi of the Competition

The modus operandi of the competition has a number of features which not only satisfy the basic aims of the competitions but has also added to the impact that the competition has had on the Australian scene.

These include:

1. Moderation of the competition papers. The competition is divided into 3 divisions (Junior, Intermediate, and Senior) spanning the 6 years of pre-college/university at school. Considerable effort is made by the Problems Committee to ensure that the first 8-10 questions in each division paper are "do-able" by most entrants from all Australian States and that all questions in each paper are of increasing difficulty. These procedures in setting the papers mean that students (and teachers) can readily relate their learning, (and teaching) of mathematical skills at school with the competition.

2. Communication strategy.

   a. February. Invitation brochures and posters are sent to all High Schools, Mathematics Department of all Universities and Colleges and to all branches of the Australian Association of Mathematics Teachers. Schools which were and were not in the previous years' competition, received different accompanying letters. National and local press releases on the competition is distributed during the 4th week of February.

   b. April. After entries close on the 31st of March, receipts and Information Bulletin #1 are sent to all the schools that have entered the competition. National and local entry statistics are released to the news media.

   c. May. Sealed bundles of question papers, Information Bulletin #2, etc. are sent to the schools.

   d. July. After processing, marking and grading the marked-sense answer sheets of all the contestants lists of results, award certificates, prizes and Information Bulletin #3 are sent to all the schools . . . all within 3-4 weeks of the students doing the papers. Often the local Bank Manager is invited to the school to present the awards. Names of national and local winners are released to the news media.

   e. September. Solutions and Statistics Booklets which also include lists of all prize winners and their grades, are available for sale. Medallists are brought from all over Australia to a special Presentation of Medals Ceremony. News media often highlight this event.

3. Recognition of high % of entrants. 45% of entrants receive some measure of recognition, ranging from credit and distinction certficates to prize money and medals. The entrants compete at a State/school grade level. For example the grade 9 entrants from Victoria compete only among themselves with the top 45% receiving awards. This competition at the local State and grade level means that entrants, generally, aren't handicapped because of syllabus differences or experience.

4. Publicity and credibility. The competition has achieved, within a couple of years, a high acceptance and credibility by the community at large. For example, in the National Capital, where the competition and its predecessors has been running for 4 years, the average entries from schools has increased from 40 to over 100 per school.

   There are many reasons for the acceptance of the competition so readily in Australia. The competition has been most fortunate in both the quality and level of sponsorship. The commercial sponsor, Australia's largest bank, organised a nationwide publicity campaign on a local basis. For example, the would send press releases to many small country town newspapers, localising both invitations and results of the competition.

   The educational sponsor is the prestigious tertiary Institution in the National Capital. The professional sponsor is the local Mathematics Teachers Association and in addition the two national mathematical professional bodies have strong associations with the competition.

   Credibility of the competition is enhanced by the high level of communication, documentation and efficient organisation associated with the competition. Much of this is due to many capable and dedicated faculty members of the educational sponsor's Institution, with encouragement being given at all times by the President of that Institution.

Other factors that helped the efficiency of the organisation include:

a. the generous guidance ad cooperation from the executives of the USA and Canadian competitions. The AMC external organisation with schools is modelled on the USA system whereas the internal administrative organisation is modelled on the Canadian system.

b. the 2 years of running the competition, firstly as a local and secondly as a pilot, in order to develop sound procedures of logistics, communications and documentation. The teachers of the local mathematics associations were most supportive during this period;

c. the considerable computer system support developed in the administrative processing and marking aspects of the competition.

Conclusion

The AMC receives good publicity in the news media over an extended period of time and as a result the community at large is aware of the competition. However, the success of the AMC could be summarised in terms of the basic marketing force - there is a community need and concern for improved numeracy, the AMC is an attractive, well designed product that satisfies that need.

PROBLEMS COMPETITIONS IN DEVELOPING COUNTRIES

Peter Sanders
University of the South Pacific,
Suva, Fiji

1. Introduction

In many developing countries, low student achievement in mathematics and science has been a continuing educational problem. This has led, in some cases, to a shortage of candidates with appropriate entry qualifications for professions which require a mathematics or science background, and hence to difficulties in manpower planning. One device for increasing pupils' interest and motivation in mathematics is the mathematics problems competition, and in this paper I describe my experience of running such competitions for a dozen years, first in Africa and, more recently in the South Pacific.

2. Background

2.1 The Eastern Africa Regional Contest

In the late 60s, several countries in East and Central Africa were running competitions independently. In 1969, a meeting in Nairobi of mathematics staff from universities in Kenya, Uganda, Tanzania and Zambia provided the occasion for the setting up of a regional mathematics contest for Forms 3 - 5 (Years 9 - 11 schooling). The arrangement was that countries should take it in turns to set the papers, and compile an international list of winners, thus sharing the extremely difficult and time-consuming task of producing suitable questions. This competition consisted of two papers. The first was a multiple choice paper, containing between 20 and 30 questions to be answered in two hours. The top 20 scorers in each country on Paper 1 were then invited to attempt the more difficult Paper 2, which consisted, usually, of 10 problem type questions (not multiple choice). Paper 1 was marked by the national co-ordinators, and the scripts from Paper 2 were all sent to the central marker, to ensure uniformity.

In spite of difficulties in communication, and a lack of continuity due to a rapid staff turn-over, the competition continued until 1978 at least, with a varying number of countries participating. In 1978, the last year for which I have any information, the competition was held in Zambia, Uganda, Lesotho, Tanzania, Kenya, Botswana, and Swaziland. (A national competition was apparently run in Tanzania in 1979.)

2.2 The South Pacific Mathematics and Logic Competition

Meanwhile in 1975, Colin Meek introduced a Mathematics and Logic Competition in Papua New Guinea, which, in 1976, was also taken by students in Fiji, Hawaii, New Zealand, Niue and the Mariannas. Soon after I arrived in Fiji, at the end of 1976, I collaborated with Colin to initiate the South Pacific Mathematics and Logic Competition, which has now run for four years. In 1979 some 60 schools in 9 countries took part, with most schools entering 15 pupils (the maximum allowed). This competition, like the African one, contains a multiple choice Paper 1 which serves as a qualifying round for the problems type Paper 2. The main differences between the two competitions are that Paper 1 is marked by the schools themselves, and there is more bias towards questions involving logical thinking.

3. Administration

The administration of the competition has always been a struggle. Few teachers have been heroic enough to volunteer for the monumental amount of extra work which it entails. In most cases the job has been taken on by enthusiastic mathematics staff in ministries of education or national universities, these institutions providing the secretarial support.

4. Prizes

Once the competition is set up, it has usually been relatively easy to obtain prizes from local companies and businesses. In Uganda, winners on one occasion appeared on television, and were taken on a tour of a game park. In Zambia, the copper mining companies regularly provided donations for the prizes, which were, on one occasion, presented by President Kaunda himself.

5. Mathematical Associations

The competitions have often been linked to the activities of mathematical associations, to their mutual benefit. Both in Zambia and Fiji prizes were presented to the winners at a ceremony during the annual conference of the local mathematics association. In

Fiji, this year, both the South Pacific competition and a senior competition (for Forms 6 and 7, or years 12 and 13) were organized through the Fiji Mathematics Association, and participating schools were required to join the FMA as corporate members.

6. Types of Question

As was stated earlier, the main aim of the competition is to increase students' interest in mathematics and to develop their problem solving abilities. Many teachers doubt whether this can be adequately assessed by multiple choice items only. This point is considered further in the next section, but, in the case of the African and South Pacific competition, it led to the evolution of the two paper exam, with the problems type paper carrying twice the weighting of the multiple choice paper. Yet multiple choice questions are the only feasible type if large numbers of pupils are to be given the opportunity of writing the paper.

7. Drawbacks of Multiple Choice Items

Books have been written on this issue. Suffice it to say that a paper which sets 30 questions in 75 minutes encourages inspired guesswork more than problem solving. Problem solving is often a long term activity. (It took Hamilton 15 years to realise that multiplication of quaternions was noncommutative). My own experience has convinced me of the unreliability of such tests in the assessment of problem solving ability. The difficulty is that if the students are allowed a reasonable time for each question (say 10 questions per hour), then the random effect of guessing is too strong, because there are too few questions, which ensures that a student who makes a superficial attempt at the question is more likely to fall into the trap of one of the distractors than a student who makes a random guess.

8. Sources of Questions

It is extremely difficult to make up good questions oneself. Consequently I have had to plagiarise. My source is a large file full of competition papers collected over the years. It is noticeable in this file, how certain stereotyped questions occur again and again, notices of copyright notwithstanding.

9. Evaluation

Research is urgently needed to establish the extent to which these competitions fulfil their objectives. My correspondence files and contacts with teachers convince me that teachers enjoy them very much - indeed on one occasion I was asked to run a competition for teachers. But as to the pupils, I am not so sure. Many of them regard it as just another exam, and try to cram for it in the same way. Perhaps we have now reached the stage where a careful evaluation of these competitions is appropriate.

SOME IDEAS ON MATHEMATICAL COMPETITIONS

Janos Suranyi
Budapest, Hungary

Different sorts of mathematical competitions have always emerged in human history but their use for raising interest for mathematical activities and stimulating mathematical abilities on the one side and for recongising mathematically gifted young people on the other, seems only to go back to less than 90 years (2). A characteristic trace of such competitions is that the use of all sorts of printed materials is usually admitted.

The Hungarian television organized quizzes in different subjects amongst them in mathematics. The mathematical ones* (held 3 times: in 1964, 1966 and 1968) proved surprisingly useful even in reducing the general repugnance against mathematics. People who hardly understood even the questions, to say nothing of the solution, watched the contesting pairs with great excitement. Seeing that normal lively children, far from the type of some dry as dust, have found the solution in 2-3 minutes, these people have lost their proudness for being ignorant in mathematics.**

Mathematical competitions, as competitions in general, have, however, their disadvantages too. Good contest results depend also on nerves, quick reaction and other psychological components besides talent, but failure in competitions discourages some from mathematics and not only from the competitions.

Another discouraging aspect of competitions, observed in Hungary, could have been eliminated easily. The first round of the two round competitions was run earlier with three problems as difficult as those in the second one. These problems had to be hard enough to select about the 10th part of the participants for the second round. This led to the complete failure of many children essentially better than average, who became disappointed and have lost their interest for mathematics. In order to avoid this effect, just contrary to the aims of the competitions, the first round is run now with more problems (with 8 at present) of different degrees of difficulty including some quite easy ones, too. The participants can then choose according to their interest. Generally, the fourth and third part of the maximal score qualifies already for the second round.

The competitions must be, and are, independent from school teaching in the sense that their result doesn't influence the marks obtained at school. Some teachers and headmasters are, however, inclined to estimate the results of their pupils in competitions as the measure of the efficiency of their work which is unjust, too. Though a low level of mathematics teaching seldom produces good mathematicians and good competitors, it is, however, not difficult and, for a sufficiently well prepared teacher, even comfortable, to train the best students and pay not much interest to the rest. The opposite case occurs, too: the teacher doesn't inform his pupils on competitions and student journals for mathematics - - which provide a good preparation for competitions - - because students then raise difficult problems.

Competitions can lead to some one sidedness, too. They enforce mainly the capacity of the problem solving, which is an essential component of mathematization, but not the only one. Moreover, they give emphasis to some topics appropriate for giving good contest problems and for some types of problems whereas other fields don't fit to competitions.

An overgrowth of competitions and generally an overfeeding of talented youngsters with mathematics can turn them in disgust towards their first interest, and it seems tome that this danger cannot be neglected either in our days.

All these are essential drawbacks of competitions and the list could even be continued. These are, however, not arguments against competitions, only warnings which must be taken attentively in consideration in order that competitions play their beneficient role which no other thing can replace.***

Notes

* A detailed description of the first two contests together with the problems and their solutions appeared in (1).

** Unfortunately, the contests of other subjects were not so successful and the contests were not repeated later on.

*** C.f. the evaluation in (2) pp. 99-100.

References

1. E. Fried, I. Lanczi, E. Gyarmati, J. Suranyi, Ki miben tudos? (Who is the expert and in what?) Budapest, 1968, 163 pp. In Hungarian.

2. I.C.M.I. Report on Mathematical Contests in Secondary Education, Olympiads, I. Editor: Hans Freudenthal, Educational Studies in Mathematics, 2, 1969, p. 80-114.

15.2 MATHEMATICS COMPETITIONS: PHILOSOPHY, ORGANIZATION AND CONTENT

MATHEMATICS COMPETITION AS A STIMULUS FOR MATHEMATICS EDUCATION

Alfred Kalfus
Babylon High School
Babylon, New York, U.S.A.

Competitions, such as mathematics leagues, fairs and contests, encourage students to research and master more topics, as well as develop more powerful techniques for problem solving, that are contained in our usual syllabi. The stimulus of these activities and the recognition given the student make the mastering of mathematics an exciting and rewarding experience.

The New York City Interscholastic Mathematics League has been in existence for almost 60 years. It has consistently stimulated, developed, and sustained a level of mathematical talent amongst the city's secondary school students that is truly outstanding. That is the past many of the city's students made notable careers for themselves in mathematics education and other mathematics-related fields is not surprising. However, at the present time, despite all the difficulties one hears about in the New York City school system, the Interscholastic Mathematics League is still turning out the largest number of exceptional mathematics students in our country. Year after year, these students earn the lion's share of awards offered by the most prestigious mathematics competitions in or country. To account for these accomplishments, one cannot ignore the training and tradition of excellence maintained to this very day by the New York City league.

If we are to judge by the number of students involved, the most successful kind of competition we have is the mathematics contest. Though there are numerous local and state contests, the most prestigious is the Annual High School Mathematics Examination, founded by Professor Charles Salkind in New York City 31 years ago (an obvious offspring of the mathematics league concept). This contest has grown to such proportions that last year in the United States and Canáda nearly 400,000 students participated. Indeed, a mathematics contest patterned after this one has been most successfully instituted in Australia. The best 100-125 student scorers in or contest are invited to participate in the U.S.A. Mathematical Olympiad Contest, from which the United States team is selected to compete in the International Mathematical Olympiad.

The mathematics fair takes many forms in our country, but the Math Fair founded in the New York Metropolitan area (Long Island) 20 years ago has proved tobe most active and successful. Its purpose is to offer the student an opportunity to develop and display a facet of mathematics education that is often overlooked in the schools. At this event the student presents a 15 to 20 minute talk on a topic he had independently researched. It may be original, and some very creative presentations have been given, but most often the talks demonstrate that the student has studied and mastered a topic well beyond its coverage in his current course. In fact, each year many of the papers are of such a noteworth caliber that a collection of them has been assembled and published by Mu Alpha Theta, the National Honor Society for mathematics students.

However, because of the space limitation, the main emphasis of this paper must be reserved for the competition known as the mathematics league. Not only is it the longest continuously existing competition in our country (59 years in New York City), but since the late 1950's this kind of activity has generated widespread enthusiasm as it continues to proliferate throughout the country. The students like it because it is run like an athletic league, with all its concomitant excitement and the opportunity to meet and compete with one's peers in an area of common interest. The teacher-coaches like it because it generates a student camaraderie built about mathematics, which leads naturally to the fostering of mathematics clubs and, in turn, to additional opportunities to enrich the student's mathematical potential.

Mathematics leagues operate on both the senior and the junior high levels. Though league may involve anywhere from 20 to 90 schools, it is divided according to small geographical areas with five or six schools per section. Each participating school, which may enter from one to three teams, takes a turn as host at one of the five or six meets held per academic year. A team is composed of five students, and when a written question is distributed to all the participants each must be prepared to work independently at his assigned seat. The coach may replace any student with another member of the squad before a new question is distributed. The strategy depends on the student's strength in a particular area of mathematics, as the odd-numbered questions may involve algebra and trigonemetry and the even-numbered ones may involve geometry. One point is alloted for each correct answer. Thus, a student may achieve a maximum of 6 points and the team a maximum of 30 points for the six questions asked.

In 1961 the two senior high mathematics leagues operating in Nassau and Suffolk counties on Long Island began a practice that continues to have significant ramifications tothis very day. Each year they held a bi-county competition to which the respective leagues brought their top scorers. This all-star event continued for some years until the various leagues in New York State united and formed the New York State Mathematics League (NYSML) in 1973. The number of teams competing in the various leagues throughout the state is quite large, though many were brought into existence with the aid and encouragement of the NYSML. Thus many hundreds of schools and thousands of students are encompassed in the present network of the NYSML operation. At this past year's annual meet 15 leagues were represented, with 225 of the state's leading mathematics students spending Friday evening and Saturday at the NYSML events at one of the state unviersities (which provided sleeping accomodations and meals, as well as other facilities for the meet).

In 1973 the New England Mathematics League was also founded, creating an inter-league competition in a six state area. The culmination of these inter-league competitions was the establishment of the Atlantic Region Mathematics League (ARML) in 1976. The NYSML, the New England Mathematics League, as well as leagues from New Jersey, Pennsylvania, Maryland, and Virginia joined together to form what has proved to be a most exciting and successful venture. Each year new states are added, and the league's all-star competition continues to grow. Some states, such as Deleware, had no league at all when they first contacted ARML. But with guidance and materials supplied by ARML a league was founded in Delaware, and an all-star team was subsequently sent to represent the state at the ARML annual meets. For the past three years additional teams have been flying to the meets, one from West Virginia, one from Michigan, and two from the Chicago area. This year North Carolina entered a team, and two teams came from Texas. This steady growth reveals that despite many logistic and financial problems, a league encompassing a large geographical area is viable.

The purpose of the ARML is to being the leading mathematics students from widely separated areas together, to meet one another and socialize in an atmosphere conducive to mathematics enrichment, to face the challenges posed at the meets, and to have fun doing mathematics. This year's meet was attended by 700 students and teachers. They arrived on a Friday evening at Rutgers University (where accomondations were provided for sleeping and for meals). The students attended a social and talks, while the coaches attended a council meeting to prepare for the meet.

The first event on Saturday is the Power Question. Each 15-member team is placed in a separate room, where all work together to produce a written solution in one hour. The problem is a difficult one, and a routine solution is not necessarily enough, as additional points are granted for originality and elegance, or if the problem is extended and generalized.

The second event is the Team Questions. Once again all members of the team are in a separate room. Ten questions are presented to each team, and they may use any strategy to apportion the problems amongst themselves, for the problems must all be done in 12 minutes.

The third event is two relays. Parallel columns of fifteen armchairs are set up in a large gynmasium. Each team is seated in a column. The team is then divided into three groups of five members each, so that when the five questions of each relay are distributed, each team member receives a different question. When the first student has finished, he may pass only a written answer back to the second student, who must then use that number in reaching the second solution. Then the second answer is passed along to the third student, and so on. There is no talking, and answers must not be passed forward. However, a student may send a corrected answer back, which begins the chain going again. The fifth student signals the team is finished by standing up with the team's final answer. Points are awarded in terms of time intervals within which the correct answer is achieved.

The last event is the Short Answer Questions. The students reassemble, after lunch, in the same large gymnasium. This time each student works independently and acquires points for himself, or herself, as well as for the team. When a question is presented, 4 to 8 minutes (depending on the difficulty) are given to find the solution. Eight questions are presented in all. At Rutgers University this year, one student was able to get all eight answers correct. As the team totals are tabulated, the results are displayed on overhead projectors, stirring up a storn of cheers and commentary, especially when two or three teams are vying for first place.

At the termination of events an awards program is held. Each student is presented a certificate of participation, and the leading students and teams are called up to receive awards. The competition is divided into two divisions: the stronger teams in Division A and the weaker teams in Division B. Parallel sets of awards are presented to the leading teams in each division. The awards are generous, sponsored by the leading national mathematical organizations (NCTM, MAA, Mu Alpha Theta, AMATYC) as well as publishers and computer firms.

Everyone goes home flushed with excitement, talking about next year's meet. The success of the ARML, which plans to keep expanding until a National Mathematics League is established, clearly demonstrates that such an operationn is not merely feasible but, because of its demonstrated attributes, eminently desirable. This, in turn, signifies that mathematics education everywhere can and should exploit the tremendous incentives that are so natural a part of these competitions.

CHAPTER 16 - Language and Mathematics

16.1 LANGUAGE AND THE TEACHING OF MATHEMATICS

LANGUAGE AND THE TEACHING OF MATHEMATICS

A.G. Howson
University of Southampton
Southampton, England

The ways in which the learning of mathematics interacts with problems of language are beginning to receive more attention than has formerly been the case. The enormous increase in the variety of children who are now introduced to mathematics has alerted us to aspects previously ignored. It is good, then, that at this Congress considerable time should be devoted to discussing language problems. We have already heard a plenary lecture by Hermina Sinclair on the role of language in cognitive development, and there has been an opportunity for members to respond to this. Further sessions are to be devoted to detailed consideration of the problems of teaching mathematics in a second language. It was my intention to attempt in this paper to survey aspects of the mathematics education/language interface which would not be touched upon in those sessions. Eventually, I rejected that idea and decided to speak on one particular aspect - - that of symbolism. Those in search of a survey will find one such in Austin and Howson (1979). Yet, even by restricting myself to the one area of symbolism, I find that all too frequently I can make only brief allusions to problems. My purpose, then, is to attempt to identify particular areas of possible research and, where possible, to mention existing work.

However, Professor D'Ambrosio has asked me to say a few words of introduction to his panel discussion on second-language problems. I am most happy to do this, for it will serve as an introduction to the particular aspect on which I wish to talk.

The difficulties of having to learn mathematics in a language that is not the mother tongue are so obvious that they are readily recognised and have given rise to considerable concern in many developing countries (see, for example, Unesco, 1974; Abidjan, 1978). It is not a new problem. In the sixteenth century, England was a developing country; its language had replaced Latin and then French as the language of government, but it still was not the language of mathematical instruction. All textbooks were written in Latin - - the international language of scholars. Then men such as Recorde and Dee saw that the result of this was to separate mathematics from the people and, in particular, the artisan - - the technician on whom the nation's prosperity so much depended. Dee argued strongly for mathematical texts in the vernacular. Books in English would make for 'better and easier learning' and would help students proceed 'more cheerfully, more skillfully and speedily forward' in their studies. (Dee, 1570). I shall not try to press this parallel too far. Yet Dee's thesis that mathematics cannot truly enter a nation's life until it enters the nation's language is, I feel certain, a valid one.

The problems of moving from teaching and publishing in an international language to the vernacular are ones which Professor D'Ambrosio's group will wish to discuss. An important issue which must not be neglected is how research can be planned so as to assist both those who make decisions and also those who teach in the classroom. Many suggestions for research activities were made in Nairobi in 1974, yet these proposals have not been acted upon. It is essential that we consider the reasons for this. Were, for example, the activities proposed too ambitious or too ill-defined? How can limited budgets and restricted manpower best be employed to produce research findings which can aid decision makers and which carry sufficient weight potentially to exert influence? These are questions that must be discussed.

Yet 'second-language' problems do not only affect the countries in Africa and Asia. It is at its most obvious where there are large communities of immigrants. Are we really aware of the size this problem can be? In the Inner London Education Authority - - Britain's largest - - 40% of school children have mothers who were born outside England and 10% come from homes in which English is not used as the everyday tongue. I am told that parts of Australia are faced with a similarly daunting situation.

Similar problems occur, of course, when children from a working-class background, with a deprived vocabulary, are suddenly placed in a milieu in which middle-class language and expressions are taken as the norm. This area has, of course, been investigated by Bernstein and others. It is necessary to emphasize that I am not here concerned with whether or not mathematical concepts can be conveyed in 'lower-class'language. There exists evidence to show that they can (Labov, 1970; Ginsberg, 1972). What I wish to draw attention to is the failure of teachers to recognise that there is a gap which somehow must be bridged. The coming of 'modern math' with its (supposed) emphasis on precision of language and its plethora of formal definitions has clearly exacerbated these problems in several countries (see, for example, Clark, 1976). Yet 'second language' problems do not end there. Recently I had to read through the transcript of a lesson given by a young teacher. Again and again I was struck by the artificiality of the language which he used. It was stilted and jargon-ridden. Eventually, the children began to learn what was required of them, and they responded in an equally unnatural way. The mathematics was not being discussed in their own language; it was not being absorbed by them as part of their own culture.

However, the most obvious difference between the language used in maths classrooms - - the language to be found on the blackboard or in texts - - and ordinary language is the preponderance of symbolism in the former. It is on this aspect that I wish to concentrate my remarks today.

The immediate stimulus to speak on this topic came from a recent correspondence with Professor Boas, editor of the American Mathematical Monthly. One paragraph of the Austin-Howson survey read "Books written for school teachers frequently emphasize the power of mathematical symbolism and its superiority over common language. How do we reconcile this advice with that given by editors of the American

Mathematical Monthly to its authors: 'Remember that most people understand words more rapidly than formulas'? Here, of course, we must remember that by 'most people' the writers meant mathematics graduates, not citizens or students!".

Professor Boas, perhaps misinterpreting our statement as a criticism of his views, wrote to me to justify his position. At this stage I should make it clear that I do believe what he says to be correct. I fear that we as mathematicians get carried away by the manner in which we can translate long verbal messages into a compact collection of symbols. It emphasizes the power of mathematical symbolism and demonstrates our personal command of it - - yet how frequently does it confuse our students or our non-mathematical colleagues? Yet whilst accepting Professor Boas's statement, there is also no doubt that the mathematician depends upon his facility to use symbols. When teaching mathematics we must always have as an important aim to introduce the learner to the use of symbolism and, more than this, to develop in him an understanding of the benefits that can accrue through the use of symbols.

Symbols do, indeed hold a peculiar position within mathematics. In it the conventional linguistic ideas of signified-signifier - - i.e., 'the thing signified' and 'that which signifies it' - - break down. In mathematics signifiers create their own signifieds. Thus, when first being formally introduced to the natural numbers, the child can be given an exemplification of the symbols '1', '2', and '3'. We do this by moving from a mathematical sign to an entity that lies outside mathematics, and we do it using ordinary language as a mediator. However, $123^{123^{123}}$ admits of no such exemplification. The symbol creates its own signified. Yet, the mathematician can readily show that the integer so designated possesses certain properties.

Helping the learner move from the one level of symbolisation to the other is, of course, a key problem of mathematical education. However, like too many others it is a process that would appear to have received very little attention from researchers and others.

Yet it is not this aspect of symbolism that we think of first when discussing symbolism and mathematics. It is, of course, the power which symbolism gives us to perform important operations without thinking about them. It is this power which allows us to manipulate algebraic symbols without having simultaneously to consider their meaning. Like all power, it can have a corrupting influence. For example, the comparative ease with which one can teach students to manipulate algebraic symbols has been one which has led to a decrease in the emphasis on pure geometry at secondary and undergraduate level and a corresponding loss in the students' powers of visualization. Interestingly, over 100 years ago Isaac Todhunter (1873) remarked on the increased use of symbolism and pondered whether this reduced the value of mathematics in a general education. More recently, Thom has pointed out the differences in the languages of algebra and geometry: that of the latter cannot so readily be divorced from meaning. At primary and secondary level we see a marked difference in children's ability to manipulate symbols and to comprehend the underlying operations. (Reports from the Chelsea College Concepts in Secondary Mathematics and Science (CSMS) Project suggest that the two abilities are to some extent independent; children can calculate without comprehending, and can comprehend without being able to calculate - - or, at least, without being able to use standard algorithms. (See Brown & Kuchermann, 1976-77; Hart, to appear). How one can simultaneously develop both an ability to use and manipulate symbols and an appreciation of their meaning is, then, a key problem. It would seem to depend upon several factors, and in the remainder of this paper I shall look briefly at a number of these.

1. On the Introduction of Symbolism

That the development of concepts in the individual might well mirror the way in which those concepts developed historically is an idea which can be traced back to the early nineteenth century. Along with that idea, there has developed the thesis that we should take account of historical developments when planning the curriculum. It is a view which, for example, Thom put forward at the Exeter ICME. What are the consequences of such a theory for the introduction of symbolism?

First we observe that prior to the sixteenth century it was common to use special words, abbreviations, and, of course, the number symbols. Then gradually, during the sixteenth and seventeenth centuries, special symbols were introduced. Many of these, for examle, +, -, x, and = we still use today; others have vanished and been replaced. Algebraic statements still did not consist solely of symbols: writers used a mixture of symbols, abbreviations, and words. Not until the end of the seventeenth century did an awareness of the power and generality of symbolism really enter mathematics. Only then did symbolic algebra arise. Yet now this process, which took centuries, is rushed through in a matter of hours or ignored completely. In the progression rhetorical - - syncopated - - symbolic (see, for example, Boyer, 1968) the first two stages are largely ignored, as examples I shall quote later will show.

The new symbolism of the seventeenth century was indeed to make possible an enormous advance in mathematics; in the words of Jacobi "with the introduction of the sign x^{12} for the 12 power of x the newer analysis started". Not even Newton, who was to introduce the idea of a general exponent, did not always make use of the new contracted forms. Side by side with expressions such as a^4, he would use the older uncontracted aaaa.

Our children, though, are expected to move straight to the contracted symbolism. Is that reasonable? The answer to this question may well be 'Yes". However, it is significant to our concerns at this Congress that after deciding to raise this question, I chanced to look again at Branford's A Study of Mathematical Education: a classic now over seventy years old. In that I found my thoughts were completely foreshadowed. Branford had, in addition, laid down as a principle of teaching:

> No symbol or contraction should be introduced till the pupil himself so deeply feels the need for such that he is either ready himself to suggest some contraction, or at least appreciate reasonably fully the advantage of it when it is supplied by the teacher.

Over seventy years later nothing would appear to have happened to invalidate Branford's principle; yet what research evidence exists which would add support to

Branford, and how many textbook authors and teachers have adopted the principle? Like so many other 'principles' in mathematics education it is still founded on 'common sense'. Can we do better than that? When discussing the problems associated with learning algebra, Freudenthal (1973) says "I do not know of any systematic investigation on school children learning the language of algebraic formulae . . . . Nobody seems to know what the original obstacles were and how they were overcome. School children would have to be observed for a long period to understand this phenomenon, short experiments would not suffice". I know of no subsequent research aimed at answering Freudenthal's questions. Certainly, there have been beginnings - - one might mention as an example Perla Nesher's work (1972) on the way in which the child in the primary grades gradually adds to his ordinary language that special language and symbolism of arithmetic. Yet, despite such pioneering efforts, our knowledge of how children learn to use symbols remains remarkably limited.

2. Diagrams and Symbolism

The three-stage progression rhetorical - - syncopated - - symbolic to which I have just referred is, of course, somewhat akin to Bruner's enactive - - ikonic - - symbolic triad. Again, we see that symbolism is the final stage and that preliminary stages of learning would appear to be essential. Bruner's three stages can, of course, be readily exemplified in terms of the learning and teaching of arithmetic and algebra. However, in the case of geometry, topology, dynamical systems, etc., the ikonic and symbolic categories would appear to merge. The diagram is no longer a stepping stone en route to a symbol. Sapir has written that 'the feeling entertained by many that they can think, or even reason, without language is an illusion . . . . No sooner do we try to put an image into conscious relation with another than we find ourselves slipping into a silent flow of words'. I question the validity of this statement so far as it relates to a mathematician reasoning with the aid of diagrams. Are then diagrams, just as symbols are, part of the language of mathematics? If so, what can we say about their syntax and semantics? One can draw attention to limited research work in this area (Hirabayashi & Katayama, 1969), but it remains a largely unexplored field.

3. Symbolic and Conceptual Structures

Considerable work has been carried out in recent years on concept formation, and various models have been proposed for cognitive growth. It is through the use of language and of symbols that we are able to communicate conceptual structures. Can we therefore obtain a theoretical model which explains the relationship between these symbolic structures used in communication and the underlying conceptual ones? One preliminary attempt to establish such a model was made by Richard Skemp (1977), who took as his starting point Chomsky's 'deep' and 'surface' structures - - corresponding in his case to conceptual and symbolic structure, respectively. As Skemp acknowledges, his ideas were merely exploratory, and it would be easy to criticise his preliminary model. Yet his suggestion that this is a topic to which further research and thought should be devoted in surely a good one.

So far I have spoken of problems of symbolism which are inherent within mathematics; they do not depend on geographical considerations, and only to a limited extent do they depend upon curricular decisions. Let us now turn to some more specific points.

4. National Problems

The symbols we use today in elementary mathematics are almost all the product of European mathematicians. As such, they reflect the particular structures of the Western European languages and, for example, the fact that it is customary to write in these from left to right. Now mathematics is taught, and many of these symbols used, in countries in which the language structures and writing practices are radically different. What special problems are likely to arise for students of mathematics in those countries, and how are they to be alleviated?

Some idea of the problems involved is given by Morris (1974). He quotes many examples, such as that in the Sinhala language relations would be more naturally expressed in the form a, b, R rather than in the mathematician's order aRb. Once again, one can point to limited research work in this area emanating from Japan, but here, too, we have a field in which further investigations would seem desirable.

This particular issue is, however, overshadowed by the more serious one of whether or not the structure of the vernacular is sufficient to 'carry' the linguistic demands of mathematics. The lack of technical terms or precise equivalents can be unfortunate, but would not seem particularly vital. The position is much more serious if, a6 would appear to be the case in many African languages, there is a lack of logical connectives and quantifiers. Calls have already been made for cooperation and systematic research to identify more precisely the nature of such difficulties and to find ways of surmounting them. However, the response to data has been poor. One hopes that matters will improve quickly.

5. What Emphasis Are We to Place on Symbolism in Our Teaching.

Weyl (1949) distinguished between the intuitionist and the formalist in the following way:

> The intuitionist does not consider symbolism as an essential characteristic of mathematics - - he sees symbols merely as a tool of communication and a support for the memory by fixation; the formalist, however, thinks of mathematics as consisting wholly of symbols, which have no meaning verifiable in sensual or mental intuition and which are manipulated according to fixed rules.

It is not our purpose here to discuss the range of validity of Weyl's characterisation. Yet we note the vital role he assigns to symbolism in defining philosophical approaches to mathematics. During the last twenty years formalist tendencies (if not an explicit underlying formalist philosophy) in school mathematics would appear to have grown.

I note, for example, that in England every year our 16-year-olds are now asked 0-level question of the type:

> 'If the operation * is defined by a*b = a+2ab, calculate 3*4 and prove * is not commutative.'

This would seem an eminently satisfactory question from a formalist point of view. To me, it would seem anti-mathematical. I accept that any mathematician ought to be able to answer such a question, but the idea of drilling students to answer such formal trivia as an end in itself appals me. Yet this is not the only kind of routine example to become established in recent years which merely tests an ability to manipulate symbols and/or to comprehend technical language, as if this by itself constituted mathematical ability.

The recent reforms through the world not only introduced to schools new symbols and language but also served to place additional emphasis on their use. Often the innovators' intentions became garbled; language that was intended to add precision and increase understanding was, as we shall see later, used so loosely and erroneously that the position is now worse than it was before. It would seem time to take stock and to consider in greater detail and depth, and with the benefit of twenty years of experiment behind us - - and one hopes many more (better regulated) ones still to come - - exactly what our aims regarding symbolism are: what are reasonable goals and how might they be attained?

6. How Do Symbols Influence our Understanding?

The great benefit of symbolism is that it allows us to perform mathematics without thinking - - we use auto-pilot when manipulating symbols. Moreover, good symbolism can actually suggest new results, prompt new mathematics. Unfortunately, symbols can also suggest incorrect results; they can easily puzzle or mislead the learner. Worries arising from a natural extension of the use of symbols in an unwarranted way have affected mathematicians as well. Hutton in his Dictionary (1815) explains how contemporary mathematicians were unsure what the product of $-a$ and $-b$ was. Euler's Algebra suggested it should be ab (but this was clearly wrong when $a = b = 1$). Emerson had suggested that it should be $-ab$. Here we see how mathematicians have formulated their own rules based on the existing symbolism and not on conceptual appreciation. At a lower level we find examples of this process in, say, Erlwanger's work (1975). It is an example of what Bob Davis (1979) has referred to as 'rules generating rules'. Deeper studies of the manner in which learners generate their own rules may well lead us better to understand the errors students make and will illuminate those symbolic/conceptual structures referred to earlier.

7. On the Duplication and Standardisation of Symbols

It is one of the marks of the professional mathematician that he can not only cope with ambiguous symbols but also switch from one set of symbols to another - - both denoting the same concepts - - with ease. Thus he can, and probably does, use a variety of symbols when dealing with partial differentiation. Here, in fact, is an example where the introduction of a unique standardised set of symbols would enfeeble mathematics. A variety of symbols has remained current simply because each has its strengths in particular circumstances. The mathematician is stronger for having a battery of alternatives at his command. Yet what is the effect on the student of mathematics? How confusing is it to him to be presented with a variety of alternative notations? Are there benefits to be gained from temporarily shielding him from alternatives? Again, are there advantages to be gained from devising interim 'pedagogial' notation for students?

Let us take as an example the 'box and 'triangle' symbols which have entered mathematics teaching in the last thirty or so years; that is, notation of the form $\square + 3 = 5$. I am willing to believe - - although I cannot recall having seen any empirical evidence to support the claim - - that this symbolism offers advantages over the traditional $x + 3 = 5$, when one first introduces the notion of equations.

Here we have an example of what I see as pedagogical symbolism. It is a symbolism which must sooner or later be cast aside for the more traditional forms. We have now had two decades in which this particular new symbolism has been in use. It is now time that there were studies - - practical and theoretical - - to determne its efficacy, its weaknesses and strengths; and suggestions concerning, say, how the transition from it to the traditional symbolism can be best effected?

As I mentioned earlier, some of the developments of the past decades have reached the schools in a garbled form. One particular sufferer has been the arrow as originally used in mappings. Perhaps the arrow is used more sanely elsewhere than in Britain. Some ideas of the horrors perpetrated there are portrayed in a book for primary school teachers which I was recently sent. In a mere five pages and in addition to the arrow's use in the section headings 'Components of numbers --> 10 and 'Addition and subtraction --> 10', the book recommended the use of the arrow with 5-7 year olds in the following circumstances:

6 --> 3,3 ; 6 --> 5,1

add / add3
4,3 --> 7 , 5 --> 8, 3 + 5 --> 8

subtract 3
7 --> 4 , 8 - 3 --> 5.

For what then is '-->' a symbol: up to, of numbers up to, goes into, is equal to, . . .? If any teacher, let alone child, could survive such an introduction without being utterly confused, I should be very surprised. This is, to me, not only an example of crazy pedagogy, but, I fear, merely symptomatic of the lack of attention which is paid in mathematical education to the problems of symbolism and the difficulties of teaching it.

Our symbolism has evolved in an unplanned way. It is a mixture of the good, the bad, and the utterly confusing. Consider, the three problems

(a) $28 - 8 =$

(b) $2\ 1/7 - 1/7 =$

(c) $2x^2 - x^2 =$

Bob Davis (1978) describes the difficulties of a typical learner with (c) and how he translated it as "2x squared 'take away' x squared leaves 2". Davis suggests that this example warns us against the danger of using 'natural language' to 'clarify' mathematical statements. I cannot altogether agree with Davis here. It is unlikely that the student would have made the error with (a); and, of course, it would not have been an erroneous argument when applied to (b)! The problem would seem to be far more complex. The puzzled student who has not comprehended the symbolism of (c) may well have reverted to those ikonic representations of subtraction to which the term 'take away' could reasonably apply.

Unfortunately, we are in such a hurry to move from 5 'take away' 3 'leaves' 2 to "5 - 3 = 2" that the symbols become associated with words or phrases that later prove misleading. (The use of the arrow would seen to do little to alleviate such problems.) This merely serves to illustrate my earlier points concerning the headlong rush to a symbolic representation. However, as we see here, the notations 2 1/7 and 2 x^2 are confusing in the extreme. In one case juxtaposition indicates addition, in the other multiplication. This is but one example of unfortunate terminology which is unlikely to be replaced. However, if we cannot 'replace', how can we 'counteract'? It is, for example, doubtful if many teachers are aware of the pitfalls that lie in store when they introduce particular symbols and if they have been given guidance on how to help students surmount the resulting difficulties. Again, further investigation of ambiguous and troublesome symbolism could prove most useful, together with detailed suggestions (which one hopes might lead to general agreement) on how particularly difficult problems might be overcome.

8. New Symbols for Old

In the previous section I referred to two fairly recent additions to the mathematician's list of symbols. As one particularly interested in the problems of curriculum development, I have a special interest in the way in which new symbols can be absorbed into our teaching, the problems they solve, and the problems they cause. Yet absorbing new symbolism would seem much easier than throwing out the old. We in Britain took well over a century to rid ourselves of the more unsatisfactory parts of Newtons notation for the differential calculus. The "pricked letters" $\dot{x}$, $\ddot{x}$, . . . were, of course, retained as super-contracted forms. More recently one has watched with some dismay as attempts to establish simpler and less ambiguous symbolism for the calculus - - mainly based on the suggestions of Menger (see, for example, 1979) - - have almost immediately foundered as a result of opposition from the users of mathematics. Clearly, one cannot supplant the notation of Leibniz, Lagrange, and Cauchy overnight. But is there no hope of our ever replacing it by anything more suitable (as the Hindu-Arabic notation gradually replaced that of the Romans)? How can the requisite curriculum development be effected? Should IMU and IMCI pay specific attention to the question of what in the present context constitutes reasonable notation? To quote Menger, "Nothing is more distasteful to an active mathematician . . . than discussions of symbolism and notation; and that dislike is perfectly understandable. After having overcome in his youth whatever difficulties the formal expression of ideas presents, the mathematician finds that certain ways of writing have become second nature and he regards any suggestion of change, even if he recognizes its merits, as nothing but a trivial nuisance." The professional mathematician by definition is one who should not be troubled by mathematical symbolism and language. It is difficult for us to imagine the problems they can cause to others.

Today I have sketched very briefly some particular problems which I believe merit further investigation. I hope that at the ICME in four or eight years time we shall be presented with papers which can cast new light on these areas and can appreciably deepen our understanding of them.

References

Abidjan, Conference on 'Mathematique et Milieu en Afrique', 1978.

Austin, J.L., & Howson, A.G. 'Language and Mathematical Education', Educ. Studies Math. 10 (1979), 161-197.

Branford, B. A Study of Mathematical Education, Oxford University Press, 1908.

Brown, M., & Kuchemann, D. 'Is It an 'Add' Miss?, Maths in School, 5 (5) (1976), 15-17; 6 (1) (1977), 9-10.

Clark, M. An Investigation into the New Zealand Forms I to IV Mathematics Syllabus, Unpublished M.Sc. Thesis, Victoria University of Wellington.

Davis, R.B., Jockusch, E., & McKnight, C.C. 'Cognitive Processes in Learning Algebra', J. Child. Math. Beh. 2 (1) (1978), 1-320.

Davis, R.B., & McKnight, C.C. 'Modeling the Processes of Mathematical Thinking' J. Child. Math. Beh. 2 (2) (1979), 91-111.

Dee, J. 'A Mathematical Preface', Euclid's Elements (trs. H. Billingsley), 1570.

Erlwanger, S.H. Case Studies of Children's Conceptions of Mathematics, Unpublished Ph.D. Thesis, University of Illinois, 1973.

Freudenthal, H. Mathematics as an Educational Task, Reidel, 1973.

Ginsberg, H. The Myth of the Deprived Child, Prentice-Hall, 1972.

Hart, K.M. Children's Understanding of Mathematics 11-16, John Murray (at the press).

Hirabayashi, I. & Katayama, K. 'Some Linguistic Aspects of Geometrical Diagrams', Reports of Math. Ed., J.S.M.E., 17 (1969), 1-14.

Hutton, C. A Mathematical and Philosophical Dictionary (2nd Ed.), 1815.

Labov, W. 'The Logic of Non-Standard English', in Language and Poverty (ed. Williams, F.), Markham, 1970.

Menger, K. Selected Papers in Logic and Foundations, Didactics and Economics, Reidel, 1979.

Morris, R.W. Linguistic Problems Encountered by Contemporary Curriculum Development Projects in Mathematics, UNESCO, 1974.

Nesher, P.A. From Ordinary Language to Arithemtical Language in the Primary Grades. (What does it mean to teach '2+3=5'?), Unpublished Ed.D. Thesis, Harvard University, 1972.

Skemp, R.R., 'Relational Mathematics and Instrumental Mathematics - - Some Further Thoughts', University of Warwick, 1977.

Thom, R. 'Modern Mathematics: Does it Exist?' in Developments in Mathematical Education (ed. Howson, A.G.), Cambridge University Press, 1973.

Todhunter, I. The Conflict of Studies, Macmillan, 1873.

UNESCO, Interactions between Linguistics and Mathematical Education, 1974.

Weyl, H. Philosophy of Mathematics and Natural Science, Princeton University Press, 1943.

16.2 THE RELATIONSHIP BETWEEN THE DEVELOPMENT OF LANGUAGE IN CHILDREN AND THE DEVELOPMENT OF MATHEMATICAL CONCEPTS IN CHILDREN

DEVELOPMENT OF LANGUAGE AND MATHEMATICAL CONCEPTS IN CHILDREN: IS THERE A RELATIONSHIP?

F.D. Lowenthal
Universite de l'Etat a Mons
Mons, Belgium

1. Introduction.

At one stage (Lowenthal, 1971, 1972) we had the opportunity to use modern techniques in teaching mathematics to problem children. These children, who refused to learn and seemed incapable of making any progress, in 18 months made up a great deal of their backwardness in mathematics (Cordier, Lowenthal, & Heraux, 1975). We inquired into the causes of this change. The subject we taught does not seem to be an essential factor; but the fact that we used in our teaching the non-verbal support described by F. and G. Papy (1970) was probably essential: this non-verbal support enabled us indeed to introduce some elements of a logically based language. We were able to show (Cordier, de Kerchove & Lowenthal, 1977) that children need these language-elements to express the logical structure of the unvierse in which they live, and also that problem children usually do not master these lingustic elements.

These first observations led us to formulate a new thesis concerning cognitive development: "one of the main factors of cognitive development is the manipulation of representations and the comparison of isomorphic representations. This helps the maturing of a metalangauge (the child is compelled to make many value judgements) as well as the simultaneous development of a logic and an object-language through which the child learns to organise the universe, to define his relationship with the rest of the world and to share his thoughts with someone else". This has several possible applications among which is an improved method of teaching mathematics and language.

To test our thesis we initated a new experimental teaching programme, similar to the previous one, but directed towards normal 8- to 10-year-olds. To assist in the development of their language and logical thought, we decided to place a formalism at their disposal. This formalism had to be non-verbal, non-ambiguous, simple and easy to handle, and flexible enough to enable the children to become aware of what they cannot communicate verbally and express to others. It seems essential to us that such a formalism should be suggestive of a logic and could be used in the frame of games; but the nature of the formalism used seems less important provided that the conditions described here are fulfilled.

2. Methodology

We described in other papers a formalism which fulfills these requirements (Lowenthal, 1978, 1980). Objects are represented by dots and relations between objects, by multicoloured arrows; each dot represents exactly one object (which can have several names) and each colour represents exactly one relation: 2 dots represent 2 different objects and 2 colours, 2 differing relations.

We also wanted to base our techniques on the use of games: this is reassuring and amusing for the children. It gives them an opportunity to handle "rules of a game". One of the best ways to analyze these rules is to connect them with a finite automaton. This can easily be done through our formalism (Kaufman, 1975) by drawing multicoloured diagrams. This representation technique thus provides the children's thought with a concrete support in the frame of the analysis of all possible plays fora given game (by definition a play is a succession of moves according to a given set of rules of a game, such that those successive moves lead to a conclusion).

We wanted to use the full power of the chosen formalism to let children learn to represent and compare their games through finite automata. For each game, 5 steps were used (Lowenthal, 1976, 1977):

1. learn a game and play it;
2. give a verbal description of what happened in the first step;
3. describe all possible plays of the game with a multicoloured diagram;
4. forget the initial game and modify the diagram arbitrarily;
5. interpret the new diagram: if it still represents a finite automaton and if you wish to view it as such, describe and play a game illustrated by the new diagram.

   or if it does no longer represent a finite automaton, or if you simply do no longer wish to view it as such, tell a coherent story which illustrates the new diagram.

One can thus associate the first 3 steps, the apprenticeship of the rules of the game, with the last step, the creation of something new, which is either action (a game) or word (a story) depending on the subject's mood, and with step 4, a concretisation of the arbitrary changes that children often introduce in the rules of a game (Kohl, 1974). The function of this iconico-symbolic bridge between apprenticeship and creation will be better understood through an example.

3. An Example: The Game of Timoleon and Nabuchodonosor.

a. The Rules.

In a garden, there are three paving stones placed in a triangle. They are called "stone A", "stone B", and "stone C". Timoleon and Nabuchodonosor spring from stone to stone. To spring they use a die with 3 green sides and 3 red sides: they play alternately and Timoleon starts. At the start, both are on stone A. The player who gets green goes one stone forward (i.e., A to B and then C), the player who gets red goes one stone backward. The first one to stop on stone C wins and the play stops then immediately. Timoleon and Nabuchodonosor must always be on one of the 3 stones (not necessarily the same!).

b. Accessories

The children draw first the circuit as it is described. They often hesitate: do they need a fourth stone "to go backward from A"? But the rules are clear: "There are only 3 stones". To go backward can only mean: "go to C".
To describe a play through a diagram, we need a code; each dot must represent 3 things: Timoleons position on the circuit, Nabuchodonosors' position, and the name of the player who must play next. These are in fact the inner states of our automaton. By convention Timoleon's position is always mentioned first. The dots are named and the names are numbers, but this does not imply any ordering of the dots.
It suffices then to use 2 "colours" to define the "next state - next position function" of our automotan. We then let one arrow of each kind start from each of the dots corresponding to an inner state which is actually in use. Dot number 1 is used as starting point (both players are on stone A and Timoleon must play). The dots corresponding to victories get special labels: no arrow starts from these dots, the play is finished. The other dots remain isolated, useless but present.

4. Results

a. The Games.

The 8- to 10-year-olds quickly learn to use multicoloured diagrams to represent finite automata. After one month they know how to use dots to represent the inner states of the structure they study: out of each dot associated with an "active" state, start as many differing arrows as there are moves to be taken care of. Only the dots associated either with a victory for one of the players or with a useless position in this game are not the origin of several arrows.

Third graders (8- to 9-year-olds) know how to use this technique of multicoloured diagrams without problems. After a few (5 or 6) training sessions, they can establish a double-entry table of "states and positions, a picture of the circuit to use, and a diagram representing the finite automaton corresponding to a given game.

After 3 months of work (one lesson a week), they willingly produce new diagrams, which "serve to play by following the arrows". But they are not yet able verbally to formulate the rules of the game thus created. It is only after 10 months of work that some (but not all) succeed in verbally expressing the rules they handle easily with arrows. Some who translate correctly into verbal language their graphical creation are astonished: how is it possible that these two different representations correspond to the same game? One must wait for 6 more months, giving the children the opportunity to compare many representations, to judge their respective merits, and thus letting the metalanguage of the youngest ones ripen, in order to let them understand that two things as different as a verbal statement and a multicoloured diagram can represent the same set of rules.

Our pupils love "to change the diagrams". We do not always know why they choose to do it one way rather than another and this is not very important, as our aim is not to teach directly something; we want to create a precise structure, a frame inside which the children can speak, think, reason. It is then possible to use diagrams easily to introduce logico-mathematical exercises using games with 8- or 10-year-olds. One could, of course, ask the child who plays Nabuchodonosor's part which colour he wishes for Timoleon, and why; this for the original game as well as for the modified games. But more complex problems arise naturally:

> There are 2 children and 3 stones. We observe experimentally that there are 9 "positions" and 18 inner states, how many would there be if there were 3 children and 3 stones? 2 children and 5 stones? 5 children and 2 stones?
>
> We can thus naturally approach (i.e., because a real problem must be solved) since the 4th grade (9-10 years) notions like "power", "combination", "permutation" without doing anything that is not (apparently) immediately needed for our games.
>
> We can also notice that Timoleon has "more chance than" Nabuchodonosor to win the original game. The children say that "the game is unfair". One can then try to better define this notion of "chance", and try also to let the children build the diagram of "a fair game".

b. Stories.

We asked the children to choose a diagram and to associate with it a coherent story by giving a name to each dot and by attaching an action or relation to each colour. This story had to be logically valid: it was produced by one child and submitted to the criticism of the whole group.

In other cases the children learned, with the help of the concrete support of a diagram, to formulate clearly their stories on their remarks in order to avoid their peer's criticisms. They also learned to use more economical formulas: our method makes it possible to bring back into the frame of games the teaching of relative pronouns, subordinates, etc., and reach simplified stories, which are produced after 10 months of work (9-10 year olds).

We thus notice in the domain of "rules of a game", as well as in the frame of "coherent stories", progress especially in the field of socialisation: the children learn to work in groups, to express their ideas with care in order to help the others to understand, to listen to the others and also never to interrupt someone who has the floor, but on the contrary to request leave to speak. We also observe progress in the use and the manipulation of representations. Parellel results have been observed with younger children (7-8 year olds) whom we trained to use the semi-concrete formalism described by Cohors-Fresenborg (1978) for older children. These advances in the field of communication have immediate consequences at the level of these children's verbal language and logic: some of our pupils approached successfully, since 2nd grade, the notion nof finite automaton (Lowenthal & Marcq, 1980). Others, trained in the use of the diagrams described in this paper, have collectively rediscovered in the 6th grade, almost without help, Pick's theorem (Lowenthal & Severs, 1980). We must especially mention the following fact: children who use multicoloured diagrams are often able from the age of 9-10 years to formulate and to analyze hypotheses.

5. Discussion.

After this observation of the results obtained by young children, as far as the pair language-rational thought is concerned, we can once again wonder which is the main factor in cognitive development.

For Piaget, there is firstly action, then imitation followed by development of logical thought, and then, but not simultaneously, development of verbal language. This does not fit our observations concerning children who work in groups and who acquire a logic through their language and vice versa. For Chomsky, there is an innate Language Acquisition Device. For him, the language seems to precede everything, since a given structure would be innate and universal. It is possible that some innate structure exists, but it is then most probably a more restricted structure than that described by Chomsky, and certainly not a universal set of rules. It is more likely a limited set of capacities to learn and to use social contacts, like those described by Bruner (1966). Bruner's theory seems to fit better with our observations: he describes three types of representations which complete one another and which develop one after the other, but which do not supersede one another. Based on these representation modes, language and logic develop concurrently, each supporting the other: the logic serves to structure the universe and the language to represent it and to communicate with others.

We want to go even further than Bruner and claim that the most important activity for the child is developing his metalanguage, learning to formulate value judgments. To help the child to do this, we think that it is useful to give him the opportunity to handle different representations of the same thing, the opportunity to notice which are the differences and why there is nevertheless invariance of the represented object. This is why we attach a great deal of importance to all manipulations and representations and to the use of formalisms which are supple, simple, changing and varying, suggestive of a logic, since these formalisms are useful tools to enable young children to handle representations.

References

Bruner, J.S.: 1966. On Cognitive Growth. (In Bruner, J.S., Greenfield, P.M. and Olver, R.R. Studies in Cognitive Growth), New York: John Wiley, 1-67.

Cohors-Fresenborg, E.: 1978. Learning problem solving by developing automata networks. Revue de Phonetique Appliquee, vol 46-47, 93-99. (Proceedings of the first Mons Conference on Language and Language Acquisition, 1979).

Cordier, J., de Kerchove, A. and Lowenthal, F.: 1977. Education du langage - innovation ou retour a la tradition? (In Le Coeur it l'Esprit, S. Reuse ed.), Bruxelles: Editions de l'Universite de Bruxelles, 749-757.

Cordier, J., Lowenthal, F. and Heraux, C.: 1975. Enseignement de la mathamatique et exercices de verbalisation chez les enfants caracteriels, Enfance, 1, 111-124.

Kaufman, A.: 1975. Introduction a la theorie des sous-ensembles flous. vol 3, 72-96. Paris. Masson.

Kohl, H.: 1974 Math, Writing and Games in the Open Classroom, New York, New York Review Book NYR 109).

Lowenthal, F.: 1972. Enseignment de la mathematique a 2 gruopes d'enfants caracteriels. NICO, 10, 69-86 (Proceedings GIRP I).

Lowenthal, F.: 1973. La mathematique peut-elle etre une therapeutique? NICO, 13, 98-104 (Proceedings GIRP II).

Lowenthal, F.: 1976. Jeux mathematiques comme base du langage. (In Langage et Pensee Mathematiques, Luxembourg: Centre Universitaire de Luxembourg, 439-441.

Lowenthal, F.: 1977. Games, graphs and the logic of language acquisition, Communication and Cognition, Vol. 10, no. 2, 42-52.

Lowenthal, F.: 1978. Logic of natural language and Games at Primary School. Revue de Phonetique Appliquee, vol. 46-47, 133-140.(Proceedings of the first Mons Conference on Language and Language Acquisition, 1977).

Lowenthal, F.: 1980. Games, logic and cognitive development: A longitudinal study of classroom situations, Communication and Cognition, vol. 13, no. 1, 43-61.

Lowenthal, F. and Marcq, J.: 1980. Dynamical mazes used to favour communication in 7-8 year olds. In Proceedings of the 4th International Conference for the Psychology of Mathematics Education, Berkeley, 1980, ed.: R. Karplus et al.).

Lowenthal, F. and Severs, R.: 1979. Langage, jeu et activite mathematique - un essai a l'ecole primaire. Educational Studies in Mathemaics. 10, no. 2, 245-262.

Lowenthal, F. and Severs, R.: 1980. Inductive and axiomatic reasoning at elementary school level. In Proceedings of the third International Conference for the Psychology of Mathematics Education, Warwick, 1979, ed.: D. Tall).

Papy, F. and Papy, G.: 1970. L'enfant et les graphes. Bruxelles. Didier.

ANALYSIS OF RECIPROCAL RELATIONSHIPS BETWEEN LINGUISTIC DEVELOPMENT AND MATHEMATICS TEACHING: A PHYCHOLOGICAL AND SOCIO-CULTURAL POINT OF VIEW

Michele Pellerey
Universita Selesiana di Roma, Italy

I would like to provide tentative answers to two related questions: a) What is the influence of the mother tongue on the language actually used in mathematics? Does this foster or impede the building up of basic concepts and the development of mathematical abilities? b) What is the influence of the language competence of a child on the learning of mathematical concepts, principles and procedures?

Mathematical education is a process through which a child enters into two basic regions of the mathematical world: concepts and theories; and mathematical activity. Such initiation, fostered and made easier by teachers, takes place within a well-knit system where as many as four main poles exert their interactive influence:

a) The culture to which the child belongs. Culture is here to be taken in an anthropological sense; it expresses itself and is communicated through the mother tongue.

b) The acts and concepts of mathematical type together with the theories that bind them into a body that is logically coherent and applicatively valid. Such a mathematical body is expressed in linguistic forms; that is, one or more systems made up of language signs according to certain usage rules that are socially determined.

c) The cognitive structure (or cognitive matrix) of the child. This is to be understood as a more or less organized set of concepts, principles, skills, and attitudes together with their interactions. Within such cognitive structure, language competence plays a central role; that is to say, the ability to pass from a system of language symbols socially accepted and organized to inner contents or meanings, and vice versa.

d) The physical world that a child encounters through his or her sensory perception.

The culture to which the child belongs and that which is instilled by the mother tongue is a comprehensive system, from which may be separated a number of cultural subsystems: representational, normative, expression, and action systems. These subsystems interact significantly, as do the elements that make up each of the subsystems: concepts, symbols, knowledge and ascertaining methods, values, behavioural norms, ways to express and single out deeper feelings, interpretation devices, technological and social mediations.

Mathematical concepts and procedures are more or less deeply embedded in this culture, either in a simple, natural form or in a rather elaborated structure. Such mathematical elements usually belong to conceptual categories whereby situations, facts and problems fit within an abstract frame for interpretation, explanation, and prediction. Categories as these may be labelled "mathematical" depending on the relationship they have with established portions of mathematics accepted by the scientific community of a given time. However, often observed in school situations are contrasts between ideas and words of the native culture, and ideas and words employed in mathematical thinking and communication.

From birth, a child grows within such culture, in which it finds its first socialization forms by communicating with its family circle, initially through non-verbal signs, then through language codes organized into a system whose usage rules are shared by a larger language community.

The child little by little masters the words it needs, the rules that belong to them, and the norms that govern the setting out of speech. At the same time it develops the ability to pass from the perception of organized sounds to the communication of contents, and conversely, to pass from inner contents or meaning to their coding in its mother tongue. We call such two-fold ability "language competence" or, to use the expression of F. de Saussure, "langage". Among the main functions of language competence (instrumental, normative, interactive, rhetorical, personal, heuristic, representative, and logical), we shall deal here primarily with the heuristic and the representative ones. That is to say, we shall look at language firstly as a tool whereby reality is studied and some notions pertaining to reality are derived, and also a meta-language, or as a speech on language itself; secondly, as a tool to communicate and express sentences that pertain to processes, persons, objects, abstractions, qualities, states, and relationships.

It is, however, necessary to consider also the different functions of language competence in its oral and written forms. Oral language is more pragmatic, as it is related to a way of representing, communicating, formulating, and exploring a knowledge directed at controlling practical activity and achieving goals which are socially and personally valuable. Written language is more explicit and logically organized. It helps to formulate ideas and judgements in an abstract and controlled form. While in the spoken language an active mood prevails, in the written language reflection and critical control predominate. Hence arises the need for an adequately developed written language competence when the mastery of reason and logic is desires, as in the conscious definition and organization of mathematical concepts.

Mathematical concepts and procedures may be regarded as a "corpus" or common background on which scholars have operated and are still operating with the aim of causing it to grow, of better organizing it to provide a finer inner cohesion, and of employing it to explain or anticipate situations and facts of both the physical and social worlds. This "corpus" is being concretized in a number of ways that include graphic forms, written language forms, more abstract symbolic forms, and so on. Leibniz' dream was to arrive at a universal symbolic form, the "characterization universalis" as he called it, whereby a logical calculus analogous to the numerical one should be invented; such a view is at loggerheads with the complex and dynamic character of mathematical thinking and communicating. However, that portion of mathematics that proved most susceptible to a mechanical formulation has been computerized.

The culture to which a child belongs exerts some influence on these forms of mathematical expression. It is enough to open a textbook written in French and another written in English to see the point: the latter more empirical, paying attention to the core of concepts, of procedures and applications; the former more abstract and formal.

We should state that a unique system of signs, or mathematical language, does not exist; there exists, instead, a set of linguistic forms that evolve according to time, geographical location, and inner regions of the discipline itself. History and geography of mathematical language are very closely knit to history and geography of scientific communities.

Comparing the child's cultural language with the complex of mathematical linquistics, both continuity and discontinuity may be seen. In order to make my point clear, it may be useful to start from the following graph:

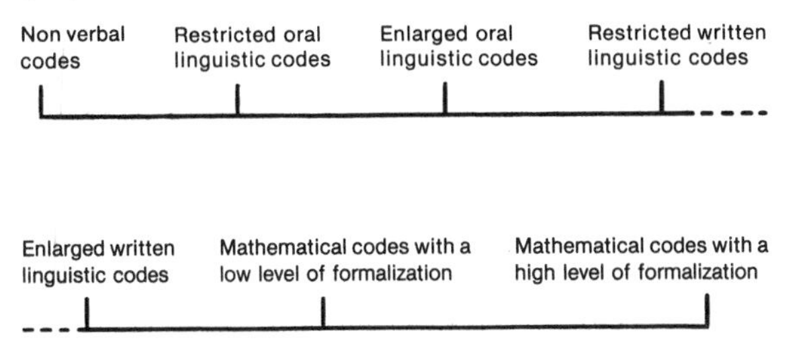

Following this schema, we may generate progressive differentiations from low-level codes to high-level codes. We pass from non-conventional linguistic forms to increasingly conventionalized forms. The rhetoric function, whose object is to foster social relationships between those engaged in dialogue, diminishes as we proceed toward abstract symbolic codes; on the other hand, the logical function, whose purpose is to specify relationships within statements as well as between different statements, increases in importance. This means that where spoken language is prevelant, textual relationships take the upper hand, especially if such language is formalized. This transition, therefore, takes a person from sharing and understanding concrete situations as well as from leading one to perform everyday activity, to delving into an abstract and coherent theory of reality as well as to organizing both concepts and procedures into a logical scheme.

The process of mathematical learning that takes place in the school must inevitably start from the code system that the child is already acquainted with through the influence of its family and social environment and with the world of ideas and know-how that make up the culture to which it belongs. The task, then, consists not only in leading the child to master more elaborated and more formalized codes, but also in leading it to grasp the different functions of such systems with regard to the development of thought and arranging suitable activity. Instead we perceive, too often, a remarkable breakdown between the child's cultural world and language, and the school cultural world and language. In the case of mathematics, this fact may become a dramatic one, because its categories and procedures are often scarcely represented in a given cultural environment and, besides, both linguistic forms and their functions are wholly ignored by the child.

In order to complete our brief sketch, we ought now to study the relationships that exist between the physical world and the child's cognitive structure under the influence of mathematical education. We have stated already that a child's cognitive structure is to be envisaged as a more or less appropriately structured set of concepts, principles, abilities, and attitudes. It is bound to develop both with the help of immediate experience and of communication. But it is chiefly through cultural communication reinforcing direct experience that things and happenings get their meaning. A certain development of child's language competence is thus implied.

When dealing with mathematical learning, we ought to distinguish between concepts and activities at cypher level and those at a higher level. In the first case, we pass initially from the world of direct experience to the world of its rational representation; in the second case, one works already with representations and schemes that are more or less abstract ones.

Mathematical concepts at cypher level are simplified reproductions of a child's direct experience. The perception process has the all-important role of selecting and structuring sensory data either according to a conceptual category somehow already present, or according to the attention being focused by outward solicitations, usually by the teacher, with the aim of isolating certain relationships or specific operations. In the second case, one or more relationships pertaining to one or more operations, physical or mental, may emerge: this allows for a wholly singular force being attached to them and adequate words to identify them being sorted out.

Two mathematical processes to the child's mind thus open. The first one applies very simple schemes of a mathematical nature to interpret and explain a portion of physical reality. The other one, under the expert guidance of a teacher or someone else, builds up a conceptual scheme and names it. To speak of building up mathematical concepts of using them when interpreting schematically some facts, phenomena, or physical situations, it is necessary that these schemes, even when still raw and unfit to settle immediately into a theory in mathematics, be open to such future settling without needing deep changes or even reversing them from their present state of rationality. This means that from a cypher level one must be able to pass to a superior level, a more abstract and more reflexive level, without meeting with obstacles or contradictions that may be insuperable. The same thing is true for the words and symbols as they are gradually being introduced.

We may draw a few simple practical orientations.

a) Cypher-level mathematization processes must bank on what a child has already acquired both conceptually and linguistically through the influence of its environmental culture. For instance, in the case of natural number, one is bound to bestow serious attention to awareness of phenomena of a recursive nature and on their main verbal representation, counting.

b) If a discrepancy should arise between the schemes and words developed in family and societal circles and those specifically mathematical, one is bound to point out clearly differences and alternative roles that are relevant to them: the former more practical, the latter more logical.

c) Ideal schematization must be achieved using more than a single representation form; those closer to a child's sensibility and practice must be given precedence.

d) In order to foster significant learning, with a view to incorporating into a child's cognitive structure the very substance of concepts and procedures, its language competence must improve in both senses: from symbols to inner meaning and vice versa; moreover, the largest possible range of codes should be used.

RELATIONS ENTRE LE DEVELOPPEMENT DU LANGAGE ET CELUI DES CONCEPTS MATHEMATIQUES CHEZ LES ENFANTS

Colette Laborde
IMAG
Grenoble, France

I. Position du Probleme

Une des specificites de la forme prise par le discours mathematique ecrit reside dans l'utilisation conjointe et parfois tres imbriquee de deux codes, la langue naturelle et ce que nous appelons l'ecriture symbolique. En effet, d'une part la designation d'un objet mathematique par un signe ou en ensemble de signes choisis en dehors de la langue naturelle est un fait courant en mathematique, qui plus qu'un simple "artifice abreviatif" peut constituer un element moteur dans l'activite mathematique; d'autre part, cette derniere ne se reduisant pas a l'application d'une suite de regles sur des expression formelles est processus de creation, et ne serait-ce que consideree comme activite humaine individuelle integre les capacites du sujet agissant. Il en resulte que l'expression de l'activite mathematique ne saurait avoir lieu sans s'appuyer sur ces deux codes. Nous pensons que cette affirmation recouvre plus qu'un seul probleme de forme; l'evolution de l'expression mathematique au cours du temps et les debats parfois passiones qu'elles a souleves chez les mathematiciens en sont revelateurs.

L'eleve de 11-12-13-14 ans se trouve particulierement confronte a ce probleme d'expression en mathematiques, des son entree en 6eme. (En France, la 6eme est la premiere classe de l'enseignement secondaire. Les eleves y sont en general ages de 11-12 ans environ.) L'activite mathematique a l'ecole primaire exige en general beaucoup moins d'activite de formulations, qui restent d'ailleurs essentiellement orales. L'expression premise est plus proche de celle quotidienne des eleves et possede une plus grande souplesse. Des le debut du premier cycle du secondaire, les textes mathematiques ecrits proposes aux eleves (en particulier les manuels) ont en general une forme assez voisine du "discours mathematique type", en particulier en ce qui concerne l'utilisation conjointe de l'ecriture symbolique et de la langue naturelle, meme si l'usage des expressions symboliques y est moins developpe qu'il peut l'etre habituellement.

Certes, des activites de codage et de designation sont proposees a l'eleve des le debut de sa scolarite, mais outre que leur finalite n'est heureusement pas d'arriver a une forme aussi "achevee" de l'expression mathematique que celle du manuel, leur domaine d'application n'est pas le meme. A l'ecole primaire, les codages interviennent essentiellement dans le domaine de la numeration, ensuite en classe de 6e et 5e l'usage de designer les objets mathematiques par des lettres, que ce soit un objet numerique ou un objet geometrique prepare la notion de variable.

Nous cherchons donc a determiner comment les eleves de cet age (11 a 14 ans) designent les objets mathematiques et expriment les relations etablies entre ces objets. Utilisent-ils le code symbolique? Comment le font-ils fonctionner dans leur discours? S'ils ne l'emploient pas et ont recours a la langue naturelle, quelles sont leurs strategies linguistiques de remplacement? Cela, non pour etudier les seules proprietes formelles de discours de l'eleve mais les apprehender en rapport avec le sujet parlant, c'est-a-dire l'eleve avec ses connaissances, et avec le contenue vehicule. L'etude du choix et de la forme du codage elabore par l'eleve n'a d'interet que par rapport aux processus cognieifs qui ont construit et mis en oeuvre ce codage (Brossard 1978).

II. Recueil de Donnees

Nous avons essaye de repondre a ces questions on deux temps:

1°. en collectant des production d'eleves en situation de classe: solutions ecrites a des problemes de recherches d'un nombre inconnu, dont l'enonce n'indiquait pas la possibilite de designer cette inconnue par une lettre et d'evoir recours a une resolution de type algebrique (eleves de 13-14 ans), echanges oraux d'eleves, travaillant en groupes sur des constructions geometriques relatives au cercle, dont ils devaient trouver le principe puis l'explique par ecrit (eleves de 11-12 ans).

2°. en creant une situation de communication precise en dehors de la classe: celle d'un "jeu de messages". Deux eleves devaient decrire une figure donnee dans un message ecrit destine a deux autres eleves devant construire la figure uniquement a l'aide de message (eleves de 11 a 13 ans). La figure ne comportait aucune lettre accompagnatrice, afin que nous puissions voir, si le recours au codage avait lieu et dans l'affirmative comment il fonctionnait et quels etaient les elements codes. Les echanges oraux entre eleves codeurs d'une part, et eleves decodeurs d'autre part, etaient observes et enregistres.

Dans la suite, nous presenterons les resultats tires de l'analyse de ces donnees qui n'est pas explicitee ici. (Des paties de l'analyse figurent dans Guillerault-Laborde (1980).

III. Choix de Code

a) Strategies linguistiques:

L'usage du code symbolique n'est pas toujours spontane chez les eleves. Ainsi, dans les problemes algebriques, aucun des eleves pourtant deja inities au calcul algebrique, ne designe de lui-meme le nombre inconnu par une lettre, ce qui les conduit a exprimer des expressions arithmetiques ou meme des relations arithmetiques par des expressions entierement en langue naturelle.

ex: "la difference entre le nombre que l'on a choisi et 10" ecrit un eleve de 14 ans pour designer x-10.

La phrase suivante d'un autre eleve de 14 ans equivant a "(10 +x) + (10 - x) = 20".

"Ma reponse est correcte car si on choisit un nombre quelconque inferieur a 10 et qu'on ajoute a 10 et ensuite on retranche ce meme nombre quelconque a 10 et qu'on ajoute les 2 resultats, cela nous donnera obligatoirement le nombre 20".

De meme dans la description de figures geometriques, 14 binomes d'eleves (sur 34) ne codent pas les elements cruciaux de la figure a l'aide de lettres.

Dans les deux cas (domaine algebrique et domaine geometrique), on peut reperer des strategies linguistiques:

- la designation de l'objet en langue naturelle est repetee integralement dans le cours du texte, a chaque reapparition, comme une lettre (x ou a...) est repetee par le mathematicien dans son discours.

- lorsque plusieurs objets de meme nature (nombres entiers, points, droites) sont en jeu, certains eleves (mais non la totalite) inserent dans la designation en langue naturelle des traits oppositifs:

ex: "le point de haut" et "le point du bas" sont les designations de deux points de la figure.

"la liste au-dessus" opposee a "la liste des 3", dans la solution d'un probleme, ou intervenaient deux ensembles de nombres, ou dans une autre copie.

"la liste des nombres" opposee a "la liste des 3 nombres".

b) lien du codage avec les procedures de resolution:

Il apparait nettement dans la description de la figure geometrique (II 2°), que l'absence ou la presence de codages symboliques sont liees a la procedure suivie pour decrire la figure. Ainsi, une procedure de description non constructive qui enumere des elements de la figure, de facon vague, sans indication de mesures n'est jamais accompagnee de codage. Parmi les procedures qui explicitent au contraire comment construire un a un les differentes parties de la figure, certaines d'entre elles commencent par donner tous les points, puis seulement apres les segments qui joignent des paires de points. Il est clair, que ces dernieres procedures exigent davantage un codage des points par des lettres, que celles qui donnent a construire chaque segment des que ses deux extremites ont ete decrites. C'est effectivement ce que nous avons pu constater dans les messages des eleves.

L'observation du travail d'elaboration des messages nous a meme permis de suivre chez certains les differentes etapes qui conduisaient de l'usage de la langue naturelle a celui d'un codage. Ainsi, certains eleves parlent du premier point, du deuxieme point (dans l'ordre de description); le troisieme point devient point n° 3 et vers le cinquieme ou le sixieme, il est seulement question du point 5 ou du point 6. D'autres ne s'arretent pas a ce codage numerique et le transforment en codage letteral A, B, C ...

Enfin, il nous parait interessant de signaler que le passage d'une description vague a une description plus precise avec indication de distances entre points a souvent ete suivi d'une decision de coder des points de la figure qui pouvaient servir d'origine de reperage pour d'autres points.

c) Fonctionnement du code symbolique

Si un codage par lettres est choisi, nous avons pu distinguer differents niveaux d'utilisation:

- des lettres ont ete choisies pour designer les objets mais ne servent plus dans la suite;

- les lettres reapparaissent dans le discours de l'eleve sans entrer dans des ecritures symboliques;

- les lettres sont utilisees dans des ecritures symboliques.

IV. De la Langue Naturelle au Symbolisme

L'examen des designations en langue naturelle d'objets mathematiques par les eleves montre, que ces designations integrent des elements exterieurs a l'objet lui-meme, a savoir:

- son ordre d'apparition dans le texte du probleme "le premier nombre", "le nombre de depart";

- sa localisation dans la feuille de papier ou au tableau;

- l'action et eventuellement l'auteur de l'action qui a produit l'objet: "le point que vous faites", "le nombre que Bernard a choise" ou meme l'absence d'action: pour distinguer une droite d'autres droites, deux eleves l'appellent "la droite qui n'a pas ete prolongee".

Ce dernier point nous avait parait tres important, car nous avons pu constater dans de nombreuses productions d'eleves correspondant a des taches differentes, combien le processus createur de l'objet, l'action du sujet y transparaissent. Le choix des termes, verbes de mouvement, d'actions, prepositions semantiquement marquees, qu'on ne trouve pas dans le discours mathematique de notre epoque en est un indice.

Nous pensons que la signification de ces objets, construite par l'eleve contient ces elements exterieurs a l'objet strict, et que c'est une phase de conceptualisation qui permet de reperer d'eliminer parmi eux les elements non pertinents pour l'usage mathematique; designer des objets par des lettres, mais surtout exprimer les relations entre objets au moyen d'ecritures symboliques sont des demarches qui passent par cette conceptualisation.

References

1. Brossard, M. "Activites cognitives et conduites verbales" in Bulletin de Psychologie T. XXXII no. 338 Annee 1978-1979 pp. 57-63.

2. Guillerault M. et Laborde, C. "Bezeichnungen mathematischer Objekte in der Sprache der Schuer" (Rapport de recherche IMAG, 1980).

3. Guillerault, M. et Laborde, C. "Une activite linguistique en geometrie" (Seminaire a la lere ecole d'ete de didactique des mathematiques. Chamrousse 1980).

ON STUDENT'S MISINTERPRETATION OF LOGICAL EXPRESSIONS INFLUENCED BY FOREIGN LANGUAGE LEARNING

Tsutomu Hosoi
Science University of Tokyo
Tokyo, Japan

When we teach college-level mathematics, we often come across a situation where logical expressions are almost unavoidable. For example, when we treat the concept of convergence in an exact way, we need the epsilon-delta method, where we meet such logical expressions as "for any" and "there exists".

As in the case of convergence, when we treat logical concepts in exact terms, we employ symbolic expressions in order to avoid such ambiguities as might be brought about by the natural language expressions. For experienced mathematicians, those symbolic expressions are quite easy to grasp and better understood than the natural language expressions. But, for students, they are real ciphers which must be decoded again into the natural language expressions. At least students must do decoding in their mind.

For the speakers of English, as well as of many other languages, the decoding might proceed in a quite direct and parallel way with respect to the symbols. But for decoding into Japanese, we have structural and semantical difficulties. And, further, the interpretation of the decoded Japanese expression very often receives some interference from the effect of foreign language education (mostly from English, because it is taught most widely in Japan). The difficulty is greatest when the logical expressions appear in the context of negation. I shall explain about these difficulties met in mathematical education in Japan with regard to language problem.

"All-Not" Interpretations by Students

First I will explain the variety of students' interpretations of Japanese "all-not" sentences, that is, those sentences containing the negated "all". Examples of such Japanese sentences are:

1. Zen-in mada kite inai.
   (all persons)(yet)(come)(not)
   (None has come yet.)

2. Sono hoteishiki no kon wa subete hu denai.
   (that)(equation)(of)(root)(p.)(all)(negative)(be not)
   (None of the roots of that equation is negative.)

3. Sono hoteishiki no kon wa doremo hu denai.
   (that)(equation)(of)(root)(p.)(every one)(negative)(be not)
   (None of the roots of that equation is negative.)
   Here p. becomes the particle indicating subject.

These sentences are not necessarily good Japanese, but they are quite common in that they, or similar patterns, quite often appear in the writings by students.

In Japanese, an "all-not" combination might mean either total negation or partial negation, depending mainly on the sentence structure. It might depend on the context, but such a case is rather rare. For the most part, "all-not" tends to mean total negation. It is rather troublesome to express the partial negation in Japanese. The above sentences all mean total negation.

I investigated how mathematics students interpret these written sentences. Table I shows the percentages of "total negation" interpretation. (The data shown in this report are gathered from 246 college students, of whom 47 are in the first year, and 49, 81, 69 are in the 2nd, 3rd, 4th year, respectively. Similar data gathered from some other groups show a similar tendency.)

Table I

	Percentage
Sentence (1)	41.8%
Sentence (2)	73.1%
Sentence (3)	91.1%

As the sentence number increases, the percentage grows rapidly. This fact reflects the situation that the sentences are getting clearer and more natural as Japanese. Only a few students claimed that the sentences were ambiguous, and hence, the remaining responses are to the partial negations.

From these data, we can see that there could be misunderstanding between teachers and students concerning "all-not" expressions as long as expressions like (1) or (2) are common among students.

Teachers must choose good "all-not" expressions as (3) in order that they should be correctly understood. But, even if they use good expressions, there remains a danger that they are transformed into non-exact expressions in students' minds. Anyway, teachers must read the inside meaning of students' "all-not" expressions like (1) or (2).

It seems that this confusion of "all-not" interpretation has not been recognized until quite recently.

Now, I would like to discuss the causes of this confusion. I see at least two big causes for this, one in the loose nature of Japanese language, and another in the interference from foreign language education.

Interference from the English Education

With the same students, I tried to see how they interpret the following two sentences in written English:

4. All the pencils are not black.

5. Not all the pencils are black.

Again, these sentences are not necessarily good English, but they are sufficient for our purpose. In contrast with Japanese, English "all-not" (including "not-all") expressions mean the partial negation. Table 2 shows the percentage of correct interpretation.

Table 2

	Percentage
Sentence (4)	45.0%
Sentence (5)	68.9%

In English classes in Japan, it is often emphasized that "not-all" means the partial negation. This might be one reason why (5) is rather well accepted. On the other hand, the structure of (4) is similar to Japanese, but the meaning is contrary. So the percentage for (4) is lower than for (5).

This comment might imply that students' English interpretation is influenced by their everyday Japanese experiences. This is quite natural for students, because many of them are thinking that every language is based on essentially the same principle. They tend to understand Japanese on the principle of English when they are intensively studying English, and conversely they understand English on the principle of Japanese when they are far from English study.

Table 3 shows evidence for one side of the above comments. I show the percentage of correct interpretation of (4) by classifying the data by students year.

Table 3

	Percentage
First year	65.9%
Second Year	57.1%
Third year	40.7%
Fourth year	27.3%
Total	45.0%

In Japan, English ability of students usually attains the highest level when they enter universities. This reflects the difficult situation of the entrance examinations. In universities, they gradually go far away from English or foreign language education. Though the percentage for the first year is almost equal to the average percentage of (5), the percentage gradually decreases as the years advance. This decrease seems to reflect the above comments.

To see the converse effect, that is, the English influence on Japanese, I would like to take up the percentages for (2) and (3), because they are mathematical materials. I would like to classify the percentages into those of the first year and those of the remainder.

	(2)	(3)
First Year	65.9%	85.1%
2, 3, 4th year	74.9%	92.5%

It seems that the first year students are a little worse for the interpretation of Japanese. And I like to see the cause in the interference from the English education as mentioned above since, for the first year students, the English influences are strongest. Anyway, the variety of interpretations gives us much trouble in Japan.

Peculiarity of Japanese Language

Next, I would like to explain the difficulty due to the peculiarity of Japanese. One day I asked the same students to write down the following symbolic expression into Japanese:

6. $\forall x \exists y : f(x, y) = 0$

About 90 percent of them wrote something like the following:

7. Subete no x ni taishite f(x, y) = 0 to naru y ga sonzairsuru.
   (for all x) (f(x, y) = 0 holds) (there exists y)

This sentence is ambiguous in Japanese and it can mean the following:

8. $\exists y \forall x : f(x, y) = 0$

The next day, I asked them to write down (8) in Japanese. Again about 90 percent of them wrote something like (7). If I had asked them to treat both (6) and (8) on the same occasion, they would have tried to express them differently. The above confusion becomes greatest if "not" is inserted somewhere.

These confusions originate in the word order in Japanese. In Japanese, a verb in a sentence comes at the last and necessarily the negation tends to be at the last part of a sentence. Hence, "there exists" or "not" must be treated as suffixal, that is, they qualify the main part from the end, while "for all" can be treated as prefixal. This fact is clearly seen in the rewriting (7) of (6) or (8).

Sometimes it is said that (6) and (8) can be distinguished in (7) by putting a comma in suitable positions. But this does not work well, because a "comma" cannot be pronounced and it is often omitted by students. At least, commas do not remain so long in students' mind.

Conclusion

Concerning the inferences from language education, it would be urgently necessary to have some consultations and arrangements among teachers of mathematics, of foreign languages, and of the mother tongue.

Concerning the peculiarity of Japanese, I think we need some contrivances in some natural way to express quantifiers as suffixal. I have some suggestions for this, which I refrain from explaining here because they are too detailed.

16.3 TEACHING MATHEMATICS IN A SECOND LANGUAGE

NATIVE LANGUAGE VS. SECOND LANGUAGE IN TEACHING ELEMENTARY MATHEMATICS: A CASE FROM THE AMAZON

Maurizio Gnerre
Unicamp
Campinas, Sao Paolo, Brazil

The main objective of this paper is to evaluate the problem of opposition between the native language and the second langauge in teaching elementary mathematics. Linguistic and sociolinguistic factors have to be taken into account in evaluating such opposition. We will analyze some aspects of the use that can be made of the lexicon of a native language without a written tradition for expressing elementary mathematical concepts. The mathematical terminology under analysis is the result of a process of "adaptation" that redefines the terms used in everyday language and rids them of major ambiguities. The terminology was elaborated recently for teaching elementary mathematics in an Amazonian language, Jivaro of Eastern Ecuador. (1) We will consider what it would mean to teach mathematics in Spanish, rather than in the native language.

The Suar Jivaros of Ecuador number approximately 25,000 and are one of the largest indigenous groups in the Amazon. Together with the neighboring Achuar, Huambiza, and Aguaruna groups, they form one ethnolinguistic family with more than 50,000 people in Ecuatorian and Peruvian territories. The Shuar language runs no risk of extinction. Bilingualism in Spanish is gaining ground but is still far from reaching high percentages, except for a few limited areas. The majority of school children are Shuar monolinguals and are now beginning to have a more systematic contact with Spanish at school, as this language is introduced in the course programs.

The Jivaros of Ecuador are organized in a Federation whose main concern is the defense of land belonging to the Indians. In order to achieve full economic independence and critical self-consciousness on the part of the Indians, the Federation operates a radio station which broadcasts in Shuar most of the time. Radio broadcasts have been used since 1972 for the education of the Indian children. In each jungle village a local teacher (referred to as "teleauxiliar") follows the program as it is broadcast and eventually adapts it to local needs. In 1972-73 there were primary schools in 31 villages. In 1975-76 there were 130 schools, with a total of 224 "teleauxiliares" teaching up to the 4th year. In 1978 the primary school program (covering 6 years) was completed and recently the 7th and 8th years have been added. The schools now reach a population of approximately 4,000 indigenous children and teenagers.

The Federation's education-by-radio program is defined as bilingual and bicultural. The Federation criticizes the definition of education as bilingual in the cases in which native languages are supposed to act as a bridge leading to the assimilation of the language and the culture of a predominant society. According to the Federation ideology, the use of the native language has the intrinsic value of leading to consciously prepared bilingual Indians, and contributes to self-confidence about their own culture and language.

In the language standardization process, the challenge is particularly evident in fields such as mathematics, which present greater difficulties for translation into the native language. Using Spanish would be tantamount to saying: "The native language is not suited for conveying these concepts".

In the last few years four primers with a total of 219 pages were written for the first grade (Nekapmarar', 1978-78). (2) The texts resulted from direct experience in teaching mathematics, with the conceptual and linguistic part composed by the Indian teacher Pedro Kunkumas. Even though the process of adapting the everyday language for expressing mathemaical concepts was realized, the problems of lexical and syntactical order continue to be present. We will consider first some lexical problems.

To express the concept of 'zero', the Indian teachers chose the word atsa', there is not, does not exist'. How does one explain to the indigenous children that the cipher associated with this name has the function of a place holder? Why is 100 larger than 99?

It was less of a problem to express the concepts of addition, subtraction, multiplication, and division. The first two concepts were expressed by two verbs, both in a form which corresponds approximately to the past participle: atakasha ikiurma 'still placed' and ju(r)ama 'withdrawn'. The concepts of multiplication and division were expressed by the nominalized forms of verb roots; these forms indicate the instrument or the location for realizing some action: ikiauntai, 'for augmenting, for multiplying'; makatai, 'for dividing'.

Besides the names of the numbers and the elementary operations, approximately 40 conceptual terms were established, terms which are indispensible for working with elementary mathematics. We will mention six of these, to give evidence of problems which are more complex than the problems of the lexical order. These are related to the grammatical characteristics of the language.

The concepts of the inclusive disjunction (V) and the exclusive disjunction ($\underline{V}$) presented difficulties in being expressed. The solution that was given was: nincha pachik, 'and naming that' and niniak pachitsuk, 'not naming that'. The second concept can also be expressed by negative verb forms, tumatskesha or turutskesha, 'not doing this way'. All the problems derived from the fact that the language does not have available a unique and explicit form for expressing a disjunction. Major problems were faced in expressing the concepts of comparative order, because in the grammar of the language there does not exist a device for expressing comparison. Thus, 'less than' and 'greater than' were translated by the adjectives that signify 'small', uchich' and 'big', uunt. In the case of the concept of 'less than or equal to' and 'greater than or equal to' the translation of the disjunction was different from the translation already indicated. In these two cases, the disjunction was omitted: uchich', meteketai, 'is small, equal to', and uunt, meteketai, 'is big, equal to'.

Another order of problems, perhaps more complex yet, concerns mathematical symbolism, constructed predominantly on the basis of western languages. Most of these languages have the basic word order subject-verb-object (SVO). The Jivaro language has the basic word order subject-object-verb (SOV). Symbolism such as 7 - 3 = 4 corresponds to the order SVO, as much in the sequence 7 - 3 as in the sequence (7 - 3) = 4. In an SOV order, the order of the indigenous language, the symbols should be put down in the linear order 7, 3 -, 4 =, which would correspond to the verbalization in the indigenous language: 'from 7, 3 taking, 4 gives'. A notational order of this type, closer to some logical notational orders, cannot be adopted, because it would mean that the indigenous children would become accustomed to a type of notation that would be valid only for their group or for other indigenous groups. It would also create enormous problems when the children or adolescents move to education or to textbooks in Spanish.

The brief considerations above are meant to stress the great difficulties brought about by the attempt to use the native lanaguage in the teaching of elementary mathematics.

On the other hand, we must consider, from a sociolinguistic point of view, what it would mean to teach mathematics in Spanish rather than in the native language. Three factors have to be taken into account in analyzing this possibility: 1) the degree of bilingualism of the indigenous group, 2) their attitudes towards the native language, and 3) their attitude toward the second language. At present we need more data. (3) Oversimplifying, however, we could say that the teaching of mathematics in Spanish, under the above circumstances, would avoid two problems, but it would also give rise to two new ones. It would avoid the linguistic problems inherent to the Shuar language, as well as those problems due to the great difference between the syntax of the native language and the syntax of mathematical symbolization. It would, however, lead to other problems, one of them a linguistic one, due to the limited knowledge of Spanish on the part of the indigenous children in the early grades, and the other, a sociolinguistic one. The Spanish required for mathematics would be learned without any need for correlating it to the regional variety of Spanish. The main problem would be the sociolinguistic one: giving up the use of the native language precisely where the challenge is greatest. The effort to use the native language has a meaning which transends analysis of the mathematical language devised, the attempt is worthwhile since we are dealing not with a system of signs, but rather with a symbol of group identification and of cultural values - the language of a people fighting for ethnic survival.

NOTES

1. Linguistic field work was carried out in 1968, 1970, 1971, and 1974.

2. These primers were not the first ones to deal with mathematics in the Shuar language. Prior to these were published Calculo (1959-1967) and the mathematical sections in Germani (1966).

3. We have no data concerning the attitude of the Jivaros towards their native language or towards Spanish. Six years ago we conducted a brief survey of these aspects among 100 Indian teachers ("teleauxiliares"). Their attitudes regarding their native language were highly positive. The teachers, as children, had studied in missionary schools, where a highly repressive policy prevailed in so far as the native language was concerned. Lately this situation has greatly improved.

REFERENCES

Calculo, Instituto Linguistico de Verano. Ministerio de Educacion Publica (6 booklets). Quito, 1959-1967.

Germani, A. Unuimiartai. Aprendamos. Texto-guia de las escuelas bilingues de los Centros Shuaras. Federacion Provincial de Centrol Shuaras, Sucua, 1966.

Nekapmarar' nakurustai, Escuelas radiofonicas. (4 booklets). Federacion de Centrol Shuar, Sucua, 1976-1978.

TEACHING MATHEMATICS IN A SECOND LANGUAGE, WITH SPECIAL REFERENCE TO JAMAICA

Althea Young
Ministry of Education
Kingston, Jamaica, West Indies

Introduction

In Jamaica, as in many other countries, educational problems are partially social, reflecting divisions and flaws in the structure of the society. These divisions run deep in Jamaica, as class is important. Once its basis was colour. With very few exceptions, this was what dictated an individual's probability of success in life. Today, there is much greater social mobility, based on money. But a person's chance of being middle-class depends, as in many other countries, on the class of his parents, especially as in Jamaica there is a wide gap between rich and poor.

The problem of upward mobility for the poorer classes and of the acceptance into the middle-class is largely a question of behaviour, and more specifically of speech. The educational problems that arise from these difficulties are outlined below.

1. Language in Jamaica

In Jamaica, there are two languages: one is official and the language of instruction, Standard Jamaican English (SJE), and the other is Jamaican Creole, thought to be a dialect of English by some linguists and called here Jamaican English (JE). There are many similarities between the two, but there are also many differences in pronunciation, vocabulary, intonation, syntax, and grammar (see, for example, Cassidy, 1964).

It is possible to make a sort of scale and to locate speakers on it according to their use of certain phonological syntactical, and logical forms (see the article by DeCamp in Hymus, 1971, p. 356). David DeCamp nd Beryl Bailey, amongst others, assert that there is a "continuum" connecting the forms of JE and SJE that differ most from each other. If point A is taken on a line to represent the position of the speaker of the socially best received form of SJE and point C to represent that of the speaker of JE whose usage is most divergent from this form, then all other speakers will be located between the two.

A B C

The line segment ABC represents such a continuum, although in real life there is not an infinite number of speakers corresponding to the number of points on the line. Any speaker not situated at B, the midpoint, would lie either in AB or BC, making his dominant language SJE or JE, respectively.

At least 60 per cent of the population lies in BC. Moreover, Frederick Cassidy estimates that JE is spoken 80 per cent of the time, because many persons of higher social standing use it in speaking to maids, workmen, office messengers, etc. (Alatis, 1970, p. 204). Many of the speakers who lie between A and B can produce forms between B and C, but the reverse is not usually true. Speakers in the positions between A and B might even be called bilingual. Sometimes it is hard to make a clear cut distinction between the two, as many people switch linguistic forms at will.

JE is sufficiently like SJE for many speakers to be able to decode more or less accurately. Most persons, for example, can get the general outline of the radio news, although sometimes the details might be misunderstood. However, they cannot recode or reproduce the language again.

In Jamaica, it is fair to say that there is a correlation between good education and acceptable English which makes it possible to assume that ability to manipulate SJE is indicative of a good education, in addition of course to birth in a higher caste or class (Bailey, 1966, p. 106). Because of this, many children, even those from JE speaking homes, are frequently urged to "speak well". The results as seen by one educator, Dennis Craig, are as follows (see Humus, 1971, p. 371):

i. A speaker aiming to produce a significantly less familiar, but socially required, system of speach usually produces a system intermediate between the two: what Craig calls the "interaction area". It is in this area, which does not represent a discrete stable speech norm of its own, that many Jamaican children find themselves.

ii. The children learn to switch linguistic forms at an early age.

iii. SJE is neither a native nor a foreign language. Therefore, the children in general are able to recognise SJE all the time but not to produce it. (Black English speakers in the USA may also have this difficulty.)

iv. For this reason, many learners of SJE feel that they "know" SJE, and this reduces their motivation to modify their language and to perceive contrasts between their own and the standard language.

Although television and radio have some impact on the language situation in Jamaica, the language used for broadcasting varies greatly; American and British Standard English for "canned" programmes, the announcer's version of American disc jockey English for programmes of this sort, SJE, and JE. Television and radio, therefore, have not helped much in stabilising the speech norms of the community. When children from homes where only JE is spoken enter school, they find the task of learning to read and write in SJE very difficult indeed - - so difficult that many never really learn to read and write fluently, nor to understand the correct use of SJE. And what makes the task of becoming functionally literate harder is that many teachers themselves do not speak SJE, "and so the Creole goes merrily on" (F. Cassidy, in Alatis, 1970, p. 207).

2. Pre-Primary and Primary Education in Jamaica.

Pre-primary and primary education in Jamaica are divided more or less along class lines. When they are four, middle-class children go to preparatory schools where they remain until they take the 11-plus examination, which is supposed to select children for academic, technical, or modern education. On an average, they read when they are about six years old. Preparatory schools are usually well-organized. Classes tend to be small. At one school, for example, although there are between thirty and thirty-five children in a class, every teacher has an assistant in the room, so that the students can get more individual attention. "Prep" schools, as they are called, will have libraries and sometimes quite an extended curriculum, with perhaps music, dancing, and a foreign language. In general, the standard of teaching is satisfactory, although since their aim is to get children through the 11-plus examination, rather too much emphasis is placed, in some schools at least, on rote learning. Fees for these schools are usually high.

At four, many, though not all, working-class children, go to basic schools, which vary widely in equipment and staff. In the main, however, equipment is poor, staff untrained, and the schools small. At seven, the children go to primary schools, where in crowded areas of Kingston a class might have ninety or a hundred students, although between forty-five and fifty is normal. Again, the standard of teaching varies widely, but it is fair to say that, in general, it is below that of the preparatory schools, which are better equipped anyway. It is in primary schools that poorer children learn to read and write, and they take the 11-plus examinations from these schools.

3. Some Symptoms of the Situation

i. Each year, all children who have not yet entered secondary school and who are 11 or 12 are supposed to take the 11-plus examination. Many poorer children from primary schools are not entered for the exam because they are adjudged by the teachers as being below standard. Usually, this is because they cannot read, comprehend, or write SJE sufficiently well. Middle-class children who have no language difficulty nearly all take the examination, and between 80 and 90% are successful. This means that as the number of spaces offered in grammar and technical schools is limited, many poorer children are excluded from this type of education almost solely on the basis of being unable to communicate effectively in a second language or dialect.

ii. Many poorer children learn to hate school early. Not surprisingly, since not much is expected of them, they do not perform well. They themselves, on the whole, do not have high expectations for themselves, and most accept it as natural that they will not succeed.

iii. Others become alienated from the mainstream of society because they know that apart from, say, the world of entertaiment, most legal routes to success are forever closed to them.

iv. The pass rates in English and mathematics in public exams are abysmal: usually between 30 and 40% of the candidates. In mathematics, perhaps more than in any other discipline, clear thinking and the tools for clear thinking are important. JE does not supply enough abstraction. Moreover, improvement in mathematics relies on the student understanding all previous steps, and many speakers of JE do not comprehend properly all the first steps and so miss succeeding ones.

4. Family Life in Jamaica for the Working Class

The traditional pattern of family life for working-class Jamaicans does not aid in solving educational problems in the country. Although there are many urban and rural working-class families in which the parents are married, this is not always the case - in marked contrast to the middle-class, whose norms are those of Western society. A law passed recently has has removed many of the disabilities of illegitimacy, at least as far as the inheritance of property is concerned, but the law has not been able to do anything about the quality of life enjoyed by many urban working-class children, particularly those of Western Kingston and the Wareika suburbs. A number of children are deprived of adult male company and often, too, of a significant amount of adult female company if the mother works. Too frequently, there is no intellectual challenge in the home, which is simply a place - - over-crowded - - to sleep.

For rural working-class children, family life is generally more stable. Sometimes the children live with the parents, or occasionally with the grandmother, for it is common for working-class mothers to send their children to their own mothers. Grandparents who are caring for six or more children of varying sizes are frequently encountered, and it is often difficult for these old people to control the children.

Although, on the whole, parents and guardians tend to be ambitious for their children, paradoxically many do not take school seriously. One reason for this may be than, given their own experiences, it is difficult for the older generation to identify with the school. Also, the children are made to help on Fridays, the traditional market day, so the four-day school week is common, even in urban areas, where many mothers sell in the markets.

Thus, to enter a school whose norms and modes tend to have a middle-class slant is a cultural shock in some ways for some students.

5. Educational Problems

The two main problems encountered in the classroom are imperfect communication and the psychological problem that users of the unofficial language often feel inferior. Let q signify that the teacher speaks SJE, and let r signify that all students speak SJE. There are four possible situations: (1) q r, (2) qr r, (3) q r, and (4) q r.

Problems of communication will not arise in cases 1 and 4. However, in case 2 the child tends to feel confused and inferior, especially if his speech is corrected in a derogatory way, and he can find it difficult if not impossible to learn. Case 3 can be very interesting from an educational viewpoint. This time it can be the teacher who feels inferior, with resultant loss of authority over his class. Sometimes children refuse to learn from a teacher "who can't even speak English."

Now consider the four cases with the additional condition that written communication (the texts or the class's written work) is in Standard English. Then case 1 becomes ideal, and good oral and written communication is possible. In case 2, the child will find such communication difficult, and in case 3, the teacher will find such communication difficult - - and may lose authority. In case 4, both the teacher and the class will find good oral and written communication difficult. If, on the other hand, the written communication were in JE (a condition not known in Jamaica), case 4 would be ideal for many Jamaican children and teachers.

Obviously, the situations outlined above are complex. Real life, in fact, is even more complex, when one remembers that the norm in Jamaican society is widely varying speech patterns, with members of a class at various points on the speech continuum.

6. Special Problems in the Teaching of Mathematics

The problems listed above occur when any discipline is taught. Other problems arise when mathematics is taught:

1. Some mistakes come about because SJE and JE have different plural forms. Since SJE does not contain the pluralizer "dem" and JE sometimes uses SJE plural forms randomly, it is sometimes difficult for JE speaker to know when a plural is intended, especially in oral work.

2. SJE speaking students come from homes where reading is normal. Many words used in mathematics may be familiar to them, so that SJE users have a link to help their learning processes. Two examples are "acute" and "pivot".

3. Generally, the language of textbooks approximates to SJE. It is hard for many speakers of JE to understand texts properly. They have to battle with language as well as to learn subject matter.

Of course, the higher up the educational ladder one goes, the less noticeable these problems become. This is because so many poorer children do not make it to high school, much less to technical college or university. Even these institutions, however, have to run courses for freshmen in the use of English.

The fact is that the problem of language is an all-pervasive one. It is not that the student does not understand specifically mathematical terms such as "less than" and "line segment". He may understand these very well but still not understand the general meaning of the sentence in which they occur. Thus, students have a negative approach to problem solving, and they will try to do purely mechanical questions on any paper. This must be related to the fact that as children they have not been allowed officially to use their own language in the classroom, and learning mathematics in what is to them a second language (SJE) has made it impossible for them to learn the language of mathematics. It also makes it hard for them to link any given problem with one which they have already encountered, as meanings are seldom exact. Effectively, this makes past experience in mathematics so much dead wood that can only be junked.

To my mind, this explains something that is peculiarly Jamaican: many mathematics students hate graphs. Assigning values arbitrarily to variables needs imagination and the ability to think clearly, and this faculty is denied them because of the language barrier.

Many speakers of JE start off knowing more about areas and volumes than do speakers of SJE. They will know very early in life what a square of land is or what a "gill" of peas is because they have had much practical experience in these matters. The traditional approach to the teaching of mathematics ignores this. The child seldom has the chance to feel that the knowledge of life that he brings with him to the class is worthwhile.

So the fact that Jamaica has a second language has different implications from those in countries, e.g. Peru, where the second language is completely different from the official language. Nevertheless, the problems are real and positive steps need to be taken if they are to be overcome.

7. Recommendations

i. Children starting school should be given an early linguistic test to see which half of the continuum they are in.

ii. Those whose language is predominantly JE should have this as their medium of instruction, with SJE as a foreign language.

iii. Every effort should be made to incorporate into the curriculum the mathematical concepts already possessed by the child when he enters school.

REFERENCES

Alatis, J.R. (Ed.). Report of the Twentieth Annual Round Table Meeting on Linguistics and Language. Washington, Georgetown University Press, 1970.

Bailey, Beryl. Jamaican Creole Syntax. Cambridge University Press, 1966.

Cassidy, Frederick. Jamaica Talk: Three Hundred Years of the English Language in Jamaica. London, MacMillan, 1964.

Humes, D. (Ed.). Pidginization and Creolization of Languages. Cambridge Universty Press, 1971.

CHAPTER 17 - Objectives

17.1 TEACHING FOR COMBINED PROCESS AND CONTENT OBJECTIVES

TEACHING FOR COMBINED PROCESS AND CONTENT OBJECTIVES

A.W. Bell
Shell Centre for Mathematical Education
University of Nottingham
Nottingham, England

THE NATURE OF MATHEMATICS

I have been concerned ever since I began to teach mathematics that a great deal of what is commonly taught and what is offered in textbooks provides a comparatively meaningless form of activity for pupils. They learn to perform certain computational processes, they perform algebraic manipulations, but those activities which would show them the purpose of mathematics are scarce.

Applied and Pure Mathematics

For me, mathematics has two main purposes. First, it is a means of gaining insight into some aspects of our environment. It may enable us to understand more about how populations grow, how many different ways we can turn the mattress on our beds, how it is possible to have a decrease in the rate of inflation, or how it is possible that the probability of getting a shattered windscreen on your car is the same the day after you've had it shattered and repaired as it was on the day before.

Those examples represent the applied side of mathematics, but we need to remember, too, that there are quite a number of people who get a lot of interest and pleasure out of solving number problems, puzzles, the kind of thing you find in magazines and newspapers, and we may take that as indicating that the possibility of enjoying playing with number puzzles and solving number problems is a capacity which a great many people have, and which we should develop.

Relationships, Concepts and Symbols

Gattegno expresses the nature of mathematics in this way:

> "To do mathematics is to adopt a particular attitude of mind in which relationships in themselves are of interest. One can be considered a mathematician when one can isolate relationships from real and complex situations . . . Teaching mathematics means helping one's pupils to become aware of their relational thought."

I think if you see mathematics as broadly as that it explains both its far-reaching usefulness and also its intrinsic difficulty. It is useful because it is abstract and applies in many different situations, and it is difficult because its power depends on being able to manipulate these relationships from their concrete embodiments. The crux of the mathematical act is the ability to manipulate symbol systems as if they were simply marks on paper to be shifted around spatially, and yet at the same time to retain the awareness of the concepts which those symbols denote.

Content and Process

As well as thinking of pure and applied mathematics, that is, looking at pattern and exploring the environment, I divide mathematics into content and process. Content represents particular ideas and skills, like rectangles, highest common factor, solution of equations. On the other side there is the mathematical process, or mathematical activity: that deserves its own syllabus to go alongside a syllabus of mathematical ideas; it includes the processes of abstraction, representation, generalisation, and proof.

EXAMPLES OF MATERIAL COMBINING PROCESS AND CONTENT OBJECTIVES

The title of this session is Teaching for Combined Content and Process Objectives, and this is the approach of the course to which I want to refer most. It is published as Journey into Maths and is a course for pupils of all abilities, aged 11-13 years; it has been developed in two Nottinghamshire comprehensive schools. This course includes process-oriented work in two ways: first, many of its units are based on the investigation of a problem involving concrete material, which develops in two directions: (a) the key mathematical concepts and skills are picked out for further development and practice, and (b) extensions and generalisations of the basic problem are suggested for investigation. Secondly, two out of the fifteen units of work in each year's course are labelled Open Investigations, and, in these, problems are suggested which are within the general field of ideas being learnt, but which have proved particularly suitable for allowing pupils to make generalisations and extensions of their own, with plenty of choice about which aspects to pursue. These units are included so that, in them, the teacher is free from anxiety about whether the pupils are mastering any particular content - there are no specific content objectives for these units.

An Open Investigation

In giving some examples I want to begin with an Open Investigation, called "The Remainder Problem". This was done by a 13-year-old girl, a pupil of a colleague of mine, Mr. Eric Love, at Wyndham School in Cumbria. The problem is:

> "A boy counts his sweets in fours, and has two left over; when he counts them in fives, he has one left over. How many sweets did he have?

This pupil's approach to the problem was as follows:

> "I thought about this problem and wrote it out like this: 4r2, 5r1." (An interesting little bit of symbolism.) "I got the answers 6, 26, 46, 66, 86 and so on."

Already this girl has generalised the problem from what was initially given to her. The answer 6 could have been the end of the story, but she looked for other solutions and has found a whole set, and has discovered the relationship connecting the differences in this sequence with the divisors, 4 and 5, given in the problem. Next she generalises the problem, considering (3r2,4r2), and later, (4r2, 6r4), which lead her to grapple with, but not quite grasp, the concepts of HCF and LCM.

> "I ended up with, if the numbers have no connection, they go up in multiples of the two numbers, and if one number is a multiple of the second the numbers go up in the largest number."

After a further generalisation, to problems with three conditions, she takes up the question of how to find the starting number. She is unable to find an explicit solution, but gives an effective algorithm, based on constructing a table.

That is an example of an Open Investigation of which the main interest is in the processes of generalisation, abstraction, representation, and proof. But it also offers an opportunity for teaching the concepts of GCD and LCM and their properties, in a situation were the need for them can be clearly seen.

An Applied Investigation

My second example is of an Applied Investigation, which practises the techiques of calculus and also requires mathematical modelling and the ability to see how to apply calculus to a practical situation. The problem is why circular filter papers are folded into a cone in the way they are, for use in filter funnels (ATM 1978). This 17-year-old boy asked the question, "Is this the most efficient way of folding a filter paper?" And what could you mean by "efficient"? You might mean, do you get the maximum amount of liquid in the paper? So he tried to investigte whether the sort of cone that is formed in this way was a maximum volume cone. He discovered that it wasn't; as his work shows, a different way of folding will give 80% more volume.

The first difficult part of that investigation was putting the variables in; you know the formula is $1/3 \quad r^2h$, but you have to decide what is varying and what is constant, and which variable you have to differentiate with respect to, and that is where the problems arise. The other region where there were problems was in working out, when the fold is made in the filter paper, what are the radius and height of the resulting cone, in terms of the radius of the original circle of paper. These are problems of modelling, problems of introducing variables and describing what are variables and what constants. It is an example of how a piece of mathematics can answer an environmental problem.

Moving Furniture

We turn now to a piece of work which starts from an investigation with a practical origin, but which is set clearly within the content area of transformation geometry, of rotations and their combinations. Entitled Table Moves, it first poses the problem shown in the diagram.

You will need spotty paper and tracing paper.

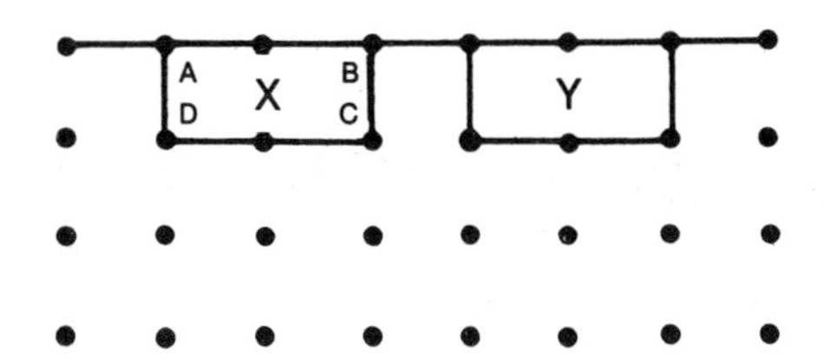

Can you move the table from X to Y using only quarter turns about corners?

Describe your moves.

Would your method work if the rectangle represented a piano instead of a table?

Answers: Yes: bc or $b\bar{a}\bar{c}\bar{a}$ or $c\bar{d}\bar{c}\bar{a}$ (see below)
No: Keys would be against the wall.

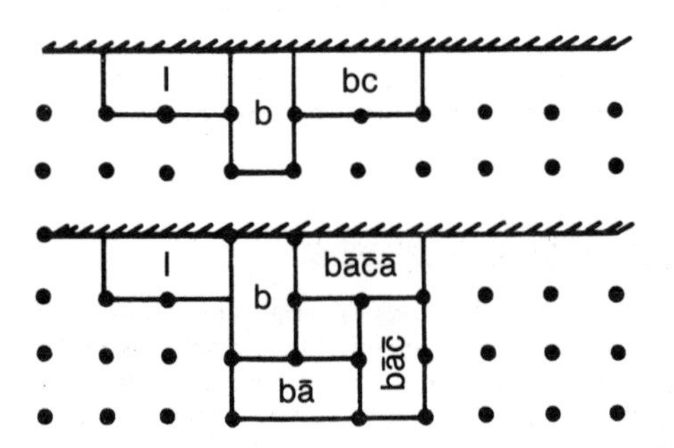

After a few specific questions to acquire familiarity with the situation, two kinds of general question can be posed, one more strongly related to general mathematical processes, the other concerning the systematisation of content. The first question could be, "Find rules for obtaining the shortest move to any given final position." This requires a combination of algebra and geometry. For example, one finds that $ac = ca$, and, in fact $xy = yx$ always, which can be proved either by checking cases and using symmetry arguments, or geometrically. One also finds that $\bar{a}b$ is a translation of $2\sqrt{2}$ in a SW direction, but $\bar{a}d$ has magnitude $\sqrt{2}$ only. Such results enable a shortest-route alogrithm to be developed. The second type of general question is, "What kinds of transformation result from combining quarter turns?" This can easily lead to the general theorems about the composites of rotations. Other material of interest in developing process aims is published by the Leapfrogs group.

EXAMINATIONS

These methods can be developed by a knowledgeable teacher in any school setting, but if teachers in general are to be encouraged to use them, it is helpful if the examination system gives full credit for attainments in process abilities, as in content. This is rare in England, but there are some schemes along these lines. The procedure adopted by the school best known to me, in which this work has been taken farthest, is that a scheme has been accepted by the appropriate examining board by which there is a written paper testing the knowledge of the content material, the particular mathematical ideas and skills, and alongside this there is the assessment of a number of pieces of individual work of the kind I have discussed here. The scheme states the list of ideas, such as decimals, fractions, percentage, algebraic formulae, matrices, graphs, geometric transformations, and so on, but also mentions the following process objectives: to be able to mathematise a situation, symbolise, quantify, and analyse it; to select and apply mathematical techiques and strategies to unfamiliar situations; to be able to argue inductively, make conjectures, verify, and generalise; to provide proof, primitive justification, counter examples, deductive argument; and to be able to formulate questions and communicate results. These are the objectives which have to be assessed from the pieces of written work submitted. The grading of these pieces is done with the aid of a scheme where there are about six headings relating to different aspects, such as the depth of mathematical ideas used, the rigour of the argument, and the clarity of the presentation; the assessment scheme describes criteria for the award of grades under each of these headings. Then a combined grade is obtained by an averaging process.

Details of these scheme are contained in ATM (1978). The British national survey of mathematical attainment, being conducted by the Assessment of Performance Unit of the Government's Department of Education and Science, also includes an assessment of mathematical activity alongside its evaluation of particular items of mathematical knowledge. The scheme for this is in process of development; it includes some use of short items and some of items which are presented and evaluated in an interview by a visiting tester.

The question is sometimes asked, is such work suitable for all levels of ability or only for the best? While it is true that the abler pupils can go very much farther along this road than the rest, the fact is that there are no pupils for whom it is right simply to teach techniques and ideas out of relation to each other and out of a context of application and use. At the school which I have just mentioned, the examination for the middle and somewhat lower range of ability of pupil also includes this assessment of extended pieces of work, but these tend to be of a more practical nature. For example, I saw one piece which involved an examination of the school registers to find out which pupils from which classes and of which sex were most prone to absenteeism. Another asked, "Which is the cheapest supermarket to shop in?" and drew up a typical week's set of groceries, compared prices in a number of shops, and drew certain conclusions. There are some similarities here with the attempts of the USMES scheme in America to develop general problem-solving abilities, and I am interested to see that the same problems of evaluation have been met by that scheme as have been met by myself and my colleagues in England. That is, we have all found that some compromise between short items testing component process skills, medium-length items which allow a measure of investigataion by the pupil but which can be standardised in that the same problems are set to all, and finally some mode of observational assessment of the actual work being done by the pupils are all necessary to produce an adequate evaluation.

EVALUATION

Some test material has been developed by our South Nottinghamshire Project for the evaluation of process attainments. The full tests are in Horton (1979). These tests have been used to evaluate the progress of pupils in the Project schools over periods of a year at a time, and to compare them with pupils in other schools. The most obvious superiority of the Project pupils was in their ability to give reasoned verbal explanations of the results obtained in the test items; other evaluations showed that the time spent by Project pupils in developing these process abilities had not resulted in any significant loss in their attainment with regard to knowledge of mathematical content.

TEACHER DEVELOPMENT

Teaching for process involves a broader range of teaching skills. Some approaches to helping teachers to develop these skills have been explored by my colleague Eric Love, who has contributed the following brief account of this work during the year 1979-80.

Introduction and Orientation

My work this year has involved working with teachers who are keen to improve their awareness of what they were doing, extending the range of objectives they were teaching for, and especially to implement teaching for the 'process' aspects.

Early discussions identified a range of issues that concerned the teacher:

> How does the teacher know when they are teaching for the process aims?
>
> What kinds of influences affect the teaching towards the process aims?
>
> How does the teacher know when students have acquired, or extended their notions of process aspects?
>
> What meaning do the activities have for the children?

I worked in the classrooms with the teachers, with one of their classes (the children were aged 11-13). Monitoring what went on was by means of tape recordings and discussions.

Promoting Teacher Awareness - I

An initial attempt was made to alter the materials used so that the teachers would be able to compare and contrast the original version and the new one.

One of these attempts concerned the Sequences and Functions Unit. Both versions use simple materials to get the children to make patterns. Some aspect of the pattern is then counted, to give a function table. The children then have to produce a formula.

In the original version the children copy and extend patterns presented by the teacher. In the new version, the children produce their own patterns and decide for themselves which aspects to count.

In some respects this revision was successful, but it threw up rather deeper issues:

> The crucial role of the teacher in interpreting the unit.

The tendency for content aims to take precedence over progress aims.

The tendency for teachers to attempt to teach directly for the process aims, the work becoming instruction in techiques.

The effect on the confidence of the teacher in trying to teach material devised 'outside'.

Promoting Teacher Awareness - 2

In the light of the issues raised above, a new orientation was given to the direction of the work.

a. Objectives I had come to feel that the very notion of objectives had become a stumbling block. We had a major aim which was to enable children to think for themselves in mathematics. What seemed to be happening was that the specific objectives pursued by the teachers shaped the directions for the children and prevented them from thinking for themselves. Instead of having the objectives as intended outcomes we attempted to specify principles of procedure for the way in which the lessons would be conducted. Thus a form of process rather that the outcomes of process was being specified.

   These principles were that children needed to be allowed to:

   1. identify and initiate their own problems for investigation.
   2. express their own ideas and develop them in solving problems;
   3. test their ideas and hypotheses against relevant evidence;
   4. rationally defend their own ideas and conclusions and submit the ideas of others to reasoned criticism.

b. Self-monitoring by the teachers How to increase the awareness of the teachers' of their own actions is crucial, for it seems that it is the teacher's ability to respond to ideas produced by the children, and in their sensitivity to the effect of things said to the children, that much of the difficulty lies. What was used here was a device produced by the Ford Teaching Project which requires the teacher to, say, refrain from interrupting a child, or to refrain from rephrasing the children's utterances in his/her own words. In attempting to put these into practice the teacher is forced into a greater awareness of the ways in which they are following, or acting against, the principles of procedure.

References

ATM (1978): Maths for Sixth Formers. Association of Teachers of Mathematics, Nelson, Lancs.

Bell, A.W., Wigley, A.R. and Rooke, D.J. (1978): Journey into Maths, The South Nottinghamshire Project. Teachers Guides 1 and 2, Number Skills 1 and 2, Pupils worksheets Packs 1 - 10. Blackie, Glasgow.

Horton, B. (1979): Tests of Mathematical Process Shell Centre for Mathematical Education, University of Nottingham.

Leapfrogs group: Leaflet from Coldharbour, Newton St. Cyres, Exeter.

THE WORK OF LAKATOS AND GATTEGNO: IMPLICATIONS FOR MATHEMATICS TEACHING

A.J. Dawson
Simon Fraser University
Vancouver, Canada

I began a doctoral program in mathematics education wanting to look at the whole process by which mathematics is generated, hoping a study of process(es) would provide some insights as to how mathematics might be presented to children with more success than had previously been the case. I did realize, however, that how children learn mathematics - - how they teach themselves mathematics - - may not be the same manner in which the mathematics was initially created. Nonetheless, the two processes may have some points of contact which would be fruitful for further investigation. The question I then had to face was: Where do I begin?

I was introduced to the writings of Sir Karl Popper, the British philosopher of science. I wrote to him asking, if anyone had applied his work to the field of mathematics. He responded with the name of Imre Lakatos (8). When I read Lakatos' work I knew I had something of immense value.

I went on to complete my doctorate producing the usual dissertation 'tomb' in which I dealt with Popper, Polya, and Lakatos and I threw in a bit of the Madison Project and the work of Robert Davis (2).

I felt I had gained a more comprehensive view of what one does when generating mathematics, but I did not feel any closer to understanding how children learn mathematics, or if there were connections between those activities. In the late 60's or early 70's no one seemed very interested in Lakatos' work, and as a result I didn't pursue his work any further, though it did effect the way I taught mathematics with prospective teachers.

In 1969, I moved to Simon Fraser University where I was immediately introduced to the work of John Trivett, and, through him, to Gattegno. I found many insights which I have subsequently found to be most fruitful in understanding how children learn mathematics.

Consequently, I come before you today as one who was thoroughly acquainted with Lakatos' ideas some ten years before they began to have an impact on the mathematics education community, and as one who has spent the better part of the last ten years trying to understand what Gattegno has been saying to mathemtics educators for over three decades (5a, 5b, 5c, 5d, 5e).

What I would like to examine with you is the relationships between what these two men have been saying to us, and to explore some of the implications their points of view have for the teaching of mathematics. I have not explored these relationships and connections in great detail, so I will invite you to join with me in beginning to do just that. Two articles particularly relevant to this are: Gattegno's article, "The Foundations of Geometry", and Agassi's article, "On Mathematics Education: The Lakotosian Revolution", from the Learning of Mathematics; which is edited by David Wheeler.

It is perhaps not surprising that Lakatos and Gattegno have similar views as to what has been (and still is, I fear) wrong with the teaching/learning process in mathematics. Back in the mid-sixties, Lakatos was saying - - and remember this was at the height of the so-called new math era (8, 142):

> Euclidean methodology has developed a certain obligatory style of presentation. I shall refer to this 'deductivist style'. This style starts with a painstakingly stated list of axioms, lemmas, and/or definitions. The axioms and definitions frequently look artificial and mystifyingly complicated. One is never told how these complications arose. There are carefully worded theorems. These are loaded with heavy-going conditions; it seems impossible that anyone should have ever guessed them. The theorem is followed by the proof.
>
> The student of mathematics is obliged, according to the Euclidean ritual, to attend this conjuring act without asking questions either about the background orabout how this sleight-of-hand is performed. If the student by chance discovers that some of the unseemly definitions are proof-generated, if he simply wonders how these definitions, lemmas and the theorem can possibly precede the proof, the conjuror will ostracise him for his display of mathematical immaturity.
>
> In deductivist style, all propositions are true and all inferences valid. Mathematics is presented as an ever-increasing set of eternal, immutable truths. Counter examples, refutations, criticisms cannot possibly enter. An authoritarian air is secured for the subject by beginning with disguised monster-barring and proof-generated definitions and with the fully-fledged theorem, and by suppressing the primitive conjecture, the refutations, and the criticism of the proof. Deductivist style hides the struggle, hides the adventure. The whole story vanishes, the successive tentative formulations of the theorem in the course of the proof-procedure are doomed to oblivion while the end result is exalted into sacred infallibility.

Agassi (1, 28) sums up Lakatos' thrust when he states:

> What Lakatos has succeeded in showing is that all mathematical presentations in standard mathematics text books are ill-conceived as to their purpose, and that mathematicians, even some of the greatest, are surprisingly vague about the rationale of mathematics, and about the aims and methods of mathematical proofs in particular.

Gattegno says things slightly differently, and implies something a bit more as far as how children might learn mathematics (4, 10). He states:

> There is a basis for rejecting the ancient geometry which I would call, from my point of view, the stultifying approach: having units of knowledge given, one after the other, and the student memorising a definition, a theorem, a proof, without really having a grasp of what the problem is. Now what mathematicians do with geometry . . . makes the difference between them and their students, is that they use a dynamic approach to the many notions, and they put things together in wholes that enlighten, that light up, the various bits. Therefore, they can, when they speak of one particular thing, put in all their experience which contains a great deal of material which doesn't appear at all to the outsider.

What is implied by both Lakatos and Gattegno is a different view of what mathematics is. Agassi writes of Lakatos' view in this manner (1, 29):

> Popper said of science, and Lakatos said of mathematics, that each is full of errors to be corrected and hence is neither nature, i.e., not only demonstrable final truths, nor convention, i.e., not truths arbitrarily nailed down and defended against criticism.

No final demonstrable truths, eh! If that is true, and if mathematics educators know it, then why is it that so many children view mathematics as if it were written in stone?

Here is what Gattegno has to say about the nature of mathematics (4, 15).

> We have come a long way from the time, round 1900, when it was thought that we could put the whole of mathematics together as a system that would be deductive. It has been recognized that this is a futile activity. What we know is injection of some definitions which are fruitful; and this is what we want to put into circulation with the teachers. An axiomatic system is only for the presentation of what I have found. I don't begin with it; I end withit. So I'll give experiences, I'll give lots of experiences to the people, and then come and say, "Let's build it in such a way that it is the least costly in terms of words and propositions and so on; so that if you know this, you know all the rest." Why? Because the mind functions; and the principle of "if you know a little you know a lot" is pedagogically sound everywhere and we can use it in any foundation of mathematics.

So the axiomatics are the end product, not the beginning! Moreover, whether or not we have the most minimal set of axioms is not important, at least not in working with children.

That is fine, but what are these experiences of wich Gattegno speaks? What is it that we are to have children experience which will hopefully provide the seeds to grow magical bean-stalks of mathematical know-how for children? What powers, capabilities, do children have which we know they can use, and which could be used to help them learn mathematics?

It is these latter questions which Lakatos' work does not address, but which Gattegno's does. Nonetheless, the answers Gattegno provides do imply a conjecture and refutation strategy being utilized by children. One point of connection, then, between Lakatos and Gattegno is that children seem to use Lakatosian type strategies when they exercise the powers of mind they all possess.

A second point of connection between Lakatos and Gattegno is their emphasis on conjecturing (to use Lakatos' term) or on intuition or insight in Gattegno's terms. Indeed, if the last few centuries may be characterized we are now entering as the age of intuition.

Let's examine these two points of connection Lakatos has proposed a proof and refutation strategy as being fundamental to the generation of mathematics. Gattegno has argued that the powers of mind possessed by us all are indispensable tools for the learning of mathematics. Both men suggest that insightful guessing - - intuition - - is part and parcel not only of the generation of mathematics, but also in the learning of mathematics. Moreover, Gattegno has suggested that children possess powers of the mind which we as teachers of children can use in helping them learn mathematics. And it is they, the children, who must do the learning - - THEY HAVE TO EDUCATE THEMSELVES. No one else can do it for them. The criticisms of how mathematics has been taught in the past which I read the teacher and the content as being paramount in the teaching/learning process. Lakatos and Gattegno are both suggesting that this is not as it should be. The teacher and the content are not paramount: THE LEARNER IS.

Let me elaborate on what Gattegno means by the powers of mind possessed by children. I don't have time to go into even a small portion of what Gattegno has written, so I will focus on just one power of mind, that power being the capability of making images in the mind, of imagination. Is there anyone here who doubts for a moment that children - - and each of you - - have fantastic powers of imaging?

What Gattegno has argued is that to tap children's minds to do mathematics we have to ensure that things like points, planes, and lines in geometry, for example, ". . . exist in the minds of our students, at the level of intuition, of imagery". (4, 11) We need " . . . to give them (our students) the dynamics, and them make them talk" (4, 11) about what they saw. Then, their " . . . words will be triggered by the images, and later the images will be triggered by the words." (4, 11)

First, we must have the images, the children's images, then we must have the children talk about their images, and only then should we attach words - - the conventional mathematical terminology, if you will - - to those images. Then, perhaps, we can formalize it all so that in the final product, the words trigger the images. What is being done is that the meaning should precede the words, the experiences precede the formalization. If this is done, then later the words, the formalizations, will activate the mind to re-collect the experiences and the meaning.

When one says that the teaching of mathematics should be meaning-full, I agree, but the meanings must be those of the children, and these meanings need not necessarily be based on real life situations, so-called. The images of the mind are also real, and they are often richer than real life. Meanings can be constructed from such images.

The initial images can be generated by perception, by looking at one of Gattegno's geometry films (7), for example, but the imaging can from there where new images are created by the self in one's own mind.

That is what I would invite you to do now, i.e., to create some images in your minds in response to some words I shall say to you. So sit back and relax; close your eyes if you wish.

> Just as we can run a finger along the edge of a coin and come back to where it started, we can conceive that a point on the circumference, of a circle could be moved around, passing through all the points of the circumference before returning to its original position.
>
> I invite you (and the readers of this article) to imagine this "running" point describing the circumference:
>
> a. at various speeds,
>
> b. alone, or with the ray which emanates from the center of the circle and passes through the point,
>
> c. with the same ray, but not showing the point . . . (6)

(Note: At this point in the presentation, members of the audience were asked to share what they had created in their minds. No records were kept of their responses. A.J.D.)

> Now, begin again but this time have the turning ray leave behind two tracks which we choose to contain two perpendicular diameters of the circle, one vertical and one horizontal.

(Again, audience response was solicited and discussed at the point, but no records were kept.)

> Finally, as the point slowly circumscribes the circle, have it, like a bird flying overhead, drop a trajectory which in one instance forms a perpendicular with the horizon axis, and in the second instance, a perpendicular with the vertical axis.

I shall leave you with the task of determining what mathematics might be generated from these images.

If we call the vertical trajectory George and the horizontal trajectory Mary, you might wish to investigate what happens to George and Mary as the point (pi, pj, pk . . .) moves around the circle.

In closing, I would like to share with you a description of mathematics given to me by Jim McDowell. A group of people were talking about a field trip, and Jim offered "Mathematics is a field trip of the mind". I like that because it draws attention to the fact that is the mind and its powers which we must tap if we are to learn mathematics, and the "we" I am talking about includes children and adults alike.

References

1. Joseph Agassi, "On mathematics education: the Lakatosian revolution", For the learning of

mathematics, Vol. 1, No. 1, Montreal, 1980. pp. 27-31.

2. A.J. (Sandy) Dawson. The Implication of the Work of Popper, Polya, and Lakatos for a Model of Inquiry in Mathematics. Unpublished doctoral dissertation, University of Alberta, 1969.

3. The journal, For the learning of mathematics is available by writing to the editor, David Wheeler, at the Department of Mathematics, Concordia University, 7141 Sherbrooke Street West, Montreal, Quebec, Canada, W4B 1R6.

4. Caleb Gattegno, "The foundations of geometry," For the learning of Mathematics, Vol. 1, No. 1, Montreal, 1980. pp. 10-15.

5. Caleb Gattegno has written many books most of which are available from Educational Solutions, Inc., 80 Fifth Avenue, New York, New York 10011. Some which may be of most interest to mathematics educators are:
   5a) The Universe of Babies
   5b) The Mind Teaches the Brain
   5c) Evolution and Memory
   5d) The Common Sense of Teaching Mathematics
   5e) What We Owe Children

6. This activity was suggested by an item in the Newsletter published by Educational Solutions, Inc., "Mathematics: visible and tangible", Vol. IX, No. 3, February, 1980, New York.

7. The geometry films developed by Gattegno are also available from Educational Solutions, Inc. They are computer animated films, relatively short in duration, but rich in their potential for helping learners to "image" in mathematics.

8. Imre Lakatos, Proofs and Refutations, Cambridge University Press, New York, 1976.

TRADITIONAL GEOMETRY AS A VEHICLE FOR TEACHING MATHEMATICAL PROCESSES

P.G. Human
University of Stellenbosch
South Africa

This paper deals with the philosophy behind the development of an alternative Geometry curriculum for South African schools. Some typical teaching-learning activities contained in the present version of the alternative curriculum are also described. The alternative curriculum is oriented towards attainment of a variety of mathematical process skills, in contrast to the traditional curriculum which mainly features geometric knowledge and deductive proof. In the traditional curriculum formal Geometry is presented as a ready-made system, i.e. the teacher supplies the definitions, axioms, theorems and proofs in a suitable order. Our present efforts are based on the assumption that for most pupils the present Geometry curriculum is of limited value, mainly because of limited opportunities for mathematical activity by pupils (excepting deductive proof for which ample opportunity is provided in the senior secondary phase) and because the material (especially the formal system) lacks meaning for the pupils.

For the purposes of the alternative curriculum geometry is viewed as a vehicle for teaching appreciation of and competence in various mathematical processes, specifically abstraction, classification, generalization of concepts and results, defining, formulation of mathematical ideas, proving, axiomatisation and mathematical reading ability. The teaching approach is characterized by active pupil participation in the production and formulation of classification schemes, hypotheses, definitions, sets of axioms and proofs.

Teaching of Classification

One activity is based on a set of 20 different quad-shaped pieces of stiff paper. Each pupil receives such a set at the beginning of the first lesson, with the instruction to sort the pieces into different sets and to motivate the classification in writing. When all pupils have completed their classifications, groups of 5 - 6 pupils compare their work and produce a documented group effort. In a second assignment pupils (first individually and then in groups) have to identify as many different properties of each type of quadrilateral as possible. Finally, individuals and then groups have to draw Venn-diagrams of the quadrilaterals. During a class discussion the teacher compares the work of the different groups, irons out differences and suggests a common classification. At this stage the teacher also supplies the standard names for the different types of quadrilaterals. During trials it became clear that the approach induced situations in which proof and precise definitions may be meaningful to pupils. This mainly happened when pupils worked in groups.

Teaching of Proving and Meanings of Proof

Proof is initially only applied to what we hope would be somewhat surprising results to the pupils, the required "intuitively obvious" results being accepted without further ado. An attempt is made to let pupils experience proof as:

> a way of convincing people of propositions that they do not readily believe (we believe that this meaning of proof is pretty close to the notion of proof pupils already have at this stage);
>
> a way of accounting for surprising results;
>
> a way of checking the validity of hypotheses.

In order to achieve experience of these meanings by the pupils, different types of classroom situations and exercises are employed, e.g.:

> the situation where the teacher strives to convince the pupils of a result they tend to doubt;
>
> the situation where pupils are led to discover a surprising result and are then required to convince somebody else;
>
> exercises requiring explanation of a result (instead of proof) e.g. "Explain why the exterior angle of a triangle is equal to the sum of the opposite interior angles." We hope that the explanatory meaning of deductive argument would in cases like this be experienced even if the teacher has to supply the argument;

discovery exercises leading to surprising inductive hypotheses (e.g. that the quadrilateral formed by joining the midpoints of the sides of any quadrilateral is a parallelogram) which the pupils are then required to check deductively (alternatively the teacher may supply a deductive check).

In this type of teaching good rapport between teacher and pupils is essential: the teacher must be aware of the pupils' real attitude (surprise, acceptance, don't care, disbelief, intrigue) towards propositions in order to judge which meaning(s) may be attached to deductive arguments.

Teaching of Defining

Another important theme is the construction of precise and economical definitions of geometric concepts (classes of figures). The possibility of various different definitions for the same concept is emphasized, thus paving the way for understanding the possibility of different axiom sets for the same set of propositions (cf. Activity 1: Economical Descriptions).

In other activities pupils have to construct "mini-deductive systems" around single concepts, e.g. parallelograms. Through these activities a different meaning of proof is emphasized: that of proof as an instrument of deductive organization

ACTIVITY 1: ECONOMICAL DESCRIPTIONS

The objective with this activity is to make pupils aware of the possibility of economizing descriptions of mathematical objects by deliberately exploiting the logical relations between properties.

Pupils are first required to formulate descriptions of some classes of geometrical figures of which they have explored the properties (e.g., rhombuses, kites - it is important that they have not encountered definitions of these concepts). They are then acquainted with the idea of economical description by means of non-mathematical exercises like the following:

J H Nel, Esq.,
"Willows Cottage"
9 Venter Avenue
P.O. Box 48639
STENNENBOSCH
7600

a. The address in unnecessarily long. Give an abbreviated version of the above address which would still enable the letter to reach Mr. Nel. (In Stellenbosch, post is delivered in boxes as well as by street address.)

b. Are there more abbreviated forms of the above address which would still enable the letter to reach Mr. Nel? Give as many shortened versions as you can. Each one should be as short as possible.

After having done this exercise satisfactorily pupils are referred back to their earlier descriptions of geometric concepts and are given the assignment to devise shorter descriptions.

Teaching of Axiomatisation

The next step in the development is to confront pupils with the multitude of geometrical assertions they have encountered so far and to suggest that these assertions be proved. (In traditional teaching formal geometry is initiated with a step like this. We hope that in the alternative development pupils would at this stage accept proof as an activity with different possible meaning, including the intellectual satisfaction that may be derived from it.) Pupils now have to learn that not all assertions can be proved in any one deductive system (i.e. that axioms are necessary) but that the choice of axioms is somewhat arbitrary. We try to achieve this by creating situations in which pupils may commit circular arguments. The fallacy of circular arguments is then pointed out employing suitable examples, and exercises are given in which pupils have to evaluate the validity of sequences of proof (i.e. they have to identify any possible circular arguments).

This is following by a simple statement on the necessity of axioms introducing exercises in which pupils have to select axioms. Possibilities for alternative choices of axioms are emphasized (compare Activity 2). Eventually, pupils are confronted with the set of axioms, theorems and definitions currently in common use in South African schools.

ACTIVITY 2: SELECTING AXIOMS

One cannot prove every geometrical proposition without arguing in circles - - some propositions simply have to be assumed; these we call axioms. Propositions which are proved are called theorems. The question, then, is what propositions are accepted as axioms. Here one has some freedom of choice.

Exercise 3

Consider the following propositions concerning two parallel lines which are cut by a transversal:

1. Alternate angles are equal
2. Cointerior angles are supplementary
3. Corresponding angles are equal

a. Select one of these three propositions as an axiom and then prove the other two propositions.

b. Now select a different proposition as axiom and then prove the remaining two propositions.

c. Are there further possibilities?

17.2 THE COMPLEMENTARY ROLE OF INTUITIVE AND ANALYTICAL REASONING

THE COMPLEMENTARY ROLES OF INTUITIVE AND REFLECTIVE THINKING IN MATHEMATICS TEACHING

Erich Wittman
Padagogisch Hochschule Ruhr
Dortmund, Federal Republic of Germany

1. Introduction

The themes "intuition", "insight", "informal mathematics", etc. on one hand and "axiomatics", "abstraction", "systematic-deductive method", etc. on the other hand have been of continuous interest at the international congresses on mathematics education - implicitly and explicitly and in varying terminology. For example at Lyon (1969) Armitage talked about "abstract and concrete mathematics at school", and Christiansen about "induction and deduction in the learning of mathematics" (Armitage 1970, Christiansen 1970). At Exeter (1972), Fischbein gave an introductory lecture to the Working Group 12 "Psychology of Mathematics Learning". The first part of this lecture dealt with "intuition", the second part with "structures" (Fischbein 1973). At Karlsruhe (1976), the Exeter Working Group 12 met again and one of the topics was "reflective and intuitive intelligence in learning mathematics" (Athen/Kunle 1977, 309-310).

The continuing interest of mathematics educators in these topics is easy to explain. Every conception of mathematics teaching rests, as we know, on a certain philosophy of mathematics. And it was just the battle of two competitive philosophic points of view which characterized the international discussion on mathematics teaching during the past two decades: one orientated towards abstract structures and the axiomatic-deductive style of presenting mathematics, and a second one considering mathematics an activity and emphasizing intuitive thinking as the source of mathematical discovery.

In the course of time extreme positions have more and more been moderated or abandoned, and one-sided or controversial considerations have been placed by attempts to reconcile the two positions. This was facilitated by understanding the axiomatic-deductive method as a continuation of more elementary forms of reflective (or analytic) thinking.

During the discussion on intuitive and reflective intelligence at Karlsruhe, Brookes stated: "The intuitive and the analytical ways of teaching are not contradictory. They must be used in a complementary manner. The analytical techniques help to correct the incorrect primary intuitions and to transform them into coherent and productive interpretations" (Athen-Kunle 1977, p. 310). It is exactly the complementary nature of intuitive and reflective thinking which will be discussed in the following sections.

2. Recent Studies of Mathematical Thinking

Mathematics education is a discipline between mathematics, psychology, philosophy, pedagogy and practice. Therefore it is natural that mathematics educators adapt results and methods from these related areas for their own use, thus enriching the variety of specific didactical approaches.

In this sense I would first like to review recent studies of mathematicians and epistemologists on the nature of intuitive and reflective thinking, which seem to be relevant to mathematics education. The available space does not permit me to discuss former contributions and their differing aspects. However, to give a first impression of the point of view in this paper, I quote from the preface to" Geometry and the Imagination" by Hilbert and Cohn-Vossen (1952, iii): "In mathematics, as in any scientific research, we find two tendencies present. On the one hand, the tendency toward abstraction seeks to crystallize the logical relations inherent in the maze of material that is being studied, and to correlate the material in a systematic and orderly manner. On the other hand, the tendency toward intuitive understanding fosters a more immediate grasp of the objects one studies, a live rapport with them, so to speak, which stresses the concrete meaning of their relations".

In 1974 Dieudonne gave a talk "Mathematical Abstraction and Intuition" at the Luxembourg-Conference on "The Mathematicians and Reality". His main points, as I see them, were:

1. The mathematicians do not only translate reality into mathematical structures, but they also construct artificial mathematical objects, which have no counterpart in reality (p. 46-47).

2. An "intuition" of these artificial objects is acquired by growing experience, familiarity and accustoming (p. 50). There are different intuitions for different fields, e.g., geometric intuition, combinatorical intuition, etc.

3. The progress of modern mathematics was essentially achieved by transfers of intuition from one mathematical field to another (p. 52ff). Examples are linear algebra (tranfser from geometry to algebra), algebraic topology (transfer from algebra to geometry), homological algebra (transfer from algebraic topology to algegra.

4. Intuition develops hand-in-hand with abstraction, i.e. the more schematically and abstractly mathematical objects are comprehended, the deeper relationships can be understood and transferred intuitively (p. 65).

A valuable contribution towards an explanation of the nature of intuition and insight is due to a controversy between two English mathematicians, the geometer Griffiths and the algebraist MacDonald. Griffiths gave a talk at the IMU-Congress Nizza 1970 (Griffiths 1971) in which he criticized the one-sided orientation of mathematics teaching to formal structural mathematics and emphasized intuition and insight. He explained intuition and insight by a series of examples. MacDonald (1978) criticized Griffiths' considerations by means of further examples and arrived at the following description of insight: "Instead of regarding insight as a mysterious identity, indefinable even if somehow knowable, we merely say that it is that understanding of a theory given by a model in physics, geometry or some familiar or accessible part of mathematics" (p. 426-427).

Let us investigate this description somewhat more thoroughly. Any piece of mathematics includes general statements on classes of special cases. It depends on the context whether these special cases are called "models" or "examples. For example, 5 + 2 = 7 is modelled by the composition of corresponding concrete objects, radioactive decay is a model for an exponential function (and, by the way, vice versa!), and the numberline is a model for the domain of real numbers. On the other hand, special polyhedra represent examples of the Euler polyhedron theorem, and the alternative group A_5 is an example of a finite, simple non-Abelian group. Referring back to Griffiths and MacDonald, "insight" into a mathematical structure is the understanding mediated by models or examples of this structure.

MacDonald points to two problems connected with the growth of knowledge by intuition and insight:

1. How do we know whether a model or a set of examples is adequate for the mathematical structure at hand?

2. How is it possible to generalize on the basis of the investigation of single examples?

Griffiths' response to the first question is as follows: we learn the suitability of a model by experience, i.e. we have to work with models or examples and test whether they fit or not. It is inevitable that single examples may give rise to wrong inferences (Griffiths 1978, p.422). With respect to (2), Griffiths (1978, p. 423) points to the "exemplary" character of "good" examples. But what does that mean? A more sophisticated answer to question (2) is given in Piaget's epistemology by the concept of reflective abstraction ("abstraction reflechissante", cf. Beth-Piaget 1961, p. 202, 211ff). Piaget distinguishes two forms of abstraction: "simple abstraction" means abstraction starting from the objects themselves, whereas "reflective abstraction" is abstraction starting from operations on objects.

I would like to illustrate reflective abstraction by us ng an example due to G. Walther (1979). The mathematical structure in question is the recursive relation

$$P_n^m = P_{n-m}^m + P_{n-1}^{m-1}, \ (n > 1, m \leq n)$$

where P_i^j is the number of partitions of the natural number i into j parts. How could this relation be proven? One could start from a series of examples

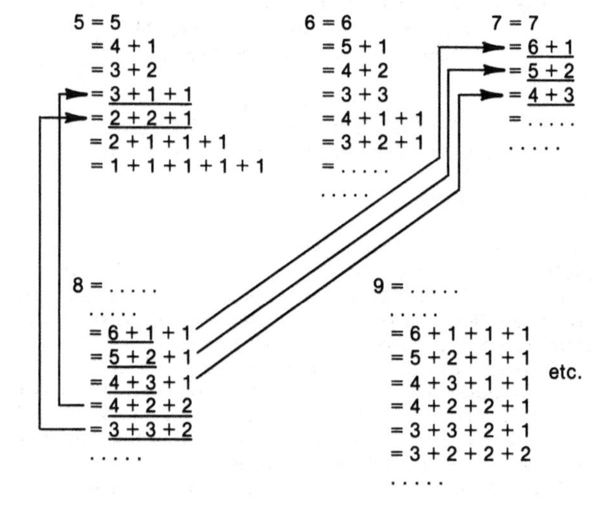

An empirical investigation will produce a lot of special relationships; in particular

$P_8^3 = 5, \ P_7^2 = 3, \ P_5^3 = 2 \qquad 5 = 3 + 2$

$P_9^4 = 6, \ P_8^3 = 5, \ P_5^4 = 1 \qquad 6 = 5 + 1$

etc.

A decisive step beyond individual examples would be provided by the perception of operative schemata establishing transitions from partitions to partitions:

> The schema "cancelling the last 1", applicable to all partitions containing 1 at least one (examples:
>
> 8 = 6 + 1 + 1 — 7 = 6 + 1, 8 = 5 + 2 + 1 — 7 = 5 + 2, 8 = 4 + 3 + 1 — 7 = 4 + 3)
>
> the scheme "diminishing every part by 1", applicable to all partitions not containing one as a part (examples:
>
> 8 = 4 + 2 + 2 — 5 = 3 + 1 + 1, 8 = 3 + 3 + 2 — 5 = 2 + 2 + 1)

What happens if we apply these schemata: Operations of the first kind reduce the sum by 1 and also reduce the number of parts by 1. Operations of the second kind leave the number of parts invariant, but reduce the sum by the number of parts. Because the schemata are reversible they establish a 1-1- correspondence between the partitions in question. We have therefore a general proof of the above relation: a proof whch is abstracted from operations, not from isolated examples.

This illustration of reflective abstraction opens the scene for the quasi-empirical interpretation of mathematics by Lakatos (1976). Although Lakatos' thesis "Proofs and Refutations" was already published in 1963 I would like to treat it as a recent contribution because its importance outside of the theory and history of science was recognized only in the last few years. Lakatos' basic idea, illustrated in detail by the historical genesis of Euler's polyhedron theorem, consists of a peculiar view of mathematical research. According to Lakatos, mathematicians work in a sense like scientists. The difference is that mathematicians study a world of artificial objects created and permanently increased by their own constructions. The constructions and operations on objects give rise to conjectures and proofs. A proof (in this paper I understand "proof" in the sense of the working mathematicians) is a kind of thought experiment, which arises from operations on examples, is applied to new examples and is thus tested. In this way weaknesses and errors are discovered which require improvement. The main source of progress is criticism. Examples do not only inspire conjectures and proofs, but they also provide criticism; and, vice-versa, general considerations are useful to criticize inadequate intuitions.

3. The Interaction of Intuitive and Reflective Thinking in Mathematics

I would like to attempt to integrate the ideas of section 2 into a rough diagram of cognitive activity separating intuitive and reflective thinking and relating them to one another. I am well aware of the shortcomings of such an attempt and of the danger of scholasticism, but I believe the diagram may facilitate further discussion.

As stated above, any mathematical investigation is related to a "universe" of concretely or artificially represented objects: the momentary reality or quasi-reality of the working mathematician. The comparison of the mathematician's activity with that of the scientist suggests to borrow from science the diagram of model building and to adapt it to mathematics

The mathematical exploration of a situation proceeds along two paths. The intuitive one works in a direct manner with the objects represented in a familiar concrete, pictorial or symbolic way. Constructions, operations or transformations are directly applied to these representations in order to discover or to establish general patterns in the behavior of the objects. The intuitive thinking is concrete, immediate, inductive and rich in non-symbolic means of representation and information processing, and is also partly unconscious. The second path, reflective thinking, aims at the explicit formulation of general relationships and procedures and at general foundations, i.e. at general statements about objects and operations with objects. The mathematical language of concepts and symbols is indispensable here and, compared with intuitive activities, marks a meta-level: No longer does the investigator simply work with objects and operations, but he reflects and talks about his activities (in his reaction to the present paper Professor Henkin put particular emphasis on modes of expression and communication as the complement of intuitive thinking).

Both paths support one another in the creation of knowledge as well as in the foundation of knowledge. Intuition is by no means restricted to the creation of knowledge - it is also important for intuitive tests of theoritical inferences, contrary to the formalistic doctrine. Let us consider an example: the "proof" of the assertion "every triangle is isosceles", which is based on a faulty diagram. This "proof" is often used as a demonstration of the failure of intuition and of the necessity to eliminate intuition from proofs. However, this view is wrong. In fact it is intuition which tells you that there are non-isosceles triangles and that the "theorem" therefore must be wrong. Moreover, in general practice intuition is indispensable in checking proofs and detecting errors (cf. Thom 1973, Manin 1979).

Vice versa, the reflective-conceptual thinking is essential also for the growth of mathematics. Single objects are uninteresting for mathematics. It is abstraction, conceptual ordering and schematisation of sets of objects which enables mathematical problems and solutions.

The results of combined intuitive and reflective activities are at first informal theories. They are directly related to intuitive experiences, therefore they need no extensive conceptual language. By further conceptual analysis and schematisation informal theories are developed to formal theories. This means another change of level. Formal inferences are dependent on tests by informal theories in the same way reflective thinking is dependent on intuitive thinking. Shibata (1973) gives an excellent analysis of the connection between informal and formal theories.

4. Some Consequences for Mathematics Teaching

The following recommendations are based on the assumption that mathematics teaching should not be modelled according tor eady-made mathematics but according to the processes of doing mathematics (cf. Freudenthal 1973, chap. 6, Wittman 1978, chap. 10). In this view the activities of students play the crucial role in teaching. I think three types of activities are necessary in order to develop a balance of intuitive, reflective and formal thinking (in the oral presentation of the paper the activities were illustrated by themes of mathematics teaching).

1. The development of intuitive thinking depends on operative studies of a rich variety of examples and models. The students must manipulate these objects and investigate the behavior of them under operations. (In Piaget's terminology such systems of objects and operations are called "groupings" The application of this concept to mathematics teaching is a main point of research at our institute, cf. Wittmann 1980). (The qualifications "objects", "operations", depend on the context. Operations in one context may well be considered as objects in another one.)

2. Students must be stimulated to reflect upon their intuitive activities. They should try to find general formulations and proof ideas for the patterns and constructions discovered by intuition. They should test, demonstrate, improve and corroborate their general ideas by means of examples.

3. The students should gradually learn to analyze concepts, constructions, theorems and proofs. Such analyses are based on a written piece of mathematics, e.g. a proof, or a small context of concepts and theorems. They aim at a deeper understanding of the assumptions of a proof, of the

form of inferences, of logical relationships and at the formulation of more systematic versions of the text at hand. Activities of this kind, which are a first step to axiomatics, have been proposed by Freudenthal (1973, chap. 16) as "local ordering of a field".

With respect to a given content, the three types of activities are not independent. According to van Hiele's theory of thinking levels (cf. Freudenthal 1973, chap. 6) (1) is a prerequisite of (2), (2) is a prerequisite of (3).

In my opinion the main task of mathematics teaching at the school level is the development of activities (1) and (2), i.e. the development of informal mathematics. I think it important to recognize that informal mathematics is genuine mathematics. Within informal mathematics intuitive activities are a natural mode of mathematical thinking. They must not be misunderstood as a concession to students not yet mature for proper mathematics. As a consequence we should resist unjustified pressure for conceptual and formal perfection and we should have more confidence in the autoregulative mechanisms of intellectual progress.

All this does not mean to neglect the achievements of the conceptual analyses due to modern structural mathematics. However, we should detach the fundamental ideas of modern mathematics from their formalistic setting and introduce them to school mathematics via the spiral approach suggested by Bruner.

5. Problems for Research and Development

We need more knowledge about students' primary intuitions and the possibilities to foster intuition by operating activities (cf. Fischbein 1975). We also need further investigations on the development of reflective thinking like the study of Bell (1976). Of particular interest are longitudinal studies.

The constructive branch of mathematics education faces the difficult problem of developing courses which base reflective thinking on intuitive activities and formal thinking on informal thinking. (An interesting approach is Griffiths, 1976). This problem is difficult because it requires a sound feeling for changing levels of rigour. It is much easier either to stick to the intuitive level - and to ignore advanced mathematics - or to rely on the formal system of mathematics - and to strive for the blessing of the watching Big Brother - mathematicians.

References

Armitage, J.V., The Relation Between Abstract and 'Concrete' Mathematics at School, Proceedings of the First International Congress on Mathematical Education, Dordrecht-Holland: Reidel 1969, 48-55.

Athen, H., Kunle, H. (ed.) Proceedings of the Third International Congress on Mathematical Education, Karlsruhe 1977.

Bell, A.W., A Study of Pupil's Proof Explanations in Mathematical Situations. Educ. Stud. in Math. 7 (1976), 23-40.

Beth, E.W., Piaget, J., Epistemologie Mathematique et Psychologie. Etudes d'epistemologie genetique XIV. Presses Universitaires de France 1961.

Christiansen, B., Induction and Deduction in the Learning of Mathematics and in Mathematical Instruction, Proceedings of the First International Congress on Mathematical Education, Dordrecht-Holland: Reidel 1969, 7-27.

Dieudonne, J., L'abstraction et L'intuition Mathematique, NICO 20 (1976), 45-65.

Fischbein, E., Intuition, Structure and Heuristic Methods in the Teaching of Mathematics, In: Howson 1973, 222-232.

Fischbein, E., The Intuitive Sources of Probabilistic Thinking in Children. Dordrecht-Holland: Reidel 1975.

Freudenthal, H., Mathematics as an Educational Task. Dordrecht-Holland: Reidel, 1973.

Giles, G., Does Teaching Inhibit Learning? Mathematics Teaching 65 (1973), 33-38.

Griffiths, H.B., Mathematical Insight and Mathematical Curricula, Educ. Stud. in Math. 4 (1971), 153-165.

Griffiths, H.B., Surfaces. London: CUP 1976.

Griffiths, H.B., Some Comments on MacDonald's Paper, Educ. Stud. in Math. 9 (1978), 421-427.

Hilbert, D., Cohn-Vossen, S., Geometry and the Imagination. New York: Chelsea 1952.

Howson, A.G., (ed.) Developments in Mathematical Education, Proceedings of the Second International Congress on Mathematical Education, Cambridge 1973.

Lakatos, I., Proofs and Refutations. London: CUP 1976.

MacDonald, I., Insight and Intuition in Mathematics, Educ. Stud. in Math. 9 (1978), 411-420.

Manin, Yu., How Convincing is a Proof?, The Mathematical Intelligencer 2 (1976), No. 1, 17-18.

Shibata, T., The Role of Axioms in Contemporary Mathematics and in Mathematical Education. In: Howson 1973, 262-271.

Thom, R., Modern Mathematics: Does it Exist? In: Howson 1973, 194-209.

Walther, G., Illuminating Examples - An Aspect of Simplification in Mathematics Teaching. Inaugural Lecture, University of Dortmund 1979.

Wittman, E., Grundfragen des Mathematikunterrichts, Braunschweig 1978[5]

Wittmann, E., Beziehungen zwischen operativen "Programmen" in Mathematik, Psychologie und Mathematikdidaktik, to appear in Journal fur Mathematikdidaktik 1980.

A version of this paper has appeared in Educational Studies in Mathematics, Vol. 12, No.3, August 1981, pp.

INTUITION AND AXIOMATICS IN MATHEMATICAL EDUCATION

Efraim Fischbein
Tel-Aviv University
Ramat Aviv, Israel

When discussing the inclusion of axiomatics in mathematical curriculums for high schools, what are we referring to? Are we considering the possibility of teaching arithmetic, geometry, or probability in an axiomatized form, or are we suggesting that pupils should learn the basic concepts related to the axiomatic method, its role in mathematical thinking, and the steps and ways of elaborating an axiomatic presentation?

Of course, both objectives can be taken into account, but in my opinion the second must be the main objective. A fully elaborated axiomatic presentation of a certain topic is possible and useful only in a much broader context; i.e., after the student has acquired a high degree of knowledge and experience in the respective domain, and after he has reached a full understanding of the meaning and techniques of the axiomatic method.

While expressing this viewpoint I am aware that my point is far from being original. Further, as far as I know, no systematic psycho-didactical research has been carried out to elucidate the role and the effects of including axiomatics in high school syllabuses. Therefore, the following lines are to be considered only as a preliminary attempt to envisage some of the psychological implications of including the topic of axiomatics in high school curriculums.

It may be supposed that a professional mathematician is able, while reading the axiomatic presentation of a piece of mathematics, to reconstruct by himself the author's ways of thinking. He may be able to reintroduce into the dry sequence of symbols the qualities of an intuitive representation, of intrinsic organization. He will probably be able to imagine for himself the moments of questioning and justification which explain the choice of axioms and primitive terms, and the trajectories followed in building the strings of deduced theorems.

But I have great doubts about the capacity of an average student to reach by his own mental resources such a meaningful re-examination of the respective piece of mathematics, presented from the beginning as a ready-made product. Moreover, I conjecture that the best way to kill genuine mathematical thinking would be to present an unprepared mind directly with an axiomatic exposition of a mathematical concept (cf. Courant & John, 1965, p. vi). I do not deny, of course, the importance for students enrolled in mathematical courses of learning axiomatic systems such as those of Peano (natural numbers), Hilbert (geometry), Kolmogorov (probability), etc. But an axiomatic presentation of a certain chapter of mathematics is no more - - and no less - - than the only certain way to verify the validity of all its statements, i.e., as items of a perfect logical organization.

Now the problem from a psycho-didactical point of view is the following: Is the adolescent's mind well enough equipped for learning successfully the concepts and the techniques required by an axiomatic elaboration? Piaget's theory might suggest an affirmative answer, but as a matter of fact, things are much more complicated. A first important difficulty refers to the adolescent's logical abilities. The qualities of formal thinking, as described by Piaget and his co-workers, constitute only virtualities if they are not especially developed in practical situations. As has been found in recent years, adolescents still have difficulties in performing logical operations, especially when they are not related to familiar, concrete contents. Truth tables and the techniques of disjunction, conjunction, implication, etc. must be learned and practiced in order to become real tools for the pupil's formal reasoning. A good knowledge of the theory of definition is indispensible for learning and using the axiomatic method, its role in mathematical thinking, and the steps and ways of elaborating an axiomatic presentation?

Of course, both objectives can be taken into account, but in my opinion the second must be the main objective. A fully elaborated axiomatic presentation of a certain topic is possible and useful only in a much broader context; i.e., after the student has acquired a high degree of knowledge and experience in the respective domain, and after he has reached a full understanding of the meaning and techniques of the axiomatic method.

While expressing this viewpoint I am aware that my point is far from being original. Further, as far as I know, no systematic psycho-didactical research has been carried out to elucidate the role and the effects of including axiomatics in high school syllabuses. Therefore, the following lines are to be considered only as a preliminary attempt to envisage some of the psychological implications of including the topic of axiomatics in high school curriculums.

It may be supposed that a professional mathematician is able, while reading the axiomatic presentation of a piece of mathematics, to reconstruct by himself the author's ways of thinking. He may be able to reintroduce into the dry sequence of symbols the qualities of an intuitive representation, of intrinsic organization. He will probably be able to imagine for himself the moments of questioning and justification which explain the choice of axioms and primitive terms, and the trajectories followed in building the strings of deduced theorems.

But I have great doubts about the capacity of an average student to reach by his own mental resources such a meaningful re-examination of the respective piece of mathematics, presented from the beginning as a ready-made product. Moreover, I conjecture that the best way to kill genuine mathematical thinking would be to present an unprepared mind directly with an axiomatic exposition of a mathematical concept (cf. Courant & John, 1965, p. vi). I do not deny, of course, the importance for students enrolled in mathematical courses of learning axiomatic systems such as those of Peano (natural numbers), Hilbert (geometry), Kolmogorov (probability), etc. But an axiomatic presentation of a certain chapter of mathematics is no more - - and no less - - than the only certain way to verify the validity of all its statements, i.e., as items of a perfect logical organization.

Now the problem from a psycho-didactical point of view is the following: Is the adolescent's mind well enough equipped for learning successfully the concepts and the techniques required by an axiomatic elaboration? Piaget's theory might suggest an affirmative answer, but as a matter of fact, things are much more complicated. A first important difficulty refers to the adolescent's logical abilities. The qualities of formal thinking, as described by Piaget and his co-workers, constitute only virtualities if they are not especially developed in practical situations. As has been found in recent years, adolescents still have difficulties in performing logical operations, especially when they are not related to familiar, concrete contents. Truth tables and the techniques of disjunction, conjunction, implication, etc. must be learned and practiced in order to become real tools for the pupil's formal reasoning. A good knowledge of the theory of definition is indispensable for learning and using the axiomatic method. But let us suppose that such instruction has been accomplished. A second type of difficulty still remains which seems to me much deeper. On that I want to insist.

The capacity to think productively does not depend only on the technical prerequisites of intelligence. We always think having in mind a certain problem, a certain objective. The nature of that objective, the representation we have of it, and the interests to which it responds determine how we proceed, guide the selection of plausible combinations, contribute to the organization and processing of the data, and finally, constitute a basic factor in ratifying our findings. In other words, the capacity anyone has to produce coherent propositional structures can be estimated only in relation to a certain content. Therefore, when considering the readiness of the adolescent's mind for understanding and managing a conceptual system, we have to take into account not only the mental mechanisms he has acquired but also the directions in which these mechanisms are prepared to work.

Thinking is basically an instrument for behavioral adaptation to reality - - to a reality that can be controlled and dominated only if we agree to be taught and guided by it. The formal thinking of the adolescent - - and of the adult as well - - is naturally oriented to reality by its objectives and by its ways of guessing, trying, and checking. The basic aim of formal procedures is not to detach us completely from reality but, on the contrary, to offer us the possibility of identifying its invariant attributes and relationships and thus to enable us to exert a more effective control over its developments. One should not confound the (virtual or actual) capacity to perform formal processes with the use we make of these mechanisms - - the objectives they are oriented to and the beliefs and interests they are supposed to serve. Our activity - - both external and mental - - is naturally oriented toward practical, behavioral objectives. And this is, I suppose, the main obstacle in teaching and learning axiomatics.

Our cognitive mechanisms are built on the implicit hypothesis that reality comes first - - in the form of given data - - and that our concepts and statements are meant to reflect, describe, and integrate these data, more or less exactly. The axiomatic approach seems to follow an exactly opposed way: We start, or are supposed to start with undefined, meaningless terms. We confer on them some properties by establishing relations between them with the help of freely chosen statements (axioms). On the basis of these primitive terms and accepted axioms we go on producing terms - - through definitions - - and inventing statements and checking their validity deductively. The whole process is supposed to be completely controlled only by the rules of logical inference.

In reality, as we very well know things do not happen that way in the mathematician's creative activity. As stressed above, it is practically impossible to produce coherent conceptual structures without having in mind a given object - - more or less defined - - that will inspire and guide the process of invention (cf. Wilder, 1952, p. 19).

In my opinion, this is the main source of difficulty with the axiomatic method: the basic contradiction between the characteristics of the investigative, creative process which is necessarily experimental, heuristic, and intuitively oriented - - even when elaborating an axiomatic presentation - - and the effort to eliminate completely from the final exposition exactly these characteristics, i.e., the intuitive sources, the heuristic approach, and the step-wise accumulation of results by trial-and-error procedures. The mathematician striving to axiomatize some piece of mathematics seems to be in a permanent struggle, in a permanent contradiction with the deepest attributes of his own thinking activity. It looks as if the main purpose of the productive thinking of the mathematician is to destroy its own sources of inspiration, its mechanisms of invention and verification, and, in spite of this, to keep on being productive, constructive, and consistent. I guess that this kind of double game, subtle and strange, cannot be found in other forms of mental activity - - though, in fact, it represents the supreme aspiration of every scientific endeavor.

A physicist, a biologist, or a psychologist is obliged in a final report of research study to relate not only the problem and the results but also the hypothesis, the methods of investigation, the obtained facts, and the reasons for supporting a certain final interpretation. In these cases the way of investigation itself is an integral part of the author's argumentation. In mathematics, and specifically in an axiomatic exposition, the mathematician has no hypothesis to report on (in the behavioral sense); no facts that have inspired, inductively, his conclusions; and no description of the methods he has used in order to gather these facts. The credibility of a mathematical statement does not depend on the quantity of results that confirm the statement. It depends absolutely only on its final proof. Therefore, when the final form of presentation is reached - - the axiomatic form - - no trace is left of its intuitive, inductive, empirical sources. All progress in approaching the final stage is, in fact, progress toward destroying the dependence of the statements on their intuitive, factual, experimental sources. Despite this, the process of thinking itself during the activity of axiomatization remains, till the end, inductive, constructive, and intuitively guided. As Felix Klein says: "The investigator . . . in mathematics as in every other science, does not work in this rigorous deductive fashion. On the contrary, he makes essential use of his imagination and proceeds inductively aided by heuristic expedients" (quoted in Kline, 1970, p. 274).

The mathematician's intuition intervenes not only in elaborating an axiomatic system, but also in his endeavors to prove its consistency or to prove the independence of its axioms. In both cases, he generally resorts to models, to interpretations. The intuitive role of such techniques is evident. As Paul Cohen (1966, p. 107) says:

> The most natural way to give an independence proof, is to establish a model with the required properties. This is not the only way to proceed since one can attempt to deal directly and analyze the structure of proofs. However such an approach to set theoretic questions is unnatural since all our intuitions come from our belief in the natural, almost physical model of the mathematical universe.

Such introspective descriptions are of great interest for the psychological analysis of the thinking process. On the other hand, we must admit that we do not possess any experimental evidence concerning the complex way by which a mathematician succeeds in producing an axiomatic system.

I feel that I can render my viewpoint more convincing by mentioning a research study carried out recently in my department concerning the concept of proof in high school students. About four hundred high school pupils were divided randomly into two groups. One group received a numerical question; the other group received a geometrical problem. The numerical example consisted, essentially, of presenting the subjects with the expression $E = n^3 - n$, and stating that it is divisible by 6 for every n (n is a natural number). After some practical checks, a proof was given: $n^3 - n = (n - 1) n (n + 1)$, etc. The pupils were asked if they accepted the proof and if they considered that the theorem was fully supported by the proof. A similar situation was produced by using a well-known geometrical theorem: "By joining the midpoints of the sides of a quadrilateral, a parallelogram is formed." In this case, too, some checks were asked for, and afterwards a complete proof was given. The students had to decide if they agreed with the proof, and if they considered that the theorem was completely demonstrated by the proof.

After that introduction, the pupils were aksed a variety of questions to check their confidence in the general validity of the theorems (i.e., the predictive capacity of the proved theorems). For instance: "Moshe claims that he has checked the divisibility of $n^3 - n$ by 6 and that he obtained $2357^3 - 2357 = 105, 513, 223$, a number which is not divisible by 6. What is your opinion?"

"How many quadrilaterals is it necessary to verify in order to be sure that the midpoints of the sides form a parallelogram?" For each of the two problems, various questions of the same type were asked.

About 40% of each group declared that they agreed with the proofs and that they considered the theorems absolutely valid as supported by the respective proofs. But only 8.1% for the geometrical example and only 14.5% for the numerical example were consistent in their correct answers (i.e., answers that demonstrated that the pupil really, intuitively, behaviorally, accepted the universal validity of a proved theorem). Most of them envisaged the possibility of counterexamples or looked forward to additional checks that would increase their faith in the predictive capacity of the statement. The pupils' basic confusion was, then, between an empirical prediction and the formal generality of a mathematical theorem. Certainly these pupils were not yet prepared for coping with an axiomatic approach. But what I want to emphasize is the fact that beyond the formal rules of logic and demonstration - - the meaning of which we may well understand - - we possess a very resistent and a very powerful background of intuitive interpretations and forms of justification. They are the sin qua non of every productive thinking process. They may productively inspire our mental endeavors, but they may also lead us to expectations, interpretations, and beliefs that contradict formally accepted truths.

So what is intuition? In my opinion, intuitions play - - at the level of intellectual activity - - a role analogous to that of perceptions at the level of practical sensory-motor behavior. Productive thinking is not reducible to a concatenation of syllogisms, i.e. to a sequence of operations in which each performed operation determines strictly the next one. A thinking activity possesses the attributes of fluency and internal continuity, but at the same time of adaptiveness and inventiveness and internal continuity, but at the same time of adaptiveness and inventiveness that characterize a practical, accomodative behavior. When we think, we run through an unknown troubled land - - following some general goals but adapting each step to the accidents of the ground and trying to learn, as promptly as possible, from the effects of each step, the next useful reaction.

My theory is that the intuitive forms of representation and interpretation play - - in the framework of intellectual activity - - a role similar to that fulfilled by perceptions at the sensory-motor level.

Like a perception, an intuition is a global, compact, direct form of knowledge. It appears to us as imposing itself with the obviousness and the intrinsic credibility of a perceived fact. A statement like "if A = B and B = C, then A = C" seems to possess the direct coerciveness and the intrinsic obviousness that are specific to a perceived reality. The role of intuitive guesses and intuitive representations or interpretations is to inspire, to guide, and to coordinate our mental actions.

In the study cited above, most of the students were not able to distinguish clearly between the general basically heuristic and intuitive nature of our ways of thinking and proving, and the formal structure of a mathematical argumentation - - after this has reached its final form. The word "distinguish" may be misleading in this context. The problem is not that of a formally stated differentiation between an empical argument and a logical proof. The confusion takes place at a much deeper level - - at the behavioral level. The pupil simply cannot accept intuitively, behaviorally, once and for all, that in every quadrilateral no matter what its form, one will obtain a parallelogram by joining the midpoints of its sides. That intuitively based mistrust appears to be much stronger than the conceptual constraint imposed by the logical-mathematical schemes. The pupil may accept the theorem formally, verbally, at the level of the propositional system - - but this acceptance is not necessarily accompanied by a feeling of intrinsic confidence in the universal validity of the statement. The student himself may not be aware of that contradiction. But as we have seen, the contradiction appears when he is asked non-standard questions. I suppose that an analogous difficulty - - but in much more complex terms - - emerges when the student has to face the task of producing (or understanding) an axiomatic structure.

In an axiomatic presentation, what is kept is exclusively the pure, logical form fo the structure considered. When a group structure is considered axiomatically, no particular category of mathematical objects is taken into account. The same group structure may correspond to geometrical transformations, to sets of numbers, or to combinatorial operations. Therefore, the contradiction is even sharper, in that case, between the intuitive attributes of the productive process of thinking and the pure logical form of the produce which is to be reached by this process. The student who strives to elaborate on axiomatic presentation of a piece of mathematics must preserve the empirical constructiveness of his mental experimentations throughout the entire process, but at the same time and at each step, he must be able to envisage his statements as if they were completely free from every possible empirical, intuitive constraint.
In my opinion, teaching axiomatics means first of all to create by way of stepwise, thoughtful instruction and practice the specific mental capabilities required for overcoming and even enjoying these specific contradictory situations. Suppes (1966, p. 70) has expressed a fundamental psychological fact: In order to develop the capacity of students to operate with formal structures, we must not strive to eliminate the intuitive mechanisms (considering that a formal structure is opposite to an intuitive argumentation). On the contrary, Suppes claims - - based on his own intuition - - that the capacity for elaborating a formal system itself needs a specific category of intuitive patterns (in addition, of course, to a good command of logical operations). The role of these intuitions is to enable the student to identify and describe coherently those systems of properties and relationships that are able to survive - - under the form of meaningful, manageable and consistent structures - - without any intuitive help.

References

Cohen, P.J. Set theory and the continuum hypothesis. New York, W.A. Benjamin, 1966.

Courant, R., & John, F. Introduction to calculus and analysis. Wiley, International Edition, New York, 1965.

Kline, M. Logic versus pedagogy. American Mathematical Monthly, March 1970, pp. 264-281.

Piaget, J. Le possible, l'impossible et le necessaire. Archives de Psychologie, 1976, 44 (No. 172), pp. 281-299.

Suppes, P. The axiomatic method in high school mathematics. In The role of axiomatics and problem solving in mathematics. Conference Board of the Mathematical Sciences, Washington, D.C., Ginn & Co., 1966, pp. 69-76.

Wilder, R.L. Introduction to the foundations of mathematics. Wiley, New York, 1952, 1955.

THE COMPLEMENT OF INTUITIVE THINKING IN MATHEMATICS TEACHING AND LEARNING A RESPONSE TO ERIC WITTMANN'S PAPER

Leon Henkin
University of California
Berkeley, California

Wittmann introduces his paper by contrasting "intuition, insight, informal mathematics, etc. on the one hand," and "axiomatics, abstraction, systematic-deductive method, etc. on the other hand." He quotes Armitage as distinguishing between "abstract and concrete mathematics at school"; Fishbein as contrasting "intuition" with "structures"; an Exeter Working Group as dealing with "refelctive and intuitive intelligence in learning mathematics"; and Brookes as referring to the "intuitive and analytical ways of teaching." Out of all these contrasts Wittmann emerges with the terminology "intuitive and reflective thinking," which forms the title and starting point of his own conribution.

Coming to this discussion from the perspective of a mathematician whose background lies primarily in the logical foundtions of the subject, it seems to me that many diverse elements of mathematics, and of mathematics instruction, have been mixed up in this citation of contrasting lists. By no means have two complementary types of thinking been even moderately well delineated. My position is that there are a multitude of components making up the thinking that goes into mathematics - - a multitude whose diversity we only dimly apprehend at present. To me it seems that the complementary component of intuitive thinking which myst be addressed, in the teaching and learning of mathematics, consists of modes of mathematical expression and communication - - something related to, but separate from, "thinking." Before discussing this thesis, I shall examine critically several of the details of Wittmann's paper.

1. Reviewing a 1974 talk by Dieudonne, Wittmann extracts what he considers the main points. Among these are that mathematicians are said to"construct artificial objects which have no counterpart in reality," and to develop an intuition of these objects "by growing experience, familiarity, and accustoming." But is it possible to gain "experience and familiarity" with objects having "no counterpart in reality?" We are led to the long-standing philosophical controversy between the nominalists and the Platonists: Are there really abstract objects, or is abstraction simply a mode of overlooking differences between "real" objects?

2. Wittmann ascribes to MacDonald the view that insight is "that understanding of a theory given by a model in physics, geometry, or some more familiar or accessible part of mathematics." It seems clear that some mathematicians may gain insight into a mathematical theory by connecting it with a biological or economic model, say, rather than with a model in physics or geometry or some more familiar part of mathematics - - providing, of course, that they are already conversant with biological or economic phenomena. Could we then make a bold leap from MacDonald's thesis and suggest that insight is the understanding of the unfamiliar in terms of what is already known to us? If such a leap is permitted us, is it a leap of intuition - - or of abstraction?

3. Wittmann cites Piaget's distinction between "simple abstraction," which is abstraction "from objects

themselves," and "reflective abstraction," which is abstractin "from operations on objects." Of course such a distinction makes sense only if one regards operations as non-objects, contrary to prevailing mathematical attitudes; or if one understands the word "objects" as used here in the non-mathematical sense of material objects. Since there are also operations on operations, and indeed a whole hierarchy of operations of higher type, must we be led to a hierarchy of forms of abstraction?

4. Wittmann attempts to "illustrate reflective abstraction" by considering in some detail the recursive relation

$$P_n^m = P_{n-m}^m + P_{n-1}^{m-1}, \quad n > 1, \quad m \leq n$$

where P_j^i is the number of ways to partition the natural number i into j parts.

Wittmann asks, "How to prove this relation?" and then leads us through a series of examples to what he calls "the decisive step" beyond such examples, which is a "perception of operative schemata." Starting with any partition X of the number n into m parts at least one of which consists of the number 1, Wittmann describes a schema A which produces a partition A(X) of the number n - 1 into m - 1 parts. And starting with any partition Y of the number n into m parts no one of which consists of the number 1, Wittmann describes a schema B which produces a partition B(Y) of the number n - m into m parts.

After describing these schemata, Wittmann writes, "Apart from further details, this is a general proof" of the relation (*), "a proof which is abstracted from operations, not from isolated examples." The "further details" that are needed are: Verifications that the schema A establishes a one-one correspondence between all type-X partitions of n into m parts; and all partitions of n - 1 into m - 1 parts, while schema B establishes a one-one correspondence between all type-Y partitions of n into m parts, and all partitions of n - m into m parts. Finally, one must observe that every partition of n into m parts is either of type X or of type Y, but not both.

By leading us through the examples of partitions, in this illustration, Wittmann seems to suggest that refelctive abstraction is a type of thinking that leads from examples to a proof. But surely the actual thinking process that leads to a proof is vastly more complicated than what is described; in particular, it is characerized by many thought paths that generate, follow, and finally discard, ideas that are wrong or do not lead to the desired proof.

Wittmann says of this illustration that the proof which is indicated for the relation (*) "is abstracted from operations." However, the proof itself is put together from a variety of components; it is not the proof, but the schemata A and B, that are abstracted from operations on examples of partitions. The role of these schemata in the proof is clear; but the way in which intuitive thought is led to focus on these schemata as a potential element of a proof is not at all illuminated by Wittmann's analysis.

5. The conception of Lakatos, that proofs are a kind of thought experiments that must be tested by examples, is discussed by Wittmann. This conception is clearly very different from the classical mathematical notion of a proof, as analyzed by logicians, according to which a proof is a sequence of sentences linked by prescribed rules of inference, the existence of which guarantees the correctness of its last line (the theorem proved). To use the same word, "proof," for two such different notions is an invitation to misunderstanding. As both notions are useful, I suggest the term "purported proof" - - which I shorten to "pproof," pronounced "puh-proof" - - for the Lakatos concept. (In another paper I have distinguished between "skills" and "sskills.")

6. Wittmann's diagram to illustrate the interaction of intuitive and reflective thinking in mathmatics is puzzling in several aspects. First, although intuitive activities are performed directly upon the objects under consideration, reflective thinking is only performed when a theory is at hand. To get from the objects to the theory, a process of "schematizing" is indicated. Is this schematizing a third kind of mathematical thinking that must be considered along with the intuitive and reflective kinds? Second, there are no arrows leading awary from "intuitive inferences" in Wittmann's diagram. How, then, do such inferences affect the final product of mathematical thought?

7. In discussing the well-known proof, based on a faulty diagram, of the proposition that every triangle is isosceles, Wittmann argues that in looking for the error "we proceed intuitively by testing the various steps and elaborating the critical ones." But what does the test consist of that we apply to each step - - is it an intuitive test as to whether the statement "fits the facts," or as to whether the statement "follows from the earlier steps?" In the former case, the "facts" are likely to be read off from the diagram, inhibiting the discovery of the error. In the latter case, it appears that we must rely on a "logical intuition" as well as on an intuition about the objects under discussion - - a distinction not mentioned by Wittmann.

8. In drawing consequences of his analysis for mathematics teaching, Wittmann proposes to help students develop intuitive thinking by manipulating objects and investigating their behavior under operations. However, to describe "the behavior" of objects under operations, one must have a store of classes and relations of objects that are considered mathematically significant, in addition to the objects and operations themselves. (For example, if a student repeatedly applies a certain function to integers and obtains in each case a value that is prime, his observation of the results will be of limited intuitive value if he has never encountered the concept of prime number.) Is he expected to construct all of these significant classes and relations *ab initio*, or is the teacher to provide the student with a supply? In the latter case, how is the teacher to choose which concepts to supply, and which ones to leave to be discovered?

I have critically examined eight details of Wittman's paper, and I wish now to return to the thesis enunciated at the beginning of these remarks. When we examine the totality of activities that enter into mathematical experience, we do indeed find observation of concrete objects by the senses and of purely ideational objects by a kind of inner observation; we find manipulation of both kinds of objects; we find expectations and guesses about the outcome of future observationa; and we find analyses of the origin of such guesses which tend to heighten or lessen our expectations. All of this activity can be subsumed under the categories of intuitive and reflective thinking propounded by Wittmann. But there is something more.

Students, teachers, and mathematicians must all communicate about their mathematical experience. To help our students achieve this dimension of mathematical activity, we must help them shape their language patterns to function efficiently in a realm quite different from the world of everyday events. To my mind, such notions as proof and axiomatic theory must be understood as ways of expressing rather than as ways of thinking. But I must leave the development of these ideas for another occasion.

The writing of this paper was supported in part by National Science Foundation Grant No. MCS 7722913.

CHAPTER 18 - Technology

18.1 THE EFFECT OF THE USE OF CALCULATORS ON THE INITIAL DEVELOPMENT AND ACQUISITION OF MATHEMATICAL CONCEPTS AND SKILLS

THE EFFECT OF THE USE OF CALCULATORS ON THE INITIAL DEVELOPMENT AND ACQUISITION OF MATHEMATICAL CONCEPTS AND SKILLS

Hartwig Meissner
Westfälische Wilhelms-Universitat
Munster, Federal Republic of Germany

There is a general public agreement that calculators will reduce number sense and computational skills. I do not know any research supporting that emotional statement, which is world-wide (USA, Canada, Russia, Europe, Asia, New Zealand). To give a fairer view of the problem, we first should analyze the role of the calculator and the role of checking in arithmetic (see Meissner, 1978).

We can use the calculator in two modes. Using the syntactical mode we work on a sequence of digits (digital data). According to the underlying algorithms we press button after button, often forgetting the preceeding ones. We work on a problem step by step without reflecting the meaning of it. (The same behaviour can be observed while doing paper-and-pencil algorithms.)

In the semantical mode we work on relations between inupt and output (analogous data). We use guess-and-test procedures to emphasize the order of magnitude of input and output. The "wrong" output gives hints for better next input. Thus we get a feeling how to vary input intervals $_1$ according to output intervals $_2$. We get " -experiences". The calculator is only the black-bos machine to give immediate feedback about our guesses (reinforcement like in Programmed Instruction).

The distinction between the syntactical and semantical use of calculators leads to two different possibilities for automatic (unconscious) error-detecting. In the syntactical mode errors may become obvious by the unconscious comparison of the calculator result with the knowledge of baseic facts (e.g., 4 x 5 = 20). In the semantical mode we have an unconscious comparison with relational or environmental experiences. (In the same way we should distinguish between approximation and estimation.)

We never have had great success in teaching our students to control their results. We see one reason for that failure in mathematics itself (all is logical and provable, therefore controls are not necessary) and two other reasons in human mentality. First, there is no real responsibility for the result and second, it is unnatural always to doubt one's own abilities. The calculator will stress this effect even more since it allows the user to identify himself with the safety of the machine.

To analyze the effect of calculator use it also is necessary to know something about the development of understanding, of abilities, and of skills. Having a concept of numbers means to have relational understanding (being convinced oneself) or communicable understanding (convincing others) of the relations in the number system (see Meissner, 1980). Without these modes of understanding there are only instrumental skills (corresponding to an instrumental understanding) like working algorithms mechanically (syntactical mode). The public fear in using calculators is to lose instrumental skills (mental arithmetic, paper-and-pencil algorithms) and higher skills or abilities (applying relational or communicable understanding, e.g. deducing, explaining, comparing, doing algorithms consciously, etc.).

Preparing this panel I sent a questionnaire to about 30 specialists all over the world asking for research and findings concerning the effect of the early use of calculators on the acquisition of number concepts and skills. In Europe there is some research going on in Sweden (projects RIMM, IAB), Great Britain (with Dataman), West Germany (project TIM), and Belgium. But still results are not available. In other parts of the world except North America there are only a few investigations. More information about the answers to the questionnaire can be obtained from the author. See also the state-of-the-art reviews by Knopf (1979) and Suydam (1979, 1980).

Here we only will give a brief description of some research from TIM:

> Three investigations in grades 3 and 4 (ages 8 - 10) on training mental arithmetic via the calculator: Computing basic facts in the syntactical mode. During the training period of 2 to 3 weeks, calculator usage was allowed for every worksheet. Many competitions in mental computing against calculator computing, were used to determine "who is quicker?" Findings: (a) Significant gains in the posttests in mental arithmetic, (b) experience (conscious insight) for the children that for mental arithmetic problems they perform better in computing mentally than in using a calculator.
>
> Training number sense by calculator games in grade 3 (a one-year study with about 900 students). During 10 to 20 minutes per week the test classes played three different calculator games (semantical mode by guess-and-test). Pre-, middle-, and posttests in mental estimation were given for the four basic operations (multiple choice, e.g., "576 + 549 is close to" (circle one of) 800, 900, 1000, 1100). Statistical data are not yet available. Observations show a better performance of the calculator group towards new or unusual problems.
>
> The development of number sense. Monitoring children aged 7 - 10 playing calculator games: Do children develop strategies by themselves or is it necessary to teach strategies? (No results before fall 1981.)
>
> Teaching percentages with the %-key of the calculator, ages 13-15. The "percent" function was used in the semantical mode via guess-and-test to develop proportional and percentage feeling.

Highly significant gains were found for the test group in traditional percentage word problems.

All results fit into the summary of Suydam (1980): "Across countries, the overwhelming majority of the data has supported the conclusion that the use of calculators does not harm achievement scores (in particular, computational scores)."

But there are two restrictions, most important ones in my opinion. Up to now there is no long-term research. And almost all of the present investigations have ignored the Hawthorne Effect. Therefore we cannot make any valid prediction without further research. We need more knowledge about the process of learning, about the role of guess-and-test procedures with the calculator and about the long-term effect of stimulus-response drill with the calculator (e.g., mental arithmetic). Only then can we make decisions about the future of paper-and-pencil algorithms. Only then we can discuss a new balance between mental and written arithmetic on the one hand and calculator arithmetic on the other hand.

References

Knopf, P.: Der Taschenrechner auf der obligatorischen Schulstufe. Schweizerische Koordinationsstelle fur Bildungsforschung. Aaurau, 1979.

Meissner, H.: Projekt TIM 5/12 - Taschenrechner im Mathematikunterricht fur 5- bis 12-Jahrige. Zentralblatt fur Didaktik der Mathematik 10, Nr. 4, 1978.

Meissner, H.: Number Sense and Computational Skills. Proceedings of the Fourth International Conference for the Psychology of Mathematics Education. Berkeley, California, 1980.

Suydam, M.N.: The Use of Calculators in Pre-College Education: A State-of-the-Art Review. Columbus, Ohio: Calculator Information Center, May 1979.

Suydam, M.N. (ed.): Working Paper on HAND-HELD CALCULATORS IN SCHOOLS. Columbus, Ohio: ERIC Clearinghouse for Science, Mathematics, and Environmental Education, March 1980.

18.2 A MINI-COURSE ON SYMBOLIC AND ALGEBRAIC COMPUTER PROGRAMMING SYSTEMS

SYMBOLIC AND ALGEBRAIC COMPUTER PROGRAMMING SYSTEMS

Richard J. Fatemen
University of California
Berkeley, California

1. The Difference Between Numeric and Algebraic Computation.

To illustrate the difference between numeric and symbolic processing, consider a computer program (in FORTRAN, say) which given the quantities a, b, and c, can apply the quadratic formula to approximate the roots of the quadratic equation $ax^2 + bx + c = 0$. The names a, b, and c must of course correspond to numerical values at run-time because the program has been written to provide numerical processing. If a had as its run-time value the expression q, b had value - pq-1 and c had value p, the FORTRAN program would not be applicable. Nevertheless by applying the quadratic formula symbolically, the two roots

$$\frac{-(-pq-1) \pm \sqrt{p^2 q^2 + 2pq + 1 - 4pq}}{2q}$$

can be represented. By further efforts, this expressionn can be reduced to the set of values p, 1/q. This substitution (in this case, into the quadratic formula) and subsequent simplification are but two of the necessary operations in an algebra system. Some of the more elaborate facilities that can be built up (and have been, in some systems), bring us to the edge of research areas in several fields of mathematics, and present us with problems in representation and communication of "knowledge" which border on the hardest problems in computer "artificial intelligence" - - natural language processing, representation of complex data, heuristic programming.

Along the way we see that problems of graphical display, parsing of programming/mathematical languages, and operating system capabilities must be solved. Furthermore, searching for the most efficient methods possible for tackling fundamental arithmetical problems (e.g. multiplying two polynomials) brings us into analysis of the complexity of computer programming systems.

This course will survey recent activity in the area of computer systems for algebraic manipulation. Among the topics to be discussed are the underlying representations for some common mathematical systems such as the ring of integers, polynomials and rational functions in several variables over a field, algebraic and transcendental extensions, matrices, combinatorial forms, etc. Algorithms for the simplification, differentiation, integration of expressions will be discussed and demonstrated using a system called MACSYMA.

Results about algorithms are of interest not only to educators (e.g. there is an algorithm for indefinite integration, even though calculus students do not learn it!), but also to researchers who require the use of a symbolic calculation engine for large scale manipulation of formulas of many types.

2. Macsyma Demonstration: A few simple examples:

(c1) /*

Note the (ci) lines are commands or expressions typed by the user, while the (d) lines are the results produced by Macsyma. "%" stands for the value of the last (d) line. The " ' " character indicates that a function is to be displayed but not evaluated.

Arithmetic

Integer arithmetic is performed in Macsyma with arbitrarily large integers. Floating point arithmetic can be performed with single word precision or with arbitrary, but specified precision. Let us illustrate with the calculation of 30 factorial. "!" is the postfix, factorial function, just as in normal notation.

30!;
Time=16 msec.

$$265252859812191058636308480000000 \quad (d1)$$

(c2) /*
Now we can factor this result. (This is really the hard way to find the primes less than 30.)
*/

factor(%)
Time=133 msec

$$2^{26}3^{14}5^{7}7^{4}11^{2}13^{2}17192329 \quad (d2)$$

(c3) /*

Indefinite Integration, Differentiation and Simplification
*/

$1/(x^3+2)$;
Time=50 msec

$$\frac{1}{x^3+2} \quad (d3)$$

(c4) /* Integrate with respect to x. */

integrate(%,x);
time=1683 msec

$$-\frac{\log(x^2-2^{1/3}x+2^{2/3})}{6\,2^{2/3}}+\frac{\operatorname{atan}\left(\frac{2x-2^{1/3}}{2^{1/3}\sqrt{3}}\right)}{2^{2/3}\sqrt{3}}+\frac{\log(x+2^{1/3})}{3\,2^{2/3}} \quad (d4)$$

(c5) /* Now let's differentiate to see if we get $1/(x^3+2)$ back*/

diff(%,x);
Time=416 msec

$$\frac{1}{\left(\frac{3\,(2x-2^{1/3})^2}{3\,2^{2/3}}+1\right)}-\frac{2x-2^{1/3}}{6\,2^{2/3}(x^2-2^{1/3}x+2^{2/3})}+\frac{1}{3\,2^{2/3}(x+2^{1/3})} \quad (d5)$$

(c6) /* That doesn't look the same! Maybe if we rationally simplify it . . . */

ratsimp(%);
Time=916 msec

$$\frac{1}{x^3+2} \quad (d6)$$

(c7) /*

Whew! Had me worried there for a minute. Looks like the factored form of expressions ought to be simpler than the expanded form. Here's a nice, compact expression.
*/

$(x+3)^{20}$;
Time=33 msec

$$(x+3)^{20} \quad (d7)$$

(c8) /* Look what happens when we expand it. */

rat(%)
Time=150 msec

$$\begin{aligned}&x^{20}+60x^{19}+1710x^{18}+30780x^{17}+392445x^{16}\\&+3767472x^{15}+28256040x^{14}+169536240x^{13}\\&+826489170x^{12}+3305956680x^{11}\\&+10909657044x^{10}+29753610120x^{9}\\&+66945622770x^{8}+123591918960x^{7}\\&+185387878440x^{6}+222465454128x^{5}\\&+208561363245x^{4}+147219785820x^{3}\\&+73609892910x^{2}+23245229340x\\&+3486784401\end{aligned} \quad (d8)$$

(c9) /* And, of course, we can differentiate it. */

diff(%,x);
Time=83 msec

$$\begin{aligned}&20x^{19}+1140x^{18}+30780x^{17}523260x^{16}\\&+6279120x^{15}+56512080x^{14}+395584560x^{13}\\&+2203971120x^{12}+9917870040x^{11}\\&+36365523480x^{10}+109096570440x^{9}\\&+267782491080x^{8}+535564982160x^{7}\\&+865143432720x^{6}+1112327270640x^{5}\\&+1112327270640x^{4}+834245452980x^{3}\\&+441659357460x^{2}+147219785820x\\&+23245229340\end{aligned} \quad (d9)$$

(c10) /* And we can factor it. */

factor (%)
Time=3066 msec

$$20(x+3)^{19} \quad \text{(d10)}$$

(c11) /* This seems like a hard way to get the same thing we could have obtained more directly. */

diff $((x+3)^{20},x)$;
Time=50 msec

$$20(x+3)^{19} \quad \text{(d11)}$$

(c12) /* Now, consider a fairly compact, fully expanded expression. */

x^6-1;
Time=0 msec

$$x^6 - 1 \quad \text{(d12)}$$

(c13) /* I wonder what happens if we factor it? */

factor (%)
Time=200 msec

$$(x-1)(x+1)(x^2-x+1)(x^2+x+1) \quad \text{(d13)}$$

(c14) /*

Hmm . . . that seems more complicated. What's simpler is in the "eye of the beholder." That's why Macsyma has a number of functions for expanding, factoring, etc. to suit the needs of a particular calculation. */
/*

Solution of Polynomial Equations

Let's find the solutions of x^6-1=0, assign them to ei labels, and store the list of ei labels in the variable, sol.
*/
sol : solve(x^6-1);
Solution:

$$x = \frac{\sqrt{3}i + 1}{2} \quad \text{(e14)}$$

$$x = \frac{\sqrt{3}i - 1}{2} \quad \text{(e15)}$$

$$x = -1 \quad \text{(e16)}$$

$$x = \frac{\sqrt{3}i + 1}{2} \quad \text{(e17)}$$

$$x = \frac{\sqrt{3}i - 1}{2} \quad \text{(e18)}$$

$$x = 1 \quad \text{(e19)}$$

Time=3250 msec

[e14,e15,e16,e17,e18,e19] (d19)

(c20) /*
Note that %i*%i = -1. Now lets pick out the first solution and substitute it back into x^6-1.
*/

trial : ev(first(sol));
Time=33 msec

$$x = \frac{\sqrt{3}i + 1}{2} \quad \text{(d20)}$$

(c21) x^6-1,trial;
Time=116 msec

$$\frac{(\sqrt{3}i + 1)^6}{64} - 1 \quad \text{(d21)}$$

(c22) /* Hmm . . . this should be zero. What if we expand it? */

expand(%);
Time=350 msec

$$0 \quad \text{(d22)}$$

(c23) /*
That's better. Let's try the first solution again, but this time as a numeric value.
*/

trial : ev(first(sol)),numer,expand;
Time=233 msec

$$x = 0.8660254037844387i + 0.5 \quad \text{(d23)}$$

(c24) x^6-1,trial,expand;
Time=183 msec

$$5.204170427930421e - 17i + 8.326672684688674e - 17 \quad \text{(d24)}$$

(c25) /*
Note that numeric evaluation is inexact while integer and symbolic evaluation is exact! (The notation "e-17" is the E-notation for 10^{-17}.)
*/

/*

Taylor Series
*/
sqrt(1+x); Time=33 msec

$$\sqrt{(x+1)} \quad \text{(d25)}$$

(c26) /* Let's give the taylor series expansion of (d25) about x=0 to fifth order. */

taylor(%,x,0.5); Time=200 msec

$$1 + \frac{x}{2} - \frac{x^2}{8} + \frac{x^3}{16} - \frac{5x^4}{128} + \frac{7x^5}{256} + \cdots \quad \text{(d26)}$$

(c27) /* If we square this series, we should get 1+x plus higher order terms.
*/

%*%; Time=216 msec

$$1 + x + \cdots \quad \text{(d27)}$$

(c28) /* If we take an infinite product. */

product(1-x^n)^{24},n,1,inf)*x; Time=183 msec

$$x \prod_{n=1}^{\infty} (1 - x^n)^{24} \quad \text{(d28)}$$

(c29) /* We should be able to find the first five terms by a Taylor expansion. */

taylor(%,x,0,5); Time=700 msec

$$x - 24x^2 + 252x^3 - 1472x^4 + 4830x^5 + \cdots \quad \text{(d29)}$$

(c30) /*

We can also work with infinite Taylor series expansions using the Powerseries function. First, let's set the verbose flag to true, so that Powerseries will tell us what it's doing internally.
*/

verbose : true$
Time=16 msec

(c31) /* note that the "$" terminator suppresses the d-line output. */

powerseries(f(x),x,0);

In first simplification we have returned:

$$f(x)$$

Time=333 msec

$$\sum_{i1=0}^{\infty} \frac{x^{i1} \left(\left. \frac{d^{i1}}{dx^{i1}} f(x) \right|_{x=0} \right)}{i1!} \quad \text{(d31)}$$

(c32) /*
That's as well as can be done right now, because we haven't defined f(x). Let's look at the infinite series expansions for some typical expressions. */

sin(z)*cos(z); Time=16 msec

$$\cos(z)\sin(z) \quad \text{(d32)}$$

(c33) powerseries(%,z,0);

In first simplification we have returned;

$$\frac{\sin(2z)}{2}$$

Time=523 msec

$$-\frac{\sum_{i2=0}^{\infty} \frac{2^{2i2+1} z^{2i2+1}}{(2i2+1)!}}{2} \quad \text{(d33)}$$

(c34) log(sin(x)/x); Time=16 msec

$$\log\left(\frac{\sin(x)}{x}\right) \quad (d34)$$

(c35) Powerseries(%,x,0); can't expand.

$$\log(\sin(x))$$

so we will try again after applying the rule:

$$\log(\sin(x)) = \int \frac{\frac{d}{dx}\sin(x)}{\sin(x)}\,dx$$

In first simplification we have returned:

$$\int \cot(x)dx - \log(x)$$

Time=1316 msec

$$\sum_{i3=0}^{\infty} \frac{x^{i3}\left(\frac{d^{i3}}{dx^{i3}}\int \cot(x)dx\Big|_{x=0}\right)}{i3!} - \log(x) \quad (d35)$$

(c36) /*
Note that the user was asked about the index, i3. Since we know that the series starts with order zero, there will be a zero value for the index of summation. In order to properly take care of this zero term, we should respond with "zero". */

/*

factoring
*/
$x^{30}+1$
Time=16 msec

$$x^{30}+1 \quad (d36)$$

(c37) factor(%);
Time=450 msec

$$(x^2+1)(x^4-x^2+1)(x^8-x^6+x^4-x^2+1) \cdot (x^{16}+x^{14}-x^{10}-x^8-x^6+x^2+1) \quad (d37)$$

(c36) /* Let's introduce a matrix (as a number of row-element lists). */

```
matrix ([x^3,x^2,x,1],[y^3,y^2,y,1],[z^3,z^2,z,1],[w^3,w^2,w,1]);
```

Time=50 msec

$$\begin{bmatrix} x^3 & x^2 & x & 1 \\ y^3 & y^2 & y & 1 \\ z^3 & z^2 & z & 1 \\ w^3 & w^2 & w & 1 \end{bmatrix} \quad (d38)$$

(c39) /* and take its determinant */

determinant(%);
Time=533 msec

$$\begin{aligned} & x(-y^2(z^3-w^3)+w^2z^3+y^3(z^2-w^2)-w^3z^2) \\ & -x^2(-y(z^3-w^3)+wz^3+y^3(z-w)-w^3z) \\ & -y(w^2z^3-w^3z^2)+y^2(wz^3-w^3z) \\ & +x^3(-y(z^2-w^2)+wz^2+y^2(z-w)-w^2z) \\ & -y^3(wz^2-w^2z) \end{aligned} \quad (d39)$$

(c40) /* This factors into a particularly simple form. */

factor (%);
Time=7283 msec

$$-(x-w)(y-w)(y-x)(z-w)(z-x)(z-y) \quad (d40)$$

(c41) /*

Array Associated Functions

As an example of defining a function, let's define an array-associated function which produces as its value new functions, Legendre polynomial functions. Note: an array-associated function is a function which produces a value for an array element if that element is referenced, but does not yet have an assigned value. This is sometimes called a "memo" function. */

```
p[n](x):=ratsimp(1/(2^n*n!)diff((x^2-1)^n,x,n));
```

Time=0 msec

$$p_n[x] := \mathrm{ratsimp}\left(\frac{1}{2^n n!}\frac{d^n}{dx^n}(x^2-1)^n\right) \quad (d41)$$

(c42) /*
This can be read as, "if array element p[n] does not have an assigned value yet, then produce a function of a single argument, given by the right-hand side and assign this function as value to the array element.

Now let's use a simple loop to display the first 5 Legendre polynomials in the variable z. */
for i:0 through 5 to display (p [i](z));

$$p_0[z] = 1$$

$$p_1[z] = z$$

$$p_2[z] = \frac{3z^2 - 1}{2}$$

$$p_3[z] = \frac{5z^3 - 3z}{2}$$

$$p_4[z] = \frac{35z^4 - 30z^2 + 3}{8}$$

$$p_5[z] = \frac{63z^5 - 70z^3 + 15z}{8}$$

Time=2316 msec

done (d42)

(c43) /*
Note that the array elements through p [5] now have assigned values and these values are functions; not simple expressions. */
p [3];
Time=0 msec

$$\lambda([x], \frac{5x^3 - 3x}{2}) \quad (d43)$$

(c44) /*

Laplace Transforms

Let's solve a pair of coupled, ordinary differential equations using the Leplace transform. First, here are the odes, eq1 and eq2. */
eq1:3*'diff(f(x),x,2)-2*'diff(g(x)=sin(x);
Time=66 msec

$$3 \frac{d^2}{dx^2} f(x) - 2 \frac{d}{dx} g(x) = \sin(x) \quad (d44)$$

(c45) eq2:a*'diff(g(x),x,2)+'diff(f(x),x)=a*cos(x);
Time=66 msec

$$a \frac{d^2}{dx^2} g(x) + \frac{d}{dx} f(x) = a \cos(x) \quad (d45)$$

(c46) /* Next, specify the boundary conditions. */

atvalue(g(x),x=0,1);
Time=0 msec

1 (d46)

(c47)atvalue('diff(g(x),x),x=0,0);
Time=33 msec

0 (d47)

(c48)atvalue('diff(g(x),x),x=0,1);

1 (d48)

(c49) /* Now Laplace transform the equations from the x domain to the s domain. */

eq1:laplace(eq1,x,s);
Time=133 msec

$$3(s^2 \mathrm{laplace}(f(x),x,s) - f(0)s) - 2(s\ \mathrm{laplace}(g(x),x,s) - 1$$

(d49)

$$= \frac{1}{s^2 + 1}$$

(c50) eq2:laplace(eq2,x,s);
Time=200 msec

$$a(s^2 \mathrm{laplace}(g(x),x,s) - s - 1) + s\ \mathrm{laplace}(f(x),x,s) - f(0)$$

(d50)

$$= \frac{as}{s^2 + 1}$$

(c51) /*
We now have a pair of coupled algebraic equations where before we had differential equations. They are linear in the unknowns, the transformed functions, laplace(f(x),x,s) and laplace(g(x),x,s), so we can use the linear equation system solver, linsolve, to solve for them. */

linsolve([eq1,eq2],['laplace(f(x),x,s)'laplace(g(x),x,s)]);
Solution

laplace(f(x),x,s)

$$= \frac{3f(0)as^4 + (f(0)(3a+2) + 2a)s^2 + 3as + 2a + 2f(0)}{3as^5 + (3a+2)s^3 + 2s}$$

3. A recent reference to the field.

Edward W. Ng, editor, Symbolic and Algebraic Computation, Lecture Notes in Computer Science #72 Springer-Verlag, Berlin, Heidelberg, New York, 1979.

18.3 THE USE OF PROGRAMMABLE CALCULATORS IN THE TEACHING OF MATHEMATICS

COMPUTER USE IN MATHEMATICS EDUCATION

K.D. Graf, K.A. Keil, H. Lothe
with the cooperation of
B. Winkelmann
Universitat Bielefeld
Bielefeld, Federal Republic of Germany

Abstract

Computers, and computer-related thinking structures, are only gradually influencing mathematics education. On the one hand, there is a discrepancy between involved teachers who already have changed their own classroom teaching to a great extent, and a majority of mathematics teachers who have not even yet taken notice of the computer for teaching purposes. On the other hand, knowledge of the computer and of algorithms is frequently merely added to the mathematical subject matter. As opposed to that, the authors argue that it is necessary to genuinely integrate such subject matter, and to include general topics such as social impact and changed attitudes toward application. With regard to implementation, they develop concrete ideas which are aligned in a differentiated manner to the specific situation and the opportunities offered in the Federal Republic of Germany. The rationale for that is that only such reference to a specific situation will provide an opportunity for readers abroad to usefully apply approaches and ideas to the situation given in their own cultural environment.

Introduction and Definitions

The topic of this paper is the influence which

subject matter and methods of informatics,

real applications of data processing in industry, business and administration.

process automats, and

the computer as an auxiliary device

are presently having on the school subject of mathematics, or will have in the future.

(In accordance with IFIP and the terminology in Western Europe we use the term "informatics" instead of the older "computer science").

For school mathematics, the questions thus are:

how the algorithmic way of thinking can be integrated into mathematics, and trained;

how the computer can be used as a medium to illustrate mathematics;

which knowledge about computers and their effects is to be taught as basic (computer literacy);

which subject matter and skills from informatics and from real data processing are to be taught in school for later real life, and which consequences for the existing mathematics curriculum ought to be considered.

For these problems, we aim to show and discuss the present status within the educational system of the Federal Republic of Germany, the intentions, trends, and consequences, as well as innovation problems as they arise in our country.

Differences in other countries, e.g. the USA, mainly consist of the specific way in which informatics as a scientific subject profoundly influences educational discussion. This influence has led to more comprehensive ideas on the effect of computers on mathematics teaching, and to broader educational objectives, which is the main reason for presenting this paper on ICME IV. On the other hand, Western Germany's bureaucratic federal system and its rigid structures of teacher training cause serious obstacles which may be less pronounced in other countries.

The Present Situation

Associations (e.g. the German Gesellschaft fuer Informatik), the Federal States, and committees associated with the Federal Ministry of Science, have demanded that the basics of computers and informatics be taught in schools.

Since the mid-seventies, increasingly efficient mini- and micro-computers have been on the market for prices no longer prohibitive to schools. The number of schools having acquired a computer in order to be able to teach informatics, or to provide vocational education with respect to data processing, has since been rapidly increasing. Inexpensive, external storage facilities with random access memory recently have permitted administrative tasks to be solved by means of small computers owned by schools, thus creating a further motive for acquisition. Model experiments in the field of informatics and school administration sponsored by federal states provided the basis for extensive experience in software production and organization.

In case a school has a computer at its disposal, there is, in most cases, the desire to use it in subjects other than informatics too, especially in mathematics, and this because mathematics teachers also frequently hold informatics courses and take care of the computer.

Some Remarks Concerning the School System

Primarstufe (grades 1 to 4)

For the grades 1 to 4, curricula have been made obligatory since 1972 which prescribe "new math" for primary school. After very important innovation problems which were predominantly due to lack of teacher training, things have now calmed down somewhat.

This innovation of "new math" which has just been weathered, and the as yet insufficiently discernible objectives, methods, and contents of computer use in primary education, do not provide the least latitude for introducing such an innovation at present. In addition to that, there is the widespread distrust of many teachers and parents toward objectives such as encouraging creativity, learning how to solve problems, acquiring a positive attitude toward learning etc. independent of the traditional subject matter of arithmetics, objectives which were proclaimed in connection with the curriculum reform of new math, but which have not been attained as yet to a recognizable degree.

The use of computers can be justified for primary education only in case teaching is done according to objectives of problem-solving rather than according to aspects of subject matter.

Hence, computers will be used in primary education, if at all during the next years, only for purposes of drill and practice and for games. While this is being done - - for microcomputers too, - - only in special schools for the disabled. Special hand-held calculators for drill and practice and games, however, which will only permit tasks of lesser complexity, are very popular in families and schools. Hence it seems that an innovation of work with the computer in primary schools could develop, by small steps, from drill and game programs.

Sekundarstufe I (grades 5 to 10)

Individual Efforts of Teachers

In the field of S I, computer use is more advanced in several respects. Generally, there is a host of opportunities to establish contacts in the curricular subject matter of mathematics (see 3.2.3) which will lead to isolated uses.

Textbooks

Individual problems and applications referring to the subject matter of mathematics are already found in textbooks, such as in the case of an algorithm used for calculating a square-root or the number pi. The flow charts symbols have taken hold as a descriptive instrument. This instrument of description is also favored for recording the student activities (e.g. "How do I transform an algebraic term?").

Syllabi

A second impediment to innovation is the fact that little can be found concerning informatics subject matter in the curricula issued by the individual states of the Federal Republic of Germany.

It is certain, however, that activities using computers are organized, by individual teachers at least, as soon as the hardware requirements are met and teachers have become familiar with the equipment. Schools at which there is a S II curriculum containing the discipline of informatics (i.e. Gymnasium and Gesamtschulen having a Sekundarstufe II), or which offer elective courses in informatics in S I are at an advantage in this.

There is a special situation in the Sekundarstufe I in Berlin because of the introduction of compulsory elective subjects for the 9th and 10th grades obliging the student to choose from a number of subjects, informatics being among these.

A course aimed at concretizing the curriculum developed by the Berlin ECIS project, for instance, shows the following structure:

algorithms and programming among other things control structures, data structures, Nassi-Shneiderman diagrams)

model construction for not too complex examples

examples of computer tuse, e.g. for simulations, polls

visit to a computer centre, discussion about data-processing vocations.

There is a strong emphasis on applications of computer science.

Availabilities of Computers

As in every other field pertaining to educationa, information as to computer density in schools can only be given in an examplary fashion owing to the federalist structure of the West German educational system. For the State of Bavaria, computer density (predominantly microcomputers, but also multi-user minicomputers) can be estimated as follows for the beginning of 1980:

below 2% of Grund- und Hauptschulen

20% of the approx. 320 Realschulen

50% of the approx. 400 Gymnasien

80% of the approx. 60 Fachoberschulen

About 30% of these, however, are only numerical computers with machine language bought before 1976.

This, however, is by no means a stable picture, but rather shows the initial stage of a development which will be strongly accelerated by the low price of minicomputers.

Sekundarstufe II

While informatics courses are relatively seldom held in the Sekundarstufe I, and the computer is used, if at all, more in other subjects, especially in mathematics, informatics, which is being offered as a basic course in most of the federal states, dominates in the last stage of secondary education. In some federal states, informatics can be selected as an examinations subject for the Abitur. There is discussion in some federal states whether informatics is to be included among the subjects eligible for advanced courses. At first, there was the possibility of selecting a two-term course within the basic mathematics course only in the federal state of Bavaria. Meanwhile, this federal state prepares to advance informatics to the 10th grade of mathematics which would confront every student in class with informatics.

Recommendations as to Computers and Languages

In 1974, a council of experts established by the Federal Ministry for Research and Technology issued recommendations to the Federal Government and the federal states concerning computer equipment and use in the Sekundarstufe II. These recommendations were then in part considered unrealistic as they could not have been financed. Meanwhile, however, the hardware has become so inexpensive as to make equipment to the extent recommended feasible. Today, however, many would prefer buying several of the microcomputers offered on the market today to the multi-user minicomputer systems suggested in 1974. Among the recommendations not implemented are the large-scale training and further education of teachers, and the production of software for educational purposes. For school administration, on the other hand, some program packages were centrally developed in some of the federal states (e.g. Bavaria, Rhineland-Palatinate). In 1976, a working group established by the above ministry examined the question which computer language is suited for teaching informatics in schools. The group recommended the languages PASCAL and ELAN which strongly assist structured programming. Because of its general availability, BASIC prevails at present, especially outside informatics, e.g. in mathematics.

Implementations in Mathematics Education

Besides the fact that the necessary knowledge is frequently still lacking, there is mainly a deficit in practical hints for computer use in teaching, in teaching models, and in didactical, methodological, and organizational helps. Textbooks for the last stage of secondary education seldom make allowance for available computers.

The most important subject matter field for the last stage of secondary education in Western Germany is analysis. Analytical geometry/linear algebra are also important, and probability calculus and statistics have become increasingly important during the last ten years. The computer, in this, may take over multiple roles, for instance

illustration of convergence behaviour

graphic assistance of curve discussions

illustration of mathematical methods, e.g. numerical integration

calculating power, e.g. in case of determining zeros of functions, value tables

simulations in probability and statistics

production of data of any given distribution

realization of algorithms developed as problem solutions by students.

Numerical methods, which in the present curricula have been rather deemphasized, would be particularly suited to computer treatment.

Conclusion

Computer use and integration of informatics subject matter into mathematics education are practically absent in primary school, and still in their beginnings in stages I and II of secondary education. During the last three years of Sekundarstufe II, however, informatics is being taught in many schools already.

Computer equipment of schools will be complete, to a certain degree, only after a few more years have elapsed. It is not yet clear, however, how many terminals or personal computers will be necessary for a school.

Teacher Education

General Structures

Teacher education, in the Federal Republic of Germany, is provided mainly according to three different teaching careers:

a. for teachers and instructors at "Grund- und Haupstschulen" (first to tenth school year);

b. for teachers at Gymnasium (fourth/sixth to 13th school year); and

c. for teachers in vocational education.

In general, teachers are trained in two or more school subjects. Teacher education is subdivided into

a. predominantly scientific studies for 7 to 10 semesters at a university or teacher college, and

b. a practical stage of teaching in school (Referendariat) lasting between 1 and 2 years.

At the end of both these stages there is a separate examination (first and secont Staatsexamen).

University and Examination Regulations with Respect to Informatics

In the present official examination regulations, subject matter fields such as "data processing", "Computer", "Computer Use", or "Informatics" are not listed as yet, neither as independent disciplines nor within other subjects of study (subjects, educational science, fundamental science). Exceptions to the rule are examination regulations in some federal states (e.g. Bavaria, Berlin) providing for the new profession of teaching informatics at vocational schools, or, in some cases, for the upper grades of generally educating Gymnasien.

University-specific study regulations for mathematics, however, are increasingly incorporating elements of study, as so-called "compulsory elective subjects", such as "Programming", "Computer Use", "Information Processing", or "Informatics". Students are allowed to select these subjects for their examinations.

At the Paedagogische Hochschule Berlin, "Informatics" is a comprehensive part (approx. 40%) of studies in the subject of cybernetics. It is studied mainly in combination with mathematics.

Informatics Courses Offered

It is almost exclusively mathematics teachers who teach subject matter and methods of informatics.

Responses to questionnaires sent out in 1977 and 1978 yielded the fact that 42 of the approximately 75 West German institutions of higher education charged with teacher education offer regular courses for the general education of prospective teachers in methods of data processing or in the subject matter of informatics.

Generally, it can be stated that there is a tendency to incorporate informatics into teacher education as an auxiliary discipline, and to prepare appropriate classroom teaching of the latter by offering didactically oriented courses. The bulk of these refers to mathematics education.

The educational activities listed in the questionnaire responses can be classified as follows:

1. Courses offered for pedagogical studies and for fundamental education.

2. Computer application in subject-specific education and teaching.

3. Education for teaching informatics.

Teaching Personnel

Least advanced, of course, is staffing higher education with teaching personnel both qualified for teaching informatics and accessible for work in teacher education. There are only 5 teaching posts known in the entire Federal Republic designated for informatics in education.

Such university teachers and assistants as get involved in the courses mentioned usually hold a teaching post for mathematics, or for didactics of mathematics, or for informatics as a profession. Mathematics teachers in higher education are usually self-taught with respect to informatics.

Intentions and Trends

To date, there are no curricular decrees issued by the governments of the federal states, nor are there recommendations of the professional associations concerning the integration of the computer into school mathematics. Or, more precisely: curriculum elements or recommendations at present still concern informatics subject matter and are at best simply attached to the school subject of mathematics. Existing curricula concernng informatics subject matter, and professoinal associations' recommendations on informatics, however, very strongly influence the curricular and didactical view taken of computer use in mathematics education.

The reason for this is that informatics is for the most part taught by mathematics teachers. Teaching informatics - be it as an independent subject or as part of mathematics education - seems to be closest to the self-understanding of mathematics teachers. Besides, schools expect the mathematics teachers to become involved in the field of computer use.

The general objectives - as they are presented below - are primarily aimed at establishing informatics as a school subject. But if informatics as an independent subject is impossible - as is true to a considerable degree in Sekundarstufe I - it will most probably be mathematics as a school subject which will be confronted with these demands and requested to find a way of integrating these elements of informatics.

This pressure on mathematics as a subject is not alleviated by understanding computer use, which has its locus in the sciences and in the social sciences, as a cross-disciplinary technology. As long as informatics is mainly understood as a methodological discipline for systematically solving and programming problems, many will say that mathematics as a school subject may very well even profit. Integration of actual data processing subject matter and of its social effects, however, is frequently seen as not belonging to school mathematics, and rejected because of the latter's little scope for innovation. The following remarks on the objectives and consequences should be considered bearing this in mind.

General Objectives

By various groups in society, and from diverse positions, school has been confronted with demands to help society to cope with data processing, and to teach basic skills required for computer use. We are going to classify these demands according to four aspects. After that, we are going to examine the consequences these general objectives for teaching computer-related subjects - be it within mathematics, or in the shape of informatics as an independent subject in school - will have on school providing general education.

Any decision as to what is important will be temporary. Until some years ago, algebra of logical circuits and programming on machine level were thought to be important, today, and probably for some time to come, it is algorithmics, i.e. the theory of defining data structures and of developing algorithms.

In future, however, and in case of further development of computer systems, it could be computer-adapted symbolizing of problem solutions. Even today, many computer applications consist of using ready-made software packages, the individual elements serving as black boxes solving complex tasks. One characteristic of the mathematical machine of the "computer" thus seems to be a hierarchy of precisely defined elements which, as a rule, will have to be accepted by the user as black boxes.

Computer Use as a Qualification

To use computers is a skill increasingly necessary for vocational and private life. School is expected to teach the basic skills and abilities required. The important thing for this education is less programming as such, but rather using the computer for processing texts, retrieving information, and as a calculating device in problem solving, etc. The ability to work with any editing or requiring system, or the ability to quickly get familiar with such a system, however, can only be taught by letting students work with machines offering suitable opportunities.

Besides this aspect of teaching basic skills, there is the objective of providing the students with insights about the opportunities and limits of the computer. Extreme confidence in computers is best reduced by practically working one. This also helps to emphasize the necessity of critically examining all data issued by the computer and of checking numerical computer results by rough estimate calculations. This aspect of de-mystifying the computer already extends to the field of social effects of computer technology.

Data Processing and its Social Impact

School should treat real computer applications in industry and administration in a model fashion in order to make social problems subject to discussion. Above all, the mechanisms of processing large amounts of data should be integrated into mathematics as one aspect of algorithmics.

Another important field of real data processing is process monitoring which is a candidate for being treated in a model fashion in school. This involves the concepts of feedback, feedback loops etc. This is how mathematics as a school subject may enhance its reference to real life, as mathematics in business and administration practically always will consist of using computers today.

Computers as Classroom Media

School mathematics should profit from the opportunity of concretisizing mathematical situations by means of the computer as a medium. Almost every subject matter field of school mathematics offers opportunities of using the computer to illustrate mathematical facts (in the widest sense of the term):

geometrical objects can be manipulated in graphic representation,

stochastic processes can be simulated with random number generators,

real processes can be represented bymeans of equations with differences instead of differentials and experienced as simulations,

data can be generated for heuristic considerations; large amounts of data can be graphically summed up,
concepts belonging to a dynamical understanding of mathematics such as iteration, recursion can be demonstrated; in particular, the process of convergence can be illustrated by means of numbers, or graphically,

languages used for the representation of algorithms (including flow charts, Nassi Shneiderman diagrams etc.) can be applied in mathematics as well.

Impact on Curricula

Among the intentions listed above, the view on the present subject matter of mathematics must change, lest activity with the computer become something alien to mathematics education, and in order to achieve genuine integration.

Data

Besides numbers, general data types are important as well: strings, Boolean variables, vectors, matrices, sets, elements of data spaces and fills. Even if the extent to which other data types than numbers can be treated in school is considered modest, the view taken of traditional subject matter must change.

Correspondingly, attention of mathematics educators should turn to the opportunity of data abstraction, of defining the selector functions and operators, and of combining into classes, modules, packets or the like offered by informatics. Expressing - say residual classes arithmetic - in a language premitting such abstractions will be more concrete and manageable for students than the concept of the algebraic structure of the quotient ring.

Algorithms

One of the important consequences of computer use in schools certainly is that students have to be taught less in executing algorithms, and more in formulating algorithmic processes.

Applications

Another consequence of computer use for the subject matter taught is that it does away with the necessity of confining selection of tasks and problems to such with simple numbers and easily solvable equations. It is only realistic numbers taken from a factual situation, complex problems, not artificially simplified ones, which will yield motivating tasks taken from real life. Beyond that, using the computer for a task may mean a decisive lowering of the abstraction level required. Just think of problems which, in the shape of differential equations, would be inaccessible for students, but manageable in the shape of difference equations, as the computer takes care of the considerable amount of numerical calculations.

Conclusion

Under the influence of informatics and computer use, mathematics as a school subject shows a more complete picture of what mathematics is in real life: proving is now more strongly accompanied by calculating, i.e. reaching the actual solution. The accessibility of the computer as a mathematical machine also has an impact on the school subject:

subject matter changes as to relative weight and appropriate views,

mathematics' application range is enhanced, and

the methods of mathematizing change under the influence of the new auxiliary device.

Changes in the Working and Teaching Style

For the teaching style, and for the working style of the individual student, far-reaching consequences may be derived from the objectives listed above. In case of didactical or methodical proceeding in school, the computer permits more experimental work with mathematical subject matter: the computer can generate data which prompt heuristic considerations, it can create a feeling for mathematical relationships by means of numerical calculations, etc.

The obvious consequence for the classroom seems to be that the computer shall be used, for illustrations, by the teacher. In this case, students may voice ideas and suggestions just as in the case of the teacher experiment in physics, but will not have immediate access.

Potential Subject Matter Elements for Mathematics in Sekunkarstufe I

For school mathematics in the Sekundarstufe I, the following areas are considered important:

a. Propaedeutics of Algorithmizing

"algorithms of everyday life" (phoning, getting up, etc.) and behaviours in the mathematics classroom such as written calculations, solving equations, evaluating algebraic terms etc.

verbally describing and algorithmizing such behaviours

representations by means of flow diagrams or other graphical systems (e.g. structograms),

student executes algorithms, student plays the computer, e.g. filling tables in according to prescription,

trying out seqences of key strokes with non-programmable or formula-programmable hand-held calculators under student control,

developing calculating schemata, game rules,

representing calculations by means of operators, assignments, developing sequences of key strokes.

This area can be treated in a loose fashion starting with the 5th grade by means of suitable classroom subject matter; it does not require additional time.

b. Applications in the Subject of Mathematics

that is algorithmizing and running suitable problems taken from school mathematics on computers. In this, the insights of the students concerning the program itself may range from no insight at all, i.e. using the program as a black box, to developing programs by themselves. Independent of that, classroom organization of computer use, and thus student access to the computer, may range from pure demonstrative teaching by the teacher to individual student exercises on the computer.

Subject matter fields suitable for applying programming in mathematics are

numerical methods, convergency, propaedeutics of infintesimal calculus,

tasks belonging to number theory,

statistics (e.g. generating data),

simulations (e.g. probability experiments, parameter variations),

strategic games,

etc.

c. Contents and Methods of Real Data Processing

Insights into techniques of

appropriate structuring of data

handling large amounts of data in files

should be treated in an examplary fashion in order to make real life applicatins of data processing and their social implications accessible to discussion.

Suitable subject matter areas are, among others

student data, staff data, community data (problem of privacy),

storage bookkeeping, seat reservations,

giro transfer business etc.

Practical work with the computer is indispensible in this field, e.g.

to have the student experience how a small program will have an important effect by working on a large file,

for the insight that objectivating and formalizing decisions concerning people will eliminate important aspects,

etc.

The suitable working mode might be project-oriented (perhaps cross-subject) work.

d. Mathematics Taught my Means of Informatics Subject Matter

The integration of informatics subject matter into mathematics education of the Sekundarstufe I permits more intensive pursuit of various objectives and intentions of mathematics education. Examples should be selected according to these criteria; a systematic course in one part area of informatics would result in too strong an impact on the present mathematics curriculum.

Examples which could be used in mathematics education are:

non-numerical algorithms (searching, sorting, etc.)

strings as mathematical objects (position systems, letter games, syntax and semantics, etc.),

translation and execution algorithms (e.g. algebraic expressions with preference of operators and brackets),

linear lists, trees (strategic games, sieve of Eratosthenes, etc.),

machines as mathematical objects (e.g. hald-held calculators),

problems of complexity and efficienty under mathematical aspects (e.g. in case of algorithms for multiplication or power calculations),

etc.

Consequences for Teacher Education

The developments listed in the above sections require adequate preparation of the teachers in the course of studies. For those teachers already teaching, an intensification of further education is generally demanded. In this connection, a comprehensive study including a conception for further education in data processing has been submitted by the German Bundesinstitut fuer Berufsbildungsforschung.

For an appropriate and successful computer use in the future however, inclusion of a regular subject of studies "basics of informatics/data processing" into the curriculum of prospective mathematics teachers, and perhaps also into that of other subjects such as physics, etc. will be crucial.

In particular, the following requirements seem to be reasonable:

1. Knowledge and skills in algorithmics, the emphasis being on their importance as a mathematical/mathematisizing method. There should be awareness of the value of the algorithmic method

   for the development of mathematical concepts and methods, and for economical problems solution,

   for a cognitively efficient representation of problems and the respective solution methods,

   for the treatment of non-numerical problems.

2. Knowledge and skills with regard to at least one higher programming language, and to the methods of structured programming. The teacher must be able to program at least the algorithms of his teaching area, and to realize them on the computer. In addition, we list familiarity with the basic structure of computer systems among this heading.

3. Knowledge and skills concerning applications of data processing relevant for practical purposes. This is important, above all, for fields removed from mathematics and the sciences. The teacher must also be able to treat such applications in the classroom in a model fashion; partly from the user's point of view, too. Besides, this includes an overall view of the social relevance of data processing, computers, and informatics.

Innovation Problems

Problems of Mathematics as a School Subject

Reform-Weariness

For the school subject of mathematics, there is no intrinsic reason to integrate the computer and informatics subject matter. There is nothing to show that an improvement of mathematics achievement due to integrating computers into mathematics education has been proved by research. Hence it is understandable that many mathematics educators do not see the need to introduce this innovation now.

Conversely, pressure to introduce this innovation is exerted in the main by the fact that microcomputers are already present in school despite the short time they have been available on the market. Teachers willing to tackle new things, or teachers prompted by the subject of informatics offered in Sekundarstufe II, will certainly use the computer in their teaching in other school levels. These individual activities, however, serve to obscure the problem how school mathematics as a whole is to be reformed.

Mathematics Teachers

The broad majority of mathematics teachers, in fact, has no experience at all with computers, and is not necessarily willing to tackle this problem on their own. Besides, such subject matter is present in teacher education only in a few instances, as most of the mathematics teachers responsible for teacher education share the attitude mentioned. Thus we are faced with a situation in which, on the one hand, very involved teachers in school and in higher education attempt to solve the problem of fruitfully introducing the computer to mathematics education, but in which, on the other, these staff resources are insufficient to reshape mathematics to a larger extent.

Intensive support of this innovation by increaed education in informatics will require much time. The urgent demand for computer knowledge for the school of today thus can only be met by further education, and by organizing an exchange of experience. Efforts made in Bavaria shall be briefly outlined as an example for that.

The Center for Programmed Instruction and Computers in Education concerned with computer assisted learning (CAL) since the beginning of the 1970's, has been charged by the Bavarian Ministry of Education to advise and to instruct teachers in the use of computers in school and to coordinate continuation courses.

Open Questions in the Discipline of Informatics

An important obstacle to innovation by means of computers in school is the fact that the requirements to be asked of computers, languages, and systems, as well as of programming methodology are not yet clear.

While informatics experts demand higher languages like PASCAL because of the support they give to structured programming, we must face the fact that schools only dispose of microcomputers using BASIC. While experts prefer in general compiler languages, teacher and students are having their first experiences with BASIC dialogue systems.

This discrepancy makes the experts lose some of their authority. As a response of school mathematicians, a "BASIC-mathematics" is beginning to emerge which may become an obstacle to the futher development of computer applications to mathematical problem formulations, and an obstacle to the further development of programming methodology, as soon as this kind of teaching has spread to a certain degree.

There are also differences of opinion as to who will decide on the subject matter to be taught to teachers. One side is taken by the mathematicians, especially the mathematics educators, who consider themselves responsible for this decision, and the other by the informatics experts who feel just as competent. Insofar, the situation may be compared to the "contact problems" mathematics has had with subjects like physics or social sciences.

An understanding is sought at present in a working group "Informatics" established by the Gesellschaft fuer Didaktik der Mathematik, and in a working group "Informatics in School" established by the Gesellschaft fuer Informatik.

Problems of the Didactics of Mathematics and of Informatics

An important reason for the reluctant introduction of computers to mathematics education are also many questions of didactics and educational theory still unresolved for teachers and for the responsible authorities.

There is a growing insight in those involved and responsible that mathematical subject matter and informatics subject matter for the classroom may look rather different, but the intentions pursued will frequently concur. Seen from the viewpoint of general eduation, it will always be a matter of teaching the skills required for systematical solving of informational problems, including justification why the solution found is correct, for both the subjects of mathematics and informatics in school.

The fundamental concerns of mathematics, calculating and proof, are to be found, in principle, in informatics as well. There is a correspondence, so to say, between the teaching method of mathematizing, and the teaching method of "computerizing", i.e. that which is generally termed "computer-adapted mathematizing" or "algorithmizing".

The principles of mathematics education normally include the logically complete justification of any method used, for instance, to calculate function values or to solve problems of differential or integral calculus. Therefore, a special didactical problem is that the computer offers the opportunity of using programs for problem solving as elements which are black boxes for the student. These could be, for instance, program packages resp. methods for approximate solutions, or problems of number theory. This recourse to something unknown and unexplained is indeed alien to the didactics of mathematics. Such a method, which has indisputable advantages, thus must be made legitimate by partial elucidation of the black boxes, or by describing the principles of the tasks allotted to such programs. Any decision about the feasibility of introducing such methods to mathematics education, and about their implementation, will require thorough discussion and prolonged trial runs.

Besides these relatively general didactical problems oriented mainly toward educational theory, there are, however, practical problems as well which must be solved when introducing the computer to the classroom, particularly to mathematics education. This will require time as well. These questions become particularly relevant for the ultimate transition from the present experimental stages, in which the students behave differently from the normal classroom situation, to regular computer use. The didactic of the subject concerned then must solve the following problems:

classroom organization for computer use, e.g. suitable equipment and social forms;

provision of teaching materials,

achievement and outcome controls,

motivating all students to work with the computer as a tool, e.g. by means of suitable subject matter and mode of work.

Other Problems

Cost

Of course, the cost of adequately equipping schools with computers for the objectives mentioned will be decisive, or at least for getting the support of the authorities. It is true that the decrease in prices during the last years has overruled many objections based on earlier cost estimates, but equipping all schools with computers remains a considerable financial problem. This is all the more true as objections have been raised against many of the present cheap personal computers from the point of view of informatics and of teaching.

Resistance Offered by Other Subjects

Naturally, it is not only subject-specific considerations which influence the decision to equip a school with a computer. An important obstacle must also be seen in the fact that curricula are overloaded in all school types and grades, and in the demand for extension as to subject matter and time allotted voiced by many disciplines simultaneously. This desire, of course, implicitly, and some times even explicitly, runs counter to changes and extensions for mathematics education. In particular, there is a strong public movement in favour of socially or social science oriented subjects and against mathematics/sciences. School is increasingly requested to make allowance for social learning at the expense of cognitive intentions. This must not necessarily be an objection against computer use, but at first glance the latter is rather subsumed among cognitive intentions.

QUELQUES EMPLOIS DU CALCULATEUR PROGRAMMABLE DANS UNE CLASSE

Guy Noel
Universite de l'Etat a Mons
Mons, Belgium

Depuis l'apparition en 1972 du premier calculateur scientifique de poche, puis en 1974 du premier calculateur programmable de poche, de nombreuses publications ont ete consacrees a des compte-rendus d'experiences d'utilisation de ces machines dans l'enseignement. La bibliographie, situee en fin d'article, est tres fragmentaire et mentionne essentiellement, a titre exemplatif, des textes en langue francaise. On peut toutefois deja repartir les utilisations possibles des calculatrices programmables en plusieurs categories dont nous retiendrons les quatre suivantes.

A. Utilisation de la Calculatrice Comme Moyen de Calcul

Cette categorie est sans aucun doute la mieux representee. Il s'agit pour l'essentiel de textes relatant l'enseignement de methodes numeriques qui n'etaient pas abordees precedemment faute de moyens de calculs. On trouve par exemple un grand nombre d'algorithmes iteratifs particulierement adaptes aux machines programmables. L'introduction de ces algorithmes rend l'enseignment des mathematiques plus concret. Les eleves acquierent la possibilite de pousser la resolution d'un probleme jusqu'au bout. Leur bagage mathematique est incontestablement plus important et plus oriente vers les applications. En meme temps ils s'initient aux principes de l'informatique. Leur attention est attiree en particulier sur l'efficacite comparee de divers algorithmes, leur rapidite d'execution, la necessite de rediger des programmes aussi economiques que possible, les problemes d'approximation, etc. C'est un mode de pensee algorithmique et informatique qui s'installe progressivement.

Il ne semble pas necessaire de developper plus en details ce point de vue (cfr. (1), (2), (4), (6), (7), (8), (9), (12), (13)).

B. Utilisation de la Calculatrice pour realiser des simulations

On utilise ici la calculatrice comme un outil premettant de remplacer des experiences concretes. Les eleves sont amenes a construire un modele algorithmique d'un phenomene dynamique. L'execution du programme leur fournit des exemples d'evolution de ce phenomene. En faisant varier les parametres de depart, il est possible d'observer la facon dont les resultats dependent de ces parametres. La technique de simulation est particulierement indiquee et efficace lors de l'observation de phenomenes probabilistes (cfr. (5), (6), (7), (9), (13), (15)). L'utilisation d'un generateur de nombres pseudo-aleatoires permet en peu de temps de repeter une experience probabiliste suffisamment soubent pour pouvoir observer le phenomene de regularite statistique. De telles experiences sont, autrement, a peu pres irrealisables en classe, alors qu'elles sont indispensables pour former l'intuition probabiliste.

Il est bien entendu possible de simuler aussi des phenomenes de type deterministe. On notera par exemple divers jeux dans (11) (12), des etudes de trajectoires dans (7), des simulations de phenomenes biologiques, physiques et chimiques dans (4). On rangera egalement dans la categorie des simulations les utilisations de tables tracantes pour representer des solutions d'equations differentielles, des transformations lineaires planes, etc. (Voir (4), (16), et l'article de A. Deledicq dans (5)).

C. Utilisation de la Calculatrice comme Support Heuristique lors de l'Introduction d'une Nouvelle Notion

Dans les activites decrites aux deux points precedents, la calculatrice fournit l'occasion de presenter aux eleves des theories ou methodes qui n'etaient pas enseignees auparavant. Nous essayerons dans les deux points suivants de montrer quelle aide la calculatrice peut apporter lors de l'introduction de points plus traditionnels. La premiere idee est d'utiliser la calculatrice comme support heuristique permettant aux eleves d'acquerir une experience et une intuition numeriques (cfr. (1), (5), (12), (14), (16), (17), (18)). La calculatrice joue alors pour le domaine numerique (algebre, analyse, arithmetique) le meme role que le dessin pour la geometrie. Et de meme que les figures geometriques supportent le raisonnement sans le remplacer, de meme la realisation d'experiences numeriques permet de mieux apprehender la portee et la validite des raisonnements numeriques. Bien entendu des precautions doivent etre prises car le calcul sur machine ne reflete que imparfaitement les proprietes du modele ideal: le calcul dans IR. C'est particulierement lors de l'introduction a l'analyse que la discordance entre la situation du modele IR et celle du calcul concret (sur machine) est la plus importante. Par exemple une serie peut etre divergente et donner, sur machine, l'impression de converger (ou vice versa). Certains en ont conclu qu'il fallait eviter l'emploi des calculatrices lors de l'initiation a l'analyse (voir par exemple l'article de Kuntzmann dans (1), et les lettres de lecteurs provoquees par (16)). Ce point de vue semble excessif. Transpose en geometrie, il reviendrait a proscrire l'utilisation de figures sous pretexte qu'il est possible d'en realiser qui induisent en erreur. Il convient de faire comprendre aux eleves que la calculatrice ne fournit que des approximations et qu'il peut en resulter des aberrations. Eventuellement ceci amene a une etude plus approfondie du calcul approche. Donnons a present deux nouveaux exemples d'utilisations du caclulateur comme support heuristique dans l'etude de notions d'analyse.

1. Introduction de la notion de limite

Limitons-nous a la formalisation de la notion de suite convergente.

L'idee intuitive de convergence correspond au phenomene de stabilisation des decimales qui apparait quand on calcule sur machine les valeurs des termes successifs d'une suite. Il faut passer de cette idee a la formalisation traditionelle en , N.

Calculons, par exemple la valeur de $(n^5 + 1/n^5)$ pour les valeurs de plus en plus grandes de n. Quelle que soit la machine que nous utilisons nous trouverons 1 si nest assez grand. Les eleves devraient admettre sans difficulte qu'en realite, $1 + 1/n^5$ n'est egal a 1 pour aucune valuer de n.

Pour quelle raison alors la machine affiche-t-elle 1? Et a partir de quelle valeur de n le fait-elle? Ici il est utile que les eleves utilisent des machines differentes, les unes travaillant avec, par exemple, 8 chiffres significatifs, les autres 10, . . .

Car la reponse depend du nombre de chiffres significatifs enregistrables dans une memoire. Si ce nombre est k (et si la machine arrondit), le nombre 1 sera afficher a la place de $1 + 1/n^5$ si $1/n^5 < 5.10^{-k}$. Voici les valeurs minimales de n pour les valeurs les plus courantes de k (en admettant que la machine affiche tous les chiffres significatifs).

k	8	10	11	13
n	29	73	115	289

La comparaison des diverses machines montre aux eleves que la convergence de $(n^5 + 1/5)$ vers 1 ne peut se deduire du fait qu'une machine determinee affiche 1 a partir d'une autre valeur de n, puisque une autre machine, travaillant avec plus de chiffres, affiche encore a ce moment un nombre different de 1.

En realite, pour pouvoir affirmer la convergence de $(n^5 + 1/n^5)$ vers 1, il faudrait que n'importe quelle machine, reellee ou hypothetique, quel que soit son nombre de chiffres significatifs, affiche 1 si n est assez grand. En formules:

$$\forall k \in \mathbb{N} \quad \exists n_0 \in \mathbb{N} \quad \forall n \geq n_0 : \left| \frac{n^5+1}{n^5} - 1 \right| < 5.10^{-k}$$

De la, on passera aisement a la formulation traditionelle et on generalisera apres eventuellement d'autres exemples choisis de facon que les erreurs d'arrondi ne provoquent pas de difficultes.

II. La formule de Taylor

La signification intuitive de la formule de Taylor est souvent perdue par les eleves de la fin de l'enseignement secondaire. On leur donne une fonction, ils calculent un polynome sans toujours voir que l'important est la majoration du reste. Ce qu'il faut leur faire comprendre, c'est que le developement taylorien de degre n est le polynome de degre n qui localement approche le mieux la fonction donnee. Il faut donc disposer d'un moyen de comparer les valeurs de deux approximations par deux polynomes de meme degre.

C'est assez facile au degre 1. Considerons par exemple la fonction $f(x) = 1/(x^2 + x + 2)$ et cherchrons la fonction affine $g(x) = mx + p$ qui approche le miwux f au voisinage de 0. On impose d'abord $f(0) = g(0)$, d'ou $p = 1/2$. Puis, pour differentes valeurs de m, on calculera $f(x) - g(x)$ (voir tableau ci-dessous). Bien entendu cette difference approche 0 en meme temps que x, ce qui ne permet pas de conclure. Mais on peut tenir compte de ce que cette difference $f(x) - g(x)$ est normalement d'autant plus grande que x est plus eloigne de 0. On est ainsi amene a relativiser cette difference en la divisant par x (qui sert en quelque sorte de fonction-etalon). On constate que $(f(x) - g(x)/x)$ ne se rappraoche de 0 que si $m = -1/4$. On a ainsi trouve une caracterisation de la meilleure approximation de f par une fonction affine: la difference $f(x) - g(x)$ doit tendre vers 0 plus vite que x.

x	f(x) − g(x)			[f(x) − g(x)]/x		
	m = −1	m = −½	m = −¼	m = −1	m = −½	m = −¼
−0,5	−0,428	−0,178	−0,054	0,857	0,357	0,107
−0,5	−0,332	−0,131	−0,032	0,830	0,329	0,079
−0,3	−0,241	−0,091	−0,016	0,804	0,304	0,054
−0,2	−0,157	−0,056	−0,006	0,783	0,282	0,033
−0,1	−0,076	−0,026	−0,001	0,764	0,264	0,014
0	0	0	0	*	*	*
0,1	0,073	0,023	−0,001	0,739	0,239	−0,011
0,2	0,146	0,046	−0,003	0,732	0,232	−0,018
0,3	0,218	0,068	−0,007	0,728	0,228	−0,022
0,4	0,290	0,090	−0,009	0,727	0,226	−0,023
0,5	0,364	0,113	−0,011	0,727	0,227	−0,023

Au degre 2, la situation est un peu plus compliquee. On cherche la meilleure approximation de f, au voisinage de 0, par un polynome de degre 2. Il est naturel d'essayer $g(x) = 1/2 - 1/4\,x + cx^2$ et de rechercher la meilleure valeur de c. Pour comparer les valeurs de c, il est tout aussi naturel, vu ce qui precede, de calculer
$(f(x) - g(x)/x)$. Et on constate que quel que soit c, cette expression tend vers 0 en meme temps que x. La fonction x est un etalon trop grand pour pouvoir distinguer les differences $f(x) - g(x)$. Il faut choisir un etalon plus petit, une fonction qui tend vers 0 plus vite que x. La plus simple d'entre elles etant x^2, on calcule $(f(x) - g(x)/x^2)$, et on constate que cette expression ne tend vers 0 que si $c = -1/8$ (construire un tableau analogue au precedent).

On est donc amene, apres cette experimentation numerique, a adopter l'idee suivante:

La meilleure approximation de f, au voisinage de x_o par des fonctions polynomiales de degre n, g, est celle (si elle existe et est unique) pour laquelle

$$\lim_{x \to x_o} \frac{f(x) - g(x)}{(x - x_o)^n} = 0$$

L'etude numerique ci-dessus doit evidemment etre completee par des representations graphiques faisant apparaitre que telle fonction polynomiale "colle mieux" a f que telle autre. Cette etude numerique n'est possible que grace a l'emploi des machines programmables. Elle a pour but de creer un terrain intuitif favorable permettant d'enoncer et de demontrer la formule de Taylor.

D. Utilisation de la Programmation pour Etablir des Formules

Jusqu'ici la programmation en elle-meme ne jouait que le role d'un outil - - un outil important, mais qui s'efface devant le resultat du calcul numerique. Mais le mathematicien manipule aussi des donnees litterales. Et il resume ses conclusions en etablissant des formules. Ainsi les solutions de l'equations $ax^2 + bx + c = 0$ sont donnees par la formule

$$x = \frac{-b +/- \sqrt{b^2 - 4ac}}{2a}$$

Nous pouvons parfaitement considerer cette formule comme un programme, ecrit dans un langage particulier que nous appellerons "langage-formule" et qui est adapte a l'execution par un calculateur humain.

Quand on dispose de calculatrices dans la classe, la tendance est d'adopter, a propos de l'equation du second degre, le schema suivant:

a) On etudie des equations particulieres, que l'on resout par la technique de "completion du carre parfait":

$$3x^2 + 2x - 1 = 0 \qquad (1)$$

$$3(x^2 + 2/3x) - 1 = 0 \qquad (2)$$

$$3(x^2 + 2 \,.\, 1/3x + 1/9) - 1 - 1/3 = 0 \qquad (3)$$

$$3(x + 1/3)^2 - 4/3 = 0 \qquad (4)$$

$$(x + 1/3)^2 = 4/9 \qquad (5)$$

$$x + 1/3 = +/- 2/3 \qquad (6)$$

$$\begin{cases} x_1 = 1/3 & (7) \\ x_2 = -1 & (8) \end{cases}$$

b) On etablit la formule en generalisant cette technique a l'equation a coefficients litteraux:

$$ax^2 + bx + c = 0$$

$$a(x^2 + (b/a)x) + c = 0$$

$$a(x^2 + 2(b/2a)x + b^2/4a^2) + c - b^2/4a = 0$$

$$a(x + b/2a)^2 + (4ac - b^2)/4a = 0$$

$$(x + b/2a)^2 = (b^2 - 4ac)/4a^2$$

$$x = -b/2a +/- \sqrt{(b^2 - 4ac)/4a^2} = (-b +/- \sqrt{b^2 - 4ac})/2a$$

On applique cette formule "a la main" a de nouvelles equations.

c) On demande aux eleves de traduire la formule obtenue en langage-machine de maniere a faire executer les calculs de facon automatique.

Dans cette organisation, la programmation apparait comme une simple operation de traduction du programme, du langage formule vers le langage machine. La phase difficile est la deuxieme, car elle fait intervenir une generalisation et une abstraction. Elle debouche sur une formule synthetique qui decrit tout un calcul en quelques symboles. Le programme ecrit en langage-machine est au contraire "analytique" car chaque instruction ne fait intervenir qu'une operation elementaire. La troisieme phase constitue en quelque sorte un retour en arriere: l'algorithme ecrit en langage machine ne se laisse pas apprehender globalement comme il le fait quand il est ecrit en langage-formule. L'activite de programmation est interessante en ce qu'elle constitue precisement une activite d'analyse. La formule reste un objectif cognitif interessant vu son caractere synthetique. Et comme toute synthese, elle devrait plutot venir en dernier lieu.

Essayons donc de permuter les phases n° 2 et 3 du schema ci-dessus: apres que les eleves aient resolu quelques equations particulieres, de maniere a avoir compris la methode de "completion du carre parfait", on leur demande de rediger un programme permettant de resoudre n'importe quelle equation de degre 2 (a ce stade, on passe sous silence le fait que certaines equations n'ont pas de solution).

Trois memoires R_0, R_1, R_2 seront utilisees pour stocker a, b, c. Analysant la suite de calculs mentionnes en a), on observe les calculs numeriques suivants:

en (2): $R_1 \leftarrow R_1/R_0$ (b divise par a)

en (3): $R_1 \leftarrow R_1/2$

en (4): $R_2 \leftarrow R_2 - R^2 . R_0$

en (5): $R_2 \leftarrow - R_2/R_0$

en (6): $R_2 \leftarrow \sqrt{R_2}$

en (7): $x_1 \leftarrow R_2 - R_1$

en (8): $x_2 \leftarrow -R_2 - R_1$

Ce programme peut tres bien etre redige et execute avant que la formule traditionelle ait ete ecrite. On pourra alors rencontrer le probleme des equations sans solution en essayant de faire executer le programme pour une equation de ce type. La machine affichera en (6) un signal d'erreur: R_2 est negatif. A quelle condition sur les coefficients a, b, c cela correspond-il? On examinera l'evolution du contenu des memoires:

	R_0	R_1	R_2
(1)	a	b	c
(2)	a	b/a	c
(3)	a	b/(2a)	c
(4)	a	$\frac{b}{2a}$	$c - \frac{b^2}{4a}$
(5)	a	$\frac{b}{2a}$	$\frac{b^2}{4a^2} - \frac{c}{a}$
(6)	a	$\frac{b}{2a}$	$\frac{\sqrt{b^2 - 4ac}}{2a}$

A ce moment, on a trouve la condition pour que l'equation admette une solution (ce qui permet de completer le programme en introduisant un test) mais de plus, on n'est qu'a un pas de la formule qui donne les racines. Cette formule apparait ici comme un sous-produit de l'activite de programmation qu'elle resumera dans l'esprit d'eleves. La phase de generalisation du schema traditionnel a ete ici coupee en deux parties: la premiere lors de la programmation, la seconde quand on examine l'evolution du contenu des memoires. Les difficultes sont donc morcelees. Le tout est supporte par l'utilisation de la machine, qui donne un caractere concret meme aux phases abstraites.

On voit qu'il est possible dans certains cas d'integrer l'activite de programmation a la construction mathematique elle-meme. On peut en avoir un autre exemple quand on veut resoudre et discuter le systeme general de deux equations lineaires a deux inconnues, par la methode de Kramer: faisons comme si ce systeme avait toujours une solution et faisons programmer celle-ci. Lors de l'execution, glissons dans la liste d'exemples a traiter des systemes n'ayant pas de solution, ou en ayant plusieurs. Cela obligera a introduire des tests pour ameliorer le programme. Quand celui-ci sera correct et complet, la discussion sera terminee et il ne restera qu'a la synthetiser.

Il faut encore noter que les idees precedentes restent valables meme si la calculatrice precedente n'est pas programmable. Car un programme est simplement la suite des touches a enfoncer et cette liste peut etre etablie pour n'importe quelle calculatrice. Ainsi etablir un programme qui permet de trouver toutes les decimales du quotient de deux naturels est possible meme si on ne dispose que d'une calculatrice elementaire. Et cela fait mieux comprendre l'algorithme de la division.

En Guise de Conclusion

Les quatre types d'emploi d'une calculatrice programmable mentionnes ci-dessus ne sont certainement pas les seuls. L'ecueil a eviter est de faire apparaitre l'emploi de la calculatrice comme une activite complementaire se juxtaposant au cours traditionnel de mathematique. La machine doit au contraire etre integree a ce cours, ce qui implique une revision de la plupart des schemas usuels de lecon. De plus la programmation doit etre maitrisee, de facon a apparaitre comme un moyen et non un but.

References

1. Quelques apports de l'informatique a l'enseignement des mathematiques, Publication de l'APMEP n° 20 (1978).

2. A.M. Brisset, A. Meyer, C. Steyaert, Quelques utilisations de l'informatique dans le cours de mathematique en terminale C, IREM de Paris-Sud - INRP (1978).

3. Calculateurs programmables et algebre de quatrieme (une recherche INTER-IREM), Publication de l'APMEP n° 24 (1978).

4. Calculateurs programmables dans les colleges et les lycees, experimentation menee par les IREM et l'INRDP, Brochure INRDP n° 75, Paris, (1975).

5. Les calculatrices et l'enseignement des mathematiques, Conferences du 4eme seminaire organise par la CIEM, Luxembourg, (1978).

6. A. Deledicq et P. Vaschalde, Le calculateur programmable de poche, Ed. CEDIC, Paris, (1978).

7. A. Deledicq, R. Feffermann, G. Mounier et J-C. Oriol, Mathemachines au stage "initiation a l'informatique", Universite de Paris 7, (1977).

8. Emploi de calculateurs programmables dans le second degre, Brochure INRDP n° 54, Paris, (1972).

9. A. Engel, Elementarmathematik von Algorithmischen standpunkt, Klettstudienbucher, Stuttgart, (1977).

10. A. Engel, Algorithmes et calculateurs dans l'enseignment des mathematiques, in "Tendances nouvelles de l'enseignment des mathematiques", vol. IV, UNESCO, (1979).

11. F. Michel, Exemples d'utilisation du calculateur programmable dans des domaines non classiques, Mathematique et Pedagogie, n° 11/12, 187-213, (1977).

12. La minicalculatrice dans les classes, Seminaire de didactique des mathematiques de l'Universite de l'Etat a Mons (Belgique), Mons, (1979).

13. Georges Noel et J. Bastier, Mathematiques et calculatrices de poche, Ed. Technique et Vulgarisation, Paris, (1978).

14. E. Maor, A summer course with the TI 57 programmable calculator, The mathematics teacher, 73, 99-106, (1980).

15. L. Rade, Tentez votre chance avec votre calculatrice programmable, Ed. CEDIC, Paris, (1978).

16. R. Johnsonbaugh, Applications of calculators and computers to limits, The mathematics teacher, 69, 60-65, (1976).

17. E. Maor, The pocket calculator as a teaching aid, The mathematics teacher, 69, 471-175, (1976).

18. M. Olson, Using calculators to stimulate conjectures and algebraic proofs, The mathematics teacher, 72, 288-289, (1979).

18.4 PERSPECTIVES AND EXPERIENCES WITH COMPUTER-ASSISTED INSTRUCTION IN MATHEMATICS

EVALUATIONS OF COMPUTER-ASSISTED INSTRUCTION IN MATHEMATICS

Donald L. Alderman
Educational Testing Service
Princeton, New Jersey, U.S.A.

There have been perhaps twenty major demonstrations of computer-assisted instruction for teaching mathematics in the United States. Four such projects also involved an independent evaluation of their educational effectiveness, and these evaluations offer an objective perspective on the impact of computer-assisted instruction apart from the strong advocacy among developers of computer systems and computer curricula.

System Descriptions and Curricula

The four projects reflected different approaches to the problems of computer delivery, curriculum design, and classroom implementation. Each demonstration, however, relied on technology which preceded micro computers and videodisks, These were mainframe computer systems which supported multiple terminals. Yet the lessons learned through the evaluations of these systems may relate to applications of other forms of technology for the teaching of mathematics. The four demonstrations discussed here cover three prominent computer systems at two levels of education: the PLATO system in elementary schools, the CCC system in elementary schools, the PLATO system in community colleges, and the TICCIT program in community colleges.

The use of the PLATO system in elementary schools involved three curriculum strands focusing on the acquisition of mathematical concepts in grades four through six. The strand covered whole numbers, fractions, decimals, graphs, variables, functions, and equations. Each lesson made available on the PLATO system followed a different format, often, often a game-like situation, according to the nature of the topic to be taught. These lessons were intended to supplement regular instruction in the classroom with four terminals per demonstration classroom. Approximately 300 students received an average of 50 hours of instruction in elementary school mathematics on the PLATO system in the 1975-76 academic year.

The CCC system supported a drill-and-practice curriculum on arithmetic facts and computation in the first through sixth grades in four elementary schools. The topics in the curriculum ranged from simple number concepts to operations with negative numbers and included equations, fractions, and decimals as well as traditional operations with positive integers (e.g., multiplication and division). Students receiving computer-assisted instruction went to a separate laboratory room with sixteen terminals in the smaller schools and thirty-two terminals in the larger schools. From 1977 through 1980 there were several hundred students exposed to this drill-and-practice mathematics curriculum in a longitudinal study of the effectiveness of computer-assisted instruction for compensatory education.

The demonstration of the PLATO system in community colleges covered several disciplines, but just one college employed the lessons for mathematics courses and participated in the evaluation. Instructors prepared their own lessons on the PLATO system, and these lessons formed a large pool from which faculty members could choose those topics appropriate for a given course. The mathematics course with the heaviest use of PLATO lessons during the evaluation was on fundamentals of mathematics, an introductory course designed for students deficient in basic mathematical skills (e.g., whole-number arithmetic, decimals, algebraic expressions).

The TICCIT program stood in sharp contrast to the preceding applications in that it offered a complete and independent alternative to full college courses rather than simply supplementary lessons. Courses content in the TICCIT program was structured in a hierarchical manner; segments of TICCIT courses essentially adhered to a rule-example-practice paradigm as an instructional strategy. The computer curricular materials represented the equivalent of three different algebra courses ranging from an intrdluctory course concentrating on a review of arithmetic skills and coverage of rational an polynomial expressions to a course which focused on functions and relations. The emphasis in the TICCIT courseware was on learner acquisition of mathematical concepts and rules for solving routine problems.

Evaluations and Results

Each evaluation followed a similar approach in assessing the impact of computer-assisted instruction on student achievement in mathematics, but the evaluations differed in their strategies for documenting other aspects of educational effectiveness.

The demonstration of an elementary school curriculum on the PLATO system had a positive impact on each aspect of educational effectiveness (see Swinton, Amarel, and Morgan, 1978). Students in the fourth through sixth grades attained higher scores on a standardized achievement test, especially on subtests for computation across grades but also on subtests for whole numbers, fractions, and applications at particular grade levels. Students in the fifth and sixth grades further demonstrated a stronger grasp of mathematical concepts in work with fractions and graphs on special tests targeted toward the curriculum. There were positive reactions to the PLATO system; students reported enjoying instruction via PLATO lessons in mathematics, and teachers preferred receiving access to a PLATO terminal to receiving financial support for a part-time classroom aide for teaching mathematics. Such consistent positive results represent an unusual pattern of findings in evaluations and even contrast sharply with the findings from a companion part of the same evaluation of the PLATO system focused on instruction in reading at elementary schools.

The longitudinal evaluation study of the CCC system, a drill-and-practice curriculum in elementary school mathematics, is still underway, but intermediate results regarding standardized tests and tests specific to the computer curriculum showed achievement differences in favor of students given daily exposure to drill-and-practice in mathematics on the computer (see Ragosta et al., 1980). These differences were especially evident on the standardized achievement subtest dealing with computation and on the curriculum-specific tests drawing heavily from exercises presented on the computer. Gains on the curriculum-specific tests could be partially attributed to the notational convention followed on the computer and to the proficiency on timed exercises promoted by extensive drill and practice on the compter (see Alderman, Swinton, and Braswell, 1979). Moreover, students lacked any meaningful iconic or concrete models for mathematical concepts despite their improved test performance.

The use of the PLATO system in mathematics courses at community colleges led to favorable student and faculty reactions to the computer system but had no effect on student achievement or on student attrition (see Murphy and Appel, 1977). The faculty had control over the computer system and accepted it as a limited supplement to their courses,received exposure to few lessons and, an average of just four hours over an entire academic term. The students, however, really had little opportunity to benefit from such scant exposure to computer-assisted instruction. Yet the PLATO system represented a powerful and attractive mechanism for learning as shown by the positive faculty and student attitudes toward it.

The TICCIT program had a significant impact on each dimension of educational effectiveness studied in the evaluation, but the direction of the impact was sometimes positive and sometimes negative depending on the outcome of interest (see Alderman, 1978). Students completing mathematics courses under the TICCIT program attained appreciably higher posttest scores than did comparable students from lecture-discussion sections of the same courses, but far fewer students actually completed all course requirements in TICCIT classes. The problem did not stem from attrition or withdrawal: students stayed in classes on the TICCIT system and simply failed to progress far enough along in the computer curriculum for obtaining full course credits.

What general inferences might be drawn from these diverse results with different applications of computer-assisted instruction to the teaching of mathematics? It would seem that each of the projects which involved persons with experience in curriculum development or teams of specialists in different fields also had positive outcomes in student achievement. Of particular importance may have been expertise in the subject matter, knowledge of educational psychology, and familiarity with computer capabilities. The assertion that any teacher can develop lessons for computer-assisted instruction seems clearly fallacious, yet there should also be real participation in developmental efforts by classroom teachers and school administrators.

Conclusion

If computer systems alone served to guarantee educational effectiveness, there would be clear and consistent results across the evaluations of computer-assisted instruction in mathematics. The results, however, varied depending on the particular approaches to curriculum design, development, and implementation (see Alderman, Appel, and Murphy, 1978). The computer does offer a broad spectrum of display capabilities for teaching mathematics and can engage students as active participants in the learning process. But the impact of computer-assisted instruction depends as much, if not more, on what the computer conveys to the student and on how it interfaces with regular classroom practices as on the chosen delivery mechanism for the instruction.

REFERENCES

Alderman, D.L. Evaluation of the TICCIT Computer-Assisted Instructional System in the Community College. (ETS PR 78-10). Princeton, NJ: Educational Testing Service, 1978.

Alderman, D.L., Appel, L.R., & Murphy, R.T. PLATO and TICCIT: An evaluation of CAI in the community college. Educational Technology, 1978, 18 (4), 40-45.

Alderman, D.L., Swinton, S.S., & Braswell, J.S. Assessing basic arithmetic skills and understanding across curricula: Computer-assisted instruction and compensatory education. Journal of Children's Mathematical Behavior, 1979, 2 (2), 3-28.

Murphy, R.T. & Appel, L.R. Evaluation of the PLATO IV Computer-Based Education System in the Community College. (ETS PR 77-10) Princeton, NJ: Educational Testing Service, 1977.

Ragosta, M., Holland, P., Juhnke, W., Woodson, R., & Jamison, D.T. Computer-assisted instruction: A longitudinal study of drill-and-practice curriculums in elementary schools. Symposium presentation at the annual meeting of the American Educational Research Association, Boston, 1980.

Swinton, S.S., Amarel, M., & Morgan, J.A. The PLATO Elementary Demonstration Educational Outcome Evaluation. (ETS PR 78-11) Princeton, NJ: Educational Testing Service, 1978.

EXPERIENCES WITH COMPUTER ASSISTED INSTRUCTION IN THE MATHEMATICS TRAINING OF HIGH-SCHOOL STUDENTS AND HANDICAPPED PERSON IN THE FEDERAL REPUBLIC OF GERMANY

Rul Guzenhaeuser
University of Stuttgart
Federal Republic of Germany

This paper reports on different examples of Computer Assisted Instruction (CAI) in mathematics training in the F.R. of Germany. Based on facts about the increasing number of computational power of micro-computers it further outlines some perspectives about the production and distribution of CAI-Teachware and its applications in schools, offices and private homes.

I. Experiences with CAI

The Center of Rehabilitation in Heidelberg (Germany) operates one of the most effective CAI-Installations in the Western World. More than 230 CAI-terminals, located in different cities of Germany and Austria, are on-line supported by an IBM 370/155 Computer System. Presently more than 600 hours of CAI-Teachware are available, most of them in the fields of elementary mathematics and technical applications. They serve more than 2,000 students each year in different courses and different levels of education. Most of the CAI-Teachware is supplementing regular work in class by providing additional drill and practice as well as tests.

The CAI-Teachware is designed for the specific needs of training handicapped people and groups of students which are extremely heterogeneous in age, learning abilities, and future professional standards. Most of the programs - like simulations or drill and practice in mathematics could also assist courses in regular high-schools. (1)

The CAI-System in Heidelberg is not based on a known CAI-author language. It is implemented in APL. All courses are administered by a single monitor program which controls a hierarchy of APL-functions and routines for the generation of individualized problems, the analysis of students' answers, or recording students' errors. A modular package system and a test item package system enable teachers to design, test and run CAI-programs without knowledge about the computer language. An easy-to-learn control-language allows students to select programs, to refer to an information system, to do calculations, or to get further help. (2)

Presently investigations are being made to implement the successful Heidelberg System on several mini-computers - each with about 20 modern terminals - which operate "stand-alone" as well as within a computer network.

To assist blind students attending high-schools and colleges a portable micro-computer based Learning and Working Environment for the Blind (LWEB) is being developed at Stuttgart University (this project is sponsored by Deutsche Forschungsgemeinschaft - DFG). A standard IBM 5110 micro-computer is supplemented by a BRAILLE-Code-Display (BD) which offers a tactile copy of the computer output. For input-information the blind student uses the standard keyboard. Drawings are plotted on an IBM 5130 printer and off-line copied by an optical-to-tactile converter (OTAC).

The following educational software is available for LWEB, implemented in APL: (3)

- Tutorial programs to learn BRAILLE and German constructed BRAILLE
- Tutorial programs to apply fractions in their specific BRAILLE-representation
- Training programs for integer arithemetic
- Training programs for the differentiation of elementary mathematical functions.
- An educational system for plotting constructions in elementary geometry.

All drawings are generated during a man-machine-dialog. The student enters his/her instructions in natural (German) language. The system also replies to different types of queries about the "constructed" drawing and about elementary geometry.

For applications in mathematics and engineering we designed and implemented as Interactive Method Bank System (IMBS) to be operated on an IBM 5110 micro-computer. IMBS is advising and training students to select and test the appropriate computational methods for given (mathematical) problems. Presently IMBS is used in the field of numerical interpolation of functions. Other applications are being investigated.

Training programs for the computer languages APL and BASIC form another part of CAI-efforts in Stuttgart. The APL-version is widely used by university students; the BASIC-version is designed for high-school students. Both versions operate on mini-computers; they apply routines to generate problems as well as to control the students' answers semantically.

2. Computers in German High-Schools

Due to recommendation of the (German) Gesellschaft fuer Informatik (GI) nearly all States of the Federal Republik of Germany developed curricula for (elective) Informatik-courses in high-schools (grades 10 to 13). The availablility of (micro-) computers at schools is required (4). Most schools use micro-computers like Commodore-PET or TRS-80; some schools operate mini-computers like DEC pdp 11 or comparable German products. Most of the teachers have a mathematical background and use the new equipment increasingly for algorithmic problem solving and doing experiments with heuristics solutions in mathematics. The use of school-computers to assist mathematics teaching will in a few years exceed the computers' use in computer science classes - - this will lead to a better integration of both subjects in high-school education.

The remaining computational power of school-computers will be used to present CAI-Teachware to students being behind in their work, being handicapped or needing extra help to cover required subjects. To provide those students with approved computer based learning aids, special efforts in producing, controlling and distributing these materials are needed. Special CAI-Centers" might be sponsored by commercial companies, university institutes or book publishing companies, because teachers cannot provide their students with CAI-programs in an effective and cost-mnimizing way.

3. Perspectives: CAI in a Public Information Network

A new way of distributing CAI-Teachware will be opened by 1982 by the Bildschirmtext-System (BT) of the Bundespost, the mail and telephone service in the F.R. of Germany. It is presently experienced in two large cities; the technical design is comparable to the British Viewdata System. Plans are made to transmit CAI-Programs as well as other personal messages via telephone-lines to home, offices and schools. The present user equipment consists of a telephone, a modem, a standard TV-set with a special keyboard. "External computers", connected with the BT-network, will provide the strategies and contents of the CAI-Programs.

Technically it will be possible to connect micro-computers to the BT-keyboard, so CAI-Teachware will be stored and presented without continuous on-line contact to the computer of an educational center.

Various system modules, implemented on the external computers, will support CAI-authors in defining, selecting, testing and improving the learnng and problem-solving strategies and their learning items. A detailed concept about the future interaction of home- and school-computers and external computers via BT is presently worked out in our group (5). For an experimental implementation of this modular system we will use micro-computers as well as a medium size computer available at Stuttgart University.

The above mentioned Heidelberg CAI System could serve as a model for an educational computer-center.

4. Remarks

Not all current applications of computers in education are pointing to useful future directions. CAI in the traditional sense - use of author languages and large time-sharing systems - has to be expanded to CAI in a wider sense which enables students to learn interactively while collecting information from their own experiences, and maintaining individualized control of content, time and sequence in using computer-based learning material.

Our experiences show that educators are just beginning to understand how to use computers effectively in mathematics education and how to prepare adequate course materials.

5. References

1. Augsburger, W.: Educational Systems, In: Interactive Systems, Lecture Notes in Computer Science, Heidelberg, 1976.

2. Lampl, G.R. and Schell-Haungs, I.: Functions in APL to assist the programming and servicing of CAI-lessons. In: APL 76 Conference Proceedings, ACM, New York, 1976.

3. Schweikhardt, W.: A computer based educational system for the blind. Proceedings of IFIP-Congress 1980, Tokyo (to be printed).

4. Brauer, W. et al: Recommendations for an Informatik-Curriculum in Secondary Schools (F.R. of Germany). Gesellschaft fuer Informatik, 1976 (in German).

5. Gunzenhaeuser, R. and Horlacher, E.: Eine bildungsorientierte Erweiterung technischer Kommunikationasnetze durch Methoden der Informatik. Techn. Report Institut fuer Informatik, Univers. Stuttgart, 1979 (in German).

18.5 COMPUTER LITERACY / AWARENESS IN SCHOOLS; WHAT, HOW AND FOR WHOM?

COMPUTER LITERACY: ASSESSING SECONDARY SCHOOL PUPIL'S GENERAL KNOWLEDGE AND UNDERSTANDING *

D.C. Johnson
University of London
London, England

(* This paper was also presented at the British Education Research Association, BERA, Annual Conference, Cardiff, Wales, September 1980.)

The notion of Computer Literacy has received considerable attention in the past decade. The optional courses in the United Kingdom in the area of computer studies have a history dating back to the mid 60's and at present approximately 40% of the 6,000 secondary schools offer some formal coursework in this area. The British Computer Society Schools Committee has provided a focal point and leadership for curriculum recommendations (e.g., see Syllabuses for the Future, 1979). The U.S. has a similar history as there have been a number of professional groups and individuals advocating school activities/curriculm orientated towards the development of pupils understanding of computing and implications of this technology (e.g., see Molnar, 1978; NCSM, 1978; Moursund, 1975; CBMS, 1972; Michael, 1968).

Today, the 1980's, the widescale availability and impact of micro-processor technology in home, school, and the world of work suggests that schools which do not incorporate the opportunity for pupils to develop an understanding of, and the capability to deal with, this revolutionary technology are remiss in fulfilling their responsibility. While wide scale availability is at present a phenomenum characteristic of developed countries it is readily apparent that developing countries must also concern themselves with the educational implications as the ever decreasing costs of microcomputers suggest that even between the time of writing this paper and publication we will see small (only in size) machines available at a cost 50-80% less than today (from 500 pounds to less than 100 pounds).

What should the educated citizen know to be able to function as a contributing member of a society which utilizes computers in the home, business, industry, and government? Obviously, to attempt an answer to this question is very much like trying to answer the similar question of "what and how much mathematics?". In education we tend to suggest that while there may be some 'minimal level' of understanding it is also the goal that each child should be educated to his or her full potential. In the case of computing and the unknowns of what the future holds this means we have the obligation to develop an awareness of the technology including a knowledge of the special vocabulary, notions of both strengths and limitations, and a understanding of the issues which accompany computerization of tasks. To be computer literate implies comprehension and the ability to discuss computing concepts, applications, and issues intelligently. At a somewhat higher level, a functional level if you wish, the phrase includes the ability to actually use or identify new uses of computers in home and profession. Obviously, as indicated previously, it should be the goal of education to provide an opportunity for all pupils to reach a level of literacy commensurate with each individual's potential.

Where are we today? What is the state of knowledge of school pupils? What is being recommended? The next section of this paper describes some of the initial work in an attempt to assess pupils' knowledge and understanding in a broadly defined area of computer literacy. The research, supported by the National Science Foundation, was conducted in Minnesota, U.S.A.

Computer Literacy and Awareness Assessment

The Minnesota research project was designed to (1) collect baseline data regarding pupil knowledge and understanding of computers, and (2) to determine the relative impact of the various computing of computer-related activities in schools on the development of computer knowledge and understanding. The research was conducted during the period 1977-1979 and the Final Report (Klassen, Anderson, Hanson, and Johnson, 1980) reports on results from a teacher survey, over 3,500 teachers (see also Anderson, et al, 1979a); a field study, assessment and background information collected on over 1100 secondary pupils; and an experimental study, 350 pupils.

Initial work of the project involved an extensive review of the available information regarding how different individuals and organizations viewed the phrase 'computer literacy' or 'computer awareness'. This enabled the research team to develop an empirical definition of the construct. This phase of the research resulted in a listing of 54 objectives in the cognitive domain under the categories of Hardware (H), 7 objectives; Software and Data Processing (S), 13 objectives; Programming and Algorithms (P), 8 objectives; Applications (A), 10 objectives; and Impact (I), 16 objectives; and 9 objectives in the affective domain in the areas of Attitude, Values, and Motivation (see Johnson, et al, 1980, for a more complete discussion of the objectives). It is interesting to contrast this listing and the general topic headings and descriptions in the British Computer Society Schools Committee, 1979, recommendations.

The Minnesota research project designed a test, The Computer Literacy Questionnaire (see Klassen, et al, 1980), to assess a subset of the objectives. (Note that the Questionnaire used in the research has been revised, the most recent version is The Minnesota Computer Literacy and Awareness Assessment (Anderson, et al, 1979b). The present version of the instrument has been used in the Minnesota Statewide assessment program (approximately 6000 pupils) and data are now being analyzed.) For purposes of this paper we will restrict the discussion to the cognitive objectives and in particular those which were included in the Questionnaire (34 or 54 objectives were assessed).

The Questionnaire was used in a field study for which complete data were available from 1106 pupils (in 51 classrooms). Of these 929 pupils were in classrooms which had some type of planned computing activity (e.g., CAL, programming, computer appreciation, etc.). The data suggest that if one accepts the objectives and corresponding items as appropriate (note again that the test has been revised), then even with these pupils who have had some exposure to computing activities there is room for improvement as average performance on the final composite cognitive computer literacy test was only 27.9 or 57 percent correct.

Epilogue

What does all this mean? Where and how should computer literacy/awareness be developed? The Minnesota research tells us that while exposure to computing in different aspects of schooling does give pupils some insights, there is still much that could be done. In particular, there is a need for more planned activity which emphasizes the specific computing objectives. These can take place in a number of disciplines (and some objectives are more appropriate for specific subject areas, e.g., programming and algorithms are a natural part of mathematics and aspects of impact and issues certainly fit into an up-to-date social studies curriculm). However, the overall program needs to be coordinated and attempts made to ensure a range of activities which will result in an education of a comprehensive and cohesive nature.

References

Anderson, R., Hansen, T., Johnson, D., Klassen, D. Acceptance and rejection of instructional computing by secondary school teachers. Sociology of Work and Occupations, May 1979a.

Anderson, R., Hansen, T., Johnson, D., and Klassen, D. The Minnesota Computer Literacy and Awareness Assessment (test). Minnesota Educational Computing Consortium, St. Paul, Minnesota (U.S.), 1979b.

Anderson, R., Klassen, D., Johnson, D., and Hansen, T., The Minnesota Literacy Tests: A Technical Report on the MECC Computer Literacy Field Study, Minnesota Educational Computing Consortium, St. Paul, Minnesota (U.S.), 1979c.

British Computer Society Schools Committee, Syllabuses for the Future, British Computer Society, London, 1979.

CBMS (Conference Board for the Mathematical Sciences). Recommendations Regarding Computers in High School Education. Washington, D.C., 1972.

Johnson D., Anderson,R., Hansen, T., and Klassen, D. Computer Literacy - what is it? Mathematics Teacher, Vol. 73 (Feb. 1980), pp.91-96.

Klassen, D., Anderson, R., Hansen, T., and Johnson, D. A study of Computer use and Literacy in Science Education, (supported by Grant No. SED 77-18658) Minnesota Educational Computing Consortium, St. Paul, Minnesota, 1980.

Michael, D. The Unprepared Society. New York, Basic Books, Inc., 1968.

Molnar, A. The next great crisis in America education: computer literacy. The Journal, July/August, 1978, pp. 35-39.

Moursund, D. What is Computer Literacy? Oregon Computing Teacher, Oregon Council for Computer Education, Vol. 2, No. 2, June 1975, p. 17.

NCSM (National Council of Supervisors of Matematics). Position statements on basic skills. Mathematics Teacher, Vol. 71, (Feb. 1978), pp. 147-52.

L'INTRODUCTION DE L'INFORMATIQUE DANS L'ENSEIGNEMENT EN FRANCE

Claudette Vieules
IREM, Université Paul Sabatier
Toulouse, France

Préliminaires

Le mot "informatique francais" n'a pas d'equivalent dans la langue anglaise.

En FRANCE, les ecoles elementaires s'adressent aux enfants de 6 a 11 ans; les colleges, a ceux de 11 a 15 ans; les lycees, a ceux de 15 a 18 ans.

1. Le Plan d'Informatique dans l'Enseignment en France (en Dehors de la Formation des Specialistes)

Des 1970, la mission a l'informatique cree aupres du Ministre de l'Education Nationale a lance ce que l'on a coutume d'appeler l'experience des 58 lycees: En 5 ans, 58 lycees ont ete equipes de mini-ordinateurs (Mitra 15 de C.I.I. ou T 1600 de Telemecanique) avec 8 consoles alphanumeriques et une teletype. Le langage LSE, defini et concu pour cette experience a ete implante.

Simultanement, 500 professeurs de toutes disciplines ont suivi une formation a temps plein durant un an alors que pres de 5000 enseignants suivaient une formation par correspondance.

Parallelement, l'INRP (Institut National de la Recherche Pedagogique) a coordonne et favorise la conception par les enseignants de programmes produits (400) et assure leur diffusion.

En particuliere, parce que le cout trop eleve des materiels ne permettait pas la generalisation de leur implantation, cette derniere a ete suspendue en 1976, la formation des enseignants n'a pas ete poursuivie et une premiere evaluation a ete entreprise.

Depuis 1979, un nouveau plan dit des 10 000 micros a ete lance sur cinq ans.

Cette anee, 416 micro ordinateurs (LX de Logabax et XI de la Societe Occitane) ont ete installes dans les lycees (8,4 ou 1 par etablissement). Une trentaine de professeurs formateurs ayant pour la plupart deja participe a l'experience precedente ont appuye l'implantation sur le terrain apres avoir suivi un state d'actualisation de leurs connaissances et de prise de contact avec ce materiel nouveau. Cette premiere annee, le seul langage disponible est be BASIC mais le LSE est en cours d'implantation. Dans chacun des lycees nouvellement equipes, les professeurs formateurs ont anime des stages de 4 fois 3 jours. Ils n'ont, en general, pas pu repondre a toutes les tres nombreuses demandes formulees dans chacun des etablissements.

Parallelement, ils ont du repondre a des sollicitations de plus en plus nombreuses et diverses d'information et de formation (inspecteurs pedagogiques regionaux, centres de formation pedagogique, ecoles normales d'instituteurs . . .).

En octobre 1980, l'operation va s'etendre aux differents secteurs de l'enseignment pre-baccalaureat:

- une centaine de lycees seront equipes chacun de 8 micros et d'une imprimante;
- l'equipement d'une cinquantaine de colleges est prevu;
- quelques ecoles normales d'instituteurs ainsi que leurs ecoles d'application seront pourvues en materiel varie.

Le nombre de professeurs formateurs est porte à 50.

II Les Objectifs de ce Plan

C'est deliberement que 8 postes de travail (plus une imprimante) sont installe dans chaque etablissement. En effet, autant dans l'experience des 58 lycees que dans le debut de generalisation actuel, l'option est prise d'introduire l'informatique a travers son apport aux differentes disciplines traditionnellement enseignees dans le secondaire: les enseignants de chacune d'entre elles ont elabore des exercices pedagogiques et c'est avec des classes entieres (ou des demi-classes suivant les effectifs) qu'ils les ont utilises.

Il est vrai qu'un grand nombre d'entre eux a eu tendance en un premier temps a faire de l'enseignement programme mais, meme dans cette voie, les aspects positifs ne sont pas negligeables: les enseignants concernes ont pris conscience autant des possibilites que des limites de cet outil moderne et leurs travaux ont souvent ete un revelateur autant pour certaines de leurs methodes pedagogiques qu'ils ont caricaturees en les programmant que pour l'actualite et la realite du contenu de leur enseignement.

D'autre part, ils ont aborde par ce biais un certain nombre de problemes de representation de l'information, d'automatisation des traitements, de modelisation et de simulation mais aussi d'interpretation de messages emis.

Conjointement, les methodes de raisonnement et d'analyse mises en oeuvre par l'informatique sont maintenant decelees ou appliquees dans un certain nombre de cours traditionnels, le developpement d'un mode de pensee structuree se faisant independamment de toute contrainte materielle, par exemple de langage.

III Le Contact des Enseignants avec l'Information

Les deux operations ont principalement touche des enseignants du secondaire.

Tout d'abord un phenomene d'impregnation s'est produit sur les volontaires formes en un an dans l'experience des 58 lycees alors que ceux qui suivaient la formation par correspondance etaient, dans l'ensemble, seulement sensibilises superficiellement. Dans leurs lycees respectifs, ils ont presente a leurs collegues interesses les differents programmes produits et propose des cours d'initiation a l'utilisation du materiel aux enseignants ou eleves benevoles mais il faut reconnaitre que d'annee en annee, les mauvaises conditions materielles (horaires en particulier) ont petit a petit decourage les bonnes volontes de part et d'autre.

Cette annee, les stages de formation organises dans les lycees ont eu une incidence sensiblement differente du fait autant des conditions sociales que materielles nouvelles. La prise de conscience de l'actualite du probleme, la dotation a l'etablissement du materiel, l'inclusion des stages dans le temps de service et leur situation sur le lieu de travail sont autant de facteurs ayant favorise la determination active des professeurs. Tres tot, ils ont essaye d'entrevoir les liens avec leur discipline et leur enseignement. Lorsqu'ils se tourneront vers les eleves, ce sera pour leur presenter une nouvelle verite ou un produit fini, ce qui peut-etre discute.

IV Le Contact des Eleves avec l'Informatique

Dans le cadre de l'experience des 58 lycees, des cours d'initiation facultatifs ont ete proposes en general aux eleves des classes de seconde ou de premiere qui n'ont pas la sanction d'un examen de fin d'annee. Il est probable qu'il en sera de meme dans le cadre de l'operation actuelle. Ceux-ci ont presente la "science informatique" en amenant les eleves a programmer: il y a cependant eu peu si ce n'est pas du tout de rigueur au niveau des methodes. Beaucoup d'eleves ont ete decourages par les contraintes materielles (autant au niveau des langages, de l'absence de graphisme, que des difficultes horaires). Un certain nombre a mene a bien des projets interessants au prix d'un investissement en temps tres important.

L'utilisation dans le cadre d'un cours traditionnel de programmes d'enseignement s'est beaucoup faite dans les 58 lycees. Cela a permis a un grand nombre d'enfants de voir de pres un ordinateur, de frapper sur un clavier, mais ceci a travers un programme deja fait.

Dans certain cas, la conception de ces programmes et l'intervention des enseignants leur ont bien montre que, par exemple, la modelisation des phenomenes reels avait ete introduite par l'homme, la machine simple automate, etc. Je pense personnellement que pour beaucoup d'entre eux la demythification n'est vraiment pas suffisante. J'ai pu meme observer que dans des cas que l'on peut souhaiter exceptionnels, l'utilisation mythique d'un ordinateur peut-etre un moyen decuplant les possibilites d'oppression de l'enseignant en occultant toute preoccupation pedagogique sous le pretexte de la presence seule de cet outil moderne.

On peut remettre en question, certes, les modalites de la formation des enseignants.

Ce que l'on peut constater, c'est la non appropriation par eux de ce qui leur a ete presente et la necessite qu'ils prennent un certain recul. Beaucoup n'en sont qu'au stade descriptif et il faut craindre qu'ils ne procedent avec les eleves qu'en termes d'utilisation aveugle d'un produit fini ou de transfert de connaissances de la science informatique ce qui est absolument contraire aux objectifs vises.

D'autre part, les ordinateurs font partie integrante au meme rang que le telephone de l'environnement des generations actuelles, et il ne faudrait tout de meme pas que les enseignants induisent sur eux une peur mythique qu'ils n'ont absolument pas.

THE COMING OF COMPUTER LITERACY: EVOLUTION OR REVOLUTION?

Andrew R. Molnar
National Science Foundation *
Washington, D.C.

* This paper is based on comments presented at the 4th International Congress on Mathematical Education, Berkeley, California, August 12, 1980. The views expressed are those of the author and do not necessarily reflect the views of the National Science Foundation.

1. Science and the Information Explosion

Discoveries of fundamental importance are occurring in all fields of science at an incredible rate. These developments have created a new body of knowledge that has affected all mankind. The computer, with its enormous power and speed, has acted as a great catalyst to scientific discovery. It has become an amplifier of human thinking, the tool for comlex problem-solving, and the repository for huge quantities of the world's data, information, and knowledge.

Many of these discoveries have been transformed into useful applications that have increased America's productivity and well-being. Today, science-based information industries account for more than one-half of our gross-national-product and over one-half of all of our jobs. It has been estimated that advances in knowledge are the largest single source of long-term economic growth, accounting for more than one-third of our economic output since World War II. Information has become a national commodity and a national resource and has altered the very nature of work. We as a nation have moved from being predominantly an industrial society to being an information (Porat, 1977; NSB, 1976).

The transition to an information society has created new national needs. While computers improve national productivity, they also make many jobs and occupations obsolute. Computers have tended to decrease the value of physical "work" and increased the value of "thinking", thereby requiring totally new occupational skills and training.

2. The Need for Computer Literacy

In education, computers have changed the very nature of many of our educational needs and have brought into question many of the basic assumptions upon which our educational process is built. Dr. Herbert Simon, Nobel Prize-winner, says that developments in science and information have changed the meaning attached to the verb "to know" from "having information stored in one's memory" to "the process of having access to information" (Simon, 1971). In short, if we are to take advantage of the ever increasing quantity of information, we must master the use of new, computer-based tools and "work smarter." If individuals are not computer literate and do not understand how these systems work, they will be unable to take advantage of information and will be unable to meaningfully participate in actions that affect their lives.

The transition to an information society has created a new dilemma. Dr. Donald Michael writes in The Unprepared Society that there is a growing separation produced by those working creatively with computers and the rest of the population. He says that "ignorance of computers will render people as functionally illiterate as ignorance of reading, writing, and arithmetic" (Michael, 1968).

In summary, if we are to have equity in our society, all citizens, not just specialists, must have access to information, and all citizens must have an understanding of computers since they are the tools that make information useful and productive. In an information society, a computer literate populace is as important as energy and rw materials are to an industrial society. Conversely, the general level of computer illiteracy may be the limiting factor to growth and productivity in an information society (Molnar, 1978, 1980).

3. Computers in Education

In the past decade, expenditures for academic computing in higher education (research, instruction, and administration) have more than doubled. Ths year, it is estimated that 1.2 billion dollars will be spent on academic computing at 2,163 institutions with an enrollment of 9.9 million students. In 1975, an estimated 350 million dollars was spent on pre-college computer use. Over 55% of all secondary schools had access to computing and 23% of all schools used computers primarily for instructional purposes. Today, a reasonable estimate of computer usage for instruction in all schools would be approximately 50% with expenditures approaching 700 million dollars. In spite of this dramatic growth, large numbers of students never come in contact with computers in the classroom. In higher education, for example, over half the students who use computers are concentrated in three departments: computer science, engineering, and business. There is a growing awareness that our educational institutions are not satisfying student and national needs to achieve comuter literacy (Molnar, 1980).

4. Computer Literacy and the Role of the National Science Foundation.

In February 1967, the President of the United States specifically directed the National Science Foundation (NSF) to work with the U.S. Office of Education in establishing an experimental program for developing effective methods for utilizing computers at all levels of education. Since then, NSF has supported a wide variety of programs designed to stimulate new innovative applications of computers in science education.

Planning Activities

Several planning activities have addressed the computer literacy question. In 1972, the Conference Board on Mathematical Sciences' Committee on Computer Education recommended to NSF that in addition to introducing computers into high school mathematics that a junior high school course in "computer literacy" be designed to provide all students with enough information about the nature of a computer so that they can understand the roles which computers may play in our society (CBMS, 1972).

As part of another long-range, planning activity for NSF, Dr. Arthur Leuhrmann of the Lawrence Hall of Science has recommended that an entire basic curriculum in computer use be developed. The courses, he says, should aim not merely at teaching the syntax of a common computer language, but should mainly show students how to perform meaningful tasks, thereby improving their analytic and problem-solving skills. Students should learn to structure problems in a logical form, to express ideas as algorithms, to simulate real systems as computer models, to process text, to construct graphs, and to search data bases among other skills (Luehrmann, 1979).

Research

Research has been supported to clarify the important dimensions of computer literacy. The Human Resources Research Organization (HumRRO) has analyzed twenty-four, locally initiated, college and university programs designed to increase the level of computer literacy among students and faculty. They define computer literacy as what a person needs to know and do with computers in order to function competently in our society. They found that these programs stress (1) skill in writing algorithms and computer programs, (2) knowledge of computer applications in one's field, and (3) understanding of computers and their impact on society. They also found that exemplary programs usually had strong administrative support from academic leaders, provided students with ready access to suitable facilities, and integraged computer-based learning into regular courses (Hunter, 1978).

Course Requirements

In the area of computer science the Association for Computing Machinery (ACM), a professional association, is completing a study of computers in society and computer literacy and will soon publish a recommended curriculum for computer literacy. The ACM maintains an on-line, annotated bibliography of several thousand books and articles in this area. ACM committees have also published recommendations for computing competencies for school teachers who are and are not teachers of computer science (Taylor, 1979; Poirot, 1979).

Needs Assessment and Evaluation

The Minnesota Educational Computing Consortium (MECC), a statewide educational agency, has developed and validated the "Minnesota Computer Literacy and Awareness Assessment Test" of computer knowledge. The test is designed to measure knowledge attitudes and skills of computer literacy. In April of 1979 the test was administered to a statewide sample in Minnesota to establish statewide norms and to evaluate those factors in the school and home that contribute to computer literacy (Klassen, 1980).

Curriculum Development

The Foundation is also supporting the development of materials for computer literacy in the schools. HumRRO is developing curricular materials that infuse computer related skills and knowledge into the traditional curriculum for science, social studies, and mathematics in elementary and junior high schools. Their goal is to enable all students, kindergarten through junior high school, to acquire minimal computer related skills they will need for their continuing education and for their roles as citizens in an information based society. MECC is developing instructional materials for computer literacy at the junior high. They are developing an integrated set of 25 learning packages and related teachers' guides for science, social science, and mathematics. The materials will focus on how the computer works, how it is used, and its impact upon the individual and society with implications for the future.

5. Computer Literacy and Mathematics

Technology and knowledge are inextricably linked. If we look at the current curriculum we find whole areas of knowledge built around technology such as the inclined plane and the pendulum. The computing metaphor provides new and useful ways of thinking about mathematics. The power and speed of the computer encourages an attack on larger and more complex problems. Computer graphics facilitate visualization of abstract mathematical ideas. And, computing in general, has increased the importance of algorithmic problem-solving.

Of special note is the work of Dr. Seymour Papert at the Massachusetts Institute of Technology. Dr. Papert has designed mathematical technologies to teach young children problem-solving. He has developed a computer-driven, electro-mechanical device called a "turtle" to teach geometry; a computer-driven "worm" to teach physiology, and a computer-driven music box to teach music composition. All of these and other technologies require an understanding of mathematics and algorithmic problem-solving.

Papert says the computer is not merely a device for manipulating symbols but is also capable of controlling real physical objects and processes. The need to intelligently control these technologies induces an immediate and practical need to understand mathematics. Children learn by doing and by thinking about what they do. So, says Papert, our task is to give them better ways to think about what they do. Papert insists that we should not teach children mathematics but we should teach children to be mathematicians (Papert, 1972).

Papert's work demonstrates that today's curriculum greatly underestimates the capacity of children to deal with complex problem-solving. Given a computer, young children are able to use mathematics in a creative way to solve complex problems. Computing provides a new way of thinking and should be introduced into the curriculum as early as possible.

6. Evolution or Revolution

Computers are a pervasive force in our society. Has the computer revolution, just as the agricultural and industrial revolutions before it, created a discontinuity in our society that our educational system is failing to meet? Do revolutionary developments in society require revolutionary changes in education?

Are we prepared for revolutionary change? Over half of the current faculty in our educational institutions completed its formal training before computers appeared on the scene. Can we train a sufficient number of computer specialists, engineers, and professionals to meet our national needs? Do we have the resources to train a whole new generation as well as retrain an older generation to be computer literate?

What are we to teach? In a field where technological change occurs literally overnight and where computer generations are measured in two to three years, what is literacy? Rapid change is not easily accepted by a profession such as education which usually measures innovative adoption by generations of teachers and decades of time.

The National Science Foundation is seeking answers to these questions through programs that emphasize (1) fundamental research (2) curriculum revision and development (3) faculty training and (4) development of microcomputers for education.

In conclusion, these questions require deep and careful thought by all segments of our society. The "pull" of these developments is great and the societal benefits high; but so are risks. Continuing research, development and evaluation is needed to maximize the benefits and reduce the risks. If education is to have social value, it must be continually adjusted to meet the needs of society. In the past, this change has been slow and continuous; however, the future may demand rapid, revolutionary change.

References

CBMS (Conference Board for the Mathematical Sciences) Recommendations Regarding Computers in High School Education, Washington, D.C. (1972).

Hunter, Beverly, "What Makes a Computer-Literate College?" Conference on Computers in the Undergraduate Curricula, University of Denver (1978).

Klassen, Daniel L., et al. A Study of Computer Use and Literacy in Science Education, Minnesota Educational Computing Consortium, 2520 Broadway Drive, St. Paul, Minnesota, 55113 (1980).

Luehrmann, Arthur, in Technology in Science Education: The Next Ten Years. National Science Foundation, Washington, D.C. 20550, (1979).

Michael, Donald N., The Unprepared Society, New York: Basic Books, Inc., (1968).

Molnar, Andrew R., "The Next Great Crisis in American Education: Computer Literacy." The Journal, July, August, 1978, pp 35-38.

"Understanding How to Use Machines to Work Smarter in an Information Society." The Computing Teacher Vol. 7, No. 5., April/May 1980, pp 68-73.

"Microcomputers and Videodisc: Innovations of the Second Kind," National Science Foundation, 1980.

NSB, (National Science Board), Science Indicators 1976. U.S. Government Printing Office, Washington, D.C. 20402 (1976).

Papert, Seymour, "A Computer Laboratory for Elementary Schools" Computers and Automation, June, 1972.

Prot, Marc Uir, The Information Economy. U.S. Government Printing Office, Washington, D.C. 20402 (1977).

Simon, Herbert A., "Designing Organizations for an Information Rich World," in Martin Greenberger (ed). Computers, Communications, and The Public Interest, Baltimore: The Johns Hopkins Press, (1971).

Taylor Robert, et al., "Computing Competencies for School Teachers: A Preliminary Projection for All but the Teachers of Computing," and Poirot, James, et al., Competencies for School Teachers: A Preliminary Projection for Teachers of Computing," Proceedings of the National Educational Computing Conference, University of Iowa (1979).

18.6 THE TECHNOLOGICAL REVOLUTION AND ITS IMPACT ON MATHEMATICS EDUCATION

THE COMPUTER AND MATHEMATICAL EXPERIENCE

Andrea A. diSessa
Massachusetts Institute of Technology
Boston, Massachusetts

"Mathematics arises from human experience." Ths innocuous phrase, nearly tautological at a literal level, nonetheless captures an essential heuristic upon which thinkers concerned with education from Dewey to Piaget (1) have dwelt: the more extensive and intimate the contact between person and mathematical structure, the more spontaneous and effective the learning process. Assuming this premise, I wish to illustrate with a couple of examples how the computer can contribute to improving our students' direct experience with mathematics.

I can sharpen the focus on what I mean by mathematical experience by asking two simple questions. How can students know that what we teach them is really so, and why should they care? These point to two separate avenues of contact between student and mathematics. The first has to do with confidence, with the salience of the mathematics as experienced by the student and with the set of channels - - including text, lectures and exercises, but going beyond to include everyday experience, common sense, and intuition - - through which students contact the structure of the subject. Moments of insight when a personal experience illumintes mathematical conent, or vice versa, are a hallmark of good contact through this channel. The second question, about caring, has to do with students feeling a want or a need for the mathemtaics. It's not so much about understanding as about students taking themselves to the mathematics and feeling its power.

These questions are particularly relevant to designing computer environments for students exploring mathematics. For, among the first few desiderata for such an environment we should list: (1) It should be simple and "known" so that students can work on their own, helping to allow personal contact and avoiding students merely following along. (2) It should be rich, complex and interesting enough to attract students' attention and get them involved. (3) As we intend it to be a learning environment, it must confront the student in a direct way with the unknown. Putting these together, a paradox about learning at least as old as Plato's Meno reappears. In its simplest sense, how can one learn something new from what one already knows? We are prompted to ask further, how can something be at the same time simple and complex?

The examples below are meant to show one way to resolve the paradox, to show how a computational environment can manage the trick of representing both halves of the dichotomies of known and unknown, simple and complex. In a nutshell, a computational environment shows the user two faces. It is first of all a world of process where how to do somethng is very prominent. Indeed the fundamental "objects" are programs, processes. Yet products are equally apparent. A calculation gives a result and, the case in point, a program produces a graphic result. Attending to one and then the other of these faces of an activity in a computational environment allows the user to see simplicity or complexity, what he understands or does not, what is pleasing or baffling.

Imagine a computer display screen on which lives a creature called a "turtle" which responds to simple commands typed at a keyboard. FORWARD 100 causes him to move 100 units in the direction he is facing. RIGHT 90 causes him to rotate in place by 90°. The number following the command is called an "input." The mathematics of FORWARD and RIGHT is called turtle geometry, and in the context of a suitable programming language becomes a very fertile area of exploration. The examples which follow are taken from turtle geometry work with high school and undergraduate students.

Suppose we define a new command called POLY which merely repeats FORWARD and RIGHT over and over with the same inputs each time. If the inputs are small, you might expect, that the continuous moving and turning of POLY might cause the turtle to go around in a circle, and that is exactly what happens.

Now suppose we make a small modification to the program which causes the input to FORWARD to be increased (e.g. add one) each time it is repeated. Intuitively, such a circle of continuously increasing size might be expected to make a spiral, and that is the case as shown in Figure 1. But the process begins to give rise in its product to some new phenomena: The distance between the arms of the spiral are approximately constant. A moment's reflection about the process of construction, constant increase in size of FORWARD input, makes the effect plausible, but not transparent. We should do a little mathematics to "prove" the effect.

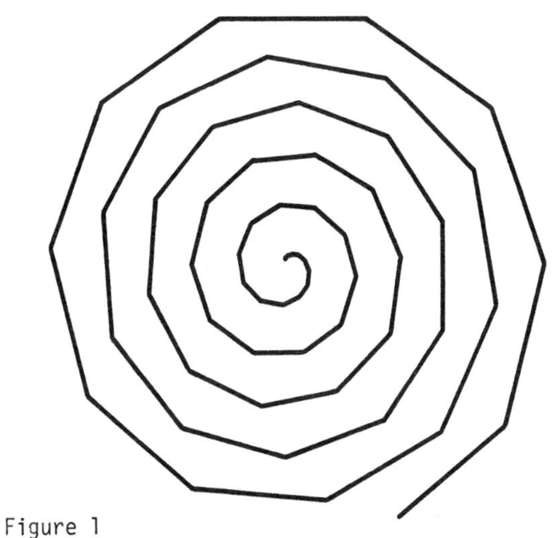

Figure 1

Notice what is going on here. A process which is utterly simple and in a certain sense totally understood makes a product which, in its own terms, is equally accessible. But these are different faces of the same thing, and how to fit them back together is an obvious question, the answer to which is mathematical and unknown. (It turns out that a little elementary turtle geometry which involves conceptualizing a turtle program as laying down successive vectors does what is needed, but those details are beside the main point.)

Suppose we make another product with the same process by marking only the vertices where the turtle turns. Figure 2a shows the result; the spiral is evident. But Figure 2b shows "the same spiral" at a different scale. Having stepped back a bit we see more. (Equivalently, one could reduce the increment to the input to FORWARD). In particular the single arm is still visible near the center, but toward the edge it magically turns into a ten arm spiral. The product of the same process has radically changed.

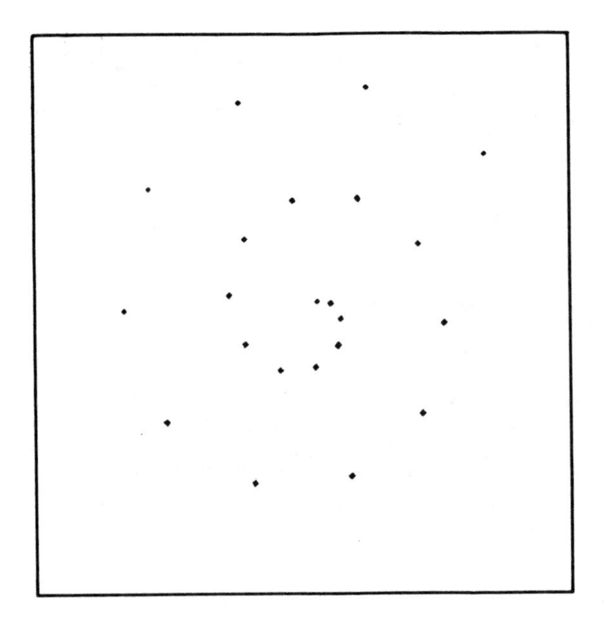

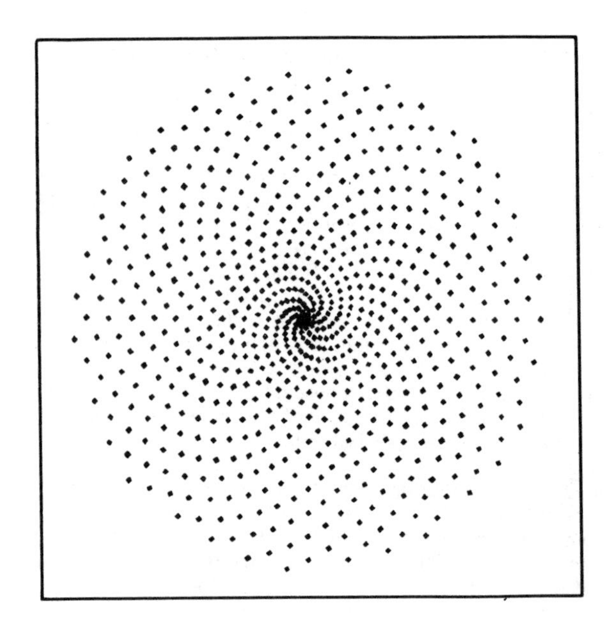

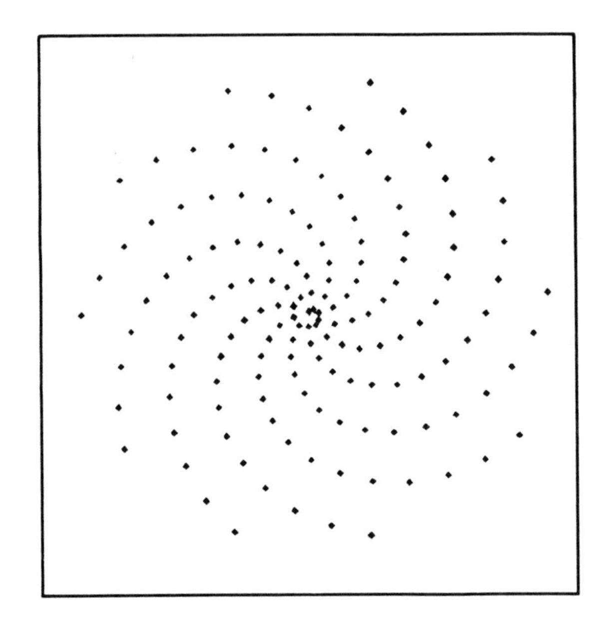

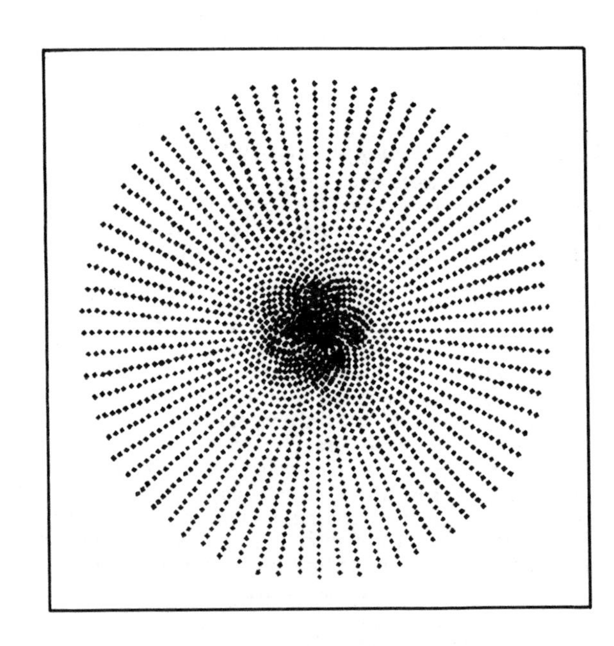

Figures 2a and 2b

We should try such a good trick again. Figure 2c shows another scale change, and more arms appear. This time simultaneously some arms turn to the left and some to the right. The ten arm spiral is still visible toward the center. Figure 2d shows one more change in scale. More, this time straight, arms appear. What would happen if the scale were changed again?

Figures 2c and 2d

We are confronted with two views, process and product, of the same thing. Each view is, in its own terms, simple and understandable. A simple program gives rise to a product which in its own static, visual terms is easy to understand. But the evident features of the two are disparate which poses directly the question of reconciling them. Thus the unknown rests squarely in the links between two knowns. Similarly, the interest, the surprise rests between the two, not in each by itself. (If the spiral pictures themselves happen to be pleasing too, that is a bonus.) It is important that the aesthetic which draws one to the problem is common sense. "Several views of the same thing should be mutually compatible." All too often "interesting mathematical problems" are interesting only to those whose mathematical aesthetics are already developed.

I can provide a glimpse into the mathematical resolution of this problem and more of its richness briefly and without depriving the reader of the pleasure of pursuit. Since the distance between one arm and the next swing around remains constant, (A in Figure 3) yet the distance from one dot to the next in the sequence of production (B in Figure 3) gets greater and greater, soon the former becomes less than the latter. One's eye will begin to associate dots from one swing around to the next rather than dots which are "adjacent" in the production sequence. One can check in Figure 2 that a set of spiral arms visually disappears just when the distance between dots within an arm becomes greater than the distance between arms (2).

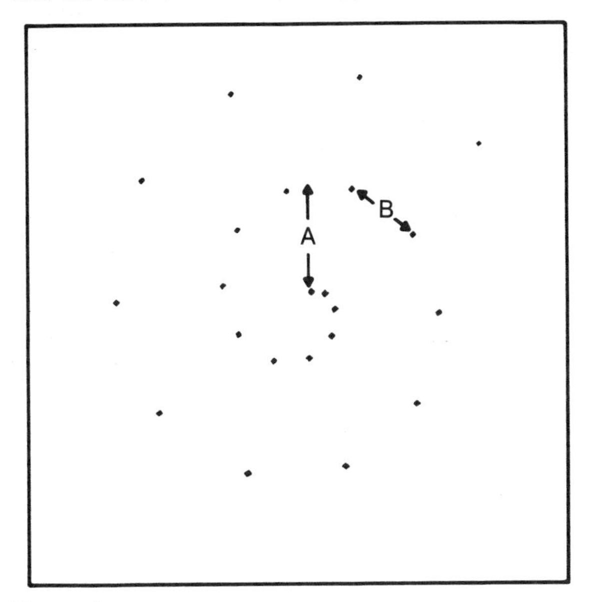

Figure 3

The second example is drawn from the work of an undergraduate student. This student came to me with a very keen interest in the tiling patterns of Arabic art. However, texts on the subject proved minimally informative, and experimentation by hand was hopelessly laborious. So he set about to make a computer tool to automate the uninteresting parts of production (3).

Figure 4 is one (of hundreds) of the products of his work. Consider its complexity for a few minutes. What kind of symmetry does it have? What are its fundamental shapes - - y's, hexagons, diamonds . . . ? One of the fascinating properties of such patterns is the many ways they can be cut up into pieces. Could you describe the pattern in such a way as to be able to reproduce it?

Figure 4

Again it turns out that this remarkable complexity is the product of an essentially trivial process, involving only the simple shape element in Figure 5. The key is to observe that that element sits nicely inside a triangle. Triangles in turn have the nice property that one can construct a larger one from 4 smaller ones. One can repeat the process to make larger and larger triangles of increasing internal complexity. This idea of the whole consisting of sub-parts identical to the whole (triangles consisting of sub-triangles consisting of . . .) is called recursion and is easy to implement in any decent computer language. All that I have left out of the description of the tiling pattern is what orientations the subtriangles have relative to the whole they construct at each level of recursion, which I leave to the dedicated reader to discover. Figure 6, however, shows how a few of the basic elements fit into the whole design.

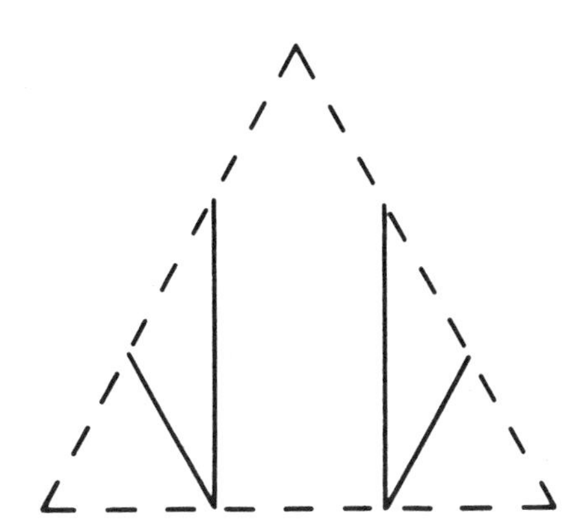

Figure 5

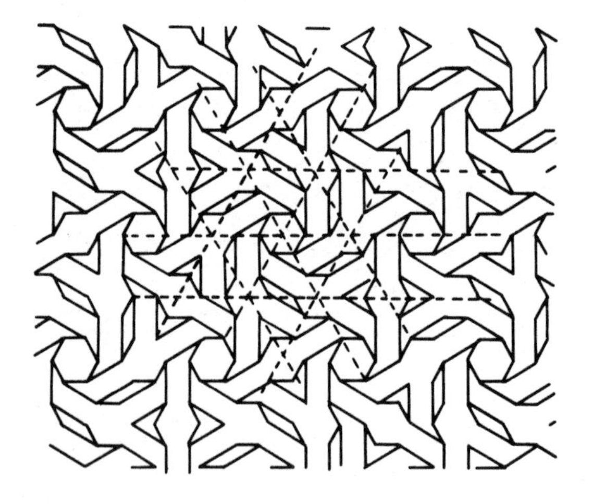

Figure 6

With the tools produced by the student, making a pattern only takes a few moments work on the computer (once the basic tile and patterns of recursion are selected). This frees one to deal with the real work, the art of design, which particularly in this case is a fine combination of realizing the mathematical constraints and principles governing construction of such figures and, on the other hand, aesthetics and inventiveness. Again we must rest with examples to make the point.

If one takes some basic tile and repeatedly applies the same composition to make larger tiles, one finds generally the pattern changes qualitatively at each stage, until suddenly things become "dull". Thereafter iterating the composition just gets more of the same, a larger patch of the same wallpaper. When and why are the obvious questions.

A second example arose when the student turned to coloring his own tilings. Seeking to color a pattern with only two colors led to repeated failures. The first turn of mind necessary to solve the problem was to see it as a mathematical problem - - it might in fact be impossible to color the pattern! This led quickly to focusing on a new feature of the patterns, the number of lines meeting at a vertex and to the conjecture that a pattern with only even vertices is colorable with two colors. The aesthetic goal which motivated the formulation of a mathematical problem, however, did not simply succumb to the fact that odd vertices mean trouble for two-color coloring. In the original pattern, odd vertices were scarce and significant parts of the whole were colorable with two colors (which also helps to explain why the reason for difficulty was not obvious). Would it be possible to color the pattern with three colors, but in pairs so as to maintain two-color coloring in different regions? Can one make a tiling in which odd vertices are extremely sparse? The answers to these questions are affirmative, and a new class of figures and colorings opens up for exploration.

The pattern of thinking in the example - - aesthetic goal meets mathematical constraint, understanding the constraint leads not just to a permanent roadblock, but offers suggestions for overcoming the immediate problem and even for new techniques and goals - - is typical of the best kinds of mathematical experience coming from personal involvement and commitment.

I have tried to show how understanding and controlling the relationships between processes of construction, and their products in a computer environment involve mathematics in an essential way and in a way which draws together seemingly incompatible elements of a learning experience, the known and unknown, the pleasing and baffling. For some, the obvious enterprise of finding more examples and fitting them into a curriculum must follow (4). But I would caution against "legislating interest" as such a task does. Selecting the few best of anything for everyone's enlightenment is at best problematic. Instead, or at least in addition, I hope computers especially in the home can carry with them myriads of examples like these as educational toys in the best sense, vastly increasing the mathematical nature of what people encounter for their own personal ends.

Footnotes and References

1. It's difficult to keep the list small. One would like to include Montessori, Freudenthal, Papert, Hawkins, etc.

2. The dedicated reader will also discover the following: Only straight arms are stable; that is, on changing scale curved arms will always dissolve into another pattern of more arms, except if the arms are straight. The number of arms at various stages makes a nice mathematical inquiry. For example, the number of arms at the last stage (straight arms) is the same as the number of vertices in the POLY program (no increment to distance) with the same angle input to RIGHT. Incidentally, this is the same spiral effect as seen in pineapple, pinecone and sunflower. Why it is that the number of whorls to the left and the number to the right in these naturally appearing spirals should be successive Fibonacci numbers (a phenomenon called philotaxis) is another fascinating exploration.

3. Doug Hi, TILER: A Program for Geometric Design, unpublished B.A. thesis, M.I.T. Department of Physics, June 1979.

4. The examples in this paper are two of many scores of similar projects to be found in: H. Abelson and A. diSessa, Turtle Geometry: The Computer as a Medium for Exploring Mathematics, M.I.T. Press, Spring, 1981.

18.7 CALCULATORS IN THE PRE-SECONDARY SCHOOL

AN INTERNATIONAL REVIEW OF CALCULATOR USES IN SCHOOLS

Marilyn N. Suydam
The Ohio State University
Columbus, Ohio

1. Introduction

Many reports at this Congress have focused on what could or should be done with calculators in schools. Now let's face reality: what is being done? My comments will synthesize a report of the Working Group on Handheld Calculators in Schools of the Second International Mathematics Study (Suydam, 1980). It contains summaries on the status of calculator use prepared by persons in 16 countries. One is struck by the similarities of issues and concerns across these countries- -although divergencies exist in how issues are being resolved.

2. Trends and Prevailing Opinions About Curriculum Implications

Calculators are in common use by scientists, engineers, economists, and other professionals; almost every household in some countries has one. Moreover, data indicate that many children can use those calculators; in the U.S., for instance, the 1978-79 National Assessment of Educational Progress indicated that 75% of 9-year-olds, 80% of 13-year-olds, and 85% of 17-year-olds in the national sample owned or had access to calculators. Data on how widely calculators are used in schools vary more, however. A March 1979 survey in Austria indicated that almost 60% of the students used calculators in grades 7-9, 95% in grades 10-11, and 80% in grade 12. However, in a 1980 survey in Belgium, only 3 of 313 elementary teachers surveyed used calculators during mathematics instruction. Many of the reports reflect this same reluctance by mathematics to use calculators, with resistance greater as grade level decreases. As Fielker notes for the U.K., "While some are willing to look into the possibilities, the majority are worried that arithmetical skills will be forgotten" (p. 56).

The reports from Hong Kong, Japan, West Germany, Canada, and the U.S. similarly reflect this concern that students master the computational skills before they use calculators. As is noted in the Canadian report, it is believed that "Calculators should be used to supplement rather than supplant the study of necessary computational skills" (p. 21).

In countries like Thailand, the issues have "not been seriously considered" yet. In fact, in many developing countries (and apparently in many eastern European countries), calculators are expensive and thus not widely available, therefore delaying the curricular issue, In some countries (e.g., New Zealand), there is little strong feeling one way or the other. Nevertheless, most countries echoed the statement from Australia that "Both individual teachers and education systems began to realize that they would have to come to terms with the calculator" (p. 5).

Over and over, similar arguments are raised for using calculators: attainment of more time for "genuine mathematics content", including concepts; emphasis on problem-solving strategies and mathematical ideas rather than routine calculations; use of more practical problems with realistic data; support for heuristic and algorithmic processess; increased motivation; enhancement of discovery learning and exploration; attainment of speed and accuracy, with relief from tedious calculation; enhancement of understanding; and lessening of the need for memorization.

The points cited for not using calculators involve: fear of dependence on calculators as a crutch; the tendency to fail to criticize calculator-given results; non-availability to all students, since they cost too much for some; the creation of a false impression that mathematics is computation; insufficient research on long-term effects; reduction of achievement in computational skills; decrease in understanding of computational algorithms; lessening of ability to think and also to memorize; and reduction of motivation to learn computational skills and mathematical principles, or to think through mathematical problems.

Predominant among these concerns is fear of loss of computational skills with paper and pencil. In fact, in several studies in the U.S., 80% of the parents and teachers responded "No" when asked, "Should children use calculators in grades 1-6" but when asked, "Should children use calculators in grades 1-6 while also learning paper-and-pencil computational skills?", 80% responded "Yes".

3. Research Activities with Calculators

The amount of evidence being accumulated on the effects of the use of calculators varies across countries. In some, there is little or no research activity (Australia, Austria, Hong Kong, Ireland, Japan, New Zealand, Switzerland, and Thailand). In several others, a limited amount of research is being conducted (Belgium, Brazil, Canada, Israel, and West Germany). In the remaining three countries (the U.K., Sweden, and the U.S.), the research is more extensive. In the U.K., the investigations have been largely exploratory and informal, with the emphasis on ascertaining what could be done with calculators and on the development of comparatively short curricular sequences. In Sweden, there is a coordinated program of research and development, studying the effect of calculators as an aid for calculation and as an aid in changing both the methods and the content of the present curriculum.

In the U.S., over 100 studies have been conducted, most independent of the others and most using an experimental design in which the achievement of calculator and non-calculator groups on various curricular topics or modes of instruction has been compared. Almost all such studies indicate either higher or comparable achievement when calculators are used than when they are not used. A handful of studies has looked at learning-oriented questions, in an attempt to ascertain how learning of mathematical ideas (rather than merely achievement) can be improved with the use of calculators. In addition, less-formal curriculum exploration studies have been underway, to develop activities. The majority of such work has involved integration of the calculator into the existing curriculum; far less attention has (thus far) been given to revising the curriculum to integrate calculators.

Across countries, the overwhelming majority of the data - from both formal experiments and information exploration - has supported the conclusion that the use of calculators does not harm achievement scores. That use of calculators can promote computational achievement, as well as the learning of other mathematical ideas, is also indicated.

In some studies, student attitudes have been ascertained. Students are generally positive about using calculators. However, the initial high level of motivation usually attained when calculators are first introduced is rarely a lasting phenomenon. It may be, however, that low achievers who have continuously failed in mathematics may find that the calculator provides a means to help them succeed. Data from the U.S. indicate, however that only 40% of the educators and 19% of the lay public surveyed would let slower students use calculators, and providing a calculator course in grade 9 for those who have not yet acquired computational skill was accepted by only 45% and 30%, respectively (PRISM, 1980).

4. Instructional Practices with Calculators

The need for curriculum development is evident in the comments from several countries; it is noted in the report from Austria, for instance, "that neither the curriculum of mathematics nor the schoolbooks are related in any way to the need and possibilities of the calculator" (pp. 11-12). Where the calculator is used, it "has been assimilated into existing curricula, and no one so far has altered the curriculum to take account of the calculator" (p. 56). The reports from the U.K., from Israel, and from Japan indicate the direction in which it is believed the curriculum should go, however: to incorporate algorithmic thinking, rather than just the learning of prescribed algorithms.

Sweden is relatively advanced in curriculum development compared with most other countries: new curricular sequences integrating calculators are being tested. In West Germany, Brazil, Argentina, Israel, the U.S., and the U.K., there are some smaller-scale efforts to develop calculator-integrated curricula. While the Swiss report indicates that "changes in curricula are for the time being not necessary because of the use of calculators" (p.47), curricular guides may include use of calculators as aids; such statements have also included in many other countries.

In most countries reporting recommendations, the use of calculators is suggested after grade 7 - - that is, after initial teaching and learning of computational skills is completed. In April 1980, however, the National Council of Teachers of Mathematics in the U.S. released its An Agenda for Action, containing recommendations for school mathematics in the 1980s. One of the eight recommendations pertains to calculators: "Mathematics programs must take full advantage of the power of calculators and computers at all grade levels." In the additional recommended actions to promote this goal, the NCTM acknowledges that computational skills are still necessary, but stresses the need to integrate calculator use at all levels, reinforces their usefulness in problem solving, notes the need for imaginative materials, and emphasizes the key component of teacher education.

The latter point may be particularly noteworthy, in view of the reports from the 16 countries. Little cohesive planning of in-service activities for teachers, to help them place the calculator into perspective and develop strategies for using it effectively, was reported.

Another matter of concern to many persons is their use on tests. Practices vary widely: for instance, in Scotland, Australia, and Hong Kong, calculators are allowed on grade 12 examinations; in New Zealand, some locally based regional examinations allow them, but not national examinations; in West Germany, calculators are allowed on tests in grades 8-13 whenever calculation is not a goal of the test; in the U.S., calculators are not allowed on standardized tests, and most teachers do not allow their use on any mathematics tests.

5. Concluding Comment

The terms "fluid situation" and "cautious approach" appear in the reports and would seem to characterize the status of calculator use in the schools of most countries. The Australian report notes that the need to preserve a reasonable arithmetical facility will continue to be argued. But, as Fielker notes:

> Unless we effect the necessary changes in educational attitudes, it could be that the classroom will be the only place where arithmetic is done by hand! (p. 58)

References

NCTM. An Agenda for Action: Recommendations for School Mathematics in the 1980s. Reston, Virginia: National Council of Teachers of Mathematics, 1980.

PRISM. Priorities in School Mathematics. Final Report to the National Science Foundation, April 1980. ERIC: ED 184 891, ED 184 892.

Sydam, Marilyn N. International Calculator Review. Working Paper on Hand-held Calculators in Schools, Second International Mathematics Study (IEA). Columbus, Ohio: SMEAC Information Reference Center, March 1980.

PRESENT AND FUTURE TRENDS IN USE OF HAND-HELD CALCULATORS AT PRIMARY AND SECONDARY LEVELS

Alexander Wynands
Pedagogische Hochschule I
Ludwigsburg, West Germany

1. State-of-the-Art of Hand-held Calculators (HHC) in Germany (FRG)

In all provinces within West German, the HHC (simple non-programmable calculators) is permitted as a calculatory aid for grades 7/8 (i.e., ages 13/14) including for homework and tests. Since the latter are not multiple-choice tests, no problems arise.

The new curriculum proposes HHCs as an alternative to the slide rule to be used from grade 7 on. If at all, school books deal with HHCs only restrictedly (3-5 pages). This does not mean that HHCs are integrated in school lessons, which might result in alteration of the problem presentation or learning aims. School books do not deal with the application of the HHC as a methodological aid to make mathematic lessons more effective.

In a project which I have conducted since 1979, attitude toward the HHC, basic skills, and the reciprocal effects between basic skills and computer application are being iwvestigated. These empirical investigations are planned for 1979-1983; however, some of the results already obtained are presented here.

- Among 170 interviewed secondary school teachers (pupils 11-16 years), the following percentages intend to use HHCs in lessons: 12% (earliest) from grade 7 (13 years), 23% from grade 8 and 47% from grade 9. 82% of all teachers are in favour of using HHCs from grade 9. A similar interview in 1976 showed that the same number of teachers were prepared to introduce the HHC a year earlier, i.e., in grade 8. Uncertainty and a dampening of initial enthusiasm could be reasons for this difference.
- 66% of the interviewed teachers do not use the HHC in grades 7-9; 15% use it occasionally; 14% often use it.
- 74% never use it for tests.
- 90% of the teachers believe pupils use HHCs for homework, but only 56% of the pupils admit having done so.
- 95% of the teachers believe that the HHC will have a (very) negative impact on mental and written calculation.
- Almost every pupil from grade 7 on has an HHC available. In 1979 the following percentages used their own HHC: 50% in grade 7, 65% in 8, 75% in 9.
- There is no indication of integrating HHCs into mathematics lessons of the first secondary stage.
- 50% of all teachers in grade 9 ignore the HHC.

In view of the increasing usage of HHC in schools and at home, will the basic skills of the pupil diminish in the future? This question can be answered by comparisons of past, present, and future calculatory efficiencies and observation of the percentage of HHCs used.

Does the pupil using the HHC calculate with paper and pencil as well as his counterpart who does not use an HHC? This is answered in my investigation of 1979/80, in which 4300 tests were analyzed by Dieter Wickmann. The test consisted of 10 problems each of basic operations, estimations, and fractions. The result was: No influence of the grade of usage of HHC on the basic skills can be proved. This may be compared to similar statements made by Weiss (1978 - USA) and Werry (1979 - New Zealand), reported in Suydam (1979).

Unfortunately, the present rating of the basic skills of pupils is regrettably low. However, a comparison with tests conducted approximately 50 years ago shows similar results (Wynands, 1980). The verdict on the long-term efects of HHCs on the basic skills must consider the actual state of these abilities.

2. Conditions for the Use of HHCs in Schools

The following points express, in my opinion, the necessary and (almost) satisfactory conditions for an effective application of HHCs:

a. Every pupil should possess his own appropriate HHC. For example, an HHC for the secondary stage is not appropriate if it can be used only for homework and not for practical daily problems.

b. Training and inservice (post-graduation) courses should give teachers more information about the HHC. Although many in-service courses for teachers in Germany deal with HHCs, this training must be intensified. Reforms in schools cannot meet with success if teachers are not sufficiently informed.

c. Suitable working material for HHCs must be prepared and integrated within the school work. In Germany these requirements are still far from fulfilled.

3. Dangers from the Use of HHCs

Five key points are:

a. Attention should be drawn to mental skills, knowledge of calculation rules, and estimatory skills as important techniques in our culture that mean human emancipation from machines.

b. "Pressing keys without thinking" must be criticized; e.g., use of the %-key. The pupil who has not understood the terms in the formulae G x p % = w and G x P % + G = E and is not aware that "$200 plus 5%" is mathematically expressed as "200 x 1.05" will solve the following problem

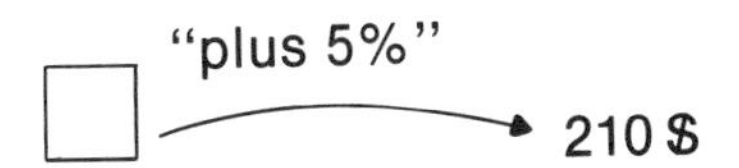

with a wrong "minus 5%" dead end unless he tries to solve the problem by the one-way-principal "trial-and-error".

Furthermore, the algorithm

200 [+] 10 [%] [=] [+] 10 [%] [=] [+] 10 [%] [=] ....
220 242 266, 2

would make a pupil believe that this is the beginning of an arithmetic progression instead of a geometric one.

Other examples could be drawn from the primary school level, where the pupil must understand an operation or the blinking HHC signal sequence has no meaning. Almost all teachers in Germany ignore the question of the place of HHCs in the curriculum of the primary level; in fact many don't approve of it at all. However, no clear statements on this issue are currently available in Germany. Some research results are to be expected from

Meissner (Universitat Munsster, FRG), leading to a project group dealing with this subject.

c. One of the dangers of balanced mathematics lessons lied in the partly unreflected transition from sturcture-oriented to algorithm-oriented lessons. The outstanding examples of algorithmic thinking propagated by A. Engel at the last ICME Congress in Karlsruhe are not to be taken for an antithesis "Algorithm versus Structure" but to be understood as an adequate supplementation of school lessons.

d. The HHC, and in particular the press-button (programable) pocket calculator, could lead to a false understanding of Informatic (Computer Science) in schools. A working group of the Gesellschaft fur Didaktik der Mathematic (GCM) in Germany is striving to establish distinguishing and common points between numerical methods and informatic theories in schools.

e. HHC results could be false! This might lead to questions of áccuracy and chain errors in digital processing and also to errors in approximation methods. If such aspects are not recognised as problems, the danger exists of the HHC being accepted as an authority that, for example, could decide about the convergence or divergence of a function.

4. Possibilities for HHC Use in Mathematics Lessons

4.1 The HHC makes mathematics lessons more effective.

A fast calculation in terms and application of formulae enables an independent processing of value tables and graphics. This makes daily work with tables and graphs easier for the pupil. Data access of HHCs forces the user to plan their forms or tables in such a manner that input-output data are properly processed. For one type of problem a whole series of exercises can quickly be calculated which gives a more lasting effect to the learning phase.

4.2 HHCs influence methods in mathematics lessons.

a. Functions

The HHC is a machine for functions, which works as an operator. The access to and the work with functions become visible. Example: Compare

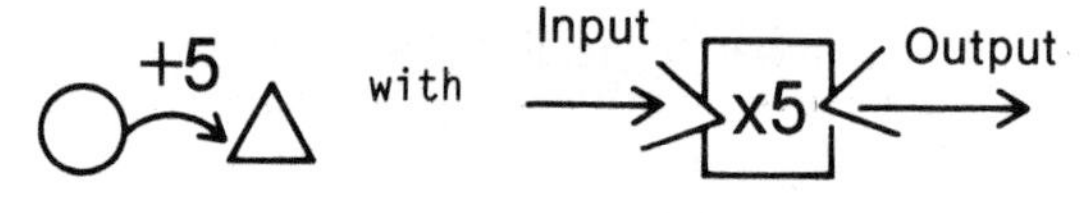

with the HHC program:

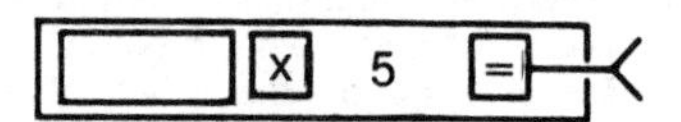

Basic operations with and compositions of functions can be presented and calculated by the HHC.

b. Rule of Proportions

The only computer-effective method of calculating several problems with a given proportional factor e.g., rates of exchange, petrol consumption per filled tank) is first to determine the factor m (i.e., an external ratio between two given quantities x_o and y_o) and then determine the y value corresponding to x or vice versa.

c. Equations and Inequalities

The trial-and-error method served as a preparation for weaker pupils; in fact, it is a substitute for closed algebraic solution-strategies. The HHC shows quickly how correctly one has guessed.

4.3 The HHC alters the time sequences of learning.

At an early stage the HHC pupil gets acquainted with "number chains that he can later interpret as negative numbers or decimal fractions.

This leads to urgent research projects with the following aims:

- What effects do the availability of HHCs have on the order of learning-sequences of negative numbers and in particular of decimal fractions?
- Does the presence of HHCs and the actual poor basic skills justify the handling of decimal fractions prior to normal fractions?

4.4 The HHC enables some part of the curriculum to be dealt with that, up to now, were unfortunately only on paper.

These areas are sufficiently described in the corresponding literature. The pupil can individually calculate numerical problems that are helpful in processing the terms "geometric progression, convergence, divergence, real numbers, exponential functions, power, roots, fix-point of a function" and are also helpful for mathematics work and proof methods, recursion, and iteration.

References

Nancy D. Cetorelli: Teaching Function Notation, The Math. Teacher, Vol. 72, Nov. 1979.

Marilyn N. Suydam; Working Paper No. 7 on Hand-Held-Calculators in School, IDM-IEA Working Group on Calculators 1979.

Bert K. Waits and James E. Schultz; An Iterative Method for Computing Solutions to Equations Using a Calculator, Mathematics Teacher, 9/1979.

Alexander Wynands and Dieter Wickmann; Taschenrechner und Rechenfertigkeit, Arbeitspapier des FEOLL, D 4790 Paderborn, Pohlweg 55, 1980.

CHAPTER 19 - Forms and Modes of Instruction

19.1 DISTANCE EDUCATION FOR SCHOOL-AGE CHILDREN

BBC SCHOOL TELEVISION

David Roseveare
BBC Television
London, England

This presentation extensively used videocassette recordings of television material. Readers interested in obtaining copies of any complete BBC Television programmes should contact BBC Enterprises, Room 503, Villiers House, The Broadway, London, W5 2PA England. (Tel: 01.743.8000 Extension 394/5). Readers in North America who would like to borrow the copy of the extracts only which were used in this presentation should write to British Broadcasting Corporation, 630 5th Avenue, New York, N.Y. 10020, U.S.A. (The videocassette is Sony 3/4" 525 lines NTSC.)

In the United Kingdom, there is no national curriculum in mathematics - - or any other subject. Each school may devise its own syllabus, although in practice there is a generally agreed body of mathematical knowledge and technique to be acquired.

There is no obligation for any teacher to use school television, but 95% of all U.K. schools are equipped with receivers and over 90% actually use BBC Television series for schools. Why, then, do U.K. teachers use broadcast television? What do they expect from it, and what are we as producers trying to provide.

Every teacher will give you a different set of criteria. Some will argue that television's principal role is to bring the real world into the classroom. It is easy to see how this applies to subjects such as geography, natural history, or careers for the early school leavers. How does it apply to mathematics at school level? Let me give you twovery different examples. The first extract is from a series on trigonemetry.

Video Sequence 2 (3'21"): from the series Math Topics, for children aged 13-16, top 25% of ability range. Trigonometry Unit, Programme 3, Sequence 2: "A Surveying Problem".

This sequence combines live-action photography with diagramatic animation and shows surveyors of the United Kingdom Ordinance Survey at work. The accompanying commentary is:

> "It's about one kilometre between the two points.
>
> "The top of the pillar is already known to be 129.21 metres above seal-level but what matters is that they know the height of the hill above mean sea level -
>
> " - and they want to find the height of the corner of the fence.
>
> "So it's only the difference in height that they need to measure - that's this distance.
>
> "They can't measure it directly because this point is deep inside the hill. The only distance they can measure is this one.
>
> "So, how do they measure, through the air, a distance of about a kilometre?
>
> "What they do is to transmit a beam of infra-red light down to the fence junction where a pair of prisms reflects it back to the transmitter. The instrument notes the travel time of the light. Knowing the speed of light, it computes the distance to the prisms - 930.387 metres.
>
> "Now they know this distance, but they don't yet know its direction.
>
> "They need to measure an angle. You can't do that with just a light beam, so they use this instrument - a theodolite. First, they get it level, using built-in spirit levels.
>
> "They're going to find the angle through which you would have to tilt the theodolite from the upward vertical until you were looking at the reflecting prisms."

Real surveyors measuring real angles, but you will have noticed some of the simplifications we had to make.

For the same age group, but at a much lower level of mathematical ability, we have also brought in the real world, but we have another reason for teachers to use television. This time, the mathematics is very simple. What matters is that it is being used in situations and by characters with whom the pupils can identify.

Video Sequence 3 (1' 48"): From the series Everyday Maths, for children aged 14-16 of low ability. Programme 1, "Pounds and Pence".

This series is entirely dramatised. In this extract, Mike is a regular character, London Cocknev, aged about 18, not very good at mathematics. Jim is his West Indian friend who works in a garage forecourt. The scene is set in the forecourt shop.

> Mike: "Dead easy. Now then - (USES CALCULATOR, WHICH WE SEE IN CLOSE-UP). One pound five pence. One . . . (HE PRESSES THE APPROPRIATE BUTTONS AS HE SPEAKS) . . . point . . . nought . . . five . . . times . . . six . . . equals . . . (CALCULATOR DISPLAYS 6.3). Six point three. Six pounds three pence. Hey, that's not right, either.
>
> Jim: That's not three pence. It's thirty.
>
> Mike: Thirty? It's three.
>
> Jim: It's point three. Point three is three tenths. Three tenths of a pound. Three of these. (HOLDING UP A 10p COIN) Thirty pence.
>
> Mike: So why hasn't it got a nought on the end?

Jim: Because a nought on the end wouldn't make any difference. (MIKE IS NONE THE WISER) The calculator doesn't know you're doing money. It could be . . . six point three . . . gallons of petrol, or six point three . . . elephants, or . . .

Mike: Elephants?

Jim: It's all the same to this thing. It just handles numbers. Only you know what the number stand for.

(The exerpt continues with more dialogue and diagrams showing the numbers.)

You will have noticed that animated diagrams are used without introduction - the characters continue the scene while we, the authors, substitute a visual explanation of what they are saying. Of course, the actors can see the diagram, but we were attempting to keep the drama itself realistic.

The BBC's Charter has, for 60 years, required it to"inform, educate and entertain". It does not say that these objectives have to be kept separate. I am a firm believer in the use of humour in educational programmes. Let me put my case with an example.

Video Sequence 5 (1'46"): From the series Mathshow, for children aged 11-12, of average ability. Programme 4, "Fair Share".

This extract starts and ends with a cartoon triangle and square acting as commentators.

Triangle: "Shouldn't you share out a remainder?

Square: "It depends what it is."

The bulk of the sequence is now a black-and-white sequence in silent movie style, with accompanying piano music. A caption appears "Eggs-actly (a bad yolk)" and we then see two characters Alpha and Beta, not unlike Laurel and Hardy, sharing out food in two equal piles. Six biscuits are no problem, nor are the first 4 or 5 sandwiches. Alpha tries to eat the 5th sandwich but Beta stops him. (Caption: "You have to share it out".) Alpha takes a knife and cuts the sandwich in half. Finally, he applies the same procedure to 3 eggs.

Triangle: "Hm, so what?" (DEVOTEES OF THIS SERIES WILL APPRECIATE THIS AND OTHER ANAGRAMS)

Square: "So you need to have a remainder sometimes. Not always.

Each Mathshow programme was intended to be a sequence of memorable scenes, dramatically unconnected - we jump from Laurel and Hardy to cartoons to Noel Coward - but with a common mathematical theme. The extract quoted above was intended for the teacher to use as a starting point for the idea that when division is not exact it may be appropriate to specify a remainder or it may be necessary to use fractions or decimals - the real-life context will normally determine which course to take.

Such pedagogical remarks are not in the television programme but are in the Notes for teachers which accompany the series. These can be bought from the BBC's Publications Department. The notes contain descriptions of the programme and comment on its mathematical content, with suggestions for classroom work related to the broadcast.

We have, then, several related media of communication. At the heart of it all is face-to-face teaching between teacher and child. Into this comes an intruder, TV programmes for pupils. And, to help the teacher use the broadcasts, printed material for pupils and notes for teachers are added. The cost of print and its distribution is currently forcing us to reduce these elements. In the past we have experimented with a more elaborate package that included radio and TV programmes for teachers and 8 mm loop films and notes for pupils, but quite apart from the cost of producing these, it was difficult to find a suitable time of day for teachers to view. The 8 mm loop films were an attempt to create a permanent kinetic image of key scenes - normally animations - from the programmes. They were never very successful, and in any case we now have a much more useful piece of technology - the videocassette recorder - which enables teachers to record and broadcast progammes and to use them selectively.

At present, 86% of U.K. secondary schools possess a videorecorder but only 12% of primary schools. There are three reasons for this. One is the obvious one of financial scale - it is easier for a large school to buy a machine costing hundreds of pounds. The second relates to the school timetable. A primary school teacher normally has the same group of children all day and can fairly easily build his timetable around the fixed times of broadcasts. A secondary school timetable is much less flexible - videorecording at last makes it possible for broadcasts to be as useable as a book. The third reason relates to the expectation which different teachers have of broadcast television. A colleague of mine once attempted to define two main types of teacher in relation to their use of broadcasts. The scavenger is probably a subject specialist who sees television as one of many possible aids, along with slides, OHP acetates, models, textbooks. He selects the material he wants to use and chooses when to use it. The depender is probably a non-specialist - and all primary teachers are non-specialists. Broadcast television appears as a voice of authority. It is there to stimulate the children, but it is also there to give the less assured teacher ideas for classroom work. It is there every week - or every fortnight - and thus can act as a pacemaker for the term's work.

Dependers almost certainly outnumber scavengers, but they form a silent majority. Our most difficult task is to obtain the views of those who are themselves uncertain what they want from school television.

We have several methods of obtaining such information. At the heart of the structure lies the School Broadcasting Council which determines the broad policy. For example, they might decide that an additional series in mathematics for future mechanical engineers was more important than a new series about Ancient Rome. But once the guidelines have been laid down - and you can see some examples (in the booklet "Broadcasting and Mathematics" distributed freely at ICME IV) at your leisure - then the detailed decisions are taken by the producer, who is expected to blend creativity and educational know-how with strict financial control and the ability to make every programme, like this paper, last more than its alloted time of 20 minutes.

19.2 TEACHING MATHEMATICS IN MIXED-ABILITY GROUPS

LEARNING TO TEACH MIXED ABILITY CLASSES AT FLEMINGTON HIGH SCHOOL

Denis Kennedy
Hadfield High School
Pascoe Vale, Australia

The Australian Aborigine has a powerful concept: "dream-time." Dream-time has familiar characteristics shared with our Genesis stories, our "when the world was young," "in the beginning," and "once upon a time" stories. But it is a deeper concept. Dream-time still lives, not in the remote past or in some far distant land but in the here and now, woven into the fabric of our real world. It is as real and as accessible as the trees, the rocks, and the water-holes. I mention dream-time becuase it is the quickest way that I can describe the impact of twelve years at Flemington High School on my wife and me, and on other members of our team. For us, it is over and two years gone. Only a shadow of the system we built exists now, but the lessons learnt live on.

During those twelve years, we manufactured our own myths. For example, like Camelot with its Round Table and Knightly Equality, we tenaciously maintained that every child has equal ability - - and it was up to us to find the best way to bring out and nurture that ability. Like Camelot, the dream lived for a time and has now faded. Teachers are human beings and not mythical heroes; there is a limit to what they can give. Worse, conditions changed. The school adopted a sub-school approach to meet the educational and social problems, and this structure would have destroyed many of the features of our individualised system. We prepared to adapt - - with bad grace, we must admit. However, the parents sprang to our defense and demanded that our system must continue. Their support was wonderful. It was moral and financial, but it proved to be fatal.

We believe that any program which cannot or does not adapt and evolve with changing circumstances is, quite rightly, doomed. We preached that lesson, of the need for evolutionary change, for hastening slowly, of not passively adapting but of cautiously making use of changing circumstance. Hence, it is ironical that we should have been caught be a sudden climatic change, frozen and killed like the mammoths in Northern Siberia.

So far I have offered you images of fabulous mythical creatures, of King Arthur, and of extinct frozen hairy elephants. It is time to talk in more concrete terms about an inner suburban high school and how its mathematics department coped with the social and educational problems of students in very mixed ability classes for twelve years.

The Growth of Individualised Instruction

Flemington High School served a "disadvantaged" community living close to the business centre of Melbourne, the capital of Victoria, a southern state and one of the more industrial states, of Australia. Flemington is part residential, part industiral area. The community is predominantly working-class, with a large ethnic component. Fourteen years ago, the community was a little more affluent and the school was academic, using streaming and setting a different year levels. As a new bright-eyed mathematics co-ordinator, I took the bottom set of the Year 10 Arithmetic group. For the first half-year the class learnt very little,, but I learnt that I had a lot to learn about teaching. In the second half-year I was desperate enough to ignore my colleagues' comments, and to use games and to structure my teaching around them. At the end of the year, the students, the critics and I were stunned by the class's academic achievement. From that beginning grew our mathematics games period that lasted five years and influenced much of what came later.

Those five years were busy and happy ones within the classrooms, but a little more tumultuous outside. People were interested, there was local and national media interest, but also the times were right. Guardian angels appeared in the form of the Mathematics Association of Victoria, our school principal, various members of the State Education Department administration, and others, who gave us the time and chance to build and prove the value of the mathematics laboratory approach. After the first four years, all mathematics classes spent 20% of their time in the laboratory, until it became absorbed into the individualised programme.

A major lesson for us from this period was something that we all know in theory but that becomes dramatic in active practice - - the importance of success in the learning process. For some students it is difficult to find something at which they can succeed and which they will recognise as a worthwhile success. One student I recall cost us six months before we found a breakthrough. We found that our students had to succeed before they really committed themselves to learning. Strangely, at the other end of the ability scale were students who hid their ability from themselves and their peers. For self-protection, they worked hard at being mediocre. These students had to be tricked into succeeding. I sometimes wonder how many students were too clever for us and evaded our traps and snares and went on convincing themselved and the world that they were strictly average. I hope that my fear is groundless. I hope that our experience is unusual and that it is not generally the case that high ability minds, consciously or unconsciously, feel the need of camouflage and the need to ape the average citizen.

The need for success in learning caused us to adopt many aspects of mastery learning. On achievement tests, each student was expected to earn a 100% mark. Students and colleagues questioned this policy, and our quick standard answer became, "How would you like to be operated on by a surgeon who could only manage 80% on his tests?" The policy proved itself in the long term by the motivation from such success. Students who considered themselves as failure since kindergarten blossomed after they had struggled through to their first

100%. The retention rate was worth the insistence on 100%. After six months and twelve months the retention rate was steady at 80%, and this became our bench mark and test when we suspected that a student was cutting corners. If he could not test out at 80% at any time, then he was not using the system properly.

The need for success also caused us to abandon set syllabuses, and to tailor each student's work to his present stage of development. In reports and assessments in Years 7 and 8, a student's achievement was compared to his assessed current ability. This last term sounds vague but it was based on objective as well as subjective measures, and I would argue that it is more meaningful than a class average, some percentage that masquerades as respectable in its garments of surplus decimals. By Year 10, we had to bow to approaching career choices and external examinations, and one of our five ratings was more like the traditional achievement mark.

The need for success in learning was one of our main principles. Every teacher is aware of it, and it is one of the main motivations for streaming and setting. There is a belief that a teacher can help students to succeed with more ease if all the students are of about the same ability. At Flemington, the mathematics department used setting but of an unusual kind.

Setting at Flemington High School

During most of the seven years of our individualised phase, all mathematics classes at each year level were time-tabled together. This suited us well although we did prefer two different half-years together. Our usual situation was six classes with seven teachers, with the extra teacher personning the resource centre. (I have to use the word "personning" as the extra teacher was usually my wife, and "manning" does not seem appropriate for her. We used to have some difficulty in describing my wife's job in the scheme. We were taken aback when we heard that, following a visit by a Christian Brother to whom we had used the term "floating teacher," my wife was being described as a "lose woman.")

The resource person, or the "floater," handed out resources beyond those standard ones kept in the classrooms. Thus, she was dealing with the problem students who might normally be described as remedial, or with students who had run their teachers out of material and needed enrichment work. Generally, she was our diagnosis and prescription expert. All students passed through her hands and used her resource room on a fairly regular basis.

The other six classes were all mixed-ability groups but tended to be set according to the personality needs of the students. If a student responded well to mothering, then we would try him in the class of a motherly-type teacher. If an older-brother-type was called for, then we would hand him over to an older-brother-type teacher. If a stern father was called for, they usually got me. Hence my classes consisted of rat-bag boys and lovely girls who used to wrap me round their little fingers. One of our best efforts was combining fifty kids in one room with two teachers where I played Daddy and a sweet first-year teacher played Mummy. That was a terrific class for the amount of work accomplished.

You will see that security was one of our principles. With the social and emotional problems of many of our students it was very important, and yet, in one of our practices, we seemed to be throwing it away. Each term the classes would be re-distributed. Often over protests from teachers and from students we would take at least one third of each class away from its teacher and place individuals with some other class. Only occasionally did teachers protest, as it tended to be their discipline problems who were moved. If a student was a discipline problem, then we assumed that we had placed him badly or that he was no longer getting what he needed from that teacher. A sure indication that it was time for a shuffle was a student calling teacher "Daddy" or showing some other sign of emotional dependence.

One of our major aims was student independence - - as learners and as people. We tried to keep the sense of security by shifting dependence - - from dependence on one teacher to dependence on two teacher (his own teacher and the resource teacher) and then to dependence on the whole team of teachers. As the breadth of dependence spread, the depth decreased. The students gained confidence in their own ability and were more able to help themselves and each other.

In describing our way of setting, I have mentioned some of the aspect of teacher influence. These came to our notice during the games phase and the first year of our individualised phase. The first discovery of the dramatic influence of teacher expectation emerged when we were testing the value of games. There were some variations in results between classes, and we were at a loss to explain them. Slowly, it dawned on us that the variations were not a "group mind" phenomenon, but simply each class reflecting its teacher's attitudes and expectations. We tried faking our attitudes in front of the classes in later tests, but the classes always brushed that aside and reflected our actual attitudes. It was unnerving.

Since then there has been much mention of teacher expectation in the literature, I will mention only one effect that was startling. In our first year of individualisation we did not shuffle our classes. One day for interest's sake, my wife constructed a histogram of each class's achievement level, and, there, as plain as a signature, each teacher had his own characteristic shape in his different class histograms. If the classes were shuffled, the teacher's signature began to emerge again about three months later. Our piece of teacher lore from these observations was that students first benefited from the strengths of any teacher but after three months began to show the effect of his weaknesses. Vague and unproven perhaps, but we felt that we had to act on it.

Up to this point I have tried not to trespass on what David Lingard may say or what can be read in the Schools Council publication, Mixed Ability Teaching in Mathematics, to which David contributed. It is a superb book and I commend it to you. The book contains a wealth of information and guidance that we had to find out the hard way, by experience. It mentions the need for variety, and we would agree and stress that point most emphatically. The teaching methods should change; one day a student might be working alone, later he is paired off with another, a week later he might be grouped with 5 other students, and at some time he might find himself in a class lesson. The teaching material should vary in media, style, and demands on students. Variety is all important, and there are pressing educational reasons for indulging, apparently, in change for change's sake.

The complexity that I have just advocated and the need for security for the students mentioned earlier make it obvious that I would not recommend that one teacher undertake individualised instruction on a long term basis on his own. You will have noticed that my referenced to the set of mathematics teachers at Flemington High School have been to a team - - not a department or a faculty but a team. For the students' sake and for his survival, each teacher needs all the support he can get in an individualised system. He must be part of a team where his strengths are used and his weaknesses covered by colleagues and by the system structure.

However, even as a member of a good team working efficiently within a well-organised structure, the teacher pays a heavy price. Nearly every advantage won for a student is purchased at the expense of the teacher. I have opened up a large topic, but let me limit it to the mention of three reactions of teachers in an individualised system. Few teachers escape the feelings of being dethroned, de-skilled and disoriented.

Dethroned: Teachers must change their role in an individualised system from the more traditional one of being king in their own classroom. They must become more like servants. They must allow and encourage other teacher's trespassing in both the physical and emotional kingdom of the classroom. And it must be more than trespass. Colleagues must share responsibilities in a way that tears away the survival device that teachers tend to build, the device that I have caricatured as "King in their own classroom."

De-skilled: This feeling attacks the experienced and skilled teacher most strongly. The tricks, devices, pat phrases, body language, pet jokes, cherished lesson plans, and seemingly the whole arsenal of successful methods and approaches not only don't work but seem counter-productive. The teacher feels that he has been thrown back beyond his first year as a teacher. He feels as though he is half trained and badly trained. I can assure you that it is a dreadful feeling, but that it does go away, that each recurrence is less savage than the previous one. However, it can be fatal in that excellent teachers have sometimes fled from an individualised system they admired because they have felt de-skilled.

Disoriented: There is no way to escape this disquieting feeling. A good individualised system has variety in all its aspects and must run to the edge of bewilderment of the team. Somewhere along that edge, any teacher must find areas where he is confused and out of his depth. In the classroom, there are many opportunities for a feeling of disorientation to arise. I found that I could conduct six conversations at once on six different mathematical topics. That, surprisingly, need not be disorienting. What is difficult is the rapid change of role - - this student needs encouragement, that student needs a stern rebuke, a third needs straight information, a fourth needs a veiled hint, and on and on it goes. It is the same problem that arises in any classroom, but in an individualised setting the problem is multiplied many times over.

The three problems - - teachers feeling dethroned, de-skilled and disoriented - -can all be overcome with team effort, but I still wonder if the individualised approach demands too much of teachers. I suggest that there is help visible on the horizon. The rescuers are not, in this case, the American cavalry but, I sugget, the American micro-computers.

With micro-computers coming on the scene, teachers in individualised system need no longer be overpowered by detailed record-keeping. Record-keeping is fundamental to a successful individualiased programme, but much of it is hack work better handled by a few silicon chips than by neurons. Further, a micro-computer could suggest options for further work and keep track of resources, leaving the humans to do the human thing - - making decisions and choices. With some caution I add that the micro-computer could take up some of the instruction, but I am not yet convinced of the practicability or value of computer assisted instruction in Australia in our present circumstances. What I do commend to you is the value of individualised instruction as a difficult, but not impossible, and extremely valuable answer to the problems of mixed-ability teaching in mathematics.

TEACHING MATHEMATICS IN MIXED ABILITY GROUPS

David Lingard
Northcliffe Comprehensive School
Doncaster, England

The original suggestion for the for the title of this panel was: "Mixed Ability Classes in Mathematics: Problems and How to Cope with Them." "What are the most crucial problems associated with mixed ability classes in mathematics?" the panel description asked. I was not happy with this title. It seemed to present mixed ability teaching in a rather negative manner. I felt that I wanted to be more positive and cheerful. However, it is with the problems that I wish to begin.

I have been teaching mathematics to mixed ability groups of 11-16 year olds now for eleven years, in three comprehensive schools, and over this period I have gradually been drawn to the following conclusion: There are no problems in teaching mixed ability groups but there are a lot if problems in teaching mathematics. To put it another way, problems associated with teaching mixed ability groups are solely problems of teaching mathematics.

What has happened is that teaching the full range of ability within one class has made some of the problems of teaching mathematics more apparent. Certain difficulties have been highlighted, and consequently teachers working in this way have focused their attention on them. For some teachers it has made them aware of problems whose existence they were previously unaware of. I would suggest three particular reasons as to how this has come about. Teaching mixed ability classes (i) teachers have been required to pay close attention to individual children, (ii) teachers have found it necessary to re-examine their aims and objectives in teaching mathematics, and (iii) teaching all abilities has encouraged teachers to offer a wider mathematical experience to all children.

The growth and development of mixed ability teaching over the last 15 years has served a most useful function. It has acted as a means of diagnosis for many of the problems relating to the teaching of mathematics. Of course, it has not necessarily always provided a therapy! Some teachers, faced with the problems that they felt were presented by mixed ability teaching, have retreated to previous practice hoping, or falsely assuming, that the problems would no longer be with them.

Mixed ability teaching has brought its own advantages in addition to this valuable diagnosis. The advantages are numerous and have been well documented by me and many others, including for example, the authors of the Schools Council report published in 1977. For the moment, perhaps I could just reiterate what, for me, are two of the major advantages:

1. Raising student motivation, mainly through countering teacher expectation of pupil performance. An end to the premature and mistaken 'labelling' of children.

2. Raising academic standards for all, but in particular for the most able. Previously these students were so often unwittingly held back by teachers imposing a ceiling on their performance - a ceiling related to what the teacher thought a child should be able to do in the fourth year or the top set or the exam group.

But perhaps in the long term, as far as mathematical education is concerned, the greatest single asset and advantage to stem from mixed ability teaching will be the focus that it has brought to some of the very many complex problems of teaching mathematics.

I would like to expand upon the general point that I have made and try to illustrate it from my own experience and that of many colleagues. The problems of teaching mathematics to mixed ability classes are not problems about suitable classrooms, adequate resources, reprographic facilities, computerised learning programmes, lack of suitable text-books, methods of assessment, etc. They are related to 3 things: (i) the teacher's view of what mathematics is; (ii) as a consequence of (i), the teacher's view about the ways mathematics can be taught and learned; and (iii) the teacher's attitude towards children and their respect for children's mathematics. Those who have taken up mixed ability teaching at secondary level have been forced (consciously or otherwise) to face these three issues. Those who have survived the challenge (and in his talk in this session Denis Kennedy clearly illustrated some of the problems for teachers in doing so) have come to know that teaching mixed ability classes is not the issue. The issue is teaching mathematics.

Let me take these three points in turn and look at their implications for teaching. First, what is mathematics? It is more than a body of knowledge, a bundle of facts and techniques. A lot has rightly been said and written in recent years about mathematising and about humanising mathematics. Mathematics is an activity. It deals with skills and processes and ideas. It is something experienced and developed by each individual. The implications for school mathematics teaching are considerable: (a) certain topics and pieces of mathematics are not sacrosanct; (b) content is not the primary concern but is subjugated to the context (context of application, of the learner, of the teacher, of the time); and (c) there is not a clearly defined linear progression of what mathematics should be taught and in what order. Achieving a coherent mathematical education for each child is important, but this is not the same thing.

The second point concerns ways mathematics can be taught and learned. It follows from the first point that you cannot simply hand over your mathematics (or anyone else's) to children. The teacher is there to promote and facilitate learning. To provide starting points for an experience. To guide and respond. I first appreciated the futility and irrelevance of much of our traditional classroom teaching when I read Caleb Gattegno's hilarious parody of the passing on of knowlegde to children, via lessons, tests, homework, examinations etc., in his book What We Owe Children. A.N. Whitehead also had similar things to say about this, particularly about "inert knowledge." (I'm not sure whether to feel encouraged or worried to discover that President Carter reads the same books as me!)

Denis Kennedy has in his talk emphasised the need for a variety of modes of learning and teaching. I support his view so strongly that would like to repeat what he said:

> The teaching methods should change; one day a student might be working along, later he is paired off with another, a week later he might be grouped with five other students, and at some time he might find himself in a class lesson.

(I think that I would have said that he should find himself in a class lesson.)

All my exprerience tells me how important this statement is, and it is a view strongly supported by the Schools Council working party. How have we got away with the monotony of class lessons for so long? In particular, I feel that I must oppose fiercely the use of individualised learning programmes where they are used to exclusion of all else - as in England they so often seem to be - and where students are left to plough their own lonely furrow for lesson upon lesson, working at "their own pace" - accurately observed by some to mean "snail's pace" all too often.

The third point concerns respect for children and their work. There has to be respect for children as doers and creators of their own mathematics. Their feelings have to be respected and their point of view listened to, developed, and encouraged. Discussion of their mathematics, writing about their mathematics are crucial and essential steps in their mathematical development. Teachers of mixed ability classes have been quick to recognise this. There is little new here. A number of teachers have accepted all that I have said. English primary schools contain many such teachers. At secondary level, however, this has rarely been the case, and still is not.

So far I have used rather general terms and may not have been as specific as some of you would have liked. I will try to remedy this. Successful teachers of mixed ability groups have been forced to use not only a variety of styles and modes of teaching but also a variety of materials and texts. They have been eclectic in their choice of textbooks or work cards. No one textbook or scheme is good enough for a single child. How foolish that we ever believed it to be so!

In the early 1960's many felt that the introduction of the "new maths" would solve some of our problems and enable us to produce a numerate population. I remember writing in 1970 that my fear was that when the dust settled after this "revolution," little would have changed and we should be no further forward along the difficult path of mathematics education. Ten years late I think that current practice, at least in the U.K., supports my view. The issue is not about content, and teachers of mixed ability groups are aware of this.

Peter Coaker, a member of British Petroleum and an enthusiastic and noted supporter and worker for mathematics education in England, recently suggested at a meeting of the Schools Council Mathematics Committee that it would be a good idea if, when the time came to finally assess pupils' competence at mathematics before they left school, we sat four children around a table, presented them with a problem, and then asked them to talk about it. He went on to suggest that such an assessment might constitute half of their final assessment. Teachers of mixed ability groups would leap at this splendid proposal! My guess is that many others would be terrified by the prospect.

I have already mentioned the need to focus on individuals. One particular development has been the attention paid by teachers to the mistakes made by children and the analysis of these mistakes. Examples of this from my classroom and others are well illustrated in two articles published recently by David Kent, a former colleague and a friend of mine, published through the National Foundation for Educational Research, entitled "Some ways in which mathematics is lost." This research was undertaken by David as just a part of his normal classroom activity as a teacher in our mixed ability classrooms at Belper and elsewhere. I was, for example, able to appreciate why Susan thought that $2x - 5 = 39$ was a stupid statement, because the classroom situation allowed us to sit and talk about it. (It wasn't because Susan couldn't accept x as a number. It wasn't because she didn't like equations. It was, in fact, because she couldn't see how you could take 5 away from twenty something and be left with thirty-nine.)

Teachers teaching mixed ability classes have again been forced to work together more closely (as Denis Kennedy has illustrated), enabling strengths to be utilised, weaknesses supported, interest developed. Chuck Allen in his talk has indicated just how wide and diverse this spectrum of interests, talents, and experience is. This has led to a much needed greater degree of professional honesty between teachers. Many of the mathematics department at my own school seem to find it difficult to cope with my regular discussions about the problems that I am having in teaching mathematics. (They have not been used to teaching mixed ability groups and working together in this way.) He is the Head, I can see them thinking, and he has all these problems and he talks about them!

But in addition to cooperation there has to be autonomy for the teacher. Individualised learning programmes threaten and sometimes remove this all together. Autonomy of the teacher in one sense has been a noted feature of the English educational system, although at present it is seriously threatened by frightened politicians, powerful employers, and others who do not understand the complexity of the problems faced by teachers. Autonomy in another sense will always be there. As I indicated earlier, the teacher's attitude, relationship with children, view of mathematics, emphasis on particular topics, etc. are all of paramount importance, and no prescription or statute from above will ever change this fact.

There must also be student autonomy. What work to do, when to do it, how long to spend on it, who to consult with, what books to refer to, how to record the work . . . these are all legitimate claims for students to make and teachers to accept. Mixed ability classes have frequently accepted this need and faced the consequent organisational re-adjustment. But the re-adjustment is not the problem. The problem was accepting the need for it. I think many of us have been (and still are) constantly amazed by the ability, persistence, resilience of children to accept responsibility for their own learning, to work on their own for periods of time, to sustain a piece of work for weeks, to accept the responsibility that goes with the freedoms offered. Teachers of mixed ability groups have regularly reaped this reward. The need for variety also applies to assessment. The foolishness of trying to assess mathematical competence by a single means of examination now seems transparently obvious. In mixed ability teaching it just isn't possible to get away with this. It isn't desirable in any sort of teaching.

Finally, one hobby horse. During the last ten years I have found one vehicle to help meet the challenge posed by some of the problems mentioned. I am speaking of the use of mathematical investigations in the teaching of mathematics. I have had children of all ages spending as much as seventy per cent of their time on investigative work. The advantages have been considerable: (i) they have encouraged children to spend more time on the processes and strategies in mathematics, (ii) they have given the children an opportunity to create and develop their own mathematics; (iii) they have enabled children (including the least able) to get a feel for pure mathematics, and to know what it is to work as mathematicians; and (iv) they have encouraged children to discuss their mathematics and to write about it.

Ten years ago there seemed to be only a handful of teachers working in this way with children. The growth of mixed ability teaching has helped to promote more use of investigative work, and I was delighted that this Congress was able to make space for a panel on this topic. The conference programme refers to "disadvantage." What can I say? It is very hard work, but you seem to learn a lot. I feel certain that the development of mathematical education has been enhanced by the contribution made by mixed ability teaching.

19.3 APPROACHING MATHEMATICS THROUGH THE ARTS

APPROACHING MATHEMATICS THROUGH THE ARTS

Emma Castelnuovo
Secondary School
Rome, Italy

In my presentation I would like to give you an idea of a particular meaning that can be given to the interaction between mathematics and art.

For many years in my Italian Junior High School classes I had noticed that 11-14 year old students could be motivated to mathematics by an aesthetic elementary feeling.

I would like to add that, as I had the opportunity of giving courses in some schools of Niger (Africa) in recent years, my observations also concern my African students.

I shall begin with some general remarks. In my opinion the aim of mathematical instruction in junior high school has two aspects: students must have an idea of the relationship between mathematics and reality, and, furthermore, they must little by little realize the power of mathematical thought also from an imaginative point of view.

Keeping in mind this aim we must "plunge" our young students in the spirit of mathematics so that they have the impression of constructing mathematics by themselves. Interest has to be motivated by dynamic problems capable of introducing them to the fundamental concepts of function, transformation and structure. Now, dynamic problems can be suggested either by questions concerning the real world and so from outside, or by aesthetic elementary pleasure and so from inside.

I will explain this last motivation with examples, also citing some comments made by the students themselves.

Aesthetic Feeling Stimulated by the Repitition of a Movement

I show the students (11-12 years old) a jointed square and ask them: "in passing from square to rhombus, does the area change?" The answer is always the same: "no, the area doesn't change because the perimiter doesn't change." Then, I incline the rhombus more and more until arriving at the limit-case; in this case - it is clear - the area is zero. It is precisely this case that leads students to realize that the area does change.

I continue with this movement returning the square once again and then inclining it towards the other side. Students are fascinated by this transformation. They say: "it is beautiful because the movement is always the same, and equal figures appear on the left as on the right of the square." "The square - they say - is the most important figure because it is the highest; it is the central figure." Students never tire of observing this movement; they are really fascinated. They repeat: "how beautiful it is!"

Now we ask ourselves: why such a reaction on the part of children? Perhaps because these symmetric figures remind them of some decorative picture as is frequently seen in Italy? No, surely that is not the reason because my African students had the same reactions, and in Niger no pictorial art exists. Clearly it is an elementary aesthetic feeling. It is precisely this aesthetic pleasure that stimulates them towards mathematical research: the area of the rhombus changes because its height changes, but the height depends on the angle. Therefore the area is a function of the angle. What is the relationship between the area and the angle? We put this problem aside for the moment because it is too difficult, but the students cannot forget it and the following year they will be enthusiastic to study it from a mathematical point of view.

Aesthetic Pleasure from the Discovery of Analogous Behaviours

We start by presenting to 11-12 year old students a very elementary operation: the addition of even and odd numbers. After some numerical examples students are led to write the composition table

+	e	o
e	e	o
o	o	e

This table becomes expressive when we discover that other operations have the same behaviour.

Let us consider the rotations and the symmetries of a square. We draw a square with its symmetry axis. Painting a figure in one of the eight triangles, students realize that they can reproduce it by folding and refolding the square on its axis. For instance we pass from figure A to B by folding on axis 1, and from B to C by folding on axis 2, and so on.

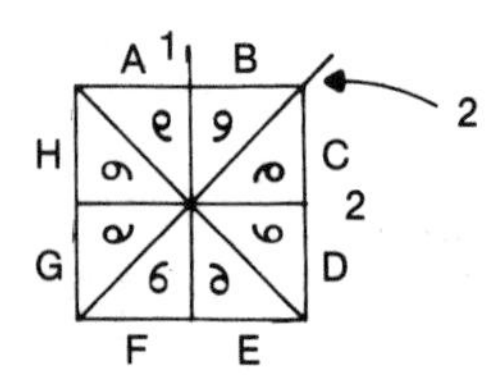

We ask them: it is possible to pass from A to C by operating only one fold? After some attempts students realize that it is impossible, but they discover that the first figure can be put on the third by rotating. Therefore the product of two symmetries is a rotation. With other attempts they arrive at constructing a composition table concerning the rotations and the symmetries of a square

o	R	S
R	R	S
S	S	R

This table impresses them very much because "it is the same thing" - they say. And their wonder increases as they discover that the same structure is valid in the multiplication table of positive and negative numbers

o	p	n
p	p	n
n	n	p

Finally, they exprience a real fascination when they discover that there is an analogous behaviour in very common linguistic rules: two negative sentences give an affirmative sentence. For instance, if I say "I don't want you not to go", this means "I want you to go". Students are able to construct by themselves this table

o	yes	no
yes	yes	no
no	no	yes

which is for them the most concrete example because it is their common language.

At this level children are not able to understand one of the greatest discoveries of mathematics: the fundamental importance of structural behaviour which considers neither the nature of elements nor the kind of operations; but they are impressed by the same theme repeating itself in so many different events, like a rhythm.

This aesthetic impression will stimulate them, later on, towards scientific research. This type of aesthetic pleasure, that is the discovery of the same behaviour in different contexts, is, without any doubt, more abstract than the aesthetic pleasure experienced from the moving square, due to a lack of any visual figure.

Aesthetic Emotion Aroused by the Transformation of a Curve by Continuity

It is even more difficult for the students (13-14 year olds) to be able to recognize equality in figures which appear different. I want to give the example of elipse, parabola and hyperbola, three so different curves at sight, which, however, belong to the same family.

In the first two years the students meet one or another of these curves in several situations. At the end of the third year we want to give them a more general view: we show these curves as sections of a cone.

Here is a cylinder and we begin by cutting this cylinder with a plane of light. We see a circle, or an ellipse or two parallel lines. Then we pass from cylinder to cone and we still proceed cutting with a plane of light. We notice that some times the curve is closed, sometimes open. We observe.

The passage by continuity from one to another curve produces a real aesthetic emotion on the students. They never tire in beholding this transformation. Then we discuss this experiement. The curve is closed if the plane cuts each generatrix, while, if the plane is parallel to one or two generatrixes, the curve opens because there is not a point of light on each generatrix.

Students observing this passage from ellipse to parabola to hyperbola say that it is as if one curve transforms itself into others. Somebody said: "it is the same thing when I change my clothes; I look different but I am always the same."

Surely this aesthetic emotion helps them to see equal curves that look different and therefore to grasp some ideas on the conics family, ideas that will lead them to an analytical study.

And now I would like to conclude with some general remarks: we know that the mathematician arrives at discovery either by seeing through his mind's eye an ideal model and its transformations or by seeing with intuition the same structure in different situations. I think that, by means of art or, more precisely, by means of an aesthetic stimulus, the student can experience this same imaginative attitude, which is the first and essential moment prior toa subsequent rational study.

APPROCHE DE QUELQUES OBJETS MATHEMATIQUES POUR LES UTILISER EN ART GRAPHIQUE

Paul Delannoy
Universite de Dijon
Dijon, France

Depuis 1977 nous programmons un calculateur couple a une table tragante pour obtenir des dessins ayant une valeur artistique. La petite taille de ce calculateur et ses fonctions de type mathematique, ainsi que notre propre formation, nous ont amenes a definir notre travail comme "recherche d'objets mathematiques et d'algorithmes simples permettant d'obtenir des dessins".

Notre groupe, dit "groupe de Recherche Esthetique et Mathematique" travaille jusqu'a present sur la machine de l'Institut de Recherche sur l'Enseignement des Mathematiques (IREM) de Dijon.

On utilise souvent en art graphique, pour l'analyse d'un dessin, d'une part la distinction entre abstrait et figuratif, d'autre part les trois notions de base que sont la couleur, la valeur et la composition.

Pour nous, la couleur reste un procede artisanal et manuel qui permet souvent une approche plus facile de notre machine aux non-inities. Pour la valeur, nous ne pouvons jouer que sur le nombre de traits que nous faisons effectuer au crayon dans une unite de longueur ou de surface fixe, et nous avons la deja l'approche d'une activite de mesure. Pour la composition enfin, nos programmes vont d'une realisation totalement controlee par l'utilisateur (composition "humaeine") a une realisation enierement laissee "au hasard" (composition "par la machine").

Dessin "figuratif"

Le probleme pose est de decrire le graphisme que l'on veut faire realiser par la calculatrice dans un langage mathematique.

La solution la plus simple est universellement utilisee; elle correspond a la notion d'ensemble fini de points decrits par exemple par leurs coordonnes nees cartesiennes dans une echelle donnee. Le dessin est ansi approche par une ligne brisee qui joint ces points.

On peut tenter une approche plus globale en recherchant un modele adequat pour differents elements du dessin a realiser. Ces modeles seront mis dans la memoire du calculateur sous forme de sous progrmmes. Chaque module pourra dependre de parametres numeriques qui permettront certaines transformations de ces modules, et l'on pourra aussi jouer sur leurs positions respectives.

Un bonhomme sera ainsi realisee avec trois sous programmes:

- un parallelogramme donnera le corps, les membres
- une secteur circulaire donnera les articulations
- une ellipse donnera la tete.

On peut enrichir un tel modele tres facilement par ajout de sous programmes.

Comment trouver un tel modele?

Nous parlerons ici de l'exemple de l'oeil humain que vous pouvez voir dans certains de nos dessins.

a) recherche les elements caracteristiques.
Pour dessiner un oeil, retenons le sourcil, le bord inferieur, l'iris, et la pupille.

b) Modeles etc) test de leur validite
La pupille est vite realisee comme un disque plein, en utilisant les fonctions trigonometriques par exemple
Le bord inferieur et le sourcil sont des lignes courbes

Une premiere tentative faite pour les realiser par des arcs de sinusoide s'est revelee frustrante, ce que nous aurions pu prevoirs en affinant le modele: les courbes recherchees doivent avoir une concavite constante, ce qui restreint l'arc de sinusoide utilisable a un intervalle entre deux points d'inflexion et ne permet plus beaucoup de deformations par les parametres d'amplitude et de periode.

Le second essai avec des paraboles s'est revele etre le bon en jouant sur la position de leur point extremal et sur leur aplatissement.

L'iris se presente comme une couronne de couleur dont nous voulions faire ressortir le caractere non uniforme par une repartition aleatoire de zones claires et sombres.

Nous avons realise cela par un modele de aryons successifs traces dans cette couronne.

Un choix aleatoire des rayons a tracee amene a des temps de traceties longs (pour chaque rayon: tirage de la position, trace, puis deplacement, eventuellement long jusqu'a la position suivante) et nous avons prefere tracer de proche en proche et le probleme pose est: determiner une serie de points successifs sur la circonference d'un circle sans donner l'impression de points bien ordonnes. Ceci est realiser actuellement par une progression geometrique de l'angle polaire du rayon, portant sur trois tours, calculee pour laisser des zones de valeur differentes et creer des interferences.

d) Parametres a fixer, parametres a laisser libres.

Outre les parametres des paraboles, de la serie geometrique, et la taille relative de l'iris par rapport a la pupille, qui sont internes a chacun des trois modeles des elements choisis, il faut etudier ceux qui decrivent leurs positions respectives, leurs tailles respectives, la position et la taille du dessin final dans la feuille, etc.

Dessin abstrait

C'est ici beaucoup plus que l'activite artistique enrichit le plus clairement l'enseignement et la comprehension des mathematiques.

Le probleme est ici "Comment regarder un objet mathematique pour en faire un dessin ou un element de dessin?"

Le premier element sur une telle route a sans doute ete donne par A. Deledicq de l'IREM de Paris lorsqu'il remarquait qu'une "courbe", lrosqu'on la trace, est en faite decrite comme un ensemble de points muni d'un ordre, et que le meme ensemble de points avec des ordres differents donne des resultats etonnants. (Nous avons realiser un film sur ce sujet qui est disponible a l'IREM de Dijon, dans lequel on voit les changements graduels d'une courbe lorsque l'ordre change.) Lorsque la courbe est formee et definie par un parametre angulaire, mesure en degres par des nombres entiers (ensemble de 360 points) les ordres differents de traces sont definis entierement par un entier si celui ci est un nombre premier avec 360.

Cette facon de "regarder" la courbe et de l'utiliser pour le dessin n'est pas la seule. Nous pouvons essayer le "tube" qui consiste a tracer non plus un point, mais un segment dont l'extremite est un point de l'ensemble. Et nous pouvons faire "tourner" ce segment sur lui meme pendant le trace. Par exemple:

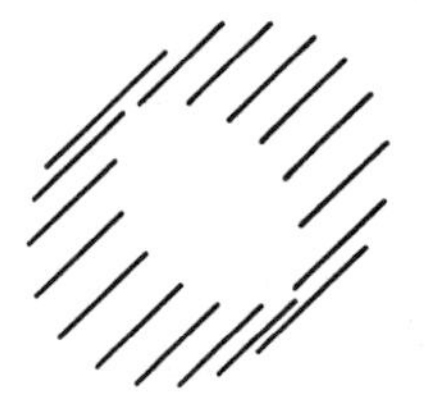

Cercle en "tube" defini par 16 points

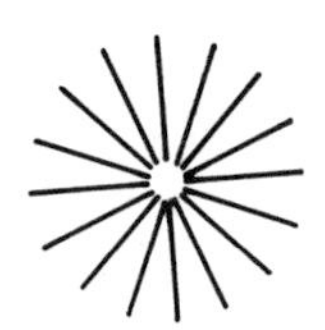

Le meme en faisant tourner le segment d'un tour par tour sur le cercle

Nous pouvons egalement tracer le segment qui joint deux points successifs en le prolongeant jusqu'aux bords de la feuille ("a l'infini"!).

Ce moyen permet d'observer des moirages interessants, ressemblant souvent a un coucher de soleil.

Nous pouvons tracer l'ensemble ordonne plusieurs fois dans des echelles variables, ce que nous appelons "perspective". En utilisant le compteur du nombre de traces a faire, et en annulant les valeurs de variation d'echelle, ce procede revient facilement a tracer une famille de courbes parametrees.

Une derniere facon de voir cette famille de courbes est de considerer, asors que l'on trace les points dans une echelle donnee, la position qu'ils auront dans la future echelle: Quelles sont les coordonnees dans l'echelle du point qui aura les memes coordonnees dans suivante? Ce procede, qui nous amene a parler de "grille" donne de tres jolis resultats avec des courbes du troisieme degre.

- Notre collegue, professeur de lycee, Christian Lalitte, nous a offert un reel exemple d'objet mathematique utilisable en art graphique.

  Nous connaissons tous des dessins de surface dans l'espace ou projetees sur un plan, tels ceux de Vasarely. L'important dans un tel objet, pour l'enseigner, est de mettre en evidence les invariants: un point, un cercle, un rectangle . . . Une transformation qui a un rectangle invariant est une chose rare, mais fort interessante pour notre table tragante. Christian en a trouve une famille dans son concours de recrutement comme professeur. Et si nous regardons ces transformations comme operant sur un quadrillage, nous pouvons obtenir des resultats etonnants.

- Une facon de faire fonctionner les fonctions trigonometriques, et d'etudier leurs pulsations et leurs dephasages, est de les utiliser comme vitesse de progression de deux points sur les bords de la feuille, que l'on joint. Deux series de segments ainsi traces donneront des moires dans toute la feuille.

- Nous pouvions citer encore de nombreux objets que nous utilisons enrichis par l'idee d'en faire des objets graphiques, dans nos programmes. Je me contentera de parler de la metamorphose d'un polygone qui se produit lorsqu'on lui applique le procede suivant, illustre ici sur un triangle: A partir du premier polygone P on en calcule un dit rP, 0 ¼ r ¼ 1 en divisant chaque cote dans le rapport r, et l'on recommence a partir de rP. Ce procede a ete imagine, au cous d'un travail realise a l'IREM de Paris Nord par Pierette Serano, par une classe d'eleves de terminale.

Conclusion

Notre demarche permet une activite ludique et creative qui peut aider l'enseignement des mathematiques parce qu'elle replace les mathematiques dans un contexte operationnel, qu'elle permet d'aborder des problemes lorsqu'ils se posent a l'utilisateur qui est alous tres motive pour les resoudre, sand poutant fixer de limites aux connaissances, ni en etercve, ni en profandeur.

Il existe de nombreuses autres manuetes rendre ka natgenatuqye vivante et active, l'avantage et l'originalite de celle ci est de ne pas opposer creativite et logique.

Nous pouvons affirmer que lors de l'utilisation de systemes informatives a possibilites graphiques etendues, comme logo par exemple, l'interet des utilisateurs pour les realisations graphiques a toujours ete preponderant et constitute un moyen efficace pour les pousser a un approfondissement de leurs connaissances mathematiques; c'est donc un bon moyen pour "apprendre a apprendre".

Un dernier mot pour signaler que les illustrations manquent dans cet expose, mais qu'on peut se procurer des copies de dessins realises par notre groupe a l'ADAO, 97 Boulevard Mansard 21100 Dijon, France.

MATHEMATICS IN THE HUMANITIES

James R.C. Leitzel
Ohio State University
Columbus, Ohio, U.S.A.

My intention to day is to describe a course I give at The Ohio State University to satisfy a distribution requirement in mathematics for students majoring in the College of Arts and Sciences. Primarily the students who elect this particular course option have current career choices that have no specific mathematics skill or content expectation. The course also services students from the College of Education who must satisfy a state certification requirement in mathematics. These latter students are usually secondary education majors in non-science related specialty areas.

The course runs for 10 weeks, has no prerequisites, and is not a prerequisite to any other offering in the Mathematics Department. The course has no fixed syllbus so instructors are free to develop their own approach. Since the course is designed as a liberal arts experience, I try to utilize as many examples from the arts as I can to illustrate the mathematical concepts I am presenting in lecture. In essence, I try to meet the students where they are. The ideas and examples are drawn, in so far as possible, from the students' major areas. The intent is to show that mathematics is not only manipulative computation, but has a much broader scope. Ideas from mathematics are utilized in the arts and the structural approach of mathematics gives structure also to aspects of the arts. The expectation is that an acquaintence with and understanding of some basic mathematical ideas will enhance the students' appreciation of their own discipline.

Since I try to tailor the experience to the audience I have found no text which uniformly works. Consequently I develop my own lectures and problems. This gives a second intent of the course. I believe students should be active participants in the development of the mathematics and not passively reading about it.

Let me turn now to a sampling of the topics I cover and briefly illustrate those topics by examples from the various arts.

I begin the course with selected topics from number theory. It is topic which is new, requires minimum computational skill, and presents questions which are easy to understand. We investigate various ways to classify numbers: even-odd; prime-composite; deficient-perfect-abundant. The discussion of perfect numbers presents various illustrations. The historical conjectures and their current status indicate that mathematics has not yet completely answered even some fairly simple questions. It also is an opportunity to discuss how the pursuit of an answer to a problem has led to significant developments. Today even some computers have their internal circuitry checked by computing known perfect numbers.

At this point I usually give a brief accounting of number mysticism and its role in literature. The early Greek separation of numbers as male and female and the special role of ten are illustrated. Because '2' is female and '3' is male, both '5' and '6' represent marriage numbers. In Plato's Republic books V and VI treat this topic. Excerpts from Dante's Divine Comedy provide further illustration. The general scope of the work is 100 cantos. These occur in three divisions with 33 cantos each, the first is a general introduction. "Three", the first male number, has in Christian tradition becoe the Trinity - Father, Son and Holy Ghost. "Four" has become associated with the earth. For example we have the four winds; the four seasons; the four ages of man (infancy, youth, manhood, old age). Also, because it is the first square number, four has been associated with justice. In Dante's work Canto XIX has Pope Nicholas III suffering on the fourth embankment. In Canto XXXII a lake of ice is found with four different types of traitor encased in a ring. Early in the work (Canto IV) Dante meets four great poets - Homer, Horace, Ovid and Lucan. When Dante and Virgil join them the number is six (and perfect). Mention is also made of gematria as it appears in Hebrew tradition and in some current literature.

While talking about the special role of 10 we also note that 1 + 2 + 3 + 4 = 10 and move into figurate numbers. The contributions of Pythagoras to music are mentioned and the capturing of the relations of octaves; fourths and fifths in a triangular way.

Another number sequence with many illustrations in the arts is of course the Fibonacci sequence: 1,1,2,3,5,8,13,21 . . . . Its occurrence in nature (pine cones, phyllotaxis) and its relation to the Goldren Ratio are examined. The architectue of Le Corbusier,, paintings of Mondrian and the music of Bela Bartok (particular the piece "Music for Celeste, Percussion and Strings") are some illustrations. Further occurrences in nature in the form of spirals is pointed out.

The discussion of architecture leads to an analysis of symmetry. Palindromes are the linguistic example: MOM, ROTOR, ETE, (French). Poetry can also serve to illustrate symmetry. One such example is Robert Browning's Meeting at Night:

> The gray sea and the long black land;
> And the yellow half-moon large and low;
> And the startled little waves that leap
> In fiery ringlets from their sleep.
> As I gain the cove with pushing prow,
> and quench its speed i' the slushy sand.
>
> Then a mile of warm sea-scented beach,
> Three fields to airs till a farm appears;
> A tap at the pane, the quick sharp scratch
> And blue spurt of a lighted match,
> And a voice less loud, through its joys and fears,
> Than the two hearts beating each to each!

Here the rhyme scheme (abc,cba) in the stanzas illustrate symmetry to simulate the water's ebb and flow.

Discussion of the seven possible patterns on a strip leads naturally to reflections and rotations. Their combinations (first investigated with simple geometric figures) give a structured approach to the analysis. For some students this is the first example of a mathematical computation system without numbers. Further examples of this "structure" approach can be drawn from music.

It is natural at this point to return to number theory and look at the structure that arises in residue class arithmetic. Aside from the investigation of the formal group properties and computational exercises that are possible in such systems, there are also string designs possible. Students can create their own art.

The course I have been discussing is a course for liberal arts students. The seven traditional liberal arts are: grammar, rhetoric, logic, arithmetic, geometry, astronomy and harmony. It is difficult to teach such a course to students who lack a foundation in liberal education. Nonetheless it is appropriate to approach mathematical ideas for these students with a strong example base from art, music and literature. In addition each offering of the course provides a new learning experience for the teacher and keeps him in touch with a variety of other disciplines.

19.4 THE USE AND EFFECTIVENESS OF MATHEMATICS INSTRUCTIONAL GAMES

STUDIES IN A CONTINUING INVESTIGATION OF THE COGNITIVE EFFECTS OF MATHEMATICS INSTRUCTIONAL GAMES

Margariete Montague Wheeler
Northern Illinois University
DeKalb, Illinois

The studies summarized here are part of a continuing, systematic investigation of the cognitive effects of instructional games on mathematics learning initiated in 1976. This investigation was initiated and continues because games seem to be important instructional activities, elementary school teachers report regular use of games in their classrooms, many elementary and middle/junior high school students indicate both that mathematics instructional games are used in their classrooms and that these games seem to help them to learn mathematics, and, prior to 1976, the very limited empirical research on the cognitive effects of mathematics instructional games had produced few substantiated effects.

A game is defined by the following seven statements:

1. A game is freely engaged in.

2. A game is a challenge against a task or an opponent.

3. A game is governed by a definite set of rules. The rules are structured so that once a player makes a move and his/her turn comes to an end, that player is not permitted to retract that move or to exchange it for another.

4. Psychologically, a game is an arbitrary situation clearly delimited in time and space from real-life activity.

5. Socially, the events of a game situation are considered in and of themselves to be of minimal importance.

6. A game has a finite state-space (Nilsson, 1971). The exact states reached during play of the game are not known prior to the beginning of play.

7. A game ends after a finite number of moves within the state-space.

An instructional game isa game for which a set of possible instructional objectives has been determined by the person(s) planning instruction and before the game is played by the students receiving instruction from the game. MULTIG (Romberg, Harvey, Moser, & Montgomery, 1974, 1975, 1976). Polyhedron-rummy (Peterson, 1971), and Equations (Allen, 1972) are examples of mathematics instructional games.

The continuing investigation seeks to identify variables which may be related to the cognitive effects of mathematics instructional games and to investigate systematically the effects of these variables upon learning. Thus far, four sets of variables which seem particularly relevant to games and game-playing situations have been identified: (a) characteristics of the game (format, constraints, responses), (b) instructional objectives (content, instructional level, cognitive level), (c) learner-game interactions (problem solving heuristics, game-playing strategies, outcomes of products, memory loading), and (d) learner-learner interactions (level of competition, amount of peer teaching). These four sets of variables are discussed in Bright, Harvey, and Wheeler (1977).

Throughout the continuing investigation, the following research strategy is being used. First, to determine which of the identified game-related variables have an effect upon learning, they are manipulated singly. Second, learning effects are investigated when two or more variables are manipulated simultaneously. Third, cognitive effects associated with games are compared to those produced by other instructional techniques. Ten studies conducted during 1976-77, 1977-78, and 1978-79 examined different game-related variables singly; 12 studies being conducted during 1979-81 are simultaneously manipulating two variables (instructional level and cognitive level); and two studies planned for 1981-83 will each manipulate a single variable (game products or level of competition), neither of which has previously been studied.

The studies all fall within the first or second phase of the three-phase research strategy. In designing studies in these two phases the following assumptions are used:

1. The thrust of each study should be to determine the learning effects which are due to the games treatments and should not be to compare the effectiveness of games with other instructional alternatives. Thus there are no control groups which receive non-game instruction designed to have the same effects on learning as does game instruction.

2. The mathematics content of the games used should be substantial and a part of the school curriculum at the grade levels at which the games are played.

3. The mathematics content of the games used should not be acquired easily or quickly by a majority of students.

The studies completed or being conducted each have a treatment which is administered in approximately eleven instructional days. On Day 1 pretests are administered; on Day 2 the game is explained; on Days 3-10 the game is played for about 20 minutes each day; and on Day 11 posttests are administered. Variations of this treatment sequence include more testing days and more or fewer game playing days. The treatment is usually spread over about six weeks, with there being at most two game-playing days per week. Pretests might include a test of formal operations, learner difference tests (e.g., spatial visualization), standardized achievement tests, tests of prerequisite skills, and game content tests. The posttests are only of game content.

In pre- and post-instructional studies, no instruction is provided on the mathematics content of the games other than that provided by the games themselves. Learning effects in these studies are generally determined by comparing the pre- and post-test scores. In co-instructional studies, game playing accompanies instruction, so part of each class is designated as a control group and receive instruction without game playing. The effect of the game is determined by comparing the performance of the game-playing and control groups and by comparing the pre-and post-test scores of both groups.

On the basis of the studies completed before mid-1979, it is possible to conclude that at the post-instructional level games can be an effective way to retrain and maintain skills with multiplication basic facts (Bright et al., 1979a, in press a) and to develop further the skill of ordering common fractions (Bright et al., in press c, submitted b); that games may not need to be played frequently in order to maintain skills with multiplication basic facts (Bright et al., in press a); that at the pre-instructional level, games can be used to instruct on the concepts of fairness (Bright et al., in press c); that the ways children are grouped to play pre-instructional fairness games or post-instructional fractions ordering game may not be critical (Bright et al., in press c); and that a game involving mathematics can be used to elicit problem-solving behavior from both naive and experienced problem solvers (Bright, in press; Kraus, 1980) and to cluster problem solvers into groups based upon the problem-solving heuristics they use (Kraus, 1980). From the studies at the post-instructional level, which have varied game constraints, it is possible to conclude that changing those constraints to increase the verbalizations of the players does not increase the achievement of those players (Bright et al., in press b); that, in a fractions ordering game, the use of additional manipulative or pictorial devices does not enhance the effect of the game alone (Bright et al., submitted b); and that in multiplication basic facts games, the incorporation of computation of those basic facts into the scoring rule does not enhance the effect of the original game (Bright et al., 1979b). It can also be reported that games do not seem to enhance a child's ability to use accurate and meaningful geometric descriptions (Bright et al., in draft) or to improve logical reasoning skills as measured by a test of those skills (Bright et al., submitted a). Finally, it can be reported that a computer simulation has been developed whose behavior significantly correlates with the game-playing behavior of human subjects when given the problem-solving heuristics used by those subjects (Kraus, 1980).

The studies being conducted in 1979-81 are investigating the 12 meaningful combinations of the two variables, instructional level and cognitive level. The instructional level variable has three values: pre-, co-, and post-instructional (Bright et al., 1977); the cognitive level variable has five values: knowledge, comprehension, application, analysis, and synthesis (Bloom, 1956). Each study investigates one of the combinations and is being conducted at two grade levels using a pre/posttest design. In the co-instructional studies, control groups within each classroom are used. It is hoped these studies can be reported together so the results can be summarized and synthesized cohesively.

Bright and Harvey have planned two studies for 1981-83. In one study the instructional objectives of the games will be held constant while the game products will be varied, principally, by changing the format of the games. In the second study the variable manipulated will be the level of competition. Initially, an operational definition of this variable will be developed; experimental manipulation will determine the cognitive effects associated with each level.

The authors believe that their continuing, systematic investigation of the cognitive effects of instructional games on mathematics learning is leading to a clear, comprehensive picture of the ways in which game playing can improve students' achievement. Then, it should be possible to design instructionn so as to take that knowleged into account.

References

Allen, L.E. EQUATIONS: The game of creative mathematics. New Haven, Connecticut: Autotelic Instructional Materials Publishers, 1972.

Bloom, B.A. (Ed.) Taxonomy of Educational Objectives. New York: McKay, 1956.

Bright, G.W., Games moves as they relate to strategy and knowledge. Journal of Experimental Education, in press.

Bright, G.W., Harvey, J.G., & Wheeler, M.M. Cognitive effects of games on mathematics learning. Madison, Wisconsin: University of Wisconsin, 1977. (ERIC Document Reproduction Service No. ED 166 007).

Bright, G.W., Harvey, J.G., & Wheeler, M.M. Using games to retrain skills basic multiplication facts. Journal for Research in Mathematics Education, 1979, 10, 103-110. (a)

Bright, G.W., Harvey, J.G., & Wheeler, M.M. Incorporating instructional objectives into the rules for playing a game. Journal for Research in Mathematics Education, 1979 10, 356-359. (b)

Bright, G.W., Harvey, J.G., & Wheeler, M.M. Using games to maintain multiplication basic facts. Journal for Research in Mathematics Education, in press a.

Bright, G.W., Harvey, J.G., & Wheeler, M.M. Game constraints, verbalization, and mathematics learning. Journal of Experimental Education, in press b.

Bright, G.W., Harvey, J.G., & Wheeler, M.M. Achievement grouping with mathematics concept and skills games. Journal of Educational Research, in press c.

Bright, G.W., Harvey, J.G., & Wheeler, M.M. Using a game to instruct on logical reasoning. Submitted a.

Bright, G.W., Harvey, J.G., & Wheeler, M.M. Varying manipulative game constraints. Submitted b.

Bright, G.W., Harvey, J.G., & Wheeler, M.M. Accurate and meaningful descriptions and the use of games. In draft.

Kraus, W.H. An exploratory study of the use of problem solving heuristics in the playing of games involving mathematics. Unpublished doctoral dissertation, University of Wisconsin-Madison, 1980.

Nilsson, N. Problem Solving Methods in Artificial Intelligence. New York: McGraw-Hill, 1971.

Peterson, W.H. Polyhedron-rummy. Glenview, Illinois: Scott, Foresman, 1971.

Romberg, T.A., Harvey, J.G., Moser, J.M., & Montgomery, M.E. Developing Mathematical Processes. Chicago: Rand McNally, 1974, 1975, 1976.

19.5 STRATEGIES FOR IMPROVING REMEDIATION EFFORTS

REMEDIATION EFFORTS IN TWO YEAR COLLEGES

Ronald M. Davis
Northern Virginia Community College
Alexandria, Virginia

Remediation in mathematics has long been a major effort of mathematics educators at two year colleges. As a faculty member at a two year college for the past twelve years I am no exception. I have devoted much of my attention to mathematic remediation. In an effort to gain an understanding of the task, I've spent much of my professional life exchanging ideas with other mathematics educators about this topic, listening to their approaches to it and reflecting upon my own instruction. I like to believe that I have gained some insight into the instructions needs for mathematics remediation in two year colleges. In this paper I will attempt to share some of those ideas with the reader.

A Brief History of Two Year Colleges

An understanding of current remediation efforts requires a brief historical development of the two year college and its purpose. In the early 1900s junior colleges were developed as local colleges providing the first two years of a baccalaureate degree. The emphasis of junior colleges was on transfer. About the same time technical institutes were organized. These emphasized preparation of students for entry into local industries with high level technical skills.

In the early 1950s the concepts of a comprehensive community college evolved in which the efforts of junior colleges and technical institute were merged. As public officials championed the concept of universal education large numbers of public comprehensive community colleges appeared on the scene. By the late 1970s there were more than 1000 public two year colleges.

With the development of public two year colleges came the emphasis on open door admissions. This meant that public two year colleges could accept students at whatever their current academic level and provide instruction that would assist them in attaining their desired level of educational development. Thus, remediation efforts became an essential part of the role of a two year college.

Efforts at Mathematical Remediation

Mathematical remediation was for many years a task unique to two year colleges. Thus, many varieties of approach have been utilized in an effort to provide effective remedial instruction. Also many different departments have been responsible for the mathematical remediation efforts. These include mathematics, developmental studies, learning laboratories, study skill centers and others.

The variety of approaches to the delivery of mathematical remediation includes lecture/discussion classes using either a college professorial format or a high school hands-on format. Another highly popular approach is the use of programmed instruction or locally produced study modules. The use of audio tapes and computer assisted instruction are also expanding.

Class structure is also a widely varying aspect in remediation efforts. Instruction for a full class has been used, as have small group formats and individual instruction. Which of these various approaches then are part of an effective mathematical remediation effort? It appears that class structure, delivery approach, and the department responsible for the remediation program are not the key factors in a successful effort. Other factors appear to be the key aspects.

Characteristics of Effective Remedial Mathematics Programs

What are these key aspects? First of all, effective programs seem to emphasize both skill and concept development. They are often developed to provide instruction for the requirements of later courses, both mathematics courses and courses in other disciplines. This development is often preceded by a careful analysis of the needs of these successor courses.

Successful programs seem to avoid being an accelerated repeat of the four years of high school mathematics curriculum. They stress the needed skills and concepts and avoid those mathematical techniques that are not needed in a particular curricular area and that are no longer essential because of advances in calculators and computers. These successful programs place a large emphasis on utilizing the life experiences of students.

The organization of effective mathematics remediation programs also has some characteristics which are common to those different programs. These characteristics might be characterized by the labels of structure, diversity, concrete experience and personalism.

Structure is the carefully controlled introduction of ideas. Ideas are introduced individually with careful attention to hierarchies of ideas and to broadening carefully an idea to show its breadth and its limitations. Ideas are presented in small blocks with much interchange occurring between students and teachers.

Diversity or variety also is carefully controlled. Alternative approaches to concepts and skills are provided but only after firm implantation of a concept. Emphasis is also placed upon providing an understanding of the context of the concept within the broad spectrum of mathematical study.

Concrete experience also looms as a key aspect of a successful program. The programs stress the relation of ideas to concrete situations and tend to provide explanations which develop from familiar ideas.

Personalism seems to be the most important of the characteristics. Successful programs are characterized by instructor interest in students, enthusiastic instructors, instructor sensitivity toward special needs and anxieties of students, and extensive interaction between students and instructors. Because of the importance of this characteristic to successful programs, it deserves a more intense examination.

Characteristics of Effective Teachers in Remedial Programs

Many of the characteristics of effective mathematics instructors of remedial programs which I have observed have been published in a pamphlet of the Mathematical Association of America. Included in that publication are the following:

A. Expertise in mathematics well beyond the remedial level,

B. Recent experience in teaching transfer and occupational-technical mathematics courses,

C. A knowledge of theories of learning processes to assist in understanding student learning patterns (e.g. Brunner, Gagne, Piaget, locus of control and developmental theory are several theories or writers considered important),

D. An ability to recognize and provide corrective instruction for misconceptions, misimpressions, and partly learned or incompletely learned concepts,

E. An ability to communicate concepts clearly and concisely

Other characteristics seem to be characteristic of effective instructors also. They are a sensitivity to student needs, a sensitivity to student anxieties and adeptness at diffusing those anxieties. Such instructors exude patience. They also express an intense enthusiasm about learning. Finally, they are adept at facilitating learning through the use of alternative explanations of ideas and by being adaptive to differing student learning styles.

Conclusions

In this paper I've attempted to provide my view of what I consider to be the important characteristics of an effective remedial mathematics program and of an effective remedial mathematics instructor. I believe these thoughts provide us with some guidelines for selecting individuals for teaching such courses. However, in this day of declining enrollments in general and increasing enrollments in remedial mathematics courses, few colleges have the luxury of hiring faculty with special appropriate qualifications but rather must be content to utilize current faculty who most likely have a focus on the transfer curriculum.

Reference

Guidelines for the Evaluation of College Mathematics Programs, The Mathematical Association of America, Washington, DC, August, 1979.

REMARKS FOR REMEDIATION PANEL

Deborah Hughes Hallett
Harvard University
Cambridge, Massachusetts

You may wonder what I'm doing here - - I who teach at a college with a selective admissions process, where the students are supposed to be well prepared and eager and certainly not in need of remediation. Let me assure you - - or perhaps shatter your illusions - - that I'm going to talk about students who have trouble with decimals, and who find l.c.d.'s impossibly mysterious, and who drop minus signs and parentheses as though they were going out of style, and who freeze at the very thought of a word problem.

For the past eleven years most of Harvard's math disaster cases have passed through my hands. Nine years ago, I managed to convince the university - - somewhat to their horror - - that Harvard does in fact have its share of students who, while they might be very bright, nevertheless can't do algebra, and, in some cases, arithmetic, properly and who, besides this, or perhaps because of it, are terrified to boot. As a result, I was allowed to start a "before calculus" course called Math Ar, and I say "before calculus" with feeling because it's not a pre-calculus course in the usual sense, because we try to do everything our students need to get them where they are going - - and that may be anything from basic arithmetic to trigonometry

Before I talk about exactly what we do, I'd like you to look at two ideas that are such an accepted part of teaching math that you may never have thought about them. First there's an idea abroad, especially in the more mathematical reaches of the land, and also, unfortunately, in some of our student's heads, that students who come to college without having learned algebra, or, worse, arithmetic, properly are hopeless cases, doomed to remain mathematical cripples forever. But just because a student is behind his classmates, does that mean that he is unteachable? Just because he is learning something for a second time - - does that mean he will be unsuccessful? I take exception to this view. I think you can always learn math - - it may take time, work, and energy but it always can be done. And don't think that you can successfully teach students with bad backgrounds unless you believe this too.

Second, I'd like you to look at our usual attitude towards students who have forgotten something they've been taught. This is a common enough phenomenon: how often have you had to stop a calculus class to tell them the volume of a sphere, or to explain in detail some algebraic step which should have been straightforward, or to give the value of log 1 or $\sin \pi$, or to howl that 1/0 is undefined, not 0? The list is endless and makes one wonder what they'll think of to forget next. What is wrong? My theory is that students very seldom forget things they genuinely understand, and that what we are dealing with here are studens who have never understood what they have learned but who have memorized their way through earlier courses. And I don't believe that you can teach students with bad backgrounds unless you are willing to confront one of their major sources of difficulty - - their ability to subtitute memorization for understanding.

Now about the "before calculus" course. In the past nine years it has expanded to several hundred students a year including undergraduates, night school students (mostly adults), and summer school students of all ages and from all parts of the country. Our students have backgrounds from the peculiar to the appalling, students who have been out of math from 4 to 40 years, and students who think that learning math is learning a collossal list of rules - - and indeed some who treat the rules rather like traffic regulations and obey them only if the policeman/instructor is watching. Just about no one is taking math because it is fun; just about all are terrified of the subject. We obviously have to try and change their attitude as well as teach them math. Surprisingly enough it can be done, and even more surprisingly, it can be fun.

Over the years we've learned a great deal by trials, errors and tribulations. First is the importance of starting at the last point at which the student is confident - - which may not be where they or we wish it was, and it certainly won't be where their last course left off. As I speak I can hear the voices of my calculus students ringing in my ears saying they "have had" algebra and of my pre-calculus students saying they "have had" arithmetic. And it is true, they have - - but all I can say is that from much bitter experience that doesn't mean they know it. And if they are to be confident you must build new material on a background of solid understanding - - otherwise you build castles in the sand. It is important to be honest and brave enough to begin at the right place - - and if that's arithmetic, then so be it.

Secondly, we fight an interminable battle against the tendency to sacrifice understanding to speed. It's so easy to be tyranized by the syllabus, and in the moments of panic when you feel that you'll never "make it" by the end of the semester, to slip into getting students to "at least memorize the method". It's so easy, when you think students are never going to do something straight, to give them practice exams enough like the real one that they end up getting a half-way decent score - - but not by understanding, but by memorizing. There are places where there's no substitute for memorization - - I've never found anything else for irregular Latin verbs - - but for math memorization is a disaster. First, it has a half-life of about a week - - people forget what they've memorized as soon as possible. Second, it contributes greatly to one's fear of a subject to know that you don't understand what you're doing, that as soon as you're given a problem that looks different, you'll have had it.

How is this course organized in practice? First we give a placement test and group students into sections of 12 to 15 or so. Different sections start at different places, some with arithmetic, some with algebra. All show algebra as a generalization of arithmetic, all do functions only after students really understand graphing, and all do a right triangle approach to trigonometrty.

Each section is taught by a teaching fellow, who, more often than not, is an undergraduate. Sometimes they are math majors, often not. When we first started using undergraduates it was considered heretical, but they have been quite superb and are now an irreplaceable part of the teaching staff in all the sciences. The sections with the lowest scores on the placement test are always taught by the most experienced T.F.'s. Promotions work downwards - - a really good job earns a teaching fellow the section with the lowest placement scores.

Over the years, we've experimented with every conceivable style of teaching - - we've team-taught, used programmed texts, and done it self-paced. The one thing that has become clear is that our teaching methods are essentially irrelevant to our students - - the only thing that really matters to them is our attitude, our spirit. They learn far more from our faith that they can and will learn mathematics than from our most lucid explanations or most brilliant innovations.

In practical terms this faith means going to some unusual lengths to reach out to students. It means taking the initiative to call people who haven't been to class for a while. It means taking the time to see math in the warm, human, logical, but non-standard, way that our students see it. It means talking to them about whatever's on their mind, mathematical or otherwise - - usually otherwise. In my case in meant learning enough about football to be able to make up word problems to convince a skeptical football star that he could do them (he could). It means having the patience to answer students' questions, and more importantly, not to answer their questions. Many students have learned to do math by a sophisticated form of guesswork. If you ask them a question they start to answer while staring hard at your forehead, as if expecting to find the answer written there. If you start to frown, they immediately stop, apologize, and try another track. And they keep trying different answers until they find one that makes you smile and nod. I've always felt the only way to teach students was with a paper bag over my head, or at least facing in the opposite direction.

On the other hand, there are students who are sufficiently scared that they feel they can't do math without checking every step of the way. Answering all their questions is unhelpful because it leads them to think they can't manage without me, though if I have them work completely on their own they get stuck. Here it is important to start off being helpful answering questions. But after a while you don't really have to answer much - - your presence will be enough to keep them going. At that stage I have students come and work at my kitchen table while I do something else. The student can then ask questions if he gets stuck, but after a little while, instead of answering, I find I can get the answer from the student. The next stage is when the student doesn't even wait for me to say anything, but goes on and answers it himself.

Our belief that people can and will learn means that there are times when we have to take a hard line. For example, I believe that people can learn to take tests, and those who don't go through our "exam immunization scheme". This belief also leads to "nostalgia problems" - - our atttempt to make sure people don't forget what they've learned: on any problem set or test there may be problems from any earlier part of the course - - for nostalgic reasons. It is improtant to be tough as well as kind and not be afraid of anger or tears. Both are part of learning math, and certainly of repairing a bad math background.

Above all, believing in your students means caring about their learning math enough not to ever give up on them, even on the not too infrequent occasions when they give up on themselves. It means never compromising your expectations - - because you will get from your students exactly what you expect of them - - no more, no less.

Let me repeat: the one crucial component in this kind of course is the one thing our students cannot provide for themselves and cannot get from any textbook: our faith and conviction that they can and will learn mathematics.

REMEDIATION METHODS IN ADULT BASIC MATHEMATICS

Gerald Kulm
Purdue University
West Lafayette, Indiana

The initial stage in a remediation effort is a careful diagnosis of the student's strengths and deficiencies. A discussion of this process is beyond the scope of this paper, but it should be noted that such a diagnosis must be sufficiently detailed to determine precisely those skills needing remediation. For example, it is not sufficient to determine that a student has difficulty with adding fractions. The diagnosis should reveal whether the deficiency is in finding common denominators; whether certain fractions are more difficult than others; whether the process is understood; and other similar details.

Once the task of diagnosis is completed, the teacher must create procedures to help a student overcome the errors, misunderstandings, or deficiencies. Before discussing the specific methods, it may help to ask a few questions about the possible remediation steps. Here are some things that should be considered.

1. Does the student need a boost in self-confidence?

2. Is the student convinced he or she is wrong?

3. Does the student have the necessary background knowledge?

4. Would it help to work out the procedure in detail?

5. Would a picture help?

6. Is my attitude positive?

(Adapted from Didactics and Mathematics, Creative Publications, 1978.)

With these questions as a guide, let us consider the major difficulties found in the four areas of computation with whole numbers, fractions, decimals, and percent.

A. Basic Facts

The basic addition, subtraction, multiplication, and division facts are essential and must be memorized. Memory takes practice, but there are many ways to memorize; for example, one can use pictures, tables, flashcards, or calculators. Many adults do well working at a computer for practice with basic facts. The non-threatening environment of the computer, the immediate feedback, and the report of percent correct combine to make automated drill a valuable tool.

B. Computational Algorithms

Computational procedures are simply a set of rules for doing arithmetic more quickly and efficiently. It is helpful for adults to understand that, given the time, almost any arithmetic problem can be done by some counting procedure. Students (and teachers) should understand that there is nothing mystical about the particular procedures we happen to use. Most of them developed historically, and slightly different ones are used in some countries.

The impact of inexpensive calculators raises difficult questions regarding the teaching of algorithms to adults. Both practical and emotional factors must be considered. In everyday application, there are few computations that could not be done with a calculator. Furthermore, few adults will go on to higher mathematics where an understanding of algorithms provides a basis for the study of more abstract notions. On the other hand, most adults feel that they aren't really learning math unless they learn to do computational algorithms. In either case, judgements must be made about which algorithms and which number domains for the algorithms should be taught. It may not make sense to teach long division by a 4-digit divisor, given the time restrictions usually imposed in adult programs. These decisions must be made on an individual basis and represent a major difference between curricula for adults and those of younger students.

There are two important ideas in learning to use any computational procedure: (1) the special notation or format in writing things, and (2) the rules to follow.

Format and Notation

Often, it is possible to correct a student's errors or to clarify a misunderstanding simply by writing things correctly, for example:

$$\begin{array}{r} \frac{2}{3} \\ + \frac{1}{6} \\ \hline \end{array} \qquad \begin{array}{r} \frac{4}{6} \\ + \frac{1}{6} \\ \hline \frac{5}{6} \end{array}$$

The vertical format for adding and subtracting fractions helps students not to confuse it with multiplying and dividing.

Questions of format and notation are really not about arithmetic, but being careful with them can often improve performance tremendously.

Computational Rules

There is often controversy about whether a student should understand <u>why</u> a rule works the way it does. Some students may do better by following the rules in a step-by-step fashion and may be confused by explanations of why it works. Other students may be frustrated by not knowing the reasons for following the steps. The teacher must know the reasons why but also must decide when and how much to explain. Here are some other examples of computational rules that have explanations.

> Invert the divisor and multiply.
>
> To change a percent to a decimal, move the decimal two places to the left.
>
> To change a fraction to a decimal, divide the numerator by the denominator.

For some of these rules, the explanations are long and tedious. Since computational procedures are to be learned in order to do computation efficiently, there is some justification for following certain rules without question because they work and give the answer quickly.

C. Teaching Computational Procedures

Perhaps the best overall practice to follow is to build computational procedures step-by-step. Breaking an algorithm into steps has been very successful with adult students. Many adults apparently were not taught in this way or did not comprehend it in lower grades. The main idea is that one rule at a time should be learned or re-learned. Many computational procedures involve so many rules in order to cover the most complex example that it is taking on the memory capacity to learn or remember them all at once.

D. Correcting Computational Errors

Many errors can be traced back to gaps or unclear understanding of fundamental concepts. Here are some important concepts to look for in various areas:

1. Whole number computation: basic facts, place value.
2. Fraction computation: meaning of fraction, "number sense."
3. Decimal computation: place value, estimation.
4. Percent computation: meaning of percent, "number sense."

In many cases, it is enough simply to point to the answer and ask the student to think about the problem and whether the answer seems reasonable. Many errors are "thoughtless" and not symptomatic of deep problems. If, however, it appears that the student really doesn't know the correct way, remediation is needed. Here is an example of remediation technique:

Example

Error

$$\begin{array}{r} 56 \\ 15\overline{)7590} \\ \underline{75} \\ 90 \\ \underline{90} \end{array}$$

Remediation

Break the problem up.

$$\begin{array}{r} 5 \\ 15\overline{)75} \\ \underline{75} \end{array} \qquad \begin{array}{r} 0 \\ 15\overline{)9} \end{array} \qquad \begin{array}{r} 6 \\ 15\overline{)90} \end{array}$$

or

Estimate: 7590 is about 7500

$$\begin{array}{r} 500 \\ 15\overline{)7500} \end{array}$$

Is 56 a reasonable answer?

It is important to decide whether the error was an oversight or not. If not, it may still be as easy as reminding the student to think about what the numbers mean. If the student doesn't know the basic meanings (e.g., 70% means .7, or 36 means 30 + 6), these must be taught. If the student attempts to memorize the computational rules without these basic concepts, the memory load will be too much.

E. Explaining Basic Concepts

There are three main sources that can be used in explaining concepts: (1) pictures and diagrams, (2) the meanings of the terms used, and (3) organizing information or ideas in useful and clear ways. Space permits only one illustration of the third resource:

Organizing Information

There are three types of percent problems and it is difficult sometimes for students to keep them straight. Here is an example of each, followed by a way to organize things to find out which is which.

1. What is 60% of 25?
2. What percent of 25 is 15?
3. 15 is 60% of what number?

Each of these can be solved the same way if the information is organized. We call this the PTA method; P (percent), T (total), A (amount). Organized the PTA way, the three problems look like this:

	P	T	A
1.	60%	of 25 is	? .
2.	?	of 25 is	15.
3.	60% of	?	is 15.

The idea is to determine which is the unknown, then fill in the other two. Next, "of" means multiply and "is" means equals, so

1. 60% x 25 = ___
2. ___ x 25 = 15
3. 60% x ___ = 15

Number 1 is easy. Numbers 2 and 3 are solved by dividing 15 by 25 and by .60 (60%). Let's try for example:

You saved $18 when you bought an appliance on sale for a 10% discount. What was the original price?

The total is unknown.

P	T	A
10% of	?	is $18
.10 x	T	= 18
	T	= 18 ÷ .10
	T	= $180

Many ideas in arithmetic seem, at first, to be different ideas. If organized, the same procedure can be used for several cases rather than learning a different rule for each.

F. Summary

This paper has presented some practical procedures that have been effective for helping adults learn arithmetic. Perhaps the most important thing to be kept in mind is that no special "tricks" or unusual aids are necessary for teaching many adults. The errors or gaps in computational skills can often be traced to a missing explanation of a rule or procedure. The most often encountered remark heard from adults is something similar to "Why didn't they (teachers) tell me that before?" The "before" may refer to 6th or 7th grade when the student was either uninterested or did not understand the explanation when it was given. In any case, a clear, well-sequenced, and detailed explanation of a computational algorithm, accompanied by a simple definition of essential concepts, is the most effective remediation procedure.

PANEL ON STRATEGIES FOR IMPROVING REMEDIATION EFFORTS

Joan R. Leitzel
The Ohio State University
Columbus, Ohio

The fact that large numbers of freshmen are entering colleges and universities in the United States without adequate preparation in mathematics has been well documented. Indeed the February 1980 Notices reports a 29% increase in course enrollments below calculus between fall 1978 and fall 1979 in the top 27 ACE ranked mathematics departments. Less formal surveys indicate that almost all two-year and four-year institutions are currently providing beginning courses that include elementary algebra and often arithmetic topics and indeed that enrollments in these courses have increased sharply since 1975.

At my own institution, which is an open-admissions university, every new student writes placement tests in mathematics. These scores together with the student's ACT or SAT score determine a mathematics placement level. Level 1 students are ready for calculus; levels 4 and 5 students need remedial courses before taking the pre-calculus courses. (Level 5 students generally show no skills in elementary algebra.) Because the University has used the same placement tests since 1966, we are able to compare the performance of incoming students over the last 14 yeras and to see the marked increase in the percentage of level 5 students.

Percent of Ohio State Autumn Quarter Freshmen with Remedial Mathematics Placement (Levels 4 and 5)

	1966	1967	1968	1969	1970	1971	1972	1973	1974	1975	1976	1977	1978	1979
Level 4	15	16	17	16	17	16	16	16	16	16	14	15	16	16
Level 5	13	16	18	17	17	21	23	24	25	26	27	27	26	27

Because the decline in mathematics skills of college freshmen and the development of remedial courses have been so widespread in this country, the Mathematical Association of American in 1978 formed a Committee on Improving Remediation Efforts in the Colleges. This Committee has spent several months studying remedial programs in many junior colleges, four-year colleges, and universities in this country. For many years the two year institutions carried the burden of remedial instruction in this country. However, five or six years ago large numbers of four-year schools began to develop new programs or to expand existing programs in response to increased numbers of under-prepared students. Many other schools are beginning now to give attention to this level of instruction as their institutions anticipate reaching our more widely for enrollments in the 80's and as colleges restore graduation requirements in mathematics that were commonly dropped in the early 70's. One audience which is presently receiving new attention is the increased number of older students entering many of our colleges.

The Committee has found many institutions where faculty feel the remediation efforts are doing little good and, indeed, their data confirm this feeling. Frequently, teachers are frustrated by the low level instruction and departments are hard pressed to absorb the costs of increased loads. On the other hand, the Committee has found in some schools considerable optimism about the outcomes of these efforts. Individual faculty members report that they can realistically set higher goals for this audience than they first thought, that they have learned a great deal about the teaching of elementary mathematics from their involvement with remedial students, and that in some cases the development of remedial courses has resulted in the modification and improvement of other undergraduate courses. These individuals welcome the opportunity to teach good mathematics (albeit elementary) to an audience who might otherwise have taken no mathematics in their college programs.

The Committee has found programs which teach the technical skills needed in later courses and at the same time provide for conceptual understanding and for analytic reasoning. Data indicate that large numbers of students are moving successfully from these programs into the mathematics courses required in degree programs. Some schools are using the instruction of remedial students as part of their program for the training of teachers and are convinced that the involvement of future teachers with students who failed to learn pre-college mathematics is very important.

Although there is a great deal of diversity among the remedial programs at various schools, the Committee has found among the successful programs certain characteristics which appear to be common. I would like to share a summary of these characteristics with you.

Characteristics of Successful Remedial Programs

- Involvement of regular faculty in the direction of the program and in the instruction.

- Evidence of institutional and departmental commitment to the program, shown both through reasonable financial support and through appropriate rewards for faculty involved.

- Requirement for strong commitment of time and effort on the part of students, with attempt to communicate honestly to students the scope and demands of the program.

- Program of study tightly structured, with the individual's pace determined by the instructor (not the student) in response to the student's ability and progress.

- Use of curricular materials often developed (or at least synthesized) by faculty working in the program.

- Strong emphasis on problem solving, perhaps with calculators, and on involving students with demanding problems at an elementary level; a de-emphasis on vocabulary.

- Content which both responds to real world experiences and provides interesting elementary applications, and at the same time strives for conceptual understanding.

- Efforts to achieve some specific short range objectives (e.g., computational skills with rational expressions) and also to get started on some longer range goals (e.g., understanding the notion of function).

- Integration of mathematical ideas both horizontally (e.g., analytic geometry with algebra) and vertically (e.g., awareness that processes are repeated in different settings).

- Awareness of the needs of different groups of students in the remedial audience (e.g., 18 year olds vs. returning adults, untaught students vs. unsuccessful students, anxious students vs. lazy students) and instructional strategies that provide for these differences.

- Emphasis on an informal instructional atmosphere that provides considerable moral support to the students and encourages peer interaction in the learning of mathematics.

- Evidence of considerable enthusiasm among teachers and strong respect for the potential of the students.

- Ratio of students to teachers not greater than 30:1 in courses that include topics in arithmetic and elementary algebra.

- Provision for some one-on-one interaction between teacher and student.

- Careful attention to the training and supervision of inexperienced teachers.

It has been heartening to find remedial programs in colleges and universities that are doing an exemplary job with under-prepared freshmen students. However, there is strong agreement among us all that the proper place for mathematics instruction at this level is in the secondary schools. The learning of elementary mathematics is best done over a period of several years. No highly concentrated program contained in a few months can substitute for the opportunity to deal with the ideas of elementary mathematics over an extended time. Several programs have been developed to encourage more adequate preparation in mathematics at the pre-college level. I would like to cite some of these efforts.

The National Council of Teachers of Mathematics and the Mathematical Association of America have made strong recommendatins on college preparatory mathematics in the publication Recommedations for the Preparation of High School Students for College Mathematics Courses. The MAA has also published The Math in High School You'll Need for College and has circulated more than 320,000 copies to teachers, guidance counselors, and students. Some colleges and universities have prepared careful statements to secondary schools in their states describing the mathematics requirements in programs at their institutions and what constitutes appropriate secondary preparation for the required mathematics. At least one college has developed a review workbook in algebra and trigonometry which students can obtain at the time they are admitted and work on over the summer months before writing the mathematics placement test in the autumn.

The discontinuation of the National Science Foundation Institutes of the 60's left secondary teachers of mathematics with too little opportunity for the continued study of mathematics. Some mathematics departments, but not yet enough, have developed credit courses for inservice teachers at both the secondary and elementary levels. Some are using their remedial courses at laboratories in the training of preservice as well as inservice teachers.

There are several organized efforts to send mathematicians and persons in math-related fields into schools to talk about the importance of early preparation in mathematics. Two of these, Women and Mathematics (WAM) and Blacks and Mathematics (BAM), have data which document significant increases in enrollment in college-preparatory courses in the schools that have been involved in their programs.

Finally I would like to mention the program initiated at Ohio State which provides for the testing in mathematics of high-school juniors. This last year more than 5,000 students were tested in 30 high schools with equivalent forms of the Ohio State placement test. All students were provided with their own scores, with a description of the mathematics courses required in their intended majors, and with a list of remedial courses that would be required if mathematics skills remained at the tested level. This program is now in its third year and will be expanded next year to 100 Ohio high schools. Two aspects of this program are very promising. First, each secondary school involved reports a sharp increase in enrollments in college preparatory mathematics courses among senior students. Secondly, the group of students tested as juniors in the first year of the program entered Ohio State last autumn as freshmen. They were retested a incoming freshmen and the results indicate considerable improvement in mathematics placement levels among those students who took proper college preparatory courses as seniors.

Efforts of these kinds suggest that it is possible to influence young people to study college preparatory mathematics in the secondary schools. We need to speak with a strong, concerted voice to our youth and their parents about the consequences of learning too little mathematics before entering college. Although we can never hope to overcome the need for remediation and review completely, we surely can hope to overcome the need for large numbers of our young people to learn elementary mathematics as beginning college students.

19.6 INDIVIDUALIZED INSTRUCTION AND PROGRAMMED INSTRUCTION

SUCCESSES AND FAILURES OF INDIVIDUALIZED INSTRUCTION IN MATHEMATICS - THE PUERTO RICAN STYLE

Noemi Alvarado
Educational Region of Ponce
Ponce, Puerto Rico

For decades the process of "educating" the child has been characterized for its great emphasis on transmitting knowledge and mathematical content to the learner through lectures, readings and mechanic drill.

The students are exposed by the teacher to large amounts of material that they are supposed to master during a fixed period of time. Often, the number of chapters and pages covered is more important for the teacher than the number of mathematical concepts and skills developed by the student. To cope with this task of absorbing more and more mathematics information, the student tries to memorize it, usually with very little comprehension of its meaning. Frequently, he is unable to apply this knowledge to handle new situations, and as a result he tends to forget it very fast.

Many teachers give to the teaching process a general group approach where the students are considered as packages and are all taught by the same methods and teaching techniques. They are also expected to learn the material in the same amount of time.

As education has become universal, the size of groups has increased and so it has been impossible for the mathematics teacher to give special attention to particular students due primarily to limitation in time. This mass education has gone through a gradual process of deterioration in quality. Many educators have pointed out that education in the 20th Century is in a real crisis. On the other hand, the principle of equal educational opportunities for all, guarantees that each individual can develop to the maximum of his abilities and human potential for his own benefit and to be useful to society.

This is the dilemma. For one thing, mathematics educators do know the kinds of learnings that we need and on the other hand, they are aware of the poor quality of the results of mathematical instruction.

What can be done?

The answer is not simple. As educational research has shown, the teaching-learning process is effected by many factors such as: teaching methods, curricular materials, learning styles and environmental conditions.

All individuals differ in the amount of knowledge they have, in their experiences, abilities, interests, attitudes and capacity to create and produce. Also, they differ in the style and rate of learning. Learning then, is an individual and humanistic process. Trying to agglomerate this process is to destroy the natural way in which it occurs.

At present, educators are looking forward for innovations that contribute to improving the quality of education. Mathematics is not an exception. Although, individualization is a teaching techique that goes back to the teachings of Plato and Socrates, at present it has been thought of as one of the educational strategies of impact in handling the teaching-learning process.

What is individualized instruction?

There are many interpretations of the term "individualization". For some teachers it means a teaching technique. For others it means a method of teaching. For others, it simply implies working with just one student at a time.

According to Piaget, Torrance, Guilford, and others, individualization of instruction is more than just a teaching technique or method.

Individualization is conceived as a teaching strategy or approach where the learner is the center of interest and teaching is adapted to each child in a humanized way. It is the instructional process by means of which the student learns at his own rate and capacity. It is the teaching based on the particular needs of the student, person to person, as a tutorial program. Its emphasis is on mastery of learning instead of memorizing content, and the student is not allowed to move to a new task until he masters the previous mathematics work. This individualized approach provides for the individual needs of the learner and also provides for continuous progress in the learning of mathematical content, attitudes and skills of a particular learning level.

The student does not move to a different learning level until he is skillful in the previous one.

Many attempts to individualize mathematics instruction include:

1. levels of learning
2. interests
3. mathematics skills
4. special assignments and projects
5. differentiated curricular materials
6. remedial classes

Individualization can be organized by means of large and small groups, independent study and laboratory work. Some factors that affect the individualization of instruction are: school organization, teaching periods, attitudes of teachers towards innovation, physical educational facilities, curricular materials, attitudes of parents, and the interrelations among teachers, students and administrators.

The role of the teacher as the organizer, planner, evaluator, guide, leader, adviser, and administrator of the process is fundamental if individualization is expected to occur.

Also, the educational materials and mathematics textbooks should provide and present the content in a way that facilitates individual use of it.

Other specialized materials such as kits, filmstrips, slides and cassettes can be used.

A progressive record of the child is required so as to have a clear and complete picture of the student. Also, a continuous regrouping in the classroom is necessary for which the teacher must evaluate constantly.

The parents should be informed and aware of the progress of their children. This can be accomplished by means of interviews, teacher-parents meetings and regular reports.

Teachers also, can use students as tutors to teach other children who need help and are temporarily slow in their learning.

Additional materials should be prepared to be used with students that need more time and additional learning experiences to master a certain number of mathematical skills. No individualization will ever take place in a classroom that lacks the necessary educational materials such as reference books, diagnostic tests, progress, and competency tests, additional exercises, and other materials.

Once, the students are properly placed, the group should work out some functional working norms to facilitate the movement of students, control of discipline and classroom management. Some examples of classroom norms are:

1. Check your exercises by using the answers sheet.
2. Put back all materials used in their assigned places.
3. Do not go on with new material until you find out how well you are doing in the previous one according to the criterion established.

The teachers also, need to plan their classes daily, make a follow up, and keep record of the progress of the student.

According to Ralph Tyler, individualizatoin of instruction should be a systematic process that can be separated into six major phases:

1. Rationale - the student recognizes the importance and utility of the topic he is going to study. This provides true motivation for learning.
2. Objectives expressed in behavioral form that establish clearly to the student what is expected from him.
3. Pre-test - to diagnose what the student already knows about the topic.
4. Learning activities - to develop the objectives established for the student.
5. Post-test - to determine the degree of progress in terms of the objectives established.
6. Evaluation - final decision of what the student should do after having the results of the post-test.

Any learning unit which the design has this organization is called an Instructional Modulus.

Based on this systematic approach the mathematics program in Puerto Rico has produced a series of curricular materials for the elementary school called "Etapas de Aprendizaje" and "Lecciones Individualizadas." For the intermediate level, a series of 56 instructional modules has been produced. At the high school level a series of Six Programmed Instruction Units have been written to be used as a remedial course in General Mathematics.

No single teaching method by itself is a panacea. The learning process is so complex that different approaches are needed to ensure high quality learning. Nevertheless, in our experience working with the individualized approach we have found many advantages of this strategy. Among them, I will point out the following:

1. The learner has an active involvement in the learning process, planning his time and been aware of his potentialities and needs. He knows exactly what is expected of him.
2. The learner has the opportunity to progress at his own rate of learning without being in competition with other students.
3. Evaluation of instruction is constant, so that the learner is always informed of his success and failures.
4. The teacher is able to provide immediate remedial work for individual students.
5. Students who master the mateial do not have to wait for others and can go on to study enrichment material or move to the following level of content.
6. The learner feels more secure as he is always informed of his progress.

The disadvantages of the individualized approach of instruction are more related to its implementation rather than to the teaching-learning process itself.

To implement this strategy adequately it is necessary to produce curricular materials that provide for individual use. Many of the curricular materials and textbooks in use lack the presentation of the content. Also, a continuous in-serve training program for teachers is absolutely necessary as many do not have the competency and the professional and academic background required to manage properly the classroom situation in an individualized setting.

Finally, it is imperative to modify the traditional system of grades in the evaluation of achievement. A more adequate system of evaluating the progress of children in a learning setting based upon individualization is necessary. This requires the thought, coordination and involvement of different individuals at the central level of the Department of Education. I look forward to reaching that day, when, individualization will have a real impact on the whole educational process.

References

1. Hillson Maurie, Ronald T. Hyman. Change and Innovation in Elementary and Secondary Organization. Holt, Rinehart and Winston, Inc. New York. 1971.
2. Johnson Stuard R., Rita B. Johnson. Developing Individualized Instructional Material. Westinghouse Learning Corp. New York. 1971.

   Assuring Learning with Self-Instructional Packages. Instructional Packages, Inc. Phillipines. 1973.
3. Kemp Jerrold E. Planeamiento Didactico. Editorial Diana Mejico, 1971.
4. Kibler Robert J., Larry L. Barker, David T. Miles. Behaviorial Objectives and Instruction. Allyn and Bacon, Inc. Boston. 1972.
5. Klaus David J. Tecnicas de Individualizacion e Innovacion de la Ensenanza. Editorial Trillas S.A. Mejico. 1972.
6. Lewis Jr. Jones. Administering the Individualized Instruction Program. Parker Publishing Company, Inc. New York.

7. Petreuin Gaynor. Individualizing Learning Through Modular-Flexible Programming. McGraw-Hill Book Company. New York 1968.

8. Popham W. James, Eva L. Baker. Systematic Instruction Prentice-Hall, Inc. New Jersey. 1970.

9. Pula Fred J., Robert J. Coft. Technology in Education: Challenge and Chance. Charles A. Jones Publishing Company. California. 1972.

CHAPTER 20 - Women and Mathematics

20.1 A COMMUNITY ACTION MODEL TO INCREASE THE PARTICIPATION OF GIRLS AND YOUNG WOMEN IN MATHEMATICS

ESTABLISHING THE EFFECTIVENESS OF PROGRAMS DESIGNED TO INCREASE WOMEN'S PARTICIPATION IN MATHEMATICS

Elizabeth K. Stage
University of California
Berkeley, California

An evaluator always faces the problem that program planners are interested in information that will help them to improvè the program (formative evaluation), while program funders are interested in information that will help them to measure the impact of the program (summative evaluation). In the case of programs of the Math/Science Network that aim to increase women's participation in mathematics, the complex origins of the underparticipation problem and the multifaceted nature of the programs combine to challenge the evaluator. Finally, the grass roots, volunteer character of many activities leads one to hesitate to spend large sums of money on data gathering when more people could be served if the same resources were allocated to staff or materials. Our resolution of this system of constraints is a flexible approach to evaluation that employs activities that provide formative and summative data, are targeted at specific and diverse goals, serve some programmatic use in addition to their assessment purpose, and are inexpensive. Examples from the evaluations of a variety of Network programs will illustrate these points and show that the planners and the funders can be equally satisfied.

The EQUALS program for inservice teacher education makes a good case study of evaluation techniques. Daily monitoring of the individual program components is handled by asking the participating educators to rate each of a given day's activities on a scale from 1 to 5 and to share any thoughts about the day on "comment cards." The staff gathers to review the day's ratings and reads the cards, enabling them to change tactics in midstream if some program component is not having a positive response. In general, the daily reports are quite favorable, with last year's 33 activites averaging a 4.2 mean rating on the 5-point scale (with a range of 3.5 to 4.9). The high satisfaction of the participants increases the likelihood of overall success, but it is only a first step toward establishing the effectiveness of a program whose ultimate goal is to influence the enrollment of young women in elective mathematics courses. Enrollment data are gathered by EQUALS participants as a way of getting them to assess their own situation, but once gathered and compared with previous years' data, they become a way of measuring the program's success on a broad scale. In some districts where teachers have been attending EQUALS and using our techniques with students for several years, there has been a notable increase in enrollment.

In most school districts and in areas where efforts have been concentrated at the elementary school levels, it will not be possible to achieve dramatic changes in elective course enrollment in high schools for several years. It is still possible to show gradual improvement, however, by examining some of the underlying causes of mathematics avoidance. Since favorable attitudes toward mathematics and awareness of the career importance of mathematics have been shown repeatedly to be important predictors of electing optional mathematics courses, the EQUALS program can take heart at the demonstrated improvement in these areas among the students of the participating teachers. The Math Attitudes Scale and the Career Awareness Scale have had both formative and summative applications, as the career component of the program was strengthened following the finding that the attitudinal improvement was superior to the career awareness gain.

The teachers of young children have presented a particular problem for evaluation, since pupil's preferences for career choices are not easily measured by a standard checklist. Using pictures eliminates the reading and vocabulary demand, but selection of preferences on the basis of gender is likely with a photographic presentation. Teacher/developer Diane Downie and psychologist Steven Pulos worked with an artist to develop The Bears, a cartoon displaying various occupations. The children were not fooled, though, and made choices that reflected a knowledge of the sex role stereotypes that are prevelant in the culture despite the artist's efforts to make "male" arms on the secretry and long eyelashes on the scientist. The Bears is a good example of something that calls the teachers' attention to their students' opinions in an entertaining but informative way. If the children fill out the form later in the year, one can assess changes in opinion and let the teacher measure his or her effectiveness. Pooling the data allows us to measure the program overall.

The final documentation measure in EQUALS is journals kept by teachers during the year. Taking notes during EQUALS sessions, they can record good or bad points of the program that seem to minor to mention. Recording their attempts at implementing EQUALS strategies during interim periods, they can remember them for discussion at subsequent gatherings.

The comprehensive scope of the evaluation for EQUALS is an unusual situation for a Network program, but it illustrates that the principle that each evaluation procedure should accomplish as many objectives as possible. Having teachers and administrators gather their own data is not only an economy measure, but also an important mechanism for raising the consciousness of district personnel, colleagues, and parents. Asking teachers to administer some of the measures to their students acts as a check on the program's effectiveness, while adding to what the teachers' learn from the program.

Other Network programs build upon and contribute to the EQUALS evaluation. The National Science Foundation has funded a curriculum development project to refine, test, and disseminate some of the materials developed in EQUALS that are designed to emphasize those areas in which female students have been found to be weaker than males. Problem solving, spatial visualization, career awareness, and favorable attitudes toward mathematics will be assessed through a combination of newly created instruments and those found useful in EQUALS.

The "Expanding Your Horizons" conferences for young women in grades 6 to 12 are held on a Saturday in March on college campuses in California and throughout the U.S.A. One measure of success is growth, from one conference in 1976, three in 1977, four in 1978, eight in 1979, to fifteen in California alone in 1980. We know that satisfaction with the conferences is high among the girls who attend them. The model of combining hands-on experiences doing science and mathematics with conversations with women who use mathematics and science in their occupations has been modified over the years by carefully noting the reactions of the girls and maintaining a list of speakers who are good at achieving rapport with teenagers. Girls' plans for further study of mathematics and science and their knowledge of careers in these areas increase over the course of the day, based on surveys administered at the start of the opening session and the close of the final session. The Network is currently seeking funds to support a study to examine the long-term effects of the conferences to see if the plans inspired by enthusiastic role models are carried out or forgotten after a year or more.

Conferences are also held for college and older women that often focus on fields in which there is currently an abundance of employment opportunities, such as engineering and computer science. Again, satisfaction with program elements is polled to add to the growing store of knowledge about dynamic speakers and capable presiders. Follow-up questionnaires used for some of these conferences have shown that a year later many women had taken action influenced by the conference experience, such as taking an academic course or looking for a new job.

Course taking is used as a primary evaluation measure for the collegiate programs. The Mills College precalculus course, which prepares students for calculus in one semester by concentrating on the essential concepts and providing a peer-taught workshop for support, has achieved such success that mathematics enrollments have doubled. The teachers of the "Math without Fear" courses at San Francisco State University and California State University at Long Beach also look for enrollment and achievement in subsequent mathematics courses as measures of their impact, though they first have to accomplish a change in attitudes towards mathematics. Math without Fear aims at eliminating the excessive reliance on rules found among poorly trained students and replacing it with a conceptual orientation to mathematics learning. This attitudinal change is fostered by a supportive classroom environment that encourages guessing and intuitive approaches to problem solving, as well as small group, cooperative activity. Observers in the Math without Fear classes help to insure that the classroom climate enhances student learning and provide qualitative evaluation data.

The Network itself has been evaluated to provide documentation to the Carnegie Corporation of New York, its funder, and to give assistance to leaders in planning future activities. One of the major procedures was the organization and compilation of data that already existed but had not been put together. While the evaluation was sparked by the need to apply for renewal funding, it helped to boost the spirits of Network members when they saw how much had been accomplished in the first two years of funding. Carnegie representiatives expressed a preference for "hard data," like the enrollment information gathered by EQUALS teachers, but the program officer also appreciated seeing The Bears and hearing anecdotes from Expanding Your Horizons conferences. The combination of quantitative and qualitative data, gathered for both formative and summative purposes from a variety of sources and dimensions, helps to assure program planners and funders alike that the Math/Science Network is accomplishing its objectives of contributing to increased participation of women in matheamtics.

EQUALS
A STAFF DEVELOPMENT PROGRAM

Kay Gilliland
Emeryville Unified School District
Emeryville, California

At the Lawrence Hall of Science, University of California, Berkeley, a group of educators in the Mathematics and Science Education Programs for Women developed direct service programs for elementary age girls and conferences for secondary school women in order to change the pattern of math avoidance among female students. It became clear that the problem of math avoidance among female students was too great for direct service activities alone to solve. The way to effect large scale change would be to work with teachers, counselors, and administrators. The EQUALS staff development program was designed to attract educators, convince them that they could help solve the problem, and keep them involved so they would return to their schools to implement EQUALS activities in the classrooms and conduct inservice presentations for their colleagues.

EQUALS was initiated in 1977 when a proposal to the Department of Education was accepted. In the first year, seven school districts were represented with 40 educators who took part in a three-week summer inservice. The inservce education was refined and intensified into a series of five day-long workshops given over a six-month period during the school year. During the 1980-81 school year, EQUALS will serve educators in many areas of the United States. In California alone, EQUALS will provide inservice education to 600 educators in 65 school districts.

EQUALS has proven effective in increasing educators' awareness of sex differences in young people's math and science education and their awareness of the inadequate preparation of women students for technical and scientific careers. EQUALS has developed methods for increasing young women's competence and confidence in doing math and for bringing information on the career options open to those who have adequate preparation in mathematics and the sciences.

One of the program activities which is designed to help the staff and the participants is the Research Project. EQUALS involves participants as researchers in their own schools. Three of the projects which EQUALS has made available to participants are "Patterns of Math Course Taking", "Typical Wednesday at Age Thirty" and "Job Picture Story".

In "Patterns of Math Course Taking", participants in schools where math courses are optional survey the classes to discover the percentage of females in each class. Most discover that the higher level math classes have a much lower percentage of female students. As soon as math becomes optional, most schools experience a drop in the percentage of female enrollment. Activities that help students develop competence and confidence in mathematics are implemented. Surveys by the same educators a year later have shown improvement in math course-taking by female students.

In "Typical Wednesday at Age Thirty", students are asked by their teachers to write essays on what will happen during a day in their lives when they are thirty years old. Participants frequently find the career aspirations of even their brightest female students to be very low. Where males look forward to high-interest activity and success, females often describe boredom and failure. The following story by an intelligent thirteen year old girl illustrates this:

> First of all I get up, brush my teeth and eat breakfast. Then I go to work. I'll be a secretary for a large company and take a letter. I go to my electric typewriter and start to type. I make a mistake and can't find any correcting fluid. Then I start all over again. I get off at nine in the night. I am tired. Then I drive home. I'm too tired to cook and I go to bed.

In "Job Picture Story", participants ask younger students to draw pictures of themselves as workers. The students then describe the picture and the teacher writes the description. Comments by the participants indicate that they are surprised at the stereotypic responses of their young pupils.

Each of the activities has a two-fold purpose: 1) to inform the participants and 2) to provide a model for use in classrooms, faculty meetings, and in other appropriate situations. These activities are designed to provide awareness of the problem of math avoidance among females, competence and confidence in doing mathematics, and encouragement toward the consideration of non-traditional careers that require a background of mathematics and science.

One awareness activity developed by EQUALS is "Going to the Workforce". Participants receive specific information on the amount of math required in preparing for various occupations. Participants are asked to rank eleven occupations according to the amount of preparatory math required to obtain a degree in the field. Having done this individually, they form groups and discuss their rankings, coming eventually to a group consensus. The catalog answers are then given, and participants can derive a score by taking the absolute difference of the ranks they assigned the degree fields, and the ranking found by consulting university and community college catalogs. Teachers are encouraged to bring catalogs and handbooks to their classrooms after this activity so students can begin learning to read them and to discover the differing requirements at various colleges and universities in their area. The activity raises awareness of math as a filter which keeps students out of certain occupations, and gives information on particular occupations which students may wish to consider.

Participants are strongly encouraged to bring women from business and industry into the classrooms as role models. A role model panel is an integral part of most EQUALS presentations. Typical panels include a woman physicist, engineer, electrician and computer scientist. Each woman talks briefly about her work, how she got into her career, and which aspects of her schooling were formative for her. Time is given for questions from the participants. The panel invariably receives extremely high ratings.

The EQUALS activities are planned to provide awareness of the problem, competence and confidence in doing math, encouragement toward non-traditional careers, and development of techniques for spreading the work. EQUALS participants find that they use the activities directly in the classrooms or modify them to meet the needs of the particular age level they teach. EQUALS participants report that they have given presentations in faculty meetings, for parents, for larger groups of colleagues, and for their school boards. Several participants have written proposals for similar projects in their districts and have received funding for them. It seems clear that EQUALS participants have recognized the importance to equity of increasing young women's participation in mathematics and science and are willing to share the responsibility and the challenge of bringing about change in the pattern of math avoidance among female students.

THE DEVELOPMENT AND GROWTH OF THE MATH/SCIENCE NETWORK

Nancy Kreinberg
University of California
Berkeley, California

Introduction

The concept of "Networks", has come into its own in the last ten years. The seventies was a decade of remarkable change for women. Women began to form alliances on issues of mutual concern.

One of the most effective alliances emerging during the seventies was the forming of Networks.

The Math/Science Network's singular force on promoting the participation of women into mathematics has been instrumental in its growth and success. It provides a setting and a framework in which people from diverse backgrounds and experiences can focus their ideas, expertise, and insights on a common goal and offer and receive assistance from each other.

Four years ago, I wrote the following about the Math/Science Network: "a new cooperative effort among women scientists, mathematicians, technicians, and educators is flourishing in the San Francisco Bay Area. Since August 1975, 54 women representing colleges and universities, school districts, and corporations have been working together to promote the participation of young women in mathematics and to encourage their entry into careers in science and technology."

The paper went on to describe the one conference we had held for 200 high school students and to detail the activities we were prepared to undertake.

It ended with a checklist for new members to indicate ways in which they would like to become involved in projected Network activities. If they were interested in producing materials, they could collect information and write copy for: a proposed career guide describing opportunities in science and technology; a handbook providing resources, information, and activities to assist math teachers in encouraging women students in mathematics.

If they wanted to provide direct services, they might: enlist workshop leaders and panelists for future conferences; develop student and educator workshops to promote awareness of career opportunities for women in science; tutor high school women in mathematics.

If they wanted to contribute to our need for financial support, they might choose to assist us in: seeking federal or private funding; helping to develop and write proposals.

Many people responded to this checklist, whch was circulated widely in the Bay Area and published in the Newsletter of the Association of Women and Mathematics, and our network grew. Four years later every one of those tasks have been accomplished or are currently underway.

The Beginnings

The history of the Network begins at a meeting held at the Lawrence Hall of Science (U.C. Berkeley) in August 1975 of educators at the elementary, secondary, and postsecondary levels, as well as researchers and students. We met to exchange information, establish contacts, and discuss research and future plans concerning women and mathematics. It became apparent at this meeting that there was a great deal of activity occurring in the Bay Area in 1975 relative to women and mathematics, and we had a desire to coordinate efforts for maximum impact.

Our first coordinated activity was a conference for high school students in which 200 young women spent the day on the Mills College campus meeting women who worked in math-based fields and being involved in math and science workshops. This one conference for 200 girls in 1976, grew to 15 in one day in 1980 for 4,000 students and 2,000 parents and teachers.

The one task that the Network continues to undertake each year is the series of conferences for young women. With the exception of the Resource Center Staff, the 15 conferences conducted in 1980 for 4,000 students were planned, conducted, and evaluated entirely by the volunteer efforts of hundreds of Network members. The benefits that accrue from this volunteer effort vary from the satisfaction of seeing 300 girls spend a day excited by ideas and people in math and science; learning about new materials or approaches for the classroom; or examining some questions about the effectiveness of intervention programs.

The structure of the Network is intended to keep the association as loose and informal as possible to foster new programs and new leadership. Currently, there are two Co-Directors: myself and Lenore Blum. We spend a large portion of our time seeking advice and guidance from Network members. We also prepare reports of Network activities, seek new funds, and provide general direction and long-range planning for the Network, based on members' suggestions. The majority of the day-to-day Networking is carried on by the coordinator of the Resource Center, Jan MacDonald Programs developed by Network members in the last few years include the EQUALS Teacher Education programs at the Lawrence Hall, the expansion of the Women in Science program at Mills, The Center for Mathematical Literacy at San Francisco State, directed by Diane Resek, "Math Without Fear" courses developed by Ruth Afflack at California State Long Beach, conferences by Sheila Humphreys for college and resuming students at Berkeley and other sites, a summer engineering program for women directed by Lee Hornberger at the University of Santa Clara, and the institutionalization of EQUAL inservice programs in the school districts of San Francisco, Emery, Novato, and Napa. Materials we've developed have included handbooks, films, and curriculum activities.

Common Approaches

In developing responses to the problems of underrepresentation of women in mathematics, Network members have concentrated on three major approaches: increasing awareness of the issues; enchancing confidence and competence in doing mathematics; and encouraging career aspirations. Whether a program serves secondary or post-secondary students, precollege teachers and administrators, or resuming women, these elements will be found in every program.

The impact of Network activities has been substantial, both in numbers of people reached and increased awareness of the issues concerning women and mathematics. A number of effective approaches have emerged that are useful at several educational levels and with students of differing educational experiences.

The Growth of the Network

Networking cannot be forced, but it can be fosterd by people who have strong but flexible leadership skills, are knowledgeable about the area in which networking is desired but are themselves seekers of new knowledge; and have an ability to encourage connections and linkages among people. Networking cannot occur simply because people have a geographic proximity to each other. They must also have an investment in exchanging ideas and resources with each other. Helping people to organize around issues that have significant and immediate meaning to them creates the possibiity for them to benefit from each other's experiences.

Several factors contribute to our Network's continued health and growth. The membership represents a cross section of occupations from business, industry, education, government, and community organizations. Access to the Network is open to anyone who is interested. Membership is free and one becomes a council member, the governing body of the Network, by attending monthly meetings. Planning and decision-making occur at these meetings, which are attended by 30-50 people, about 1/3 of whom are new at any one meeting. Committees are formed to work on projects and issues as needed, and disbanded when done.

In analyzing the qualities that define our Network, four have been pointed out to me as being evident: informality, generosity, humor, and productivity.

Informality in this case exists on a number of levels. It is a powerful nonorganization. Projects develop informally by members who form subgroups based upon common interests in the project.

Generosity is evident in the ways we encourage people to take leadership roles and believe that they can make important contributions. Members see that their professional development can be enhanced by Network participation, which benefits both the group and the individual. We can thus encourage bonds between individuals and help to reduce the tendencies to compete for scarce resources that can negate our efforts and sever Network connections.

Humor. We enjoy the process as much as the end result. While we are a diverse group, we have an ease of communication that characterizes people who welcome new ideas and interactions. The open style of discussion in our Network meetings allows us to banter ideas back and forth without a loss of self-confidence.

Finally, there is productivity. The Network provides a structure that enables people to become involved in purposeful activity that succeeds because of the wealth of talent that each individual contributes and the institutional resources that he or she brings to any cooperative venture. It is this crossing of institutional lines for collective action that makes the Math/Science Network a unique model. The alliance of people from all levels and types of education, industry, government, and community organizations is the source of our strength and substantial impact. We are eager to share our strength and multiply the impact of our network that is transforming issues into action.

REFERENCE

Sarason, S.B., and Lorentz, E. The Challenge of the Resource Exchange Network. San Francisco: Jossey-Bass, 1979, pp. 227-271.

WOMEN AND MATHEMATICS IN THE UNITED STATES: THE NEW MYTHOLOGY

Elizabeth Fennema
University of Wisconsin
Madison, Wisconsin

Ten years ago, the study of sex-related differences was a respectable, if dull, area of investigation for many psychologists, sociologists, and even a few educators. Such investigators described and documented sex differences and theorized as to why differences existed. There was little or no concern about the elimination of the differences except in the case of underachieving males. The "feminized" school was attacked as not being appropriate for boys, and many studies were done to attempt to find ways to help boys and young men to achieve at their full potential. Little, if any, of this concern was in the area of mathematics. It was accepted, almost without question, that males were better at mathematics than were females starting at least as early as kindergarten. This believe in male superiority was not only accepted as true, but also accepted as appropriate.

Within the last two decades, there has been increasing examination of the role of women in American society. A portion of that examination has been concerned with the eduational preparation of females which either enables or prohibits them from entering a variety of occupations and professions. One of the main components of this concern has been a focus on women and their learnng of mathematics. Within the last 7-8 years, a tremendous amount of money, time, and energy has been spent on various aspects concerned with women and mathematics. This current emphasis on mathematics and sex differs in at least two ways from earlier emphasis on sex-related differences: (1) the appropriateness or inevitability of differences is being strongly questioned, and (2) there is a tremendous commitment to affecting change and eliminating sex-related differences in mathematics.

There have been a variety of components of the emphasis on women and mathematics. Intervention programs at the university level have been developed. For example, the Tobias math anxiety clinic at Wesleyan University emphasizes psychological aspects, while the Blum emphasis at Mills College is on the mathematical aspects. The National Science Foundation has funded short workshops designed to encourage women returning to school to update math/science skills. Intervention programs designed to encourage high school females to continue to elect mathematics courses have a variety of foci. Some emphasize providing incentives for females to change their career aspirations by providing successful role models and information about the usefulness of mathematics. Other interventions aimed at high school students attempt to change not only the females' beliefs, but also the attitudes of significant others toward women and mathematics.

Research studies dealing with identification of variables related to women and mathematics have been another component of the women and mathematics concern. The National Science Foundation (NSF) led the way by funding a few studies in 1974, and the National Institute of Education (NIE) funded ten major studies which have just been completed. Both NIE and NSF are continuing their funding of research in this area. Private foundations are beginning to fund in this area. At least two books totally focused on women and mathematics have been published in the last few months. Another component of the concern for women and mathematics is the development of materials for preservice elementary and secondary teachers as well as materials for elementary school children. There are organizations of women mathematicians and those concerned with the education of females in mathematics.

I have been actively involved in both doing research in the area and in developing, implementing, and evaluating intervention programs designed to eliminate sex differences in mathematics, and let me share with you a question that is facing me, and causing me major concern. It has to do with whether I, and others like me, are helping females to achieve true equity in mathematics education, or whether I am helping to perpetuate the myth that there are large and non-changeable sex-related differences in mathematics. Are we, indeed, creating a new mythology of female inadequacy in the learning of mathematics? Research over the last few years has given us some knowledge. Let me share with you what I believe is known about women and mathematics in the United States.

1. There are still sex-related differences in electing to study mathematics in high schoo. While the differences are not as dramatic as was once suggested, females tend not to study, as much as do males, the most advanced mathematics courses and courses peripheral to math, such as computer

science, statistics, and physics. It appears that the size of the differences varies tremendously by school and by region of country. At the post-high-school levels, differences are still large.

2. Even when the amount of mathematics studied is controlled, females appear not to be learning math as well as males in some instances. The mathematics portion of the second National Assessment of Educational Progress (NAEP) indicated a trend that should be of concern to us all. When females excelled, it was in lower level cognitive tasks. Even when females and males reported they had been enrolled in the same mathematics courses, males performed better on more difficult and complex tasks. One positive note from NAEP II is that overall differences between females and males in achievement appear to becoming smaller.

3. There are psychological variables which may help in understanding sex-related differences. Females, as a group, more than males as a group, have less confidence in learning mathematics, believe their success in mathematics is due to other people, and perceive mathematics to be less useful to them.

4. There is some evidence that the classroom learning environment is different for females and males in a variety of ways.

5. There is some evidence, and probably will soon be more, that sex-related differences are stronger in certain specific mathematics areas such as measurement and geometry.

Let me expand a bit on this difference in geometry and measurement skills. Spatial visualization has been suggested as a reason why females are not learning mathematics as well as males. Starting at about adolescence, male superiority on tasks involving spatial visualization is found. It appears logical that the differences found in geometry and measurement learning are related to these differences in spatial visualization skills.

Currently, my colleagues and I are engaged in gathering data about how mathematics learning is dependent upon spatial visualization. Let me share with you my current thinking and hypothesizing. It appears evident to me that tasks which measure spatial visualization skills have components which can be mathematically analyzed or described. From such an examination, one could hypothesize a direct relationship between mathematics and spatial visualization. An item from the Space Relations portion of the Differential Aptitude Test requires that a two-dimensional figure be folded mentally into a three-dimensional figure. The Form Boards Test requires that rigid figures be rotated and translated to specific locations. The Cubes Test requires rotation of a three-dimensional shape. The activities required by those tests can be described as mathematical operations. Yet this set of operations is only a minute subset of mathematical ideas which must be learned, and indeed, one could go a long way in the study of mathematics without these specific ideas.

The hypothesis currently under investigation is that the critical relationship between mathematics and spatial visualization is not direct, but quite indirect. It involves the translation of words and/or mathematical symbols into a form where spatial visualization skills can be utilized.

We know that females tend to score lower than do males on spatial visualization tests. What we do not know is if females differ from males in their abiity to visualize mathematics, i.e., in the translation of mathematical ideas and problems into pictures. Also not known is if good spatial visualizers are better at these translations than are poor spatial visualizers. However, I am increasingly convinced that there is no direct causal relationship between spatial visualization skills and the learning of mathematics in a broad general sense. While I am continuing to investigate the impact of spatial visualization skills, I am less convinced than I was that spatial visualization is important in helping understand sex-related differences in the studying and learning of mathematics.

In addition, since I have been here, I have heard a number of other statements, some of which are agreed upon, but some of which are not:

1. Females prefer to learn their mathematics in classroom discussions; males prefer to work individually.

2. Classroom interactions are more important to females than they are to males.

3. Females are passive; males are active.

4. Females do well in computational tasks; males do well in problem solving tasks.

5. Female teachers teach mathematics more poorly than do male teachers.

6. Females fear success; males fear failure.

While there are major disagreements that some of these differences do exist, we could get general consensus on some of them. Since I can identify some fairly specific ways in which sex-related differences exist and variables which are quite closely related to these differences, it appears that focusing on these areas should do no harm, but can only help to eliminate the problem. Why then am I becoming concerned that a new mythology is being created?

I am concerned because, while there are still sex differences in a variety of mathematics-related variables, the magnitude of each of those differences is usually not large and appears to be diminishing. In fourth year high school mathematics classes, there is about 2:3 female to male ratio. However, under 10% of all males take fourth year mathematics. Almost as many males as females do not take fourth year mathematics. In the NAEP achievement data, the mean differences between the sexes were usually small. Obviously, many females are achieving at much higher levels than are many males. Many females are extremely confident of their ability to do mathematics. Only when the mean score of a large group of females is compared to the mean score of a large group of males are signficant differences found. Intra-sex differences are in reality much larger than inter-sex differences.

So, how have I answered my personal question about working in the area of women and mathematics. I shall continue to work in the area until equity is achieved.

Women need to learn mathematics so that many options are open to them. However, my focus in both studying the problem and working to alleviate it is going to change. Rather than focusing on females, I shall focus on those attributes which seem to be of particular importance to females' learning of mathematics. For example, I shall study, or emphasize, confidence in mathematics. In so doing, the focus is on the attribute of confidence, not the attribute of femaleness. Sex becomes an irrelevant attribute in some respects. I shall work with teachers, parents, and counselors so that they understand that mathematics is useful for all learners and that females must be included.

Let us all monitor our conversations and actions so that we do not create a new mythology that women and mathematics do not mix. Let us not say women are anxious about learning mathematics, but talk of the necessity to help all those who are math anxious. Let us continue our work and our study because the problem is far from solved. At the same time, let us not perpetuate the idea that females are inadequate in mathematics.

20.2 CONTRIBUTIONS OF WOMEN TO MATHEMATICS EDUCATION

CONTRIBUTIONS OF WOMEN TO MATHEMATICS EDUCATION

Kristina Leeb-Lundberg
City College of the
City University of New York
New York, New York

I believe in more women mathematics educators than presently exist - many more - but of a kind that are not copies of men, that are truly female in their approach, especially to the pupils. We need to find our true roles, both as men and women. We do not need to become like each other.

I am here in an effort to be impartial, to understand the connection of the true women educators who were interested in the meaning of children and humanity. A teaching position can be filled by either a man or a woman, but certainly the impulse, the nature of the teacher, carries different vibrations - - to be taught by both male and female teachers allows for compassion and reason at the same time. I really wish to emphasize children, not male or female.

Historically, Lore Rasmussen, who developed the first mathematics laboratory for elementary school children in the United States in the mid-nineteen fifties, fulfills, as an example, much of the role I envision for a woman mathematics educator who is a creative innovator: her work was conducted with a brilliant mind and the feeling of a woman. It is, in my view, only with this combination that a woman can make her true innovative contribution to mathematics education - a contribution sorely needed, particularly for elementary school children, who in most parts of the world are still being exposed to a rigid, traditionally male constructed curriculum, with little true understanding of children's psychological needs.

When I am studying women's contribution to mathematics eduation, it is this double role, of approaching mathematics teaching with your mind and your feeling, that I shall be looking for. The women who can fulfill this double role are strong, and their work must, seen over a long period of time, be truly lasting since they are working along with their role in nature. They will be able to nudge at the edge of the intellectual age of technocracy which we are suffering from and help us in the direction of a solar tomorrow where mathematics and technology will have a place as man's servant instead of his master. Technocracy is a male invention, and teachers have far too often become stooges of technocracy.

As part of an approach to the subject of women's contributions to mathematics education, I sent a questionnaire to some women innovators in this field. Edith Biggs of England, the most widely travelled and widely known woman mathematics educator of our time, answered my question, "What do you consider as your contribution to mathematics education, in regard toideas and methods?" in two parts. In the first part she states that her contribution has been:

> Working with pupils (5-18 years) to help them enjoy mathematics and to learn by means of investigation.

This short answer of Edith Biggs strikes me as simple and profound. It is practical and non-intellectual. Her answer does not even begin in mathematics - - it is about children and doing and enjoyment.

Is Edith Biggs' spontaneous answer typically female? Is she, as perhaps the leading woman mathematics educator of our time, more concerned with children than with mathematics and, above all, more concerned with children than men generally are?

My answer is yes! I have gone through a great deal of material in relation to this question, and I found that not only Miss Biggs, but the circle of the other women I have questioned, do represent this quality for which I, from a broader viewpoint was looking and searching. These women do mathematics with children, with mind, and with compassion. Edith Biggs' contribution is indeed the uniquely woman's contribution to mathematics education.

The second part of Edith Biggs' answer concerns teachers. Still, her child-centered focus is clear. She says that her contribution has been

> Helping teachers to enjoy mathematics themselves and to teach the subjects so that children: (a) understand what they are doing through firsthand experience and investigation, (b) are able to discuss with their peers and exchange ideas, (c) enjoy mathematics and appreciate its applications in the world today, and (d) create some mathematics themselves, at however humble a level.

Here we again hear Edith Biggs talk about understanding and feeling good about mathematics at the level where you are and being allowed to peek into the structure of mathematics from the bottom up, not from the top down - in fact, creating your own mathematics. This contribution, of course, also deals, as a sideline, early and efficiently with the widespread fear of and block on mathematics in both boys and girls.

Surprisingly, Miss Biggs gives credit for her innovative approach to a woman outside the field of mathematics, the art educator Marion Richardson. She opened up, in the twenties, a new and freer way of drawing and painting with children in Great Britain. Her methods were a radical departure from the rigid nineteenth century approach that had encompassed geometric drawing and copying and that gave children static impressions of art that were incompatible with their perception and reason.

We still live in this dilemma of rigid versus creative education. But we, as women, know more about how things should be for children. Yet, we do not always seem able to speak up strongly enough, or to the right people. As Madame Frederique Papy, a woman innovator from Belgium, says in her answer to my question whether she as a woman has had problems in making her voice heard: She feels it has been better heard by the teaching profession at the botton of the educational system, which finds itself in direct contact with children, than by the official commissions, which generally consist of a great majority of men who are more occupied with administration and organization than with the effective and intellectual life of children. Perhaps this is, in fact, what Edith Biggs refers to in her answer to the question whether she as a woman has had difficulties in making her voice heard. Her answer is, "No - - except when it comes to being an idealist!"

How does one defend the right kind of education for children? How does one, as a woman, defend one's views in front of a man, when one's emphasis on reasoning is not the same as his or, perhaps, not of the same type as his?

Perhaps this is so difficult because an innovative idealist sets her goals high, and others can't follow because they are not creative, nor do they put a high value on creativity. The women elementary teachers who are afraid of mathematics are not really teachers of mathematics. They *may* be trying to teach mathematics like a man and therefore hate it. A real woman cares for the child first and sees the needs of the curriculum second, while a man generally sees the needs of the curriculum first and cares for the child second. That is why we need a child-centered mathematics that begins with the child, and not with mathematics.

In the circle of experimentation and courage our search for a meaning of their life as true women in the teaching of mathematics, there are many anonymous women. The most noteworthy here are the unsentimental and dedicated primary classroom teachers in England whose ideas and common sense became the basis for the practical work of the Nuffield Mathematics Teaching Project. The male mathematicians working on this project never made any bones about the fact that their approach was mostly based on the work of female classroom teachers' experience and intuition. It is to the eternal credit of these male mathematicians that they respected and listened to these teachers, whose work is documented in the most beautifully illustrated books for teachers in the history of not only mathematics education, but also general education.

How many men really understand the implications of this child-centered approach? Are men teachers aware - as these women teachers in England are - of how, for example, a right teaching of mathematics can explode into other areas of learning and also revitalize other parts of the elementary curriculum - - that a right teaching of mathematics can be such a powerful change agent for a whole curriculum and a whole school? Do they, in general, know enough about this general curriculum, or are they content with approaching their work through a narrow and isolated view of mathematics? Women are interested in, and experienced with, the *whole child* and the *whole curriculum*, and this is *where men need to listen and learn from* gifted women, not least in the construction of the mathematics curriculum.

Women innovators' interest in the whole child has made then center around how mathematics is a transition of an experience from the real world into the minds of children, a basic concern which is not always, but often, overlooked by a too male approach. In my research, I have found that it is the details of the transition of the impressions from the real world which is the overwhelming concern of innovative women in different countries. They all have this in common. their emphasis only varies with the level they are working at. What else is, for example, Anna Zofia Krygowska's ardent interest in and search for provision for a creative, pupil-centered "mathematics in the act" on the secondary level in Poland? Or Emma Castelnuovo's discussion in Italy of the "operative concrete" and the "methodology as changer of attitudes? on the same level?

There is a great need for the growth of mathematics through speech and language as well, an expansion of a more mobile, imaginative, self-satisfying education through a more three-dimensional concept of mathematics, with language. I return to England, since it has been a center of innovative women mathematics educators for decades now, without a real parallel in any other country. Dora Wittaker's high valuation of *linguistic/mathematical interaction* has long been *particularly well formulated.* She deals the traditional textbook a formidable blow when she says that teachers have become imbued with the idea that symbolic arithmetic is the only constituent of mathematics. The present widespread "decoding" practice is time wasting, like learning a foreign language through analysis rather than speaking it, and the practice of line-by-line answers has its harmful sequel of finding the answer the "teacher wants." Linquistic expression leads to refinement of understanding. Dora Whittaker's concern with a more organic approach to mathematics learning is related to other ideas such as relatedness, the keynote of mathematical education: problem solving should be treated as an ongoing contextual activity rather than a series of unrelated puzzles. She shares Edith Biggs' recognition of the absolute need for a *sensitive classroom organization* for anyone involved *with the functioning of the whole child.* Such an innovative room provides a practical balance between activities and recorded work in mathematics and is a room where the teacher, through the ingenuity of this new type of classroom, has time to do her own assessment by observing and listening to children.

Elizabeth Williams of Britain talks about women's natural harmonizing and pacifying role, a more old-fashioned term for what now goes under the name of the affective domain. As the first president of the British

Mathematical Association who was a mother, she issued a primary school report, in 1955, which appeared at about the same time as Edith Biggs began her wider work, and which had an important influence in making materials available for children to have direct practical experience working with mathematics, thus laying a groundwork for subsequent innovations in England. And long before her, another English woman by the name of Mary Boole (widow of the initiator of Boolean algebra) had tried to reform the teaching of arithmetic and published a book about it, Lectures on the Logic of Arithmetic, in 1903; and her contemporary, Edith Somervell, had inspired teachers to offer children experiences of curve stitching.

Now, if a woman feels both the child and the mathematics, then she is closer to the problem than a man. Italy's leading woman innovator in mathematics education, Emma Castelnuovo, went already in the 1930's beyond using mathematical models for beginning secondary students' understanding of the transition from the real world into mathematics. She writes that the use of the concrete should not be reduced to showing the student a cube, a pyramid, or a cupboard. Children should, from the beginning, be plunged into mathematical situations which are right under their nose, for example, by observing shadows and their geometric properties, an image in the mirror, or something they themselves made. Such a truly "operative concrete" excites the intuition, the imagination, and therefore the mathematical activity of the student: it fosters intellectual operations in global situations leading to analysis, comparison, grouping, classifying, and synthesis. She calls it false integration to do a "little arithmetic" in chemistry or biology. True integration is integration of content, when the student discovers the same rule and the same structure in different disciplines. This, then, becomes an integration of method, a concrete mathematics directed toward understanding the analytic and synthetic method.

The relationship between Emma Castelnuovo and Anna Zofia Krygowska in Poland is clear. Madame Krygowska has, however, been an instrument of change on a large scale, not only in her own country, but through the Eastern European world and further, into the world at large, through UNESCO. No man has, since the beginning of her influence in the 1940s, had a similar influence in mathematics education in Eastern Europe. Yet, as the grand old dame of a constructive mathematics education on the secondary levels, she does not stop her questioning about the passage from the real world of pupils into mathematical language and structure. She is against the many so-called "useful" pieces of mathematical subject matter taught through endless exercises in traditional textbooks. True mathematical applications can only be learned by the students themselves, while they are going through a rich and systematic variety of processes of mathematization in unusual situations. Students need, just as professional mathematicians, to experience mathematical adventures with all their dangers, successes, and failures. Students' reasoning, imagination, and intuition must not be blocked by the school - - although a real problem is that many secondary teachers, who by habit move freely in the comfortable domain of pure mathematics, resent the feeling of uncertainty they may encounter (almost incompetence) when it comes to crossing the borderland which surrounds pure mathematics. For the secondary level, Madame Krygowska's contribution is a rare one in the history of mathematics education.

Frederique Papy's solution in Belgium to the problems younger children have in entering the land of mathematics is, again, by incorporating manipulation, color, strings, and through teaching based on "situations" with non-verbal means, a mathematical language of dots, arrows, strings, and stories. The aim is that the mathematical language the children learn through this approach will become so natural to them that they one day, as adults, will be like the character Jourdain in Le Bourgeois Gentilhomme, who one day woke up and discovered he had been speaking prose all his life. One of the common errors, says Madame Papy, is to underestimate young children's tendency towards abstraction. Children are not afraid of abstraction, she says, but some adults are, and they do children much harm by communicating their dread to children. Madame Papy - at times together with her husband, the mathematician George Papy, the creator of the minicomputer which she incorporates in her work - had influence in several European countries besides her own. Because of her five years of participation in a United States government sponsored mathematics project, CEMREL, here ideas may eventually spread also in the Americas.

The case for a reappraisal of our attitude to mathematics in relation to children has, however, perhaps not been made more eloquently than by Madeleine Goutard of France, most of whose writings have been related to the Cuisenaire rods. In a completely unconventional approach this material, she says, is to provide the mind with the experiences from which it will elaborate its own structures, in order to generate the intuitions which permit children's minds to penetrate into the structure, thus freeing them from memory and from the concrete material which they transcend. Madeleine Goutard is yet another woman who certainly has carried the vibrations of caring and feeling into the methodology of teaching mathematics. She displays an unusual interest in language and notation. There must be, says she, a time lag between action, verbalization, and writing. The first two should always be well in advance of the other two. She asks how can it be, since we never would even get the idea to attempt to teach written forms of speech before a child can speak, that in mathematics we do must this and ask the child to write, long before he has been given a chance to develop a mathematical language by using it and mastering its construction.

All these women bring me to a question: Nowhere in the intuitive, creative wish of these women innovators and master teachers do I basically find anything mechanical, and I ask myself; Does the present day use of the calculator cut out the true, intuitive, creative exchange that comes between teacher and at least the young child? I am concerned that 200, or even 500, years from now we may be judged by posterity as having made too fast decisions about computers while forgetting about the organic, and that computers are being put in at too young an age at the expense of a developing child's brain.

My observation is that the human brain is equivalent to electrical circuits in a certain way. When circuits are not used, oxidation sets in - - circuits don't work. It takes continual use and a flow of energy to keep circuits clear, for the message to flow. In a child that is getting new impressions, the cells of the brain work in the same way. Impressions work with connections from the flow of energy that relates to the flow of electricity in electric circuits. The sequence of impressions that make logic in mathematics is a gradual connecting of these circuits that gives the impression of order and reason. Leave out a set of circuits in a house, and there is a room without lights. Use a computer with a too young child: it blots out the sensitivity, and it becomes a dark room in the child's mind. If this becomes permanent, it can do permanent damage to a sequence of reasoning that encompasses mathematical order - - an order that every adult needs to find his way in life.

A woman's voice, at best, has always been an individual's. Maria Montessori was a woman individual who in the 1890's took classes in engineering before she entered medical school and became Italy's first woman doctor. She had a life-long interest in mathematics, although she eventually became a pioneer in auto-education, self-teaching through the use of sensory materials and situations. This led her to the center of children's thinking, the relationships which are the subject matter of mathematics. With her deep feeling for children she felt them and mathematics at the same time and became closer to the problem than most men of her era. Her observations of children's responses stimulated the production of apparatus which involved not only such ideas as sequency and matching, but also materials embodying the structure of arithmetic. Her bead sticks, squares, and cubes preceded the so-called Dienes' materials by some 50 years. Still, in the Montessori classrooms, a part of the standard equipment is boxes with beautifully painted puzzles, which the intiated observer can recognize as embodiments of algebraic theorems. She advocated, from her observations of children's needs, as her special road from a real world into the land of mathematics of school children, work with mathematical models, which she called "materialized abstractions." As the children worked with these sensory materials, their fears were alleviated and abstraction arrived at through "the law of least resistance and least effort." The importance of this sensory approach through the use of organic materials - - there still are no plastic mathematics materials in the Montessori schools, but they are all made of wood and beautifully finished - - is, in my view, not sufficiently researched. In modern language, Maria Montessori may have said, "We need an organic education, just like organic gardening."

And organic in mathematics means to connect numbers to reason with feeling! I feel a deep question about the interjection of calculators as a mechanical wedge into children that are too young. In Maria Montessori I see another example of a most interesting and prolific woman educator and innovator, who, if alive today, would not likely have even considered electronic computers for young children, but who would have insisted on sensitive and organic materials instead.

The subject of women and mathematics education is vast. When one searches one finds many intelligent and dedicated women teachers and innovators. I would like to dwell longer on all of them, and include more - - such as Catherine Stern, Caroline Pratt, and Lucienne Felix — women who have added their innovations and understanding to the meaning of what I call organic mathematics, intuition, feeling, listening, and patience.

My wish is to hear deeper this message from these women so that I myself can lead pupils towards the solar age with a memory that "my teacher had a heart."

20.3 THE STATUS OF WOMEN AND GIRLS IN MATHEMATICS: PROGRESS AND PROBLEMS

THE STATUS OF WOMEN AND GIRLS IN MATHEMATICS: PROGRESS AND PROBLEMS

Marjorie Carrs
University of Queensland
Australia

There is now an extensive literature available on sex differences in participation and performance in school mathematics. The most recent report of a study funded by the U.S. National Institute of Education confirmed the finding of most previous studies that at the end of primary (or elementary) education there are no differences in the achievement of boys and girls in mathematics. At the same time, large-scale studies of student performance over the past twenty years have generally reported superior performance in mathematics by males at the secondary school level, a trend which continues through adulthood. These findings, together with the significant decline in the participation of females in senior mathematics classes relative to males, must be noted with some concern. This is not a new phenomenon, but the identification of factors that contribute to the relatively low place of females in mathematics classes in the senior high school has become more critical with the increased participation of women in the work force for a longer term, together with increased demands for mathematical knowledge and skills as a prerequisite for entry to many occupations.

The decision of many girls not to pursue the study of mathematics at the senior level may result from their junior high school experience in this subject. It is in the junior high school that differences in the performance of boys and girls in mathematics begin to appear. Girls are unlikely to persist in participating in mathematics classes at senior high school when their performance level falls during the period in junior high. Even when they perform well at this level, they sometimes still move out of mathematics. In some countries, early poor performance usually means the student is guided to take a course at year ten which, because it provides an inadequate background, denies access to further study in school mathematics (e.g., CSE courses in the United Kingdom, "general mathematics: in the United States and Australia).

It is also noteworthy that differences in mathematics performance begin to appear during a period which usually coincides with the onset of puberty. Studies indicate girls reach puberty earlier than boys, yet biological, cultural, cognitive, and environmental considerations that might be taken into account in planning and organization in response to the differential growth and maturity of students appear to have little impact on school mathematics curricula. It is in the early years of junior high school that we begin to see the alienation of girls to mathematics. Much closer examination of all factors that influence student/school relationships in this period is required if intervention to increase female participation and performance in mathematics is to be effective.

As the years of junior high school encompass the years of transcience from childhood to adulthood, so they also encompass the years when the views of peers, counsellors, teachers, and other significant adults are likely to have the greatest influence on students' views, and in particular their view of mathematics. There is considerable evidence to suggest that girls are concerned about negative effects of success in mathematics. Where there is conflict between the two images of academic success or being viewed as feminine, they often choose the latter.

Studies which have examined the relationship between sex-role stereotyping and the participation of women in mathematics suggest there is a conflict for females who plan to pursue a career, between that career and the traditional role of homemaking. The subtle and hidden assumptions are often hard to identify. While the participation of women in the work force has increased markedly in most Western countries in the past thirty years, counselling both by parents and professionals does not appear to have taken cognizance of the increased demands for demonstrated mathematics performance in actual jobs (such as nursing and secretarial work) and as a basis for most areas of further (adult) education. Despite some attention to the contributions of women to the work force, they still tend to be employed in nonprofessional and low status occupations with limited opportunities for advancement and job satisfaction.

This view of sex-role is confirmed in many instances in the sex bias in problems in mathematics texts, especially at the junior high school level. Only recently have we seen the appearance of mathematics texts for primary and junior secondary children in which a conscious effort has been made, firstly, to included within problems examples of female participation in interesting activities, non-traditional occupations, and a variety of situations, and secondly, to give approximately equal time to male and female activities. The failure of the majority of mathematics text writers to do this is a subtle and pervasive reflection of a view of life roles.

Many of the careers traditionally viewed as male domains (especially science and engineering) have mathematics as a major component. The need for early identification of proposed careers in these areas so that appropriate prerequisite courses are chosen has assured male participation in school mathematics classes and helped confirm the view of mathematics as a male domain. While the study of mathematics has been perceived as useful for most traditionally male occupations, the opening of these occupations to equal female participation is unlikely unless females can be persuaded to actively pursue studies in mathematics from the beginning of secondary schooling.

At the same time, the increased demands for background knowledge and experience in mathematics in careers and occupations traditionally regarded as non-mathematical (e.g., librarian, sociologist, psychologist) must also affect a student's perception of mathematics as a useful subject for study and therefore help in developing a positive attitude towards mathematics. Even occupations which do not require a high level of proficiency in mathematics often use mathematics performance as a "filter" to those occupations.

In Australian schools, as elsewhere in the western world, the tendancy is to exacerbate the different expectations of boys and girls as they enter secondary schooling and reinforce sex-role stereotyping through differential educational experiences for males and females rather than to act as an agency for social change. Girls far more frequently than boys are offered courses in home economics, typing, and business studies, while boys are more frequently offered woodwork and technical drawing. While we can encourage equal participation of both sexes in all courses offered by a school, much more subtle factors than course offerings need to be identified if the relatively low performance of girls in mathematics is to be accounted for.

There are significant differences in the organization of primary and secondary schooling, and these affect, among other things, the quality of interactions between students and teachers and between students and curriculum materials. Language and communication in mathematics classes seem to be important factors for girls and a rich area in which to investigate both the obvious and subtle, yet pervasive, interactions that have an important influence on views of subject, of self, and of the expectations and values of others.

Personal experiences and discussions with talented girls participating in an annual summer school confirmed views that female students are more unhappy than males with teachers' explanations and the extent of discussions in mathematics classes. This general dissatisfaction with classroom verbal interaction was confirmed in a recent study of sex differences in adolescents in an Australian city in which it was reported that girls thought participation in classroom discussion more important than boys; more boys than girls were inclined to agree that teachers were good at getting their ideas across (58% male to 44% female); and 74% of girls said teachers rarely listened to students, while 43% of boys felt they did. The kind and quality of teacher-student interaction appears to be more important to girls than boys, especially in junior high school.

In Queensland, a preliminary study to examine the language of the mathematics classroom more closely was undertaken. Pairs of boys and girls of average and high ability were taped while working in mathematics classrooms, and the classroom was independently observed for teacher-student interactions. For both high and average ability groups, girls spent more time discussing the meaning of problems and puzzling over the processes involved than did boys. Girls seemed to be more concerned with understanding the problem than getting the correct answer. Bright boys had more interactions with teachers than the other three groups. These observations were made in classes where students worked at their own rate through the same text series. In traditional classrooms it is not as easy to investigate student language; hence teacher introductions to sections of content will be made and student explanations of how to do problems in that content area recorded at the end of the period of instruction.

Some paper-and-pencil tasks were used to investigate sex differences in identifying the mathematical meaning of words used in junior high school texts that also have several meanings in everyday English; in identifying the symbol or abbreviation for a mathematical concept and giving a written explanation or definition; and in identifying precise mathematical meaning from a list of synonyms. On all these tasks, girls were more successful than boys.

The study was very limited, but results indicate it is worth pursuing as there clearly are sex differences in the use of language and in the level of satisfaction with the quality of teacher/text/student communication. Explanation that facilitates understanding appeared to be more important for girls than for boys. Yet most texts (and teachers) tell how to do problems, not why you do them that way, or that they can be solved by different methods.

These subtle influences in mathematics classrooms appear to be fruitful avenues for further research and a basis for the development of effective intervention programs to encourage and assist more girls to pursue studies in mathematics at least to senior high school level. Programs are also needed for both girls and the public at large to dispel the myth that mathematics is unfeminine and to emphasize the importance of pursuing the study of mathematics so that options for career and occupational choice are kept open. As mathematics educators we need to be concerned to identify all factors that contribute to the disparity of performance and participation of males and females in mathematics and help teachers develop programs that ensure equal opportunity.

REFERENCES

Armstrong, Jane M.: Achievement and Participation of Women in Mathemics: An Overview. Denver, Colorado: Education Commission of the States, March 1980.

Fox, Lynn H., Fennema, E., & Sherman, J.: Women and Mathematics: Research Perspectives for Change. Washington, D.C., National Institute of Education, November 1977.

Fennema, Elizabeth, & Sherman, A.: Sexual Stereotyping and Mathematics Learning. Arithmetic Teacher, May 1977.

Poole, Millicent E.: La Trobe 15 to 18 Year Old Project: Sex Differences in the Response of a Sample of Melbourne Adolescents. Melbourne, VIER Bullentin, No. 42, June 1979.

Carss, Marjorie C.: Sex Bias in Junior High School Mathematics Texts. Report to the Schools Commission, 1979.

Carss, Marjorie C., & Barnes, E.: Some Perspectives on Sex Differences in the Language and Learning of Mathematics in Early Secondary School. Report to the Schools Commission, 1980.

THE STATUS OF WOMEN AND GIRLS IN MATHEMATICS: PROGRESS AND PROBLEMS

Eileen L. Poiani
Saint Peter's College
New Jersey

> Nothing can be more absurd than the practice which prevails in our country of men and womennot following the same pursuits with all their strength and with one mind, for thus the state, instead of being a whole, is reduced to a half.
> Plato
> THE DIALOGUES, (The Laws - Book 7)

Plato's quote rings true today, well over 2300 years later. If I were a pessimist, I would dwell on the fact that women still do not follow the same pursuits as men in proportionate numbers. Of the 222 million people in the United States, more than half are women. However, these women comprise barely 3% of the United States engineering force, 10% of the doctors and 3% of the dentists, 1% of the school superintendents, and hold fewer than 10% of the nation's elective offices; while they continue to capture 99% of the nursing and secretarial positions.

But being an optimist, I prefer to focus on the progress that has been made and on the action that needs to be taken to restore the nation to a whole.

We are here because we recognize that one pursuit in which women have continually fallen behind men is the study of mathematics and, in particular, in calculus. Nearly 300 years after the discovery of the calculus, sociologist Lucy Sells identified mathematics as the "critical filter" in career access for both college and non-college bound students. Without a solid background in mathematics, career doors are closed to all but a handful of occupations.

Women and minorities have traditionally been "math avoiders." Even the American response to the launching of Sputnik in 1957 did not bring women into the mathematical mainstream. The nation was mobilized for a technological revolution and the nation's youth was sensitized to attend college and seek math and science careers. However, the overwhelming majority of young women persisted in taking only the minimum high school mathematics requirement - usually one or two years. They continued to enter traditionally female occupations.

Perhaps the most frequently cited statistics - already classic - that support this claim are those uncovered on this very campus: 57% of the men entering the University of California at Berkeley in 1972 had had four full years of high school mathematics, but only 8% of the women. Thus, 92% of the entering women were not qualified to enter the mathematics track required of all majors except librarianship, elementary education, the social sciences, and humanities - all typically female career fields.

But the 70's brought a new era of social consciousness. A growing wealth of literature is now probing the reasons for this "math avoidance" - reasons ranging from the masculine stereotype of mathematics and the shape of children's toys to the phenomenon called "math anxiety" and the lack of encouragement by teachers, parents, and counselors. The mere fact that these reasons have been identified and publicized has focused public attention on the "women and math" issues. Also, the national climate is ripe for giving special attention to mathematics because poor performance of students in "basic skills" has led to a national outcry for better preparation.

These factors have indeed brought progress in improving the participation of women in mathematics. I will mention other evidence of progress and then comment on something that challenges every mathematician - namely, unsolved problems.

A 1980 Census Bureau report compared college majors in 1978 with those in 1966. Four to six times as many women were majoring in the traditionally male-dominated fields of business and engineering in 1978 than in 1966. The shift is slow but sure.

A national study conducted earlier this year by the Education Commission of the States also gives cause for optimism. Based on a survey of about 1500 thirteen-year-olds and 1800 high school seniors, results show that sex differences in participation in mathematics courses have diminished noticeably in recent years (41% men, 37% women had 4 years of math). In terms of achievement, the thirteen-year-old females began high school with at least the same mathematical abilities as males and were even better in computation and spatial visualization. Unfortunately, by senior year the males surpassed them in problem-solving and were on a par in spatial and computational areas.

Previous studies also confirmed that young women often outperformed men in mathematics tests at the elementary and junior highlevels, but are outdistanced by young men in high school. In fact, national data shows that sex difference in the College Board Mathematics Scholastic Aptitude Test scores still hovers around 50. The gap is narrowing slightly each year with 1980 senior women scoring 48 points less than their male counterparts.

High marks go to several new programs which were developed in the 70's to help reverse the female "math avoidance" trend and to thereby prevent the decline in mathematics achievement. These programs will be discussed at various sessions of the Congress. The one which I direct is called WAM: Women and Mathematics.

WAM is a national lectureship program established in 1975 by the Mathematical Association of America under a grant from IBM. Presentations are aimed at ninth and tenth graders as well as at those who influence course and career choice - including parents, teachers at all levels, counselors, curriculum designers, administrators, and legislative leaders.

The idea for WAM was sparked when IBM representatives hosted a reception for high achievers on the U.S.A. Mathematics Olympiad, and they noticed that there were no women among the winners. The U.S.A. Olympiad is an annual contest for invited high school students who have excelled in previous mathematics competitions. From the high achievers on the U.S.A. Olympiad, a team is chosen to participate in the summer International Olympiad. Two young women did participate for the first time in the training sessions in 1978, although none has yet made the team.

The absence of women from the Olympiad symbolized their absence from a host of fields requiring sound mathematical preparation. To encourage ninth and tenth graders to elect more than the minimum requirement and to keep their career doors open, IBM agreed to support the WAM secondary school lectureship program. Other corporations have also joined with IBM to contribute to the support of WAM, including the John Hancock Mutual Life Insurance Company and the Polaroid Foundation.

WAM speakers are all women and are drawn from a variety of fields which use mathematics both directly and indirectly. The emphasis is on the usefulness of mathematics - be it for the practicality of taking calculus in a pre-med or accounting major or for the day-to-day demands for mathematics in engineering, physics, economics and computer science. Speakers in a myriad of occupational roles represent the worlds of business, medicine, academe, research, industry, government, and skilled trades.

By conservative estimates, WAM has reached more than 54,000 students, 660 schools, and 6,400 parents, teachers, and counselors since it began in the Fall of 1975. WAM has active regions in New York/New Jersey, Connecticut, Boston, Chicago, San Francisco Bay, Southern California, Oregon, South Florida, and Seattle, Washington. Plans are underway for expansion to other areas of the country.

A typical WAM school visit consists of the following elements: a formal talk on how the speaker uses mathematics or on a mathematical application; an informal discussion "for women only;" and a conversation with teachers and counselors.

Since schedule conflicts frequently prevented counselors from participating in school visits, WAM designed a one day conference called MODE: Math Opens Doors Everywhere, especially for them. Such conferences have been held in New Jersey, Chicago, and Portland. Faculty in-service workshops, career days, professional meetings, parents nights - all are events which WAM services to increase public awareness of the importance of mathematics.

The program undergoes continual scrutiny with tripartite evaluations by the hosts, the students, and the speaker herself. A recently completed doctoral dissertation analyzed the short-term effects of WAM by pre-testing all tenth graders at sample schools, then presenting a WAM talk to half the students, and finally post-testing. Conclusions show that sophomores in urban schools who had participated in WAM were more likely to plan to take more and more appropriate mathematics than those who had not. Furthermore, those sophomores who heard the WAM talk were more likely to perceive the career usefulness of taking more than the minimum mathematics requirement than those who had not. Some high schools have attributed to WAM an increase in the number of their female students who are taking advanced mathematics courses.

A logical corollary to the success of programs like WAM is that they should put themselves out of business. Such programs have the goal of instilling a commitment on the part of school authorities and parents, in particular, and society, in general, to encourage proportionate participation of men and women in mathematics and consequently in the pursuits which rely on mathematics.

But before we go out of business, we must tackle a number of unsolved problems. Here are a few:

1. Women still fear success: men fear failure. One of the most frequent questions asked in informal WAM sessions is: "If I excel in math or science, the boys won't date me. What should I do?"

2. Women's colleges have often provided supportive environments for women in mathematics and science. Their number is dramatically dwindling. New support systems need to be developed.

3. Public confusion and misunderstanding of the goals of the so-called "women's movement" have led to resentment, indifference, and even a backlash against any program designated "for women."

4. Influential decision-makers continue to be mostly men. In a period of inflation, budget cuts, and unemployment, the purse-string holders are inclined to cut back on special interest programs which may be perceived as frills. Limited resources signal the need for greater cooperation and networks.

5. The public image of mathematicians and mathematics persists in being negative. Attitudes are slow to change, but like it or not, mathematics opens career doors, so the improvement of public perceptions of mathematics remains a top priority.

Simplistic as it sounds, we live in a technological age. The demand for advanced technology in the next two decades and throughout the twenty-first century is predicted to grow exponentially. to meet that demand we can no longer tolerate, as Plato said, "state . . . reduced to a half."

Indeed we have made progress - significant progress - in bringing American women into the mathematical mainstream. But, to update the words of American poet, Robert Frost, from the poem "Stopping by Woods on a Snowy Evening":

> "We have promises to keep,
> And (kilometers) to go before we sleep,
> And (kilometers) to go before we sleep."

THE HUMANNESS OF MATHEMATICS

Nancy Shelley
Canberra College of T.A.F.E.
Australia

Other participants have led us through the problems they have encountered, and the progress they have made, in the field of women and girls in mathematics. I am going to take you on a different journey, borrowing more from the skills of the poet than those of the social scientific analyst. I invite you to share with me the type of understanding that I attempt with the adult women I teach.

As any good mathematician knows, it is the premises we start with which set the stage, and which, largely determine our outcomes. So, we must take care with our premises! Where do you start from?

I started by taking a walk in the park, and found myself delighted with the rose-pink-breasted galahs which swept against the cloudless sky of an autumn day. They came to rest in the branches of a leafless tree, they grey backs merging into the anonymity of the branches; their task done. I pondered upon that delight. An everyday path; no particular spring in my gait; and then, unforeseen, my spirit soaring with that flock of birds and the possibility they offered. It was limitless: calling on my willingness to expand at the prospect. With quick intake of breath, I followed the flight and awaited its outcome. It came in a harmony of natural orderliness, and, joy in my heart--the uphill tread now lighter, inexplicably affected by an ordinary event offering its beauty. It grew cold and I was forced to return to the warmth of my study to record that experience and reflect upon where it led me. It led to this my first premise: the opportunity for the imagination to soar in limitless vein with the experience of an exhilarating touch-down.

The title of the session will serve to articulate my second premise: The Status of Women and Girls in Mathematics: Progress and Problems. Let me not forget that my assignment under that heading is to incorporate something of my interest in a theoretical model within which to study the factors which prevent women from achieving in mathematics. That's a task!

Now would you believe, that while I was writing that last paragraph, at separate intervals, three different aircraft from the Fairbairn Air Base flew overhead, piercing the solitude of my study and the sky with an ear-splitting presence? They have demanded their place among these premises of mine. (Three more have come and gone as I write of them! It is clearly a morning of maneauvres.)

Before I leave the list, another premise needs stating. It is perhaps the hardest of all. Taking its strength from the bones of mathematics itself, its face is masked by its presence in the very marrow of our being, so that we perceive it not, and ignore, nay, will deny, that it resides there. Having enmeshed itself within that attribute with which mathematics most prides itself, this premise lurks in the alleys of mathematicians' minds and derives its power by being submerged. When from the 'objectivity' of that god-like realm I dare to wrest this premise and whisper its name, I shall tread upon the susceptibilities of those most affected by its interpenetration and so anticipate their scorn. The premise is none other than the politics of mathematics and its learning.

There you have them: four. Have you decided upon yours yet? I'll spell mine out: The flight of the spirit, the status of women in mathematics, the manoeuvres of aeroplanes, and the politics of mathematics and its learning.

Now we must look to the rules of the game. Those rules should employ a logic which will, in turn, permeate the whole structure we erect. So, when we stand aside to admire our completed handiwork, I shall hope to do so, not only with pride, but with job, remembering that pride entails the backward glance at the task accomplished, while joy holds the potential of life in its experience. So, the rules of my game are those which incorporate and propagate life-giving properties.

Next, the consistency I shall expect, and have you carefully monitor, will be that which adheres rigorously to the life-experience of persons and harnesses that experience solely to extend and expand it.

One further aspect we should think of is the hidden expectation which we cherish, and in the best circles ought not to disclose, for fear of laying ourselves open to the charge of being influenced in our pursuit by that end we hoped for. Since we all possess an expectation, I think it only right and proper to proclaim it. You may confess it to yourself, I shall tell mine aloud, that you may know my secret from the start.

Are you hoping for a system which is complete and entire unto itself? Or will you join me in looking for one that's open and fluid, a system which not only admits of change, one which does alter when it alteration finds, and one which thrives for its lifeblood upon the creativity and ingenuity of human experience?

So, there you have it: I have declared my premises. I have stated the rules by which I shall operate, I have defined what I take to be consistency, and I have admitted to what system I hope will be built. Now let us proceed with the architecture.

Beginning with the second premise, recall it was: the status of women and girls in mathematics, both progress and problems, and from the confines of my stated ground, what phoenix doth arise to challenge and enlighten us?

First, women: no difficulty there, for woman necessarily incorporates and propagates life-giving properties. That is, as long as she affirms herself, and is affirmed, as woman. Now you may tell me that except she become as men, she shall not enter the kingdom of mathematics. And I will answer: that of such is the kingdom of darkness; and you will instantly revealed one of the problems relating to the status of women in mathematics, and one of the chief barriers to their progress in it.

If a woman, an ordinary one, needs to deny her experience as woman, and put on the mantle of pre-characterized man, in order to understand the complexities of that kingdom, she must falter at the denial of herself as person and at the irrelevance of her life experience. She must doubt, too, that the kingdom, as you depict it, propagates and incorporates any life-giving theses. What status would she find there? For us, no apeing of man can be countenanced, for if she would deny herself, she stultifies her life, and in my thesis has no place.

What of mathematics? This is less direct. Here, my secret expectation for an open system begins to have its influence. An analogy may serve to point it up.

Consider some of the maginificent Greek statues carved in marble, where the form is enclosed within the shape selected by the sculptor, and therefore, taken by the stone. There is no question of beauty, but what of the system it embodies: It is a completed one and exists within closed boundaries. In contrast, a spring of water arising from the ground makes for itself a shape as it first ascends and then falls to the rocks beneath. The drops of water forming the shape are not static, as are the atoms of marble, and indeed, can be said to pass through the shape. This is an open system.

If one moves one's hand through the water's shape it demonstrates the power to restore itself imemdiately. No so the marble. And here is highlighted a very important difference: the closed system does not hold within itself the power to restore its grandeur and beauty: witness the chipping of a statue with time and the loss of some of its features. The open system on the other hand, contains within itself the power of renewal.

How do you see mathematics? Open or closed?

Before I leave the analogy, if I should try to move myhand through the marble of the closed system, it is certain I shall be bruised; I may not recover. Many women will witness to a similar experience when brought in contact with a closed system which their initiators called mathematics.

But what phoenix arises out of the juxtaposition of the two words **women** and **mathematics**? Those two words thus incorporated, in title and in premise, announce that men have strode the domain too long alone. Yet men have delighted in expression of their natural craetivity and had much say in the formation of the beauty, richness, power that there abounds. In this respect they're players within my rules.

Alas, it is not all. For some have sought to compass that domain with fences insurmountable, observing not how thus they moved it from its openness to a system closed in aspect and in access; how prohibitions on the entry affects alike its beauty and the aspirations of those humans it encounters - - both those within and those excluded. It once the fluid, life-generating power which permeates a domain requiring creativity for its maintenance is transformed to one of stone. And the austere beauty of a monolith inspires two forms of action: either to attempt to scale it, or to stand in awe, remaining at its feet.

This picture serves to illustrate how humans are affected in this change. These men have turned themselves, and others who would scale, into conquerors, with power to wield over new initiates and those who opt for different terms of entry. And all at once, I call upon both fourth and third premises to crystallize the scene.

The fourth enunciated the politics of mathematics and its learning. For when one says that some are chose, others fall - - no matter how respectable, or learned, or hopefully substantiated, those claims and justifications - - one makes a political statement and we must recognize it is so. It remains, of course, for one to say and live by it, but not to deny its nature.

It often is the case, those self-same men, ignoring how creative, intuitive, and inspirational in origin was that art they have transformed, regard only the form which later came: arranged, and placed in frigid, pristine form, exhibiting its grandeur through constancy, logic, and consistency. With this static aspect they then dictate the manner of the learning, to leave the goddess in its unscathed form.

'Tis here the politics again is clear. For learning incorporates within its acquisition the power to handle the world as we encounter it. Indeed, it is the perceived guarantee of that power which ensures that learning will take place, and determines the manner of it. We are able to depart from this basis of our learning only when there is a more than reasonable expectation that should we take on the structures of another's world view, a high possibilityresides, indeed a probability, that we shall gain access to that other's power. If that expectation is not there, as is the cast for most of womankind, it is an affirmation of her humanness when she rejects seducing overtures for embarkation, even though she reaps the tag "unable", and walks the road to powerlessness.

And what of premise three? Recall first how those men, imposing a closed, pristine image upon that erstwhile open system, became conquerors in their own eyes, demanding that intent of all who wished to know and share their power. In that same act, they set in motion so many of the attributes which men feel bound to, in order to establish their "maleness", thus defined.

My premise spoke of planes manoeuvring overhead and is the symbol of the war game, which that same "maleness", thus defined, inevitably is heir to. Let me but signpost it that you may glimpse the way.

It moves through logic, consistency, and deduction - - forgetting their secondary nature - - to give them primacy. The world is classed in terms "mechanical"; the tools are named "objective"; all life becomes subjected to analysis - - those parts resisting being defined out of existence; what once appeared one mode to view the world, gives way to be the only mode.

That "only mode" proclaims rigidity in cast of mind and ultimately demands one way to think. The power of mind is judged according to one's aptitude, proficiency, and speed of application. A gnawing doubt that life holds more is treated as a weakness to be quelled. To stay on top is what's required to prove just what? That we posses what's needed to excel in this mastering race?

The system's closed, as is the mind, and of its nature must decay. Once creativity gives way to maintenance of place, limitation upon access results, ambition dictates, "And Art's made tongue-tide by Authority.

Possessors of the knowledge are self-proclaimed protectors of the realm and live accordingly. Their skills become enlisted to that state which guards their place, defends their cause. Both place and cause become identified with an outlook that requires a structure of defense, while fear and vulnerability cement the thrust to war.

These men, obsessed with both rigidity and power, no longer know the attributes of humanness, nor how to respond to inspiration or creativeness within themselves or others. This power and its obsession are rooted in a mode of thinking which admits no rivals, is embedded in the myth of objectivity, incarcerated deep within a monolith. For some, the monolith's inscribed, **mathematics**, and their attitudes descend directly from my premise three: the war game.

How comes it then as premise in my thesis which designated rules pertaining to properties of life, and desired consistency in terms expanding the life experience of persons? It has its place because it is a fact we are immersed in, and needs must come to terms with.

Clearly, we must acknowledge a contradition and move to resolution. Here too, we rest upon the fourth premise: since choice for one to be involved in either war or peace is also choice political.

For me, the resolution lies in premise one with rules applied, consistency maintained, and system-hoped-for kept before our eyes. In it we'll find the progress both of women and of mathematics, the two entwined, if such we seek. Let me repeat: An everyday path; no particular spring in my gait; and then unforeseen, my spirit soaring with that flock of birds and the possibility they offered. It was limitless: calling on my willingness to expand at the prospect. With quick intake of breath, I followed the flight and awaited its outcome. It came in a harmony of natural orderliness, and, joy in my heart - the uphill tread now lighter, inexplicably affected by an ordinary event offering its beauty.

For in that premise is embodied what one needs to link the art of mathematics with one's life, to keep alive its inspiration and one's own, and in such openness to find a joy empowering all who tread the daily path uphill. For women, then, the progress must be sought in first the nature of mathematics and then attributes of humanness which can be utilized, enhancing the persons and the art, in learning and in use.

1. I am indebted to Jonathan Miller in his B.B.C. television programme, "The Body in Question", for this example.

2. Shakespeare, William. Sonnet LXVI.

THE STATUS OF WOMEN AND GIRLS IN MATHEMATICS: PROGRESS AND PROBLEMS

Dora Helen Skypek
Emory University
Atlanta, Georgia, U.S.A.

Many of us here are affiliated with schools or departments of teacher education. As specialists in mathematics education we are seldom involved, however, in the more general education program. My own courses in the professional training sequence at Emory make up 8% of the undergraduate requirements and 8 to 23% of the graduate requirements for elementary or secondary school teachers of mathematics. I am increasingly convinced that to ignore that other 75 to 90 percent of the teacher training experience is to ignore a significant source of problems and a significant resource for progress in the mathematics education of girls and women.

Consider a recent analysis (1) of 24 teacher education textbooks that are widely used in the USA. The textbooks, with copyright dates 1973 to 1978, are in the areas of educational psychology, foundations of education, and methods in teaching elementary school mathematics, science, social studies, reading and language arts. The findings were these:

> Twenty-three of the 24 books give less than 1% of book space to the issue of sex equity. (The one exception gives 1.7% of its content to discussions of the issue.)
>
> Eight of the 24 books make no mention of the issue at all.
>
> Two of those eight are survey or introduction to education texts. The two other introductory or foundational texts devote less than 1/2 of one percent to the issue of sexism in education. None of the four tells the history of women in American

education; the contributions of women to education are ignored.

The three methods books in mathematics completely ignore the issue of sex equity. Only one of the three science methods books takes note of the problem. It devotes six sentences to the issue in a section entitled "A Special Handicap." I quote the first two sentences:

"Girls at all socioeconomic levels act with respect to science as though they were handicapped. They know less, do less, explore less and are prone to be more superstitious than boys."

There is no diagnosis of the special handicap and no discussion of how to treat it.

In the books that attend to sex differences, the treatments are for the most part stereotypic, even discriminatory. For example a methods book in the language arts advises the teacher that, while girls will read books about boys, boys will not read about girls. Therefore, in the classroom library collections, the ratio of "boy books" to "girl books" should be about two to one.

It is not surprising that students of these materials, the graduates of our teacher education programs, continue to reinforce biased attitudes and behaviors that erode the self-esteem of girls, that inhibit their independence and that limit their aspirations.

Studies in achievement motivation and attribution confirm the critical influence of teachers on student attitudes and behaviors related to educational and career aspirations. For instance, investigators have found that boys attribute their successes in mathematics to ability and their failures to luck or to not trying hard enough. For girls, the pattern of attribution is reversed. They attribute their successes in mathematics to hard work or luck and the failures to lack of ability. The usual result is that when boys are faced with difficult tasks, they assume that they have the ability to do them, that they just have to try harder. Girls often opt out of the difficult tasks when given the chance.

Carol Dweck (2), co-investigator in several studies involving 4th and 5th graders, explains these sex differences in terms of teacher evaluations. The effect of negative feedback is of particular interest. She writes that boys are given negative feedback for a variety of reasons - - their social behavior, their dress, their lack of effort, their academic behavior. On the other hand, girls, usually well-behaved, properly dressed, and assumed to be working at full potential, receive negative feedback for rather specific reasons, usually having to do with the intellectual quality of their work. Boys come to disregard negative feedback because it is so diffuse, while girls attach high regard to negative feedback because it is so specific. Carol Jacklin (3) summarized both negative and positive teacher evaluations this way: Boys receive more negative reinforcements for social behaviors and more positive reinforcements for academic behaviors. Girls receive more positive reinforcements for social behaviors and more negative reinforcements for academic behaviors.

Dweck points out that attitudes students acquire about themselves and their abilities are crucial in junior and senior high school when academic achievement involving new skills comes into conflict with academic demands. Mathematics, in particular, involves new units of conceptualization and information as one moves beyond elementary school arithmetic. Increasingly more specific and differentiated areas must be learned. At each new level a large dose of persistence is called for. Boys have in general learned to be persisters. Many girls have not.

Still other studies suggest that teachers at every grade level from kindergarten on are frequently and unwittingly sexist in their interactions with students. Teachers too have been taught the attitudes and beliefs that govern their behaviors. Their own educational experiences have not challenged the general invisibility of women or the stereotypic treatment of the sexes. Teachers, the majority of whom are working women, even engage in the paradoxical behavior of denigrating working women. They (we) frequently attribute a student's deviant behavior to the fact that his or her mother works - - the student doesn't do his homework, lacks motivation, is disruptive, etc. because his mother isn't at home. She works. Father works too, of course, but that fact has nothing to do with it. The message to both girls and boys is that mothers are not supposed to work. It follows then that if women are not supposed to work, the do not need to persist, to learn independence, to take hard courses, to keep educational and career options open.

These are the problems. The progress to date is minimal. The most promising effort I know in the arena of general teacher education is the set of materials (5) developed by the Non-Sexist Teacher Education Project, directed by Sadkers, and field-tested last year at ten teacher education institutions in the USA. The revised materials, currently in press, consist of six modules designed to fill the gap in current programs. One module provides a history of women's contributions to American education; another looks at the research on sex differences and similarities in school achievement. Still another examines research on sex bias in teacher expectations and interaction patterns. All of them suggest discussion topics and activities that help teachers to counteract the effect of bias in their own classrooms.

Those of us in mathematics education must, of course, continue our efforts and advances in our special fields, but let us also keep in mind that a large part of our problem and the potential for significant progress are rooted in general education - - in particular in the general education of teachers.

References

Myra and David Sadker. Beyond Pictures and Pronouns: Sexism in Teacher Education. Published and disseminated by the Women's Educational Equity Act Program, U.S. Department of Health, Education and Welfare. Available from Education Development Center, 55 Chapel Street, Newton, Massachusetts 02160.

Carol Dweck, William Davidson, Sharon Nelson and Bradley Enna. Sex Differences in Learned Helplessness: II. The Contingencies of Evaluative Feedback in the Classroom, and III. An Experimental Analysis. Developmental Psychology, 14, no. 3 (1978): 268-276.

Carol Jacklin, In keynote address at the annual meeting of the Research Council on the Diagnostic and Prescriptive Teaching of Mathematics, Tampa, Florida, April 1979.

Non-Sexist Teacher Education Project Materials. For information on availablility of the modules, write: Drs. David and Myra Sadker, School of Education, The American University, Washington, D.C. 20016. The titles of the modules are: Sexism in American Education; The Impact of Women on American Education; Boys and Girls in School: A Psychological Perspective; Between Teacher and Student: Overcoming Sex Bias in the Classroom; Beyond the Dick and Jane Syndrome: Confronting Sex Bias in Instructional Materials; and Promoting Sex Equity in School Organizations.

20.4 SPECIAL PROBLEMS OF WOMEN IN MATHEMATICS

SPECIAL PROBLEMS OF WOMEN IN MATHEMATICS

Erika Schildkamp-Kundiger
University of Saarbrucken
West Germany

1. Introduction

"Mathematics and gender" is a topic whose relevance is of increasing importance. This is especially true for the U.S.A., where a vivid discussion is going on in newspapers and magazines; moreover, there are many research projects, of which extensive reviews are available. Some of the most recent reviews are by Fox, Fennema, and Sherman (1977) and Ernest (1976). Where does this interest come from? The reason is not - - as might be hypothesized - - that the problems of girls in mathematics have become greater and greater, thus increasing the necessity to be engaged in this topic. On the contrary: if there have been changes with time, the problems of girls in mathematics seem to have diminished. The interest in "mathematics and gender" has increased remarkably just at the moment when the promotion of sex equity for students and professionals has become a central public demand. When mathematical competence was identified as a critical skill directly related to admission to many colleges and most professional occupations, in both technical and non-technical fields (Sells, 1976), the National Institute of Education (USA) in 1978 started a research grant program aimed at understanding more about sex differences in mathematical achievement.

Interest in this topic increased not only in the U.S.A. but world-wide. This can readily be recognized by the fact that at the ICME IV (in 1980) there are several activities in this area, whereas at the ICME III (in 1976) there were none.

Some selected aspects of the topic will be considered in the following exposition.

2. The Appearance of Sex-Related Achievement Differences

The situation can be summarized as follows: Differences begin to appear around adolescence and then continue into adulthood. If sex-related differences in mathematical achievement tests appear, they often are in favor of boys, although mathematics marks need not necessarily differ between the sexes and even favor girls sometimes (Aiken, 1976; Astin, 1974; Robitaille & Sherrill, 1979; NAEP, 1975). Among those pupils good in mathematics, more males than females enroll in advanced courses; moreover, Fennema (1977) suggests that sex differences in mathematics achievement are not as great today as they were in earlier years.

Differences in favor of males seem to appear not only in America but in many countries (Husen, 1967; Keeves, 1973). But these differences vary considerably from country to country. Concerning the extent to which the achievement of boys exceeds that of girls, Keeves (1973) summarizes:

The range of differences across countries is to great for a simple explanation to be advanced as to why such sex differences should have been observed. Furthermore, no combination of factors, operating together, comes readily to mind. Nevertheless, it is clear from this evidence that girls tend to be less well prepared to enter occupations and careers that require a prior knowledge of mathematics and science. To this extent inequalities between males and females are built into educational systems.

3. Explanations for Sex-Related Achievement Differences

In the following, some central aspects which have been investigated will be considered. They will be summarized in three groups according to the groups of variables that are looked at to explain behavior. The three groups can be characterized in a simplified manner as follows:

a. Achievement differences are explained by sex-role perceptions.

b. Achievement differences are explained by differences in personality traits.

c. Achievement differences are the result of a differentiated interaction process of environment and personality variables between which cognitions bridge the gap.

3a. Sex-Role Perceptions. In Western culture mathematics and natural sciences are widely looked upon as being male domains. Newer studies suggest that the sex-typing of mathematics is not as prevalent as it once was (Fox, 1977; Schildkamp-Kundiger, 1973), but the evidence shows that male prejudice against girls' mathematical engagement still exists, or at least that girls believe it to exist and have accepted the stereotype that mathematics is a male pursuit (Fennema & Sherman, 1977; Tobias, 1978; Westoff, 1979). The opinion that girls are not as able in mathematics as boys, or that they should not waste their time by choosing advanced mathematics courses, is held by school counselors - - male and female - - more often than might be thought (Casserly, 1979a).

To explain the learning of the sex role, all general psychologicaltheories explaining change of performance can be considered: Psychoanalytic theories based on Freud; stimulus-response theories; social learning and modeling; and cognitive learning theories. These approaches not only give models of how sex roles are learned, but also try to explain what has happened when not-sex-appropriate behavior appears, e.g., interests and achievement in mathematics. These aspects will not be considered in further detail here (see Maccoby & Jacklin, 1974; Fox, 1977; Schildkamp-Kundiger, 1980).

In speaking of sex role, one might forget that sex role is not a set of well-defined habits that govern the behavior of an individual and the behavior expectations of others; sex role is a concept, the perception of which differs not only between persons but also within a person depending on situations and personal development. Misconceptions may lead to using the concept "the sex role" quickly as a global explanation for sex-related differences, without trying to get a deeper understanding, e.g., girls are not interested in mathematics because mathematics belongs to the male area, and so they might be right.

There is great empirical evidence supporting the hypothesis that the sex typing of mathematics is related to achievement differences in such a way that the linking of mathematics to the male domain can explain a great deal of any inferior achievements and engagement of women in mathematics. The following is a small selection of results based on empirical research.

Parallels between the learning of one's sex role and the appearance of achievement differences are found; differences in favor of men mostly begin to appear around adolescence and then continue into adulthood (Stein, 1977). Emmerich (reported in Westoff, 1979, p. 11 f.) found that certain sex stereotypes appear to be breaking down, and corresponding to this, achievment differences seem to diminish (Fennema, 1977 .

The degree to which a girl has adopted the concept that mathematics is a male domain significantly depends on the sex-role conceptions of her environment. Casserly (1979a) and Fox (1977) extensively report studies that demonstrate the influence of parents, peers, and teachers. Even in textbooks and in tests this bias is found (Rogers, 1975; Glotzer, 1979). When a child perceives a school subject as appropriate for his or her own sex, then this influences most the achievement in this subject, more than liking the subject or even whether the child is male or female (Dwyer, 1974). Obviously, what is looked upon as sex-role-education behavior is of great relevance in our context - - especially sex-role-adequate achievement behavior.

Not only do sex-role perceptions differ from country to country and between individuals, as already mentioned, but there is an individual development over time. By entering new situations people develop variations of their sex roles as they grow older. If a woman enters an area looked upon as a male domain, she may redefine her activity as suitable for the female role (Stochard & Johnson, 1980). There seem to be some developments that have loosened the linkage male/mathematics. Senechal (1974) extensively discusses that through this process not only sex roles are altered but also the connotative meaning of mathematics.

The question occurs whether mathematics has always been looked upon as being related to the male domain. Perl (1979) tires to get an answer to this question by analyzing the Ladies Diary, one of the first popular magazines that appeared in England from 1704 to1941. She concludes that in the late seventeenth and eighteenth centuries women were not considered less capable of learning mathematics than men. In the eighteenth through the nineteenth centuries the attitudes towards women changed; they were looked upon as weaker, less capable of engaging in intellectual areas like mathematics. Perl sees the stereotype of mathematics as a male domain as an effect of changing social roles in the eighteenth century accompanied by a growth of mathematical knowledge that for understanding necessarily presupposes an intensive training.

3b. Personality Traits Starting from factorial models of intelligence, it seems reasonable to try to find intelligence factors relevant to mathematics, and after that to look and see if differences in achievement can be explained by differences in these factors (Werdelin, 1958; Very, 1967; Aiken, 1971; Treumann, 1974). Problems arise, as there obviously do not exist two or three well-defined factors totally explaining achievement in mathematics. The major intellectual factor related to mathematics learning is general intelligence. This does not help in explaining sex-related differences in mathematics performance (Armstrong, 1975).

Sex-related differences appear relatively consistently in spatial visualization and verbal ability - - the former in favor of men, the latter in favor of women. Verbal ability correlates hghly with mathematical performance (Aiken, 1971), and if relations are found between spatial visualization and mathematical performance, they are positive too - - although there are some authors who neglect relationships between the last two variables (see Very, 1967, p. 171; Smith, 1964). Thus, the results lead to an inconsistent picture.

In the last ten years the relevance of cognitive style variables for achievement in mathematics has been considered (see Robinson & Gray, 1974; Radatz, 1976). Compared with the more static factorial models of intelligence, the cognitive style construct is more dynamic, allowing hypotheses about learning processes. Fennema (1977) supposes that especially the spatial-verbal cognitive style dimension will help in understanding sex-related differences. Beside a possible better understanding of sex-related differences, research on the relevance of cognitive styles will lead to a greater knowledge of learning processes in mathematics in general, and this is an essential demand considering not only that girls have problems, but that mathematics per se is a difficult subject to learn.

In relation to achievement differences attitudes have been of interest. For this, the first IEA Study gives a good survey (Husen, 1967; Keeves, 1973). In most countries and age levels boys show a greater interest in mathematics than girls (Keeves, 1973, p. 59). Moreover, an analysis of variance shows that there is a significant effect between countries and between sexes and a significant interaction for the lower secondary school populations and for the mathematics specialists (Husen, 1967, Vol. II, p. 245). The common concept that ability on the one side and attitudes and interests on the other side explain together a great deal of students' achievement does not, in this generality, resist empirical check (Suydam, 1975). Besides a better definition of the attitudes considered (e.g., liking mathematics, believing it is important), and a more elaborated theoritical framework is needed linking attitudes and achievements.

The perceived usefulness of mathematics is one aspect that may help to understand differences in achievement and in course-taking behavior. Empirical research has found that sex-related achievement differences only appear when there are differences concerning this attitude (Fennema & Sherman, 1977; Hilton & Berglund, 1974). Moreover, this attitude was a good predictor of course-taking behavior in mathematics (Haven, 1972). In relation to the achievement motive this phenomenon becomes understandable. Only if an area is achievement relevant for an individual will he or she make efforts toreach an achievement goal. Achievement striving is embedded in overarching projects, e.g., the desired profession. Special sex-role perception can cause a girl to perceive mathematics as a non-achievement-relevant area (Fox, 1977).

The relevance of self-esteem of one's mathematical abilities and math anxiety for achievement and achievement differences becomes understandable in a broader theoretical context, in the center of which is a cognitive view of the achievement motive. This will be considered next.

3c. Differentiated Interaction Processes Research about the learning of mathematics and its sex-typing is strictly related to areas considered with the achievement motive. During their lessons pupils have to solve tasks (achievement situation) that, especially in mathematics, are looked upon as intellectual problems. The relevance of intellectual achievement means that failure may burden the pupil as not having reached the standard of performance the school and the society have set. A young girl sees herself confronted, on the one side, with the intellectual achievement norms of the school and, on the other side, with sex-role perceptions that may be in contrast to the school norms. Heilbrun (1963) could verify that some female students already perceive the general intellectual demands of a college as being incompatible with their female sex role. As a consequence of the sex-typing of mathematics, mathematically gifted girls may fear negative consequences for their relationships with boys when accelerating their progress in mathematics (Casserly, 1975; Fox, 1976; Fennema & Sherman, 1977).

Stein and Bailey (1973) discuss studies concerned with the achievement motive of women. As forthe effects of sex role on female achievement striving, there is empirical evidence that the attainment value for a given area of achievement is a good predictor both of female efforts as well as of their performance. Schildkamp-Kundiger (1974) found that girls having lower achievements in mathematics than would be predicted on the basis of their intelligence (underachievers) see intellectual achievement as appropriate for men only; whereas female mathematical overachievers believe intellectual achievement to be suitable for women too. The persistence a student develops to master learning difficulties depends on the one hand on his or her experience of the efficiency of the efforts and on the other hand, on the relevance the subject matter has for the student (Raynor, 1974). This becomes obvious - - as already mentioned - - by the importance the variable "perceived usefulness of mathematics" has for achievement and course-taking. The conflicts caused by sex-role perceptions that may arise for a young woman in developing career interests in general, and especially in mathematics and scientific fields, are discussed by Fox (1977).

If a student has not reached the antecedent achievement goals, this has two different impacts for the learning processes to come: Firstly, one sees a more subject matter directed impact: the student's deficiency lessens his or her chance of reaching the following goal because of fragmentary antecedent knowledge. An accumulating deficiency will appear the more the subject matter is hierarchical in nature and is taught in this way. Secondly, a more motivation directed impact can be described when cognitive achievement motivation models (Heckhausen, 1974; Gorlitz et al., 1978) are adapted to school situations (Fuchs, 1975; Kornadt, 1975). It is assumed that during the years at school a student has developed a subject matter related achievement motive associated with a self-concept of his or her abilities and casual attributions of success and failure (e.g., good luck, effort, ability, easiness of the task).

Failure can lead to different reactions. It is possible that the student makes more efforts to compensate for the deficiency, and if these efforts lead to success, the learning potential for the next learning steps is strengthened and the achievement motivation too. If the efforts fail and if this occurs over and over again, a failure cycle is settled (Shapito, 1962). The negative motivational development and the lack of knowledge relevant for the next learning step are affecting each other. In analogy to this failure cycle a success cycle can be developed. The developmental process theoretically leads to certain types of students. The extremes may be characterized as

> The achievement motivated student: good knowledge, positive self-esteem of his mathematicalabilities; success will be attributed tohis own ability and efforts, failure due not to lack of ability but to internal or external variable causes, e.g., lack of efforts or bad luck; and
>
> the fear-of-failure student: poor knowledge, negative self-esteem of his mathematical abilities; success in mathematical tasks is attributed not to ability but primarily to external reasons such as good luck and easiness of the task, failure to lack of ability.

Sex differences in self-confidence in mathematics are found in quite a lot of studies (Lorenz, 1979; Robitaille, 1977). There is some evidence that women, having a more traditional sex-role perception in intellectual achievement situations, show an attribution pattern and have a self-esteem of their ability similar to the fear-of-failure student (Stein & Bailey, 1973; Deaux, 1976). The effects of failure may even lead to a special math anxiety (Dreger & Aiken, 1957; Richardson & Suinn, 1972). As to the hypothesis that girls score higher on math anxiety, the empirical results are not consistent (Suydam, 1975; Westoff, 1979; Caserly, 1975, 1979). But as sex role perceptions were not considered, this might not be surprising.

The development of a student's self-concept of his mathematical abilities and his casual attributions is a process which is remarkably influenced by the teacher, seen as significant other. By direct or indirect causal attribution of the achievement results of a student, the teacher delivers information to the student about the achievement standards the student can or should reach and by this, influences the self-concept and attribution of the student; e.g., if a student has success in solving a problem and the teacher reacts "Very nice that you gave us the right solution; you worked well at home, but the question was not so very difficult," this may mean that the teacher thinks that the student is not very able (he praised him for solving an easy task) and that the student could give the right solution only because he made great efforts. It can easily be assumed that a girl will make no special effort to solve a mathematical problem when the teacher delivers to her the information that he does not expect her to be successful because of lack of ability or because mathematics is not appropriate for her. There are quite a lot of empirical studies showing that the teacher-student interaction is different for boys and girls (e.g., Brophy & Good, 1970; Sikes, 1972). For mathematics classes, the study of Becker (1979) demonstrates how different teachers' expectations based on sex for student performance and behavior are transformed into a sex-biased interaction pattern. Teachers' expectations closely followed traditional sex-role stereotypes; correspondingly, they gave, e.g., more encouragement to males than to females, and the latter were even discouraged in some cases. If a boy gave a wrong answer, teachers were more likely to give hints or to ask a new question and by this made it possible for the student to solve the problem.

4. Possibilities of Supporting Change

Considering the reported results, it is obvious that one essential changeable cause for special problems of women in mathematics lies in the stereotyping of mathematics as a male domain. This stereotyping does not belong to the denotative meaning of mathematics, that is, it is not a constitutive critical attribute of mathematics.

There are great differences between countries and between individuals in the extent to which this stereotype is still held. By changing perceptions about the female sex role, the linkage "mathematics/male" has already been diminished. But there are still a lot of teachers, counselors, parents, and further "significant others" who should think over their sex-role perception about female mathematicians, espeically because of their great influence on the developing sex-role perceptions of young women and, combined with this, on their career plans. A further aspect is that the sex-typing of mathematics and science in learning materials has to disappear. Moreover, the relevance of mathematical knowledge should be demonstrated not only for occupations belonging to the field of mathematics and natural and technical sciences but also for occupations belonging to fields more often chosen by women, e.g., social sciences and economics. By ths means, a girl perceives the usefulness of mathematics, and accordingly course-taking behavior may be altered.

In analyzing high school mathematics and science programs that attract and hold high proportions of girls, Casseerly (1979a) found some organizatinal aspects that seem to be helpful, e.g., homogeneous grouping of able students (boys ad girls) as in Advanced Placement. Encouragement to engage in mathematics should start as early as possible and can be facilitated not only by teachers and counselors but also by older female students (Casserly, 1979b).

Proposals to facilitate learning processes in mathematics cannot be made for lack of empirical results. The relationships between sex, cognitive style, teaching style, and mathematical achievement should be investigated intensively. Research is needed that is based on elaborated theoretical backgrounds. This is true not only in the areas considered (Travers, 1963).

What can be done for those men and women who do not recognize that mathematics is useful and necessary for a lot of different occupations and studies before they enter college or university? These persons not only have a lack of mathematical knowledge but also, because of failure experiences, have negative feelings about mathematics; they may even have math anxiety. In the U.S.A., so-called "Math Clinics" have been established which not only offer courses to fill in knowledge gaps but also offer different therapies to overcome math anxiety (Tobias, 1977, 1978). Of course, encouraging them to participate (Tobias, 1980).

Literature

Aiken, L.R.: Intellective variables and mathematics achievement: Directions for research. Journal of School Psychology, 1971, 9, 201-209.

Aiken, L.R.: Update on attitudes and other affective learning in mathematics. Review of Educational Research, 1976, 46, 293-311.

Armstrong, J.R.: Factors in intelligence which may account for differences in mathematics performance between the sexes. In E. Fennema (Ed.): Mathematics learning: What research says about sex differences. Columbus, Ohio, Ohio State University, 1975.

Astin, H.S.: Sex differences in mathematics and scientific precocity. In Stanley, J.C., Keating, D.P., & Fox, L.H. (Eds.), 1974, 70-86.

Becker, J.B.: A study of differential treatment of females and males in mathematics classes. Dissertation, University of Maryland, 1979.

Brophy, J.E. & Good, T.L.: Teacher's communication of differential expectations of children's classroom performance: some behaviorial data. Journal of Educational Psychology, 1970, 61, 365-374.

Casserly, P.L.: An assessment of factors affecting female participation in advanced placement programs in mathematics, chemistry, and physics. National Science Foundation Grant GY-11325, 1975.

Casserly, P.L.: Helping young women take math and science seriously in school. In Colangelo-Zaffrann: New voices in counseling gifted, 1979 a.

Casserly, P.L.: Factors related to young women's persistence and achievement in Advanced Placement Mathematics. Paper presented at AERA, 1979b.

Casserly, P.L.: Factors leading to success - present and future. In Jacobs, J.E. (Ed.): Perspectives on women and mathematics. ERIC/SMEAC Clearinghouse for Science, Mathematics and Environmental Eduation, 1978, 119-124.

Deaux, K.: Ahh, she was just lucky. Psychology Today, 1976, 10(6), 70-75.

Dreger, R.M. & Aiken, L.R.: The identification of number anxiety in a college population. Journal of Educational Psychology, 1957, 48, 344-351.

Dwyer, C.A.: Influence of children's sex role standards on reading and arithmetic achievement. Journal of Educational Psychology, 1974, 66, 811-816.

Ernest, J.: Mathematics and sex. American Mathematical Monthly, 1976, 83, 595-614.

Fennema, E.: Influence of selected cognitive, affective, and educational variables on sex-related differences in mathematics learning and studying. In Fox, L.H., Fennema, E., & Sherman, J. (Eds.), 1977, 79-135.

Fennema, E. & Sherman, J.A.: Sex-related differences in mathematics achievement, spatial visualization and affective factors. American Educational Research Journal, 1977, 57, 51-71.

Fox, L.H.: Sex differences in mathematical precocity: Bridging the gap. In Keating, D.P. (Ed.): Intellectual talent: Research and development. Baltimore, Maryland, 1976, 183-214.

Fox, L.H.: The effects of sex role socialization on mathematics participation and achievement. In Fox, L.H., Fennema, E. & Sherman, J. (Eds.), 1977, 1-77.

Fox, L.H., Fennema, E., & Sherman, J. (Eds.): Women and mathematics: Research perspectives for change (NIE papers in education and work, No. 8). National Institute of Education., U.S. Department of Health, Education, and Welfare, Washington, D.C. 20208, 1977.

Fuchs, R.: Lehrziele und Lernziele als Determinanten des Lehrer und Lernerverhaltens. In Kornadt, H.-J., 1975, 162-176.

Glotzer, J.: Rollenfixierung. In: Kritische Stichwortere zum Mathematikunterricht, 1979, 234-249.

Gorlitz, D., Meyer, W.-U., & Weiner, B. (Eds.): Bielefeldr Symposium uber Attribution. Stuttgart, 1978.

Haven, E.W.: Factors associated with the selection of advanced academics courses by girls in high school (Dissertation, University of Pennsylvania, 1971, RB-72-12). Princeton, N.J.: Educational Testing Service.

Heckhause, H.: Motive and ihre Entstehung. In Weinert, F.E. et al. (Eds.): Padagogische Psychologie I. Frankfurt, 1974, 135-171.

Heilbrun, A.B.: Sex role identify and achievement motivation. Psychological Reports, 1963, 12, 483-490.

Hilton, T.L., Berglund, G.W.: Sex differences in mathematics achievement: A longitudinal study. Journal of Educational Research, 1974, 67, 231-237.

Husen, T. (Ed.): International study of achievement in mathematics: A comparison of twelve countries, Vol. I and II. New York 1967.

Keeves, J.: Differences between the sexes in mathematics and sciences courses. International Review of Education, 1973, 19(1), 47-63.

Kornadt, H.-J.: Lehrziele, Schulleistung und Attribuierungen im Mathematikunterricht. Dissertation. Bielefeld, 1979.

Maccoby, E.E. & Jacklin, C.: The psychology of sex differences. Stanford, 1974.

NAEP, National Assessment of Educational Progress. The first assessment of mathematics: An overview (Mathematics Report No. 04-MA-00). Washington, D.C.: U.S. Government Printing Office, 1975.

Perl, T.: The Ladies Diary. Historia Mathematica, 1979, 6, 36-53.

Radatz, H.: Individuum und Mathematikunterricht. Hannover, 1976.

Raynor, J.O.: Future orientation in the study of achievement motivation. In Atkinson, J.W. & Raynor, J.O. (Eds.): Motivation and achievement. Washington, 1974, 121-153.

Richardson, F.C. & Suinn, R.M.: The mathematics anxiety rating scale: Psychometric data. Journal of Counseling Psychology, 1972, 19(6), 551-554.

Robinson, J.W. & Gray, J.L.: Cognitive style as a variable in school learning. Journal of Educational Psychology, 1974, 66, 793-799.

Robitaille, D.F.: A comparison of boys' and girls' feelings of self-confidence in arithmetic computation. Canadian Journal of Education, 1977 2(2), 15-22.

Robitaille, D.F. & Sherrill, J.M.: Achievement results from the B.C. mathematics assessment. Canadian Journal of Education, 1979, 4(1), 39-53.

Rogers, M.A.: A different look at word problems. Mathematic Teacher, 1975, 68, 285-288.

Schildkamp-Kundiger, E.: Geschlechtsrollenvorstellungen und Mathematikleistung bei Madchen. Dissertation. Saarbrucken, 1973.

Schildkamp-Kundinger, E.: Mathematics and gender. In: Steiner, H.-G. (Ed.): Comparitive studies of mathematics curricula - - change and stability 1960-1980. Proceedings of a conference jointly organzed by the IDM and the IMC of the 2nd International Mathematics Study of the IEA, Osnabruck FRG, Jan 7-11, 1980. Materialien und Studien Bd 19 Institut fur Didaktik der Mathematik, Unversitat Bielefeld FRG, 1980, 601-622.

Sells, L.W.: The mathematics filter and the education of women and minorities. Paper presented at the annual meeting of the American Association for the Advancement of Science, Boston, Mass., Feb. 1976.

Senechal, B.: A woman or a mathematician? In: La Gazette SMF, 1974.

Shapiro, E.W.: Attitudes towards arithmetic among public school children in the intermediate grades. Dissertation Abstracts 1962, 22, 3927 f.

Sherman, J.: Effects of biological factors on sex-related differences in mathematics achievement. In: Fox, L.H., Fennema, E., Sherman, J. (Eds.), 1977, 136-221.

Sikes, J.W.: Differential behavior of male and female teachers with male and female students. Dissertational Abstracts International, 1972, 33, 217A, (University Microfilms No. 72-19670).

Smith, I.M.: Spatial ability. San Diego, 1964.

Stamp. P.: Girls and mathematics: Parental variables. British Journal of Educational Psychology, 1979, 49, 39-50.

Stockard, J. & Johnson, M.: Sex roles: Sex inequality and sex role development. Englewood Cliffs, N.J. 1980.

Stanlay, J.C., Keating, D.P. & Fox, L.H.: The socialization of achievement orientiation in females. Psychological Bulletin, 1973, 80, 345-366.

Suydam, M.N.: Research on some key non-cognitive variables in mathematics education. In: Schriftenreihe des Instituts fur Didaktik der Mathematik. Bielefeld, 4/1975, 105-135.

Tobias, S.: Counseling the math anxious. In: Zentralblatt fur Didaktik der Mathematik, 1977, 3.

Tobias, S.: Overcoming math anxiety, W.W. Norton, 1978.

Tobias, S.: The problem: Math anxiety and math avoidance; the solution: reentry mathematics. In: Information of the Institute for the Study of Anxiety in Learning 1980.

Travers, R.M.: An introduction to educational research. New York, 1969.

Treumann, K.: Dimensionen der Schulleistung. Teil 2: Leistungsdimensionen im Mathematikunterricht. Stuttgart, 1974.

Very, P.S.: Differential factor structures in mathematical ability. Genetic Psychology Monographs, 1967, 75, 169-207.

Westoff, L.A.: Woman - - in search of equality. Focus (Educational Testing Service) Whole No. 6, 1979.

Wendelin, I.: The mathematical ability: Experimental and factorial studies. Lund, 1958.

CHAPTER 21 - Special Groups of Students

21.1 CURRICULUM ORGANIZATIONS AND TEACHING MODES THAT SUCCESSFULLY PROVIDE FOR THE GIFTED LEARNER

THE GIFTED CHILD IN AUSTRALIA

A.L. Blakers
University of Western Australia
Nedlands, Australia

After many years of virtual neglect, the last two or three years have witnessed a remarkable surge of interest and activity in the area frequently referred to as "gifted and talented children." This development in Australia is partly indigenous, but to a considerable degree it seems to parallel (and to draw inspiration from) similar movements in other countries. In a very short period of time we have moved from a situation where concern for gifted children was generally dismissed an unnecessary or elitist, to a situation where the education of gifted children is now occupying the attention of all major educational authorities.

The current resurgence of interest in giftedness in Australia has led to many attempts to define such terms as "giftedness" and "talented student," and there has been a great diversity of frequently incompatible views. It is my experience that most mathematicians and mathematics teachers who are trying to do something about mathematically gifted children are not greatly concerned with the precise description and measurement of the concept of mathematical giftedness. If pressed, they would probably settle for some such definition as "the upper 1% (or some other figure) in mathematical ability," recognising both the arbitrariness of the percentage and the dubious underlying assumption that mathematical ability can be measured on a linearly ordered scale. Most mathematicians of my acquaintance are satisfied to assume that there is some underlying reality to the concept of mathematical giftedness, and that they, and mathematics teachers, can recognise this with sufficient accuracy to justify special attention to those identified.

I will attempt to give some indication of what has been and is now being done in Australia to help the mathematically gifted. Until quite recently the answer to this would have been "very little, and at the official level almost nothing." What had been done in this area was almost entirely due to the interest and dedication of individual teachers (at all levels, primary, secondary, and tertiary) and to the efforts of professional organisations of teachers; but this situation is rapidly changing.

One type of activity which has existed for some time in almost every state is the holding of mathematical competitions. In some places these are run by branches of the Australian Association of Mathematics Teachers, and in others by university departments of mathematics. Some are run at only one level for all secondary students, while others are run at two or even three levels. In all cases the objective is to detect students with special mathematical abilities, particularly those abilities which lead to success in the solving of the sort of problems which are set in the competitions. (For those interested in more detail I could arrange to send copies of some of the competition problems which have been used.) The prizes which are awarded to the leading candidates are usually donated by business organisations. Closely related to these competitions are regional "Talent Searches," in which the entrants submit reports on mathematical investigations or projects. Sometimes the organisers provide a list of suggestions for entrants sometimes the students have to find their own problems under the guidance of their teachers. From my observation, these mathematics talent searches are not as successful as the problem competitions in attracting entrants or in detecting gifted students, but they certainly have a useful educational role, particularly as a means for generating interest in mathematics and for encouraging team work.

One particularly mathematical competition of recent origin has had such a spectacular growth that it deserves special mention. This is the Australian Mathematics Competition, which began as a local competition in Canberra three years ago and has rapidly developed into an Australia wide competition, operating at three secondary school levels *years 7/8, 9/10/11/12). Sponsorship and logistic support is provided by Australia's largest commercial bank, and there are now well over 100,000 entrants each year. The questions are multiple choice and they are machine marked. With such numbers, computers are used extensively in the organisation of the competition and in the recording and analysis of the results. While this particular competition is designed to generate interest rather than to discover those with special mathematical gifts, there is no doubt that it can be used to give a general (and sometimes early) indication of special mathematical ability.

A quite recent development in the competition area is a move to enter an Australian team in the International Mathematical Olympiad. It is expected that an Australian team (not necessarily a full team of eight) will be entered in the Olympiad which is to be held in the United States next year. Efforts are already under way to identify (through all sorts of channels including state competitions and teacher recommendations) potential members of such an Olympiad team, and to sharpen their problem solving skills through voluntary participation in a problems-by-mail activity. In this respect it has to be realised that Australia is a big country - - as big as the continental United States - - but the population is only about 14 million; it would be extremely difficult (and costly) to assemble all potential team members together in one place for competition training. It is far too early to say what effect Olympiad participation might eventually have on the advancement of the mathematically gifted in Australia, but thought is already being given to finding ways of providing some spin off to a larger group than those few who seek Olympiad selection.

Another activity aimed at the recognition and stimulation of the mathematically gifted is the National Mathematics Summer School. This Summer School (on which I reported at the Karlsruhe Congress) has taken up much of my time for over twelve years. It is a residential summer school for over seventy high school students from years 11 and 12, selected solely on the basis of their achievement and potential in mathematics, without regard to their intentions regarding future careers. Selection is decentralised to the states and territories on a population basis, and a variety of procedures are used, including specific selection tests in some states. It is most unlikely that those selected are in fact the top seventy mathematics students of their age in Australia, but in any reasonable sense they are certainly mathematically gifted. Our task in running the summer school is to use the short period of two weeks (in residence at the Australian National University) to enlarge their intellectual horizons and to open up (for most of them) entirely new perspectives of what mathematics is about and what creative mathematicians and users of mathematics are like and what they do. We push them hard, but we watch and counsel them quite carefully so as to avoid discouragement. Staff and students reside together so that they can interact freely at a social as well as at a mathematical level. By the end of the program we are all good friends - - partners in a common intellectual endeavour. Over the twelve years of operation, we believe that we have progressively increased the effectiveness of this summer school, most especially in the last five years through the efforts of Arnold Ross, who has an extraordinary talent for inspiring mathematically gifted students. By the end of the two weeks most of them are truly surprised at what they have achieved, and this growth in the awareness of their own capabilities is one of the most valuable outcomes of the program. Most of them have never been seriously pushed before and then tend to underestimate their own potential. A unique feature of this summer school is the opportunity which it gives for the development of peer-group stimulation at the highest level of student mathematical ability - - something which is not possible for these students in their normal environment, as they are usually the leading mathematics students (frequently by a considerable margin) in their own schools.

The students in the Summer School are selected without regard to their financial means. Each is asked to contribute about 20% of the total cost per student, and the rest of the money has been obtained from a variety of sources. Initially it all came from business and industry, plus one-off encouragement grants from such bodies as the increasingly difficult to retain business support, but for five of the past six years some assistance has been given by the Ausent of peer-group stimulation at the highest level of student mathematical ability - - something which is not possible for these students in their normal environment, as they are usually the leading mathematics students (frequently by a considerable margin) in their own schools.

The students in the Summer School are selected without regard to their financial means. Each is asked to contribute about 20% of the total cost per student, and the rest of the money has been obtained from a variety of sources. Initially it all came from business and industry, plus one-off encouragement grants from such bodies as the increasingly difficult to retain business support, but for five of the past six years some assistance has been given by the Australian Schools Commission. This is no longer available, but the Summer School will continue in 1981 through a large number of relatively small contributions, mainly from past students and their schools. It is hoped that the current rapid expansion of interest and activity in the gifted student area will lead to the provision of governmemt assistance through a suitable funding agency.

There is a conventional wisdom which says that most mathematically able students are self-sufficient and bound to succeed without special encouragement; as a result of my experience with a considerable number of mathematically gifted students, from their own comments, and from reports from their teachers, I believe that this is false. I find that most of them do indeed need encouragement if they are to achieve anything like their real potential, and that they respond to encouragement in a most gratifying way.

The National Mathematics Summer School is not the only "Summer Camp" for mathematics students. Two such camps have been held in Victoria for over twenty years, but these have been aimed at interested students rather than the gifted. Another summer camp, with similar objectives to the National Mathematics Summer School, has been held annually in Western Australia for nearly twenty years; the success of that activity was the stimulus which persuaded me to seek the suport of the Australian Association of Mathematics Teachers, twelve years ago, in the launching of the National Mathematics Summer School Project. Several other centres have similar activities for area students - - usually based on interest rather than on giftedness. In saying this, I do not intend to imply that there is no connection between programs designed to stimulate interest and the needs of the gifted. In fact I am confident that many mathematically gifted students have received significant stimulation through activities whose primary purpose was to create interest.

Another activity, which has played a significant role in the stimulation of mathematcal interest amongst gifted students in the Canberra region, has been a program of Friday evening seminars that has run over fifteen years. This program is sponsored by the Canberra branch of the Australian Association of Mathematics Teachers, and its effectiveness has been due to the efforts of a few dedicated mathematicians.

In recent years there have been a number of new initiatives taken by individuals and organisations to assist mathematically gifted children. For example:

> In one state there is a mentor scheme, which brings together mathematicians and high school students on a one to one basis.
>
> Closely related to this is a program at the University of New South Wales which attaches High School students to mathematics staff members as "vacation scholars."
>
> In other centres professional associations and Education Departments have prepared enrichment materials and have run workshops designed to assist teachers in providing special activities for their mathematically gifted students.
>
> In Canberra the local educational authority has established, with Schools Commission funds, a Mathematics Centre for students from Year 4 to Year 10. The emphasis is on problem solving with the aid of models and concrete materials. The long-term objective is to generate interest and to develop more positive community attitudes to mathematics, but some of the problem material available challenges the more able students. A

similar centre has been established in at least one other city.

Many schools have established mathematics clubs.

In one large city there is a Mathematics Circle which draws interested (and able) students from all over the city and from nearby rural areas to its periodic meetings; most of the talks are given by the students themselves, under the guidance of members of the local association of mathematics teachers.

At official levels mathematics seems to have been given some priority. In part this is the result of a common view that mathematical talent is more easily recognised and more easily provided for than talents in other directions, but it is also due, in part, to the fact that during the lean years mathematics teachers and mathematicians, and their professional organisations, have remained more active in the gifted student area than their counterparts in most other disciplines. During this period many individual teachers have gone out of their way (without waiting for official support and frequently without knowledge) to encourage and foster students with special mathematical gifts. In fact one of the most interesting letters which I received during my search for background information for this paper was from an upper secondary teacher in a small private school. Over a long period of time he had devoted much thought and effort to the development of strategies aimed to capture the interest of all students in mathematics, and to challenge the best of them. His methods were sometimes unusual and essentially personal, but what stands out from his letter is that his efforts have been a labor of love - - a love of mathematics and a love of children. I know that he has been successful because at different times I have had two of his students at the National Mathematics Summer School. Many mathematicians (and also of course many whose gifts lie in other directions) recall with gratitude and affection a particular teacher who gave them the first indication that they had a special gift, and that it was something precious that should be carefully nurtured. No matter what official programs are developed, this special role of the individual teacher will retain its significance.

The situation of gifted children in Australia is much more encouraging than it was only a year or two ago, but it would be naive to assume that the final battle for the rights of gifted children has been won. There is still much to be done, but what has already been achieved reflects a growing acceptance that gifted children do indeed have needs and that the state has an obligation to consider these needs. As our former Federal Minister for Education, Kim Beazley, wrote (in a forthright article in the Sydney Morning Herald on July first of this year) "no education system has the right to neglect, frustrate or refuse to identify the gifted." It is a time for cautious optimism.

THE GOVERNOR'S SCHOOL FOR THE GIFTED IN VIRGINIA

Isabelle P. Rucker
The Governor's School
for the Gifted in Virginia
Richmond, Virginia

In introducing the chapter, "The Music of the Spheres," from his book, The Ascent of Man, Jacob Bronowski writes:

> Mathematics is in many ways the most elaborated of the sciences - - or so it seems to me, as a mathematician. So I find both a special pleasure and constraint in describing the progress of mathematics, because it has been part of so much human speculation: a ladder for mystical as well as rational thought in the intellectual ascent of man. However, there are some concepts that any account of mathematics should include: the logical idea of proof, the empirical idea of exact laws of nature (of space particularly), the emergence of the concept of operations, and the movement in mathematics from a static to a dynamic description of nature.

I open my remarks with this passage because I believe that Bronowski captures the ultimate in what we would all like to do for our gifted young people. The passage is somewhat descriptive of harmony. It is a way of saying that mathematics is elegance, it is beauty, it is rhythm, it is simplicity. These and more are the attributes of mathematics for bright young people at whatever stage of their development. To be sure, the basics must be at one's command whether that one can be gifted or not! (Someone, I do not recall who, said, "Practice not only makes perfect - - it makes permanent!") But to deny bright young students the ethereal element of mathematics is nothing short of committing an educational crime.

For years, educators have ignored the needs of gifted students primarily, I think, because of several misconceptions. For too long it was thought that the gifted need no special programs, no special attention - - because they will get along anyhow; for too long it was thought that giving special attention to the intellectually gifted engendered an elite and snobbish society; for too long attention has been focused and untold millions of dollars have been spent on students at the opposite end of the spectrum from the gifted. This is not to condem programs for those students, because they both need and deserve help. But, so do the gifted. Our gifted children are our most valuable natural resource.

Give gifted students an opportunity to learn, give them freedom to progress at their rate, give them exciting alternatives and freedom to choose, give them challenges so that they will be eager to learn for the sake of learning, and they, in turn, will assume their share of responsibility for their own education. Eventually, they will be the group to make the greatest contribution to society; they will provide the leadership and the brain power that will keep their country strong. Deny them these opportunities and challenges, and they will likely turn away from learning and devote their efforts to the fulfillment of selfish and disruptive motives.

Why are the gifted different from the "average" or the "below-average" students (whoever they are)? Let's look at an hierarchy of intellectual operations: 1) cognition (comprehension, discovery, rediscovery, awareness); 2) knowledge (retention, recall); 3) convergent thinking (a single correct or best answer); 4) divergent thinking (imaginative, spontaneous self-expression); 5) analysis; 6) synthesis (a high degree of original thinking); 7) evaluation (ability to make projections "based on . . . "). Where do the gifted fit into this hierarchy? Certainly at numbers 4, 5, 6, and 7. If you read the comic strip, "Dennis the Menace," perhaps you remember the time that Dennis went into the house dripping wet and dirty and said to his mother, "You can't tell how deep a puddle is from the top!" That's an example of evaluation, based on . . . !

Gifted children possess special characteristics, some of which are: an insatiable curiosity, reading at ages two to four years, prolonged span of attention, tendency to be mavericks, imagination, sense of humor, capacity for commitment, concern for their fellow man, logic in the sense of ordering the chaos they see about them, eagerness to know themselves and their world, critical thinking, and superiority in abstract subject.

Of course, standardized test scores are often used in the identification process, and I certainly have no quarrel with this. Indeed, a high IQ can produce a Gauss or an Einstein, but a high IQ, in itself, does not assure achievement. Other than the concrete evidence which tests are supposed to produce, we must look for the characeristics that cannot be measured by any instruments known today.

Many organizational patterns can be employed in providing special services for the gifted. Some of these ways are: enrichment (within and without the classroom), acceleration, special schools, college courses taken in high school, early college admission, independent study, homogeneous grouping, correspondence courses, and college credit by examination.

In Virginia, as in some other places, startling strides have been made in regard to differentiating the curiculum for our gifted and talented youth. (We still have a long way to go!) Every school division is now mandated by our State Legislature to provide for the speical needs of the gifted. And there is money to back this mandate - - not a lot of money, only aboout 1.5 million annually. But it is a step. In addition to providing funds for programs during the school year, our legislature has made a separate appropriation in each of the past eight years for the operation of a residential summer program, the Governor's School for the Gifted. The foregoing comments serve as a springboard for a discussion of this exciting activity for some of our gifted students.

The purpose of the Governor's School is to provide intellectually challenging and enriching experiences for 420 high school students who are rising juniors and seniors and who are academically gifted or artistically inclined. The school is of four week's duration and is held on three college campuses with 140 students at each. Students are nominated by the school division superintendents (public schools) and headmasters (private schools) according to criteria provided by the Virginia Board of Education. Each high school may nominate one student for every 500 enrolled in grades 10 to 12 in that school. The maximum number of nominees from any one school is five. At the state level, the approximately 650 annual nominees are carefully screened, based on information they have provided on a 13-page nomination form, by a state-wide committee of about twenty educators, lay people, parents, and former GS participants. This committee makes the final selections.

The curriculum includes classes, seminars, workshops, and independent studies in a variety of subjects and topics from the general areas of the fine and performing arts, the humanities, the natural and physical sciences, and history and the social sciences. Each participant chooses categorically, from a list of about twenty, two specific subjects which comprise his or her major and minor emphasis during the four-week period. Some of you may have heard Mrs. Casey Humphreys at this Congress on Monday afternoon discussing the exciting mathematics course that she teaches at one center of the Governor's School.

No grades or credit are offered; however, each participant does receive, at the close of the session, a very impressive Certificate of Commendation.

All costs of tuition, room and board, field trips, and other activitied planned for the school are paid from state funds. The only expense to students is their transportation (one round-trip) and their spending money.

Teachers are carefully selected from high schools and colleges. Each must be impeccably prepared in his or her own discipline, must be able to deal successfully with bright teenagers, must have the physical stamina to withstand the grueling though highly motivating experience, and must be willing and prepared to accept some responsibilities other than for his or her own classes.

The following excerpts from students' spontaneously written letters attest to the success of the Governor's School:

> Governor's School unscrewed the lid! . . . Besides giving me a much-needed jump in botany and biology, GS gave me a sense of worth and a sense of just how little I know. There's so much more to learn. GS took me, a rather frustrated person, jaded at seventeen (a fearsome thing in itself), and instilled a love of learning and association with my peers. Since GS, the only television I watch is a half-hour of Monty Python every Sunday night. I am so busy learning, picking up where GS left me off, that there's little time for idleness . . .

> Before I went to Governor's School I thought that I would be bored out of my skull, trapped like a fish out of water with 140 would-be Werner Von Brauns whose idea of a good time on Friday night was reading Spinosa in the original language. HA! Never before have I met 140 more happy, normal, good-natured, and intelligent people. Where I live (and I guess where everyone else lives) the stereotyped image of individuals with above-average intelligence prevails. Now, however, I know that that's a ridiculous generalization. I learned three important things at GS: (1) There's nothing abnormal about sitting down and discussing "heavy" topics such as man's purpose on earth. (2) Hath not a gifted person eyes? If you prick him, doth he not bleed? If you tickle him, doth he not

I regret that I have but one summer to give to my Governor's School!

In my opening remarks, I referred to the harmony one finds in mathematics. I close on the general idea of harmony as seen through the mind and pen of a Governor's School participant. Each year, nominees for the GS must write an essay on a given situation. The essay must be imaginative, concise, of less than 500 words, and the student's own work. For 1979, the situation was:

> By the twenty-first century, significant changes are likely to have taken place in the United States and around the globe. Let us assume that changes will occur in government, the environment, life styles, the culture. What changes do you predict in any one or two of these areas? Based on your prediction, what preparation will you need to take a leadership role in society early in the 21st century?

One student wrote:

The Creation is finished.
A new world has evolved;
Twenty-nine cents for a
bottle of bubbles
Any my living experience
comes into form.

Bubbles, bubbles, bubbles,
The circles of existence,
Life in spheres,
No purpose, no reason, just
bubbles occupying space.

They are formed by the
Supreme Being,
The Almight Wand,
Releasing them out into
the world
To survive on their own.

Each bubble's existence
is unique.
They all receive a dif-
ferent form
And time allotment.
They are like no other
being.

The time they have
is limited.
It depends upon the group.
Harmony is the key.
The card can always run out.

Ruling the bubbles
Is the bank of hours,
Dealing out time
As it sees fit.

Bubbles form bonds,
Bonds form groups,
Groups expand to communities,
Communities form larger spheres.

Bubbles can break
If the harmony is broken.
Harmony is the test for survival.
The tranquil sphere outlives.

The sphere of bubbles has a leader,
A dominant figure that stands
above all.
This figure will break the
harmony first
And the sphere disbands.

A sphere is also unique.
It cannot reform.
The community is gone forever.
Life if broken down.

The Wand works again.
Life is brought back to space.
Again, time and harmony will
tell the story
Of the twenty-nine cent
bubble world.

* * * * * * *

In order to become a leader in the society I have created, I would have to become a master of the harmony on which each bubble depends. I would train my mind in the creative process and design a language of creativity that would ennoble the moment of birth and give breadth of the moment of death. The impulse to destroy would not be present because death would be understood as the ultimate stroke of one's imagination and, therefore, of one's renewal.

FOR GIFTED MATHEMATICS STUDENTS: PROJECT MEGSSS

Burt Kaufman and Gerald R. Rising
CEMREL, Inc.
St. Louis, Missouri, U.S.A.

The Mathematics Education for Gifted Secondary School Students (MEGSSS) Project was until recently a major activity of the Comprehensive School Mathematics Program (CSMP) in St. Louis, Missouri, but Project MEGSSS has now separated and incorporated as an independent activity with participant parents providing direction. In this paper we shall consider the Elements of Mathematics (EM) series, the textbooks used in the MEGSSS instructional program, the Gifted Math Center (the locus of St. Louis area MEGSSS instruction), the newly established Gifted Math Program MEGSSS satellite in Buffalo, and the elementary CSMP program.

The Elements of Mathematics

Starting in 1966, an international team of authors under the direction of current MEGSSS director and until 1979 CSMP director Burt Kaufman has carried on the development of the EM text series for gifted students in secondary school grades 7-12. The first edition of these books was completed in 1976. Currently these texts are undergoing revision based on teacher comments on their use in classroom trials in the Gifted Math Center and other locations, most notably the Baltimore County, Maryland, Schools. They are available from CSMP (3120 59th Street, St. Louis, MO 63139).

The EM texts are designed to communicate mathematics of the highest quality to very able (top 1-2%) and motivated students. The mathematics of the full program is essentially all of that included in school and undergraduate university. In response to the greater ability of the gifted student to master abstractions, much of the content is formal, focusing on careful language, underlying concepts, and proof. In order to provide a bridge to this exposition of serious mathematics in EM Books I - XII, a preliminary "Book" 0 (whose 16 chapters are themselves each of book length and are separately published as such) was written for use in grades 7-9 in parallel with the early books of the more formal series. An EM Problem Book also provides students opportunities to address original challenging problems at the difficulty level of international school mathematics competitions.

The unified development of the EM texts provides an overall continuity of presentation, a monotonic increase of content sophistication and corresponding demand on students, and an opportunity to introduce with care and to employ to great advantage consistent and powerful formal notation. Book 0 provides a "zoo" of examples of these "zoo animals" and their conceptual "cages" are studied informally in Book 0, and they provide an ongoing resource for the more serious considerations of the later books.

The EM development of mathematics is not simply a focus on acceleration of the trip through the standard curriculum (as is, for example, Julian Stanley's program at Johns Hopkins University), but is instead a total reorganization of the point at which students complete school mathematics and start on college mathematics; rather they study a rich mix of school and college content over the full span of the program.

The MEGSSS Gifted Math Center

Over the years as the EM series has been developed and revised, CSMP has continuously maintained an instructional program utilizing these texts. But because it was moved first from Fort Lauderdale, Florida, to Carbondale, Illinois, and then on to St. Louis, only a small number of students were carried through the full program. The quality of the achievements of these students in universities and graduate schools, however, supports the reasonableness of the hypothesis that this unique program does serve: (1) to retain gifted students longer in educational programs, and (2) to lead them to higher achievement levels in mathematics.

As a response to the identified need for fuller trial, the Gifted Math Center was organized in St. Louis in early 1978, with student testing and selection carried out in the spring and seventh grade EM instruction commencing in September of that year. During the first academic year of operation, 1978-79, 47 students were enrolled in two classes. Students attended class twice each week in the late afternoon for two hours each day. Agreements with local school systems provided that students would be excused from their regular school mathematics classes in order to provide work time for the replacement MEGSSS program. Students commuted as many as 50 miles, with parents providing transportation, often in carpools. Attrition from the program totaled eight during this first year, this loss due almost entirely to outmigration, scheduling difficulties, or transportation problems. Substantial attrition is expected for the full six-year program, however, despite the development and continued refinement of the entry testing program. It is to be expected that there will be (a) changes in student goals as sixth graders mature, (b) increased competition for their time, and even (c) misidentifications. In addition and in particular, the staff is concerned about the toil taken by extreme peer pressures against continuing placed upon junior high school girls. The current projections are for about 35% completing the full six-year program. (Note that this apparently low proportion compares very favorably with that of any college mathematics program and in fact with many general college programs.)

In each subsequest year and in addition to the continuing students, about 100 new students are processed and start the seventh grade course in St. Louis. Webster College of St. Louis offers college credit as an option to these MEGSSS students for their work in Books I-XII.

The SUNY Buffalo Gifted Math Program

A new satellite of MEGSSS is this year being organized at the State University of New York (SUNY) at Buffalo. To give an idea of the student selection procedures for MEGSSS programs here are the steps followed for identification of the first class to start the Buffalo program in September.

During January and February nominations of highly gifted and interested sixth grade students were solicited from schools in Erie and Niagara Counties of New York - - that is, within about 40 miles of the university - - and about 180 were received. Parent information meetings were held, and 125 students enrolled for the screening tests in March.

The MEGSSS test battery includes the mathematics sections of the SAT, the so-called "College Board" examination used in the United States to screen high school eleventh and twelfth graders for college admission. The average score of the 125 Buffalo sixth gaders taking this test was better than 18% of college bound seniors taking the same exam, and of those finally enrolling in the university class, better than 30%. Other tests in the battery are: the Watson-Glaser Critical Thinking Appraisal, a College Entrance Examination Board sylogism test, and a National Longitudinal Study of Mathematics Learning aptitude test. The full three-hour battery imposes the further burden of sustained concentration on these sixth graders.

The students were then ranked on the basis of a composite score, and the first 38 interviewed together with their parents. This interview was conducted not as a further screening device but rather to clarify the student, parent, and school expectations. Students must plan to spend at least an hour per day studying program content and completing assignments. (Note that this means that these gifted students must spend about the same amount of study time that would be required of an average student seeking to achieve well in a regular program.) They must be prepared to compete, ususally for the first time, with students as able as or even more able than they. They must give up two afternoons each week. Parents must provide transportation and support in the home. They must also pay a textbook and grading fee. (To date, there is no MEGSSS tuition in either St. Louis or Buffalo, the program support deriving from other sources.) Parents must also see that the required school arrangments are worked out: release from the regular math class to provide study time and schedule rearrangement if necessary to allow students to reach class promptly. The schools with participating students must also nominate a staff member to serve as liaison with the program. In response to all of these preconditions 36 of the 38 students are enrolled for September.

While the curriculum and entry testing are the same and the two staffs work in close cooperation, there are several important differences between the St. Louis and Buffalo programs. As the administrative and creative center, the St. Louis program employs full time staff; Buffalo enjoys savings based upon the use of part-time and volunteer help. The fact that the Buffalo program is based at a university provides several benefits. Members of the university mathematics department, and in particular the chairman, Lewis Coburn, are offering good cooperation and will staff the fourth through sixth years of the course instruction. Professor Rising will teach the first section of students for the first three years, but high quality local school mathematics teachers will take over those grade levels after that. These teacher must intern in the program through the grade level that they will later teach. If present plans mature, these teachers will enjoy part-time appointments in the university's College of Continuing Education.

In response to the complete absence of urban minority students and the low (2:7) ratio of girls to boys enrolled for the first year, activities are planned for 1980-81 to encourage and to sponsor greater participation by these groups. These activities will not, however, by-pass the regular examination screening and ranking procedures.

We point out here that it is the simple arithmetic of the numbers of gifted that constitutes the major reason for the almost complete absence of such programs in the United States. It is our estimate in fact that with MEGSSS, Julian Stanley's program, and the few other local developments, less than one in 100 of our gifted secondary school students receives mathematics instruction that is suitably challenging. What is this arithmetic? It is a simple matter of percent. If we identify the gifted as the top 1%, then our largest schools and even many school systems do not have enough students to form special classes. Out of an age cohort of 1000, 1% would identify only ten students, not enough to form a class in these times of serious fiscal concerns and especially since attrition is to be expected from a demanding instructional regimen.

An additional problem with in-school gifted programs is staffing. It is our observation that virtually no school by itself can staff a program stronger than the U.S. AP course, a one-year acceleration curriculum minimally suitable for perhaps the top 10-15%. They can do no more because of the accumulative demands on teachers of a stronger program. If the program does more than what is normally done at one grade level, teachers of subsequent grades must take into account that additional instruction. Thus it becomes necessary to provide an interschool program like MEGSSS to answer the special needs of this important intellectual resource.

CSMP Elementary Mathematics

CSMP Elementary Mathematics is an instructional series for grades K-6. This program for all students responds to some of the needs of the gifted in the elementary schools in ways quite different from those of the secondary program. In this program the pedagogical languages of Frederique and George Papy -- strings, arrows, and the Papy minicomputer - - are utilized to bring children into early contact with significant and challenging mathematics. And within this program for all, more talented children are provided opportunities to extend these ideas in interesting ways.

At the same time the delightful accompanying stories of Frederique associate the concepts of mathematics with the kind of literary creativity that has before always been too carefully excised from the curriculum, leaving only those dry bones that have turned away so many. It is our belief that this methaphorical extension of mathematics especially encourages very bright and creative students to see mathematics as more than a sterile collection of algorithms and facts.

Some work is currently being undertaken at CSMP headquarters by Clare Heidema and Ron Ward to identify specific activities in the program suitable to gifted students. While we note this activity, it is our belief that in these grades the general enrichment of the total program is itself an excellent contribution to the needs of talented and gifted youngsters.

A SCHOOL DISTRICT - UNIVERSITIES COOPERATIVE PROGRAM FOR GIFTED MATHEMATICS STUDENTS

Dorothy S. Strong
Chicago Public Schools
Chicago, Illinois

It is well known that in order to prove a theorem, you should begin by having faith in its truth. I am going to tell you about a faith that I have and a theorem that I have proved: If we work together, then nothing is impossible. The Chicago Comprehensive Program for Mathematically Talented Students is an example of the total mathematics academic community joining hands to nurture our most valúable resource - - our mathematically precocious students.

As the director of mathematics for the Chicago Public School System, I have the responsibiilty to develop and give leadership to the mathematics curriculum in a large urban school system serving approximately 450,000 children from kindergarten to twelfth grade. Mathematics ability and achievement levels go from one extreme to the other. We have a commitment and a responsibility to provide an academic program that meets the needs of every age level and every ability level in the system.

In 1977, William George, associate director of the Study of Mathematically Precocious Youth at the Johns Hopkins University, introduced us to the work of Julian C. Stanley regarding the Mathematics Talent Search. Immediately we realized that they had a workable solution to the problem of identifying and meeting the needs of mathematically talented students early enough to do something about it. The Johns Hopkins Program gave birth to the Chicago Comprehensive Program for Mathematically Talented Students. This program has been a godsend for mathematically precocious seventh and eighth year students. I am going to tell you about this program that the Lord dropped in my lap and how he gave me everything that I needed for its success - - even a coordinator. The program is one of early identification of mathematically precocious elementary school students followed by intensive instruction and evaluation.

Identification Stage

Our first job was to find all of our mathematically talented students and to determine their degree of giftedness. Test scores from the Iowa Test of Basic Skills were used for initial screening. This gave every student in the system an opportunity to participate since all students must take this test annually. Sixth, seventh, and eighth year students who scored in the 95th percentile or higher were invited to participate in the second stage of the testing program.

The instrument used in the second level screening of these elementary school students was the School and College Ability Test (SCAT) - - a test that is usually administered to students in the fourth year of high school. A conversion table was used to report the scores in terms of the Scholastic Aptitude Test (SAT) both verbal and non-verbal. The results of the testing have been most gratifying.

Securing Participants

Invitations were mailed to all eligible students, inviting them to participate in the talent search.

Dissemination of Results

After testing was completed and tests were scored, a detailed letter was sent to each participant. The letter explained the results and suggested action that parents might take to further their children's advancement in mathematics.

All participants were invited to an awards assembly that was held in December, 1979. All participants were awarded certificates. Special prizes were awarded to top scorers on the test. Over 2,000 people attended and saw Zal Usiskin hold the students spell bound with a speech on misconceptions about mathematics. He even played the piano.

Program Phase

Identification is easy. The hard job comes when you must now meet the needs of these students you have identified with NO money. Unlike Johns Hopkins, we were not a university that could just offer a course. We also knew that a program for urban students could not have a price tag for students. This would eliminate the low income families in the district.

Within the City of Chicago, there are over a dozen colleges and universities. We felt it wise to attempt to tap this vast reservoir of knowledge and resources at our doorstep. Every college and university mathematics department was invited to send a representative to a meeting held in our office. More than half of the institutions responded, forming an advisory committee. The suggestions and offers for assistance were overwhelming. They saw this as an opportunity to fulfill their urban mission. (I am always looking for missionaries.)

Three universities - - The University of Illinois at Chicago Circle, Chicago State University, and Northeastern Illinois University - - volunteered facilities and staff at no charge. The committtee agreed to begin accelerated "Fast-Paced Mathematics" classes in the three universities for these elementary school students. The program follows to a degree the model implemented at Johns Hopkins University. It was decided to invite students who scored 430 or above to participate in the university classes. This includes all students in the 75th percentile or higher on the math portion of the SAT.

Net Value

Thus, this one meeting netted a program worth over $100,000 that could help us to clear our consciences regarding neglecting our gifted students.

Team Teaching

The staff for each class included a university professor and a high school mathematics teacher. The high school teacher performs the following duties:

1. Assist in instruction.
2. Serve as liaison between the Bureau of Mathematics and the university.
3. Serve as liaison between the program and the local school.
4. Coordinate the pace of instruction between classes.
5. Handle all business matters for the classes.
6. Visit home schools.

The university professor is primarily responsible for instruction. A close working relationship has developed, with each gaining new respect for the other. They truly complement each other.

As we planned and organized at an unbelievable pace with limited staff, community pressure was mounting. Parents and schools were now demanding that we should do something about the problem. Classes were finally scheduled to begin January 29, 1979. However, we were not really ready. As you may recall, the Lord sent the big snow, which gave us the extra week we needed for planning.

Letters were sent to students inviting them to participate in the classes. Parents were told that they had to provide transportation for their children. Classes were scheduled for 9:00 a.m. to 11:30 a.m. to take advantage of daylight for traveling. Each class met one morning per week. Parents were informed immediately if students were absent. Parents were invited to first session. Over 90% of the parents attended.

Fortunately, there were very few problems. The routine was in general similar from class to class with some variation from instruction to instructor. Besides the traditional lecture, there were also small group discussions, individualized instruction, some drill and practice, and testing. Instructors assigned a fair amount of homework to be completed for the next class.

An enjoyable rapport existed in each of the classes. Each teacher gave the students their home and office telephone numbers to call them if they encountered difficulties. Students used this privilege but did not abuse it.

Chicago is presently under pressure to integrate its schools. The first year minimum minority population was not acceptable. We felt a need to do something before someone began to apply pressure, so we went to PUSH EXCEL and convinced them to offer enrichment for those identified minority students who did not get into the classes because of their SAT scores. They agreed and offered Saturday enrichment classes.
For the 1979-80 school year we had a similar situation: gifted students and lack of money. For the second year of our program, the University of Illinois at Chicago Circle and Chicago State University offered assistance. (Northeastern Illinois University was unable to continue.) The curriculum for the first year of the program was refined and expanded to include the use of computers.

In the original program design, it was our desire to accelerate these eighth year students in algebra and to continue their acceleration in high school eventually into Advanced Placement Calculus. After completing AP Calculus, but while still in high school, the students would attend appropriate mathematics classes at a university or college of their choice.

For the 1979-80 school year, we had a small group of students who were in a state of limbo. These were the students who were in the seventh year for the 1978-79 program. Now they had successfully completed algebra and completed elementary school mathematics. A geometry class was formed for this group at UICC. (Since we only include seventh year students for testing after 1979, this situation will not occur again.) The geometry class also used the computer and studied program-solving strategies in algebra and geometry.

An enrichment program open to all the participants in the Talent Search (and indeed open to the general public) was also developed in conjunction with the universities. This was a Saturday morning lecture series conducted at the Museum of Science and Industry. Leading mathematicians conducted self-contained lectures at a level of understanding of our talented students. Attendance averaged close to 100 people per lecture for the first two years.

Evaluation

At the end of the year the students were evaluated in a variety of ways. Of course the instructors had monitored the performance throughout the year, keeping records of the homework, special projects, and puzzles completed. Teacher-made tests were administered to each class. In addition, the Educational Testing Service Cooperative Mathematics Test in Algebra was administered during the last week of class. For the 1979 classes, 91.7 percent of the students scored in the 99.6 percentile. Over half, 50.7 percent, scored in the 90th percentile, and 83 percent of the total scored in at least the 75th percentile. The 1980 test results were equally impressive, with 51 percent scoring at or above the 90th percentile, and 77.6 scoring at or above the 75th percentile. There was no direct correlation between SAT score and the course grade. This can be accounted for in part by motivation.

Summary

To summarize my remarks, nothing is impossible if we cooperate with one another. This program is the result of the cooperation between a large urban school system and several universities. This cooperation was responsible for serving the very special needs of some of the world's most precious resources, gifted students - - hopefully, the leaders of tomorrow.

But it is also a lesson in faith. The situation I have described may appear at first glance as possible only in a large city; however, most school systems have some university, college, or two-year college within proximity to themselves. If you have faith, nothing is impossible.

WHAT MATHEMATICS FOR GIFTED YOUNG PEOPLE: THE PROBLEMS OF SELECTION OF CONTENT AND OF BRINGING ABOUT DEEP STUDENT INVOLVEMENT

Arnold E. Ross
The Ohio State University
Columbus, Ohio, U.S.A.

In attempting to provide worthwhile mathematical experience for the very young, one quickly recognizes the inadequacy of teacher's eloquence as a surrogate for students experience. In our efforts to provide appropriate mathematical experience for the student we became keenly aware of the special qualities of mind characteristic of the various stages of development of very young individuals. Also, we have found that is is the better part of wisdom to provide a means of sharing the newly gained experience by borrowing from our linguistic friends the technique of introducing language through usage (1).

The opportunities and the challenges presented by the special qualities of mind of the very young are often very surprising and ever thought-provoking. Let me discuss two pages of the classroom logbook of a class of twelve and thirteen year old children who came to us, in part, from the inner city (2).

This particular class already had some experience with Euler graphs. It was my intention to experiment with graphs that morning until the class would discover conditions for a graph to be a Euler graph. To my surprise, each time I drew a graph on the blackboard there developed keen competition as to who would draw first the required complete path if such would exist or who would declare this to be impossible. So successful was this competition both in speed and in unfailing accuracy that I could not get anyone in the class to take a searching look at the graph itself and seek the reason for success or failure. We sidestepped further discussion of graphs that day and proceeded to do a partition problem.

When we returned to the graph problem on another day we concentrated on determining the degrees of vertices, the number of vertices, the number of edges and the relations between these. After some discussion about the hazards presented by vertices of odd degree, the correct conjecture about the criteria for an Euler graph was produced by the class.

The class was not experienced enough to seek a proof of sufficiency. We gave the essential part of the argument by taking advantage of the constructive nature of the proof and of the fact that everyone in the class was throwing away each initial unsuccessful attempt to draw a desired path. We suggested that one child could build on each unsuccessful attempt by starting again from a vertex of the first crossed path by a suitable cut and by the pasting together of the resulting pieces. Everyone liked this construction as assuring success and we used in on a number of examples.

The lack of space will not permit a detailed account of other mathematical ideas which proved to be very accessible to the very, very young.

Let me conclude this part by disputing the prevalent misconception that there is magic of success in combining applications with relevant mathematical ideas when one deals with an audience which lacks appropriate phenomenological experience. I have found that in order to achieve a really effective blending of the two kinds of experience one must take equally great pains in developing appreciation of the subtleties of the concerns of the experimentalist as one is willing to take in making accessible the subtleties of mathematical concepts.

In a science fiction industrial democracy such as ours the quality of life, indeed our very survival depends upon our resources of developed significant talent.

Although any effort to encourage worthwhile interest in the very young is preferable to neglect, we must, I believe, give adequate attention to the task of nurturing carefully, intensively, and persistently individuals with a very high potential for achievement. We do this in sports and we do this in music.

In sports, winning in competition is the popularly accepted objective of training. In music the end result is performance judged through many subtleties and in its deeper aspects not involving competition among individuals. Even greater subtleties and greater variations of intellectual temperament must be recognized in the development of creative individuals in mathematics, in science, and in the related disciplines.

Thus, whatever is the range of mathematical ideas with which we are concerned at any given time, the effective education of a creative scientist (mathematician! engineer!) demands that we encourage the kind of involvement which develops the student's capacity to observe keenly, to ask astute questions and to recognize significant problems. This last is important in scientific education for two reasons. First, the progress of every science depends upon the capacity of its practitioners to ask penetrating questions and to identify important problems. Second, we believe that personal discovery is a vital part of the learning process for every individual eager to gain deep insight into his subject.

Selection and organization of subject matter depends very much upon the experience and the ultimate destination of the young audience. In my experiment (1973 -) with a two-year mathematics honor sequence at the university, we included as many ideas as possible which every young scientist (and mathematician) should master as early as possible. We began with an intensive and problem oriented discussion of vectors in the analytic geometry of two and three dimensions. We used this experience to make the student feel at home with complex numbers and their geometric applications. We moved on to the study of derivatives of vector as well as scalar functions of the scalar variable. This permitted us to introduce elementary differential geometry. Students' familiarity with complex numbers made it possible not to separate completely the study of real and the study of complex calculus (such separation is an unfortunate current practice, I believe). We tried not to be glib about limits but to emphasize approximations and error estimates. We did a great deal with integration before we turned to power series. We studied the geometrical (qualitative) theory of nonlinear differential equations and used computers, digital as well as analog, to investigate (after Euler) the behavior of trajectories in a two dimensional phase space. We did linear algebra reaching canonieal forms more directly and more quickly than is usual. We did multivariate calculus and some geometrical determinant theory in connection with that. We learned to be precise and concise, avoided abstraction for mere pleasure, and laid considerable emphasis on problem solving.

In the limited time alloted to me in this session I would like to discuss in greater detail our work in the summer with mathematically gifted pre-college students whose ages range from 15 to 18. The duration of this program has varied from eight weeks at The Ohio State University in the U.S.A., to four weeks at the Mathematisches Instut of the Heidelberg University in Germany, to three weeks in Bangalore, India, to two weeks at the Australian National University in Canberra (3). In all these programs Number Theory has been used as the basic vehicle for the development of the student's capacity for observation, invention, the use of language, and all those traits of character which constitute intellectual discipline. (See plates 1 and 2.)

This has been a happy choice since number theory not only abounds in deep but accessible mathematical ideas but it is rich in opportunities for acquiring the kind of experience which is vitally needed in science and technology.

The traditional use of problems develops, in some instances to a very high degree, the student's capacity to resolve difficult problems provided they are clearly formulated.

For many years we have used suitably designed problems to develop also (4) the student's capacity for observation, to encourage the spirit of adventure in conjecturing and to develop their staying power in the exploration of what have been to them (as well as to the historical originators) some very difficult questions. Abstract reasoning is a very important tool in creative work in all the sciences and in technology. To master the use of this tool it is important, I believe, to give the student an opportunity to participate in the process of generalization. The design of problems must respond to this need as well (5).

It has been our practice in the above mentioned summer programs to group the problems in each Problem Set under subtitles which are meant to emphasize dramatically the purpose which the problems are meant to serve. In our Ohio State University program participants are given 40 sets of problems with the total number of problems exceeding 400.

The rather demanding studies during the first summer create a significant momentum for further studies in mathematics, science, or engineering. Those of our program participants who are not ready to enter college and who return to us for the second and possibly the third summer are able, as a rule, to study interesting and useful mathematics and science taught in a stimulating manner by our accomplished colleagues.

Math. Inst. Schülerseminar – Praktikum – Zahlentheorie, Aufgabe 7, 12/9/78
Heidelberg

ARNOLD E. ROSS

Exploration – Erforschung.

P1. In P10, Aufgabe 6 we discovered that if u_1, u_2 are elements of U_m, if n_1 is the order of u_1 and n_2 is the order of u_2 and if $(n_1, n_2) = 1$, then $n_1 n_2$ is the order of $u_1 u_2$. What useful observation can you make if $(n_1, n_2) > 1$? Perhaps the following example may help: Consider U_{19}. 12 is of order 6; 16 is of order 9; 3 is of order 18; 18 is the l.c.m. of 6 and 9; $3 = 16 \cdot 12^3$ and the order of 16 is 9 and of 12^3 is 2. We note that $(9,2) = 1$.

Prove or disprove and salvage if possible.

P2. If $u \in U_m$, if n is the order of u and if a is a positive integer such that $u^a = 1$, then $n | a$.

P3. 1) $\mathbb{Z}_p$ is a field if p is a prime. 2) $(\mathbb{Z}_3[x])_{x^2+1}$ also is a field. Explain carefully.

P4. A polynomial $f(x) = a_0 + a_1 x + \cdots + a_n x^n$, $a_n \neq 0$, of degree n with coefficients $a_0, a_1, \ldots a_n$ in a field F (F may be $\mathbb{Z}_p, \mathbb{Q}, \mathbb{R}, \mathbb{C}, \ldots$) cannot have more than n distinct roots. Hint: A field has no divisors of zero.

P5. If $u \in U_m$ and n is the order of u in U_m, then by P3 Aufgabe 6 (Prof. Jahno's lectures), $n | \varphi(m)$ and hence $u^{\varphi(m)} = 1$ in U_m. In particular $u^{p-1} = 1$ in U_p if p is a prime.

P6. $\mu(ab) = \mu(a)\mu(b)$. True in $\mathbb{Z}$. True in $\mathbb{Z}[i]$. True in $\mathbb{Z}_p[x]$.

Technique of Generalization – Die Kunst von Verallgemeinerung.

P7. On the basis of your experience with $\mathbb{Z}$, develop the arithmetic in $\mathbb{Z}_3[x]$. Would a similar discussion yield the arithmetic of $\mathbb{Z}_p[x]$, where p is a positive prime in $\mathbb{Z}$? Would such a discussion yield the arithmetic of $\mathbb{Q}[x]$? Of $\mathbb{R}[x]$? Of $\mathbb{C}[x]$? Justify your assertions.

Numerical Problems (Some Food for Thought).

P8. Find the g.c.d. (f,g) of $f(x) = x^4 - 3x^3 + 2x^2 + 4x - 1$ and $g(x) = x^2 - 2x + 3$ in $\mathbb{Z}_5[x]$.

P9. Use the results in P8 to find a solution $\{X, Y\}$ of the "Diophantine" equation

$$f(x)X(x) + g(x)Y(x) = (f(x), g(x))$$

in $\mathbb{Z}_5[x]$. Does our algorithm in $\mathbb{Z}$ apply here?

P10. Find the g.c.d. of $7 + 11i$ and $3 + 5i$ in $\mathbb{Z}[i]$.

P11. Out of the ring $\mathbb{Z}_3[x]$ construct a ring $(\mathbb{Z}_3[x])_{x^2+x+2}$ of remainders in division by x^2+x+2. Do this in a manner similar to that used to construct $\mathbb{Z}_m$ out of $\mathbb{Z}$. How many distinct elements are there in $(\mathbb{Z}_3[x])_{x^2+x+2}$? Calculate the number of units in $(\mathbb{Z}_3[x])_{x^2+x+2}$. Is the group of units cyclic? Is $\sqrt{-1}$ in our new "ring of remainders"? Is our ring a field? Justify your assertions.

Counting techniques – Die Kunst von Rechnung.

P12. Let $n = p_1^{\alpha_1} p_2^{\alpha_2} \cdots p_e^{\alpha_e}$ be the canonical factorization of a positive integer n into positive primes in $\mathbb{Z}$. Use the formulas in P18, Aufgabe 2 to obtain a formula for $\varphi(n)$.

P13. Calculate the number of units in $(\mathbb{Z}[i])_{3+i}$ by any method. Could one generalize the results in P12 and introduce a function $\varphi(a_1 + a_2 i)$ with $a_1, a_2 \in \mathbb{Z}$ which would give the number of units in the ring $(\mathbb{Z}[i])_{a_1+a_2 i}$ of remainders with respect to the modulus $a_1 + a_2 i$?

The Ohio State University, Columbus, Ohio 43210, U.S.A. Plate #1.
1980 Summer Program.

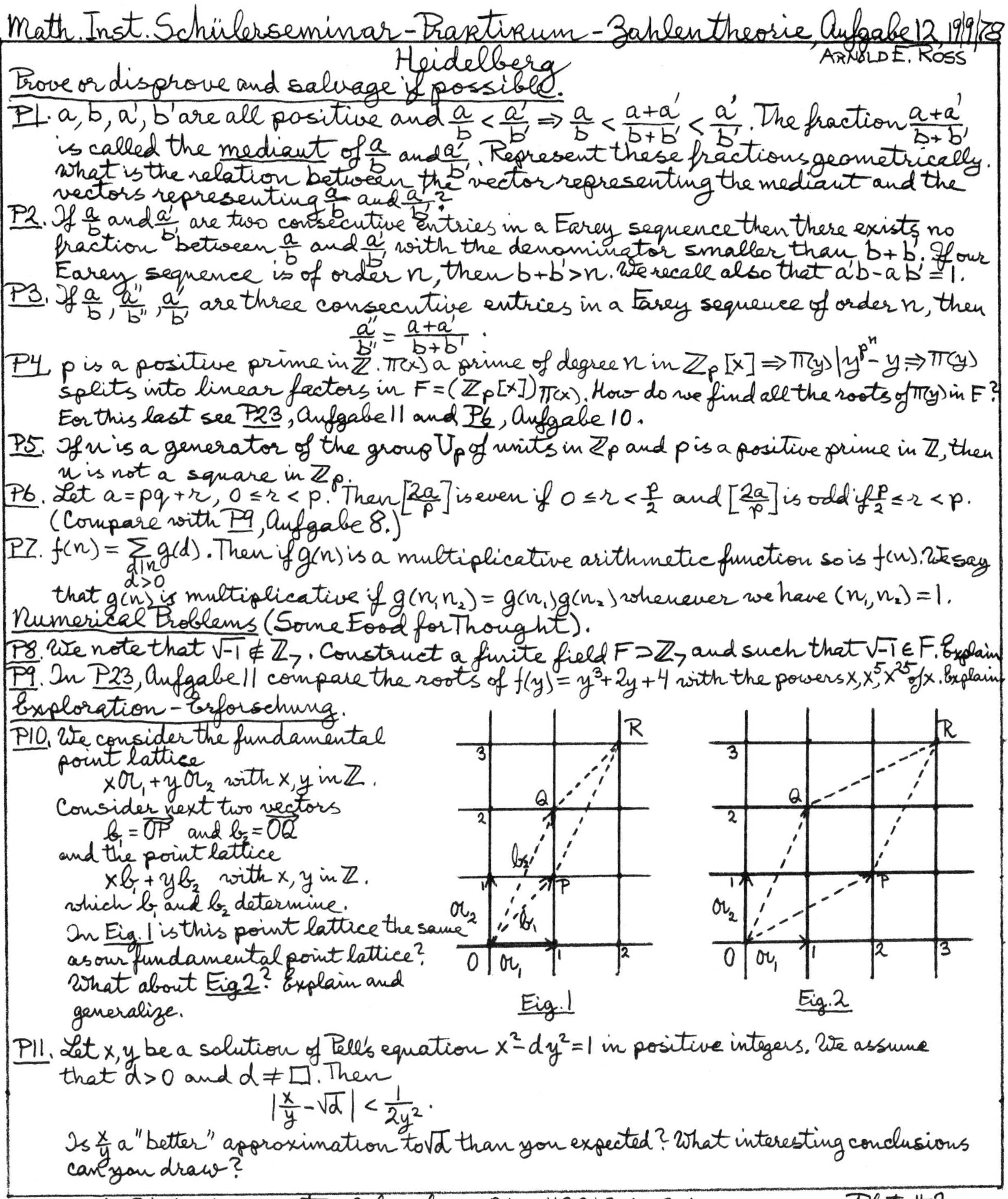

Math. Inst. Schülerseminar – Praktikum – Zahlentheorie, Aufgabe 12, 19/9/78
Heidelberg
ARNOLD E. ROSS

Prove or disprove and salvage if possible.

P1. a, b, a', b' are all positive and $\frac{a}{b} < \frac{a'}{b'} \Rightarrow \frac{a}{b} < \frac{a+a'}{b+b'} < \frac{a'}{b'}$. The fraction $\frac{a+a'}{b+b'}$ is called the mediant of $\frac{a}{b}$ and $\frac{a'}{b'}$. Represent these fractions geometrically. What is the relation between the vector representing the mediant and the vectors representing $\frac{a}{b}$ and $\frac{a'}{b'}$?

P2. If $\frac{a}{b}$ and $\frac{a'}{b'}$ are two consecutive entries in a Farey sequence then there exists no fraction between $\frac{a}{b}$ and $\frac{a'}{b'}$ with the denominator smaller than $b+b'$. If our Farey sequence is of order n, then $b+b'>n$. We recall also that $a'b-ab'=1$.

P3. If $\frac{a}{b}, \frac{a''}{b''}, \frac{a'}{b'}$ are three consecutive entries in a Farey sequence of order n, then
$$\frac{a''}{b''} = \frac{a+a'}{b+b'}.$$

P4. p is a positive prime in $\mathbb{Z}$. $\pi(x)$ a prime of degree n in $\mathbb{Z}_p[x] \Rightarrow \pi(y) \mid y^{p^n} - y \Rightarrow \pi(y)$ splits into linear factors in $F = (\mathbb{Z}_p[x])_{\pi(x)}$. How do we find all the roots of $\pi(y)$ in F? For this last see P23, Aufgabe 11 and P6, Aufgabe 10.

P5. If u is a generator of the group U_p of units in $\mathbb{Z}_p$ and p is a positive prime in $\mathbb{Z}$, then u is not a square in $\mathbb{Z}_p$.

P6. Let $a = pq + r$, $0 \le r < p$. Then $\left[\frac{2a}{p}\right]$ is even if $0 \le r < \frac{p}{2}$ and $\left[\frac{2a}{p}\right]$ is odd if $\frac{p}{2} \le r < p$. (Compare with P9, Aufgabe 8.)

P7. $f(n) = \sum_{\substack{d \mid n \\ d>0}} g(d)$. Then if $g(n)$ is a multiplicative arithmetic function so is $f(n)$. We say that $g(n)$ is multiplicative if $g(n_1 n_2) = g(n_1)g(n_2)$ whenever we have $(n_1, n_2) = 1$.

Numerical Problems (Some Food for Thought).

P8. We note that $\sqrt{-1} \notin \mathbb{Z}_7$. Construct a finite field $F \supset \mathbb{Z}_7$ and such that $\sqrt{-1} \in F$. Explain.

P9. In P23, Aufgabe 11 compare the roots of $f(y) = y^3 + 2y + 4$ with the powers x, x^5, x^{25} of x. Explain.

Exploration – Erforschung.

P10. We consider the fundamental point lattice
$x\mathfrak{a}_1 + y\mathfrak{a}_2$ with x, y in $\mathbb{Z}$.
Consider next two vectors
$\mathfrak{b}_1 = \overline{OP}$ and $\mathfrak{b}_2 = \overline{OQ}$
and the point lattice
$x\mathfrak{b}_1 + y\mathfrak{b}_2$ with x, y in $\mathbb{Z}$.
which $\mathfrak{b}_1$ and $\mathfrak{b}_2$ determine.
In Fig. 1 is this point lattice the same as our fundamental point lattice? What about Fig. 2? Explain and generalize.

Fig. 1

Fig. 2

P11. Let x, y be a solution of Pell's equation $x^2 - dy^2 = 1$ in positive integers. We assume that $d > 0$ and $d \ne \square$. Then
$$\left|\frac{x}{y} - \sqrt{d}\right| < \frac{1}{2y^2}.$$
Is $\frac{x}{y}$ a "better" approximation to $\sqrt{d}$ than you expected? What interesting conclusions can you draw?

The Ohio State University, Columbus, Ohio 43210, U.S.A.
1980 Summer Program.

Plate #2.

In the summer of 1980 we introduced a triple track course in analysis for our returnees. The elementary calculus of p-adic variable (Track 1) was taught by Professor Kurt Mahler and contained material from the second edition of his Cambridge Tract: On p-adic numbers and their functions. The basic real analysis (Track 2) was taught by Professor Bogdan Baishanski. A problem seminar pointing up the similarities and the differences between the properties of real valued and p-adic valued functions was directed by Professor Ranko Bojanic.

All of our advanced participants studied p-adic and real calculus. The more advanced among them also participated in one of the following three courses: A course in geometry taught by Professors Hans Zassenhaus and Jill Yaqub, a course in Algorithmic Methods taught by Professor Hans Zassenhaus, and a university course in Combinatorics taught by Professors Thomas Dowling and Neil Robertson was open to our advanced participants.

References

1. The Shape of our Tomorrows. A.E. Ross, American Mathematical Monthly, Vol. 77, 1970, pp. 1001-1007.

2. Horizons Unlimited. A.E. Ross, AAAS Education of Minorities Symposium, 1972, a preprint.

3. Telent Search and Development. A.E. Ross, Mathematical Scientist, 1978, 3, pp. 1-7.

4. Arthur Engel, Mathematische Schulerwettbewerbe, Jahrbuch Uberblicke Mathematik, 1979, Wissenschaftsverlag Bibliographisches Institut, Mannheim.

5. Towards the Abstract. A.E. Ross, Mathematical Spectrum, 1977-78, Vol. 10, No. 3.

GUIDANCE FOR TEACHERS IN THE IDENTIFICATION OF AND PROVISION FOR MATHEMATICALLY GIFTED CHILDREN

Graham T. Hoare
Dr. Challoner's Grammar School
Buckinghamshire, England

In a section concerning the provision of Primary and Secondary Education, the 1944 Education Act of England and Wales charged the local authorities with the responsibility of ensuring that their schools should provide education for all pupils according to 'their different ages, abilities and aptitudes'. It has become increasingly clear, however, that many with marked mathematical ability are insufficiently challenged and stretched in the classroom. In his fascinating book, 'Adventures of a Mathematician', Stanislaw Ulam, referring to mathematical talent, thinks 'there is an almost continuous passage from mediocrity to the highest levels of people like Gauss, Poincare and Hilbert'. Accordingly, below the potential genius level, we would expect to find a considerable reservoir of talent, and much evidence accrues to justify our belief in its existence. In a recent book of essays, entitled 'Mathematics Today', Allen Hammond refers to experimental programmes in which practicing mathematicians have taught very young ghetto children advanced algebra and similar subjects. The rapidity with which the students absorbed concepts far advanced beyond those normally taught at their grade level, their enthusiasm, and the rapport between them and mathematicians astounded professional educators. To the charge that, by showing particular concern for the gifted, we are creating an elite, we would reply that this minority already exists, whether we decide to ignore or frustrate it, or to provide appropriate facilities and opportunities for nurturing it.

Who are the gifted and how can we identify them? Although it is hard to quantify mathematical talent, an initial screening could be made by conducting, for example, a non-verbal reasoning group test at quite an early age. Choosing, administering and judging the purpose of tests, and interpreting the scores on them, require expertise which teachers generally do not have, and they would be strongly advised to seek the guidance of a trained psychologist in these tasks. We might add that we know of no written test which, unaided by skillful interviewing, could be used to isolate deep aspects or levels of mathematical understanding for individuals.

It is, however, classroom teachers who are the most important agents in the identification process and they need, therefore, to be aware of those characteristics which distinguish the mathematically able. The list below, which owes much to V.A. Krutetskii, can be used to guide and refine the teacher's subjective observations and assessments of individual children. By studying the processes, rather than the results, of mathematical thinking, Krutetskii's approach to the subject of mathematical ability differs markedly from that employed by factor analysts.

1. Facility for generalisation; verbally and symbolically.

2. Capacity and facility for abstract thinking.

3. Flexibility in thinking, ingenuity; inventiveness.

4. Striving for clarity, simplicity, conciseness.

5. Quickness in comprehending new ideas and the ability to apply them readily.

6. Logical acumen - a grasp of the notion of proof and of an algorithm.

7. Mathematical memory.

8. "Mathematical cast of mind" - a tendency to interpret the world mathematically.

These characteristics interweave and influence each other, although they may emerge and be recognised at different times in a child's development. Some of them may not be intrinsically mathematical qualities, but mathematically gifted pupils reveal them especially when dealing with numerical and spatial concepts and relationships. Number itself is an abstraction, and a facility for and a delight in handling numbers might be an early indication of mathematical talent, but care is needed to distinguish this from the more limited ability to do calculations.

In addition to the qualities already mentioned, there are personal factors which seem to be crucial in determining whether a person reaches his full potential, given that he has the required intellectual mastery of his subject. We regard the following, which depend greatly upon habits mostly acquired in childhood or early youth, as especially important in mathematics.

1. Persistence; stubbornness; endurance.

2. Independence of thought.

3. Self-confidence; venturesomeness.

The functioning of high mathematical ability is a formidable phenomenon and, to the trained eye, is difficult to miss. Most teachers will be aware of the pupil in a class whose mathematical abilities are markedly different from those of other pupils. Ulam relates how Banach's talent was said to have been accidentally discovered by Steinhaus when he overheard a mathematical conversation between him and another student as they sat together on a park bench! G.H. Hardy, who 'discovered' Ramanujan, made the following assessment of his outstanding intellectual attributes: "With his memory, his patience, and his power of calculation, he combined a power of generalisation, a feeling for form, and a capacity for rapid modification of his hypotheses, that were often really startling, and made him, in his own field, without a rival in his day'.

In view of the model we have given of the intellectual and personal qualities to be found in the mathematicaly gifted, what provision can be made for them? Unfortunately, classrooms have frequently failed to accomodate them. Ulam relates how he found his high school mathematics classes generally unsatisfying and preferred reading on his own. He did encounter a stimulating teacher when he was about sixteen, however, and having studied two of Sierpinski's books in depth, was able to engage him in discussions of set theory after the classes. Professor Lipman Bers of Columbia University, Latvian by birth, having been rebuffed by his teacher when he approached him about a problem that was puzzling him involving infinite sets, tells of how he felt lost until, at the age of seventeen, he chanced upon a book by Kamke in a bookstore which inspired him to take up mathematics.

Mathematicians speak of the power, beauty, unity, range and abstractness of their subject, and the universality of its language. Those who are capable of eventually appreciating mathematics in these terms need to be stimulated by intellectually challenging problems and ideas, in which the subject abounds, and to experience the joy and satisfaction of sudden insight, perhaps after sustained mental effort. It is precisely this kind of effort that those who find mathematics an intensely enjoyable and rewarding occupation are prepared to expend. Teachers, however, are usually preoccupied with those who can achieve but a limited understanding of the subject, with the result that very capable students can be left languishing. Again, it can be uncomfortable to encounter such pupils for they are often more clever than ourselves. They tend to ask awkward questions, make unexpected responses, miss out steps in logical arguments, and shun the rather pedestrian expositions offered. Bearing in mind that mathematical talent often appears early, the unusual insights and shortcuts of the gifted are apt to be particularly unwelcome to the junior school teacher whose own competence in mathematics may be limited. We admire Gauss' teacher, who readily admitted that his pupil had outstripped him, but the great mathematician-to-be was fortunate to find encouragement from Bartels, a pupil teacher. Galois and Cantor, by contrast, were either misunderstood, ignored or riduculed in their time, but their ideas were eventallly accepted by the mathematical community.

Mathematically gifted children, who race through the usual offerings, need opportunities for independent study and accelerated learning. Providing these requires considerable organisational flexibility. Challenging them with problems which require mulling over, or encouraging them to write essays and articles, will be greatly more beneficial than setting more routine work. If the work is investigative or requires reading mathematical material, they may need guidance in acquiring appropriate study skills.

Polya has said that mathematics is not a spectator sport, and Emil Post recorded in his Diary that he studied mathematics as a product of the human mind and not as absolute. Here, mathematics is viewed as a process rather than as a body of knowledge or a set of skills. To be effective a teacher needs to encourage pupils to formulate and tackle problems, to make generalisations, model situations, make conjectures and attempt proofs. An alert teacher will recognize and exploit opportunities pupils have for making mathematical rediscoveries. Acquiring the techniques appropriaate to the level of the work involved presents few difficulties for the gifted, but even here there are possibilities of sudden insights. Professor Penrose, of Oxford University, writes how he became intested in particular cases of the relation, $a + b = ab$, and how he became fascinated by the power of algebraic manipulation when his older brother showed him how to deduce: $1/a + 1/b = 1$.

Maintaining an atmosphere of spontaneity helps to sustain interest. Ulam tells how Banach's lectures were none too well prepared and, though he invariably pulled through, watching him struggling at the blackboard was most stimulating. Recently, I was discussing various proof strategies with a class of bright sixteen-year-olds. They were making conjectures about the Fibonacci sequence, which I had chosen to illustrate these strategies when, at the end of the session, a strikingly beautiful one was proposed which I had not seen before. We agreed to attempt a proof when we next met, and I was leaving the room when one student made the perceptive remark: 'you should not look at this before next time; then we shall get a better idea how to go about proofs!' Needless to say, they were ready with a correct general formulation of the conjecture at the next lesson.

If a school is short of mathematically qualified teachers it is important that, once identified, able pupils should get the best mathematical teaching available at least once a week. It would be valuable to assemble an easily accessible bank of resources, to which each teacher should be encouraged to contribute, and to designate someone to coordinate the work with able children and ensure they are recognised by colleagues.

An HMI report on 'Gifted Children in Middle and Comprehensive Schools' in England and Wales, published in 1977, states that: 'In the experience of HMI, in almost all cases where the mathematical attainments in a school are unusually high, there is some form of voluntary activity outside normal lesson hours'. A mathematics club, computer activities, problem competitions, can all play their part in augmenting classroom activities. Computers can be imaginatively exploited by teachers and pupils alike to illuminate important mathematical concepts and problem-solving procedures and to effect simulations.

Beyond school confines many activities, such as the International Olympiad, National Competitions, Saturday clubs, the Russian Pioneer Clubs, the 'Study of Mathematically Precocious Youth', (SMPY), at the Johns Hopkins University, Baltimore, exemplify the regard educators have for mathematics in general, and their concern for the mathematicaly gifted in particular. Mathematicians have not sold their subject well and have tended to work in isolation, but there are indications of greater cooperation, essential for continuity, between those who work in various educational institutions. What is required is a climate of public opinion which values excellence and appreciates the importance of 'transmitting to the young mind the beauty of intellectual work', and mathematicians have a responsibility in helping to create it.

21.2 DISTANCE EDUCATION FOR ADULTS

DISTANCE TEACHING FOR ADULTS: THE PROBLEMS OF THE DISTANCE LEARNER

Michael Crampin
Open University
Milton Keynes, United Kingdom

(This is an edited version of the presentation given at Berkeley. The actual presentation was copiously illustrated with overhead projector slides, tape recordings, and film of teaching material from two Open University courses: the Mathematics Foundation Course (Code M101), an introductory mathematics course; and 'Developing Mathematical Thinking' (Code EM235), a course in mathematical education.)

Introduction

We discuss the problems that face students of mathematics and mathematics education, in distance teaching programmes, and how these problems are tackled at the Open University of the United Kingdom. These problems arise because such students are geographically isolated from their teachers and fellow students. But many students of mathematics, not just distance learners, are isolated in another sense: they feel a sense of isolation which comes from the difficulties they experience in learning the subject. The special problems of distance teaching force one to face this second problem of isolation, too, and to find some new ways of tackling it. We believe, therefore, that we have something of value to share with all teachers of mathematics.

Background to the Open University

The Open University was established just over ten years ago for adult students who study part-time in their own homes. The University operates on a national scale. It employs about 2,000 people; the Mathematics Faculty and the Educational Studies Faculty each have over 40 academic staff. The Open University Production Centre of the British Broadcasting Corporation has about 350 staff engaged exclusively in the production of broadcasting and audio-visual material for the University. The mainstays of the University's teaching activities are correspondence tuition - about 4 million packages of correspondence material are mailed to students each year - and broadcasting - 1300 hours of educational television programmes are broadcast by the national broadcasting network per year. The Open University has some 62,000 students at present, and produces about 6,000 graduates a year. About 10,000 students will be taking a course with a significant element of mathematics at any one time. The University operates a totally open admissions system, with one proviso: students must be over 21 years of age. The median age for students on admission is about 32. The students are drawn from a wide variety of backgrounds. Teachers (21%) form a significant proportion of the student population,, but not a dominant one: professional people (managers, scientists, engineers) and technicians (computer programmers, draftsmen, laboratory assistants) are also strongly represented. A rather different but important category is disabled students.

There is clearly a danger that students will find such a large-scale organisation dispiritingly impersonal. A great deal of thought has gone into a human touch to all interactions between the student and the University. The necessity of providing some personal contacts for the students has long been recognised. Each student is assigned to a tutor and a counsellor, who normally work part-time for the University. There are various opportunties for the more normal kind of face-to-face teaching and discussions: for example, students can meet tutors, counsellors, and fellow students at local 'study centres', of which there are 300 dotted around the country.

However, most of the student's time is spent working on his own at home. He studies from correspondence material, from broadcasts and other audio-visual material, and by doing assignments or homework. A half, or more, of the 10 to 12 hours he spends studying during a typical week will be devoted to correspondence material, the remaining time being divided between the other elements (including possibly a class tutorial at the study centre), so that for example an hour or so might be taken up with viewing a television programme and doing related work.

The Open University degree is credit based, six credits making up the ordinary degree. The student has to start his programme by taking a Foundation Course in any one of five areas of study: Arts, Social Science, Mathematics, Science, or Technology; there is no Foundation Course in Educational Studies, though there are higher level courses in this area. Having taken his first Foundation Course - - a normal programme, without exemptons, must contain two Foundation Courses - - the student goes on to take courses in whatever subjects he choses. There is no compulsion to specialize. So a student may spedialize to some extent in Mathematics, by taking one and a half credits, say, in

Mathematics and a half credit in Technology; but there is nothing to prevent his taking in addition courses in Science and Educational Studies, if such a pattern of courses suits his particular needs - - if he is a teacher, for example. By not dictating to students which courses to take, but allowing them to make their own choices, we feel that we treat them as human; beings and not as cogs in the machine.

The Teaching Process

Once a student has enrolled for a course, he begins to receive the course materials. The correspondence text for each week's work is an attractively printed and bound booklet containing, in the case of the summary of the television programme for that week, and a section to be studied in conjunction with an audio-tape. In addition, there is a booklet which contains extra practice exercises and another containing the assignment that the student has to submit as part of the continuous assessment for the course.

One of the major problems we have to face is how to humanise the teaching. The use of audio-tapes has been very successful in this respect. The idea is quite simple: the student has a set of visuals in the correspondence text, which the voice on the tape guides him through. The student experiences something not unlike individual personal instruction, In the production of these tapes we have tried to ensure that each speaker adopts a friendly, informal manner of presentation.

The expository part of a correspondence text cannot be like a textbook. We try to emulate in print the same kind of approach that we use for audio-tapes by adopting a pleasant, conversational style. The material is properly designed and printed, which conveys to the student the feeling that care and thought are being devoted to his needs.

Television, or film, is sometimes used in a very impersonal way. We believe that, valuable as television is as a means of showing pictures which could not be shown in any other way, it also has an important point to play in providing psychological support for the student. Thus, although our television programmes are a long way from being merely televised lectures, they almost almost always make extensive use of lecturers, or presenters, who address the student audience directly.

Fostering the Private Rewards of Study

For the isolated learner, keeping going is more difficult than it is for the ordinary student: it is very easy to drop out. There are two motives a student has for keeping going: the public rewards such as the value he expects to get from his degree, or the praise of his family and friends; and the private rewards of satisfying his curiosity, his desire to find out about things. We have to make these motives work for us, we have to foster them.

We look first at the private rewards. One important function of the television programmes is to intrigue the student. It is widely recognised that getting and keeping students' interest is one the most difficult problems in teaching mathematics. It has often been said that the way to do it is to make the mathematics relevant - - to provide lots of real world examples. In the past we have tried making television programmes which showed, for example, industrial applications of the mathematics we were teaching. When applications was the main topic of the course, that was fine. But for basic mathematics courses not essentially concerned with applications, this approach did lead to failures. We now look for ways of intriguing the students within the topic we are actually trying to teach. One example of this is a programme for the Mathematics Foundation Course which is concerned with the integral definition of the natural logarithm function. The programme begins with a physical demonstration of the logarithmic property of the areas under the graph of the reciprocal function, in which pieces of card cut to fit under the graph are weighed against each other on a chemical balance. Being intrigued by the unexpected fact that the pieces balance if their end points satisfy the appropriate product rule, the viewer is invited to examine evidence that the areas add like logarithms. The rest of the programme introduces in visual terms the concept of integration and makes copious use of animated diagrams to demonstrate how areas are affected by scaling. This leads to a final animation depicting the proof of the logarithm property.

The same programme is an example of how incidental learning is built into the course. Though it may not be apparent from the description above, the main purpose of the programme for the students is to review ideas that they met earlier in the course: ideas such as logarithms and powers, graphs, transformations of the plane such as scalings, and nests of intervals given by upper and lower sums. The students had met nests several times before in simpler contexts: in the bisection method of numerically solving equations, where they gain concrete experience of the idea through the use of a calculator; and in the definition of a^x for the irrational x. And the discussion of how areas are calculated by integration is itself incidental to the main purpose of the programme.

So learning becomes an exciting and dynamic process for the student, with each stage rehearsing ideas from past studies, but raising questions in his mind to be answered in later study.

It is easy to give a false impression of the role of television in Open University courses. Our television time is limited to at best one programme for each 12 hours of study. Our courses cannot be described as television courses. It is quite the reverse: the text carries the course material; we use television to pick out certain ideas so that in the act of viewing our students can rehearse newly studied concepts. We try to devise programmes which will at the same time lead to new ideas which themselves are developed in the correspondence texts. We regard it as essential to provide both pre-programme and follow-up work linking the television to the other study sections of the course.

Teaching the Teachers of Mathematics

So far we have concentrated on the problems of teaching mathematics: but among our students are many teachers, some of whom are teachers of mathematics. These teachers may take mathematics courses, but they may also study, in education courses, teaching techniques that may be applied to their own teaching - - the teaching of children. Teaching the teachers of mathematics is an important part of our programme. The difficulties of the isolated learner are not just confined to the isolated learner of mathematics.

In the United Kingdom there are large numbers of teachers of mathematics who are not in any sense mathematical specialists: Mathematics is only one of the many things that they teach, and they are often much more confident in other areas of the curriculum. A large proportion of the mathematics teaching that children from 5 to 14 years old receive is provided by such teachers, and it is them that we are most anxious to help. One of the two courses we are developing which is intended for this audience is called 'Developing Mathematical Thinking'; the comments below relate to this course.

As with the Mathematics Foundation Course, we aim to use television in the mathematics education course to keep the student constantly involved with what he is studying, though the subject matter is very different. Our students see short extracts from a number of mathematics lessons, recorded in the classroom; they are to think about them and analyse them. It is hard for students to do his adequately if they can watch only once, which would be the case with a transmitted programme. So for this course we are making video cassettes, which students can view in small groups at teachers centres or in schools.

Many of the problems children encounter in learning mathematics arise from a premature rush to abstraction. We believe that this can be avoided by teaching according to the following model. To begin with, there should be a long period of discussion about practical situations, perhaps involving the manipulation of apparatus, or pattern making, or some similar activity. When the children begin to sove problems without much recourse to the apparatus, it is time for them to make a record of what they are doing, in their own terms, often in words and pictures. Gradually this record making becomes more and more abbreviated, until a conventional notation is reached. Then and only then is it appropriate for them to practice 'sums' in this standard notation.

This approach has two main advantages. First, because the path from concrete experience to standard algorithm is securely built, the algorithm can be reconstructed if partially forgotten. Second, because the skill stems from a rich immersion in varied practical situations, expressed both physically and verbally, it has become a generalised idea, and hence more readily applicable in new contexts.

The videotapes give the students of the course the opportunity to view activities in the classroom and discuss them in terms of the model. One videotape is concerned with the concept of area. The children's activities progress from filling up a sheet of paper with irregular shapes (potato prints, in fact), via a similar activity using regular shapes such as triangles and rectangles, to discovering how to find the area of a relatively awkward shape such as a parallelogram by dissection. We provide one or two key questions for the students of the course as a focus for their discussion and analysis of the activities.

Our testing of material like this reveals that our students do get very involved in discussing the children's activities and do a lot of useful thinking. But of course, viewing sessions like these are not enough in themselves: the ideas they trigger off need consolidating and amplifying, and this is done in the course work which follows. At this stage, the student is working at home on his own, mistly with printed material. But he will have a collection of vivid experiences which can be constantly used in the correspondence text.

Moreover, much incidental learning occurs here too. Our students gradually get more and more understanding of the role of discussion, and of practical experience in general, although their attention is primarily upon the particular content involved - - area, subtraction, elementary algebra, or whatever. In addition, they get a feeling for ways of organising a large class so as to make working in this way a practical proposition.

Fostering the Public Rewards of Study

We mentioned earlier the importance of public rewards as a motive for keeping going. To harness this to our purpose of encouraging the student to keep studying, we have to provide this kind of reward pretty frequently. This is why we use continuous assessment. The student has to do assignments about once a month through his course, which are marked by his tutor, and the marks count, half and half with the examination at the end of the course, towards deciding whether or not he passes the course. The tutors are encouraged to comment freely, in writing, on the student's assignment. This correspondence tuition is a very important part of the teaching process.

We are not ashamed to teach directly to the achievement of good scores on the assignments and examination, by teaching techniques, for example. Audio-tape can be very useful for this purpose. For example, in the calculus section of the Mathematics Foundation Course we decided to devote all the audio-tapes to helping students with techniques. We concentrated on two topics: differentiation, especially the chain rule; and on that traditional stumbling block, integration by substitution.

In designing audio-tapes of this kind we have found it necessary to devote a lot of attention to three things: correlating voice and visual so that the student's eye would always be at the right place in the frame; secondly, providing a clear work structure for solving the problem; and lastly, encouraging the student to do progressively more of the steps for himself.

Each of the various media at our disposal can be used to help the student keep going by offering the frequent reward of knowing that he is making progress towards getting his degree. They can be more effective if used in conjunction: for example, problems about how the geometrical properties of graphs relate to the analytical properties of the functions they portray may be posed in print, but are demonstrated more effectively on television. The reward of knowing that one has correctly answered the problem comes when one reads the solution in the correspondence text.

The Course Team Approach to Teaching

One of the peculiarities of the Open University is that, although the student is isolated, the teachers work together as a team. This team approach to teaching is forced on us by the need to employ skills not usually held by academics, but it is also a distinctive contribution to teaching practice.

A typical course team is a collection of academics, specialist staff, external consultants, and members of the BBC. The course is prepared co-operatively by this group, which may be as large as 25 or as small as 5 or 6. The advantage of the course team approach is that it gives us the chance of extracting the best of people's specialisms and expertise.

It is our experience of the team approach that small working groups charged with developing specific parts of a project will generate new ideas, now innovations, that an individual would not generate alone.

Summer School

One important feature of many courses at the Open University, which we have left till last because it seems to be like ordinary teaching, is Summer School. Students of the Mathematics Foundation Course, and indeed of about a quarter of the courses the University offers, have to attend a week's residential course as part of their studies. This takes place in the middle of our academic year, which coincides with the calendar year. Since the students are together for such a short period, we put a lot of effort into planning Summer Schools so that they get the maximum benefit; it is a very intensive experience for them.

The least important part of the Summer School programme is the lectures, of which there are comparatively few. The thing that students can do at Summer School, which is difficult for them in the rest of the year, is working together, so this is what we concentrate on. A large part of the programme is devoted to what we call investigations, in which the students work together in groups of four to eight, with a little guidance but no direct assistance from a tutor, on open-ended problems we have devised. It is stressed that if is more important for the students to explain their thinking than to reach some final conclusion. One of the assignments for the course is based on this activity. The point of interest here is that we give credit for imagination and perseverence, and particularly for making an honest attempt to reflect on and record one's own thought processes in investigating a mathematical situation, rather than for correctly answering the question: indeed, the questions posed do not have single correct answers. One of the questions we use is this: road repairs on a two-lane road reduced it to single file traffic - - how should the temporary traffic lights be timed? Another, more abstract, is this: a polygon is drawn using points of the unit grid in the plane as vertices - - how is its area related to the number of grid points involved? These investigations are perhaps the clearest expression of the desire, which runs throughout our work, to get the students actively involved with mathematics - - with doing mathematics, rather than simply learning it.

Conclusion

Our purpose was to demonstrate what has been learned, in the ten years of the Open University's existence, about how one should tackle some problems which are particularly severe for the teaching of mathematics, and mathematics education, at a distance, but which are relevant in general: the problem of emphasising the human element; the problem of fostering the personal aspect of motivation, of capturing and developing the student's interest; and the problem of using to best advantage the public rewards for successful study. The most convincing evidence for our success in dealing with these problems is provided by the teaching materials that we produce.

References

The following audio-visual materials were used to illustrate the talk.

Audio-vision extracts from:-
M101 Mathematics Foundation Course
- Block I Unit I "Computation"
- Block III Unit 3 "Integration"

Television extracts from:-
M101 Mathematics Foundation Course
- Block II Unit I "Functions and Graphs"
Block II Unit 5 "x 1/x, Areafor Revision"
- Block VI Unit I "Complex Numbers"

EM235 Developing Mathematical Thinking
Four extracts from location recordings in schools not yet edited into programmes.

21.3 ADULT NUMERACY - PROGRAMMES FOR ADULTS NOT IN SCHOOL

ADULT NUMERACY - - ADULTS NOT IN SCHOOL

Anna Jackson
BBC Television
London, England

I am a producer in the Continuing Education Department of BBC Television, London, England. Recently, with Chris Jelley, I produced a series of television programmes designed to help adults not in school to improve their basic mathematics at a practical, everday level. Peter Kaner was a consultant to the series; he was then Inspector for Mathematics in the Inner London Education Authority and is now working writing mathematics books.

The theme of the session is : "Adult learners need adult methods of teaching: What are the key characteristics of successful learning material for adult innumerates?" With our recent experience on this series of programmes in mind I will address this question with particular reference to distance learning by television.

I believe myself that nothing can replace the direct relationship between a student and a good teacher; however, as we know, this is not always possible. Adults may have difficulty in attending classes, whether daytime or evening; ot there may be no classes available. Also, television may by its very nature be able to provide learning aids may be adapted for class use, and I hope that those television aids not open to the class teacher. Nevertheless, many of these television aids may be adapted for class use, and I hope that those among you who are teachers may also find something of interest in the points I shall raise.

I should like to take in turn a number of characteristics that we found to be crucial to our audience. (In the Congress presentation each one was illustrated with a relevant extract from the programmes.)

There have been so far three series of television programs in Britain directed towards the less numerate adult; that is , adults who have difficulty with basic mathematics. Yorkshire Television's 13-part series "Make it Count" was shown in 1978 and aimed to teach elementary arithmetic - - simple addition, subtraction, multiplication, and division, and went on to cover fractions and decimals. The BBC series "It Figures," which I produced, consists of ten 25-minute programmes which have now been transmitted three times since 1979 and aim to help the home-based learner who can manage with the four rules, but who needs revision and help with decimals, fractions, percentages, graphs, and/the application of arithmetic skills in everyday life. Yorkshire Television followed with "Numbers at Work," which covered similar ground, using a variety of situations where mathematics was required to solve a problem at work.

The criteria which follow, and on which the "It Figures" programmes were based, were partly the result of some formative research commissioned by the BBC and carried out, while the series was in preparation, by Professor Percy Tannenbaum of the University of California at Berkeley. Professor Tannenbaum spent several months with us in London preparing the research and evaluating the results. His conclusions are presented in this report.

There were enormous limitations of time, financial resources, personnel, and therefore sampling. I made four pilot programmes, each aiming to teach the same content, but in different ways, and with different combinations of production method and presentation. These programmes were taken all over the country to show to sample groups of adult learners. About a week previously, they had all been given a test based on the content of the programme. Immediately after viewing one of these four pilot programmes, they were asked to do another test to see whether they had learned anything! They were also asked to answer a simple questionnaire to grade their responses to various elements in the programme.

Due to the sampling limitations I have mentioned, evaluation was difficult. For me, the most valuable aspect was to sit in on the piloting sessions, watching the responses of the adults watching the programmes and hearing their comments afterwards - - especially if I could draw them out individually over a cup of tea. "In tea, veritas."

This was particularly important because I have my doubts about the possibility of teaching, say, fractions, in one 25-minute programme to a wide range of unseen adults who bring various types of half-knowledge or despair to each section of the topic.

What I was aiming at above all was to cultivate a positive attitude to mathematics: that it could be pleasant, even fun; that failure was not inevitable; that it could be amusing, relevant, alive, and accessible.

So it was a result of these observations, a mixture of formal evaluation and product-testing, which led to the presentation methods and format of the final programmes and which I will illustrate.

1. Encouragement

The first characteristic I should like to discuss is encouragement: many adults approach the whole business of learning basic mathematics with a built-in sense of failure, and sometimes even a positive fear of anything to do with numbers. To overcome this we wanted to show that we are all sharing a common difficulty, and most programmes begin with a selection of quick street interviews, so that the viewer can identify with other real people confronting similar problems. For example, Programme One of "It Figures," which deals with estimation, has street interviews after the front titles and the introduction. Jimmy Young, the presenter is not a teacher, but a well-known disc-jockey, who has his own two-hour radio show and attracts seven million listeners every morning. This part of our plan, to give viewers a feeling that having difficulties with basic mathematics is not something to be ashamed of, because well-known people are prepared to be involved with putting it over.

The use of these street interviews was very much liked by our viewers, and Jimmy Young generally went down well - - more with the students than with the teachers!

2. Participation of "Real" People

Following from this identification with other people, the second point we considered essential was the participation of "real" people in the programmes, and we decided to have a panel of "real" people in the studio learning and tackling the problems along with the viewer. I say "real" as opposed to "actors" - - which I shall come to later. In fact, our panel are reasonably well-known television personalities of different kinds, none of them particularly good at mathematics, none of them primed beforehand. For example, in "Going Metric," Programme 5 in the series, we had a panel measuring their hands.

The panel was not considered altogether successful by some of our viewers,who felt patronised by celebrities whom they thought were sometimes pretending to have difficulties when they were not. Other viewers, however, liked the variety of people and voices. They were glad that Jimmy Young was not teaching all the time and that answers were being given by other people and then reinforced by the teacher.

3. Real-Life Situations in Documentary Film

The third characteristic, which seems key to all of us, was the use of real-life situations. Adults are motivated to learn by their practical need for mathematics in everyday life; even more than with children, examples can relate to their own experience. Some adults will have vague memories of having to calculate how much water is running out of the bath if the speed of the train going by is so many miles per hour - - which did not seem to mean much in terms of their own needs. We wish to confirm the adult's desire to learn by showing how simple mathematics is genuinely needed in a wide range of different situations, and that we are dealing not with abstractions but with reality.

Of course, the difficulty with reality is that its complexity obscures the fundamentals of the problem: an important element in coping with real situations is to know which calculation to use to solve a problem. In Yorkshire Television's "Numbers at Work," for example, several different situations are shown that lead to the same calculation - - twenty-one take away nine, how many left?

4. Use of the Media - Entertainment and the Wider World

Leading on from everyday life, I occasionally used material from the media which might stimulate the adult to be aware of numbers used in the wider world, in the context of sport, or in entertainment. In the "It Figures" programme "Going Metric" I showed a sequence of sporting events in the hope that they would entertain the viewer, thereby encouraging him or her to accept the reality of metric measurement in a pleasurable way. Not every country has this problem, of course, but in Britain many adults are psychologically reluctant to accept metrication. This item was designed not as formal teaching "so many metres is the same as so many yards," but to give an impression of metric distance in an entertaining way.

5. Dramatised Material

Another problem with reality is that it does not always fit into neat packages which suit the teacher. In the making of these programmes I have had to reject real-life situations over and over again which confused a teaching point by introducing others, not yet dealt with. So the next characteristic which seemed to very successful with our adult viewer was the real-life situation tailored to make a particular point, in a dramatised form. Every "It Figures" programme had a short dramatised sequence, using the same three characters - - sometimes only two of them. Sometimes they were measuring a room in metres; sometimes making a temperature graph for a sick wife; sometimes estimating the amount of money spent on shopping. In the programme "Pictures and Charts," which dealt with reading pictograms and bar charts, our two friends were in their work canteen for a British tea-break. From time to time we tried to put in a bit of humor.

6. Simple Strong Images

Another key characteristic is the use of simple, strong images to convey a teaching point. Cartoon film animations can be an ideal way of making something memorable - - and it can be humorous as well. We used an animation to put over concepts of fractions and to illustrate the use of co-ordinates. In the first programme we used a cartoon sequence with fish to teach grouping numbers in tens. The response to this little piece of film was overwhelming. Students told us that they understood the grouping in tens idea for the first time. We went on to reinforce this by taking it on into decimals in a later programme. On the other hand, we found we had to take care in our use of cartoon animation. During our pre-programme research, when piloting the series, we included in the pilot programme another cartoon making a teaching point. Some of the viewers complained that cartoons were "childish," and they clearly felt patronised; others, however, enjoyed it. We tried to take these responses into account when making the programmes.

7. Straight Teaching

Of course it would be wrong to ignore the main element of successful learning to a motivated audience - - straight teaching. This is obvious, but I do not want to make it seem as if it has got lost for that reason. Adults who wish and need to improve their basic skills are prepared to listen to straight instruction, although, as I have suggested, we found it advisable to vary the mixture to take into account the attention span of the viewer. The pattern we followed was teaching/ reinforcement/ practice.

I mentioned at the beginning that nothing can replace the one-to-one relationship of the student and the teacher. Although distance learning by television at home may be the only way open to the adult, it is not enough for a subject like basic mathematics. Both the BBC and YTV provide back-up material in the form of books that provide reinforcement and materials for practice, two more essentials for successful learning. In addition, each of the "It Figures" programmes ended with an interview with someone - - and a "real" person again - - who was attending classes. This was designed to show adults with a sense of failure dating from their schooldays that these classes could be different. The interviewees were all of different ages and types, and the viewer could identify with their fears and be encouraged by their success.

The Adult Literacy Support Services Fund provided the funding for a telephone referral service after each programme. Members of the public were invited to phone in if they needed help. Their names and addresses were taken and passed to their nearest numeracy co-ordinator, who then contacted them about classes in their area. The response was better than expected, but still not as hoped. However, many viewers may have felt unable or unwilling to go to classes, and bought the book to study at home.

The "Make it Count" programme sold 8,000 workbooks.

The "It Figures" book has sold, up to May this year, over 15,000 copies and has gone into a second edition. The programmes are still being repeated from time to time during the mornings on television, and we expect the book to continue to sell as new viewers discover the series. The BBC is currently planning a two-year CSE/0-level equivalent course in mathematics linked to a flexi-study system.

IT FIGURES A BOOK FOR ADULT INNUMERATE PEOPLE

Peter Kaner
Inspector of Schools
London, England

Adults with numeracy problems are not necessarily non-readers, so the notion of teaching from a book is quite realistic. There are, however, some difficult problems to face if the book is to function without the support of TV programmes. (Even people who are watching the programmes may miss some programmes or start after the first programme.)

Some of the problems are:

1. The variation in education skills of the learners.

2. The wide structure of situations in which the book must be used.

   a. numeracy classes with teacher

   b. groups without teacher

c. isolated learners who have purchased the book from a bookshop.

3. The wide range of motivation of the students

   a. trying to improve so that they can get a better job

   b. learning to cope with the numerical aspects of everyday life (particularly financial)

   c. wanting to understand an important subject in which they failed at school

   d. wanting to help their children or wanting to avoid seeming stupid to their children.

Features of the book which attempt solutions of the problems.

1. A helpful design which gives examples, explanations, and practise questions for everyday situations.
2. Use of a relaxed style of pictures and cartoons.
3. Emphasis on the use of calculator in solving everyday problems but also offering simple techniques in pencil and paper arithmetic and estimating.
4. Tabloid style, avoiding long words and technical definitions. Short sentences. Use of topical problems (although this can make the book out of date).
5. A wide view of numeracy as the ability to solve quantity problems but also to read graphs and tables and interpret statistics in diagram form.

Copies of the book and information may be obtained from BBC Publications, 35 Marylebone High Street, London W1, England.

21.4 PROBLEMS OF DEFINING THE MATHEMATICS CURRICULUM IN RURAL COMMUNITIES

PROBLEMS OF DEFINING THE MATHEMATICS CURRICULUM IN RURAL COMMUNITIES

Desmond Broomes and P.K. Kuperes
University of the West Indies
Bridgetown, Barbados

Introduction

Emphasis of the development of rural communities focuses our attention on the 70-95% of the population of developing countries who live in rural areas, especially the 85% of all children for whom the primary school is the sum total of their formal schooling. To show ways through which mathematics educators have grappled with problems of rural development, we propose to explore certain aspects of mathematics education at the primary level. The methodology used allows for easy extrapolation to the secondary level. We will explicate, very briefly, some concepts associated with the ruralisation of primary education and, in so doing, discover what contributions mathematics education can make towards realising these concepts.

Primary mathematics as a set of experiences for terminal primary school leavers.

Ruralisation of primary education implies, among other things, that primary mathematics education should be degined as ensuring at least a threshold level of mathematical learning required for effective participation in the economic, social, and political life of rural communities. Learning experiences such as organising the water supply in a rural village; participating in a cooperative marketing society; building a hen-house, a fence, or a school latrine; and pricing and marketing goods and services, not merely represent interesting applications of mathematical skills, but also define the minimum levels of mathematical understanding and skills to be achieved at the end of terminal course in primary mathematics. And most important, these learning experiences are organised as genuine instruments of reconstruction of rural life.

The objectives of such a mathematics course have therefore to be expressed in terms of (1) knowledge and skills to understand and undertake the keeping of farm records and household accounts, to read the plan of a village water-supply scheme, etc.; and (2) attitudes to participate in making common decisions and plans and to work cooperatively.

The content of a primary mathematics course concerned with effective social and economic participation has to be derived mainly from task analyses of common everyday experiences.

The environment in which the primary school leaver lives may be analysed into at least three domains: economic, technical-professional, and social-environmental. An adequate and valid content of primary mathematics may be specified by doing task analyses within each of the three domains. This may be complemented by a participatory approach that allows more information about local needs, problems, and experiences to become available for effective planning and implementation, and also ensure closer integration of activities from below rather than from above, within rather than from outside. These approaches to defining the content of primary mathematics require that within each domain (1) the learning needs of the primary school leaver be identified and (2) the learning process itself be brought into closer contact with the realities of everydaylife in rural communities. They also allow the teaching and the instructional materials used to exploit the interdisciplinary character of the three domains.

Primary mathematics as a set of experiences leading to secondary mathematics

The emphasis on a terminal primary mathematics programme for rural communities must not lead us to ignore the interests of those who have the capacity and the opportunity to continue their schooling at secondary level. Educational policy in rural communities should allow for at least three basic functions of the primary school: (1) to teach basic skills of communication and arithmetic (literary); (2) to prepare the better pupils for secondary schooling (propaedutic); and (3) to serve as a vehicle for rural reconstruction (developmental).

All too often mathematics is seen only as a stepping stone towards secondary mathematics, which in turn is seen as a preparation for a university education. Such a view tends to emphasize mathematics as a discipline, as an organised and structured body of knowledge. To introduce at the primary school level certain mathematical concepts merely because they have been found to have great organising and explanatory force at a highly abstract level is educationally unsound.

But mathematics is not a discipline alone. It contributes to the training of thought, underlies activitiy, and is an approach to problem-solving. We are therefore committed to using approaches for teaching and learning mathematics in which mathematics is viewed essentially as a process:

> We learn mathematics by doing it.
>
> The nature of mathematicas thought is best discovered when we perform experiments, design symbolisms, and recognize structures.
>
> We cannot become completely familiar with the processes of mathematical thought through a study of its end-product: we have to do mathematics in order to get to know the processes and be able to use them in analysing our environs and creating new syntheses.

Thus, the propaedeutic function of the primary mathematics programme (to prepare the better pupils mathematically for secondary schooling) may be described as an attempt by curriculum developer, textbook writer, teacher, and examiner to challenge the pupil, wherever possible, to recognize similarities and differences, to proceed systematically, to symbolise, to generalise, to find formulae for generalisations, to use models, to schematise models leading to algorithms, to recognize isomorphisms, to reason on the basis of symmetry, to proceed inductively, and so on.

But good mathematics education is not only about objectives, content, and teaching methods reflecting the nature of mathematics and the economic and social realities of a community. We also have to take into account the ways by which boys and girls, men and women learn and are motivated to learn mathematics. The insights and understandings gained from thinking about the three major functions of primary mathematics education (literacy, developmental, and propaedeutic) and the didactical aspects of the mathematics programme in rural communities combine to facilitate the search for answers to the question: what mathematics is valid and is of most worth to those who live in rural communities.

Three examples in the instructional design of mathematics education in rural communities

1. Village water-supply scheme

Water comsumption of different consumers such as villages in remote areas, hospitals, schools, etc. Average daily consumption. Graph of cumulative percentage of water consumption in a rural water supply. Capacity of water storage tank sufficient for one day's supply. Tables of monthly rainfall. Water-supply situation plan. Standard design of a public wash-place.

2. The school farm plot

Observations of school farm work. Farm records. The farm diary. Labour distribution on the school farm. Student's harvest-records. Methods of recording. Bar-graphs of yields. Pie charts of labour distribution. Profit and loss in trading farm produce. The use of percentages. Calculating the yield of a given farm plot. Incomes derived from yields. Frequency tables and graphs. Average yields and incomes.

3. The hen-house

Constructing a hen-house. Interpretation of diagrams and plans. Designing and constructing a paper model of a hen-house. A playground activity as an introduction to map reading. Calculating the building costs of a hen-house. The bill of materials. The woodmill. Cost price and retail prices of wood. Transportation charges. Calculating profit and loss. Investments and the introduction to comparative advantage. Estimating the area of a cross section of a tree. Calculating the volume of a tree trunk. Pricing of a log. The use of grids, upper and lower limits. Scale drawing. Map reading. Table of distances. Transporting logs.

Three major strategies for achieving aims of mathematics education in rural communities

Our study of the ways through which mathematics education has grappled with the problems of rural education across different countries has led us to identify three major strategies which seem most promising for developing, implementing, and evaluating an appropriate mathematics curriculum for rural communities. The first strategy emerges from a search for answers to the question: what part should mathematics play in education and in the curriculum of all children? Our analyses suggested that the general aims of education should include (a) intellectual, physical, emotional, social and spiritual development of persons; (b) providing each person with a cultural frame of reference; (c) encouraging persons to ask questions and search for their own answers; and (d) realising the full potential of each person as he makes his way through life.

Though not exhaustive, this list gives a sense of the scope of any education programme. the aims of mathematics should therefore be concerned with persons

(a) acquiring certain basic skills, knowledge and attitudes necessary for everyday life;

(b) acquiring further skills and knowledge for particular courses and careers;

(c) developing the ability to formulate problems in mathematical terms, and hence enabling them to appreciate the role of mathematics in a wide variety of disciplines and settings;

(d) acquiring the facts, ideas and processes of mathematics as a language which may be used to express ways of thinking about a phenomenon and communicating these ways to others; and

(e) developing an appreciation of mathematics as a social activity, and as the queen and servant of the sciences.

This strategy seems to be guided by the Panoramic View of Education Strategy, under which curriculum planning in schools should take into account the wider social and economic aspects of teaching. Schools therefore have a fundamental responsibility to demonstrate and interpret the content and methods of the mathematics curriculum, the unit of head, hand, and heart - - that is, of theory, practice, and commitment. Schools in developing countries must eschew a purely theoretical approach in conducting their curriculum, as this type of approach tends to produce an academic elite rather than a community of doers whose activities are informed by empiricism.

The second major strategy is associated with the nature of the motivations and the way the motivational forces operate within a community and a culture. Under this Community Involvement Strategy latent abilities of rural persons are activated through a reward system that allows them to become involved in (planning, carrying out, and evaluating) the affairs of their communities as they pursue whatever they do in schools, in homes, in factories, and in the fields.

This strategy seeks to ensure that knowledge is acquired by persons as a living force, and not as isolated facts and, above all, not as dogma; that mathematics is treated as an effort to present the quantitative aspect of reality and to analyse (and sometimes create) reality in quantitative terms rather than as an ability to do "sums" in aritificial, abstract contexts. Thus in the developing countries the task of building minds and personalities and the taks of catering to a changing economic, social, and political scene must be closely interrelated, and must not be age-bound,time-bound, place-bound, nor form-bound. Hitherto, education

> ". . . was provided for children for a certain fixed period of their lives in primary, secondary and tertiary stages clearly delimited in terms of the age of the child. It was provided in institutions and only in these institutions, by professional teachers and only by them. Certain forms of instruction were appropriate and certain forms were not." (Hawes, 1975, p. 23-24)

Thus the activities which have traditionally been organized piecemeal as responses to particular needs within the community must now be conceived as part of an educational philospohy which should undergird any community effort.

Therefore, the nature and direction for educational regenration within developing countries would suggest activities that

(a) bring the curriculum of schools closer to the activities of community life and to the needs and aspirations of individuals;

(b) integrate educational institutions, vertically and horizontally, into the community so that outputs of such institutions are beter adopted to the life and work in the community;

(c) redistribute teaching in space, time, and form and so include in the education process certain living experiences that are found in the community and in the lives of persons;

(d) broaden the curriculum of schools that it includes, in a meaningful way, socioeconomic, technical, and practical knowledge and skills, that is, activities that allow persons to combine mental and manual skills to create and maintain and promote self sufficiency as members of their community.

The third major strategy may be described as an Evaluation Strategy. Activities under this strategy are motivated by a view of education as a process which changes the learner in desirable ways within a community. This viewpoint sets the activities of learning firmly within a cultural frame of reference, allowing us to conceptualize the problems of mathematics curriculum differently.

First, what students learn and how they learn it are markedly affected by who are the teachers; by the interaction among learners, teachers and subject matter; and by the source and locus of the experience that contains the subject matter.

Second, the fundamental curriculum decisions to be made through the pooled wisdom of the society (What is taught? Why is it taught? When should it be taught?) compel the entire community to conceive of curriculum development "in terms and activities that may enable schools to become directly in tune with the world as it is" (Broomes, 1976).

Third, it seems easy to apply an important curriculum development principle; that is, every curriculum change requires political and professional ccoperation which thrives best when there is quick, easy flow of information about students' learning, teachers' teaching, and the effectiveness of school activities upon the lives and living of persons within the community. Or put more strongly, the quality of cooperation is best when it ensures that there is a dialectical relationship, back and forth between theory and practice, between classroom learning and living abundantly.

The Evaluation Strategy therefore allows three useful statements to be made about the role of evaluation in curriculum development within developing countries:

1. Evaluation is a set of strategies for monitoring new programmes, instrinsically and extrinsically. Many innovations in developing countries are characterized by (a) trial and error and (b) lack of relevant theoretical bases. Such innovations requrie careful evaluation of trials in order to identify errors and weaknesses. What is needed most is feedback information and correctives at each stage of the curriculum process.

2. Evaluation is a set of procedures for finding answers to certain questions or for realising certain purposes. Curriculum development has different stages and different phases and each stage or phase has different purposes. Evaluation activities must ensure what questions need answers and how best to collect and process the information for the answers.

3. Evaluation is a multivariate analysis of a research design involving independent, intervening,and dependent variables. Under this formulation, the hyposthesis of curriculum development may be

stated as follows: By changing A, through a planned programme of activities, the process B will be affected in such a way that the probability of producing effect C will be increased.
This formulation allows curriculum workers to lay bare the differences between failure of a curriculum project due to a technical flaw (programme does not contain the theory adequately, or programme is badly implemented and failure due to a theoretical weakness (the theory is invalid).

Nothing has been written to suggest that the three major strategies are mutually exlusive: indeed, the activities they define tend to overlap and complement each other. Nevertheless, the analysis gained in sharpness and focus by identifying and discussing the three strategies independently and individually.

References

Broomes, Desmond. "Structure, form and Organisation for the Evaluation of Curriculum". Caribbean Journal of Education, Vol. 3, No. 1, 1976.

Harbison, Federick H. "Human resources planning in modernizing economics", Internations Labour Review, Vol. LXXXV, No. 5 (Geneva, May 1962).

Hawes, H.W.R. Lifelong Education, Schools and Curricula in Developing Countries. Hamburg: UNESCO Institute for Education, 1975.

Malassis, Louis. The Rural World: Education and Development. Paris: The UNESCO Press, 1976.

21.5 PARTICIPATION OF THE HANDICAPPED IN MATHEMATICS

LE LANGAGE DES FLECHES

Robert Deischbourg
Institute Pedagogique
Walferdange, Luxembourg

Personne ne contestera que la mathematique d'aujourd'hui est essentiellement relationnelle et s'interesse davantage aux relations entre objets qu'aux objets eux-memes.

Dans toute initiation a cette mathematique au niveau scolaire se posera le probleme pedagogique de rendre accessible a de jeunes enfants le langage abstrait des mathematiciens qui parlent et pensent en termes de relations et de fonctions.

Tout langage s'apprend d'autant plus vite et d'autant mieux qu'on y est confronte etant jeune. Le langage des fleches en couleur ou papygrammes que nous preconisons a l'ecole primaire n'echappe sans doute pas a cette regle. Aussi conviendra-t-il d'y initier les enfants des leur entree dans les classes prescolaires ou, a defaut, des la premiere annee primaire.

L'apprentissage de ce langage peut se faire dans le cadre de situations non mathematiques (11), (12). Donnons un exemple (lecon faite dans une classe HMC):

Voici les enfants de la classe prets a se rendre dans la cour de recreation

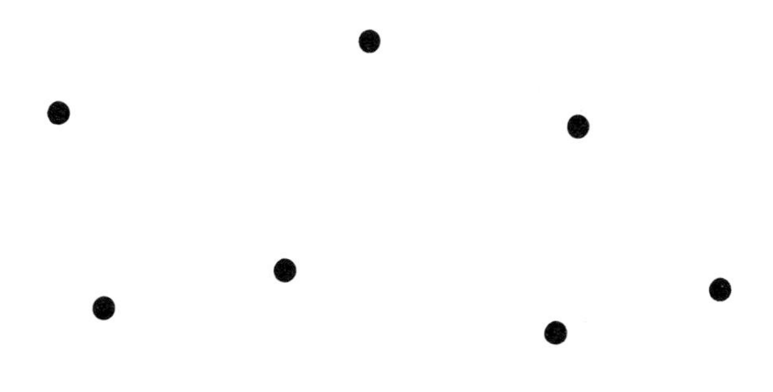

Chaque enfant montre l'eleve a cote duquel il va se mettre en quittant la salle de classe.

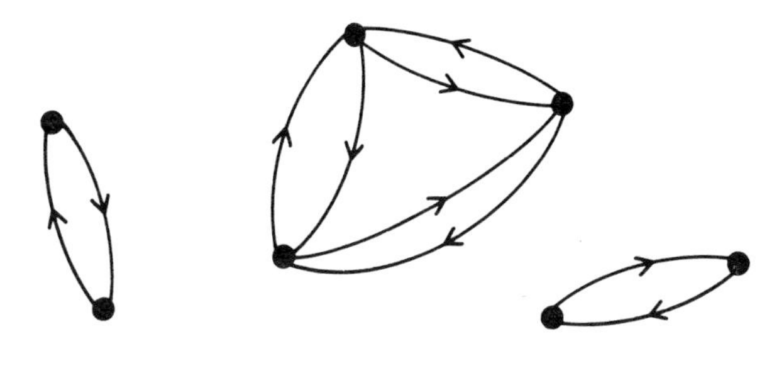

D'emblee, les "personnages" sont representes par des points, les "actions" de ces personnages par des fleches!

Le jour ou nous avons presente cette situation, l'un des grands etait malade. Les eleves ont insiste pour que le graphe ait l'aspect ci-apres

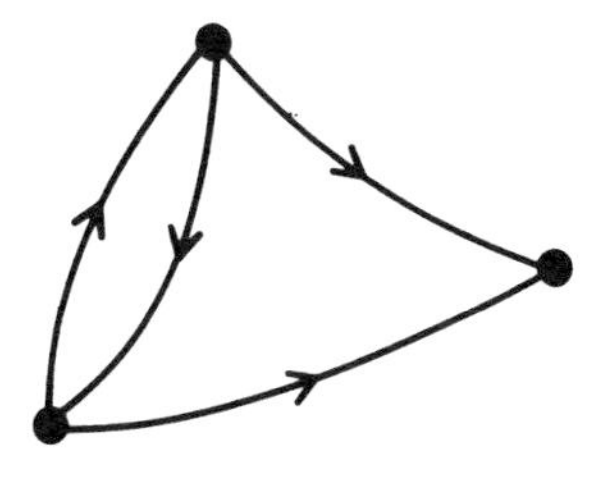

l'enfant absent ne pouvant "evidemment" pas montrer ses deux compagnons habituels. Ce n'est que le lendemain, en presence de l'eleve en question, que l'on a complete le dessin.

Cette attitude des enfants nous semble bien montrer combien ce graphe abstrait est reel a leurs yeux et l'avantage qu'on a de ne pas raisonner dans une situation purement concrete, puisque par des points on peut representer des enfants absents et les faire participer au jeu dans une certaine mesure.

Il s'agit d'une premiere ebauche mathematique d'algebraisation que l'on utilisera consequemment dans la suite.

Dans notre exemple nous avons traduit dans le langage des fleches une situation vecue. On pourra de meme presenter aux eleves un graphe et leur demander de decrire une situation compatible avec ce graphe.

L'etape suivante consistera a faire traduire dans le langage des fleches de petits textes du livre de lecture:

Denis	joue avec	Capi
Colette	joue avec	Papa
Aline	joue avec	Denis
Papa	montre	Capi

On observera les deux regles evidentes d'ailleurs:

1. Jamais deux fleches de meme couleur entre deux points
2. Actions differentes — Fleches de couleurs differentes

Nous avons aborde ensuite des situations plus numeriques:

La comparaison de la taille d'un eleve avec celle des autres donne lieu au graphe:

Les points vont finir par representer des nombres, concretises par les reglettes Cuisenaire.

Dans les deux problemes suivants, il s'agissait de trouver des nombres compatibles avec les graphes donnes et de dessiner par apres les fleches manquantes.

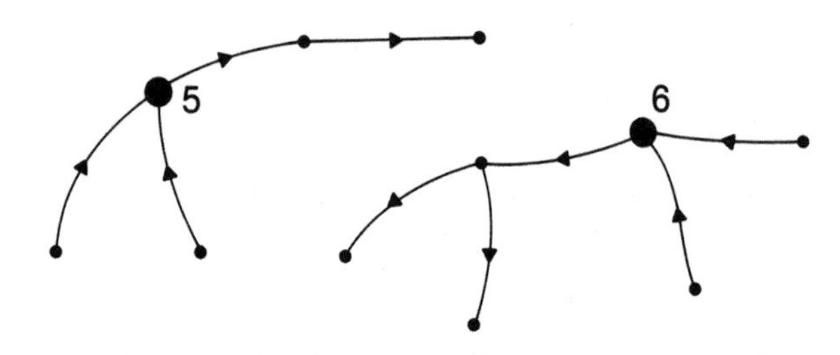

Nous avons aborde ensuite les fonctions additives et multiplicatives

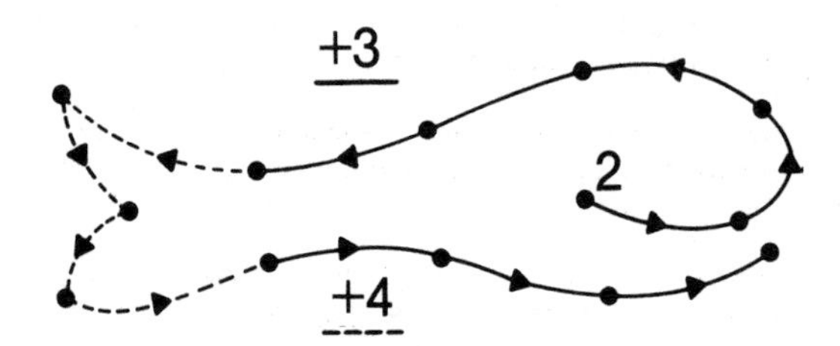

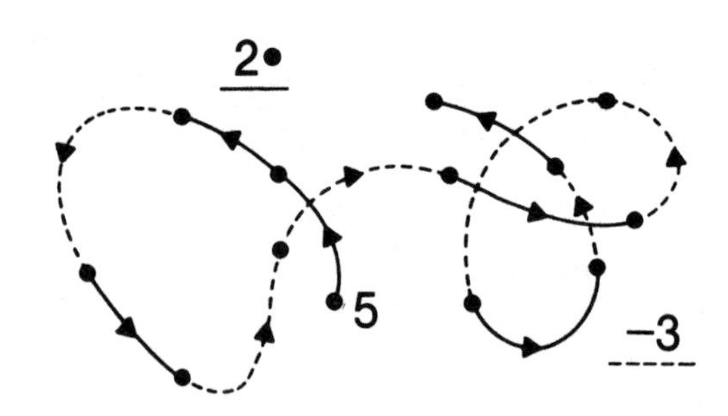

remplacant avantageusement les ennuyeuses colonnes de calcul de jadis.

Le langage des fleches trouve enfin a cote d'autres langages son application dans la resolution des problemes arithmetiques.

Donnons un exemple:
Dans un parking sont garees 5 voitures. Quatre nouvelles voitures arrivent. Combien de voitures y a-t-il au parking?

Représentation des données

dans le langage des cordes

dans le langage des flèches

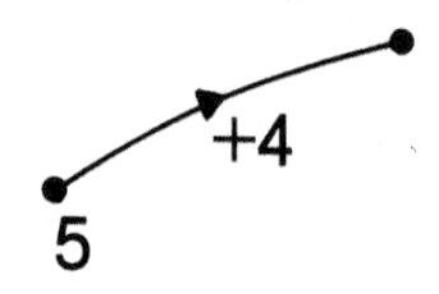

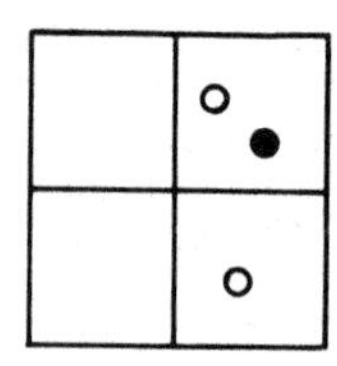

dans le langage de Minicomputer

Il est clair que dans la suite le langage des fleches s'imposera pour la traduction du probleme, Minicomputer se chargeant de l'aspect purement numerique (1), (13).

Donnons encore un exemple:

Maman fait des achats. Elle depense 256 F chez le boucher, puis encore 132 F pour des legumes. Enfin, elle achete 4 bouteilles de vin a 42 F la bouteille. A la sortie du magasin, elle a encore 327 F dans sa bourse. Combien d'argent Maman avait-elle importe?

1ere phase: Traduction du probleme dans le langage des fleches.

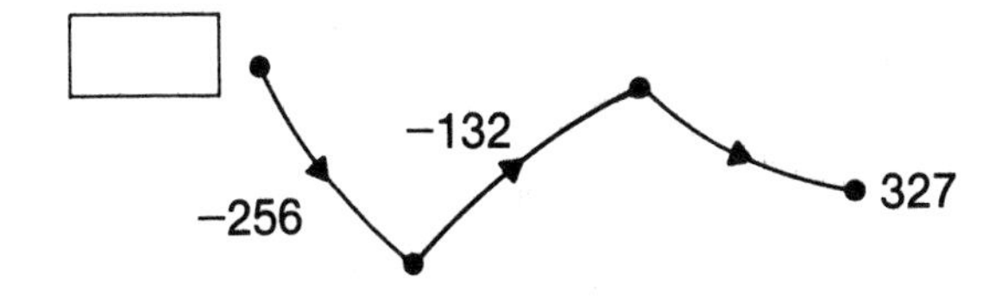

Les fleches indiquent les depenses successives de Maman: le dernier point represente donc la somme restee dans la bourse. Il n'est pas inquietant qu'on ne connaisse pas la derniere fleche! La traduction du probleme n'est pas encore terminee:

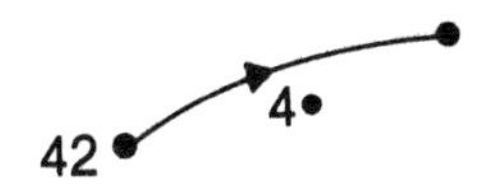

La fleche signifie donc: -168

En fait, la partie la plus ardue de la resolution du probleme est terminee. Partant du point connu, il suffira de revenir au premier point en renversant le sens des fleches (c'est-a-dire utiliser les fonctions reciproques).

2eme phase:

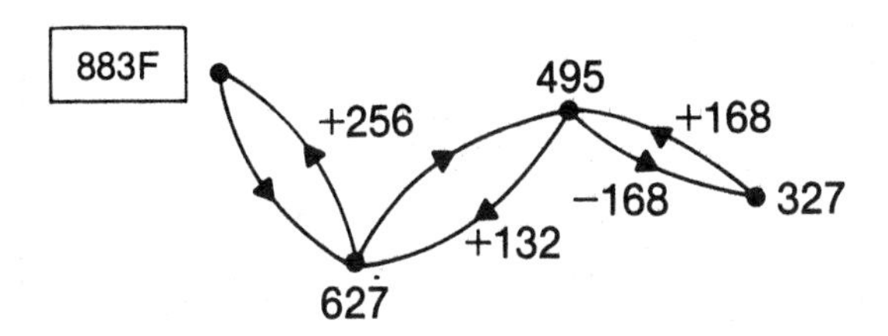

La partie purement numerique peut, au besoin, etre confiee a Minicomputer.

Dans nos deux exemples, chaque etat etait represente par un point (voitures en stationnement, somme contenue dans la bourse) chaque action par une fleche (arrivee de voitures, depenses de Maman).

Qu'en est-il des problemes ou ne se passe pas d'action?

Voici un exemple:

Jean possede 32 F. Paul en a 26. Combien d'argent ont-ils ensemble?

En general, les enfants proposent de representer par deux points l'avoir de Jean et celui de Paul. Il sera des lors naturel de representer la somme totale par un nouveau point.

32 • • 26

Comment obtenir a partir de l'avoir de Jean la somme totale? On prendra l'avoir de Paul et on l'ajoutera a celui de Jean. On aurait pu ajouter l'avoir de Jean a celui de Paul! On aura, dans les deux cas, le meme resultat, a savoir 58 F.

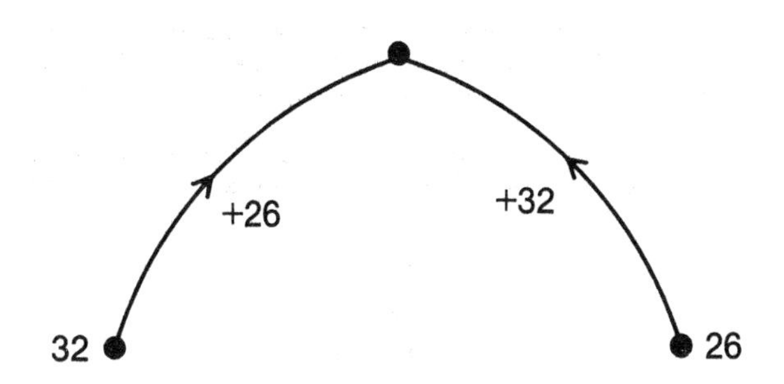

A noter que, dans le graphe, chacun des nombres 32 et 26 est represente par un point, mais aussi par une fleche (additive).

Dans d'autres problemes, les nombres seront representes par des fleches multiplicatives (cf. plus loin).

Dans bien des problemes, cette double representation est possible et permet de simplifier la demarche de resolution: toute inconnue pourra etre representee par un point, procede qui facilite un raisonnement direct!

Donnons deux exemples qui, bien que semblables, presentent pour les eleves des degres de difficulte differents:

1. Quinze bouteilles sont arranger dans 3 caisses. Combien de bouteilles aura-t-on dans chaque caisse?

2. Quinze bouteilles sont a arrangar en caisses de 5 bouteilles. Combien de caisses remplira-t-on?

Les enfants hesitent a diviser des bouteilles par des bouteilles!

Dans la langage des fleches, les deux problemes donnent lieu au meme graphe.

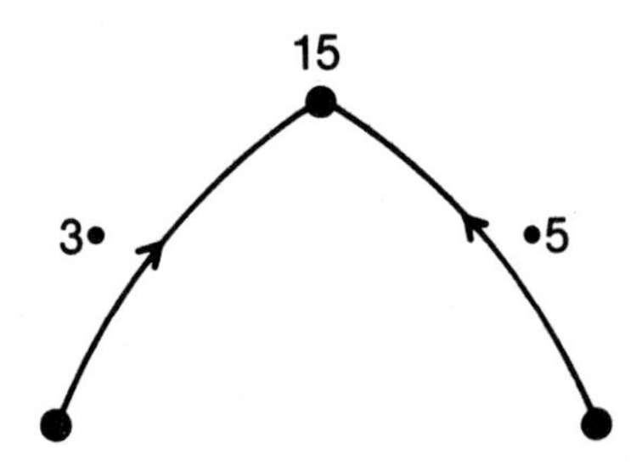

On y arrive par un raisonnement direct: representant dans le premier probleme le nombre de bouteilles par caisse a l'aide d'un point, il est clair que la multiplication par 3 donnera le nombre total de bouteilles.

De meme dans le deuxieme probleme: partant du point representant le nombre de caisses, une multiplication par 5 donnera le nombre total de bouteilles.

Dans les deux problemes il s'agira, partant du point connu, de calculer le point inconnu en utilisant les fonctions reciproques!

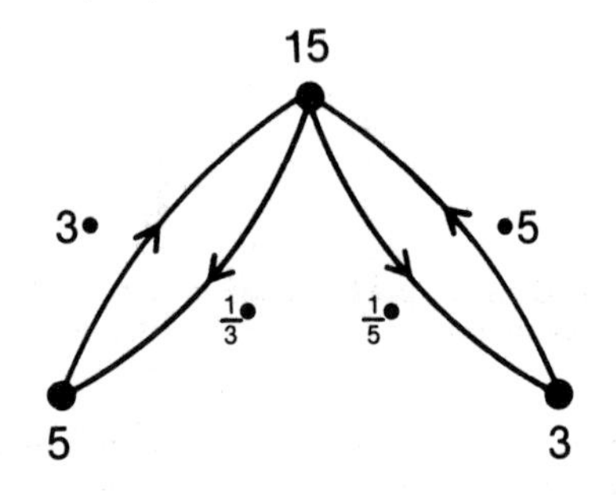

Ainsi, la traduction du probleme dans le langage des fleches etant achevee, il suffira que les eleves connaissent les diagrammes sagittaux suivants pour se tirer d'affaire.

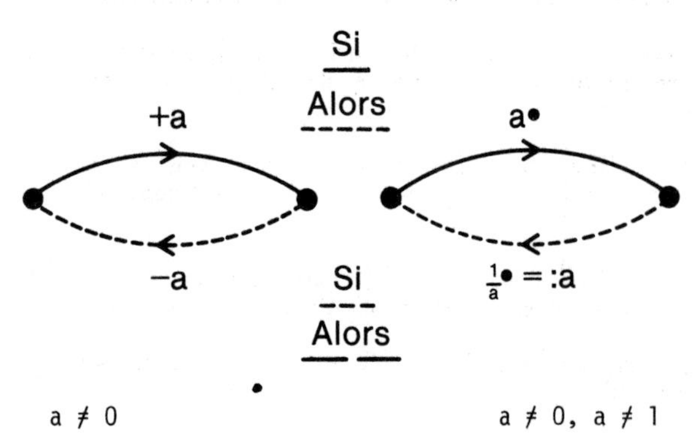

a ≠ 0 a ≠ 0, a ≠ 1

Lors de l'etude de quadrillages, nos eleves ont eu un premier contact avec les notions carre, rectangle et les unites de longueur et d'aire: cm et cm^2. Les concepts perimetre et aire ont ete mis en evidence simultanement et les enfants ont facilement pu arriver aux formules qui permettent de calculer ces deux grandeurs. (Les originaux des dessins d'enfants reproduits au long de cet article sont en couleurs. On s'est efforce d'en respecter la saveur malgre les exigences du noir et blanc.) Se reporter à l'illustration, page suivante.

Nous leur avons demande de traduire ces formules dans le langage des Papygrammes.

Pour le calcul de l'aire, une seule representation est possible

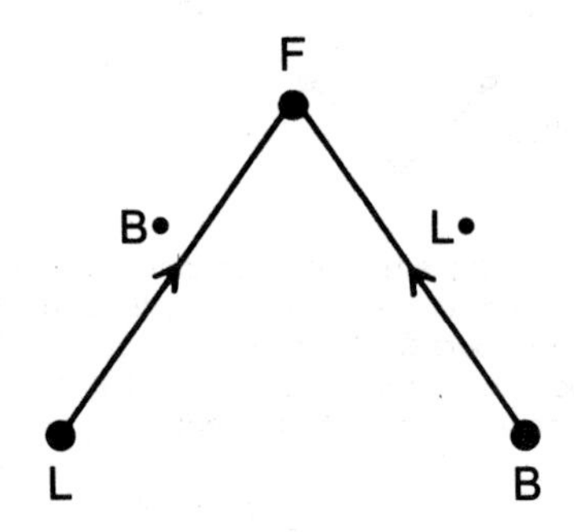

Pour le calcul du perimetre, on peut s'y prendre de deux manieres,

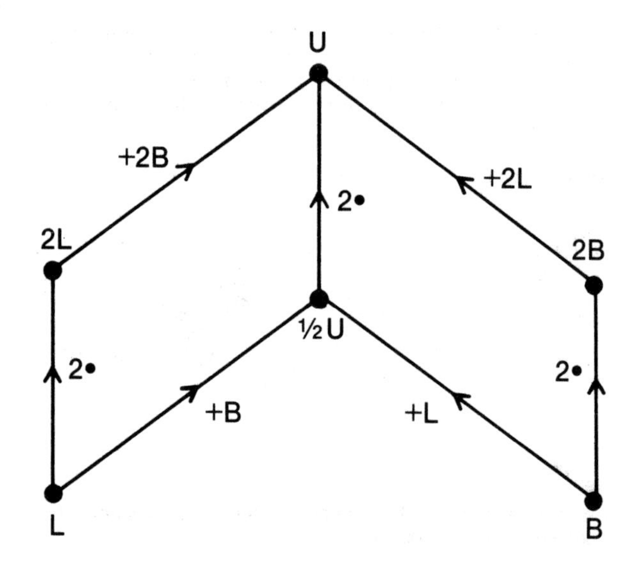

On notera que, dans ces graphes, la longueur (L) et la largeur (B) apparaissent sous l'aspect "point" et sous des aspects fonctionnels (+L, +B resp L.,B.).

Calculer l'aire et le perimetre d'un rectangle (d'un carre) connaissant ses dimensions ne presentait pas de difficultes pour nos eleves.

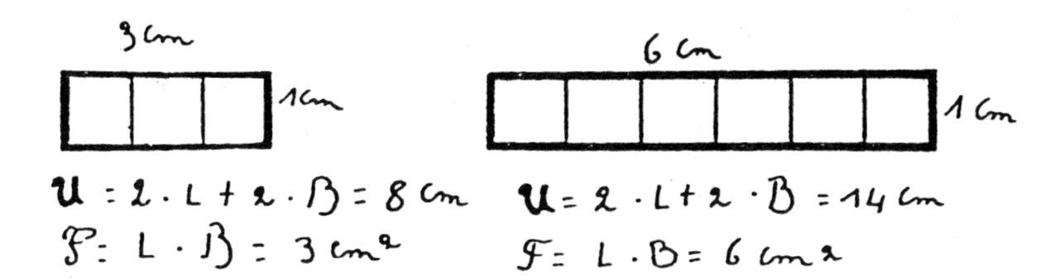
3 cm
1 cm
U = 2 · L + 2 · B = 8 cm
F: L · B = 3 cm2
6 cm
1 cm
U = 2 · L + 2 · B = 14 cm
F = L · B = 6 cm2

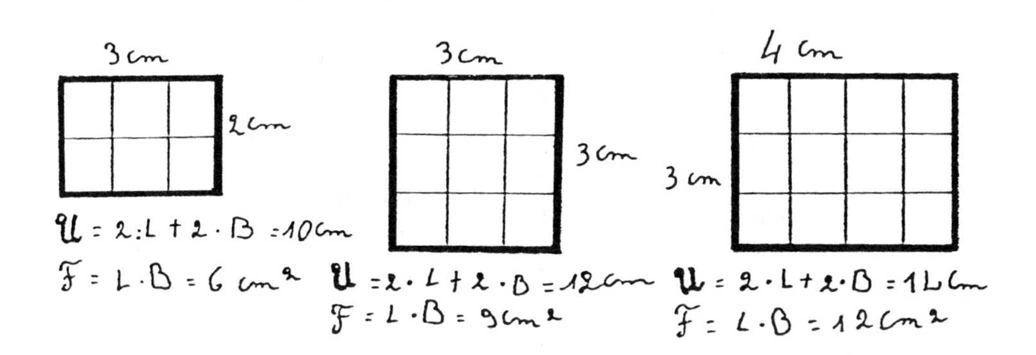
3 cm
2 cm
U = 2 : L + 2 · B = 10 cm
F = L · B = 6 cm2
3 cm
3 cm
U = 2 · L + 2 · B = 12 cm
F = L · B = 9 cm2
4 cm
3 cm
U = 2 · L + 2 · B = 14 cm
F = L · B = 12 cm2

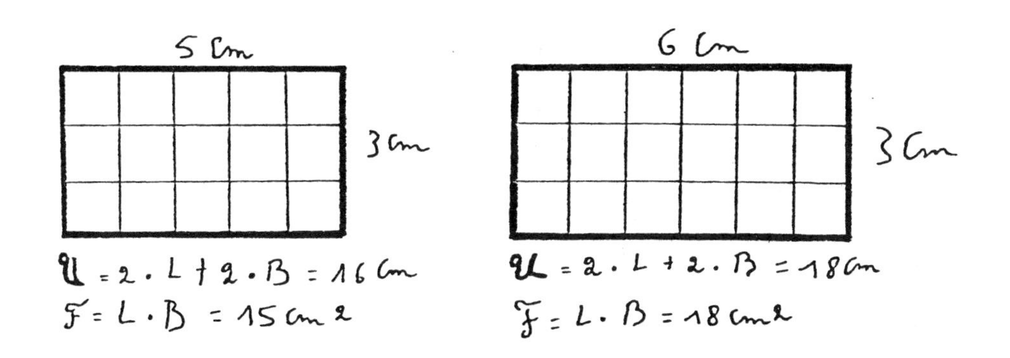
5 cm
3 cm
U = 2 · L + 2 · B = 16 cm
F = L · B = 15 cm2
6 cm
3 cm
U = 2 · L + 2 · B = 18 cm
F = L · B = 18 cm2

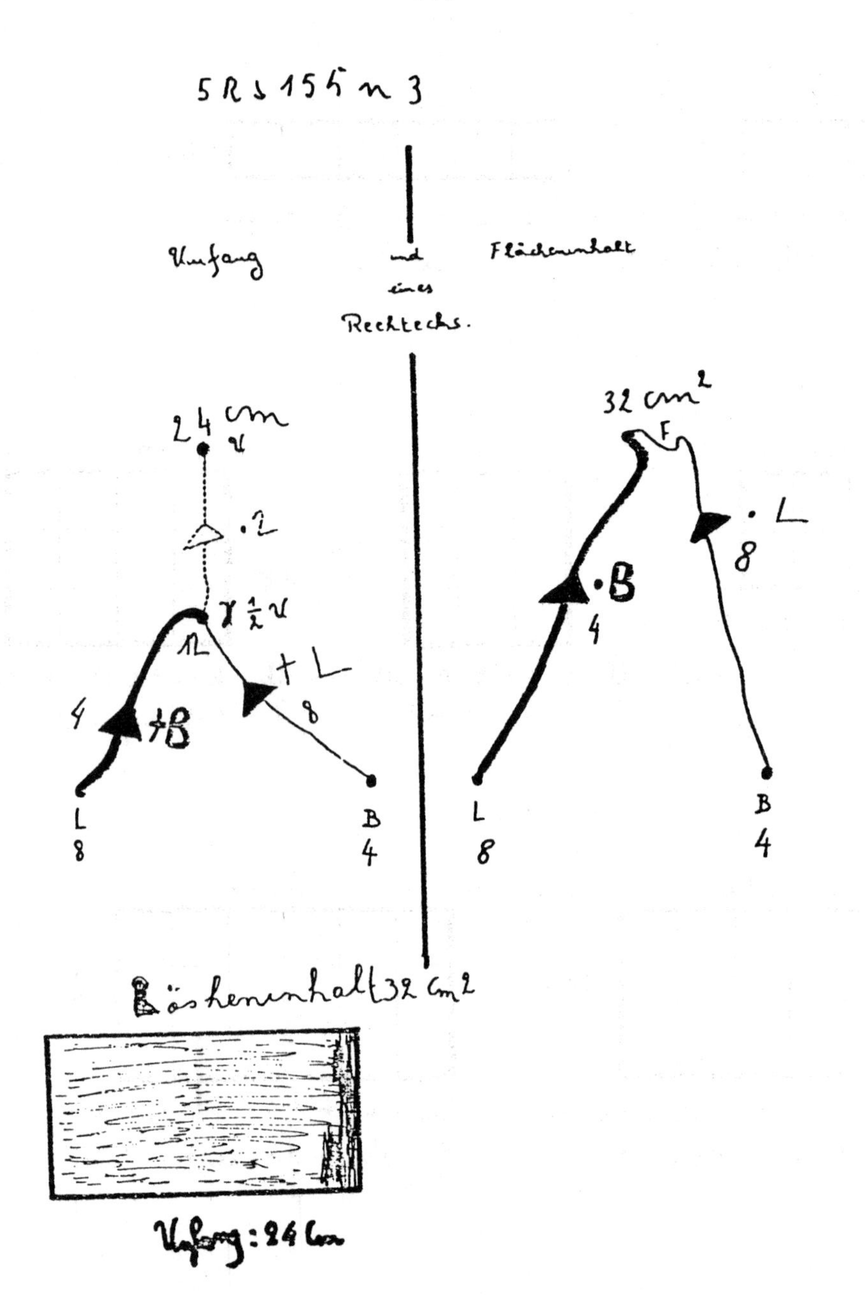

Mais avantage decisif: pour les enfants de notre classe, rompus au maniement du langage des fleches, les problemes: "connaissant l'aire (le perimetre) et une dimension d'un rectangle, calculez l'autre" sont du meme degre de difficulte que les precedents. Nul besoin d'une nouvelle formule! On utilise le meme graphe, tout au plus faut-il renverser des fleches.

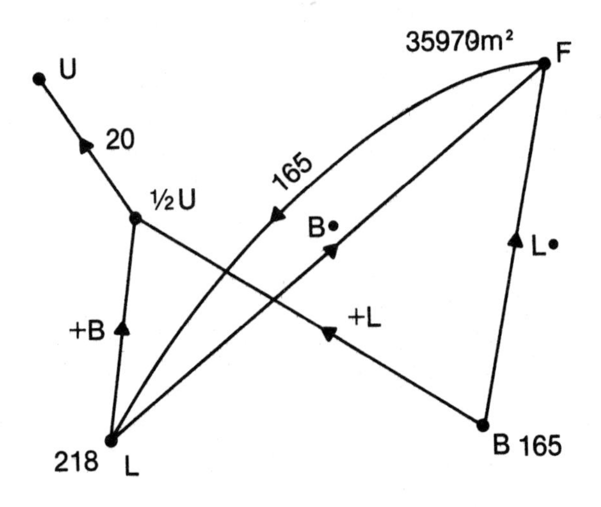

Du fait que nous avons superpose les figures 2 et 3, des problemes plus complexes se laissent, grace a la representation sagittale, resoudre sans peine.

A titre d'exemple: Le perimetre d'un rectangle mesure 780 cm. Sa longueur est de 240 cm. Calculez son aire.

Sur le graphe il suffit de mettre en evidence les donnees, puis de trouver un "chemin" du point representant le perimetre au point representant l'aire du rectangle, en retournant certaines des fleches connues

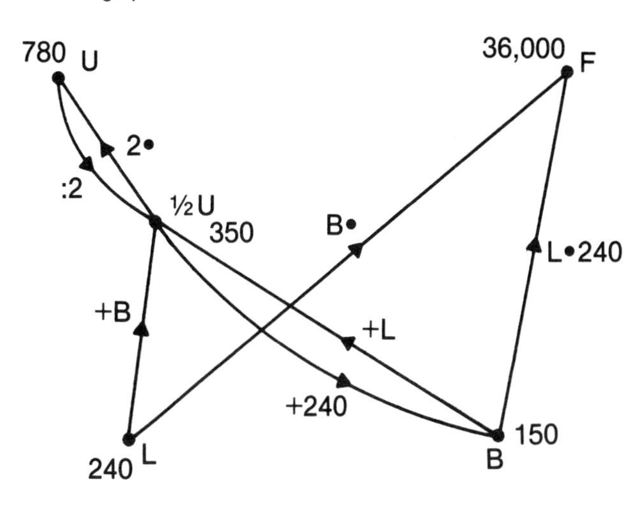

et nous avons pu sans difficultes traiter avec les eleves les problemes habituels sur les aires de ces surfaces.

Donnees: base du triangle: 13,3 m
hauteur: 4,2 m
Calculer l'aire

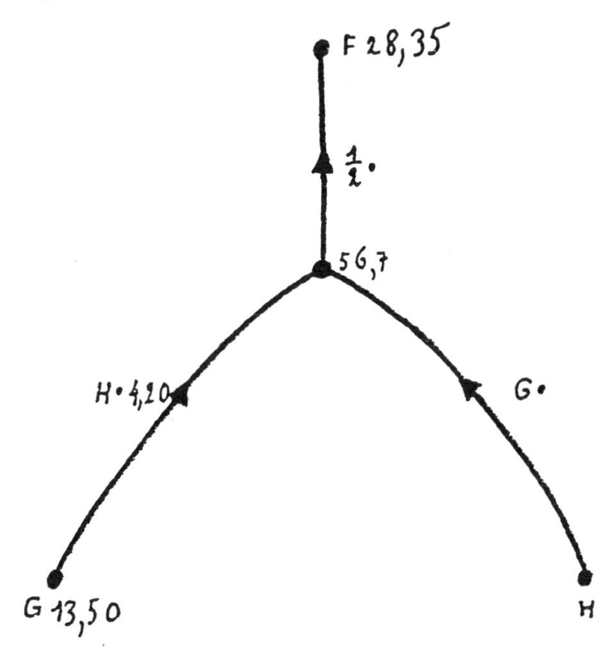

Donnees: aire du triangle 168cm^2
base 28 cm
Calculer sa hauteur

Antwort: Höhe des Dreieckes = 12 cm

Le probleme a resoudre etait devenu du meme ordre de difficulte que les precedents!

A noter que souvent pour ce genre de probleme, nous permettions aux enfants d'effectuer tous les calculs sur une calculatrice electronique.

De nouveau par voie experimentale sommes-nous arrives a la formule donnant l'aire d'un trapeze.

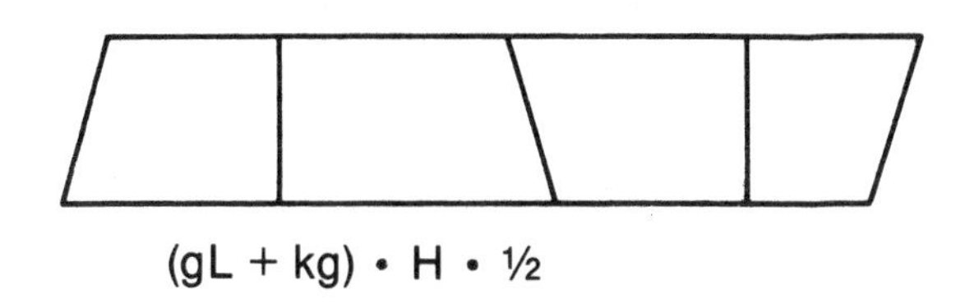

A noter que nous avons prefere ne pas faire intervenir la notion de base moyenne.

Comme pour les autres surfaces, pas de difficultes pour la resolution des problemes classiques.

6RS43 Nº 1

1 Antwort: Flächeninhalt des Feldes 9588 m^2

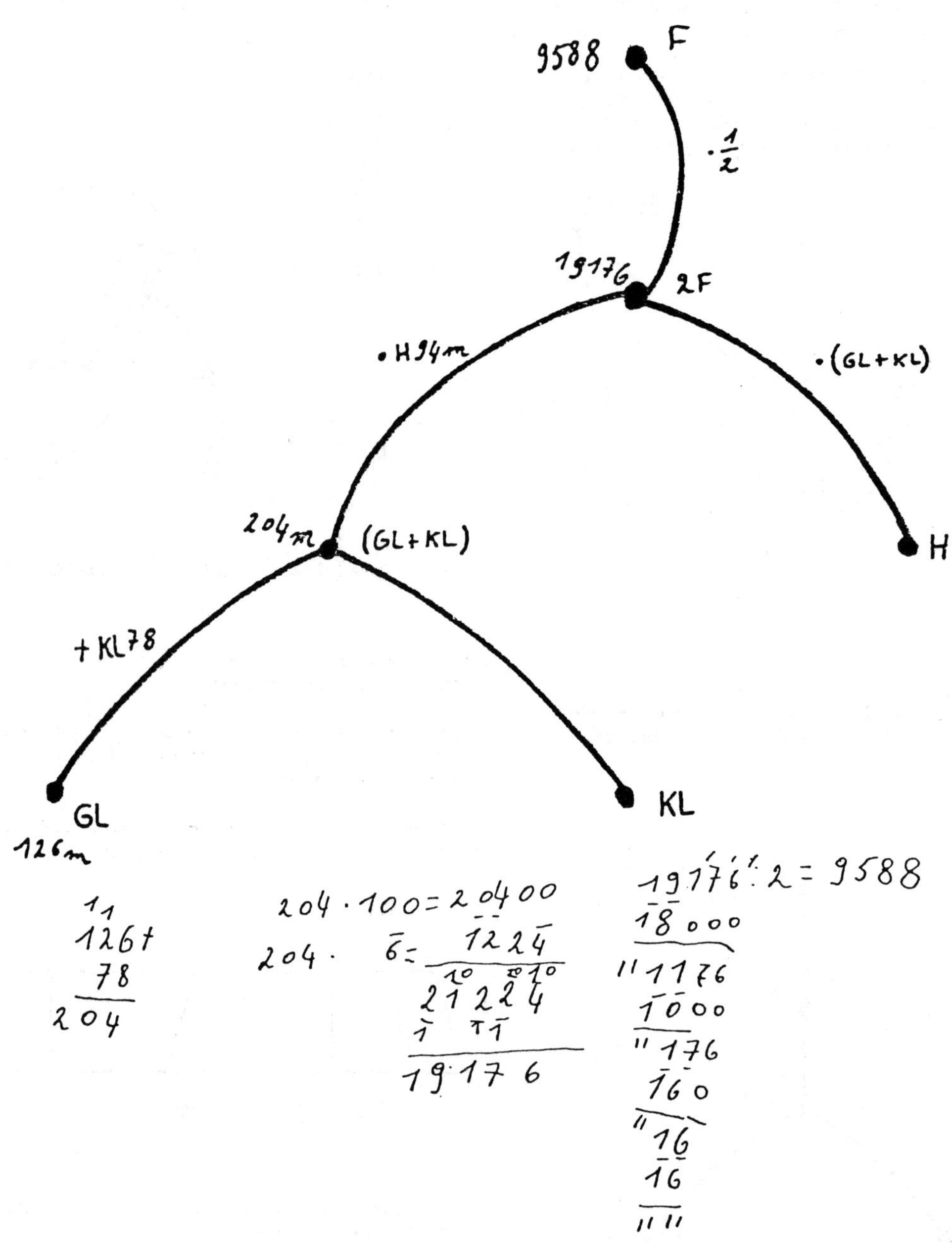

Pour la transformation des unites d'aire (de longeur) nos enfants disposaient de fiches.

km²		hm²		dam² (Ar)		m² (Quadrat Meter)		dm²		cm²		mm²
			1		1							
				1	2	3						
				1	6	7						
		3	4	8	7	5						
			1	6	0	0						
		1	3	4	0	0						
		4	8	4	6	5						
						1	1	1	1			
						0,	4	5				
						1,	6	7				
						0,	2	3	6	7		
						0,	0	0	5	6		
						2,	3	6	2	3		
	1	2	2									
	2	4	3	0	0	0						
		2,	8	9	0	0						
	5	6	9	4	0	0						
		3	6	7	0	0						
	8	7	8	0	0	0						

Pour illustrater les progres realises par nos eleves, terminons par le probleme suivant, resolu en equipe par la classe.

On echange une prairie ayant la forme d'un trapeze (grande base 126 m, petite base 84 m et hauteur 70 m) et valant 6 5000,- F l'aire contre un champ rectangulaire de meme prix revenant a 750 000,- F le ha. La longueur du champ est de 127,4 m. Calculez l'aire et le prix de la prairie ainsi que l'aire et la largeur du champ.

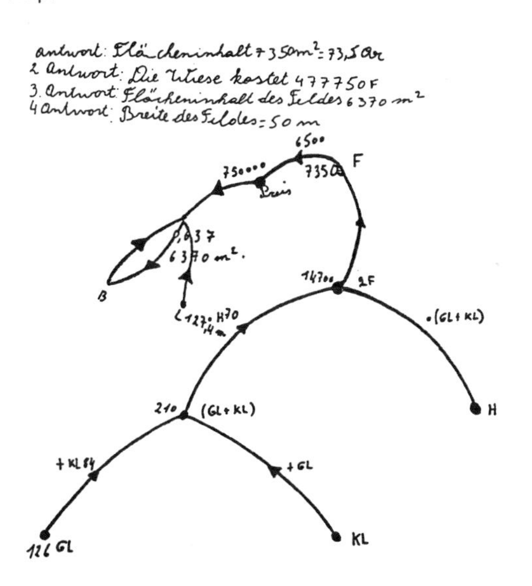

PARTICIPATION OF THE HANDICAPPED IN MATHEMATICS: AN OVERVIEW

Carole E. Greenes
Boston University
Boston, Massachusetts.

Since the beginning of compulsory education, various philosophical positions related to the education of handicapped children have been advocated. For many years, a distinction was made only between "learning" and "non-learning" children. "Non-learners" who were either brain injured, emotionally disturbed, auditorially handicapped, visually impaired, intellectually subnormal, suffering from motor deficits, or hampered by any combination of these handicaps were grouped together for instruction. Their instruction, whether in schools dedicated to special needs or in special classes within public schools, focused on the development of self-help skills.

Advances in medicine and technology in the 1950s and early 1960s contributed greatly to the integration of many special needs students into the regular classroom. Drug therapy for the emotionally disturbed and the hyperactive, prosthetic devices for the physically handicapped, and teaching aids for the sensorially handicapped permitted many of these students to participate in normal learning environments.

In the mid 1960s, interest focused on other students with learning difficulties, difficulties resulting from impairments of the central nervous system. These were students who for many years had been labelled as mentally deficient; students who were not deaf, but count not process auditory stimuli; students who were not blind, but had visual perception problems; and students who had difficulty in learning but were not mentally retarded. These students have come to be known as the "learning disabled," the children with symbolic language disorders (i.e., aphasia, a defect or loss in language, loss of expression in speech or writing, or comprehension of spoken or written language; dyslexia, an inability to attain language skills of reading, writing, and spelling, agraphia, an inability or extreme difficulty with the formation of letters, numerals, and forms in handwriting; dyscalculia, an inability to do or to learn mathematic.

Initially, special differentiated diagnosis and treatment or remediation of students with learning disabilities did not focus on children with learning difficulties in mathematics. There are several reasons for this neglect.

First, throughout history there has always been a certain mystique surrounding mathematics. While we assume that everyone can learn to read, we do not believe that everyone can learn to do mathematical manipulations. Mathematical ability has been thought to be a "God-given" talent - - a talent that is inherited and unalterable.

Second, the development of arithmetic skills and concepts has been relegated to a position of secondary importance in the educational program. Society has endorsed the teaching of reading as a primary goal of education because of its obvious relevance to the demands of living in a literate society. There has been less concern for mathematics because of its supposed lack of relevance to the daily living needs of the "average" citizen.

Third, the learning of mathematics has been recognized as the acquisition of a complex set of integrated skills and concepts, a set that cannot be easily segmented for the purpose of diagnosis and remediation. Learning to read, on the other hand, has been segmented into the skill of decoding and the process of comprehension. Thus remediation programs have dealt primarily with reading and language disorders.

When educational concern for the characterization, diagnosis, and remediation of mathematical difficulties did develop, the concern was sparked by legislation, funding, and research in education, psychology, and neurology.

In order to insure equal rights for all handicapped children, the Congress of the United States passed the Education for All Handicapped Children Act, Public Law 94-142, in 1975. This law required that each state establish policies for the identification and evaluation of all handicapped children; that procedures be established to safeguard the rights of handicapped children and their parents in decisions relating to identification, evaluation, and educational placement; and that a free appropriate public education be made available to all handicapped children from three to eighteen years of age. This legislation resulted in the "mainstreaming" of special needs students into the regular classroom.

Federal funds available to effect mainstreaming required schools to design individual educational programs IEP's for each special needs student. This required plan created an urgent need for diagnostic tools and appropriate curriculum materials in a variety of subject areas including mathematics.

For many years, the diagnosis and remediation of learning difficulties in mathematics was hampered by the commonly held belief that poor mathematical skills arises from low intelligence. Many children, however, who are sufficiently intelligent and highly motivated are not making progress in school mathematics. Research studies have demonstrated that many of these mathematical inadequacies are not due to lack of number concepts or subnormal intelligence, but are psychoneurogenic in nature, related to concentration, approach to tasks, visual processing, and temporal sequencing. A child may have problems in the learning of mathematics because of poorly developed concepts of space and time, figure-ground difficulties, linguistic failures, overstimulation difficulties, or memory (particularly auditory memory) failure. Results of these studies suggest a remediation-treatment program that is not content specific or dependent on the decomposition of the complex set of mathematical skills and concepts, but focuses on the development of some of the deficient perceptual, linguistic, and memory skills.

During the past 30 years, neurologists in the United States, Canada, and Russia have investigated the functions of the right and left hemispheres of the brain. They explain differences in cognition as assymmetries in brain fuction of hemisphere specialization. It appears that each side of the brain becomes specialized for different cognitive functions; the left hemisphere for logical-analytical operations and the right hemisphere for gestalt-synthetic activity. There is speculation that some students have difficulty learning mathematics because one hemisphere is not contributing to the development of mathematical thought. Studies in education, psychology, and neurology are just beginning to explore cognitive styles of learning and learning modalities. In the future, studies such as these may have significant implications for the design of programs for students with special needs in mathematics.

Currently, we are in a diagnosis-remediation gap. We have identified many different types of learning difficulties in mathematics, but have not designed specialized programs to help students overcome these difficulties.

PROBLEMATICA DE LA EDUCACION MATEMATICA PARA NINOS DE CLASES TRABAJADOR Y PROLETARIA EN LATINOAMERICA

Esther Pillar Grossi
Grupo de Estudos sobre o Ensina de Matematica de Porto Alegre
Porto Alegre, Brasil

En esta panel, que se titula "Matematicas a Nivel Elemental: Objetivos Generales y Contenidos, en Contextos Sociales y Culturales Diferentes", mi participacion se relaciona al hechoo de que soy latinoamericana, trabajo en investigacion y educacion matematica hace diez anos y, muy especialmente, nuestro Grupo de Estudios en Port Alegre estudia el aprendizaje de ninos de clases trabajadora y proletaria. Es en esa investigacion que quiero fundamentarme para hablar de los Objetivos Generales de la Matematica a nivel Elemental en Contextos Sociales y Culturales Diferentas.

Presentare algunos datos brasilenos y espero que en la discusion mis colegas de LatinoAmerica incorporen los de sus paises para que se caracterice aqui el contesto social y cultural de esta gran parte del continente americano.

En Brasil, somos 113 millones de habitantes de los cuales apenas 23 millones asisten a la escuela desde el Jardin de Infantes a la Universidad, mientras que mas de 25 millones entra 7 y 14 anos quedan duera de la escuala.

Pero lo que es mas grave aun es que de los ninos que estan en la escuela tenemos las siguientes datos, recibidos por telefone ayer, de un profesor de la "Fundacao Getulio Vargas", en Rio de Janeiro:

De la poblacion escoar que se inscribe en el primer ano primario el 59% no se inscribe en segundo y de los alumnos que ingrean en al primario mas del 90% abandonan la escuela a nivel de quinto grado, permaneciendo por lo tanto semi analfabetos y siendo obligados en su vida profesional a conformarse con un subempleo.

Estos datos son globales; esto es, mezclados con los de las clases media y alta que son los ninos que tienen exito enn la escuala, teniendo presente que las clases trabajadora y proletaria en Brasil representan el 75% de la poblacion.

Es evidente la imortancia que tiene para nosotros esa problamatica constituida por el contexto social y cultural en que vivimos.

Basados en los datos relativos a las dificultades de los ninos de las clases trabajadora y proletaria en la escuela, es que afirmamos que no vale la pena intentar lla asistencia de esta poblacion a la escuela en las condiciones actuales.

Ademas, es una idea aceptada en los circulos educacionales y en la sociedad en general, que la capacidad de aprendizaje de esos ninos es muy baja, proque son ninos que fueron y estan malnutridos y que viven en un ambiente cultrual considerado muy deficiente. En function de ello, son concebidos como normales los resultados obtenidos en su aprendizaje.

En este sentido, estamos haciendo una invetigacion , para detectar las posibilidades reales de aprendizaje de esos ninos que viven en un contexto cultural especial. Nuestro objetivo es intentar conocer sin prejuicios la realidad de vida de esos ninos para crear una propuesta didactica adaptada a ella, sin querer tomar como patron de referencia a los ninos de otras clases sociales.

A continuacion, describire rapidamente la metodologia de investigacion y analizare algunos resultados parciales del estudio en proceso.

El grupo de investigacion es compuesto por tres equipos. El primaro tiene un sociologo, un antropologo, un psicologo, un medico y un especialista en aprendizaje.

El segundo equipo esta formado por professores en cada una de las siquientes disciplinas: matematicas, lengua nacional, ciencias naturales y sociales, artes plasticas, musica, educacion fisica y, como intentamos trabajar con ninos que deben a aprender a leer y escribir, un especialista en este area.

El tercer equipo esta formado por observadores y auxiliares de investigacion.

Elegida una villa miseria, entre las 150 existentes en Porto Alegre, mediante observaciones, contactos personales y utilizacion de instrumentos de sondeo, se intenta el conocimiento de su realidad. A partir de estos datos, se plantean secuencias didacticas que son experimentadas en clases de alumnos de una escuela de esa villa. "Feed-backs" parciales son realizados con la participacion de todos los integrantes del grupo de investigacion y, con base en esto, se estan determinando algunas lineas directrices sobre el aprendizaje y la ensenanza para estos ninos.

Algunas diapositivos podran dar una breve idea de como es la villa miseria y como se realiza el trabajo de investigacion.

Josefina Cosentino, de Argentina, a quien reemplazo en este panel, porque no pudo venir, escribio que "La fijacion de nuevos objetivos, apunta a actualizar y educuar los concimientos matematicos a los cambios operados en la cultura de la humanidad en los ultimos anos."

Ella continua: "Si bien es cierto que el realizador directo, para cumplir los objetivos deseables, sera el maestro, este poco puede hacer sin el apoyo del sistema politico al cual pertenece."

Y sigue escribiendo sobre las grandes resistencias a la realizacion de cambios, sobretodo en educacion, y termina refereiandose el importante rol que pueden tener Reuniones Internacionales como fuerza de presion en favor de los importantes cambios que se hacen necesarios en nuestros paises.

En la investigacion que estamos en proceso de realizacion, algunas tendencias aparecen como norteadoras de los cambios que esta sociedad esta exigiendo para que sea efectivo el aprendizaje de esa gran parte de la poblacion en LatinoAmerica.

El primer punto, es la confirmacion de la necesidad de organizar una nueva manera de ensenanza para esos ninos.

Este nueva manera parece tener que considerar:

1. Que lo manual y lo intelectual deben marchar juntos en las actividades para esos ninos.

2. Que en vista de la falta de motivaciones primarias para aprendizajes del tipo escolar en funcion del ambiente no intelectual en que viven esos ninos, resulta prioritario el que las actividades de clase produzcan placer porque asi sera posible despartar el interes de esos ninos.

3. Que se deben caracterizar las habiidades especificas que esos ninos presentan, las practicas locales que construyen a partir de las actividades que realizan en lo cotidieno y utilizarlas en la escuela como ligazon entre lo que ya saban y lo nuevo que tendremos que solicitar de ellos.

4. Que para esos ninos con relacion a las matematicas, es mas importante el proceso de matematizacion que el aprendizaje de tecnicas de calculo u otras heramientas en terminos de contenido matematico. Porque constatamos que aprenden solos estas tecnicas, cuendo en su vida se les presente la necesidad repectiva; tales como, problemas de calculo, medida, etc.

5. Hay que tener en cuenta muy especialmente, la integracion social de esos ninos en la clase, ya que al trabajo individualista por lo general realizado en la escuaela, es rechezado frontalmente por ellos. Trabajar en equipos de ninos, les facilita su busqueda de identidad, la cual puede presantar dificultades en las numerosas familias y en el contexto de una villa miseria.

6. Hay que considerar desde el punto de vista antropologico, por ejemplo, la importancia que tiene para ellos la religion, no impora la que sea y no intentar una ensenanza, como es posible realizar en otras clases.

7. Parece muy importante incrementar la ejecucion de tareas experimentales en quimica o fisica para conduzir esos ninos a una lectura objetiva de la realidad y ayudarles a establecer correlaciones valederas entre causa y efecto. Eso les permite que se alejen del mundo magico en que viven, donde las enfermedades son producidas por malos deseos de enimigos y las curas son tambien debidas a la accion de un curandero, o pensar que su mala situacion socio-economico-cultural es voluntad de Dios debiendo por lo tanto rasignarse a ella.

8. Para que sea fructifera la accion educativa de la escuala sobre esos ninos, nos parace necesario empezar la labor didactica conellos antes de los 7 anos, esto es, antes de la transicion del periodo pre-operatorio al operatorio. Ademas, hasta los 7 anos, las experiencias de vida de los ninos de nuestra villa miseria se presentan como mas ricas que las de ninos de clase alta y media.

9. Podemos pensar que un largo y amplio trabajo en el sentido que proponemos, puede conducir a un cambio muy grande respecto a contenidos matematicos de los curriculums para esta poblacion y, quizas, par todo el contexto brasileno.

Por primera vez se podria tomar como base para la organizacion del curriculum, una investigacion cientifica con la colaboracion de expertos en otras areas del conocimiento, como sociologos, antropologs, psicologos, maestros, medicos, artistas, etc. . . . De esa forma, conceremos mejora los ninos de clases trabajodora y proletaria, auscultando sus verdaderos interses y posibilidades. Por ahora, ademas de los aspectos especiales y temporales en geometria asi como de la aritmetica, no inclinamos muy particularmente a incluir juegos aleatorios. Pensamos que, con esto, se pueda iniciar la formacion de una mentalidad probabilistica, tan importante para las matematicas de hoy, como para el hombre moderno. Sumamos a estas dos razones, una tercera, ligada a la necesidad de provocar el pesaje del pensamiento magico al pensamiento cientifico.

Una pregunta, dentro de tantas que no provoca esta investigacion, es la siguiente: Cuales son los efectos de la situacion tipica del punto de vista economico-social, en la capacidad de aprender de esta poblacion, que se caracteriza por la dominacion? Es posible hacer algo legitimo en la educacion, si esta situacion perdura, o se necesita irremediablemente un cambio en esta problematica?

Por ahora, lo mas importante que aprendimos con estos ninos es que ellos no parecen corresponder a la idea accptada de que tienen muy baja capacidad de aprendizaje alrededor de los 7 anos; sino que esto depende de la forma como las actividaes escolares se seleccionen y les sean presentadas.

Por todo esto, concluimos que es necesaria la elaboracion de un nuevo curriculum adaptado a la realidad, a esta realidad que acerca una extension muy significativa de la poblacion brasilena actuel.

AUTHOR INDEX

Page numbers do not always refer to the first page of individual articles but to the first page of the section in which an individual article appears.